PROBLEMS ON GENETICS MOLECULAR GENETICS EVOLUTIONARY GENETICS AND IMMUNOGENETICS

PROBLEMS
ON
GENETICS
MOLECULAR GENETICS
EVOLUTIONARY GENETICS
AND
IMMUNOGENETICS

[For the Undergraduate and Postgraduate Students of Biological Sciences and Other Competetive Examinations (NET, SLET, etc.)]

Dr Pranab Kr Banerjee MSc [CU], PhD [CU], FZS, FZSEI

Associate Professor
Department of Zoology [UG & PG]
Serampore College, Serampore, Hooghly

Department of Microbiology, R K Mission Vidyamandira
Belur Math, Howrah

Department of Zoology, R K Mission
Vivekananda Centenary College, Rahara

New Central Book Agency (P) Ltd

Co-Published by

Bijay Publications

E-mail : bijaypublications@gmail.com

Exclusively Distributed by

Bijay Pustak Bhandar (P) Ltd

90/6A M. G. Road, Bhawani Dutta Lane, Kolkata-700007

✆ : (033) 4605-7856, 8777898576, 8274916104

E-mail : bijaypustakbhandar@gmail.com

NCBA

REGD OFFICE
8/1 Chintamoni Das Lane, Kolkata 700 009, India
email: ncbakolkata@gmail.com

CORPORATE OFFICE
963/8 Govind Puri, Kalkaji, New Delhi 110 019
email: delhincba@gmail.com

BIJAY PUBLICATIONS
90/6A M. G. Road, Bhawani Dutta Lane, Kol-7
E-mail : bijaypublications@gmail.com

BIJAY PUSTAK BHANDAR (P) LTD
90/6A M. G. Road, Bhawani Dutta Lane, Kol-7
E-mail : bijaypustakbhandar@gmail.com

PROBLEMS ON GENETICS, MOLECULAR GENETICS, EVOLUTIONARY GENETICS AND IMMUNOGENETICS
• Dr Pranab Kr Banerjee

First Published: 2006
Reprinted: 2009
Thoroughly Revised, Enlarged and Updated Second Edition: 2011
Reprinted: 2013, 2015, 2016, 2018
Thoroughly Revised, Enlarged and Updated Third Edition: 2021
Reprinted: 2023, 2026

PUBLISHER • TYPESETTER
New Central Book Agency (P) Ltd
8/1 Chintamoni Das Lane, Kolkata 700 009

PRINTER
New Central Book Agency (P) Ltd
Web-Offset Division, Dhulagarh, Sankrail, Howrah

CO-PUBLISHED BY
Bijay Publications
90/6A M G Road, Bhawani Dutta Lane, Kolkata 700007

TECHNICAL EDITOR
Pradip Kumar Chowdhury

PROJECT SUPERVISOR
Ashit Ghosh

ISBN: 978 81 7381 498 3

Price ₹ 1250.00

Dedicated

To The Holy Mother

Sri Sri Ma Sarada

and

To The Departed Soul of my Father

Panchanan Banerjee

Contents

Foreward

Telephone
2654 9999 / 2654 9181
2654 9632 / 2654 6081

Fax
2654 9999

E-mail :
rkmveri@gmail.com
vidyamandira@vsnl.net

RAMAKRISHNA MISSION VIVEKANANDA EDUCATIONAL AND RESEARCH INSTITUTE

(A Deemed University under UGC Act)

P.O. Belur Math, Dist. Howrah
West Bengal 711202, India

It gives me great pleasure to have been called upon to write a foreword to Dr Pranab Kumar Banerjee's book Problems on Genetics, Molecular Genetics and Evolutionary Genetics.

We are celebrating the year 2005 as the International Year of Physics in commemoration of Einstein's great discoveries that revolutionized our conceptions of space, time and matter. Human life, science included, is nothing but "man's dialogue with Nature". In this dialogue, the human mind that takes up the challenge of unravelling Nature's mysteries through the path of investigation is ultimately stilled into silence by being overwhelmed by a sense of great wonder at Nature's variety and charm, power and beauty, order and abandon. Einstein spoke of himself as being engaged in a "humble adoration of an Illimitable Intelligence that our dull faculties could comprehend only in the most primitive forms", choosing to "wonder and stand in rapt awe" rather than pry, prove and penetrate into Nature's secret shrine through the dry intellectual exercise of investigation.

When we transcend the world of matter that physical scientists deal with to enter the world of life and consciousness which is the abode of bio-scientists, the mystery and the wonder become more pronounced. Consciousness is a word that is shrouded in greater mystery than life. After the dialogue with Nature in the form of matter, the human mind is now entering into a greater dialogue with Nature in the form of life. Rightly has the twenty-first century been called the "century of bio-science" just as the twentieth century was christened the "century of physical science". Man's dialogue with Nature began at the dawn of the twentieth century with the quantum theory, relativity theory, quantum mechanics, elementary particles, fields and forces. And the last century ended with the Human Genome Project, encountering on the way the path-breaking discovery of the DNA structure. The 'bio' word has become ubiquitous, entering into technology and compounded with 'molecular', 'micro' and an endless procession of such words, throwing up new sciences and new technology—bio-technology, molecular biology, microbiology, etc. It is no wonder that there is a near craze in the colleges and universities to learn and to explore this fascinating world of bio-sciences and bio-technology.

Mathematics, till recently considered the unquestioned partner of only physical sciences, has now become wedded to bio-sciences also. Bio-statistics has emerged as an important subject. Students of bio-sciences cannot any longer ignore mathematics and statistics as being unnecessary to the pursuit of life sciences. The myth of bio-science students being proverbially weak in mathematics is being exploded. Bio-sciences are now in a state of transition, from being descriptive to being mathematical in approach.

In this period of transition, it is urgently necessary that textbooks containing bio-statistical and bio-mathematical problems and their solutions appear on the stage. It is in this context that the present book by Dr Pranab Kumar Banerjee gains great relevance and importance. Dr Banerjee is a painstaking teacher who has been in the profession for a couple of decades. He understands the students' needs, their fears and bugbears, their strengths and weaknesses. This book, I believe, contains the stamp of this understanding. I hope, therefore, it would prove extremely useful to the students as well as to the teachers who would guide the students in this area.

Swami Atmapriyananda

Vice-Chancellor

Ramakrishna Mission Vivekananda
Educational and Research Institute

Belur Math
18 November 2005

Preface to the Third Edition

The popularity of this book has gained over the years as per commitment to bring this book in excellence. Therefore, I feel great pleasure to place before students and academics this thoroughly revised enlarged, elegant and appropriately updated version, to maintain its hallmark currency and continue to provide frame work for understanding this exciting and rapidly advancing subject, without making the book too voluminous and cumbersome.

Genetics is not just about genes, but much more or rather less than nature's great achievement — the miracle of DNA. Today science of genetics, stripped down to sub-molecular level is poised to make a quantum leap to solve the most elusive mystery of all the origin of life.

While retaining much of the flavour of earlier editions, style, content, several alterations and additions with new advancement of genetics, molecular genetics, evolutionary genetics, immunogenetics, including new colour illustration of matter have been incorporated and expanded in order to improve the quality of the content as well as to reflect the modern trend in this field.

Emphasis has been given in concept rather than excessive details. All the matters are written clearly and directly for the benefit of the student in order to provide understandable complex example of analytical topics along with solved problems. Chapters are written to be as independent of one another as possible.

Part 1. Genetics: Gene interaction epistasis, non-mendelian inheritance, genomic imprinting, somatic cell genetics, QTL mapping have been incorporated along with suitable problems in this section.

Part 2. Molecular Genetics: Special attention has been given on cot curve analysis along with suitable numerical problems. Concept of genetic code and related problems are incorporated. Microbial genetics has been enriched with experimental problems. Complementation mapping has been elaborated with several problems. Gene regulation along with epigenetic gene regulation has been introduced in this edition. In genetic engineering and biotechnology section, analysis of the transposable element, AC-DC elements and P elements have been illuminated. Modernization of cutting edge topic like restriction mapping and DNA sequencing are well designed and beutifully executed with numerical problems.

Part 3. Evolutionary Genetics and Immunogenetics: Population genetics is the study of alleles in population. Neutral and synthetic theory, molecular phylogeny have been incorporated in this edition. Extension and application of population genetics includes inbreeding, heritability and coverage of an emerging discipline i.e. conservation genetics have also been incorporated.

Immunology has remained as a discipline completely isolated from rest of the biology. It covers a wide area. Recently it has been treated as the discipline concerned chiefly with the preparation of vaccine. It plays a key role in the production of monoclonal antibody by the development of hybridoma technique. In this section, numerical problems are also incorporated.

I hope, this revised, enlarged and updated edition will open a new vista and will be a good guide for both under graduate and post graduate students in appearing various competitive examination in national level.

In preparing this book, I have taken profuse help from several books. I have expressed my acknowledgements in the Bibliography. I convey my respect and pronams to Swami Atmapriyanandji Maharaj (Vice-chancellor, Ramakrishna Mission Vivekananda Educational and Research Institute). I extend my respect and gratitude to my teacher, Professor Ranbindranath Chatterjee (Genetics Research Unit, Dep. of Zoology, University of Calcutta) and Dr. Tarit Kumar Banerjee (RPM college Uttarpara) for their enthusiastic inspiration. I also extend my thanks to my departmental colleagues especially Dr Chiranjib Dey for his constant inspiration as well as help in the preparation of manuscript for the chapter immunogenetics.

My heartful thanks are due to Sri Amitabha Sen, New Central Book Agency (P) Ltd for their contribution towards thoroughly revised, enlarged and elegant edition of this book in painstakingly suitable form. Thanks are due to Mr. Ashit Kr. Ghosh of New Central Book Agency (P) Ltd. for his continuous help in composing, drawing and overall supervision throughout the progress of this book.

Last but not the least, I extend my sincere thanks to Mrs Mandira Banerjee (Headmistres Baidyabati Charusila Bose Balika Vidyalaya) and Debdatta Banerjee (daughter) for their endurance, active assistance and cooperation during the preparation of third edition of this book.

Dr Pranab Kr. Banerjee

29th January 2020
Sri Saraswati Puja

Dept. of Zoology, Serampore College

Preface to the First Edition

This book entitled "Problems on Genetics, Molecular Genetics and Evolutionary Genetics" is based on my experience of teaching Genetics, Molecular Genetics and Evolutionary Genetics to the students of Biological Sciences.

Genetics—the science of heredity—is a core subject in Biological Science. Most of the Genetic principles can be precisely expressed in mathematical terms. Basic Genetics can easily be expressed through mathematical illustrations and problems.

Molecular Biology is a modern dynamic biological subject in the 21st century. Biotechnology is one of the facets of Science. It has been introduced in the syllabus of most of the Indian Universities at Undergraduate and Postgraduate levels. Therefore, problems regarding Molecular Genetics including Biotechnology have been incorporated in this book for the growing interest in this subject.

Population Genetics, Natural Selection, Molecular Evolution are current topics of Evolutionary Biology as well as Evolutionary Genetics. Evolutionary theory can easily be analysed through mathematical expression. Therefore, numerous problems regarding Evolutionary Genetics have been added in this book.

I claim no originality for the matter presented in this book but the method of presentation and illustration is my own. This book has been written in clear, lucid language to cover theoretical and mathematical approaches of the matter. It will help the undergraduate and post-graduate students of Biological Sciences of different colleges and universities for academic and competitive examinations.

In preparing this book, I have taken profuse help from several books. I have expressed my acknowledgement in the bibliography. I am indebted to my respected teacher Prof. R. N. Chatterjee (Genetics Research Unit, Dept. of Zoology, Calcutta University) for his constant inspiration and encouragement. I convey my respect and pranam to Swami Atmapriyanandaji Maharaj [Vice Chancellor, Ramakrishna Mission Vivekananda Educational Research Institute (Deemed University under U. G. C. Act)] for his enthusiastic inspiration which has driven me to prepare this book. I extend my respect and gratitude to Prof. A. K. Dutta Gupta (Retired) (Genetics Research Unit, Dept. of Zoology, Calcutta University) for his help and co-operation.

I also extend my thanks to Prof. C. Chakrabarty (Head of the Dept. of Zoology, Burdwan University), Dr T. K. Banerjee (R. P. M. College, Uttarpara), Dr K. K. Mukherjee (Serampore College) and my best friend Swaraj for their inspiration and encouragement for the preparation of this book.

I take the opportunity to extend my thanks to my students R. Surana and A. Das for their active co-operation and encouragement during the preparation of this book. Comments, positive criticisms and healthy suggestions from students as well as from erudite teachers for corrections and improvement of this book will be highly appreciated and incorporated in the future edition of this book.

My heartful thanks are due to Sri Amitabha Sen, Director, New Central Book Agency (P) Ltd. I also extend my sincere thanks to entire production unit of New Central Book Agency (P) Ltd for their contribution towards the publication of this book in nice form.

Last but not the least, I extend sincere thanks to Mrs. Mandira Banerjee and Miss Debdatta Banerjee (daughter) for their endurance and active assistance during the preparation of this book.

Dr Pranab Kr. Banerjee

Dept. of Zoology, Serampore College

10th November 2005
Jagatdhatri Puja

SECTION-I

GENETICS

Extensions of Mendelian Genetics 1

INCOMPLETE DOMINANCE

The situation in which phenotype of a heterozygote is intermediate between the two homozygotes on some quantitative scale of measurement is called **incomplete dominance.**

Characteristics

1. In the absence of complete dominance, every genotype has a distinguishable phenotype.
2. Sometimes incomplete dominance is also termed as **semi-dominance** because alleles may produce the same product but in lesser quantities as compared with the dominant allele.
3. Here in the heterozygous condition, the total product is intermediate between that of the dominant and recessive alleles.
4. The phenotypic ratio for the monohybrid cross between two heterozygote results 1 : 2 : 1 ratio, instead of 3 : 1 for dominant alleles.

Example: In snapdragons, heterozygous have pink flower in contrast to red and white for the dominant and recessive homozygous respectively.

	Red		White
	RR		rr
Gametes ⟶	R		r
F_1 ⟶		Rr (Pink)	
F_2 ⟶	Pink	×	Pink
	Rr		Rr
Gametes ⟶	R, r		R, r

♀ \ ♂	R	r
R	RR	Rr
r	Rr	rr

Result:

Genotype	Genotypic ratio	Phenotype	Phenotypic ratio
RR	1	Red	1
Rr	2	Pink	2
Rr		Pink	
Rr	1	White	1

Molecular Explanation of Incomplete dominance

1. In incomplete dominance only one allele in a heterozygote is expressed to produce a product.
2. (a) A homozygote for the expressed allele, then has double doses of the gene product and as a result full phenotypic expressions (Red in snapdragon) of gene is observed.

   (b) On the other hand, a homozygote for the other allele that is not expressed, no phenotypic expression (White in snapdragon) of gene is observed.

   (c) In a heterozygote the single gene (allele) is expressed and result only enough product for an intermediate phenotype, (pink in snapdragon).
3. In contrast in a heterozygote for a gene showing normal dominance, half the amount of a protein produced by the homozygote may be sufficient for normal cell function. In this case, the gene is said to be **haplosufficient.**

   Alternatively, the expression of the normal gene (allele) in the heterozygote may be increased to produce protein levels that give normal cell function.

Genetic basis for incomplete dominance

Phenotype	Genotype	Amount of Gene product
Red	$\frac{+}{+}$	$2x$
Pink	$\frac{+}{w}$	x
White	$\frac{w}{w}$	0

1. The allele + is completely dominant over white. Differences among the phenotypes could be due to differences in the amount of the product specified by '+' allele.
2. + allele specifies the product but 'w' allele does not. $\frac{+}{+}$ homozygote will have twice as much of the product as $\frac{+}{w}$ heterozygotes do and will therefore be shown deeper (red) colour.
3. When the heterozygote's phenotype is mid-way between the phenotypes of the two homozygote. Therefore, it is sometimes called '**partial**' or **semidominant**.

Problem 1

In *Mirabilis* (the four o'clock), a plant hybrid for red flowers (+) and white (w) is pink (w/+); one pink plant is crossed to a red, another to white. Give the genotypic and phenotypic ratios in which the offsprings are produced in each case.

Solution

Pink × Red

$\frac{w}{+}$ × $\frac{+}{+}$

Gametes: w, $+$, $+$

F_1

♀ \ ♂	w	$+$
$+$	$\frac{w}{+}$	$\frac{+}{+}$

Genotype	Genotypic ratio	Phenotype	Phenotypic ratio
$\frac{w}{+}$	1	Pink	1
$\frac{+}{+}$	1	Red	1

Pink × White

$\frac{w}{+}$ × $\frac{w}{w}$

Gametes: w, $+$, w

F_1

♀ \ ♂	w	$+$
w	$\frac{w}{w}$	$\frac{+}{w}$

Result:

Genotype	Genotypic ratio	Phenotype	Phenotypic ratio
$\frac{w}{w}$	1	White	1
$\frac{+}{w}$	1	Pink	1

Problem 2

There is a race of chicken known as "Andalusians", some of which have black feathers, others white. Consider black as the normal (+) and white as mutant (w). When black is crossed by white ($\frac{+}{+} \times \frac{w}{w}$), the F_1 hybrid ($\frac{+}{w}$) is a *slaty blue* (neither black nor white being dominant). Give the F_2 genotypic and phenotypic ratios when the F_1 are interbred ($\frac{+}{w} \times \frac{+}{w}$).

Solution

Slaty blue × Slaty blue

$\frac{+}{w}$ $\frac{+}{w}$

Gametes: $+$, w, $+$, w,

F_2

♀ \ ♂	w	$+$
w	$\frac{w}{w}$	$\frac{+}{w}$
$+$	$\frac{+}{w}$	$\frac{+}{+}$

Genotype	Genotypic ratio	Phenotype	Phenotypic ratio
$\frac{w}{w}$	1	White	1
$\frac{+}{w}$, $\frac{+}{w}$	2	Slaty blue, Slaty blue	2
$\frac{+}{+}$	1	Black	1

Problem 3

In radishes, flower colour may be red, purple or white. The edible portion of the radish may be long or oval. When only flower colour is studied, no dominance is evident and red × white yields purple. If red long plant is crossed with white oval and F_1 plants are interbred, determine F_1 and F_2 phenotypes, genotypes and their ratios.

Solution

Red long × White oval

RR LL × rr *ll*

Gametes: RL r*l*

F_1 Rr L*l* (purple long)

$F_2 = F_1 \times F_1$

Purple long × Purple long

Rr L*l* × Rr L*l*

Gametes: RL, rL, R*l*, r*l*; RL, rL, R*l*, r*l*

♀ \ ♂	RL	rL	R*l*	r*l*
RL	RR LL	Rr LL	RR L*l*	Rr L*l*
rL	Rr LL	rr LL	Rr L*l*	rr L*l*
R*l*	RR L*l*	Rr L*l*	RR *ll*	Rr *ll*
r*l*	Rr L*l*	rr L*l*	Rr *ll*	rr *ll*

Result:

Genotype	Genotypic ratio	Phenotype	Phenotypic ratio
RR LL	1	Red long	
RRL*l*	2	Red long	3, i.e., $\frac{3}{16}$
RRL*l*		Red long	
RR *ll*	1	Red oval	1, i.e., $\frac{1}{16}$
RrLL	2	Purple long	
RrLL		Purple long	
RrL*l*		Purple long	
RrL*l*	4	Purple long	6, i.e., $\frac{6}{16}$
RrL*l*		Purple long	
RrL*l*		Purple long	
Rr*ll*	2	Purple oval	2, i.e., $\frac{2}{16}$
Rr*ll*		Purple oval	
rrLL	1	White long	
rrL*l*	2	White long	3, i.e., $\frac{3}{16}$
rrL*l*		White long	
rr *ll*	1	White oval	1, i.e., $\frac{1}{16}$

Problem 4

When chickens with splashed white feathers are crossed with black-feathered birds, their offsprings are all slate blue (Blue Andalusian). When Blue Andalusians are crossed among themselves, they produce splashed white, blue and black offsprings in the ratio of 1 : 2 : 1 respectively.

(a) How are these feather traits inherited?

(b) Using any appropriate symbols, indicate genotypes for each phenotype.

Solution

Let the genotype of splashed white feather is F^SF^S and black feather is F^BF^B and slate blue is F^SF^B.

Splashed white × Black

F^SF^S F^BF^B

Gametes: F^S F^B

F_1 F^SF^B (Blue Andalusian)

$F_2 = F_1 \times F_1$ F^SF^B × F^SF^B

Gametes: F^S, F^B F^S, F^B

♀ \ ♂	F^S	F^B
F^S	F^SF^S White	F^SF^B Blue
F^B	F^SF^B Blue	F^BF^B Black

Result: White : Blue : Black

1 : 2 : 1

(a) Single pair of codominant alleles.

(b) White = F^SF^S, Blue = F^SF^B, Black = F^BF^B.

Problem 5

Yellow coat colour in guineapig is produced by the homozygous genotype C^yC^y, cream colour by the heterozygous genotype C^yC^w and white by the homozygous genotype C^wC^w. What genotypic and phenotypic ratios are expected from matings between cream-coloured individuals?

Solution

Phenotype	Genotype
Yellow	C^yC^y
White	C^wC^w
Cream	C^yC^w

Cross: Cream × Cream

C^yC^w C^yC^w

Gametes: C^y, C^w C^y, C^w

♀ \ ♂	C^y	C^w
C^y	C^yC^y	C^wC^y
C^w	C^wC^y	C^wC^w

Result:

Genotype	Genotypic ratio	Phenotype	Phenotypic ratio
C^yC^y	1	Yellow	1, i.e. $\frac{1}{4}$
C^yC^w } C^yC^w	2	Cream } Cream	2, i.e. $\frac{2}{4}=\frac{1}{2}$
C^wC^w	1	White	1, i.e. $\frac{1}{4}$

Problem 6

Palomino horse is a hybrid exhibiting a golden colour with lighter mane and tail. A pair of codominant alleles (D^1 and D^2) is known to be involved in the inheritance of these coat colours. Homozygous genotypes for the D^1 alleles are chestnut-coloured (reddish), heterozygous genotypes are palomino-coloured, and genotypes homozygous for the D^2 alleles are almost white and are called cremello.

(a) From matings between palominos, determine the expected palomino and non-palomino ratio among the offspring.

(b) What percentage of the non-palomino offspring in part (a) will breed true?

(c) What kind of mating will produce only palominos?

Solution

Phenotype	Genotype
Chestnut	D^1D^1
Palomino	D^1D^2
White	D^2D^2

Palomino × Palomino

D^1D^2 D^1D^2

Gametes: D^1, D^2 D^1, D^2

♀ \ ♂	D^1	D^2
D^1	D^1D^1	D^1D^2
D^2	D^1D^2	D^2D^2

Result:

Genotype	Genotypic ratio	Phenotype	Phenotypic ratio
D^1D^2 } D^1D^2	2	Palomino	1
D^1D^1 } D^2D^2	2	Non-Palomino	1

(a) Palomino : Non-palomino 1 : 1

(b) 100% (i) D^1D^1 × D^1D^1 (Chestnut)

Gametes: D^1 D^1

F_1 D^1D^1 (Chestnut)

(ii) D^2D^2 × D^2D^2 (Cremello)

Gametes: D^2 D^2

F_1 D^2D^2 (Cremello)

(Chestnut)   (Cremello)

(c) D^1D^1 × D^2D^2

Gametes: D^1 D^2

F_1 D^1D^2 (Palomino)

Problem 7

In snapdragons, red flower colour (C^R) is incompletely dominant to white (C^w); the C^R/C^w heterozygous are pink. A red coloured snapdragon is crossed with a white flowered one. Determine the flower colour of (a) the F_1 snapdragons (b) the F_2 snapdragons (c) the progeny of the cross of the F_1 snapdragons to the red parent and (d) the progeny of the cross of the F_1 snapdragon to white parent.

Solution

(a) Red coloured snapdragon   White flowered one

C^RC^R × C^wC^w

Gametes: C^R C^w

F_1 C^RC^w (Pink) Colour flower

(b) $F_1 \times F_1$ – F_2 progeny

F_1: Pink × Pink

C^RC^w C^RC^w

Gametes: C^R C^w C^R C^w

♀ \ ♂	C^R	C^w
C^R	C^RC^R	C^RC^w
C^w	C^RC^w	C^wC^w

Result:

Genotype	Genotypic frequency	Phenotype	Phenotypic frequency	
C^RC^R	1	Red	1	$\frac{1}{4}$
C^RC^w } C^RC^w	2	Pink } Pink	2	$\frac{2}{4}=\frac{1}{2}$
C^wC^w	1	White	1	$\frac{1}{4}$

One red, 2 pink and 1 white flower is found. Half of the progeny have parents coat colour.

(c) F_1 × Red parent

C^RC^w C^RC^R

Gametes: C^R C^w C^R

C^RC^R C^wC^R

Red Pink

Result: one is red other is pink, i.e., 1 : 1 ratio.

(d) F_1 × White

C^RC^w × C^wC^w

Gametes: C^R C^w C^w

C^RC^w C^wC^w

Pink White

Result: one is pink other is white colour.

Problem 8

In horses, there are three coat colour patterns termed Cremello (beige) Chestnut (brown) and palomino (golden with light more and tail). If two palomino horses are mated, they produce about $\frac{1}{4}$ cremello, $\frac{1}{4}$ chestnut and $\frac{1}{2}$ palmino offspring. In contrast, cremello horses and chestnut horses breed true. (In other words, two cremallo horses will produce only cremello offspring and two chestnut horses will produce only chestnut offspring). Explain this pattern of inheritance.

Solution

1. This is an example of incomplete dominance.
2. Heterozygous horses are palomino.

C represents chestnut, c represents cremello. Chestnut horses are CC, palominos are Cc, cremellos are cc.

(a) Palomino × palomino, i.e., Cc × Cc → $\frac{1}{4}$ (CC) chestnut, $\frac{1}{4}$ (cc) cremello and $\frac{1}{2}$ (Cc) palomino.

(b) Cremello × cremello → cc × cc → produces (cc) (cremello); chestnut × chestnut → CC × CC $\xrightarrow{\text{produces}}$ (CC) chestnut.

Problem 9

In snapdragon, Tall (T) is dominant to dwarf (t) and red flowers (R) are incompletely dominant to white (w), the hybrid being pink. The F_1 flies are test crossed. Give F_2 ratio.

Solution

Parental phenotypes

Tt RR × tt rr

TR tr

F_1 Tt Rr (Tall pink)

Test cross Tt Rr × tt rr

Gametes TR, tR, Tr, tr tr

♀ \ ♂	TR	tR	Tr	tr
tr	$\frac{TR}{tr}$	$\frac{tR}{tr}$	$\frac{Tr}{tr}$	$\frac{tr}{tr}$

	Genotype	Genotypic ratio	Phenotype	Phenotypic ratio
I.	$\frac{TR}{tr}$	1	Tall pink	1
II.	$\frac{Tr}{tr}$	1	Tall white	1
III.	$\frac{tR}{tr}$	1	Dwarf pink	1
IV.	$\frac{tR}{tr}$	1	Dwarf red	1

Partial dominant

- **Semi-dominant**

I. When the heterozygote phenotype is midway between the phenotypes of the two homozygotes is called semidominant.

II. Bar eye *Drosophila*, Normal eye is BB, Bar eye is B^+B^+ in heterozygous (B^+B) is kidney shaped eye.

BB (Round eye) → B

B^+B^+ → B^+

→ BB^+

CODOMINANCE

Alleles that lack dominant and recessive relationship and both have capability to express themselves phenotypically in the heterozygous conditions are termed as **codominance**.

Characteristics

1. It exhibits a unique pattern of expression with heterozygotes being phenotypically distinguishable from both of the homozygotes.
2. Here two alleles appear to contribute independently to the phenotype of heterozygote.
3. Here both alleles are expressed equally and simultaneously.
4. In general, two products are the same with respect to function but different in exact amino acid sequence.
5. Codominant alleles are represented by superscripts on the symbol for the gene.

Example:

1.(i) The coat colour of the short horn breed of cattle is the best example of codominance. The genotype of red coat colour is C^RC^R and white coat colour is C^wC^w. The heterozygous or hybrid is roan coat, i.e. C^RC^w.

(ii) In roan coat red and white hairs occur in definite patches, but no hair has intermediate colour of red and white.

Genotype	Phenotype
C^RC^R	Red
C^RC^w	Roan
C^wC^w	White

2.(i) In human, MN blood group antigens are good example.

Allele L^M for M-type antigen, L^N for N-type antigen. The heterozygote (L^ML^N) expresses the characteristics of both M and N antigens (MN blood) that is L^ML^N heterozygote possesses red blood cells that carry both M and N antigens on their surface.

Genotype	Phenotype
L^ML^M	M blood group
L^ML^N	MN blood group
L^NL^N	N blood group

Genotype	Blood type (antigen present)	Reaction with antisera agglutination (+) Non-agglutination (–)	
		Anti-M serum	Anti-N serum
L^ML^M	M (M)	+	–
L^ML^N	MN (M & N)	+	+
L^NL^N	N (N)	–	+

Detection of the M and N antigens on blood cells by agglutination with specific antisera (anti-M and anti-N sera).

Molecular explanation of Codominance

1. In codominanace two alleles appear to contribute independently to the phenotype of the heterozygote, i.e., products result from both alleles in a heterozygote, so that often (but not always) both homozygote phenotypes are observed.
2. Codominance implies independence of allele function.

Example:

L^ML^N individual express both M and N blood group antigen.

Human ABO blood group system is an ideal example of codominance

[C.U.(Zoo H.) 1998]

Alleles that lack dominant and recessive relationship and both have capability to express themselves phenotypically in the heterozygous conditions are termed as codominance.

1. ABO locus is concerned with the production of **glycosyl transferase enzymes** that mediate the synthesis of polysaccharides.
2. That polysaccharides attach with lipid (associated with cell membrane) and form cell surface glycolipid.
   Glycosyl transferase presence in cell which is influenced by ABO locus for glycolipid production.

3. I^A allele of ABO locus-synthesize glycosyl transferase, i.e., αN acetyl **galacto-saminyl transferase.**

I^B allele of ABO locus-synthesize **αD galactosyl transferase.**

There I^AI^A have αN **acetyl galactosaminyl transferase** chain.

I^BI^B have **αD glactosyl transferase** chain.

I^O has lack of **glycosyl transferase** enzyme.

Crossing: A × B [Both homozygous]

$I^A I^A$ × $I^B I^B$

Gametes: I^A I^B

F_1 I^AI^B (Phenotype AB blood group)

1. Here both alleles are expressed simultaneously which are manifested through AB blood group.
2. Biochemically in AB blood group both the glycosyl transferase enzymes are found.

Therefore, ABO blood group is codominant.

I gene is said to be polymorphic:

All three alleles (I^A, I^B and i) are found at appreciable frequencies in human population. Thus, the I gene said to be polymorphic.

Problem 1

In short horn cattle, coat colour may be red, white or roan. Roan is an intermediate phenotype expressed as a mixture of red and white. Red is governed by the genotype C^RC^R, roan C^RC^w and white by C^wC^w.

(a) When roan short horns are crossed among themselves, what genotypic and phenotypic ratios are expected among their progeny?

(b) If red, short horns are crossed with roans and the F_1 progenies are crossed among themselves to produce F_2, what percentage of F_2 will probably be roan?

Solution

(a) Roan × Roan

C^RC^w C^RC^w

Gametes: C^R, C^w, C^R, C^w

♀ \ ♂	C^R	C^w
C^R	C^RC^R	C^RC^w
C^w	C^RC^w	C^wC^w

Genotype	Genotypic ratio	Phenotype	Phenotypic ratio
C^RC^R	1	Red	1
C^RC^w } C^RC^w	2	Roan } Roan	2
C^wC^w	1	White	1

(b) Red × Roan

C^RC^R C^RC^w

Gametes: C^R C^R, C^w

F_1 = C^RC^R C^RC^w

Ratio = 1 : 1 (Red) (Roan)

i.e. Red × Red, Red × Roan

Roan × Roan

$C^RC^R \times C^RC^R$, $C^RC^R \times C^RC^w$, $C^RC^w \times C^RC^w$

There are three types of mating possible for the production of F_2, viz. Red × Red, Red × Roan, Roan × Roan. $F_2 = F_1 \times F_1$

Mating table:

	C^RC^R♀ $(\frac{1}{2})$	$(\frac{1}{2})$ C^RC^w♀
$(\frac{1}{2})$ C^RC^R	C^RC^R♂ × C^RC^R♀ $(\frac{1}{2} \times \frac{1}{2})$	C^RC^w♀ × C^RC^R♂ $(\frac{1}{2} \times \frac{1}{2})$
$(\frac{1}{2})$ C^RC^w♂	C^RC^w♂ × C^RC^R♀ $(\frac{1}{2} \times \frac{1}{2})$	C^RC^w♂ × C^RC^w♀ $(\frac{1}{2} \times \frac{1}{2})$

1. The mating red × red ($C^RC^R \times C^RC^R$) produces only red (C^RC^R) progeny, i.e., $\frac{1}{2} \times \frac{1}{2} = \frac{1}{4}$. Therefore, $\frac{1}{4}$ of all the F_2 should be red.
2. The mating red × roan ($C^RC^R \times C^RC^w$) are expected to produce red and roan progeny, i.e., $\frac{1}{2} \times \frac{1}{2} = \frac{1}{4}$. Therefore, $\frac{1}{4}$ of all the F_2 should be red and $\frac{1}{4}$ should be roan from this cross.
3. The mating roan × roan ($C^RC^w \times C^RC^w$) produces red (C^RC^R), roan (C^RC^w) and white (C^wC^w) progeny. The mating type constitutes $\frac{1}{2} \times \frac{1}{2} = \frac{1}{4}$ of all crosses. Therefore, F_2 should be Red $\frac{1}{4} \times \frac{1}{4} = \frac{1}{16}$, Roan $= \frac{1}{4} \times \frac{1}{2} = \frac{1}{8}$ and White $\frac{1}{4} \times \frac{1}{4} = \frac{1}{16}$.

The F_2 contributions from all three types of mating:

Result:

Types of mating	Frequency of mating	F_2 progeny		
		Red	Roan	White
Red × Red	$\frac{1}{4}$	$\frac{1}{4}$	0	0
Red × Roan	$\frac{1}{2}$	$\frac{1}{4}$	$\frac{1}{4}$	0
Roan × Roan	$\frac{1}{4}$	$\frac{1}{16}$	$\frac{1}{8}$	$\frac{1}{16}$
Totals		$\frac{9}{16}$	$\frac{3}{8}$	$\frac{1}{16}$

Roan should be $\frac{3}{8} \times 100 = 37.5\%$.

Problem 2

In a ICMR laboratory in Delhi, a scientist has discovered a new blood typing system for human beings. The system involves two antigens, P and Q, each determined by different allele of a gene named N. The allele for these antigens are about equally frequent at the general population. If he N^P and N^Q alleles are codominant, what antigens should be detected in the blood of N^PN^Q heterozygotes?

Solution

Both P and Q antigens should be detected because codominance implies that both the alleles will be expressed in heterozygotes.

Problem 3

In short horn cattle, the heterozygous condition of the alleles for red colour (C^R) and white coat colour (C^W) is roan coat colour. If two roan cattle is mated what proportion of the progeny will resemble their parents in coat colour?

Solution

(a) Roan × Roan

C^RC^w C^RC^w

Gametes: C^R, C^w C^R, C^w

♀ \ ♂	C^R	C^w
C^R	C^RC^R	C^RC^w
C^w	C^RC^w	C^wC^w

Genotype	Genotypic ratio	Phenotype	Phenotypic ratio	
C^RC^R	1	Red	1	$\frac{1}{4}$
C^RC^w	2	Roan	2	$\frac{1}{2}$
C^RC^w		Roan		
C^wC^w	1	White	1	$\frac{1}{4}$

Therefore, half $(\frac{1}{2})$ of the progeny resemble the parents in coat colour.

Contrast Incomplete dominance and Codominance

Incomplete dominance	Codominance
1. It can be viewed more as a **quantitative** phenomenon where heterozygote exhibits an intermediate phenotype between the limits set by the homozygotes.	1. It can be viewed more as a **qualitative** manner where both of the alleles in the homozygote are expressed.
2. In incomplete dominance, only one allele in a heterozygote is expressed to produce a product.	2. In codominance, products result from both alleles in a heterozygote, so that often (but not always) both homozygote phenotypes are observed.
3. There is partial dominance of one allele and hence an intermediate phenotype.	3. Both alleles are equally dominant and hence both express themselves fully.
Example: Plumage colour in chicken. Black (BB) × white (bb) results bluish gray (Bb) plumage.	*Example*: In human M-N blood group system MM × NN-results MN blood group type.

OVERDOMINANCE

(SUPERDOMINANCE, HETERODOMINANCE)

The phenomenon in which a heterozygote has greater reproductive success or fitness may or exceed the phemotypic measurement compared to either of the corresponding homozygotes is called **overdominance** or **heterozygote superiority** or **heterozygote advantage.**

1. A well-documented example involves a human allele that causes sickle cell disease in homozygous individuals.
2. It is an autosomal recessive disorders.
3. Most individuals carry the Hb^A allele and make haemoglutin A. Their genotype is Hb^A/Hb^A.
4. Individual affected with sickle cell anaemia are homozygous for Hb^S allele and produce only haemoglobin S. Their genotype Hb^S/Hb^S. This causes their red blood cells to deform into a sickle shape under condition of low oxygen concentration. For this reasons Hb^S/Hb^S individual has usually shortened life span compared to normal Hb^A/Hb^A individual.
5. However, red blood cells of heterozygoes Hb^A/Hb^S are likely to rupture when infected by *Plasmodium* parasite, thereby preventing the parasite from propagating. People who are heterozygous have better resistance to malaria compared to homozygous Hb^A/Hb^A.
6. Therefore, even though homozygous Hb^S/Hb^S condition is detrimental, the greater survival of the heterozygotes has selected for the presence of Hb^S allele within population where malaria is prevalent.

Heterozygous × Heterozygous

$\frac{Hb^A}{Hb^S}$ $\frac{Hb^A}{Hb^S}$

Gametes: Hb^A, Hb^S Hb^A, Hb^S

F_1

♀ \ ♂	Hb^A	Hb^S
Hb^A	$\frac{Hb^A}{Hb^A}$	$\frac{Hb^A}{Hb^S}$
Hb^S	$\frac{Hb^A}{Hb^S}$	$\frac{Hb^S}{Hb^S}$

Result:

Genotype	Genotypic frequency	Phenotype	Phenotypic frequency
$\frac{Hb^A}{Hb^A}$	1	Normal	$1(\frac{1}{4})$
$\frac{Hb^A}{Hb^S}$, $\frac{Hb^A}{Hb^S}$	2	Sickle cell trait, Sickle cell trait	$2(\frac{2}{4})$
$\frac{Hb^S}{Hb^S}$	1	Sickle cell anemia	$1(\frac{1}{4})$

Molecular explanation

There are three possible explanations for overdominance at molecular level.

1. **First molecular explanation**

   There are two alleles that produce proteins with slightly different amino acid sequences.

   Pathogen can succesfully propagate.

   (A, A_1) Normal homozygote (Sensitive to infection).

   Pathogen cannot successfully propagate.

   (A, A_2) Heterozygote (resistant to infection).

2. **Second molecular explanation**

   (i) It is related to subunit composition of protein.

   (ii) In some cases, proteins function as a complex of multiple subunits; each subunit is composed of one polypeptide.

   (iii) A protein composed of two subunits is called **dimer.** When both the subunits are encoded by the same gene, the protein is a **homodimer.**

   (iv) Here a gene exists in two alleles that encode polypeptides designated A_1 and A_2. Homozygous individuals can produce only A_1A_1 or A_2A_2 homodimer whereas heterozygous can produce A_1A_2 homodimer. Thus, heterozygous can produce three forms of homodimers and homozygous can produce only one.

(v) For same proteins A_1A_2 homodimers may have better functional activity, because they are more stable or able to function under wide range of conditions.

$Hb^S/Hb^A \longrightarrow$ (A_1A_1) (A_2A_2) (A_1A_2) [Homodimer formation]

3. **Third molecular explanation**

(i) Proteins encoded by each allele exhibit difference in their functional activity.

(ii) Suppose, a gene encodes a metabolic enzyme that can be found in two forms (corresponding to the two alleles) one that functions better at low temperature and other functions optimally at higher temperature.

(iii) The heterozygote, which makes a mixture of both enzymes, may be at an advantage under a wider temperature range than either of the corresponding homozygotes.

(E_1)	(E_2)	
27-32°C	30-37°C	[Variation in functional activity]
(Optimum temperature range)	(Optimum temperature range)	

Problem 1

Sickle cell anaemia shows incomplete dominance (Cellular or Cell shape level) and Codominance (at molecular, i.e., haemoglobin level). Explain.

Solution

1. There are 3 genotypes with different phenotypes:

Genotype	Phenotype
Hb^A/Hb^A	(i) Normal, (ii) They contain one type of haemoglobin, i.e., Hb^A, (iii) Red blood cells are round not sickled.
Hb^S/Hb^A	(i) No anaemia, (ii) They contain both types of haemoglobin, i.e., Hb^A and Hb^S, (iii) Red blood cells are sickle-shaped under abnormally low oxygen concentration.
Hb^S/Hb^S	(i) S ever often fatal anaemia, (ii) They contain one type of haemoglobin, i.e., Hb^S, (iii) Red blood cells are sickle-shaped.

2. The gene pair concerned Hb^A (for haemoglobin A) and Hb^S (for haemoglobins) affects oxygen transport molecule haemoglobin.

3. In **regard to anaemia** the Hb^A allele is dominant. Hb^A/Hb^A is disease-free and homozygous dominant. In regard to blood cell (RBC) shape, it shows **incomplete dominance**, i.e., RBC are sickled shaped. In respect to haemoglobin, it is cis **dominant**. Here both sickle **shape and normal haemoglobin** are present.

4. **The Hb^S allele in homozygous** (Hb^S/Hb^S) condition acts as lethal. It causes severe anaemia, results in death of its bearer.

5. In heterozygous condition (Hb^A/Hb^S) it is carrier, i.e., sickle cell trait, both types are present.

Problem 2

A true breeding male fly with eosin eyes is crossed to a white eyed female that is hetrozygote for wild type (C) and cream allele (c^a). What are the expected proportions of their offspring?

Solution

Eosin eyed male × White eyed of male [Cheterozygote for wildtype (C) and Cream allele (c^a)]

$CC\,X^{w-e}Y$ × $Cc^a\,X^wX^w$

Gametes: CX^{w-e} CY, CX^w c^aX^w

♀ \ ♂	CX^{w-e}	CY
CX^w	$CCX^{w-e}X^w$	CCX^wY
c^aX^w	$Cc^aX^{w-e}X^w$	Cc^aX^wY

Result:

Genotype	Phenotype		Ratio
$CCX^{w-e}X^w$	Light eosin	female	2
$CCX^{w-e}X^w$	Light eosin		
CCX^wY	White	male	2
CcX^wY	White		

Inference: (i) C is dominant, the phenotypic ratio two light eosin female and two white males. Eosin eye, i.e., pale orange colour. (ii) Females that have one eosin allele and one white allele have light eosin eyes. Where as females with two eosin alleles or males with one eosin allele have eosin eyes, which are bit darker than light eosin allele.

LETHAL ALLELES

The phenotypic manifestations of some mutant genes result in the death of the individual either in the prenatal or postnatal period prior to sexual maturity. Such mutant genes are called **lethal genes.**

Characteristics

1. Here genetic defect causes 100% mortality.
2. They are not limited to the X chromosome.
3. They are of more frequent occurrence in the autosome.
4. It involves the interaction of many genes.
5. It might interfere with the development of some vital organs such as heart or it might upset some vital processes, chemical or physiological early in development.

A. Dominant lethal:

1. Individual with dominant lethal dies before they can leave progeny.
2. Therefore, the mutant dominant lethal is removed from the population in the same generation in which it arose.
3. More lethal effects occur in heterozygous condition.

Example: "*Epiloia*"—in humans causes abnormal skin growth, severe mental defects and multiple tumours, Huntington's disease.

B. Recessive lethal:

1. It kills only in homozygotes.
2. It may be two kinds
   (a) One which has **no obvious phenotypic** effect in heterozygous.

Genotype	Phenotype
LL, Ll	Normal viability
ll	Lethal

   (b) One which exhibits a **distinctive phenotype** when heterozygous.

Example: In snapdragons (*Antirrhinum*), the amount of chlorophyll is controlled by an incompletely recessive gene which exhibits a lethal effect when homozygous and a distinctive phenotypic effect when heterozygous.

Genotype	Phenotype
CC	Green (normal, with chlorophyll)
Cc	Golden/Auria (yellowish green with carotenoids)
cc	White (lethal and without chorophyll)

Auria × Auria

Cc × Cc

Gametes: C c × C c

♀ \ ♂	C	c
C	CC	Cc
c	Cc	cc lethal

Result:

Genotype	Genotypic ratio	Phenotype	Phenotypic ratio
CC	1	Green	1
Cc, Cc	2	Auria	2
cc	1	White (lethal)	×

C. Completely lethal:

1. It usually causes death of the zygote, later in embryonic development or after birth.
2. Here no individual of a certain genotype attains sexual maturity. Here lethal genes have a high degree of penetrance and expression.
3. However, in many cases lethal genes permit larger proportions of affected genotyps of survive.
4. Such lethal genes are called *subvital, sub-lethal* or *semilethal.*

D. Semilethal:

Genes that kill fewer than 100%, but more than 50% of their victims are called Semilethal.

- **Subvital:**

Genes that cause less than 50% mortality are known as subvital.

- **Supervital:**

Some mutant alleles enhance the survival of those individuals that carry them in the appropriate genotype as compared to that of wild type allele, such genes are known as supervital.

- **Some lethal genes**

1. 'Yellow' lethal in mice: The yellow gene is a lethal (homozygous), since it causes early death. But it is also a visible since it expresses itself as yellow in heterozygotes.
2. In chickens a certain mutant gene interferes with the development of skeleton. In heterozygotes, they have deformed skeleton and are known as **"creeper fowls"** because of the shortness of their legs and their gait.

   In homozygotes, they have badly deformed skeleton and die before hatching. Therefore, the mutant gene is lethal in homozygote but visible in **heterozygote.**

   Manx individual are heterozygous genotype. Mm × Mm result $\frac{1}{3}$ normal progeny the lack of tail (taillessness) in the *Manx* cat.
3. In *Drosophila,* **curly wing** (upward bendings of wing), **plum** (eye colour) eyes, **Dichaete** (absence of certain bristles and spread apart wings).
4. In human—**congenital ichthyosis** (lethary skin with deep fissure); **Thalassaemia** (Cooley's anaemia), **Amouretic idiocy**, etc.
5. Many genes have alleles that affect rate of migration but are not lethal, these are termed as detrimental alleles.

- **Balanced lethal**

1. Mutations sometime have a dominant visible effect and a recessive lethal effect.
2. In *Drosophila,* '**Beaded**' mutation (indentations in the margin of the wing leaving a bead-like remnant of the margin in between the indentation). A fly with pure beaded $\left(\frac{Bd}{Bd}\right)$ cannot develop because of lethal effects. But hybrid $\left(\frac{+}{Bd}\right)$ can develop.
3. Now a fly which has Beaded (Bd) in one chromosome and ordinary recessive lethal (*l*) at some other locus in the homologous chromosome.
4. We can represent the fly $\frac{Bd+}{+l}$(Bd for beaded and *l* for lethal and + for normal allele). Assume further that **+*l* chromosome contains an inversion which act as crossover suppressor (c)** between Bd and *l*. Then the genotype (formula) of the fly becomes $\frac{Bd+}{+cl}$.
5. $\frac{Bd+}{+cl}$ × $\frac{Bd+}{+cl}$

Gametes: Bd +, + c*l*, Bd +, + c*l*

♀ \ ♂	Bd+	+ c
Bd+	$\frac{Bd+}{Bd+}$	$\frac{+cl}{Bd+}$
+ c*l*	$\frac{+cl}{Bd+}$	$\frac{+cl}{+cl}$

Result:

Genotype	
$\frac{Bd+}{Bd+}$	pure-beaded dies ×
$\frac{+cl}{Bd+}$	survive
$\frac{+cl}{Bd+}$	survive
$\frac{+cl}{+cl}$	dies ×

Comment: Here only the heterozygous offsprings live and stock is constant in composition. It is due to the following:

1. The presence of lethal at different locus in each chromosome.
2. The crossover suppressor prevents crossing-over.
3. Thus, the two lethals are held opposite each other or balanced from generation to the next. We can refer it as **balanced lethal.**

Solved Problems

Problem 1

The Pelzer anomaly of rabbits involves abnormal white blood cell nuclear segmentation. Pelzers are heterozygous (Pp), normal individuals are homozygous (PP). The homozygous recessive genotypes (pp) have grossly deformed skeleton and usually die before or soon after birth. If Pelzers are mated together, what phenotypic ratio is expected in the adult F_2?

Solution

Phenotype	Genotype
Pelzer	Pp
Normal	PP
Grossly deformed skeleton (lethal)	pp

Pelzer × Pelzer

Pp   Pp

Gametes: P p   P p

♀ \ ♂	P	p
P	PP	Pp
p	Pp	pp

Result:

Genotype	Genotypic ratio	Phenotype	Phenotypic ratio
PP	1	Normal	1
Pp, Pp	2	Pelzer	2
pp	1	Lethal	1 ×

Normal 1 and Pelzer 2, i.e. 1 : 2.

Problem 2

Thalassemia is a hereditary disease of the blood of humans resulting in anaemia. Severe anaemia (thalassemia major) is found in homozygotes (T^MT^M) and a milder form of anaemia (thalassemia minor) is found in heterozygotes (T^MT^N). Normal individuals are homozygous T^NT^N. If all the individuals with thalassemia major die before sexual maturity, (a) what proportion of the adult F_1 from marriages of thalassemia minors by minors would be expected to be normal (b) what fraction of the adult F_1 from marriages of minors by normal would be expected to be anaemic?

Solution

Phenotype	Genotype
Thalassaemia minor	T^MT^N
Thalassaemia major	T^MT^M
Normal	T^NT^N

(a) Thalassaemia minor × Thalassaemia minor

T^MT^N T^MT^N

Gametes: T^M, T^N T^M, T^N

♀ \ ♂	T^M	T^N
T^M	T^MT^M	T^MT^N
T^N	T^MT^N	T^NT^N

Result:

Genotype	Genotypic ratio	Phenotype	Phenotypic ratio
T^MT^M	1	Thalassaemia major	1 (die)
T^MT^N, T^MT^N	2	Thalassaemia minor	2
T^NT^N	1	Normal	1

(b) Minor × Normal

T^MT^N T^NT^N

Gametes: T^M, T^N T^N

♀ \ ♂	T^M	T^N
T^N	T^MT^N	T^NT^N

Result:

Genotype	Genotypic ratio	Phenotype	Phenotypic ratio
T^MT^N	1	Minor	1
T^NT^N	1	Normal	1

Anaemic $\frac{1}{2}$ = 50%.

Problem 3

In *Drosophila*, a dominant gene (D) for a phenotype "dichaete" alters the bristles and also makes the wings remain extended from the body while the fly is at rest. It is homozygous lethal. (a) Design a cross between two dichaete (Dd) flies and summarise the expected results. (b) Design a cross between dichaete and wild type and summarise the expected results.

Solution

(a) Dichaete (Dd) × Dichaete (Dd)

Gametes: D d   D d

♀ \ ♂	D	d
D	$\frac{D}{D}$ lethal	$\frac{D}{d}$
d	$\frac{D}{d}$	$\frac{d}{d}$

Result:

Genotype	Genotypic ratio	Phenotype	Phenotypic ratio
$\frac{D}{D}$	1	Lethal	0
$\frac{D}{d}$, $\frac{D}{d}$	2	Dichaete, Dichaete	2
$\frac{d}{d}$	1	Wild type	1

(b) Dichaete × Wild type

$\frac{D}{d}$ $\frac{d}{d}$

Gametes: D, d   d

♀ \ ♂	D	d
d	$\frac{D}{d}$	$\frac{d}{d}$

Result:

Genotype	Genotypic ratio	Phenotype	Phenotypic ratio
$\frac{D}{d}$	1	Dichaete	1
$\frac{d}{d}$	1	Wild type	1

Problem 4

In *Drosophila*, the mutant allele responsible for upward bending of wing curly (cy) is dominant over its wild type alleles (+) and is homozygous lethal. (a) Design a cross between two curly flies and summarise the expected results. (b) Design a cross between curly and wild type flies and summarise the expected results.

Solution

Genotype	Phenotype
$\frac{cy}{+}$	Curly
$\frac{+}{+}$	Wild type

(a) Curly × Curly

$\frac{cy}{+}$ $\frac{cy}{+}$

Gametes: cy +   cy +

♀ \ ♂	cy	+
cy	$\frac{cy}{cy}$ lethal	$\frac{+}{cy}$
+	$\frac{+}{cy}$	$\frac{+}{+}$

Result:

Genotype	Genotypic ratio	Phenotype	Phenotypic ratio
$\frac{cy}{cy}$	1	Lethal	0
$\frac{cy}{+}$, $\frac{cy}{+}$	2	Curly, Curly	2
$\frac{+}{+}$		Wild type	1

(b) Curly × Wild type

$\frac{cy}{+}$ $\frac{+}{+}$

Gametes: cy +   +

♀ \ ♂	cy	+
+	$\frac{cy}{+}$	$\frac{+}{+}$

Result:

Genotype	Genotypic ratio	Phenotype	Phenotypic ratio
$\frac{cy}{+}$	1	Curly	1
$\frac{+}{+}$	1	Wild type	1

Problem 5

In *Drosophila*, two eye-colour mutations brown (bw) and cinnabar (cn) are recessive to the

wild type alleles (+) and responsible for brown and cinnabar colouration respectively. Interestingly, when both mutants are present in homozygous condition, colouration of eye will be white. The mutant allele responsible for upward bending of the wing, curly (cy) is dominant over wild type allele (+) and is homozygous lethal. From mating between curly and white flies the following progenies are obtained:

Curly male: 108, Curly female: 102, White male: 106, White female: 104.

(a) Determine the genotype of the parent and offsprings.

(b) Design the cross.

Solution

Here the phenotypic ratio between curly and white is 1 : 1, i.e., curly 108 + 102 = 210; white 106 + 104 = 210.

Phenotype	Genotype
White	$\frac{bw\ cn+}{bw\ cn+}$
Curly	$\frac{+\ +\ cy}{bw\ cn+}$

Parent: Curly × White

Genotype: $\frac{+\ +\ cy}{bw\ cn+}$ $\frac{bw\ cn+}{bw\ cn+}$

Gametes: $\underline{+\ +\ cy}$ $\underline{bw\ cn+}$; $\underline{bw\ cn+}$

♂ \ ♀	+ + cy	bw cn +
bw cn +	$\frac{+\ +\ cy}{bw\ cn+}$	$\frac{bw\ cn+}{bw\ cn+}$

Result:

Genotype	Genotypic ratio	Phenotype	Phenotypic ratio
$\frac{+\ +\ cy}{bw\ cn+}$	1	curly	1
$\frac{bw\ cn+}{bw\ cn+}$	1	white	1

Problem 6

In *Drosophila,* two eye-colour mutations brown (bw) and cinnabar (cn) are recessive to the respective wild type alleles (+) and responsible for brown and cinnabar colour respectively. When both mutants are present together in homozygous condition eye colour is white. The mutant allele responsible for upward bending of the wing, curly (cy) is dominant over its wild type alleles (+) and is homozygous lethal. From a cross between curly and curly flies the following progenies are recovered:

Curly male: 214, Curly female: 216, White male: 110, White female: 105.

Design the cross and determine the genotype of the parent and offsprings.

Solution

Brown and cinnabar colour genes occupy separate loci on the same strand of the chromosome. Curly males and females are heterozygous $\left(\frac{+}{cy}\right)$ in genetic configuration. White males and females are homozygous for brown and cinnabar and recessive eye colour. Here curly trait 214 + 216 = 430 and white trait 110 + 105 = 215. The phenotypic ratio is 430 : 215 = 2 : 1.

$$\text{White} = \frac{bw\ cn\ cy}{bw\ cn+} \qquad \text{Curly} = \frac{+\ +\ cy}{bw\ cn+}$$

Parents: Curly $\frac{+\ +\ cy}{bw\ cn+}$ × Curly $\frac{+\ +\ cy}{bw\ cn+}$

Gametes: $\underline{+\ +\ cy}$, $\underline{bw\ cn+}$; $\underline{+\ +\ cy}$, $\underline{bw\ cn+}$

♂ \ ♀	+ + cy	bw cn +
+ + cy	$\frac{+\ +\ cy}{+\ +\ cy}$ lethal	$\frac{bw\ cn+}{+\ +\ cy}$
bw cn +	$\frac{+\ +\ cy}{bw\ cn+}$	$\frac{bw\ cn+}{bw\ cn+}$

Result:

Genotype	Genotypic ratio	Phenotype	Phenotypic ratio
$\frac{bw\ cn+}{+\ +\ cy}$, $\frac{bw\ cn+}{+\ +\ cy}$	2	Curly Curly	2
$\frac{bw\ cn+}{bw\ cn+}$	1	White	1

Problem 7

Two fruit flies of wild type phenotype were crossed, and in the progeny there were 202 females and 98 males.

(a) What is unusual about this result?

(b) Provide a genetic explanation.

(c) Provide a test of your hypothesis.

Solution

(a) The sex ratio is expected to be 1:1.

(b) The female parent was heterozygous for an X-linked lethal allele.

Female (wild) Male (wild)

$$♀\frac{\ell}{+} \quad \times \quad \frac{+}{\rightharpoondown}$$

Gametes: $\underline{\ell}$ $\underline{+}$ $\underline{+}$ $\rightharpoondown$

♂ \ ♀	ℓ	+
$\underline{+}$	$\frac{\ell}{+}$	$\frac{+}{+}$
$\rightharpoondown$	$\frac{\ell}{\rightharpoondown}$	$\frac{+}{\rightharpoondown}$

Result:

Genotype	Genotypic ratio	Phenotype	Phenotypic ratio
$\frac{\ell}{+}$	1	Wild type } Female	2
$\frac{+}{+}$	1	Wild type } Female	
$\frac{\ell}{\rightharpoondown}$	(dead)	(Male)	1
$\frac{+}{\rightharpoondown}$	1	Wild type male	

(c) (i) Half of the female progeny should be heterozygous for the lethal allele and half should be homozygous for the non-lethal (wild type) allele.

(ii) Individually mate the F_1 females and determine the sex ratio of their progeny are as follows:

If $(♀)\frac{\ell}{+} \times \frac{+}{\rightharpoondown}$ (♂) → results will be 2 : 1

If $(♀)\frac{+}{+} \times \frac{+}{\rightharpoondown}$ (♂) → results will be 1 : 1.

Problem 8

(a) If two *Manx* cats are mated (all *Manx* are genotype Mnx) what proportion of the progeny should be tailless? (b) If a *Manx* male cat mates with female cat with normal trait what proportion of the progeny should have tail?

Solution

(a) This trait is homozygous dominant lethal (tailless).

Manx × *Manx*

$$\frac{M}{m} \qquad \frac{M}{m}$$

Gametes: $\underline{M}$ $\underline{m}$ $\underline{M}$ $\underline{m}$

♂ \ ♀	$\underline{M}$	$\underline{m}$
$\underline{M}$	$\frac{M}{M}$	$\frac{m}{M}$
$\underline{m}$	$\frac{M}{m}$	$\frac{m}{m}$

Result:

Genotype	Genotypic ratio	Phenotype	Phenotypic ratio
$\frac{M}{M}$	1	Tailless (lethal)	1
$\frac{M}{m}$ }	2	Tailless }	2
$\frac{M}{m}$ }		Tailless }	
$\frac{m}{m}$	1	Tailless (normal)	1

Tailless progeny will be $\frac{2}{3}$.

(b) Male *Manx* × Female with normal tail

$$\frac{M}{m} \qquad\qquad \frac{m}{m}$$

Gametes: $\underline{M}$, $\underline{m}$ $\underline{m}$

♂ \ ♀	$\underline{M}$	$\underline{m}$
$\underline{m}$	$\frac{M}{m}$	$\frac{m}{m}$

Tailed progeny = $\frac{1}{2}$

Result:

Genotype	Phenotype
$\frac{M}{m}$	*Manx* (tailless)
$\frac{m}{m}$	Normal tail

Problem 9

In the genetics laboratory, a researcher mated two yellow heterozygous mice and obtained the following progeny 56 yellow and 31 wild type. Are these data consistent with 3 : 1 ratio of Mendelian dominant gene inheritances?

Solution

It is not consistent with 3 : 1 ratio, because yellow gene is dominant and homozygous lethal. Therefore, crossing of two heterozygous yellow results:

Yellow × Yellow

$$\frac{Y}{y} \qquad \frac{Y}{y}$$

Gametes: $\underline{Y}$ $\underline{y}$ $\underline{Y}$ $\underline{y}$

♂ \ ♀	$\underline{Y}$	$\underline{y}$
$\underline{Y}$	$\frac{Y}{y}$	$\frac{Y}{y}$
$\underline{y}$	$\frac{Y}{y}$	$\frac{y}{y}$

Result:

Genotype	Genotypic ratio	Phenotype	Phenotypic ratio
$\frac{Y}{y}$	1	Yellow lethal	×
$\frac{Y}{y}$, $\frac{Y}{y}$	2	Yellow	2
$\frac{y}{y}$	1	Wild type	1

Observed result was yellow & wild (56 + 31= 87)

Expected 2 : 1, i.e., 58 : 29, i.e., 2 : 1

Therefore, it is the result of 2 : 1.

Problem 10

Heterozygous $\frac{CP}{cp}$ chicken expresses a condition called creeper, in which the leg and wing bones are shorter than normal $\left(\frac{cp}{cp}\right)$. The dominant CP allele is lethal when homozygous. Two alleles of an independently segregating gene determine white (W) versus yellow (w) skin colour. From matings between chickens heterozygous for both these gene. What phenotypic classes will be represented among the viable progeny, and what are their expected relative frequencies?

Solution

Creeper white (heterozygous) × Creeper white (heterozygous)

$\frac{CP}{cp}\ \frac{W}{w}$ $\frac{CP}{cp}\ \frac{W}{w}$

Gametes: (i) $\underline{CP}\ \underline{W}$, $\underline{CP}\ \underline{w}$, $\underline{cp}\ \underline{W}$, $\underline{cp}\ \underline{w}$

(ii) $\underline{CP}\ \underline{W}$, $\underline{CP}\ \underline{w}$, $\underline{cp}\ \underline{W}$, $\underline{cp}\ \underline{w}$

♂ \ ♀	$\underline{CP}\ \underline{W}$	$\underline{CP}\ \underline{w}$	$\underline{cp}\ \underline{W}$	$\underline{cp}\ \underline{w}$
$\underline{CP}\ \underline{W}$	$\frac{CP}{CP}\ \frac{W}{W}$	$\frac{CP}{CP}\ \frac{w}{W}$	$\frac{cp}{CP}\ \frac{W}{W}$	$\frac{cp}{CP}\ \frac{w}{W}$
$\underline{CP}\ \underline{w}$	$\frac{CP}{CP}\ \frac{W}{w}$	$\frac{CP}{CP}\ \frac{w}{w}$	$\frac{cp}{CP}\ \frac{W}{w}$	$\frac{cp}{CP}\ \frac{w}{w}$
$\underline{cp}\ \underline{W}$	$\frac{CP}{cp}\ \frac{W}{W}$	$\frac{CP}{cp}\ \frac{w}{W}$	$\frac{cp}{cp}\ \frac{W}{W}$	$\frac{cp}{CP}\ \frac{w}{W}$
$\underline{cp}\ \underline{w}$	$\frac{CP}{cp}\ \frac{W}{w}$	$\frac{CP}{CP}\ \frac{w}{w}$	$\frac{cp}{cp}\ \frac{W}{w}$	$\frac{cp}{cp}\ \frac{w}{w}$

Result:

Genotype	Genotypic ratio	Phenotype	Phenotypic ratio
1. $\frac{CP}{CP}\ \frac{W}{W}$	1		
2. $\frac{CP}{CP}\ \frac{W}{w}$, 3. $\frac{CP}{CP}\ \frac{W}{w}$	2	Creeper white lethal	(lost)
4. $\frac{CP}{CP}\ \frac{w}{W}$	1		
5. $\frac{CP}{cp}\ \frac{W}{W}$	1		
6. $\frac{CP}{cp}\ \frac{W}{W}$, 7. $\frac{CP}{cp}\ \frac{W}{w}$	2	Creeper white	5
8. $\frac{CP}{cp}\ \frac{W}{w}$			
9. $\frac{CP}{cp}\ \frac{W}{w}$	1		
10. $\frac{CP}{cp}\ \frac{W}{W}$	1	Creeper white	3
11. $\frac{CP}{cp}\ \frac{W}{W}$, 12. $\frac{CP}{cp}\ \frac{W}{W}$	2		
13. $\frac{cp}{cp}\ \frac{W}{W}$	1	Normal white	3
14. $\frac{cp}{cp}\ \frac{W}{w}$, 15. $\frac{cp}{cp}\ \frac{W}{w}$	2	Normal white, Normal white	
16. $\frac{cp}{cp}\ \frac{w}{w}$	1	Normal yellow	1

Problem 11

***Huntington's* disease is a rare neurodegenerative human disease determined by a dominant allele, HD. The disorder is usually manifested after the age of 45. A young man has learned that his father has developed the disease.**

(a) What is the probability that the young man will later develop the disorder?

(b) What is the probability that a child of the young man carries the HD allele.

Solution

(a) This trait is rare and it is assumed that the father is heterozygous $\frac{HD}{hd}$. $\frac{1}{2}$ of his gamete contain is HD allele. So the probability is

$\frac{1}{2}$ that his son will receive the allele and will develop the disorder.

(b) We do not know whether the man is heterozygous $\frac{Hd}{HD}$ or homogyzous $\frac{HD}{HD}$; therefore, $\frac{1}{2}$ of his gamete will carry HD allele. So the overall probability that his child will carry HD allele is $\frac{1}{2} \times \frac{1}{2} = 25\%$.

Problem 12

An allele in *Drosophila* produces a "star eye" trait in the heterozygous individual. However, stareye allele is lethal in homozygotes. What would be the ratio and phenotypes of surviving flies of star eyed flies were crossed to each other.

Solution

$$\frac{Str}{+} \times \frac{Str}{+}$$

Gametes: $\underline{Str}$ $\underline{+}$ $\underline{Str}$ $\underline{+}$

♂ \ ♀	Str	+
Str	$\frac{Str}{Str}$	$\frac{+}{Str}$
+	$\frac{+}{Str}$	$\frac{+}{+}$

Result:

	Genotype	Phenotype	Frequency
1.	$\frac{Str}{Str}$	Lethal (l)	lost
2.	$\frac{+}{Str}$	Star eye	2
3.	$\frac{+}{Str}$	Star eye	
4.	$\frac{+}{+}$	Normal	1

- **Huntington Disease**
  1. It is an autosomal dominant disorder, which affects the central nervous system of both sexes and appears after the age of 40.
  2. The patients with H.D. suffer from involuntary movements such as twitching of the limbs and face.
  3. The gene (huntington) responsible for H.D. is located on **chromosome 4**. The gene is expressed in many cells in the C.N.S.
  4. The gene contains a repeated **CAG sequence** near the 5′ end. In normal individual, the sequence is repeated **26 times** or less. However in individual with H.D. the repeat sequence is expanded and occurs more than **40 times.**

Phenotype		Number of CAG repeat
(i) Normal	=	11 to 34 average 18
(ii) Mutant I	=	40 to 86
(iii) Muant II	=	33 to 49

  5. All affected persons are heterozygous (Hd), homozygous dominant (HD) are rare. Non-affected persons have hd allele.

- **Congenital Ichthyosis**
  1. It occurs in homozygous condition for its recessive lethal gene.
  2. At birth children are afflicted with this disease. They have dry, thickened scaly, crusted leathery skin with deep fissures; fissures lead to bleeding, infection and death.

- **Genocopy and Phenocopy:**

Genocopy: The term genocopy is used when the same phenotype is observed due to different genetic loci. When the same phenotype is observed because of environmental factor, it is referred to as phenocopy.

Phenocopy: An organism whose phenotype (but not genotype) has been changed by the environment to resemble the phenotype of a different (mutant) organism.

(i) A phenocopy is not a type of mutation. It is an environmentally induced, nonhereditary phenotypic modification, that matches a similar phenotype (mutant) which is determined by genetic factor while mutations are inherited.

(ii) Example: Tetraptera in *Drosophila.* This fly has single pair of wings and second pair is reduced to knob-like structure known as **haltere (balancing organ)**. The recessive gene in tetraptera in homozygous condition leads to the development of the balancers into wing-like structure when a wild type embryo is treated with **heat shock** or ether a phenocopy of tetrapteron condition is produced.

MULTIPLE ALLELES

Two or more kinds of genes occupying the same locus in a given pair of homologous chromosome or arisen as a result of mutation from the normal gene and affecting the same character and mapping together, are termed as **multiple alleles.**

Characteristics

1. They are always at the **same locus** of the chromosome.
2. There is **no crossing-over** within the alleles of a same multiple allelic series.
3. The wild type allele is usually **dominant over** all others at the same locus.
4. When the two of the mutant-multiple alleles are crossed, the phenotype is **mutant type** and not the **wild type.**
5. Multiple alleles always influence the **same character.**
6. It controls the same character but each of them is characterized by **different manifestation.**

Example 1 (*Drosophila* eye colour)

(i) w^+ allele results in wild type brick red eyes and the recessive 'w' allele when homozygous result in white eyes. w^e is the recessive mutant eosin (reddish orange) allele.

(ii) Like white (w), eosin is x linked recessive to the wild type colour.

(iii)

(a) When eosin eyed female is crossed with white eyed male, all F_1 females are eosin eyed.

Eosin(♀) × White(♂) F_1

$\frac{w^e}{w^e} \times \frac{w^e}{\rightharpoonup} \longrightarrow \frac{w^e}{w}$ (females) eosin eyes

Here w^e is dominant over w.

(b) F_1 females (eosin eyed) $\frac{w^e}{w}$ is crossed with red eyed male ($\frac{w^+}{\rightharpoonup}$), all the female progenies are heterozygous and red eyes, because they contain w^+ allele. Half the male progeny are eosin ($\frac{w^e}{\rightharpoonup}$) eyed and other half are white eyed ($\frac{w}{\rightharpoonup}$).

Eosin(♀) × Red(♂) F_1

$\frac{w^e}{w} \times \frac{w^+}{\rightharpoonup} \rightarrow$ Females ($\frac{w^+}{w}$ and $\frac{w^+}{w^e}$) are red,

Males ($\frac{w}{\rightharpoonup}$) white & ($\frac{w^e}{\rightharpoonup}$) eosin

Sturtevant (1913) concluded that red is dominant to eosin and white, eosin is dominant to white, but recessive to wild type (Red). He also concluded that eosin and white are both mutant allele of a single gene. In other words, they are multiple alleles of the white eye gene.

(iv) Many known alleles of white genes are distinguisable because they produce different eye colour ranging from red to white. Eye colour phenotypes depend on the amount of pigment deposited in the eye cells. Mutant white (w) has least amount of pigment.

Example 2 (Rabbit coat colour)

(i) The coat colour of rabbit is referred to as 'agouti' or full colour, i.e., wild type. The gene for this colour is represented by capital letter 'C'.

(ii) The coat colour lacks yellow pigment and appears silvery grey, i.e., Chinchilla. The gene for Chinchilla is represented by C^{ch}.

(iii) The Himalayan type coat colour is white. The gene for it is represented by C^h.

(iv) Coat colour lacking the pigment is called albino. The gene for it is represented by c.

The dominance hierarchy may be as follows: $C > C^{ch} > C^h > c$

Phenotypes	Possible genotypes
Agouti	CC, Cc^{ch}, Cc^h, Cc
Chinchilla	$C^{ch}C^{ch}$, $C^{ch}c$
Himalayan	C^hc^h, c^hc
Albino	cc

Example 3 (Peppered moth: *Biston betularia*)

(i) It has three alleles for body colouration.

(ii) The pale typical allele (m) is recessive. The second allele, the darkly mottled insularia (M′) is dominant to typical (pale). The third allele is nearly black melanic (M) is dominant to both of the other two alleles.

(iii) There are six genotypes but only three phenotypes.

(iv) The dominance hierarchy may be as follows: M > M′ > m

Possible genotypes	Phenotype
MM, MM′, Mm	Melanic
M′M′, M′m	Insularia
mm	Typical (Pale)

Example 4 (ABO blood group system of human)

(i) Here the allele I^A for the A antigen is codominant to allele I^B for B antigen. Both I^A and I^B are completely dominant over allele i which fails to specify any detectable antigenic structure.

(ii) The hierarchy of dominance relationship is as follows: $I^A = I^B > i$

Genotype	Phenotype (Blood group)
I^AI^A, I^Ai	A
I^BI^B, I^Bi	B
I^AI^B	AB
ii	O

- **Multiple allele and Molecular genetics**

(i) The base pair sequence of a gene specifies the amino acid sequence of a protein and the function of a protein depends on its amino acid sequence.

(ii) An amino acid change at one or many places in the protein affects its function adversely and the position and type of the change determine the extent of loss of function of the protein.

Genotype number of Multiple alleles

Number of alleles	Kinds of genotypes	Kinds of homozygotes	Kinds of heterozygotes
n	$n\frac{(n+1)}{2}$	n	$n\frac{(n-1)}{2}$
1	1	1	0
2	3	2	1
3	6	3	3
4	10	4	6
5	15	5	10

ABO blood group is a multiple allelic system
C.U.(Zoo H.) 1991

Two or more kinds of genes occupying the same locus in a given pair of homologous chromosome or arisen as a result of mutation from the normal gene and affecting the same characteristic and mapping together are termed as multiple alleles.

ABO blood group

1. Land Steiner (1900) observed that agglutination (clumping of RBC) is due to antigen and antibody reaction.
2. Antibody is present in serum and antigen on the surface of RBC or in the body fluid such as saliva.
3. Antigen in saliva is called secretors (Sc). Secretor (Sc) is dominant over nonsecretors (sc). It is controlled by different gene pair than the ABO blood group gene.

Blood group	Antigen	Antibody	Agglutination
A	A	b	B, AB
B	B	a	A, AB
AB	AB	×	×
O	×	ab	A, B, AB

4. O is universal donor as it has no **antigen** and AB is universal recipient as it has no **antibody**.

Genetics of ABO blood group

1. ABO blood group represents a multiple allelic series. Its genes are located at the same locus (long arm of chromosome 9) and affect the same phenotypic character (formation of antibody).
2. Genes of A and B are dominant, i.e., I^A and I^B and of 'O' type gene I^O is recessive to both A and B types.
3. When A and B present together, i.e., I^A and I^B, both the genes are expressed simultaneously, i.e., blood contains both A and B antigens, i.e. **codominant**.

Biochemistry of ABO blood group

1. A, B, O substances have been isolated and their structures are determined. They contain

H-substance, i.e., precursor of A and B substances and the blood group substances of O type.

2. Precursor H-substances (**galactose and N-acetyl glucosamine**) is converted into H-substance (**fucose-gal-N-acetyl glucosamine**) by adding **fucose** with the help of **fucosyl transferase** enzyme which is synthesized by H-gene.
3. A-blood group has A-substance (Fucose-gal-Ac-Glu-NH).
   |
   (Ac-Gal-NH)

   It is produced from H-substances by adding ***N-acetyl galactosamine*** with the help of ***N-acetyl* galactosaminyl transferase** enzyme which is synthesized by I^A allele.
4. B-blood group has B-substance (Fucose-gal-Ac-Glu-NH).
   |
   Gal

   It is produced from H-substances by adding **galactose** with the help of **galactosyl transferase** enzyme which is synthesized by I^B allele.
5. O-blood group has only H-substance. I^O allele has no such terminal transferase enzyme for incorporation of such group to H-substances. This is why 'I^O' is called **null allele.**

Mendelian inheritance

ABO blood group genes are inherited according to the law of Mendelian segregation.

Phenotype	Genotype
A	I^AI^A, I^AI^O
B	I^BI^B, I^BI^O
AB	I^AI^B
O	I^OI^O

(i) Table indicates that there are 4 possible phenotypes and 6 possible genotypes for ABO locus.

(ii) Therefore, in ABO locus there are three possible alleles. One person at a time can carry only two of the possible alleles, such gene which has multiple different (more than two) alleles are called **multiple alleles.**

Pattern of inheritance

Cross	Result	Phenotype
i) A × O = $I^AI^O \times I^OI^O$ =	I^AI^O, I^OI^O	A & O
ii) B × O = $I^BI^O \times I^OI^O$ =	I^BI^O, I^OI^O	B & O
iii) A × O = $I^AI^A \times I^OI^O$ =	I^AI^O	A (only)
iv) B × O = $I^BI^B \times I^OI^O$ =	I^BI^O	B (only)
v) A × A = $I^AI^O \times I^AI^O$ =	I^AI^A, I^AI^O, I^OI^O	A:O = 3:1
vi) B × B = $I^BI^O \times I^BI^O$ =	I^BI^B, I^BI^O, I^OI^O	B:O = 3:1
vii) AB × AB = $I^AI^B \times I^AI^B$ =	I^AI^A, I^AI^B, I^BI^B	A:AB:B=1:2:1
viii) O × O = $I^OI^O \times I^OI^O$ =	I^OI^O	O (only)

(i) Table indicates in heterozygous condition of A × O and B × O results in two different phenotypes. It reveals O is recessive.

(ii) A × O and B × O (A and B homozygous) results show all phenotypes are A and B respectively. It indicates O is recessive.

(iii) A × A (heterozygous) and B × B (heterozygous) result in 3 : 1 ratio.

(iv) AB × AB results in 1 : 2 : 1 ratio. It indicates A and B are codominant.

(v) O × O results in O.

All the crosses reveal that O is recessive to both A and B. A and B are codominant.

Inference

1. I^O is alternative to I^A.
2. I^O is alternative to I^B.
3. I^A and I^B genes are also alleles of the same gene.

The heterozygous of I^AI^B shows codominance and appears as AB blood group.

Mendelian segregation

Marriage between A × A (both heterozygous) results in $\frac{3}{4}$A and $\frac{1}{4}$O.

In case of B × B (both heterozygous) same ratio is observed, i.e., $\frac{3}{4}$B and $\frac{1}{4}$O.

Inference

This indicates

(i) A gene is allele of 'O' and dominant over 'O'.

(ii) B gene is allele of 'O' and dominant over 'O'.

Hence, A is the allele of B. Therefore, A, B and O are three different alleles of the same locus or are multiple alleles.

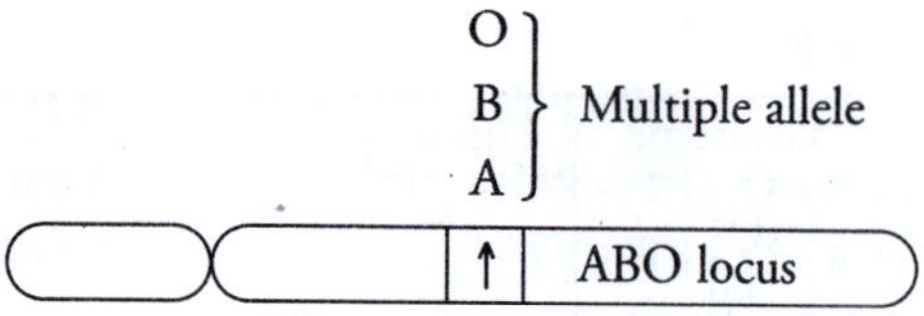

Chromosome '9'

Conclusion

1. 3 gene control ABO blood group (I^A, I^B & I^O).
2. All three genes occupy same locus, i.e., they are multiple alleles.
3. A is dominant over O and B is dominant over O. A and B are codominant.
4. 4 phenotypes and 6 genotypes are possible, i.e., A, B, AB, O and I^AI^B, I^AI^A, I^AI^O, I^BI^B, I^BI^O & I^OI^O.

H-SUBSTANCE AND ABO SYSTEM

The **H-substance** has been isolated and its structure is determined. It is the precursor of both the A and B substances and is the blood group substance found in persons of type O.

Characteristics

1. H-substance itself consists of three sugar molecules—galactose (Gal), N-acetyl glucosamine (AcGluNH), and fucose. These are chemically linked together.
2. H-substance itself is formed by the action of a **fucosyl transferase** which catalyzes the addition of the terminal fucose on to the terminal Gal residue (α 1-2 linkage).
3. The H-locus codes for this **fucosyl transferase.** The h allele of the H-locus codes for an **inactive fucosyl transferase.** The gene 'h' is termed 'amorph' because it produces no demonstrable product. The individual with 'hh' genotype cannot generate H-substance, the precursor of A and B antigens. Thus individual of 'hh' genotype will have red blood cells of type 'O', even though they may possess the enzyme necessary to make the A and B substances.
4. The I^A allele is responsible for an enzyme (N-acetyl galactosaminyl transferase) that can add the terminal sugar N-acetyl galactosamine to the H-substance. The I^B allele is responsible for modified enzyme (galactosyl transferase) that cannot add N-acetyl galactosamine but instead can add terminal galactose.

Heterozygous (I^AI^B) add either one or other entities (sugar) at many sites (substrate) available on the surface of the red blood cells.

AB posseses both enzymes and thus has two oligosaccharide chains, one terminated by GalNac (N-Acetyl glactosamine) and other by Gal.

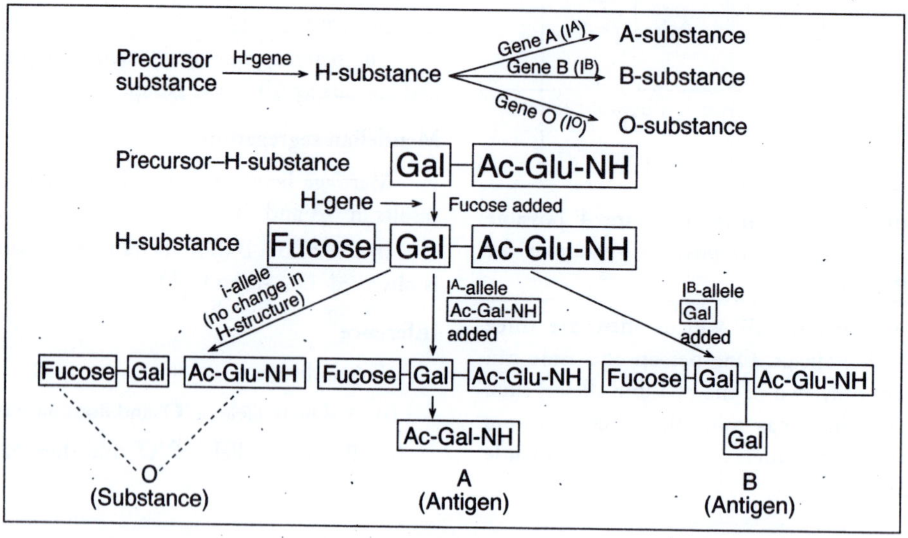

BOMBAY PHENOTYPE

A rare variant of ABO system in which affected individuals do not have A or B antigen and thus, appear as blood type O, even though their genotype may carry unexpressed alleles for the A or B antigen.

It is so named because it was first described in a family from **Bombay metropolitan** (i.e., one case in 13,000 persons).

Characteristics

1. Antigens A and B of A, B and O blood phenotypes are synthesized from a precursor mucopolysaccharide in the presence of dominant alleles H and h.
2. With genotype HH or Hh the precursor is converted into H antigen.
3. In presence of I^A and I^B alleles, H antigen is converted into A and B antigens respectively. Therefore, only individuals with H genotype (HH/Hh) can form A and B antigens.
4. As a result individual with genotypes I^A or I^B allele who are also hh configuration can express the type 'O' phenotype.

$I^A I^B$ Hh × $I^A I^B$ Hh

I^A I^B I^A I^B

♀ \ ♂	I^A	I^B
I^A	I^AI^A	I^AI^B
I^B	I^AI^B	I^BI^B

Result:

Genotype	Phenotype	Phenotypic ratio
I^AI^A	A	$1 = \frac{1}{4}$
I^AI^B	AB	$2 = \frac{2}{4}$
I^AI^B	AB	
I^BI^B	B	$1 = \frac{1}{4}$

Hh × Hh

Gametes: H, h   H, h

♀ \ ♂	H	h
H	HH	Hh
h	Hh	hh

Result:

Genotype	Phenotype	Phenotypic ratio
HH	H	$3 = \frac{3}{4}$
Hh	H	
Hh	H	
hh	h	$1 = \frac{1}{4}$

Considering both gene-pairs together

Of all offsprings	Of all offsprings	Final probabilities
$\frac{1}{4}$A	$\frac{3}{4}$ forms H-substance	$\frac{1}{4} \times \frac{3}{4} = \frac{3}{16}$ = type A
	$\frac{1}{4}$ do not form H-substance	$\frac{1}{4} \times \frac{1}{4} = \frac{1}{16}$ = type O
$\frac{2}{4}$A	$\frac{3}{4}$ forms H-substance	$\frac{3}{4} \times \frac{2}{4} = \frac{6}{16}$ = type AB
	$\frac{1}{4}$ do not form H-substance	$\frac{1}{4} \times \frac{2}{4} = \frac{2}{16}$ = type O
$\frac{1}{4}$B	$\frac{3}{4}$ forms H-substance	$\frac{1}{4} \times \frac{3}{4} = \frac{3}{16}$ = type B
	$\frac{1}{4}$ do not form H-substance	$\frac{1}{4} \times \frac{1}{4} = \frac{1}{16}$ = type O

Group A	Group B	Group AB	Group O
$\frac{3}{16}$	$\frac{3}{16}$	$\frac{6}{16} = \frac{3}{8}$	$\frac{4}{16} = \frac{1}{4}$

- **Molecular explanation of ABO blood group types**
  (i) DNAs of the I^A and I^B alleles were compared; four consistent nucleotide substitutions were found.
  (ii) It is assumed that the changes in the amino acid sequence of **glycosyl transferase** gene products leads to the different modifications of the **H-substance.**
  (iii) Individuals of 'O' blood group type, homozygous for I^OI^O allele lack **glycosyl transferase** activity and fail to modify H-substance.
  (iv) Analysis of the DNA of this allele show one consistent change, i.e., the deletion of a single nucleotide early in the coding sequence causing **frame shift** mutation. As a result polypeptide chain terminates prematurely and produce nonfunctional product.

- **Secretor and non-secretor trait**
  (i) Among some A and B blood group type persons, the antigens A and B, also occur in a aqueous secretion of the eyes (tears), nose and saliva of their body.
  (ii) The persons with this trait is called **secretors.** They produce water soluble antigen.
  (iii) They are homozygous (SS) or heterozygous (Ss)

  The person lacking this trait is called **non-secretors.** They are homozygous (ss).

Solved Problems

Problem 1

In rabbits, C = agouti colour, c^{ch} = chinchilla, c^h = Himalayan and c = albino. The four alleles constitute a multiple allelic series. The agouti C is dominant over the three other alleles, c is recessive to all three other alleles and chinchilla is dominant over Himalayan. Determine the phenotypes of progeny from the following crosses:

(a) CC × cc, (b) Cc^{ch} × Cc, (c) Cc × Cc, (d) Cc^h × c^hc, (e) Cc^h × cc, (f) $c^{ch}c$ × c^hc, (g) Cc^h × Cc, (h) Cc^h × Cc^{ch}, (i) c^hc × cc.

Solution

(a) CC × cc

Gametes: C c

F_1 Cc = (Agouti). All are agouti.

(b) Cc^{ch} × Cc

Gametes: C , c^{ch} C , c

F_1

♂ \ ♀	C	c^{ch}
C	CC	C c^{ch}
c	Cc	$c^{ch}c$

Result:

Genotype	Phenotype	Phenotypic ratio
CC	Agouti	3
Ccch	Agouti	
Cc	Agouti	
$c^{ch}c$	Chinchilla	1

(c) Cc × Cc

Gametes: C, c C, c

F_1

♂ \ ♀	C	
C	CC	Cc
c	Cc	cc

Result:

Genotype	Phenotype	Phenotypic ratio
CC	Agouti	3
Cc	Agouti	
Cc	Agouti	
cc	Albino	1

(d) Cc^h × c^hc

Gametes: C, c^h c^h, c

F_1

♂ \ ♀	C	c^{ch}
c^h	Cc^h	c^hc^h
c	Cc	cc^h

Result:

Genotype	Phenotype	Phenotypic ratio
Cch	Agouti	2
Cc	Agouti	
c^hc^h	Himalayan	2
c^hc	Himalayan	

(e) Cc^h × cc

Gametes: C, c^h c

F_1

♂ \ ♀	C	c^{ch}
c	Cc	c^hc

Result:

Genotype	Phenotype	Phenotypic ratio
Cc	Agouti	1
c^hc	Himalayan	1

(f) $c^{ch}c$ × c^hc

Gametes: c^{ch}, c c^h, c

F_1

♂ \ ♀	c^{ch}	c
c^h	$c^{ch}c^h$	cc^h
c	$c^{ch}c$	cc

Result:

Genotype	Phenotype	Phenotypic ratio
$c^{ch}c^{h}$	Chinchilla	2
$c^{ch}c$	Chinchilla	
$c^{h}c$	Himalayan	1
cc	Albino	1

(g) Cc^{h} × Cc

Gametes: C, c^{h} C, c

F_1

♂ \ ♀	C	c^{h}
C	CC	Cc^{h}
c	Cc	$c^{h}c$

Result:

Genotype	Phenotype	Phenotypic ratio
CC	Agouti	3
Cc	Agouti	
Cc^{h}	Agouti	
$c^{h}c$	Himalayan	1

(h) Cc^{h} × Cc^{ch}

Gametes: C, c^{h} C, c^{ch}

F_1

♂ \ ♀	C	c^{h}
C	CC	c^{ch}
c^{ch}	Cc^{ch}	$c^{ch}c^{h}$

Result:

Genotype	Phenotype	Phenotypic ratio
CC	Agouti	3
Cc^{h}	Agouti	
C^{ch}	Agouti	
$c^{ch}c^{h}$	Chinchilla	1

(i) $c^{h}c$ × cc

Gametes: c^{h}, c c

F_1

♂ \ ♀	c^{h}	c
c	$c^{ch}c$	cc

Result:

Genotype	Phenotype	Phenotypic ratio
$c^{ch}c$	Himalayan	1
cc	Albino	1

Problem 2

The inheritance of coat colours of cattle involves a multiple allelic series with a dominance hierarchy as follows: S > s^{h} > s^{c} > s. The S allele puts a band of white colour around the middle of the animal and is referred to as Dutch belt; the s^{h} allele produces Hereford-type spotting; solid colour is a result of the s^{c} allele, and Holstein-type spotting is due to the s allele. Homozygous Dutch belted males are crossed to Holstein-type spotted females. The F_1 females are crossed to a Hereford type spotted male of genotype $s^{h}s^{c}$. Predict the genotypic and phenotypic frequencies in the progeny.

Solution

Homozygous Dutch belted male × Holstein-type spotted female

SS ss

Gametes: S s

F_1 Ss (Dutch-belted)

Dutch-belted female × Hereford-type male

Ss $s^{h}s^{c}$

Gametes: S, s s^{h}, s^{c}

F_1

♂ \ ♀	S	s
s^{h}	Ss^{h}	$s^{h}s$
s^{c}	Ss^{c}	$s^{c}s$

Result:

Genotype	Phenotype	Phenotypic ratio
Ss^{h}	Dutch	$\frac{2}{4} = \frac{1}{2}$
Ss^{c}	Dutch	
$s^{h}s$	Hereford	$\frac{1}{4}$
$s^{c}s$	Solid	$\frac{1}{4}$

Problem 3

Flower colour in a garden plant is under the control of a gene with multiple allele. The phenotypes of the homozygotes and heterozygotes of this gene are as follows:

- **Homozygotes**

WW	Red
ww	pure white
W^SW^S	white stipped with red
W^PW^P	white with red patches

- **Heterozygotes**

$Ww/Ww^S/Ww^P$	Red
W^PW/W^PW^S	white with red patches
W^sw	white stipped with red

Arrange the alleles in a dominance hierarchy.

Solution

(i) W is dominant to all the other allele, W^P is dominant to W^s and W^s is dominant to w.

(ii) The hierarchy of dominance is as follows:

$W > W^P > W^s > w$.

Problem 4

A man of blood group O marries a woman of blood group A. The wife's father is of blood group O. What are the probabilities that their children will belong to blood group O?

Solution

Blood group	Genotype
O	ii
A	i^Ai^A/i^Ai

(i) A man of blood group O, his genotype is ii.

(ii) He marries a woman of blood group A, her father has blood group O; therefore, her genotype will be i^Ai (as i gene is donated by her father).

	Man	Woman
	O	A
	ii	i^Ai
Gametes:	i	i^A, i

F_1

♂ \ ♀	i^A	i
i	i^Ai	ii

Result:

Genotype	Phenotype	Phenotypic ratio
i^Ai	A	1
ii	O	1

The probability of children belonging to 'O' blood group is 50%.

Problem 5

A female *Biston betularia* moth having typical pale phenotype is mated to male melanic. If half of the progeny are melanic and half are insularia, what are the genotypes of the two parents.

Solution

In this moth three alleles are involved in the development of colouration. Melanic (M) is dominant to insularia (M′) and pale (m); Dominant hierarchy M > M′ > m.

	Female: Typical pale		Male: Melanic
	$\frac{m}{m}$	×	$\frac{M'}{M}$
Gametes:	m		M′, M

F_1

♀ \ ♂	M′	M
m	$\frac{M'}{m}$	$\frac{M}{m}$

Result:

	Genotype	Ratio	Phenotype	Ratio
(i)	$\frac{M'}{m}$	1	Insularia	1 $(\frac{1}{2})$
(ii)	$\frac{M}{m}$	1	Melanic	1 $(\frac{1}{2})$

Problem 6

If two peppered moth a melanic (Mm) and an Insularia (M′m) are mated, what types of F_1 progeny could you expect?

Solution

	Melanic	×	Insularia
	Mm		M′m
Gametes:	M, m		M′, m

F_1

♂ \ ♀	M′	m
M	M′M	Mm
m	M′m	mm

Result:

Genotype	Phenotype	Phenotypic ratio	
MM′	Melanic	2	$\frac{2}{4}=\frac{1}{2}$
Mm	Melanic		
M′m	Insularia 1	1	$\frac{1}{4}$
mm	Pale 1	1	$\frac{1}{4}$

Problem 7

In mice, a series of five alleles determine for colour. In order of dominance, these alleles are A^Y (yellow fur) but homozygous lethal. A^L (agouti with light belly); A^T(agouti) wild type a^t (black tan) and a (black). From the following crosses give the coat colour of the parents and the phenotypic ratios expected among the progeny.

(a) $A^YA^L \times A^YA^L$ (b) $A^Ya \times A^La^t$ (c) $a^ta \times a^Ya$ (d) $A^La^t \times A^LA^L$ (e) $A^LA^L \times A^YA^t$ (f) $A^ta^t \times a^ta$ (g) $a^ta \times aa$ (h) $A^Ya^L \times A^YA^t$

Solution

Parent	Progeny
(a) $A^YA^L \times A^YA^L$ = yellow × yellow	2 yellow and 1 agouti with light belly
(b) $A^Ya \times A^La^t$ = black & tan × yellow	2 yellow, 1 light belly, 1 black and tan
(c) $a^ta \times a^Ya$ = black & tan × yellow	2 yellow, 1 black tan and 1 black
(d) $A^La^t \times A^LA^L$ = light belly × light belly	All light belly
(e) $A^LA^L \times A^YA^t$ = light belly × yellow	1 yellow and 1 light belly
(f) $A^ta^t \times a^ta$ = agouti × black & tan	2 agouti and 2 black tan
(g) $A^YA^L \times A^ta^t$ = yellow × agouti	2 yellow, 2 light belly
(h) $a^ta \times aa$ = black & tan × black	black tan and black
(i) $A^Ya^L \times A^YA^t$ = yellow × yellow	2 yellow, 1 light belly

Problem 8

If a given island population of diploid organisms contain only three alleles of a particular gene (W^1, W^2 and W^3) how many different diploid genotypes are possible in the population? List all possible genotypes of diploids (consider only three given alleles).

Solution

Three alleles are viz. W^1, W^2 and W^3. Therefore the number of genotype will be $\frac{n(n+1)}{2} = \frac{3(3+1)}{2} = 6$ kinds of homozygote will be n = 3 and heterozygote will be $\frac{n(n-1)}{2} = \frac{3(3-1)}{2} = 3$.

Kinds of genotypes	Homozygote	Heterozygote
6	3	3
6+3+3	W^1W^1	W^1W^2
	W^2W^2	W^2W^3
	W^3W^3	W^1W^3

Problem 9

If a gene has three alleles, how many genotypes are possible? (b) If another gene has five alleles, how many homozygotes and how many heterozygotes are possible?

Solution

(a) Here n = 3, i.e., $\frac{n(n+1)}{2} = \frac{3(3+1)}{2} = 6$

(b) n = 5 = Homozygote will be n = 5 and Heterozygote will be $\frac{n(n-1)}{2} = \frac{5(5-1)}{2} = \frac{20}{2} = 10$.

Problem 10

In humans a series of alleles has been associated with the ABO blood type as follows: I^A A type, I^B B type, I^O O type. I^A and I^B are codominant. I^AI^B heterozygotes have AB type blood. I^O is recessive to both I^A and I^B. What phenotypes and ratio might be expected from the following matings:

(a) $I^AI^A \times I^BI^B$, (b) $I^AI^B \times I^OI^O$, (c) $I^AI^O \times I^BI^O$, (d) $I^AI^O \times I^OI^O$, (e) $I^AI^O \times I^AI^O$, and (f) $I^AI^O \times I^AI^B$.

Solution

(a) $I^AI^A \times I^BI^B$

Gametes: I^A I^B

F_1 I^AI^B (AB)—All AB phenotype

(b) $I^AI^B \times I^OI^O$

Gametes: I^A I^B I^O

F_1 I^AI^O (A) I^BI^O (B)

(c) I^AI^O × I^BI^O

Gametes: I^A, I^O I^B, I^O

F_1

♂ \ ♀	I^A	I^O
I^B	I^AI^B	I^BI^O
I^O	I^AI^O	I^OI^O

Result:

Genotype	Phenotype	Phenotypic ratio
I^AI^B	AB	1
I^BI^O	B	1
I^AI^O	A	1
I^OI^O	O	1

(d) I^AI^O × I^OI^O

Gametes: I^A, I^O, I^O

F_1

♂ \ ♀	I^A	I^O
I^O	I^AI^O	I^OI^O

Result:

Genotype	Phenotype	Phenotypic ratio
I^AI^O	A	1
I^OI^O	O	1

(e) I^AI^O × I^AI^O

Gametes: I^A , I^O I^A , I^O

F_1

♂ \ ♀	I^A	I^O
I^A	I^AI^A	I^AI^O
I^O	I^AI^O	I^OI^O

Result:

Genotype	Phenotype	Phenotypic ratio
I^AI^A	A	3
I^AI^O	A	
I^AI^O	A	
I^OI^O	O	1

(f) I^AI^O × I^AI^B

Gametes: I^A , I^O I^A , I^B

F_1

♂ \ ♀	I^A	I^O
I^A	I^AI^A	I^AI^O
I^B	I^AI^B	I^BI^O

Result:

Genotype	Phenotype	Phenotypic ratio
I^AI^A	A	2
I^AI^O	A	
I^AI^B	AB	1
I^BI^O	A	1

Problem 11

In the ABO blood grouping system in human beings, alleles I^A and I^B are codominant, and both are dominant over the i allele. In a paternity dispute, a AB type woman claimed that one of four men, each with different blood types, was the father of her type A child. Which of the following could be the blood type of the father of the child on the basis of the evidence given?

(a) Type—A, (b) Type—B, (c) Type—O, (d) Type—AB.

Solution:

Since the child was type A, it must have received I^A allele from its mother. The other allele in the child is either I^A or i,

Type	Genotype
A	I^AI^A/I^Ai
B	I^BI^B/I^Bi
O	ii
AB	I^AI^B

Sl. No.	Father	Probability	Genotype of the child
1.	I^AI^A/I^Ai	May donate either I^AI/i	I^AI^A/I^Ai
2.	I^Bi	May donate i	
3.	I^AI^B	May donate I^A	
4.	ii	May donate i	

Therefore, all may donate their gene. Their father may be A, B, AB, O blood type.

Problem 12

From the following problems state whether the child mentioned can actually be produced from the marriage. Explain your answer:

(a) An 'O' child from the marriage of two 'A' individuals

(b) An 'O' child from the marriage of an 'A' and a 'B'

(c) An 'AB' child from the marriage of an 'A' and an 'O'

(d) An 'O' child from the marriage of an 'AB' and an 'A'

(e) An 'A' child from the marriage of an 'AB' and a 'B'.

Solution:

(a) A × A

Genotype: I^AI^O I^AI^O

Child: O, i.e. I^OI^O.

Therefore, parents must be A with I^AI^O genetic configuration.

(b) O child has genotype I^OI^O.

Marriage between A and B produces 'O' child. The parents must be I^AI^O and I^BI^O genotype.

(c) An 'AB' child having genotype I^AI^B cannot produce AB child.

(d) An 'O' child has genotype I^OI^O.

Marriage between AB(I^AI^B) & A(I^AI^O/I^AI^A) cannot produce 'O' child.

(e) An 'A' child has genotype I^AI^O/I^AI^A.

Marriage between AB(I^AI^B) & B(I^BI^B/I^BI^O) can produce A(I^AI^O) child.

Problem 13

In a maternity ward of Uttarpara General Hospital, four babies were accidentally mixed up. The ABO types of the four babies are known to be O, A, B and AB. The ABO types of the four sets of parents are determined. Indicate which baby belongs to each set of parents:

(a) AB × O, (b) A × O, (c) A × AB, (d) O × O.

Solution:

(a) AB × O, i.e., $I^AI^B \times I^OI^O$. It produces baby of B (I^BI^O) blood group.

(b) A × O, i.e., $I^AI^O \times I^OI^O$. It produces baby of A(I^AI^O) blood group.

(c) A × AB, i.e., $I^AI^O \times I^AI^B$. It produces baby of AB(I^AI^B) blood group.

(d) O × O, i.e., $I^OI^O \times I^OI^O$. It produces baby of O(I^OI^O) blood group.

Problem 14

A man is suing his wife for divorce on the grounds of infidelity. Their first and second child whom they both claim are blood groups O and AB respectively. The third child whom the man disclaims is blood type B.

(a) Can this information be used to support the man's case?

(b) Another test was made in the MN blood group system. The third child was group M, the man was group N. Can this information be used to support the man's case?

Solution:

(a) (i) The first baby has 'O' blood group. His genotype is ii. Therefore, each parent must have recessive gene 'i'.

(ii) The second child is AB, so the genotype is I^AI^B.

It indicates that one of his parents has I^A allele and other has I^B allele.

Therefore, any four blood groups can appear among the children whose parents are $I^Ai \times I^Bi$.

The information given on ABO blood groups is no use in supporting the man's claim.

(b) The MN blood groups are governed by a pair of codominant alleles where M and N are produced by homozygous genotypes, i.e., L^ML^M and L^NL^N.

N group father must pass L^N allele to his offspring. All offsprings have N antigen on their red blood cells.

As the child is group M, i.e., L^ML^M, the man (L^NL^N = N) could not be the father of a group M child.

Problem 15

A man of blood group B is being used by a woman of blood group A for paternity. The woman's child is blood group O.

(a) Is this man the father of this child? (b) If this man actually is the father of this child, specify the genotypes of both parents. (c) If it is impossible for the group B man to be the father of a type O child, regardless of the mother's genotype, specify his genotype. (d) If the man is of

blood group AB, could he be the father of a group O child?

Solution:

(a) Man of blood group B, i.e. I^BI^B/I^BI^O.
As the child is of 'O' blood group, his genotype is I^OI^O. Therefore, the man could be the father but paternity cannot be proved by blood type.

(b) If the man is father, then the genotype of the parents would be

$I^BI^O \times I^AI^O$ (woman)

(c) If the genotype of the father is I^BI^B, it would be impossible for him to be the father of 'O' child, regardless of mother's genotype.

(d) If the man is of blood group AB, he could not be the father of the child of 'O' blood group.

Problem 16

A man belongs to AB blood group while his wife is B blood group. Predict the probable genotypes and phenotypes of their offsprings.

C.U.(2000 Hons.) Part I, 1991, '97

Solution:

(e) Man (AB) × Wife (B)

I^AI^B $\qquad$ I^BI^B/I^BI^O

Gametes: I^A, I^B $\qquad$ I^B, I^B, I^O

(i)

♀ \ ♂	I^A	I^B
I^B	I^AI^B	I^BI^B

(ii)

♂ \ ♀	I^A	I^B
I^B	I^AI^B	I^BI^B
I^O	I^AI^O	I^BI^O

Result:

(i) If wife is homozygous, i.e. I^BI^B (B blood group)

Genotype	Phenotype	Phenotypic ratio
I^AI^B	AB	1
I^BI^B	B	1

(ii) If wife is heterozygous I^BI^O (B blood group)

Genotype	Phenotype	Phenotypic ratio
I^AI^B	AB	1
I^BI^B	B	
I^BI^O	B	2
I^AI^O	A	1

Problem 17

The A and B antigens in human may be found in water-soluble form in secretions, including saliva, of some individuals (Se/Se and Se/se) but not in others (Se/Se). The population thus contains "secretors" and "nonsecretors".

(a) Determine the proportions of various phenotypes (blood type and ability to secrete) in matings between individuals that are blood type AB and type O, both of whom are Se/se.

(b) How will the results of such matings change if both parents are heterozygous for the gene controlling the synthesis of H substance (Hh)?

Solution:

Secretor = Se $\qquad$ Nonsecretor = se

"AB" secretor × "O" secretor

i^Ai^B Se se $\qquad$ i^Oi^O Se se

Gametes: i^ASe, I^BSe, I^Ase, I^Bse; i^OSe, i^Ose

	I^ASe	I^BSe	I^Ase	I^Bse
I^OSe	I^AI^OSeSe	I^BI^OSeSe	I^AI^OSese	I^BI^OSese
I^Ose	I^AI^OSese	I^BI^OSese	I^AI^Osese	I^BI^Osese

Result:

Genotype	Phenotype	Phenotypic ratio
I^AI^OSeSe	A secretor	
I^AI^OSese	A secretor	$3 = \frac{3}{8}$
I^AI^OSese	A secretor	
I^AI^Osese	A nonsecretor	$1 = \frac{1}{8}$
I^BI^OSeSe	B secretor	
I^BI^OSese	B secretor	$3 = \frac{3}{8}$
I^BI^OSese	B secretor	
I^BI^Osese	B nonsecretor	$1 = \frac{1}{8}$

Problem 18

A woman who is blood type A and a man who is blood type B have a child who is blood type O. What are the genotypes of three individuals? What is the probability of that a second child would again be blood type O? Blood type AB?

Solution

	Woman		Man
	Blood type A	×	Blood type B
	$\frac{iA}{i}$		$\frac{iB}{i}$
Gamete:	iA i		iB i

Child Blood type O (ii) = $\frac{i}{i}$

	Parent	Child
Genotype:	(i) Woman $\frac{iA}{i}$ (A),	(ii) Child $\frac{i}{i}$ (O)
	(iii) Man $\frac{iB}{i}$ (B)	

From the above cross the probability of offspring having blood type are as follows:

	iA	i
iB	$\frac{iA}{iB}$	$\frac{iB}{i}$
i	$\frac{iA}{i}$	$\frac{i}{i}$

	Genotype	Phenotype		Probability
1.	$\frac{iA}{iB}$ =	AB	$= \frac{1}{4}$	25%
2.	$\frac{iA}{i}$ =	A	$= \frac{1}{4}$	25%
3.	$\frac{iB}{i}$ =	B	$= \frac{1}{4}$	25%
4.	$\frac{i}{i}$ =	O	$= \frac{1}{2}$	25%

Therefore, probability of blood type (O) is 25% and blood type AB is 25%.

Problem 19

A woman with blood type B has a child with O type blood. What are the possible genotypes and blood types of the father?

Solution

(i) Here father could not be I^AI^B, I^BI^B or I^AI^A.

(ii) He must contribute the O allele to his child. Therefore genotypically he is i^Ai, i^Bi or ii and lower type A, B or O blood group respectively.

B (i^Bi) × A B O: i^Ai i^Bi ii

i — i

ii = O blood type

Problem 20

A woman with blood type M had a child with MN type blood. She said that a certain man who had blood type MN was the father and he claimed he was not.

(a) With this information can you exclude this man as the father?

(b) Further investigation showed that the woman was blood type B, the child blood type O, and the man blood type AB. Can the man be excluded as the father with this information? Explain your answer.

Solution

(a) Women has blood type M and child has MN blood. Therefore, his father has N or MN blood type. So the man with MN type blood can not be excluded.

(b) Woman was blood type i.e. i^Bi^B or i^Bi and her child has O (ii) blood type. The man blood type AB i.e. i^Ai^B. The child (ii) had neither of the alleles of the man at ABO locus. The man can not able to contribute 'i' allele to the child.

Problem 21

Pigment in mouse fur is only produced when the C allele is present. Individuals of the cc genotype are white. If colour is present, it may be determined by the A, a alleles. AA or Aa results in agouti colour, while aa results in black coats.

(a) What F_1 and F_2 genotypic and phenotypic ratios are from a cross between AACC and aacc mice?

(b) In three crosses between agouti females whose genotypes were unknown and males of the aacc genotypes, the following phenotypic ratios were obtained:

(i)	(ii)	(iii)
8 agouti	**9 agouti**	**4 agouti**
9 white	**10 black**	**5 black**
		10 white

What are the genotypes of these female parents?

Solution:

(a) In a cross

AACC × aacc

Gametes AC × ac

F_1 AaCc — (Agouti)

(i) All the F_1 offsprings are agouti because the C allele allows pigment to be deposited in the hair and when it is, it will be agouti.

(ii) F_2 offspring would have the following simplified genotypes with the corresponding phenotypes:

AACC and others: agouti = $\frac{9}{16}$

Aacc (colourless because cc is epistatic to A) = $\frac{3}{16}$

aaCC and others: black = $\frac{3}{16}$

aacc = $\frac{1}{16}$ (colourless because cc is epistatic to aa).

(iii) The two colourless classes are phenotypically indistinguishable. Therefore, the final ratio is 9 : 3 : 3.

(b) Results of crosses of female agouti (AC) × aacc (males)

(i) To produce an even number of agouti (8) and colourless or white (9) offspring, the female parent must have been AACc, so that half of the offspring are able to deposit pigment because of C and when they do, they are all agouti (having received only A from the female parent)

AACc × aacc → AC, Ac and ac → AaCc and Aacc (white)

(ii) To produce even number of agouti (9) and black (10) offspring, the mother must have been Aa and so that no colourless offsprings were produced, the female must have been CC. Her genotype must have been AaCC.

♂ aacc × ♀ AaCC

Gametes ac AC aC

F_1 AaCc (agouti) aaCc (black)

(iii) Agouti = 4

Black = 5

White = 10

From the result, it is noticed that half of the offspring are colourless; therefore, the female must have been Cc. Half of the pigmented offspring are black and half are agouti; therefore, the female must have been Aa. Overall, the genotype of the female is AaCc.

♂ aacc × ♀ AaCC

Gametes ac AC Ac aC ac

F_1 AaCc (agouti) Aacc (white) aaCc (black) aacc (black)

Problem 22

In *Drosophila* the mutant allele bw^{dts} causing brown eyes (normal eyes are red) is temperature sensitive. In flies reared at 29°C the mutant allele is dominant but this reared at 22°C the mutant allele is recessive. In a cross of $bw^{dts}/+$ × $bw^{dts}/+$, where + sign denotes the wild type allele of bw^{dts}, what is the expected ratio of brown eyed flies to red eyed flies if the progeny are reared at 29°C ; at 22°C.

Solution

Wild type red eyes × Wild type red eyes

$\frac{+}{bw^{dts}}$ × $\frac{+}{bw^{dts}}$

Gamete: $+$ bw^{dts} $+$ bw^{dts}

	$+$	bw^{dts}
$+$	$\frac{+}{+}$	$\frac{bw^{dts}}{+}$
bw^{dts}	$\frac{+}{bw^{dts}}$	$\frac{bw^{dts}}{bw^{dts}}$

Result:

	Genotype	Phenotype	(29°C)	Probability	(22°C)
1.	$\frac{+}{+}$	Red	1	Normal (Red)	
2.	$\frac{+}{bw^{dts}}$	Brown eyed		Normal (Red)	3
3.	$\frac{+}{bw^{dts}}$	Brown eyed	3	Normal (Red)	
4.	$\frac{bw^{dts}}{bw^{dts}}$	Brown eyed		Brown eyes	1

Problem 23

In Uttarpara general hospital, in the choice of donors for blood transfusion, a patient's brother or sister is often selected. Would these be more likely to be successful donors if both parents belong to group AB or if both belonged to group O. Explain with proper genotype.

Solution

(a) Parents are both AB, their genotype would be i^Ai^B.

(Father) i^Ai^B × i^Ai^B (Mother)

Gametes: i^A i^B i^A i^B

♂ \ ♀	i^A	i^B
i^A	$i^A i^A$	$i^A i^B$
i^B	$i^A i^B$	$i^B i^B$

$i^A i^A$ = A $i^B i^B$ = AB $i^A i^B$ = AB $i^B i^B$ = B

Brothers and sisters would be A, B or AB. So when they are selected they likely to be successful but not every case.

(b) When both parents are O blood group

(♂) i i × ii (♀) → i i (O)

Their all children would be O (i i) blood group. Then the brother and sister become successful to donate blood at every case.

Problem 24

In what sense is vertebrate haemoglobin gene a "luxary" to the red blood cells? Why is it not a luxary to the whole organism?

Solution

(i) The red blood cell itself can live without haemoglobin, that is why it is luxary to red blood cells.

(ii) But the whole organism would die without this product which is needed to carry oxygen to all its tissues.

Some multiple alleles:

(a) (i) Wing size in *Drosophila*:

L^+ = for normal wing, L^{vg} = for vestigial wing, L^a = Antelered wing

(ii) L gene locus in *Drosophila* consists of three allelic form $L^+ > L^{vg} > L^a$.

(b) Eye colour in *Drosophila*.

(c) (i) $\frac{w^+}{w^+}$ = Red

(ii) $^w/_w$ = white

(iii) $\frac{w^t}{w^t}$ = tinged

(iv) $\frac{w^a}{w^a}$ = apricot = orange yellow

(v) $\frac{w^e}{w^e}$ = eosin : reddise orange.

PSEUDO ALLELES

These are closely linked allele, arranged linearly in a complex locus (consist of different sub-loci) within which crossing over can occur, they are structurally non-allelic and functionally allelic and they show cis-trans position effect.

Characteristic of Pseudoalleles:

1. These are closely linked allele within which crossingover can occur. Hence, it is structurally **nonallelic in nature.**
2. They affect the same character. So they are **allelic** in **function.**

Example: Red eye colour in *Drosophila* has different mutants like white and apricot. They affect eye pigmentation, i.e., **affect the same character.** So they are allelic.

They can undergo **recombination**, i.e., they are **nonallelic.**

3. These alleles are arranged in **linear order** within **complex loci.** This complex locus divisible into **several sublocus.**
4. Pseudoalleles of the same sublocus do not undergo recombination. Those (alleles) occupying **different** sublocus **undergo recombination.**
5. Pseudoalleles have "**cis-trans**" position effect. When they are in **cis** configuration, they show **wild type**; when they are in **trans** configuration, result is **mutant phenotype.**

Example: Garnet locus in *Drosophila* (Chovnick 1961) new wild type may arise through recombination among garnet alleles $\left(\frac{g^1}{g^2}\right)$. Recombination is 0.003%.

(i) Female heterozygote of genotypes sable (s) and garnet (g^1) on the X chromosome and garnet (g^2) and pleated (pl) on other X chromosome.

(ii) Sable is at locus 43.0 and lies to the left of garnet (locus 44.4) while pleated lies to the right of garnet at locus 47.9.

(iii) Since heterozygous female also carry wild type allele for sable and pleated $\frac{sg^1+}{+g^2pl}$ crossing-over results female wild type, i.e., s + + pl.

(iv) Due to crossing-over between g^1 and g^2 result is wild type offspring.

$$\frac{s+g^1+}{sg^2+pl} \longrightarrow \left.\begin{matrix} s++pl \\ sg^2g^1+ \end{matrix}\right\} \text{Recombinant}$$

$$\frac{g^1\,g^2}{+\,+}\ \text{(wild type), cis position} \qquad \frac{g^1\,+}{+\,g^2}\ \text{(mutant type) trans position}$$

Lozen locus (Lz)

In 1940 and later Oliver and his student Green conducted experiments using different mutants and heterozygotes at the locus **lozenge** (*lz*).

(i) The lozenge (*lz*) locus (x-linked) of *Drosophila melanogester* is responsible for smaller, darker and more elliptical (narrow ovoid) eyes.

(ii) M. M. Green and K. C. Green showed that the complex locus lozenge consists of four subloci (groups), each containing one or more different alleles.

(iii) Recombination distances between these subloci are on the order of approximately 0.03 to 0.09% or frequencies of less than one per thousand chromosomes.

(iv) Since all these mutant alleles are recessive, they show no phenotypic effect when two mutants at two sub-loci are present in cis position and they have their wild type allele on the other homologue.

(v) However, two mutants are placed in trans-position any heterozygous combination of different lozenge alleles do show mutant effects.

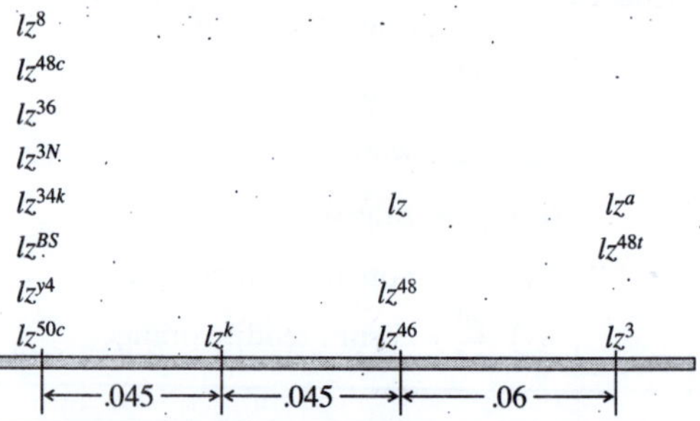

Fig. 1.1: Some alleles of lozenge occupying different subloci within the lozenge locus of *Drosophila melanogaster* (based on data of M. M. and K. C. Green; and M. M. Green).

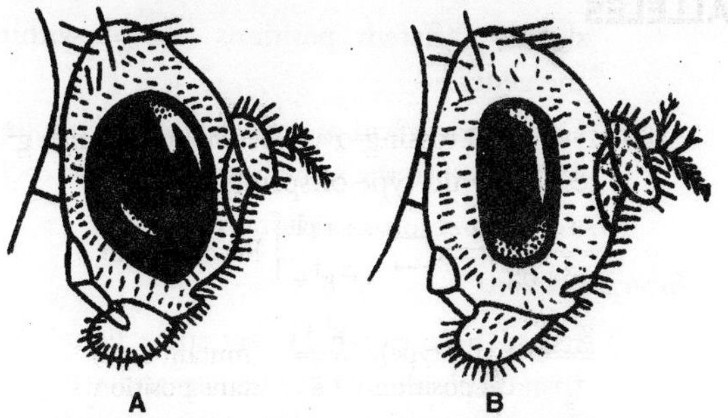

Fig. 1.2: Lozenge eye in *Drosophila*: A. Wild type eye, B. Lozenge eye (after Gardner, 1968)

Experiment:

(i) Two mutant ***Drosophila*** having spectacle eye (Lz^s) and glossy eye (Lz^g) at lozenge locus were taken for experiment.

(ii) They were presumed to be alleles which had been represented in exactly corresponding position on a chromosome pair Lz^s/Lz^g.

(iii) The heterozygous female (lz/Lz^s) were crossed with either Lz^s or Lz^g males.

$$\frac{Lz^s}{Lz^g} \times \frac{Lz^s}{\rightarrow} \;/\; \frac{Lz^g}{\rightarrow}$$

$F_1 \rightarrow$ (♀) $\frac{Lz^g +}{+ Lz^s}$ (Mutant) × $\frac{Lz^g +}{\rightarrow}$ or $\frac{Lz^s}{\rightarrow}$

(♂) $\frac{Lz^g\ Lz^s}{\rightarrow}$ (mutant) $\frac{+\ +}{\rightarrow}$ [wild type ♂]

Cross:

♂ \ ♀	Lz^g +	+ lz^s	$Lz^g\ lz^s$	+ +
Lz^g +	$\frac{Lz^g +}{Lz^g +}$ glossy	$\frac{+ Lz^s}{Lz^g +}$ (intermediate) mutant	$\frac{Lz^g Lz^s}{Lz^g +}$ (mutant)	$\frac{+\ +}{Lz^g +}$ wild *
$\rightarrow$	$\frac{Lz^g +}{\rightarrow}$ glossy	$\frac{+ Lz^s}{\rightarrow}$ (spectacle)	$\frac{Lz^g\ Lz^s}{\rightarrow}$ mutant	$\frac{+\ +}{\rightarrow}$ *wild type

Observation:

(i) All progeny from this cross were expected to have lozenge phenotype.

(ii) But few expressed wild type phenotype with a low frequency.

Explanation:

(i) Generally, a fly heterozygous for two lozenge mutant alleles (Lz^g & Lz^s) will not produce wild type progeny because wild type alleles are absent. This indicated that these two alleles (Lz^g & Lz^s) were not true alleles, i.e., two genes were not in identical positions on homologous chromosomes.

(ii) But these two mutants were located in slightly different positions $\left[\frac{lz^g +}{+ Lz^s}\right]$ within the genes.

(iii) Crossing over had presumably occurred, placing lz^g and Lz^s on the same homologue and each in heterozygous condition $\left[\frac{Lz^g Lz^s}{+\ +}\right]$.

(♀) Lozenge eye $\frac{Lz^g +}{Lz^g +}$ Lozenge eye (♂) $\frac{+ Lz^s}{\rightarrow} \;/\; \frac{Lz^g +}{\rightarrow}$ ♂ / ♂

Interpretation:

(i) It had been observed from the crosses that a few red eyed male and female *Drosophila* flies were appeared.

(ii) Although their rate of appearance was very low.

(iii) In the cis position the alleles lz^g and lz^s will express wild type (red) phenotype).

$\frac{Lz^g Lz^s}{+\ +}$ (red eye), when they are in transposition, i.e. $\frac{lz^g +}{+ Lz^s}$ will express mutant phenotype).

(iv) The two mutants lz^g and lz^s are functionally allelic as they produce **mutant phenotype**, i.e., the different expression of same phenotype.

But, they are structurallly non-allelic, because they are separated by low order of crossing over and produce wild type phenotype (red eyes). They also show cis-trans position effect.

(v) Such functionally related closed linked alleles which express same phenotypic trait (same mutant phenotype) and consist of different subloci within which crossing over may occur that produce wild type phenotype are termed as **pseudo-allele.**

So lozenge locus genes are **pseudo-allele.**

White apricot: Pseudo-allele

Experiment:

(i) Two mutant *Drosophila* having apricot eye colour (w^{apr}+) of female and white eye colour (+w^a) of male were taken for experiment.

(ii) They are presumed to be alleles which had been represented in exactly corresponding position on a chromosome pair $\frac{w^{apr}+}{+w}$.

(iii) The F_1 females were light eyed or of pale apricot $\left(\frac{w^{apr}+}{+w}\right)$ and male were of apricot eye $\left(\frac{w^{apr}+}{\rightarrow}\right)$.

(iv) Light eyed (pale apricot) females were crossed with white eyed male ($\frac{+\,w}{\longrightarrow}$).

Cross:

Apricot eye colour female × White eye colour male

$$(♀)\ \frac{w^{apr}\,+}{w^{apr}\,+} \quad \times \quad (♂)\ \frac{+\,w}{\longrightarrow}$$

$$F_1 = (♀)\ \frac{w^{apr}+}{+\ w}\left(\text{Pale apricot}\right) \quad \times \quad (♂)\ \frac{w^{apr}+}{\longrightarrow}\ (\text{apricot})$$

$$F_2 = (♀)\ \frac{w^{apr}+}{+\ w}\left(\text{Pale apricot}\right) \quad \times \quad (♂)\ \frac{+\,w}{\longrightarrow}\left(\text{white eyed}\right)$$

Gametes: $\frac{w^{apr}\,+}{}$, $\frac{+\ w}{}$, $\frac{w^{apr}\,w}{}$, $\frac{+\ +}{}$; $\frac{+\ w}{}$, $\longrightarrow$

♂ \ ♀	w^{apr} +	+ w	w^{apr} w	+ +
+ w	$\frac{w^{apr}+}{+\ w}$	$\frac{+\ w}{+\ w}$	$\frac{w^{apr}w}{+\ w}$	$\frac{+\ +}{+\ w}$ *
$\longrightarrow$	$\frac{w^{apr}+}{\longrightarrow}$	$\frac{+\ w}{\longrightarrow}$ *	$\frac{w^{apr}w}{\longrightarrow}$	$\frac{+\ +}{\longrightarrow}$ *

Result:

	Genotype	Phenotype	Phenotypic frequency
1.	$\frac{w^{apr}+}{+\ w}$	(♀) light eyed (pale apricot)	1
2.	$\frac{+\ w}{+\ w}$	(♀) white eye	1
3.	$\frac{w^{apr}w}{+\ w}$	(♀) pale apricot eye	1
4.	$\frac{+\ +}{+\ w}$	(♀) *Red (wild) eye	1
5.	$\frac{w^{apr}+}{\longrightarrow}$	(♂) apricot eye	1
6.	$\frac{+\ w}{\longrightarrow}$	(♂) White eye	1
7.	$\frac{w^{apr}w}{\longrightarrow}$	(♂) pale apricot	1
8.	$\frac{+\ +}{\longrightarrow}$	(♂) *Red eye	1

Observation:

(i) All progeny from this crosses were expected to be apricot or white.

(ii) But few expressed wild type phenotype, i.e., red eye progeny with low frequency.

Explanation:

(i) Generally, a heterozygous for two mutant alleles ($\frac{w^{apr}+}{}$ and $\frac{+\,w}{}$) will not produce wild type (red) progeny because wild type alleles were absent.

(ii) This indicated that these two alleles ($\frac{w^{apr}+}{}$ and $\frac{w}{}$) are not true alleles, i.e., two alleles or genes are not in identical position on homologous chromosome.

(iii) But these two mutants were located just slightly different positions ($\frac{w^{apr}+}{+\ w}$) within the genes.

(iv) Crossingover had presumably occurs placing w^{apr} and w in the same homologue and each in heterozygous condition ($\frac{w^{apr}w}{+\ +}$).

Interpretation:

(i) It had been observed that some light-eyed (pale apricot), red-eyed male and red-eyed females appeared. Although the rate of appearance is less than 0.001%. This is above the mutational background rate.

(ii) If the wild type alleles ($\frac{+\ +}{}$) are in the cis position ($\frac{+\ +}{+\ w}$, $\frac{+\ +}{\longrightarrow}$) red eyes will appear. If they are in trans configuration ($\frac{w^{apr}+}{+\ w}$) the phenotype is light (pale apricot). This is the **cis-trans effect.**

(iii) The genes w^{apr} and w are **functionally allelic,** i.e., they produce different expression of same phenotypic trait but they structurally **non-allelic** because they are separated by low order of crossing over and they exhibit **cis-trans** position effect.

Such functionally related closely linked alleles containing clusters of sub-alleles is known as **pseudo-alleles.**

✱ Pseudoallele and position effect

Green and Lewis tried to demonstrate intragenic or interallelic recombination of a heterozygote for two mutant alleles.

(i) If two mutant alleles occupy different positions showing close linkage rather than same position, it may be expected that F_1 heterozygous ($\frac{Lz^{g}\,+}{+\ Lz^{s}}$) would express wild type phenotype.

(ii) But it had been found that F_1 exhibits mutant phenotype.

(iii) Close linked alleles which have similar phenotypic effect but can recombine each other are considered to occupy complex loci. It can be divided into subcoci between which recombination can occur.

(iv) Lozenge (*Lz*) is a complex loci and consists of four subloci which contain different lozenge alleles. Recombination distances between the groups (subloci) are 0.03 to 0.09%.

(v) When both mutant alleles ($\frac{Lz^g\ lz^s}{}$) are on the same homologue and their counterpart ($\frac{+\ +}{}$) on the other homologue, they are in the **cis arrangement** $\left[\frac{Lz^g\ Lz^s}{+\ +}\right]$.
Two mutant alleles (Lz^g & Lg^s) are on two different homologues. They are in **trans arrangement** $\left[\frac{Lz^g\ +}{+\ lz^s}\right]$.

(vi) In cis position **(coupling configuration)** they will express wild type phenotype and in transposition **(repulsion configuration)**, they will show mutant phenotype.

(vii) This may be due to the **position effect** where mutant does not allow adjust region to express wild type phenotype. This difference in phenotypic expression is due to **cis-trans** arrangement which is known as **cis-trans effect.**

$\frac{Lz^g\ Lz^s}{+\ +}$ (wild type) (cis arrangement) $\frac{Lz^g\ +}{+\ lz^s}$ (mutant phenotype) (trans arrangement)

Remarks:

(i) E.B. Lewis hypothesized the sequential nature of a gene product synthesis.

(ii) According to him if two mutations (Lz^g & Lz^s) are in the same chromosome and other chromosome is with wild type allele (+ & +) they are able to produce wild type substance. So the phenotype is **wild type**, the genes are in **cis position.**

(iii) A gene being linear, produces linear complex substances which can be interrupted at a specific step by a mutant allele. In trans position gene function is interrupted by mutant allele on both of the two homologues, which result **mutant phenotype.**

Pseudoallele and complementation

The functional allelism of any two recessive mutation is determined experimentally by the cis-trans or complementation test.

(i) Two mutations (Lz^g and Lz^s) in the trans-heterozygote are usually **allelic** if they produce **mutant phenotype**, and **non-allelic** if they form a **wild type product.**

$\frac{Lz^g\ +}{+\ Lz^s}$ → → mutant (allelic)
(Lz^g and Lz^s are in same gene)

$\frac{Lz^g\ +}{+\ Lz^s}$ → → wild type (nonallelic)

(ii) The ability of two recessive mutations to restore wild type phenotype either partially or completely in the transposition has been termed as **complementation.**

(iii) Two lozenge pseudoalleles are *Drosophila*. Lz^k and Lz^{50e} are complementary and represent mutation at different subloci, since the trans-combination $\frac{Lz^k\ +}{+\ lz^{50e}}$ is wild type, Lz^k and Lz^{50e} are two lozenge pseudoallele and are complementary.

(iv) Complementation suggests that different products or portions of products are affected by each allele, while lack of complementation indicates both alleles affect the same gene product or same portion of the gene product.

Example:

$\frac{Lz^k\ +}{+\ Lz^{50e}}$ → complement product of Lz^k complement product of Lz^{50e} → wild type phenotype

Isoallele

The alleles which express the same phenotypic range are called **isoalleles.**

(i) Isoalleles within the phenotypic range of abnormal character, i.e., mutant isoalleles.

(ii) Some isoalleles within the phenotypic range of wild type, i.e., normal isoalleles.

In reality, multiple allelic system, therefore, be quite complex and may consist of various subsidiary isoallelic systems.

Example: Eye colour gene in *Drosophila*—One recessive mutant is white associated with complete absence of eye colour. Many alleles of white have been found which have a quantitative effect between white and wild type (red).

Combination	Eye pigment colour	Relative amount of pigment
w/w	white	0.0044
w^a/w^a	apricot	0.0197
w^e/w^e	eosin	0.0324
w^{ch}/w^{ch}	cherry	0.0410

GENE INTERACTION AND EPISTASIS

• Intraallelic or Allelic interaction

(i) Interaction between the two alleles of a single gene is known as intraallelic interaction.

(ii) Such interaction gives a typical Mendelian ratio 3 : 1 in a monohybrid cross and 9 : 3 : 3 : 1 in a dihybrid cross.

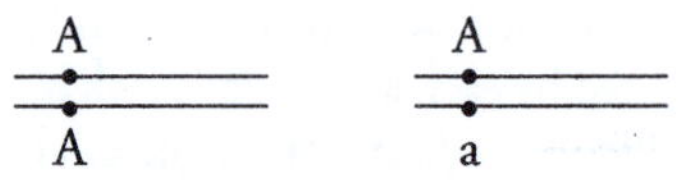

Interaction between AA or Aa (intraallelic interaction)

• Interallelic interaction

(i) Interaction between the non-homologous independent genes located on the same or on different chromosomes is known as interallelic interaction.

(ii) **Epistasis** is a type of interallelic interaction.

(iii) The gene which suppressed or masked the action of a gene at another locus is termed as **epistatic gene** and the gene which is suppressed by an **epistatic gene** is called **hypostatic gene.**

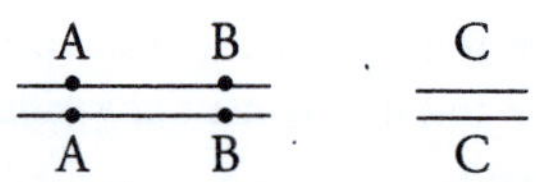

Interaction between two genes AA and BB or between AA and CC (interallelic interaction).

• Epistasis (Greek word "Stand above")

It is a phenomenon when the presence of one gene suppresses the expression of another gene at different locus.

(i) It is an example of a single character controlled by two or more pairs of genes.

(ii) The gene which suppresses the effect is called **epistatic gene.** The gene whose effect gets suppressed is called **hypostatic gene.**

Pleotropy or Pleotrophism:

(Greek word "to take many turns")

It is a phenomenon when a single pair of genes affect more than one phenotypic traits. Such genes are called **pleotropic gene.**

(i) Mendel proposed law of unit character which means each character is controlled by one **gene** (factor).

(ii) A **pleotropic** gene along with controlling the manifestation of its phenotypic trait, also influences some other related or unrelated character, i.e., a single gene influences more than one **phenotype.**

Mechanism:

The phenotypic effect is produced by a gene owing to a cascade of reactions during some metabolic pathway which are influenced from the original gene product and contribute to different phenotypic effect.

Example:

1. **PKU (Phenylketonuria)**

   (i) It is an autosomal recessive character controlled by mutant gene present in **12th chromosome.**

   (ii) The mutant gene fails to code for the enzyme **phenylalanine hydroxylase** (PAH) needed for normal metabolism of amino acid **phenylalanine to tyrosine.**

   (iii) As a result **phenylalanine** accumulates in body fluids (blood, sweat and cerebrospinal fluid) and appearance of an abnormal breakdown of toxic product **phenylketone** in urine.

2. In ***Drosophila***: White eye mutation leads to depigmentation in many other parts of the body.

3. In **transgenic organisms**: The introduced gene can produce different effects depending on where the gene has introgressed.

Molecular basis of gene interaction, i.e., Biochemical Explanation.

(i) The expression of a particular character is the net result of a series of enzyme mediated biochemical reactions.

(ii) The production of each enzyme depends upon a specific gene. Each reaction step is controlled by a single gene, i.e., 1 : 1 correspondence of gene, and biochemical reaction exists.

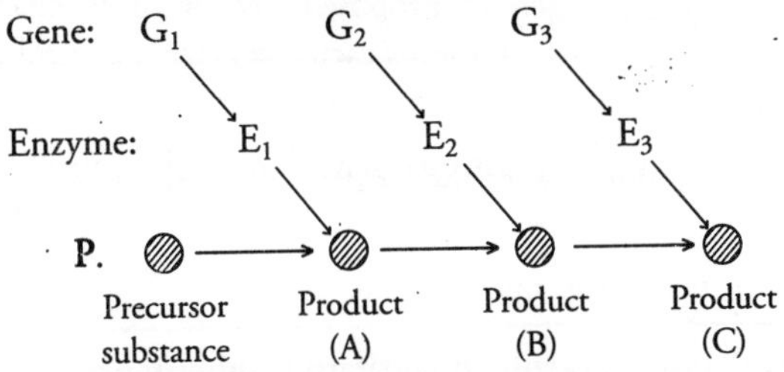

(iii) E_1, E_2 and E_3 are specific enzymes controlled by genes G_1, G_2 and G_3 respectively.

(iv) P is the precursor substance, A and B are the intermediate product and C is the end or final product.

(v) Now suppose, G_2 becomes mutant, the product 'B' will not be converted into final product C and hence intermediate product 'B' could accumulate. Gene G_3 would not have any effect on the phenotype.

(vi) Thus, G_2 can mask the phenotypic expression of alleles at G_3 locus. G_2 is known as epistatic gene and G_3 as hypostatic.

Types: (a) **Dominant epistasis** (Here a dominant gene does not allow the expression of the non-allelic gene) Ratio — 12:3:1

(b) **Recessive epistasis** (Here homozygous recessive condition of a gene determine the phenotype, irrespective of the alleles of other gene pairs, i.e., recessive alleles mask or hides the effect of the other gene pair) Ratio — 9:3:4.

Monogenic (qualitative) inheritance

Classical Mendelian traits which have two contrasting expressions and are controlled by a single pair of genes, are monogenic traits. The inheritance of monogenes or qualitative characters is called **qualitative** or **monogenic** inheritance.

Polygenic (quantitative) inheritance

A single phenotypic character governed by more than one pair of genes is called **polygenic** character or **quantitative** character.

(i) Their inheritance is called **polygenic** inheritance.

(ii) Polygenes are also called **quantitative** genes or **cumulative** genes or multiple factors.

Example: Skin colour in man.

	Dark Red	White
Parents:	$R_1R_1R_2R_2$	$r_1r_1r_2r_2$
Gamete:	R_1R_2	r_1r_2

F_1 → $R_1r_1R_2r_2$ (Medium red)

F_2 → Dark red, medium red, light red and red progenies.

- *Positive allele*: one of the alleles which produces seed colour and is called positive allele. It is denoted by a capital cells e.g. R, R_1 etc.
- *Negative allele*: The other allele which dies not produce any clour is known as negative allele. It is denoted by the corrsponding smal letter, e.g., r_1, r_2 etc.

- **Transplantation**

Replacement of diseased tissue or organs by healthy one is called transplantation.

- **Autograft**

(i) Tissue is grafted from one area to another on the same individual.

(ii) Rejection is not problem.

- **Isograft**

(i) It is a graft between two genetically identical individuals such as identical twins.

(ii) Rejection is not problem.

- **Allograft**

Tissue is grafted from one individual to a genetically different individual of the same species.

- **Xenograft**

It is a graft between individuals of different species such as pig to human.

Table 1.1: Difference or Disimilarities between Dominance and Epistasis

Dominance	Epistasis
(i) It involves intra-allelic gene suppression or masking effect of one allele.	(i) It involves inter-allelic gene suppression or masking the action of a gene to another locus.
(ii) Here interaction of two alleles of the same gene occur. Thus, it involves single locus.	(ii) Here interaction of two or more genes occur. Thus, it involves two or more loci.
(iii) It always refers to heterozygotes.	(iii) It may involve both homo- and heterozygotes.
(iv) It is known as intragenic or intralocus gene action.	(iv) It is known as intergenic or interlocus gene interaction.

Table 1.2: Similarities between Dominance and Epistasis

(i) Dominant gene always affect the same character.	(i) Epistatic genes always affect the same character.
(ii) It involves nuclear gene.	(ii) It involves nuclear gene.

Table 1.3: Differences between Monogenic (qualitative) and Polygenic (quantitative) inheritance

Monogenic	Polygenic
1. It deals with qualitative character.	1. It deals with quantitative character.
2. Each trait is controlled by one pair of contrasting character.	2. Each trait is controlled by more than one pair of non-allelic genes (polygenes).
3. Each trait or character exhibits two distinct phenotypes.	3. Each trait or character has an integrating range of phenotype.
4. It exhibits discontinuous pattern of inheritance.	4. It represents continuous pattern of inheritance.
5. Individual exhibits dominant expression of parent.	5. F_1 individual exhibits intermediate expression betwen the two parents.
6. Phenotypic expression is not influenced by environment.	6. Phenotypic expression is influenced by environment.

Transgressive segregation

The appearance of F_2 individual with higher or lower intensity of a character than those present in both parents involved in the cross is termed transgressive segregation and such individuals are called transgressive segregrants.

They affect quantitative trait.

Solved Problems

Problem 1

What is the difference between dominance and epistasis? Why did not any pairs of genes used by Mendel show epistasis?

Solution

(i) Dominance concerns the interactions of two alleles at a given gene to determine phenotype.

(ii) Epistasis involves the interaction of alleles at different genes.

All of the traits examined by Mendel were determined by a single gene.

Problem 2

In domestic chickens, a dominant allele C is required for coloured feathers but a dominant allele I of an unlinked gene is an inhibitor of colour that overrides the effect of C. White leghorns have genotype CC II, whereas white Wyandottes have genotype cc ii. Both breeds are white, but for different reasons.

(a) In the F_2 generations of a cross between white leghorns and white Wyandottes, what is the phenotypic ratio of white coloured?

(b) Among F_2 chicks that are white, what is the proportions of Cc Ii?

Solution

(a)

(i) The initial cross CC II × cc ii will yield F_1 chickens with genotype Cc Ii.

(ii) Because the genes are not linked, the offspring into F_2 generation will have 9 : (C-I-) : 3(C-ii) : 3(cc-I-) : I(cc ii).

(iii) Only the chickens with genotype C-ii will be coloured because C is required for coloured feathers but I inhibits its effect of C. This type of interactions of gene if **epistasis** will lead to phenotypic ratio 13 white : 3 colour.

(b)

(i) In any dihybrid cross with unlinked genes i the probability that an F_2 progeny is heterozygous for both genes equals $(\frac{1}{2} \times \frac{1}{2}) = \frac{1}{4}$. Because only $\frac{13}{16}$ of the chicks in this example are white, the proportions of Cc Ii among white chicks is $\left(\frac{1}{4} / \frac{13}{16}\right) = \frac{1}{4} \times \frac{16}{13} = \frac{4}{13}$.

(ii) A total of 13 genotypes yield white chicks and among these four are heterozygous for both genes.

Problem 3

In Durco-Jersey pigs, animals with genotype R–S– are red, those with genotype R–SS or rr S– are sandy coloured and those with genotype rr ss are white. The gene show independent assortment. In the F_2 generation of a cross between the genotypes RR ss and rr SS.

(a) What is the phenotypic ratio of red : normal

(b) Among F_2 piglets that are red, what is the proportion of Rr Ss.

Solution

(a) (i) The initial cross RR ss × rr SS will yield F_1 pigs with genotype Rr Ss.

(ii) The progreny of the F_2 generation will have the genotypic ratio 9(R–S–) : 3(RR–ss) : 3(rr S–) : 1(rr ss).

(iii) Only the pigs with genotype R–S– will be red while all the others will be non red (some sandy and some white in the ratio (6 : 1).

(iv) This type of interaction of two genes (**epistasis**) will lead to a phenotypic ratio of 9 red : 7 non red).

(b) (i) The proportion of Rr Ss is $\frac{4}{9}$.

(ii) The proportion of pigs with the genotype Rr Ss is the frequency of those heterozygous for both the genes which equals to $\frac{1}{2} \times \frac{1}{2} = \frac{1}{4}$.

(iii) Because $\frac{7}{16}$ of pigs are red, the proportion of Rr Ss among them is $(\frac{1}{2} \times \frac{1}{2}) / \frac{9}{16} = \frac{1}{4} \times \frac{16}{9} = \frac{4}{9}$.

(iv) A total 9 genotypes will yield red pigs among which 4 are heterozygous for both genes.

Problem 4

In *Drosophila*, red eyes is the wild type phenotype, there are several different genes (with each gene existing in two or more alleles) that affect the

eye colour. One allele causes purple eyes, and a different allele causes sepia eyes. Both of these alleles are recessive compared to red eyes. When the F_1 offsprings were allowed to mate with each other, the following data were obtained.

Purple eyes 146, sepia eyes 151, parplish sepia eyes 50 and red eyes 444.

Explain the pattern of inheritance. Conduct a Chi-square analysis to see if the experimental data fit with your hypothesis.

Solution

(i) If eye colour is affected by two different genes that exist in dominant and recessive alleles, the result obtained by crossing of the F_1 offspring appearing to be 9 : 3 : 3 : 1.

(ii) Neither gene is X-linked. Let pr^+ represent the red allele of the first gene and pr the purple allele.

(iii) Let sep^+ represent the red allele of the second gene and sep the sepia allele.

First crosses

$pr^+ pr^+$ sep sep × pr pr $sep^+ sep^+$

Gamet pr^+ sep   pr sep^+

F_1

pr^+ pr sep^+ sep (Red eyes)

All F_1 are red because red is dominant for each gene cross of F_1 offspring.

pr^+ pr sep^+ sep × pr^+ pr sep^+ sep

Gametes:

pr^+sep^+, pr^+sep, prsep^+ prsep × pr^+sep^+, pr^+sep, prsep^+ prsep

♀ \ ♂	pr^+ sep^+	pr^+ sep	pr sep^+	pr sep
pr^+ sep^+	pr^+ sep^+ pr^+ sep^+ Red	pr^+ sep pr^+ sep^+ Red	pr sep^+ p^+ sep^+ Red	pr sep pr^+ sep^+ Red
pr^+ sep	pr^+ sep^+ pr^+ sep Red	pr^+ sep pr^+ sep Sepia	pr sep^+ pr sep Red	pr sep pr^+ sep Sepia
pr sep^+	pr^+ sep^+ pr sep^+ Red	pr^+ sep pr sep^+ Red	pr sep^+ pr sep^+ Purple	pr sep pr sep^+ Purple
pr sep	pr^+ sep^+ pr sep Red	pr^+ sep pr sep Sepia	pr sep^+ pr sep Purple	pr sep pr sep pur/sepia

Analysis:

(i) In this case, one gene exists as the red (dominant) or purple (recessive) allele and the second gene exists as the red (dominant) or sepia (recessive) allele.

(ii) If an offspring is homozygous for the purple allele, it will have purple eyes. Similarly, if an offspring is homozygous for sepia gene, it will have sepia eyes.

(iii) If offspring is homozygous for both recessive alleles, purple and sepia will be purplish sepia eyes.

To have red eyes it must have one copy of dominant and others for both genes.

(iv) Based on as expected 9(Red) : 3(Purple) : 3(sepia) : 1(purplish sepia) the observed and expected answers are as follows:

Observed	Expected
Red eyes = (444)	$\frac{9}{10} \times 791 = 445$
Purple = (146)	$\frac{2}{10} \times 791 = 148$
Sepia = (151)	$\frac{9}{15} \times 791 = 148$
Purplish sepia = (50)	$\frac{1}{10} \times 791 = 49$

Chi-square

Observed	Expected	job–Ex1 = D	$\frac{D^2}{E}$	X^L
444	445	1	$\frac{1}{445} = 0.002$	
146	148	2	$\frac{1}{148} = 0.027$	= 0.109
151	148	3	$\frac{9}{148} = 0.060$	= 0.11
50	49	1	$\frac{1}{49} = 0.020$	
			0.109	

Tabulated value of 3 df at 5% is 7.82, but our data revealed that $x^2 = 0.11$.

Therefore, we cannot reject our hypothesis, i.e., purple and sepia alleles are in two different genes that these recessive alleles are epistatic to each other.

Problem 5

In Arambagh hatchery, the chickens that came both the alleles for Rose comb (R) and Pea comb (P) produce walnut comb. On the otherhand chickens that lack both of these alleles (i.e., they are genotypically pprr) have single combs. Determine the expected phenotype and proportions of the offspring from the following crosses.

(a) RRPp × rrPp (b) RrPp × Rrpp (c) rrPP × RrPp (d) RrpP × rrpp (e) Rrpp × RrPp

Solution

(a) RRPp × rrPp

Gamete: RP, Rp rP, rp

♂ \ ♀	RP	Rp
rP	$\frac{RP}{rP}$	$\frac{Rp}{rP}$
rp	$\frac{RP}{rp}$	$\frac{Rp}{rp}$

	Genotype	Phenotype		Frequency
1.	$\frac{RP}{rP}$	Walnut comb	1	Walnut = 3
2.	$\frac{RP}{rp}$	Walnut comb	1	
3.	$\frac{Rp}{rP}$	Walnut comb	1	rose comb = 1
4.	$\frac{Rp}{rp}$	Rose comb	1	

(b) RrPp × Rrpp

Gametes: RP, rP, Rp, rp; Rp, rp

	RP	Rp	rP	rp
Rp	$\frac{RP}{Rp}$	$\frac{Rp}{Rp}$	$\frac{rP}{Rp}$	$\frac{rp}{Rp}$
rp	$\frac{RP}{rp}$	$\frac{Rp}{rp}$	$\frac{rP}{rp}$	$\frac{rp}{rp}$

	Genotype	Phenotype	Phenotype frequency	
1.	$\frac{RP}{Rp}$	Walnut	3	$\frac{3}{8}$
2.	$\frac{RP}{rp}$	Walnut		
3.	$\frac{rP}{Rp}$	Walnut		
4.	$\frac{Rp}{Rp}$	Rose comb	3	$\frac{3}{8}$
5.	$\frac{rp}{Rp}$	Rose comb		
6.	$\frac{Rp}{rp}$	Rose comb		
7.	$\frac{rP}{rp}$	Pea comb	1	$\frac{1}{8}$
8.	$\frac{rp}{rp}$	Single comb	1	$\frac{1}{8}$

(c) rrPP × RrPp

Gametes: rP ; RP, Rp, rP, rp

	RP	Rp	rP	rp
rP	$\frac{RP}{rP}$	$\frac{Rp}{rP}$	$\frac{rP}{rP}$	$\frac{rp}{rP}$

	Genotype	Phenotype	Frequency	
1.	$\frac{RP}{rP}$	Walnut	2	$\frac{2}{4} = \frac{1}{2}$
2.	$\frac{Rp}{rP}$	Walnut		
3.	$\frac{rP}{rP}$	Pea comb	2	$\frac{2}{4} = \frac{1}{2}$
4.	$\frac{rp}{rP}$	Pea comb		

(d) Rrpp × rrpp

Gametes: Rp, rp ; rp

♂ \ ♀	RP	rP
rp	$\frac{Rp}{rp}$	$\frac{rp}{rp}$

	Genotype	Phenotype	Phenotype frequency
1.	$\frac{Rp}{rp}$	Rose comb	$\frac{1}{2}$
2.	$\frac{rp}{rp}$	Single comb	$\frac{1}{2}$

(e) RrPP × RrPp

Gametes: RP, rP; RP, rP, Rp, rp

♂ \ ♀	RP	rP	Rp	rp
RP	$\frac{RP}{RP}$	$\frac{rP}{RP}$	$\frac{Rp}{RP}$	$\frac{rp}{RP}$
rP	$\frac{RP}{rP}$	$\frac{rP}{rP}$	$\frac{Rp}{rP}$	$\frac{rp}{rP}$

	Genotype	Phenotype	Phenotype frequency
1.	$\frac{RP}{RP}$	Walnut	$\frac{6}{8} = \frac{3}{4}$
2.	$\frac{rP}{RP}$		
3.	$\frac{Rp}{RP}$		
4.	$\frac{rp}{RP}$		
5.	$\frac{rP}{RP}$		
6.	$\frac{Rp}{rP}$		
7.	$\frac{rP}{rP}$	Rose comb	$\frac{1}{8}$
8.	$\frac{rp}{rP}$	Pea	$\frac{1}{8}$

NON-MENDELIAN INHERITANCE

Extranuclear inheritance

Inheritance of genetic material located outside the nucleus is termed as extranuclear inheritance.

Titbit of extranuclear or Non-Mendelian inheritance

(i) The pattern of inheritance shown by genes located outside the nucleus (organelles) differs strikingly from the pattern shown by nuclear genes, which is also termed as **non-Mendelian inheritance.**

(ii) Extra chromosomal inheritance, **extranuclear inheritance, somal inheritance** and **maternal inheritance** are all synonyms.

(iii) The genes controlling cytoplasmic inheritance are present outside the nucleus and in the cytoplasm they are known as **plasma genes, cytoplasmic genes, extranuclear genes or extra chromosomal genes.**

(iv) The sum total genes present in the cytoplasm of a cell is known as **plasmon.**

All the genes present in a plastid are known as **plastoms.** All the genes present in a mitochondrion are known as **chondrioms.**

Basic features of Extranuclear or Extrachromal inheritance

1. The results of reciprocal crosses indicate that one parent (usually maternal) exerts greater influence than the other on a particular trait.
2. Cytoplasmic factor, i.e., organelles in the cytoplasm of maternal reproductive cell containing extranuclear DNA are associated with the extrachromosme or cytoplasmic inheritance. It is an evidence of **maternal transmission** or **inheritance.**
3. Absence of Mendelian segregation and Mendelian ratios in the product is an evidence of extrachromosomal inheritance.
4. Absence of linkage to known chromosomal genes in the product is an evidences of extrachromosomal inheritance.

Main characteristics of Non-Mendelian inheritance

1. Mendelian segregation (typical ratios) is not found here because meiosis based Mendelian segregation is not involved.
2. In multicellular eukaryotes hereditary traits are transmitted by cytoplasm and do not show Mendelian segregation. The result of reciprocal crosses involving extranuclear genes are not same as reciprocal crosses involving nuclear genes because meiosis based Mendelian segregation is not involved.
3. Mitochondrial and Chloroplast genes usually show **uniparental** inheritance from generation to generation. It may be paternal (transmission through father) or maternal (transmission through mother).

   (i) In uniparental inheritance, all progeny (both males and females) have the phenotype of only one parent.

   (ii) For multicellular eukaryotes, the mother's phenotype is expressed exclussively—the phenomenon is called **maternal inheritance.**

   It occurs, because the amount of cytoplasm in the female gamete (ovum) usually greatly exceeds that in the male gamete (sperm). Therefore, the zygote receive most of its cytoplasmic factors from the female parent and a negligible amount from the male parent. It is also known as **transovarion transmission.**

   In contrast, reciprocal crosses between a wild type and a mutant strain are identical if the genes are located in the nuclear chromosome.
4. Extra nuclear genes cannot be mapped to the chromosme in the nucleus.
5. Non-Mendelian inheritance is not affected by substituting a nucleus with a different genotype.

Maternal inheritance

It means the inheritance controlled by cytoplasmic factors that are transmitted to the succeeding generation through the egg of female organism.

They show the following characters:

- Reciprocal crosses result differences in F_1.
- In most cases that result F_1 dissappear in F_2.
- A smaller differences in F_2 as compared to that in F_3.

Maternal effect

This is a phenomena in which mother's nuclear genotype directly influences the phenotype of the off-spring and may be mediated by the substances present in the cytoplasm of the egg or through nurturing.

Table 1.4: Distinction between Maternal inheritance and Maternal effect

Maternal inheritance	Maternal effect
1. The hereditary determinants of a trait are extra nuclear and genetic transmission is only through maternal cytoplasm.	1. The hereditary determinants are nuclear genes transmitted both sexes. Here the nuclear genotype of the mother determines the phenotype of the progeny.
2. Mendelian segregation is not observed.	2. Mendelian segregation is observed through suitable crosses.
	Ex. Intrauterine environment of female mammals and their ability to produce milk.

Non-Mendelian inheritance and Nuclear gene inheritance

(i) In eukaryotic species, most genes within the nucleus obey Mendelian pattern of inheritance because homologous pairs of chromosomes segregate during gamete formation, except for sex linked genes.

(ii) Extra-nuclear genes does not display Mendelian pattern. Sorting of chromosome during meiosis explains the inheritance pattern of nuclear genes.

Non-Mendelian inheritance in *Chlamydomonas*

Ruth Sagar and her colleagues worked on the genetics of *Chlamydomonas reinharditii.*

- **Basic features**

(i) The motile haploid, unicellular algae (*C. reinharditii*) has two flagella and a single large chloroplast containing 75-80 copies of duplex cpDNA molecule approximately 195 kb in size.

(ii) There are two mating types designated as mt^+ and mt^- are of equal size and appear to contribute same amount of cytoplasm to the zygote.

(iii) The opposite mating type (mt^+ and mt^-) fuse (syngamous mating), to form diploid zygote. A thick-walled cyst develops around the zygote.

(iv) The diploid zygote undergoes meiosis to form a tetard of four haploid cells. As the mating type is determined by nuclear gene, a 2 : 2 segregation of $mt^+ : mt^-$ mating type results.

Cytoplasmic inheritance in *Chlamydomonas*

Requirement

(i) R. sagar isolated two strains of *chlamydomonas.* One strain is *streptomycin resistant* (Str^r) and other one is *streptomycin sensitive* (Str^s).

(ii) The trait of streptomycin resistance (Str^r) is believed to be located in its cp DNA.

Experiment and Observation

A. (i) Mating was made between streptomycin resistance (Str^r) and streptomycin sensitive (Str^s) [$mt^+ Str^r \times mt^- Str^s$] strain. It results about 95% of the progeny which are streptomycin resistant.

(ii) This is a clear example of uniparental inheritance because the resistance or sensitivity comes from the mt^+ parent.

(iii) In contrast, the mt^+ allele in the nucleus behaves strictly in Mendelian fashion, yielding the 1 : 1 progeny ratio expected of nuclear genes.

B. (i) The reciprocal cross or mating between streptomycin sensitive and streptomycin resistance ($mt^+ Str^s \times mt^- Str^r$). The result also showed that 95% of the progeny are streptomycin sensitive.

(ii) This also shows an uniparental inheritance. Here streptomycin sensitive phenotype is inherited.

(a) (♀) $mt^+ Str^r \times mt^- Str^s$ (♂)

↓

All (95%) are streptomycin resistant (Str^r) mt^+ allele segregated in 1 : 1 ratio.

(b) (♀) $mt^+ Str^s \times mt^- Str^r$ (♂)

All (95%) are streptomycin sensitive and mt^+ allele segregated in 1 : 1 ratio.

Inference

(i) By comparison Sagar and her colleagues discovered that Str^r gene was inherited from mt^+ parent but not from the mt^- parent. Therefore, Str^r gene was not inherited in a Mendelian manner.

(ii) This pattern occured because only the mt^+ parents chloroplast transmitted to daughter progeny and the Str^r gene was found in the chloroplast genome.

Cytoplasmic inheritance in *Paramoecium*

Kappa particle

(i) Intracellular bacteria (*Caedibacter taeniospiralis*) in *paramoecium aurelia* that releases a substance called *paramecin* which kills some but not all strain of *paramecia* are called **kappa** particles.

(ii) It is 0.4μ long and has its own DNA. A dominant nuclear gene **K** within the kappa particle encode the **paramecin** toxin.

(iii) The kappa bacteria possess a refactile protein containing 'R' body and are called **bright's** because they are infected with a virus that controls the synthesis of viral protein as well as **R protein body** in kappa bacterium.

Mu particles

Bacteria (*Caedobactor conjugatus*) like particles found in the cytoplasm of *Paramoecium*, that imparts the male killer phenotype is called **Mu particle**.

Killer strain of *Paramoecium*

(i) Some strains of *Paramoecium aurelia* containing kappa particles are known as **killer strain.** Its character has a nuclear as well as cytoplasmic basis.

(ii) The existance of kappa particle is determined by presence of a **nuclear dominant** gene.

Sensitive strain of *Paramoecium*

(i) The strain of *paramoecium* in which the kappa particles are absent is called sensitive strain.

(ii) They are killed by killer strain.

Conjugation of killer and sensitive strain of Paramoecium

A. Killer strain with kappa particle and nucleus with gene K. Sensitive strain without kappa particle and nucleus with gene k.

(killer) KK × kk (sensitive)
[conjugation] ↓ [without cytoplasmic exchange]
Kk

Kk (killer) Kk (sensitive)
Ex-conjugant cells

killer (Kk) kk (sensitive)

KK (sensitive)
Kk (sensitive)

B. killer (KK) × sensitive (kk)
[conjugation] ↓ [without cytoplasmic exchange]
Kk

killer (KK) × killer (kk)
(conjugation) ↓
Kk

Kk (killer) Kk (killer)

killer (Kk) sensitive (kk)

KK (killer)
kk (sensitive)

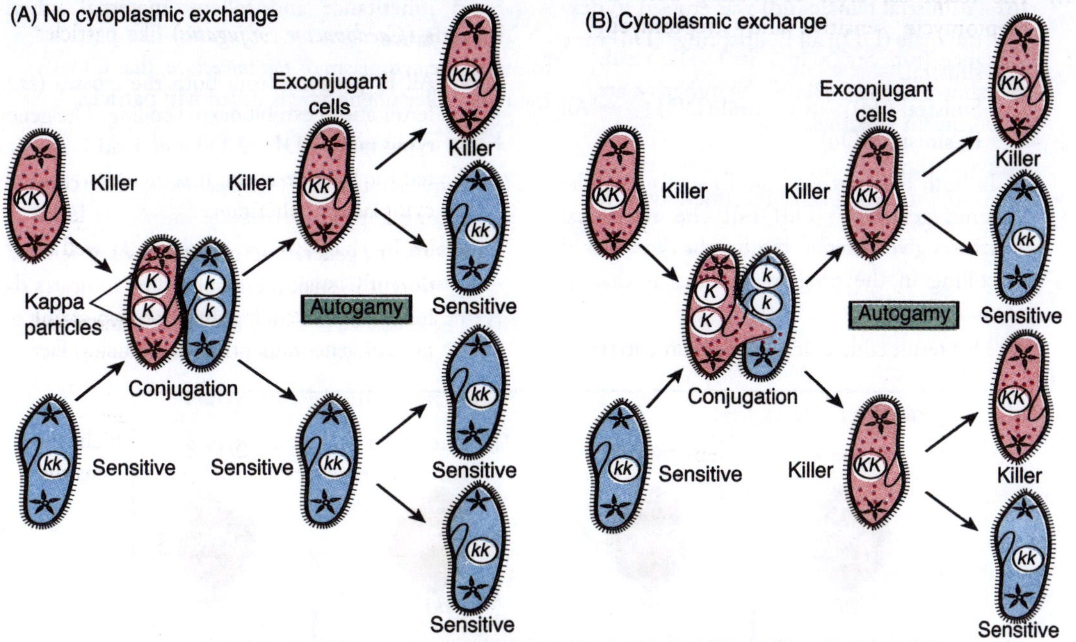

Fig. 1.3: Crosses between killer (KK) and sensitive (kk) *Paramoecium*. (A) The result when no cytoplasmic exchange takes place in conjugation. (B) The result when cytoplasmic exchange occurs. Cells containing kappa are shown in yellow. The genotype and phenotype of each cell are indicated.

Table 1.5: A comparison of phenotypes following conjugation of killer (KK) cells and sensitive (kk) cells with and without cytoplasmic mixing

KK × kk (cytoplasmic exchange)	KK × kk (No cytoplasmic exchange)
1. Here both exconjugants are killer cells because each cell has kappa particles derived from the cytoplasm of the killer.	1. Here 1 : 1 ratio of killer to sensitive exconjugant cells are observed. Kappa particle is not transferred.
2. Autogamy of both exconjugants Kk (heterozygote) produces homozygotes KK (killer) and kk (sensitive) cells in the 1 : 1 ratio as expected from Mendelian segregation. kk cell cannot maintain kappa particle and so becomes sensitive.	2. Autogamy of both exconjugants Kk (heterozygote), i.e., one is killer and other is sensitive, produce one killer (KK) and one sensitive (kk) from killer one. On the other hand, sensitive (Kk) exconjugant produces all sensitive but KK and kk forms.

• Maternal effect in Snail shell coiling

Maternal transmission is not only mediated by cytoplasmic organelles but also mediated through the cytoplasmic transmission of the products of nuclear genes.

(i) In water snail (*Lymnaea peregra*), the shell and internal organ are arranged either in a right-handed (**dextral**) or left-handed (**sinistral**) direction.

(ii) The dextral orientation is more common and is **dominant** to the sinistral orientation.

(iii) Many combinations of crosses results that could not be explained by Mendelian pattern of inheritance.

(a) A dextral female (DD) was crossed to a sinistral male (dd), all F_1 offsprings (Dd) were sinistral.

Dextral (DD) ♀ × Sinistral (dd) ♂ → All F_1 dextral (Dd)

(b) A sinistral female (dd) was crossed to dextral male (DD); all F_1 offsprings (Dd) were sinistral.

Sinistral (dd) ♀ × Dextral (DD) ♂ → All F_1 sinistral (Dd)

(iv) In both these crosses, the F_1 snails have the same genotype (Dd) but the reciprocal crosses give different results, the direction of coiling in the progeny is same as that of their mother.

This result contradicts **Mendelian pattern** of inheritance and follows **maternal inheritance.**

(v) All F_2 progeny from both the crosses (self-fertilisation) exhibit dextral coiling. The genotypic ratio 1 DD, 2 Dd and 1 dd is predicted for F_2 generation. It is inconsistent with cytoplasmic inheritance.

(vi) The F_3 generation, obtained by self-fertilization of F_2 snail (hermaphrodite) indicates the inheritance of coiling direction depending on nuclear genes rather than extranuclear factors.

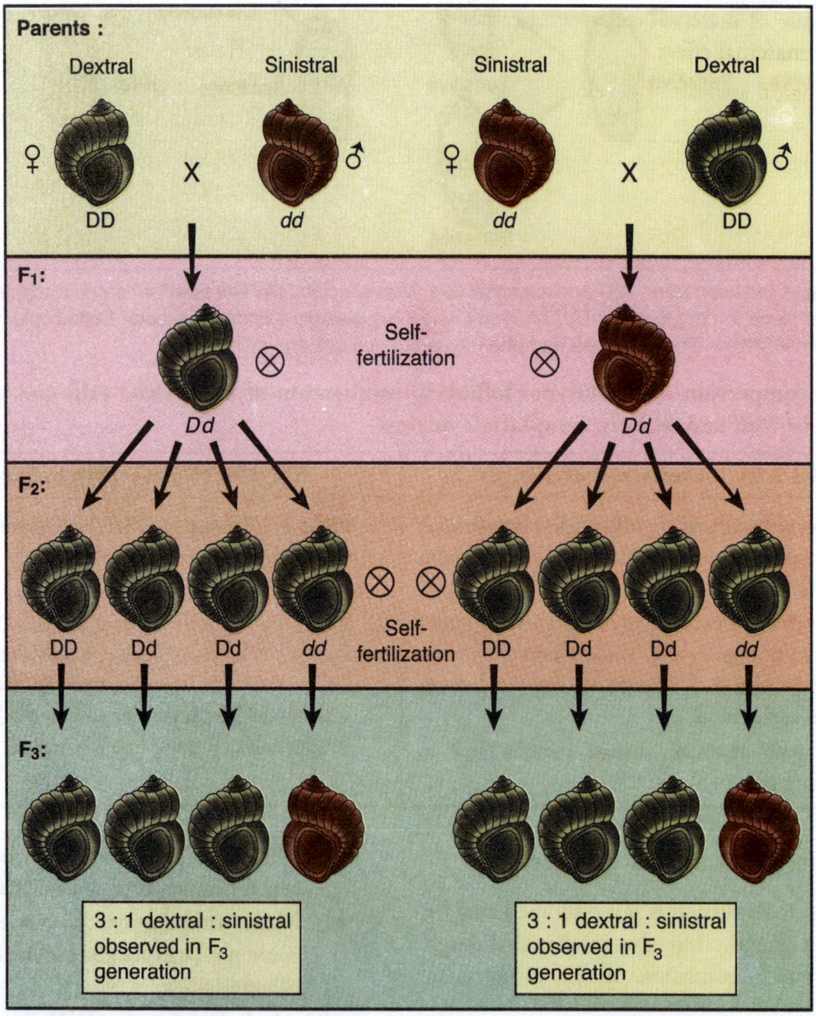

Fig. 1.4: Inheritance of the direction of shell coiling in the snail *Limnaea*. Sinistral coiling is determined by the recessive allele d, dextral coiling by the wild-type D allele. However, the direction of coiling is determined by the nuclear genotype of the mother, not by the genotype of the snail itself or by extranuclear inheritance. The F_2 and F_3 generations can be obtained by self-fertilization, because the snail is hermaphroditic and can undergo either self-fertilization or cross-fertilization.

(vii) $\frac{3}{4}$ of the F_3 progeny have dextrally coiled shells and $\frac{1}{4}$ have sinistrally coiled shells. This is typically 3 : 1 ratio with dextral coiling (DD & Dd) dominant over sinistral coiling (dd). The ratio indicates Mendelian segregation but it is not observed in F_2 generation.

The Mendelian ratio is delayed for one generation.

(viii) The coiling phenotype of an individual snail is determined by the genotype of the mother. This means that direction of shell coiling is not a case of **maternal inheritance** rather a case of **maternal effect**.

The cytological analysis of developing eggs has provided this explanation.

The genotype of the mother determines the orientation of the spindle in the initial mitotic division after fertilisation and thus in turn controls the direction of shell coiling of the offspring.

- **Shell coiling is not a case of maternal inheritance but rather a case of maternal effect**

(i) Cytological analysis of developing egg provides this explanation.

(ii) The genotype of the mother's egg determines the orientation of the spindle in the initial mitotic division after fertilization and this in turn controls direction of shell coiling of the offspring.

(iii) It begins during the second cleavage division in zygotes.

(iv) In the next phase of the cleavage spiral patterns are established either in dextral, i.e., right-ward coiling or sinistral, i.e., left-ward coiling. The sperm genotype is irrelevant because the expression of sperm's gene would occur too late.

- **The mechanism of maternal effect on snail coiling.**

1. Genes → Traits:
   Nurse cells are present in the egg. Gene products are transferred from nurse cell to an egg.
2. If nurse cells are homozygous or heterozygous (DD/Dd), they will transfer D gene product to the egg, thereby resulting offspring will be **dextral.** If the nurse cells are dd, only the d gene product will be transferred to the egg, so the resulting offspring will be **sinistral.**
3. A **dd** mother contribute only d gene product that promote a left-handed, i.e., sinistral body, even if the egg is fertilized by a sperm carrying a 'D' allele. The sperm's genotype is irrelevant, because the expression of sperm's gene would occur too late.

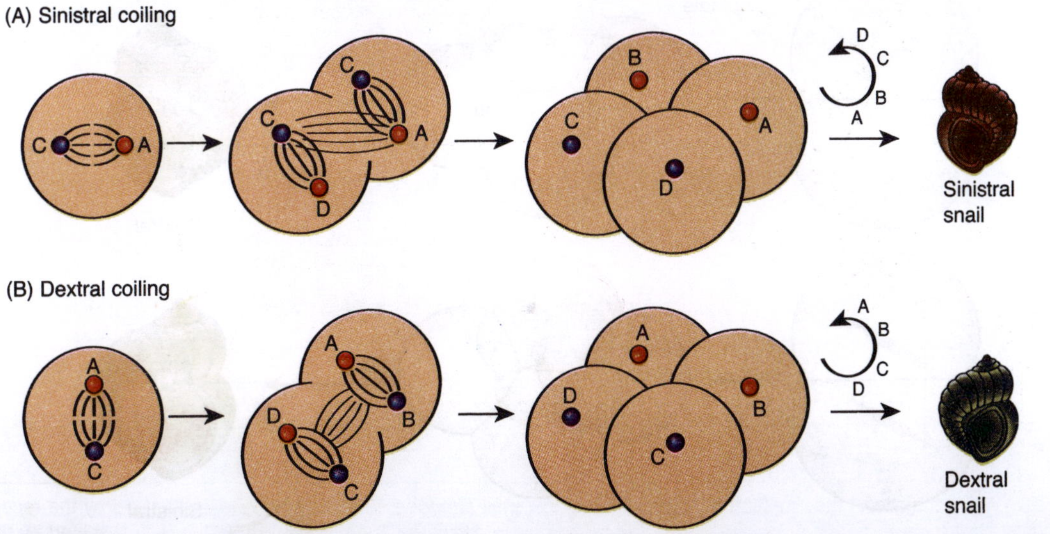

Fig. 1.5: The direction of shell coiling in *Limnaea*, is determined by the orientation of mitotic spindles during the second cleavage division in the zygote which is determined by the Egg cytoplasm as a result of mother cytoplasm. (A) Sinistral left orientation, (B) Dextral right orientation.

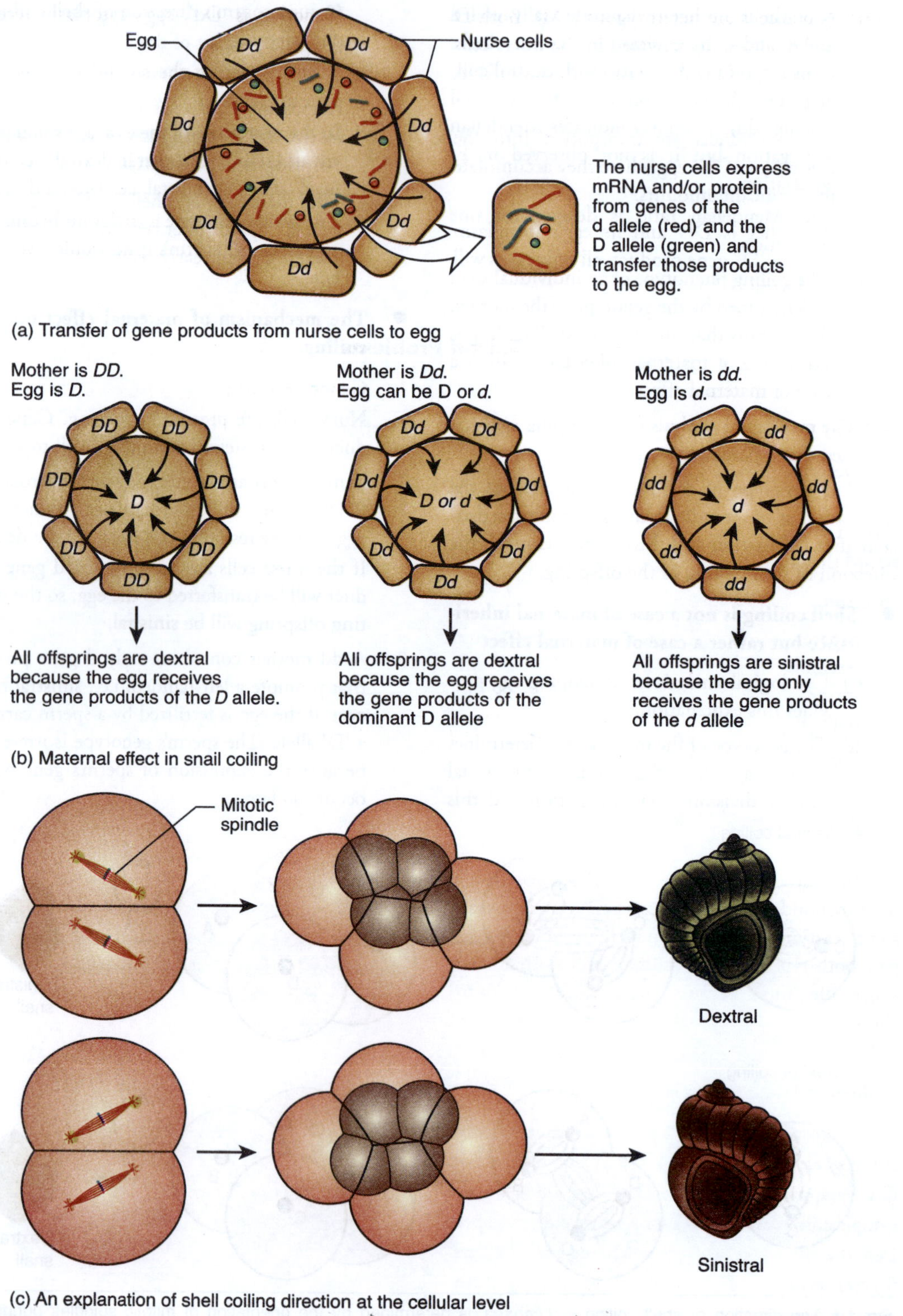

Fig. 1.6: The mechanism of maternal effect in snail coiling

(i) Nurse cells are heterozygous (Dd). Both D and **d** alleles are activated in the nurse cells to produce D and d gene products (mRNA or protein).

(ii) These products are transported into the cytoplasm of the egg where they accumulate in significant amounts.

(iii) If mother is DD and its egg's cytoplasm receives 'D" gene product, all offsprings will be **dextral**.

If mother is 'Dd', egg's cytoplasm receives the gene product of dominant gene 'D'. All offsprings will be dextral.

If mother is 'dd', it's egg's cytoplasm will receive the gene products of 'd' allele and all offsprings will be **sinistral**.

(iv) The direction of shell coiling is determined by differences in the cleavage planes during early embryonic development.

Solved Problems

Problem 1

Describe how biparental pattern of extra-nuclear inheritance would resemble a Mendelian pattern of inheritance for a particular gene. How would they differ?

Solution

1. Biparental extranuclear inheritance would resemble Mendelian inheritance if that offspring could inherit alleles of given genes from both parents.
2. For a Mendelian trait, the law of segregation tells us that a heterozygote passes one allele for a given gene to an offspring but not both.

In contrast, if a parent has a mixed population of mitochondria (e.g. some carrying mutant gene z, some carrying normal gene) that parent could pass both types of gene (mutant and normal) to a single offspring, because more than one mitochondrian could be contained within a sperm or egg cell.

Problem 2

A cross between the killer strain and sensitive strain *of paramoecium* produced F_1 that were killers (kk + kappa) and sensitive (kk) depending on the cytoplasmic exchange. What result would you expect from the following: (a) A cross between two F_1 killers. (b) A cross between F_1 killer and F_1 sensitive. (c) Autogamous reproduction of the F_1 sensitive. (d) Crosses of a kk sensitive (produced in c) to a F_1 killer.

Solution

Killer — Nonkiller

KK (kappa) — kk

F_1 = killer (Kk + kappa) — Sensitive (Kk)

(a) Killer (Kk + kappa) × (Kk + kappa) killer
→ 3 killer (1 KK : 2 Kk) : 1 kk) (sensitive)

(b) Killer (Kk + kappa) × Sensitive (kk)
Killer exconjugant produces 3 killers and one sensitive. Descendent of sensitive exconjugant will be senstive since they do not have kappa).

(c) All sensitive.

(d) (kk) (sensitive) × (Kk) (killer)
Descendent of killers will be killers. They have all kappa and at least one dose of dominant gene K.

Problem 3

Determine which of the three *Paramecial* phenotypes (killer, unstable or sensitive) is produced by the following combination of genotypes and cytoplasmic state.

	Genotype	Cytoplasm
1.	KK	kappa
2.	Kk	no kappa
3.	kk	kappa
4.	KK	no kappa
5.	Kk	kappa
6.	kk	no kappa

Solution

(a) Killer (1, 5); (b) Unstable (3), (c) Sensitive (2, 4, 6).

Problem 4

In *Paramoecium*, the maintenance of kappa particles requires the dominant nuclear gene K. A Kk killer cell conjugates with a sensitive cell of the same genotype without cytoplasmic exchange. Predict the genotypes and phenotypes, so that result of each exconjugant then undergoes autogammy.

Solution

1. Conjugation produces exconjugants with same genotype, but which haploid micronucleus survives in each cell is random.
2. Therefore, 25% genotypes of the exconjugants are expected to be KK, 25% are to be kk and 50% are to be Kk.
3. The sensitive exconjugant remains sensitive, regardless of which of the genotypes is present.
4. Autogamy does not affect the genotype of homozygote conjugants, but it does affect the heterozygote. Among the killer exconjugants we expect the following:

Killer exconjugants (with kappa)	Autogamous products	Phenotypic ratio
$\frac{1}{4}$KK	All KK	$\frac{1}{4}$killer } $\frac{1}{2}$killer
$\frac{1}{2}$Kk	$\frac{1}{2}$KK	$\frac{1}{4}$killer }
	$\frac{1}{2}$kk	$\frac{1}{4}$sensitive (kappa are lost)
$\frac{1}{4}$kk	All kk	$\frac{1}{4}$sensitive (kappa are lost)

Problem 5

From the matings of *Paramoecium*, predict the phenotypic ratios among exconjugants under normal conditions of cytoplasmic matings.

Conjugating *Paramecium*

	Genotype	Cytoplasm	×	Genotype	cytoplasm
(a)	KK	kappa		KK	no kappa
(b)	KK	no kappa		KK	no kappa
(c)	KK	kappa		kk	no kappa
(d)	Kk	kappa		kk	no kappa

Solution:

							Normal conjugation	Cytoplasm mixing
(a)	KK	kappa	×	KK	no kappa	→	1 killer, 1 sensitive	Both killers
(b)	KK	no kappa	×	KK	no kappa	→	Both sensitive	Both sensitive
(c)	KK	kappa	×	kk	no kappa	→	1 killer, 1 sensitive	Both killers
(d)	kk	kappa	×	kk	no kappa	→	1 killer, 1 unstable 2 sensitive	1 killer, 1 unstable

Problem 6

Mention the percentage of exconjugants from the following *Paramoecium* matings will possess and be able to maintain kappa through subsequent asexual division?

	Mating type	
	Kappa present ×	**Kappa absent**
(a)	**KK**	**KK**
(b)	**KK**	**Kk**
(c)	**KK**	**kk**
(d)	**Kk**	**Kk**
(e)	**Kk**	**kk**
(f)	**kk**	**kk**

Solution:

	Normal conjugation	**Cytoplasmic mixing**
(a)	$\frac{1}{2}$ = 50%	Both 100%
(b)	$\frac{1}{2}$ = 50%	Both 100%
(c)	$\frac{1}{2}$ = 50%	Both 100%
(d)	$\frac{1}{4}$ = 25%	$\frac{3}{4}$ = 75%
(e)	$\frac{1}{4}$ = 25%	$\frac{1}{2}$ = 50%
(f)	– 0%	0 = 0%

Problem 7

Most strains of *Chlamydomonas* are sensitive to streptomycin. A strain is found which require streptomycin in culture medium for its survival. How couild it be determined where streptomycin dependence is due to a chromosomal gene or a cytoplasmic element.

Solution:

SD = Streptomycin dependent

SS = Streptomycin sensitive

(i) SDmt^+ (female) × SSmt^- male

If chromosomal $\frac{1}{4}$ (25) SDmt^+, $\frac{1}{4}$ (25%) SDmt^-, $\frac{1}{4}$ (25%) SSmt^+, $\frac{1}{4}$ (25%) SSmt^-;

(ii) If cytoplasmic, all should follow maternal line, i.e., streptomycin dependent.

Problem 8

A streptomycin-resistant strain of *Chlamydomonas* carrying resistant factors located both in the nuclus and in cytoplasms was crossed to a streptomycin sensitive strain.

What result would you expect (a) if the resistant strain was the plus parent and the sensitive strain the minus parent? (b) if the reciprocal cross was made?

Solution:

(a) (♀) mt^+ Str^r × mt^- Str^s (♂)

All are streptomycin resistance, mt allele in the nucleus segregates in 1 : 1 ratio as Mendelian fashion.

(b) (♀) mt^+ Str^s × mt^- Str^r (♂)

All are streptomycin sensitive, mt allele segregates in 1 : 1 ratio as Mendelian fashion.

Problem 9

When chloroplast DNA from *Chlamydomonas* is digested with a particular restriction enzyme and then hybridized with a particular probe, two bands are detected. Some strains type (1) yield bands of 1.5 kb and 3.7 kb, other strains type (2) yield bands of 2.5 kb and 6.0 kb. For the following crosses predict the progeny.

(a) mt^+ strain 1 × mt^- strain 2

(b) mt^+ strain 2 × mt^- strain 1

Solution:

(a) mt^+ strain-1 × mt^- strain-2

All are type 1, i.e., 1.5 and 3.7 kb

(b) mt^+ strain-2 × mt^- strain-1

All are type 2, i.e., 2.5 and 6.0 kb.

Chloroplast DNA does not appear in the present.

Problem 10

If a and b are chloroplast markers in *Chlamydomonas*, what genotypes of progeny are expected from the cross a^+ b^- mt^+ × a^- b^+ mt^-?

Solution:

a^+ b^- mt^+ × a^- b^+ mt^-

(i) mt^- allele is preferentially lost and the mt allele segregate in 1 : 1 ratio.

(ii) The expected result is $\frac{1}{2}$ a^+ b^- mt^+; $\frac{1}{2}$ a^+ b^- mt^-.

Problem 11

Consider three types of antibiotic resistance (r) versus sensitivity (s) in *Chlamydomonas*.

(a) Cpl results from a mutation in chloroplast DNA. (b) mit-r results from a mutation in mitochondrial DNA. (c) muc-r results from a mutation in nuclear chromosomal DNA.

If a cross of resistant mt^+ × sensitive mt^- is carried out for each of these three types of resis-

tances individually, what types of progeny are expected, and in exact proportions? Explain your answers.

Solution:

(a) Resultant$^+$ × Sensitive mt$^-$
All progeny will be cpl-r because chloroplast came from the mt$^+$ parent.

(b) All progeny will be mt-r because mitochondria comes from the mt$^+$ parent.

(c) muc-r is present in chronosomal DNA, i.e., in nucleus.
Therefore, result will be $\frac{2:2}{1:1}$. It obeys Mendelian segregation.
muc-r mt$^+$ × muc-s mt$^-$
muc-r mt$^+$, muc-r mt$^-$, muc-s mt$^+$, muc-s mt$^-$
resistant 2 , sensitive 2 or 1 : 1.

Problem 12

What possible phenotypes and genotypes could the female parent of a sinistrally coiled snail have?

Solution:

(i) If a snail is sinistrally coiled, its mother must have had the dd genotype, since sinistrality is recessive.

(ii) If the female parent is recessive homozygote, its mother must have contributed a recessive 'd' allele.

(iii) Therefore, its mother (the grand mother) could have had either a Dd or dd genotype.

(iv) Its daughter could, therefore, be either dextrally or sinistrally coiled.

Thus, a sinistrally coiled snail could have had a mother that was either dextrally or sinistrally coiled, but only of the dd genotype.

Problem 13

A dextral snail is self-fertilized and produces only sinistral progeny. What is the probable genotype of the snail and its parents?

Solution:

(i) Since the selfing produces only sinistral progeny, individual must be homozygous for sinistral coiling (dd).

(ii) Individual must get one 'd' allele from each parent. So, each parent must be at least heterozygous.

(iii) Since a dextral snail is self-fertilized, the mother must be heterozygous. The father could be either D/d or d/d.

(iv) Female parent (dextral) × Male parent

Dd × Dd/dd

Gamete: D d; D, d / d

F_1 Sinistral (dd)

Problem 14

The direction in which the shell coils in hermophroditic snail *Limnaea peregra* is a maternal effect. They can reproduce either by crossing or by self-fertilization. A homozygous dextral snail is fertilized with sperm from a homozygous sinistral snail. The heterozygous F_1 individual undergoes two generation of self-fertilisation.

(a) Diagram the parents F_1 and two selfing generations showing phenotypes and genotypes and their expected ratio.

(b) What are the phenotypes of the parental individual.

Solution:

(a)

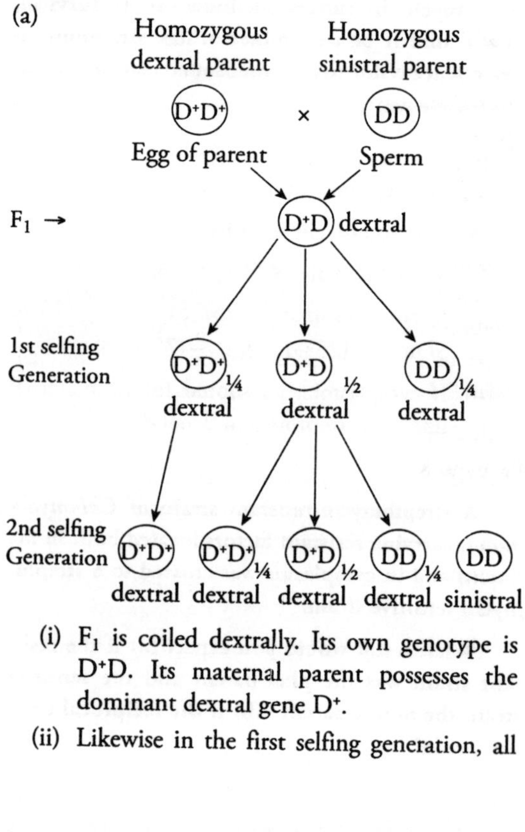

(i) F_1 is coiled dextrally. Its own genotype is D$^+$D. Its maternal parent possesses the dominant dextral gene D$^+$.

(ii) Likewise in the first selfing generation, all

First selfing Generation	Second selfing Generation	Summary: Genotypes	Summary: Phenotypes
D^+D^+ $(\frac{1}{4}) \times \rightarrow$	D^+D^+ (All) $\frac{1}{4}$	$\frac{1}{4} + \frac{1}{8} = \frac{3}{8}$	$\frac{5}{8}$ Dextral
D^+D $(\frac{1}{2}) \times$	D^+D^+ $(\frac{1}{4} \times \frac{1}{2}) = \frac{1}{8}$; D^+D $(\frac{1}{2} \times \frac{1}{2}) = \frac{1}{4}$; DD $(\frac{1}{4} \times \frac{1}{2}) = \frac{1}{8}$	D^+D $\frac{1}{4} = \frac{1}{4}$	
DD $(\frac{1}{2}) \times$	DD (All)	DD $\frac{1}{8} + \frac{1}{4} = \frac{3}{8}$	Sinistral

are phenotypically dextral regardless of their own genotype because the F_1 was D^+D.

(iii) In the second selfing generation the phenotypes are dextral and sinistral.

(b) (i) We know the genotypes of the parents, we have no information concerning the genotypes of the immediate maternal ancester, which was responsible for the organization of the egg cytoplasm from which the parental individuals are developed.

(ii) Therefore, we are unable to determine what phenotypes those individuals exhibit.

Problem 15

In Sagar island, it was found that a male snail coils to the right. However, in the same area several shells (not containing living snails) are obtained that coil to the right. Therefore, it is not possible to predict the genotype of the male snail on a different island near to Sagar, where a large colony of snails of the same species was found. All of these snails coil to the right, every snail that has been found on the Sagar island coils to the right. With regard to the maternal effect of gene that determines coiling pattern, how would you determine the genotype of the male snail obtained from the Sagar island? Describe the expected results.

Solution:

A. (i) We assume that the snails in the large colony found in the other island are true breeder with DD genotype.

(ii) Let the male snail of Sagar island mate with the female snail from the large colony.

(iii) Let the F_1 snails mate with each other to produce F_2 generation.

Then let F_2 generation mate with each other to produce F_3 generation.

- Female DD × Male DD
  D D

  $F_1 \rightarrow$ DD (coil to right, i.e., dextral)
- $F_2 = F_1 \times F_1$ DD × DD → DD (Dextral) (coil to the right)

  $F_3 \rightarrow$ All (DD) coil to right, i.e., dextral

B. Female DD × Male DD

$F_1 \rightarrow$ (DD)/DD = Coil to the right

F_2 ♀ Dd × Dd ♂ → Snails coil to the right

$F_3 \rightarrow DD(\frac{1}{4}), Dd(\frac{1}{2}), dd(\frac{1}{4}) = \frac{1}{4} + \frac{1}{2} = \frac{1+1}{4} = \frac{3}{4}$ (coil to the right)

C. Female DD × Male dd

$F_1 \rightarrow$ All Dd (Coil to the right)

F_2 Female DD × Male Dd

$DD(\frac{1}{4}) + Dd(\frac{1}{2})$ & $dd(\frac{1}{4}) = \frac{3}{4}$(coil to the right)

$\frac{1}{4}$(coil to the left)

Problem 16

The inheritance of shell coiling direction in the snail *Limnea seregra* has ben studied extensively. A snail produced by a cross between two individuals has a shell with a right-hand twist (dextral coiling). This snail produces only left-hand (sinistral) progeny on selfing. What are the genotypes of the F_1 snail and its parents.

Solution:

(i) The F_1 snail gives sinstral offspring when selfied. So, it is dd. Therefore both parents had a 'd' allele.

(ii) Since, the F_1 has a dextral pattern, its maternal parent was Dd. The paternal parent could have been either Dd or dd.

P ♀ (Maternal) × ♂ (Paternal)

Dd × Dd/dd

Gamete: D d   D, d or only d

F_1 D d (Dextral)

$F_2 = F_1 \times F_1 =$ D d × D d

D d   D d

F_2 = Sinistral =   d d (Sinistral)

Problem 17

The conjugation between a KK killer strain of *Paramoecium* and kk nonkiller strain, why is cytoplasmic exchange necessary to obtain 1:1 ratio of killer : no n-killer after autogamy of the exconjugants has taken place? What is the ratio of killer : non-killer in the absence of cytoplasmic exchange?

Solution:

A. Consider the situation without cytoplasmic exchange.

(i) In this case the KK killer exconjugant becomes Kk (kappa) which means that the kappa bacterium is present in the cytoplasm and the kk non-killer exconjugant becomes Kk (no kappa).

(ii) Immediately after autogamy, the ratio becomes $\frac{1}{4}$ KK (kppa). $\frac{1}{4}$ kk (kappa) $\frac{1}{4}$ KK (no kappa) $\frac{1}{4}$ kk (no kappa).

(iii) The kappa bacterium cannot be maintained in kk (kappa) strain, so the overall ratio of killer : non-killer is that of (kappa) : (no kappa) which is 1 : 3.

B. (i) On the other hand, with cytoplasmic exchange, the $\frac{1}{4}$ KK (no kappa) and $\frac{1}{4}$ kk (no kappa) become $\frac{1}{4}$ KK (kappa) and $\frac{1}{4}$ kk (kappa).

(ii) The latter cannot maintain kappa, but the former can. So the overall ratio of killer : non-killer with aptoplanic exchange is 1 : 1.

(iii) The cytoplanic exchange is necessary to obtain the 1 : 1 ratio, because without the K allele necessary to maintain kappa, it cannot create kappa.

(iv) The kappa bacteria must be transferred for the killer partner.

Problem 18

A snail of the species *limnaea peregra* in which the cell coils to the left undergoes self-fertilization, and all of the progeny coil to the right. What is genotype of the parent? If the progeny are individually self-fertilized, what coiling phenotypes are expected?

Solution:

(i) In this species shell coiling is determined by the genotype of the mother. Here the snail is left coiling, so its mother had the genotype Ss. Therefore, it must carry at least 's' allele.

(ii) On the other hand, after selfing it produces right coiling progeny, so its genotype must be S^+s.

(iii) The progeny genotype would be $S^+s \times S^+s$

	S^+	s
S^+	S^+S^+	S^+s
s	S^+s	ss

$S^+S^+ = \frac{1}{4}$

$\left.\begin{matrix} S^+S \\ S^+S \end{matrix}\right\} = \frac{1}{2}$

$ss = \frac{1}{4}$

(iv) The S^+S^+ and S^+s yield rightward coiled progeny and ss individuals yield leftward coiled progeny.

Problem 19

A scientist carried out reciprocal crosses between two species A and B of the evening *Primrose* known to have same chromosome constitution, when the seed parent was A, the plastids of the progeny were green and when the seed parent was B, the plastid of the progeny were yellow. How might these difference in the results of reciprocal crosses be explained?

Solution:

(i) It is given that during the reciprocal crosses between two species A and B of the evening primrose which have same chromosome constitution, when the seed parent was A, the plastid of the progeny were green and when the seed parent was B, the plastid of the progeny were yellow.

(ii) The result indicates that the plastids of the offsprings are similar to the female parent, which indicates the presence of extranuclear inheritance as the maternal inheritance is involved.

Linkage and Chromosome Mapping

LINKAGE

When two or more genes located in the same chromosome are said to be linked for the obvious reason that they are physically linked or joined together by the segment of chromosome between them, it indicates

- Genes are linked
- Alleles of the genes are linked
- Gene loci are linked.

Genes present in the same chromosome have the tendency to be inherited together, from one generation to the other. This phenomenon is known as **linkage**.

• LINKAGE PHASE

There are two possible arrangements of two pairs of linked genes:

1. The genes on the same chromosome are said to be in **coupling phase**, while those on the opposite chromosomes of the pair are in the **repulsion phase**.
2. The arrangement of genes in particular genotype is called **linkage phase**.

Both the linked genes have their normal (dominant) alleles in one chromosome and mutant (recessive) alleles in the other, e.g.,

$$\frac{\text{A B}}{\text{a b}}.$$

(i) A and B in one chromosome and a and b on the other chromosome of the same homologous pair,

$$\frac{\text{A B}}{\text{a b}}.$$

Heterozygote with this configuration is called **cis-heterozygote.**

(i) The normal alleles (also mutant ones) lie in opposite chromosome of the homologous pair, e.g., $\frac{\text{A b}}{\text{a B}}$

The heterozygote of such configuration is called **trans-heterozygote.**

• LINKAGE MAP

1. The diagrammatic graphical representation of relative distances between linked genes of a chromosome is called **linkage map** or **genetic map.**
2. It is also known as **crossover map** because this map is generated by estimating the number of times of crossover event occurred in a particular segment of chromosome out of all meiosis examined.

Significance:

1. This map displays the exact locations, arrangement and combinations of genes in a linkage group or chromosome.
2. It shows the number of linkage groups of an organism.
3. It shows the relative distance and chances of crossing-over between the genes.
4. They are useful in predicting the results of dihybrid and trihybrid crosses.

• DETERMINATION OF MAP DISTANCE

1. The distance between two given genes is calculated according to the percentage of crossing-over because crossover frequency is directly proportional to distance between the genes.
2. This distance is measured in **map units** (m.u.). A crossover frequency of 1 per cent

between two genes is defined as **map unit.** It is also expressed as **morgan unit**, i.e., 100% crossing-over.

The map unit is sometimes referred to as **centimorgan** (cM), i.e. one percent crossing-over in honour of geneticist, **T. H. Morgan**, the first geneticist to win the Nobel Prize. 10% crossing-over referred to one **deci-morgan**.

3. Although the map unit is not a physical distance, it is a relative distance that makes it possible to know the order of and relative separation between loci on a chromosome.

• CYTOLOGICAL OR PHYSICAL MAP

It is the physical location of a gene in a chromosome which is accomplished by identifying a gene locus with relation to some visible landmark, viz., chromomere, or cross band is termed as physical or cytological map.

• LINKAGE MAP VS. PHYSICAL MAP

1. The frequency of crossing-over usually varies in different segments of the chromosome, but it is highly predictable event between the two gene loci.
2. The actual physical distances between linked genes bear no direct relationship to the linkage map distances.
3. The linear order of gene is identical in both cases.

• LINKAGE GROUP

All the genes which are linked with one another constitute a **linkage group** and tend to be inherited together. The number of linkage group in a species is equal to the number of homologous pairs of chromosome or haploid number of chromosome of that species.

Name of the organism	Number of pairs of homologous chromosome	Number of linkage group
Drosophila	4 pairs	4
Human	23 pairs	23
Maize	10 pairs	10

• TYPES OF LINKAGE

Linkage 〈 Sex-linked / Autosomal 〈 Complete / Incomplete

(a) Complete Linkage

In this type of linkage genes are closely associated and tend to transmit together upon coming from the same parent for two or more generations in a continuous and regular fashion.

Example:

The genes for bent wings (bt) and shaven bristles (svn) of the fourth chromosome mutant of *Drosophila melanogaster* show complete linkage.

P: Normal fly $\frac{bt^+\ svn^+}{bt^+\ svn^+}$ × Bent shaven $\frac{bt\ svn}{bt\ svn}$

Gametes: $bt^+\ svn^+$, $bt\ svn$

F_1: $\frac{bt^+\ svn^+}{bt\ svn}$ × $\frac{bt\ svn}{bt\ svn}$ (Back-cross)

Gametes: $bt^+\ svn^+$, $bt\ svn$ and $bt\ svn$

F_2:

♂ \ ♀	$bt^+\ svn^+$	$bt\ svn$
$bt\ svn$	$\frac{bt^+\ svn^+}{bt\ svn}$	$\frac{bt\ svn}{bt\ svn}$

Result:

Genotype	Phenotype
$\frac{bt^+\ svn^+}{bt\ svn}$	Wild type (normal)
$\frac{bt\ svn}{bt\ svn}$	Bent shaven

Inference:

(i) Parental combination restored.

(ii) Complete absence of independent assortment.

(b) Incomplete Linkage

In this type of linkage genes are widely located in the chromosome and have chances of separation by crossing-over. Hence, the parental combination is more than that of recombinant.

Example:

It is found in female *Drosophila* and various other organisms.

When F_1 flies of grey-body long wing (heterozygous) were test crossed to double recessive black-body vestigial wing, four types of progenies were obtained.

P Grey-body long wing × Black-body vestigial wing

$$\frac{b^+\ vg^+}{b^+\ vg^+} \times \frac{b\ vg}{b\ vg}$$

Gametes: $\underline{b^+\ vg^+}$ $\underline{b\ vg}$

F_1 $\frac{b^+\ vg^+}{b\ vg}$ (Grey-body long wing)

$$\frac{b^+\ vg^+}{b\ vg} \times \frac{b\ vg}{b\ vg} \text{ (test crossed)}$$

Gametes: $\underline{b^+\ vg^+}$, $\underline{b\ vg}$, $\underline{b^+\ vg}$, $\underline{b\ vg^+}$; $\underline{b\ vg}$

{Non-crossover}{Recombinants}

♂ \ ♀	$b^+\ vg^+$	$b\ vg$	$b^+\ vg$	$b\ vg^+$
$b\ vg$	$\frac{b^+\ vg^+}{b\ vg}$	$\frac{b\ vg}{b\ vg}$	$\frac{b^+\ vg}{b\ vg}$	$\frac{b\ vg^+}{b\ vg}$

Result:

	Genotype	Phenotype	Ratio	
(i)	$\frac{b^+\ vg^+}{b\ vg}$	Grey long	41%	Parental combination
(ii)	$\frac{b\ vg}{b\ vg}$	Black vestigial	41%	
(iii)	$\frac{b^+\ vg}{b\ vg}$	Grey vestigial	9%	Recombinant combination
(ii)	$\frac{b\ vg^+}{b\ vg}$	Black long	9%	

Inference: (i) Here parental combination is more than 50%.

(ii) Recombinant is less than 50%.

CROSSING-OVER

Recombination of genes on the same chromosome is accomplished by crossing-over, a process by which parts of homologous chromosomes are interchanged.

• IMPORTANT FEATURES OF CROSSING-OVER

1. The loci of the genes on a chromosome are arranged in a linear sequence.
2. At any point of crossing-over two non-sister chromatids of a **bivalent** are involved.
3. Crossing-over involves the breakage of each of two **homologous chromosomes** (actually chromatids) and the exchange of parts.
4. Crossing-over occurs at pachytene after the synapsis of the homologous chromosome has occurred in prophase. Crossing-over occurs at **4-chromatid stage**.
5. The probability that crossing-over will occur between two loci increases with increasing distances between the two loci on chromosome.
6. Crossing-over is always **less than 50%**.

• FACTORS CONTROLLING CROSSING-OVER

1. **Temperature**: High and low temperatures increase the frequency of crossing-over.
2. **Maternal age**: Bridges showed that when maternal age increases, crossing-over tends to decrease.
3. **Radiation**:
   (i) X-ray irradiation increases crossing-over (Muller).
   (ii) Radium irradiation increases frequency of crossing-over.
4. **Sex**:
   (i) Crossing-over in *Drosophila* **male** is negligible.
   (ii) Males of mammals also show little crossing-over.
   (iii) In silk moth crossing-over does not occur in **female**.
   (iv) A third chromosome gene (**C 3G**) in *D. melanogaster* completely suppresses crossing-over when homozygous is female.
5. **Chromosome structure**:
   (i) Near centromere the tips of the chromosome crossing-over are less frequent.
   (ii) **Inversion** of chromosome suppresses C.O. (crossing-over suppressor/C factor). In heterozygous condition, '**C' factor** suppresses crossing-over.

6. **Mutation:**
   (i) Certain mutant genes called **minutes** (small bristles) increase the frequency of crossing-over.
   (ii) Some mutations decrease crossing-over.
7. **Distance:** Chiasma formation at one point discourages the chiasma formation at the nearest vicinity due to strain on the chromatid. This is known as **chiasma interference.**

- **CHIASMA FREQUENCY AND GENETIC MAP DISTANCE**

Sl. No.	No. of Chiasmata (A)	No. of cells observed (B)	Product A × B
1.	0	10	0 × 10 = 00
2.	1	25	1 × 25 = 25
3.	2	30	2 × 30 = 60
4.	3	15	3 × 15 = 45
5.	4	15	4 × 15 = 60
6.	5	5	5 × 5 = 25
		100	215

1. Recombination between the genetic marker is an outcome of crossing-over.
2. Each chiasma is thought to represent the resolution of a crossing-over.
3. The average number of crossovers occuring on a chromosome can be estimated by counting the chiasmata.
4. It has the use as an estimate of genetic map.

- Average number of chiasmata per cell $= \frac{215}{100} = 2.15$
- Average number of chaismata per chromatid $= \frac{2.15}{2} = 1.07$
- Genetic length of chromosome = 1.07 Morgans = 107 centimorgan.
- Relationship between genetic length and average number of chiasma $= \frac{107}{2.15} = 50$ cM/chiasma.

We can conclude that average each chiasma corresponds to 50 cM on the genetic map.

- **Three-point crosses are a valuable approach to learn about the nature of linkage of genes.**
  (i) The use of three-point crosses confirms the linearity of genes on chromosomes predicted from two locus crosses.
  (ii) In addition, one three-point cover can give as much information as three crosses involving two genes each.
- **Generally genetic map distance is similar to physical distance on the chromosome, but in some cases it is not justified.**

Recombination may be localized into certain regions.

Recombination:

(i) Recombination involves physical exchange of material between duplex DNAs

(ii) Three different types of events share this feature: Homologous recombination, Site specific recombination and Transposition.

(iii) (a) Homologous recombination involves exchange of precisely corresponding sequences between homologous DNA duplexes.

(b) Site specific recombination occurs between specific pairs of DNA sequences, since the enzymes involve can act only on particular pair of target sequences.

(c) In transposition breakage reunion of DNA strands allows one DNA sequence to be inserted into another without reliance on sequence homology.

INTERFERENCE

Interference means the tendency of one crossing-over to interfere in the occurence of another crossing-over in its nearest vicinity or adjacent region.

The phenomenon of chiasma interference or chromosome interference was first observed by H. J. Muller (1916).

Characteristics

1. The degree of interference may vary in different regions of the chromosome.
2. It is found to be **inversely proportional** to the distance between the points of crossing-over.

   It means interference is maximum over a short distance and decreases as distance increases. It is maximum near centromere and telomere.

   In *Drosophila,* interference prevents D.C.O. between the genes of **10-15 units** apart.
3. The degree of interference is measured by **coefficient of coincidence.**

$$\left\{ \text{c.c.} = \frac{\text{observed D.C.O.}}{\text{expected D.C.O.}} \right\}$$

4. Sometimes the level of interference is given as the **coefficient of interference** which equals 1-minus **coefficient of coincidence.**

TYPES

1. **Positive interference:**
   (i) Positive interference indicates that one crossing-over interferes with the occurrence of second crossing-over nearby.
   (ii) Interference value between 0 and 1 indicates that positive interference (partial) has occurred.
   (iii) Interference is cent percent (absolute) when double crossovers are altogether absent.
   (iv) If the frequency of double crossovers is equal to the expected value, it means that there is no interference (none).
   (v) It is generally observed in **eukaryotes.**
2. **Negative interference:**
   (i) Negative interference indicates that the occurrence of one crossover increases the likelihood of additional crossovers occurred in the nearest vicinity (so-called **clustering of crossing-over**).
   (ii) It occurs in certain microorganism particularly in **bacteriophages, fungus** (*Aspergillus*).

Table 2.1: Difference between linkage and crossing over

Linkage	Crossing over
1. It keeps two or more genes to stay together during inheritance.	1. It leads exchange and separation of linked genes.
2. It involves individual genes.	2. It involves between non-sister chromatids of homologous chromosome.
3. It reduces variation.	3. It increases variations by forming new gene combination.
4. Strength of linkage increases as the distance between the two linked genes decreases.	4. Frequency of crossing over decrease with the decrease of distance between genes.
5. Linkage groups can never be more than haploid number.	5. Frequency of crossing over always less than 50 percent.
6. It provides higher frequency of parental types than recombinant progeny among test-cross progeny.	6. It provides equal frequency of parental and reombinant types among test cross progeny.

Table 2.2: Difference between positive and negative interference

Positive interference	Negative interference
1. One crossing over reduces the chance of another crossing-over in its nearest vicinity.	1. One crossing-over enhances the occurrence of another crossing over in its nearest vicinity.
2. It is observed both in eukaryotes and prokaryotes.	2. It is observed in lower organisms (bacteria and fungus).
3. Here coefficient of coincidence is less than one.	3. Here coefficient of coincidence is always more than one.

Table 2.3: Difference between Linkage and Independent assortment

Linkage	Independent assortment
1. Here two pairs of genes are located in the same chromosome, i.e., the genes are like $\frac{A\ B}{a\ b}$.	1. Here two pairs of genes (A and B) located in the different chromosome $\frac{A}{a}\ \frac{B}{b}$.
2. At the time of gamete formation during meiosis the linked genes fail to separate, because there is no crossing-over.	2. The genes (A & B) assorted independently at the time of gamete formation during meiosis.
3. In absence of crossing-over only two types of gametes are formed with equal proportions. Only parental types are produced [AB(50%) ab(50%)]. When they have undergone crossing-over, four types of gametes are produced.	3. Here four types of gametes are produced in equal proportion.

COINCIDENCE

The coincidence or **coefficient of coincidence** is the inverse measure of interference and is expressed as the ratio between the observed (actual) number (frequency) of double crossover and the expected number (frequency) of double crossover.

Coefficient of coincidence =

$$\frac{\text{observed double crossover frequency}}{\text{expected double crossover frequency}}$$

Characteristics

1. It is the complement of interference, i.e., coincidence + interference = 1.0.
2. Coincidence becomes **zero** when interference is complete (1.0) and therefore, no **double crossing-over.** Coincidence values ordinarily vary between **0 and 1.**
3. Coefficient of coincidence larger than one indicates the occurrence of double crossing-over more frequent than expected.

Coefficient of Coincidence	Frequency of D.C.O.	Type of Interference
0	If D.C.O. = 0%	absolute
<1	D.C.O less frequent than expected	partial
1	Observed D.C.O. = Expected D.C.O.	none
>1	D.C.O. more frequent than expected	negative

Cytological basis of Crossing over:

In 1931 C. Stern showed the direct cytological evidence of crossing over in ***Drosophila*** where homologous chromosomes exchange parts during crossing over.

1. Normally the two chromosomes of any homologous pair are morphologically indistinguishable. **Stern, Creighton** and **Mc Clintock** identified homologous chromosome that were morphologically distinguishable, i.e., they were not entirely homologous.

2. They paired and segregated normally during meiosis.

- **Requirements:**

Stern's special type of *Drosophila* female:

(i) Two X-chromosomes differ from each other.

(ii) One X-chromosome was shorter than normal and was composed of two approximately equal fragments, each with its own centromere. This X-chromosome had recessive gene **car** (carnation, i.e., light red colour) and the dominant gene B (Bar eye shape).

(iii) The other X-chromosome was of normal length but a segment of Y-chromosome was translocated into its short arm. This chromosome had the dominant gene car⁺ (wild-type, i.e., red-eye

Experimental design:

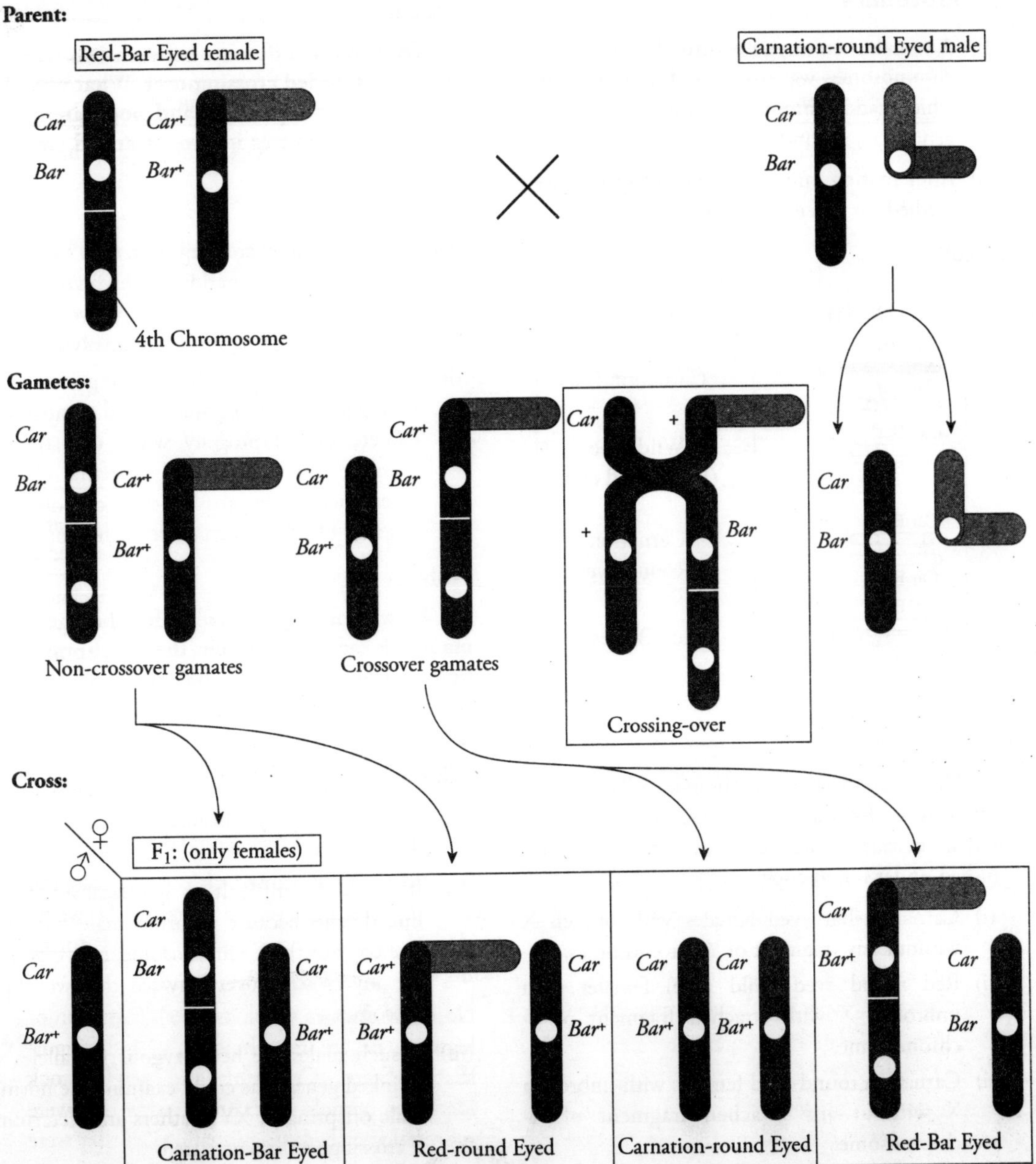

colour) and recessive gene B⁺ (wild-type, i.e., ovate shape). Thus both aberrant X-chromosomes of female heterozygote were cytologically detectable.

(iv) This female was heterozygous for their morphologically distinguishable X-chromosomes (easily detected in microscope).

(v) A male fly had car (carnation) and B⁺ (round eyes).

- **Procedure**

(i) A female heterozygote with the aberrant X-chromosomes was mated with a normal male which had X-chromosome with both recessive genes, i.e., car and B⁺ alleles.

(ii) After mating only the female offsprings were studied for this experiment.

Result

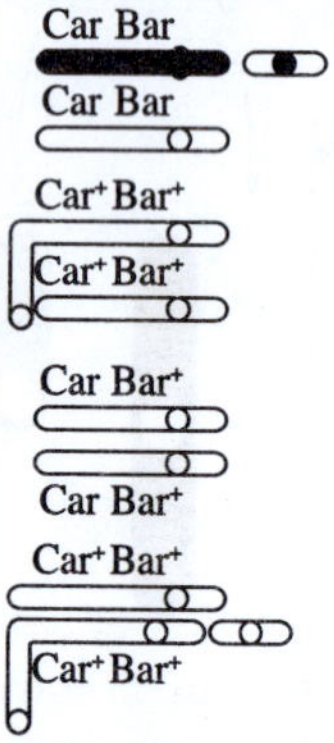

	Genotype	Phenotype
1.	Car Bar / Car Bar	Carnation Bar
2.	Car⁺ Bar⁺ / Car⁺ Bar⁺	Wild type (Red round eye)
3.	Car Bar⁺ / Car Bar⁺	Carnation Round eye
4.	Car⁺ Bar⁺ / Car⁺ Bar⁺	Red Bar eye

Observation

The chromosomes of observed offspring are examined under microscope. Y-chromosome was used as marker so that cross-over can be distinguished under microscope.

(i) Carnation-Bar eyed females with broken X without any fragment of Y-chromosomes.

(ii) Red round eyed (wild type) females with unbroken X with attached fragment of Y-chromosome.

(iii) Carnation round eyed females with unbroken X without any attached fragment of X-chromosome.

(iv) Bar eyed females with broken X with attached fragment of Y-chromosome.

Inference

(i) Genetic crossing-over, i.e., recombination was accompanied by cytological crossing-over.

(ii) Actual exchange of chromatin material between homologous chromosome was observed phenotypically as well under microscope, i.e., cytologicallly.

Problem

Discuss the difference between two-stranded and four-stranded crossing-over. What would be the proportions (parental and non-parental) of the resulting gametes in two- stranded crossing-over?

Solution

(i) In four-stranded crossing-over, only two of the four present chromatids are involved in any recombination event while in two-stranded crossing-over all (in both) are involved.

(ii) As a result four-stranded crossing-over generally will give both parental and nonparental gamete in the progeny while two-stranded crossing-over will give only parental (100%) when there is no crossing-over, or only non-parental (100%) when there is crossing-over.

Problem

How would you determine that genes in mammals are located on the Y chromosome linkage group? Is it possible to conduct crosses (Let's say in mice) to map the distances between genes along the Y chromosome? Explain.

Solution

(i) A gene in the Y chromosome in mammals would be transmitted only from father to son.

(ii) It would be difficult to genetically map Y-linked genes because a normal male has only one copy of the Y chromosome, so we do not get any crossing over between the two chromosomes.

(iii) If such males are heterozygous for alleles of Y-linked genes, one could examine the normal male offspring of XYY fathers and determine if crossingover has occurred.

Solved Problems

Problem 1

In a three-point test-cross $\frac{ABC}{abc} \times \frac{abc}{abc}$, following data are obtained (only phenotypes are given):

ABC	abc	aBc	AbC	ABc	abC	aBC	Abc	Total
230	240	96	104	138	142	12	8	970

Find out the correct linear order of the genes. Calculate the map distance between the genes and the coefficient of coincidence.

Solution:

(i) With the three gene pairs, eight phenotypic classes are expected and eight are observed.

(ii) The most frequent reciprocal pairs of classes (phenotypes) which arise from non-crossover gametes, tell us the genotypes of the original parents.

Here ABC and abc are the parental classes, i.e., 230 + 240 = 470.

(iii) The order of the genes can be determined by inspecting the least frequent reciprocal pairs of phenotypic classes that represent the results of double crossing-over.

Here aBC and Abc are the double crossover classes, i.e., 12 + 8 = 20.

So, the frequency of D.C.O. $= \frac{20}{970} \times 100 = 2.06\%$.

The gene pair that has changed its position relative to the other two pairs of genes (alleles) is the central gene which in this case is A/a. Therefore, the order of genes is BAC (or CAB).

Diagram of F_1 cross is as follows:

$$\frac{BAC}{bac} \times \frac{bac}{bac}$$

(iv) A single crossing-over between B and A genes gives the recombinants Bac and bAC which have 96 and 104 members respectively for a total of 200 (96 + 104).

Therefore, frequency of S.C.O. between B and A is $\frac{200}{970} \times 100 = 20.61\%$.

(v) Similarly, other single crossing-over between A and C genes gives the BAc and baC recombinants which have 138 and 142 members respectively for a total of 280 (138 + 142).

Therefore, frequency of S.C.O. between A and C is $\frac{280}{970} \times 100 = 28.86\%$.

(vi) The map distance between B and A is the all crossover between B and A. So we must add D.C.O.

B – A = S.C.O. + D.C.O. = 20.61 + 2.06 = 22.67%.

Similarly, the map distance between A and C = S.C.O. + D.C.O. = 28.86 + 2.06 = 30.92%.

(vii) B $\underline{\quad 22.67\% \quad}$ A $\underline{\quad 30.92\% \quad}$ C (Map)

The coefficient of coincidence =

$$\frac{\text{Frequency of observed double crossing-over}}{\text{Frequency of expected double crossing-over}}$$

Expected D.C.O. $= \frac{22.67}{100} \times \frac{30.92}{100} = 0.2267 \times 0.3092 = 0.07$.

Observed D.C.O. $= \frac{2.06}{100} = 0.02$.

C.C. $= \frac{0.02}{0.07} = 0.2857 = 0.29$.

So, the coefficient of coincidence is 0.29 or 29%.

Problem 2

Determine the constitution of heterozygous parent, gene sequence and map distance from the following data:

+b+	abc	a+c	++c	a++	+bc	ab+	+++
104	180	109	21	5	5	31	191

Solution:

(i) With the three-gene pairs, eight phenotypic classes are expected and eight are observed.

(ii) The most frequent reciprocal pairs of classes (phenotypes) which arise from non-crossover gametes, tell us the genotypes of the original parents.

Here +++ and abc are the parental classes, i.e. 191 + 180 = 371.

(iii) The order of the genes can be determined by inspecting the least frequent reciprocal pairs of phenotypic classes that represent the results of double crossing-over.

Here a++ and +bc are the double crossing-over classes.

Therefore, frequency of D.C.O. = $\frac{5+5}{646} \times 100$ = 1.54 = 1.54%.

The gene pair that has changed its position relative to the other two pairs of genes (alleles) is the central gene which in this problem is a.

Therefore, the order or sequence of gene is bac or cab.

Constitution of heterozygous parent is $\frac{+++}{bac} \times \frac{bac}{bac}$.

(iv) A single crossing-over between b and a produces the b++ and +ac recombinants, which have members 104 and 109 respectively for a total of 213 (104 + 109).

Therefore, frequency of S.C.O. between b and a is $\frac{104+109}{646} \times \frac{213}{46} \times 100 = 32.97\%$.

(v) Similarly, other single crossing-over between a and c produces ++c = 21 and ab+ = 31 progeny.

The frequency of S.C.O. between a and c is $\frac{21+31}{646} \times 100 \times \frac{52}{646} = 8.049\% = 8.05\%$.

The map distance between b and a = S.C.O. + D.C.O. = 32.97 +1.54 = 34.51.

The map distance between a and c = S.C.O. + D.C.O. = 8.05 + 1.54 = 9.59%.

The above-mentioned data reveal the following map:

$$b \xrightarrow{34.51\%} a \xrightarrow{9.59\%} c.$$

Problem 3

In *Drosophila*, Lyra (Ly) and Stubble (Sb) are dominant mutations at loci 40 and 58 respectively on chromosome 3. A recessive mutation with bright red eyes was discovered and shown also to be on chromosome 3. A map was obtained by crossing a female who was heterozygous for all three mutations to a male homozygous for the bright red mutations (which here refer to as br). The following data are obtained:

	Phenotypes			Number
1.	**Ly**	**Sb**	**br**	**404**
2.	**+**	**+**	**+**	**422**
3.	**Ly**	**+**	**+**	**18**
4.	**+**	**Sb**	**br**	**16**
5.	**Ly**	**+**	**br**	**75**
6.	**+**	**Sb**	**+**	**59**
7.	**Ly**	**Sb**	**+**	**4**
8.	**+**	**+**	**br**	**2**
			Total	**1000**

Determine the location of the br mutation on chromosome 3.

Solution:

(i) With the three-gene pairs, eight phenotypic classes are expected and eight are observed.

(ii) The most frequent reciprocal pairs of classes (phenotypes) which arise from non-crossover gametes tell us the genotype of the original parents. Here these are phenotypes (1) and (2). Each one represents the arrangement of alleles (genes) on one of the homologous. Therefore, the arrangement is

$$\frac{\text{Ly Sb br}}{+\ +\ +}$$

(iii) The order or correct sequence of the genes can be determined by inspecting the least frequent reciprocal pairs of phenotypic classes that represent the results of double crossing-over. Here these phenotypic classes are (7) and (8).

Therefore, frequency of double crossing-over is $\frac{4+2}{1000} \times 100 = \frac{6}{1000} = 100 = 0.6\%$.

The gene pair that has changed its position relative to the other two gene pairs (alleles) is the central gene which is br/+. Therefore, the sequence of gene is as follows:

$$\frac{\text{Ly br Sb}}{+\ +\ +}$$

(iv) Inspection shows single crossing-over between Ly and br yields phenotypes of classes (3) and (4), i.e., Ly ++ and + br Sb.

Therefore, crossing-over frequency between Ly and br is $\frac{18+16}{1000} \times 100 = \frac{34}{1000} \times 100 = 3.4\%$.

(v) Similarly, single crossing-over between br and Sb yields the phenotypic classes (5) and (6), i.e., Ly br + and + + Sb.

Therefore, crossing-over frequency between br and Sb is $\frac{18+16}{1000} \times 100 = \frac{134}{1000} \times 100 = 13.4\%$.

(vi) For mapping distance between Ly and br, we must add D.C.O. with S.C.O.

Therefore, Ly – br = 3.4 + 0.6 = 4%.

Similarly, the distance between br and Sb is 13.4 + 0.6 = 14%.

The final map is

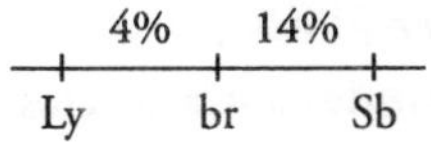

It shows that br is located at the locus 44, since Lyra and stubble are known.

Problem 4

In the Chinese primrose, slate-coloured flower(s) is recessive to blue flower (S); red stigma (r) is recessive to green stigma (R); and long style (*l*) is recessive to short style (L). All these genes are located in the same chromosome. The F_1 of a cross between two breeding strains, when test-crossed, gave the following progeny:

	Phenotypes	Number
1.	slate flower, green stigma, short style	27
2.	slate flower, red stigma, short style	85
3.	blue flower, red stigma, short style	402
4.	slate flower, red stigma, long style	977
5.	slate flower, green stigma, long style	427
6.	blue flower, green stigma, long style	95
7.	blue flower, green stigma, short style	960
8.	blue flower, red stigma, long style	27
	Total	3000

(a) What are the genotypes of the parents in the cross of the two true breeding strains?

(b) Make a map of these genes, showing gene order and distance between them.

(c) Derive the coefficient of coincidence for interference between these genes.

Solution:

(a) (i) With the three-gene pairs, eight phenotypic classes are expected and eight are observed.

(ii) The most frequent reciprocal pairs of classes (phenotypes) which arise from non-cross-over gametes, tell us the genotypes of the original parents. Here the two classes (4 and 7) are slate flower, red stigma, long style and blue flower, green stigma short style (977 + 960).

Thus, the F_1 triple heterozygous parent of this generation must have been SRL/srl. So the true breeding parents were $\frac{SRL}{SRL}$ (blue, green, short) and $\frac{srl}{srl}$ (slate, red, long).

(b) (i) The order of genes can be determined by inspecting the least frequent reciprocal pairs of phenotypic classes that represent the results of double crossing-over.

Here the double-crossover classes are 1 and 8, i.e., slate plus green plus short (sRL) and blue plus red plus long (Srl), which have 27 and 27 members respectively for a total of 54 (27 + 27). Therefore, frequency of double-crossing-over (D.C.O.) is $\frac{54}{3000} \times 100 = 1.8\%$.

(ii) The gene pair that has changed its position relative to the other two pairs of genes (alleles) is the central gene. Here S/s has changed its position. Therefore, the order of gene is RSL (or LSR). We can diagram the F_1 test-cross as follows:

$$\frac{RSL}{rsl} \times \frac{rsl}{rsl}$$

(iii) A single crossing-over between R and S gives the green plus slate plus long (Rsl) and red plus blue plus short (rSL) which have 427 and 402 members respectively for a total of 829 (427 + 402). Therefore, the frequency of S.C.O. between R and S is $\frac{829}{3000} \times 100 = 27.63\%$.

(iv) Similarly, a single crossing-over between S and L produces green plus blue plus long (RSl) and red plus slate plus short (rsL) which have 95 and 85 members respectively for a total of 180. Therefore, frequency of S.C.O. (between S and L) is $\frac{180}{3000} \times 100 = 6\%$.

(v) The map distance between R and S is S.C.O. + D.C.O. = 27.63 + 1.8 = 29.43%.

The map distance between S and L is 6 +1.8 = 7.8%.

Map: R ——— S ——— L
←— 29.4 —→←— 7.8 —→

(c) The coefficient of coincidence = $\frac{\text{frequency of obserbed D.C.O.}}{\text{frequency of expected D.C.O.}}$

Expected D.C.O. = $\frac{29.4}{100} \times \frac{7.8}{100} = 0.023$,

Observed D.C.O. = 1.8% = 0.018,

Coefficient of coincidence = $\frac{0.018}{0.023} = 0.78 = 78\%$.

Problem 5

The following numbers were obtained from test-cross progeny in *Drosophila* (Phenotypes):

	Phenotypes			Number
1.	+	m	+	218
2.	w	+	f	236
3.	+	+	f	168
4.	w	m	+	178
5.	+	m	f	95
6.	w	+	+	101
7.	+	+	+	3
8.	w	m	f	1
			Total	1000

Construct a genetic map.

Solution:

(i) The most frequent reciprocal pairs of classes (phenotypes) which arise from non-crossover gametes tell us the original parental progeny. Here (1) and (2), i.e., + m + and w + f are the parental classes.

(ii) The order of genes can be determined by inspecting the least frequent reciprocal pairs of phenotypic classes that represent the results of double crossing-over. Here (7) and (8), i.e., + + + and wmf are the double crossover progeny, which have 3 and 1 members respectively for a total of 4. Therefore, the frequency of double crossing-over is $\frac{4}{1000} \times 100 = 0.4\%$.

(iii) The gene pair that has changed its position relative to the other two pairs of genes (alleles) is the central gene. Here +/m has changed its position. Therefore, the order of gene is wmf (or fmw). We can diagram the F_1 test-cross as follows:

$$\frac{\text{wmf}}{+\,+\,+} \times \frac{\text{wmf}}{\text{wmf}}$$

(iv) A single crossing-over between w and m gives the progeny (5 and 6), i.e., +mf and w++ which have 95 and 101 members respectively for a total of 196(95 + 101). Therefore, the frequency of S.C.O. between w and m = $\frac{196}{1000} \times 100 = 19.6\%$.

(v) Similarly, a single crossing-over between m and f gives the progeny 3 and 4, i.e., ++f and wm+ which have 168 and 178 members respectively for a total of 346(168 + 178). Therefore, the frequency of S.C.O. between m and f is $\frac{346}{1000} \times 100 = 34.6\%$.

(vi) The map distance between w and m = S.C.O. + D.C.O. = 19.6 + 0.4 = 20%.

The map distance between m and f = S.C.O. + D.C.O. = 34.6 + 0.4 = 35%.

w ——— m ——— f
←— 20% —→←— 35% —→

Problem 6

Assume that in *Drosophila* there are three pairs of alleles, $\frac{+}{x}$, $\frac{+}{y}$ and $\frac{+}{z}$. As shown by symbols, each mutant gene is recessive to its wild type allele. A cross between female heterozygous at these three loci and wild type males yielded the following progeny:

	Phenotypes			Number
Females:	+	+	+	1010
Males:	+	+	+	39
	+	+	Z	430
	+	Y	Z	32
	X	+	+	27
	X	Y	+	441
	X	Y	Z	31
			Total	2010

Draw the appropriate linkage map for these data showing the order of the three markers and the map distances for each marked interval.

Calculate the coefficient of coincidence for these data.

Solution:

(i) These genes are sex-linked, thus male progenies are hemizygous for X chromosome markers that are transmitted through eggs (from their female parents).

(ii) All female progenies will necessarily carry the wild type alleles of all three genes, namely, wild type alleles transmitted through the sperm (from their male parents).

(iii) Therefore, linkage relationship can be determined only by analyzing male progeny of the cross. Only six genotypes are represented among the progeny. No double crossover types are present among the progeny analyzed.

However, the missing double crossover genotypes can be determined by elimination.

Cross: Female heterozygous Male wild type

$$\frac{YX+}{++Z} \times \frac{+++}{\longrightarrow}$$

Gamets: YX+, ++Z, Y+Z, +X+, +++, YXZ

Females		Males		
(i) $\frac{YX+}{+++}$	wild type	(i) $\frac{YX+}{\longrightarrow}$		430
(ii) $\frac{++Z}{+++}$	wild type	(ii) $\frac{++Z}{\longrightarrow}$		441
(iii) $\frac{Y+Z}{+++}$	wild type	(iii) $\frac{Y+Z}{\longrightarrow}$		32
(iv) $\frac{+X+}{+++}$	wild type	(iv) $\frac{+X+}{\longrightarrow}$		27
(v) $\frac{YXZ}{+++}$	wild type	(v) $\frac{YXZ}{\longrightarrow}$		31
(vi) $\frac{+++}{+++}$	wild type	(vi) $\frac{+++}{\longrightarrow}$		39
		Total	=	1000

(i) The most frequent reciprocal pairs of classes which arise from non-crossover gametes tell us the genotypes of the parents.

Here wild type females are $\frac{YX+}{+++}, \frac{++Z}{+++}$; Males are $\frac{YX+}{\longrightarrow}$ and $\frac{++Z}{\longrightarrow}$, their numbers are 1010 females and 871 (430 + 441) males.

(ii) Single crossing-over between Y and X gives the Y + Z and + X + which have 32 and 27 members respectively for a total of 59. Therefore, the frequency of S.C.O. between X and Y is $\frac{59}{2000} \times 100 = 5.9\%$.

(iii) Similarly, other single crossing-over between X and Z produces YXZ and + + + which have 31 and 39 members respectively for a total of 70. Therefore, the frequency of S.C.O. between X and Z is $\frac{70}{1000} \times 100 = 7\%$.

(a) Map: $Y \underset{5.9 \text{ m.u.}}{\text{———}} X \underset{7 \text{ m.u.}}{\text{———}} Z$

(b) Coefficient of coincidence = $\frac{\text{Observed D.C.O.}}{\text{Expected D.C.O.}}$

Observed D.C.O. = 0, Expected D.C.O. = $\frac{5.9}{100} \times \frac{7}{100} = 0.059 \times 0.07 = 0.00413$

Coefficient of coincidence = $\frac{0}{0.00413} = 0$. Therefore, interference is 100%.

Problem 7

Three of the many recessive mutations in *Drosophila melanogaster* that affect body colour, wing shape or bristle morphology are black (b) body versus grey in wild type, dumpy (dp), obliquely truncated wing versus long wing in the wild type and hooked (hk) bristles at the tip versus not hooked in the wild type. From a cross of a dumpy female with a black and hooked male, all the F_1 are wild type for all three characters. The test-cross of an F_1 female with a dumpy, black, hooked male gave the following results:

	Phenotypes	**Number**
1.	**wild type**	**169**
2.	**black**	**19**
3.	**black hooked**	**301**
4.	**dumpy hooked**	**21**
5.	**hooked**	**8**
6.	**hooked, dumpy, black**	**172**
7.	**dumpy black**	**6**
8.	**dumpy**	**304**
	Total	**1000**

(a) Construct linkage map of the linkage group (or groups) these genes occupy. If applicable, show the order and give the map distances between the genes.

(b) (i) Determine coefficient of coincidence for the portion of chromosome involved in this cross. (ii) How much interference is there?

Solution:

(i) With the three-gene pairs eight phenotypic classes are expected and eight are observed.

(ii) The most reciprocal pairs of classes (phenotypes) which arise from non-crossover gametes tell us the genotypes of the original parents. Here the two classes (3 and 8), i.e., black hooked and dumpy (301 + 304) represent the parents.

Cross: dumy female × black hooked male

$$\frac{+dp+}{+dp+} \times \frac{b+hk}{\longrightarrow}$$

$$F_1 \quad \frac{+dp+}{b+hk} \times \frac{b\,dp\,hk}{\longrightarrow}$$

Thus, F_1 triple heterozygous parent of this generation must have been $\frac{+dp+}{b+hk}$.

(iii) The order of genes can be determined by inspecting the least frequent reciprocal pairs of phenotypic classes that represent the results of double crossing-over. Here the double crossing-over classes 5 and 7 are hooked and dumpy black have 8 and 6 members respectively for a total of 14(8 + 6).

Therefore, the frequency of double crossing-over is $\frac{14}{1000} \times 100 = 1.4\%$.

(iv) The gene pair that has changed its position relative to the other two pairs of genes (alleles) is the central gene. Here +/b has changed its position. Therefore, b is the central gene. Therefore, F_1 genotypic configuration is $\frac{dp++}{+b\,hk} \times \frac{dpbhk}{\longrightarrow}$.

(v) A single crossing-over between dp and b produces dumpy black hooked and wild type progeny which have 172 and 169 members respectively for a total of 341(169 + 172). Therefore, the frequency of S.C.O. between dp and b is $\frac{341}{1000} \times 100 = 34.1\%$.

(vi) Similarly, the single crossing-over between b and hk gives black and dumpy plus hooked which have 19 and 21 members respectively for total of 40(19 + 21). Therefore, the frequency of S.C.O. between b and hk is $\frac{40}{1000} \times 100 = 4\%$.

(vii) The map distance between dp and b is S.C.O. + D.C.O. = 34.1 + 1.4 = 35.5% or 35.5 mu.

The map distance between b and hk is S.C.O. + D.C.O. = 4 + 1.4 = 5.4% or 5.4 mu.

(a) The order of genes in the chromosome is dp – b – hk, with 35.5 mu between dp and b and 5.4 m.u. between b and hk.

```
dp                    b             hk
|---------------------|-------------|
<------- 35.5 ------><--- 5.4 ----->
```

(b) (i) The observed D.C.O. is 1.4%, i.e., 0.014 and expected D.C.O. is 35.5% × 5.4%, i.e., $\frac{35.5}{100} \times \frac{5.4}{100} = 0.355 \times 0.054 = 0.0191 = 0.019$.

Coefficient of coincidence $= \frac{\text{Observed D.C.O.}}{\text{Expected D.C.O.}} = \frac{0.014}{0.019} = 0.7368 = 0.74 = 74\%$.

(ii) The interference value is 1 – coincidence = 1 – 0.74 = 0.26 or 26%.

Problem 8

In the tomato mutant genes o (oblate = flattened fruit), p (peach = hairy fruit), and s (compound inflorescence = many flowers in a cluster) were found to be on chromosome 2. From the following data (test-cross mating of an F_1 heterozygote for all three genes × homozygous recessive for all three genes), determine: (a) the sequence of these three genes in chromosome 2, (b) the genotypes of the homozygous parents used in making the F_1 heterozygote, (c) the recombination distance between the genes, (d) the coefficient of coincidence:

	Phenotypes			Number
1.	+	+	+	73
2.	+	+	s	348
3.	+	p	+	2
4.	+	p	s	96
5.	o	+	+	110
6.	o	+	s	2
7.	o	p	+	306
8.	o	p	s	63
			Total	1000

Solution:

(i) The most frequent reciprocal pairs of classes which arise from non-crossover gametes, tell us the genotypes of the original parents. Here classes (2) and (7), i.e., + + s and o p + are parental classes (348 + 306).

Thus, F_1 genotypic configuration is $\frac{++s}{op+}$.

(ii) The order of genes can be determined by inspecting least frequent reciprocal pairs of phenotypic classes that represent the result of double crossing-over. Here the double crossing-over classes are (3) and (6), i.e., + p + and o + s which have 2 and 2 members respectively for a total of 4. Therefore, the frequency of double crossing-over is $\frac{4}{1000} \times 100 = 0.4\%$.

(iii) The gene pair that has changed its position relative to the other two pairs of genes (alleles) is the central gene. Here +/o has changed its position. So the order of genes is p o s (or s o p). Diagram of the F_1 test-cross is as follows:

$$\frac{++s}{po+} \times \frac{pos}{pos}$$

(iv) A single crossing-over between p and o produces the + o + and p + s which have 110 and 96 members respectively for a total of 206 (110 + 96).

Therefore, the frequency of S.C.O. between p and o is $\frac{206}{1000} \times 100 = 20.6\%$.

(v) Similarly, a single crossing-over between o and s gives + + + and p o s which have 73 and 63 members respectively for a total of 136 (73 + 63).

Therefore, frequency of S.C.O. between o and s is $\frac{136}{1000} \times 100 = 13.6\%$.

(vi) The map distance between p and o is S.C.O. + D.C.O. = 20.6 + 0.4 = 21%.

The map distance between o and s is S.C.O. + D.C.O. = 13.6 + 0.4 = 14%.

(a) The sequence of three genes is

p —— o —— s

|← 21 mu →|← 14 mu →|

(b) Genotypes of the homozygous used in making F_1 heterozygotes

P_1: $\frac{++s}{++s} \times \frac{po+}{po+}$ or, $\frac{s++}{s++} \times \frac{+op}{+op}$

↓

F_1: $\frac{++s}{po+} \times \frac{pos}{pos}$

(c) The recombination distance:

(i) p–o distance—21% or 21 mu

(ii) o–s distance—14% or 14 mu

(iii) p–s distance 20.6 + 13.6 = 34.2% = 34.2 mu

(d) Expected double-crossing-over is $\frac{21}{100} \times \frac{14}{100} =$ 0.21 × 0.14 = 0.0294

Observed D.C.O. = 0.4% = $\frac{4}{100}$ = 0.004

Coefficient of coincidence = $\frac{\text{Observed D.C.O.}}{\text{Expected D.C.O.}} = \frac{0.004}{0.029} = 0.1379 = 0.14$.

Therefore, coefficient of coincidence = 0.14 × 100 = 14%.

Problem 9

Two normal looking *Drosophila* are crossed and yield the following phenotypes among the progeny:

	Phenotypes			Number
Females:	**+**	**+**	**+**	**2000**
Males:	**+**	**+**	**+**	**3**
	a	**b**	**c**	**1**
	+	**b**	**c**	**839**
	a	**+**	**+**	**825**
	a	**b**	**+**	**86**
	+	**+**	**c**	**90**
	a	**+**	**c**	**81**
	+	**b**	**+**	**75**
			Total	**4000**

Give parental genotypes, gene arrangement in the female parent, map distances, and the coefficient of coincidence. **CU. 2013**

Solution:

(i) These genes are sex-linked, thus male progenies are hemizygous for X chromosome markers that are transmitted through egg (from their female parents)

(ii) All female progenies will necessarily carry the wild type alleles of all three genes, namely, wild type alleles transmitted through the sperm (from their male parents)

(iii) Therefore, linkage relationship can be determined only by analyzing male progeny of the cross.

Cross: Normal male × Normal female

$$♂\ \frac{+\ +\ +}{\longrightarrow} \quad \times \quad \frac{b\ +\ e}{+\ a\ +}\ ♀$$

Gametes: (i) $\frac{+\ +\ +}{}$, ——→; (ii) b + c, + a +, b a +, + + c, b + +, + a c, b a c, + + +

Females		Males	
(i) $\frac{b\ +\ c}{+\ +\ +}$	wild type	(i) $\frac{b\ +\ c}{\longrightarrow}$	839
(ii) $\frac{+\ a\ +}{+\ +\ +}$		(ii) $\frac{+\ a\ +}{\longrightarrow}$	825
(iii) $\frac{b\ a\ +}{+\ +\ +}$		(iii) $\frac{b\ a\ +}{\longrightarrow}$	86
(iv) $\frac{+\ +\ c}{+\ +\ +}$		(iv) $\frac{+\ +\ c}{\longrightarrow}$	90
(v) $\frac{b\ +\ +}{+\ +\ +}$		(v) $\frac{b\ +\ +}{\longrightarrow}$	75
(vi) $\frac{+\ a\ c}{+\ +\ +}$		(vi) $\frac{+\ a\ c}{\longrightarrow}$	81
(vii) $\frac{b\ a\ c}{+\ +\ +}$		(vi) $\frac{b\ a\ c}{\longrightarrow}$	1
(vii) $\frac{+\ +\ +}{+\ +\ +}$		(vi) $\frac{+\ +\ +}{\longrightarrow}$	3
		Total =	2000

(i) The most frequent reciprocal pairs of classes which arise from non-crossover gametes tell us the genotypes of the parents. Here wild type females are $\frac{+\ +\ +}{b\ +\ c}$ and $\frac{+\ +\ +}{+\ a\ +}$, and males are $\frac{b\ +\ c}{\longrightarrow}$ and $\frac{+\ a\ +}{\longrightarrow}$. Their numbers are 2000, 1664 (839 + 825).

(ii) Single crossing-over between b + a gives the b a + and + + c which have 86 and 90 numbers respectively for a total of 176. Therefore, frequency of S.C.O. between b and a is $\frac{176}{2000}$ = 100 = 8.8%.

(iii) Similarly, other single crossing-over between a and c gives rise b + + and + a c which have 75 and 81 members respectively for a total of 156. Therefore, frequency of S.C.O. between a and c is $\frac{156}{2000}$ = 100 = 7.8%.

(iv) The order of genes can be determined by inspecting the least frequent reciprocal phenotypic classes that represent the result of double crossing-over. Here + + + and a b c are the double crossing-over classes and have 3 and 1 members respectively for a total of 4. Therefore, the frequency of D.C.O. is $\frac{4}{2000}$ = 100 = 0.2%.

(v) The map distance between b and a is the all crossover between b and a. So we must add D.C.O. b – a = S.C.O. + D.C.O. = 8.8 + 0.2 = 9.0 mu.

Similarly, map distance between a and c is S.C.O. + D.C.O. = 7.8 + 0.2 = 8.0 mu.

Parental genotype $\frac{+\ +\ +}{\longrightarrow}$ (male) & $\frac{b + c}{+ a +}$ (female)

(c) The map distance = a —— 9.0 —— b —— 8.0 —— c

$$\text{Coefficient of coincidence} = \frac{\text{Freq. of observed D.C.O.}}{\text{Freq. of expected D.C.O.}}$$

Expected D.C.O. = $\frac{9}{100} \times \frac{8}{100}$ = 0.09 × 0.08 = 0.0072

Observed D.C.O. = 0.2% = $\frac{0.2}{100}$ = 0.002.

Coefficient of Coincidence (CC) = $\frac{0.002}{0.0072}$ = 0.27777 = 0.278 = 27.8%.

So the coefficient of coincidence is 27.8%.

Problem 10

Genes a, b, c are recessive. Females heterozygous at these three loci are crossed to phenotypically wild type male. The progeny are phenotypically as shown:

	Phenotypes				Number
Daughters:	+	+	+		
Sons:	+	+	+	=	23
	a	b	c	=	26
	+	+	c	=	45
	a	b	+	=	54
	+	b	c	=	427
	a	+	+		424
	a	+	c	=	1
	+	b	+	=	0
			Total	=	**1000**

(a) What is known about the genotype of the female parent with respect to these three loci? Give gene order and the arrangement in the homologous.

(b) What is known of the genotype of the male parent ?

(c) Map the three genes and coefficient of coincidence.

Solution:

(i) These genes are sex-linked, thus male progenies are hemizygous for X chromosome markers that are transmitted through eggs (from their female parents).

(ii) All daughter progenies will necessarily carry the wild type alleles of all three genes, namely, wild type alleles transmitted through sperm (from their male parents).

(iii) Therefore, linkage relationship can be determined only by analyzing male progeny of the cross.

Cross: Wild type male × Wild type female (heterozygous)

$$\frac{+\ +\ +}{\longrightarrow} \male \quad \times \quad \frac{+\ c\ b}{a\ +\ +} \female$$

Gametes: (i) + c b, a + +, + + +, a c b, + c +, a + b, + + b, a c, or (a + c)

(ii) $\frac{+\ +\ +}{}$, ——→

Daughter (females):		Son (males):	
(i) $\frac{+\ c\ b}{+\ +\ +}$	wild type	(i) $\frac{+\ c\ b}{\longrightarrow}$	427
(ii) $\frac{a\ +\ +}{+\ +\ +}$	wild type	(ii) $\frac{a\ +\ +}{\longrightarrow}$	424
(iii) $\frac{+\ +\ +}{+\ +\ +}$	wild type	(iii) $\frac{+\ +\ +}{\longrightarrow}$	23
(iv) $\frac{a\ c\ b}{+\ +\ +}$	wild type	(iv) $\frac{a\ c\ b}{\longrightarrow}$	26
(v) $\frac{+\ c\ +}{+\ +\ +}$	wild type	(v) $\frac{+\ c\ +}{\longrightarrow}$	45
(vi) $\frac{a\ +\ b}{+\ +\ +}$	wild type	(vi) $\frac{a\ +\ b}{\longrightarrow}$	54
(vii) $\frac{a\ +\ c}{+\ +\ +}$	wild type	(vi) $\frac{a\ +\ c}{\longrightarrow}$	1
(vii) $\frac{+\ b\ +}{+\ +\ +}$	wild type	(vi) $\frac{+\ b\ +}{\longrightarrow}$	0
		Total =	1000

(i) The most frequent reciprocal pairs of classes which arise from non-crossover gametes tell us the genotypes of the parents. Here wild type daughters (females) are $\frac{+\ c\ b}{+\ +\ +}$ and $\frac{a\ +\ +}{+\ a\ +}$, and sons (males) are $\frac{+\ c\ b}{\longrightarrow}$ and $\frac{a\ +\ +}{\longrightarrow}$. Their numbers are, including all daughters and sons 851(427+424).

(ii) Single crossing-over between a and c gives + + + and a c b which have 23 and 26 members respectively for a total of 49. Therefore, frequency of S.C.O. between a and c is $\frac{49}{1000} \times 100 = 4.9\%$.

(iii) Similarly, other single crossing-over between c and b gives + c + and a + b which have 45 and 54 members respectively for a total of 99. Therefore, frequency of S.C.O. between c and b is $\frac{99}{1000} \times 100 = 9.9\%$.

(iv) The order of genes can be determined by inspecting the least frequent reciprocal phenotypic classes that represent the result of double crossing-over. Here a + c and + b + are double crossing-over classes and have 0 and 1 members respectively for a total of 1. The frequency of D.C.O. is $\frac{9}{1000} \times 100 = 0.1\%$.

(v) The map distance between a and c is all the crossover between a and c. So we must add D.C.O. a–c = S.C.O.+D.C.O. = 4.9%+0.1%=5.0 mu.

Similarly, the map distance between c and b is the all crossover between c and b. So we must add D.C.O. c – b = S.C.O. + D.C.O. = 9.9% + 0.1% = 10.0% = 10 mu.

(a) Genotype of female parent: $\frac{+\ c\ b}{a\ +\ +}$

(b) Genotype of male parent: $\frac{+\ +\ +}{\longrightarrow}$

(c) The map distance = a —— 5 mu —— c —— 10 mu —— b

$$\text{Coefficient of coincidence} = \frac{\text{Freq. of observed D.C.O.}}{\text{Freq. of expected D.C.O.}}$$

$$\text{Expected D.C.O.} = \frac{5}{100} \times \frac{10}{100} = 0.05 \times 0.1 = 0.005$$

$$\text{Observed D.C.O.} = 0.1\% = \frac{0.1}{100} = 0.001.$$

Coefficient of Coincidence $= \frac{0.001}{0.005} = 0.2$, i.e., $0.2 \times 100 = 20\%$.

Problem 11

The following three recessive markers are known in lab mice: h, hot foot; o, obese and wa, waved. A trihybrid of unknown origin is test crossed, producing the following offspring:

Phenotypes	Number
hot foot, obese, waved	357
hot foot, obese	74
waved	66
obese	79
wild type	343
hot foot, waved	61
obese, waved	11
hot foot	9
Total =	1000

(a) If the genes are linked, determine the relative order and map distances between them.

(b) What was the cis-trans allele arrangement in the trihybrid parent?

(c) Is there any crossover interference? If yes, how much?

Solution:

(i) With the three gene pairs eight phenotypic classes are expected and eight are observed.

(ii) The most frequent reciprocal phenotypic classes which arise from non-cross-over gamete tell us the genotypes of the original parents.

Here hot foot, obese, waved and wild type classes are parental classes, i.e., 357 + 343 = 700.

(iii) The order of gene can be determined by inspecting the least frequent reciprocal pairs of phenotypic classes that represent the result of double crossing-over. Here obese, waved and hot foot are double crossover classes which have 11 and 9 members respectively for a total of 20 (11 + 9). Therefore, the frequency of double crossover is $\frac{20}{1000} \times 100 = 2\%$.

(iv) The gene pair that has changed its position relative to the other two pairs of genes (allele) is the central gene. Here +/h has changed its position. Therefore, the order of gene is o h wa or wa h o. We can diagram the test cross as follows:

$$\frac{\text{o h wa}}{+\,+\,+} \times \frac{\text{o h wa}}{\text{o h wa}}$$

(v) A single crossing-over between o and h gives the progeny obese and hot foot wave which have 79 and 61 members respectively for a total of 140 (79 + 61). Therefore, the frequency of S.C.O. between o and h is $\frac{140}{1000} \times 100 = 14\%$.

(vi) Similarly, another single crossing-over between h and wa gives the progeny hot foot obese and waved which have 74 and 66 members respectively for a total of 140(74 + 66). Therefore, the frequency of S.C.O. between h and wa is $\frac{140}{1000} \times 100 = 14\%$.

(vii) The map distance between o and h is the all crossover between o and h. So, we must add D.C.O. Therefore, map distance between o – h = S.C.O. + D.C.O. = 14 + 2 = 16%.

Similarly, the map distance between h – wa = S.C.O. + D.C.O. = 14 + 2 = 16%

(a) Relative map distance: o $\overset{16\%}{\text{———}}$ h $\overset{16\%}{\text{———}}$ wa

(b) Parents genotype: $\frac{\text{o h wa}}{+\,+\,+}$ and $\frac{\text{o h wa}}{\text{o h wa}}$

(c) Coefficient of coincidence

$= \frac{\text{Freq. of observed D.C.O.}}{\text{Freq. of expected D.C.O.}}$

Observed D.C.O. = 2% = $\frac{2}{100}$ = 0.02.

Expected D.C.O. = 16% × 16% = $\frac{16}{100} \times \frac{16}{100} = \frac{256}{1000}$ = 0.0256.

Coefficient of coincidence = $\frac{0.02}{0.0256}$

Interference = 1 – coincidence = 1 – 0.78 = 0.22 = 22%.

Problem 12

The following three recessive genes are found in corn: bt, brittle endosperm; g117, glossy leaf; rg, ragged seedling. A trihybrid of unknown origin is test crossed, producing the following offspring:

Phenotypes	Number
brittle glossy, ragged	236
brittle glossy	241
ragged	219
glossy	23
wild type	224
brittle ragged	17
glossy ragged	21
brittle	19
Total =	1000

(a) If the genes are linked, determine the relative order and map distances.

(b) Reconstruct chromosome of trihybrid.

(c) Is there any crossover interference? If yes, how much?

Solution:

(i) With the three gene pairs eight phenotypic classes are expected and eight are observed.

(ii) The most frequent reciprocal phenotypic classes which arise from non-crossover gametes tell us the genotype of original parents.

Here brittle glossy ragged and wild type are parental classes which have 236 and 224 members of total 460 (236 + 224).

(iii) The order of gene can be determined by inspecting the least frequent reciprocal pairs of phenotypic classes that represent the result of double cross-over. Here glossy and brittle ragged are double crossover classes having 23 and 17 members of total 40 (23 + 17). Therefore, the frequency of double crossing-over is $\frac{40}{1000} \times 100 = 4\%$.

(iv) The gene pair that has changed its position relative to the other two pairs of genes (alleles) is the central gene. Here gl/+ has changed its position.

Therefore, the order of gene is bt gl rg. We can diagram the test-cross as $\frac{bt\,gl\,rg}{+++} \times \frac{bt\,gl\,rg}{bt\,gl\,rg}$.

(v) A single crossing-over between bt and gl produces the progeny brittle and glossy ragged which have 19 and 21 members respectively for total of 40(19 + 21). Therefore, the frequency of S.C.O. between bt and gl is $\frac{40}{1000} \times 100 = 4\%$.

(vi) Similarly, another single crossing-over between gl and rg gives rise to the progeny brittle glossy and ragged which have 241 and 219 members respectively for a total of 460(241 + 219). Therefore, the frequency of single crossing-over between gl and rg is $\frac{460}{1000} \times 100 = 46\%$.

(vii) The map distance between bt and gl is the all crossover between bt and gl. So we must add D.C.O. Therefore, map distance between bt and gl is S.C.O. + D.C.O., i.e., 4% + 46% = 50%.

Similarly, the map distance between gl and rg is S.C.O. + D.C.O., i.e., 4 + 4 = 8%.

(a) Relative order and map distance =

bt —— 50% —— gl —— 8% —— rg

(b) $\frac{bt\,gl\,rg}{+\,+\,+} \times \frac{bt\,gl\,rg}{bt\,gl\,rg}$

(c) Expected D.C.O. = $50\% \times 8\% = \frac{50}{100} \times \frac{8}{100} = \frac{400}{1000} = 0.04$.

Observed D.C.O. = $4\% = \frac{4}{100} = 0.04$.

Coefficient of coincidence = $\frac{\text{observed D.C.O.}}{\text{expected D.C.O.}} = \frac{0.04}{0.04} = 1$.

Interference = 1 – CC = 1 – 1 = 0.

No, there is no interference.

Problem 13

In the house mouse, the autosomal alleles Trembling and Rex (short hair) are dominant over not-trembling (normal) and long haired, respectively. Heterozygous Trembling, Rex females were crossed with normal, long-haired males and yielded the following offspring:

Phenotypes	**Number**
Trembling Rex	**42**
Trembling long-haired	**105**
Normal Rex	**109**
Normal long-haired	**44**
Total	**300**

(a) Are the two genes linked? How do you know?

(b) In the heterozygous females where is Trembling and Rex—in cis or transposition? Explain.

(c) Calculate the percentage of recombination between the two genes.

Solution:

(a) The two genes are linked. It is a test-cross. If the two genes are not linked, there would be 1 : 1 : 1 : 1 ratio. But here (data) no such ratio is found.

(b) The most frequent reciprocal pairs of phenotypic classes which arise from non-crossover gametes, tell us the genotype of the original parents.

Here Trembling long-haired and normal Rex are the parental phenotypic classes which have 105 and 109 members respectively for a total of 214(105 + 109).

Let us take T for Tremb**i**ng, R for Rex.

Genotype will be $\frac{Tr}{tR} \times \frac{tr}{tr}$.

Gametes: Tr, tR, TR, tr; tr

♀ \ ♂	Tr	tR	TR	tr
tr	TTrr	ttRr	TtRr	ttrr
	Tremb Long	Normal Rex	Tremb Rex	Normal long
	105	109	42	44

(c) Crossing-over classes are Trembling Rex and normal long-haired which have 42 and 44 members respectively for a total of 86 (42 + 44). The frequency will be $\frac{86}{300} \times 100 = 28.666$ = 28.7%.

Problem 14

In *Drosophila* kidney-shaped eye (k), cardinal eye (cd) and ebony body (e) are three recessive genes. If the homozygous kidney, cardinal females are crossed with homozygous ebony males, the F_1 offsprings are all wild type. If heterozygous F_1 females are mated with kidney, cardinal, ebony males, the following 2000 progeny appear:

Phenotypes	Number
kidney cardinal	**880**
ebony	**887**
kidney ebony	**64**
cardinal	**67**
kidney	**49**
ebony cardinal	**46**
kidney, ebony cardinal	**3**
wild type	**4**
Total	**2000**

(a) Determine chromosomal composition of the F_1 females.

(b) Derive the map of the three genes.

Solution:

(i) With the three-gene pairs eight phenotypic classes are expected and observed.

(ii) The most frequent reciprocal phenotypic classes which arise from non-cross-over gametes, tell us the genotype of the original parents. Here kidney, cardinal and ebony are the parental classes which have 880 and 887 members respectively for a total of 1767 (880 + 887).

(iii) The order of genes can be determined by inspecting the least frequent reciprocal pairs of phenotypic classes that represent the results of double crossing-over. Here kidney ebony, cardinal and wild type classes have 3 and 4 members respectively, for a total of 7(3 + 4). The frequency of D.C.O. is $\frac{7}{2000} \times 100 = 0.35$.

(iv) The gene pair that has changed its position relative to the other two pairs of genes (allele) is the central gene. Here +/e is the middle gene. Therefore, the order of genes is cd e k.

We can diagram the test cross as follows:

$$\frac{cd + k}{+ e +} \times \frac{cd\,e\,k}{cd\,e\,k}$$

(v) A single crossing-over between cardinal and ebony produces cardinal ebony and kidney which have 46 and 49 members respectively for a total of 95 (46 + 49). Therefore, the frequency of S.C.O. between cd and e is $\frac{95}{2000}$ × 100 = 4.75%.

(vi) Similarly, another crossing-over between ebony and kidney gives cardinal and ebony kidney which have 67 and 64 members respectively for a total of 131(67 + 64). Therefore, single cross-over frequency between e and k is $\frac{131}{2000} \times$ 100 = 6.55%.

(vii) The map distance between cd to e is the all crossing-over between cd and e. So we must add D.C.O.

The map distance between cd – e = S.C.O. + D.C.O. = 4.75 + 0.35 = 5.10%.

Similarly, the map distance between e and k is S.C.O. + D.C.O. = 6.55 + 0.35 = 6.9%.

(a) Composition of F_1 females $\frac{cd + k}{+ e +}$ or $\frac{k + cd}{+ e +}$

(b) Map distance

cd —— 5.1% —— e —— 6.9% —— k

Problem 15

In *Drosophila* a cross was made between females expressing the three X-linked recessive traits, scute (sc) bristles, sable body (s) and vermilion eyes (v) and wild type males. In the F_1, all females are wild type, while all males expressed all three mutant traits. The cross was carried to the F_2 generation and 1000 offsprings were

counted with the result shown below. No determination of sex was made in the F_2 data.

Phenotypes			Offspring number
sc	s	v	314
+	+	+	280
+	s	v	150
sc	+	+	156
sc	+	v	46
+	s	+	30
sc	s	+	10
+	+	v	14

(a) Determine the genotypes of P_1 and F_1 parents using proper nomenclature.

(b) Determine the sequence of three genes and the map distance between them.

(c) Are there more or fewer double cross-over than expected? Calculate the coefficient of coincidence. Does this represent positive or negative interference?

Solution:

(a) Genotypes of P_1 and F_1

$$P_1\ ♀\ \frac{sc\ v\ s}{sc\ v\ s} \quad \times \quad \frac{+\ +\ +}{\longrightarrow}\ ♂$$

$$F_1\ ♀\ \frac{sc\ v\ s}{+\ +\ +} \quad \times \quad \frac{sc\ v\ s}{\longrightarrow}\ ♂$$

(b) (i) With the three gene pairs eight phenotypic classes are expected and observed.

(ii) The most frequent reciprocal classes (phenotypes) which arise from non-cross-over gametes, tell us the genotype of the original parents. Here sc s v and + + + are the parental classes which have 314 and 280 members respectively for a total of 594 (314 + 280).

(iii) The order of genes can be determined by inspecting the least frequent reciprocal pairs of phenotypic classes that represent the double crossover classes. Here sc s + and + + v are the double crossover classes which have 10 and 14 members respectively for a total of 24 (10 + 14). Therefore, the frequency of D.C.O. is $\frac{24}{1000} \times 100 = 2.4\%$.

(iv) The gene pair that has changed its position relative to the other two pairs of genes (alleles) is the central gene. Here +/v is the central gene. Therefore, the order of genes is sc v s or s v sc.

(v) The single crossing-over between sc and v produces sc + + and + v s which have 150 and 156 members respectively for a total of 306 (150 + 156).

Therefore, the frequency of S.C.O. between sc and v is $\frac{306}{1000} \times 100 = 30.6\%$.

(vi) Similarly, other single crossing-over between v – s is sc v + and s + + which have 46 and 30 members respectively for a total of 76 (46 + 30). Therefore, the frequency of S.C.O. is $\frac{76}{1000} \times 100 = 7.6\%$.

(vii) The map distance between sc – v is the all crossover between sc – v. So we must add D.C.O. Therefore, map distance between sc – v = 30.6 + 2.4 = 33.0%. Similarly, map distance between v – s = 7.6 + 2.4 = 10.0%.

The final map is sc —— 33% —— v —— 10% —— s

(c) Expected D.C.O. = $\frac{33}{100} \times \frac{10}{100} \times \frac{330}{10000} = 0.033$.

Among 1000 progeny, the expected D.C.O. is $0.033 \times 1000 = 33$.

Observed D.C.O. = 24 and expected D.C.O. is 33, it is fewer than expected.

Observed D.C.O. = 2.4%, i.e. $\frac{2.4}{100} = 0.024$.

Coefficient of coincidence = $\frac{\text{observed D.C.O.}}{\text{expected D.C.O.}} = \frac{0.024}{0.033} = 0.727 = 72.7\%$.

Interference = 100 – 72.7 = 27.3%.

Problem 16

A cross in *Drosophila* involved the recessive, X-linked genes yellow (y), white (w) and cut (ct). A female that was yellow-bodied and white-eyed with normal wings was crossed to a male whose eyes and body were normal but whose wings were cut. The F_1 females were wild type for all three traits while the F_1 males expressed the yellow body, white traits. The cross was carried to an F_2 and only male offsprings were tallied. On the basis of the data shown below a genetic map was constructed:

	Phenotypes			Male offspring
(i)	y	+	ct	9
(ii)	+	w	+	6
(iii)	y	w	ct	90
(iv)	+	+	+	95
(v)	+	+	ct	424
(vi)	y	w	+	376
(vii)	y	+	+	0
(viii)	+	w	ct	0
			Total =	1000

(a) Diagram the genotypes of the F_1 parents.

(b) Construct a map, assuming that white is at locus 1.5 on the X chromosome.

(c) Were any double crossover offsprings expected?

(d) Could F_2 female offspring be used to construct the map? Why or why not?

Solution:

(a) (i) With the three gene pairs, eight phenotypic classes are expected and eight are observed.

(ii) The most frequent reciprocal pairs of phenotypic classes which arise from non-crossover gametes, tell us the genotypes of the original parents. Here + + ct and y w + which have 424 and 376 members respectively for total of 800 (424 + 376).

(iii) The order of genes can be determined by inspecting the least frequent reciprocal pairs of phenotypic classes that represent the double crossing-over. Here y + + and + w ct are the double crossover classes which have 0 and 0 members respectively for total of 0. Therefore, the frequency of D.C.O. is 0.

(iv) The gene pair that has changed its position relative to the other two gene pairs (alleles) is the central gene. Here +/w is the central gene. Therefore, the order of genes is y – w – ct.

(v) The diagram of the F_1 is as follows:

$$F_1 \ ♀ \frac{y\ w\ +}{+\ +\ ct} \times ♂ \frac{y\ w\ +}{\longrightarrow}$$

[∵ As male of P_1 was $\frac{+\ +\ ct}{\longrightarrow}$]

(vi) A single crossing-over between y and w gives the progeny y + ct and + w + which have 9 and 6 members respectively for a total of 15(9 + 6). Therefore, the frequency of S.C.O. between y – w is $\frac{15}{1000} \times 100 =$ 1.5%.

(vii) Similarly other single crossing-over between w – ct produces y w ct and + + + which have 90 and 95 members respectively for a total of 185 (90 + 95). The frequency of S.C.O. between w and ct is $\frac{185}{1000} \times 100 =$ 18.5%.

(viii) Therefore, the map distance between y – w is S.C.O. + D.C.O. = 1.5% + 0 = 1.5% and map distance between w – ct is S.C.O. + D.C.O. = 18.5% + 0 = 18.5%.

(ix) For construction of genetic map, assuming w is 1.5, so ct is 18.5 + 1.5 = 20.0.

(b) Map distance

y	w		ct
0	1.5	18.5 mu	20.0

(c) Expected D.C.O. = $\frac{1.5}{100} \times \frac{1.5}{100} \times 100 = 1.5\%$.

In 1000 flies it would be 0.002775 × 1000 = 2.775 = 2.8.

Here only 2.8, i.e., 3 D.C.Os. are expected.

(d) No, as in F_2 females are only yellow white $\left(\frac{y\,w\,+}{y\,w\,+}\right)$ and wild type $\left(\frac{+\,+\,ct}{y\,w\,+}\right)$ are found.

Problem 17

In *Drosophila* heterozygous female for the X-linked recessive traits a, b and c was crossed to a male that was phenotypically a, b, c. The offspring occured in the following phenotypic ratios:

Phenotypes			Number
+	b	c	460
a	+	+	450
a	b	c	32
+	+	+	38
a	+	c	11
+	b	+	9
		Total =	1000

No other phenotypes were observed.

(a) What is the genotypic arrangement of the alleles of these genes on the X-chromosome of the female?

(b) Determine the correct sequence and construct the map of these genes on the X-chromosome.

(c) What progeny phenotypes are missing? Why?

Solution:

(i) The most frequent reciprocal pairs of phenotypic classes which arise from non-crossover gametes tell us the genotype of the original parents. Here + bc and (a + +) are the parental classes which have 460 and 450 members respectively for a total of 910 (460 + 450).

(ii) A single crossing-over between a – b produces + + + and a b c which have 38 and 32 members respectively for a total of 70(38 + 32). Therefore, the frequency of S.C.O. between a – b is $\frac{70}{1000} \times 100 = 7\%$.

Similarly, other single crossing-over between b and c produces + b + and a + c phenotypic classes which have 9 and 11 members respectively for a total of 20(9 + 11). Therefore, the frequency of S.C.O. between b – c is $\frac{20}{1000} \times 100 = 2\%$.

(a) Genotype of F_1 female is $\frac{+\,b\,c}{a\,+\,+}$.

(b) The map distance a —— 7% —— b —— 2% —— c

(c) Missing progeny phenotypes are a b + and + + c which have members $\left(\frac{7}{100} \times \frac{2}{100} \times 1000 = 0.07 \times 0.02 \times 1000 = 1.4\right)$ among 1000 would be 1.4 or 2 expected.

Problem 18

In *Drosophila*, dichaete (D) is a chromosome-3 mutation with dominant effect on wing shape. It is lethal when homozygous. The genes ebony (e) and pink (p) are chromosome-3 recessive mutation affecting the body and eye colour, respectively. Flies from a dichaete stock were crossed to homozygous ebony, pink flies and the F_1 progeny with a dichaete phenotype were back-crossed to the ebony pink homozygous. The results of this back-cross were as follows:

[P.G., M.G., E.G. & I.G. 6]

Phenotypes	Number
i. dichaete	401
ii. ebony, pink	389
iii. dichaete, ebony	84
iv. pink	96
v. dichaete, pink	2
vi. ebony	3
vii. dichaete, ebony pink	12
viii. wild type	13
Total =	1000

(a) Diagram the cross, showing the genotypes of the parents and offspring of both crosses.

(b) What is the sequence and interlocus distance between these three genes?

Solution:

(i) With the three gene pairs eight phenotypic classes are expected and observed.

(ii) The most frequent reciprocal pairs of phenotypic classes which arises from non-crossover gametes tell us the original genotypes of the parents. Here dichaete and ebony pink phenotypic classes are parental classes which have 401 and 389 members respectively for a total of 790 (401 + 389).

(iii) The order of gene can be determined by inspecting the least frequent reciprocal pairs of phenotypic classes that represent the result of double crossing-over. Here dichaete pink and ebony are double-crossover classes having 2 and 3 members for a total of 5 (2 + 3). Therefore, the frequency of double crossing-over is $\frac{5}{1000} \times 100 = 0.5\%$.

(iv) The gene pair that has changed its position relative to other two pairs of genes (alleles) is the central gene. Here $\frac{+}{p}$ has changed its position. Therefore, the order of genes is D p e. Diagrams of the crosses are as follows:

$$P = \frac{D\,+\,+}{+\,p\,e} \times \frac{+\,p\,e}{+\,p\,e}$$

$$F_1 \; \frac{D\,+\,+}{+\,p\,e} \text{ (Dichaeta)} \times \frac{+\,p\,e}{+\,p\,e} \text{ (ebony pink)}$$

(v) A single crossing-over between dichaete and pink produces dichaete pink ebony and wild type progeny which have 12 and 13 members respectively for a total of 25(12 + 13).

Therefore, the frequency of S.C.O. between D and p is $\frac{25}{1000} \times 100 = 2.5\%$.

Similarly, another single crossing-over between pink and ebony gives dichaete ebony and pink progeny which have 84 and 96 members respectively for a total of 180 (84 + 96).

Therefore, the frequency of S.C.O. between p and e is $\frac{180}{1000} \times 100 = 18\%$.

(a) P: Dichaete female × ebony pink male

$$♀ \frac{D\,+\,+}{+\,p\,e} \times \frac{+\,p\,e}{+\,p\,e} ♂$$

F_1: Dichaete female × ebony pink (homozygous)

$$\frac{D\,+\,+}{+\,p\,e} \times \frac{+\,p\,e}{+\,p\,e}$$

F_2: Phenotype	Genotype
1. Dichaete	$\frac{D\,+\,+}{+\,p\,e}$
2. Pink ebony	$\frac{+\,p\,+}{+\,p\,e}$
3. Dichaete ebony	$\frac{D\,+\,e}{+\,p\,e}$
4. Pink	$\frac{+\,p\,+}{+\,p\,e}$
5. Dichaete pink ebony	$\frac{D\,p\,e}{+\,p\,e}$
6. Wild type	$\frac{+\,+\,+}{+\,p\,e}$
7. Dichaete pink	$\frac{D\,p\,+}{+\,p\,e}$
8. Ebony	$\frac{+\,+\,e}{+\,p\,e}$

(b) The map distance between Dichaete and pink is S.C.O. + D.C.O. = 2.5% + 0.5% = 3.0%.

Similarly, the map distance between p and e is S.C.O. + D.C.O. = 18% + 0.5% = 18.5%.

The sequence and interlocus distance among these genes are as follows:

D —— 3% —— P —— 18.5% —— e

Problem 19

The following offspring phenotypes were obtained from a cross between *Drosophila* females heterozygous for three alleles, approximated vein, curled wing and striped thorax (+ app, + cu, + sr) and male homozygous for these alleles:

Phenotypes			Number
+	+	+	337
app	cu	sr	348
+	cu	sr	58
app	+	+	49
+	+	sr	55
app	cu	+	48
+	cu	+	3
app	+	sr	2
		Total =	**900**

Calculate the coefficient of coincidence.

Solution:

(i) The most frequent reciprocal pairs of phenotypic classes which arise from non-crossover gametes, tell us the genotype of the original parents. These are + + + and app cu sr parental classes which have 337 and 348 members respectively for a total of 685 (337 + 348).

(ii) The order of the genes can be determined by inspecting the least frequent reciprocal pairs of phenotypic classes that represent the double crossover classes. Here it is + cu + and app + sr which have 3 and 2 members respectively for a total of 5(3 + 2). Therefore, the frequency of D.C.O. is $\frac{5}{900} \times 100 = 0.5555 = 0.56\%$.

(iii) The gene pair that has changed its position relative to the other two gene pairs (allele) is the central gene. Here +/cu is the central gene.

app —— cu —— st

(iv) A single crossing-over between app – cu produces the progeny app + + and + cu sr, which have 49 and 58 members respectively for a total of 107(49 + 58). Therefore, the frequency of S.C.O. between app and cu is $\frac{107}{900} \times 100 = 11.8888 = 11.89$.

(v) A single-crossing-over between cu – sr produces + + sr and app cu + for a total of 103 (55 + 48). Therefore, the frequency of cu and sr is = 11.45%.

(vi) The map distance between ap – cu is S.C.O. + D.C.O. = 11.89 + 0.56 = 12.45%. The map distance between cu – sr is S.C.O. + D.C.O. = 11.45 + 0.56 = 12.01.

(vii) The expected D.C.O. is $\frac{12.45}{100} \times \frac{12.01}{100} = 0.1245 \times 0.1201 = 0.01495$.

Observed D.C.O. is $\frac{0.56}{100} = 0.0056$.

Coefficient of coincidence $\frac{0.0056}{0.0149} = 0.3758 = 0.38 = 38\%$.

Problem 20

In a plant there are three linked genes, all of which control fruit phenotypes. A recessive allele 'p' of the first gene determines purple fruit colour in contrast to the yellow colour of the wild type. A recessive allele 'r' of the second gene results in a round fruit shape, as compared with an elongated fruit in the wild type. A recessive allele 'j' of third gene gives a juicy fruit instead of the wild type dry fruit. From a test-cross of a triple heterozygous ($p^+r^+j^+/prj$) [yellow, elongated and dry (wild type)] with a triple homozygous recessive (prj/pj) [purple, round and juicy] the following results were found:

Phenotypes	Number
1. Yellow elongated, dry ($p^+ r^+ j^+$)	**179**
2. Purple, round, juicy (p r j)	**173**
3. Purple, elongated, dry ($p\ r^+ j^+$)	**52**
4. Yellow, round, juicy ($p^+ r\ j$)	**46**
5. Purple, elongated, juicy ($p\ r^+ j$)	**22**
6. Yellow, round, dry ($p^+ r\ j^+$)	**22**
7. Yellow, elongated, juicy ($p^+ r^+ j$)	**4**
8. Purple, round, dry ($p\ r\ j^+$)	**2**
Total	**500**

(a) Determine the order of the genes on the chromosome.

(b) Determine the map distance between the genes.

(c) Determine coefficient of coincidence and interference for the portion of chromosome involved in this cross.

Solution:

(i) With the three gene pairs, eight phenotypic classes are expected and eight are observed.

(ii) The most reciprocal pairs of classes (phenotypes) which arise from non-cross-over gametes tell us the genotypes of the original parents. Here the two classes (1 and 2), i.e., yellow elongated dry and purple round juicy are (179 + 173).

(iii) The order of genes can be determined by inspecting the least frequent reciprocal pair of phenotypic classes that represent the result of double crossing-over. Here the double crossing-over classes (7 and 8), i.e., yellow elongated juicy and purple round dry are respectively for a total of 6 (4 + 2). Therefore, the frequency of double crossing-over is $\frac{6}{500} \times 100 = 1.2\%$.

(iv) The gene pair that has changed its position relative to the other two pairs of alleles (genes) is the central gene. Here j^+/j has changed its position. Therefore, j is the middle gene. Therefore, F_1 genotypic configuration is $\frac{p\,j\,r}{p^+j^+r^+} \times \frac{p\,j\,r}{p\,j\,r}$.

(v) A single crossing-over between p – j produces $p\ r^+\ j^+$ and $p^+\ r\ j$ which have 52 and 46 members respectively for a total of 98 (52 + 46). Therefore, frequency of S.C.O. between p and j is $\frac{98}{500} \times 100 = 19.6\%$.

(vi) Another single crossing-over between j – r produces $p\ r^+\ j$ and $p^+\ r\ j^+$ which have 22 and 22 members respectively for a total of 44 (22 + 22). Therefore, frequency of S.C.O. between j – r is $\frac{44}{500} \times 100 = 8.8\%$.

(vii) The map distance between p – j = 19.6 + 1.2 = 20.8%. The map distance between j – r = 8.8 + 1.2 = 10.0%.

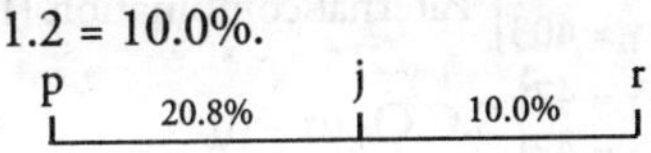

Expected D.C.O. = $\frac{20.8}{100} \times \frac{10.0}{100} = 0.208 \times 0.1 = 0.0208$.

Observed D.C.O. = $1.2 = \frac{1.2}{100} = 0.012$.

Coefficients of coincidence = $\frac{\text{observed D.C.O.}}{\text{expected D.C.O.}}$

$= \frac{0.012}{0.0208} = 0.5769 = 0.577$.

Interference = 1 – 0.577 = 0.423, i.e., 42.3%

Coefficient of coincidence = 0.577 = 57.7%.

Problem 21

The linkage map for three recessive genes located on the same chromosome in a certain species is measured in percent recombination frequencies.

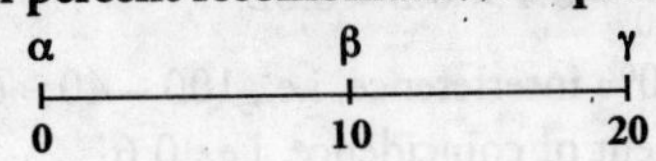

If the coefficient of coincidence is 0.60 in this case, determine the frequencies of phenotypes expected among 1000 offsprings of a cross $\frac{\alpha\beta\gamma}{+++} \times \frac{\alpha\beta\gamma}{\alpha\beta\gamma}$.

Solution:

(i) Here $\alpha - \beta = 10\%$ and $\beta - \gamma = 10\%$ (20 – 10).

Therefore, expected D.C.O. = $\frac{10}{100} \times \frac{10}{100} = 0.1 \times 0.1 = 0.01$.

Among 1000 progenies expected D.C.O. would be $0.01 \times 1000 = 10.00 = 10$.

(ii) Coefficient of coincidence = 0.6

$$0.6 = \frac{\text{observed D.C.O.}}{\text{expected D.C.O.}} = \frac{\text{observed D.C.O.}}{10}.$$

Observed double crossing-over 6.0 = 6

The members are $+ \beta +$ and $\alpha + \gamma$.

(iii) The number of recombinants between $\alpha - \beta$, i.e., 10% is $\frac{10}{100} \times 1000 = 100$, the members are 100 6 = 94.

They are $+ \beta \gamma$ and $\alpha + +$.

(iv) The number of recombinants between $\beta - \gamma$ = 10% is $\frac{10}{100} \times 1000 = 100$. It would be 100 – 6 = 94.

They are $+ + \gamma$ and $\alpha \beta +$.

(v) The number of parental progeny is 1000 – (94 + 94 + 6) = 1000 – 194 = 806.

$+++ = 403$, $\alpha\beta\gamma = 403$ } Paternal combination (PC)

$+\beta\gamma = 47$, $\alpha++ = 47$ } S.C.O. ($\alpha - \beta$)

$++\gamma = 47$, $\alpha\beta+ = 47$ } S.C.O. ($\beta - \gamma$)

$+\alpha+ = 3$, $\beta+\gamma = 3$ } D.C.O.

Problem 22

Given a segment of map a —10 mu— b —20 mu— c with 40% interference, calculate the observed D.C.O. types.

Solution:

(i) Given a – b = 10 mu and b – c = 20 mu.

Therefore, expected D.C.O. would be $\frac{10}{100} \times \frac{20}{100} = 0.1 \times 0.2 = 0.02$.

(ii) 40% interference, i.e., 100 – 40 = 60% coefficient of coincidence, i.e., 0.6.

Coefficient of coincidence = 0.6

$$0.6 = \frac{\text{observed D.C.O.}}{\text{expected D.C.O.}} = \frac{\text{observed D.C.O.}}{0.02}.$$

Observed D.C.O. = $0.6 \times 0.02 = 0.012 = 1.2\%$.

Problem 23

In *Drosophila* three autosomal genes have the following:

a —20 mu— b —20 mu— c

Provide the data, in terms of the expected number of flies in the following phenotypic classes when $a^+b^+c^+$/abc females are crossed to abc/abc males. Assume 1000 flies were counted:

a^+	b^+	c^+
a	b	c
a^+	b	c
a	b^+	c^+
a^+	b^+	c
a	b	c^+
a^+	b	c^+
a	b^+	c

Solution:

(i) Here a – b = 20 mu and b – c = 10 mu.

Therefore, expected D.C.O. is $\frac{20}{100} \times \frac{10}{100} = \frac{200}{10000} = 0.02$.

Among 1000 flies expected D.C.O. would be $0.02 \times 1000 = 20$.

Therefore, the D.C.O. progeny, i.e. $a^+ b c^+$ and $a b^+ c$ are 20.

(ii) S.C.O. between a – b results in $a^+ b c$ and $a b^+ c^+$.

Recombinants are $\frac{20}{100} \times 1000 = 200$.

The S.C.O. (a and b) progeny numbers are 200 – 20 = 180.

(iii) Another S.C.O. between b – c results in the progeny $a b c^+$, $a^+ b^+ c$.

Recombinants are $\frac{10}{100} \times 1000 = 100$.

The S.C.O. (b and c) progeny numbers are 100 – 20 = 80.

(iv) The number of parental progeny is 1000 – (180 + 80 + 20) = 1000 – 280 = 720.

Phenotype of progeny	Numbers	
$a^+ b^+ c^+$	360	Parental types
a b c	360	
$a\ b^+ c^+$	90	C.O. between a–b
$a^+ b\ c$	90	
$a\ b\ c^+$	40	C.O. between a–b
$a^+ b^+ c$	40	
$a\ b^+ c$	10	D.C.O.
$a^+ b\ c^+$	10	

Problem 24

In the following test-cross, genes a, b are 20 map units apart and genes b and c are 10 map units apart:

$$\frac{a + c}{+ b +} \times \frac{a\,b\,c}{a\,b\,c}$$

If the coefficient of coincidence is 0.5, how many triple homozygous recessive individuals would you expect to find among 1000 progenies from a cross?

Solution:

(i) Here a – b = 20% and b – c = 10%.

Therefore, expected D.C.O. is $\frac{20}{100} \times \frac{10}{100} = \frac{200}{10000} = 0.02$.

Coefficient of coincidence = 0.5

$$0.5 = \frac{\text{observed D.C.O.}}{\text{expected D.C.O.}} = \frac{\text{observed D.C.O.}}{0.02}.$$

$\therefore$ Observed D.C.O. = 0.5 × 0.02 = 0.010 = 0.01.

Among 1000 progenies it would be 0.01 × 1000 = 10.

D.C.O. members are + + + and a b c, and total 10 progenies. Therefore, a b c would be 5.

Problem 25

Suppose that you have determined the order of three genes to be a c b and that by doing two-point crosses you have determined map distances as a – c = 10 and c – b = 5. If the interference is –1.5 and the three-point cross is

$$\frac{A\,C\,B}{a\,c\,b} \times \frac{a\,c\,b}{a\,c\,b},$$

what frequency of double crossover do you expect?

Solution:

Here a – c = 10 mu; c – b = 5 mu.

$\therefore$ Expected D.C.O. = $\frac{10}{100} \times \frac{5}{100} = \frac{50}{10000} = 0.005$.

Interference is –1.5 and therefore, coefficient of coincidence is 1 – (–1.5) = 1 + 1.5 = 2.5.

Coefficient of coincidence = 2.5

$$2.5 = \frac{\text{observed D.C.O.}}{\text{expected D.C.O}} = \frac{\text{observed D.C.O.}}{0.005}.$$

Observed D.C.O. would be = 2.5 × 0.005 = 0.0125.

Problem 26

Map location of genes a, b and c are respectively 3, 13 and 25 mu. Assuming 10 map units is equal to 10% recombination approximately, calculate the frequency of observed D.C.O., when coefficient of coincidence is 28%.

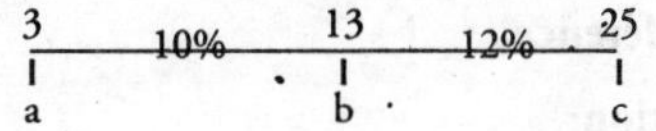

Solution:

Expected D.C.O. is $\frac{10}{100} \times \frac{12}{100} = \frac{120}{10000} = 0.012$.

Coefficient of coincidence = 28% = 0.28.

$$0.28 = \frac{\text{observed D.C.O.}}{\text{expected D.C.O}} = \frac{\text{observed D.C.O.}}{0.012}$$

Observed D.C.O. would be 0.00336 = 0.0034 = 0.34%.

Problem 27

In a series of two-point map crosses involving three genes linked on chromosome 3 in *Drosophila*, the following distances were calculated:

cd – sr = 13 mu

cd – ro = 16 mu

Determine the sequence and construct the map of three genes.

Solution:

It is impossible to do so. There are two possibilities based on these limited data:

Case 1: cd $\overset{13}{\text{———}}$ sr $\overset{3}{\text{———}}$ ro [16–13]

or

Case 2: ro $\overset{16}{\text{———}}$ cd $\overset{13}{\text{———}}$ sr

(i) The map distance is determined by crossing-over between ro and sr. If case 1 is correct, it should be 3 mu and if case 2 is correct, it should be 29 mu. In fact, this distance is 13 + 16 = 29 mu, demonstrating that the case 2 is correct.

(ii) There the sequence will be

$$\text{ro} \xrightarrow{16} \text{cd} \xrightarrow{3} \text{sr}$$

or

$$\text{sr} \xrightarrow{13} \text{cd} \xrightarrow{16} \text{ro}$$

Problem 28

The accompanying diagram summarizes the recombination frequencies observed in a large experiment to study three linked genes:

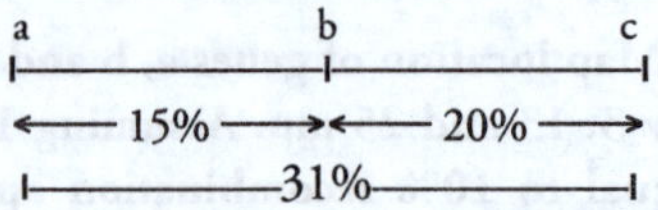

What was the observed frequency of double crossing-over in this experiment? Calculate the interference.

Solution:

(i) Let x be the observed frequency of double crossing-over. Then single crossing-over in the a-b interval occurs with the frequency 0.15 – X and single crossing-over in the b-c interval occurs with frequency 0.20 – X.

(ii) The observed recombination frequency between a-c equals the sum of the single crossing-over frequencies (because the double crossing-overs are undetected) which implies that 0.15 – X + 0.20 – X = 0.31 or, 0.35 – 2X = 0.31 or, –2X = 0.31 – 0.35 = –0.04 or, –2X = –0.04 or, X = 0.02.

(iii) The expected frequency of double crossing-over equals = 0.15 × 0.20 = 0.0300 = 0.03.

The coefficient of coincidence

$$= \frac{\text{Observed D.C.O.}}{\text{Expected D.C.O.}} = \frac{0.02}{0.03} = 0.1379 = 0.14.$$

The interference = 1 – 0.67 = 0.33 = 33%.

Problem 29

There are three genes in the chromosome 9 of corn. Determine shrunken sh versus plump Sh kernels, waxy wx versus nonwaxy Wx endosperm, and glossy gl versus nonglossy Gl leaves. The genetic map of this chromosomal region is sh — 30 — wx — 10 — gl. From a plant of genotype Sh wx Gl/sh Wx gl, what is the expected frequency of sh wx gl gamete,

(a) in the absence of interference?

(b) assuming 60 percent interference?

Solution:

(a) In the absence of interference, expected D.C.O. 0.30 × 0.10 = 0.03. sh wx gl is one of the D.C.O., the expected frequency = $\frac{0.03}{2}$ = 0.015.

(b) If interference is 60%, then coincidence = 40%. The observed frequency of double crossover is 40% as great as with independence. So the expected frequency of sh wx Sl = 0.015 × 0.40 = 0.00600 = 0.006.

Problem 30

In *Drosophila* the genes cn (cinnabar eyes) and bw (brown eyes) are located in the same chromosome but so far apart that they appear unlinked. Compare the possible offspring produced by the crosses

$$♀\,\frac{\text{cn bw}}{+\ +} \times \frac{\text{cn bw}}{\text{cn bw}}\,♂ \text{ and } ♀\,\frac{\text{cn bw}}{\text{cn bw}} \times \frac{\text{cn bw}}{+\ +}\,♂$$

The + symbols denote the wild type alleles. Both cn and bw are recessive. Flies with genotype $\frac{\text{cn}}{\text{cn}}$ have cinnabar (bright red) eyes, $\frac{\text{bw}}{\text{bw}}$ flies have brown eyes and the double mutant $\frac{\text{cnbw}}{\text{cn bw}}$ has white eyes.

Solution:

(i) There is no crossing-over in *Drosophila* males. When male is $\frac{\text{cn bw}}{+\ +}$, his gametes are $\frac{1}{2}$ cn bw and $\frac{1}{2}$ + +. So the progenies $\frac{\text{cnbw}}{\text{cn bw}}$ (white eyes) and $\frac{\text{cn bw}}{+\ +}$ (wild type eyes) are in equal proportions.

(ii) Crossing-over occurs in female *Drosophila* but cn and bw are so far apart that they undergo independent assortment. When the female is $\frac{\text{cn bw}}{+\ +}$, her gametes are $\frac{1}{4}$ cn bw, $\frac{1}{4}$ + +, $\frac{1}{4}$ + bw and $\frac{1}{4}$ cn +. So the progenies $\frac{\text{cnbw}}{\text{cn bw}}$ (white eyes), $\frac{\text{cn}\ +}{\text{cn bw}}$ (bright red), $\frac{+\ \text{bw}}{\text{cn bw}}$ (brown) and $\frac{+\ +}{\text{cn bw}}$ (wild type) are in equal proportions.

Problem 31

Among 3000 progeny from a trihybrid testcross the following phenotypes are observed. [a = causing absence of anthocyanin pigment, hi = causing hairless plant, j = causing jointless fruit in tomato plant]

Phenotypes	Number
1. Hairless	259
2. Jointless, hairless	40
3. Jointless	931
4. Normal	260
5. Anthocyaninless, jointless, hairless	268
6. Anthocyaninless, hairless	941
7. Anthocyaninless	32
8. Anthocyaninless, jointless,	269

(a) Find out genotype of the parent.

(b) Estimate distance between the genes.

(c) Calculate the value of interference.

Solution:

(a) Parents are jointless (931): A j Hi

Anothocyaninless and hairless (941): a J hi

D.C.O. jointless hairless (140): j hi A

Anthocyaninless (32): J Hi a

From parent hi gene is replaced. Therefore, middle gene is hi. (A — Hi — J):

Parent: (i) $\frac{\text{A Hi j}}{\text{A Hi j}}$; (ii) $\frac{\text{a hi J}}{\text{a hi J}}$ (original stock)

(b) A —20 CM— Hi —20 CM— J

|——35.2——|

$$\frac{259+269+268+260}{3000} = 35.2;$$

$$\frac{32+40+268+260}{3000} \times 100 = 20\%$$

C.O.C. = 0.20 × 0.20 = 0.04 × 3000 = 120

(expected 100 – 60 = 40% = $\frac{72}{100}$ = 0.6 = 60%).

Problem 32

In corn, a trihybrid tunicate (T), glossy (G) liguled (L) plant was crossed with a nontunicate, nonglossy, linguleless plant, and the following results are obtained.

Phenotypes	Number
Tunicate, liguleless, glossy	58
Tunicate, liguleless, nonglossy	15
Tunicate, liguled, glossy	55
Tunicate, liguled, nonglossy	13
Nontunicate, liguled, glossy	16
Nontunicate, liguleless, nonglossly	53
Nontunicate, liguleless, glossly	14
Nontunicate, liguleless, nonglossly	59

(a) Determine the genotype of the heterozygote.

(b) Calculate the map distances between the linked genes.

Solution:

(a) Parental combination:

Tunicate, ligule, glossy (TLG) = 55

Nontunicate, liguleless, nonglossy (tlg) = 59

D.C.O. = 13 + 14 = 27

Tunicate, liguled, nonglossy (TLg) = 13

Nontunicate, liguless, glossy (tlG) = 14

$\frac{\text{TGL}}{\text{tgl}}$ Genotype of the heterozygote.

(b) T—G = $\frac{15+16}{283} = \frac{31}{283} \times 100 = 10.95\%$,

G—L = $\frac{58+53}{283} = \frac{111}{283} \times 100 = 39.22\%$

D.C.O. = $\frac{13+14}{283} = \frac{27}{283} \times 100 = 0.095 \times 100 =$ 9.5%

T—G = 10.95 + 9.5 = 20.45 = 20.5%

G—L = 39.22 + 9.5 = 48.72 = 48.8%

T —20.5— G —48.8— L

Expected D.C.O. = $\frac{28.19}{100} \times \frac{48.72}{100}$ = 0.2819 × 0.4872 = 0.137

Observed D.C.O. = 0.095

C.O.C. = $\frac{0.095}{0.137}$ = 0.69 = 69%.

Problem 33

***Drosophila* females heterozygous for three recessive X-linked markers y (yellow body) ct (cut wings) and m (miniature wings) and their wild type alleles were crossed to y, ct, m males.**

The following progenies were obtained

	Phenotypes	Number
1.	Yellow, cut, miniature	30
2.	Wild type	33
3.	Yellow	10
4.	Cut, miniature	12
5.	Miniature	8
6.	Yellow cut	5
7.	Yellow, miniature	1
8.	cut	1

(a) Which classes are parental types?

(b) Which classes represent double crossover?

(c) Which gene is the middle of the other two?

(d) What was the genotype of the heteroygous females used in the cross? (Show the correct linkage phase as well as the correct order of the markers along the chromosomes)

Solution:

(a) Yellow, cut miniature (30) and wild type (33) are most numerous and therefore represent parental class.

(b) The double crossover classes are the least number, i.e., yellow miniature (1) and cut wings (1).

(c) In this data 'ct' allele is separated from y and m in double crossover class, therefore, 'ct' is the middle gene, i.e., $\underline{\text{y ct m}}$.

(d) Genotype of heterozygous female was $\frac{+\,+\,+}{\text{y ct m}}$.

Problem 34

Assume that in *Drosophila* there are three genes x, y & z, with each mutant allele recessive to the wild type allele. A cross between females heterozygous for these three loci and wild type males yielded the following progeny.

	Phenotypes			Number
Females:	+	+	+	1515
Males:	+	+	+	55
	+	+	Z	630
	+	Y	Z	65
	X	+	+	43
	X	Y	+	640
	X	Y	Z	52
			Total	3000

Using these data construct a linkage map of the three genes ad calculate the coefficient of coincidence.

Solution:

(i) These genes are sex-linked, thus male progenies are hemizygous for X-chromosome markers that are transmitted through eggs (from thier female parents).

(ii) All daughter progenies will necessarily carry the wild type allele of all three genes namely (X Y Z). Wild type alleles transmitted through sperm (from their male parents).

(iii) Therefore, linkage relationship can be determined by analyzing male progeny of the cross.

Male wild type × Female heterozygous

$\frac{+\,+\,+}{\longrightarrow}$ ♂ × $\frac{+\,+\,Z}{Y\,X\,+}$ ♀

Gametes:
(♀) + + Z, Y X +, + + +, Y X Z
+ X +, Y + Z, + X Z, Y + +
(♂) + + +, ⟶

Daughter (females):	Son (males):
(i) $\frac{+\,+\,Z}{+\,+\,+}$	(i) $\frac{+\,+\,+}{\longrightarrow}$ = 55
(ii) $\frac{Y\,X\,+}{+\,+\,+}$	(ii) $\frac{Y\,X\,Z}{\longrightarrow}$ = 52
(iii) $\frac{Y\,+\,Z}{+\,+\,+}$	(iii) $\frac{+\,X\,+}{\longrightarrow}$ = 43
(iv) $\frac{+\,X\,+}{+\,+\,+}$	(iv) $\frac{Y\,+\,Z}{\longrightarrow}$ = 65
(v) $\frac{+\,+\,+}{+\,+\,+}$	(v) $\frac{+\,+\,Z}{\longrightarrow}$ = 630
(vi) $\frac{Y\,X\,Z}{+\,+\,+}$	(vi) $\frac{Y\,X\,+}{\longrightarrow}$ = 640
(vii) $\frac{+\,X\,Z}{+\,+\,+}$	
(vii) $\frac{Y\,+\,+}{+\,+\,+}$	

C.O. between Y – X = 43 + 65 = 108

$\frac{108}{3000} \times 100 = 3.6\%$.

C.O. between X – Z = 55 + 52 = 107

$\frac{107}{3000} \times 100 = 3.57\%$.

Distance among genes are:

Y – X = 3.6 + 0 = 3.6%

X – Z = 3.57 + 0 = 3.57%

Genetic map: Y $\overset{3.6\%}{\rule{1cm}{0.4pt}}$ X $\overset{3.57\%}{\rule{1cm}{0.4pt}}$ Z

Here D.C.O. is missing because distance between genes are less than 20%.

$$\text{Expected D.C.O.} = \frac{3.6}{100} \times \frac{3.57}{100} = 0.036 \times 0.0357$$

$= 0.001285$, i.e. 0.00129.

$$\text{Coefficient of coincidence} = \frac{\text{observed D.C.O.}}{\text{expected D.C.O.}}$$

$$= \frac{0}{0.00129} = 0.$$

Problem 35

A cross in corn is carried out with three linked genes in a triply heterozygous genotype for the allele pairs (A, a), (D, d) and (R, r). The dominant A allele results in red leaves; aa plants have green leaves. The dominant D allele results in tall plants. dd plants are dwarfed. The dominant R allele dd plants are dwarfed. The dominant R allle results in ragged leaf margin; rr plants have smooth leaf margins. The triply heterozygous is crossed with homozygous recessives for all three alleles, and the following phenotypes of progeny are observed. Assume that interference in each interval is complete, so that the map distance in centimorgans (cM) corresponds to the percent recombination.

Phenotypes	Number
Red, tall, ragged	265
Red, tall, smooth	24
Red, dwarf, ragged	120
Red, dwarf, smooth	90
Green, tall, ragged	70
Green, tall, smooth	140
Green, dwarf, ragged	16
Green, dwarf, smooth	275
Total	1000

(a) What is the order of the gene?

(b) What is the genotype of the triply heterozygous parent with genes written in correct order?

(c) What is the shortest map distance between adjacent genes?

(d) What is the longest map distance between adjacent genes?

(e) What is the value of the interferene across the interval?

Solution:

(a) Parents A D r and a d R

D.C.O. = A D R and a d r. Here R and r are interchanged, therefore, order of the genes A R D or D R A.

(b) Triply heterozygous parent $\frac{A\,R\,D}{a\,r\,d}$.

(c) The shortest map distance between the genes R and D equals to $\left(\frac{90+70+16+24}{1000}\right) = 20$ cM.

(d) The longest map distance between the genes A and R equals to $\left(\frac{120+140+16+2+}{1000}\right) = 20$ cM.

(e) The expected D.C.O. = 0.20 × 0.30 = 0.6

Among 1000 it is 0.6 × 1000 = 60

Observed D.C.O. = 24 + 16 = 40.

$\therefore$ coefficient of coincidence $= \frac{40}{60} = \frac{2}{3} = 0.666$

$= 0.67$, i.e., 67% and interference 100 – 67 = 33%.

Problem 36

R. K. Sokai, K. Akhtar and C. J. Dubash reported data from a set of test-cross with mosquito *Anopheles culicifacies*, a vector for malaria in Southern Asia. The data involved three mutations bw (brown eyes), c (colourless eyes) and Blk (black body). In each cross repulsion heterozygous were mated to mosquitoes homozygous for the recessive alleles of the genes and the progeny were scored as having either parental or a recombinant genotype. Are any of the three genes studied in this crosses linked? If so, construct a map of the linkage relationship.

Repulsion Cross Heterozygote	Progeny Parental	Progeny Recombinant	Percent Recombinant
1. $\frac{bw+}{+\,c}$	850	503	37.2
2. $\frac{bw+}{+\,Blk}$	750	750	24.0
3. $\frac{C\,+}{+\,Blk}$	629	183	22.5

Solution:

(i) In each cross, the frequency of recombination is less than 50 percent, so all the three loci are linked.

(ii) To place them in a linkage map, we estimate the distances between each pair of genes from the observed recombination frequencies.

bw —24.0— B/k —22.5— C

|——— 37.2 ———|

(iii) Recombination frequency between bw and c 37.2% for Cross I, i.e., less than the actual distance 46.5 (24 + 22.5). This shows that widely separated genes, recombination frequencies underestimates the true map distance.

Problem 37

The frequencies of gametes of different genotypes determined by test-crossing a triple heterozygous are shown in the following table:

Gamete genotypes			Percentage
+	+	+	12.9
a	b	c	13.5
+	+	c	6.9
a	b	+	6.5
+	b	c	26.4
a	+	+	27.2
a	+	c	3.1
+	b	+	3.5
		Total	100.0

(a) Which gametes are known to have been involved in double crossing-overs?

(b) Which gamete types have not been involved in any exchanges?

(c) The order shown is not necessarily correct which gave locus in the middle.

Solution:

(a) a + c (3.1) and + b + (3.5) are D.C.O. progeny

(b) + b c (26.4) and a + + (27.2) are non-crossover product.

(c) c is the middle gene, i.e., order $\frac{\text{a c b}}{}$.

Problem 38

A geneticist has estimated the number of exchanges that occurred during meiosis on each of 100 chromatids that were recovered in gametes. The data are as follows:

Number of exchange	0	1	2	3	4	Total
Frequency	18	20	40	16	6	100

What is the genetic length in centimorgans of the chromosomes analysed in this study.

Solution:

The genetic length of the chromosome is the average number of exchanges on a chromatid at the end of meiosis. For the data at hand the average is $\left(0\times\frac{18}{100}\right)+\left(1\times\frac{20}{100}\right)+\left(2\times\frac{40}{100}\right)+\left(3\times\frac{16}{100}\right)+\left(4\times\frac{6}{100}\right)$ = 0.00 + 0.20 + 0.80 + 0.48 + 0.24 = 1.72 morgan or 172 centimorgan.

Problem 39

The accompanying diagram shows the linkage map of the three genes a, b and c, where the distances are given in centimorgans (map units).

a —10— b —15— c

Assume that double crossovers across these intervals show a coefficient of coincidence 0.40. Among thousand gametes from an individual of genotype ABC/abc, what are the expected numbers of each of the possible allele combinations?

Solution:

Coefficient of coincidence = 0.4

$$\text{Coefficient of coincidence} = \frac{\text{observed D.C.O.}}{\text{expected D.C.O.}}$$

(i) Observed D.C.O. = 0.4 × 0.10 × 0.15 = 0.006, these are two types A b C and a B c $\left[\frac{\text{A B C}}{\text{a b c}} = \text{A b C, a B c}\right]$.

Among 1000 gametes, the number will be 0.006 × 1000 = 6, i.e., 3 in each type.

(ii) In a – b, crossover product will be 0.10 – 0.006 = 0.094.

Among 1000 gametes 0.094 × 1000 = 94. There are two types A b c, a B C, $\frac{94}{2}$ = 47 of each.

(iii) Similarly, in b – c, crossover product will be 0.15 – 0.006 = 0.144.

Among 1000 gametes 0.144 × 1000 = 144; There are two types, i.e., A B c and a b C; $\frac{144}{2}$ = 71 of each.

(iii) Non recombinant, i.e., parental progenies will be 1000 – (6 + 94 + 144) = 1000 – 244 = 756; they are also two types A B C and a b c, i.e., 378 of each.

Problem 40

The genes for vermilion (v) eyes, sable (s) body colour and forked (f) bristles are on the X-

chromosome of *Drosophila melanogaster*. They map in the order vsf with 10 map units betwen v and s ad 14 map units between s and f. Suppose that you cross a wild type male with a vsf female and then test-cross the F_1 females.

(a) Indicate the progeny classes and their expected numbers among 2000 total progeny, assuming that the interference is i = 1.

(b) Make the same calculation assuming an interference value of i = 0.5.

(c) Assume again that i = 1 and suppose that you cross homozygous wild type female to Vsf male and then mate the progeny among themselves. What are the possible classes of progeny and their expected numbers among 2000 progeny animals?

Solution:

(a) Assuming interference 1 the coefficient coincidence is close to zero. It occurs when map distance is less than 20 cM and here D.C.O. seldom occur.

Therefore, in this case, there will be no D.C.O. products here.

$♂ \frac{+\,+\,+}{\longrightarrow}$ (wild type male) × $\frac{v\,s\,f}{v\,s\,f}$ (female) ♀

F_1 (female) $\frac{v\,s\,f}{+\,+\,+}$ (wild type) & $\frac{v\,s\,f}{\longrightarrow}$ (male)

$\frac{v\,s\,f}{+\,+\,+} \times \frac{v\,s\,f}{\longrightarrow}$

Female			Male	
1. $\frac{v\,+\,+}{v\,s\,f}$	vermilion	(100)	1. $\frac{v\,s\,f}{\longrightarrow}$	(vermilion scute froked)
2. $\frac{+\,s\,f}{v\,s\,f}$	scute forked	(100)	2. $\frac{+\,+\,+}{\longrightarrow}$	(wildtype)
3. $\frac{v\,s\,+}{v\,s\,f}$	vermilion scute	(140)		
4. $\frac{+\,+\,f}{v\,s\,f}$	forked	(140)		
5. $\frac{v\,s\,f}{v\,s\,f}$	vermilion scute forked	(760)		
6. $\frac{+\,+\,+}{v\,s\,f}$	wild type	(760)		

$$2000 - (200 + 280) = 2000 - 480 = \frac{1520}{2} = 760$$

(b) $i = 0.5 = \text{C.O.C.} = 1 - 0.5 = 0.5 = \frac{\text{observed D.C.O.}}{\text{expected D.C.O.}}$

Observed D.C.O. = 0.5 × 0.10 × 0.14 = 0.007

Among 2000 it will be 0.007 × 2000 = 14.000

Female

	Genotype	Phenotype	Number	
(i)	$\frac{v\,+\,+}{v\,s\,f}$	vermilion =	43	$[100-14 = \frac{86}{2} = 43$ each]
(ii)	$\frac{+\,s\,f}{v\,s\,f}$	scute forked	43	
(iii)	$\frac{v\,s\,+}{v\,s\,f}$	vermilion scute	93	$[200-14 = \frac{186}{2} = 93$ each]
(iv)	$\frac{+\,+\,f}{v\,s\,f}$	forked	93	
(v)	$\frac{v\,+\,f}{v\,s\,f}$	vermilion forked	7	
(vi)	$\frac{+\,s\,+}{v\,s\,f}$	scute	7	
(vii)	$\frac{+\,+\,+}{v\,s\,f}$	wild type	857	[2000−(86+186+14) =1714]
(viii)	$\frac{v\,s\,f}{v\,s\,f}$	vermilion scute forked	857	

(c) $\frac{+\,+\,+}{+\,+\,+}$ (Homozygous) wild type female) × $\frac{v\,s\,f}{\longrightarrow}$ (♂)

F_1 (female) $\frac{+\,+\,+}{v\,s\,f}$ × $\frac{+\,+\,+}{\longrightarrow}$ (wild type male)

F_2

$\frac{+\,+\,+}{v\,s\,f}$, $\frac{+\,+\,+}{+\,+\,+}$ (wildtype female)

	Genotype	Phenotype	Number
1.	$\frac{+\,+\,+}{\longrightarrow}$	wildtype	760
2.	$\frac{v\,s\,f}{\longrightarrow}$	vermilion scute forked	760
3.	$\frac{+\,s\,f}{\longrightarrow}$	scute forked	100
4.	$\frac{v\,+\,+}{\longrightarrow}$	vermilion	100
5.	$\frac{+\,+\,f}{\longrightarrow}$	forked	140
6.	$\frac{v\,s\,+}{\longrightarrow}$	vermilion scute	140

As interferene is 1, so no D.C.O.

Problem 41

In *Drosophila* females heterozygous for three recessive mutations a, b, c were crossed to males homozygous for all three mutation. The cross yielded the following results.

Phenotype			Number
+	+	+	75
+	+	c	348
+	b	c	96
a	+	+	110
a	b	+	306
a	b	c	65

Construct a linkage map showing the correct order of these genes and estimate the distances between them.

Solution:

+ + c =	348	Parental classes
b a + =	306	
+ + + =	75	140 = 14% C.O between a – c
a b c =	65	
a + + =	96	206 = 20.6% C.O between b – a
+ b c =	110	

b —20.6%— a —14.0%— c
(20.6 cM, 14.0 cM)

Here D.C.O is missing (b + +, + a c)

Problem 42

Female *Drosophila* heterozygous for three recessive mutations e (ebony body), st (scarlet eyes), and ss (spineless bristles) were test-crossed and the following progeny were obtained.

Phenotype	Number
wild type (+ + +)	67
ebony (e + +)	8
ebony scarlet (+ e st)	68
ebony spineless (e + ss)	347
ebony scarlet spineless (e st ss)	78
scarlet (+ st +)	368
scarlet spineless (+ st ss)	10
spineless (+ + ss)	54
	1000

(a) What indicate that the genes are linked?

(b) What are the genotypes of the original heterozygous female?

(c) What is the order of the genes?

(d) What is the map distance between e and st?

(e) What is the coefficient of coincidence?

(f) Diagram the crosses in this experiment?

Solution:

(a) Parental classes (ebony spineless – 347 and scarlet – 368) are more. The other recombinant classes are less number.

(b) $\frac{st + +}{+ \; ss \; e}$ (Heterozygous female).

(c) S.C.O. between st – ss = 78 + 67 = 145 = $\frac{145}{1000} \times 100 = 14.5\%$.

S.C.O. between ss – e = 68 + 54 = 122 = $\frac{122 \times 100}{1000} = 12.2\%$.

D.C.O. = 10 + 8 = 18 = $\frac{18}{1000} \times 100 = 1.8\%$.

st – ss = 14.5 + 1.8 = 16.3, ss – e = 12.2 + 1.8 = 14.

st —16.3%— ss —14%— e

(d) Map distance between e and st = 16.3% + 14% = 30.3% or 30.3 CM.

(e) Expected D.C.O. = $\frac{163}{1000} \times \frac{14}{100} = 0.163 \times 0.14 = 0.228$

Observed D.C.O. = $\frac{1.8}{100} = 0.018$.

$$\text{C.O.C.} = \frac{\text{observed D.C.O.}}{\text{expected D.C.O.}} = \frac{.018}{.0228} = .7894 = .789.$$

(f) $\frac{st + +}{+ \; ss \; e}$ (female) × $\frac{st \; ss \; e}{st \; ss \; e}$ (male)

Problem 43

In *Drosophila*, an allele causing vestigial wings is 12.5 mu away from another gene that causes purple eyes. A third gene that affects body colour has an allele that causes black body colour. This third gene is 18.5 mu away from the vestigial wings gene and 6 mu away from the gene causing purple eyes. The alleles causing vestigial wings, purple eyes and black body are all recessive. The dominant traits are long wings, red eyes and gray body. A researcher crossed wild type flies to flies with vestigial wings, purple eyes, and black bodies. All F_1 flies were wild type. F_1 females were then crossed to male flies with vestigial wings, purple eyes and black bodies. If 1000 offsprings were obsserved, what are the expected numbers of the following types of flies?

1. **Long wings, red eyes, gray body**
2. **Long wings, purple eyes, gray body**
3. **Long wings, red eyes, black body**
4. **Long wings, purple eyes, black body**
5. **Short wings, purple eyes, gray body**
6. **Short wings, red eyes, gray body**
7. **Short wings, red eyes, black body**
8. **Short wings, purple eyes, black body.**

Which kinds of flies can be produced only by a double crossing over event?

Solution:

(i) Vestigial wing allele (vg) is 12.5 mu away from purple (pr) eyes gene.

(ii) Black body colour gene (b) is 18.5 mu away from vestigial wing (vg) gene and 6 mu away from the purple eye gene (pr).

The distance between vg and pr is 12.5 mu. The distance between b and vg is 18.5 mu and the distance from b to pr is 6 mu.

As the distance between black (b) to purple (pr) is 6 mu, the distance between vestigial (vg) to purple is 12.5. Then distance between black to vestigial will be 12.5 + 6 = 18.5 mu which is written in the problem. Therefore, purple (pr) is the middle one.

$$b \xrightarrow{6\text{ mu}} pr \xrightarrow{12.5\text{ mu}} vg$$
$$\vdash\!\!-\; 18.5\text{ mu} \;-\!\!\dashv$$

Expected D.C.O. will be $\frac{6}{100} \times \frac{12.5}{100} = 0.06 \times 0.125 = 0.0075$.

1000 offsprings are observed, therefore, expected D.C.O. will be 0.0075 × 1000 = 7.5 = 8 (approx).

F_1 females are all wild type.

$P_1 \quad \frac{+++}{+++} \times \frac{b\ pr\ vg}{b\ pr\ vg}$

Gamete: $\underline{+++} \quad \underline{b\ pr\ vg}$

$F_1 \quad \frac{+++}{b\ pr\ vg}$ (wildtype) × $\frac{b\ pr\ vg}{}$

D.C.O. will be + pr + (Gray body purple eyes long wings) and b + vg or (black body, red eyes and vestigial or short wing)

+ pr + (gray body purple eyes & long wing) }
b + vg (black body red eyes & vestigial wing) } 8

Problem 44

Three autosomal genes are linked along the same chrmosome. The distance between genes A and B is 7 mu, the distance between B and C is 11 mu and the distance between A and C is 4 mu. An individual who is AAbbCC was crossed to an individual who is aaBBcc to produce heterozygous F_1 offspring. The F_1 offsprings were then crossed to homozygous aabbcc individuals to produce F_2 offsprings.

A. Draw the arrangement of alleles on the chromosomes in the parents and in the F_1 offsprings.

B. Where would a crossover have to occur to produce F_2 offspring that was heterozygous for all three genes?

C. If we assume that no double crossovers occur in this region, what percentage of F_2 offsprings is likely to be homozygous for all three alleles.

Solution:

A – B = 7 mu, B – C = 11 mu, A – C = 4 mu

$$B \xrightarrow{7\text{ mu}} A \xrightarrow{4\text{ mu}} C$$
$$\vdash\!\!-\; 11\text{ mu} \;-\!\!\dashv$$

A. bbAACC × BBaacc

$\frac{b\,A\,C}{b\,A\,C} \times \frac{B\,a\,c}{B\,a\,c}$

Gametes: $\underline{b\,A\,C} \quad \underline{B\,a\,c}$

$F_1 \quad \frac{b\,A\,C}{B\,a\,c}$ (offspring)

B. $\frac{b\,A\,C}{B\,a\,c} \times \frac{b\,a\,c}{b\,a\,c}$

Heterozygous for all three genes.

$\frac{b\,A\,C}{B\,a\,c}$ C.O. between b – a

(i) b a c, (ii) B A C are crossed with $\underline{b\,a\,c}$

1. $\frac{b\,a\,c}{b\,a\,c}$ all are homozygas recessive.

2. $\frac{B\,A\,C}{b\,a\,c}$ heterozygous for all three genes.

C. Expected D.C.O. = 0.07 × 0.04 = 0.0028 = 0.28% i.e. 0

Therefore, 50% will be heterozygous for all three genes.

Problem 45

A hairy winged (h) *Drosophila* female is mated with yellow bodied (y), white eyed (w) male; the F_1 females are all wild type. The F_1 progenies are then crossed with hairy male and the F_2 emerge flies as follows:

Sex	Phenotypes	Number
Females	**Wild type**	**757**
	Hairy	**243**
Males	**Hairy**	**390**
	Wild type	**130**
	Yellow	**4**
	White	**3**
	Hairy, yellow	**1**
	Hairy, white	**2**
	Yellow white	**360**
	Hairy yellow white	**110**
	Total	**2000**

Give genotypes of the parents and the F_1 and show the linkage relations and distances where possible.

Solution:

Hairy winged female × Yellow white male

$\frac{h++}{h++}$ × $\frac{+yw}{\longrightarrow}$

Gametes: h + +   + y w ⟶

F_1 $\frac{h++}{+yw}$(wild type) × $\frac{h++}{\longrightarrow}$ (male)

Gametes (♀): h + +, + y w, h y w, + + +, h + w, + y +, h y +, + + w

Gametes (♂): h + +, ⟶

	Male		Number		Female	Number
1.	$\frac{h++}{\longrightarrow}$	Hairy	390	750	$\frac{h++}{h++}$ Hairy	243
2.	$\frac{+yw}{\longrightarrow}$	Yellow white	360		$\frac{h+w}{h++}$ Hairy	
3.	$\frac{hyw}{\longrightarrow}$	Hairy yellow white	110	240	$\frac{h++}{h++}$ Hairy	
4.	$\frac{+++}{\longrightarrow}$	Wildtype	130		$\frac{hyw}{h++}$ Hairy	
5.	$\frac{h+w}{\longrightarrow}$	Hairy white	2	6	$\frac{+yw}{h++}$ Wildtype	757
6.	$\frac{+y+}{\longrightarrow}$	Yellow	4		$\frac{+++}{hy+}$ Wild	
7.	$\frac{hy+}{\longrightarrow}$	Hairy yellow	1	4	$\frac{+y+}{h++}$ Wild	
8.	$\frac{++w}{\longrightarrow}$	White	3		$\frac{++w}{h++}$ Wild	
			1000			1000

S.C.O between h – y = $\frac{240}{1000}$ × 100 = 24%

S.C.O between y – w = $\frac{6}{1000}$ × 100 = 0.6%

D.C.O (h – y and y – w) = $\frac{4}{1000}$ × 100 = 0.4%

Distance between h – y = 24 + 0.4 = 24.4

Distance between y – w = 0.6 + 0.4 = 1.0

h —24.4%— y —1.0— w

Problem 46

A homozygous claret (ca = claret eye colour) curled (cu = curved wing), fluted (fl-creased wigs) fruit fly is crossed with a pure breeding wild type fly. The F_1 females are test-crossed with the following results.

1. **fluted (fl + +) = 4**
2. **claret (ca + +) = 173**
3. **curled (cu + +) = 26**
4. **fluted claret (fl + ca) = 24**
5. **fluted curled (fl + cu) = 167**
6. **claret curled (ca + cu) = 6**
7. **fluted claret curled (fl cu ca) = 298**
8. **wild type (+ + +) = 302**

(a) Are the loci linked? (b) If so, give the gene order map distance and coefficient of coincidence.

Solution:

(i) Highest (Parental) group fluted curled claret = (fl cu ca) = 298 and wild type (+ + +) = 302

(ii) Genes are in cis position.

(iii) D.C.O. is fluted (fl + +) = 4 and claret and curled (ca cu +) = 6. The middle gene is fluted (fl), the order of gene $\frac{ca\,fl\,cu}{+++}$.

(iv) claret (ca + ++) = 173, fluted curled (fl cu +) = 167, Total 340.

(v) curled (cu + +) = 26, fluted claret (fl + ca) = 24, Total 50.

(vi) Total progeny = 298 + 302 + 173 + 167 + 24 + 26 + 6 + 4 = 1000.

C.O. product betwen ca – fl = ca + + = 173 and + fl cu = 167 = 340 = 34%.

C.O. product betwen fl – cu = ca fl + = 24 and curled + + cu = 26, Total 50, i.e., 5%.

D.C.O. = fl + + = 4 and + ca cu = 6 = 10 = 1%.

Distance between claret fluted = 34 + 1 = 35%

fluted claret = 5 + 1 = 6%.

(a) Genes are linked $\frac{+\ +\ +}{ca\ fl\ cu}$.

(b) Order

$ca \underline{\quad 35\,mu \quad} fl \underline{\quad 6\,mu \quad} cu$

$$\text{C.O.C.} = \frac{\text{observed D.C.O.}}{\text{expected D.C.O.}}$$

$$\text{Expected D.C.O.} = \frac{35}{100} \times \frac{6}{100} = \frac{210}{1000} = 0.021$$

$$\text{Observed D.C.O.} = 1\% = \frac{1}{100} = 0.01$$

$$\text{C.O.C.} = \frac{0.01}{0.021} = 0.47619 = 47.6\% = 48\%.$$

Problem 47

The following is a linkage map for three recessive genes located in the same chromosome of a certain species measured in percent recombination frequencies.

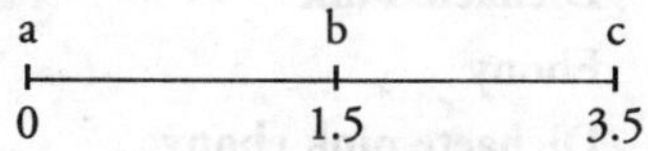

If the coefficient of coincidence is 0.7 in this case, determine the number of phenotypes among 1000 offsprings of a cross $\frac{a\,b\,c}{A\,B\,C} \times \frac{a\,b\,c}{a\,b\,c}$.

C.U. 2007

Solution:

a – b = 15%, b – c = 35 – 15 = 20%

$$\text{Expected D.C.O.} = \frac{15}{100} \times \frac{15}{100} = \frac{300}{10000} = 0.03$$

Coefficient of coincidence = 0.7

$$0.7 = \frac{\text{observed D.C.O.}}{\text{expected D.C.O.}} = \frac{ob}{.03}.$$

∴ Observed D.C.O. = 0.03 × 0.7 = 0.021.

Among 1000 progenies it would be 0.021 × 1000 = 21.

D.C.O. would be = 21.

S.C.O.:(i) $\frac{15}{100} \times 1000 = 150$; 150 – 21 = 129

(ii) $\frac{20}{100} \times 1000 = 200$; 200 – 21 = 179

1. A B C =, a b c = } 671

A b C =, a B C = } 129

A B c =, a b C = } 179

Parental Combination

PC = 1000 (S.C.O. + S.C.O. + D.C.O.)

= 1000 (129 + 179 + 21)

= 1000 – 329 = 671

A b C =, a B c = } 21

Problem 48

In *Drosophila*, Dichaete is a 3rd chromosome mutation with a dominant effect on wing shape and pink ebony are 3rd chromosome recessive mutation which affect eye colour and body colour respectively. Flies from Dichaete 'stock' were crossed to homozygous 'pink ebony' flies and the F_1 flies with Dichaete phenotype were back crossed to pink ebony homozygotes. The F_2 results were as follows:

Phenotypes	Number
Dichaete	776
pink ebony	744
Dichaete ebony	205
pink	188
Dichaete pink	12
ebony	17
Dichaete pink ebony	25
wild type	33

(i) What are the recombination distances and linkage order between each of these genes?

(ii) What is the coefficient of coincidence?

Solution:

(i) Nonrecombinant, i.e., parental progeny Dichaete (D + +), 776 and pink ebony (D p e) = 744.

(ii) D.C.O. progeny Dichaete pink (D p +) = 12 and ebony (+ + e) = 17, Total = 29.

So, the middle gene is p.

$$\text{D.C.O. percentage} = \frac{29}{2000} \times 100 = 1.45\%.$$

C.O. product betwen Dichaete and pink are:

(i) Dichaete pink ebony (D p e) = 25, Wild type(+++) = 33 } 58, i.e., $\frac{58}{2000} \times 100 = 2.9\%$

(ii) C.O. products between pink and ebony are

Dichaete ebony (D + e) = 205, Pink (+ p +) = 188 } 393, i.e., $\frac{393}{2000} \times 100 = 19.65\%$

Distance between dichaete and pink = 2.9 + 1.45 = 4.35%.

Distance between pink and ebony = 19.65 + 1.45 = 21.15%.

Expected D.C.O. = $\frac{4.35}{100} \times \frac{21.1}{100} = \frac{91.780}{10000}$ = 0.009178.

Observed D.C.O. = 1.45 = $\frac{1.45}{100}$ = 0.0145.

C.O.C.= $\frac{\text{observed D.C.O}}{\text{expected D.C.O.}} = \frac{0.0145}{0.00917} = \frac{0.045}{0.0092}$

= 1.57.

Gene order D $\xrightarrow{4.35 \text{ mu}}$ p $\xrightarrow{21.1 \text{ mu}}$ e

Problem 49

In *Drosophila*, the sex linked genes cut (ct) lozenge (lz) and froked bristles (f) are with the following map distance apart.

Cut to lozenge = 8.2 units lozenge to forked bristles = 28.0 units. They have an interference of 0.3; work out

(a) the expected number of genotypes out of 2000 flies recovered from the cross $\frac{\text{ct lz f}}{\text{+ + +}}$ × ct lz f

(b) What will be the result, if there is no interference?

Solution:

ct —— 28.0 —— Lz —— 0.3 —— f

Expected D.C.O. = $\frac{280}{100} \times \frac{0.3}{100}$ = 0.28 × 0.003 = 0.00084.

Interference 0.3

∴ coefficient of coincidence = 1 – 0.3 = 0.7.

0.7 = $\frac{\text{observed D.C.O.}}{\text{expected D.C.O}} = \frac{\text{observed D.C.O.}}{0.00084}$.

∴ Observed D.C.O. = 0.00084 × 0.7
= 0.000588.

Among 2000 flies = 0.000588 × 2000
= 1.176000 = 1.17.

(i) 28% C.O. between ct and lz.

Recombinant will be $\frac{28}{100}$ × 2000 = 560 – 1 = 559.

(ii) 0.3% C.O. between lz and f

Recombinant will be $\frac{0.3}{1000}$ × 2000 = 6 – 1 = 5

Total recombinant = 559 + 5 + 1 = 565

Parental progeny = 2000 – 565 = 1435.

(b) If there is no interference, the coincidence will be 100%.

Problem 50

In *Drosophila*, dichaete is a 3rd chromosome mutation with a dominant effect on wing shape. Pink (p) and ebony (e) are 3rd chromosome recessive mutation, which affects eye colour and body colour respectively. Flies from 'Dichaete' stock (D) were crossed to homozygous 'pink-ebony' flies, and the F_1 female flies with 'Dichaete' phenotype were back-crossed to "pink ebony" homozygous mates. The F_2 results were as follows:

Phenotypes	**Number**
Dichaete	**824**
Pink ebony	**878**
Dichaete ebony	**214**
Pink	**187**
Dichaete Pink	**11**
Ebony	**18**
Dichaete pink ebony	**68**
Wild type	**71**

(a) What are linkage distances between these genes? Mention their linkage order with proper diagram.

(b) What is the interference in this case?

(i) With the three gene pairs of phenotypic classes which arises from non-cross-over gametes, tell us the original genotypes of the parents. Here Dichaete and pink ebony are phenotypic classes which have 824 and 878 members respectively for a total 1702 i.e. (824 + 878).

(ii) The order of gene can be determined by inspecting the least frequent reciprocal pairs of phenotypic classes that represent the result of double crossing-over. Here Dichete pink and ebony are double cross-over classes having 11 and 18 members of a total 11 + 18 = 29. Therefore, frequences of D.C.O. is $\frac{29}{2271}$ × 100 = 1.276%.

(iii) The gene pair which has changed its position relative to other two pairs of genes (alleles) is the middle gene. Here $\frac{+}{p}$ has changed its position. Therefore, order of genes are Dpe.

(iv) Diagram of the crosses are as follows:

$$P: \quad \frac{D\,+\,+}{+\,p\,e} \times \frac{+\,p\,e}{+\,p\,e}$$

$$F_1 \quad \frac{D\,+\,+}{+\,p\,e}\text{(Dichaete)} \times \frac{+\,p\,e}{+\,p\,e}\text{(pink ebony)}$$

(v) A single crossing-over between Dichaete and pink produces Dichaete pink ebony and wild-type progeny which have 68 and 71 members respectively for a total of 68 + 71 = 139.

Therefore, S.C.O. between D and p is $\frac{139}{2271} \times 100 = 6.12\%$.

Similarly other single crossing-over between pink and ebony gives dichaete ebony and pink progeny which have 214 and 187 members respectively for a total of 214 + 187 = 401.

Therefore, frequency of S.C.O. between p and e is $\frac{401}{2271} \times 100 = 17.65\%$.

(vi) The map distance between D – p = 6.12 + 1.276 = 7.396 = 7.4 and map distance between p – e = 17.65 + 1.276 = 18.926 = 18.9.

Linkage distance and order of genes are

$$D \underline{\quad 7.4\% \quad} p \underline{\quad 18.9\% \quad} e$$

$$\text{C.O.C.} = \frac{\text{observed D.C.O.}}{\text{expected D.C.O.}} = \frac{1.276\%}{7.4\% \times 18.9\%}$$

$$= \frac{\frac{1.276}{100}}{\frac{7.4}{100} \times \frac{18.9}{100}} = \frac{0.01276}{0.074 \times 0.189} = \frac{0.01276}{1.3986}$$

$$= 0.09, \text{ i.e., } 0.09 \times 100 = 9.12\%.$$

Interference = 100 – 9.12 = 90.88%.

Problem 51

In the Genetic Research unit of Calcutta University, one scientist has estimated the number of chiasmata that occurred during meiosis on each of hundred chromatids which were recovered in gametes.

The data are as follows:

Number of Chiasmata	Frequency
4	6
0	18
1	20
3	16
2	40

What is the genetic length in centimorgan of chromosome analyzed in this research study?

Solution:

(i) The number of chromatid is 100.

(ii) The number of chiasmata on chromatid is 4 × 6 + 0 × 18 + 1 × 20 + 3 × 16 + 2 × 40 = 24 + 0 + 20 + 48 + 80 = 172.

The average number of chiasmata per chromatid is $\frac{172}{100}$ = 17.2 morgans, i.e., 172 centimorgans.

Problem 52

The total length of *Drosophila* genome is about 270 cM. How many chiasmata will occur in an Oocyte going through meiosis?

Solution

One chiasmata during meiosis is equivalent to 50 centimorgan of genetic map unit.

50 centimorgan = 1 chiasmata.

270 centimorgan = $\frac{1}{50} \times 270 = 5.4$.

Problem 53

A chromosome is 150 cM long. On the average how many chiasmata will occur when it goes through meiosis?

Solution

One chiasma is equivalent to 50 centimorgan (cM) of genetic map unit.

50 cM = 1 chiasmata.

150 cM = $\frac{1}{50} \times 150 = 3$.

EXERCISE

1. Singed bristle (sn), cross veinless wings (cv) and vermilion eye colour (v) are due to recessive mutant alleles of three sex-linked genes in ***Drosophila melanogaster***. When a female heterozygous for each of three genes was test-crossed with a singed, cross veinless, vermilion male, the following progenies were obtained:

Phenotypes	Number
Singed cross-veinless vermilion	3
Cross-veinless vermilion	392
Vermilion	34
Cross-veinless	61
Singed cross-veinless	32
Singed vermilion	65
Singed	410
Wild type	3
Total	1000

(a) What is the correct order of three genes on the chromosome?

(b) What are the genetic map distances between sn and cv, sn and v?

(c) What is the coefficient of coincidence?

2. In corn, a plant heterozygous for three gene pairs, A, a; B, b; C, c is test-crossed to the triple recessive with the following results:

ABC	142	ABc	22
abc	148	abC	18
Abc	235	AbC	95
aBC	255	aBc	85

(a) What is the parental cross?

(b) What is the genotype of the heterozygous plant?

(c) Give the single and double crossover progeny.

(d) Determine the correct order of the genes without mapping them.

(e) Map the genes.

(f) Calculate the coefficient of coincidence.

3. In *Drosophila*, a fly that is a triple recessive for three sex-linked genes—ruby, vermilion, cut (rb, v, ct) is crossed to a wild type fly. The F_1 heterozygous daughters are crossed to triple recessive males. The offsprings fall into the following eight classes:

+ + +	729	rb + ct	129
rb v ct	701	+ v +	151
rb + +	138	rb v +	13
+ v ct	122	+ + ct	17

(a) Map these genes.

(b) What is the coefficient of coincidence value?

4. Alfred Sturtevant (1921) reported the following data from crosses made with *Drosophila simulans*. The genes were yellow body (y), carmine eyes (c) and forked bristles (f), all three recessive to wild type and sex-linked. Yellow carmine females were mated to forked males and the F_1 females were mated to yellow carmine forked males to produce the following offsprings:

y c +	725	y c f	134
+ + f	719	+ + +	109
y + f	419	y + +	34
+ c +	383	+ c f	32

(a) Map these genes.

(b) What is the coefficient of coincidence value?

5. Assume a three-point test-cross in which F_1 kk^+ ll^+ mm^+ was crossed to kk ll mm and the offsprings had the following genotypes:

kk^+ ll^+ mm^+	619	kk^+ ll mm	109
kk^+ ll^+ mm	3	kk ll^+ mm	57
kk^+ ll mm^+	64	kk ll mm^+	7
kk ll^+ mm^+	103	kk ll mm	608

Construct a genetic map of these loci including their order and the map distance between them. Calculate the magnitude of interference.

[C.U. 1995]

6. A male *Drosophila* having cut wings (ct), vermilion eyes (v) and a yellow body (y) is crossed to a female heterozygous with respect to these three characters. The following classes of phenotypes are observed in the offsprings:

Yellow body	440
Yellow body, vermilion eyes	50
Wild type	1780
Vermilion eyes, cut wings	470
Vermilion eyes	300
Yellow body, vermilion eyes, cut wings	1710
Yellow body, cut wings	270
Cut wings	55

Construct the genetic map of the three loci involved indicating both map distance and the correct sequence of genes. What is the coefficient of coincidence involved here? What classes represent the occurrence of double-crossover events? [C.U. 2003]

7. *Drosophila* stubble (Sb) is a third chromosome mutation and dominant. Two other mutations pink (p) and ebony (e) are also on the third chromosome and are recessive. Homozygous Sb is lethal. Sb flies were crossed with homozygous pink and ebony flies. The F_1 flies were backcrossed to homozygous pink ebony flies. The results in the F_2 are as follows:

Stubble	900	Stubble pink	10
Pink ebony	925	Stubble pink ebony	25
Stubble ebony	200	Ebony	25
Pink	180	+ + +	27

(a) What are the parental F_1 and F_2 genotypes?

(b) What is the present recombination between the three genes?

(c) What is the sequence of the three genes in the chromosome?

(d) What is the coefficient of coincidence and is there any interference? [C.U. '93]

8. In male *Drosophila*, recombination does not occur, F_1 males of a cross were heterozygous for the three alleles dachs (d), vestigial (vg) and black (b). These flies when test-crossed following F_2 progenies:

wild type	9	vg b	79
b	339	d b	63
vg	71	d vg	373
d	85	d vg b	3

From the data above, write down the genotypes of parental flies; and determine the order of the genes. Calculate the map distance between the three loci, and comment on the interference. What will be the phenotypes of the progenies if the F_1 males are test-crossed? [C.U. 1999]

9. 'Star' is a dominant mutation in the second chromosome of *Drosophila*. 'Aristaless' and 'dumpy' are two recessive mutations in the same chromosome. A cross is made between 'star' and homozygous 'aristaless' and 'dumpy'. The F_1 'star' females were then backcrossed to homozygous 'aristaless' 'dumpy males' and the following phenotypes were obtained as a result:

Star	958	Aristaless dumpy	916
Aristaless star	8	Star dumpy	99
Aristaless	158	Dumpy	4

From the above data determine the order of loci for the three genes. Find out the recombination distance between them. What classes of phenotypes are missing and why? [C.U. 1997]

10. In a cross that includes three loci in *Drosophila* the non-wild type alleles sc (scute, or loss of certain thoracic bristles), ec (echinus, or roughned eye surface) and vg. If we cross sc ec vg flies with homozygous wild type and then test-cross the F_1 females which are sc/+ ec/ + vg/+, we obtain the following results:

sc ec vg	235	sc + vg	12
+ + +	241	+ ec +	14
sc ec +	243	sc + +	14
+ + vg	233	+ ec vg	16

Determine the order in which the three loci occur in the chromosome and prepare a chromosome map.

11. A male *Drosophila* having echinus eyes (ec), crossveinless wing (cv) and cut wings (ct) is crossed to a female heterozygous with respect to these three characters. The following classes of phenotypes are observed in the offspring:

Crossveinless	2207
Echinus cut	2125
Echinus crossveinless	273
Cut	265
Echinus	217
Crossveinless	223

Wild type	5
Echinus, crossveinless, cut	3

The above characters are due to recessive mutant allele of three x-linked genes in *Drosophila*.

(a) Construct the genetic map of the three loci involved indicating both map distance and correct sequence of genes.

(b) What are coefficient of coincidence and interference involved here? [C.U. 2006]

12. Female *Drosophila* heterozygous for ebony (e^+/e), scarlet (st^+/st) and spineless (ss^+/ss) were test-crossed and the following progenies (Phenotype) were obtained:

Progeny phenotypes	Number
wild type	67
ebony	8
ebony scarlet	68
ebony spineless	347
ebony scarlet spineless	78
scarlet	308
scarlet spineless	10
spineless	54

(a) Write the genotypes of the flies involved in the parental cross and test cross.

(b) Construct the genetic map of the three loci involved indicating the map distance and correct sequence of the gene.

(c) Calculate the coefficient of coincidence. [C.U. 2008]

13. You have a *Drosophila* line that is homozygous for autosomal recessive alleles a, b and c, linked in that order. You cross females of this line with males homozygous for the corresponding wild type alleles. You then cross the F_1 heterozygous males with their heterozygous sisters. You obtain the following F_2 phenotypes (where the letters denote recessive phenotypes and pluses denote wild type phenotypes)

+ + + = 1364, abc = 365, ab + = 87, + + c = 84, a + + = 47, + bc = 44, a + c = 5 and + b + = 4.

(a) What is the recombinant frequency between a and b, between b and c? (Remember, there is no crossing-over in *Drosophila* males.)

(b) What is coefficient of coincidence?

14. Chromosome 3 of corn carries three loci (b for plant-colour booster, v for virescent and lg for liguleless). A test-cross of triple recessives with F_1 plants heterozygous for three genes yields progeny having the following genotypes:

305 = + v lg, 275 = b + +, 128 = b + lg, 112 = + v +, 74 = + + lg, 66 = b v +, 22 = + + + and 18 = b v lg.

Give the gene sequence on the chromosome, the map distances between the genes and the coefficient of coincidence.

15. Assume that three pairs of alleles are found in *Drosophila*: x^+ and x, y^+ and y, and z^+ and z. As shown by the symbols, each non-wild type allele is recessive to its wild type allele. A cross between females heterozygous at these three loci and wild type males yields progeny having the following genotypes:

	Phenotypes	Number
Females:	$x^+ y^+ z^+$	1010
Males:	$x\ y^+ z$	430
	$x^+ y\ z^+$	441
	$x\ y\ z$	39
	$x^+ y^+ z$	32
	$x^+ y^+ z^+$	30
	$x\ y\ z^+$	27
	$x^+ y\ z$	1
	$x\ y^+ z^+$	0

(a) On what chromosome of *Drosophila* are three genes carried?

(b) Draw the relevant chromosomes in the heterozygous female parent, showing the arrangement of the alleles.

(c) Calculate the map distances between the genes and the coefficient of coincidence.

16. From the five sets of data given in the following table, determine the order of genes by inspection—that is, without calculating recombination values. Recessive phenotypes are symbolized by lower case letters and dominant phenotypes by pluses:

Phenotypes observed in 3-point test-crosses			Data sets				
			1	2	3	4	5
+	+	+	317	1	30	40	305
+	+	c	58	4	6	232	0
+	b	+	10	31	339	84	28
+	b	c	2	77	137	201	107
a	+	+	0	77	142	194	124
a	+	c	21	31	291	77	30
a	b	+	72	4	3	235	1
a	b	c	203	1	34	46	265

17. From the phenotype data given in the following table for two three-point crosses for (1) a, b and c and (2) b, c and d, determine the sequence of the four genes a, b, c and d, and the three map distances between them. Recessive phenotypes are symbolized by lowercase letters and dominant phenotypes by pluses:

1		2	
+ + +	669	b c d	8
a b +	139	b + +	441
a + +	3	b + d	90
+ + c	121	+ c d	376
+ b c	2	+ + +	14
a + c	2280	+ + d	153
a b c	653	+ c +	65
+ b +	2215	b c +	141

18. The father of Mr Spock, first officer of the Starship Enterprise, came from planet Vulcan; Spock's mother came from Earth. A Vulcan has pointed ears (determined by allele P), adrenal absent (determined by A) and a right-sided heart (determined by R). All these alleles are dominant to normal Earth alleles.

The three alleles loci are autosomal, and they are linked as shown in this linkage.

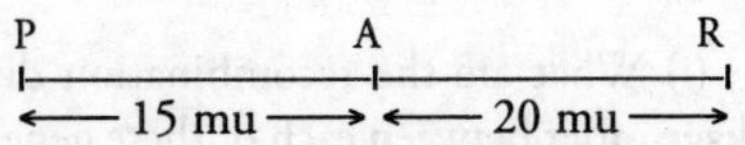

If Mr Spock marries an Earth woman and there is no (genetic) interference, what proportion of their children will have

(a) Vulcan phenotypes for all three characters?
(b) Earth phenotypes for all three characters?
(c) Vulcan ears and heart but Earth adrenals?
(d) Vulcan ears but Earth heart and adrenals?

19. In a certain diploid plant, the three loci A, B and C are linked as follows:

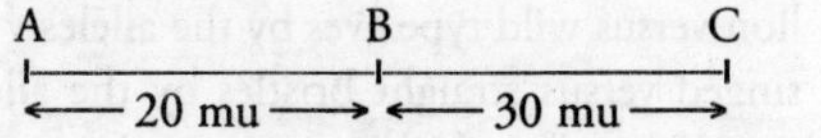

One plant is available to you (call it the parental plant). It has constitution $\frac{\text{Abc}}{\text{aBC}}$.

(a) Assuming no interference, if the plant is selfed, what proportion of the progeny will be of the genotype $\frac{\text{abc}}{\text{abc}}$?
(b) Again assuming no interference, if the parental plant is crossed with the $\frac{\text{abc}}{\text{abc}}$ plant, what genotypic classes will be found in the progeny? What will be their frequencies if there are 1000 progenies?
(c) Repeat part (b), this time assuming 20 percent interference between the regions.

20. In corn (*Zea mays*), the genetic map of part of chromosome 4 is as follows, where w, s and e represent recessive mutant alleles affecting the colour and shape of the pollen:

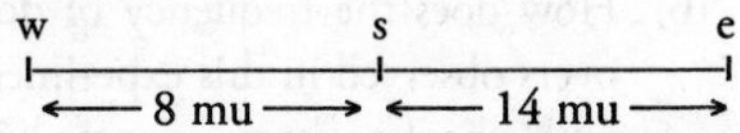

If the following cross is made $\frac{+++}{+++} \times \frac{\text{wse}}{\text{wse}}$ and the F_1 is test-crossed to $\frac{\text{wse}}{\text{wse}}$, and if it is assumed that there is no interference on this region of this chromosome, what proportion of progeny will be of genotypes (a) +++, (b) wse, (c) +se, (d) w++, (e) ++e, (f) ws+, (g) w+e, (h) +s+?

21. The recessive mutations b (black body colour), st (scarlet eye colour) and hk (hooked bristles) identify three autosomal genes in *D. melanogaster*. The following progeny were obtained from a test-cross of females heterozygous for all three genes.

black scarlet	243
black	241
black hooked	15
black hooked scarlet	10
hooked	235
hooked scarlet	226
scarlet	12
wild type	18

What conclusions are possible concerning the linkage relation of these three genes? Calculate any appropriate map distances.

22. Yellow versus gray body colour in *D. melanogaster* is determined by the alleles y and y^+, vermilion versus wild type eyes by the alleles v and v^+ and singed versus straight bristles by the alleles sn and sn^+. When female heterozygous for each of these X-linked genes were test-crossed with yellow vermilion, singed males, the following classes and progenies were obtained:

yellow, vermilion singed	53
yellow, vermilion	108
yellow singed	331
yellow	5
vermilion singed	3
vermilion	342
singed	95
wild type	63

(a) What is the order of the three genes? Construct a linkage map with the genes in their correct order, and indicate the map distances between the genes.

(b) How does the frequency of double cross-overs observed in this experiment compare with the frequency expected if crossing-over occurs independently in the 2 chromosome regions? Determine the coefficient of coincidence and the interference.

23. In a corn, the alleles C and c result in coloured versus colourless seeds, Wx and wx in nonwaxy versus waxy endosperm, and Sh and sh in plum versus shrunken endosperm. When plants grown from seeds heterozygous for each of these pairs of alleles were test-crossed with plants from colourless, waxy, shrunken, the progeny seeds are as follows:

colourless nonwaxy shrunken	84
colourless nonwaxy plum	974
colourless waxy shrunken	20
colourless waxy plum	2349
coloured waxy shrunken	951
coloured waxy plum	99
coloured nonwaxy shrunken	2216
coloured nonwaxy plum	15

Determine the order of three genes and construct a linkage map showing the genetic distances between adjacent genes.

24. In *Drosophila* the genes ct (cut wing margin), y (yellow body) and v (vermilion eye colour) are x linked. Female heterozygous for all three markers were mated with wild type males and the following male progenies were obtained. As in conventional *Drosophila* genetics the wild type allele of each gene is designated by a + sign in the appropriate column:

ct	y	v	4
ct	y	+	93
ct	+	v	54
ct	+	+	349
+	y	v	331
+	y	+	66
+	+	v	97
+	+	+	6
		Total	1000

(a) What was the genotype of the female parents?

(b) What is the order of genes?

(c) Calculate the recombination frequencies and the interference between the genes.

(d) Draw the genetic map of the genes.

25. In *Drosophila*, Dichaete is a 3rd-chromosome mutation with a dominant effect on wing shape, and pink, ebony are 3rd-chromosome recessive mutations which affect eye colour and body colour respectively. Flies from a 'Dichaete' stock were crossed to homozygous 'pink ebony' flies, and the F_1 flies with Dichaete phenotype were backcrossed to the pink ebony homozygotes. The F_2 results were as follows:

dichaete—776, pink ebony—744, dichaete ebony—205, pink—188, dichaete pink—12, ebony—17, dichaete pink ebony—25, wild type—33.

(i) What are the recombination distances and linkage order between each of these genes? (ii) What is the coefficient of coincidence?

[C.U. Zoology (Hons.) 2009]

26. A wild type strain of flies was crossed to another strain homozygous for the recessive X-linked mutants m, r and v. The F_1 females were back-crossed to recessive strain males, and the following numbers of pogeny were observed.

+	+	+	=	331
m	+	+	=	2
+	r	+	=	73
+	+	v	=	14
m	r	+	=	10
m	+	v	=	81
+	r	v	=	0
m	r	v	=	309
				820

(a) What is the correct order of three genes?

(b) Construct the genetic map.

(c) What is coefficient of coincidence?

27. Assume a three-point test-cross in which F_1 kk^+ ll^+ mm^+ was crossed to kk ll mm and the offspring had the following genotypes.

kk^+	ll^+	mm^+	=	621
kk^+	ll^+	mm	=	3
kk^+	ll	mm^+	=	64
kk	ll^+	mm^+	=	103
kk^+	ll	mm	=	109
kk	ll^+	mm	=	57
kk	ll	mm^+	=	7
kk	ll	mm	=	608

Construct the genetic map of these three ions including the order and the map distance between them.

28. Three recessive alleles are lz (for lazy or prostrate growth habit), gl (for glossy leaf), and su (for sugary endosperm). F_1 heterozygous parent is crossed with homozygous recessive for all three genes and the following results are obtained.

Phenotype of test cross progeny	Number
Lazy (lz + +)	33
Glossy (+ gl +)	59
Lazy glossy sugary (lz gl su)	272
Glossy sugar (+ gl su)	40
Lazy sugary (lz + su)	44
Normal (wild type) (+ + +)	286
Sugary (+ + su)	4
Lazy glossy (lz gl +)	2

(a) Construct genetic map.

(b) Calculate coefficient of coincidence.

29. In the tomato, three genes are located in the same chromosome. Tall is dominant to dwarf, skin that is smooth is dominant to skin that is peachy, and fruit with a normal tomato shape is dominant to oblate shape A plant that is true breeding for the dominant trait was crossed to a dwarf plant with peachy skin and oblate fruit. The F_1 plants were then test-crossed to dwarf plants with peachy skin and oblate fruit. The following results were obtained.

Phenotypes	Number
1. Tall smooth normal (+ + +)	152
2. Tall smooth oblate (+ + o)	33
3. Tall peach normal (+ p o)	11
4. Tall peach normal (+ p +)	2
5. Dwarf peach oblate (d p o)	156
6. Dwarf smooth normal (d + +)	13
7. Dwarf peach normal (d p +)	31
8. Dwarf smooth oblate (d + o)	2

Construct a genetic map that describes the order of these three genes and the distances between them.

30. The following X-linked recessive traits are found in fruit flies. Vermilion eyes are recessive to red eyes, miniature wings are recessive to long wings and grey body is dominant over sable body. A cross was made between wild type males with red eyes, long wings and grey bodies to females with vermilion eyes, miniature wings and sable bodies. The heterozygous females from this cross, which had red eyes, long wings and grey bodies were then crossed to males with vermilion eyes, miniature wings and sable bodies. The following outcome was obtained.

Phenotypes	Number
Vermilion eyes, miniature wings, sable body	1320
Red eyes, long wings grey body	1346
Vermilion eyes, miniature wings, grey body	102
Red eyes, long wings, sable body	90
Red eyes, miniature wings, sable body	48
Vermilion eyes, long wings, grey body	42
Vermilion eyes, long wings, sable body	2
Red eyes, miniature wings, grey body	1

(a) Calculate map distance among the three genes.

(b) Is positive interference occurring?

31. In mice, a trait called snubnose is recessive to a wild type nose, a trait called pintail is dominant to a normal tail and a trait called Jerker (a defect in motor skills) is recessive to a normal gait. Jerker mice with a snubnose and pintail were crossed to normal mice and then the F_1 mice were crossed to Jerker mice that have snubnose and normal tail, the outcome of this cross was as follows:

Phenotypes	Number
Jerker, snubnose, pintail (j sn p)	557
Normal gait, normal nose, normal tail (+ + p)	547
Jerker, snubnose, normal tail (j sn p)	105
Normal gait, normal nose, pintail (+ + p)	103
Jerker, normal nose, normal tail (j + p)	77
Normal gait, snubnose, pintail (+ sn p)	71
Jerker, normal nose, pintail (j + p)	11
Normal gait, snubnose, normal tail (+ sn p)	9

Construct the genetic map that describes the order and distance between these genes.

32. Three recessive traits in the garden of pea plants are as follows: Yellow pods are recessive to green pods, bluish green seedlings are recessive to green seedlings creeper (a plant cannot stand up) is recessive to normal. A true breeding normal plant with green pods and green seedlings was crossed to a creeper with yellow pods and bluish green seedlings. The F_1 plants were then crossed to creepers with yellow pods and bluish green seedlings. The following results were obtained.

2059	Green pods, green seedlings, normal
151	Green pods, green seedlings, creeper
281	Green pods, bluish green seedlings, normal
15	Green pods, bluish green seedlings, creeper
2041	Yellow pods, bluish green seedlings, creeper
157	Yellow pods, bluish green seedlings, normal
282	Yellow pods, green seedlings, creeper
11	Yellow pods, green seedlings, normal.

Construct a genetic map that describes the map distance among these three genes.

33. The data in the accompanying table show 1000 gametes observed from a triply heterozygous parent in a three-point test-cross to determine the genetic map of three linked genes. Neither the parental genotype of the heterozygous parent nor the order of the genes is known.

F	G	H	4
F	G	h	410
F	g	H	393
F	g	h	50
f	G	H	64
f	G	h	413
f	g	H	33
f	g	h	2

(a) What is the genotype of the heterozygous parent?

(b) What is the correct linear order of the genes?

(c) What is the map distance between the middle gene and its nearest neighbour?

(d) What is the map distance between the middle gene and its farthest neighbour?

(e) What is the expected number of double crossovers?

(f) What is the coincidence (c) and interference (i)?

34. In maize, the coloured aleurone (c), full endosperm (sh), starchy endoperm (w) genes are dominant over white (e), shurken (sh) and waxy (w) endosperm respectively. These genes are on the same chromosome. F_1 plants heterozygous for all three pairs of alleles were test-crossed and the following phenotypes are obtained.

Phenotypes	Number
White, shurken, starchy	113
Coloured, full starchy	4
Coloured, shurken, starchy	2538
Coloured, shurken, waxy	601
White, full starchy	626
White, full waxy	2708
White, shurken, waxy	2
Coloured, full waxy	116

(a) Construct genetic map.

(b) Calculate coefficient of coincidence.

35. A *Drosophila* female heterozygous for the sex-linked recessive traits a b and c was crossed to a male which was phenotypically a b c. Offsprings were of six phenotypes which occurred in the following ratio. [C.U. 2012 (2015)]

+	b	c	450
a	+	+	450
a	b	c	32
+	+	+	38
a	+	c	11
+	b	+	9

(i) Find out genotype of the female.

(ii) Analyse correct gene order.

(iii) Construct a linkage map of a b c.

(iv) Which progeny phenotypes are missing? Explain their absence.

(v) Determine coefficient of coincidence.

36. In *Drosophila,* Dichaete is a third chromosome dominant wing shape mutation, and pink and ebony are third chromosomal recessive mutation affecting the eye colour and body colour respectively. *Drosophila* males from a Dichaete stock were crossed to homozygous pink, ebony females. The F_1 flies with Dichaete phenotype were back-crossed to pink ebony homozygotes. The following results were obtained.

Phenotype	Number
Dichaete	810
Pink ebony	845
Dichaete ebony	201
Pink	198
Dichaete pink	10
Dichaete pink ebony	32
Ebony	17
Wild	37

(a) Provide the gamete cross that gives rise to the above progeny phenotypes.

(b) Analyse correct gene order.

(c) Find the distance between the three genes and construct genetic map.

(d) Calculate coefficient of coincidence and interference, if any. [C.U. 2014]

37. A *Drosophila* melanogaster female heterozygous for three 3rd chromosomal genes namely glass (gl, eyes are glassy), giant (gt, body larger) and kidney (k, eyes kidney shaped) was test-crossed and obtained following progeny.

Phenotype	Number
glass	460
giant-kidney	450
giant	16
glass-kidney	14
giant-glass	32
kidney	28

(a) Estimate recombination frequencies among the genes and draw a linkage map for these three genes.

(b) Determine the genotypes of the parental classes.

(c) Which two classes are absent and why?

(d) What would be the minimum total of progeny number if you want to have nine of each of missing classes of phenotype? [C.U. 2018]

38. In *Drosophila* a stock carrying the dominant eye mutation star on 2nd chromosome were crossed to a stock homozygous for the 2nd chromosome recessive mutations aristalers and dumpy. The F_1 females were then back-crossed to homozygous aristalers and dumpy males and the following phenotypes were obseved.

Phenotype	Number
Aristalers, dumpy	915
Star	959
Aristalers, star	7
Dumpy	5
Aristalers	128
Star, dumpy	104

(i) What are the recombination distances and the order of loci for these three genes.

(ii) What classes of phenotypes are missing and why? **[C.U. 2010]**

39. In the Genetics Research Laboratory of Kolkata, *Drosophila* female heterozygous for three recessive X-linked markers Y (yellow body), Ct (cut wings), m (miniature wings) and thier wild type alleles were crossed to $\frac{y\ ct\ m}{\longrightarrow}$ males. The following progenies were obtained.

Phenotype	Number
1. y ct m (yellow cut miniature)	30
2. y + + (yellow)	10
3. ct + m (cut miniature)	12
4. + + + (wild type)	30
5. ct + + (cut)	1
6. + + y (yellow)	1
7. + m + (miniature)	8
8. + y ct (yellow cut)	5

(a) What classes are parental types?

(b) Which classes do represent double crossovers?

(c) Which gene is in the middle of the other two?

(d) What was the genotype of the heterozygous females used in the cross? (Show correct linkage phase and correct order of the marker along the chromosome).

40. The following numbers were obtained from test-cross progeny in *Drosophila* (phenotypes).

Phenotype	Number
1. b^+ pr^+ vg^+ (Grey body red eyes and long wing)	411
2. b^+ pr^+ vg (Grey body red eyes and vestigial wing)	61
3. b^+ pr^+ vg^+ (Grey body purple eyes and long wing)	2
4. b^+ pr vg (Grey body purple eyes and vestigial wing)	30
5. b pr^+ vg^+ (black body red eyes and long wing)	28
6. b pr^+ vg (black body red eyes and vestigial wing)	1
7. b pr vg^+ (black body purple eyes long wing)	60
8. b pr vg (black body purple eyes and vestigial wing)	412

Construct genetic map and C.O.C.

41. A *Drosophila* geneticist in BHU crossed females homozygous for three X-linked mutations (y, yellow body, B-bar eye shape v and vermilion eye colour) to wild type males. The F_1 females which had grey bodies and bar eyes with dark pigment were then crossed to y B^+ v males yield in the following results.

Phenotype	Number	
1. yellow bar vermilion	289	581
2. wild type	292	
3. yellow	102	200
4. bar vermilion	98	
5. yellow vermilion	87	173
6. bar	86	
7. yellow bar	22	46
8. vermilion	24	

Determine the order of these three loci in the X-chromosomes.

42. In the following test-cross, genes a and b are 30 cM apart and genes b and c are 20 cM apart $\frac{a + c}{+ b +} \times \frac{a\ b\ c}{a\ b\ c}$. If the coefficient of coincidence is 0.5 over this interval on the linkage map, how many triply homozygous recessive individuals are expectd among 1000 offsprings?

43. In *Drosophila*, a fly that is triple recessive for three genes – cut wings (ct), crossveinless wing (cv) and vermillion eyes (v) is crossed to a wild type fly. The F_1 heterozygous daughters are crossed to triple recessive males. The following classes of phenotypes are observed.

Progeny Phenotype & Genotype	Number
1. Wild type (+ + +)	592
2. Vermilion eyes, crossveinless, cut wings (V ct cv)	580
3. Crossveinless (cv + +)	45
4. Vermilion eyes, cut wings (v ct +)	40
5. Vermilion eyes (v +)	89
6. Crossveinless, cut wings (cv + ct)	94
7. Cut wings (+ ct +)	3
8. Vermilion eyes, crossveinless (v + cv)	5
	1448

(a) Construct the genetic map of three loci involved indicating both map distance and correct sequence of genes.

(b) Calculate coefficient of coincidence and interference.

44. A three-point cross in *Drosophila* was performed in Genetic Engineering College, Badu and involved three recessive sex-linked genes, crossveinless (cv), cut wing (ct) and vermilion eye (v). The F_1 female from this cross was $\frac{+\ +\ ct}{cv\ v\ +}$ and she was test-crossed to the recessive $\frac{cv\ v\ ct}{\Longrightarrow}$ male. The following progenies were obtained

Phenotype	Number
1. cut wing (+ + ct)	759
2. crossveinless, vermilion eye (cv v +)	766
3. vermilion eye (+ v +)	73
4. crossveinless, cut wing (cv ct +)	80
5. vermilion eye cut wing (+ v ct)	140
6. crossveinless (cv + +)	158
7. wild type (+ + +)	3
8. crossveinless, cut wing vermilion eye (cv v ct)	1

(a) What is the order of the three genes? Construct a linkage map with the genes in their correct order and indicate the map distances between the genes.

(b) Determine the coefficients of coincidence and interference.

45. In a test-cross of an individual heterozygous for each of three linked genes, the most frequent classes of progeny were ABc/abC and abC/abc, and the least frequent classes were ABC/ab and abc/abC. What was the genotype of the triple heterozygous parent and what is the order of the genes?

46. Chromosome 5 of corn carries three genes v (virescent seedling), pr (red aleurone) and bm (brown midrib). A test-cross of triple recessives with F_1 plants heterozygous for these genes yields 1000 progenies with the following phenotypes.

Phenotypes	Number
1. v^+ pr bm	209
2. v pr^+ bm^+	213
3. v^+ pr bm^+	175
4. v pr^+ bm	181
5. v^+ pr^+ bm	69
6. v pr bm^+	76
7. v^+ pr^+ bm^+	36
8. v pr bm	41

(a) Determine the recombinant frequency betwen the adjacent genes.

(b) Construct genetic map of three genes. Calculate coefficients of coincidence and interference.

(c) Explain why in this experiment, the recombination frequencies are not good to estimate the gene map?

47. In *Drosophila*, recessive genes scarlet eyes (sc), spineless bristles (ss) and ebony body (e) are linked and their map distances from the end of the chromosome are 44, 58 and 70 respectively.

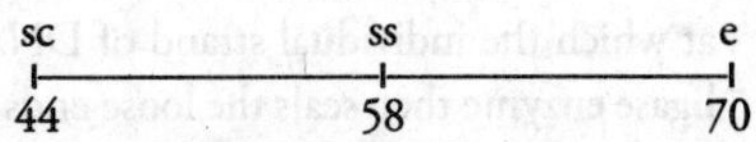

If the C.O.C is 0.70 in this case, calculate the frequencies of double crossing-over phenotypes among 1000 fly offsprings.

48. *Aedes albopictus*, a dominant gene for DDT (dichloro diphenyl trichloroethane, an insecticide) resistant and a dominant gene Dl for dieldrin resistant (dieldrin—long lasting insecticide) are known to be in the same chromosome. A cross was carried out between a DDT resistant strain and a dieldrin resistant strain, and female progeny resistant to both insecticides were test-crossed with wild type males.

The progenies were 99 resistant to both insecticides, 88 resistant to DDT only, 89 resistant to Dieldrin only and 106 sensitive to both insecticides.

(a) Are DDT and Dl alleles of the same gene? How can you tell?

(b) Are DDT and Dl linked?

(c) What can you say about the position of DDT and Dl along the chromosome?

49. A cross between female *Drosophila* heterozygous for three third chromosomal mutation Ly, Sb and br and recessive mutant male yielded following F_1 progeny:

Ly	Sb	br	—	404
+	+	+	—	422
Ly	+	+	—	18
+	Sb	br	—	16
Ly	+	br	—	75
+	Sb	+	—	59
Ly	Sb	+	—	4
+	+	br	—	2

(a) Find the correct gene order.

(b) Estimate recombination frequency between the gene pairs.

(c) Draw a linkage map of the genes.

(d) Estimate coefficients of coincidence and interference.

(e) What would be the expected number of total progeny if someone wants to have 210 D.C.O. classes? **[C.U. 2017]**

50. Ebony body colour (e), rough eye (ro) and brevis bristle (bv) are three recessive mutation in *Drosophila* melanogaster and the map is as follows:

e ——— 20 mu ——— ro ——— 12 mu ——— bv

(a) If the interference between the gene pairs is 0.4, find out the expected genotypes and number of progeny from a cross between $\frac{e^+ ro^+ bv^+}{ero\ bv} \times \frac{ero\ bv}{ero\ bv}$, when the total progeny number is 10,000.

(b) If there is no interference what would be the expected number of D.C.O. progeny classes from the same cross, when the total number of progeny is 50,000?

HOLLIDAY MODEL

1. Recombination between chromosome occurs through breakage and reunion of parts between two non-sister chromatids (each contain duplex DNA). An endonuclease enzyme creates a single strand nick at an identical point in the two homologous DNA duplex and allows movement of free end. The ends of the strands are then displaced by proteins (SSB and helicases) and subsequently pair with their complement strand on the opposite DNA duplex by the help of rec A protein. The exchange creates a connection between the two DNA duplexes. The connected pair duplex (**joint molecule**) and the point at which the individual strand of DNA crosses from one duplex to other is known as **recombinant joint**. A ligase enzyme then seals the loose ends and creates **heteroduplex** or **hybrid molecules** consisting of one strand from each of parental DNA molecule.
2. The exchange creates a planar or cross bridge structure i.e. chi (χ). One end of this figure rotates 180° and generates **Holliday structure**. The position of the cross bridge structure can then move down the chromosome by a process known as **branch migration** which occur by zipper like action as hydrogen bonds are broken and are reformed between complementary bases of displaced strands of each duplex. The branch point migrates in either direction. The joint molecule formed by strand exchange must be resolved into two separate duplex molecules. Resolution or separation requires further pair of nicks. Resolution may be horizontal (left to right cut) or vertical (up and down cut).
   - **Splice recombinant DNA duplexes:** The duplex one parent strand is covalently linked to the duplex of other parent strand a stretch of heteroduplex or hybrid DNA.
   - **Patch recombinant duplexes:** The duplex is largely unchanged except for a DNA sequence on one strand that came from the homologous chromosome.

Answers of the Exercise (Problems)

1. (a) The order of genes cv —— sn —— v
   (b) The map distance cv $\overset{13.2\%}{\text{——}}$ sn $\overset{7.2\%}{\text{——}}$ v
   (c) The coefficient of coincidence = 0.72
2. (a) The order of genes a —— b —— c
   (b) The map distance a $\overset{33\%}{\text{——}}$ b $\overset{22\%}{\text{——}}$ c
   (c) The coefficient of coincidence (C.O.C.) = 0.56
3. (a) The order of gene rt —— ct —— v
   (b) The map distance rt $\overset{14.5\%}{\text{——}}$ ct $\overset{15.5\%}{\text{——}}$ v
   (c) The C.O.C. = 0.67
4. (a) The order of gene y —— c —— f
   (b) The map distance y $\overset{33.9\%}{\text{——}}$ c $\overset{12.0\%}{\text{——}}$ f
   (c) The C.O.C. = 0.63
5. (a) The order of gene kk^+ —— mm^+ —— ll^+
   (b) The map distance kk^+ $\overset{14.14\%}{\text{——}}$ mm^+ $\overset{8.34\%}{\text{——}}$ ll^+
   (c) The C.O.C. = 0.54
6. (a) The order of gene y —— ct —— v
   (b) The map distance y $\overset{19.97\%}{\text{——}}$ ct $\overset{13.29\%}{\text{——}}$ v
   (c) The C.O.C. = 0.78
7. (a) The order of gene Sb —— p —— e
   (b) The map distance Sb $\overset{3.79\%}{\text{——}}$ ct $\overset{18.10\%}{\text{——}}$ e
   (c) The C.O.C. = 2.2
   (d) No, positive interference is lacking.
8. (a) Parents:
   Male (♂) $\frac{d+vg}{+b+} \times \frac{dbvg}{dbvg}$ Female (♀)
   (b) The order of gene d —— b —— vg
   (c) The map distance d $\overset{14.28\%}{\text{——}}$ b $\overset{17.22\%}{\text{——}}$ vg
   (d) The C.O.C. = 0.47 and the interference = 0.53 = 53%.
9. (a) The order of gene as —— st —— dp
   (b) The map distance as $\overset{0.56\%}{\text{——}}$ st $\overset{11.9\%}{\text{——}}$ dp
   (c) As the genes are very closed, so D.C.O. is missing
10. (a) The order of gene sc —— ec —— vg
    (b) The map distance sc $\overset{48.99\%}{\text{——}}$ ec $\overset{5.5\%}{\text{——}}$ vg
11. (a) The order of gene ec —— cv —— ct
    The map distance ec $\overset{10.26\%}{\text{——}}$ cv $\overset{8.42\%}{\text{——}}$ ct
    (c) The C.O.C. = 0.174
12. (a) Parental genotype: ♂ $\frac{st++}{st++} \times \frac{st\,ss\,e}{st\,ss\,e}$ ♀
    $F_1 = \frac{st++}{st++} \times \frac{st\,ss\,e}{st\,ss\,e}$ (test cross)
    (b) Genetic and map distance st $\overset{16.3\%}{\text{——}}$ ss $\overset{14.0\%}{\text{——}}$ e
    (c) The C.O.C. = 0.78
    The coefficient of interference = 0.826.
13. (a) We cannot distinguish between $\frac{+++}{+++}$ and $\frac{+++}{abc}$. Therefore, use the frequency of $\frac{abc}{abc}$ to estimate the frequency of + + + (parental) gametes from the female.

    Parental $\frac{abc}{abc}$ = 365 = 365 × 2 = 730

    C.O. between a & b = a + + = 47, + b c = 44 } = 91

    C.O. between b & c = a b + = 87, + + c = 84 } = 171

    Double crossing-over (a – b and b – c) + b + = 4, a + c = 5 } = 09

    Total = 1001

    Recombinant frequency between a – b = 91 + 9 = $\frac{100}{1001}$ × 100 = 9.99% = 10 mu.

    Recombinant frequency between b – c = 171 + 9 = $\frac{180}{1001}$ × 100 = 17.98 = 18 mu.

    (b) Coefficient of coincidence = $\frac{\text{observed D.C.O.}}{\text{expected D.C.O.}}$

    Expected D.C.O. = $\frac{9.99}{100} \times \frac{17.98}{100}$
    = 0.0999 × 0.1798 = 0.017962.

    Observed D.C.O. = $\frac{9}{1001}$ = 0.08991.

    Coefficient of coincidence = $\frac{0.08991}{0.017962}$ = 0.5.
14. (a) Gene sequence v —— b —— lg
    (b) Map distances v ← 28% → b ← 14% → lg
    [Expected D.C.O. = 0.28 × 0.14 = 0.05; Observed D.C.O. = 4% = 0.04]

C.C.O. = $\frac{0.04}{0.05}$ = 0.8.

15. (a) Genes are located on the X-chromosome of *Drosophila*.

(b) Heterozygous female $\frac{x\,z\,y^+}{x^+z^+y}$

(c) x —— z —— y
← 6 mu → ← 7 mu →

Coefficient of coincidence = $\frac{\text{observed D.C.O.}}{\text{expected D.C.O.}}$

= $\frac{0.001}{0.0042}$ = 0.238 = 23.8%

16. (a) Map distances. Gene order is deduced by inspection, without recombinant frequency analysis.

(b) It is generally true that double recombinant classes are the smallest ones. Only one order is compatible with the smallest classes and therefore formed by double crossovers.

(i) The gene order for (1) = b – a – c
(ii) The gene order for (2) = b – a – c
(iii) The gene order for (3) = b – a – c
(iv) The gene order for (4) = a – c – b
(v) The gene order for (5) = a – c – b

17. Sequence of genes

(i) In case (1): a —— c —— b
← 4.36 → ← 21.81 →

(ii) Map distances c —— b —— d
← 24.52 → ← 13.73 →

18. (i) Pointed ears = P, Adrenal absent = A, Right-sided heart = R

(ii) The genotype of Mr. Spock is $\frac{+++}{PAR}$

The genotype of his wife is $\frac{+++}{+++}$

As there is no interference, their expected D.C.O. = observed D.C.O., i.e., 0.15 × 0.20 = 0.0300 = 0.03 = 3%.

(iii) C.O. between (A–R) = $\frac{+\ \ +\ \ +}{P\ \ A\ \ R}$ = + A R = 6%, = P + + = 6%

[15% – 3% = 12%]

C.O. between (A–R) = $\frac{+\ \ +\ \ +}{P\ \ A\ \ R}$ = + P A = 8.5%, = + + R = 8.5%

[20% – 3% = 17%]

C.O. among (P–A–R) i.e. D.C.O. (3%) =

$\frac{+\ \ +\ \ +}{P\ \ A\ \ R}$ = + A + = 1.5%, = P + R = 1.5%

(a) Vulcan phenotypes for all 3 characters (PAR)
{100 – (12 + 17 + 3)} = 100 – 32 = 68%
PAR = 34%

(b) Earth phenotypes = 34%
(+ + +)

(c) Vulcan ears and heart but Earth adrenal (P + R) = 1.5%

(d) Vulcan ears but Earth heart and adrenal (P + +) = 6%.

19. Plant genotype $\frac{A\,b\,c}{a\,B\,C}$;

(a) Assuming no interference; C.O. between A – B = 20% and C.O. between B – C = 30%; expected D.C.O. = observed D.C.O.

20% × 30% = 0.20 × 0.30 = 0.006 = 6%

$\frac{A\ \ b\ \ c}{a\ \ B\ \ C}$ = A B C, = a b c (20 – 6 = 14%) abc = 7% = 0.07

As it is selfed, i.e. proportion will be 0.07 × 0.07 = 0.0049

(b) $\frac{A\,b\,c}{a\,B\,C} \times \frac{a\,b\,c}{a\,b\,c}$ Number of progeny = 1000

Expected D.C.O. = observed D.C.O. (as no interference).

1. A b c = 280
2. a B C = 280 } 560

A – B = 20 – 6 = 14% = $\frac{14}{100}$ × 1000 = 140

3. A B C = 70
4. a b c = 70 } 140

B – C = 30 – 6 = 24 = $\frac{24}{100}$ × 1000 = 240

5. A b C = 120
6. a B c = 120 } 240

D.C.O = 6% = $\frac{6}{100}$ × 1000 = 60

7. A B c = 30
8. a b C = 30 } 60

Parental frequency = 1000 – (240 + 140 + 60) = 1000 – 440 = 560

(c) For 1000 progenies and assuming 20% interference.

Coefficient of coincidence = 80%

80% = $\frac{\text{observed D.C.O.}}{\text{expected D.C.O.}}$

Observed D.C.O. = 0.20 × 0.30 × 0.8 = 0.048

Among 1000 progenies it is 48000

C.O. between A – B = 20%, among 1000 progenies, the number will be

$\frac{20}{100} \times 1000 = 200; 200 - 48 = 152 = \frac{A\quad bc}{a\quad BC}$ (crossover between A and bc) = A

1. A B C = 76 } 152
2. a b c = 76
3. A b C = 126 } 252
4. a B c = 126

$\left[\frac{30}{100} \times 1000 = 300;\ 300 - 48 = 252\right]$

5. A B c = 24 } 48 (D.C.O.)
6. a b C = 24
7. A b c = 274 } 548
8. a B C = 274

$[1000 - (152 + 252 + 48) = 1000 - 452]$

20. As there is no interference, expected D.C.O. = observed D.C.O.

Expected D.C.O. $= \frac{8}{100} \times \frac{14}{100} = .08 \times .14 = .0112$ = 11200 = 1.12%

C.O. frequency between w – s = 8 – 1.1 = 6.9%

C.O. frequency between s – e = 14 – 1.1 = 12.9%

Parental progenies = 100 – (6.9 + 12.9 + 1.1) = 100 – 20.9 = 79.1

(a) + + + } 79.1%
(b) w s e
(c) + s e } 6.9%
(d) w + +
(e) + + e } 12.9%
(f) w s +
(g) w + e } 1.1%
(h) + s +

21. (i) b and st are unlinked as are hk and st. The evidence is the 1:1:1:1 segregation observed.

(ii) For b and st, the comparisons are black scarlet = 253(243 + 10); versus black nonscarlet = 256(241 + 15), versus nonblack scarlet = 238(226 + 12), versus nonblack nonscarlet = 253(235 + 18).

(iii) For hk and st, the comparisons are hook scarlet = 236(226 + 10), nonhook scarlet = 255(243 + 12), versus hook nonscarlet = 250(235 + 15), versus non-hook nonscarlet = 259(241 + 18).

On the other hand, b and hk are linked and the frequency of recombination between them $\left(\frac{15+10+12+18}{1000}\right) \times 100 = 5.5\%$.

22. (a) (i) The non-recombinant gametes (the most frequent classes) are y v^+ sn and y^+ v sn^+; and the double recombinant (the least recombinant classes) y v^+ sn^+ and y^+ v sn indicate sn is the middle gene. Therefore, the correct order of gene is y–sn–v.

(ii) The recombination frequency between

$y - sn = \left(\frac{108+95+5+3}{1000}\right) \times 100 = 21.1\%$

The recombination frequency between

$sn - v = \left(\frac{53+63+5+3}{1000}\right) \times 100 = 12.4\%.$

Genetic map y $\underline{\ 21.1\%\ }$ sn $\underline{\ 12.4\%\ }$ v.

(b) Expected D.C.O. = 0.211 × 0.124 = 0.0262 and observe D.C.O. $= \frac{8}{1000} = 0.008.$

Coefficient of coincidence $= \frac{\text{Observed}}{\text{Expected}} = \frac{0.008}{0.0262}$ = 0.31%; Coefficient of interference = 100 – 0.31 = 0.69, i.e., 69%.

23. (i) The non-recombinant classes are c wn sh and C Wx sh; and double recombinant classes are C Wx sh and c wx sh (15 + 20), indicate sh is the middle gene.

(ii) The parental genotype is, therefore, c sh wx and C sh Wx.

The frequency of recombination between C – sh is $\left(\frac{84+99+20+15}{6708}\right) \times 100 = 3.25\%$ and between sh – wx is $\left(\frac{974+951+20+15}{6708}\right) \times 100$ = 29.2%.

The genetic map is c $\underline{\ 3.25\%\ }$ sh $\underline{\ 29.2\%\ }$ wx

24. (a) The most common classes are non-recombinant.

Here ct + + (349) and + y v (332) represent the parents.

Genotype of female parent is $\frac{+\ ct\ +}{y\ +\ v}$.

(b) y —— ct —— v or v —— ct —— y

(c) Recombinant frequencies between y – ct = 20%.

Recombinant frequencies between ct – v = 13%.

Expected D.C.O. = 0.20 × 0.13 = 0.026, Observed D.C.O. = 0.01.

Coefficient of coincidence = 0.385, Coefficient of interference = 1 – 0.385 = 0.615.

(d) Genetic map y —20%— ct —13%— v.

25. (i) Recombination distance between d–p = 21.1% and p–e = 4.35%.

Linkage order d —21.1%— p —4.35%— wx

(ii) Coefficient of coincidence = $\frac{\text{observed D.C.O.}}{\text{expected D.C.O.}}$

$= \frac{0.0145}{0.00918} = 1.5797.$

26. (a) v —— m —— r

(b) y —3.17— m —19%— r

(c) Expected D.C.O. = 4.9 = 5; Observed D.C.O. = $\frac{2}{49}$ = 0.40, C.O.C. = 0.08.

27. kk ←14.12%→ mm ←8.33%→ *ll*

28. (a) lz —— su —— ?

← 10.7 mu → ← 14.7 mu →

(b) $\frac{6}{11.6}$ = 0.51; C.O.C. = 0.51.

29. d ←7%→ p ←17%→ o

30. (a) v ←3.2%→ m ←6.6%→ s

(b) Expected D.C.O. = 0.066 × 0.032 = 0.002; 2951 × 0.002 = 5.9

C.O.C. = $\frac{3}{5.9}$ = 0.508 = 0.51

[1 – 0.51 = 0.49, i.e., 49% interference]

31. j ←11.35%→ sn ←15.4%→ p

32. s ←11.8%→ G ←6.7%→ C

33. (a) $\frac{\text{FgH}}{\text{fGh}}$; (b) FGH or HGF; (c) F – G = 0.08; (d) G – H = 0.12, or, 12%, or, 12 cM; (e) 0.08 × 0.12 × 1000 = 9.6; (f) C.O.C. $\frac{6}{9.6}$ = 0.625; [interference = 1 – C = 1 – 0.625 = 0.375.]

34. (a) Map distance: C ←3.5 cM→ Sh ←18.4 cM→ w

(b) 14.41%.

35. (i) Female $\frac{+\,b\,c}{a\,+\,+}$ and Male $\frac{a\,b\,c}{\quad}$

(ii) $\left(\frac{+\,b\,c}{a\,+\,+}\right)$

(iii) a ←7.0%→ b ←2.0%→ c

(iv) Missing progeny a b + and + + c which have $\frac{7}{100} \times \frac{2}{100} = \frac{14}{100}$ = 0.0014 × 990 = 1.38 = 1.4

Among 990 it would be 1.4, i.e., 2 only expected.

36. d ←19.82%→ p ←4.46%→ c

37. (i) $\frac{+\,b\,c}{a\,b\,c}$ and $\frac{a\,+\,+}{a\,b\,c}$

(ii) a – b – c

(iii) a ←7%→ b ←2%→ c

(iv) D.C.O. is missing because distances among the genes (abc) are less than 20 cM (7 cM and 2 cM)

(v) 0.14% and 0.86%

38. (a) art ←10.95%→ S ←0.57%→ dm

(b) Double crossing-over classes are missing. Interference is strong because map distances less than 20 cM. Here distances are 10.95% and 0.5+%. Therefore, D.C.O.s are missing.

39. (a) Parental class: yellow cut miniature and wild type (30 and 33).

(b) Yellow miniature and cut (1 and j)

(c) Double crossover classes tell us the middle marker. Here it is ct (cut) which must lie between y and m.

(d) Genotype of heterozygous $\frac{y\,ct\,m}{+\,+\,+}$.

40. b – pr = 6.1 and pr – rg = 12.3

b ←6.1 mu→ pr ←12.3 mu→ vg

C.O.C. = $\frac{0.00298}{0.00750}$ = 0.397 = 39.7 = 40%

41. y ←24.6 mu→ v ←21.9 mu→ B

42. $\frac{20}{100} \times \frac{30}{100} = \frac{600}{104}$ = 0.06

Observed D.C.O. = 0.5 × 0.06 = 0.030

Among 1000: 0.030 × 10^3 = 30

So triply homozygous recessive will be $\frac{30}{2} = 15$.

43. (a) cv —6.4%— ct —13.2%— v

(b) $\text{C.O.C.} = \frac{\text{Observed D.C.O.}}{\text{Expected D.C.O.}} = \frac{0.0055}{0.0084} = 0.65$

Coefficient of interference $= 1 - 0.65 = 0.35$.

44. (a) cv —7.9%— ct —15.3%— v

(b) C.O.C. = 0 = 0.166 = 16%, coefficient of interference = 100 – 16 = 84%.

45. (i) The most frequent of classes of gametes are the nonrecombinant. So the parental genotype was ABc/abC.

(ii) The least frequent gametes are double recombinants. The allele that distinguish them from the nonrecombinant (in this case C and c) is the middle.

Therefore, the order of genes A——C——B.

Genotype of heterozygote parent $\frac{\text{ACB}}{\text{acb}}$.

46. (a) Recombinant frequency between pr–v = 14.5%

Recombinant frequency between v–bm = 35.6%

(b) Genetic map distance between pr–v = 14.5 + 7.7 = 22.2%

Genetic map distance between v–bm = 35.6 + 7.7 = 43.3%

pr —22.2%— v —43.3%— bm

(c) $\text{Coefficient of coincidence} = \frac{\text{Observed D.C.O.}}{\text{Expected D.C.O.}} =$

$$\frac{\frac{7.7}{100}}{\frac{22.2}{100} \times \frac{43.3}{100}} = \frac{0.077}{0.222 \times 0.433} = \frac{0.077}{0.096} = 0.80 = 80\%$$

Coefficient of interference $= 100 - 80 = 20\%$.

The true map distances are certainly larger than 20 cM (22.2 cM and 43.3 cM). The frequency of recombination between the genes are so large that there are undoubtedly many undefected D.C.O.s in each region.

47. ss – sc = 58 – 44 = 14 cM and e – ss = 70 – 58 = 12 cM

$$\text{Expected D.C.O.} = \frac{14}{100} \times \frac{12}{100} = 0.0168$$

$$\text{Coefficient of coincidence} = \frac{\text{Observed D.C.O.}}{\text{Expected D.C.O.}} =$$

or, observed D.C.O. = 0.7 × 0.0168 = 0.01176

Among 1000, it will be 0.01176 × 1000 = 11.76.

So, the double crossing-over progeny will be 11 among flies.

48. (a) (i) No, they are not alleles. If they were alleles, they would undergo segregation and the progeny would consist equally of either DDT-resistant or Dieldrin resistant animals.

(ii) There would be no progeny resistant to both insecticides and no progeny sensitive to both.

(b) (i) The test for linkage is whether the parental and non-parental chrmosomes occur in equal proportion, so the reciprocal classes must be pooled. This yield 177 (89+88) parental and 205 (106+99) nonparental.

(ii) The chi square 2.05 with 1 df for who in P is approximately 0.15.

The genes are, therefore, unlinked. They show independent assortment, even though they are in the same chromosome.

(c) The conclusion is that DDT and Dl are so far apart in the chromosome that one or more crossover always occur between them.

49. (a) The correct gene order: Ly——br——sb

(b) Recombination frequency between Ly—br = 3.4% and between br—sb = 13.4%.

(c) Linkage map: Ly —4 mu— br —14 mu— sb

(d) $\text{C.O.C.} = \frac{\text{Observed D.C.O.}}{\text{Expected D.C.O.}} = \frac{0.006}{0.4 \times 0.14} =$

$\frac{0.006}{0.0056} = 1.07$ or $\frac{0.006}{0.006} = 1$

then coefficient of interference = 1 – 1 = 0

(e) When D.C.O. is 6, total progeny is 1000, when D.C.O. is 210, total progeny would be $\frac{1000}{6} \times 210 = 35000$.

50. Expected D.C.O. = 0.20 × 0.12 = 0.0240

Interference 0.4, i.e., coincidence $= 1 - 0.4 = 0.6$.

Observed D.C.O. = C.O.C. × Expected D.C.O. × total progeny = 0.6 × 0.024 × 10,000 = 0.144.

If there is no interference, then coincidence = 1, i.e., observed D.C.O. and expected D.C.O. are same.

Expected D.C.O. × total progeny = Observed D.C.O. = 0.20 × 0.12 × 50,000 = 1200.

Twin spot

In *Drosophila* female, the result of mitotic crossing-over within the somatic tissue from a heterozygous individual was observed as **twin spot** (yellow patch and singed bristles). Here a pair of adjacent cell groups genetically belongs to different region of tissues differing in genotype and phenotype.

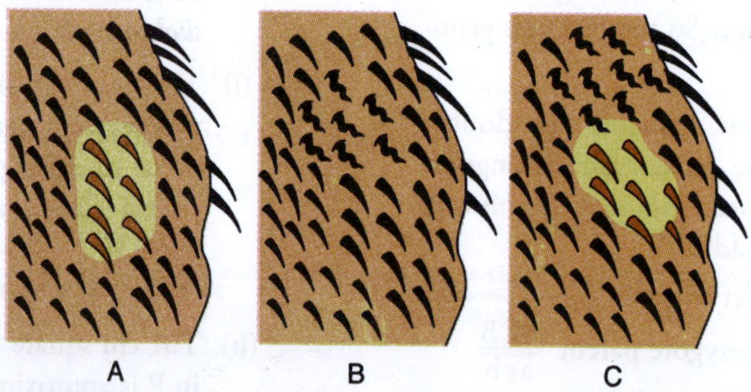

Fig. 2.1: Showing (A) Single yellow spot, (B) Single singed (twisted) bristle spot, (C) Twin spot: yellow spot and singed bristle spot from $\frac{y^+ \; sn}{y \; sn^+}$ *Drosophila* strain.

Retinoblastoma (a human tumer) and Mitotic Recombination

(i) Retinoblastoma is a dominant trait. The retinoblastoma (RB) normally prevent retinoblastoma located in the long arm of chomosome 13 ($13q_{14.2}$). The most people have two normal copies of RB gene in cell.

(ii) Khudson proposed that both familial and sporadic cases of retinoblastoma two copies of the genes (RB) have been inactivated (rb). Person with familial retinoblastoma have one inactivated (rb) and one normal RB) copy of each cell including retinal cell. On the other hand in sporadic retinoblastoma, both the inactivated mutation of retinoblastoma (RB/RB) occurs during developments.

(iii) These result uncontrolled proleferation and tumor to develop. In familial tumour occur in both eyes (bilateral). In sporadic tumours occur in one eye (unilateral).

(iv) Cavenee W. K. and his colleagues proposed several mechanism for the inactivation/ mutation of wild type retinoblastoma (RB) gene.

(a) Loss of whole chromosome by mittic nondisjunction (b) Mitotic recombination proximal to the RB locus, (c) Deletion of wild type allele. Retinoblastoma gene some how replaced by a duplicated copy of the homologous chromosome region carrying mutant (rb) allele. Mitotic recombination is the prime case of it.

LOD SCORE

The logarithm (base 10) of the odds (LOD) is the lod score and is symbolized as Z. Newton Morton (1955) demonstrated the most efficient statistical method for evaluating linkage analysis in a family (pedigree) studies.

In genetics LOD score is a statistical estimate whether two genes or a gene and a diseased gene are likely to be located close to each other on a chromosome and are, therefore, likely to be inherited together as a package.

Criteria for LOD score analysis:

(i) A large data sample is needed to determine a Lod score. In this type of statistical analysis, the bigger the sample is better.

(ii) Huge samples tend to yield more statistically meaningful information, because they reduce the risk of flukes and clusters which twist the data.

(iii) The analysis would be less ambiguous if the families are with large number of children.

(iv) For pedigree analysis at least three generations are to be taken.

(v) The method simply calculates two different probabilities for obtaining a set of results in a family.

(vi) The first probability is calculated assuming independent assortment and the second assuming a specific degree of linkage. Then the two ratios (odds) of the two probabilities are calculated and the logarithm of this number gives the LOD value.

(vii) Human genetics calculate a LOD score to confirm or reject cosegregation.

Calculation of LOD score:

This method is based on the ratio of the probability of observing cosegration if the genes are linked and the probability of observing cosegregation if the genes are not linked.

(i) NR denotes the number of non-recombinant progeny and R denotes the number of recombinant progeny.

(ii) The reason of using 0.5 in the denominator is that any alleles that are completely unlinked have a 50% chance of recombination due to independent assortment.

(iii) θ is the recombinant fraction, i.e., the fraction of births in which recombination has happened between the studied genetic marker and the putative gene associated to R/NR+R.

Significance of LOD score:

1. If genes are on different chromosomes, for instance, the number is expressed logarithmically usually inherited independently, because a great deal of crossover can occur during meiosis (the process of cell division by which sperm and egg cells are formed).

2. If genes are on the same chromosomes but far apart, they are also inherited independently of each other. Although, when LOD scores are in close proximity on the same chromosome, they tend to be inherited together. This information is a root of the LOD score.

3. If the LOD score is high (Positive LOD), it means that the traits are closely linked, and therefore, usually inherited together. LOD scores (Negative LOD), on the other hand, indicate a low linkage.

4. As a general rule, if the LOD score is 3 or higher, it means that there is a high probability of genetic linkage. Lower number indicates less likelihood of genetic linkage, even though they can still be useful in the process of elimination. An LOD score of +3 indicates 1000 to 1 odd that the linkage being observed did not occur by chance.

5. (a) $z(\theta) \geq 3.0$—significant evidence for linkage;

$$\text{LOD score} = Z = \log_{10}\frac{\text{Probability of observating cosegregation if the genes are linked}}{\text{Probability of observing cosegregation if the genes are not linked}} = \frac{(1-\theta)^{NR} \times \theta^{R}}{0.5^{(NR+R)}}$$

(b) $z(\theta) \le -2.0$—significant evidence for non-linkage; (c) $2.0 \le z(\theta) \le 3.0$—insufficient data.

6. The hypothesis of linkage is typically extended to consider a variety of sub-hypotheses about how tight the linkage is between the disease and marker locus. LOD scores are routinely calculated for 0%, 5%, 10%, 15%, 20%, 30%, and 40% recombination to allow a thorough consideration of potential relationships between the disease and marker locus. The highest LOD score will be obtained at the best estimate of the recombination fraction.

Difficulties in applying LOD score (Z):

(i) Problems with locus heterogeneity.

(ii) Vulnerability to errors.

(iii) Computational limits regarding the analysis of types of pedigree.

(iv) It is necessary to specify a precise genetic model and mode of inheritance, penetrance of each genotype and gene frequencies.

Problem 1

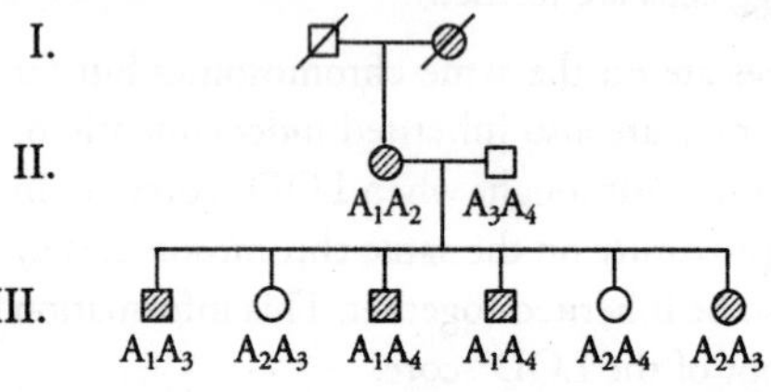

Pedigree of a family with an autosomal dominant disease, typed for a marker with alleles A_1-A_6. In generation II-1 (mother) could have inherited either marker allele A_1 or A_2 with disease. Her phase is unknown as her father and mother are diseased (death). Among her children III generation, either five (III_1–III_5) are non-recombinant and one recombinant III_6, or it may be five (III_1–III_5) recombinant and one non-recombinant (III_6).

Solution

I. The mother's (II-1) phase is unknown, i.e., both parents are dead and therefore, not available for genotyping.

II. There are two possibilities: (a) if she inherited A_1 with the disease, there are five non-recombinants, one recombinant (b) if she inherited A_2 with the disease, or five recombinants and one non-recombinant.

III. The overall likelihood is

$$\frac{1}{2}\left\{\frac{(1-\theta)^5\times\theta^1}{\theta^6}\right\}+\frac{1}{2}\left\{\frac{(1-\theta)\times\theta^5}{\theta^6}\right\}$$

θ = recombinant fraction, i.e., $\frac{1}{6} = 0.167$.

$$Z = \log_{10}\left[\frac{1}{2}\left\{\frac{(1-0.167)^5\times 0.167}{(0.5)^6}\right\}+\frac{1}{2}\left\{\frac{1-0.167\times(0.167)^5}{(0.5)^6}\right\}\right]$$

$$= \log_{10}\left[\frac{1}{2}\left\{\frac{(0.833)^5\times 0.167}{0.15625}\right\}+\frac{1}{2}\left\{\frac{0.833\times 0.0001299}{0.15625}\right\}\right]$$

$$= \log_{10}\left[\frac{1}{2}\left\{\frac{0.401\times 0.167}{0.15625}\right\}+\frac{1}{2}\left\{\frac{0.0001082}{0.15625}\right\}\right]$$

$$= \log_{10}\left[\frac{1}{2}\left\{\frac{0.066967}{0.15625}\right\}+\frac{1}{2}\left\{0.0006925\right\}\right]$$

$$= \log_{10}\left\{\frac{1}{2}\times 0.42858+\frac{1}{2}\times 0.0006925\right\}$$

$$= \log_{10}(0.214294 + 0.00034626)$$

$$= \log_{10}(0.2146)$$

$$Z = -0.668.$$

Here LOD scores give negative result, so the genes are not linked.

Problem 2

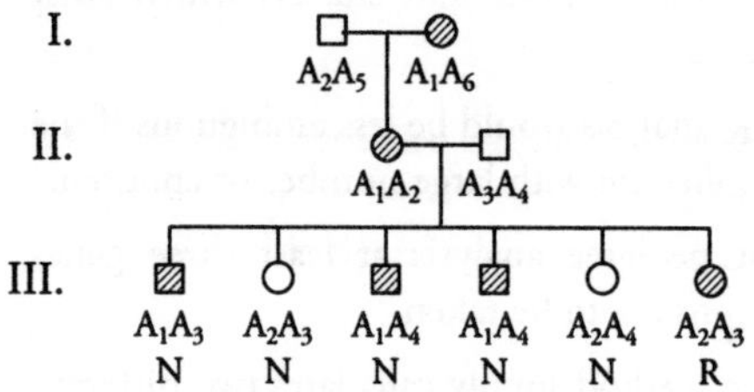

Pedigree of a family with an autosomal dominant disease, typed for a marker with alleles A_1-A_6. Generation III, III_1–III_5 are non-recombinant (N) and III_6 is recombinant (R).

Solution

I. If the loci are truly linked, the recombination fraction is θ and non-recombinant is $(1 - \theta)$.

II. If the loci are not linked (unlinked) recombinant or non-recombinant fraction is 0.5.

III. In this pedigree, in III generation six members are present. There are five non-recombinant $(1 - \theta)$ and one recombinant (θ) members in this family, i.e., $\frac{1}{6} = 0.167$.

IV. Here $\theta = 0.5$ and $1 - \theta = 1 - 0.5 = 0.5$.

$$Z = \log_{10}\frac{\text{Probability of observating co-segregation of gene (within given linkages)}}{\text{Probability of co-segregation of gene with no linkage}}$$

$$= \log_{10} \frac{(1-\theta)^{NR} \times \theta^{R}}{0.5^{(NR+R)}} \qquad R = \tfrac{1}{6} = 0.167$$

$$= \log_{10} \frac{(1-0.167)^5 \times (0.167)^1}{(0.5)^6}$$

$$= \log_{10} \frac{(0.833)^5 \times 0.167}{0.015625}$$

$$= \log_{10} \frac{0.401 \times 0.167}{0.015625}$$

$$= \log_{10} \frac{0.066967}{0.015625}$$

$$= \log_{10} 4.28 = 0.63.$$

Here LOD score is positive, so the genes are linked.

Problem 3

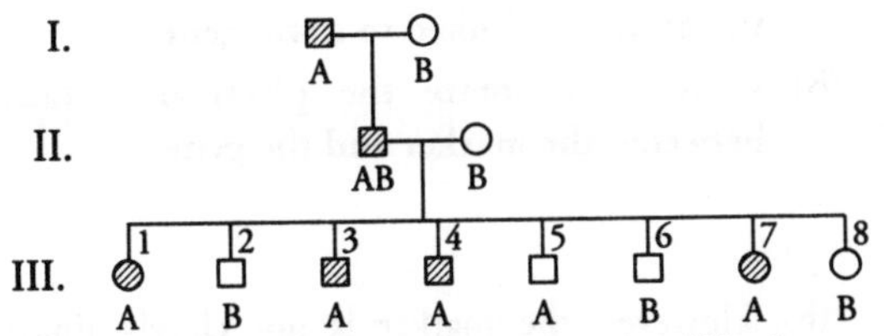

Inheritance of nail-patella syndrome (dominant) and blood group.

The pedigree shows the inheritance of nail-patella syndrome (poorly developed nails and knee caps) and ABO blood group. Calculate LOD score.

Solution

I. The above pedigree shows that there are three generations.

II. The two genes nail-patella syndrome (NPS). A dominant gene and ABO blood group gene is inherited.

III. LOD score analysis is made to know the coinheritance of blood group A gene with NPS gene.

IV. In generation II, male individual is AB with nail-patella disease.

V. In generation III, the individual III-5 is possible recombinant, i.e., A blood group without the syndrome.

VI. To assess the likelihood of linkage of the two genes, LOD scores are calculated.

$$Z = \log_{10} \frac{\text{Probability of observating co-segregation if the genes are linked}}{\text{Probability of observing co-segregation if the genes are not linked}}$$

θ = recombination fraction.

N = number of informative progeny

R = number of recombinant

considering III generation in the pedigree, recombinant (R) III_5, i.e., $\frac{1}{8}$ (Total N = 8) = 0.125.

$$\theta = 0.125$$

$$Z = \log_{10} \frac{(1-\theta)^{n-r} \times \theta^{r}}{(0.5)^{n}} = \log_{10} \frac{(1-0.125)^{8-1} \times (0.125)^1}{(0.5)^8}$$

$$= \log_{10} \frac{(0.875)^7 \times (0.125)^1}{(0.5)^8} = \log_{10} \frac{0.392696 \times 0.125}{0.00390625}$$

$$= \log_{10} \frac{0.0490869}{0.00390625} = \log_{10} 12.566268 = 1.0992$$

Here LOD score is positive, so the genes are linked.

I. In this set of calculation the maximum LOD score is 1.099 at recombination frequency (θ) 0.125.

II. It indicates that the possibilities of linkage is 10 times more likely than the two genes assorting independently.

III. In this case of nail-patella syndrome and ABO blood groups, combined LOD scores produce value greater than 3 indicating a strong likelihood linkage between two genes. They are linked about 10 cm apart.

Problem 4

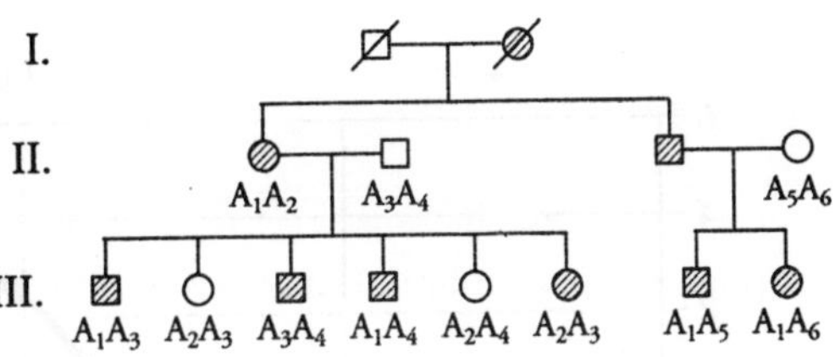

Autosomal dominant inheritance along with the marker alleles A_1–A_6. Among 8 members 2 are recombinant.

Solution

I. Members of the pedigree III-7 and III-8 have inherited marker alleles A_1, along with the disease from their father but we cannot sure whether their father's alleles A_1 is identical by descent to the allele A_1 in the sister II_1.

II. There may be two copies of alleles A_1 among four grand parental marker alleles.

III. The recombinant frequency is $\frac{2}{8} = \frac{1}{4} = 0.25$, i.e., θ = 0.25.

$$Z = \log_{10} \frac{(1-0.25)^8 \times (0.25)^2}{(0.5)^8} = \log_{10} \frac{(0.75)^8 \times (0.25)^2}{(0.5)^8}$$

$$Z = \log_{10}\frac{0.1001 \times 0.0625}{0.003906} = \log_{10}\frac{0.006256}{0.003906}$$

$$Z = \log_{10}160 = 0.204.$$

Here LOD score is positive.

It indicates that genes may be linked.

Problem 5

Some cats love water, while others avoid it. A cat that loved water was mated to a cat that avoided it and thier F_1 progeny were interbred to develop F_2. The parental F_1 and F_2 progenies were evaluated by DNA typing and LOD score method was applied to assess the linkage between DNA markers and genes for water affection (waf genes). Suppose that the following data were obtained for one marker, where θ gives the value of the recombination frequency between the marker and a waf gene used in calculating LOD score.

Sl. No.	θ	Lod score
1	**0**	–α
2	**0.05**	–12.51
3	**0.10**	–2.34
4	**0.15**	–1.32
5	**0.20**	2.66
6	**0.25**	4.01
7	**0.30**	3.21
8	**0.35**	2.14
9	**0.40**	1.56
10	**0.50**	0

(a) Graph the LOD scores and evaluate whether the marker is linked to a waf gene.

(b) If it is, estimate the physical distance between the marker and the gene.

Solution

(a) A graph θ versus LOD score reveals a maximum LOD score of 4.01 at a map distance about 25 mμ. Since the LOD score is greater than 3 the marker is linked to the waf gene.

(b) However, the marker is not closely linked. Estimate from humas that 1 mμ corresponds in average to 1 mb, the markers about 25 mb away from the waf gene.

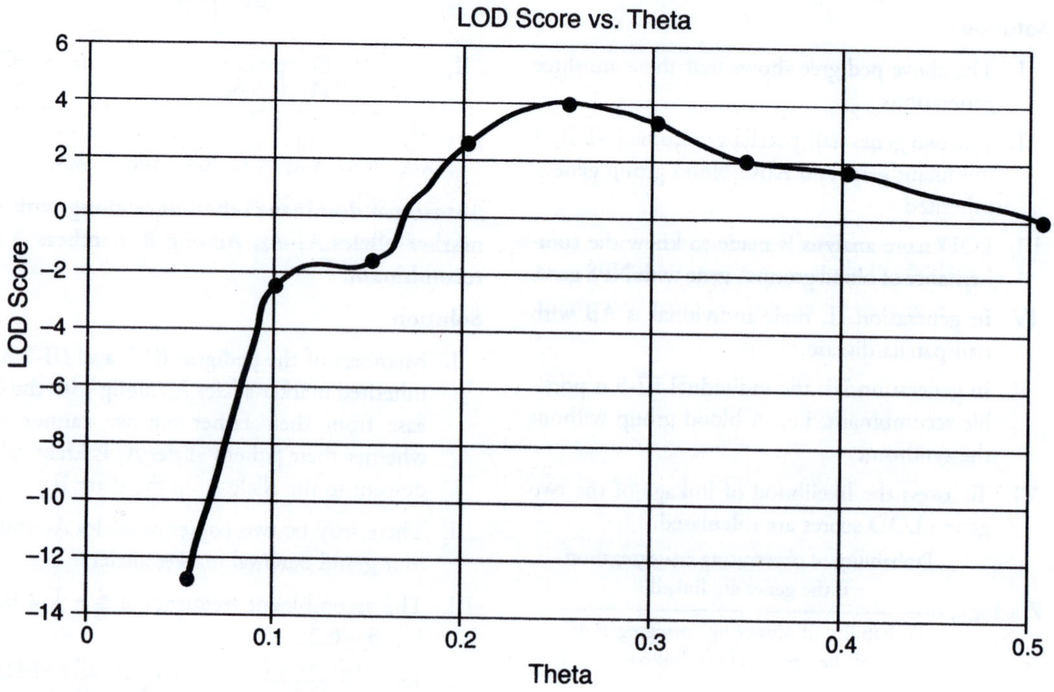

MAPPING FUNCTION

The relationship between the recombination fraction and the genetic map distance is defined by the mapping function:

(i) It is required in **multilocus mapping.**

(ii) The simplest map function, i.e., **Haldane** function is

$$w = -\frac{1}{2}\ln(1-2\theta) \text{ or, } \theta = \frac{1}{2}[1-\exp(-2w)],$$

where w = map distance

θ = the recombination fraction.

In the logarithm to the base e, exp means 'e to the power of'.

Features:

I. This approach was worked out by J. B. S. Haldane.
Mapping function relates recombination frequency (RF) value to "real physical" distance.

II. An accurate measure of physical distance is the mean number of crossover that occurs in that segment per meiosis. Let us call this **m.** Hence, our aim to find a function that relates **RF** to **m.**

III. Genes that are closely linked and located in the same chromosome are usually unlikely to experience more than one crossing-over. Hence, small map distance are likely to reflect that actual amount of crossing over.

IV. However, as the genetic *distance increases* the opportunity for multiple crossing-over increases tending to follow a *Poison distribution.* **The occurence of crossovers in a specific chromosomal region is analysed by a statistical method known as Poisson distribution.**

V. Any finite number of crossing-over per meiosis produces an observed recombination frequency (**RF**) of 50% among their products. Theoretically they cannot give **RF** more than 50%. No matter how far apart two loci are linked.

VI. The further *apart two genes* are, the greater the error of estimate of true linkage distance provided by the recombinant progeny.

VII. The proportion of meiosis in which one or more crossovers are equal to one minus the fraction of zero crossovers, i.e., **RF = (1–e^{-m}) where 50 percent recombinant chromatids will occur. Therefore, RF = $\frac{1}{2}(1-e^{-m})$ where m is the actual number of zero crossing-over event (not observed number).**

VIII. If there is no zero crossing-over events, then RF = $\frac{1}{2}(1-0) = \frac{1}{2}$, because expected zero class frequency is (RF) = $\frac{e^{-m}m^i}{\lfloor i}$

[e = the base of natural logarithm approximately 2.7.
m = the number of crossing-over.
i = actual number of crossing-over.
RF = Recombination frequency.]

The Poisson distribution tell us that i=0 class.

RF = $e^{-m}\frac{m^0}{\lfloor 0}$;

since, m^0 and $\lfloor 0$ = 1, the formula reduces to e^{-m}.

IX. Now we can write a function that relates RF to m. The frequency of the class with any non-zero number crossovers will be 1 – e^{-m} and in these meiosis $\frac{1}{2}$ of the product will be recombinant.
So, RF = $\frac{1}{2}(1-e^{-m})$.

- **Recombination fraction:**

At the end of the first meiotic division, each daughter cell will have a 50% chance of receiving each locus. This recombination fraction will be 0.5 (i.e., recombination percentage is 50%). Thus, the recombination fraction is the index of the distance between two loci.

- **Kosambi's function:**

$$w = \frac{1}{4}\ln\left[\frac{(1+2\theta)}{(1-2\theta)}\right]$$

$$\text{or, } \theta = \frac{1}{2}\left[\exp(4w) - \frac{1}{\exp(4w)+1}\right]$$

- **Physical map distance to correct map distance**

I. In very small genetic regions, recombination frequency (RF) is expected to be an accurate measure of physical distance because there are not any multiple crossing-over.

II. In fact, meiosis will show either no crossover or one crossover.

III. The frequency of crossover (m) will then be translatable into 'correct' recombinant fraction of $\frac{m}{2}$ because the recobinants will be $\frac{1}{2}$ of the chromatids arising from the single crossover class.

IV. This defines a *general relationship between m and a corrected recombinant fraction.* Thus, the corrected recombinant fraction can be thought as $\frac{m}{2}$.

Distribution of event	Probabilities
0	e^{-m}
1	me^{-m}
2	$\frac{m^2}{2(e^{-m})}$
3	$\frac{m^3}{6(e^{-m})}$
etc.	

Example:

From one test-cross, it is found that RF (recombination fraction) is 27.5%, i.e., 0.275.

$$RF = \frac{1}{2}(1 - e^{-m})$$

or, $0.275 \times 2 = 1 - e^{-m}$

$$e^{-m} = 1 - 2 \times 0.275 = 1 - .550 = 0.45$$

$$\frac{1}{e^{-m}} = \frac{1}{0.45}$$

$$m = \ln\left(\frac{1}{0.45}\right) = \ln 2.22$$

$$= 0.7979 = 0.8.$$

Thus, on average there are 0.8 crossovers per meiosis in that chromosomal region.

The corrected recombinant fraction is $\frac{0.8}{2} = 0.4$, i.e., 40% or 40 map units.

Solved Problems

Problem 1

If two linked genes produce 33% recombinants, estimate the probable amount of crossing-over that actually occurred between the loci, assuming that probabilities of 0, 1, 2, 3, ...n exchange occur during meiosis according to Poisson distribution.

Solution:

Recombinants 33% = 0.33

I. $RF = \frac{1}{2}(1-e^{-m})$ [m = number of crossing-over]

$0.33 = \frac{1}{2}(1-e^{-m})$

or, $0.66 = 1-e^{-m}$

or, $e^{-m} = 1 - 0.66 = 0.34$

or, $\frac{1}{e^m} = 0.34$.

or, $e^m = \frac{1}{0.34}$

$\Rightarrow m = \ln\left(\frac{1}{0.34}\right) = \ln 2.94 = 1.078.$

value

e = 1 = 0.36786

e = 2 = 0.13534

Therefore, $e^{-m} = 0.34$ lies between e = 1 and e = 2.

Thus, $\frac{0.34}{0.36788} = 0.9242.$

II. 0.9242 corresponds to m = 1.08

Therefore, $RF^1 = \frac{m}{2} = \frac{1.08}{2} = 0.54$ or, 54 map units.

III. The error under estimate of actual crossing-over is $\frac{54}{33} = 1.636 = 1.64$.

Problem 2

If two linked genes produe 27.5% recombinants, estimate probable amount of crossing-over that actually occurred between these loci, assuming that probabilities of 0, 1, 2, ...n exchange occur during meiosis accordance to Poisson distribution.

Solution:

Recombinant = 27.5% = 0.275

$RF = \frac{1}{2}(1-e^{-m})$

or, $e^{-m} = \frac{1}{2}(1-e^{-m}) = 0.45$

or, $\frac{1}{e^m} = 0.45.$

or, $e^m = \frac{1}{0.45}$

$\Rightarrow m = \ln\left(\frac{1}{0.45}\right) = \ln 2.22 = 0.0798 = 0.8.$

I. From the table, value of e = 1 = 0.36788, e = 2 = 0.13536, e = 0 = 1.000.

Thus, $\frac{0.45}{0.36788} = 1.223.$

$RF' = \frac{RF}{2} = \frac{m}{2} = \frac{0.8}{2} = 0.4 = 0.4 \times 100$

= 40% = 0.40 map units.

Sex-linked Inheritance

The genes which occur exclusively on X-chromosome (mammals, *Drosophila*, etc.) or on X-analogous Z-chromosome (birds and other) are called **X-linked genes.** The genes which exclusively occur in Y chromosome are called **holandric genes.** The inheritance of X- or Z-linked and holandric genes is called **sex-linked inheritance.**

• CHARACTERISTICS OF SEX-LINKED INHERITANCE

The genes which are located in the nonhomologous sections of X-chromosome and have no counterpart on the other kind of sex chromosome (Y-chromosome) are commonly known as **sex-linked** (or **X-linked**) genes.

(i) These genes, dominant or recessive, show their effects in the male phenotype. Genes in the nonhomologous section is called **hemizygous.**

(ii) The X-linked recessive genes show the **criss-cross** pattern of inheritance (X-linked recessive gene is transmitted from P_1 male parent (father) to F_2 male progeny (grandsons) through F_1 heterozygous female (daughter), i.e., **carrier**, and difference in F_1 and F_2 results in reciprocal crosses.

• PATTERN OF INHERITANCE

(i) More sex-linked genes in male heterogametic animal are on the X-chromosome.

(ii) Some animals may carry a few genes on the Y-chromosome, that produces visible effect on phenotype of the organism.

(iii) Y-linked (holandric genes) would be transmitted from **father to son.**

(iv) In *Drosophila* studied with attached X-chromosome with X-linked genes can be transmitted directly from **father to son.**

(a) Attached X-chromosomes are **aberrant compound X-chromosomes** possessing a fused centromere when attached X females are mated with normal XY males.

(b) The XX female gametes fertilized with Y male gametes result in XXY females (Y is not a male sex determiner in *Drosophila*).

(c) Male X-chromosome gametes that are fertilized with female gametes lacking X-chromosome result in XO sterile males that express the X-linked genes of the father.

(v) In the absence of dominant alleles, a recessive allele (w) responsible for white eyes in *Drosophila* can express itself.

(vi) The **criss-cross pattern of inheritance** is the characteristic of sex-linked genes. It means that traits appearing in males are transmitted through their daughters to the males (grandsons) in the next generation.

(vii) In *Drosophila* X-chromosome is rod-shaped with centromere near one end and Y is **hook-like** having a long and short arm.

Y is composed of heterochromatin and devoid of genes. One small part of the short arm in *D. melanogaster* has a homologous section on the X-chromosome.

(viii) The allele (bb) for bobbed bristles on the X-chromosome or its wild type allele (b^+b) may occur in the short arm of Y-chromosomes. Genes such as bb with a locus on the Y-chromosome as well as on a homologous part of the X-chromosome are said to be **incompletely sex-linked.**

Criss-cross inheritance

A type of gene transmission is passed from a male parent to a female child to a male grand child.

• NON-DISJUNCTION

The failure of disjunction or separation of homologous chromosome or sister chromatid in anaphase (mitosis or meiosis) resulting in too many chromosomes in some daughter cells and too few in others.

Example

(i) In meiosis both members of a pair of chromosomes go to one pole, so that the other pole does not receive either of them.

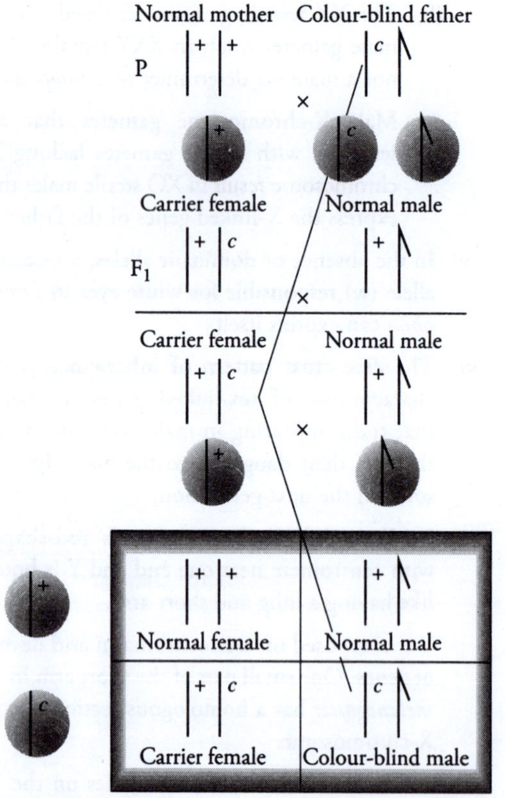

Fig. 3.1: Criss-cross inheritance from father to daughter to grandson (of original father) that is characteristic of a sex-linked gene. Genes are shown on the chromosomes illustrating a cross between a woman with normal vision and a man with colour vision deficiency. The symbol c represents the sex-linked recessive gene for colour vision deficiency.

(ii) In mitosis both sister chromatids go to the same pole.

A. Primary non-disjunction

In 1916, C. B. Bridges reported that white-eyed females crossed to red-eyed males occasionally produced exceptional offspring (1 per 2000 flies) that were phenotypically exactly like their parents rather than red-eyed females and white-eyed males normally expected from such a cross.

White-eyed female × Red-eyed male

$\frac{w}{w}$ × $\frac{w^+}{\rightharpoondown}$

Normal gametes: $\underline{w}$ $\underline{w^+}$ $\rightharpoondown$

Non-disjunctional egg: $\frac{w}{w}$ Nullo

(a) Normal sperm

♀ \ ♂	w^+	$\rightharpoondown$
Normal egg $\underline{w}$	$\frac{w^+}{w}$ red-eyed ♀	$\frac{w}{\rightharpoondown}$ white-eyed ♂

(b) Normal sperm

♀ \ ♂	w^+	$\rightharpoondown$
Non-disjunctional egg $\frac{w}{w}$	$\frac{w}{w}$ $\underline{w^+}$ super ♀	$\frac{w}{w}$ $\rightharpoondown$ white-eyed ♀
$\underline{o}$	$\frac{w^+}{o}$ Red-eyed ♂	$\frac{o}{\rightharpoondown}$ dies

Results:

Genotype		Phenotype
(i) $\underline{\underline{w^+}}$	(xo)	red-eyed male
(ii) $\frac{w}{w}$ $\rightharpoondown$	(xxy)	white-eyed female
(iii) $\frac{w}{w}$ $\underline{w^+}$	(xxx)	super female usually dies
(iv) $\frac{o}{\rightharpoondown}$	(yo)	dies

(i) The result indicates that xxy female is fertile and quite normal in appearance except that it is white-eyed, because both X-chromosomes are **maternal origin.**

(ii) The male which has xo genotype, i.e. red-eyed, because X-chromosome is **paternal origin.**

This is due to the phenomenon of **primary non-disjunction**. Here two XX-chromosomes migrate to one pole and other pole contains only autosome. Such non-disjunction with normal female parent would be called **primary non-disjunction.**

B. Secondary non-disjunction

(i) The exceptional **white-eyed female** (xxy) was obtained due to primary non-disjunction.

(ii) This female was again mated with red-eyed male, 96% female progeny had red eyes and only 4% **female had white** eyes. These indicate that 4% **secondary was non-disjunction.**

(iii) xxy would normally disjoin into x and xy but secondary non-disjunction will result into gametes carrying xx and y.

(iv) Result obtained xxy ($x_w x_w y$) white-eyed female and xy ($x_{w+}y$) red-eyed male.

Partial Sex linkage

(i) The human X- and Y-chromosomes are morphologically distinct but they pair during meiosis in male cell.

(ii) The pairing occurs in the two telomeric regions (pseudoautosomal regions, i.e., **PAR**).

(iii) The pseudoautosomal regions are known as PAR-1 and PAR-2.

(a) PAR-1. (i) It is 2.6 Mb long major pseudoauosomal region and located at the tip of the short arm of the X- and Y-chromosomes.

(ii) It is the site of obligate crossing-over during male meiosis.

(iii) The frequency of recombination in PAR-1 is 70 times more than normal recombination frequency.

(iv) PAR-1 region has around a dozen genes (Ex: SHOX, XE7, ANT-3 Tramp, MIC2, etc.)

(b) PAR-2. (i) It is 320 kb long minor pseudo-autosomal region and located at the tips of the long arm of X- and Y-chromosomes.

(ii) Crossing-over is not so frequent.

(iii) It contains at least two genes (IL9R and SYBL-1).

Explanation: (i) The genes PAR-1 and 2 are present in both X- and Y-chromosomes, therefore, they do not show the typical inheritance for sex linkage.

(ii) Their inheritance pattern resembles that of autosomal gene, this phenomenon is called **partial sex linkage** and the chromosomal region involved on it is referred to as pseudo-autosomal region.

- **Cause of Sex linkage**
  (i) The locatiobn of a gene in the X chromosome.
  (ii) The absence of its allele in the Y chromosome.

Sex linked character	Sex limited character
• Located on sex or X chromosome.	• Located on sex chromosomes and autosomes.
• Express on both sexes.	• Express on one sex only.
• Include character not related to sex.	• Include primary and secondary sex character.

- **People suffering from haemophilia may acquire AIDS?**
  (i) The people suffering from haemophilia also acquire AIDS, which is a threat to the victims of the disease.
  (ii) It is arisen due to their continuous dependence upon blood transfusions.
  (iii) Extensive surveillance of donor blood supplies is required to protect haemophilious and all other requiring transfusion.

Solved Problems

Problem 1

In *Drosophila* the gene for bobbed bristles (recessive allele bb; wild type allele bb⁺ normal bristles) is located on the X-chromosome and on a homologous segment of the Y-chromosome. Give the genotypes and phenotypes of the offspring from the following crosses:

(a) $x^{bb}x^{bb} \times x^{bb}y^{bb^+}$ (b) $x^{bb}x^{bb} \times x^{bb^+}y^{bb}$

(c) $x^{bb^+}x^{bb} \times x^{bb^+}y^{bb}$ (d) $x^{bb^+}x^{bb} \times x^{bb}y^{bb^+}$

Solution:

(a) $x^{bb}x^{bb} \times x^{bb}y^{bb^+}$

Gametes: x^{bb} x^{bb} y^{bb^+}

F₁:

♀ \ ♂	x^{bb}	y^{bb^+}
x^{bb}	$x^{bb}x^{bb}$	$x^{bb}y^{bb^+}$

Results:

Genotype	Phenotype	Ratio
$x^{bb}x^{bb}$	bobbed female	$\frac{1}{2}$
$x^{bb}y^{bb^+}$	wild type male	$\frac{1}{2}$

(b) $x^{bb}x^{bb} \times x^{bb^+}y^{bb}$

Gametes: x^{bb} x^{bb^+} y^{bb}

F₁:

♀ \ ♂	x^{bb^+}	y^{bb}
x^{bb}	$x^{bb^+}x^{bb}$	$x^{bb}y^{bb}$

Results:

Genotype	Phenotype	Ratio
$x^{bb^+}x^{bb}$	wild type female	$\frac{1}{2}$
$x^{bb}y^{bb}$	bobbed male	$\frac{1}{2}$

(c) $x^{bb^+}x^{bb} \times x^{bb^+}y^{bb}$

Gametes: x^{bb^+} x^{bb} x^{bb^+} y^{bb}

F₁:

♂ \ ♀	x^{bb^+}	x^{bb}
x^{bb^+}	$x^{bb^+}x^{bb^+}$	$x^{bb}x^{bb^+}$
y^{bb}	$x^{bb^+}y^{bb}$	$x^{bb}y^{bb}$

Results:

Genotype	Phenotype	Ratio
$x^{bb^+}x^{bb^+}$, $x^{bb^+}x^{bb}$	wild type female	$\frac{1}{2}$
$x^{bb^+}y^{bb}$	wild type male	$\frac{1}{4}$
$x^{bb}y^{bb}$	bobbed male	$\frac{1}{4}$

(d) $x^{bb^+}x^{bb} \times x^{bb}y^{bb^+}$

Gametes: x^{bb^+} x^{bb} x^{bb} y^{bb^+}

F₁:

♂ \ ♀	x^{bb^+}	x^{bb}
x^{bb}	$x^{bb^+}x^{bb}$	$x^{bb}x^{bb}$
y^{bb^+}	$x^{bb^+}y^{bb^+}$	$x^{bb}y^{bb^+}$

Results:

Genotype	Phenotype	Ratio
$x^{bb^+}x^{bb}$	wild type female	$\frac{1}{4}$
$x^{bb}x^{bb}$	bobbed male	$\frac{1}{4}$
$x^{bb^+}y^{bb^+}$, $x^{bb}y^{bb^+}$	wild type male, wild type male	$\frac{1}{2}$

Problem 2

The recessive incompletely sex-linked gene, called "bobbed" bb, causes the bristles of *Drosophila*. It is shorter and of smaller diameter than the normal bristles produced by its dominant wild type alleles bb⁺. Determine the phenotypic expectations of the F₁ and F₂ when bobbed females are crossed with each of the two possible heterozygous males.

Solution:

Incompletely sex-linked gene has an allele on homologous portion of the Y-chromosome in male.

First: Bobbed females Heterozygous male (wild type)

$x^{bb}x^{bb} \times x^{bb^+}y^{bb}$

Gametes: x^{bb} x^{bb^+} y^{bb}

♀ \ ♂	x^{bb^+}	x^{bb}
x^{bb}	$x^{bb^+}x^{bb}$	$x^{bb}y^{bb}$

F₁: Wild type females Bobbed males

$x^{bb^+}x^{bb} \times x^{bb}y^{bb}$

Gametes: x^{bb^+} x^{bb} x^{bb} y^{bb}

F₂:

♀ \ ♂	x^{bb^+}	x^{bb}
x^{bb}	$x^{bb^+}x^{bb}$	$x^{bb}x^{bb}$
y^{bb}	$x^{bb^+}y^{bb}$	$x^{bb}y^{bb}$

Results:

Genotype	Phenotype	Ratio
$x^{bb^+}x^{bb}$	wild type female	½
$x^{bb}x^{bb}$	bobbed female	½
$x^{bb^+}y^{bb}$	wild type male	½
$x^{bb}y^{bb}$	bobbed male	½

Second: Bobbed females × Wild type male (heterozygous)

$x^{bb}x^{bb}$ × $x^{bb}y^{bb^+}$

F_1: Bobbed female × Wild type male

$x^{bb}x^{bb}$ × $x^{bb}y^{bb^+}$

Gametes: x^{bb} x^{bb} y^{bb^+}

F_2:

♀ \ ♂	x^{bb}	y^{bb^+}
x^{bb}	$x^{bb}x^{bb}$	$x^{bb}y^{bb^+}$

Results:

Genotype	Phenotype	Ratio
$x^{bb}x^{bb}$	bobbed female	½
$x^{bb}x^{bb^+}$	wild type male	½

Problem 3

A female fruit fly with a yellow body is discovered in wild type culture. The female is crossed with a wild type male. In the F_1 generation, the males are yellow-bodied and the females are wild type. When these flies are crossed among themselves, the F_2 produced are both yellow-bodied and wild type, equally split among males and females. Explain the genetic control of this trait.

Solution:

P_1: Yellow female × Wild type male

$\frac{y}{y}$ × $\frac{y^+}{\rightharpoonup}$

Gametes: $\underline{y}$ $\underline{y^+}$ $\rightharpoonup$

F_1:

♀ \ ♂	$\underline{y^+}$	$\rightharpoonup$
$\underline{y}$	$\frac{y^+}{y}$ wild type ♀	$\frac{y}{\rightharpoonup}$ yellow ♂

wild type ♀ × yellow ♂

$\frac{y^+}{y}$ × $\frac{y}{\rightharpoonup}$

Gametes: $\underline{y^+}$ $\underline{y}$ $\underline{y}$ $\rightharpoonup$

♀ \ ♂	$\underline{y^+}$	$\underline{y}$
$\underline{y}$	$\frac{y^+}{y}$ wild type ♀	$\frac{y}{y}$ yellow ♀
$\rightharpoonup$	$\frac{y^+}{\rightharpoonup}$ wild type ♂	$\frac{y}{\rightharpoonup}$ yellow ♂

Results:

Genotype	Phenotype	Ratio
$\frac{y^+}{y}$	wild type female	1
$\frac{y}{y}$	yellow female	1
$\frac{y^+}{\rightharpoonup}$	wild type male	1
$\frac{y}{\rightharpoonup}$	yellow male	1

(i) Since the results in the F_1 generation differ between the two sexes, we suspect a sex-linked gene is responsible for the body colour.

(ii) If we assume, yellow body colour is recessive in nature, then the female parent must be homozygous in nature and male must be hemizygous wild type.

(iii) Let yellow body colour allele is y, then female parent is $\frac{y}{y}$ and wild type male is $\frac{y^+}{\rightharpoonup}$.

(iv) In F_1, y^+ allele is segregated to female, then female offspring is wild type and male becomes yellow bodied.

(v) Similarly, in F_2 crosses wild type and yellow trait in both male and female appear as 1:1:1:1.

Thus, a **recessive x-linked gene** controls the yellow body colour of fruit fly.

Problem 4

An autosomal recessive gene (tra) when homozygous, transforms a *Drosophila* female (x/x) into a phenotypic male. All such "transformed" males are sterile. The gene is without effect in males ($^x/_y$). A cross is made between a female heterozygous at the tra locus and a male homozygous recessive at the same locus. What is the expected sex ratio in the F_1 and F_2?

Solution:

Parent: Normal female (Heterozygous) × Normal male (Homozygous recessive)

$\frac{x}{x}\ \frac{+}{tra}$ × $\frac{x}{\rightharpoonup}\ \frac{tra}{tra}$

Gametes: $\underline{x}\ \underline{+}$, $\underline{x}\ \underline{tra}$ $\underline{x}\ \underline{tra}$, $\rightharpoonup\ \underline{tra}$

F_1:

♀ \ ♂	$\underline{x}\ \underline{+}$	$\underline{x}\ \underline{tra}$
$\underline{x}\ \underline{tra}$	$\frac{x}{x}\ \frac{+}{tra}$	$\frac{x}{x}\ \frac{tra}{tra}$
$\rightharpoonup\ \underline{tra}$	$\frac{x}{\rightharpoonup}\ \frac{+}{tra}$	$\frac{x}{\rightharpoonup}\ \frac{tra}{tra}$

Results:

Genotype	Phenotype	Ratio
(i) $\frac{x}{x}$ $\frac{+}{tra}$	female normal	$\frac{1}{4}$ female
(ii) $\frac{x}{x}$ $\frac{tra}{tra}$	transformed male (sterile)	$\frac{3}{4}$ male
(iii) $\frac{x}{\rightharpoonup}$ $\frac{+}{tra}$	normal male	
(iv) $\frac{x}{\rightharpoonup}$ $\frac{tra}{tra}$	normal male	

(i) The transformed F_1 males are sterile, so they do not contribute gametes to the offsprings.

(ii) Two kinds of mating must be considered:

(a) $\frac{x}{x}$ $\frac{+}{tra}$ (female) × $\frac{x}{\rightharpoonup}$ $\frac{+}{tra}$ (male)

(b) $\frac{x}{x}$ $\frac{+}{tra}$ (female) × $\frac{x}{\rightharpoonup}$ $\frac{tra}{tra}$ (male)

♂ \ ♀	x tra	x +
x +	$\frac{x}{x}$ $\frac{tra}{+}$	$\frac{x}{x}$ $\frac{+}{+}$
x tra	$\frac{x}{x}$ $\frac{tra}{tra}$	$\frac{x}{x}$ $\frac{+}{tra}$
$\rightharpoonup$ tra	$\frac{x}{\rightharpoonup}$ $\frac{tra}{tra}$	$\frac{x}{\rightharpoonup}$ $\frac{+}{tra}$
$\rightharpoonup$ +	$\frac{x}{\rightharpoonup}$ $\frac{tra}{+}$	$\frac{x}{\rightharpoonup}$ $\frac{+}{+}$

Results:

Genotype	Phenotype	Ratio
(i) $\frac{x}{x}$ $\frac{tra}{tra}$	female	
(ii) $\frac{x}{x}$ $\frac{+}{tra}$	female	$\frac{3}{8}$ female
(iii) $\frac{x}{x}$ $\frac{+}{+}$	female	
(iv) $\frac{x}{x}$ $\frac{tra}{tra}$	transformed male	
(v) $\frac{x}{\rightharpoonup}$ $\frac{tra}{tra}$	male	
(vi) $\frac{x}{\rightharpoonup}$ $\frac{+}{tra}$	male	$\frac{5}{8}$ male
(vii) $\frac{x}{\rightharpoonup}$ $\frac{+}{tra}$	male	
(viii) $\frac{x}{\rightharpoonup}$ $\frac{+}{+}$	male	

(a) (i) Thus, F_2 offspring from this mating type appears in the proportion $\frac{3}{8}$ female, $\frac{5}{8}$ male.

(ii) But this type of mating constitutes only half of all possible matings.

Therefore, the contribution to the total F_2 from this mating is $\frac{1}{2} \times \frac{3}{8} = \frac{3}{16}$ female and $\frac{1}{2} \times \frac{5}{8} = \frac{5}{16}$ male.

(b) In $F_1 \frac{x}{x}\frac{+}{tra} \times \frac{x}{\rightharpoonup}\frac{tra}{tra}$ result $\frac{1}{4}$ female and $\frac{3}{4}$ male. It also constitutes $\frac{1}{2}$ of all possible matings, i.e., $\frac{1}{2} \times \frac{3}{4} = \frac{3}{8}$ males and $\frac{1}{2} \times \frac{1}{4} = \frac{1}{8}$ females.

Summary of the F_2 from both matings:

males $= \frac{5}{16} + \frac{3}{8} = \frac{11}{16}$; females $= \frac{3}{16} + \frac{1}{8} = \frac{5}{16}$.

Problem 5

Vermilion eye colour in *Drosophila* is determined by the recessive allele v of an x-linked gene; the wild type eye colour produced by the v⁺ allele is brick-red. What phenotypic ratios are expected from the crosses (a) vermilion female × wild type male (b) homozygous wild type female × vermilion male and (c) heterozygous female × vermilion male?

Solution:

(a) Vermilion female × Wild type male

$\frac{v}{v}$ $\frac{v^+}{\rightharpoonup}$

Gametes: $\underline{v}$ $\underline{v^+}$ $\rightharpoonup$

Cross:

♀ \ ♂	v^+	$\rightharpoonup$
v	$\frac{v^+}{v}$ wild type ♀	$\frac{v}{\rightharpoonup}$ vermilion ♂

Results:

Genotype	Phenotype	Ratio
$\frac{v^+}{v}$	wild type female	1
$\frac{v}{\rightharpoonup}$	vermilion male	1

(b) Wild type female (homozygous) × Vermilion male

$\frac{v^+}{v^+}$ $\frac{v}{\rightharpoonup}$

Gametes: $\underline{v^+}$ $\underline{v}$ $\rightharpoonup$

Cross:

♀ \ ♂	v	$\rightharpoonup$
v^+	$\frac{v^+}{v}$ wild type ♀	$\frac{v^+}{\rightharpoonup}$ vermilion ♂

Results:

Genotype	Phenotype	Ratio
$\frac{v^+}{v}$	wild type female	1
$\frac{v}{\rightharpoonup}$	vermilion male	1

(c) Heterozygous female × Vermilion male

$\frac{v^+}{v}$ $\frac{v}{\rightharpoonup}$

Gametes: $\underline{v^+}$ $\underline{v}$ $\underline{v}$ $\rightharpoonup$

Cross:

♀ \ ♂	v	$\rightharpoonup$
v^+	$\frac{v^+}{v}$	$\frac{v^+}{\rightharpoonup}$
v	$\frac{v}{v}$	$\frac{v}{\rightharpoonup}$

Results:

	Genotype	Phenotype	Ratio
(i)	$\frac{v^+}{v}$	normal female	$\frac{1}{4}$
(ii)	$\frac{v}{v}$	vermilion female	$\frac{1}{4}$
(iii)	$\frac{v^+}{\rightharpoonup}$	normal male	$\frac{1}{4}$
(iv)	$\frac{v}{\rightharpoonup}$	vermilion male	$\frac{1}{4}$

Problem 6

In *Drosophila* white eye is a sex-linked character. The mutant allele for white eye (w) is recessive to the wild type allele for brick-red eye colour (w^+).

A white-eyed female is crossed with a red-eyed male, and an F_1 female from this cross is mated with her father and an F_1 male is mated with her mother. What will be the appearance of the offspring of these last two crosses with respect to eye colour?

Solution:

(a) P_1 White-eyed female × Red-eyed male

$\frac{w}{w}$ $\frac{w^+}{\rightharpoonup}$

Gametes: $\underline{w}$ $\underline{w^+}$ $\rightharpoonup$

F_1: ♀ $\frac{w^+}{w}$ (Red-eyed) ♂ $\frac{w}{\rightharpoonup}$ (White-eyed)

Cross: 1

F_1: Female (Red) $\frac{w^+}{w}$ Male (P_1) $\frac{w^+}{\rightharpoonup}$

Gametes: $\underline{w^+}$ $\underline{w}$; $\underline{w^+}$ $\rightharpoonup$

F_1:

♀ \ ♂	$\underline{w^+}$	$\rightharpoonup$
$\underline{w^+}$	$\frac{w^+}{w^+}$ red ♀	$\frac{w^+}{\rightharpoonup}$ red ♂
$\underline{w}$	$\frac{w^+}{w}$ red ♀	$\frac{w}{\rightharpoonup}$ white ♂

Results:

Genotype	Phenotype	Ratio
$\frac{w^+}{w^+}$	Red-eyed } female	$2 = \frac{1}{2}$
$\frac{w^+}{w}$	Red-eyed } female	
$\frac{w^+}{\rightharpoonup}$	Red-eyed male	$1 = \frac{1}{4}$
$\frac{w^+}{\rightharpoonup}$	White-eyed male	$1 = \frac{1}{4}$

Cross: 2

F_1 White-eyed female × White-eyed male (P)

$\frac{w}{w}$ $\frac{w}{\rightharpoonup}$

Gametes: $\underline{w}$; $\underline{w}$ $\rightharpoonup$

F_2:

♀ \ ♂	$\underline{w}$	$\rightharpoonup$
$\underline{w}$	$\frac{w}{w}$ (white ♀)	$\frac{w}{\rightharpoonup}$ (white ♂)

All progenies are white.

(b) White-eyed (♀) × Red-eyed (♂)

$\frac{w}{w}$ × $\frac{w^+}{\rightharpoonup}$

Result: White-eyed (♂) $\frac{w}{\rightharpoonup}$; Red-eyed (♀) $\frac{w^+}{w}$

Problem 7

A normal woman whose father was colour-blind has a son. Nothing is known of the colour vision phenotype of the father. What is the probabilities that the son will be colour-blind?

Solution:

As a normal woman's father was colour-blind, so her genotype will be $x_c^+ x_c$, i.e., she is carrier.

(i) She has a son. As father always donate his Y-chromosome to his son, so there is no need to know the phenotype of the father.

(ii) As the woman has two X-chromosomes, i.e., x_c^+ and x_c, so the probability of colour-blind son will be $\frac{1}{2}$, i.e., 50%.

Problem 8

If a mother carried the sex-linked gene for green defective colour vision and the father was normal, would their sons or daughters be defective in colour vision?

Solution:

Mother carried sex-linked gene for green defective colour vision and father was normal. The woman was carrier, i.e., $x_c^+ x_c$ (x_c = for colour blindness). Half of the sons will be colour-blind, as mother carried 50% colour-blind gene (x_c). All the daughters will be normal.

Problem 9

A normal woman, whose father had haemophilia, married a normal man. What is the chance of haemophilia in their children?

Solution:

(i) Normal woman's father had haemophilia. So the woman will be carrier (heterozygous) normal. Her genotype is $\frac{x_h^+}{x_h}$.

(ii) She married a normal man $\frac{x_h^+}{x_h} \times \frac{x_h^+}{y}$.

Gametes: $\underline{x_h^+}$ $\underline{x_h}$ $\underline{x_h^+}$ $\overline{y}$

F₁:

♀ \ ♂	x_h^+	y
x_h^+	$\frac{x_h^+}{x_h^+}$	$\frac{x_h^+}{y}$
x_h	$\frac{x_h^+}{x_h}$	$\frac{x_h}{y}$

Results:

Genotype	Phenotype	Probability
$\frac{x_h^+}{x_h^+}$	Normal child (female)	$\frac{1}{4}$ } $\frac{1}{2}$
$\frac{x_h^+}{x_h}$	Normal child female (carrier)	$\frac{1}{4}$
$\frac{x_h^+}{y}$	Normal child (male)	$\frac{1}{4}$
$\frac{x_h}{y}$	Affected child (male)	$\frac{1}{4}$

$\frac{1}{4}$ child (male) will be affected with haemophilia. Among the sons $\frac{1}{2}$ child (male) will be affected.

Problem 10

A girl of normal vision marries a boy of normal vision. Fathers of both girl and boy are colour-blind. What type of vision you expect in their offspring?

(Note the gene for colour-blind is sex-linked and recessive.)

Solution:

Here the girl of normal vision is heterozygous or carrier as her father is colour-blind ($x_c y$). The boy is of normal vision and hemizygous for dominant allele x_c^+ as he gets Y-chromosome from his father.

Boy (normal vision) × Girl (normal vision)

$x_c^+ y$ $x_c^+ x_c$

Gametes: x_c^+ y x_c^+ x_c

♀ \ ♂	x_c^+	y
x_c^+	$x_c^+ x_c^+$	$x_c^+ y$
x_c	$x_c^+ x_c$	$x_c y$

Results:

Genotype	Phenotype
$x_c^+ x_c^+$	Normal girl
$x_c^+ x_c$	Normal girl (carrier)
$x_c^+ y_c$	Normal boy
$x_c y$	Colour-blind boy

Problem 11

A husband and his wife have normal visions, although both of their fathers are red green colour-blind, an inherited x-linked recessive condition. What is the probability that their 1st child will be a (a) normal son? (b) normal daughter? (c) colour-blind son? (d) colour-blind daughter?

Solution:

Here the female (wife) is heterozygous or carrier ($x_c^+ x_c$) as her father is colour-blind ($x_c y$).

$x_c^+ y$ × $x_c^+ x_c$

Gametes: x_c^+ x_c x_c^+ y

♀ \ ♂	x_c^+	y
x_c^+	$x_c^+ x_c^+$	$x_c^+ y$
x_c	$x_c^+ x_c$	$x_c y$

Results:

Genotype	Phenotype	Probability
(i) $x_c^+ x_c^+$	Normal vision female	$\frac{1}{4}$ } $-\frac{1}{2} = 25\%$
(ii) $x_c^+ x_c$	Normal vision female (carrier)	$\frac{1}{4}$
(iii) $x_c^+ y$	Normal vision male	$\frac{1}{4}$
(iv) $x_c y$	Colour-blind male	$\frac{1}{4}$

(a) The probability of first child (son) will be normal $= \frac{1}{4} = 25\%$

(b) The probability of normal daughter $= \frac{1}{4} + \frac{1}{4} = \frac{2}{4} = \frac{1}{2} = 50\%$

(c) The probability of colour-blind son $= \frac{1}{4} = 25\%$

(d) The probability of colour-blind daughter = nil, i.e., zero.

Problem 12

In *Drosophila*, an x-linked recessive mutation, scalloped (sd) causes irregular wing margins. Diagram the F₁ and F₂ results if (a) a scalloped female is crossed with a normal male, (b) a scalloped male is crossed with a normal female.

Compare these results with those that would be obtained if the scalloped were not x-linked.

Solution:

(a) Scalloped female × Normal male

$\frac{x^{sd}}{x^{sd}}$ × $\frac{x^+}{\rightharpoondown}$

Gametes: $\underline{x^{sd}}$ $\underline{x^+}$ $\rightharpoondown$

F$_1$:

♀ \ ♂	X^+	$\rightharpoonup$
X^{sd}	$\frac{X^+}{X^{sd}}$	$\overset{X^{sd}}{\rightharpoonup}$

Result: Normal ♀ Scalloped ♂

$$\frac{X^+}{X^{sd}} \quad \overset{X^{sd}}{\rightharpoonup} = \frac{1}{2}:\frac{1}{2} = 1:1$$

Scalloped male × Normal female

$$\overset{X^{sd}}{\rightharpoonup} \qquad \frac{X^+}{X^{sd}}$$

Gametes: X^{sd} $\rightharpoonup$ X^+ X^{sd}

♂ \ ♀	X^+	X^{sd}
X^{sd}	$\frac{X^+}{X^{sd}}$	$\frac{X^{sd}}{X^{sd}}$
$\rightharpoonup$	$\overset{X^+}{\rightharpoonup}$	$\overset{X^{sd}}{\rightharpoonup}$

Results:

Genotype	Phenotype	Ratio
(i) $\frac{X^+}{X^{sd}}$	Normal female	$\frac{1}{4}$
(ii) $\frac{X^{sd}}{X^{sd}}$	Scalloped female	$\frac{1}{4}$
(iii) $\overset{X^+}{\rightharpoonup}$	Normal male	$\frac{1}{4}$
(iv) $\overset{X^{sd}}{\rightharpoonup}$	Scalloped male	$\frac{1}{4}$

(b) Scalloped male × Normal female

$$\overset{X^{sd}}{\rightharpoonup} \times \frac{X^+}{X^+}$$

Gametes: X^{sd} $\rightharpoonup$ X^+

F$_1$: $\frac{X^{sd}}{X^+}$ $\overset{X^+}{\rightharpoonup}$

Result:

Genotype	Phenotype	Ratio
(i) $\frac{X^{sd}}{X^+}$	Normal female	$\frac{1}{2}$
(ii) $\overset{X^+}{\rightharpoonup}$	Normal male	$\frac{1}{2}$

F$_2$ Normal male × Scalloped female

$$\frac{X^{sd}}{X^+} \times \overset{X^{sd}}{\rightharpoonup}$$

Gametes: X^{sd}, X^+ ; X^{sd}, $\rightharpoonup$

♀ \ ♂	X^{sd}	$\rightharpoonup$
X^{sd}	$\frac{X^{sd}}{X^{sd}}$	$\overset{X^{sd}}{\rightharpoonup}$
X^+	$\frac{X^{sd}}{X^+}$	$\overset{X^+}{\rightharpoonup}$

Results:

Genotype	Phenotype	Ratio
(i) $\frac{X^{sd}}{X^{sd}}$	Scalloped female	$\frac{1}{4}$
(ii) $\frac{X^{sd}}{X^+}$	Normal female	$\frac{1}{4}$
(iii) $\overset{X^{sd}}{\rightharpoonup}$	Scalloped female	$\frac{1}{4}$
(ii) $\overset{X^+}{\rightharpoonup}$	Normal male	$\frac{1}{4}$

If the scalloped were not x-linked, then all of the F$_1$ offsprings would be wild type (phenotypically) and 3:1 ratio of normal and scalloped would occur in the F$_2$.

Problem 13

In *Drosophila*, the x-linked recessive mutation, vermilion (v) causes bright red eyes which is contrast to brick-red eyes of wild type. A separate autosomal recessive mutation, suppressor of vermilion (su-v) causes flies homozygous or hemizygous for v to have wild type eyes. In the absence of vermilion alleles su-v has no effect on eye colour. Determine the F$_1$ and F$_2$ phenotypic ratios from a cross between a female with wild type alleles at the vermilion locus, but who is homozygous for su-v with a vermilion male who has wild type alleles at the su-v locus.

Solution:

Female × Male

$$\frac{X^+}{X^+}\ \frac{su\text{-}v}{su\text{-}v} \times \overset{X^v}{\rightharpoonup}\ \frac{su\text{-}v^+}{su\text{-}v^+}$$

Gametes: X^+ $su\text{-}v$; X^v $su\text{-}v^+$, $\rightharpoonup$ $su\text{-}v^+$

F$_1$

♀ \ ♂	X^v $su\text{-}v^+$	$\rightharpoonup$ $su\text{-}v^+$
X^+ $su\text{-}v$	$\frac{X^v}{X^+}\ \frac{su\text{-}v^+}{su\text{-}v}$	$\overset{\rightharpoonup}{X^+}\ \frac{su\text{-}v^+}{su\text{-}v}$

Results:

Genotype	Phenotype	Ratio
1. $\frac{X^+}{X^v}\ \frac{su\text{-}v^+}{su\text{-}v}$	Normal female	$\frac{1}{2}$
2. $\overset{X^+}{\rightharpoonup}\ \frac{su\text{-}v^+}{su\text{-}v}$	Normal male	$\frac{1}{2}$

F$_2$ = F$_1$ × F$_1$

Normal female Normal male

$$\frac{X^v}{X^+}\ \frac{su\text{-}v^+}{su\text{-}v} \qquad \overset{X^+}{\rightharpoonup}\ \frac{su\text{-}v^+}{su\text{-}v}$$

Gametes:

Female	Male
(i) X^v $su\text{-}v^+$	(i) X^+ $su\text{-}v^+$
(ii) X^+ $su\text{-}v$	(ii) $\rightharpoonup$ $su\text{-}v$
(iii) X^v $su\text{-}v$	(iii) X^+ $su\text{-}v$
(iv) X^+ $su\text{-}v^+$	(iv) $\rightharpoonup$ $su\text{-}v^+$

♂ \ ♀	X^v $su\text{-}v^+$	X^+ $su\text{-}v$	X^v $su\text{-}v$	X^+ $su\text{-}v^+$
X^+ $su\text{-}v^+$	$\frac{X^v}{X^+}\ \frac{su\text{-}v^+}{su\text{-}v^+}$	$\frac{X^+}{X^+}\ \frac{su\text{-}v}{su\text{-}v^+}$	$\frac{X^v}{X^+}\ \frac{su\text{-}v}{su\text{-}v^+}$	$\frac{X^+}{X^+}\ \frac{su\text{-}v^+}{su\text{-}v^+}$
$\rightharpoonup$ $su\text{-}v^+$	$\overset{X^v}{\rightharpoonup}\ \frac{su\text{-}v^+}{su\text{-}v^+}$	$\overset{X^+}{\rightharpoonup}\ \frac{su\text{-}v}{su\text{-}v^+}$	$\overset{X^v}{\rightharpoonup}\ \frac{su\text{-}v}{su\text{-}v^+}$	$\overset{X^+}{\rightharpoonup}\ \frac{su\text{-}v^+}{su\text{-}v^+}$
X^+ $su\text{-}v$	$\frac{X^v}{X^+}\ \frac{su\text{-}v^+}{su\text{-}v}$	$\frac{X^+}{X^+}\ \frac{su\text{-}v}{su\text{-}v}$	$\frac{X^v}{X^+}\ \frac{su\text{-}v}{su\text{-}v}$	$\frac{X^+}{X^+}\ \frac{su\text{-}v^+}{su\text{-}v}$
$\rightharpoonup$ $su\text{-}v^+$	$\overset{X^v}{\rightharpoonup}\ \frac{su\text{-}v^+}{su\text{-}v^+}$	$\overset{X^+}{\rightharpoonup}\ \frac{su\text{-}v}{su\text{-}v^+}$	$\overset{X^v}{\rightharpoonup}\ \frac{su\text{-}v}{su\text{-}v^+}$	$\overset{X^+}{\rightharpoonup}\ \frac{su\text{-}v^+}{su\text{-}v^+}$

Result:

Genotype	Sex	Phenotype	Ratio
i. $\frac{x^v}{x^+}\ \frac{su\text{-}v^+}{su\text{-}v}$	Female	Wild = 6 type	$\frac{6}{16}$
ii. $\frac{x^+}{x^+}\ \frac{su\text{-}v}{su\text{-}v^+}$	Female		
iii. $\frac{x^v}{x^+}\ \frac{su\text{-}v}{su\text{-}v^+}$	Female		
iv. $\frac{x^v}{x^+}\ \frac{su\text{-}v^+}{su\text{-}v^+}$	Female		
v. $\frac{x^v}{x^+}\ \frac{su\text{-}v^+}{su\text{-}v}$	Female		
vi. $\frac{x^+}{x^+}\ \frac{su\text{-}v^+}{su\text{-}v}$	Female		
vii. $\frac{x^+}{x^+}\ \frac{su\text{-}v}{su\text{-}v}$	Female	Wild type (suppression of vermilion)	$\frac{2}{16}$
viii. $\frac{x^v}{x^+}\ \frac{su\text{-}v}{su\text{-}v}$	Female		

Result:

Genotype	Sex	Phenotype	Ratio
i. $\frac{x^v}{\rightharpoondown}\ \frac{su\text{-}v^+}{su\text{-}v}$			
ii. $\frac{x^v}{\rightharpoondown}\ \frac{su\text{-}v^+}{su\text{-}v}$	Male	Vermilion male	$\frac{3}{16}$
iii. $\frac{x^v}{\rightharpoondown}\ \frac{su\text{-}v}{su\text{-}v^+}$			
iv. $\frac{x^v}{\rightharpoondown}\ \frac{su\text{-}v}{su\text{-}v}$	Male	Wild type (Vermilion gene is suppressed)	$\frac{1}{16}$
v. $\frac{x^+}{\rightharpoondown}\ \frac{su\text{-}v^+}{su\text{-}v}$	Male	Wild type	$\frac{4}{16}$
vi. $\frac{x^+}{\rightharpoondown}\ \frac{su\text{-}v}{su\text{-}v^+}$			
vii. $\frac{x^+}{\rightharpoondown}\ \frac{su\text{-}v^+}{su\text{-}v^+}$			
viii. $\frac{x^+}{\rightharpoondown}\ \frac{su\text{-}v}{su\text{-}v}$			

(i) $\frac{8}{16}$ wild type females (none of the females is homozygous for the vermilion gene).

(ii) $\frac{4}{16}$ wild type males ($\frac{4}{16}$ have no vermilion gene and $\frac{1}{16}$ x-linked, hemizygous vermilion gene is suppressed by $\frac{su\text{-}v}{su\text{-}v}$).

(iii) vermilion males ($\frac{x^v}{\rightharpoondown}$).

Problem 14

In a cross in *Drosophila*, involving the x-linked recessive eye mutation white and the autosomally linked recessive eye mutation sepia (resulting in dark eye), predict the F_1 and F_2 results of crossing true breeding parents of the following phenotypes:

(a) White females × Sepia males

(b) Sepia females × White males.

Note that white is epistatic to the expression of sepia.

Solution:

(a) White females × Sepia males

$$\frac{w}{w}\ \frac{se^+}{se^+} \quad \times \quad \frac{w^+}{\rightharpoondown}\ \frac{se}{se}$$

Gametes: $\underline{w}\ \ \underline{se^+}$, × $\underline{w^+}\ \ \underline{se}$, $\rightharpoondown\ \ \underline{se}$

F_1 $\frac{w}{w^+}\ \frac{se^+}{se}$ (wild type female) × $\frac{w}{\rightharpoondown}\ \frac{se^+}{se}$ (white male)

$F_2 = F_1 \times F_1$

$\frac{w}{w^+}\ \frac{se^+}{se}$ (wild type female) × $\frac{w}{\rightharpoondown}\ \frac{se}{se}$ (white male)

Gametes: (i) $\underline{w}\ \underline{se^+}$ (ii) $\underline{w^+}\ \underline{se}$ (iii) $\underline{w}\ \underline{se}$ (iv) $\underline{w^+}\ \underline{se^+}$

(i) $\underline{w}\ \underline{se^+}$ (ii) $\rightharpoondown\ \underline{se}$ (iii) $\underline{w}\ \underline{se}$ (iv) $\rightharpoondown\ \underline{se^+}$

♂ \ ♀	w se⁺	w⁺ se	w se	w⁺ se⁺
$\underline{w}\ \underline{se^+}$	$\frac{w}{w}\ \frac{se^+}{se^+}$	$\frac{w^+}{w}\ \frac{se}{se^+}$	$\frac{w}{w}\ \frac{se}{se^+}$	$\frac{w^+}{w}\ \frac{se^+}{se^+}$
$\rightharpoondown\ \underline{se}$	$\frac{w}{\rightharpoondown}\ \frac{se^+}{se}$	$\frac{w^+}{\rightharpoondown}\ \frac{se}{se}$	$\frac{w}{\rightharpoondown}\ \frac{se}{se}$	$\frac{w^+}{\rightharpoondown}\ \frac{se^+}{se}$
$\underline{w}\ \underline{se}$	$\frac{w}{w}\ \frac{se^+}{se}$	$\frac{w^+}{w}\ \frac{se}{se}$	$\frac{w}{w}\ \frac{se}{se}$	$\frac{w^+}{w}\ \frac{se^+}{se}$
$\rightharpoondown\ \underline{se^+}$	$\frac{w}{\rightharpoondown}\ \frac{se^+}{se^+}$	$\frac{w^+}{\rightharpoondown}\ \frac{se}{se^+}$	$\frac{w}{\rightharpoondown}\ \frac{se}{se^+}$	$\frac{w^+}{\rightharpoondown}\ \frac{se^+}{se^+}$

Result:

Genotype	Sex	Phenotype		Ratio
1. $\frac{w^+}{\rightharpoondown}\ \frac{se^+}{se}$	Male	Wild type	3	$\frac{3}{16}$
2. $\frac{w^+}{\rightharpoondown}\ \frac{se}{se^+}$	Male			
3. $\frac{w^+}{\rightharpoondown}\ \frac{se^+}{se^+}$	Male			
4. $\frac{w^+}{\rightharpoondown}\ \frac{se}{se}$	Male	Sepia	1	$\frac{1}{16}$
5. $\frac{w}{\rightharpoondown}\ \frac{se}{se^+}$	Male	White	4	$\frac{4}{16}$
6. $\frac{w}{\rightharpoondown}\ \frac{se^+}{se}$	Male	White		
7. $\frac{w}{\rightharpoondown}\ \frac{se^+}{se^+}$	Male	White		
8. $\frac{w}{\rightharpoondown}\ \frac{se^*}{se}$	Male	White		
9. $\frac{w^+}{w}\ \frac{se}{se^+}$	Female	Wild	3	$\frac{3}{16}$
10. $\frac{w^+}{w}\ \frac{se^+}{se^+}$	Female	Wild		
11. $\frac{w^+}{w}\ \frac{se^+}{se}$	Female	Wild		
12. $\frac{w}{w}\ \frac{se^+}{se^+}$	Female	White	4	$\frac{4}{16}$
13. $\frac{w}{w}\ \frac{se}{se^+}$	Female	White		
14. $\frac{w}{w}\ \frac{se^+}{se}$	Female	White		
15. $\frac{w}{w}\ \frac{se^*}{se}$	Female	White		
16. $\frac{w^+}{w}\ \frac{se}{se}$	Female	Sepia		$\frac{1}{16}$

(b) Sepia females × White males

$\frac{w^+}{w^+}\ \frac{se}{se}$ $\qquad\qquad$ $\frac{w}{\rightarrow}\ \frac{se^+}{se^+}$

Gametes: $\underline{w^+}\ \underline{se}$, $\qquad$ $\underline{w}\ \underline{se^+}$, $\rightarrow\ \underline{se^+}$

F_1 $\quad$ $\frac{w^+}{w}\ \frac{se}{se^+}$ (wild female) $\qquad$ $\frac{w^+}{\rightarrow}\ \frac{se}{se^+}$ (wild male)

$F_2 = F_1 \times F_1$

$\frac{w^+}{w}\ \frac{se}{se^+}$ $\qquad\qquad$ $\frac{w^+}{\rightarrow}\ \frac{se}{se^+}$

Gametes:

	♀		♂
(i)	$\underline{w^+}\ \underline{se}$	(i)	$\underline{w^+}\ \underline{se}$
(ii)	$\underline{w}\ \underline{se^+}$	(ii)	$\rightarrow\ \underline{se^+}$
(iii)	$\underline{w^+}\ \underline{se^+}$	(iii)	$\underline{w^+}\ \underline{se^+}$
(iv)	$\underline{w}\ \underline{se}$	(iv)	$\rightarrow\ \underline{se}$

♂ \ ♀	$\underline{w^+}\ \underline{se}$	$\underline{w}\ \underline{se^+}$	$\underline{w^+}\ \underline{se^+}$	$\underline{w}\ \underline{se}$
$\underline{w^+}\ \underline{se}$	$\frac{w^+}{w^+}\ \frac{se}{se}$	$\frac{w}{w^+}\ \frac{se^+}{se}$	$\frac{w^+}{w^+}\ \frac{se^+}{se}$	$\frac{w}{w^+}\ \frac{se}{se}$
$\rightarrow\ \underline{se^+}$	$\frac{w^+}{\rightarrow}\ \frac{se}{se^+}$	$\frac{w}{\rightarrow}\ \frac{se^+}{se^+}$	$\frac{w^+}{\rightarrow}\ \frac{se^+}{se^+}$	$\frac{w}{\rightarrow}\ \frac{se}{se^+}$
$\underline{w^+}\ \underline{se^+}$	$\frac{w^+}{w^+}\ \frac{se}{se^+}$	$\frac{w}{w^+}\ \frac{se^+}{se^+}$	$\frac{w^+}{w^+}\ \frac{se^+}{se^+}$	$\frac{w}{w^+}\ \frac{se}{se^+}$
$\rightarrow\ \underline{se}$	$\frac{w}{\rightarrow}\ \frac{se}{se}$	$\frac{w}{\rightarrow}\ \frac{se^+}{se}$	$\frac{w^+}{\rightarrow}\ \frac{se^+}{se}$	$\frac{w}{\rightarrow}\ \frac{se}{se}$

Result:

Genotype	Sex	Phenotype		Ratio
1. $\frac{w^+}{\rightarrow}\ \frac{se}{se^+}$	Male	Wild	3	$\frac{3}{16}$
2. $\frac{w^+}{\rightarrow}\ \frac{se^+}{se^+}$	Male	Wild		
3. $\frac{w^+}{\rightarrow}\ \frac{se^+}{se}$	Male	Wild		
4. $\frac{w}{\rightarrow}\ \frac{se^+}{se^+}$	Male	White	4	$\frac{4}{16}$
5. $\frac{w}{\rightarrow}\ \frac{se}{se^+}$	Male	White		
6. $\frac{w}{\rightarrow}\ \frac{se^+}{se}$	Male	White		
7. $\frac{w}{\rightarrow}\ \frac{se}{se}$	Male	White		
8. $\frac{w^+}{\rightarrow}\ \frac{se}{se}$	Male	Sepia		$\frac{1}{16}$
9. $\frac{w}{w^+}\ \frac{se}{se^+}$	Female	Wild	6	$\frac{6}{16}$
10. $\frac{w^+}{w^+}\ \frac{se^+}{se}$	Female	Wild		
11. $\frac{w^+}{w^+}\ \frac{se^+}{se^+}$	Female	Wild		
12. $\frac{w}{w^+}\ \frac{se^+}{se^+}$	Female	Wild		
13. $\frac{w^+}{w^+}\ \frac{se^+}{se^+}$	Female	Wild		
14. $\frac{w^+}{w^+}\ \frac{se}{se^+}$	Female	Wild		
15. $\frac{w^+}{w^+}\ \frac{se}{se}$	Female	Sepia	2	$\frac{2}{16}$
16. $\frac{w}{w^+}\ \frac{se}{se}$	Female	Sepia		

Problem 15

In the human disorder sickle-cell anaemia, many phenotypic traits are evident beside those where the red blood cells undergo sickling and clumping under low oxygen tension and have a shorter half-life than normal, causing anaemia. The overall phenotype includes episodes of severe abdominal pain, weakness and fatigue, lengthened long bones, heart enlargement, kidney failure and respiratory difficulties, including pneumonia.

(a) What term describes cases of multiple phenotypic manifestations of a single gene?

(b) Relate each of these phenotype responses to the sickling and anaemia.

(c) Perform a web, search (or jump ahead) and determine the molecular basis of disorder.

Solution:

(a) The term **pleotropy** is used when there are multiple phenotypic manifestations of a single gene.

(b) (i) Each phenotypic response is a result of the inability of the red blood cells to adequately carry oxygen and sickle shape of the red blood cells, blocks their passage to various tissues.

(ii) As a result of vaso-occlusive crisis, tissue necrosis occurs and a variety of health problems result.

(c) The molecular basis of sickle-cell anaemia is a substitution of valine for glutamic acid at the sixth position of β chain.

Problem 16

In *Drosophila*, the lozenge phenotype, caused by a sex-linked recessive allele (Lz) is narrow eyes. Diagram the F_2 generation across a lozenge male and homozygous normal female. Diagram the reciprocal cross.

Solution:

(a) Lozenge male × Normal females (homozygous)

$\frac{Lz}{\rightarrow}$ $\qquad\qquad$ $\frac{+}{+}$

Gametes: $\underline{Lz}$, $\rightarrow$; $\qquad$ $\underline{+}$

F_1 $\quad \frac{Lz}{+}$ (Normal female) $\quad \frac{+}{\rightharpoondown}$ (Normal male)

$F_2 = F_1 \times F_1$

$$\frac{Lz}{+} \quad \times \quad \frac{+}{\rightharpoondown}$$

Gametes: $\underline{Lz}$, $\underline{+}$; $\quad \underline{+}$ $\rightharpoondown$

F_2

♀ / ♂	$\underline{Lz}$	$\underline{+}$
$\underline{+}$	$\frac{Lz}{+}$	$\frac{+}{+}$
$\rightharpoondown$	$\frac{Lz}{\rightharpoondown}$	$\frac{+}{\rightharpoondown}$

Result:

Genotype	Sex	Phenotype	Ratio
(i) $\frac{Lz}{Lz}$	Female	Lozenge	1
(ii) $\frac{+}{Lz}$	Female	Normal	1
(iii) $\frac{Lz}{\rightharpoondown}$	Male	Lozenge	1
(iv) $\frac{+}{\rightharpoondown}$	Male	Normal	1

Problem 17

In *Drosophila*, cut wing is controlled by a recessive sex-linked allele (ct) and fuzzy body is controlled by a recessive autosomal allele (fy). When a fuzzy female is mated with a cut male, all the members of the F_1 generation are wild type. What is the proportion of F_2 phenotypes by sex?

Solution:

Fuzzy female $\times$ Cut males

$$\frac{fy}{fy}\ \frac{+}{+} \quad \times \quad \frac{fy^+}{fy^+}\ \frac{ct}{\rightharpoondown}$$

Gametes: $\underline{fy}$, $\underline{+}$; $\quad \underline{fy^+}$ $\underline{ct}$, $\underline{fy^+}$ $\rightharpoondown$

F_1 $\quad \frac{fy}{fy^+}\ \frac{+}{ct}$ (Wild type female) $\quad \frac{fy}{fy^+}\ \frac{+}{\rightharpoondown}$ (Wild type male)

$F_2 = F_1 \times F_1$

$$\frac{fy}{fy^+}\ \frac{+}{ct} \qquad \frac{fy}{fy^+}\ \frac{+}{\rightharpoondown}$$

Gametes:

$\underline{fy}\ \underline{+}$, $\underline{fy^+}\ \underline{ct}$, $\underline{fy^+}\ \underline{+}$, $\underline{fy}\ \underline{ct}$; $\underline{fy}\ \underline{+}$, $\underline{fy^+}\ \rightharpoondown$, $\underline{fy}\ \rightharpoondown$, $\underline{fy^+}\ \underline{+}$

♀ / ♂	$\underline{fy}\ \underline{+}$	$\underline{fy^+}\ \underline{ct}$	$\underline{fy^+}\ \underline{+}$	$\underline{fy}\ \underline{ct}$
$\underline{fy}\ \underline{+}$	$\frac{fy}{fy}\ \frac{+}{+}$	$\frac{fy^+}{fy}\ \frac{ct}{+}$	$\frac{fy^+}{fy}\ \frac{+}{+}$	$\frac{fy}{fy}\ \frac{ct}{+}$
$\underline{fy^+}\ \rightharpoondown$	$\frac{fy}{fy^+}\ \frac{+}{\rightharpoondown}$	$\frac{fy^+}{fy^+}\ \frac{ct}{\rightharpoondown}$	$\frac{fy^+}{fy^+}\ \frac{+}{\rightharpoondown}$	$\frac{fy}{fy^+}\ \frac{ct}{\rightharpoondown}$
$\underline{fy}\ \rightharpoondown$	$\frac{fy}{fy}\ \frac{+}{\rightharpoondown}$	$\frac{fy^+}{fy}\ \frac{ct}{\rightharpoondown}$	$\frac{fy^+}{fy}\ \frac{+}{\rightharpoondown}$	$\frac{fy}{fy}\ \frac{ct}{\rightharpoondown}$
$\underline{fy}\ \underline{+}$	$\frac{fy}{fy}\ \frac{+}{+}$	$\frac{fy^+}{fy}\ \frac{ct}{+}$	$\frac{fy^+}{fy}\ \frac{+}{+}$	$\frac{fy}{fy}\ \frac{ct}{+}$

Result:

Genotype	Sex	Phenotype	Ratio
1. $\frac{fy^+}{fy}\ \frac{ct}{+}$	Female	Wild type	$\frac{3}{8}+\frac{1}{8}=\frac{4}{8}=\frac{1}{2}$
2. $\frac{fy^+}{fy}\ \frac{+}{+}$	Female	Wild type	
3. $\frac{fy^+}{fy^+}\ \frac{ct}{+}$	Female	Wild type	
4. $\frac{fy^+}{fy^+}\ \frac{+}{+}$	Female	Wild type	
5. $\frac{fy}{fy^+}\ \frac{ct}{+}$	Female	Wild type	
6. $\frac{fy}{fy^+}\ \frac{ct}{+}$	Female	Wild type	
7. $\frac{fy}{fy}\ \frac{+}{+}$	Female	Fuzzy	$\frac{1}{16}=\frac{1}{8}$
8. $\frac{fy}{fy}\ \frac{ct}{+}$	Female	Fuzzy	
9. $\frac{fy}{fy^+}\ \frac{+}{\rightharpoondown}$	Male	Wild type	$3=\frac{3}{16}$
10. $\frac{fy^+}{fy^+}\ \frac{+}{\rightharpoondown}$	Male	Wild type	
11. $\frac{fy^+}{fy}\ \frac{+}{\rightharpoondown}$	Male	Wild type	
12. $\frac{fy^+}{fy}\ \frac{ct}{\rightharpoondown}$	Male	Cut	$3=\frac{3}{16}$
13. $\frac{fy}{fy^+}\ \frac{ct}{\rightharpoondown}$	Male	Cut	
14. $\frac{fy^+}{fy}\ \frac{ct}{\rightharpoondown}$	Male	Cut	
15. $\frac{fy}{fy}\ \frac{+}{\rightharpoondown}$	Male	Fuzzy	$1=\frac{1}{16}$
16. $\frac{fy}{fy}\ \frac{ct}{\rightharpoondown}$	Male	Fuzzy cut	$1=\frac{1}{16}$

Problem 18

In *Drosophila*, white eye is an x-linked recessive trait, and ebony body is an autosomal recessive trait. A homozygous white-eyed female is crossed with a homozygous ebony male.

(a) What phenotypic ratio do you expect in the F_1 generation?

(b) What phenotypic ratio do you expect in the F_2 generation?

(c) Suppose, the initial cross was reversed. Ebony female × White-eyed male. What phenotypic ratio would you expect in F_2 generation?

Solution:

(a) White-eyed female × Ebony males

$$\frac{w}{w}\ \frac{+}{+} \times \frac{+}{\rightharpoonup}\ \frac{e}{e}$$

Gametes: $\underline{w}$, $\underline{+}$; $\underline{+}\ \underline{e}$, $\underline{\rightharpoonup}\ \underline{e}$

F_1 $\frac{w}{+}\ \frac{+}{e}$ (Wild type female) $\frac{w}{\rightharpoonup}\ \frac{+}{e} = 1:1$ (White-eyed male)

(b) $F_2 = F_1 \times F_1$

Wild type female × White-eyed male

$$\frac{w}{+}\ \frac{+}{e} \qquad \frac{w}{\rightharpoonup}\ \frac{+}{e}$$

Gametes:

	Female	Male
(i)	w +	w +
(ii)	w e	w e
(iii)	+ +	⇀ +
(iv)	+ e	⇀ e

♂ \ ♀	w +	w e	+ +	+ e
w +	$\frac{w}{w}\ \frac{+}{+}$	$\frac{w}{w}\ \frac{e}{+}$	$\frac{+}{w}\ \frac{+}{+}$	$\frac{+}{w}\ \frac{e}{+}$
w e	$\frac{w}{w}\ \frac{+}{e}$	$\frac{w}{w}\ \frac{e}{e}$	$\frac{+}{w}\ \frac{+}{e}$	$\frac{+}{w}\ \frac{e}{e}$
⇀ +	$\frac{w}{\rightharpoonup}\ \frac{+}{+}$	$\frac{w}{\rightharpoonup}\ \frac{e}{+}$	$\frac{+}{\rightharpoonup}\ \frac{+}{+}$	$\frac{+}{\rightharpoonup}\ \frac{e}{+}$
⇀ e	$\frac{w}{\rightharpoonup}\ \frac{+}{e}$	$\frac{w}{\rightharpoonup}\ \frac{e}{e}$	$\frac{+}{\rightharpoonup}\ \frac{+}{e}$	$\frac{+}{\rightharpoonup}\ \frac{e}{e}$

Result:

Genotype	Sex	Phenotype		Ratio
1. $\frac{w}{w}\ \frac{+}{+}$	Female	Wild type		
2. $\frac{+}{w}\ \frac{e}{+}$	Female	Wild type	3	$\frac{3}{16}$
3. $\frac{+}{w}\ \frac{+}{e}$	Female	Wild type		
4. $\frac{w}{w}\ \frac{+}{+}$	Female	White-eyed females		
5. $\frac{w}{w}\ \frac{e}{+}$	Female	White-eyed females	3	$\frac{3}{16}$
6. $\frac{w}{w}\ \frac{+}{e}$	Female	White-eyed females		
7. $\frac{w}{w}\ \frac{e}{e}$	Female	White-eyed ebony	1	$\frac{1}{16}$
8. $\frac{+}{w}\ \frac{e}{e}$	Female	Ebony body	1	$\frac{1}{16}$
9. $\frac{+}{\rightharpoonup}\ \frac{+}{+}$	Male	Wild type		
10. $\frac{+}{\rightharpoonup}\ \frac{e}{+}$	Male	Wild type	3	$\frac{3}{16}$
11. $\frac{+}{\rightharpoonup}\ \frac{+}{e}$	Male	Wild type		
12. $\frac{w}{\rightharpoonup}\ \frac{+}{+}$	Male	White-eyed male		
13. $\frac{w}{\rightharpoonup}\ \frac{+}{e}$	Male	White-eyed male	3	$\frac{3}{16}$
14. $\frac{w}{\rightharpoonup}\ \frac{+}{\rightharpoonup}$	Male	White-eyed male		
15. $\frac{w}{\rightharpoonup}\ \frac{e}{e}$	Male	White ebony	1	$\frac{1}{16}$
16. $\frac{+}{\rightharpoonup}\ \frac{e}{e}$	Male	Ebony	1	$\frac{1}{16}$

P_1 Ebony female × White-eyed males

$$\frac{+}{+}\ \frac{e}{e} \times \frac{w}{\rightharpoonup}\ \frac{e}{+}$$

Gametes: $\underline{+}\ \underline{e}$, $\underline{w}\ \underline{+}$, $\underline{\rightharpoonup}\ \underline{+}$

F_1 $\frac{+}{w}\ \frac{e}{+}$ (Wild type female) $\frac{+}{\rightharpoonup}\ \frac{e}{+}$ (Wild type male)

$F_2 = F_1 \times F_1$

Wild type female × Wild type male

$$\frac{+}{w}\ \frac{e}{+} \times \frac{w}{\rightharpoonup}\ \frac{e}{+}$$

Gametes:

	Female	Male
(i)	+ e	+ e
(ii)	w +	⇀ +
(iii)	+ +	+ +
(iv)	w e	⇀ e

♂ \ ♀	+ e	w +	+ +	w e
+ e	$\frac{+}{+}\ \frac{e}{e}$	$\frac{w}{+}\ \frac{+}{e}$	$\frac{+}{+}\ \frac{+}{e}$	$\frac{w}{+}\ \frac{e}{e}$
⇀ +	$\frac{+}{\rightharpoonup}\ \frac{e}{+}$	$\frac{w}{\rightharpoonup}\ \frac{+}{+}$	$\frac{+}{\rightharpoonup}\ \frac{+}{+}$	$\frac{w}{\rightharpoonup}\ \frac{e}{+}$
+ +	$\frac{+}{+}\ \frac{e}{+}$	$\frac{w}{+}\ \frac{+}{+}$	$\frac{+}{+}\ \frac{+}{+}$	$\frac{w}{+}\ \frac{e}{+}$
⇀ e	$\frac{+}{\rightharpoonup}\ \frac{e}{e}$	$\frac{w}{\rightharpoonup}\ \frac{+}{e}$	$\frac{+}{\rightharpoonup}\ \frac{+}{e}$	$\frac{w}{\rightharpoonup}\ \frac{e}{e}$

Result:

Genotype	Sex	Phenotype	Ratio
1. $\frac{w}{+}\ \frac{+}{e}$	Female	Wild type	
2. $\frac{+}{+}\ \frac{+}{e}$	Female	Wild type	
3. $\frac{+}{+}\ \frac{e}{+}$	Female	Wild type	$\frac{6}{16} = \frac{3}{8}$
4. $\frac{w}{+}\ \frac{+}{+}$	Female	Wild type	
5. $\frac{+}{+}\ \frac{+}{+}$	Female	Wild type	
6. $\frac{w}{+}\ \frac{e}{+}$	Female	Wild type	
7. $\frac{+}{+}\ \frac{e}{e}$	Female	Ebony body	$\frac{2}{16} = \frac{1}{8}$
8. $\frac{w}{+}\ \frac{e}{e}$	Female	Ebony body	
9. $\frac{w}{\rightharpoonup}\ \frac{+}{+}$	Male	White	
10. $\frac{w}{\rightharpoonup}\ \frac{e}{+}$	Male	White	$\frac{3}{16}$
11. $\frac{w}{\rightharpoonup}\ \frac{+}{e}$	Male	White	
12. $\frac{+}{\rightharpoonup}\ \frac{e}{+}$	Male	Wild type	
13. $\frac{+}{\rightharpoonup}\ \frac{+}{+}$	Male	Wild type	$\frac{3}{16}$
14. $\frac{+}{\rightharpoonup}\ \frac{+}{e}$	Male	Wild type	
15. $\frac{+}{\rightharpoonup}\ \frac{e}{e}$	Male	Ebony body	$\frac{1}{16}$
16. $\frac{w}{\rightharpoonup}\ \frac{e}{e}$	Male	White ebony	$\frac{1}{16}$

Problem 19

A woman who is heterozygous for both phenyl ketonuria mutation and X-linked haemophilia mutation has a child who is phenotypically normal and is also heterozygous for a phenyl ketonuria mutation. What is the probability that the child will be affected by both diseases? (Assume that they are equally likely to have a boy or a girl)?

Solution:

(i) Since the two genes are in different chromosomes (mutation for phenyl ketonuria is autosomal, that for haemophilia X-linked), they will assort independently and so we consider them separately.

(ii) As both parents are heterozygous for an autosomal recessive phenyl ketonuria mutation there is $\frac{1}{4}$ chance that their child will be homozygous recessive. For X-linked haemophilia, only a boy could be affected, a girl would have inherited a normal X-chromosome for her father. So she could not have haemophilia.

(iii) So the probability of a girl being affected with both the diseases is nil (0) and the probability of a boy being affected with both diseases is $\frac{1}{4} \times \frac{1}{2} = \frac{1}{8}$.

Overall the probability is $\left(\frac{1}{2}\right)(0) + \left(\frac{1}{2} \times \frac{1}{8}\right) = \frac{1}{16}$.

Problem 20

A recessive mutation is X-linked gene which results in haemophilia (prolonged time needed for blood to clot). Suppose that two phenotypically normal parents produce three normal daughters and a son affected with haemophilic.

(a) What is the probability that all the daughters are heterozygous carriers?

(b) If one of the daughters mates with a normal male and produces a son, what is the probability that the son will be affected?

Solution:

(a) (i) In this problem phenotypically normal parents have an affected son, who gets his only X-chromosome from his mother? The mother must be carrier of mutation.

(ii) Therefore, the probability that any particular daughter is a carrier is $\frac{1}{2}$. The probability that all 3 daughters are carrier is $\left(\frac{1}{2}\right)^3 = \frac{1}{8}$ because their birth is independent.

(b) (i) If the daughter is not carrier, the probabilities of an affected son is 0 and if the daughter is a carrier; the probability of an affected son is $\frac{1}{2}$.

(ii) In part (a) the probabilities of a daughter being carrier is $\frac{1}{2}$. The overall probability of an affected son is $\left(\frac{1}{2}\right) \times (0) + \left(\frac{1}{2}\right) \times \left(\frac{1}{2}\right) = 0 + \frac{1}{4} = \frac{1}{4}$.

Problem 21

A true breeding male fly with eosin eyes is crossed to a white-eyed female that is heterozygous for wild type (C) and cream alleles (i). What are the expected proportions of that offspring?

Solution:

The eosin gene is X-linked.

Wild type (C) is dominant.

Eosin-eyed male × White-eyed female

$$\frac{C}{C}\ \frac{X^{w.e}}{\rightharpoondown} \quad \times \quad \frac{C}{c^a}\ \frac{X^w}{X^w}$$

Gametes: $\underline{C}\ \underline{X^{w.e}}$, $\underline{C}\ \underline{\rightharpoondown}$, $\underline{C}\ \underline{X^w}$, $\underline{c^a}\ \underline{X^w}$

♀ \ ♂	$\underline{C}\ \underline{X^{w.e}}$	$\underline{C}\ \underline{\rightharpoondown}$
$\underline{C}\ \underline{X^w}$	$\frac{C}{C}\ \frac{X^{w.e}}{X^w}$	$\frac{C}{C}\ \frac{X^w}{\rightharpoondown}$
$\underline{c^a}\ \underline{X^w}$	$\frac{C}{c^a}\ \frac{X^{w.e}}{X^w}$	$\frac{c^a}{C}\ \frac{X^w}{\rightharpoondown}$

Results:

Genotype	Ratio	Phenotype	Ratio
1. $\frac{C}{C}\ \frac{X^{w.e}}{X^w}$	1	Wild type eosin female	2
2. $\frac{C}{c^a}\ \frac{X^{w.e}}{X^w}$	1	Wild type eosin female	
3. $\frac{C}{C}\ \frac{X^w}{\rightharpoondown}$	1	Wild type white-eyed male	2
4. $\frac{c^a}{C}\ \frac{X^w}{\rightharpoondown}$	1	Wild type white-eyed male	

Two light eosin female and two white-eyed male.

Note: Female that have one eosin and one white allele have light eosin eyes, whereas female with two eosin allele or males with one eosin allele have eosin eyes which are a bit darker than light.

Problem 22

In *Drosophila*, vermi lion eye colour is due to recessive allele (v) located on the X-chromosome. Curved wings is due to a recessive allele (cu) located on one autosome and ebony body is due to a recessive allele (e) located on another autosome. A vermilion male is mated to a curved ebony female and the F_1 males are phenotypically wild type. If these males were back-crossed to curved ebony females, what proportion of the F_2 offspring will be wild type males?

Solution:

The genotype of vermilion male is $\frac{v}{\rightharpoondown}\ \frac{+}{+}\ \frac{+}{+}$ and genotype of curved ebony female is $\frac{+}{v}\ \frac{cu}{cu}\ \frac{e}{e}$. After mating they produce wild type male.

vermilion male × curved ebony female

(♂) $\frac{v}{\rightharpoondown}\ \frac{+}{+}\ \frac{+}{+}$ × $\frac{+}{v}\ \frac{cu}{cu}\ \frac{e}{e}$

Gametes: $\underline{v}\ \underline{+}\ \underline{+}$, $\rightharpoondown$ + +; $\underline{+}\ \underline{cu}\ \underline{e}$, $\underline{v}\ \underline{cu}\ \underline{e}$

F_1: $\frac{+}{\rightharpoondown}\ \frac{cu}{+}\ \frac{e}{+}$ (wild type male)

(♂) $\frac{+}{\rightharpoondown}\ \frac{cu}{+}\ \frac{e}{+}$ × $\frac{+}{v}\ \frac{cu}{cu}\ \frac{e}{e}$ (♀) (back-crossed)

Gametes: (♂) $\underline{+}\ \underline{cu}\ \underline{e}$, $\rightharpoondown$ + +, $\underline{+}\ \underline{+}\ \underline{+}$, $\rightharpoondown\ \underline{cu}\ \underline{e}$, $\underline{+}\ \underline{cu}\ \underline{+}$, $\rightharpoondown\ \underline{+}\ \underline{e}$, $\underline{+}\ \underline{+}\ \underline{e}$, $\rightharpoondown\ \underline{cu}\ \underline{+}$

(♀) $\underline{+}\ \underline{cu}\ \underline{e}$, $\underline{v}\ \underline{cu}\ \underline{e}$

	1	2	3	4	5	6	7	8
	+ cu e	⇀ + +	+ + +	⇀ cu e	+ cu +	⇀ + e	+ + e	⇀ cu +
+ cu e	$\frac{+}{+}\ \frac{cu}{cu}\ \frac{e}{e}$	$\frac{\rightharpoondown}{+}\ \frac{+}{cu}\ \frac{+}{e}$	$\frac{+}{+}\ \frac{+}{cu}\ \frac{+}{e}$	$\frac{\rightharpoondown}{+}\ \frac{cu}{cu}\ \frac{e}{e}$	$\frac{+}{+}\ \frac{cu}{cu}\ \frac{+}{e}$	$\frac{\rightharpoondown}{+}\ \frac{+}{cu}\ \frac{e}{e}$	$\frac{+}{+}\ \frac{+}{cu}\ \frac{e}{e}$	$\frac{\rightharpoondown}{+}\ \frac{cu}{cu}\ \frac{+}{e}$
v cu e	$\frac{+}{v}\ \frac{cu}{cu}\ \frac{e}{e}$	$\frac{\rightharpoondown}{v}\ \frac{+}{cu}\ \frac{+}{e}$	$\frac{+}{v}\ \frac{+}{cu}\ \frac{+}{e}$	$\frac{\rightharpoondown}{v}\ \frac{cu}{cu}\ \frac{e}{e}$	$\frac{+}{v}\ \frac{cu}{cu}\ \frac{+}{e}$	$\frac{\rightharpoondown}{v}\ \frac{+}{cu}\ \frac{e}{e}$	$\frac{+}{v}\ \frac{+}{cu}\ \frac{e}{e}$	$\frac{\rightharpoondown}{v}\ \frac{cu}{cu}\ \frac{+}{e}$

F_2 = males:

Genotype	$\frac{\rightharpoondown}{+}\ \frac{+}{cu}\ \frac{+}{e}$	$\frac{\rightharpoondown}{+}\ \frac{cu}{cu}\ \frac{e}{e}$	$\frac{\rightharpoondown}{+}\ \frac{+}{cu}\ \frac{e}{e}$	$\frac{\rightharpoondown}{+}\ \frac{cu}{cu}\ \frac{+}{e}$	$\frac{\rightharpoondown}{v}\ \frac{+}{cu}\ \frac{+}{e}$	$\frac{\rightharpoondown}{v}\ \frac{cu}{cu}\ \frac{e}{e}$	$\frac{\rightharpoondown}{v}\ \frac{+}{cu}\ \frac{e}{e}$	$\frac{\rightharpoondown}{v}\ \frac{cu}{cu}\ \frac{+}{e}$
Fenotype	wild type	curved ebony	ebony	curved	vermilion	vermilion ebony curve	vermilion ebony	vermilion curved

Only $\frac{1}{8}$ wild type male.

Problem 23

A human female with turner syndrome also expresses the sex-linked trait haemophilia, as her father did, which parent underwent non-disjunction during meiosis giving rise to the syndrome?

Solution:

(i) Human female with turner syndrome (44 + X_0) expresses sex-linked haemophilic traites, i.e., 44 + X_hO.

(ii) She received X-chromosome from her father, i.e., X_h (22 + X_h).

(iii) Mother must have undergone nondisjunction during meiosis and thus contributed to the gamete (ovum) devoid of X-chromosome (22 + 0).

Problem 24

Colour blindness in human beings is an X-linked trait. A man who is colour blind has a tumer syndrome (44 + X_0 = 45). His daughter is also colour blind. Did the nondisjuntion that led to the

44 + X_0 (45) child occur in father or mother? How can you tell?

Solution:

(i) Colour blindness in human beings is an X-linked trait. It is given that the daughter having karyotype 45 X_0, contains colour blind allele.

(ii) She must have received X-chromosome from father through X_c-chromosome bearing sperm.

(iii) Nondisjunction must occur in mother which results in an egg cell deficient of X-chromosome.

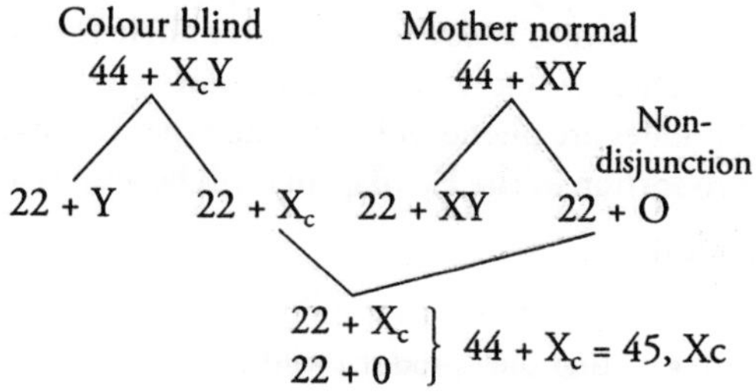

- **Sex-linked inheritance**
  - (i) The pattern of inheritance showing by the trait determined by a gene locating on the non-homologous part of the sex chromosome, (X-chromosome) is known as **sex-linked inheritance.**
  - (ii) It obeys **cris-cross** inheritance.

  Ex.: Colour blindness in man.
- **X-linked recessive**
  - (i) Usually more male than female are affected.
  - (ii) No offspring of an affected male are affected, making the trait skip generations in the pedigree, always with an unaffected female in the intermediate generation. **(An exception to this pattern occurs in the rare instance when the affected male mates with female carrier, produces affected female offspring).**
- **X-linked dominant**
  - (i) Affected male produces all affected female offsprings and no affected male offsrping.
  - (ii) Approximately half of the offsprings of affected females are affected, regardless of their sex.
- **Y-linked (holandric) inheritance**
  - (i) The pattern or mode of inheritance showing by the trait and determined by a gene locating on the nonhomologous part of the Y-chromosome is termed as Y-linked inheritance.
  - (ii) Transmission occurs from male to male.
  - (iii) It is commonly known as **holandric inheritance.**

  Ex.: Hairy pinna.
- **Sex-limited inheritance**
  - (i) The pattern or mode of inheritance shows by a trait which is expressed in only one sex not in other, even though the trait may not be X-linked.
  - (ii) Although the genes are present in both sexes and located on any of the autosomes but they manifest phenotypically on one sex not in other. The phenotypic expression is determined by the presence or absence of one of the sex hormones.
  - (iii) The sex limited trait in males could be transmitted through females to males in subsequent generation.
  - (iv) The sex limited genes are mainly responsible for **secondary sex character.**

  Ex.: Breast and ovary formation in female and facial hair distribution and sperm production in male.
- **Sex-influenced inheritance**
  - (i) The mode of inheritance shows by the trait whose phenotypic expression depends on the sex of the individual.
  - (ii) A heterozygote may express one phenotype in one sex/male and alternate phenotype in other sex/female (two recessive genes).
  - (iii) It has a dominant expression in male but recessive in female and depends on the sex of the individual.

  Ex.: The pattern of baldness in human.

Genotype	Phenotype	
	Female	Male
BB	Bald	Bald
BB	non-bald	Bald
bb	non-bald	non-bald

Sex Determination

- **Sexual Orientation**
- **Monoecious:**
  (i) Individual of one species produces both eggs and sperms.
  (ii) All individuals look alike.
  (iii) They carry both types of gonads (**testes** and **overies**), i.e., *hermaphrodite.*
- **Dioecious:**
  (i) Individual of one species exists in two sexual forms.
  (ii) One produces sperm and other produces egg
  (iii) Sperm producing individuals carry only testes and are called **male.** On the other-hand, egg producing individuals carry only ovaries and are called **female.**
  (iv) Generally males and females differ in morphology, i.e., shape, colour and ornamentations, etc. and this condition is called **sexual dimorphism.**

Sex differentiation:

The process by which individual of dioecious species becomes either male or female is called sex differentiation.

(i) The first and most crucial stage is the establishment of genetic sex of the zygote that occurs during fertilization. Genetic sex is determined by the nature of the chromosomes contributed by both sperm and egg.

(ii) Genetic sex differentiation establishes gonadal sex.

(iii) The next stage is the differentiation of somatic sex in embryo, i.e., development of potential gonadal ducts. Under the influences of hormone of differentiated gonad-genital ducts and specific sex gland develop.

(iv) The last and final stage of sexual differentiation is the overall body configuration with the appearance of male and female gonad specific sex appearance including development of secondary sexual characters which require input of specific hormones.

Mechanism of Sex determination

(i) In chromosomal or genotypic sex determination (CSD) the maternal and paternal chromosome complements unite, and zygote inherits a combination of gene.

(ii) In environmental sex determination (ESD) sex is determined after fertilisation by the influence of environmental factors like temperature, or location of development.

	Group of Animals	Mechanism	Sexes	
1.	Bonelia	ESD/Female attachment	Attached male	No attachment of female
2.	*C. elegans*	CSD/X : A ratio	XO male	XX Hermaphrodites
3.	*Drosophila*	CSD/X : A ratio	XY male	XX female
	Musca	CSD/M Dominant	M/m male	m/m female
	Apis	CSD/Haplodiploidy	Haploid male	Diploid female
4.	Turtle	ESD/Temperature	Cool male	Warm female
	Alligator	ESD/Temperature	Warm male	Cool female
5.	Bird	CSD/W dominant	ZZ male	ZW female
6.	Mammal	CSD/Y dominant	XY male	XX female

SEX DETERMINATION IN C. ELEGANS

Caenorhabditis elegans

The nematode worm *C. elegans* has become a popular organism for genetic studies. It has two sexual phenotypes, hermaphrodites and males. Hermaphrodites are essentially female animals that produce sperm during larval development and oocytes during adulthood. Males have only testes and females have both testes and ovaries.

Self-fertilization always produces more hermaphrodites (99%) and less male offspring (<1%). Outcome of cross-fertilization between hermaphrodites and male produces hermaphrodite (50%) and male (50%).

(i) Generally hermaphrodites are XX and males are XO with respect to X-chromosome, both have five pairs of autosomes. They lack Y-chromosome altogether.

(ii) The genetic signal that determines maleness in contrast to hermaphroditic development is provided by genes located on both the autosome and sex chromosomes.

(iii) It is believed that the ratio of X-chromosome to the number of sets of autosomes ultimately determines the sex of these worms.

(iv) When this ratio is 1 or greater, it results female or hermaphrodite phenotype, whereas when the ratio is 0.5 or less it develops male phenotype.

Hermaphrodite
↓
Hermaphrodite (>99%) (Self-fertilization) Male (<1%)
↓
(Cross-fertilization)
↓
Hermaphrodite (50%) Male (50%)

- **Process of sex determination in *C. elegans***

(i) *C. elegans* has two sexes: hermaphrodites and males. During larval development hermaphrodites produce sperm that are stored in spermatheca. In adulthood hermaphrodites are anatomically female and produce oocytes.

(ii) The sexual identity of *C. elegans*, hermaphrodite versus male, is a trait determined very early embryonic development. It is controlled by the activities of many genes that interact in a regulatory cascade that determines the sex of the worm and dosage compensation.

(iii) In *C. elegans*, XX genotype is hermaphrodite and are believed to be female in its soma (somatic tissue) and mixed in germ line (both male and female). XO genotype is male.

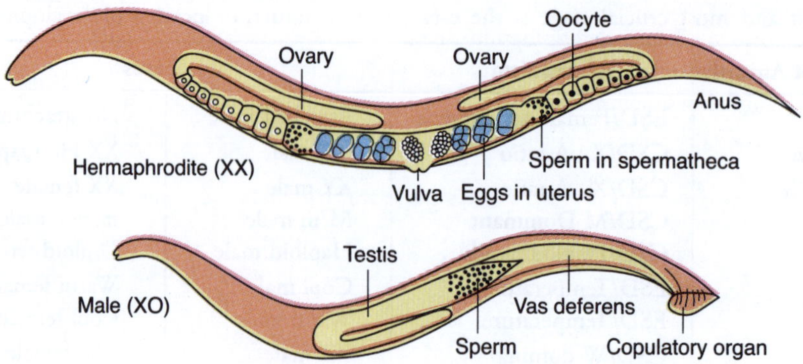

Fig. 4.1: Hermaphrodite and male *C. elegans*. The hermaphrodite has a 'two-armed' gonad and makes both eggs and sperm. The eggs are fertilized internally. The male makes sperm only.

(iv) In *C. elegans*, the expression of genes on the X-chromosome in the hermaphrodite, which carries two X-chromosomes, is decreased to 50% compared to males which carry one X-chromosome.

(v) Like *Drosophila* with minor variation, a similar pattern in X : A ratio was observed in this worm.

In C. elegans		***In Drosophila***	
Ratio	**Sex**	**Ratio**	**Sex**
$\frac{X}{A} = <0.67$	Leads to male	$\frac{X}{A} = 0.5$	Male
$\frac{X}{A} = <0.75$	Leads to female or hermaphrodite (female)	$\frac{X}{A} = 0.5$	Female

1. **$\frac{X}{A}$ ratio, sex-linked master control gene.**

(i) The primary signal for sex-determination is the ratio of X-chromosomes to sets of autosomes which causes XX animals to become hermaphrodites and XO animals to become males.

(ii) Early in development X : A ratio regulates the activity of XOL-1 (XO-lethal) an X-linked gene, i.e., a key developmental switch gene that controls both **sex determination** and **dosage compensation.**

(iii) XOL-1 encodes a noble protein. During early embryogenesis (during gastrulation) high levels of XOL-1 protein turn off the pathway of hermaphrodite development and thereby turning on the pathway of male development. XOL-1 genes appears to accomplish this by repressing the SDC (sex determination and dosage compensation) gene whose activity makes the animal hermaphrodite.

(iv) The male specifying XOL-1 transcript is not needed after the end of gastrulation.

2. **Autosomal regulatory genes for somatic and germ line sexual phenotype**

(i) A cascade of regulatory genes are located on the autosome of *C. elegans.* They also have dual effect on soma and germ line.

(ii) There are seven genes in this cascade, her-1 (hermaphrodite), tra-1, tra-2 and tra-3 (transformer), fem-1, fem-2 and fem-3 (feminization).

(iii) In hermaphrodite (XX), higher X:A ratio (X:A >0.75) enhances the expression of fox-1 and sex-1 genes, which inhibit the expression of XOL-1 genes.

(iv) The protein product of XOL-1 gene inhibit, the expression of SDC-1, SDC-2 and SDC-3 genes. In hermaphrodite the activity XOL-1 gene is inhibited. This permits the expression of SDC genes.

(v) The expression of SDC-1, SDC-2 and SDC-3 has two effects:

(a) These are necessary for **dosage** compensation

(b) They are necessary to permit downstream expression tra-1 gene, which is needed for **hermaphrodite** development. These occur through chain of events.

(vi) The **SDC** (1, 2, 3) gene products inhibit the expression of her-1. As a result, it (her-1) permits the expression of **tra-2** and **tra-3** genes.

(vii) When **tra-2** is active it inhibits the expressions of **fem-1, fem-2** and **fem-3** genes.

When the expression of **fem gene** is inhibited, it permits the expression of tra-1 gene.

Tra-1 gene products are essential for female somatic developments and promote hermaphrodite developments. They inhibit male development.

(viii) In XO (male) individual, lower X:A ratio (X:A = <0.67) the expression of fox-1 and sex-1 is insufficient to inhibit the expression of XOL-1 gene.

(ix) When XOL-1 gene is expressed, its protein product inhibits the expression of **SDC** (1, 2, 3) genes. This prevents dosage compensation and permits the expression of **her-1.**

(x) The her-1 gene product then inhibits the expression of **tra-2, tra-2** genes. This in term permits the expression of **fem**(1, 2, 3) genes. The active **fem**(1, 2, 3) gene inhibits the expression of tra-1 gene. As a result worm develops into male instead of hermaphrodite.

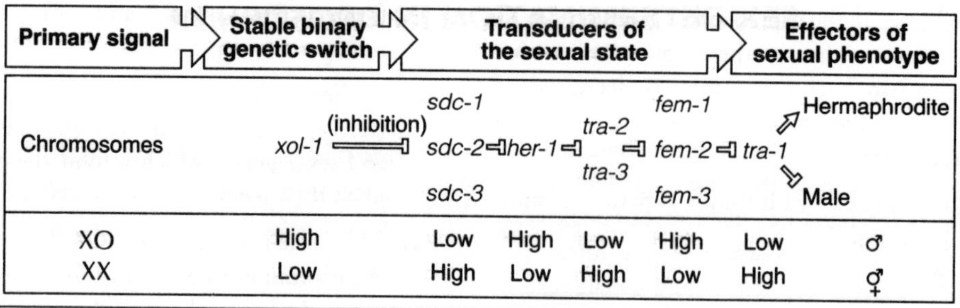

Fig. 4.2: **Outline of the somatic sex determination pathway in *C. elegans*.** The primary signal for sex determination is set by the number of X-chromosomes. When two X-chromosomes are present, the expression of the gene XO lethal (XOL-1) is low, leading to hermaphrodite development, whereas XOL-1 is expressed at a high level in males. There is a cascade of gene expression starting from XOL-1 that leads to the gene transformer-1 (tra-1), which codes for a transcription factor. If tra-1 is active, development as a hermaphrodite occurs, but if it is expressed at low levels, males develop. The product of the hermaphrodite-1 (her-1) gene is a secreted protein, which probably binds to a receptor encoded by transformer-2 (tra-2), inhibiting its function.

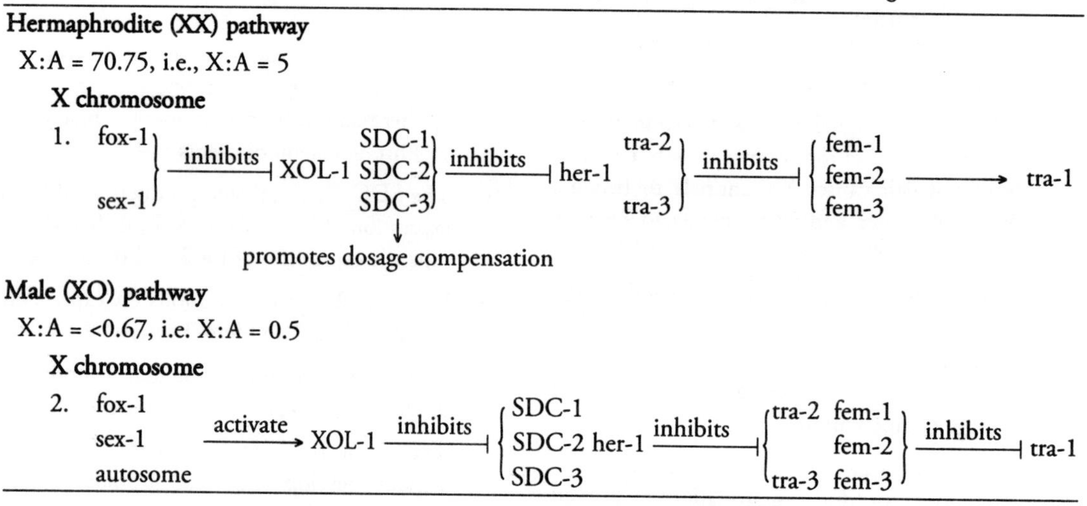

Fig. 4.3: Sex determination pathway for *C. elegans*

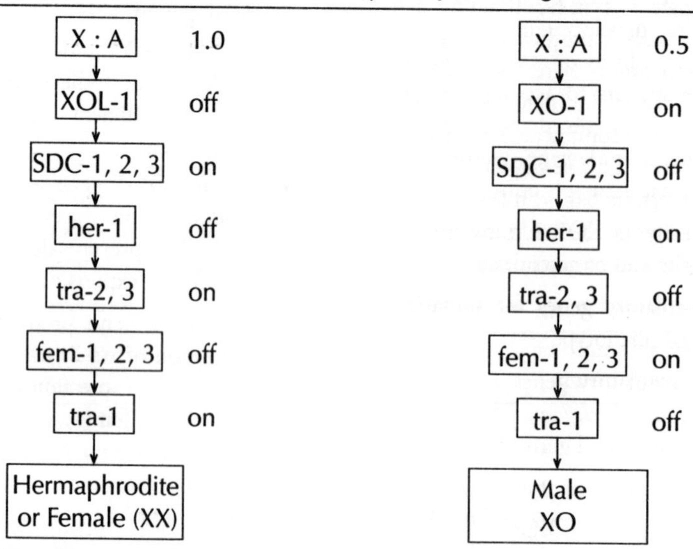

Fig. 4.4: Sex determination in *C. elegans* (a simplified scheme)

SEX DETERMINATION IN *DROSOPHILA*

Members of all species are often divided into two sections according to the kind of gamete or sex cells produced by them, i.e., male sex or female sex. The sex cells and reproductive organs form the primary sexual characters of male and female sexes.

(i) Besides these primary sexual character, male and female sexes differ from each other in many somatic characters known as secondary sexual characters. The phenomenon of morphological, physiological, molecular or behavioural differentiation between male and female sexes is called **sexual dimorphism**.

(ii) In *Drosophila molanogaster*, XX-XY types of sex determination is observed. The female possesses two homomorphic X-chromosomes in their body cell (hence referred to as XX) and they being **homogametic**, produce one kind of eggs, each with one X-chromosome. The male of these organism possesses one X-chromosome and one Y-chromosome (hence referred to as XY). The males having two heteromorphic sex chromosome (XY) and being **heterogametic**, produces two kinds of sperm half with X-chromosome and half with Y-chromosome.

(iii) The sex of an embryo depends on the kind of sperm. An egg fertilized by 'X' bearing sperm produces a female but if fertilized by a Y-bearing sperm a male is produced. Soon after the sex chromosomes were identified it becomes obvious that sex determination in *Drosophila* was more complicated than preliminary observation. In *Drosophila,* the presence of Y-chromosome has been found essential for the fertility of the male sex but has nothing to do with determination of male sex.

- **Bridges proposed that both X-chromosome and autosomes together play a critical role in sex-determination.**

Bridges' work can be divided into two phases:

(a) A study of offspring resulting from non-disjunction of the X-chromosome during meiosis in females.

(b) Subsequent work with progeny of triploid (3n) females.

(i) Non-disjunction is the failure of paired chromosome to segregate or separate during anaphase stage of the first or second meiotic divisions. It results the production of two abnormal gametes, one of which contains an extra chromosome (n + 1) and the other which lacks a chromosome (n − 1). Fertilization of such gametes produces aneuploid zygotes (2n + 1)(2n − 1).

(ii) Bridges found that the XXY flies were normal females and XO flies were sterile males. The presence of Y-chromosome in XXY flies did not cause maleness and its absence in the XO flies did not produce females. From this data, he concluded that Y-chromosome in *Drosophila* lacks male-determining factors but apparently contains genetic information essential to male fertility as the XO males were sterile.

(iii) Bridges was able to clarify the mode of sex determination in *Drosophila* by studying the progeny of triploid females (3n) which have three copies each of the haploid complement chromosome.

Effect of Sex Chromosome Dosage on Somatic Sexual Phenotype in Diploid *Drosophila* and Humans

Sex-chromosome constitution		Somatic sexual phenotype	
		Drosophila	Human
Euploidy:	XX	Female (♀)	Female (♀)
	XY	Male (♂)	Male (♂)
Aneuploidy:	XXY	Female (♀)	Male (♂)
	XO	Male (♂)	Female (♀)

Experiment

Requirements:

1. Triploid female (3A + XXX)
   (i) It has heavy-set bodies, coarse bristles and coarse eyes, and they may be fertile.
   (ii) Because of the odd number of each chromosome (three) during meiosis a wide range of chromosome complements is distributed into the gametes that give rise to offspring with a variety of abnormal chromosome constitution.
2. Diploid male (2A + XY).

Procedure:

1. He crossed the experimentally produced triploid (3A + XXX) female *Drosophila* to diploid males (2A + XY).
2. The results obtained from such crosses are as follows:

3A + XXX ♀ × 2A + XY ♂

Functional gametes (female)	Gametes (male): A + X	Gametes (male): A + Y
2A + XX	3A + XXX triploid female	3A + XXY triploid inter-sex
A + X	2A + XX diploid female	2A + XY diploid male
2A + X	3A + XX triploid inter-sex	3A + XY super or meta-male
A + XX	2A + XXX super or meta-female	2A + XXY diploid female

Chromosome composition	Chromosome formulation: Autosome set		Sex chromosome	Ratio of X-chromosomes to autosomes set	Sexual morphology
	2A	:	3X	$\frac{3X}{2A} = 1.5$	Meta-female
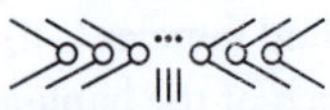	3A	:	3X	$\frac{3X}{3A} = 1.0$	Female
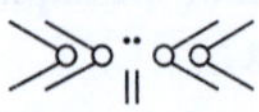	2A	:	2X	$\frac{2X}{2A} = 1.0$	Female
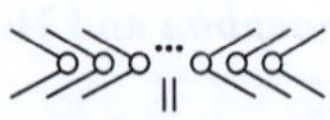	3A	:	2X	$\frac{2X}{3A} = 0.67$	Intersex
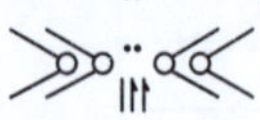	2A	:	XY	$\frac{XY}{2A} = 0.50$	Male
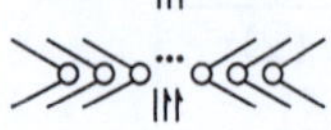	3A	:	XY	$\frac{XY}{3A} = 0.33$	Meta-male
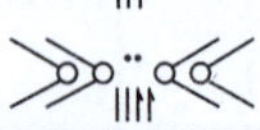	2A	:	2XY	$\frac{2XY}{2A} = 1.0$	Female

Inference:

Bridges realized that critical factor in determining sex is the ratio of X-chromosome to the number of haploid sets of autosomes present.

(i) Normal (2X : 2A) and triploid (3X : 3A) females each has a ratio equal to 1.0 and both are fertile.

(ii) As the ratio exceeds unity (3X : 2A or 1.5) what was originally called a **superfemale** is produced. Because this female is rather weak and infertile and has lowered viability, this type is now more appropriately called a **metafemale.**

(iii) Normal (XY : 2A) and sterile (XO : 2A) males each has a ratio of 1 : 2 or 0.5.

(iv) When the ratio decreases to 1 : 3 or 0.33 as in the case of an XY : 3A male, infertile **metamales** result.

(v) Other flies recovered by Bridges contained an X : A ratio intermediate between 0.5 and 1.0. These flies were generally larger and they exhibited a variety of morphological abnormalities and **rudimentary bisexual gonads** and **genitalia.** They were invariably sterile and were designated as **inter-sexes.**

Metamale	Male	Intersex	Female	Meta female
0.33	← 0.5 →	0.67	← 1.0 →	1.5

Key features regarding sex determination in *Drosophila*:

(i) Ratio between the number of X-chromosome and number of complete set of autosome will determine the sex.

(ii) The X-chromosome is believed to carry female tendency genes and male-determining factors are found in autosome.

(iii) Both these sets of genes start functioning and there has to be a balance between them for an individual to become male or female.

(iv) In one complete set of chromosome (A + X), if female tendency genes are more $\left(\frac{X}{A} = 1\right)$, it will be a female.

(v) A threshold of maleness reached when the X : A ratio is 1 : 2 (X : 2A) but the presence of an additional X(XX : 2A) alters the balance and results in female differentiation.

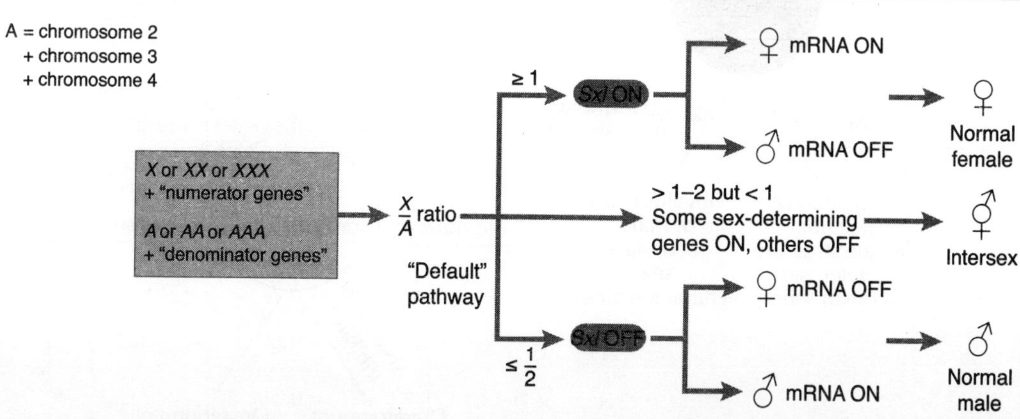

Early steps in the genetic control of sex determination in *Drosophila* through the activity of the Sex-lethal protein and ultimately through the "numerator" and "denominator" genes that signal the ratio of X-chromosomes to sets of autosomes (the X/A ratio).

SEX DETERMINATION IN MAMMALS (HUMAN)

In most sexually dimorphic species, the primary sexual phenotype is determined genetically by sex chromosomes. Secondary sex characteristics may also be determined using sex chromosomes or a alternatively by hormonal signalling.

Gonadal sex determination in mammals:

The expression of a mammal's *gonadal sex*—whether it will form testes or ovaries—necessitates the expression of the zygote genome and occurs relatively late in development. Genes on the sex chromosomes determine whether the gonads of the embryo will develop into normal testes or ovaries.

(i) The gonad primordium region in human embryo remains in an undifferentiated "inter-sex" state for about **five weeks**. During this period, the embryo develops both **Mullerian ducts** (the precursors of oviduct) and **Wolffian ducts** (the precursors of sperm ducts).

(ii) Up to sixth week, there are pairs of rudimentary gonads in the embryo which are neither those of a female nor those of a male but which are capable of developing into either ovaries or testes.

(iii) About sixth week, the presence of **Y-chromosome** (XY-embryos) triggers the embryonic gonads in the male pathway. Differentiation is then initiated in the direction of testes, and seminiferous tubules develop in their gonadal primordia.

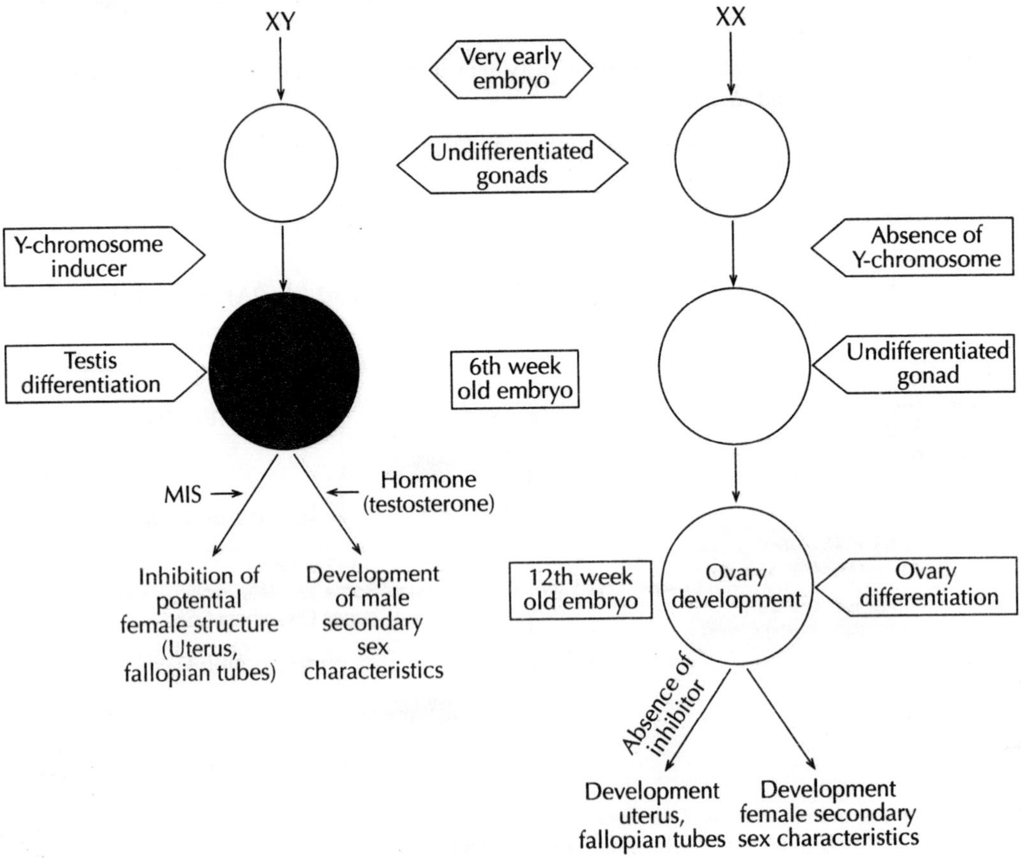

Fig. 4.5: Developmental pathway of testis and ovary.

(iv) In absence of Y-chromosome, gonads of XX embryo retain their undifferentiated gonadal state. Since there is no corresponding female inducing substance produced by the XX condition.

(v) In the absence of male inducer, the gonad will proceed into ovaries about twelfth week. In human, two X-chromosomes are required for normal ovarian development.

(vi) The genital-duct sex of a mammal normally includes a penis and sperm duct in the male and a vagina, uterus and oviducts in the female. The Wolffian and Mullerian primordia for these organs exist in the embryo before its gonadal sex is determined.

(vii) After gonadal sex is determined, two events must occur; the "appropriate" primordia must go on to develop into fully formed structures and the "inappropriate" primordia must degenerate.

(viii) After the development of testis at least two signals testosterone and Mullerian inhibiting substance (MIS) are released from the testis and direct male specific development of genito-urinary system. **Testosterone** regulates male secondary sex characters. **Mullerian inhibiting substance** (MIS) is required to ensure degeneration of the Mullerian duct.

(ix) In the absence of male inhibitor, i.e., **MIS**, the embryonic parts can develop into uterus and fallopian tube.

Mechanisms of Sex determination

The mechanisms of sex determination may be classified into the following three main groups:

1. **Environmental**

(i) In some cases, sexual differentiation is determined by environmental factors. For example, larvae of the sea worm Bonelia are sexually undifferentiated.

Only females develop at high temperature (30-35°C) and only males develop at low temperatures (23°C). In some turtles incubation of egg at intermediate temperatures (28-30°C) produces both male and females. This situation is found in most species of turtles.

(ii) In other animals (crocodiles, alligators and lizard) develop only males at high temperature (>33°C) and only females, at low temperature (<30°C). Intermediate temperatures (30-33°C) produce both males and females.

(iii) Australian crocodile, snapping turtle and some other turtle species. Only females develop both at high temperature (30-35°C) and low temperature (<25°C). At intermediate temperatures, both males and females are produced, but at some intermediate temperatures, only males are produced.

- The effect of temperature on estrogens, androgens and inhibitors of the enzymes controlling their synthesis are involved in the sexual diferentiation of ovaris and testes.
- The enzyme aromatase converts male hormone (testosterone) to estrogen (estradiol) and its activity of this enzyme is correlated with gonodal differentiation pathway.
- The thermosensitive factor mediates transcription of the reptilian aromatase gene, which leads to temperature dependent sex determinations.

2. **Chromosomal**

3. **Genic mechanisms of Sex determination.**

HUMAN Y-CHROMOSOME

It has been observed that only a very small part of the short arm of the Y-chromosome is endowed with the gene(s) for sex determination and gonadal differentiation.

Regions of human Y-chromosome:

(i) Both ends of the Y-chromosome are called **pseudoautosomal regions (PARS)** that share homology with the regions of X-chromosome and which synapse and recombine with it during meiosis.

(ii) The remainder of the chromosome (about 95%) does not synapse or recombine with X-chromosome. It is designated as nonrecombining regions of the Y (NRY). It is also referred to as **male specific region of the Y (MSY).**

(iii) The MSY is divided about equally between euchromatic regions containing functional genes and heterochromatic regions lacking genes.

(iv) Within euchromatin, adjacent to the PAR of the short arm of the Y-chromosome, is a critical gene that controls male sexual development, called the **sex-determining region Y (SRY).**

(v) SRY encodes a gene product that somehow triggers the undifferentiated gonadal tissue of the embryo to form testes. This product is called **testis-determining factor (TDF).**

(vi) SRY is the gene responsible for the human male sex determination. In human, other autosomal genes are believed to be part of a cascade of genetic expression initiated by SRY. It also includes **SOX 9** gene and **WT-1** (on **chromosome 11**). Another **SFI (orphan steroid factor-1)** gene is involved in the regulation of enzyme affecting steroid metabolism.

(vii) The MSY consists of about 23 million base pairs (23 Mb) and can be divided into three regions:

(a) **X-transposed region:**

(i) It contains 15% of MSY.

(ii) Two genes, both with X-chromosome homologs, are present in this region.

(b) **X-degenerative region:**

(i) It contains about 20% of the MSY.

(ii) It contains 27 single copy genes including numerous **pseudogenes.**

(iii) Within 27 genetic units, 14 are capable of being transcribed. SRY is one of the units of its expressed only in the testes.

(c) **The ampliconic region:**

(i) It contains about 30% of the MSY.

(ii) Most of the genes are closely associated with testes development.

(iii) There are 60 transcription units divided among 9 gene families in this region, most represented by multiple copies.

(iv) Each repeat unit is an **amplicon** and is contained within seven segments scattered across the euchromatic

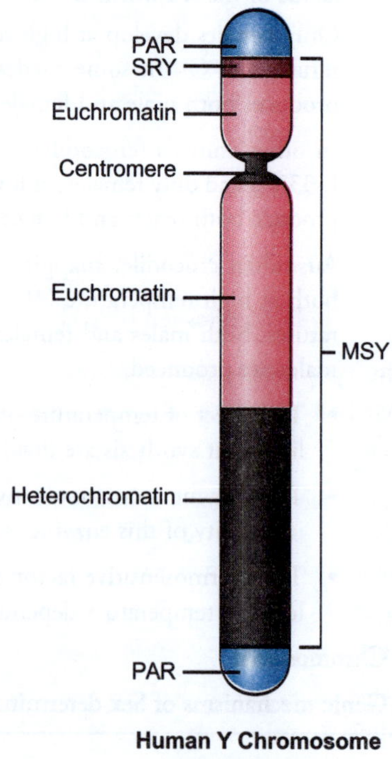

Fig. 4.6: The various regions of the human Y-chromosome

regions of both long and short arms of the Y-chromosome.

(v) Genes on the amplicon encode protein are specific to the development and function of testes. Many of these genes are directly related to fertility.

(vi) **SRY** is the **key gene** that switches the bipotential genital ridge towards testis determination.

SRY gene has been located in the sex-determining region of the Y-chromosome. It is located on the Y-chromosome and has **250 bp** and code for **80 amino acids.** The SRY protein is a transcription factor. Its role may be to activate genes encoding other transcription factors including **SOX-9** and **SF-1** which are also expressed in the male gonadal ridge and in whose absence the XY male is **feminized.**

(vii) **SOX-9** has an important role in testis differentiation as well as subsequent organogenesis of the male genital system.

SF-1 transcription factor regulates the gene encoding **Mullerian inhibiting substance (MIS)** as well as genes involved in the testosterone synthesis.

SF-1 and **SOX-9** bring about the activation of MIS gene whose product leads to regression of Mullarian duct.

Highlight the molecular mechanism of Y-chromosome in Sex determination of Mammals

In most mammals, such as humans, mice and marsupials, the presence of SRY gene on the Y-chromosome determine maleness.

(i) The SRY gene located on the Y-chromosome regulates the male sex determination pathway.

(ii) It encodes a protein that contains a DNA binding domain HMG box (**High Mobility Group**). HMG box is known to cause DNA bends.

(iii) It may act as transcription factor and promote changes in DNA architecture.

(iv) Several genes in mammals have been identified that are expressed very early in embryonic development and may be directly or indirectly involved in turning on the **SRY gene.**

(v) Two genes **SF-1** (stereoidogenic factor-1) and **WT-1** (Wilm's tumour gene-1) are expressed in the early embryo prior to sexual differentiation and may regulate SRY gene expression.

(vi) SRY gene expression activates the other gene to promote male development.

(vii) SOX-9 (**SRY-like HMG box**), DMRT-1 (**double sex and mab-3 related transcription factor**) and DAX-1 (**dosage-sensitive sex reversal, adrenal hypoplasic congenital × Chromosome**) genes may also play significant role in the sex determination.

(viii) SOX-9 protein contains two transcriptional activation domain. SOX-9 expression is necessary for male development because a loss of function allele of SOX-9 results in female developmnt.

It has been reported from research that in XX mice with three copies of the SOX-9 gene develop into male.

(ix) DMRT-1 gene in mammals is evolutionary related to the mab - 3 gene in *C. elegans* and the DSX gene in *Drosophila.* These genes encode transcription factors involved in differentiation of gonad.

In mammals DMRT-1 expression is specific and it is expressed in higher levels in testes. XY mice lacking DMRT-1 are infertile.

(x) The DAX-1 gene is X-linked and its gene product is thought to prevent male development.

In XX animals its expression remains high in contrast to XY males.

XY animals with two copies of the DAX-1 gene can develop into females when SRY gene expression is low. This suggest that DAX-1 can inhibit the effects of SRY gene.

(xi) DAX-1 gene encodes a hormone receptor protein.

Sex determination pathway for mammals

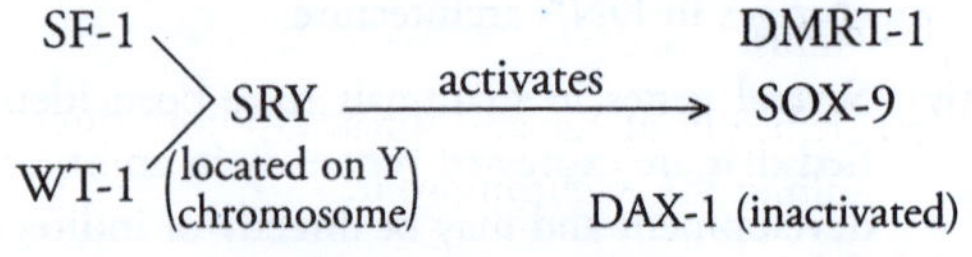

(a) Male pathway (XY)

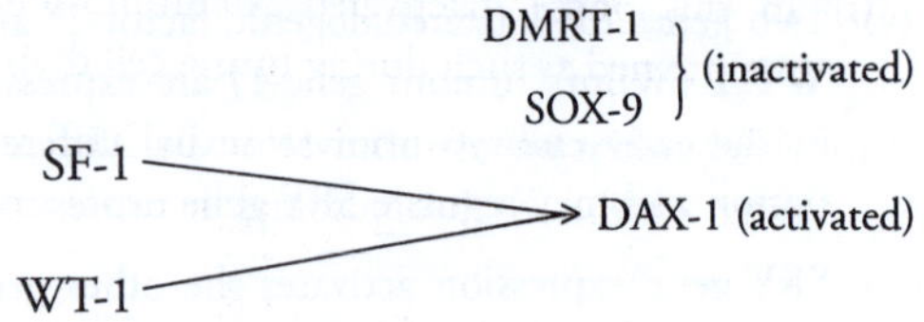

(b) Female pathway (XX)

Human X-Chromosome

(i) It has long (q) arm and short (p) arm. Its two ends are known as **pseudo autosomal regions** (PARp and PARq).

(ii) PARp is a 2.7 Mb region at the terminas of short arm and PARq is a 3.0 Mb region at the terminas of the long arm.

(iii) In normal females two X-chromosomes are counted and one is inactivated while in male one X-chomosome is counted and none is inactivated.

(iv) The gene inactivation in the inactive X-chromosome affects individual genes and blocks of genes. Some genes in the inactive X are not silenced. Tips of the long arm (PARq) and short arm (PARp) of X are escaped from inactivation.

1. XIC (X-Inactivation Centre)

(i) The constant expression of XIC (Xic) and X-chromosome inactivation supports the conclusion that this region is an important genetic component in the inactivation process.

(ii) XIC (Xic) is about 1 Mb (10^6bp) in length and is known to contain several putative regulatory units and four genes.

2. XIST (Xist) [X-Inactive Specific Transcript]

(i) It is believed to represent the critical locus within XIC.

(ii) The RNA product of this gene is quite large and lacks what is called an extended open reading frame (ORF).

(iii) An ORF includes the information necessary for translation of the RNA product into protein. Thus the RNA is not translated but instead serves a structural role in the nucleus, presumably in the mechanism of chromosome inactivation.

(iv) This finding has led to believe that the RNA products of XIST spread over and coat the X-chromosome bearing the gene that produced it, creating some sort of molecular cage that entraps it, leading to its inactivation.

(v) Inactivation is, therefore, cis acting.

(vi) Transcription of XIST occurs initially at low levels on all X-chromosomes. As the inactivation process begins, however, transcription continues and is enhanced only in the X chromosome(s) that becomes inactivated.

3. Xce (X-chromosomal controlling element)

(i) It is located very close to the end of XIC or may even lie within Xic.

(ii) It serves as binding site for proteins that regulate the expression of genes in Xic, such as Xist of Tsix.

(iii) Genetic variation in Xce that enhances Xist expression would tend to promote Barr body formation, whereas Xce variation that enhances Tsix, an expression would tend to prevent X inactivation.

(iv) The Tsix is regulated by Xce region and is expressed only during early embryonic development. It encodes an RNA which is complementary to the Xist RNA but does not synthesize protein. Therefore, it is called *anti-sense RNA*. It transcribes in opposite direction.

(v) The Xist RNA is sense RNA, becase it promotes inactivation of X-chromosome. On the other hand, Tsix antisence RNA binds for Xist RNA to inhibit its inactivator function.

- **Process of Chromosome inactivation**

The process of X inactivation occurs in three phases: **Initiation, Spreading** and **Maintenance**.

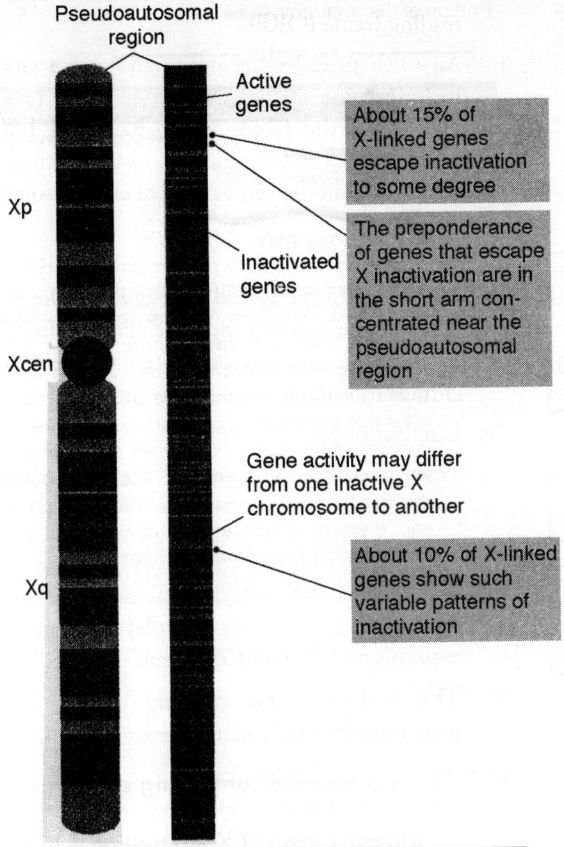

Fig. 4.7: Distribution along the human X-chromosome of the approximately 15 percent of genes that escape complete transcriptional silencing in the inactive X-chromosome [Data from L. Carrel and H. F. Willarm, *Nature* 434 (2005): 400-404]

A. **Initiation:**

(i) In this phase, one of the X-chromosomes is targeted to remain active and other is chosen to be inactivated.

(ii) X-chromosome inactivation begins at the X-inactivation centre (Xic) located near the centromere of the long arm between $Xq^{11.2}$ and $Xq^{21.1}$ which contains **Xist** gene.

B. **Spreading:**

(i) In this phase, the chosen X-chromosome is inactivated.

(ii) This spreading requires the expression of Xist gene.

(iii) The Xist RNA coats the inactivated X-chromosome and recruits protein that promote compaction through modification of histone proteins.

(iv) It begins at Xic and spreads in both directions along the X-chromosome.

C. **Maintenance:**

(i) In this phase inactivated X-chromosome is maintained as such during future cell division.

(ii) After initiation of inactivation by Xist, silencing of the X-chromosomal gene is maintained by two ways:

(a) In the active X-chromosome, Xist locus becomes methylated. On the other hand, Xist locus is unmethylated in the inactive X-chromosome. Promoter regions of numerous genes are methylated in the inactive X-chromosome but in nonmethylated in active X chromosome.

(b) Inactive X-chromosome shows depletion of acetylated histone H4.

Xist RNA is sense RNA and promotes inactivation of X-chromosome. On the other hand, Tsix RNA is antisense, i.e., complementary to Xist RNA can able to bind to Xist RNA to inhibit its inactivation function, which leads to degradation of Xist RNA (Fig. 4.8).

Xist and Tsix

Xist

(i) Xist (X-inactive specific transcript) gene is located in the Xic of inactivated X-chromosome.

(ii) They are active, but the other genes in the inactivated X-chromosomes are silenced.

(iii) The Xist gene product is an RNA molecule that does not encode protein.

(iv) The role of Xist-RNA is to coat the X-chromosome and inactivate it. After coating other protein associates with Xist RNA it promotes chromosomal compaction into a **Barr body**.

Tsix

(i) This gene is also found within Xic. It transcribes in opposite direction of Xist (the name Tsix is Xist spelled backwards).

(ii) It function appears to decrease the expression of Xist gene.

(iii) In inactive X-chromosome Xist gene is expressed and the Tsix gene is not. On the other hand, in active X-chromosome Tsix is expressed but Xist gene is not expressed.

(iv) Xist RNA act as sense RNA and Tsix RNA acts as antisence RNA.

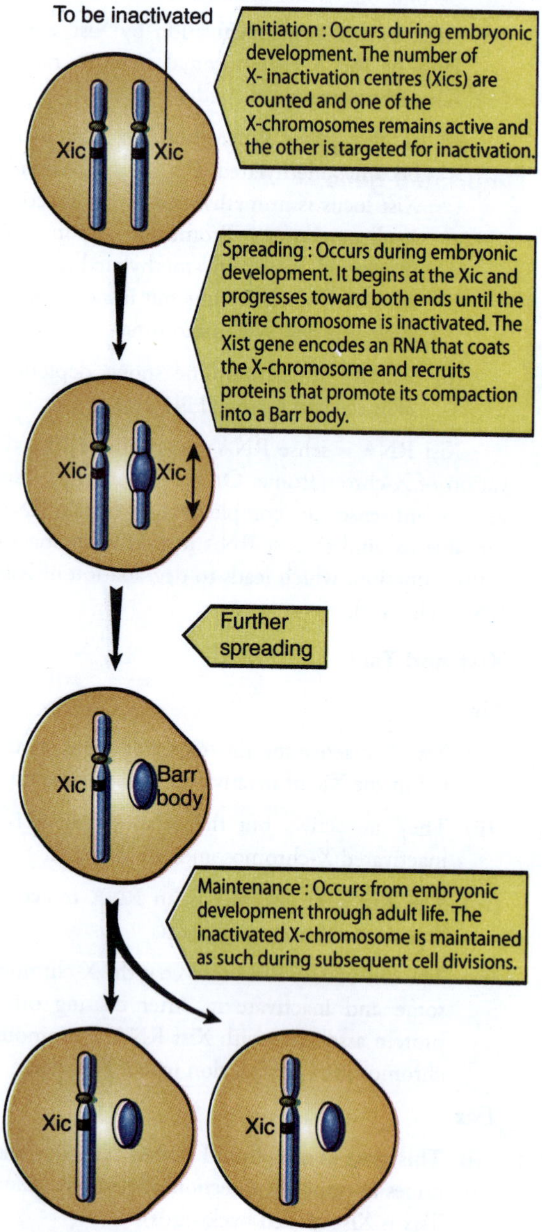

Fig. 4.8: The function of the Xic during X-chromosome inactivation.

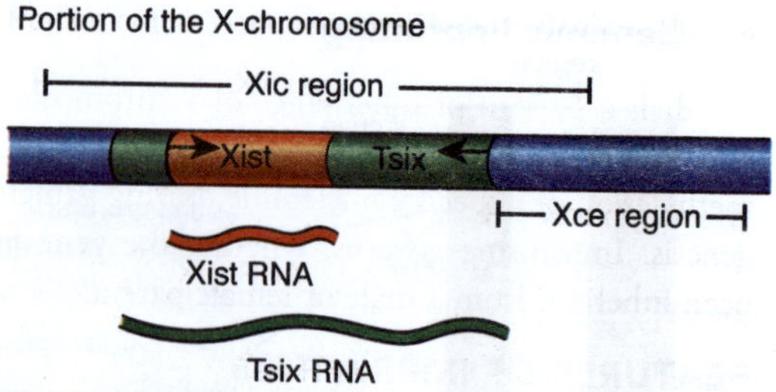

Fig. 4.9: The X inactivation centre (Xic) of the X-chromosome. The Xist gene is transcribed into RNA from the inactive X-chromosome but not from the active X-chromosome. This Xist RNA binds to the inactive X-chromosome and recruits proteins that promote its compaction. The precise location of the Xce is not known, but it may be wthin the Xic or adjacent to it. Xcc may regulate the transcription of the Xist and/or Tsix genes and thereby influences the choice of the X-chromosome that remains acitve.

(v) The choice of which X-cromosome is to inactivate, involves a complex interplay between the expression of Xist and Tsix gene.

(vi) The X-chromosome carrying mutant Tsix gene is preferentialy inactivated.

XCC (X-chromosomal controlling elements)

Molecular basis of X inactivation

(1) Counting, (2) Initiation, (3) Propagation (spreading) and (4) Maintenance

	Mammal	Human
Inactivation centre =	Xic	XIC
Inactivation key gene =	Xist	XIST

- ## Genetic Anticipation:

It is a phenomenon of heritable disorders which exhibit a progressively earlier age of onset, and the severity of the symptoms of the disorder increases in each successive generation.

It is caused by the expansion of trinucleotide repeats within or near the gene.

Example:

Myotonic dystrophy (MD)

Individual afflicted with this autosomal dominant disorder exhibits extreme variation in the severity of symptoms.

- **Genomic Imprinting:**

It is a pattern of inheritance of a chromosome or a gene in nucleotide sequene often associated with methylation of a specific nucleotide, during gametogenesis. Imprinting governs, whether the gene has been inherited from a male or female parent.

FEATURES OF IMPRINTING

(i) It is also termed as monoallelic expression or allelic exclusion and is consistently maintained in specific tissue.

(ii) At an early stage of development, the allele form of one parent is selectively inactivated.

(iii) Imprinting step is thought to occur before or during gamete formation, leading to differentially marked genes (or chromosomal region) in sperm-forming versus egg-forming tissues.

(iv) DNA is methylated in gametes but between 6 to 8 hours of fertilization, the pronucleus from the sperm is actively demethylated. The maternal DNA is not methylated after replication. It is passively demethylated (demethylated by dilution) at each mitosis.

(v) Imprinting is controlled at imprinting centre. At least twenty (20) imprinted genes have been identified in mice and humans. The majority of imprinted genes are mapped in three clusters. There is evidence that specific imprinting boxes in each cluster are required in cis to regulate parent specific expression.

(vi) The imprinting boxes are differentially methylated in the male and female gametes and direct the parent specific gene expression patterns.

In human, there are two large imprinted regions, one on chromosome $11p^{15.5}$ and other on chromosome $15q^{12}$.

Example:

Inactivation of one of the X-chromosome in mammals

(i) In mice (prior to development of the embryo, imprinting occurs in tissue such that) the X-chromosomes of paternal origin is genetically inactivated in all cells while the genes of the maternal X-chromosome remain genetically active.

(ii) Methylation is controlled by three enzymes two **methylases** and one **demethylase.** First methylase is known as denovo methylase which recognise unmethylated DNA at the region DMRs to make it hemimethylated state.

(iii) Second one is **maintenance methylase** which convert hemimethylated sites to fully methylated form.

(iv) The third enzyme **demethylase** which eliminate methylated portion of cytosine in DNA.

Imprinted gene

The gene for insulin like growth factor type two (IGF-2) located in region $11p^{15.5}$, is active on the paternal chromosome.

A gene for insulin like growth factor receptor (IGFR) removes and counteracts. IGF_2 is active in maternal chromosome.

Key features of Imprinting:

- Imprinting occurs in germ line.
- It affects almost a few hundred genes (many of them are located in clusters).
- It is accompanied by heavy methylation.
- Imprinted genes are differentially methylated in the female and male germ lines.
- Once imprinted and methylated, a silenced gene remain transcriptionally inactive during embryogenesis.
- Imprints are erased early in germ line development. Then later they are re-established according to sex-specific patterns.

Stages of Imprinting:

- Establishment phase (at fertilisation)
- Maintenance of imprint (throughout the developmental period)
- Erasure and reestablishment (during gametogenesis)

Molecular mechanism:

Nearer the imprinted genes on there are some regions called differentially methylated region (DNR).

Table 4.1: Comparison—Sex determination in *Drosophila* and Man

Drosophila	Man
1. Autosome and X-chromosome determine sex.	1. Autosomes are not responsible for the determination of sex.
2. Sex is determined by the ratio of X-chromosomes to autosomes as proposed by Bridges (1916). X-chromosome is believed to possess the **feminizing** genes and autosomes possess **masculine** genes (genic balance theory).	2. The ratio of autosomes and sex chromosomes does not determine sex. Studies on various non-disjunction types of sex chromosome indicate that X-chromosome possesses **feminizing genes** and Y-chromosome possesses the **masculine genes.**
3. Y-chromosome has **no role** in sex determination.	3. Y-chromosome plays a **vital role** in sex determination, Y determines male sex.
4. Increase or decrease of X-chromosome (besides autosome) does not help in determination of female sex.	4. Increase or decrease of X-chromosome helps in determination of female sex.
5. tra-genes located on 3rd autosome in homozygous condition convert normal female into sterile male. $\frac{tra}{tra}\ \frac{x}{x} \rightarrow$ Sterile male.	5. tra-gene has no such role.
6. Barr body has no role in determination of sex.	6. The number of Barr bodies determine male or female sex.
7. Individuals with two sets of autosomes and one X-chromosome (2A + X) develop into males without a Y-chromosome $\left[\frac{X}{2A} = \frac{1}{2} = 0.5\right]$. However, they are **less fertile** than XY male.	7. Individuals with only X-chromosome (44A + XO) develop in **Turner syndrome** with feminine appearance. Chromosome no. 45 (44 + XO). Sex organ does not develop. Menstrual cycle is absent. Short stature with webbed neck and infertile.
8. Individuals with two sets of autosomes, two X-chromosomes and a Y-chromosome (2A + XXY) are **normal females** with an additional Y-chromosome $\left(\frac{2X}{2A} = 1\right)$.	8. Individuals with XXY (i.e., 44 + XXY = 47) become **Klinefelter syndrome** with **masculine appearance.** Gonadal dysgenesis, testis small, spermatogenesis absent, some develop female-like breast (**Gynecomastia**).
9. Individuals with two sets of autosomes, three X-chromosomes, a Y-chromosome (2A + XXXY) are super-females with an additional Y-chromosome $\left(\frac{3XY}{2A} = 1.5\right)$.	9. Individuals with XXXY (i.e., 44 + XXXY) become (tetrasomic) **klinefelter syndrome** with masculine appearance, mentally retarded with small atrophied testis.
10. Individuals with two sets of autosomes and three X-chromosomes develop into **super-female** or **meta-female** with poor viabilities $\left(\frac{XXY}{2A} = \frac{3}{2} = 1.5\right)$.	10. Individuals with XXX (44 + XX) become meta-female or triple X-female with normal gonads and show no somatic abnormality but intelligence is slightly impaired.
11. Individuals with two sets of autosomes, X-chromosome and double Y-chromosomes are males (i.e., 2A + XYY = 0.5).	11. Individuals with XXY are males (44 + XYY) normal with or without fertility (Swanson et. al., 1967).
12. Sex determination is called **polygenic.**	12. Sex determination in man is **monogenic.**

Contd.

Contd. Table 4.1:

Drosophila	**Man**
13. TDF has no role in sex determination.	13. TDF (testis-determining factor) gene is the master sex determining locus in the human Y-chromosome.
14. Sex determination seems to be **quantitative type.** No role of Y-chromosome.	14. Sex determination is of **qualitative type.** Y-chromosome has male-determining (TDF) gene.

A. Comparison of the Features of Sex Determination in Drosophila and Mammals

Aspect of Sex determination	***Drosophila***	**Mammals**
1. Chromosomal basis of XX versus XY.	X : A ratio.	Presence or absence of Y.
2. Molecular basis of XX versus XY.	Numerator: denominator transcription factor sub-unit interactions.	Transcription factor for testis determination located on Y-chromosome.
3. Decision-making process.	Cell by cell.	Testis acts as source of male-determining organ instructing other tissues.
4. Switch.	Transcription of Sxl (sex lethal) gene in early embryo.	Presence or absence of SRY (human) or Sry (mouse) in gene.
5. Level of downstream regulation.	Alternative mRNA splicing.	Transcription.
6. Positive feedback mechanism development.	Positive feedback loop of 'SXL' protein on Sxl mRNA splicing.	Continued presence of testis throughout.
7. Elaboration of sex specific gene expression.	(i) Presence of male or female specific form of dsx (double sex) encoded protein. (ii) Generally DSX-F is a repressor of expression of male specific genes and DSX-M is a repressor of female specific genes. (iii) For a few genes, DSX-F or DSX-M may act as a activators of female specific or male specific gene expression.	(i) Testosterone-mediated activation of androgen receptor in target tissues. (ii) Testosterone-androgen receptor regulates transcription by binding to testosterone responsive elements (enhancers and silencers).

Twins: Two or more offsprings (babies) are born by the same pregnancy. Such babibies are called twins or multiple births.

- **Identical or Monogenetic twin:**

  Twins arise from splitting of a single fertilized egg, and have same genetic make up, and same sex.

- **Non-identical or fraternal:**

  Twins arise from two fertilized eggs and have different genetic make up and of same in different sex.

Concordance: Presence of same trait in both the members of a pair of twin.

Discordance: One express the trait and other does not.

EXERCISE

1. Discuss briefly the regulatory cascades for the sex determination in *Drosophila*.

Solution:

Sex in *Drosophila* is determined by the ratio of X-chromosomes to autosomes (X : A). When the ratio is 0.5 (1X : 2A), males are produced, even when no Y-chromosome is present, when the ratio is 1.0 (2X : 2A) females are produced. At intermediate ratios, i.e., 0.67(2X : 3A), inter-sexes are produced.

(i) The chromosomal ratios are interpreted by a small number of genes that initiate a cascade of developmental events resulting in the production of male or female somatic cells and corresponding to male or female phenotype.

(ii) The genes in this pathway are sex lethal (Sxl), transformer (tra), transformer-2 (tra-2), double sex (dsx) and inter-sex (ix).

(iii) In females, the chromosomal ratio (X : A = 1) activates Sxl gene and the production of SXL protein. This activation involves expression of several intermediate genes sis-a, sis-b (sisterless) and da (daughterless).

(iv) In males, the chromosomal ratio (X : A = 0.5) does not activate feminizing switch gene Sxl (sex lethal) and the production of SXL protein.

(v) When the SXL protein is present, it binds to and controls the splicing of the pre-mRNA for the next gene in the pathway, transformer (tra). The pre-mRNA of tra includes a terminator codon (UGA) in exon 2. The female specific SXL protein binds to the tra pre-mRNA and directs the splicing to remove exon 2 from the mature mRNA.

(vi) In males, where no SXL protein is present, splicing proceeds along the ground state or default pathway, and the stop codon is incorporated into the mature mRNA. In males, translation of tra mRNA is prematurely terminated by the stop codon (**UGA**), resulting in an inactive gene product.

(vii) Whereas in female translation produces a functional gene product. The result is a female specific tra protein that acts in conjunction with the tra-2 gene to produce female phenotype.

(viii) The **tra-2 gene** is active in both males and females and contains an RNA binding domain that binds to the pre mRNA of the dsx gene, but only in the presence of tra protein (TRA).

(ix) The **dsx gene** is a critical control point in the development of sexual phenotype, for it produces functional mRNA and protein for both males and females. However, the pre mRNA is processed in a sex specific manner to yield different transcripts.

(x) In females, the tra protein and the tra-2 protein bind to the dsx pre-mRNA and it directs its processing in an alternative, female specific manner. The female **dsx protein (DSX-F)** causes female sexual differentiation in somatic tissues by repressing male sexual development in conjunction with the protein encoded by **ix (inter-sex) gene.**

(xi) In males, where only the tra-2 protein is present, default splicing of the dsx pre-mRNA results in a male specific mRNA and protein product. The male dsx **protein (DSX-M)** brings about the male sexual development by suppressing the female pathway.

(xii) In the absence of **dsx function**, flies develop into phenotypic inter-sex.

2. Role of Sxl gene in sex determination of *Drosophila*.

Solution:

In *Drosophila* sex determination, the master switch is the transcriptional regulation of the Sxl gene during early embryogenesis.

(i) In a fly with an X : A ratio of 1, SXL protein is synthesized and the fly develops as a female. In a fly with an X : A ratio of 0.5, no functional SXL protein is synthesized and consequently, the fly develops as a male.

(ii) The Sxl gene is essentially a "toggle switch" that is permanently locked into "on" position in females and "off" position in males.

A. In case of female:

(i) To set the Sxl switch in the "on" position, the level of the active X : A NUM (numerator elements, i.e., genes on X-chromosome, viz. sis a, sis b) transcription factors must be high (owing to an X : A ratio = 1). With high (female) levels,

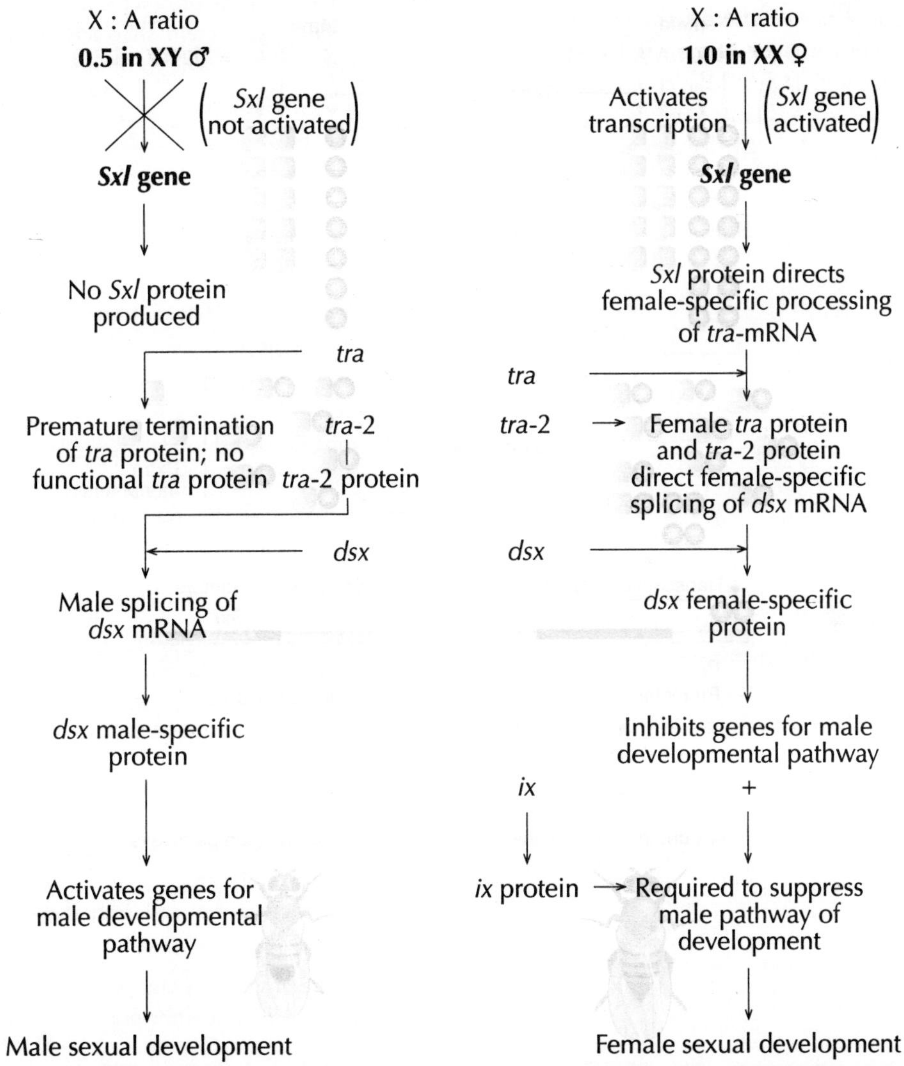

Fig. 4.10: Hierarchy of gene regulation for sex determination in *Drosophila*. In males, the X : A ratio does not activate Sxl. The result is a male-specific processing of tra pre-mRNA, resulting in no functional tra protein. This is turn leads to male-specific processing of dsx transcripts, resulting in male-specific dsx protein that activates the pathway for male sexual development. In females, the X : A ratio activates transcription of the Sxl gene. The product of this gene binds to pre-mRNA of the tra gene and directs its splicing in a female-specific fashion. The female-specific tra protein, in combination with the tra-2 protein, directs female-specific splicing of dsx pre-mRNA, resulting in a female-specific dsx protein. This protein, in combination with the ix protein, suppresses the pathway of male sexual development and activates the female pathway.

the X:A transcription factors present in the early embryo bind to enhancers of the Sxl gene, activating its transcription from the Sxl early promoter (P_E) shortly after fertilization. The early Sxl promoter (P_E) produces active SXL protein.

In contrast, if the levels of NUM factors are too low (when X : A ratio is 0.5), then there is insufficient transcription factor to activate Sxl transcription and so no SXL protein is produced.

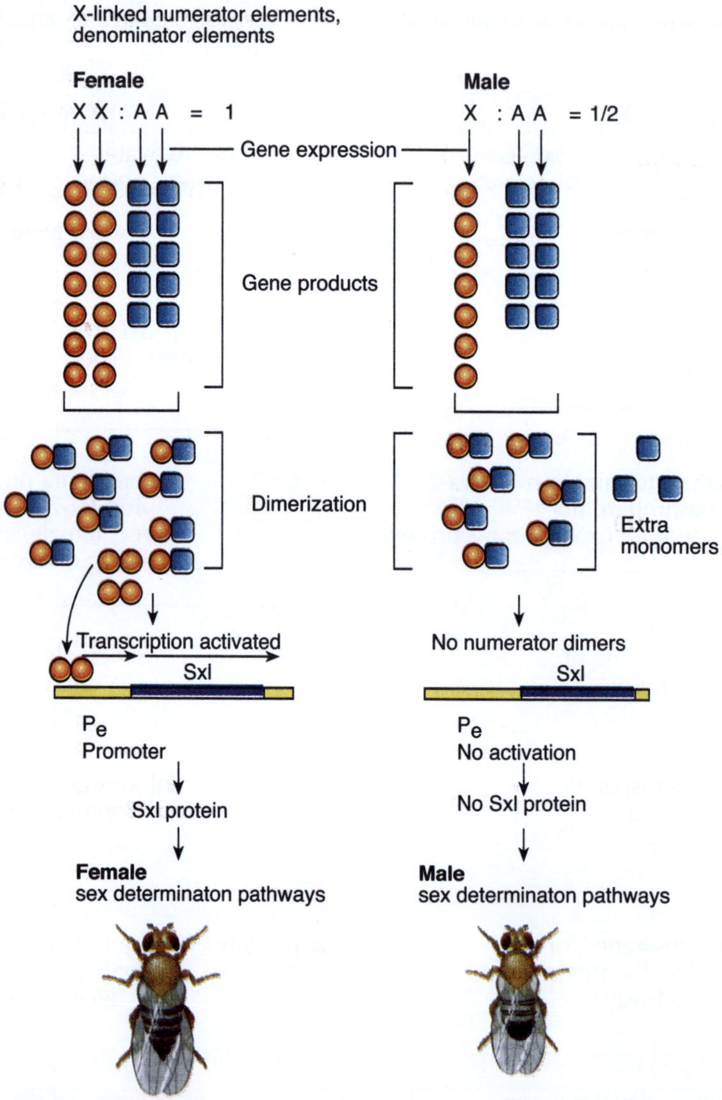

Fig. 4.11: The X-to-autosome ratio determines the expression of Sxl, the first gene in the sex determination pathway. Numerator elements are produced by the X-chromosome at a slightly higher level than denominator elements produced by autosomes. Denominator elements cannot form homodimers but they bind tightly to, and sequester, numerator elements. When the X : A ratio is one (in females), there are too many numerator elements to be occupied by denominators, and those not sequestered can form homodimers which act as activators of the Sxl gene. When the X : A ratio is ½ (in males), there are fewer numerators than denominators, and all the numerators get sequestered. Thus, males are unable to produce numerator homodimers, and Sxl remains turned off.

(ii) The Sxl gene has two promoters, viz. **early promoter** (P_E) and **late promoter** (P_L). The early promoter is the only one that is activated by NUM-NUM transcription factors. The early promoter (P_E) is only active in early embryogenesis. The later in embryogenesis and for the remainder of the life cycle, the Sxl gene is transcribed from the late promoter (P_L) regardless of the X : A ratio.

(iii) After early embryogenesis, the Sxl (P_E) is **shut off**, with a concomitant activation of constitutive Sxl transcription from the late promoter (P_L) on the Sxl gene.

(iv) Sxl gene from P_L now transcribes in both males and females. However, the primary transcript produced by Sxl transcription from the late promoter (P_L) is much larger than the primary transcript from the early promoter (P_E) and is subject to alternative mRNA splicing, depending on the presence or absence of pre-existing active SXL protein in the cell.

(v) Sxl mRNA in females differs from that in males, in the result of differential RNA processing. SXL protein is an RNA binding protein that on binding to its own mRNA precursors (Pre-mRNA) can alter splicing of the

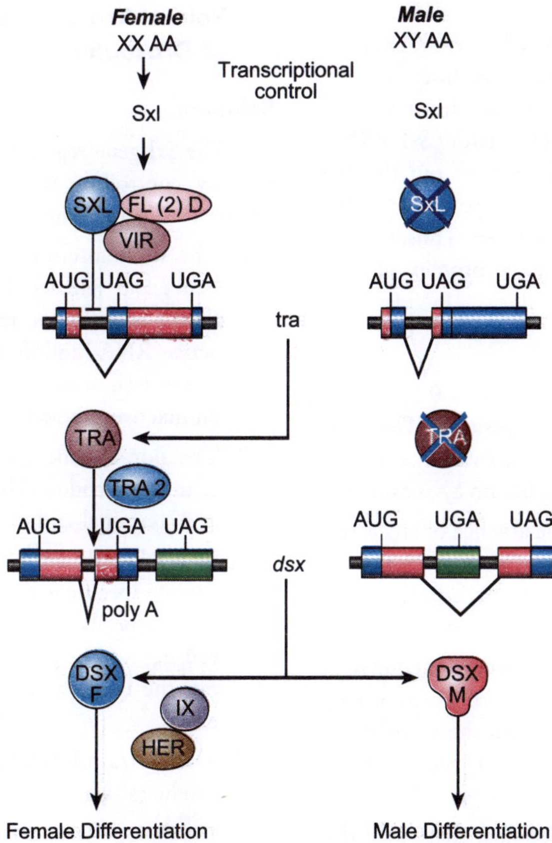

Fig. 4.12: Hierarchy of gene regulation for sex determination in *Drosophila*. In females, the X : A ratio activates transcription of the Sxl gene. The product of this gene binds to pre-mRNA of the tra gene and directs its splicing in a female-specific fashion. The female-specific TRA protein, in combination with the TRA-2 protein, directs female-specific splicing of dsx pre-mRNA, resulting in a female-specific protein (DSX-F). This protein, in combination with the IX protein, suppresses the pathway of male sexual development and activates the female pathway. In males, the X : A ratio does not activate Sxl. The result is a male-specific processing of tra pre-mRNA, resulting no functional TRA protein which in turn leads to a male-specific processing of dsx transcripts which results in a male-specific (DSX-M) protein that activates the pathway for male sexual development.

nascent Sxl transcript coming from this late promoter (P_L).

(vi) The female-specific splicing pattern of Sxl produces an mRNA that makes active SXL protein that is functionally identical with that produced from early promoter.

(vii) This SXL protein in turn binds to more Sxl primary transcript from the late promoter, creating the spliced form of the mRNA that encodes functional SXL protein. Thus, a feedback, or autoregulatory, loop controlled at the level of RNA splicing maintains SXL activity throughout the development in flies with an X : A ratio of 1.0.

(viii) The SXL protein produced in early embryogenesis from mRNA transcript from Sxl (P_E) can splice the Pre-mRNA came from Sxl (P_L) to mature Sxl mRNA. The mature Sxl mRNA comprises 7 exons and exercises out the 3rd exon (in the form of a large intron) that contained the stop codon UGA. Thus, female-specific mRNA encodes a protein of 354 amino acids.

B. In case of male:

(i) In contrast, when X : A ratio is 0.5, the Sxl switch is set in the 'off' position. The early promoter (PE) is not activated early in embryogenesis and thus has no SXL protein.

(ii) As a consequence, in the absence of any active SXL protein, the primary Sxl transcript of the late, constitutive Sxl promoter is processed in the "default" mRNA splicing pattern.

(iii) This "default" Sxl mRNA is nonfunctional, in the sense that it encodes a terminator or **stop codon (UGA)** shortly after the translation-initiation codon of its protein coding region (after 48 amino acids).

(iv) Differential RNA processing that puts this termination codon (UGA) into male-specific Sxl mRNA that thus cannot synthesize full functional SXL protein.

(v) In male necessary transcript is spliced in a manner that yields 3 exons and the terminator UGA codon is within exon-3 of mRNA and this polypeptide chain synthesis terminates at stop codon (UGA) in the exon-3.

(vi) Thus in *Drosophila* with a low level of active NUM-NUM tanscription factor $\left(\frac{X}{A} : 0.5\right)$ the absence of active SXL protein early development predestines that there will be no Sxl activity throughout the remainder of the development.

While in female RNA processing yields 7 exons because of splicing out the male-specific exon 3 (as a large intron) that contained the terminator codon (UGA). Female-specific mRNA lacking the terminator codon can produce full complement of active SXL protein (354 amino acids).

3. Role of Tra gene is sex determination of *Drosophila*:

Solution:

The Sxl gene regulates somatic sex determination by controlling the processing of the transformer (tra) gene transcript.

(i) The tra transcript is alternatively spliced in males and females. In females it produces a female-specific tra mRNA that encodes an active RNA-binding protein and in males it produces non-specific tra mRNA that encodes an inactive product.

(ii) The non-specific tra pre-mRNA, contains a terminator codon (UAG) early in the message at 3′ terminal end of its second exon, making the tra protein (TRA) non-functional.

(iii) Female-specific active SXL protein activates a female-specific 3′ splice site in tra pre-mRNA, causing it to be processed in a way that splices out the terminator codon containing second exon.

(iv) Mature tra mRNA lacking terminator codon produces active female-specific tra protein (TRA).

(v) Active TRA protein in association or conjunction with another protein tra-2 (a product of tra-2 gene) controls female-specific splicing of the pre-mRNA of the transcript of the final gene (dsx) and it also helps to generate female phenotype.

(vi) In males only the tra-2 protein is present; default splicing of the dsx pre-mRNA results in a male-specific mRNA and protein product. The male dsx protein (M) brings about male sexual development.

4. Role of double sex (dsx) gene in sex determination of *Drosophila*:

Solution:

dsx is a kind of the switch gene of sex determination.

(i) The dsx is active in both males and females but its primary transcript (Pre- mRNA) is processed in a sex-specific manner followed by alternative RNA splicing.

(ii) Alternative splicing of the primary transcript (Pre-mRNA) of dsx gene (6-exons) is done by active tra protein in association with tra-2 protein, that splices out the terminal 5th and 6th exons from pre-mRNA to process it to a mature female dsx mRNA.

(iii) This female-specific dsx mRNA produces a female-specific protein. In conjunction with the protein encoded by inter-sex (ix), gene DSX-F activates the female-specific genes (e.g., genes of yolk, protein, etc.) to transcribe constitutively and repress transcription only of male-specific genes, thereby conferring a female phenotype.

(iv) In absence of functional tra protein default splicing occurs leading exclusion of exon 4 but to inclusion of exons 5 and 6 to a male-specific transcript of DSX gene and leads to the production of a DSX-M transcript factor that represses the female-specific gene expression. Thus, it promotes male-specific gene, thereby conferring a male phenotype.

(v) Both the DSX proteins (F and M) bind to enhancer upstream of promoter for YPI gene which encodes a yolk protein; females make this protein in their fat body organs and then transfer it to developing eggs. The binding of DSX-F stimulates transcription of the YPI

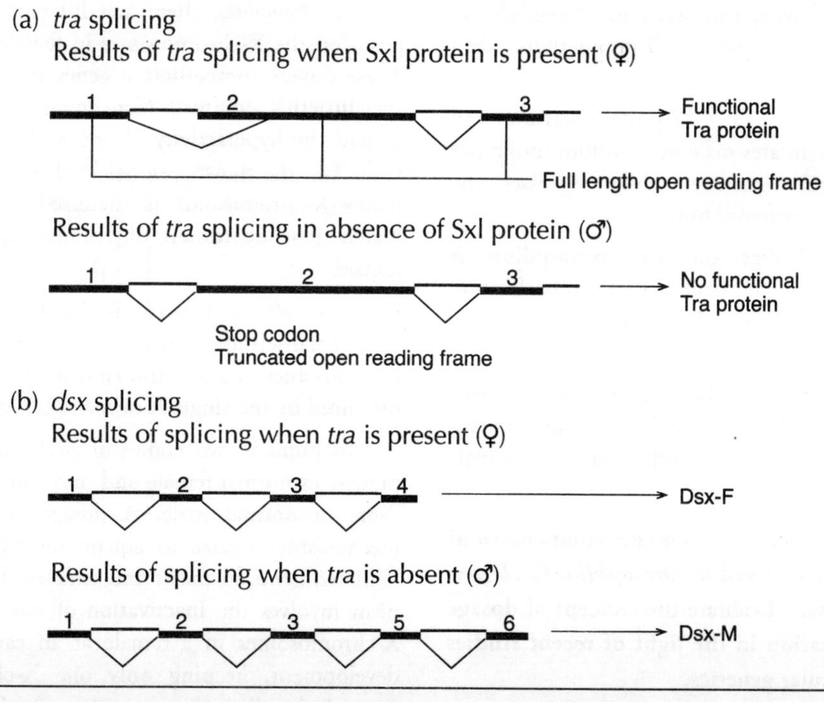

Fig. 4.13: The expression or non-expression of *Sxl* leads to downstream differences in expression of products that also play a role in the sex determination pathway. (a) The presence of *Sxl* alters the splicing of *tra* mRNA. Female transcripts produce functional Tra protein, while male transcripts have a truncated open reading frame and are unable to produce Tra. (b) Tra protein in turns plays a role in altering the splicing pattern of the *dsx* mRNA. A different Dsx product results in males (Dsx-M) than in females (Dsx-F).

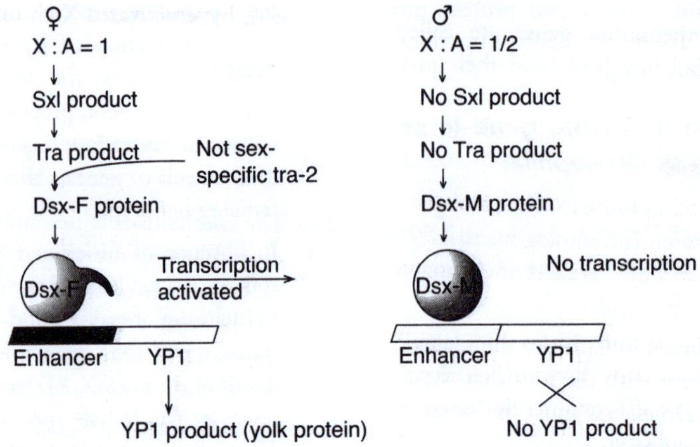

Fig. 4.14: Alternative forms of the Dsx gene product both bind to the same YP1 enhancer but they have opposite effects on the expression of the YP1 gene. Dsx-F acts as a transcriptional activator, while Dsx-M acts as a transcriptional repressor.

gene in females, the binding of DSX-M to the same enhancer region inactivates (in conjunction with other transcription factors) transcription of YPI in males.

(vi) Male and female genitalia in *Drosophila* are derived from separate cell populations called **primordia**.

  (a) In XY flies DSX-M protein activates and differentiates male primordium into adult genital structure, while represses the female primordium.

  (b) In XX flies, the male primordium is repressed and female primordium is differentiated into adult female genital structures by DSX-F protein.

  (c) In absence of dsx gene, intersexual genitalia and developed due to simultaneous development of both male and female primordium.

5. What is meant by dosage compensation? How is it achieved in *Drosophila, C. elegans* and human? Evaluate the concept of dosage compensation in the light of recent studies in molecular genetics.

Solution:

The compensatory mechanism of sex determination for **equalization** of the total gene expression from a single X (i.e., one dosage) in one sex to the two X-chromosomes (i.e., double dosage) in the other sex is referred to as **dosage compensation.**

In *Drosophila*, there are different dosages of genes on the X-chromosomes in female and male. These dosage compensation genes are called **male-specific lethal loci (msl).** Here dosage compensation is made by **hyperactivity** of one X-chromosome in male, i.e., the transcriptional level of genes on the male's X-chromosome is increased **two folds to** match gene expression of X-linked gene in the female.

In *C. elegans*, hermaphrodites (XX) **halves** the level of transcripts produced by each of their two X-chromosomes to attain the same level of transcripts produced by the single X of males (XO).

In human, two copies of X-chromosome are present in normal female and only one X-chromosome in normal male. A dosage compensation mechanism operates to equate the expression of X-linked genes in males and females. This mechanism involves the **inactivation** of one of the two X-chromosomes in a female at an early stage of development, keeping only one X-chromosome transcriptionally active, as in the case of males. The **inactivation** process involves the transcription of **XIST gene** on the chromosome to be inactivated. Inactive X is manifested as **Barr body.**

In *Drosophila*:

(i) The dosage compensation genes are called male-specific lethal loci (**msl**) and their products are called MSL proteins. Each of the MSL proteins binds specifically to the X-chromosome in males.

(ii) One of these protein products of the msl gene, called **maleless (mle)**, is homologous to **DNA** and **RNA helicases** which are enzymes capable of unwinding nucleic acids.

(iii) The putative helicase function of the maleless protein is consistent with the idea that the X-chromosome of *Drosophila* must be "opened" for vigorous transcription.

(iv) MSL protein complex also contains specific types of RNA. Two different RNAs, roX1 and roX2 (for RNA on the X-chromosome), colocalize on the male's X-chromosome with MSL protiens.

(v) Both of these RNAs are transcribed from X-linked genes and both are spliced and polyadenylated. However, they do not contain long open reading frames and therefore, probably not translated into proteins. roX RNAs seem to be the end product of the genes in *Drosophila*. The expression of the roX genes depends on the proper functioning of the msl genes. Furthermore, this expression is turned off when Sxl, the master control regulator gene of the *Drosophila* dosage compensation and sex determination system, is turned on, as it naturally is in *Drosophila* females. The **roX RNAs** are, therefore, produced only in *Drosophila* males.

(vi) The current models proposed that MSL proteins from a complex is joined by the roX RNAs. This complex then binds to 30 to 40 sites along the male X-chromosome including the loci that contain the two roX genes.

(vii) From each of these entry sites, the MSL/rox complex spreads bidirectionally until it reaches all the genes on the male X-chromosome that needs to be hyperactivated. The process of hyperactivation may involve chromatin remodelling by the MSL/roX complex.

(viii) One of the MSL proteins is a **histone acetyl transferase** (HAT) and a particular acetylated version of histone H_4 is exclusively associated with **hyperactivated** X-chromosome.

- ***C. elegans*:**

(i) In *C. elegans*, dosage compensation involves the **partial repression** of X-linked genes in the somatic cells of hermaphrodites.

(ii) The mechanism is not fully understood, but the products of several genes are involved. Like MSL genes in *Drosophila*, the proteins encoded by these genes bind specifically to the X-chromosome. However, they bind only when two X-chromosomes are present.

(iii) The proteins apparently do not bind to the single X-chromosome in males or hermaphrodites. Therefore, dosage compensation in *C. elegans*, hermaphrodites occur by **hypoactivation** of the two X-chromosomes mediated by a protein complex which binds to the X-chromosomes and represses rather than enhances transcription of their genes.

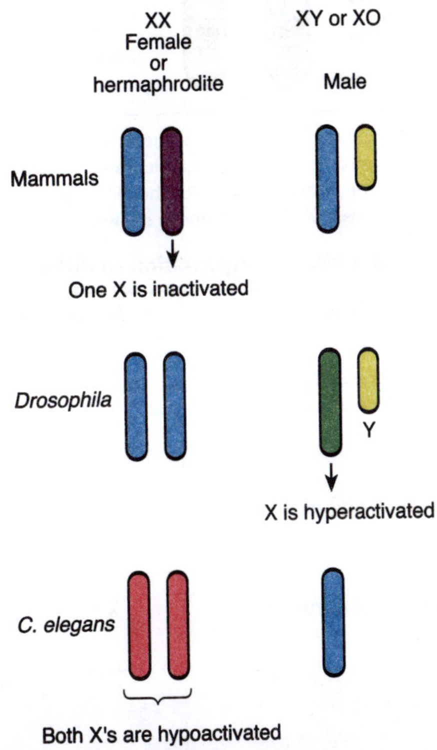

Fig. 4.15: Dosage compensation—X-linked genes: (i) In-activation, (ii) Hyper activation, (iii) Hypo activation.

In mammals:

(i) In human (mammals), X-chromosome inactivation begins at a particular site called the **X-inactivation centre** (XIC) and then spreads in opposite directions towards the ends of the chromosome.

(ii) Curiously, not all genes on an inactivated X-chromosome are transcriptionally silent. One that remains active is called **XIST** (for X inactive specific transcript); this gene is located within **XIC.**

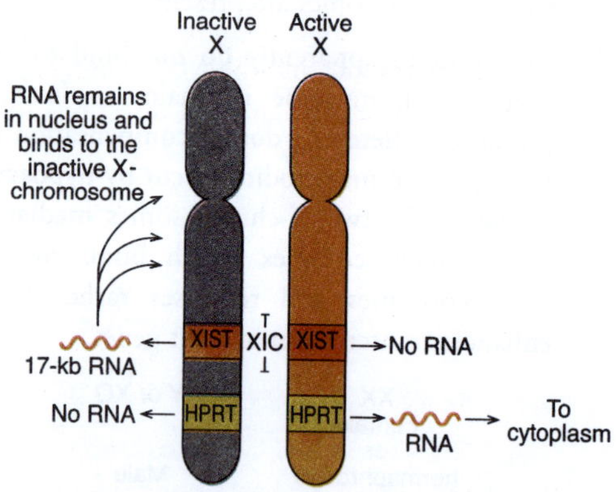

Fig. 4.16: Expression of the XIST gene in the inactive X-chromosome of human females. For comparison, the expression of the HPRT gene on the active X-chromosome is shown.

(iii) It has been shown that, in human beings, the XIST gene codes a 17 kb transcript devoid of any significant open reading frames. Therefore it seems unlikely that the XIST gene codes for a protein. Instead, the RNA itself is probably the functional product of the XIST gene.

(iv) Though polyadenylated, this RNA is restricted to the nucleus and is specifically localized to inactivated X-chromosomes; it does not appear to be associated with active X-chromosomes in either males or females.

(v) Therefore, dosage compensation of human (mammals) is made through **inactivation** of an X-chromosome (in XX female and is mediated by a noncoding RNA) transcribed from XIST gene on that chromosome.

(vi) **Acetylated H_4** is associated with all the chromosomes of human genome. However, on the inactive X it seems to be restricted to three fairly narrow bands, each corresponding to a region that contains some active genes. Acetylated H_4 is also depleted in areas of heterochromatin on the other chromosome. These findings suggest that **depletion of acetylated H_4** is a key feature of **inactive X-chromosome.**

6. There are dominant mutations of the Sxl gene called Sxl^M alleles, that transform XY animals into females but do not affect XX

Table 4.2: Dosage compensation in different animals

Animals	Sex chromosome		Process of dosage componentation
	Male	Female	
1. *C. elegans*	XO	XX	**Hypoactivation** of two X-chromosomes in hermaphrodite (female) to attain same level of activity of X of male (XO).
2. *Drosophila melanogaster*	XY	XX	**Hyperactivity** of male X-chromosome (i.e., two-fold increase activity) to equalize with female XX chromosome.
3. Marsupial mammal	XY	XX	Paternally derived X-chromosome is inactive in somatic cells of female.
4. Placental mammal	XY	XX	(i) One of the X-chromosomes, i.e., paternal chromosome is inactivated in some species. (ii) In other species-like human paternal or maternal X-chromosome is randomly inactivated throughout the female body in the form of barr body.

animals. Reversions of these SxlM alleles can be readily induced with mutagen treatment. These reversions, called Sxlf alleles, yield normal individuals in XY animals, but in homozygotes, they transform XX animals into phenotypic males. Provide a possible explanation for these observations, keeping in mind that Sxl is ordinarily dispensable for male development.

Solution:

(i) The concentration of Sxl is crucial for female development and dispensable for male development.

(ii) The dominant SxlM male lethal mutation may not actually kill all males but simply produces an excessive amount of gene product, so only females (fertile XX and XY) result.

(iii) The reversions may eliminate all gene products resulting in XX (sterile) and XY males. The reversions would be recessive because presumably, a single normal copy of the gene may produce enough gene products to "toggle the switch" in development to female.

7. **Name the noncoding RNA's that have been implicated in dosage compensation in *Drosophila* and mammals. Explain how X-chromosome specific proteins MSL bind to the male X-chromosome and play a role in remodelling the chromatin structure for dosage compensation in *Drosophila*.**

Solution:

Noncoding RNAs are followings:

Drosophila—ROX-1 and ROX-2

Mammals—XIST, TSIX

(i) In *Drosophila*, transcriptional activity is increased two folds in males to match that in females, who have twice the number of X-chromosome. Key male-specific lethal genes are mle (maleness), msl-1 (male-specific lethal), msl-2, msl-3 and mof (male absent on the first). The products of these genes are collectively called **male specific lethal (MSL)** proteins. They bind to X-chromosome and play a role in remodelling chromatin structure which is involved in regulating transcriptional activity for dosage compensation.

(ii) The SXL late protein plays a key role in dosage compensation. In females, SXL late protein binds to the transcript of MSL-2, blocking its translation, hence no MSL protein is produced.

(iii) In male, MSL-2 transcript can be translated because SXL late protein is absent. MSL-2 forms a complex with other MSL proteins MLE, MSL-1, MSL-3 and MOF.

(iv) This MSL complex binds to about 35 chromatin entry sites (CES) on the *Drosophila* male X-chromosome and the MSL complexes spread from those sites in both directions in the flanking chromatin.

(v) The MOF protein of the MSL complex is a **histone acetyl transferase** and its chromatin remodelling activity spreading along the X-chromosome is responsible for the two-fold higher level of transcription of X-chromosome genes in males than females.

Acetylation is associated with **activation** and **deacetylation** is correlated with **repressions.** The optimum balance of **acetylation** and **deacetylation** of histone, regulates gene expression.

(vi) In females, the MSL proteins other than MSL-2 are produced. However, since MSL-2 is essential for the binding of the MSL complex to the X-chromosomes no chromatin remodelling can occur in XX females.

8. **Distinguish between the concepts of sexual differentiation and sex determination.**

Solution:

(i) Sexual differentiation is the response of cells, tissues and organs to signals provided by the genetic mechanisms of sex determination.

(ii) On the other hand, genes are present in that signal developmental pathways whereby the sexes are generated.

(iii) Sexual differentiation is the complex set of responses to those genetic signals.

(iv) Sexual differentiation (in multi-cellular granism) includes both primary (related to gonad) and secondary (related to mammary gland and external genitalia) differentiation.

9. What is the major difference between the sex determination of *Drosophila* and humans?

Solution:

In *Drosophila,* it is the balance between the number of X-chromosomes and the number of haploid sets of autosomes that determine the sex.

In humans, there is a small region in the Y-chromosome (SRY) that determines the maleness.

10. What is sex switch? What genes serve as switches in human beings and *Drosophila*?

Solution:

The gene which directs the developmental pathway of undeterminate gonads towards the development of ovary (in *Drosophila*) or testes (in mammals) is called **sex switch**.

In human, a single gene (**SRY**), a **testis-determining factor (TDF)** located on the Y-chromosome, acts as a sex switch to initiate male development.

In *Drosophila,* the gene **Sxl** (sex lethal) is located on the X-chromosome that acts as a sex switch to initiate female development.

Sxl has two states of activity, when it is '**on**', it directs **female development**; when it is '**off**', **maleness ensures.**

Other genes located on the X-chromosome and the autosome regulate this sex switch gene.

(i) Genes on the X-chromosome that act to regulate Sxl into the **on state** (female development) are called **numerator element,** because they act on the numerator of the $\frac{X}{A}$ genic balance equation.

The numerator genes are **sis a, sis b** and **sis c.**

(ii) Genes on the autosomes that act to regulate Sxl into the **off state** (male development) are called **denominator elements.**

The denominator gene is **deadpan** (dpn).

11. In developmental pathways, the crucial events seem to be the activation of master switches that set in motion a programmed cascade of regulatory responses. Identify the master switches and explain how they operate in the two examples of sex determination and differentiation, viz., (a) sex determination of *Drosophila*; (b) sex determination in mammals.

Solution:

(a) Sex determination in *Drosophila*:

(i) Here the master switch is the transcriptional regulation of the **Sxl gene** during early embryogenesis.

(ii) The properties of the Sxl protein and the existence of a constitutive late promoter for Sxl ensure that when active SXL protein is produced, an auto-regulatory loop will continue to maintain SXL protein activity in the cell.

(iii) SXL protein will also initiate downstream set of regulated RNA-splicing events, culminating in the production of dsx-F mRNA.

(iv) In the absence of SXL, the default RNA-splicing machinery will produce dsx-mRNAs.

(v) The key then is whether or not the Sxl switch is **flipped on** or **stays off**. This is controlled by the level of active numerator-encoded transcription factors.

(vi) The higher level of these transcription factors conferred by an X : A ratio of 1 is necessary to activate early Sxl transcription and sets the female developmental pathway in motion.

(vii) In the absence of these higher levels of encoded transcription factors, development continues along the default pathway leading to male development.

(b) (i) In mammalian sex determination, the master switch is the presence or absence of the **SRY-Sry gene** which is ordinarily located on the Y-chromosome.

(ii) In the presence of the protein product of this gene, which acts as a DNA-binding protein, certain cells of the gonad (Leydig cell) synthesize androgen, male-inducing steroid hormones.

(iii) These hormones are secreted into the blood stream and act on target tissues to induce the transcription factor activity of the androgen receptors.

In the absence of androgen-receptor activation, development proceeds along the default pathway leading to female development.

(iv) The factors that activate SRY-Sry expression in the testis are not understood. Because the master switch here is the actual presence or absence of **SRY-Sry gene** itself.

(v) It is likely that the regulatory molecules that activate SRY-Sry are present in the indifferent gonad early in development.

12. In *Drosophila*, you have identified a new mutation, mll (male-like), that causes flies with an X : A ratio of 1 to develop as phenotypic males. You want to understand how the product of the mll gene acts in the developmental pathway for Drosophila sex determination. You measure the presence of functional SXL, TRA, DSX-F and DSX-M protein in animals with an X : A ratio of it. In each of the following cases propose a role for MLL protein in the sex determination and differentiation pathway:

(a) You observe that the functional early SXL, late SXL, TRA and DSX-M proteins are produced.

(b) You observe that functional early SXL and DSX-M are produced. No functional late SXL or TRA is produced.

(c) The only functional protein that you observe is DSX-M.

Solution:

It is actually through experiments like these that the sequential pathway of sex determination and differentiation in *Drosophila* was elucidated. More generally, it is through observations like these, where the effects of a mutant in one gene on the mRNA or protein expression of another gene are studied, that developmental pathway are pieced together.

(a) (i) It is given that **SXL** and **TRA** proteins are still operating normally in the context of X : A ratio of 1.

(ii) Whereas the alternative splicing of **dsx mRNA** is leading to the male-specific form of the protein.

(iii) It is likely that **mll** contributes to **dsx RNA-splicing** regulation.

(b) (i) Here, the block seems to be between early and late SXL proteins production.

(ii) This finding may indicate that **mll** plays a role in the **auto-regulation** of late **Sxl** alternative splicing.

(c) (i) In this example, the block is before **Sxl** expression in early embryo.

(ii) Here, the possible explanation is that MLL protein contributes to the numerator function in interpreting the X : A ratio.

13. In *Drosophila*, sex type is determined by the X : A ratio.

(a) How is this detected early in development and how does it lead to sex-specific transcription at Sxl?

(b) A mutation in Sxl affects P_E so that early transcription of Sxl does not occur. The upstream P_L promoter is unaffected, so constitutive transcription from this late promoter occurs in all cells regardless of their X : A ratio. What phenotype do you expect in this mutation to have in animals with an X : A ratio of 1 : 2; in animals with X : A ratio of 2 : 2?

(c) A tra mutant has a nonsense mutation into exon 2. What phenotype do you expect this mutant to have in animals with an X : A ratio of 1 : 2, in animals with an X : A = 2.2?

(d) The TRA protein targets the dsx pre-mRNA for alternatively splicing. However, if TRA is not present, male differentiation ensues. Why then do animals with knock-out mutations in dsx have both male and female characteristics?

Solution:

(a) (i) The X : A ratio is detected by interactions of the protein products of three X-linked **numerator** genes (sis-a, sis-b, sis-c) and one autosomal **denominator** gene (dpn).

(ii) The numerator gene products can form either homodimers or heterodimers with the denominator gene product.

(iii) When the X : A ratio is 2 : 2, an excess numerator gene product leads to the formation of many homodimers. These serve as transcription factors to activate Sxl transcription from P_E.

(iv) When the X : A ratio is 1 : 2, most numerator subunits are found in heterodimers, so Sxl transcription is not activated.

Therefore, activation of Sxl at P_E by the homodimers serves to detect the X : A ratio and leads to the early sex-specific synthesis of SXL protein.

(b) (i) Transcription at P_E is essential to generate a functional SXL protein in animals with an X : A ratio of 2 : 2.

(ii) It is not used in individuals with an X : A ratio of 1 : 2. So these animals will be unaffected and differentiated as males.

(iii) In individual with an X : A ratio of 2 : 2, SXL initiates a cascade of alternative mRNA, splicing at Sxl, tra and dsx that lead to implementation of female differentiation.

(iv) If there is no transcription from P_E, no functional SXL protein will be produced in these animals and a default set of splice choices at Sxl, tra and dsx will be used.

(v) In principle, this would lead to male differentiation in individuals with X : A ratio of 2 : 2. However, SXL also prevents the translation of msl-2 transcripts so that the dosage compensation does not normally occur in individuals with an X : A ratio of 2 : 2.

(vi) Without SXL, msl-2 transcripts will be translated; so dosage compensation will occur leading to four doses of X-linked gene products.

Due to imbalance in X and autosomal gene product dosage is likely to be lethal to animals.

(c) (i) This tra mutant will eliminate functional TRA protein. Since the functional TRA is not normally present in animals with an X : A ratio of 1 : 2, this mutation will have no effect on these animals—they will differentiate normally into male.

(ii) TRA is normally present in individuals with an X : A ratio of 2 : 2, where it functions to regulate alternative splicing at dsx and produces DSX-F, which implements female differentiation by repressing male-specific gene expression.

(iii) Without TRA default splicing will occur at dsx and produce *dsx-M* which implements male differentiation by repressing female-specific gene expression. Thus, animals with X : A ratio of 2 : 2 will be males.

(d) (i) Animals with **knock-out mutations** at dsx will produce neither dsx-M which represses female-specific gene expression nor *dsx-F* which represses male-specific gene expression.

(ii) Therefore, neither male- nor female-specific gene expression will be repressed, and both male and female differentiation pathways will proceed.

14. Campomelic dysplasia (CMD1) is a congenital human syndrome, featuring malformation of bone and cartilage. It is caused by an autosomal dominant mutation of a gene located on chromosome 17. Consider the following observations in sequence and in each case, draw whatever appropriate conclusions are warranted:

(a) Of these with the syndrome who are karyotypically 46, XY, approximately 75 percent are sex-reversed, exhibiting a wide range of female characteristics.

(b) The nonmutant form of the gene, called SOX-9, is expressed in the developing gonad of the XY male, but not in XX female.

(c) The SOX-9 gene shares 71 amino acid coding sequence homology with the Y-linked SRY gene.

(d) CMD1 patients who exhibit a 46, XX karyotype, develop as females, with no gonadal abnormalities.

Solution:

(a) Something is missing from the male-determining system of sex determination either at the level of the genes, gene products or receptor, etc.

(b) The SOX-9 gene or its gene product is probably involved in male development. Perhaps, it is activated by SRY.

(c) (i) There is probably some evolutionary relationships between the SOX-9 gene and SRY.

(ii) There is considerable evidence that many other genes and pseudogenes are also homologous to SRY.

(d) Normal female sexual development does not require the SOX-9 gene or gene product(s).

- **Gynandomorph:**

(i) A sexual mosaic (XX/XO chromosome mosaic) occurs in *Drosophila*. Here one part of the body of an individual is female and other part is male phenotype.

(ii) Gynandomorphs represent one kind of mosaic or an organism made up of tissues of male and female genotypes.

(iii) Here XX cells develop in the female direction and XO cells develop in the male direction.

(iv) A bilateral gynandomorph of *Drosophila* is male on one side (left or right) and female on other. A fertilized *Drosophila* egg, i.e., XX represents as female. One X-chromosome carries the recessive white eye gene and miniature wing alleles, while the other X-chromosome carries the wild type alleles.

(v) The X-chromosome carrying the wild type alleles was lost from one of the cells during first mitotic division (one XX and XO cells). XO cells develop as male with white eye miniature wing and XX cells develop as female.

- **Free martin:**

In cattle, when twin calves of different sexes occur, the female member usually a sterile intersex is called a free martin. It has external females genitalia but internal sex organs are more or less like those of male. The male twin is usually normal.

It is due to the fusion of the foetal membrane of the twin calves when they are in the uterus of female.

- **State briefly the Sex determination in hymenoptera:**

(i) Fertilization of egg produces diploid zygote that develop into larvae (2n). It gives rise to worker (XX) (sterile female), but the larvae (2n) taking royal jelly develop into fertile female (XX).

(ii) Unfertilized egg parthenogenetically develop into fertile male (XO) called drone.

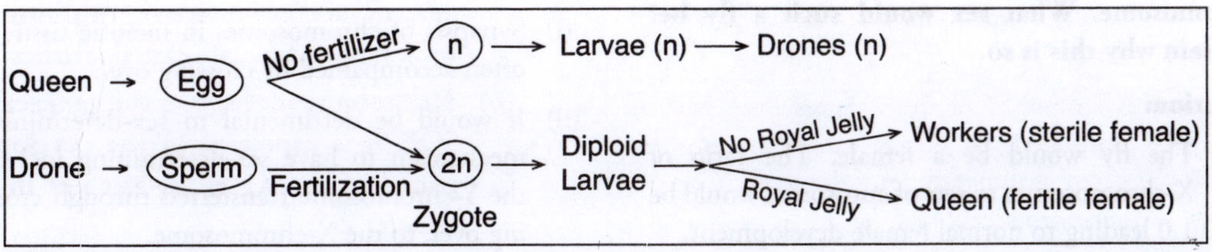

PROBLEMS ON SEX DETERMINATION

Problem 1

The autosomal recessive double sex (dsx) gene in homozygous converts males and females into developmental intersexes. Two fruit flies, both heterozygous for the double sex (dsx) allele, are mated. What are sexes of their offspring?

Solution:

Heterozygous for double sex

$$♀\ \frac{+}{+}\ \frac{+}{dsx} \quad \times \quad ♂\ \frac{+}{\rightharpoonup}\ \frac{+}{dsx}$$

Gametes:

i. $\underline{+}\ \underline{+}$, ii. $\underline{+}\ \underline{dsx}$; i. $\underline{+}\ \underline{+}$, ii. $\rightharpoonup\ \underline{dsx}$, iii. $\underline{+}\ \underline{dsx}$, iv. $\rightharpoonup\ \underline{+}$

♀ \ ♂	$\underline{+}\ \underline{+}$	$\rightharpoonup\ \underline{dsx}$	$\underline{+}\ \underline{dsx}$	$\rightharpoonup\ \underline{+}$
$\underline{+}\ \underline{+}$	$\frac{+}{+}\ \frac{+}{+}$	$\frac{\rightharpoonup}{+}\ \frac{dsx}{+}$	$\frac{+}{+}\ \frac{dsx}{+}$	$\frac{\rightharpoonup}{+}\ \frac{+}{+}$
$\underline{+}\ \underline{dsx}$	$\frac{+}{+}\ \frac{+}{dsx}$	$\frac{\rightharpoonup}{+}\ \frac{dsx}{dsx}$	$\frac{+}{+}\ \frac{dsx}{dsx}$	$\frac{\rightharpoonup}{+}\ \frac{+}{dsx}$

Results:

	Genotype	Sex and Phenotype	Ratio
1.	$\frac{+}{+}\ \frac{+}{+}$	Female Normal	$3 = \frac{3}{8}$
2.	$\frac{+}{+}\ \frac{+}{dsx}$	Female Normal	
3.	$\frac{+}{+}\ \frac{+}{dsx}$	Female Normal	
4.	$\frac{\rightharpoonup}{+}\ \frac{dsx}{+}$	Male Normal	$3 = \frac{3}{8}$
5.	$\frac{\rightharpoonup}{+}\ \frac{+}{+}$	Male Normal	
6.	$\frac{\rightharpoonup}{+}\ \frac{+}{dsx}$	Male Normal	
7.	$\frac{+}{\rightharpoonup}\ \frac{dsx}{dsx}$	Inter-sex	$2 = \frac{2}{8} = \frac{1}{4}$
8.	$\frac{+}{+}\ \frac{dsx}{dsx}$	Inter-sex	

Problem 2

In *Drosophila*, the X-chromosomes may become attached to one another (XX) such that they always segregate together. Some flies contain both an attached X-chromosome and Y-chromosome. What sex would such a fly be? Explain why this is so.

Solution:

(i) The fly would be a female. The ratio of X-chromosomes to sets of autosomes would be 1.0 leading to normal female development.

(ii) The Y-chromosome has no influence on the sex determination in *Drosophila*.

Problem 3

Predict the sex of the offspring that would occur in a cross between an attached X fly and a normal one of the opposite sex.

Solution:

Attached X fly (female) Normal male

Gametes:

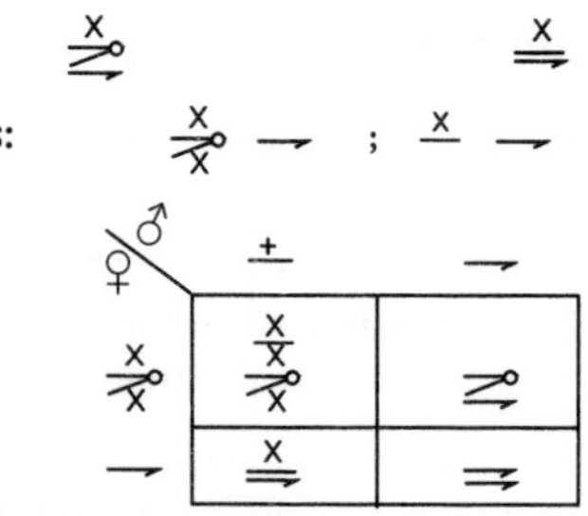

Results:

	Genotype	Sex and Phenotype	Ratio
1.	($\widehat{XX}$Y)	Female (meta) Trisomic	2
2.	($\widehat{XX}$Y)	Female (attached X)	
3.	(XY)	Male Normal	1
4.	(YY)	No development (dies)	x

Problem 4

It has been suggested that any male-determining genes contained in the Y-chromosome in humans cannot be located in the limited region that synapses with the X-chromosome during meiosis. What might be outcome if such genes were located in this region?

Solution:

(i) Synapsis of chromosomes in meiotic tissue is often accompanied by crossing over.

(ii) It would be detrimental to sex-determining mechanism to have sex-determining loci on the Y-chromosome transferred through crossing over, to the X-chromosome.

Problem 5

An attached X female fly $\widehat{XX}Y$, expresses the recessive X-linked white eye phenotype. It is crossed to a male fly that expresses the X-linked recessive miniature wing phenotype. Determine the outcome of this cross regarding the sex, eye colour and wing size of the offspring.

Solution:

White eye female (with attached X) Male-miniature wing

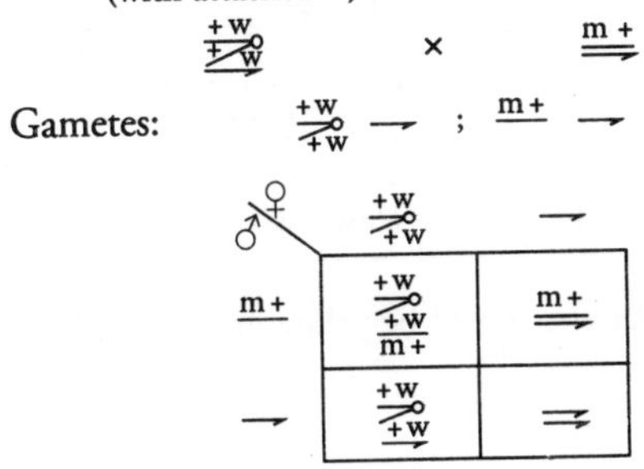

Results:

	Genotype	Sex and Phenotype	Ratio
1.	+w/+w/m+ [($\widehat{XX}$)X]	Super-female (white trisomic)	1
2.	+w/+w/→ [($\widehat{XX}$)Y]	White-eyed female (attached X and Y)	1
3.	m+/→ (XY)	Normal miniature-winged male	1
4.	→/→ (YY)	No development (dies)	X

Daughters are white-eyed and sons are miniature-winged phenotype.

Problem 6

Can the Lyon hypothesis be tested in human female who is homozygous for one allele of the X-linked G6PD gene? Why or why not?

Solution:

(i) Unless other markers, cytological or molecular, are available, one cannot test the Lyon hypothesis with homozygous X-linked genes.

(ii) The test requires identification of allelic alternatives to see differences in X-chromosome activity.

Problem 7

Cat-breeders are aware that kittens expressing the X-linked calico-coat pattern and tortoise-shell pattern are almost invariably females. Why?

Solution:

(i) Dosage compensation and the formation of Barr bodies occur only when there are two or more copies of X-chromosomes.

(ii) Males normally have only one X-chromosome; therefore, such mosaicism cannot occur.

(iii) Females normally have two X-chromosomes. Therefore, cases of male calico cats are XXY.

Problem 8

Devise as many hypotheses as you can explain why so many human male conceptions than human female conceptions occur.

Solution:

(i) One could account for the significant departures from 1 : 1 ratio of males to females by suggesting that at anaphase 1 of meiosis, the Y-chromosome more often goes to the pole that produces the more viable sperm cells.

(ii) One could also speculate that the Y-bearing sperm has a higher likelihood of surviving in the female reproductive tract, or that the egg surface is more receptive to Y-bearing sperm. But, still the mechanism is unclear.

Problem 9

What are the process involved in X-chromosome inactivation in mammals.

Solution:

(i) X-chromosome inactivation begins at the X-inactivation centre (**XIC**) which contains **XIST** gene.

(ii) During early embryogenesis in XX-mammals, the XIST genes produce unstable 17 kb long RNAs that remain closely associated with XIC region.

(iii) Transcripts from one of the two XIST genes stabilize and eventually coat the entire X-chromosome containing that gene. Transcript from the other XIST gene disintegrate and further transcription from that gene is repressed by **methylation** of nucleotides in the gene's promoter.

(iv) The chromosome that becomes coated with XIST RNA becomes inactive X-chromosome, it condenses and forms a **Barr body**.

Problem 10

Differences between male and female *Drosophila*.

Solution:

Characters	Male	Female
1. Sexcomb	Present	Absent
2. Abdomen	Blunt (rounded) and more darker	Pointed and less darker
3. Band	Thicky pigmented band-like structure at terminal tergite	Absent

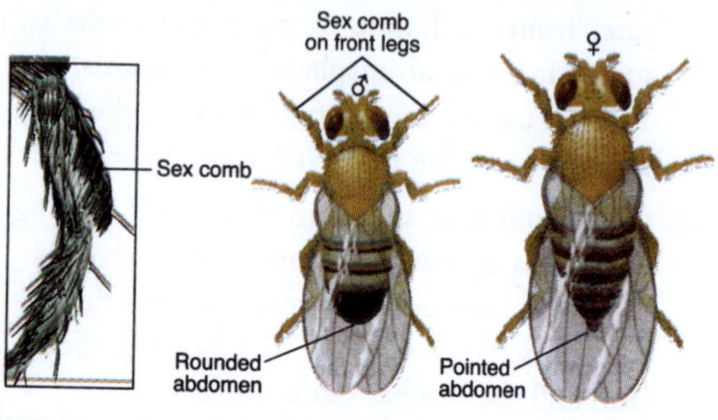

Fig. 4.17: Figures showing differences in male and female *Drosophila*.

Problem 11

A yellow bodied *Drosophila* female with attached X-chromosome was crossed to a white eyed male. Both the parental phenotypes are caused by X-linked recessive mutation. Predict the phenotype of the progeny.

Solution:

(i) Yellow body with attached X female = $\widehat{X^yX^y}$

(ii) White eyed male = $X^w/\rightharpoonup$

$$♀\ \widehat{X^yX^y} \quad \times \quad ♂\ \frac{X^w}{\rightharpoonup}$$

Gametes: $\widehat{X^yX^y}$, O ; X^w, $\rightharpoonup$

♀ / ♂	$\widehat{X^yX^y}$	O
X^w	$\widehat{X^yX^y}X^w$	O X^w
$\rightharpoonup$	$\widehat{X^yX^y}\rightharpoonup$	$\rightharpoonup$ O

Results:

	Genotype	Phenotype
1.	$\widehat{X^yX^y}\ X^w$	= Die
2.	$\widehat{X^yX^y}\ \rightharpoonup$	= Yellow bodied female
3.	O $\rightharpoonup$	= die
4.	X^w O	= White eyed male

All the daughters will be yellow bodied and all sons are white eyed.

Problem 12

A *Drosophila* female homozygous for recessive X-linked mutation causing yellow body was crossed to a wild type male. Among the progeny, one fly had sectors of yellow pigment in an otherwise gray body. These yellow sectors were distinctly male, whereas grey areas were female. Explain the peculiar phenotype of this fly.

Solution:

(i) The fly (yellow sectors are male and grey areas are female) is a gynandromorph, i.e., sexual mosaic.

(ii) The yellow tissue is $X_{(y)}/O$ and the grey tissue is $X_{(y)}/X^+$.

(iii) This mosacism must have arisen through loss of the X-chromosome that carries wild type allele (X^+) presumably during one of the early embryonic cleavage divisions.

Problem 13

In the Geneticist laboratory of Calcutta University obtained *Drosophila* females that carry attached X-chromosome homozgous for a recessive mutation yellow (y) body colour instead of grey body colour. In one experiment some of

these females are crossed with wild type males and in another these females were crossed to wild type males that have their X- and Y-chromosomes attached to each other, i.e., they carry a compound XY-chromosome. Predict the phenotypes of the progeny from these two crosses and indicate which, if any, will be sterile.

Solution:

(i) First we have to know their genotypes.

(ii) In the first cross, yellow body attached X females $\widehat{X^yX^y}$ with wild type males (X^+Y).

(iii) Females produce two kinds of gametes—XX and nullo.

Males also produce two types of gametes—X and Y.

$\widehat{X^yX^y}$ × X^+Y

(Females) (Males)

Gametes: $\widehat{X^yX^y}$, O ; X^+, Y

♂ \ ♀	$\widehat{X^yX^y}$	O
X^+	$\widehat{X^yX^y}$ X^+	X^+ O
Y	$\widehat{X^yX^y}$ Y	O Y

Results:

	Genotype	Phenotype
1.	$\widehat{X^yX^y}$ X^+	= die
2.	$\widehat{X^yX^y}$ Y	= yellow female
3.	X^+ O	= wild (grey) male
4.	O Y	= die

Yellow body females are produced like their mother except that they carry Y-chromosome and XO zygote develops into grey body male like their father, but they have deficiency of Y-chromosome. Y-chromosome in females will have no effect of fertility, but absence of Y-chromosome in male will cause them to be sterile.

Second cross

Yellow bodied attached X female = $\widehat{X^yX^y}$

Male with compound XY chromosome = $\widehat{X^+Y}$

Female gametes $\widehat{X^yX^y}$, O

Male gametes $\widehat{X^+Y}$, O

Gametes:

♂ \ ♀	$\widehat{X^yX^y}$	O
$\widehat{X^+Y}$	$\widehat{X^yX^y}$ $\widehat{XY}$	$\widehat{X^+Y}$ O
O	$\widehat{X^yX^y}$ O	O O

Results:

	Genotype	Phenotype
1.	$\widehat{X^yX^y}$ $\widehat{XY}$	dies
2.	$\widehat{X^yX^y}$ O	Yellow female
3.	$\widehat{X^+Y}$ O	Grey male
4.	O O	

Result indicates that yellow bodied females with attached X ($\widehat{X^yX^y}$) and grey bodied males with compound $\widehat{X^+Y}$-chrmosome flies will be viable and fertile.

Problem 14

In a human Genetics Research Laboratory in Delhi, cytological examination of the sex chromosomes in a man revealed that he carried an insertional translocation. A small segment was deleted from the Y-chromosome and inserted into the short arm of the X-chromosome. This segment contained the gene responsible for male differntiation (SRY). If this man marries a karyotypically normal woman, what types of progeny will the couple produce?

Solution:

Male (X-chromosome with SRY) Female (Normal)

SRY

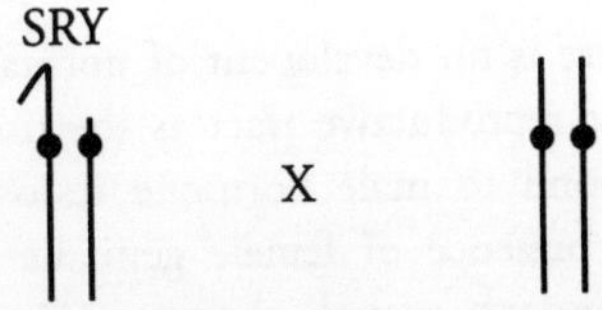

XY (deleted SRY) XX

Gametes:

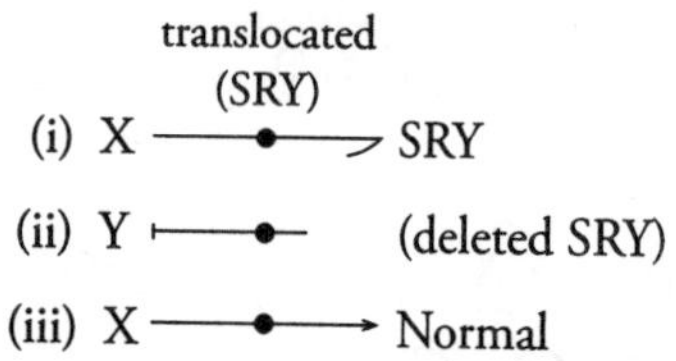

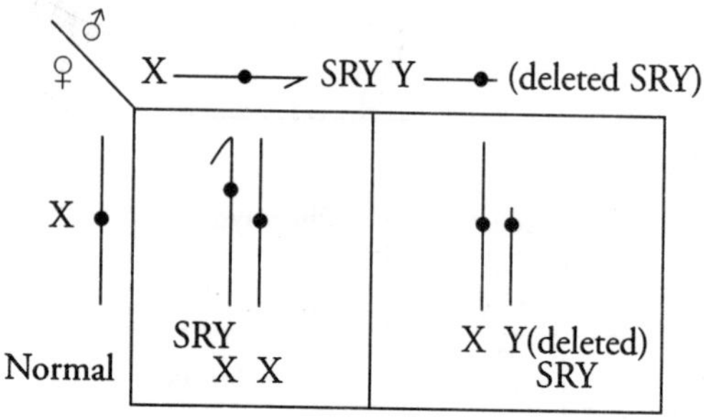

Results:

	Genotype	Phenotype
I.	X (SRY) X	Male
II.	Y (deleted) SRY X	Female

(i) XX_{SRY} zygote develops into male.

(ii) XY (deleted SRY) zygote develops into female.

Problem 15

If the diagnosis is testicular feminizations, that will be the genotype and phenotype of an individual?

Solution:

(i) The testicular feminization indicates that the genotype of the individual is XY while the phenotype is female despite of the presence of the testis.

(ii) There is no developent of normal functional male reproductive tract as the tissues do not respond to male hormone testosterone, but the presence of female genitalia and female secondary sexual characters indicates the female phenotype.

Problem 16

What are numerator and denominator elements? State the role of numerator and denominator elements during sex-determination of *Drosophila*. **C.U. 2015**

Solution:

Numerator elements

Genes on the X-chromosome of *Drosophila* that act on the numerator of $\frac{X}{A}$ genic balance equation and that regulate the sex switch gene (Sxl) to the on condition (female development) are called *numerator elements*.

Example: Sisterless a, (sis a), sisterless b (sis b) sisterless c (sis c) and runt.

Denominator elements

Genes on the autosomes of *Drosophilas* that act on the denominator of $\frac{X}{A}$ genic balance equation and that regulate the sex switch gene (Sxl) to the off conidition (male, development) are called **denominator element.**

Example: deadpan (dpn) & Extramacrochaetae; daughterless (da).

(i) Numerator elements (protein) are produced by X-linked genes at slightly higher level than denominator elements (protein) produced by autosomes.

(ii) Denominator proteins antagonize numerator proteins. Denominator proteins can not form homodimer but they bind tightly to and sequester numerator proteins.

(iii) In XX-embryos where X : A ratio is 1.00 (female), there are too many numerator protein to be occupied by denominators and those not sequestered can form **homodimer** which acts as activator of the **Sxl gene,** i.e., binds to the early promoter of Sxl (P_E) gene.

(iv) In XY embryos where X : A ratio is 0.5 (male), there are fewer numerators than denominators and all the numerators get sequenstered, thus males are unable to produce numerator homodimers, and Sxl remains turned off.

(v) In female, X : A ratio 0 : 1 activates the transcription of Sxl gene. The product of this gene binds to pre-mRNA of tra gene and directs its splicing in a female specific fashion.

(vi) The female-specific TRA protein, in combination with the TRA-2 protein directs female-specific splicing of dsx pre-mRNA, resulting in a female-specific protein (dsxF). This protein combination with IX protein suppresses the pathway of male development and activates female pathway.

(vii) In male, X : A ratio 0.5 does not activate Sxl gene. This result is male-specific processing of tra pre-mRNA, resulting no functional TRA protein which in turn leads to male specific proessing of dsx transcripts, which produce male specific protein (dsx-M) that activates pathway of male sexual development (Fig. 4.11).

Problem 17

Comment on the role of MSL protein in dosage comensation of *Drosophila*. [C.U. 2015]

Solution:

(i) In *Drosophila*, transcriptional activity is increased two folds in males to match that in females, which have twice the number of X-chromosome?

(ii) Key male-specific lethal genes are mle (**male ness**) msl-1 (**male specific lethal**), msl-2, 6, msl-3 and mof (**male absent on the first**). The products of these genes are collectively called *male-specific lethal* (msl) loci and their products are called MSL proteins.

(iii) MSL proteins bind to X-chromosome and play a role in remodelling chromatin structure which is involved in regulating transcriptional activity for dosage compensation.

(iv) One of these protein products of the msl gene, called maleless (mle), is homologous to DNA and RNA helicases, which are enzymes capable of unwinding nuleic acids. The putative helicase function of the maleless protein is consistent with the idea that the X-chromosome in a *Drosophila* male must be 'opened' for vigorous transcription. It is also consistent with the observation that polytene X-chromosomes in males appear bloated and have a difference banding patten perhaps because they are *hyperactivated.*

(v) Five MSL proteins apparently form a complex that binds to a single X-chromosome in *Drosophila* male. MSL protein complex also contains two different RNAs, viz. rox-1 and rox-2 (RNA on X). They are co-localized on the male's X-chromosome with MSL proteins.

(vi) Both these RNAs are transcribed from X-linked genes and both are spliced and polyadenylated. However, they do not contain long open reading frame and are, therefore, probably not translated into protein. The expression of rox genes depends on proper functioning of *msl* gene. Their expression is *turned off* when Sxl, the master control regulatory gene for dosage compensation and sex determination in *Drosophila,* is ***turned on.*** roX RNAs are produced only in *Drosophila* male.

(vii) The SXL late protein plays a key role in dosage compensation. In females, SXL late protein binds to the transcript of msl-2, blocking its translation and no MSL protein is produced.

(viii) In male, msl-2 transcript can be translated because SXL late protein is absent. MSL-2 formed a complex with other MSL proteins MLE, MSL-3 and MOF. The MSL protein complex also form a large complex with roX RNAs.

(ix) This new MSL complex then binds to 30 to 40 (average 35) chromatin entry sites (CES) along the *Drosophila* male X-chromosome including the loci that contains two roX genes.

(x) From each of these entry sites, the MOF protein of the MSL complex (MSL/roX RNA) is a histone **acetyl transferase** and its chromatin remodelling activity spreads bidirectionally until it reaches all the genes on the male's X-chromosome, which is responsible for the two-fold higher level of transcription of X-chromosome genes in males than females.

(xi) *Acetylation* is associated with *activation* and *deacetylation* is correlated with *repression*. The optimum balance of acetylation and deacetylation of histone H_4 is exclusively associated with **hyperactivity of X-chromosome.**

(xii) In females, the MSL-protein other than MSL-2 are produced. However, since MSL-2 is essential for binding of the MSL complex to the X-chromosomes, no chromatin remodelling can occur in XX females.

Problem 18

Role of alternative or differential splicing in sex determination of *Drosophila*.

Solution:

The process for generating different functional mRNAs from a single precursor mRNA by incorporating different combinations of exons in the mature mRNA is called *alternative splicing*. As a result, varieties of mRNAs are produced from a single pre-mRNA transcript.

The sex in *Drosophila* is determined by X : A ratio and is followed by an alternative splicing cascade. It results a cascade of regulatory gene that determines the sex of the flies.

1. The first gene in this cascade is Sxl (X-linked sex lethal) which is the master regulator of the sex determination process in *Drosophila*.
2. A. *In female* (XX):
   - (i) The pre-mRNA transcript of Sxl gene consists of 8 exons. The mature Sxl mRNA comprises 7 exons and exercises 3rd exon (**contained UGA stop-codon**) by alternative splicing. Thus, in female a regulatory protein, i.e., Sxl with 354 amino acid is produced. It regulates splicing and processing of tra gene which has four exons.
   - (ii) Through alternative splicing second exon (**contained UAG stop codon**) is exercised by splicing factor U2AF. Thus, a female specific tra protein (TRA) is produced.
   - (iii) Active TRA protein in association with another protein tra-2 (A product of tra-2 gene) controls female specific splicing of dsx (double sex) gene. The dsx pre-mRNA transcript has 6 exons. Through alternative splicing 5th and 6th exons are exercised. Thus, DSX-F protein is produced which inhibits male development and helps to generate female phenotype.

   B. *In male* (XY):
   - (i) Here Sxl genes are transcribed from late promoter (P_L). RNA splicing does not able to splice out 3rd exon (**containing UGA stop codon**) from Sxl pre-mRNA which includes all the 8 exons (incuding 3rd). As a result, truncated SXL protein with 48 amino acid is produced. It is **non-functional**.
   - (ii) The transcript of tra gene consists of 4 exons in which 2nd exon contains nonsense codon (**UAG**). Absence of functional SXL protein default splicing includes all the four exons within the spliced tra mRNA. As a result truncated tra protein is produced due to the presence of stop codon in second exon. It is also **inactive**.
   - (iii) In absence of active TRA and tra 2 protein default splicing of dsx pre-mRNA transcript leads to exclude exon 4 and include exon 5 and exon 6. As a result a male-specific DSX-M protein is produced which represses female development and promotes development of male phenotype.

Remark:

(a)

(i) If the *X : A ratio is* 1.0, through female specific splicing, active regulatory SXL protein with 354 amino acid is produced.

(ii) Active Sxl protein regulates female-specific splicing of tra pre-mRNA and produces active TRA protein.

(iii) Active tra and tra 2 protein regulate splicing of DSX pre-mRNA to form active DSX-F protein. It results female development.

Sxl ⟶ tra ⟶ DSX F

(b)

(i) If X : A ratio is 0.5, male-specific splicing results truncated SXL protein.

(ii) Default splices of tra pre-mRNA transcript is due to absence of active SXL protein. As a result truncated TRA protein is formed.

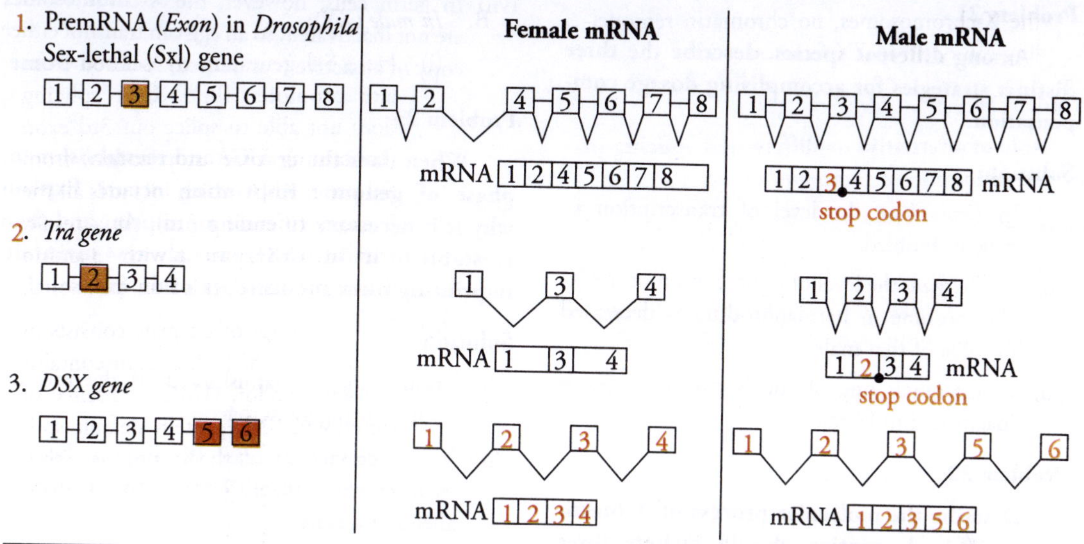

Fig. 4.18: Differential splicing of Sex-regultory gene in *Drosophila*.

(iii) Absence of active TRA and tra-2 protein results default splicing of DSX pre-mRNA, which develops DSX-M protein. DSX-M protein helps in expression of male phenotype.

SXL (inactive) ⟶ tra (inactive) ⟶ DSX-M

Problem 19

Explain briefly the role of 'Xic' in dosage compensation in mammals. Why 'Xist' is only expressed in eutherian mammals.

C.U. 2014

Solution:

There are two copies of X-chromosome in female (XX) mammals and single copy of X-chromosome in male (XY). The compensatory, mechanism of sex determination for equalization of total gene expression from single X-chromosome (one dose) in one sex to the two X-chromosomes (double dose) in other sex is known as **dosage comensation.**

(i) In normal female, two X-chromosomes are counted and one is inactivated and compacted into Barr body. On the other hand, in male, one X is counted and none is inactivated,

(ii) **Xic** (X inactivate centre) a short region in X chromosome play a key role in X inactivation.

(iii) **Xist** (X inactive specific transcript) gene on the inactivated X-chromosome is active. Its gene product is an Xist RNA (**sense RNA**) which coats the X-chromosome and inactivates it.

After coating, other proteins associate with Xist RNA and promote chromosomal compaction into Barr body.

(iv) The number of Barr body = Total number of X-1. Therefore, only one X is active and others remain inactive. The active X-chromosomes repress the Xist gene expression.

In this way male : female, i.e., 1 : 1 X-chromosome dose persists.

Problem 20

With regard to the number of sex chromosome, explain why dosage compensation is necessary?

Solution:

(i) Dosage compensation refers to the Phenomenon that the level of expression of genes on the X-chromosome is equal in both male and female; even though they have different number of sex chromosome.

(ii) In many species it seems necessary, so that the balance of genes expression between the autosomes and sex chromosome is similar between two sexes.

Problem 21

Among different species, describe the three distinct strategies for accomplising dosage compensation.

Solution:

(i) In *Drosophila*, the level of transcription in male is doubled.

(ii) In *C elegans* the level of transcription of the X-chromosome in hermaphrodites is decreased by 50% of that male.

(iii) In mammals one of the X-chromosomes is inactivated in females.

Problem 22

Describe the molecular process of X inactivation. This description should include three phases of inactivation and the role of the Xic. Explain what happens to X-chromosome during embryogenesis in adult somatic cells and during oogenesis.

Solution:

(i) X inactivation begins with counting of Xics. If there are two X-chromosomes, in the process of initiation, one is targeted for inactivation.

(ii) During embryogenesis, this inactivation begins at the Xic locus and spreads to both ends of the X-chromosomes until it becomes a highly condensed Barr body.

(iii) The *Tsix* gene may play a role in the choice of the X-chromosome that remains active.

(iv) The *Xist* gene which is located in the *Xic* region remains transcriptionally active on the inactivated X-chromosome.

(v) It is thought to play an important role in the X inactivation by coating the inactive X-chromosome.

(vi) After X inactivation is established, it is maintained in the same X-chromosome in the somatic cells during subsequent cell divisions.

(vii) In germ cells, however, the X-chromosomes are not inactivated, so an egg can transmit either copy of an active (condensed) X-chromosome.

Problem 23

When does the erasure and reestablishment phase of genomic imprinting occur? Explain why it is necessary to erase an imprint and then reestablish if in order to always maintain imprinting from the same sex of the parent.

Solution:

(i) Erasure and reestablishment of the imprint occurs during gametogenesis.

(ii) It is necessary to erase the imprint because each sex will transmit either active or inactive alleles of a gene.

(iii) In somatic cells the two alleles for a gene are imprinted according to the sex of the parent from which the allele was inherited.

Problem 24

How is the process of X inactivation is similar to genomic imprinting? How is it different?

Solution:

(i) In some species, such as marsupials, X inactivation depends on the sex. This is similar to imprinting.

(ii) Once X inactivation occurs during embryonic development, it is remembered throughout the rest of the life of the organism, which is similar to imprinting.

(iii) X inactivation in mammals is different from genomic imprinting. It is not sex dependent.

(iv) The X-chromosome that is inactivated could be inherited from the mother or the father.

(v) There was no marking process on the X-chromosome that occured during gametogenesis.

(vi) In contrast, genomic imprinting always involves a marking process during gametogenesis.

Cell Cycle and Cancer

5

CELL CYCLE

Genetics of Cell Cycle (Yeast)

Yeasts are tiny single celled fungi. Two species, viz. **fission yeast** (*Schizosaccharomyces pombe*) and **baker's yeast** (*Saccharomyces cerevisiae*) are generally used in studies of cell cycle.

- **Yeast as a genetic model organism:**

[Brief idea about life cycle in yeast]

(i) Yeast cells undergo both haploid (n) and diploid (2n) phases in the life cycle. During this phases cell division occurs by budding, a mitotic process by which a smaller genetically identical daughter cells bud from the mother.

(ii) Haploid cell occurs in two mating types—a and α. Both a and α cells grow mitotically by budding until a chemical signal known as pheromone stimulates them to mate.

(iii) The fusion of one 'a' with one 'α' cell is followed by fusion of their nuclei and formation of a diploid yeast cell.

(iv) The diploid cell either continues to bud or undergoes meiosis and sporulates depending upon the availability of nutrients. The process of meiosis results in genetic reshuffling and formation of four haploid spores (Fig. 5.1).

- **Highlight the importance in studying of cell cycle genetics in yeast**

(i) Alternating haploid (n) and diploid (2n) nature of the yeast life-cycle is particularly useful for genetic analysis.

(ii) It is easy to detect recessive mutations in haploid cells, as the recessive allele is not masked by a wild type allele.

(iii) Recessive **lethal mutation** can be maintained in a diploid yeast strain that carries the mutation on one chromosome and the wild type allele on the other chromosome.

(iv) The major advantage of yeast as a genetic model is the availability of DNA sequence information and collection of mutants and deletion of strain. In *S. cerevisiae*, genome have sequenced.

(v) Bioinformatics analysis of the yeast **genome** reveals that there are 6000 yeast genes (approximately).

- **State briefly the cell cycle regulation in yeast:**

The progression of cells through the division cycle is regulated by extracellular signals from the environment, as well as by internal signals that monitor and co-ordinate the various processes that take place during different cell cycle phases.

(i) A major cell cycle regulatory point in many types of cells occurs late in G_1 and controls progression from G_1 to S. This regulatory point is known as "START" in budding yeast (*Saccharomyces cerevisiae*).

(ii) Once cells have passed 'START', they are committed to entering 'S' phase and undergoing one cell division cycle.

(iii) In the yeast cell cycle, the passage through 'START' is controlled by the availability of nutrients, mating type and cell size.

(iv) 'START' represents a decision point at which the cell determines whether sufficient

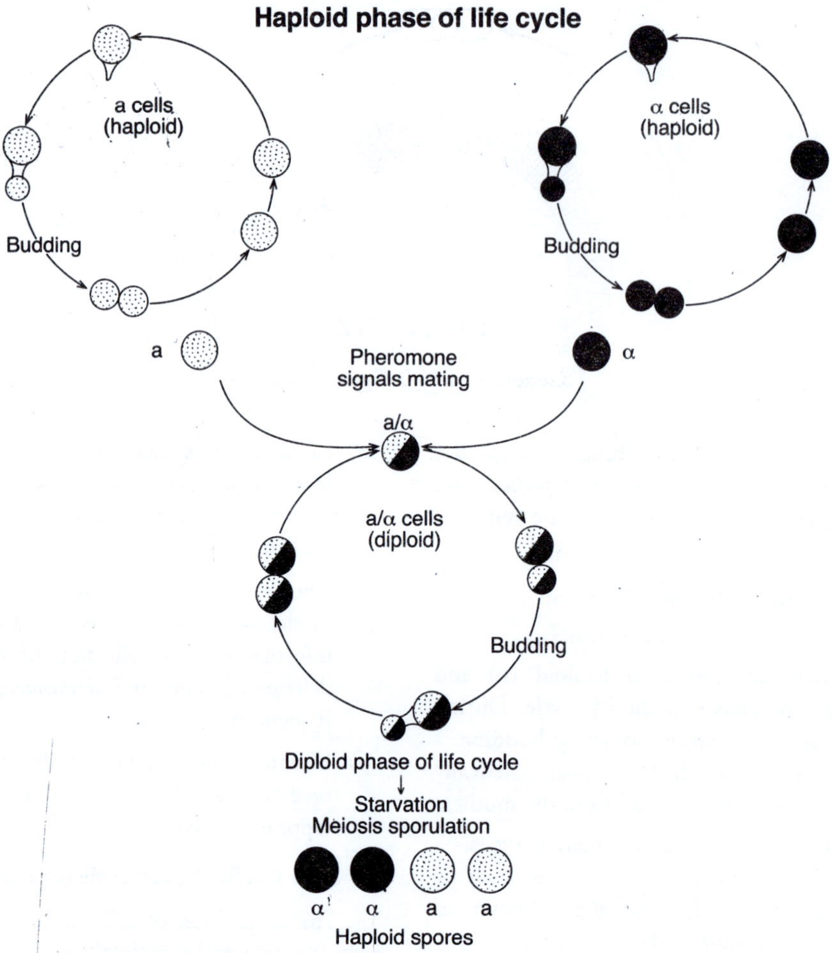

Fig. 5.1: Life cycle of the budding yeast, *S. cerevisiae*. Yeast can grow as either haploid or diploid cells. Both haploid and diploid cells divide by budding. Haploid cells when stimulated by pheromones, fuse to form diploids. In response to starvation, diploid cells undergo meiosis and sporulation, yielding four haploid spores.

nutrients are available to support progression through the rest of the division cycle. Mating factors also arrest cell cycle at 'START', allowing haploid yeast cells to fuse with one another instead of progression to 'S' phase.

(v) 'START' is the point at which cell growth is coordinated with DNA replication and cell division. In yeast, the small daughter cell must grow more than the large mother cell before they divide again. Thus, cell size must be monitored in order to coordinate cell growth with other cell cycle events. The small daughter cell growth with other cell cycle events. The small daughter cell spends a longer time in G_1, and grows prior to reach 'STAR point.

(vi) The proliferation of most cells are regulated primarily in G_1. Example: *Saccharomyces cerevisiae*. On the other hand, some cell cycles are controlled principally in G_2. Example: *Schizosaccharomyces pombe*.

A. Cell cycle regulation in fission yeast:

Fission Yeast (*Schizosaccharomyces pombe*)

(i) It is a rod-shaped cell and grows by elongation at its end.

(ii) Division occurs by the formation of septum or cell plate in the centre.

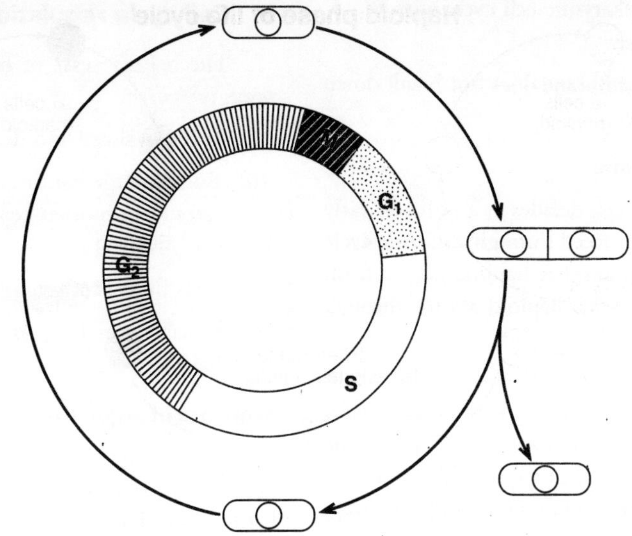

(a) Cell cycle of fission yeast (*Schizosaccharomyces pombe*)

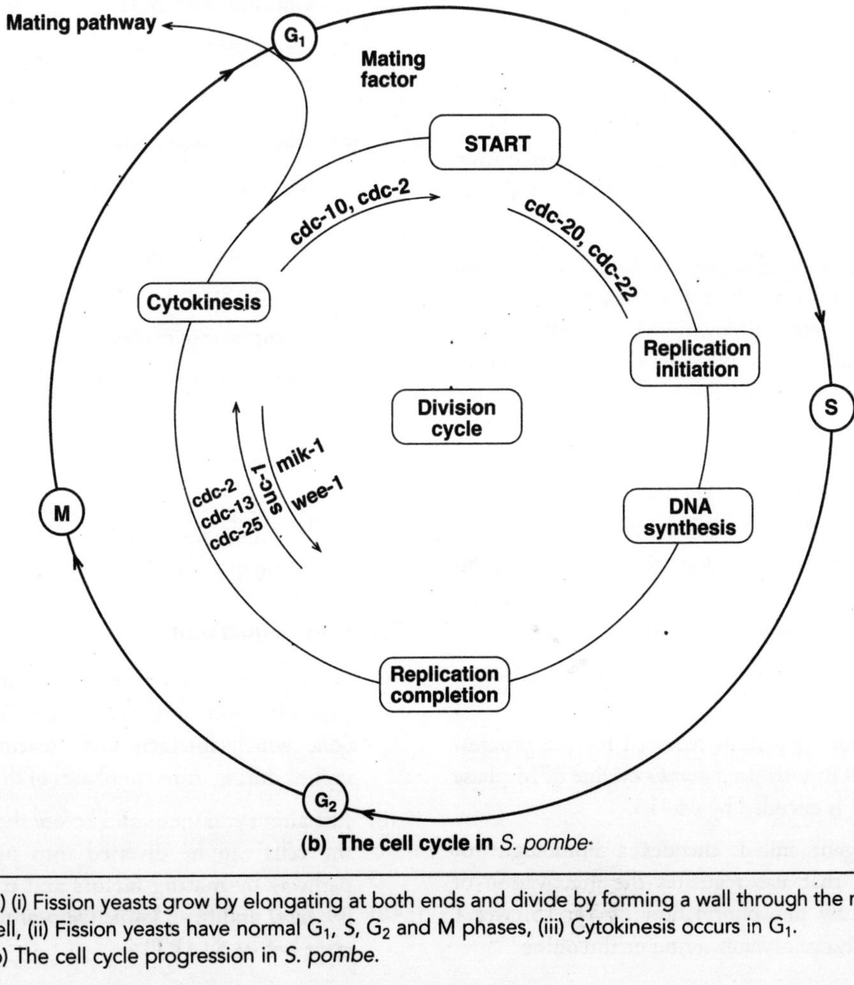

(b) The cell cycle in *S. pombe*

Fig. 5.2: (a) (i) Fission yeasts grow by elongating at both ends and divide by forming a wall through the middle of the cell, (ii) Fission yeasts have normal G_1, S, G_2 and M phases, (iii) Cytokinesis occurs in G_1.
(b) The cell cycle progression in *S. pombe*.

(iii) It has typical eukaryotic cell cycle with G_1, S, G_2 and M phases.

(iv) Here nuclear membrane does not break down during M phase.

Sequence of regulation:

(i) A haploid yeast cell decides at a point in early G_1 whether to proceed through a mitotic cycle or to mate with another haploid cell to form diploid which forms haploid spores through meiosis.

(ii) 'Start' is the point in early G_1 at which cells become committed to the mitotic cell cycle. Cells which have completed this point are committed to the mitotic cycle. Those are not able to complete it, undergo mating pathway.

(iii) In fission yeast, two genes cdc-2 and cdc-10 have been identified whose functions are required for start. The cdc-2 gene also participates prior to M phase of the cell cycle.

(iv) In G_1, the other two genes cdc-20 and cdc-22 are likely to be required for initiation of DNA replication. The factor Rum-1 is active during G_1 and G_2, and inhibitor of M phase kinase. It controls entry to S phase.

(v) The control regulating the initiation of mitosis at late G_2 has been investigated. The wee-1 mutant genes initiate mitosis and cell division. Normally, wee-1 (wild type) inhibits mitosis. The cdc-2 gene product induces mitosis. During mitosis cdc-2, cdc-13 dimer are in active state.

(vi) The cdc-25 gene has been identified. The product of cdc-25 is needed for activation of cdc-2. The cdc-25 and wee-1 gene functions are antagonistic and independent in their regulation initiation of mitosis.

(vii) The suc-1 gene product acts as a regulatory component of cdc-2.

(viii) The cdc-13 gene is required for the progression of mitosis. In *S pombe* cycline of M phase kinase is encoded by cdc-13.

(ix) The gene mik-1 encodes a homologue of wee-1 that also regulates the inactivation of cdc-2 by phosphorylation of Tyr-15. wee-1 also phosphorylates serine or threonine.

B. Cell cycle regulation in budding yeast:

The baker's yeast or budding yeast (*Saccharomyces cerevisiae*)

(i) It is a oval cell and divides by forming buds.

(ii) Buds enlarge continually and ultimately separates from mother cell.

(iii) It first appears in G_1 and grows until it separates from mother cell after mitosis.

(iv) It has normal G_1 and S phase but does not have a normal G_2.

Sequence of regulation:

The cell cycle of *S. cerevisiae* consists of three cycles that separate after start and join before cytokinesis. The cells may be diverted to mating pathway before the point of start, i.e., early G_1.

(a) Chromosome cycle

(b) Cytoplasmic cycle

(c) Centrosome cycle

(a) Chromosome cycle:

(i) Duplication and separation of chromosome.

(ii) Completion of S phase and nuclear division.

(b) Cytoplasmic cycle:

Bud emergence and nuclear migration into the buds.

(c) Centrosome cycle:

(i) Duplication and separation of spindle polar body (SPB).

(ii) Microtubules are formed.

Steps of regulation

(i) In budding yeast (*S cerevisiae*) a single cyclin dependent kinase (Cdk) encoded by CDC-28 gene which interacts with several different cyclins during different phases of the cell cycle.

(ii) Just after cytokinesis and before the 'START', the cells can be diverted into mating type pathway by **mating factors** and cdc-36, cdc-39, far-2 and fus-3 (which appear to block the cycle before START).

(iii) After cytokinesis both the mother cell and the bud cell remain in the G_1 period. With proper environmental condition cells are able to pass 'START'. Prior to bud formation spindle polar body (SPB), duplication and DNA replication occur and nutrients are exhausted.

(iv) In G_1 stage, with sufficient growth medium, the cells reach the critical size and committed to complete cell cycle.

(v) The crucial gene CDC-28 is responsible in passing the cells through 'START'. CDC-28 is homologous to cdc-2 in *S pombe.* The three cyclins cln-1, cln-2, cln-3 encoded by CLN-1, CLN-2, CLN-3 gene respectively are active in G_1 phase.

(vi) The complexes are formed between cdc-28 and the three G_1 cyclins (cln-1, cln-2 and cln-3) and have protein kinase activity. They constitute a hypothesized 'S' phase promoting factor (SPF). cln-3 is expressed throughout the cell cycle, and cln-1 and cln-2 are expressed at the late stage of G_1. When their accumulation reaches a threshold level, that triggers the passage of 'START' to 'S' phase.

(vii) cdc-28, cln-3 phosphorylate and activate SPF and MPF. These induce transcription of CLN-1 and CLN-2 genes as well as other genes required for DNA replication.

(viii) cdc-28-cln-1 and cdc-28-cln-2 phosphorylate APC (anaphase promoting complex) in late G_1 and inactivate it.

(ix) Two B types cyclin genes CLB-5 and CLB-6 are also regulated by MBF (*Microtubules Binding Factor*) and transcribed in late G_1. The corresponding proteins Clb-5, Clb-6 accumulate to inactivate APC.

(x) At late G_1, cdc-28, clb-5 and cdc-28, Clb-6 heterodimers accumulate and they are inactivated by Sic-1 which is an 'S' phase inhibitor but has no effect on cdc-28-cln complex. When Sic is being degraded following ubiquitination by E-2 associated with E-3, then cdc-28, clb-5 and cdc-28, clb-6 kinases induce DNA replication. Sic-1 is a cdk inhibitor protein (CKI) which binds M-Cdk in late G_1. Sic is inhibited by M-Cdk.

(xi) Initiation of DNA replication requires both assembly of cdc-28-cln complex. Another heterodimeric protein kinases Cdc-7, Dbf-4 expressed in G1 and trigger initiation of DNA replication. Mcm (**Minichromosome maintenance complex**) and cdc-45 move away from the origin.

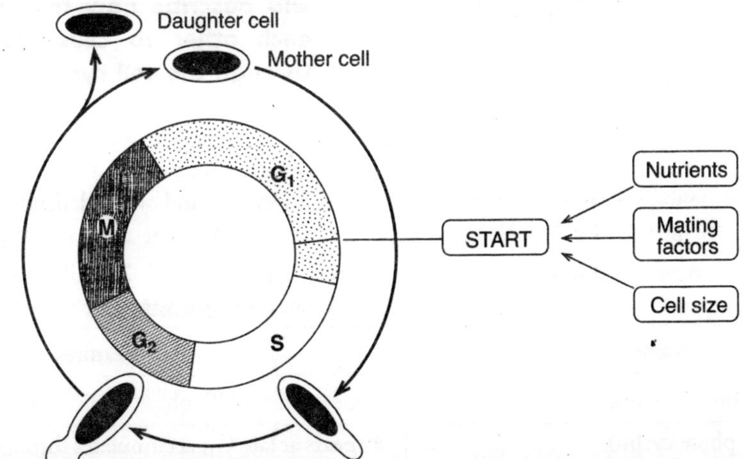

Fig. 5.3: Regulation of cell cycle in budding yeast (*S. cerevisiae*). Regulatory point 'START' is at the late G_1. START is controlled by nutrients, mating factors and cell size. Yeast divides by budding. Buds form just after 'START' and continue growing until they separate from mother cell after mitosis. The daughter cell is formed from the bud and is smaller than the mother cell and therefore, requires more time to grow during G_1 phase of the next cell cycle. Here lacks distinct G_2 phase.

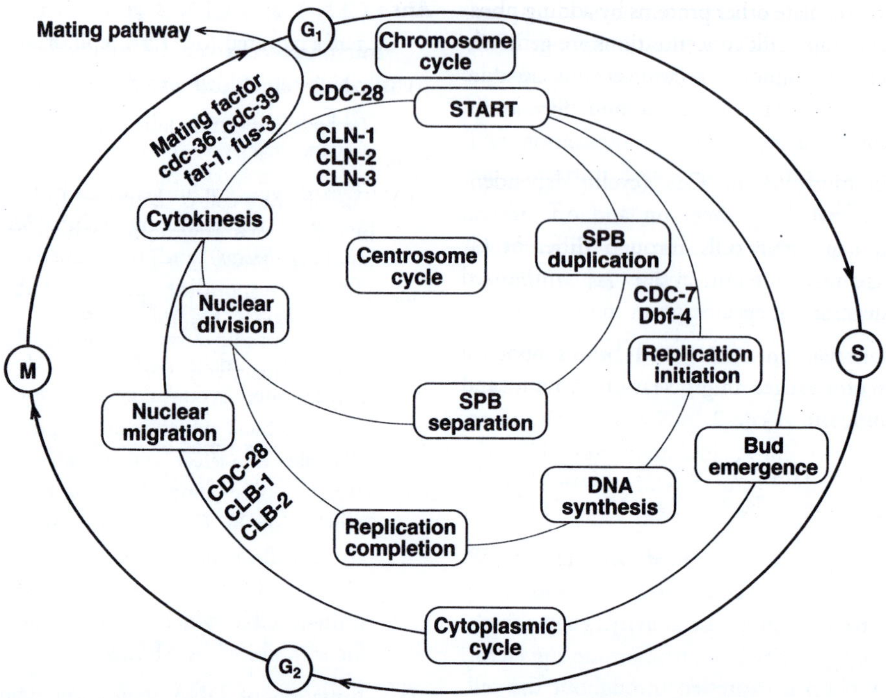

Fig. 5.4: The cell cycle in *S. cerevisiae* consists of three cycles separate after start and cytokinesis.

(xii) Later in S phase transcription of CLB-3 and CLB-4 gene occurs that encodes two B cyclins clb-3 and clb-4 which form heterodimeric protein kinase with cdc-28. These cdc-28, clb-3 and clb-4 form a complex that initiates spindle formation at the beginning of mitosis.

(xiii) After replication cells enter into G_2 phase, two more B cyclins clb-1 and clb-2 encoded by CLB-1, CLB-2 are synthesized. They form a complex with cdc-28 that is required for chromosome segregation and nuclear division.

(xiv) All the B cyclins are degraded in late anaphase.

- **Progression through the cell cycle depends on the interaction of two types of regulatory proteins, kinases and cyclins. List the function of each and describe how they interact with each other to cause cells to move through the cell cycle.**

Ans.:

(i) Progression through the cell cycle is controlled by cyclins and cyclin dependent (Cdk) kinases and regulated at a series of checkpoints.

(ii) Cdk-1 and Cdk-2 regulate entry into mitosis and S phase respectively.

Names	Features
1. Cyclin B (Mitotic cyclin)	Appears in mid 'S' phase, peaks at G_2/M transition.
2. Cyclin A (S phase cyclin)	Appears at late G_1, accumulates through S and peaks at G_2.
3. Cyclin C, D1, D2, D3 and E (G1—cyclin)	C—accumulates at G_1 and declines at S. D_1—accumulates at early G_1 and remains constant. D_2, D_3 and E—accumulate at last half of G_1 and reach peak at S and decline at G_2.

(iii) Kinases regulate other proteins by adding phosphate; group. Cdk concentrations are generally constant throughout the cell cycle but Cdks are active only when they are bound by a cyclin protein.

(iv) Cyclin binds to the Cdk (cyclin dependent kinase), switching them on and off. Several cyclin can move cells through different G_1 checkpoints. Different cyclins are synthesized and degraded at specific point in cell cycle.

(v) Phosphorylation occurs and brings about a series of changes in nuclear membrane and cytoskeleton.

- **Mention the different stages and checkpoint of cell cycle. What is G_0?**

Stages of cell cycles are G_1—S—G_2.

G_1 = Begins after mitosis and is involved in the synthesis of many cytoplasmic elements.

S = DNA synthesis occurs.

G_2 = Begins after S phase. Growth and preparation occurs for mitosis.

- **The G_1/S checkpoint:**

(i) A cell that passes the G_1/S boundary is committed to mitosis.

(ii) It is controlled by Cdk/cyclin E.

- **The G_2/M checkpoint:**

Entry into mitosis depends on the activation of Cdk 1/cyclin B by the phoshatase Cdc 25 C.

- **The spindle checkpoint**: Separation of chromatids at anaphase of mitosis is triggered by the anaphase promoting complex (APC) or **cyclosome**.

G_0 = (i) It is a point in the G_1 phase at which cells withdraw from the cell cycle and enter a prolonged nondividing but metabolically active state.

(ii) They can become terminally differentiated.

(iii) Some G_0 cells do not become terminally differentiated but are **quiescent**.

- **With the help of experimental evidence, explain the methods used to estimate the duration of cell cycle.**

Ans.:

Interphase cells can be subdivided into three parts, which can be delineated in two ways.

The first is percent labelled mitosis technique (Fig. 5.5):

(i) The dividing cell is incubated for a short time in radioactive thymidine, which is incorporated into DNA as it is synthesized.

(ii) Following this pulse, the cells are "chased" in excess nonradioactive thymidine. The fate of these radioactively tag cells can now be followed.

(iii) After various times of chase, cells are sampled for autoradiography, and the percentage of mitotic figures that are labelled is scored.

Step I:

(i) Initially (pulse) there are labelled interphase cells but no label mitosis. This shows that DNA must be made during interphase but not during mitosis.

(ii) On examination, interphase nuclei show only a fraction of them labelled, implying that DNA synthesis (S synthesis phase) occupies only part of interphase.

Step II:

(i) As the time of chase prior to autoradiography lengthens, labelled mitosis is observed. These mitotic cells were in 'S' phase during the labelling pulse, then passed through a non-DNA synthesis G_2 (gap) phase before mitosis.

(ii) The chase time for 50% mitotic labelling is in an estimate of G_2.

Step III:

(i) The S phase is the time between the 50% mitotic labelling points on the rise and fall of the wave of labelled mitosis (Fig. 5.6).

(ii) The total cell cycle time (T_c) is the interval between successive waves of labelled mitosis.

(iii) Calculation of G_2, S and T_c coupled with the estimate of M, usually does not account for all of the time spent in the cell cycle.

(iv) The remaining portion, another period of interphase when DNA is not synthesized, is

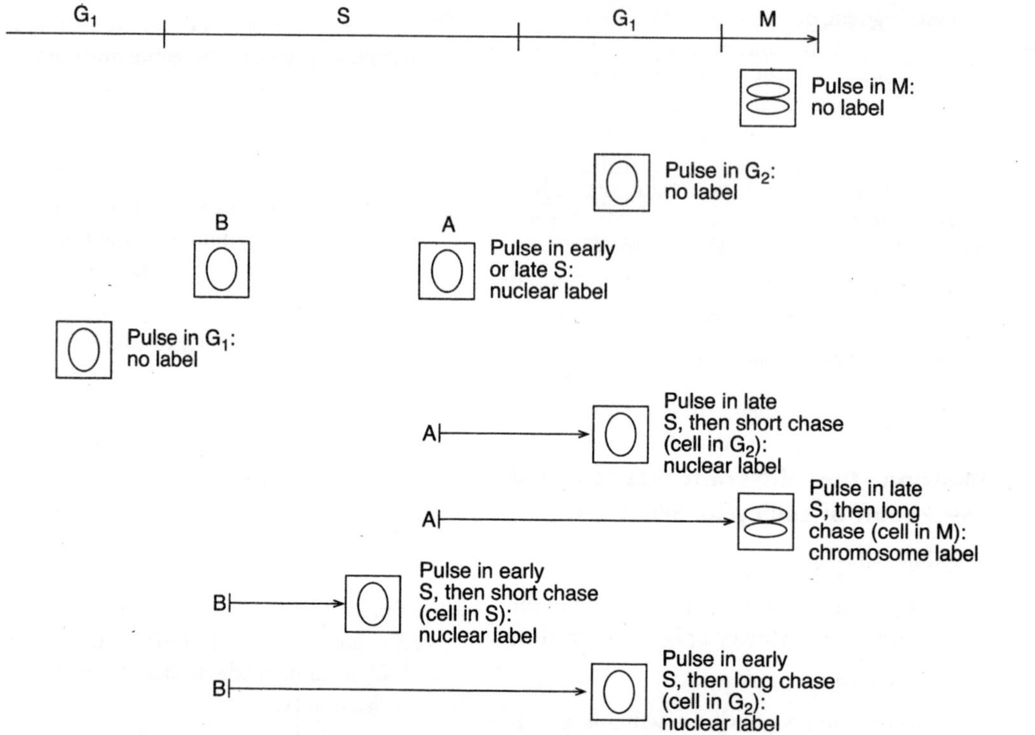

Fig. 5.5: The labelled mitosis method to determine the lengths of cell cycle phases. Cells are incubated in radioactive thymidine (pulse) and then in nonradioactive thymidine (chase).

calculated by subtraction and is termed as G_1. Thus, cell cycle is

$$G_1 \longrightarrow S \longrightarrow G_2 \longrightarrow M$$

S = The average S period is taken as the time between the two points in the first wave where 50% of mitosis are labelled.

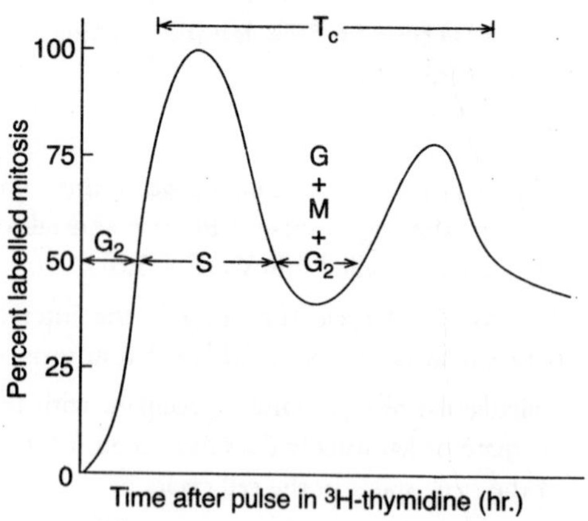

Fig. 5.6: Typical results using percent labelled method

G_2 = (i) The time between the start of this experiment and the first of these point (the point where 50% of the labelled mitosis were scored) is taken as $G_2 + \frac{1}{2}M$.

(ii) Half of the mitotic time is added to the G_2 because mitosis is normally scored in metaphases, which means the cells have to go through not only G_2 but also prophase before they are scored and prophase lasts roughly $\frac{1}{2}$ of the time of mitosis.

The total cycle time (T_c)

The total cycle time is roughly between two similar points (e.g., the peaks) in the first and second cycles.

G_1 period = $G_1 + \frac{1}{2}M$ is obtained by difference,

i.e., $G_1 + \frac{1}{2}M = T_c - S - G_2 - \frac{1}{2}M$

$$G_1 = T_c - S - G_2 - \frac{1}{2}M - \frac{1}{2}M$$

$$= T_c - S - G_2 - M = T_c - (S + G_2 + M)$$

- **ONCOGENE:**

(i) It is a gene whose protein products promotes cell proliferation as well as tumour formation and can transform a cell cancerous which is found in active state in **retrovirus** and **transformed cell** and inactive state in **non-transformed cell**.

(ii) It is a mutant form of **proto-oncogene**.

(ii) It may be viral or **v-onc** (present in virus) and cellular or **c-onc** (present in host cell).

- **PROTO-ONCOGENE:**

It is a gene in normal cell whose protein products usually concerned with the regulation of cell proliferation and which can be converted into a cancer promoting oncogene by mutation.

- **PRODUCTS OF PROTO-ONCOGENE:**

Gene	Products
CCD1	Cyclin
CDK 2 & 4	cdK

- **TUMOUR SUPPRESSOR GENE:**

It is a gene in normal cell whose protein products inhibit normal cell proliferation. Some tumor suppressor gene products prevent in appropriate cell cycle progression. When both copies of the tumour suppressor genes are mutated cellular transformation or cancer results.

Class	Gene	Chromosome
Oncogene	K-ras	12
Oncogene	myc	8
Tumor suppressor	p53	17 p13.1
Tumor suppressor	Rb	13q14.1 q14.2

- **Why are the most tumour suppressor genes expected to be recessive?**

(i) Since tumor or cancers represent a significant threat for survival.

(ii) Therefore, Darwinian fitness, strong evolutionary forces would favour a variety of co-evolved and perhaps complex conditions in which mutations in these suppressor genes would be recessive.

(iii) Tumour suppressor gene regulate cell cycle by producing a protein product (p^{53}). Cellular conditions have evolved in such a way that sufficient quantities of such gene product are made from just one gene (two for diploid) to provide normal function.

- **Characteristics of Oncogene:**

Oncogenes are mutant forms of normal cellular genes (proto-oncogene), which have important role in regulating cell division and cell proliferation.

(i) When oncogenes are carried by virus it is called V-oncs. If they reside in the host chromosome, it is called **C-oncs**.

(ii) A transducing retrovirus, then carries a significantly altered form of cellular proto-oncogene (now—a **V-oncs**).

When the transducing retrovirus infects a normal cell the hitchhiking oncogenes transform the cell into a **cancer cell**.

"Cell cycle Dysregulation result in Cancer"—Justify

(i) Dysregulation of the cell cycle is the cardinal feature of cancer cells. These cells progress relentlessly through round after round of mitosis, even when it would be more appro-

Table 5.1: Difference between Proto-oncogene or Cellular oncogene and Viral oncogene **C.U. 2019**

Cellular oncogene (C-oncs)	Viral oncogene (V-oncs)
1. Cellular oncogene and proto-oncogene have multiple exons separated by introns.	1. Viral oncogenes have single exon and no introns.
2. The chicken cellular src proto-oncogene contains 12 exons and 11 introns.	2. In Rous sarcoma virus–V-src gene has single uninterrupted coding sequence.
3. C-src protein is 533 amino acid long.	3. V-src protein is 526 amino acid long.
4. It regulates cell division and cell proliferation.	4. These are transduced to normal host cell and induced oncogenesis.

priate to pause exist from the cycle, or commit suicide by **apoptosis.**

(ii) Not surprisingly, the genes that generate and interpret the signals controlling cell cycle progression feature strongly in list of oncogenes and tumour suppressor genes.

(iii) Progression through cell cycle is controlled by cyclins and cyclin dependent kinase and regulated at a series of check points.

(iv) Progression is controlled by fluctuating levels of a series of cyclin proteins which are required to activate cyclin dependent kinases.

(v) The upstream of the Cdks is a crowd of activators and inhibitors that act on the cyclin-Cdk complexes.

(vi) The whole network of kinases, cyclins, inhibitors and activators integrates many different signals, ensuring a flexible and responsive control of the cell cycle.

(vii) In G_1 phase, the cell spends most of their time and does the most of the work. Controls in progression through G_1 are especially important targets of oncogenic mutations.

(viii) Progression from one phase of the cell cycle to the next is controlled at a series of check points.

Check points	Control
G_1/S check point	Cdk-2/Cyclin E
G_2/M check point	Cdk-1/Cyclic B
Spindle check point	Anaphase promoting complex (APC)

In cancer cells, all the checkpoint mechanisms are defective, cells replicate their DNA despite damage, enter mitosis with unrepaired damage and become inefficient at segregating their chromosome correctly. All of these destabilize the genome and lay their ground for evolution toward full blown malignancy.

• CHARACTERISTIC FEATURES OF ROUS SARCOMA VIRUS:

In 1910 Francis Peyton Rous first inferred the existence of specific genes associated with transformation of normal cell into **cancerous cell.**

He studied tumour in chickens known as **sarcoma.** The infectious agent which causes connective tissue tumour, i.e. sarcoma of chicken is known as **Rous sarcoma virus** (RSV).

(i) It is a single-stranded oncogenic virus.

(ii) It infects cells and reproduces within them. Once inside the single-stranded RNA genome of RSV is transcribed by the enzyme **reverse transcriptase**, converting the RNA genome into a single-stranded DNA molecule.

(iii) Single-stranded DNA molecule is used as a template to synthesize a complementary strand, creating a double-stranded DNA molecule that integrates into host genome forming a **provirus.**

(iv) Later DNA is transcribed into RNA which is translated into viral proteins. Packaging RNA molecules into these proteins form new virus particles. Because the replication cycle of like RSV reverse the flow of genetic informations; they are called **retrovirus.**

(v) In RSV, tumours result from a single gene src present in the viral genome. This gene is **oncogene.**

  (a) The retroviruses that carry oncogenes are known as **acute transforming virus.**

  (b) The retroviruses that do not carry oncogenes but can induce transcription of cellular genes that cause tumour formation are known as **nonacute viruses.**

• Structure of a Retrovirus:

(i) RNA viruses have an RNA genome which is extended at either end by a long terminal repeat (LTR). The LTR contains many of the signals that allow retrovirus to function.

(ii) Retroviruses use their genomic RNA as a template to make DNA with the help of **reverse transcriptase.** This DNA is then integrated into host's DNA as **DNA provirus.**

(iii) The DNA provirus is transcribed to yield genome length RNA. Provirus directed transcription involves a promoter—a sequence that directs the RNA polymerase to a specific initiation site and an enhancer—a sequence that facilitates transcription.

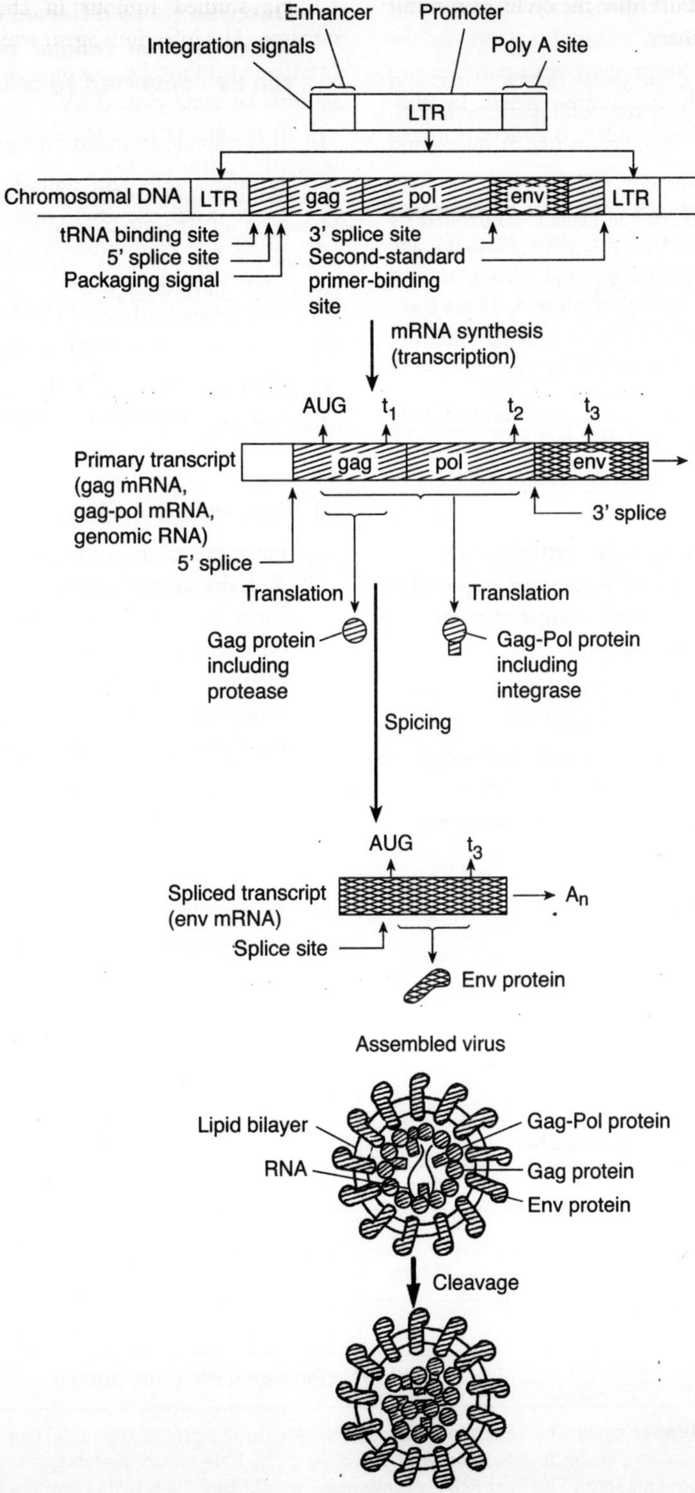

Fig. 5.7: Genetic elements of proviral DNA and the corresponding gene products.

(iv) The promoter and enhancer are located in the LTR. The primary transcripts serve as the Genomic RNA for progeny virus particles and as mRNA for the gag and pol genes. In addition, the full length RNA is spliced to yield mRNA for env.

(v) The **gag** gene encodes the viral **protease and structural proteins** of the virus particles, **pol** encodes **reverse transcriptase** and **integrase** and **env** encodes **envelop glycoprotein.** These three genes are only required for viral replication but play no role in cell transformation.

(vi) This type of retrovirus causes tumour only when any mutation results at the time of integration of proviral DNA within or adjacent to host's genome.

(vii) But there are some other **retroviruses** which contain specific genes which are responsible for the induction of **cell transformation** and acts as a potent **carcinogens.**

The first cancer-causing gene is found in retrovirus, called **Rous sarcoma virus** (RSV) that produces sarcomas in chicken. It was later named as **src** gene. Genes like src which are capable of inducing malignant transformation are referred to as **oncogenes.**

- **Describe three different mechanisms by which normal cellular proto-oncogene can be converted to cellular oncogene.**

In normal cells, expressions of proto-oncogene is tightly controlled so that cell growth and division occur only as appropriate for the cell type involved. However, when proto-oncogenes are changed into oncogenes the tight control can be lost, and unregulated cell proliferation can take place.

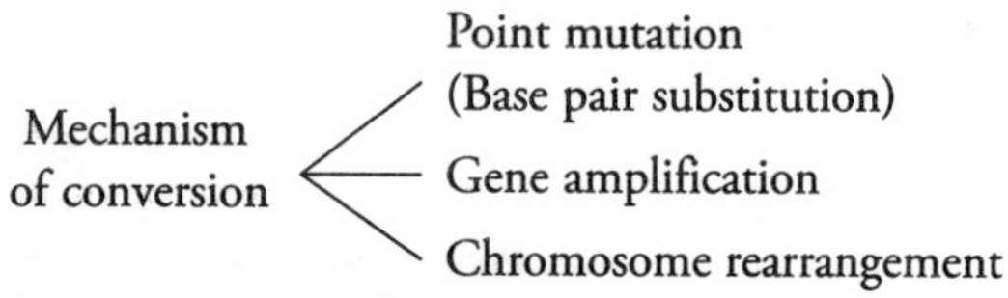

(a) Point mutation (Base pair substitution):

(i) Point mutation in the coding region of a gene or in the controlling sequences (promoter, regulatory elements, enhancer) can change proto-oncogene into oncogene by causing an increase in either the activity of the gene product or the expression of the gene, leading in turn to an increase in the amount of gene product.

(ii) For example, the genomes of all vertebrates contain three distinct but closely related **ras**

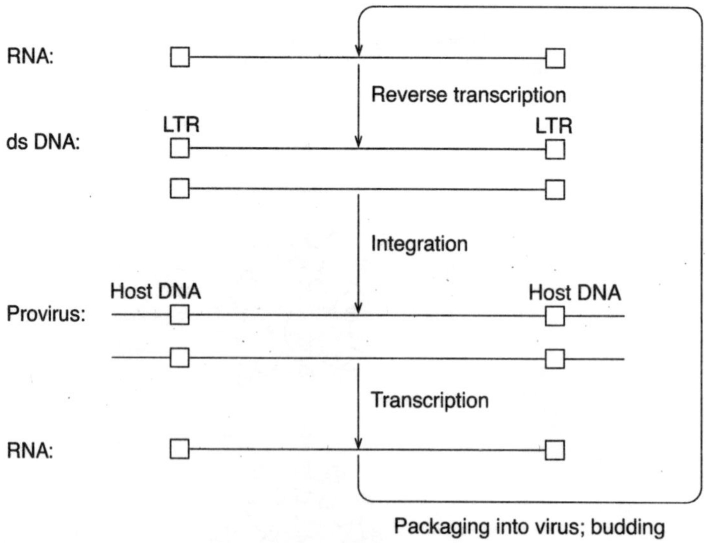

Fig. 5.8: **Retroviral replication cycle**: The viral genome is an RNA with long terminal repeat (LTR) at each end. Reverse transcriptase makes a linear double-stranded DNA copy of the RNA which then integrates into the host DNA creating the provirus form. The host RNA polymerase II transcribes the provirus forming genomic RNA. The viral RNA packaged into a virus particles which buds out of the cell and infects another cell, starting the cycle over again.

proto-oncogenes. Two of these are **c-H-ras** and **c-k-ras**, and are closely related to the **v-ras** oncogenes of the Harvey and Kristen strains respectively of the murine virus of the rat. The third one is designated as **N-ras**. All three of these cellular ras proto-oncogenes are known to encode very similar GTP binding protein with GTPase activity.

(iii) The **ras** gene codes for a protein p21 is made up of 188 amino acid residues. Single-point mutation in codons 12, 59, 61 result in activation of mammalian **ras** gene, which results the formation of mutant protein that can transform normal cells into malignant cells. Here GGC (glycine) is converted into GTC (valine). These types of mutations are found in the malignant cells of bladder cancer, lung cancer, colon cancer, etc.

(iv) The effect of mutations is to cause the G-proteins to lose their ability to be regulated. So that constitutive growth signals are transmitted into the cell. As a result unregulated cell proliferation can commence.

(b) Gene amplification:

(i) Some tumours have multiple copies of proto-oncogenes. These have probably occurred by random over replication of small segments of the genomic DNA.

(ii) One of the few instances wherein gene amplification is definitely correlated with the cancer of neuroblastoma. In these tumours, the gene **N-myc** (with sequence homology to **c-myc**) is amplified and is associated with the tumour phenotype.

(iii) The **c-myc** oncogene is also amplified in a majority of cases of one form of small cell lung cancer.

(iv) In mouse adrenocortical tumours multiple copies of **ras** are found.

(v) In general extra copies of the proto-oncogene in the cell result in an increased amount of gene product, thereby inducing or contributing unscheduled cell proliferation.

(c) Chromosome rearrangement:

(i) Chromosomal abnormalities of various kinds occur in human cancer. Several well-characterized translocations correlate with **tumorigenesis.**

(ii) **Burkitt's lymphoma** (cancer associated with the Epstein-Barr virus) affecting humans.

Here, **c-myc** gene is present near the end of chromosome 8 when it is translocated to chromosome 14, **c-myc** is placed contiguous with the immuno-globulin heavy chain (**IgH**) locus. This gene is very active in lymphocyte. Hence, **c-myc** is now transcribed at much higher rate than normal.

(iii) The well-characterized **Philadelphia chromosome** (Ph′) is found in chronic **myelogenesis leukaemia.** The reciprocal translocations in this case result in the **c-abl** gene being moved from

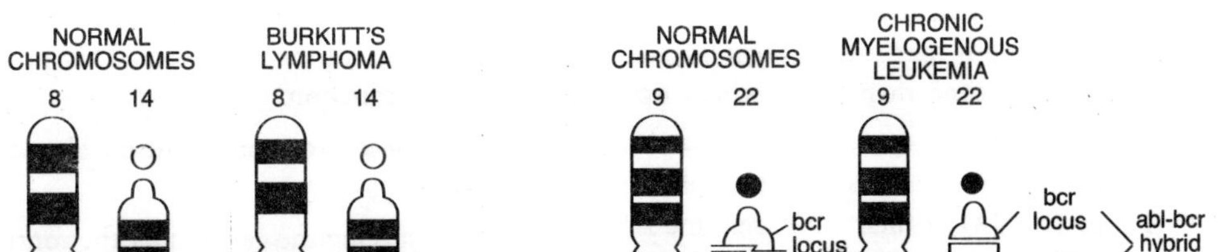

Fig. 5.9: The chromosomal translocations and associated oncogenes in Burkitt's lymphoma and chronic myelogenous leukaemia.

chromosome 9 to chromosome 22; concomitantly the **c-sis** gets translocated in the opposite direction (22 → 9). In this translocation, the breakpoint always occurs within a region of chromosome 22 known as **breakpoint claster region** which corresponds to a gene known as **bcr.** This results in enhanced transcription of **c-abl** (however, c-sis is not expressed) and the formation of abnormal **bcr-abl** fusion protein (210 kb) and to oncogenesis (Fig. 5.9).

- **State briefly the role of insertion mutation and retroviral transduction in the activation of proto-oncogene.**

- **Insertion mutation:**

(i) Not all retroviruses are pirated cellular proto-oncogene. Instead, some retroviruses may activate proto-oncogene by a process known as **insertion mutation.**

(ii) In this process a retrovirus may insert a DNA copy of its RNA genome into the chromosomal DNA of host cell.

(iii) If the proto-oncogene lies close to the site of insertion it may come under the powerful replicating influence of the viral genome.

(iv) Such an event may increase the protein product of the oncogene or create an abnormal protein product or turn on a silent proto-oncogene.

(a) Promoter insertion:

(i) Certain retroviruses lack oncogene (e.g., avian leukaemia virus) but may cause cancer over a longer period of time than those which do contain oncogenes.

(ii) When particular retroviruses infect a cell, a DNA copy (cDNA) of their RNA genome is synthesized by **reverse transcriptase** and the cDNA is integrated into the host genome. The integrated double-stranded cDNA is called a "**provirus**".

(iii) The cDNA copies of retroviruses are flanked at both ends by sequences named **long terminal repeat (LTR)** as certain transposons.

(iv) The long terminal repeat sequences appear to be important in the mechanism of proviral integration and they can act as promoters of transcription.

Example:

Infection of chicken B lymphocyte by certain avian leukaemia virus. The provirus became integrated near myc gene. The myc gene is activated by an upstream, adjacent viral LTR acts as a promoter resulting in transcription of both the corresponding myc mRNA and translation of its product.

(b) Enhancer insertion:

In some cases:

(i) The provirus is inserted downstream from the myc gene, or upstream from it, but oriented in the reverse direction; nevertheless, the myc gene becomes activated.

(ii) The activation cannot be due to promoter insertion. Since a promoter sequence must be upstream of the gene whose transcription increases and the sequence must be in the correct 5′ to 3′ direction. Instead, enhancer-sequences present in the long terminal repeat-sequences of the retroviruses under consideration appear to be involved.

Promoter and enhancer insertion mechanisms commonly operate in viral carcinogenesis. They can be classified as examples of insertional mutagenesis. Proto-oncogenes other than myc may also be involved (see Table 5.2 on next page).

- **Retroviral transduction:**

Many retroviruses pick up proto-oncogenes from host cells during infection. This process is known as **transduction.**

A. A transduced proto-oncogene may be activated in two ways:

   (a) By placing the gene under the control of a viral gene, the pirated gene may become **tumourogenic** because of sustained expression and mutation.

      (i) Here retroviral DNA (provirus) integrates randomly into cellular DNA. Sometimes there occurs a genetic rearrangement by which the transcriptional unit of the provirus connects to nearby cellular genes.

Table 5.2: Process of Insertion Mutations

(a) **Promoter insertion**

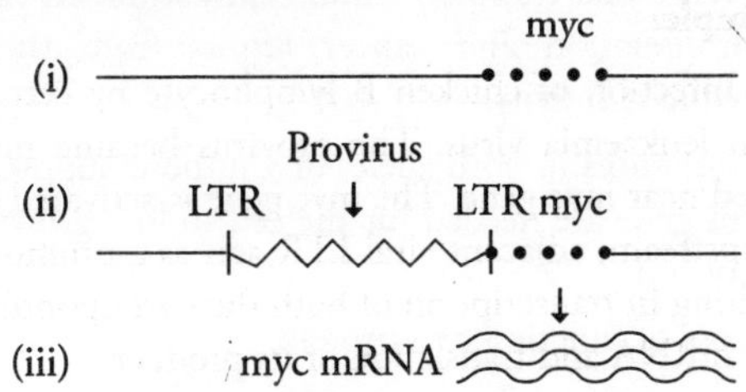

(a) Schematic representation of promoter insertion may activate proto-oncogene.

(i) Inactive myc gene.

(ii) Avian leukaemia virus (ALV) has integrated in the proviral form adjacent to myc genes. Presence of strong promoter at LTR results transcription of mRNA.

(b) **Enhancer insertion**

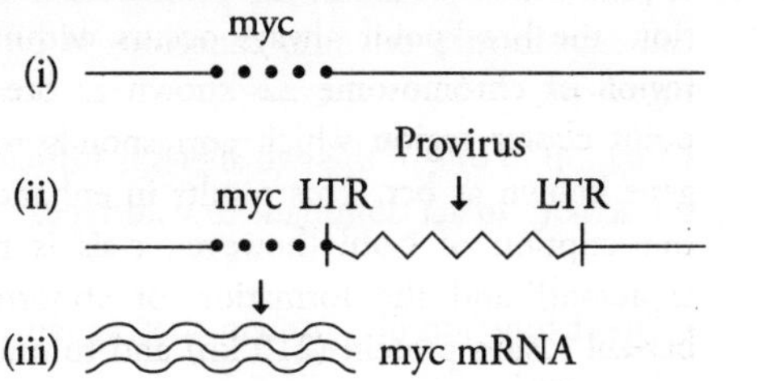

(b) Schematic representation of enhancer may activate proto-oncogene.

(i) myc inactive gene in chicken.

(ii) ALV has integrated in the pro-viral form, adjacent to myc gene.

(iii) Site of integration is just down-stream of myc gene. It cannot act as promoter, instead it acts as enhancer and produces mRNA.

(ii) Here often deletion events involving the loss of some or all of gag, pol, env genes. In this way viral RNA contains all or part of a cellular gene.

(iii) All viral progenies then carry cellular gene and under the influence of viral promoter in the LTR express cellular protein in infected cell.

(iv) If the cellular gene picked-up was a proto-oncogene, the modified retrovirus will be **oncogenic**.

(b) By fusing proto-oncogene with a structural gene whose product may stimulate cell division.

• CARCINOGENESIS: THE MOLECULAR BASIS OF CANCER

(a) Non-lethal genetic damage lies at the heart of carcinogenesis:

Such genetic damage (or mutation) may be acquired by the action of environmental agents such as chemical radiation or viruses or it may be inherited in the germ line.

(b) Three classes of normal regulatory genes:

The growth promoting **proto-oncogenes**, the growth-inhibiting genes and **cancer suppressor genes** (anti-oncogenes) that regulate programmed cell death or **apoptosis** are the principal target of genetic damage.

(i) **Mutant alleles of proto-oncogenes** are considered **dominant** because they transform cells despite the presence of their normal counterpart.

(ii) In contrast, both normal alleles of the **tumour suppressor genes** must be damaged for transformation to occur, so this family of genes is sometimes referred to as **recessive oncogene**.

(iii) Genes that regulate apoptosis may be dominant as proto-oncogenes or they may behave as cancer suppressor genes.

(c) A fourth category of genes that regulate repair of damaged DNA, are also pertinent in carcinogenesis.

(i) The DNA repair genes affect cell proliferation or survival indirectly by influencing the ability of the organism to repair non-lethal damage in other genes, including proto-oncogenes, tumour suppressor genes and genes that regulate apoptosis.

(ii) A disability in the DNA repair genes can predispose to widespread mutation in the genome and hence to **neoplastic** transformation.

Oncogene	Tumour suppressor gene
1. A gene which promotes cell proliferation and can transform a cell into cancerous is called oncogene.	1. Genes that normally inhibit un-controlled cell proliferation and cancer results from their reduced activity.
2. Mutation in one of the two alleles is sufficient for activity to act dominant to wild type.	2. Mutations in both alleles of a tumour suppressor gene are needed for the cell to lose growth control.
3. Gain of function of a protein that signals cell division.	3. Loss of function of a protein.
4. Mutation arises in somatic tissues, not inherited.	4. Mutation present in germ cell (can be inherited) or somatic cell.
5. Some tissue preference.	5. Strong tissue preference (e.g., effect of RB gene in the retina).
6. Example: H-ras, K-ras, abl, myc.	6. Example: p^{53}, RB.

Sporadic cancer	Hereditary cancer
1. It is not associated with germ line mutation.	1. It is associated with inheritance of germ line mutation.
2. This cancer does not appear to be inherited and is known as sporadic cancer.	2. This cancer runs in the families and is known as familial or hereditary cancer.
3. Sporadic cancers are more frequent.	3. Hereditary cancers are less frequent.
4. Sporadic retinoblastoma occurs in 60% cases. Here both alleles are inactivated somatically.	4. Hereditary retinoblastoma occurs in 40% cases.
5. Here both alleles are mutated somatically (RB/RB).	5. In hereditary retinoblastoma the patient inherits one copy of retinoblastoma gene through germ line (RB^+/RB). Single additional mutation required to produce retinoblastoma (RB/RB).

- **Some properties of p^{53} gene:**
- This gene is located on short arm of chromosome 17 ($17p^{13\text{-}1}$).
- The gene is about 10 kb and with 11 exons.
- Product (p^{53}) is nuclear phosphoprotein of 53 kDa. p^{53} binds specific sequences in ds DNA.
- p^{53} acts as transcriptional regulator, presumably regulating genes that participates in cell division.
- p^{53} binds various viral proteins (e.g., SV40 large T-antigen) forming inactive oligometric complexes.
- p^{53} is a tumour suppressor gene which controls the passage of the cell from G_1 to S phase of cell cycle.
- If p^{53} is inactivated DNA damage would accumulate in the cell and the cell would be genetically less stable. Thus, p^{53} has popularly acted as "**guardian of the genome**" or "**molecular police man**".
- Mutation in p^{53} gene is most common genetic alterations resulting in human cancer (lungs, colon and breast).
- The mutations are usually found at highly conserved codons.

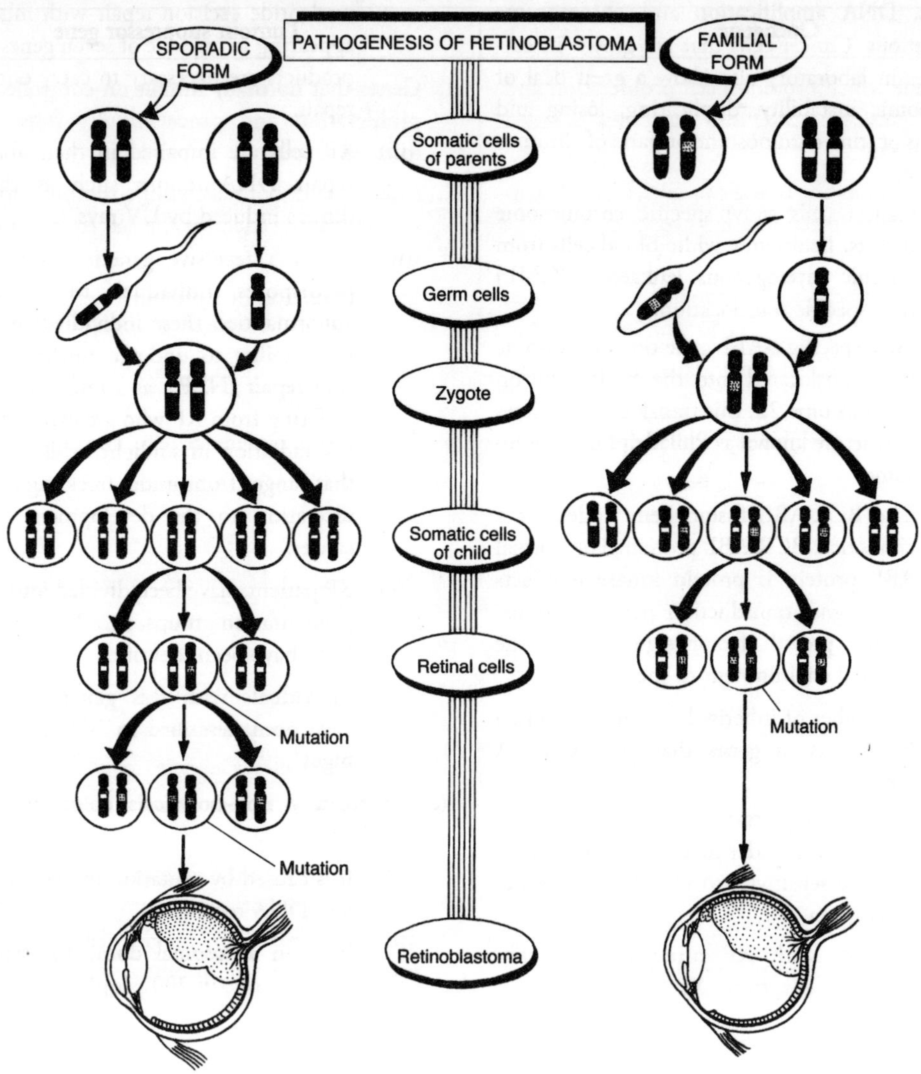

Fig. 5.10: Pathogenesis of retinoblastoma. Two mutations of the Rb locus on chromosome 13q14 lead to neoplastic proliferation of the retinal cells. In the familial form, all somatic cell inherit one mutant Rb gene from a carrier parent. The second mutation affects the Rb locus in one of the retinal cells after birth. In the sporadic form, on the other hand, both mutations at the Rb locus are acquired by the retinal cells after birth.

- **"Cancer cells contain genetic defects—affecting genetic instability and DNA repair." Discuss.**

(i) Cancer cells show higher normal rates of mutation, chromosomal abnormalities and genomic instability.

The fundamental defect in cancer cell is a derangement of the cell's normal ability to repair DNA damage. This loss of genomic integrity leads to mutations in specific genes that control various aspects of cell proliferation, programmed cell death and cell to cell contact.

(ii) In turn, the accumulation of mutation in genes controlling these processes leads to cancer. The high level of genomic instability seen in cancer cells is known as the **mutator phenotype.**

(iii) The genomic instability in cancer cells manifests itself in the presence of gross defects, such as **translocations, aneuploidy, chromosome**

[P.G., M.G., E.G. & I.G. 13]

loss, DNA amplification and **chromosome deletions**. Cancer cells that are grown in cultures in laboratory also show a great deal of **genomic instability**, duplicating, losing and translocating chromosome or parts of chromosome.

(a) (i) Cancer cells show specific chromosome defects. Leukaemic white blood cells from **chronic myelogenous leukaemia (CML)** bear specific translocation.

Here specific c-ABL gene on chromosome 9 is translocated into the BCR gene on chromosome 22, this translocation creates a structure known as **Philadelphia chromosome.**

(ii) The BCR-ABL fusion gene codes for a chimeric c-BCR-ABL protein. The normal ABL protein is protein kinase that acts with signal transduction pathways transferring growth factor signals from the external environment to the nucleus.

(b) (i) A number of inherited cancers are caused by defects in genes that control DNA repair.

Xeroderma pigmentosum (XP) is a rare hereditary disorder that is characterized by extreme sensitivity to UV light and other carcinogen.

(ii) Patients with XP often develop skin cancer. Cells from XP are defective in nucleotide excision repair with mutations appearing in any one of seven genes whose products are necessary to carry out DNA repair.

(iii) XP cells are impaired in their ability to repair DNA lesions such as thymine dimers induced by UV rays.

(iv) XP is a recessive genetic disorder that predisposes individuals to severe skin abnormalities, these individuals have lost their ability to undergo **nucleotide excision repair (NER)**; as a result individuals suffering from XP who are exposed to the UV-radiation in sunlight exhibit reaction that ranges from initial freckling and skin ulceration to the development of skin cancer.

(v) XP patients have been divided into 7 complementation groups, i.e., 7 genes XPA to XPG have been identified.

(vi) Individuals with this genetic condition have been described as **"Children of the night"**.

(c) **Hereditary non-polyposis colorectal cancer (HNPCC)**

(i) It is caused by mutation in genes controlling DNA repair.

(ii) It is an **autosomal dominant syndrome** affecting one in 200 people.

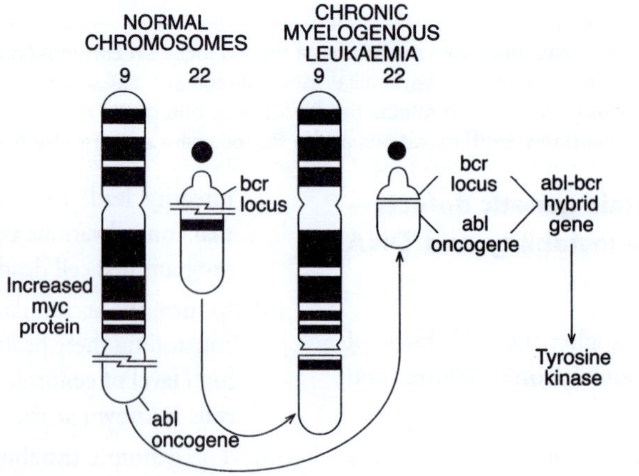

Fig. 5.11: The chromosomal associated oncogenes in Burkitt's lymphoma and chronic myelogenous leukaemia.

(iii) At least eight (8) genes are associated in HNPCC and four (4) of these genes are involved in DNA mismatch repair.

(iv) Inactivation of any of these four genes MSH-2, MSH-6, MLH-1 and MLH-3 causes rapid accumulation of genome wide mutation and subsequent development of colorectal and other cancer.

Table 5.3: Difference between Benign and Malignant tumour

Features/Symptoms	Benign tumour	Malignant tumour
1. Mode of onset	Insidious.	Sudden.
2. Rate of growth	Slow, chronic process by expansion.	Rapid by infiltration into the tissue.
3. Age of incidence	Usually younger age.	Adult or old.
4. Surface	Smooth well-capsulated compact, removal is possible.	Irregular, noncapsulated, vascular congestion is prominent, but removal is not possible.
5. Cut surface	(i) Fibrous capsule at periphery. (ii) Capsule developed by compressed normal parenchyma.	(i) No capsule at periphery. Sometimes pseudocapsule is present. (ii) Heterogenous surface with haemorrhage and necrosis are noticed.
6. Infiltration of surrounding medium	The tumor is fully demarcated from the surrounding tissue.	Surrounding tissue shows infiltration and invasion by the proliferating tumour cells.
7. Metastasis	No distant spread of the tumor cells takes place.	Distant spread of the tumour cell with involvement of the draining lymph node and vital organ of the body is common.
8. Histological		
(a) Cell character	Cells are almost normal in shape and size.	Cells are pleomorphic.
(b) Nucleus	Normal.	Proportionally larger, hyperchronic, often eccentric.
(c) Reproduction	The parent tissue from which the tumour develops almost always retained.	Sometimes lack of reproduction. No resemblance with the parental tissue.
(d) Mitosis	Nuclear division insignificant.	Evidences of mitosis and giant cell formation are present.
(e) Differentiation	Fully differentiated like adult cell.	Undifferentiated or anaplastic.
(f) Loss of polarity	Absent.	Present.
(g) Blood vessel	(i) Adult type. (ii) Haemorrhage and necrosis usually absent.	(i) Embryonic type. (ii) Haemorrhage and necrosis usually present.

Table 5.4: Difference between Carcinoma and Sarcoma

Features/Symptoms	Carcinoma	Sarcoma
1. Origin	A malignant turnover of epithelial origin.	A malignant turnover of connective tissue origin.
2. Age	Common in later life, above forty years.	Common in early life in second decade.
3. Growth	Rapidly growing.	Slowly growing.
4. Metastasis	(i) Usually spreads through lymphatics. (ii) Local draining lymph nodes are frequently involved except in adenocarcinoma of prostate, kidney, liver.	(i) Usually spread through haematogenous route. (ii) Lungs, brain, bone are involved.
5. Clinical features	(i) The turnover is fixed to the surrounding structure. (ii) The superficial overlying skin is not hyperaemic. (iii) In superficial part of the body, the turnover usually does not show fungation as haemorrhage from surface.	(i) The turnover is adhered to surrounding structure. (ii) Overlying skin is hyperaemic. (iii) The superficial part shows fungation or haemorrhage from the surface.
6. Microscopical features	(i) (a) Malignant cells are arranged in clumps or groups. (b) In between clumps of cells presence of connective tissue. (ii) Evidence of haemorrhage and necrosis are minimum. (iii) Stroma is less vascular but lymph vessels are more in number. (iv) Cells are less pleomorphic. (v) Nuclear division and mitosis and giant cells are only a few.	(i) (a) Cells are arranged roughly. (b) In between individual cells connective tissue strand is present (fibrillar structure). (ii) Haemorrhage and necrosis are frequent. (iii) Stroma is highly vascular, lymph vessels are ill found. Presence of open blood sinuses are characteristics. (iv) Cells are highly pleomorphic. (v) Mitosis is a significant feature and giant cells are frequently present.
7. Macroscopical features	(i) The turnover is soft pale, white fleshy in appearance. (ii) Consistency is homogeneous. (iii) Area of haemorrhage and necrosis is minimum.	(i) Surface show areas of haemorrhage and necrosis. (ii) Heterogenous consistency. (iii) Area of haemorrhage and necrosis is maximum.

	Tissues	Benign tumour	Malignant tumour
1.	Bone	Osteoma	Osteosarcoma
2.	Nerve	Neuroma	Neuroblastoma, Neurocytoma
3.	Cartilage	Chondroma	Chondrosarcoma
4.	Epithelia	Papilloma, Adenoma	Carcinoma
5.	Germinal cell	Dermatocyst	Teratoma
6.	Vascular tissue	Angioma	Angiosarcoma

- **Apoptosis:**

It is a genetically controlled process of programmed cell death or suicide and subsequent elimination of damaged and potentially harmful cells. It may be induced by intracellular damage or by signals from neighbouring cells.

- **Caspases:**

A member of a family of related proteases that have a cysteine at their active site and cleave their target proteins at specific aspartic acids which take part in initiating and carrying out the cell death response. These proteases are sometime called **executioners** of the cell.

There are two types of caspases, viz. **initiator and executioners.**

- **Characteristic changes during Apoptosis:**

(i) The cell membrane forms bubble-like blebs on the outer surface.

(ii) Chromatin becomes condensed and the DNA of the chromosome fragmented.

(iii) Organelle structure is disrupted and the cell loses its normal shape and becomes spherical.

(iv) The cell breaks up into small fragments called **apoptotic bodies.**

(v) Breakdown of mitochondria with release of **cytochrome** C. Fragmented cells are engulfed by phagocytic cells.

- **Characteristics of Apoptotic Cell:**

(i) Nuclear fragmentation,

(ii) Chromatin contendensation,

(iii) Chromosomal DNA fragmentation.

- **Mechanism of Apoptosis:**

Triggered by internal signals: Protein BCL-2—holds apoptotic protease-activating factor (Apaf-1)	Triggered by external signals: Some receptor proteins (FAS and TNF)	Apoptosis Inducing Factor (AIF)

- **Genetic control of Apoptosis:**

1. Some genes, responsible for apoptosis, have been identified, viz. ICE (Interleukin 16 converting Enzyme) and P^{53}.

2. **Some other factors**—viz.,

   (a) Signal protein which is released either due to some cell injury or cytokinin-mediated pathway.

   (b) Some critical proteins or modulating factors which determine whether a cell will be repaired or undergo death.

3. These genes or factors may initiate some stimuli for cell death or induce cellular susceptibilities or apoptosis.

- **Suicide gene therapy:**

Suicide gene therapy is the specific expression of a toxic gene in cancer cells to inhibit their proliferation through killing the affected cells.

- **Cancer gene therapy (a molecular approach):**

Methylation leads to silence the genes (DNA/RNA) as well as the cells. **Hypermethylation** patterns have been found in many cancer cells. Therefore, attempt has been made to devise a way to reverse **methylation**, allowing the reexpression of silenced genes, present in cancer cells.

Solved Problems

Problem 1

The nucleotide pair sequences of cellular proto-oncogenes are highly conserved in large number of species in the animal kingdom. Does the conservation of proto-oncogene structure tell us anything about the function of proto-oncogene?

Solution:

(i) The conservation of the sequence of proto-oncogenes across the animal kingdom indicates that these genes encode products that perform essential function that must be preserved in all species.

(ii) Proto-oncogene products play a central role in cellular metabolism and intercellular communication.

(iii) They provide functions that are essential for normal growth and development of the host organism.

Problem 2

How might the amplification of a proto-oncogene lead to oncogenesis?

Solution:

(i) The amplification of a proto-oncogene might result in over-production of its product—each copy of the gene contributing a certain quantity of product.

(ii) If the proto-oncogene is a positive regulator of cell division, its over-production might stimulate precocious cell division.

Problem 3

What evidence is there that the proto-oncogenes found in chromosomes of normal cells and normal animals are not simply integrated retroviral oncogenes?

Solution:

(i) The proto-oncogenes and retroviral oncogenes have very different structures.

(ii) The viral oncogenes contain a single uninterrupted coding sequence or exon, whereas the proto-oncogenes contain several coding sequences interrupted by several introns.

Example: Chicken cellular src proto-oncogene contains 11 introns separating 12 coding sequences, in RSV, v-src oncogene has a single-uninterrupted coding sequence.

Problem 4

Which is most likely: (1) that cellular proto-oncogenes originated from viral oncogenes or (2) that viral oncogenes are derivatives of cellular proto-oncogenes?

Solution:

(i) Comparisons of the sequences of cellular proto-oncogenes and related viral oncogenes indicate that viral oncogenes have evolved from proto-oncogenes.

(ii) Many of the viral oncogenes have nucleotide sequences suggesting that they have been derived from processed mRNA products of the related cellular oncogenes.

Problem 5

The majority of c-ras oncogenes have alteration in one of three codons in the c-ras mRNA that lead to substitution of amino acids in the c-ras proteins at position 12, 59 or 61 from amino terminus. How can the clustering of mutational events to these three sites in the three c-ras genes be explained?

Solution:

(i) The most of the mutations that convert c-ras proto-oncogenes to c-ras oncogenes alter three specific codons causing amino acid substitution at position 12, 59 and 61 of the c-ras proteins.

(ii) It indicates that amino acids at these sites (glycine, alanine and glutamine respectively) are critical to the normal functioning of the c-ras proto-oncogene products.

(iii)

Tissue source	ras-allele	Codon 12 59 61
Normal: Human cells	c-H-ras	GGG GCC CAG gly ala gln
Normal: Human cells	c-K-ras	GGG GCA CAA gly ala gln
Normal: Human cells	N-ras	GGG GCA CAG gly ala gln

It seems that the observed amino acid substitutions at these sites (above-mentioned table) cause the proteins to be "locked" in a form that stimulates cell division.

Problem 6

What do you mean by Gatekeeper and Caretaker genes?

Solution:

Gatekeeper genes:

(i) These genes normally inhibits cell growth.

(ii) In different cell types only one or few genes serve as gatekeeper. If both copies of the gatekeeper mutate a specific cancer may develop (retinoblastoma).

(iii) In general tumor-suppressor gene is also gatekeeper gene.

(iv) FAP (familial adenomatous polyposis) gene is a gatekeeper.

Caretaker genes:

(i) These genes (viz., MSH-2 and MLH-1) maintain the integrity of the genome including DNA repair.

(ii) These genes repair DNA damage caused by environmental agents such as ultra-violet light, mismatches that occurs during DNA replication.

(iii) Mutation of a caretaker gene does not promote tumour formation directly but leads to genetic instability, that increases the mutation rate of all genes including gatekeeper.

Problem 7

What is tumour markers? Mention its application.

Solution:

(i) Many cancers are associated with abnormal production of enzymes, proteins and hormones, which can be measured in plasma or serum.

(ii) These molecules are known as tumour markers.

Marker	Associated cancer
Carcino-embryonic antigen (CEA)	Colon, lung, breast, pancreas,
Alpha fetoprotein (AFP)	Liver, germ cell
Human chorionic gonadotrophin (HCG)	Trophoblast germ cell
Prostatic acid phosphatase (PAP)	Prostate

- **Application:**
  - Detection = Screening in asymptomatic person.
  - Diagnosis = Differentiating malignant from benign.
  - Staging = Defining extent of disease.
  - Localization = Nuclear scanning of injecting radioactive antibodies.
  - Classification = Choosing therapy and prognosis.
  - Monitoring = Predicting effect of therapy.
  - Therapy = Cytotoxic agent directed to marker containing cells.

Problem 8

Explain why HIV-1, the causative agents of AIDS, is considered a non-oncogenic retrovirus.

Solution:

(i) The retrovirus HIV-1 remains surrounded by a viral envelop in which gp120, a viral encoded glycoprotein, remains embedded.

(ii) The gp120 glycoprotein is recognised by CD4 receptor of helper T-cells (TH cell) of our immune system.

(iii) Once the virus is bound to a receptor of the cell surface virus particles enter the cell and the viral replication causes the death of the TH cells.

(iv) Thus, repetitive infection of HIV-1 retrovirus to TH cells, causes steady destruction of this cells. The decrease of TH cell population effectively disables the responses of B-cells and macrophages, and thus the entire immune system becomes less functional.

(v) As a result, an HIV-1-infected person becomes very much susceptible to various types of cancers.

This is why though HIV-1 is a non-oncogenic retrovirus, does not contain any oncogene, various types of cancers are still found in AIDS (caused by HIV retrovirus) particle.

Problem 9

Why is cell cycle progression regulated?

Solution:

(i) Regulation at checkpoints ensures that the cell cycles progress to the next stage, when proper preconditions have been fulfilled.

(ii) The cell cycle is wired into the intercellular signalling system, proliferation can be halted until the proper signals for it received.

Problem 10

Carcinogenesis is a multistep process at both the phenotypic and genetic levels.

Solution:

(i) A malignant neoplasm has several phenotypic attributes—such as excessive growth, local invasiveness and the ability to form distant metastasis.

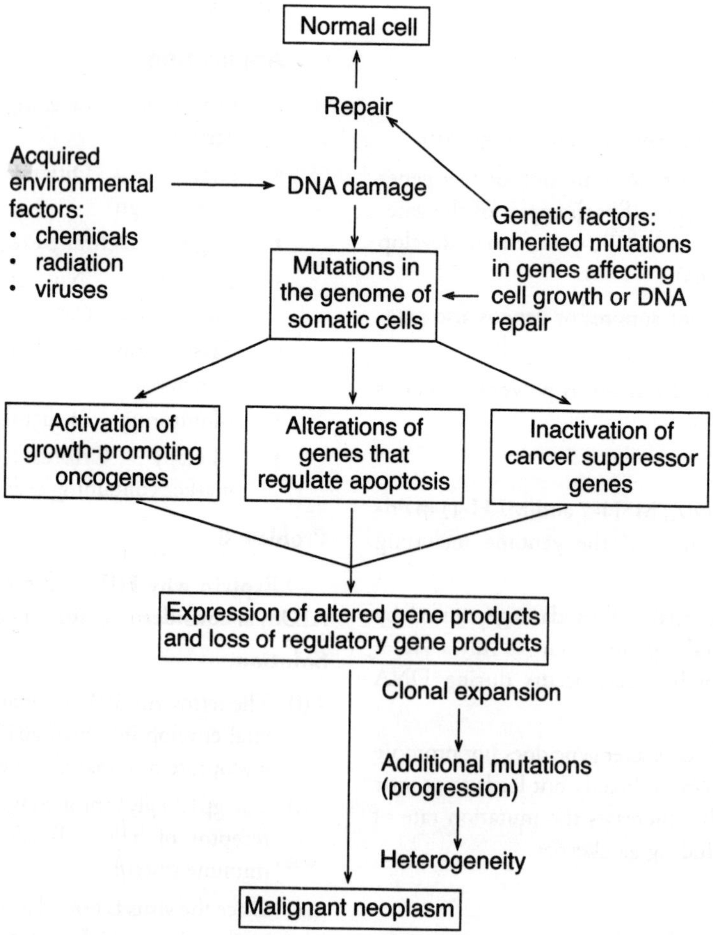

Fig. 5.12: Flow chart depicting a simplified scheme of cancer pathogenesis.

(ii) These characteristics are acquired in a stepwise fashion, a phenomenon called **tumour progression.**

(iii) At the molecular level, progression results from accumulation of genetic lesions, that ensure, instances are favoured by defects in DNA repair.

Problem 11

Why is programmed cell death necessary?

Solution:

(i) Apoptosis is a normal self-destruction mechanism that eliminates damaged and potentially harmful cells as well as cells that have only temporary functions in cell development.

(ii) Cell survival requires direct survival signals that inhibit activation of the self-destruction programme.

(iii) The cost of replacing discarded cells is far less than the cost of compromising the health of the entire organism.

Problem 12

How do neighbouring cells influence cell proliferation and cell death?

Solution:

(i) Proteins secreted by neighbouring cells act as signal.

(ii) These proteins bind to transmembrane receptor proteins on the cells receiving the signals, activating signal-transduction programme pathway that ultimately changes the expression of key genes in the cell cycle or apoptosis pathways.

Problem 13

Why is cancer considered a genetic disease of somatic cells?

Solution:

(i) Cancers are caused by the accumulation of a set of mutations in a single somatic cell.

(ii) These mutation can be induced by environmental factors, viz., tobacco, smoke, chemical pollutants, radiation, ultraviolet ray, etc.

(iii) These mutations undermine the basic control mechanisms that keep a cell from proliferating out of control.

Problem 14

How do mutations promotes tumours?

Solution:

(i) Tumour-producing mutations either release normal inhibitory controls on the cell cycle or block cell death.

(ii) In addition, mutations that increase the mutation rate of the cells are tumour promoting, presumably because mutations increase in the cell proliferation or **apoptosis** pathways.

Problem 15

Describe three mechanisms used to control the activities of the proteins in the cell cycle and cell death pathway.

Solution:

The activities of proteins controlling the cell cycle or cell death pathways are controlled by **phosphorylation/dephosphorylation**, protein interactions and proteolysis.

Problem 16

What are three major categories of mutations that lead to development of cancer?

Solution:

Point mutations, gene fusions or deletions are the key-regulatory domains.

Problem 17

How is the activity of caspasses controlled?

Solution:

(i) As the **caspasses** (cysteine-containing aspartate specific proteases) are first translated, they contain sequences that prevent their activity.

(ii) These inactive versions are called **zymogen.**

(iii) Part of the polypeptide then must be removed by enzyme through the process of **cleavage.**

Problem 18

Give an example of tumour-suppressor gene. Why is the mutation recessive?

Solution:

Examples: p^{53} and **Rb** gene.

(i) Tumour-promoting mutant alleles of tumour-suppressor gene inactivate the proteins that they encode.

(ii) Only when both copies of one of these genes are inactivated, then tumours are observed.

(iii) These losses of function mutations are the definition of recessive.

Problem 19

Suppose that you had the ability to introduce normal copies of a gene into tumour cell in which mutations in the gene caused it to promote tumour growth.

(a) If the mutations were in tumour-suppressor gene, would you expect these normal trans-genes to block the tumour-promoting activity of the mutations? Why or why not?

(b) If the mutations were of the oncogene type, would you expect the normal trans-genes to block their tumour-promoting activity? Why or why not?

Solution:

(a) (i) Mutations in tumour-suppressor gene are recessive due to loss of function.

(ii) That functions can be restored by the introduction of wild type allele.

(b) (i) Mutations in an oncogene are dominant and due to gain of function (mis-expression or over-expressions).

(ii) The normal function will not inhibit these mutants and the introduced gene would be ineffective in restoring the normal phenotype.

Problem 20

Many of the proteins that participate in the progression pathway of the cell cycle are reversibly modified, whereas in the apoptosis pathway, the modification events are irreversible. Rationalize these observations in relation to the nature and end result of the two pathways.

Solution:

(i) Once apoptosis is initiated, a self-destruct switch has been thrown, endonuclease and protease are released, DNA is fragmented, and organelles are disrupted.

(ii) This state is a "terminal", a state from which the cell will not have a need or chance to reuse the machinery of destruction.

(iii) On the other hand, the various proteins needed for the regulation and execution of the cell cycle will be needed again if the cell continues to be divided.

(iv) By recycling many of these proteins, the cell conserves its resources (proteins are energetically expensive to make) and recycling allows for rapid divisions because the cell does not have to spend time remaking all the pieces.

Problem 21

The wild type Rb protein functions to sequester E2F in the cytoplasm. At the appropriate time in the cell cycle, E2F is released, so that it can act as a functional transcription factor.

(i) In cells homozygous for a loss of function of rb mutation, where do you expect to find E2F in the cell?

(ii) If a cell were doubly homozygous for loss of function mutations in both the RB and E2F genes, would you expect tumour growth? Explain.

(iii) How would a mutation in RB protein that irreversibly binds E2F affect the cell cycle?

Solution:

(i) In the absence of functional Rb protein, E2F will be in the nucleus.

(ii) No. The absence of functional E2F would be epistatic to the absence of functional Rb protein.

(iii) A mutation that causes permanent sequestering E2F in the cytoplasm will likely inhibit the cell cycle.

Problem 22

What is apoptosis? Why is the cell death associated with apoptosis desirable, and how is it regulated?

Solution:

Apoptosis is a genetically controlled programme cell death or suicidal cell death that is triggered by intracellular damage (e.g., DNA lesions) or by external signals from neighbouring cells.

(i) Cells targeted for **apoptosis** are those that have high levels of DNA damage and so are at a greater risk for neoplastic transformation.

(ii) **Apoptosis** is regulated by P^{53}, among other proteins. In cells with large amounts of DNA damage, p^{53} accumulates and functions as transcription factor to activate transcription of DNA repair genes and WAF1 whose product p^{21} leads cells to arrest in G1.

(iii) If very high levels of DNA damage exists, p^{53} does not induce DNA repair genes and WAF1, but activates **BAX gene** whose product blocks the BCL-2 protein from repressing the apoptosis pathway.

(iv) By blocking BCL-2 function, the **apoptosis** pathway is activated.

Problem 23

Distinguish between direct-acting carcinogens, procarcinogens and ultimate carcinogens in the induction of cancer.

Solution:

Direct acting **carcinogens** are chemicals that bind to DNA and act as mutagens.

Procarcinogens are chemicals that must be converted by normal cellular enzymes to become active carcinogens.

These products, most of which also bind DNA and act as mutagens, are referred to as **ultimate carcinogens.**

Problem 24

In the inherited form of retinoblastoma, an affected child is heterozygous for an rb mutation, which is either passed on from a parent or has newly arisen in the sperm or the oocyte nucleus that gave rise to the child. The heterozygous RB/rb cells are nonmalignant, however. The RB allele of the heterozygote must be knocked out in the developing retinal tissue to create a tumourous cell. Such knockout can occur through an independent mutation of the RB allele or by mitotic crossing-over such that original rb mutation is now homozygous.

(a) If retinoblastoma is passed on to other siblings as well, could we determine whether the original mutation was derived from mother or the father? How?

(b) Could we determine whether the rb mutation was maternally or paternally derived if it arose de novo in a germ cell of one parent?

Solution:

(a) (i) If the trait is inherited, we can determine from which parent it came.

(ii) The most straight forward approach is to identify DNA polymorphisms, such as restriction fragment length polymorphisms (RFLPs), that map within or near RB gene.

(iii) RB has the curious property of being inherited as an **autosomal dominant**, even though it is recessive on a cellular basis.

(iv) Given the dominant pattern of inheritance, by finding DNA differences in the parental genomes that map near each of the four parental alleles, we should be able to determine which allele has been passed on to all affected offsprings. That allele is the mutant gene.

(b) (i) If the trait arose de novo in the sperm or the oocyte, we can still possibly determine from which parent it came, but with considerable difficulty.

(ii) One way, to do so would be to use recombinant DNA-cloning techniques to isolate each of the two copies of the RB gene from normal cells. Only one of these alleles should be mutant.

(iii) When the gene has been cloned by DNA sequencing of the two alleles we may be able to identify the mutation that inactivates RB. If it arose de novo in the sperm or the oocyte, this mutation would not be present in the somatic cells of the parents.

(iv) If sequencing also reveals some polymorphisms that distinguish the alleles, we should then be able to go back to the parent's DNA to find out whether the mother or the father carries the polymorphisms that were found in the cloned mutant allele.

(v) If the mutation arose by mitotic crossing-over, additional tools are available. In this case the entire region around the rb gene will be homozygous for the mutant chromosome.

(vi) By examining DNA polymorphisms known to map this region, we should be able to determine whether this chromosome derived from the mother or from the father.

Problem 25

What are the roles of Apaf and Bcl proteins in apoptosis?

Solution:

(i) Cytochrome C, a mitochondrial protein, is one of the switches for the activation of apoptosis.

(ii) Cytochrome 'C' that has leaked into cytoplasm is thought to bind to another protein called **Apaf** (apoptotic protease-activating factor).

(iii) Cytochrome **C-Apaf** complex then binds to the initiator **caspase** and activates it.

(iv) Under normal conditions, **apoptosis** pathway remains "**off**". Proteins such as **Bcl-2** and **Bcl-x** in mammals act as **road blocks** to prevent **apoptosis** from being triggered.
Bcl proteins block the release of cytochrome C from mitochondria and bind to **Apaf** which prevent its interaction with the initiator **caspase**.

Problem 26

The protein product of the BAX gene negatively regulates the protein product of the BCL-2 gene that is BAX protein interfere with the function of the BCL-2 protein. Predict the phenotype of cells homozygous for a loss of functions mutation in the BAX gene. Would this gene be classified as a proto-oncogene or as a tumor-suppressor gene?

Solution:

(i) Cells homozygous for loss of function mutation in the **BAX** gene would be unable to prevent repression of the programmed cell death pathway by the **BCL-2 gene** product.

(ii) Consequently, these cells would be unable to execute that pathway in response to DNA damage induced by radiation treatment.

(iii) Such cells would continue to divide and accumulate mutations; ultimately they would have a good chance of becoming **cancerous**. The **BAX gene** would, therefore, be classified as a tumour-**suppressor gene.**

Problem 27

p^{53} is a guardian of the genome or molecular policeman or role of p^{53} in cell cycle.

[C.U. 2019, CC-4]

Solution:

A. Role of wild type p^{53}:

(i) The p^{53} gene is located at chromosome 17p13-1. It is a phosphoprotein of 393 amino acid long and act as transcription factor. It has three domains viz. N-terminal transcription activation domain (**TAD**), a central DNA binding core domain (**DBD**) and a C-terminal homo-oligomerization domain (**O.D.**).

(ii) Normally p^{53} is highly unstable, the **MDM-2** (or Mdm-2—murine double minute chromosome clone number 2) protein binds its amino terminal end and ubiquinates it, leading to rapid degradation of p^{53} in the proteasome within several minutes. As a result, in normal cell the amount of p^{53} is very low.

(iii) When DNA damages (due to radiation or other mutagen) p^{53} initiates a cascade of events leading to arrestation of cell at G_1. DNA damage results phosphorylation of both p^{53} and Mdm-2 on the domain where they normally interact.

(iv) Through phosphorylation, p^{53} becomes stable and active. Activated p^{53} turns on the transcription of gene WAF-1 whose product is p-21. p^{21} is an inhibitor of CDK 4/Cyclin-D-1 complexes. It inactivates CDK 4/Cyclin-D_1 complexes and the cell cycle is arrested at G_1.

(v) During this timeout, the damaged DNA can be repaired. If it fails to repair, it activates other set of gene (bax) whose protein BCL-2 (repressor of apoptotic pathway) and promotes the cell to commit suicide.

(vi) Thus p^{53} is responsible for activating a brake on the cell cycle and this brake allows the cell to maintain a genetic integrity. Due to the wide spread activity and supervision of the p^{53} gene over the genome, it is called the **guardian of the genome** or **molecular policeman.**

bcl (B – cell) leukemia/lymphoma

bax (bcl – 2 associated × gene)

B. Role of mutant p^{53}:

(a) ***Loss of function mutation***

(i) Most of the mutation of p^{53} is located in the **DBD**.

(ii) These mutation abolishes the ability of p^{53} to bind to specific DNA sequence that are embedded in its target genes, thereby preventing the transcriptional activity of these genes.

(iii) Thus, mutations in DBD are typically recessive loss function mutation.

(b) ***Gain of function mutation***

(i) Mutation occur in **OD** portion.

(ii) p^{53} missense mutation stimulates cell division rather than suppresses it.

(c) ***Trans dominant mutation***

(i) p^{53} missense mutation inactivates the function of wild type p^{53} in a trans-dominant fashion, i.e., the mutant polypeptide dimerizes with wild type polypeptide and prevents wild type polypeptide from functioning as transcriptional activators.

(ii) Therefore, mutant p^{53} acts as **dominant negative.**

Problem 28

Genotoxic carcinogen and its role in carcinogenesis

Solution:

The carcinogens that initiate tumour formation by chemically interacting with DNA molecules are termed genotoxic carcinogens.

(i) The chemical carcinogen covalently binds with DNA and may damage DNA (by oxidative alteration and strand breakage). In most cases these lesions are repaired, or injured cells are eliminated.

(ii) If neither events occur, a lesion in parental DNA strand may induce a heritable alteration or mutation in the daughter strand during replication.

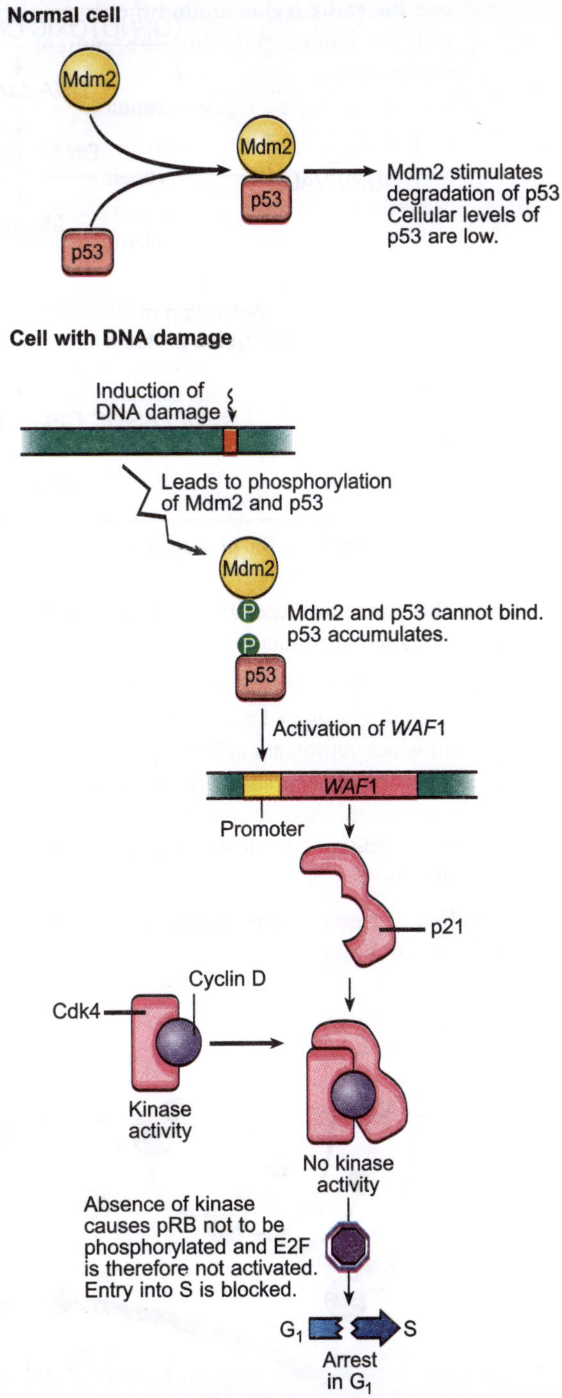

Fig. 5.13: Role of p^{53} gene in cell cycle.

(iii) The mutation may remain silent if it does not alter the protein encoded by the mutant gene

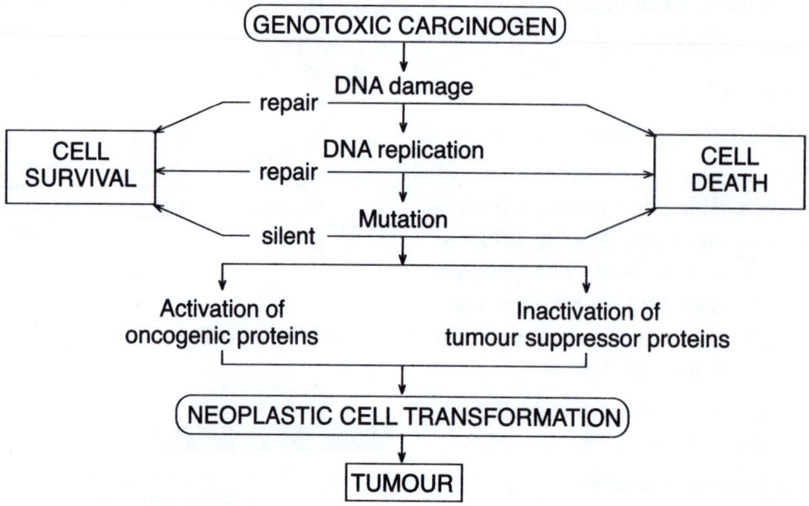

Fig. 5.14: The process of carcinogenesis initiated by genotoxic carcinogens.

or if the mutation causes an amino acid that does not affect the function of a protein.

(iv) Genetic alteration may be incompatible with cell survival. The altered genes express mutant protein which reprogram cells for growth and multiplication. When such cells survive and undergo mitosis, their descendants also have a similar propensity for uncontrolled proliferation of cells.

(v) The final outcome of this process is a nodule, followed by a tumour.

Problem 29

pRB a key regulator of cell cycle regulation.

Solution:

(i) RB is a tumour suppressor gene, whose protein product is denoted by p^{RB} and is involved in cell cycle regulation.

(ii) Early in G_1 phase of the cell cycle pRB binds to E2F protein (a family transcription factor) that controls the expression of several genes whose products move the cell through the cell cycle.

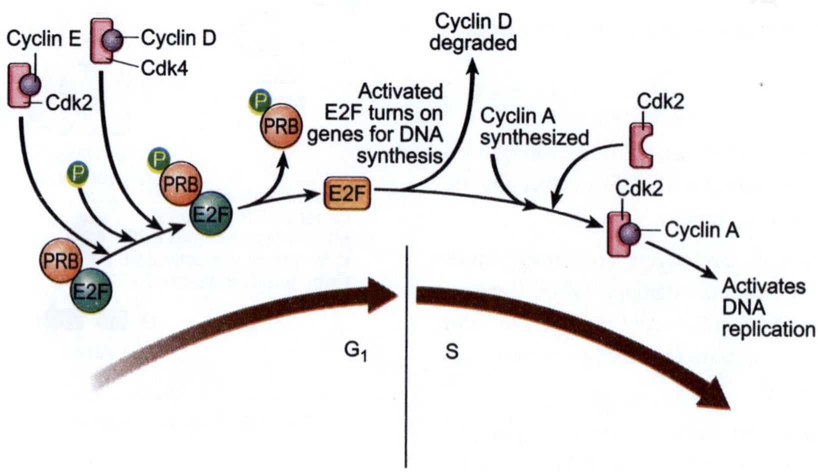

Fig. 5.15: Role of pRB in cell cycle.

(iii) The bound E2F transcription factors are unable to stimulate transcription of their target genes. Thus, the cell cycle factors encoded by these genes are not produced. As a result, DNA synthesis and the cell division remain quiescent.

(iv) In G_1, p^{RB} is phosphorylated through the activation cyclin and Cdk [Cyclin E/Cdk-2; Cyclin D/Cdk-4 or Cdk-6]. The phosphorylated p^{RB} releases the bound E2F transcription factors. The E2F stimulate the transcription of a vaiety of the target genes which encode proteins that induce the cell to progress through the S phase and into mitosis.

(v) After mitosis p^{RB} is dephosphorylated and each of the daughter cells enters the quiescent phase.

(vi) In any type of cancer, both the copies of p^{RB} have been inactivated either by deletion or by other mutation.

Problem 30

Draw the diagram of Cells cycle, its phases and components.

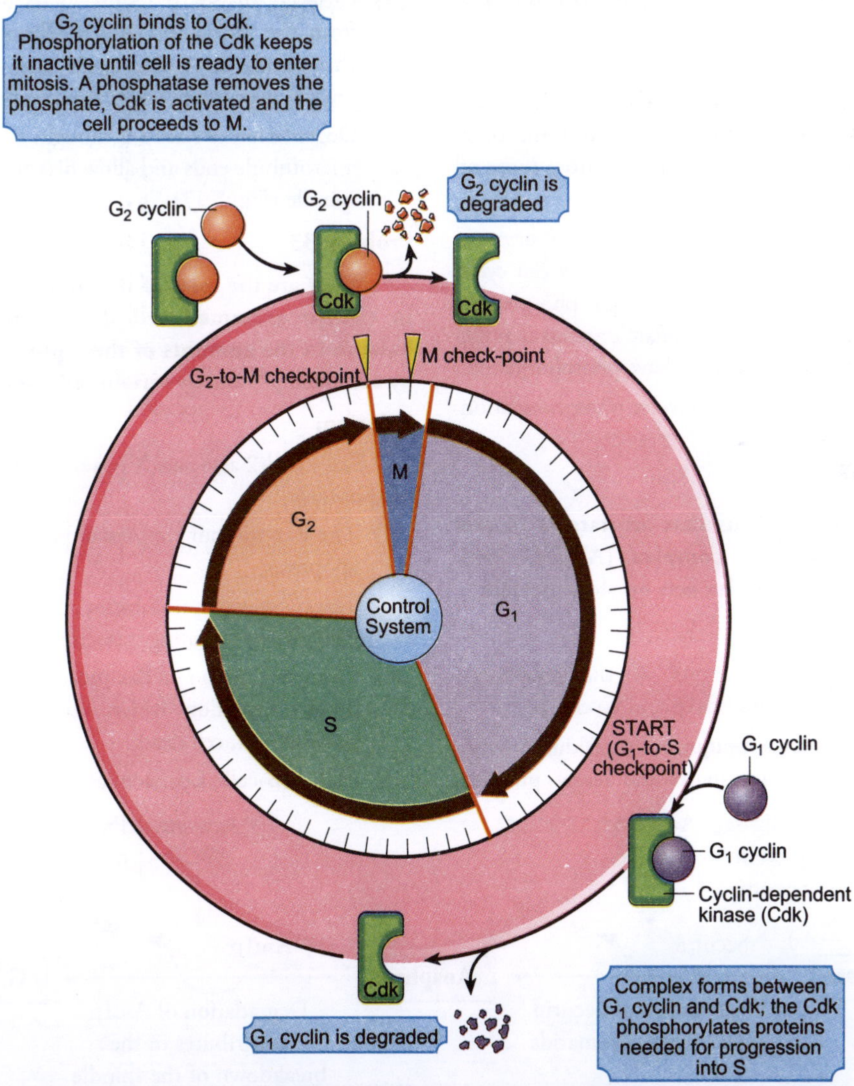

Fig. 5.16: Phases of cell cycle and different types of cyclin.

Problem 31

How do Cyclin-CDk complexes exert their effect on cells? What counterpart class enzymes reverses the effects of Cyclin-CDk complexes.

Solution:

(i) Through their protein kinase activity, they add phosphate groups to the hydroxyl groups present in the amino acids (Serine, threonine, and tyrosine) of their targeted proteins.

(ii) The cyclin components of the complex bind to the protein substrate and tether it, allowing CDk component to phosphorylate tethered substrate.

(iii) Once the targeted protein is phosphorylated it dissociates from the cyclin-CDk complex. The activities of the phosphorylated form of a protein often differ dramatically from those of the unphosphorylated form.

(iv) Phosphorylation may activate one enzyme but inactivate another. Another way cell cycle regulation is mediated through phosphatase enzymes that dephosphorylate proteins that the cyclin-dependent kinases have phosphorylated.

(v) The counterparts that reverse phosphorylation are **Phosphatase.**

Problem 32

Name two proteins that are targets of the anaphase promoting complex (APC/C) and explain how this complex acts on these protein.

Solution:

Two important targets of the anaphase-promoting complex are Pds1p and Cyclin B.

(i) This complex is ubiquitin 76-residue protein ligase that uniquitinates the target proteins and tags them for destruction in the proteosome (large multifunctional complex is responsible for cytoplasmic proteolysis).

(ii) Destruction of Pds1p allows disjunction of the sister chromatid.

(iii) **Securin** is another substrate for the APC/C which inhibits a protease called **separase** and Ase1p (microtuble associated protein). As the securin is degraded, separase protein becomes free to degrade Scc1p (glue-protein) a component of cohesin complex that condenses chromosome and holds sister chromatids.

(iv) Cleavage of Scc1p results in its dissociation from the chromatin which allows the sister chromatids to come apart and be pulled toward opposite poles of the spindle.

(v) Degradation of Ase1p is thought to release the microtubule ends and allow disassembly of the spindle (Fig. 5.17).

Problem 33

What are the roles of Bax and BCl_2 proteins in the programmed cell death? How is the balance in the amounts of these protein affected by activated p^{53} and by certain cellular oncogenes.

Solution:

Bax = (bcl-2 associate X gene) positive regulator of apoptosis.

(i) The Bax-tumour suppressor protein promotes apoptosis.

(ii) The p^{53} transcription factor activates apoptotic pathway by activating transcription of Bax gene.

(iii) Over production of Bax protein results homo-dimer formation which activates apoptotic pathway, i.e., cell death.

(iv) BCl_2 (B-cell Leukaemia/Lymphoma-2) an

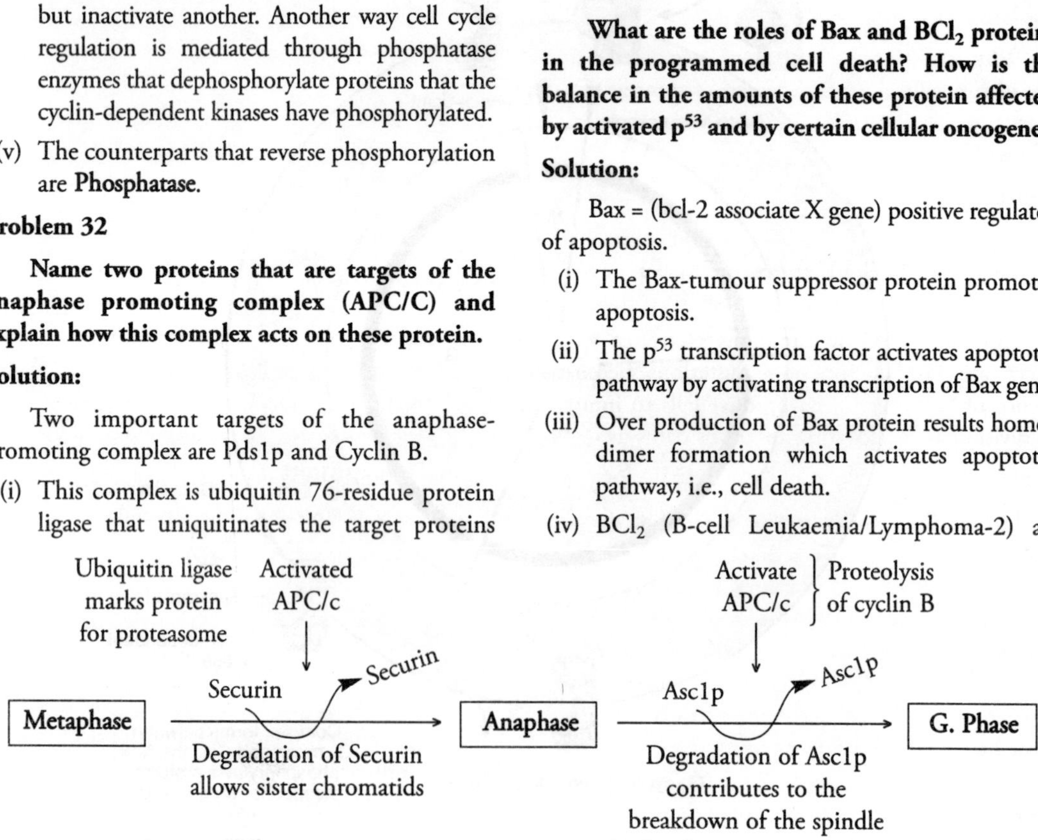

Fig. 5.17: Role of APC/c

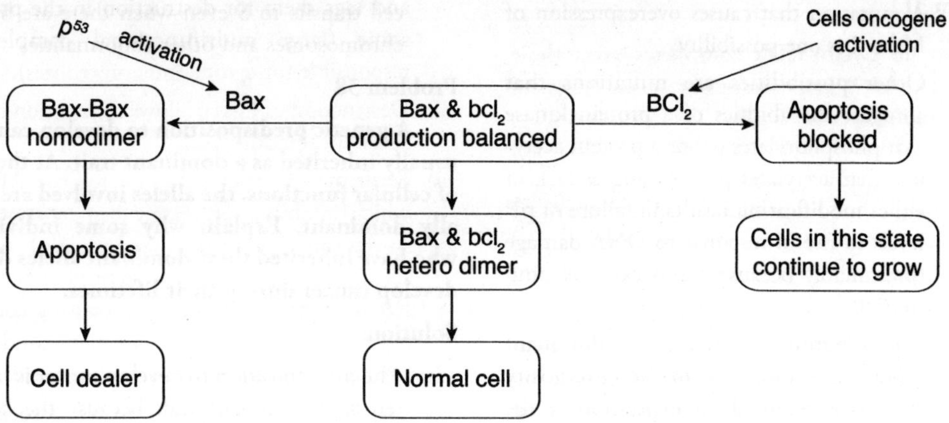

Fig. 5.18: Bax and bcl_2 interaction

oncogene is present in 18 chromosome. Activation of certain cellular oncogenes results in over-production of bcl_2 which counteracts the effect of Bax over-production by sequestering Bax subunits in Bax/bcl_2 heterodimer.

Problem 34

Why are oncogene mutations associated with gain of function mutation, whereas tumour suppressor mutations associated with cancer are loss of function mutation?

Solution:

Oncogenes are gain of function mutations because their enhanced expression makes uncontrolled cell division or prevents apoptosis pathways.

Tumour suppressor mutations associated with cancer are loss of function mutations because absence of functional gene product fails to inhibit cell division or in activating apoptosis pathway.

Problem 35

Li-Fraumeni syndrome is a familial cancer syndrome in which affected relatives show easy onset of a diverse set of tumours, including brain, breast, lung cancer and sarcomas. Nonsense mutation, missense mutation and mutation altering splice donor sites of the p^{53} gene have been found segregating in diverse Li-Fraumeni families. In about half of the families shading the syndrome, no mutation in the p^{53} coding regions or in splice donor or acceptor sites has ben detected. The question is whether in these families, the syndrome is due to change in p^{53} or due to some other cause.

[P.G., M.G., E.G. & I.G. 14]

(a) How would you determine whether the mutations in such families are actually associated with p^{53} gene in any way?

(b) If the diseased gene is p^{53} associated, what kinds of changes would you look for?

(c) Suppose that the evidence indicates that the segregating mutation is not linked to the p^{53} gene, what other possible causes of the syndrome might you suggest? (Remember the inheritance of the syndrome is done).

Solution:

(a) (i) Test for the genetic linkage of the segregating cancer gene reveals that DNA markers are very near the p^{53} gene.

(ii) If there is tight linkage between the diseased gene and a DNA marker at p^{53}, then it is likely that the disease is p^{53} related.

(iii) If the diseased gene is not genetically linked to p^{53}, another gene is probably involved.

(b) Possible site for changes:

(i) Promoter changes that reduce expression of the p^{53} gene.

(ii) Changes in introns that correspond to essential splice sites or a deletion that removes the whole gene (so that the sequence of p^{53} from a heterozygous individual is that of the wild type p^{53+} homolog).

(c) Anything that can mimic loss of p^{53} function could underline the phenotype.

(i) A mutation that causes overexpression of **Mdm 2** is one possibility.

(ii) Other possibilities are mutations that abrogate the abilities of a protein kinase that phosphorylates p^{53} or a protein acetylase that acetylates p^{53}, so long as lack of either modification results in failure of p^{53} levels to rise in response to DNA damage (presumably because Mdm can still bind to p^{53}).

(iii) The syndrome should show dominant inheritance but loss of heterozygosity should be required for expression at the cellular level.

Problem 36

There is an average of three retinal tumours per heterozygous carrier of mutation, assuming that the number of retinal cells at risk is 2 × 10^6^ in each eye and that each tumour results from an independent loss of heterozygosity.

What is the probability that an individual who inherits an RB mutation will not develop any retinoblastomas? What is the 'penetrance' of familiar retionoblastoma?

Solution:

Since the probabilities of loss of heterozygosity per cell is 7.5×10^{-7} and there are 4×10^6 cells at risk in both eyes together, the probability of no loss of heterozygosity in any cell is $(1 - 7.5 \times 10^{-7})$ raised to the power of 4×10^6 which is equal to 0.05 or 5%.

The penetrance of familiar retinoblastoma is therefore 95%.

Problem 37

Many types of cancer cells have defects in the G_1/S checkpoint, these also tend to have abnormalities in chromosome number or structure. Why would chromosome abnormalities be expected in such cases.

Solution:

(i) Chromosome abnormalities are expected because the normal G_1/S checkpoint serves to allow DNA damage (such as broken chromosome) to be repaired prior to entering the S period.

(ii) In the absence of normal G_1/S checkpoint, the cell transits to S even when there are broken chromosomes and other abnormalities.

Problem 38

A genetic predisposition to develop cancer is usually inherited as a dominant trait. At the level of cellular functions, the alleles involved are actually dominant. Explain why some individuals who have inherited these dominant alleles do not develop cancer during their lifetimes.

Solution:

(i) The predisposition to develop cancer is inherited as a dominant trait because the heterozygote has the higher predisposition.

(ii) The mutant allele is actually recessive at the cellular level. On the other hand, we have so many cells in our bodies, it becomes relatively likely that a defective mutation will occur in a single normal gene and lead to a cancerous cell.

(iii) Some heterozygous individuals may not develop the disease as a matter of chance. They may be lucky and not get a defective mutation in the normal gene, or perhaps their immune system is better at destroying cancerous cells once they arise.

Problem 39

A p^{53} knockout mouse in which both copies of p^{53} are defective has been produced by researchers in the Cancer Research Institute of Kolkata. This type of mouse apears normal at birth. However, it is sensitive to UV light. Based on our knowledge of p^{53}, explain the normal appearance at birth, and the high sensitivity to UV light.

Solution:

(i) The role of p^{53} protein is to sense DNA damage and prevent damaged cells from proliferating.

(ii) Perhaps, prior to birth, the foetus is in a protected environment, so the damage may be minimal.

(iii) In other words, the foetus may not really need p^{53}.

(iv) After birth, agents such as UV light may cause DNA damage. At this point, p^{53} is important.

(v) A p^{53} knockout is more sensitive to UV light because it cannot repair its DNA properly in response to this DNA damaging agent, and it cannot kill cells that have become irreversibly damaged.

Problem 40

Relatively few inherited forms of cancer involve the inheritance of mutant oncogenes. Instead most inherited forms of cancer are defects in tumour-suppressor genes or DNA repair genes. Give two or more reasons why we seldom see inherited forms of cancer involving activated oncogenes.

Solution:

(i) If an oncogene was inherited, it may cause uncontrolled proliferation cell at an early stage of development which results improper development of an embryo.

(ii) This could lead to an early spontaneous abortion.

(iii) Another possibility is that inherited oncogene may adversely affect gamete survival which would make it difficult for them to be passed from parent to offspring.

(iv) Third possibility would be that oncogenes could affect fertilised zygote in a way that would prevent the correct implantation in the uterus.

Problem 41

In familiar retinoblastoma, there is an average of three retinal tumours per heterozygous carrier of the mutation. Assuming that the number of retinal cells at risk is 2×10^{-6} in each eye and that each tumour results from an independent loss of heterozygosity, what is the estimated rate of loss of heterozygosity per cell? Should this be regarded as a mutation rate? Why or why not?

Solution:

(i) The number of retinal tumour = 3.

(ii) The number of retinal cells at risk is 2×10^{-6} in each eye. Therefore, in two eyes it will be $2 \times 2 \times 10^{-6}$ cells.

(iii) Three tumours 4×10^{-6}. Therefore, risk of tumour for one cell = 4×10^{-6} = 3 tumour.

$\therefore$ Rate is $\frac{0.75}{10^6} = 7.5 \times 10\%$.

(iv) This is not exactly mutation rate, because it includes such events as chromosome loss, mitotic recombination and so forth which are strictly spoken as mutation.

Problem 42

Cancer is a disease involving heritable defects in cellular control mechanisms that result in the formation of invasive tumours capable of releasing cells that can spread the disease to distant sites in the body. Justify the problems.

Solution:

(i) Normal cells proliferate and form a single layer (monolayer) over the bottom one.

(ii) Cancer cells continue to grow in culture, pilling on top of one another to form clump.

(iii) The cancer cells include an abnoral number of chromosomes.

(iv) The cancer cells has an ability to divide continuously and indefinitely.

Problem 43

Normal cells can be converted to cancer cells by treatment with a wide variety of chemicals, ionizing radiations and a variety of DNA-RNA containing viruses. They act by causing changes in the genome of the transformed cells: Explain

Solution:

(i) Analysis of the cells of a cancerous tumour almost always shows that the cells have arisen by growth of a single cell (tumour is said to be monoclonal).

(ii) The development of a malignant tumour is multi-step process charactrized by a progression of genetic alterations that make the cells increasingly less responsive to the body's normal regulatory machinery and invade normal tissues.

(iii) The genes involved in carcinogenesis constitute a specific subset of the genome whose products are involved to control the cell cycle, intercellular adhesion and DNA repair.

(iv) The level of expression of a particular genes in different types of cancers can be determined by using DNA microarrys.

(v) The growth of tumour cells is also influenced by both nongenetic and epigenetic influences that allow the cells to express its malignant phenotype.

Problem 44

Tumour suppressor genes and oncogenes play a significant role in carcinogenesis—state briefly.

Solution:

(i) Tumour suppressor genes encode protein that restrains cell growth and prevents cells from becoming malignant.

(ii) Tumour suppressor genes act recessively. Since both copies must be deleted or mutated before their protective function is lost.

(iii) In contrast oncogene encodes protein that promote the loss of growth control and malignancy.

(iv) Oncogene arises from the mutation of proto-oncogenes that encode protein having a role in a cell's normal activities. Mutation of proto-oncogene that alters the protein or its expression which causes proto-oncogene to act abnormally and promotes tumour formation.

(v) Oncogene act, dominantly, i.e., single copy of oncogene will cause the cell to express its altered phenotype.

(vi) Most tumours contain alterations in both tumour-suppressed genes and oncogenes.

(vii) As long as a cell retains at least one copy of all its tumour suppressor genes, it should be protected from the consequences of oncogene formation.

(viii) Conversely the tumour suppressor function should not be sufficient to convert the cell to become malignant.

Problem 45

Genes that encode proteins involved in DNA repair system play an important role in carcinogenesis.—Explain.

Solution:

(i) The genome of patient with most common hereditary form of colon cancer called hereditary **nonpolyposis colorectal cancer** (HNPCC) contains microsatellite sequences of abnormal nucleotide number.

(ii) Changes in the length of a microsatellite sequence arise as an error during replication which is normally recognized by mismatch repair enzymes.

(iii) It indicates that defects in repair system may be responsible for carcinogenesis.

(iv) It has been reported that extracts from HPCC tumour cells display DNA repair deficiencies. Cells that contain such deficiencies would be expected to display greatly elevated level of secondary mutations in tumour suppressor genes and oncogenes.

(v) It would lead to a greatly increased risk of malignancy.

(vi) Certain micro RNAs have also been implicated as oncogenes or tumour suppressor gene.

Problem 46

Role of Telomere in carcinogenesis, or Telomerase activity is not normally present in differentiated cells, but it almost always present in malignant cell. Explain.

Solution:

(i) Each cell has an intrinsic limitation on the number of cell division in its lifetime.

(ii) After having reached the stipulated division (50 to 70 times) the cell becomes senescent and stops division.

(iii) During DNA replication, DNA polymerase cannot replicate the tips of the chromosome. Therefore, following each cell division chromosome gradually shortens.

(iv) Once telomere shortens to a critical length, cells have to reach senescent stage. On the other hand, tumour cells overcome the intrinsic limitation to cell division by activating the gene that encodes **telomerase.**

(v) This enzyme is a type of **reverse transcriptase** that replaces the telomeric segments normally lost during cell division. The tumour cell is thus freed of the restriction of telomeric shortenings and continues to divide without becoming senescent.

(vi) The telomerase enzyme is normally absent in normal cells, but is found 85-90% in malignant cells whose activation induces unregulated uncontrolled cell division.

(vii) DNA replication continues and tumour increases in size.

6

Concept of Gene

ONE-GENE ONE-ENZYME HYPOTHESIS

1. All the biological processes are under genetic control.
2. Biochemical processes occur in series of stepwise reaction.
3. Each reaction step is controlled by one gene, i.e., 1:1 correspondence of gene and biochemical reaction exists.
4. Mutation of a single gene alters one reaction step through alternation of one enzyme.

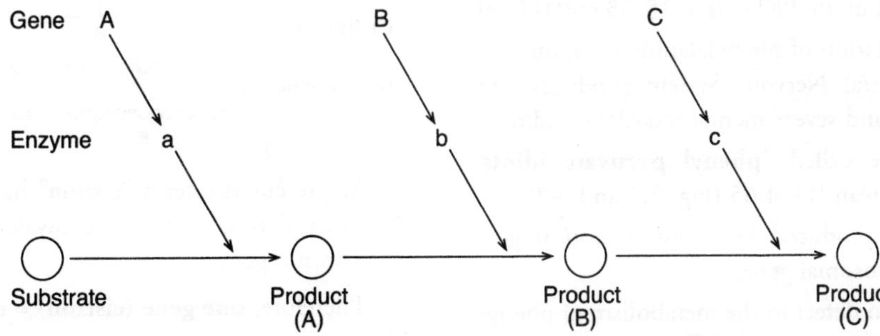

Product C is produced in three reaction steps. Each step is controlled by one enzyme (a/b/c) controlled by one gene A/B/C).

ONE-GENE ONE-POLYPEPTIDE HYPOTHESIS

1. The *one-gene*: *one-enzyme* concept developed in the early 1940s was not immediately accepted by all geneticists. This was not surprising since it was not even then clear how mutant enzymes could cause variation in many phenotypic traits (altered eye size, wing shape, etc. in *Drosophila*).
2. Two factors soon modified the **one-gene: one-enzyme hypothesis.**
   (a) First, while nearly all enzymes are proteins, not all proteins are enzymes. As the study of biochemical genetics proceedes, it becomes clear that all proteins are specified by the information stored in genes, leading to the more accurate phraseology one-gene one-protein.
   (b) Second, proteins were often shown to have a subunit structure consisting of two or more polypeptide chains.

   Each distinct polypeptide chain is encoded by a separate gene, a more modern statement of Beadle and Tatum's basic principle is **one-gene one-polypeptide.**

Experimental Evidences:

Phenylketonuria (PKU): A hereditary condition in human associated with the instability to metabolize the amino acid **phenylalanine.** Persons having autosomal homozygous recessive genes or alleles (pp) fail to produce an enzyme *phenylalanine hydroxylase* which is required by the body to convert phenylalanine to tyrosine. It results the accumulation of *phenylalanine* in the blood.

Characteristic features:

(i) PKU is produced by the defects in enzyme *phenylalanine hydroxylase.*

(ii) In PKU individual phenylalanine is converted to phenyl derivatives (phenyl pyruvate, phenyl acetate, phenyl lactate and others) which appear and excreted in urine.

Normal persons have phenylalanine 1-2 mg/100 ml, but in PKU, it is 15-63 mg/100 ml.

(iii) Accumulation of phenylalanine compound in the Central Nervous System produces irreversible and severe mental retardation idiocy.

(iv) They are called "**phenyl pyruvate idiots**" having mean IQ of 65 (Fig. 6.1 and 6.2).

(v) PKU is produced by recessive homozygous (p/p) autosomal gene.

(vi) In human defect in the metabolism of phenylalanine and tyrosine at different steps produced "**inborn error of metabolism**".

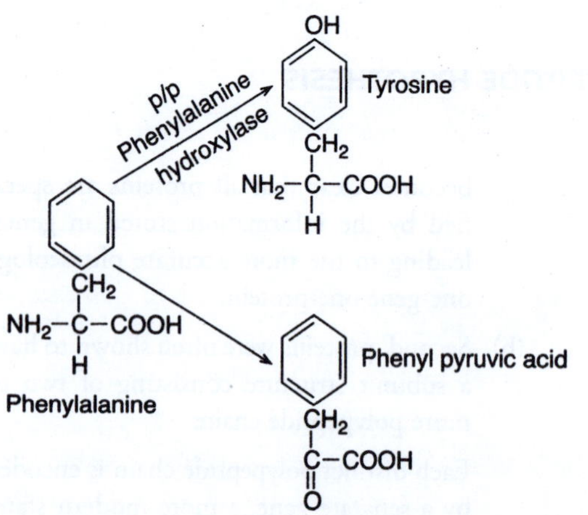

Fig. 6.1: The normal breakdown of phenylalanine into tyrosine and into phenyl pyruvic acid by p/p phenyl ketonuric individual.

Discussion:

(i) One enzyme may consist of more than one kind of polypeptide and as one gene controls the synthesis of only one kind of polypeptide, i.e., 1 : 1 relation between gene : enzyme do not exist in those cases where one enzyme = more than one polypeptide.

(ii) Thus, one enzyme more than one polypeptide or one gene = one enzyme = more than one polypeptide, i.e., one gene ≠ one function.

(iii) So latter scientist, in order to restore **one gene one function**, proposed **one cistron = one polypeptide** indicating that cistron is the smallest functional unit, performing unit function (one polypeptide). This indicates that one gene may consist of more than one cistron.

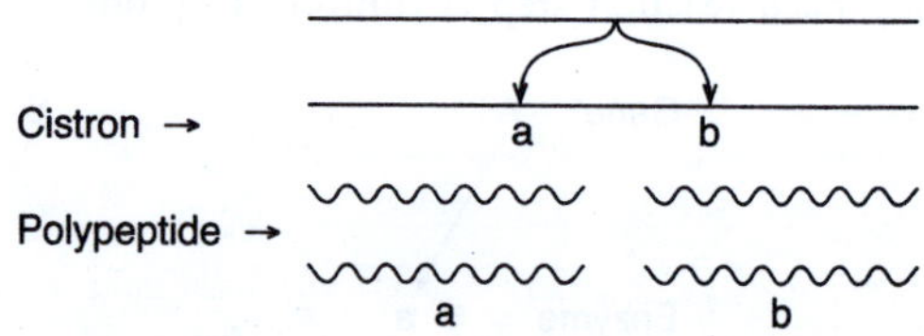

At present the term "cistron" has replaced by gene, considering "gene" as equivalent to cistron, i.e., cistron = gene.

Therefore, **one gene (cistron) = one polypeptide.**

- **Sickle-cell anaemia:**

A genetic disease in humans caused by an autosomal recessive gene, fatal in the homozygous condition, if untreated, causes by an alternation in the amino acid sequence of the β-chain of haemoglobin.

(i) The term sickle cell refers to the peculiar **sickle** or **crescent shape** of the RBC of affected person.

(ii) It is a heritable disease and the people who are homozygous for sickle-cell gene not only suffer from anaemia but also from kidney damage, spleen enlargement and skin lesions.

(iii) Genotype: $\frac{Hb^S}{Hb^S}$

- **Sickle-cell trait:**

The phenotype exhibited by individual heterozygous for sickle-cell gene.

(i) The RBC shows moderate sickling.

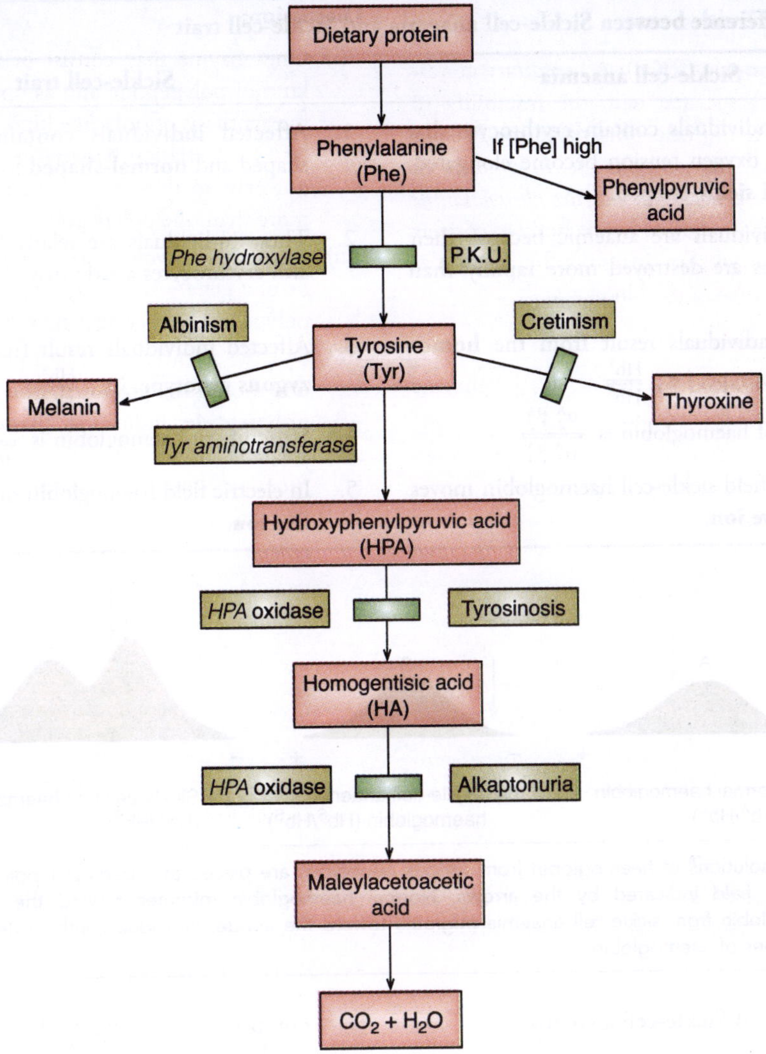

Fig. 6.2: A section of the phenylalanine metabolic pathway, including diseases associated with enzyme blockages. The disease PKU is produced when the enzyme phenylalanine hydroxylase malfunctions. Accumulation of phenylalanine results in an increase in phenylpyruvic acid, which interferes with the development of the nervous system. [After I. M. Lerner and W. J. Libby, Heredity, Evolution, and Society, 2nd ed., Copyright 1976 by W. H. Freeman and Company].

(ii) Such heterozygotes are "carrier" of the defective gene.

(iii) Genotype: $\frac{Hb^S}{Hb^A}$

- **Sickle-cell anaemia is a molecular disease:**

(i) The molecular disease is a kind of disease which is caused by the change of molecular architecture.

(ii) A change of amino acid (sixth position) of β-chain results sickle-cell anaemia. Here amino acid **glutamine** is replaced by **valine**.

(iii) In sickle-cell anaemia a point mutation, i.e., GAA (glutamine) to GUA (valine) takes place.

This is why it is a molecular disease.

Table 6.1: Difference between Sickle-cell anaemia and Sickle-cell trait

Sickle-cell anaemia	Sickle-cell trait
1. Affected individuals contain erythrocyte that under low oxygen tension become elongated, curved and **sickle-shaped**.	1. Affected individuals contain both sickle-shaped and **normal-shaped** haemoglobin.
2. These individuals are anaemic because their erythrocytes are destroyed more rapidly than normal.	2. These individuals are relatively less anaemic and erythrocytes are destroyed slowly.
3. Affected individuals result from the **homozygous** genotype, i.e., $\frac{Hb^S}{Hb^S}$.	3. Affected individuals result from the **heterozygous** genotype, i.e., $\frac{Hb^S}{Hb^A}$.
4. Structure of haemoglobin is $\frac{\alpha_2^A\ \beta_2^A}{\alpha_2^A\ \beta_2^A}$.	4. Structure of haemoglobin is $\frac{\alpha_2^A\ \beta_2^{A(glu)}}{\alpha_2^A\ \beta_2^{A(val)}}$.
5. In electric field sickle-cell haemoglobin moves as a **positive ion**.	5. In electric field haemoglobin moves as a **negative ion**.

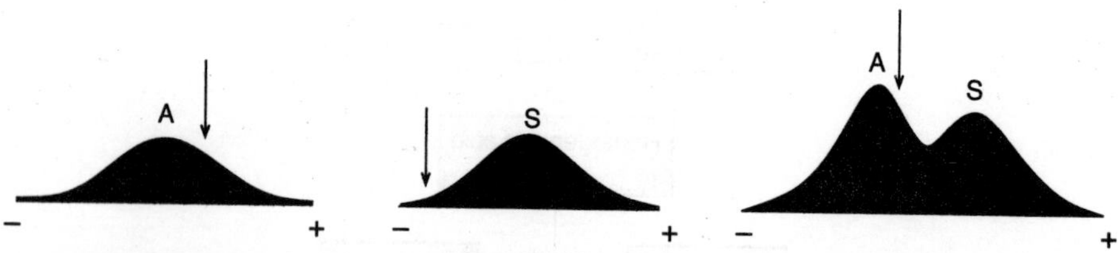

(a) Normal haemoglobin (Hb^A/Hb^A) (b) Sickle-cell anaemia haemoglobin (Hb^S/Hb^S) (c) Sickle-cell trait haemoglobin (Hb^A/Hb^S)

Fig. 6.3: In (a-c) solutions of haemoglobin from various genotypes are placed at a particular point in an electrophoretic field indicated by the arrows. Normal haemoglobin migrates toward the cathode, while haemoglobin from sickle-cell anaemia migrates toward the anode. Individuals with sickle-cell trait show both types of haemoglobin.

Molecular basis of Sickle-cell anaemia:

(i) L. Pauling (1952) and Ingram (1966) observed that haemoglobin molecules contain **globin** which is a tetrameric protein with 600 amino acids arranged in four (4) polypeptide chains with two (2) identical alpha (α) chains and two (2) identical β (beta) chains.

(ii) Amino acid sequence of the polypeptide alpha (α) and beta (β) is controlled by two structural genes (cistron), viz., alpha (α) and beta (β) genes respectively. The alpha gene is located on the long arm of chromosome number 16 (16q) and the beta gene is located on the long arm of chromosome number 11 (11q) (Fig. 6.4).

(iii) Sickle-cell condition is due to change in the amino acid, in the sixth (6th) position of the β (beta) chain.

(iv) In normal haemoglobin (Hb^A) in the 6th position (from N terminal) there is an amino acid **Glutamine** in β (beta) chain. On the other hand, in sickle-cell haemoglobin (Hb^S) amino acid glutamine is replaced by **valine** (Ingram, 1966).

	1	2	3	4	5	6	7	8
Normal (Hb^A)	*val*	*his*	*leu*	*thr*	*pro*	*glu*	*glu*	*lys*
Sickle-cell (Hb^S)	*val*	*his*	*leu*	*thr*	*pro*	***val***	*glu*	*lys*

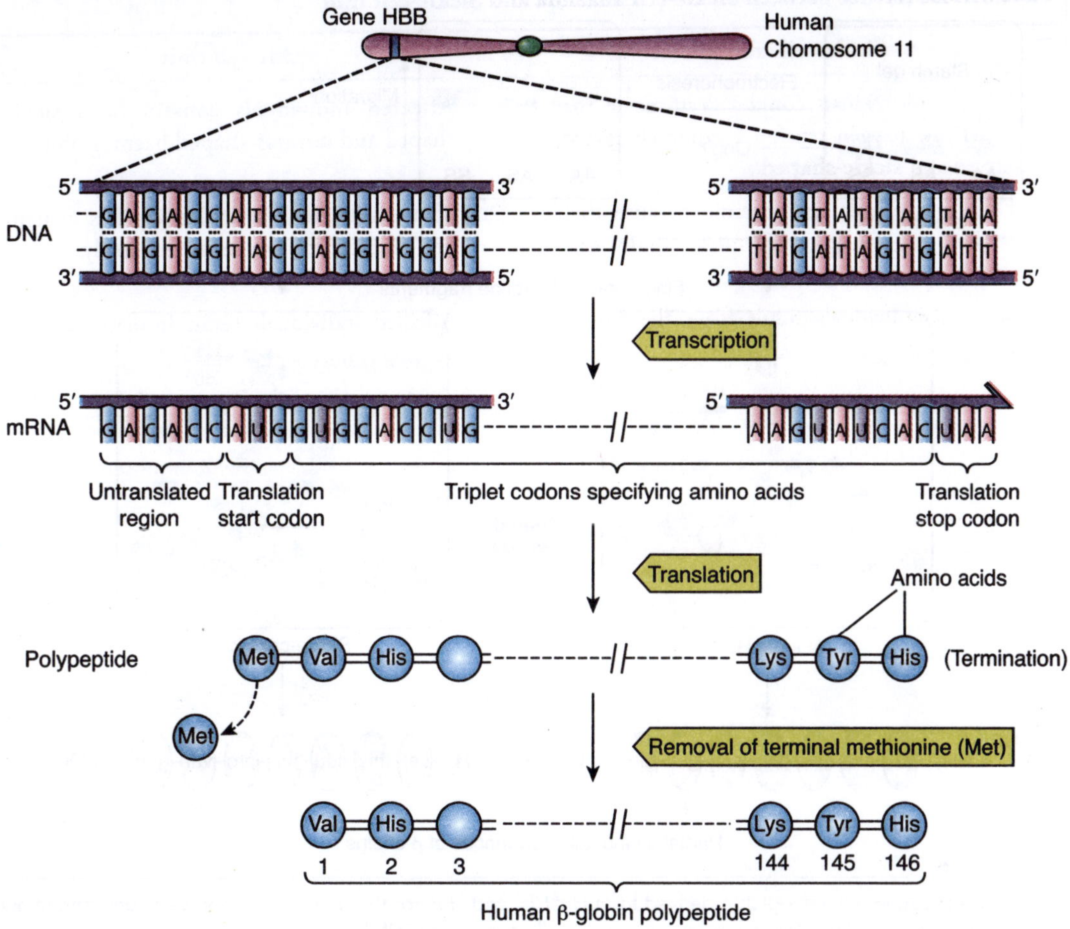

Fig. 6.4: Expression of the human gene *HBB* coding for the β-globin polypeptide of haemoglobin. During transcription (step 1), one strand of the *HBB* DNA (here the bottom strand shown highlighted) serves as a template for the synthesis of a complementary strand of RNA. After undergoing modifications the resulting mRNA (messenger RNA) is used as a template to synthesize the β-globin polypeptide. This process is called translation (step 2). During translation each triplet codon in the mRNA specifies the incorporation of an amino acid in the polypeptide chain. Translation is initiated by a start codon, which specifies the incorporation of the amino acid methionine (Met), and it is terminated by a stop codon, which does not specify the incorporation of any amino acid. After translation is completed, the initial methionine is removed (step 3) to produce the mature β-globin polypeptide.

(v) Sickle-cell haemoglobin shows different electrophoretic behaviours. It migrates towards the positive pole slowly than normal (Fig. 6.5). Heterozygous shows one migrates rapidly towards the positive pole and other slowly, indicating that both Hb^A and Hb^S are present.

(vi) Since haemoglobin protein is composed of two independent gene products. Ingram (1963) proposed that "**one-gene: one-polypeptide**" rather than **one gene: one enzyme**, which defines gene as a section of DNA that determines the amino acid sequence of a polypeptide.

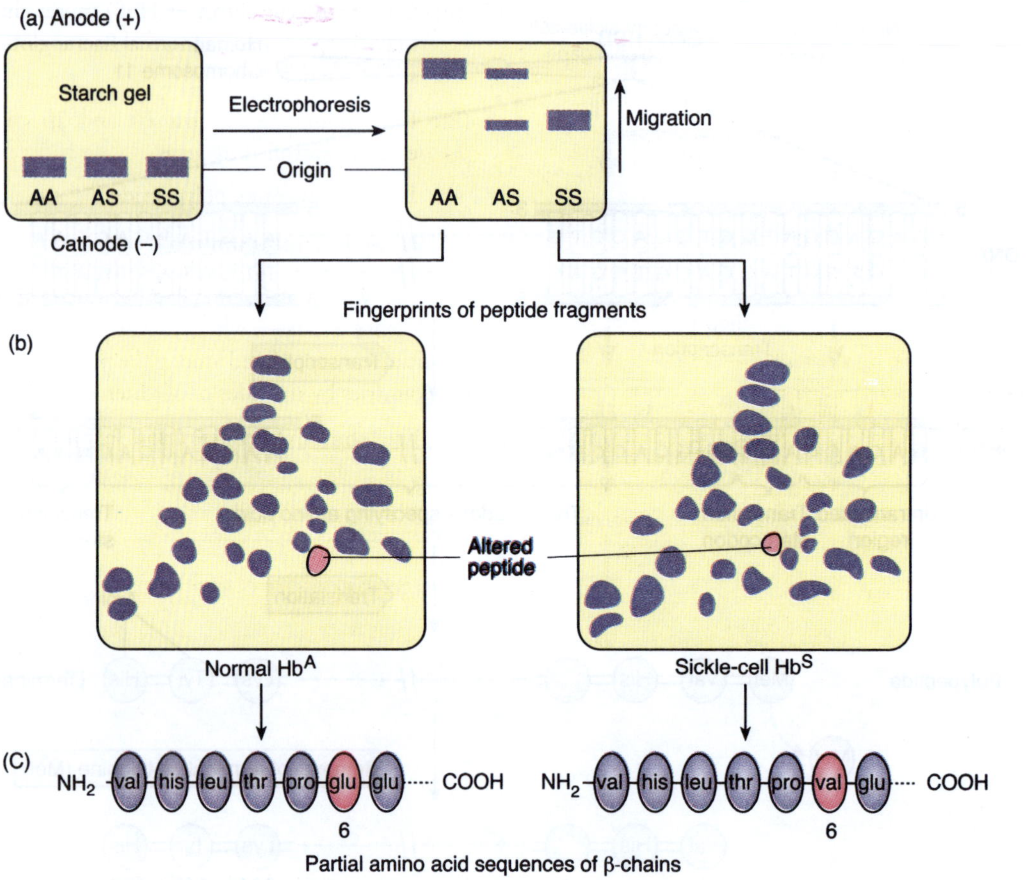

Fig. 6.5: Investigation of haemoglobin derived from Hb^AHb^A and Hb^SHb^S through electrophoresis fingerprinting and amino acid analysis. Haemoglobin with sickle-cell anaemia (Hb^SHb^S)
(a) migrates differently in an electrophoretic field,
(b) is an altered peptide in fingerprint analysis, and
(c) an altered amino acid, **valine**, at the sixth position in the β-chain. During electrophoresis, heterozygotes (Hb^AHb^S) reveal both forms of haemoglobin.

Genetic change:

It has been proposed:

(i) A 'transversion' type of base substitution produced sickle-cell condition (one base such as thiamine changes to adenine).

(ii) So Hb^S is a point or gene mutation. The genetic change and its sequence are shown in the next page:

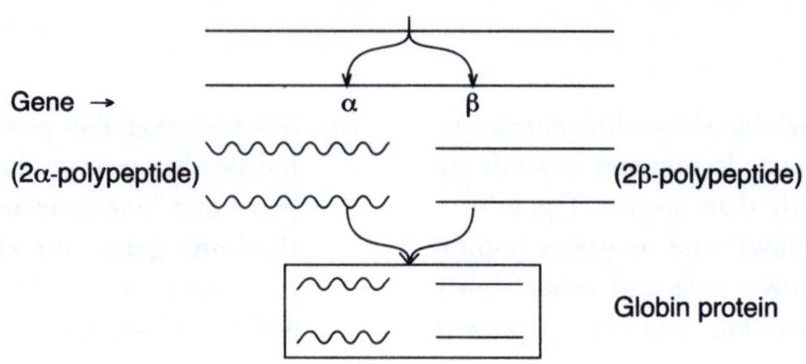

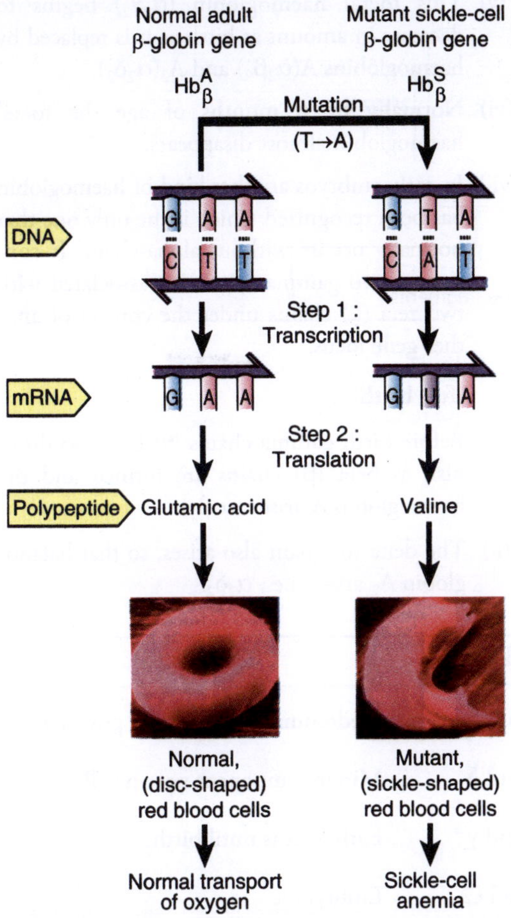

Fig. 6.6: The nature and consequence of a mutation in the gene for human β-globin. The mutant gene (Hb^S_β; top right) responsibe for sickle-cell anaemia resulted from a single base-pair substitution in the adult β-globin gene (Hb^A_β; top left). Transcription and translation of the mutant gene produce a β-globin polypeptide containing the amino acid valine (centre right) at the position where normal β-globin contains glutamic acid (centre left). Thus single amino acid change results in the formation of sickle-shaped red blood cells (bottom right) rather than the normal disc-shaped cells (bottom left). The sickle-shaped cells cause a severe form of anaemia.

Conclusion:

(i) The above observations reveal that change of morphology is related with the change of a particular polypeptide.

(ii) Change in gene (HbA → HbS) causes change in polypeptide nature (normal haemoglobin to sickle haemoglobin).

(iii) Here globin protein has 2α and 2β chains, each is controlled by gene.

So one gene = one polypeptide.

- **What is Thalassaemia?**

The word thalassaemia had been derived from Greek word **Thalassa** which means ocean. Initially this disease was confined among the people living in the countries by the sides of Mediterranean Sea.

The abnormal constitution of polypeptide chains of globin protein of haemoglobin for which a serious disease is developed and inherited called **thalassaemia**.

Molecular structure of Haemoglobin:

(i) Haemoglobin resides in the human RBC as carrier of oxygen.

(ii) Haemoglobin has two parts—one part is haem and other part is globin protein.

(iii) Globin protein of adult human consists of two alpha (α) and two beta (β) chains. But the embryonic haemoglobins contain alpha (α), beta (β), gamma (γ), delta (δ), epsilon (∈) and zeta (ζ).

(iv) The globin protein is controlled by gene. The gene responsible for controlling α-chain is located in the long arm of the chromosome **number 16**. On the other hand, the gene controlling β-chain is located in the small arm of the chromosome **number 11**.

(v) Haemoglobin < Haemoglobin A / Haemoglobin A_2

Haemoglobin A:

(i) It is composed of two identical alpha (α) and two identical beta (β) chains. It is represented by $\alpha_2\beta_2$.

(ii) It is very common and found in majority of human beings.

Haemoglobin A_2:

(i) It is composed of two identical alpha (α) and two delta (δ) type chains. It is represented by $\alpha_2\delta_2$.

(ii) A_2 is very uncommon and found in small proportion in human beings.

Steps of Haemoglobin development:

- **Before birth**

(i) The haemoglobin is composed of two alpha (α) and two epsilon ($\in$) polypeptide chains.

(ii) It is represented by $\alpha_2\in_2$.

(iii) This embryonic haemoglobin is soon (after 12 weeks) replaced by two alpha (α) and two gamma (γ) polypeptide chains. Gamma may be two types—γG and γA.

(iv) Foetal haemoglobin is represented by $\alpha_2\in_2$.

(v) This foetal haemoglobin ($\alpha_2\gamma_2$) begins to decrease in amount at birth as it is replaced by haemoglobins A($\alpha_2\beta_2$) and A_2($\alpha_2\delta_2$).

(vi) Normally by 6 months of age the foetal haemoglobin almost disappears.

(vii) In early embryos another kind of haemoglobin has been recognized which is the only one that normally occurs without alpha-chain. It consists of two gamma (γ) chains associated with two zeta (ζ) chains under the control of another gene locus.

- **After birth**

(i) Before birth gamma-chains (γ) begin to diminish as beta (β) chains are formed and the haemoglobin A arises ($\alpha_2\beta_2$).

(ii) The delta (δ) chain also arises, so that haemoglobin A_2 arises, i.e., $\alpha_2\delta_2$.

Type of haemoglobin	Chain present	Controlling gene	Occurrence
1. HbA ($\alpha_2\beta_2$)	$\frac{\alpha}{\alpha}\ \frac{\beta}{\beta}$	α and β	Predominant component post-natally
2. HbA$_2$ ($\alpha_2\delta_2$)	$\frac{\alpha}{\alpha}\ \frac{\delta}{\delta}$	α and δ	Minor component post-natally
3. HbF ($\alpha_2\gamma_2$)	$\frac{\alpha}{\alpha}\ \frac{\gamma}{\gamma}$	α and γ	Early foetus until birth
4. Embryonic ($\alpha_2\in_2$)	$\frac{\alpha}{\alpha}\ \frac{\in}{\in}$	α and $\in$	Embryonic
5. Embryonic ($\gamma_2\zeta_2$)	$\frac{\gamma}{\gamma}\ \frac{\zeta}{\zeta}$	γ and ζ	Embryonic

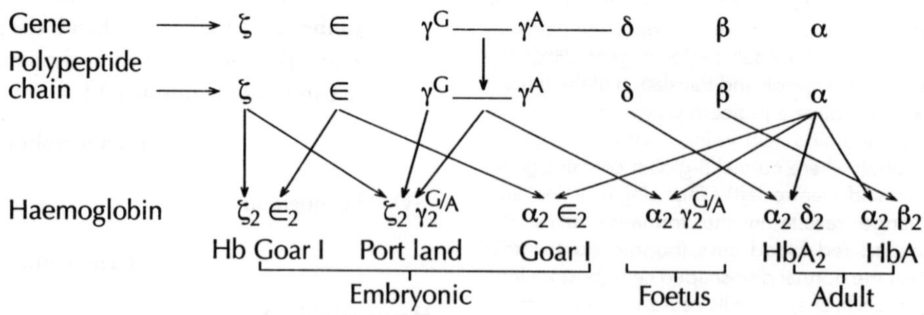

Globin polypeptide and Gene

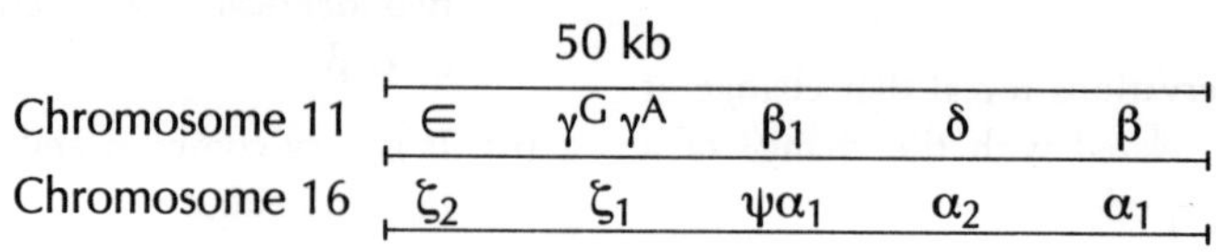

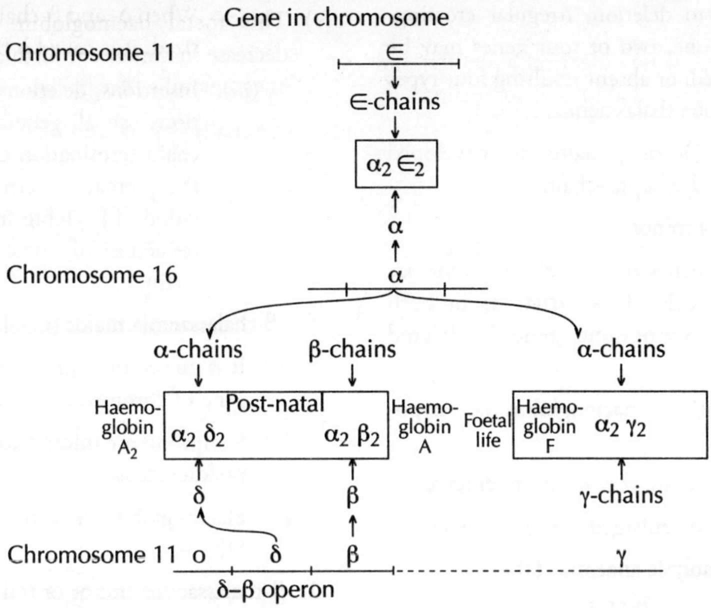

Summary of the genetic loci and their expression in the formation of human haemoglobin.

- **Features of thalassaemia:**

(i) The differential and defective rate of globin synthesis is the principal factor of this disease.

(ii) The defective globin synthesis is inherited.

(iii) As a result of defective synthesis of globin, the quantity of haemoglobin becomes far less than necessity.

(iv) This leads to reduce the life span of RBC and the child carrying this defect gradually succumbs to death.

(v) Normally globin consists of two alpha-(α_2), two beta-chains (β_2). The synthesis of α-chain is controlled by a gene located at the long arm of **16 chromosome** (16q) and the synthesis of β-chain is controlled by β-gene located at the long arm of **11 chromosome** (11q).

(vi) Due to the deletion or other mutations of the gene, normal chains (polypeptide) are not synthesized which leads to **defective globin** production.

- **Types of thalassaemia:**

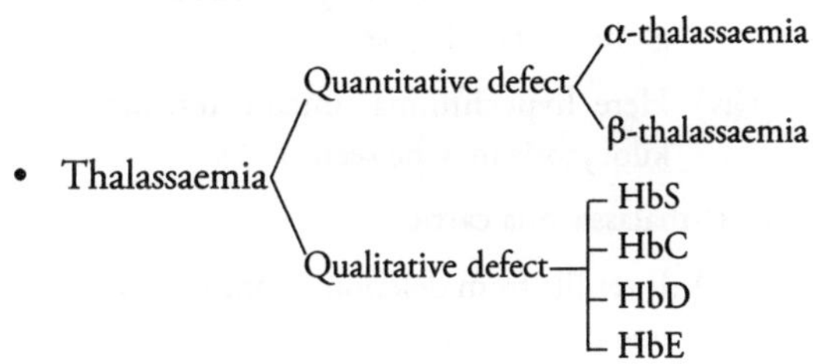

- **Quantitative defect:**

(i) It is known as thalassaemia syndrome.

(ii) Here one of the two globin polypeptides are not synthesized.

(iii) It is of two types like alpha and beta thalassaemia.

(a) Alpha thalassaemia:

(i) The alpha gene located on the short arm of 16 chromosome. Total four alpha-chains producing genes are present.

(ii) Normal genotype is αα/αα.

(iii) Due to deletion, irregular crossing-over one, two or four genes may be deleted, or absent resulting four types of alpha thalassaemia.

(iv) Here β- or γ-chains are developed instead of alpha-chain.

1. **α-thalassaemia minor**

   (i) Here two genes of a 16 chromosome are absent or deleted (– –/αα) or in both 16 chromosome one gene is deleted (– α/– α).

   (ii) Low quantity of haemoglobin, i.e., **hypochromia.**

   (iii) RBC may be short size, i.e., **microcytes.**

   (iv) RBC may be enlarged, i.e., **poikilocytosis.**

2. **Chronic haemolytic anaemia (HbH)**

   (i) It is known as HbH disease.

   (ii) It may be developed by deletion of two genes of one 16 chromosome and one gene from other 16 chromosome, i.e., total loss of function of three genes (– –/– α).

   (iii) Here only β-chain is synthesized. Haemoglobin is HbH type.

   (iv) Here hypochromia, microcytes and poikilocytosis may be seen.

3. **α-thalassaemia carrier**

   (i) It results from deletion of one α gene from one of the 16 chromosome.

   (ii) Here some HbF ($\alpha_2\gamma_2$) may be present.

4. **Hydrops foetalis**

   (i) It results from the deletion of 4α genes, i.e., (– –/– –).

   (ii) RBC may not carry O_2.

   (iii) Death occurs in embryonic condition.

(b) Beta thalassaemia:

(i) This is a process in which defective β-chains are produced.

(ii) When β-chains are less produced, it is known as **β^+ thalassaemia**; when β–chains are never produced, it is called **β^o thalassaemia.**

When δ- and β-chains are less synthesized, it is called δ β thalassaemia.

(iii) Insertions, deletions and point mutations can all generate a nonsense or chain termination codon resulting in the premature termination of translation of β-globin mRNA, that result reduction of translation of β-globin chain.

1. **β-thalassaemia major** (Cooley's anaemia)

   (i) It is due to the mutation of the regulatory gene of chromosome number 11 (11q).

   (ii) Symptoms are **microcytosis, hypochromia, poikilocytosis.**

   (iii) Haemoglobins may be HbA_2 ($\alpha_2\delta_2$) and HbF ($\alpha_2\gamma_2$).

2. **β-thalassaemia minor or trait**

   (i) This type of thalassaemia develops due to mutation of any one gene of 11 chromosomes.

   (ii) Iron accumulated in intestine results haemochromatosis.

3. **δ–β thalassaemia**

   (i) Here δ- and β-chains are less synthesized.

   (ii) In blood, hemoglobins are HbA_2 ($\alpha_2\delta_2$) or HbF ($\alpha_2\gamma_2$).

4. **Hereditary persistence of foetal haemoglobin**

   Here HbF ($\alpha_2\gamma_2$) type of haemoglobins are found in blood.

- **Qualitative defect:**

(i) It is also known as **haemoglobinopathy.**

(ii) Here defects are observed in α and β polypeptide chains.

(a) **HbS disease:** Sickle-cell anaemia. Here in polypeptide chain **glutamic** acid is replaced by **valine.**

(b) **HbC disease:** Here in polypeptide chain **glutamic acid** is replaced by **lysine.**

(c) **HbD disease:** He in β-chain in 121st **glutamic acid** is replaced by **glutamine.**

(d) **HbE disease:** Here in β-chain 26th **glutamic acid** is replaced by **lysine.**

Genetics of thalassaemia:

1. Due to unequal crossing-over between the repeated nucleotides of the α-globin gene, chromosome with more α or without α is produced.

   Example: Normal man: α-globin genotype: αα/αα, genotype with (α–/α–) or αα/– – or α–/– – may be developed. Genotype like – –/– – may also form but usually in a **lethal** condition.

2. For some thalassaemia patients, temporary hemoglobin mRNA or **faulty** proteins are considered.

Hereditary transmission of thalassemia:

I. Thalassaemia minor × Normal

$$♂\ \frac{\alpha - thal^+}{\alpha - thal} \times ♀\ \frac{\alpha - thal^+}{\alpha - thal^+}$$

Gametes: $\underline{\alpha - thal^+}$, $\underline{\alpha - thal}$; $\alpha - thal^+$

F_1 $\frac{\alpha - thal^+}{\alpha - thal^+}$ (Normal) 50% $\frac{\alpha - thal^+}{\alpha - thal}$ (Carrier) 50%

II. Thalassaemia minor × Minor

$$♂\ \frac{\alpha - thal}{\alpha - thal^+} \times ♀\ \frac{\alpha - thal}{\alpha - thal^+}$$

Gametes: α – thal, $\underline{\alpha - thal^+}$; $\underline{\alpha - thal}$, $\underline{\alpha - thal^+}$

F_1

	α – thal	α – thal⁺
α – thal	$\frac{\alpha - thal}{\alpha - thal}$	$\frac{\alpha - thal^+}{\alpha - thal}$
α – thal⁺	$\frac{\alpha - thal}{\alpha - thal^+}$	$\frac{\alpha - thal^+}{\alpha - thal^+}$

Results:

	Genotype	Phenotype	Ratio
(i)	$\frac{\alpha - thal}{\alpha - thal}$	Thalassaemia major	1 = 25%
(ii)	$\frac{\alpha - thal}{\alpha - thal^+}$	Thalassaemia minor	2 = 50%
(iii)	$\frac{\alpha - thal}{\alpha - thal^+}$	Thalassaemia minor	
(iv)	$\frac{\alpha - thal^+}{\alpha - thal^+}$	Normal	1 = 25%

- ## Molecular diagnosis of Sickle-cell anaemia:

(i) The gene for a human disease has been cloned and 6 sequenced and the nature of the mutations which cause the disorder is known as **molecular test** for the mutant alleles.

(ii) These test can be performed on small amount of DNA by using PCR to amplify DNA segment of interest.

(iii) Some molecular diagnoses involve simply testing for the presence or absence of a specific restriction enzyme cleavage site in DNA.

A. By Southern blotting:

(i) Sickle cell anaemia is caused by a single amino acid substitution in β-globin gene.

(ii) This change is caused in gene by nucleotide substitution, coincidentally removes cleavage site for the restriction enzyme Mst II.

(iii) The Hb^S_β (sickle-cell) allele can be distinguished from the normal β-globin allele (Hb^A_β) by amplifyng part of the β-globin gene by PCR.

(iv) Cutting the amplified DNA with restriction enzyme Mst II. Then restriction fragments are separated by agarose gel electrophoresis.

(v) Then preparing Southern blot of the separated fragments and hybridising the DNA on the blot to probe spanning the site of the mutation.

(vi) The probe will hybridize with two small fragments from normal β-haemoglobin gene, but with only one fragment from sickle cell β-globin gene. This difference in RFLP pattern is used to diagnose any defect of heterozygotes as well as homozygotes of sickle cell anaemia.

Alternately

(i) PCR primers can be synthesized, that are complementary to DNA sequences flanking the sickle-cell mutation in the Hb^S_β gene and used to amplify this segment from genomic DNA.

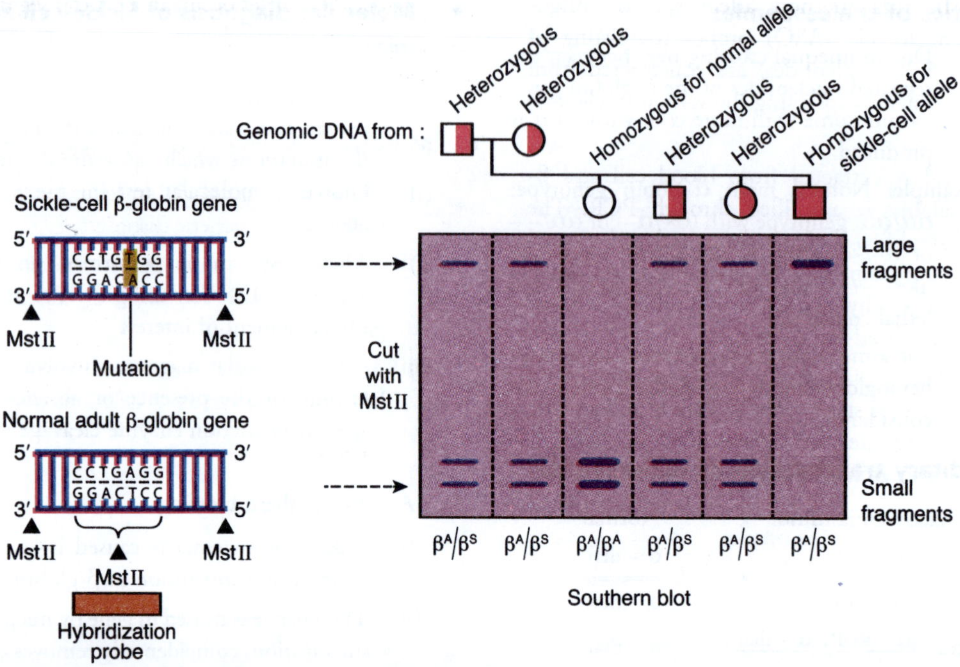

Fig. 6.7: Detection of the sickle-cell haemoglobin mutation by Southern blot analysis or genomic DNAs cut with restriction enzyme Mst II.

(ii) The amplified DNA can be treated with MstII and the products of the reaction examined by agarose gel electrophoresis to see whether or not the amplified fragment was cleared by MstII.

B. By Dot-blot hybridization employing alleles specific oligonucleotide (ASO) probe:

(i) Application of dot-blotting involves distinguishing between alleles that differ by even a single nucleotide substitution.

gene	β^A-allele	5′......GTG CAC CTG ACT CCT GAG GAG AAG TCT GCC......3′
sequence	β^S-allele	5′......GTG CAC CTG ACT CCT GTG GTG AAG TCT GCC......3′
allele-specific	β^A-ASO	TG ACT CCT GAG GAG AAG TC
oligonucleotides (ASO)	β^S-ASO	TG ACT CCT GTG GTG AAG TC

dot-blots

β^A-ASO probe

β^S-ASO probe

genotype: $\beta^A\beta^A$ $\beta^A\beta^S$ $\beta^S\beta^S$

Fig. 6.8: Testing for the sickle-cell mutation with dot-blot hybridization. The sickle-cell mutation is a single nucleotide substitution (A→T) at codon 6 in the β-globin gene, resulting in a GAG (Glu)→GTG (Val) substitution. Allele-specific oligonucleotides (ASOs) were designed to match the nucleotide sequence flanking the substitution: β^A-ASO is specific for the normal β-globin allele and β^S-ASO for mutant β-globin allele. The schematic dot-blots show the results of probing with these two ASOs under stringent hybridization conditions. With the normal probe (β^A-ASO), the results are positive (filled circle) for normal individual and for heterozygotes, but negative for sickle-cell homozygotes (unfilled circle). With the mutant probe (β^S-ASO), the results are positive for the sickle-cell homozygotes and heterozygotes, but negative for normal individuals.

(ii) In this method allele specific oligonucleotide (ASO) probes (consisting of 15-20 nucleotides) are constructed from sequences spanning the variant nucleotide site.

(iii) DNA extracted from blood cell and β-globin is amplified through PCR. The amplified DNA is denatured and spotted onto strips in DNA binding filters. Each strip is hybridized to a specific ASO and visualized on X-ray.

The DNA duplex between probe and target is stable only if there is perfect base complimentarily between them. A single mismatch between probe and target sequence is sufficient to render the short heteroduplex unstable.

(iv) In distinguishing the sickle-cell allele from normal β-globin alleles, two ASOs are designed to represent the two alleles, so that the single nucleotide difference between alleles occurs in a central segment of the oligonucleotide sequence to maximize the thermodynamic instability of a mismatched duplex.

- **Molecular diagnosis of Thalassaemia:**

(i) Mutation in the β-globin locus leads to a wide range of genetic disorders.

(ii) β thalassaemia is an autosomal disorder associated with decreased or absent β-globin production.

(iii) At the molecular level, this condition can be caused by a variety of deletions or point mutations.

(iv) One such deletion covers about 600 nucleotides and includes third exon of the gene.

(v) Restriction enzyme digestion and Southern blot hybridization to a cloned β-globin probe can be used to diagnose the genetic status of a foetus and other members of an affected family.

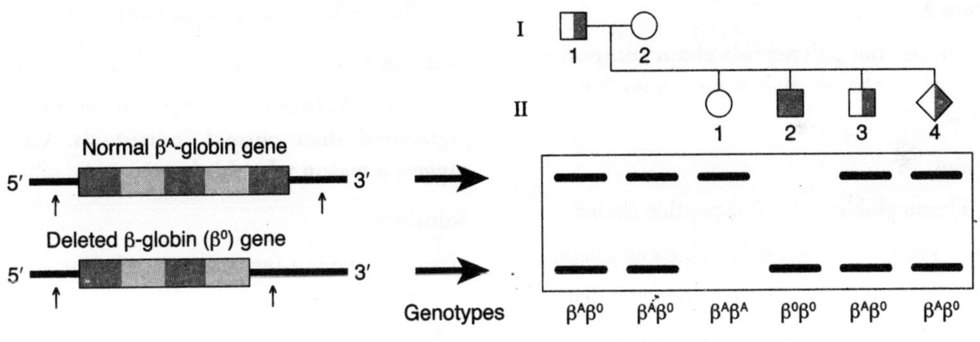

Fig. 6.9: Diagnosis of β-thalassaemia caused by a partial deletion of the β-globin gene. The family pedigree is shown above each individual's genotype on a Southern blot. The normal β-globin allele (β^A) contains three exons and two introns. The deleted β-globin allele (β^0) has the third exon deleted. Arrows indicate the cutting sites for restrictiojn enzymes used in this analysis. The normal gene produces a larger fragment (shown as the top row of fragments on the Southern blot); the smaller fragments produced by the deleted gene are represented at the bottom of the gel. The genotypes of each individual in the pedigree can be determined from the pattern of bands on the blot and is given below the blot.

[P.G., M.G., E.G. & I.G. 15]

Solved Problems

Problem 1

Explain why the one-gene one-enzyme concept is not considered totally accurate today.

Solution:

(i) The fact that enzymes are a subclass of the general term protein, a one-gene one-protein statement is more appropriate.

(ii) However, some proteins are made up of subunits, each different type of subunit (polypeptide chain) being under control of a different gene.

(iii) Under the circumstances one-gene one-polypeptide might be more reasonable. However, many additional complexities exist and simple statement regarding the relationship of a stretch of DNA to its physical product is difficult to justify.

Problem 2

Contrast the polypeptide chain components of each of the haemoglobin molecules found in humans.

Solution:

Haemoglobin	Polypeptide chains
1. HbA	$2\alpha\ 2\beta$ (alpha and beta)
2. HbA_2	$2\alpha\ 2\delta$ (alpha and delta)
3. HbF	$2\alpha\ 2\gamma$ (alpha and gamma)
4. Gower 1	$2\zeta\ 2\in$ (zeta and epsilon)

The alpha and beta chain contain 141 and 146 amino acids, respectively.

The zeta-chain is similar to the alpha chain while the other chains are like the beta chain.

Problem 3

Contrast the contribution of Pauling and Ingram to our understanding of the genetic basis for sickle-cell anaemia.

Solution:

(i) In the late 1940's, Pauling demonstrated a difference in the electrophoretic mobilities of HbA and HbS (sickle-cell haemoglobin) and concluded that the difference had a chemical basis.

(ii) Ingram determined that the chemical change occurs in the primary structure of the globin portion of the molecule using fingerprinting technique. He found a change in the 6th amino acid in the β chain.

Problem 4

Discuss the potential difficulties involved in designing a diet to alleviate the symptoms of the phenylketonuria.

Solution:

(i) Phenylalanine is an amino acid, that like other amino acid is required for protein synthesis.

(ii) While too much phenylalanine and its derivatives cause PKU in phenylketonuria, too little will restrict protein synthesis.

Problem 5

Phenylketonurics are often more lightly pigmented than normal individuals. Can you suggest a reason why this is so?

Solution:

(i) Tyrosine is a precursor to melanine, skin pigment.

(ii) Individuals with PKU fail to convert phenylalanine to tyrosine and even though tyrosine is obtained from the diet, at the population level, individuals with PKU have a tendency for less skin pigmentation.

Problem 6

Thalassaemia is a type of human anaemia rather common in Mediterranean populations but relatively rare in other peoples. The disease occurs in two forms, minor and major, the latter is much more severe. Persons with thalassaemia major are homozygous for an aberrant recessive gene; mildly affected persons (with thalassaemia minor) are heterozygous; persons, normal in this regard are homozygous for the normal allele.

(a) A man with thalassaemia minor marries a normal woman. With respect to thalassaemia, what types of children and in what proportions, may they expect (let 't' be the allele for thalassaemia minor and T its normal allele)?

(b) Both father and mother in a particular family have thalassaemia minor. What is the chance of their baby will be severely affected, mildly affected, normal?

Solution:

(a) Thalassaemia minor × Normal woman
(heterozygous)

$\frac{T}{t}$ × $\frac{T}{T}$

Gametes: T, t; T

♀ \ ♂	T	t
T	$\frac{T}{T}$	$\frac{T}{t}$

Result:

Genotype	Phenotype	Ratio
TT	Normal	1
Tt	Thalassaemia minor	1

(b) Father (Thalassaemia minor) Mother (Thalassaemia minor)

$\frac{T}{t}$ × $\frac{T}{t}$

Gametes: T, t; T, t

♀ \ ♂	T	t
T	$\frac{T}{T}$	$\frac{T}{t}$
t	$\frac{T}{t}$	$\frac{t}{t}$

Result:

Genotype	Phenotype	Phenotypic	Ratio
TT	Normal	1	(25%)
Tt Tt	Minor	2	(50%)
tt	Major	1	(25%)

25% baby will be severely affected.
50% baby will be mildly affected.
25% baby will be normal.

Problem 7

Human haemoglobin C is a variant in which a lysine in the β haemoglobin chain is substituted for a particular glutamic acid. What single base substitution can account for haemoglobin C-mutation?

Solution:

(i) Possible lysine codons are AAA and AAG, and the possible glutamic acid codons are GAA and GAG.

(ii) Therefore, the mutation results from AT → GC transition in the first position of the codon.

Lysine: A A A A A G
↓ ↓
Glutamic acid: [G]A A [G]A G – Transition

Problem 8

Assuming that the β-globin chain and the α globin chain shared a common ancestor, what mechanism might explain the differences that now exist in these two chains? What changes in DNA and mRNA codons would account for the differences that have resulted in unlike amino acids at corresponding positions.

Solution:

Mutations: Transitions, transversions and frameshifts.

Problem 9

If CTT is a DNA triplet (transcribed strand of DNA) specifying glutamic acid, what DNA and mRNA base triplet alterations could account for valine and lysine in position 6 of β-globin chain?

Solution:

Amino acid	mRNA	DNA
Glutamic acid	— G A A →	— G A A
		↑
		C T T ← Transcribed strand
		↓ Mutation T-A
Valine	— G U A →	C A T
		G U A
		↓ Mutation C-T
Lysine	— A A A →	T T T
		A A A

Problem 10

Consider sickle-cell anaemia is an example of a devastating disease, that is the result of an autosomal recessive genetic mutation on a specific chromosome. Explain what molecular or

genetic disease is similar and different to this disease with a disease caused by an invading microorganism such as a bacterium or virus.

Solution:

(i) A genetic disease such as sickle-cell anaemia is caused by a change in DNA that alters the levels or forms of one or more gene products.

(ii) In turns, the altered forms or levels of gene products result in changes in cellular functions which in turn cause a diseased state.

(iii) Genetic disease (sickle-cell anaemia) associated with mutation in a single gene that affects the protein products.

(iv) Sickle-cell anaemia is caused by mutation lead to amino acid substitution that causes β-globin polypeptide to fold incorrectly. This in turn leads to sickle red blood cells and anaemia.

(v) The environment can have a significant effect on the severity of many genetic diseases. Many genetic diseases are treatable. For example, PKU can be treated by altering the diet. Unlike diseases caused by an invading microorganism or other external agent that are subject to the defence of the human immune system and that generally have short-lived clinical symptoms and treatments.

(vi) Genetic diseases are caused by heritable changes in DNA that are associated with chronically altered levels or forms of one or more gene products.

Problem 11

	Amino acid position							
Haemoglobin (Hb)	**1**	**2**	**16**	**30**	**57**	**58**	**68**	**141**
(Normal & Chain):	**Val**	**Leu**	**Lys**	**Glu**	**Gly**	**His**	**Asn**	**Arg**
Variant (Hb norfolk):	**Val**	**Leu**	**Lys**	**Glu**	**ASP**	**His**	**Asn**	**Arg**

Haemoglobin β chain normal	**1**	**2**	**3**	**6**	**7**	**26**	**63**	**67**	**121**	**146**
(Hb)	**Val**	**His**	**Leu**	**Glu**	**Glu**	**Glu**	**His**	**Val**	**Glu**	**His**
Variant										
Hb^S	**Val**	**His**	**Leu**	**Val**	**Glu**	**Glu**	**His**	**Val**	**Glu**	**His**
Hb^C	**Val**	**His**	**Leu**	**Lys**	**Glu**	**Glu**	**His**	**Val**	**Glu**	**His**

From the above human haemoglobin variants, what would you expect the phenotype to be in people heterozygous for the following two haemoglobin mutations?

(a) Hb norfolk and Hb^S, (b) Hb^C and Hb^S.

Solution:

(a) (i) Hb norfolk affects the α chain and on the other hand, Hb^S affects the β-chain of haemoglobin.

(ii) Since each chain is encoded by a separate gene, there remains one normal allele at the genes for each of α- and β-chains in double heterozygote.

(iii) Thus, some normal haemoglobin molecules form and double heterozygotes do not have severe anaemia.

(iv) However, unlike double heterozygotes for two different, completely recessive mutations that lie in one biochemical pathway, these heterozygotes exhibit an abnormal phenotype. This is because some mutations in the α- and β-chains of haemoglobin show **partial dominance**.

(v) In particular, $Hb^S/+$ heterozygotes show symptoms of anaemia if there is a sharp drop in oxygen tension. So these double heterozygotes exhibit mild anaemia.

(b) (i) Both Hb^C and Hb^S affect the sixth amino acid of the β-chain.

(ii) The Hb^C mutation alters the normal glutamate to lysine.

(iii) On the other hand, Hb^S mutation alters its valine, since both affect the β-chain, no normal haemoglobin molecules are present.

(iv) Accordingly only one type of β-chain is found in any one haemoglobin molecule. Therefore, on Hb^C/Hb^S heterozygote has two types of haemoglobin. Those with Hb^C β-chains and those with Hb^S β-chain.

Problem 12

The gene for a globin is present in all vertebrate species. Over millions of years the DNA sequence of this gene has changed in the lineage of each species. Consequently, the amino acid sequence of a globin has also changed in these lineages. Among the 141 amino acid positions in this polypeptide, human α-globin differs from shark α-globin in 79 positions, it differs from carp α-globin in 68 and from cow α-globin in 17. Do these data suggest an evolutionary phylogeny for these vertebrate species?

Solution:

1. Human α-globin differs from;
   (i) Shark α-globin in 79 positions.
   (ii) Carp α-globin in 68 positions.
   (iii) Cow α-globin in 17 positions.

   The human and cow α-globins are least different.

   Therefore, on the assumption that differences in a globin reflect the degree of phylogenetic relationship among the organisms.

   The human and the cow are most closely related organisms among those mentioned.

   The next closest relative of human is the carp and the most distant relative is shark.

   Human → Cow → Carp → Shark
   (17) (68) (79)

Problem 13

Sickle-cell anaemia is caused by a mutation in one of the codons in the gene for β-globin; because of this mutation the sixth amino acid in the β-globin polypeptide is a valine instead of a glutamic acid. A less severe type of anaemia is caused by a mutation that changes this same codon to one specifying lysine as the sixth amino acid in the β-globin polypeptide. What word is used to describe the two mutant forms of this gene? Do you think that an individual carrying these two mutant forms of the β-globin gene would suffer from anaemia? Explain.

Solution:

(i) The two mutant forms of the β-globin gene are properly described as alleles.

(ii) Because neither of the two alleles can specify a normal polypeptide.

(iii) Thus, an individual who carries each of them would probably suffer from anaemia.

Problem 14

Pathogenic DNA variants—causes sickle anaemia.

Solution:

Genomes of healthy individuals have huge number of sequence variants. The great majority of these are completely harmless and have no known effect on the phenotype. But, those DNA variants which either make us ill or make us susceptible to illness, are called **pathogenic** DNA variants.

(i) A single nucleotide substitution within the coding sequence of a gene may or may not alter the sequence of the encoded protein as genetic code is degenerate. Thus, same codon changes do not alter the amino acid. They are **silent** or **synonymous.**

(ii) When codon change does result in a change of amino acid, the effect depends partly on the chemical differences of old and new amino acids.

(iii) Some amino acids are crucial to the functioning of a particular protein. Others may be important for maintaining the protein structure.

(iv) Globular proteins tend to have uncharged nonpolar amino acids in the interior and charged ones on the outside. Any substitution that changes this may disrupt the three dimensional folding.

(v) The sickle-cell mutation is **pathogenic** because it replaces a polar amino acid with a non-polar one on the outside of the globin molecule. This makes the molecules tend to stick together to

protein aggregation, the result of abnormal protein having sticky external area has emerged as pathogenic mechanism.

(vi) The mutation in β-globin gene replaces glutamic acid (GAG), a **hydrophilic** charged amino acid with valine (GUG), a **hydrophobic** non-polar amino acid. This change in the surface of the globin protein allows adhesive interactions between haemoglobin molecules.

(vii) The result is sickle-cell anaemia in which haemoglobin molecules clump together distorting the shape of the red blood cells.

- **Phenogenesis:**

  It is a mechanism by which phenotype of an organism is produced under the control of DNA in a given environment.

- **Field inversion gel electrophoresis:**

  An electrophoresis technique for the separation of large molecules of DNA.

- **Orthogonal field-alternation gel electrophoresis (OFAGE):**

  A gel electrophoresis technique that employs a pulsed electric field to achieve separation of very large molecules of DNA.

- **Gel retardation:**

  A technique that identifies a DNA fragment that has a bound protein by virtue of its decreased mobility during gel electrophoresis.

- **Gene cluster:**

  A group of related genes grouped together on a eukaryotic chromosome.

- **Gene therapy:**

  The treatment of inherited disease by introducing wild type copies of the defective gene causing disorder into the cells of affected individuals.

  1. Germ line or heritable gene therapy: Here reproductive cells are modified.
  2. Somatic cell or nonheritable gene therapy: Here somatic cells are modified.

- **Genetic counselling:**

  Analysis of risk for genetic disorders in a family and presentation of options available to avoid or ameliorate possible risks.

- **Gene tree:**

  A phylogenetic tree based on the divergence observed within a single homologous gene is called Gene tree.

- **Gene amplification:**

  The process by which gene sequences are selected and differentially replicated either extra-chromosomally or intrachromosomally is termed as Gene amplification.

- **Homeotic gene:**

  Genes that control the fate and identity of segments along the anterior posterior axis of higher animals is called Homeotic gene.

  Example: Antennapedia (*Drosophila*).

- **Heat shock genes:**

  Genes which are switched on in response to environmental insult (heat) are called Heat shock genes.

- **Gene conversion:**

  A non-reciprocal recombination process in which one allele in a heterozygote is changed to the other allele, thus converting a heterozygous genotype to a homozygous genotype is called Gene conversion.

- **Gene replacement:**

  The incorporation of a transgene into a chromosome at its normal location by homologous recombination, thus replacing the copy of the gene originally present at the locus is termed as Gene replacement.

7 Mutation

Definition

Any detectable, inheritable qualitative or quantitative changes in the genetic material of an organism not caused by genetic recombination are termed as **mutation.**

Features:

1. Mutations not only means changes in the genetic material but also the **method** or **process** by which the change occurs.
2. **Mutant**: An organism exhibiting a novel phenotype as a result of a mutation is referred to as **mutant.**
3. Mutation is a process by which a change in the **chromosome** or change in **DNA base pair** occurs.
4. Mutations at the level of chromosome are called **chromosomal mutations.** Mutations in the sequences of genes at the level of base pair are called **gene mutations** or **point mutations.**
5. **Mutagen:** Any physical or chemical agent that significantly increases the frequency of mutational events above the spontaneous mutation rate is called mutagen.
6. Mutation is the ultimate source of all **genetic variations.** It provides the raw material for evolution.
7. **Mutation rate:**
   (i) The frequency with which mutation takes place at a given locus or in a population is called **mutation rate.**
   (ii) It is the probability that a gene undergoes mutation in a single generation or in forming a single gamete.

Different types of mutation:

Nature of Mutation	Type
1. According to the size and quality	(i) Chromosomal mutation — Variation in number; Variation in structure
	(ii) Point mutation
2. According to direction	(i) Forward
	(ii) Reverse
3. According to magnitude and phenotype	(i) Dominant
	(ii) Recessive
	(iii) Lethal
4. According to origin	(i) Spontaneous
	(ii) Induced
5. According to cell type	(i) Somatic
	(ii) Gametic

- **Lethal mutation:**

The mutation of gene which brings death to an organism is called **lethal mutation.**

Example: Curly wing mutation becomes lethal in homozygous conditions. $\frac{Cy}{Cy}$ is a lethal condition in *Drosophila*.

- **Balanced lethal mutation:**

When one or more lethal genes remain in one organism, the stock may be kept viable keeping the lethal alleles in a heterozygous condition.

Example: Curly lobe and plum alleles can be maintained in heterozygous form $\frac{Cy + L}{+ \; pm \; +}$.

- **Conditional lethal mutation:**

This is a condition when one of two sexes becomes affected due to lethality.

Example: Notch (N) is a mutant allele which is lethal for male in hemizygous condition but a heterozygous female can survive.

- **Dominant mutation:**

The mutation which produces an allele having dominant effect on the normal allele.

Example: Mutant allele for bar-shaped eye is dominant over its normal eye-shaped allele.

Phenotype:	**Bar eye**	**Normal eye**
Genotype:	$\frac{B^+}{B^+} / \frac{B^+}{B}$	$\frac{B}{B}$

- **Genetic screen:**

(i) A mutagenesis procedure in which essentially mutagenized progenies are recovered and are individually evaluated for mutant phenotype.

(ii) The desired phenotype is marked in some way to enable its detection.

- **Genetic selection:**

Some mutagenesis protocols are cleverly designed so that only the desired phenotype survives or replicates. Such protocols are called **genetic selection.**

- **Leaky mutation:**

A mutation that confers a mutant phenotype but still retains a low but detectable level of wild type function.

- **Genetic dissection:**

The use of recombination and mutation to piece together the various components of a given biological function.

- **Paramutation:**

An epigenetic phenomenon in plants in which genetic activity of a normal allele is heritably reduced by virtue of that allele having been heterozygous with a special 'paramutagenic' allele.

- **Phenocopy**

An environmentally induced phenotype that resembles the phenotype produced by mutation.

DELETION (or Deficiency)

It is a chromosomal mutation involving the loss or removal or deletion of a chromosomal segment from a chromosome set.

Characteristics:

(i) The deletion can occur either near one end or from the interior end of the chromosome. These are called **terminal** or **intercalary** deletion respectively.

(ii) Terminal deletion requires only one break and intercalary deletion requires double breaks.

(iii) Chromosomes with deletion can never **revert** to normal condition.

(iv) Smaller deletions, present in heterozygote condition can be tolerated by an organism. Such individual at meiosis (synapsis to occur between deficiency and normal chromosome) will form (unpaired region of the normal homolog must loop out of the linear structure) a **deletion loop** or **compensation loop.**

(v) It helps to find out the relative position of genes in the chromosome.

(vi) A small deletion within a gene, called as **intragenic deletion**, inactivates the gene and has the same effect of **null mutation** of that gene.

(vii) The phenotypic expression of recessive gene on one homolog caused by the deletion of dominant gene on the other homolog is known as **pseudo-dominance.**

Deletion can be formed by two major ways:

1. **Chromosome breakage and reunion**
   (i) Chromosome breaks result from double-stranded breaks in the DNA backbone.
   (ii) Chromosome breaks occur spontaneously at low rate, but they also can be induced by X-rays and certain chemicals that cause double-stranded breaks in DNA.
   (iii) A deletion is created when a chromosome arm is broken in two places. When the broken ends bearing the centromere and the telomere fuse and the part left out remaining as an **acentric fragments** that is lost.
2. **By Ectopic recombination**
   (i) Deletion can also be created by homologous recombination between repeated DNA sequences present at different sites along the DNA, the process is known as **ectopic recombination.**
   (ii) In this process, each copy of repeated DNA sequences is indicated by coloured gradient. Note that the gradient runs from left to right in both copies which indicates that both copies of the repeated DNA sequences have the same orientation along the DNA, a configuration known as **direct repeat.**
   (iii) If the direct repeats undergo pairing and homologous recombination, the result is deletion of the material between the direct repeats, because the small circular acentric fragment containing this material is lost.

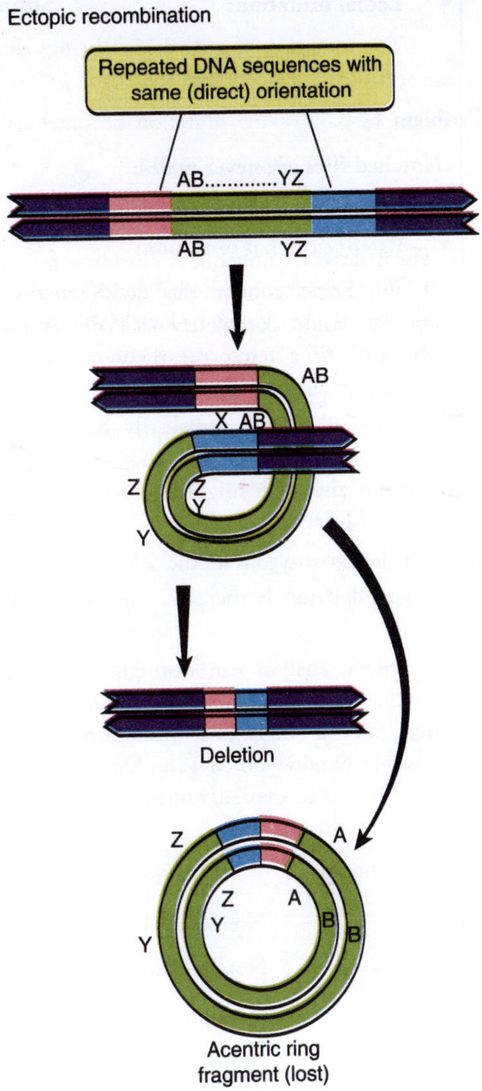

Fig. 7.1: Ectopic recombination between direct repeats in the same DNA molecules results in deletion of the material between the repeats.

Example:

(a) (i) 'Minute' bristles in *Drosophila.*
   (ii) A **Notch wing** in *Drosophila* is an X-linked dominant trait and homozygous lethal.
   (iii) **Deletion loop** was found along X-chromosome from **3C2–3C11.**

(b) In human "**Cri du chat**" syndrome deletion of tip of the short arm of 5 chromosomes ($5p^{15.2}$ to $5p^{15.3}$).

(c) Prader-Willi syndrome—deletion of one or several genes ($15q^{11-13}$).

Notch:

(i) In *Drosophila* many deletions of a gene ($3c^{2-11}$) is called **notch**, which results notch wing margin.

(ii) It is a **sex-linked dominant** trait and homozygous lethal.

Solved Problems

Problem 1

Notched flies are never males:

Solutions:

(i) The males carry only one X-chromosome and if this should contain the 'notch' deletion, the fly would completely lack the deleted segment. As a result the fly would fail to develop or survive.

(ii) On the otherhand, female fly has two X-chromosome and if one should have the deletion, the other might be normal and the fly would develop.

(iii) Females heterozygous for the deletion are notch $\left(\frac{+}{N}\right)$ so deletion is therefore regarded dominant.

(iv) A genetic analysis can distinguish between notched flies carrying deletion versus those that carry a single gene mutation. This is possible because notch gene happens to be located next to the red/white eye colour gene on the X chromosome.

Genotype	Phenotype
$\frac{N}{+}$	Notch female
$\frac{N}{N}$	Notch female (lethal)
$\frac{N}{\rightharpoonup}$	Notch male (lethal)
$\frac{+}{\rightharpoonup}$	Normal male

Drosophila with this phenotype have discovered that some of the mutant flies are due to small deletion that includes the notch gene as well as few genes on either side of it. Other notch mutations are confined within notch gene itself.

Problem 2

How would you distinguish between a notched phenotype due to a deletion that included the notch gene and the adjacent eye colour gene versus a notch phenotype due to a small mutation only within the notch gene itself.

Solutions:

(i) If the notch mutation is due to a deletion, red eyed females with the notched phenotype can be crossed to white eyed males.

(ii) Only the daughters with notched wings need to be analyzed.

(iii) If they have red eyes, this means, the notch mutation has not deleted the red allele from the X-chromosome.

(iv) Alternatively, it they have white eyes, this indicates the red eye allele has been deleted from the X-chromosome that carries the notch mutation. In this case, the white allele is expressd because it is preserved in a single copy in a female fly with two X-chromosomes. This phenomenon is called **pseudodominance.**

Red eyed female with notch × White eyed male

P_1 ♀ $\frac{+\ N}{+\ +}$ × $\frac{w\ +}{\rightharpoonup}$ ♂

Gametes $\frac{+\ N}{}$ $\frac{+\ +}{}$; $\frac{w\ +}{}$, $\rightharpoonup$

F_1 1. $\frac{+\ N}{w\ +}$ ♀,   2. $\frac{+\ +}{w\ +}$ ♀

Notched red eyes   Normal red eyes

(i) Here $\left(\frac{+\ N}{w\ +}\right)$ no deletion of red allele.

(ii) If they have white with notch gene, i.e., $\left(\frac{d\ N}{w\ +}\right)$ this situation will occur.

Problem 3

The loss of Genetic material in a deletion tends to be detrimental to an organism.

Solutions:

(i) Chromosomal deletion occurs when a chromosome breaks in one or more places and the fragment of the chromosome is lost.

(ii) If a norml chromosome is broken into two separate pieces, the piece without centromere (acentric) will be lost and degraded. This event produces a chromosome with terminal deletion.

(iii) If a normal chromosome is broken into two pieces to produce three chromosomal fragments, the central fragment will be lost and

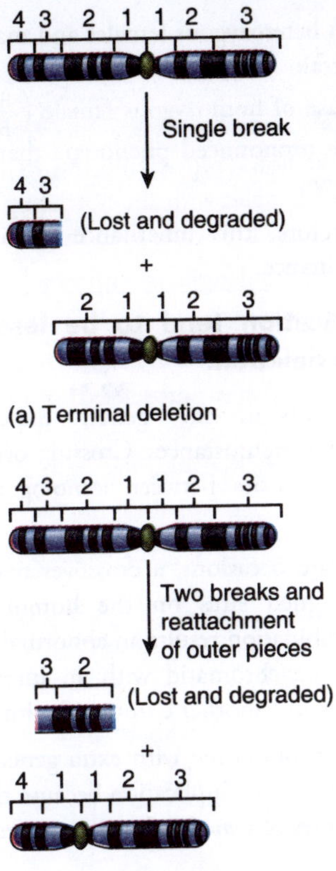

Fig. 7.2: Production of terminal and interstitial deletions. This illustration shows the production of deletions in human chromosome 1.

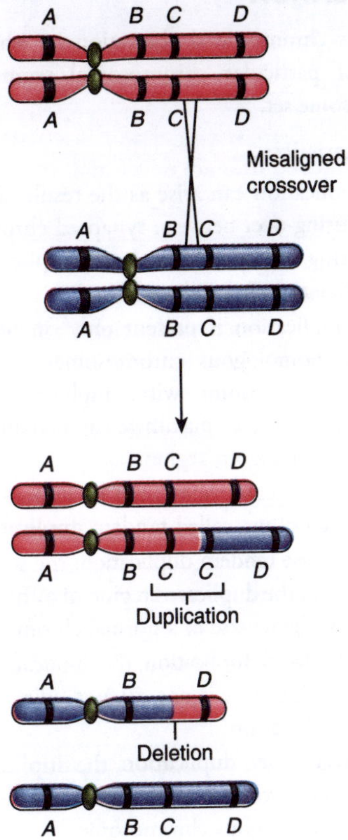

Fig. 7.3: Abnormal crossing-over, leading to a duplica tion and a deletion. A crossover has occurred a sites between genes C and D in one chromatic and between genes B and C in another chroma tid. After crossing-over is completed, one chro matid contains a duplication, and the othe contains a deletion.

the two outer pieces reattach to each other. This process creates **interstitial deletion.**

(iv) Deletion can also be created when recombination takes place at incorrect location between two homologous chromosomes. The product of this type of aberrant recombination event is one chromosome with deletion and other chromosome with duplication.

(v) Deletion heterozygotes are genetically unbalanced, because they have only one dose of every gene in the region of the deletion as compared to two elsewhere.

(vi) In case, the deletion is large, the heterozygote cannot develop due to genic unbalance. Offsprings that are homozygous for a large deletion cannot develop because they completely lack the genes in the region of the dele tion and any large deletion as a rule contain one or more genes necessary for life.

(vii) Small deletions might not prevent develop ment but often they cause the adult to hav some mutant trait.

Example: Small deletion in certain part of X-chromo some of *Drosophila* creates 'Notch' wing.

Again in *Drosophila*, small deletions often caus a reduction in bristle size (**minute bristle**) and a known as "Minutes". Homozygous minutes can n live. The heterozygotes appear Minute, so that th deletion is **dominant**.

DUPLICATION

It is chromosomal aberration in which repetition of particular chromosomal segment in a chromosome set.

Characteristics:

(i) Duplication can arise as the result of **unequal crossing-over** between synapsed chromosomes during meiosis. Here both duplication and deficiency are produced.
(ii) If duplication is present only on one of the two homologous chromosomes, at meiosis, the chromosome with duplicated segment forms a loop to maximize the juxtaposition of the homologous regions.
(iii) The duplicate regions can be located adjacent to each other, called **tandem duplication**.
(iv) In **reverse tandem duplication**, the sequence of genes in the duplicated region of a chromosome is just the reverse of a normal chromosome.
(v) In **displaced duplication**, the duplicated portion of chromosome is not situated adjacent to the normal section.
(vi) In **transposed duplication**, the duplicated portion of chromosome becomes attached to the nonhomologous chromosome.

Example: **Bar eye** in *Drosophila*

(i) X-linked dominant mutation
(ii) Narrower, oblong bar-shaped eye with few facets
(iii) It is associated with the duplication of 16A section of X-chromosome.

Significance

(i) It may result in **gene redundancy**.
(ii) As with deletion, duplication can produce **phenotypic variation**.
(iii) Duplication is an important source of **genetic variability** during evolution.

	Genotype	X chromosome	Mean No. of Facet	Phenotype
(i)	$\frac{B^+}{B^+}$	$\frac{16A}{16A}$	779-800	Wild type 
(ii)	$\frac{B}{B^+}$	$\frac{16A\cdot16A}{16A}$	350-358	Heterozygous bar 
(iii)	$\frac{B}{B}$	$\frac{16A\cdot16A}{16A\ 16A}$	68-70	Homozygous bar 
(iv)	$\frac{B^D}{B^+}$	$\frac{16A16A16A}{16A}$	45	Double bar 

Bar is semidominant

(i) Both heterozygous females and males exhibit the trait.
(ii) In case of homozygous female $\left(\frac{B}{B}\right)$ exhibits more pronounced phenotype than either of the two.
(iii) Therefore, this inheritance is called **semi-dominance.**

- **Duplication tend to be less harmful than deletion.**

(i) It results in extra genetic material under normal circumstances. Crossing-over occurs at analogous sites between homologous chromosomes.
(ii) On rare occasions, a crossover may occur at misaligned sites on the homologous and recombination results an abnormal condition, i.e., one chromatid with an **internal duplication** and another chromatid with a **deletion.**
(iii) The chromosome with extra genetic material carries a gene duplication because the number of copies of gene has been increased from one to two.
(iv) In most cases, gene duplications happens as rare sporadic events during evolution of species.
(v) Like deletion, the phenotypic consequences of duplications tend to be correlated with size. Duplications are more likely to have phenotypic effects if they involve a large piece of the chromosome.
In gerneral, small duplications are less likely to have harmful effects than the deletions of comparable size.
(vi) The consequences of a chromosomal deletion and whether it includes genes or portion of genes that are vital to the development of the organism.
(vii) When deletions have a phenotypic effect, they are usually detrimental. Larger deletion tends to be more harmful because more genes are missing.

Example: **Cri-due-chat** syndrome is created by a deletion in a segment of the short arm of human chromosome 5.

TRANSLOCATION

It is a chromosomal aberration/mutation associated with the relocation or transfer of a chromosomal segment from one chromosome to another non-homologous chromosome.

Characteristics:

(i) There is no loss or addition of genes during translocations, only **rearrangement**.

(ii) In **simple translocation**, a single break occurs in chromosome and broken segment gets attached to one end of a nonhomologous chromosome.

(iii) Non-reciprocal intrachromosomal translocation where chromosome segment changes position within the same chromosome.

In nonreciprocal interchromosomal translocation, a chromosome segment is transferred from one chromosome to another i.e. one way transfer.

(iv) **Reciprocal translocation** or segmental interchanges which involve mutual exchange of chromosome segment between two nonhomologous chromosomes.

(v) In **shift translocation**, broken segment of one chromosome gets inserted interstitially in a nonhomologous chromosome.

Example:

(a) **Robertsonian translocation** or **centric** fusion

(i) It involves breaks at the extreme ends of the short arms of two nonhomologous **acrocentric chromosomes** (13, 14, 15, 21 and 22).

(ii) The small acentric fragments are lost and the larger chromosomal segments fuse at their centromeric region, producing a new larger **submetacentric** or **metacentric** chromosome.

(b) **Philadelphia chromosome (Ph')**: The distal part of longer arm of **22 chromosome** ($22q^-$) is deleted and the missing piece is translocated to the end of long arm of **chromosome 9**.

This deleted 22 chromosome is known as **Philadelphia chromosome.**

Semisterility:

(i) An organism that is heterozygous for a reciprocal translocation usually produces only about half as many offspring as normal—a condition that is called **semisterility.**

(ii) The reason for the semisterility is the difficulty in chromosome **segregation** in meiosis.

(iii) This semisterility has an impact in the **reproductive fitness** of an organism, thus playing a role in evolution.

- **Events of Translocation**

(i) During meiosis translocated (translocation heterozygote) chromosome would be expected to pair with untranslocated homologous in a **cruciforms** or **cross** like pattern.

(ii) The two translocated chromosomes face each other opposite the centre of the cross and the two untranslocated chromosomes do likewise.

(iii) To maximize pairing the translocated chromosomes and untranslocated chromosomes remain alternate with each other forming the arms of cross.

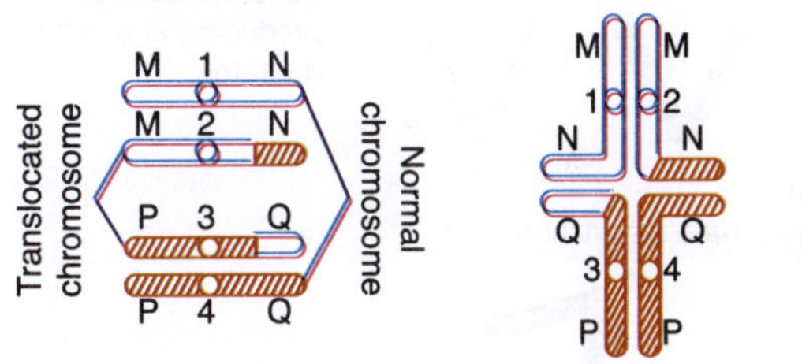

Structure of chromosome in translocation heterozygote

Pairing of chromosome in translocation heterozygote

(iv) This pairing configuration is diagnostic of a **translocation heterozygote.**

- **Features of translocation:**

(i) Translocation can change both the size of chromosome and the position of centromere.

(ii) Even though chromosomal segments have been exchanged between chromosomes in a reciprocal translocation, the affinity of the homologous regions results in pairing during meiosis I.

(iii) If nearly equal parts of the chromosomes are exchanged or not exchanged, the paired chromosomes in a translocation heterozygote have a cross appearance in metaphase I.

(iv) During anaphase I, three types of segregation occur:

(a) Alternate centromeres go to the same pole.

(b) Adjacent nonhomologous centromeres migrate to the same pole.

(c) Adjacent homologous centromeres migrate to the same pole.

(a) **Alternate segregation**

(i) Here alternate centromeres go to the same pole.

(ii) In early anaphase I, the chromosomes often form a figure eight shape.

(iii) This type of segregation is called **alternate disjunction** or **segregation.**

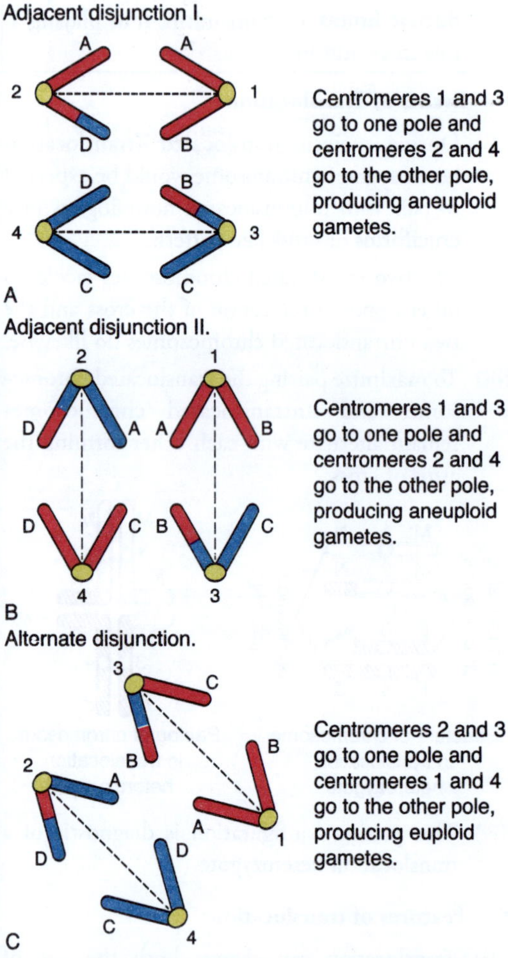

Fig. 7.4: Types of disjunction in a translocation heterozygote during meiosis I. For simplicity, only one after chromatid of each duplicated chromosome is shown: (A) One form of adjacent disjunction in which homologous centromeres go to opposite poles during anaphase. (B) Another form of adjacent disjunction in which homo-logous centromeres go to the same pole during anaphase. (C) Alternate disjunction in which homologous centromeres go to opposite poles during anaphase.

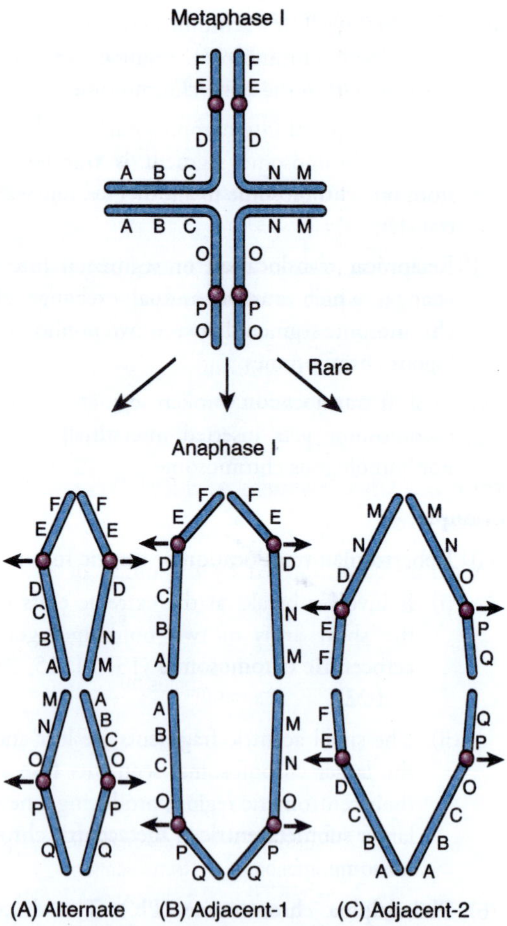

Fig. 7.5: A heterozygote for a reciprocal translocation and the outcomes of meiotic division. (A) Alternate segregation produces balancd chromosomes. (B) and (C) Adjacent-I and adjacent-2 segregation results in duplication and deletions.

(iv) All the gametes are balanced (euploid) which means none has duplication or deletion or gamete has full complement of chromosomes either two untranslocated or two balanced translocated.

(b) **Adjacent I segregation (Nondisjunctional segregation) (Figs. 7.4 & 7.5)**

(i) Here adjacent centromeres segregate together, i.e., homologous centromeres go to the opposite poles, but each normal chromosomes goes with one part of the reciprocal traslocation.

(ii) Chromosomes appear as **ring** at metaphase I or **cruciform** pattern.

(iii) All gametes formed from adjacent I segregation are **aneuploid** because some chromosome segments will be deficient for genes and others will be duplicated.

(c) **Adjacent II segregation or disjunctional segregation (Figs. 7.4 & 7.5)**

(i) Here homologous centromeres migrate to the same pole at anaphase I.

(ii) In this case, all genes have a large duplication and deficiency of the proximal part of the translocated chromosome.

(iii) Adjacent II segregation seldom occurs.

Compound Chromosome and Robertsonian translocation:

A. A chromosome formed by the union of two separate chromosomes from the same pair is known as compound chromosome.

Example: Attached X-chromosomes or attached X-Y-chromosome.

(i) A compound chromosome can exist stably in a cell as long as it has a single functional centromere.

(ii) Sometimes one chromosome fuses with its homologue, or two sister chromatids become attached to each other, forming a single genetic unit.

(iii) Compound chromosome may also be formd by the union of homlogous chromosome segments.

Example: Right arms of the two second chromosome in *Drosophila* might detach from their left arms and fuse and the centromere creating **compound half-chromosome.** Sometimes cytogneticists call this strucure an **isochromosome** because two arms are equal.

B. A special type of non-reciprocal translocation in which two non-homologous acrocentric chromosomes undergo fusion of their short arms for the formation of a single chromosome with single function at centromere. Here the loss of some materials from the ends of each of the fusing chromosomes occurs.

It is specially a risk factor for familial type of **Down syndrome,** when chromosome 21 occurs with any other aerocentric chromosome (14).

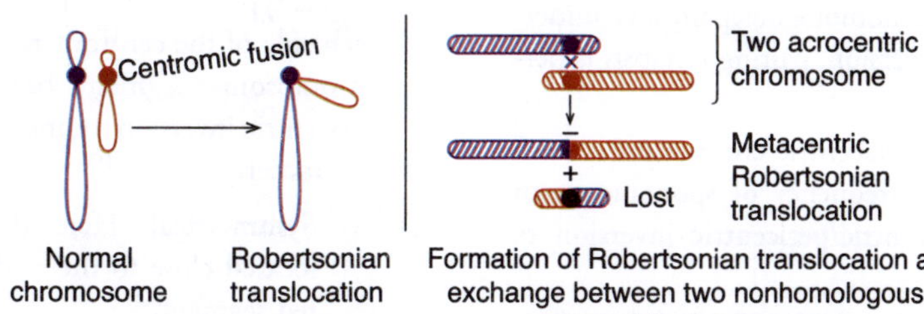

Formation of Robertsonian translocation as exchange between two nonhomologous acrocentric chromosome

INVERSION

It is complex type of chromosomal aberration in which parts of the chromosome become detached and reinserted in such a way that the orders of the chromosomal segment has been reversed, i.e., the sequence of genes are altered by the rotation of gene block with a chromosome by 180°.

Features of inversions:

1. Inversion always involves **two breaks** in a single chromosome and intercalary segments reunited in **reverse order.**
2. It is always **intercalary**, rarely terminal.
3. It causes a distinctive **looped configuration** in polytene or meiotic cell.
4. Inversion occurs in a single chromosome is called **homosomal.** Inversion occurs in both the members of a homologous pairs is called **allelosomal.**
5. Allelosomal inversions occurring one homologous arm of the two chromosomes are called **allelobranchial.** Inversions occurring in two nonhomologous chromosomes are called **heterosomal.**
6. Inversion alters the **order of linkage.** If one member of a homologous pair undergoes inversion, the inversion heterozygote fails to form **viable cross-over** gametes, pairing and hence crossing-over between homologous chromosomes are difficult.

   In inversion homozygote, where both the members of the homologous pair have undergone similar inversion, pairing synapsis is perfectly normal.
7. In the case of metacentric chromosomes (arm ratio 1 : 1), the frequency of spontaneous or induced, **paracentric/pericentric** inversion to be **equal.**
8. On the other hand, as the arm ratio gets smaller, **paracentric** inversions will become **commoner** and **pericentric** one **rare.**
9. In case truly acrocentric chomosome with very minute short arm breaks through the latter, it will necessarily be very rare and pericentric inversions will correspondingly be infrequent.
10. The primary effect of inversion is **position effect** or a change in the **relative positions of the genes.** It results in phenotypic changes, sometimes involving **variegation.**

- ## Paracentric and Pericentric Inversion

- **Paracentric inversion:**

A chromosomal inversion where inverted segment does not include centromere. In this type of inversion both the breaks occur in the same side of the chromosome arm or centromere. Paracentric inversion means inverted "beside' (para) the centromere.

(i) In this type of inversion only **one chromosome arm** is involved.

(ii) There is no change in the **shape** of the chromosome concerned.

(iii) This type of inversion is **more frequent.**

(iv) If a **single** paracentric inversion occurs, the inversion is known as **intraradial** or **homobranchial.**

(v) If **two** paracentric inversions occur on either side of centromere, the inversion is known as **interradial** or **heterobranchial.**

- **Pericentric inversion:**

A chromosomal inversion where inverted segment **includes centromere.**

(i) In this type of inversion, **one break** occurs on each side of the centromere. This involves the chromosome segment bearing centromere. Pericentric inversion means "around" (Peri) the centromere.

(ii) (a) **Symmetrical**: Here the centromere is located close to the middle of the inverted segment.

(b) **Asymmetrical**: The centromere is not located in the middle. It results in the change in the chromosome morphology.

(iii) These are **less frequent** than paracentric type (except Grasshopper).

Table 7.1: Difference between Paracentric and Pericentric inversion

Paracentric inversion	Pericentric inversion
1. This inversion does not include centromere.	1. This type of inversion includes centromere.
2. Only one chromosome arm is involved in this type of inversion.	2. Both the chromosome arms are involved here.
3. Here two breaks occur on one side of the centromere, i.e., on the same arm of the chromosome.	3. Here two breaks occur in each side of the centromere, i.e., in both the arms of the chromosome.
4. The arm ratio of the chromosome cannot be changed by this type of inversion.	4. This type of inversion transforms acrocentric chromosome into metacentric chromosome and metacentric chromosome into submetacentric chromosome.
5. In this type of inversion, crossing-over generates acentric and dicentric chromosome.	5. In this type of inversion crossing-over generates duplication and deficiency in chromosome.
6. The acentric and dicentric chromosomes are not inherited, i.e., lost, so it is known as **crossing-over suppressor.**	6. Chromosome with duplication or deficiency may be lost. So crossing-over **suppression seems to be present.**

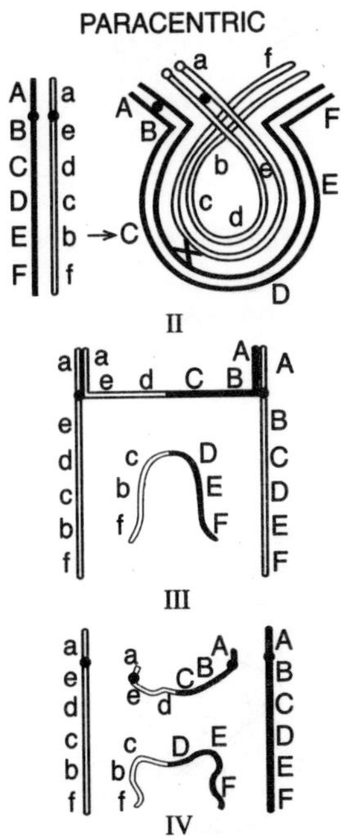

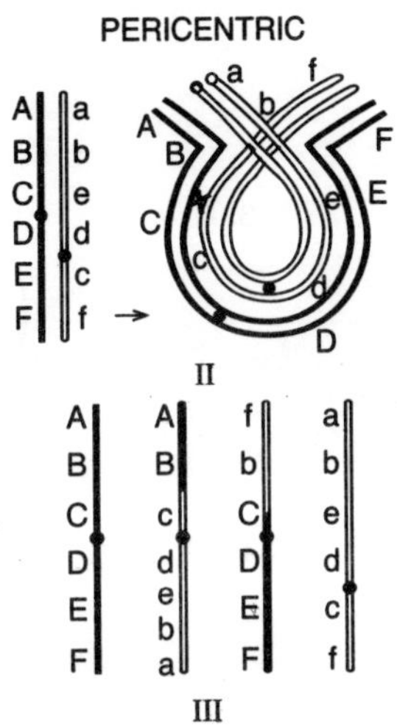

Inversion Heterozygotes:

Zygotes having one inverted chromosome and the other normal chromosome are known as inversion heterozygotes.

(i) Suppose, a normal egg has normal chromosome (a b c d) and sperm has inverted chromosome A C B D.

(ii) The offspring has $\frac{\text{A C B D}}{\text{a b c d}}$ genotype. It is heterozygous both for inversion and genes at loci a b c d.

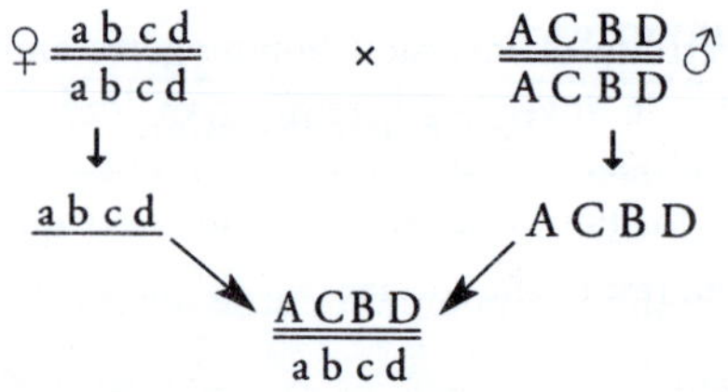

inversion heterozygotes

- **Dicentric Bridge:**

In case of paracentric inversion, if crossing-over takes place, crossover products, i.e. **dicentric** and **acentric** chromatids will be observed at anaphase 1 in the form of bridge and fragments. Here non-crossover strands are connected by means of the **dicentric chromosome.** This connection is known as **dicentric bridge** (Fig. 7.6).

Discussion:

(i) Taking inversion heterozygotes (paracentric) $\frac{\text{ACB D}}{\text{a b c d}}$.

Bridge formation:

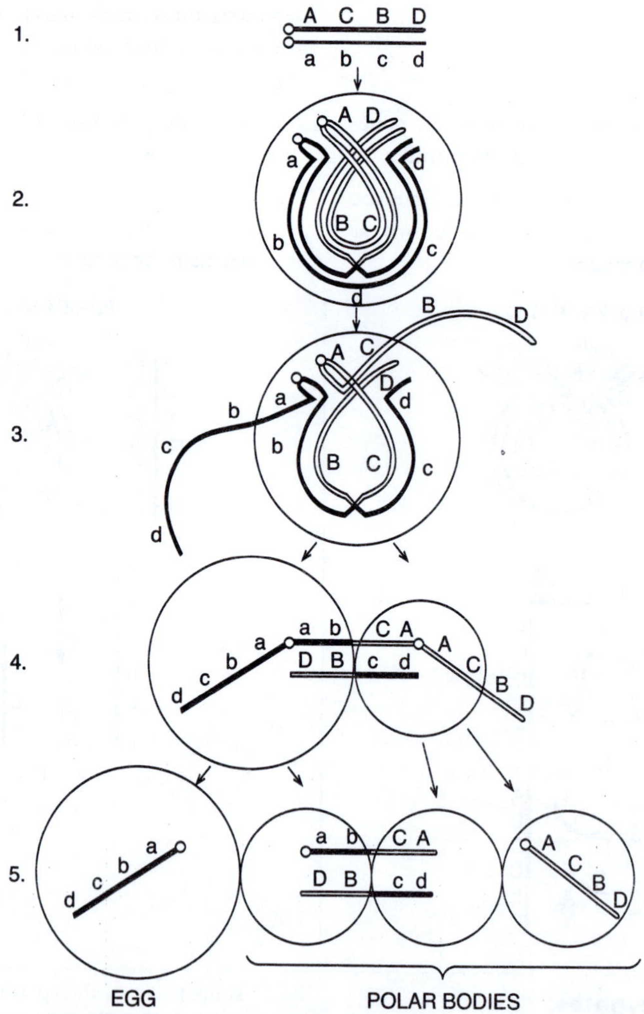

Fig. 7.6: The apparent suppression of crossing-over by an inversion. Stage 1, the tetrad before crossing-over; Stage 2, crossing-over between two of the strands (nonsister); Stage 3, the two non-crossover strands separated from their sister strands (for the sake of clarity); Stage 4, first meiotic division completed with non-crossover strands at the end of a chromatid tie; Stage 5, the second meiotic division completed. The egg contains a non-crossover strand (as does also the end polar body).

(ii) Before crossing over, each chromosome splits to form four chromatids, i.e., tetrad.

(iii) Sister strands are held together by centromere, crossing-over now takes place between one of the strands A C B D and one of a b c d.

(iv) One of the crossover strand is a b C A. This has both centromere, i.e., **dicentric** and hence both the non-crossover strands are connected with it, i.e., **dicentric bridge form.**

(v) Non-crossover strand a b c d is connected with **a end** of a b C A and non-crossover strand A C B D is connected with **A end** of a b c A.

(vi) At the completion of first meiotic division, two non-crossover strands are connected by means of the dicentric crossover chromosome. This is known as **chromatid tie.**

(vii) At second meiosis division, the centromere at each end of chromatid tie divides, and the non-crossover strands are drawn into the outer two of the four cells formed.

(viii) As the egg is one of these outer cells, it will always receive one of the non-crossover strands. Thus, it happens that in an inversion heterozygotes, the crossovers get lost in the polar bodies and the eggs always receive the **non-crossovers** when single crossing-over takes place within the limits of the inversion.

Remark:

(i) Non-crossover products are inherited in the next generation.

(ii) There is no recombinant opportunities.

• Effect of Inversion:

(i) Inversion heterozygotes often have mechanical pairing problems in the area of the inversions that **reduces crossing-over and recombination frequency** (Suzuki et al 1986).

(ii) Recovery of crossing-over product is suppressed in an inversion heterozygotes, when chromosome exchange occurs within the inverted region. This is why inversion is called **crossover suppressor.**

(iii) Major effect of inversion is to keep alleles of different genes together on the same chromosome, provided that they are contained within the inversion.

(iv) Fertility of inversion homozygotes and sterility in inversion heterozygotes lead to establishment of two groups or varieties which are mutually fertile, but do not breed well with the rest of the species.

Problem

A wild type *Drosophila* specimen from a natural population is crossed with a homozygous recessive laboratory stock dp cn/dp cn (the dp allele codes for dumpy wing and cin codes for cinnabar eyes. The two genes are known to be 45 map units apart on chromosome 2). The F_1 generation is wild type, when an F_1 female is crossed with the recessive parent, the progenies are

Wild type (+ +/dp cn) = 250

Dumpy cinnabar (dp cn/dp cn) = 246

Dumpy (dp +/dp cn) = 5

Cinnabar (+ cn/dp cn) = 7

Solution:

(i) This cross is effectively a dihybrid test cross, the total number of progeny = 508 (250 + 246 + 5 + 7). 45 percent of the progeny are expected to be dumpy or cinnabar (they constitute the crossover classes).

(ii) But only 12 out of 508, about 2.36 percent, are obtained. Something is reducing crossing-over in this region, and a likely explanation is an inversion spanning most of the dp cn region. Because the expected RF (recombination frequency) was based on measurements made on laboratory strains, the wild type fly from nature was the most likely source of the inverted chromosome. Hence, chromosome 2 in the F_1 can be represented as follows:

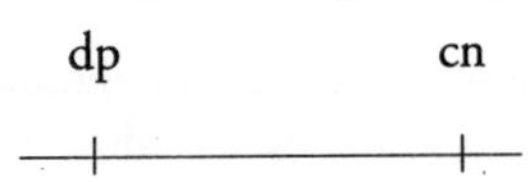

(iii) Pericentric inversions also can be detected microscopically through new ratios. Consider the following pericentric inversion.

Normal ———●— Arm ratio, long : short ~ 4 : 1

Inversion —(●)— Arm ratio, long : short ~ 1 : 1

Note that the length ratio of the long arm to the short arm has been changed from about 4 : 1 to about 1 : 1 by inversion.

(iv) Paracentric inversions do not alter the arm ratio, but they may be detected microscopically by observing changes in banding or other chromosomal landmarks, if available.

(The main diagnostic features of heterozygous inversions are **inversion loops**, reduced recombinant frequency and **reduced fertility** because of unbalanced meiotic product.)

- **Balancer**

A chromosome with multiple inversions, used to retain favourable allele combinations in the inverted homolog is termed as **balancer**.

(i) In some model experimental systems (*Drosophila, Caenorhabditis elegans*) **inversions** are used as **balancer.**

(ii) A balancer chromosome contains **multiple inversions** so that when it is combined with the corresponding wild type chromosome, there can be no viable crossover products.

(iii) In some analysis, it is important to keep the stock with all the alleles on one chromosome together. Combining such chromosome with a balancer eliminates crossovers, and **only parental** combination survive.

(iv) For convenience, balancer chromosomes are marked with a **dominant** morphological **mutation.** The marker allows the geneticist to track the segregation of the balancer or its normal homolog by following the presence or absence of the marker.

Aneuploidy:

A variarion in chromosome numbers that are not multiple of the basic haploid (n) number of chromosome.

It involves changes of chromosome number by addition or deletion of less than a whole set of chromosome.

Features:

(i) Loss of chromosome from a set, i.e., **hypoploid.**

(ii) Gain of chromosome from a set, i.e., **hyperploid.**

(iii) A diploid cell missing a single chromosome is called **monosomic.**

(iv) A cell missing both copies of that chromosome is known as **nullisomic.**

(v) A cell missing two nonhomologous chromosome is called **double monosomic.**

(vi) A diploid cell with an extra chromosome is known as **trisomic.**

- **Patau Syundrome** (Trisomy 13), 47, XX or XY, +13 (13 has three copy)
- ***Karyotype***: 2n + 1, i.e., 46 + 1 = 47, + 13
  - *Features*:

    (i) Occurs in about 2 in 10,000 live births.

    (ii) Small eyes and small brain.

    (iii) Cleft palate and cleft lip.

    (iv) Polydactyly is noticed.

    (v) Severe mental retardation with deafness.

    (vi) Congenital heart diseases.

    (vii) Mortality is very high.
  - *Origin*:

    (i) In this syndrome three copies of chromosome no. 13 are present.

    (ii) Due to addition of extra chromosome it becomes 47 + 13, i.e., 46 + 1 (Extra one 13 chromosome).
- **Edward Syndrome** (Trisomy 18), 47, XX or XY, +18 (18 has three copy)
- ***Karyotype***: 22 + 1, i.e., 46 + 1 = 47, + 18
  - *Features*:

    (i) Occurs in about 1 in 80,000 live births.

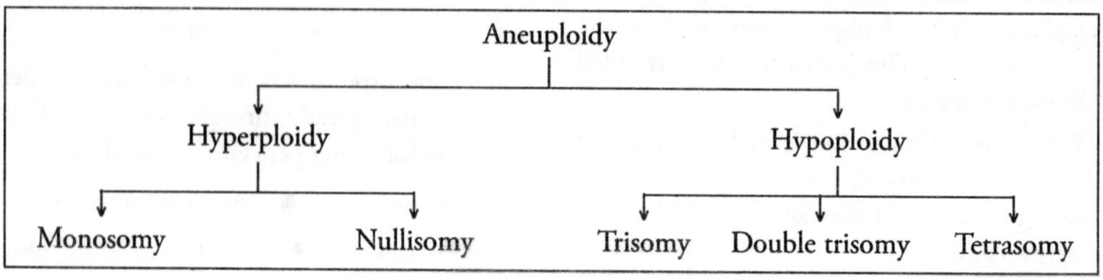

(ii) The infant usually has an elfin appearance with small nose and mouth.

(iii) Short sternum.

(iv) Elongated skull with low set malformed ears.

(v) Horse-shoe shaped kidney.

(vi) Most affected individuals are female.

- *Origin*:

(i) In this syndrome, extra chromosomes no. 18 are present in the original set, i.e., trisomy 18.

(ii) The karyotype is 47, + 18.

- **Down Syndrome** (Trisomy 21)
- ***Karyotype***: 27 + 1 i.e. 46 + 1 = 47, + 21
- *Features*:

(i) Occurs in about 1,430 per million live births.

(ii) Mentally retarded child.

(iii) Short stature, stubby hands and feet.

(iv) A round full face with "**epicanthaic folds**" in their eyes. Here upper lids cover the lower eyes lids; the term "**Mongolism**" is applied for this character.

(v) On the palm, the two normal more or less diagonal main creases may be replaced by a single transverse crease, i.e., "so decalled **simian crease**" is present.

(vi) Short and broad skull.

(vii) Average lifespan 16–20 years.

(viii) No sexual maturity.

- *Origin*:

(i) In this syndrome trisomy 21 i.e. three copies of chromosome 21 preserve, i.e. 47, + 21.

(ii) In this syndrome a translocation joins the long arm of chromosome-21 with another chromosome (mostly the chromosome 14) in the same complement.

- **Cri-du-chat Syndrome** (46, XX, 5p⁻)
- ***Karyotype***: 46, XX, 5p⁻.
- *Features*:

(i) Small and weak infants the cat-like mewing cry (cri=cry chat-cat).

(ii) Severe mental retardation.

(iii) Small head, broad face and sadle nose.

(iv) Widely spaced eyes.

(v) Epicanthic fold present.

(vi) Both physical and mental retardation.

(vii) Congenital heart disease.

- *Origin*:

(i) In this, syndrome develops due to deletion or deficiency on the short arm of cromosome 5 (5 p⁻)

(ii) The karyotype is written as 46 XX (5 p⁻).

- **Turner Syndrome** (45, XO)
- ***Karyotype***: 2n – 1 = 45 = (44 + XO).
- *Features*:

(i) Short stature (140–145 cm).

(ii) Shield-like chest, underdeveloped breast with widely spaced nipples.

(iii) Webbed neck.

(iv) Low set ears.

(v) Secondary sexual characters are poorly developed.

(vi) Adult almost without ovary, i.e., primary amenorrhoea.

(vii) Swelling of hands and feets.

(viii) Congenital heart diseases.

(ix) Presence of barr body (2x – 1 = 0).

- *Origin*:

(i) Turner people are monosomic for one x chromosome.

(ii) Individuals are phenotypical female and are lacking of Barr body.

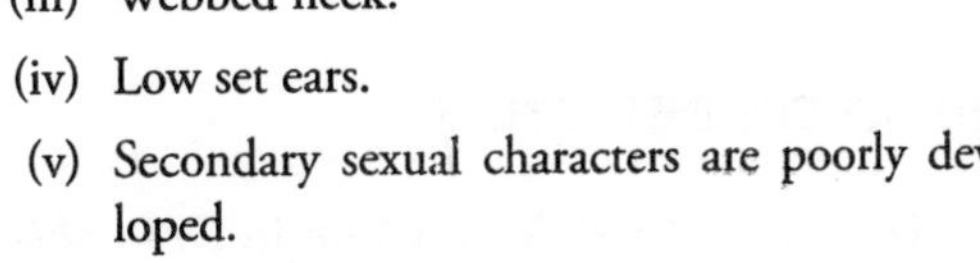

(Origin of monosomy at fertilization)

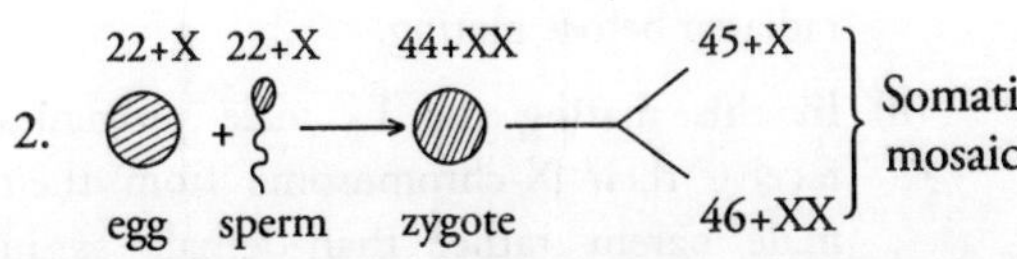

(Origin of monosomy in the cleavage after fertilization)

- **Klienfelter Syndrome** (47, XXY)
- ***Karyotype***: 44 + XXY.
  - *Features*:

(i) They are phenotypically male but showing tendency towards female.

(ii) They are tall with long limbs.

(iii) Almost sterile with small testis alongwith azospermia.

(iv) Enlargement of breast (Gynecomastia).

(v) Underdevelopd body hair.

(vi) Mental retardation more.

- *Origin*:

(i) Fertilization of exceptional XX egg (formal due to non-disfunctions by a Y sperm) results 44 + XXY.

(ii) Fertilization of an X egg by an exceptional sperm XY (formed due to nondisjunction).

MUTATION DETECTION

- **Detection of visible mutation in *Drosophila* with attached X stock:**

(i) Attached X is a special stock of *Drosophila* in which two X-chromosomes remain attached to each other with their centromere. Moreover, the female also possesses an Y-chromosome.

(ii) It is a simpler, special technique to detect recessive morphological mutation, because it requires only one generation.

- **Requirement:**

(i) An attached X female fly; (ii) X-radiated male.

- **Procedure:**

(i) Attached X ($\widehat{XX}Y$) female *Drosophila* was crossed to normal male which is X-radiated before mating.

(ii) In this mating the F_1 male progenies receive their X-chromosome from their male parent rather than female as in normal mating.

- **Experimental design:**

Mutagenic treatment (X-ray)

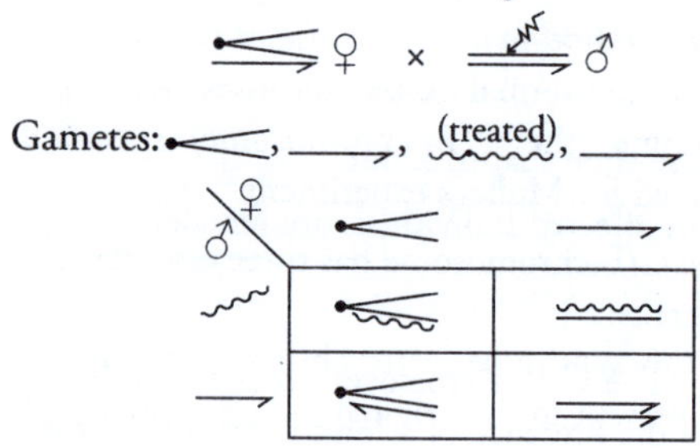

Observation:

Genotype	Phenotype	Ratio
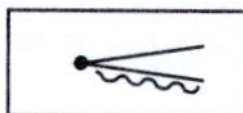	triplo-X	0
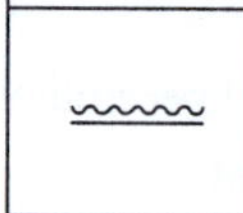	Male induced mutation	1
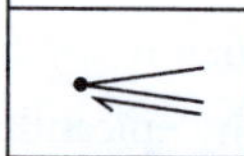	attached X Female	1
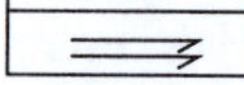	inviable	0

F_1: Triplo-X (♀), Attached X-Y (♀), Dies XY (♂) (Express induced mutation).

- **Results and inferences:**

(i) In F_1, four types of offsprings are produced, two abnormal or inviable offsprings, viz. $\widehat{XX}X$ female and YY males.

(ii) Here XY males receive X from their father. As father is treated with X-ray, therefore, any mutation which treated with X will be found in this male.

(iii) If the males appear as normal and show 1 : 1 ratio with females, then the inference would be that there was no mutation in F_1 males.

(iv) If the F_1 males are much less than female, it would be due to **semilethal effect** of X-radiation.

(v) If the F_1 males are completely absent, then it would indicate a case of **complete lethality**.

(vi) If the F_1 males show any new phenotype, then it would be due to induction of **visible mutation** as a result of X-radiation.

- **Detection of lethal mutation in the X-chromosome of *Drosophila* by C*l*B method:**

C*l*B method involves the use of female heterozygous for a normal X-chromosome and an X-chromosome (the C*l*B chromosome) specifically constructed for Muller's experiment.

The C*l*B chromosome has three essential components:

(a) The **C** (for crossover suppressor) refers to the presence of a long inversion (chromosomal level) that prevents recombination of genetic markers on the C*l*B chromosome and alleles on the normal X-chromosome. In homozygous condition **C** allele containing *Drosophila* cannot survive.

(b) *l* refers to lethal which can kill the individual if it becomes homozygous. The *l* gene is recessive.

(c) The B refers to the presence of the **partially dominant** mutation that causes the **Bar eye** phenotype which is kidney-shaped or narrow slit-shaped eye. Bar is a kind of duplication where a specific band of **16A** in its polytene chromosome becomes duplicated.

Requirements:

(i) Heterozygous C*l*B females $\left(\frac{+++}{C/B}\right)$ are taken.

(ii) Normal males ($\frac{+++}{\longrightarrow}$) treated with X-ray are taken.

(iii) F_1 Bar female that is heterozygous for such a lethal mutation is taken.

Procedure:

(i) Wild type males are treated with X-ray (mutagenic agent). C*l*B heterozygous females are mated with that treated male.

(ii) In the F_1 generation females expressing Bar eye are selected. They receive one X from their mother (C*l*B chromosome) and other X from their father.

(iii) Some of the parental X-chromosomes bear an induced X-linked lethal mutation.

(iv) F_1 Bar female, i.e., heterozygous for such lethal mutation is then back-crossed to a wild type male.

Experimental design:

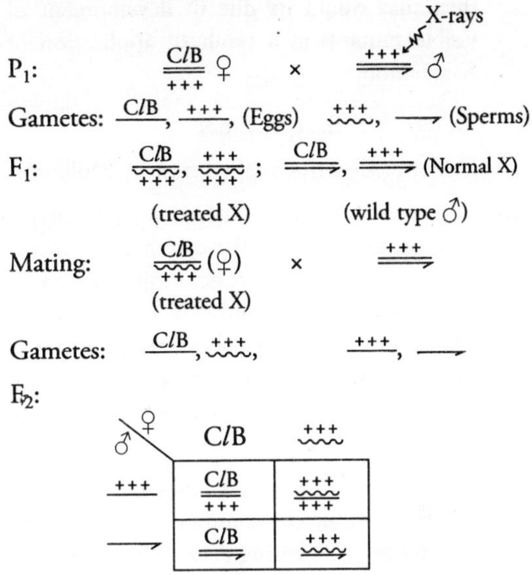

Results:

Genotype	C*l*B / +++ ♀	+++ / +++ (treated X) ♀	♂ C*l*B	♂ (treated X)
Phenotype	Bar eye female (C*l*B—heterozygous female)	Normal female	C*l*B male dies	Male dies* X-linked lethal induced

*or may have special features

Observation:

(i) Two types of females are produced and one type of male is developed.

(ii) Bar-eyed female and normal female are scored. One male is lost due to C*l*B gene in X-chromosome.

Inference:

(i) If F_2 males showed 2 : 1 ratio with the females, at the same time male appeared as normal, then the result would indicate X-radiation could not produce any mutation among the male progeny.

(ii) If the F_2 males appeared as much less than females, then it would be due to some **semilethal effect** of X-ray induced mutation.

(iii) If no male is found in F_2, then the interpretation would be that males were absent due to **complete lethality** caused by X-radiation.

(iv) If any mutants are found among F_2 males, then that would be due to development of **visible mutants** as a result of application of X-radiation.

- **Muller's-5 or Basc method:**

(i) Muller made a stock called **Basc** or **Muller-5**.

(ii) This stock of *Drosophila* has two phenotypically visible character. **Bar eye** shape is one of the two visible character, determined by an X-chromosomal **semi-dominant** mutant allele. The other phenotypically visible character is **apricot colour** (w^a). This is **recessive** in nature and is expressed in homozygous condition in female.

(iii) The third is an inversion involving the **scute** (sc) region of the chromosome which prevents occurrence of crossing-over. Therefore, the female heterozygous for **Basc**, produce Bar eyes assume to carry intact Basc chromosome).

Requirements:

(i) Muller's-5 stock (**Bar apricot**) female (Basc/Basc)

(ii) Wild type male (+++ →)

Procedure:

(i) Muller's-5 stock (Bar apricot) females are crossed with wild type male. Males are treated with X-ray before mating.

(ii) F_1 females in which c.o. do not happen, F_1 females are crossed with Bar apricot male.

Experimental design:

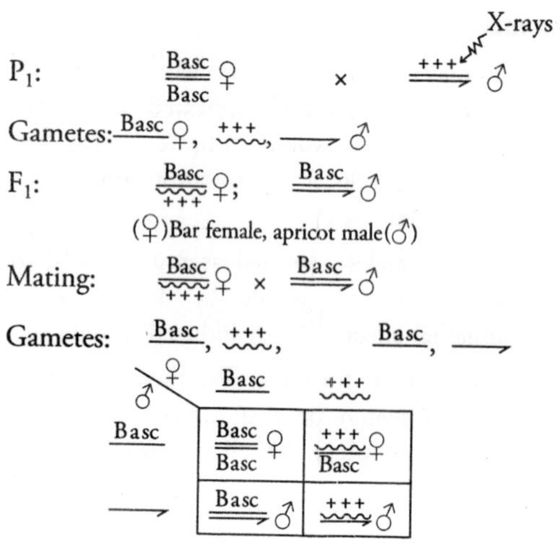

Results:

	Genotype		Phenotype
(i)	Basc/Basc	Female	Bar apricot
(ii)	~~~/+++	Female	Normal
(iii)	~~~→	Male	Normal/induced
(iv)	Basc→		Bar apicot

Observations:

(i) In F_2 generations four types of progenies are obtained, those are two females and two males.

(ii) Females are bar apricot and bar-eyed. Males are bar apricot and normal (wild-type).

Inference:

(i) If the normal males show 1 : 1 ratio with the bar females, then it would be considered as a case of **no mutation** by X-radiation.

(ii) If the F_2 normal males are much less than the bar apricot male, then it would be a case of **semilethal effect** of X-rays.

(iii) If F_2 normal males are totally absent, it would be a case of **complete lethal effect** of X-rays.

(iv) On the other hand, if the normal male shows any other phenotypic character, then it would be considered as a case of **visible mutation** due to X-radiation.

- **Detection of mutation of autosomal gene:**

Curly plum lobe-mutant stock of *Drosophila* is used for detection of autosomal mutation in *Drosophila* (presence in new chromosome).

(i) **Curly** (Cy) is a second chromosome autosomal dominant mutant allele which is responsible for curly wing character of *Drosophila*. Chromosomally curly is an **inversion** which can prevent the occurrence of crossing-over. Therefore, it is known as **crossing-over suppressor**. Curly (Cy) stock is maintained in heterozygous condition because it is **homozygous lethal**.

(ii) **Plum** (pm) is also a **dominant** mutant allele, responsible for reddish eye colour. This allele is maintained as a **heterozygous**.

(iii) **Lobe** (L) is also a second **autosomal dominant** mutant allele. It produces a modification in the normal eye shape. It is maintained as **heterozygous**.

(iv) **Curly plum-lobe** stock is considered as a **balanced lethal** stock. Because lethal genes are kept in such a way that the stock can survive due to balance effect of the genes.

Requirements:

(i) A curly plum-lobe female *Drosophila* $\left(\frac{Cy+L}{+pm+}\right)$ is taken.

(ii) Normal male.

(iii) X-ray radiation.

Procedure:

(i) A curly plum-lobe *Drosophila* is crossed to a normal male.

(ii) The males may or may not be X-radiated before mating.

(iii) If mutation can be induced of X-radiation, it would be considered as X-radiated mutant or induced mutant.

Experimental design:

(Curly lobe plum female) × (Normal male)

X-ray

P_1: $\frac{Cy+L}{+pm+}$ ♀ × $\frac{\widetilde{+++}}{+++}$ ♂

Gametes: $\underline{Cy+L}$, $\underline{+pm+}$ and $\widetilde{+++}$, $\underline{+++}$

F_1: ♂/♀ $\frac{Cy+L}{\widetilde{+++}}$ $\frac{+pm+}{\widetilde{+++}}$ ♂/♀ X-ray

(Curly lobe) (Normal)

Curly lobe male *Drosophila* × Curly lobe female *Drosophila*

$F_2 = F_1 \times F_1$: (♀) $\frac{Cy+L}{+++}$ × $\frac{Cy+L}{\widetilde{+++}}$ (♂) X-ray

Gametes: $\underline{Cy+L}$, $\underline{+++}$; $\underline{Cy+L}$, $\widetilde{+++}$

♀/♂ — ♂/♀	$\underline{Cy+L}$	$\underline{+++}$
$\underline{Cy+L}$	$\frac{Cy+L}{Cy+L}$ (dies)	$\frac{+++}{Cy+L}$ (Curly lobe)
$\widetilde{+++}$	$\frac{Cy+L}{\widetilde{+++}}$ (Curly lobe)	$\frac{\widetilde{+++}}{+++}$ (Normal)

Results/Observation

	Genotype	Phenotype
(i)	$\frac{Cy+L}{Cy+L}$	dies
(ii)	$\frac{Cy+L}{\widetilde{+++}}$	Curly lobe
(iii)	$\frac{+++}{Cy+L}$	Curly lobe
(iv)	$\frac{\widetilde{+++}}{+++}$	Normal

Inference:

(i) Among the F_2 progenies only curly lobe and normal *Drosophila* are produced.

(ii) If 1 : 2 the ratio heterozygous curly lobe flies, with normal indicates no mutation has developed after X-radiation.

(iii) If the number of normal *Drosophila* is much less than the number of curly lobe flies, then the result would indicate a case of **semilethal mutation** induced by X-radiation.

(iv) If no normal males are produced in F_2, it would be a case of **complete lethality.**
If new mutant character is expressed among the normal flies, it would be due to appearance of visible mutation after X-radiation.

Comparison of Different Methods of Detection of Mutation in *Drosophila*

	Attached X Method	C/B Method	Muller 5 Method	Curly plum lobe Method
1.	Sex-linked visible mutations are detected	Sex-linked lethal mutations are detected	Sex-linked recessive lethal mutations are detected	Visible autosomal mutations are detected
2.	Attached X females (⋛) are fused	Heterozygous C/B female $\left(\frac{+++}{C/B}\right)$ are used	Homozygous Bar apricot females $\left(\frac{Basc}{Basc}\right)$ are used	Heterozygous curly plum lobe $\left(\frac{Cy+L}{+m+}\right)$ females are used
3.	YY individuals die due to lethal effect	C/B males die due to lethal effect	No lethal effects are observed	Homozygous curly lobe individuals die.
4.	Involves one cross only	Involves two crosses	Involve two crosses	Involve three crosses
5.	Males with visible mutation are detected.	Mutation is detected by the absence of males in the progeny of second cross.	Mutation is detected by the absence of normal males in the progeny of second cross.	Individuals with homozygous treated autosomes show visible mutation.

POINT MUTATION

A mutation that can be mapped to a single specific locus and is associated with the alternation of one or a very small number of contiguous bases (R. F. Weaver, 2006).

- **Base substitution:**

(i) When one base pair is replaced by another base pair, the changes are often referred to as base substitution.

(ii) Example AT to GC.

(iii) These mutations alter the **phenotype** of an organism.

(iv) It is of two types, viz. **transition** mutation and **transversion** mutation.

(a) **Transition mutation:**

(i) When a **purine** (e.g., adenine) base of a triplet codon of a cistron is substituted by another **purine** (guanine) base or **pyrimidine** (thymine) is substituted by another **pyrimidine** (cytosine), such kind of substitution is called **transition.**

(ii) The four types of transition mutations are AG to GA, TC to CT.

(iii) It may arise through tautomerism.

(b) **Transversion mutation**:

(i) When the base pair substitution or replacement involves a change from a **purine** to **pyrimidine** or a **pyrimidine** to **purine**, it is called transversion.

(ii) The four types of transversion mutations are AT to TA, GC to CG, AC to CA and GT to TG.

(iii) It may arise through **tautomerism** and **rotation.**

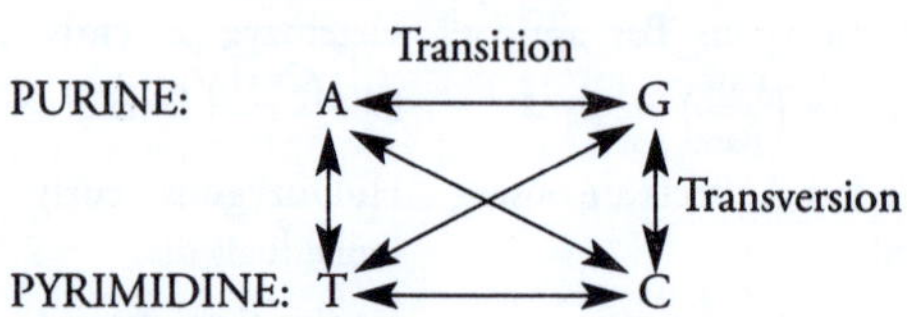

Chemical Mutagenesis:

(i) Some chemicals act by accelerating the rate of mutations that occur spontaneously.

(ii) In some cases, chemicals that are not mutagenic can be altered to a mutagenitically active form after they have been ingested into the body.

(iii) Cellular enzymes such as oxidases have been shown to activate some mutagens.

(iv) Certain foods contain chemicals that act as antioxidants.

(v) Many scientists are investigating whether a particular antioxidants found in food can counteract the effects of oxidases and thereby lower the cancer rate.

(vi) Colourful fruits and vegetables, including grapes, blue barries, citrus fruits, beets, beans, red peppers and carrots are usually high in antioxidants.

Chemical Mutagens	Effects on DNA
Nitrous acid	Deaminate bases
Nitrogen mustard	Alkylating agent
Ethyl methanosulfonate	Alkylating agent
Pro flavin	Intercalates within DNA complex
5-bromucracil	Base analogue
2-aminopurine	Base analogue

- **Deamination:**

(i) Some chemical substances such as nitrous acid (HNO_2) is a very potent mutagen that acts directly on either replicating or nonreplicating DNA by oxidative deamination of the bases that contain amino (NH_2) groups—adenine, guanine and cytosine.

(ii) Treatment of guanine with HNO_2 produces Xanthene, but since the purine base has the same pairing properties as guanine, no mutation results.

- However when cytosine is treated with HNO_2, uracil is produced which pairs with adenine instead of guanine.
- Deamination of adenine produces hypoxanthine which base pairs with cytosine rather than thymine.

(iii) Since deamination of adenine leads to AT → GC transitions and the deamination of cytosine results in GC → AT transition, nitrous acid induces transition in both directions, AT ↔ GC. Nitrous acid mutation can thus also be induced to revert with nitrous acid (Fig. 7.7).

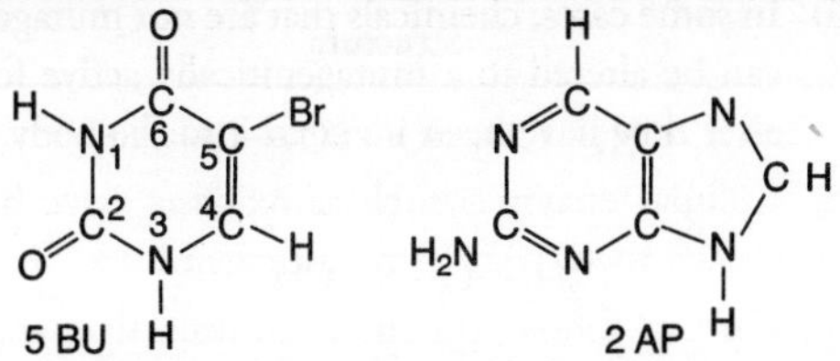

Fig. 7.7: 5-Bromouracil and 2 aminopurine are base analogue.

- **Base analogs:**

(i) Some chemical compounds have molecular structure sufficiently similar to the normal bases of DNA so that they are occasionally metabolized and incorporated into DNA during replication, such compounds are called base analogs.

(ii) Many of these analogs have pairing properties unlike those of the normal bases; thus they can produce mutations by causing incorrect nucleotides to be inserted during replication.

(iii) **Example**: 5-bromouracil (5-BU) and 2-aminopurine (2-AP).

5-Bromouracil (5-BU)

(i) 5-bromouracil has a bromine residue instead of methyl group of thymine. On its normal state, 5-BU resemble thymine and pairs only with adenine in DNA.

(ii) In rare state it pairs only with guanine.

(iii) 5-BU induces mutations by switching between its two chemical states once the base analog has been incorporated into the DNA.

(iv) During replication bromouracil pairs with adenine. However, at a relatively high rate, 5-BU will undergo a tautomeric shift and base pair with guanine. When this occur during replication, 5-BU causes a mutation in which TA base pair is changed to a 5-BU base pair. This is a transition because the adenine has been changed to guanine both of which are purines.

- **Base Alteration:**

Some mutagens are not incorporated into the DNA but instead alter a base causing specific mis-pairing.

Example: Certain alkylating agent Ethylmethano sulphonate (EMS) nitrosoguanidine (NG), etc.

- **Alkylating agents:**

Some chemicals, viz., **diethyl sulphonate** (DES), **dimethyl sulphonate** (DMS), **methyl methane sulphonate** (MMS), **Ethyl methane sulphonate** (EMS) are alkylating agents. They donate or add an alkyl group (methyl or ethyl) to amino (NH_2) or keto (C=O) groups in nucleotides and alter their base pairig potentials.

(i) They add **ethyl** (EMS) or **methyl** (MMS) groups to guanine. This makes guanine the base analog of adenine and thus produces **pairing errors.**

(ii) Loss of alkylated guanine (depurination) producing gap in the DNA chain may interfere DNA replication leading to mutation.

(iii) The gap may also produce deletion causing mutation.

Remark:

Alkylation results the formation of 7 ethyl guanine, i.e., G : C — A : T transition in the next round replication.

Hydroxylamine (NH_2OH)

(i) Hydroxylamine (NH_2OH) is a hydroxylating mutagen that reacts specifically with cytosine, modifying it by adding a hydroxyl (OH) group.

(ii) The resulting hydroxylaminocytosine base-pairs with adenine, leading to G : C → A : T transitions.

(iii) Mutation induced by hydroxylamine can only be CG to TA transitions, so hydroxylamine induced mutation cannot be reversed by a second treatment with this chemical.

- **Intercalating agent**

Intercalating mutagens, viz., **proflavin, acridine orange, ethidium bromide** (commonly used to stain DNA in electrophoresis) act by inserting themselves between adjacent bases in one or both strands of the DNA double helix.

(i) These agents are the flat plannar molecules that mimic base pairs and are able to insert or sandwich or slip themselves in (intercalate) between the stacked nitrogen bases at the core of the DNA double helix.

Chemical name	Common name	Structure
(i) Ethylmethane sulphonate ($C_3H_8SO_3$)	EMS	$CH_3—CH_2—O—SO_2—CH_3$, or 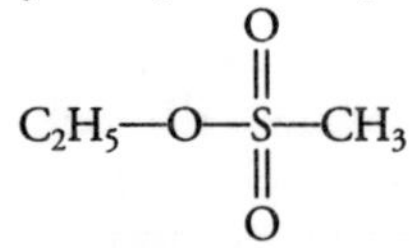
(ii) Nitrosoguanidine ($C_2H_5N_5O_3$)	NG	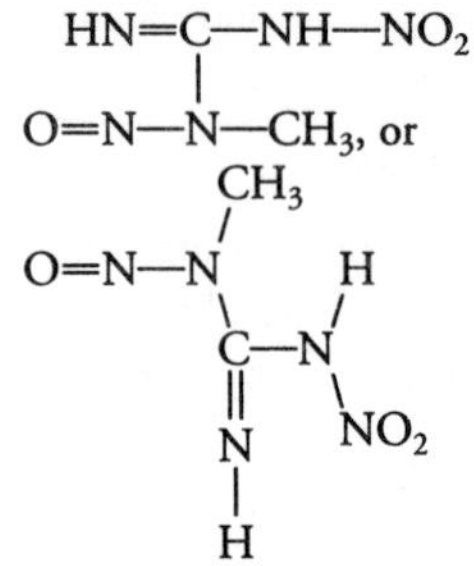

Chemically induced mutation:

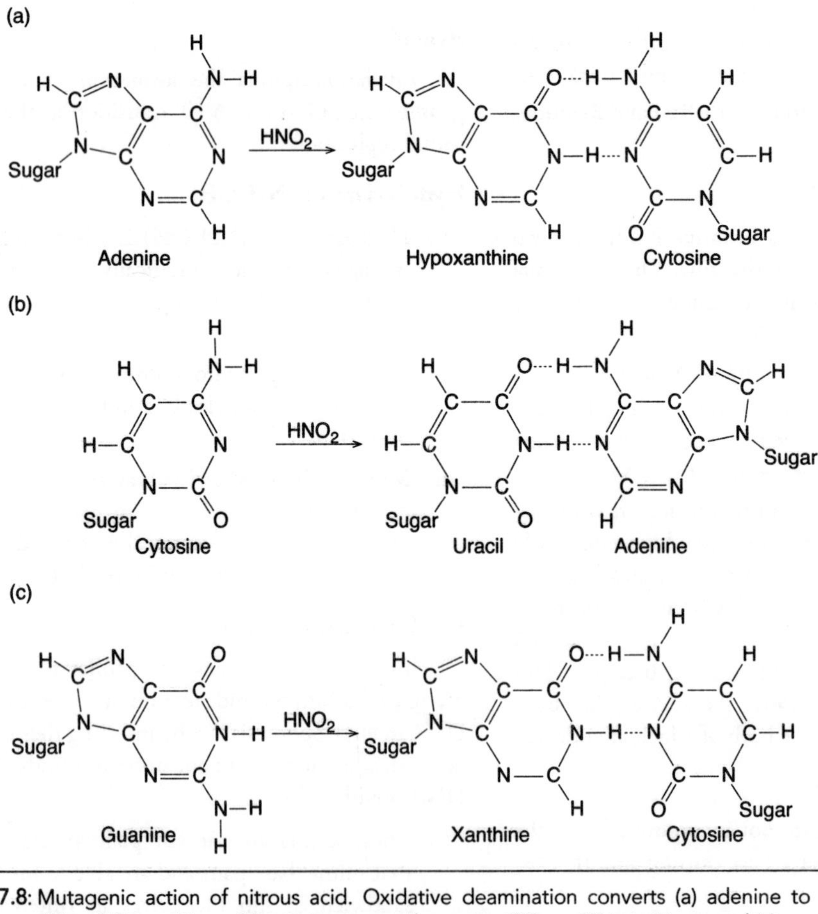

Fig. 7.8: Mutagenic action of nitrous acid. Oxidative deamination converts (a) adenine to hypoxanthine, causing AT → GC transitions; (b) cytosine to uracil, causing GC → AT transitions; and (c) guanine to xanthine, which is not directly mutagenic. The combined effects of nitrous acid on adenine and cytosine explain its ability to induce transitions in both directions, AT ↔ GC.

(ii) In this intercalated position, an agent can cause simple nucleotide pair insertions or deletion.

(iii) **Acridine** dyes such as **proflavin, acridine orange** are roughly the same dimensions as nitrogenous base pair and are known to intercalate or wedge between the purines and pyrimidines of intact DNA inducing contortions in the DNA helix and causing deletion or insertions.

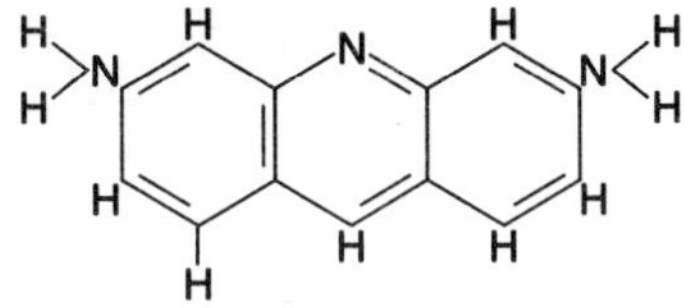

Proflavin

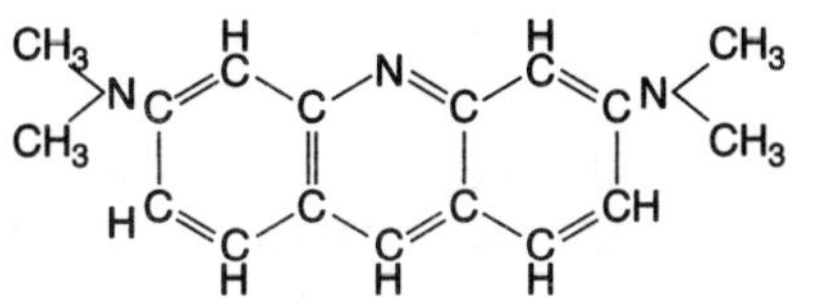

Acridine orange

Mutation (Step)

(a) Mutation by addition:

Molecule of intercalating agent

↓

(i) 5′ATCAG□TTACT3′ Template strand

3′TAGTC[G]AATGA5′ New DNA strand

| A randomly chosen base is inserted opposite to intercalating agent, the base is G

Replication

↓

(ii) 5′ATCAG[C]TTACT3′ Template strand

3′TAGTC[G]AATGA5′ New DNA strand

Result: frameshift mutation due to insertion of one bp (CG)

(b) Mutation by deletion:

(i) 5′ATCAGTTACT3′ Template strand

3′TAGTC□ATGA5′ New DNA strand

| Intercalating agent

Replication of new strand after intercalating agent is lost

↓

(ii) 5′ATCAGTACT3′

3′TAGTCATGA5′

The consequences of point mutations on polypeptide structure:

- **Silent mutation** (Synonymous/Same sense):

  (i) Here a base pair change in a gene that alters a codon in the mRNA such that the same amino acid is inserted into the protein.

  (ii) Silent mutations are possible because of the degeneracy of the genetic code.

  ATT → ATC (both code isoleucine)

  AAA → AAG (both code lysine)

  5′ ... CAAAA[A]TTT ... 3′ ⟶ 5′ ... CAAAA[G]TTT ... 3′

  3′ ... GTTTT[T]AAA ... 5′ ⟶ 3′ ... GTTTT[C]AAA ... 5′

  Gln Lys Phe ⟶ Gln Lys Phe

  AT to GC transition mutation produces same amino acid.

Effects:

1. Silent or synonymous mutations are usually **neutral.**
2. They may generate a phenotype (i) if there is **codon bias** because of a rare tRNA species, (ii) if the mutation **disrupts** an **internal regulatory element** or (iii) if the mutation happens to generate **cryptic splice site.**

- **Neutral mutation:**

  (i) Here a base pair change in the gene that changes a codon in the mRNA so that there is no change in the function of the protein translated from that message.

  (ii) Neutral mutation is a **subset** of **missense mutations** and is the case where new codon codes for a different amino acid that is chemically equivalent to the original and hence does not affect the protein's function.

5′ ... AAA 3′ → 5′ AGA 3′
(Lys) (Arg)

Arginine and lysine are basic amino acid and sufficiently similar in properties.

5′ ... CAAA [A] ATTT ... 3′
3′ ... GTTT [T] TAAA ... 5′
Gln Lys Phe

⟶

5′ ... CAAA [G] ATTT ... 3′
3′ ... GTTT [C] TAAA ... 5′
Gln Arg Phe

Here, AT → GC transition mutation changes the codon from lysine to arginine.

- **Missense mutation: Nonsynonymous**

(i) Here a base pair change in the DNA causes a change in mRNA codon so that a different amino acid is inserted into the polypeptide in place of the one specified by the wild type codon—resulting altered phenotype.

5′ ... AAA 3′ (Lys) → 5′ GAA 3′ (Glutamic acid)

5′ ... CAA [A] AATTT ... 3′
3′ ... GTT [T] TTAAA ... 5′
Gln Lys Phe

⟶

5′ ... CAA [G] AATTT ... 3′
3′ ... GTT [C] TTAAA ... 5′
Gln Glu

- **Nonsense mutation:**

(i) Here a base pair change in the DNA results in the change of an mRNA codon from one that specifies an amino acid to a chain terminating (nonsence) codon **ochre (UAA), amber (UAG), opal (UGA).**

(ii) **UAC** codes for **tyrosine.** If it undergoes base substitution (C-G), it becomes **UAG**, i.e., **terminator codon.**

5′ ... CAA **A** AATTT ... 3′
3′ ... GTT **T** TTAAA ... 5′
Gln Lys Phe

⟶

5′ ... CAATAATTT ... 3′
3′ ... GTTATTAAA ... 5′
Gln Lys Phe
↓
5′ ... GTTUAAUUU ... 3′
Stop

Effect:

(i) This type of mutation results in synthesis of **incomplete polynucleotide** chain which remains inactive.

(ii) The nonsense mutation brings about **drastic change** in expression of **phenotypic characters** because in this mutation the structure and function of enzymes are changed.

- **Frameshift mutation:**

(i) A mutational event leading to insertion or deletion of a base pair or 3n ± 1 nucleotides in a gene, shifting the codon reading frame in all codons of an mRNA following the mutational site.

5′ ... CAAAAATTT ... 3′
3′ ... GTTTTTAAA ... 5′
Gln Lys Phe

⟶

5′ ... CAA [G] AAATTT ... 3′
3′ ... GTT [C] TTTAAA ... 5′
Gln Glu Phe

Effect:

(i) It generates new codons resulting in a **shortened protein** or they read through of the normal **stop codon**, resulting in longer than normal proteins.

(ii) Frameshift mutation usually results in a **nonfunctional protein.**

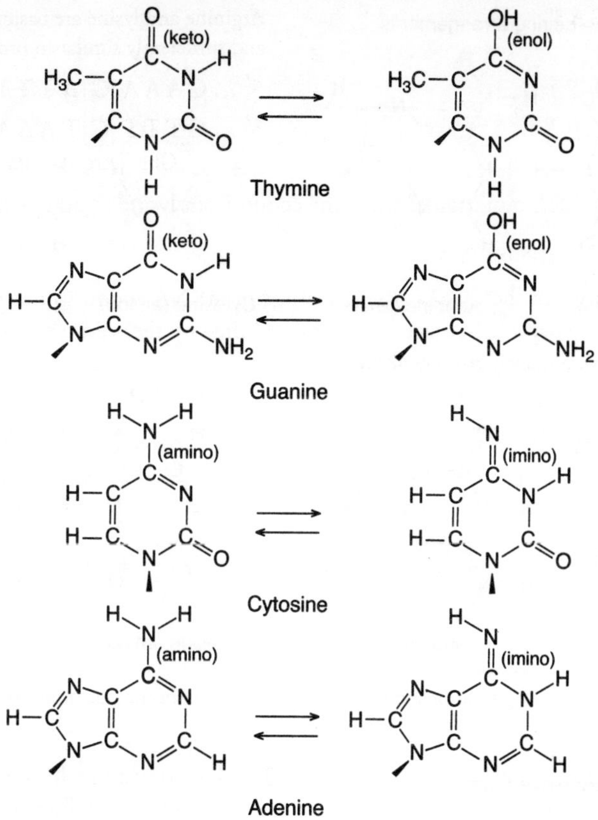

Fig. 7.9: Rare tautomeric shifts that occur in the chemical structure of the four nitrogenous bases of DNA. The dense arrow-heads indicate the point of bonding to the pentose sugar. Prior to bonding, an H atom exists at each of these points.

- **What is tautomerism? How does it occur? Mention its consequences.**

- **Tautomerism:** It is a phenomenon of existence of two distinct isomeric forms (one common and another rare) of a chemical compound because of its internal arrangement of atoms. Here hydrogen atom can move from one position in a purine or pyrimidine to another positon, i.e., amino form (NH_2) to imino form (NH) or keto form (C=O) to rare enol (C—OH) form.

Process:

(i) In stable DNA molecule nitrogen bases remain in amino (NH_2) or keto (C=O) forms.

(ii) Guanine and thymine are found in keto (C=O) form while adenine and cytosine remain in amino (NH_2) form.

(iii) After **tautomerism** the functional group (NH_2) of adenine and cytosine changes into imino (NH) group.

The functional group (C=O) of guanine and cytosine changes into enol (C—OH) group.

(iv) In stable forms, adenine always pairs with thymine (A=T) and guanine always pairs with cytosine (G≡C).

(v) After **tautomerism**, the bases remain less stable imino (NH) and enol (C—OH) state can form adenine-cytosine (A—C) and guanine-thymine (G—T) base pairs.

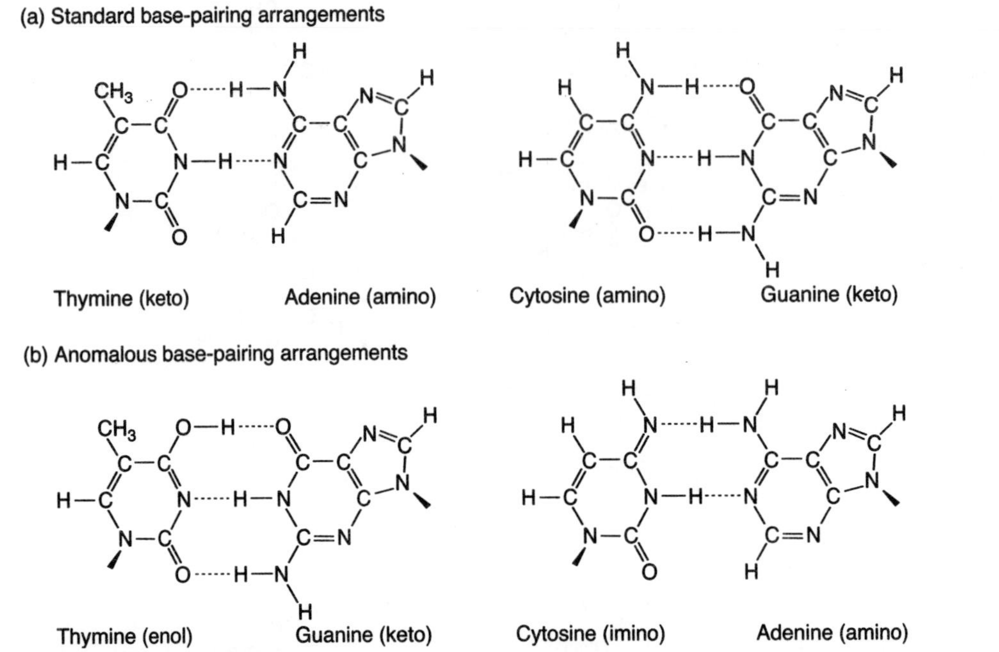

Fig. 7.10: The standard base-pairing relationships compared with two anomalous arrangements occurring as a result of tautomeric shifts.

Consequences of tautomerism

1. This type of pairing is the formation of DNA molecule that contains an exceptional base pair. This situation is not stable and at the next replication cytosine is expected to return to its common state and pair with guanine. Adenine is introduced into the complementary strand due to the tautomeric shift in cytosine and would then pair thymine.
2. Thus, there would be formed two kinds of DNA molecule, one that is identical to the original DNA and other has undergone base pair substitution of T : A for C : G.
3. This transitionally substituted DNA molecule has altered coding at a point and results in recognizable mutation. Such mutations which formed during DNA replication are called copy error mutation.

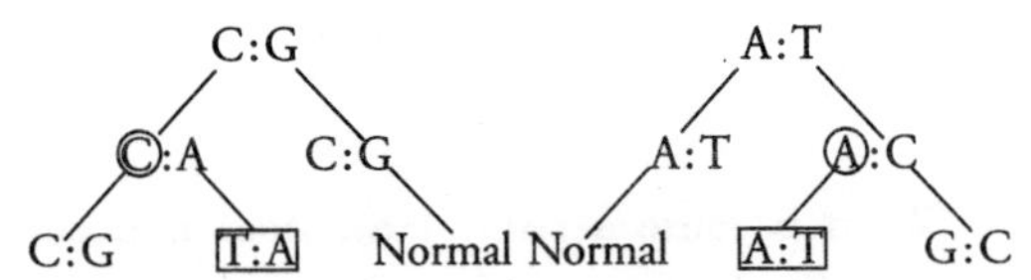

A. Tautomerism of cytosine B. Tautomerism of Adenine
Copy error mutation due to tautomerism.

INDEL MUTATION

A mutation in which insertion (addition) or deletion of one or more nucleotide base pairs occurs, is termed as **indel mutation.**

(i) These mutations add or substract a number of bases not divisible by three, they produce **frameshift mutations** in the protein coding regions.

(ii) The nucleotide sequence at frameshift mutation hot spots was determined in the lysozyme-encoding gene of phage T_4. These mutations often occur at repeated bases.

(iii) Indels arise when loops in single-stranded regions are stabilized by the "**slipped mispairing**" of repeated sequences in the course of replication. This mechanism is sometimes called **replication slippage.**

(iv) In the *E. coli* Lac I gene containing **hotspots** results from repeated sequences, just as predicted by this model.

(v) In lac I, the major indel hotspot is a four-base pair sequence (CTGG) repeated three times in tandem with the wild type (for simplicity, only one strand of the DNA is shown).

(vi) The majority of the mutations at this site (represented here by the mutations FS5, FS25, FS45 and FS65) result from the addition of one extra set of the four bases CTGG. A minority (represented here by the mutations FS2 and FS84) results from the loss of one set of the four bases CTGG.

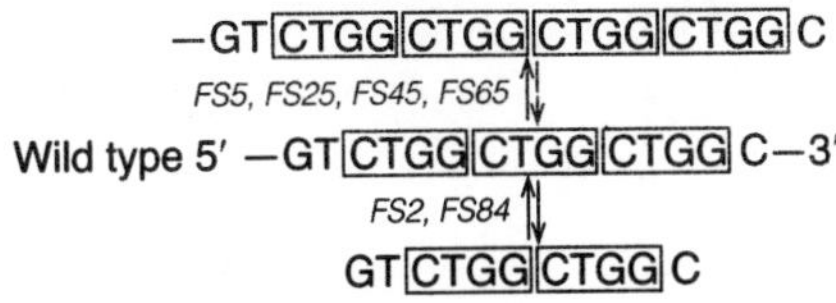

(vii) The model predicts that the frequency of a particular indel depends on the number of base pairs that can form during the slipped mispairing of repeated sequences.

(viii) The wild type sequence of Lac I gene can slip out one CTGG sequence and stabilize the structure by forming nine base pairs (see model). Whether deletion or an insertion is

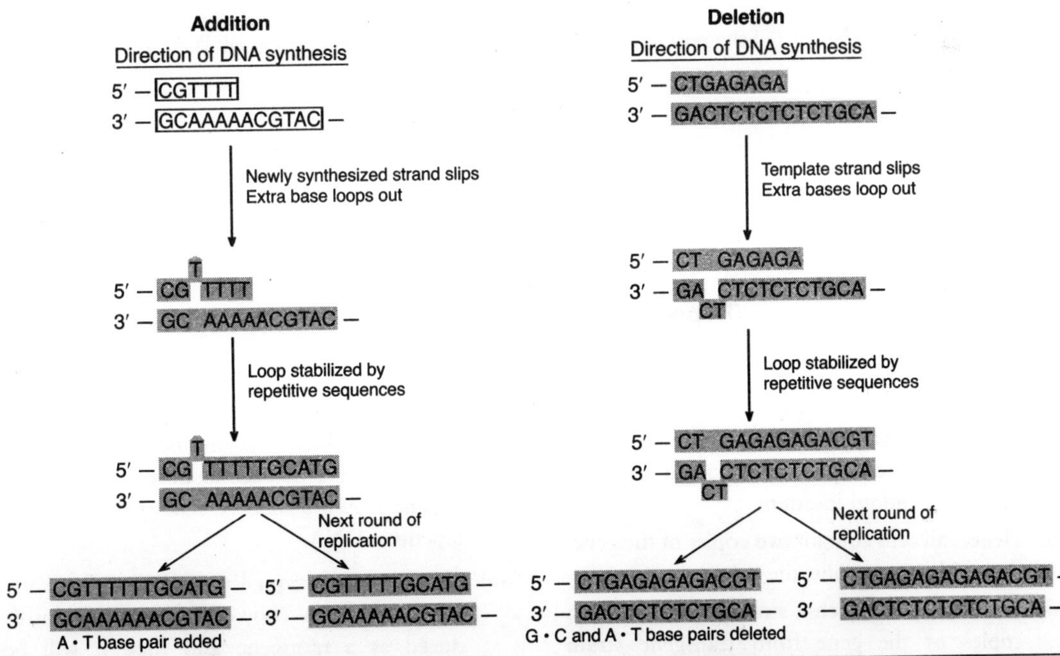

Fig. 7.11: A model for indel mutations resulting in frameshifts; dr = deoxyribose.

generated depends on whether the slippage is on the template or on the newly synthesized strand, respectively.

(ix) Larger deletions (more than a few base pairs) constitute a sizable fractions of observed spontaneous mutations. Most, although not all, of the deletions are repeated sequences.

(x) 9 deletions are analyzed at the DNA sequence level in the Lac 1 gene of *E. coli*. The results of further studies have shown that the longer repeats constitute **hotspots** for deletions.

(xi) Duplications of DNA segments have also been observed in many organisms. Like deletions, they often occur at sequence repeats.

S74, S112 75 bases

CAATTCAGGGTGGTGAATGTGAAACC......CGCGTGGTGAACCAGG

Site (no. of bp)	Sequence repeat	No. of bases deleted	Occurrences	
020 to 095	GTGGTGAA	75	2	S74, S112
146 to 269	GCGGCGAT	123	1	S23
331 to 351	AAGCGGCG	20	2	S10, S136
316 to 338	GTCGA	22	2	S32, S65
694 to 707	CA	13	1	S24
694 to 719	CA	25	1	S56
943 to 956	G	13	1	S42
322 to 393	None	71	1	S120
658 to 685	None	27	1	S86

Fig. 7.12: Analysis of deletions in *lac* 1 in regions containing repeated sequences. For S74 and S112, one of two repeated sequences and all of the intervening sequences are deleted, producing the same final sequence [from P. J. Farabaugh, U. Schmeissner, M. Hofer, and J. H. Miller, *Journal of Molecular Biology* 126, 1978, 847]

RIP
(Repeat-Induced Point Mutation)

(i) When a transgene is introduced into a haploid cell, it does not replace the resident gene but inserts at random locations.

(ii) Hence, all cells contain two copies of the gene, the **resident copy** plus the **transgene.**

(iii) However, when such a strain is crossed, both copies of the gene from transgene strain emerge from the cross with multiple GC-AT transition, which acts just before meiosis. It seems to be a defence against pesky "**transposons**". *Neurospora* is remarkably free of repetitive DNA.

(iv) Geneticists can capitalize on RIP because any cloned wild type copy of a gene can be introduced as a transgene and then it will be rendered null by RIP process.

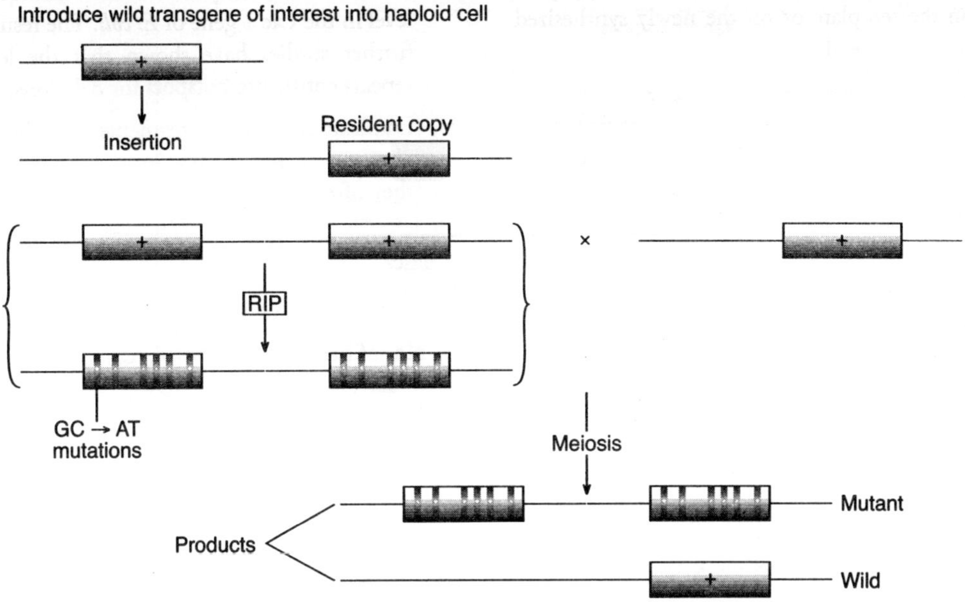

Fig. 7.13: Using the RIP process to obtain targeted mutations in *Neurospora*.

Notes on

- **Mismatch repair:**

Enzyme mediated DNA repair process that replace base pairs that are not properly hydrogen bonded.

(i) Mismatch repair systems have to do at least three things

(a) Recognise mismatched base pairs.

(b) Determine which base in the mismatch is the incorrect one.

(c) Excise the incorrect base and carry out repair synthesis.

(ii) In this repair system, four gene products (proteins) are involved—**Mut H, Mut S, Mut L** and **Mut U** (= uvrD) encoded by gene **mut H, mut S, mut L** and **mut U**.

(iii) In the first step, the Mut S protein recognizes mismatches and binds to initiate repair process of the Mut H and Mut L proteins, then joins the complex.

(iv) The second step involves the recognition of a specific sequence of the template which is GATC in *E. coli* in which A (adenine) is methylated in N-6 position. The proteins Mut L and Mut H bind to the unmethylated GATC sequence of the new strand forming a complex with Mut S which is bound to the mismatch pair.

(v) Thereby mismatch pair is brought close to the GATC sequences. The Mut H protein nicks the unmethylated DNA strand at the GATC site and the mismatch is removed by an exonuclease.

(vi) After the excision process has removed the mismatched nucleotide from the unmethylated strand. DNA polymerase III ligase seals the nick.

- **SOS repair:**

SOS (**Save Our Soul**) regulation is an induced emergency bypass repair system in which a whole battery of DNA repair recombinant and replication protein are synthesized in order to prevent cell death in the presence of significant DNA damage. It is the last ditch effort to save the genome from death and allow the cell to survive.

(i) In *E. coli* two genes **lex A** and **rec A** are key regulator of SOS system. Cells with mutant lex A and rec A gene have their SOS response

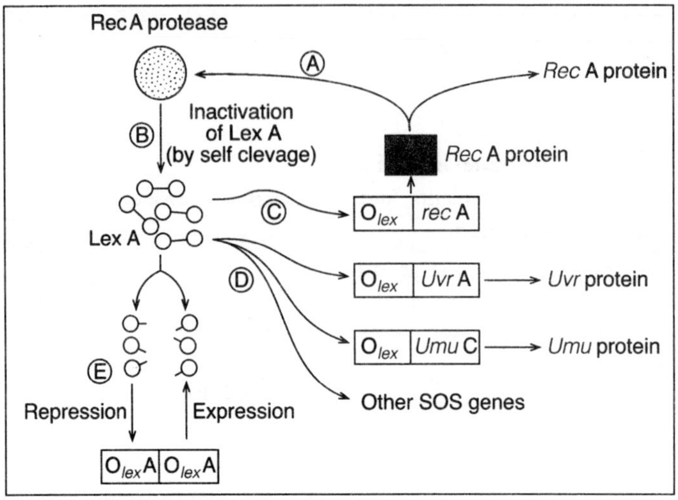

Fig. 7.14: Mechanisms of SOS repair system, Olex-lex operator, Lex A-protein expressed by gene lex A; Rec A-protein expressed by gene rec A; Uvr A-protein expressed by gene Uvr A, Umu C-protein expressed by Umu C gene.

permanently turned on. In the uninduced state (no DNA damage) the lex A encoded protein Lex A repressor binds to an operator near each gene and prevents transcription of about 17 or 18 genes whose protein products are involved in repairing and dealing with various kinds of DNA damage. Both Lex A and Rec A are synthesized at low level in the absence of DNA damage.

(ii) The Rec A protein plays two different complementary roles in the SOS response. When there has been sufficient damage to DNA, the rec A encoded **Rec A protein** binds to single strand region of DNA caused by the inability of DNA polymerase III to replicate the damaged regions. A conformational change in Rec A protein occurs and Rec A becomes activated. Active Rec A then triggers the autocatalytic properties of **Lex A.** Lex A becomes inactive by proteolytic self-cleavage which in turn relieves the repression of DNA repair genes including Rec A, Lex A, Umu C, Umu D, Uvr A and others having SOS box. As a result SOS specific genes are transcribed and DNA repair proceeds.

After DNA repair, Rec A protein will once again lose its proteolytic activity. This will alow Lex A repressor levels to increase, shutting off the expression of the genes involved in the SOS response and allowing pol III to resume its normal proof reading function.

- **Mutation induced by Radiation**

The portion of electromagnetic spectrum with wave length of shorter and of higher energy than visible light (380 nm – 750 nm) is subdivided into ionizing radiation (X-rays, Gamma rays and Cosmic rays) and non-ionizing radiation (ultraviolet rays).

The unit of radiation of all types are *rems.*

A. **Ionizing radiation:**

Ionizing radiation are the portion of electromagnetic spectrum that carries high energy to release free electron from atoms or molecules, thus create positively charged free radicals or ions.

(i) Normal atoms are electrically neutral. When ionizing radiations pass through matter, they dissipate their energy in part through the ejections of electrons from the outer shell of atoms and the loss of these balancing, negatively charged particles (electrons) leave atoms and which are no longer neutral but are positively charged atoms are called *ion* or *free radicals.*

(ii) The ions, in turn, collide with other molecules and cause the release of other electrons.

(iii) Ionizing radiations are of high energy and are useful for medical diagnosis because they penetrate living tissues for substantial distances.

(iv) Chromosomal mutations such as break, deletion, addition, inversion and translocation.

B. **Non-ionizing radiation:**

(i) Ultraviolet rays (2600 Å) having lower energy than ionizing radiations penetrate only the surface layer of cells in higher plants and animals. It does not cause ionization.

(ii) UV-rays dissipates thir energy to the atoms they encounter, raising the electrons in the outer orbitals to the higher energy level, a state referred to as **excitation.**

(iii) The excited molecules become reactive and mutated and is called *photoproduct.*

(iv) UV radiation produces several effects on DNA. In this reaction abnormal chemical bonds are formed between adjacent pyrimidines on the opposite strands, the chemical bonds between adjacent thymine residues, i.e., two thymine residues associate or *dimerise* to form thymine dimer ($\hat{TT}$). This *dimerisation* interferes with proper base pairing of thymine with adenine and results in thymine pairing with guanine. This will produce a **T-A** to **C-G** transition.

- Dimers perturb the structure of DNA double helics and interfere with accurate DNA replication.
- Error occurs during cellular process that repair defects in DNA such UV induced thymine dimer.

- **Sources of Radiations**
  - **Natural:** It is called background radiation. **Example:** Cosmic rays, radioactive elements in soil and terrestrial product (wood and stones) and atmosphere.
  - **Artificial:** Medical testing (X-ray) Nuclear testing and powerplants.

Dose	Range
Sub-lethal	100–250 rems
lethal	350–450 rems

ULTRAVIOLET RAYS (UV):

It is less energetic and therefore, nonionizing but its wavelengths are preferentially absorbed by bases of DNA.

Category	Wavelength range	Mutagenic action
UV-A	320 nm–visible	deleterious effect
UV-B	290 nm–320 nm	major lethal mutagen
UV-C	180 nm–290 nm	most energetic and lethal

- **SOS Box**

The region of promoter of various genes that the Lex A repressor recognizes and each of these genes has a consensus sequence in its promoter, called Sox box: 3′ CTGX10 CAG 5′ (where X10 refers to any ten base). Release of repressor results in induction of SOS response.

- **Depurination**

(i) It involves the loss of one of the purine base either adenine or guanine in an intact double helical DNA molecule.

(ii) It occurs when the N glycosidic bond breaks between the 1′ C of the deoxyribose and the number 9 position of the purine base.

(iii) It leads to the creation of apurinic (AP) site on one strand of the DNA.

Antimutator mutation:

Mutation of DNA polymerase that decreases the overall mutation rate of a cell or of an organism.

Acridine dyes:

(i) A class of positively charged polycyclic molecules that binds to DNA and intercalate into the double stranded structure.

(ii) It induces frameshift mutation.

(iii) Examples: Proflaven and Acridine orange.

Solved Problems

Problem 1

How are mutations characterized?

Solution:

(i) They can be classified as gain of function, loss of function or new function.

(ii) Then their sequence and cell expression patterns can be characterized.

Problem 2

What is the difference between forward and reverse mutation?

Solution:

A forward mutation is any change away from the wild type allele, whereas reverse mutation is any change back to the wild type allele.

Problem 3

Neomorphic mutations can be reverted by treatment with standard mutagen. When the revertants are examined, they typically turn out to be recessive loss of function mutations. Explain this observation.

Solution:

(i) Neomorphic mutations result in novel gene activity and are dominant.

(ii) Reversion of the dominant phenotype will commonly be the result of introducing another mutation in the already mutant gene that will now eliminate its function completely.

(iii) Most genes knock-out mutations are recessive and so most "revertants" likely to actually be recessive are loss of function mutation.

Problem 4

How can (a) paracentric, (b) pericentric inversions act as crossover suppressor? (c) Is crossing over really suppressed?

Solution:

(a) (i) When crossing over occurs in the area of paracentric inversion, acentric and dicentric chromosome occur.

(ii) They do not separate properly to the poles during meiotic division.

(iii) Gametes carrying crossover chromatids are often abnormal and viable.

(b) Crossing over within pericentric inversions also results in unbalanced chromosome arrangements that make the crossover gametes or zygote inviable.

(c) Crossing over is reduced to some extent, but the main "suppression" results from inviable gametes or zygotes.

Problem 5

What is the significance of chromosome inversion and translocation in evolution?

Solution:

(i) Inversions and translocation act as isolating factors among individuals in populations.

(ii) Thus, it prevents normal gene exchange and promotes speciation.

Problem 6

Inversions are said to "suppress crossing over". Is this terminology technically correct? If not, restate the description accurately.

Solution:

(i) While there is the appearance that crossing over is suppressed in inversion "heterozygotes", the phenomenon extends from the fact that the crossover chromatids end up being abnormal in genetic content.

(ii) As such they fail to produce viable (competitive) gametes or lead to zygote or embryonic death.

Problem 7

Ebony body (e) in flies is an autosomal recessive trait. A true breeding colony female $\left(\frac{e}{e}\right)$ is mated with a true breeding wild type male that has been irradiated. Among the wild type progeny is a single ebony male. Explain this observation.

Solution:

The cross is $\frac{e}{e} \times \frac{+}{+}$ (X-ray treatment)

Gametes: $\underline{\ e\ }$; $\underline{\ +\ }$ – – (irradiation)

F$_1$: $\frac{+}{e}$ $\qquad \frac{- -}{e}$ or $\frac{e}{e}$

(wild type ebony male) (mutation from either + to e, or, $\underline{+}$ to ——)

(i) The use of irradiation alerts us to the possibility of chromosomal breaks as well as simple mutations.

(ii) What type of chromosomal aberration would allow a recessive trait to appear unexpectedly? A deletion, which creates **pseudodominance** when there is no second allele, is a good possibility.

(iii) The male in question could have got the ebony allele from its mother and no homologous allele from its father.

(iv) Alternatively, the wild type allele from the father could have mutated to an ebony allele.

Problem 8

Does crossover suppression occur in an inversion homozygote? Explain.

Solution:

(i) No crossover suppression occurs in an inversion homozygote.

(ii) No inversion loops are formed in homozygotes.

Problem 9

White eye colour in *Drosophila* is an X-linked recessive trait. A wild type male is irradiated and mated with a white-eyed female. Among the progeny is a white-eyed female.

(a) Why is this result unexpected, and how could you explain it?

(b) What type of progeny would you expect if this white-eyed female is crossed with normal non-irradiated male?

Solution:

(a) White-eyed female × wild type (irradiated) male

$\frac{w}{w} \quad \times \quad \frac{+}{\rightarrow}$ (irradiation)

Gametes: $\underline{w}$; $\underline{+}$ $\rightarrow$

F$_1$ $\frac{- -}{w}$(white-eyed female) [+ – –]

All female should get a wild type X-chromosome from their father. Irradiation breaks the chromosome, so a deletion of part of X is possible, producing a situation of **pseudo-dominance.**

Alternatively the offspring could have got a mutant X^w from the father.

(b) White-eyed female (deleted X-chromosome) × Normal male (Non-irradiated)

$\frac{- -}{w} \quad \times \quad \frac{+}{\rightarrow}$ (– – = deleted portion)

↓

Gametes: $\underline{w}$, – – ; $\underline{+}$, $\underline{y}\rightarrow$

♂ \ ♀	– –	w
+	$\frac{- -}{+}$	$\frac{w}{+}$
$\rightarrow$	$\frac{- -}{\rightarrow}$	$\frac{w}{\rightarrow}$

Restult/Observation:

Genotype	Phenotype	Ratio
$\frac{- -}{+}$	wild type female	2
$\frac{w}{+}$	wild type female	
$\frac{w}{\rightarrow}$	white-eyed male	1
$\frac{- -}{\rightarrow}$	male but dead	×

Interference:

$\frac{- -}{+}$ (Wild type female) $\quad \frac{w}{+}$ (Wild type female) $\quad \frac{w}{\rightarrow}$ (White-eyed male) $\quad \rightarrow$ (Male but dead)

We expect all wild type females, and all white-eyed males, but in a ratio 2 females and one male (2 : 1).

Problem 10

In *Drosophila*, the recessive gene for white eyes is located near the tip of the X-chromosome. A wild type male is irradiated and mated with a white-eyed female. One red-eyed male is found among the progeny. How can you explain the red-eyed male and how could you test your hypothesis?

Solution:

(i) A translocation from the tip of the normal X in the male to the y. We expect all males to recessive an X-chromosome with the white-eyed allele from the female.

(ii) For the male to be wild type, we still must have a part of the wild type X-chromosome.

Male (wild type) (irradiated) × Female (white-eyed)

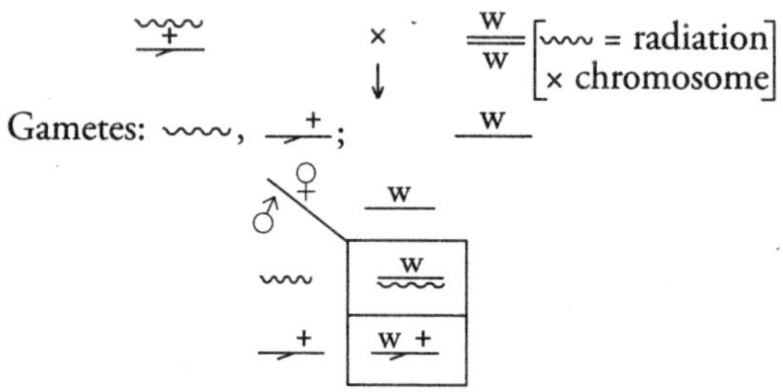

Result:

Genotype	Phenotype
w / ∿∿∿	white eyed female
w +	red-eyed male

(iii) To test the hypothesis, cross this wild type male with white-eyed females. All the female progeny should be white-eyed and all the male progeny red-eyed.

Wild type male × White-eyed female

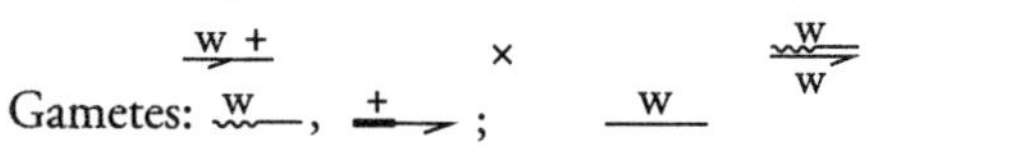

(iv) Cytological examination of the chromosome should reveal the translocation.

Wild type male × White-eyed female

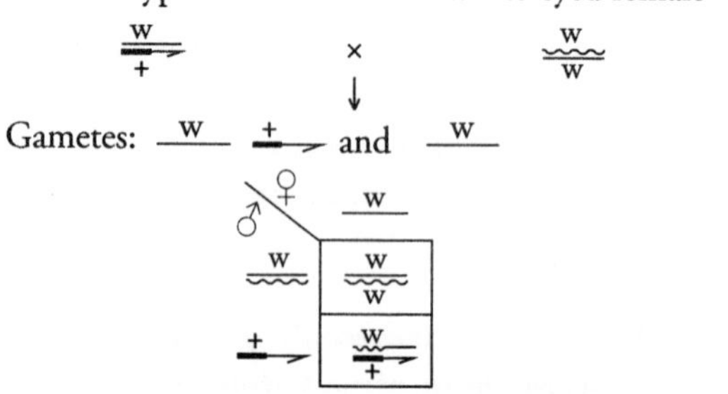

Genotype	Phenotype
$\frac{w}{w}$	white eyed female
w	red-eyed male

Problem 11

A chromosome has the gene sequence ABCDEFG. What is the sequence following a C through E inversion and following a C through E deletion? Two chromosomes with the sequences ABCDEFG and MNOPQRSTUV undergo a reciprocal translocation after breaks in E–F and S–T. What are the possible products? Which products are genetically stable?

Solution:

(i) The chromosome has the sequence ABCDEFG. After inversion it will be ABEDCFG and following ABFG through R deletion (C to E) it will be ABFG.

(ii) ABC DEFG ↓ break   MNOPQRSTUV ↓ break

(Translocation reciprocal)

(a) ABCDETUV and MNOPQRSFG

(b) ABCDESRQPONM and VUTFG

(iii) One of these possibilities includes two monocentric chromosomes, other includes a dicentric and an acentric chromosomes.

Only translocation with two monocentric chromosomes is genetically stable.

Problem 12

Four strains of *Drosophila melanogaster* are isolated from different localities. The banding patterns of particular region of salivary chromosome 2 have the following configurations (each letter denotes a band):

(a) a b f e d c g h i j

(b) a b c d e f g h i j

(c) a b f e h g i d c j

(d) a b f e h g c d i j

Assuming that part (c) is the ancestral sequence, deduce the evolutionary ancestry of the other chromosome.

Solution:

1. Starting with the chromosome (c) compare the chromosome pairwise to find those that differ by a single inversion.

(c) a b f e h g i d c j — Inversion (i-d-c) results–(d)

(d) a b f e h g c d i j —Inversion region (h g c-d) results–(a)

(a) a b f e d c g h i j — Inversion region (c d e f) results–b

(b) a b c d e f g h i j

(c) is the ancestral sequence, evolutionary ancestry is (c) → (d) → (a) → (b).

Problem 13

Two species of Australian grasshopper co-exist side by side. In meiosis, each has 8 pairs of chromosomes. When the species are crossed, the chromosomes in the hybrid form 6 pairs of chromosomes and one group of 4. What alternation in chromosome structure could account for this result?

Solution:

(i) The group of four synapsed chromosomes implies that a reciprocal translocation has taken place.

(ii) The group of four includes both parts of reciprocal translocation and their non-translocated homologs.

Problem 14

Why are the translocation heterozygotes semisterile? Why are translocation homozygotes fully fertile? If a translocation homozygote is crossed with an individual that has normal chromosome, what fraction of that F_1 is expected to be semisterile?

Solution:

(i) Translocation heterozygotes are semisterile because adjacent segregation from the four synapsed chromosomes in meiosis produces aneuploid gametes which have large parts of chromosomes missing or present in excess.

(ii) The aneuploid gametes fail to function in plants and in animals. It results in zygotic lethality.

(iii) The translocation homozygote is fully fertile, because the translocated chromosomes undergo synapsis in pairs, and segregation is completely regular.

(iv) In the cross of translocation homozygote × normal homozygote, all progenies are expected to be semisterile.

Problem 15

How can a deletion mutant be distinguished from point mutation?

Solution:

(i) The most reliable way to determine, if mutation is a deletion, is to perform a reverse mutation analysis.

(ii) Deletion do not revert to wild type but point mutations do.

Problem 16

Normal chromosome has the following gene sequences:

A B C D°E F G H

Determine the chromosomal mutation illustrated by each of the following chromosomes:

(a) A B C F E°D G H

(b) A D°E F B C G H

(c) A B C D°E F E F G H

(d) A B C D°E F F E G H

(e) A B D°E F G H

Solution:

(a) Here pericentric inversion (centromere lies within the inverted region, i.e., D°EF) is taken place.

Normal (i) A B C D°E F G H

(ii) A B C F °(E) D G H (Inversion)

(iii) A B C F E°D G H

(b) Here non-reciprocal translocation takes place. BC moved from left to right arm.

(i) A B C D°E F G H

(ii) A D°E F B C G H

(c) Here tandem duplication occurs. EF duplicated.

(i) AB CD o EF G H

↓

(ii) AB CD o EFEFGH

(d) Here reverse tandem duplication occurs. EF duplicated and inserted in reverse order.

(i) AB CD o EF G H

↓

(ii) ABCD o EFFEGH

(e) Deletion occurs. C gene is deleted.

(i) ABCD o EFGH

↓

(ii) ABD o EFGH

(ii) ● AB CDA o (Dicentric, duplication for A and deletion for E)

(iii) EB CD E (Acentric, duplication for E and deletion for A)

(iv) o ADCB E (Inverted order)

Problem 17

Inversions are known to affect crossing over. The following homologs have the indicated gene order (the filled and open circles are homologous centromere):

● ABCD E

o ADCB E

(a) Considering the position of the centromere, what is this sort of inversion called?

(b) Diagram the alignment of these chromosomes during meiosis.

(c) Diagram the results of a single crossing over between homologous genes B and C inversion.

Solution:

(a) This is a paracentric inversion, because the centromere is not included in the inverted DNA segment.

(b)

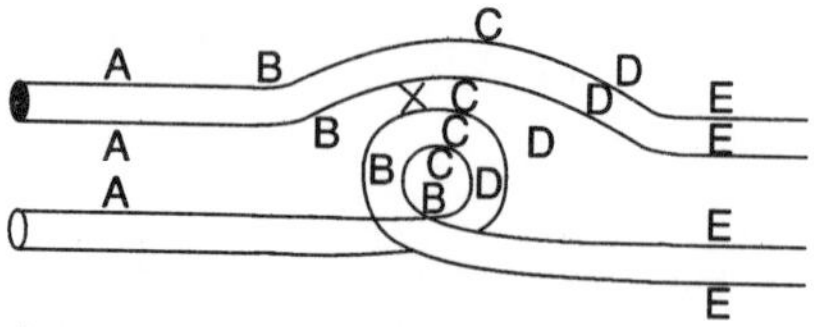

(c) A crossover between B and C results in the following chromosome:

(i) ● AB CDE (Normal order)

Problem 18

A corn plant is obtained that is heterozygous for a reciprocal translocation and therefore, semisterile. This plant is crossed to a chromosomally normal strain that is homozygous for the recessive allele brachytic (b) located on chromosome 2. A semisterile F_1 plant is then backcrossed to the homozygous brachytic strain. The progenies obtained show the following phenotypes:

Nonbrachytic		Brachytic	
Semisterile	**Fertile**	**Semisterile**	**Fertile**
334	27	42	279

(a) What ratio would you expect to result if the chromosome carrying the brachytic allele is not involved in the translocation?

(b) Do you think that chromosome 2 is involved in the translocation? Explain your answer, showing the conformation of the relevant chromosome of the semisterile F_1 and the reason for the specific numbers obtained.

Solution:

(a) (i) We should start with the methodical approach and simply restate the data in the form of a diagram, where

= The 2 chromosome involved in the translocation.

= Chromosome two which carries the brachytic allele.

To simplify the diagram, we do not show the chromosomes divided into chromatids (although they would be at this stage of meiosis). We then show the first cross.

(ii) All the progeny from this cross will be heterozygous for the chromosome carrying the brachytic allele, but what about the chromosome involved in the translocation? Only alternate segregation products survive and that half of these survivors will be chromosomally normal and half will carry the two **rearranged chromosomes.**

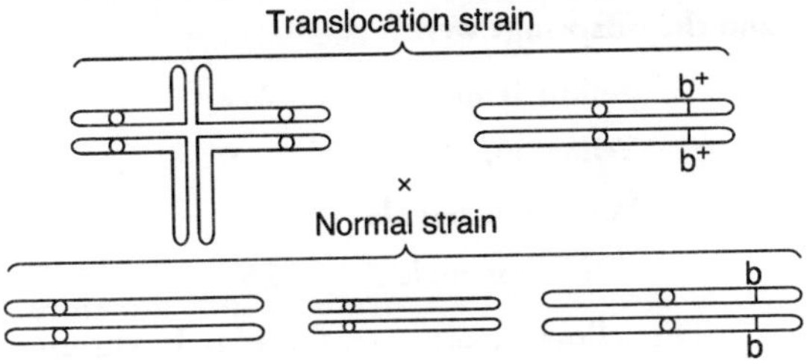

The rearranged combination will regenerate a translocation heterozygote when it combines with the chromosomally normal complement from the normal parent. These latter types are the semisterile F_1.

(iii) This semisterile F_1 plant is then backcrossed to the homozygous brachytic strain.

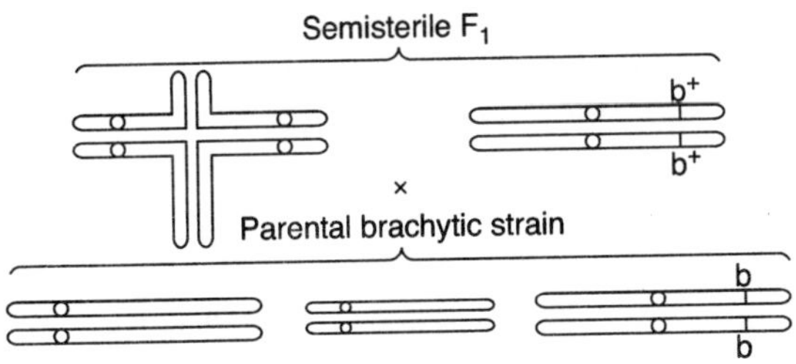

(iv) In calculating the expected ratio of phenotypes from these crosses, we can follow or treat the behaviour of the translocated chromosome independently of the behaviour of chromosome 2. Hence, we can predict that the progeny will be

$\frac{1}{2}$ translocation heterozygotes (semisterile)

↓

$\frac{1}{2}\,\frac{b^+}{b}$ → $\frac{1}{4}$ semisterile nonbrachytic

$\frac{1}{2}\,\frac{b}{b}$ → $\frac{1}{4}$ semisterile brachytic

$\frac{1}{2}$ normal (fertile)

↓

$\frac{1}{2}\,\frac{b^+}{b}$ → $\frac{1}{4}$ fertile nonbrachytic

$\frac{1}{2}\,\frac{b}{b}$ → $\frac{1}{4}$ fertile brachytic

This predicted 1 : 1 : 1 : 1 ratio is quite different from that obtained in the actual cross.

(b) (i) Chromosome 2 is likely to be involved in the translocation. Here we observe a departure from the expected ratio based on the independence of the brachytic phenotype and semisterility.

(ii) Let us assume that the brachytic locus(b) is on the shaded chromosome. But where? For the purpose of the diagram, it does not matter where we put it, but it does matter genetically because the position of the 'b' locus affects the ratio of the progeny. If we assume that the 'b' locus is near the tip of the piece that is translocated.

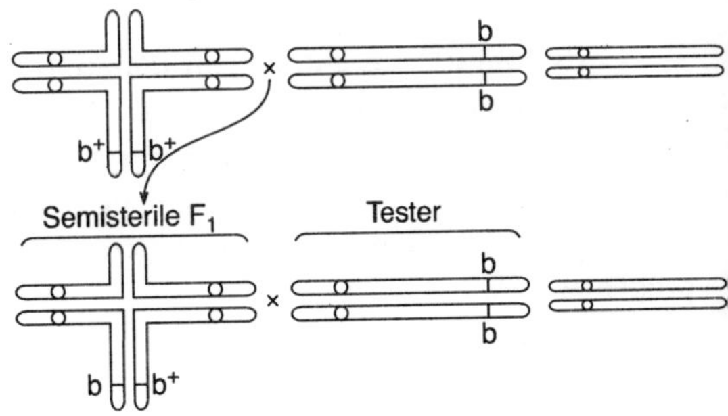

If the chromosomes of the semisterile F_1 segregate as diagrammed here (above) we could then predict

$\frac{1}{2}$ fertile, brachytic

$\frac{1}{2}$ semisterile, non-brachytic

(iii) Most progenies are certainly of this type, so we must be on the right track. How are the two less frequent types produced? Somehow, we have to get the b^+ allele on to the normal shaded chromosome and the b allele onto the translocated chromosome. This must be achieved by crossing over between the translocation break point

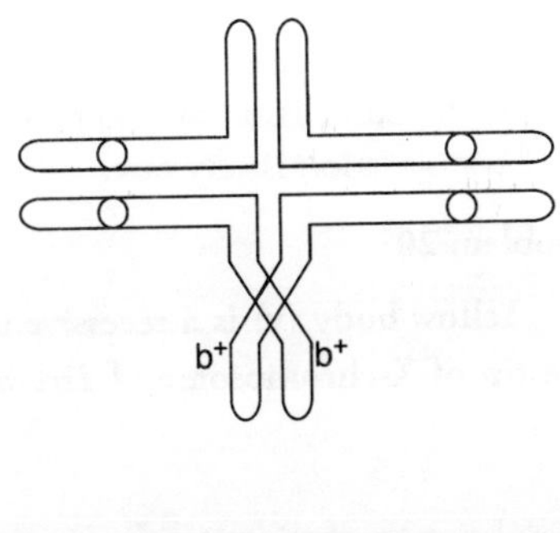

(the centre of the cross-shaped structure) and the brachytic locus.

(iv) The recombinant chromosomes produce some progenies that are fertile and non-brachytic and some that are semisterile and brachytic (these two classes together constitute 69 progeny of a total 682 or a frequency of about 10% or $\frac{69}{682} \times 100$.

(v) We can see that this frequency is really a measure of the map distance (10 mu) of the brachytic locus from the break point.

(The same basis result would have been obtained if we had drawn the brachytic locus in the part of the chromosome on the other side of the break point).

Problem 19

Curly wings(Cy) is a dominant mutation in the second chromosome of *Drosophila*. A Cy/+ male was irradiated with X-rays and crossed with +/+ females and the Cy/+ males were mated individually with +/+ females. From one cross, the progenies were

Curly males	**146**
Wild type males	**0**
Curly females	**0**
Wild type females	**163**

What abnormality in chromosome structure is the most likely explanation for these results? (Remember that crossing over does not take place in male *Drosophila*).

Solution:

(i) In this male, there was a reciprocal translocation between the curly-bearing chromosome and the Y-chromosome.

(ii) Because there is no crossing over in the male, all Y-bearing sperms that give rise to viable, progenies contain Cy, and all X-bearing sperms that give rise to viable progeny contain Cy⁺. Curly is a inversion, so occurrence of translocation is very rare.

Problem 20

Yellow body (y) is a recessive mutation near the tip of X-chromosome of *Drosophila*. A wild type male was irradiated with X-rays and crossed with $\frac{y}{y}$ females and one y^+ male progeny was observed. This male was mated with $\frac{y}{y}$ females, and the offsprings were

Yellow female	**242**
Yellow male	**0**
Wild type female	**0**
Wild type male	**260**

The yellow females were found to be chromosomally normal, and y^+ males were found to breed in the same manner as their father. What type of chromosome abnormality could account for these results?

Solution:

(i) Yellow females $\frac{y}{y}$

(ii) In this male, the tip of the X-chromosome containing y^+ become attached to the Y-chromosome giving the genotype Y.

Wild type male (y^+ male)

$\frac{x}{y}\ \frac{y}{y^+}$

(iii) $\frac{y}{y^+}$ (wild type male) × $\frac{y}{y}$ (yellow female)

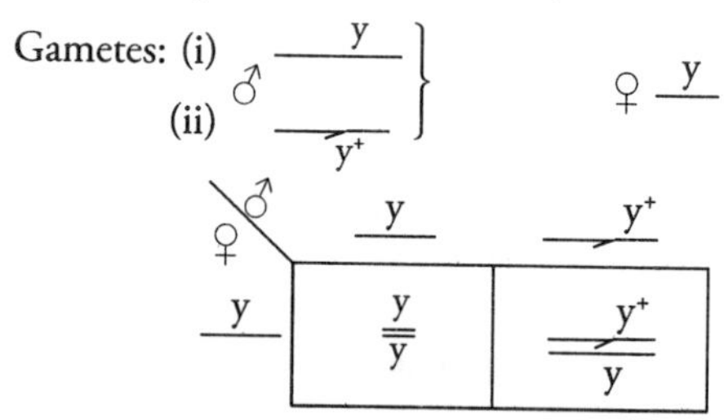

Genotype	Phenotype	Ratio
$\frac{y}{y}$	yellow female	1
$\frac{y^+}{y}$	wild type male	1

So all the offsprings are either wild type of males or yellow females.

Problem 21

A phenotypically abnormal individual has a phenotypically normal father with an inversion on one copy of chromosome 7 and a normal mother without any changes in chromosome

structure. The order of genes along chromosome 7 in the father is as follows:

RTDM centromere PUXZC (normal chromosome 7)

RT DUP centromere M X ZC (inverted chromosome 7)

The phenotypically abnormal offspring has a chromosom 7 with the following order of genes

RTDM centromere PUDTR

with a sketch, explain how this chromosome was formed. Explain where the crossover occurence (i.e. between which two genes).

Solution:

RTDM$_{\circ}$PUXZC (Normal chromosome 7)

RTDUP$_{\circ}$MXZC (Inverted chromosome)

↓

Step 1:

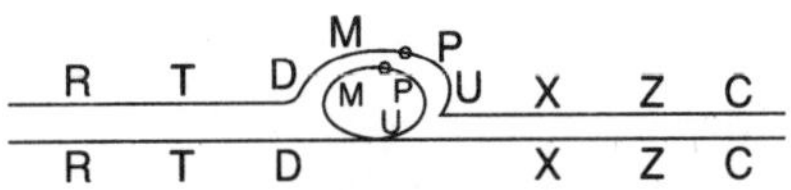

Step 2: Crossing over between P & U

(i) RTDM$_{\circ}$PU DTR

(ii) CZXUP$_{\circ}$MXZC

Problem 22

Genes a, b, c, d and f are closely linked in a chromosome, but their order is unknown. Three deletions in the region are found to uncover recessive alleles of the genes as follows: Deletion 1 uncovers a, b, and d. Deletion 2 uncovers a, d, c and e. Deletion 3 uncovers e and f. What is the order of the genes?

Solution:

(i) The genes uncovered by a single deletion must be contiguous.

(ii) The gene order is deduced from the overlaps between the deletions.

(iii) Deletion 1 uncovers a, b and d.

Deletion 2 uncovers a, d, c and e.

Deletion 3 uncovers e and f.

The overlaps are a and d between the 1st and 2nd deletions and e between the 2nd and 3rd deletions.

(iv) The gene 'b' is the far left and then a and d unknown order (either a, d or d, a), then c, e and f. Gene c must be left of e.

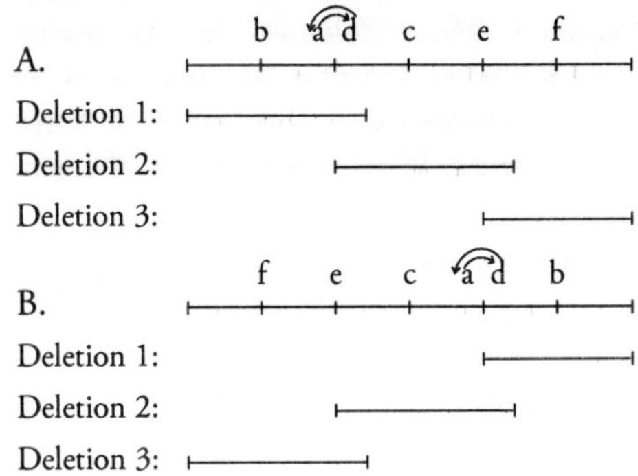

Problem 23

Recessive genes a, b, c, d, e and f are closely linked in a chromosome, but their order is unknown. Four deletions in the region are examined. One deletion uncovers a, c, and f; another uncovers b and c; the third uncovers d and f and the fourth uncovers d, e and f. What is the order of the genes?

Solution:

(i) The genes uncovered by a single deletion must be contiguous.

(ii) The gene order is deduced from the overlaps between the deletions.

(iii) 1st deletion uncovers a, c and f.

2nd deletion uncovers b and c

3rd deletion uncovers d and f

4th deletion uncovers d, e and f

(iv) The overlaps f among 1st, 3rd and 4th and d between 3rd and 4th.

The gene b is the far left and then other. The e also the left end.

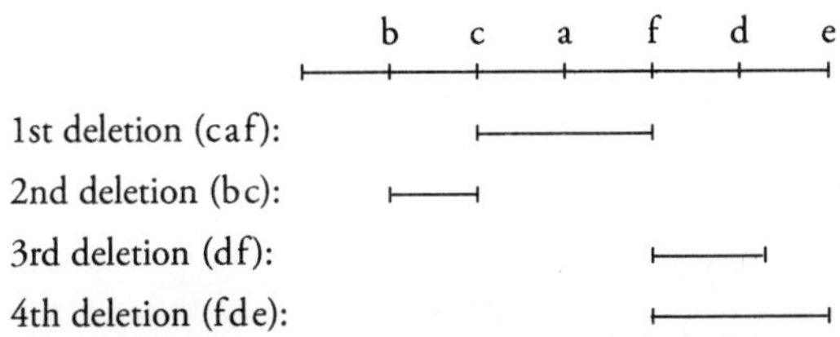

Order of gene:

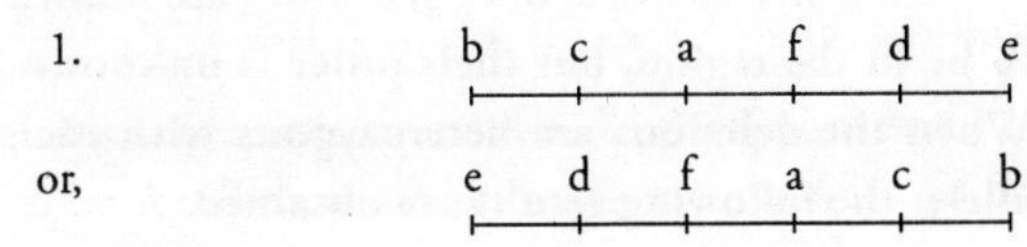

Problem 24

Recessive genes a, b, c, d, e and f are closely linked in a chromosome, but their order is unknown. Three deletions in the region are examined. One deletion uncovers a, d and e, another uncovers c, d and f and the third uncovers b and c. What is the order of the genes?

Solution:

(i) The genes uncovered by a single deletion must be contiguous.

(ii) The gene order is deduced from the overlaps between the deletions.

(iii) 1st deletion uncovers a, d and e
2nd deletion uncovers c, d and f
3rd deletion uncovers b and c
Here gene e overlaps between 2nd and 3rd deletions.

(iv) The gene b is far left. The gene a, e are unknown order (i.e., a e or e a)

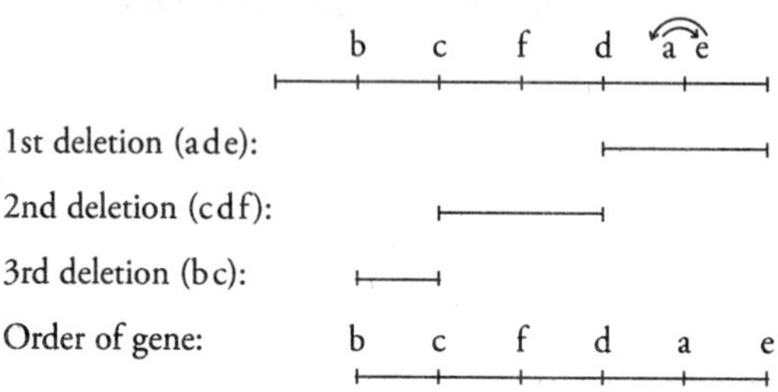

Problem 25

Six bands in a salivary gland chromosome of *Drosophila* are shown in the accompanying figure, along with the extent of five deletions (Del 1 – Del 5).

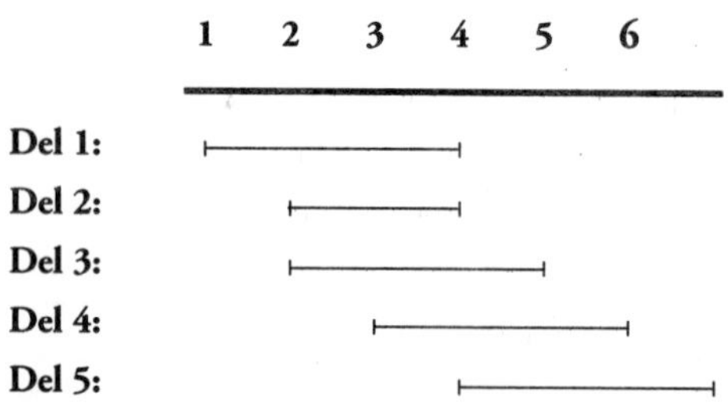

Recessive alleles a, b, c, d, e, and f are known to be in the region, but their order is unknown. When the deletions are heterozygous with each allele, the following results are obtained:

	a	b	c	d	e	f
Del 1:	–	–	–	+	+	+
Del 2:	–	+	–	+	+	+
Del 3:	–	+	–	+	–	+
Del 4:	+	+	–	–	–	+
Del 5:	+	+	+	–	–	–

In this table, the sign – means that the deletion is missing the corresponding wild type allele (the deletion uncovers the recessive allele) and sign + means that the corresponding wild type allele is still present. Use these data to infer the position of each gene relative to the salivary gland chromosome bands.

Solution:

(i) Genes b, a, c, e, d and f are located in bands 1, 2, 3, 4, 5 and 6 respectively.

(ii) Deletion 1 uncovers any gene in band 1 but the other deletion do not. So the pattern – + + + + is observed for gene b puts in band 1.

(iii) Deletions 1–3 uncover band 2, but deletions 4–5 do not; so the pattern – – – + + is observed for gene a localizes gene a to band 2.

(iv) Genes in band 3 are uncovered by deletions 1–4, but not 5. So the pattern – – – – + is observed for gene c in band 3.

(v) Band 4 is uncovered by deletions 3–5, but not 1–2, so the pattern + + + – – is observed for gene e in band 4.

(vi) Genes in band 5 is uncovered by deletion 4–5, but not 1–3, so the pattern + + + – – implies that gene d is in band 5.

(vii) Genes in band 6 is uncovered by deletion 5, but not 1–4 and so the pattern + + + + – puts the gene f in band 6.

Gene:	b	a	c	e	d	f
Band:	1	2	3	4	5	6

Problem 26

A semisterile maize plants heterozygous for a reciprocal translocation between chromosomes 1 and 2 were crossed with chomosomally normal plants homozygous for a recessive allele in chromosome 2 that result in virescent leaves. When semisterile F_1 plants were test crossed, the following data were obtained.

Semisterile virescent	**36**
Fertile virescent	**248**
Semisterile nonvirescent	**284**
Fertile nonvirescent	**44**

(a) What is the frequency of recombination between the gene for virescent leaves and the translocation breakpoint in chromosome 2?

(b) What phenotypic ratio in the test cross progeny would be expected if the gene for virescent leaves had not been in one of the chromosomes involved in the translocation?

Solution:

(i) The frequency of recombination between the gene for virescent leaves and the translocation break point in chromosome 2 equals

$$\frac{80\,(36 + 44)}{612\ (36 + 44 + 246 + 284)} = \frac{80}{612} = 0.1307 \times 100$$

= 13.07% map unit.

(ii) If there were no translocation, the expected ratio would be 1 : 1 : 1 : 1.

Problem 27

A fruit fly was found to be heterozygous for a paracentric inversion. However, it was impossible to obtain flies that were homozygous for the inversion even after many attempts. What is the most likely explanation for this inability to produce a homozygous inversion?

Solution:

The data suggest that one or both break points of the inversion are located within the essential gene causing a recessive lethal mutation.

Problem 28

A diploid species with 44 chromosomes (i.e., 22 set) is crossed to anothr diploid species with 38 chromosomes (i.e., 19 set). What is the number of chromosome in an allo-diploid or allo-tetraploid produced from this cross? Would you expect the offspring to be sterile or fertile?

Solution:

(a) An allodiploid would have 22 + 19 = 41 chromosomes. This individual would likely be sterile, because all the chromosomes would not have homologous partners to pair with during meiosis. This yields aneuploidy, which usually causes sterility.

(b) An allotetraploid would have 44 + 38 = 82 chromosomes. In this individual, each chromosomes would have homologous partner to pair with during meiosis. So the allotetraploid would likely to be fertile.

Problem 29

A person is born with one X-chromosome zero Y-chromosome, trisomy 21 and 2 copies of the other chromosomes. How many chromosomes does this person have altogether?

Explain whether this person is euploid or aneuploid.

Solution:

(i) This person has a total of 46 chromosomes. However, this person would be considered aneuploid rather than euploid because one of the sets is missing a sex chromosome.

(ii) One set has an extra copy of chromosme 21.

Problem 30

What are homeologous chromosome?

Solution:

(i) Homeologous chromosomes are chromosomes from two species that are evolutionarily related to each other.

(ii) For example, chromosome 1 in the sable antelope (*Hippotragus equinus*) and the roan antelope (*Hippotragus niger*) are homeologous.

(iii) They carry many of the same genes.

Problem 31

Contrast the Notch locus with the Bar locus in *Drosophila*. What phenotypic ratios would be produced in a cross between Notch females and Bar males.

Solution:

Notch female	×	Bar male
$\frac{N\,+}{+\,+}$	×	$\frac{+\,B}{\longrightarrow}$

Gametes: $\frac{N\ +}{}$ $\frac{+\ +}{}$; $\frac{+\ B}{}$, $\longrightarrow$

♀ \ ♂	$\frac{+\ B}{}$	$\longrightarrow$
$\frac{N\ +}{}$	$\frac{+\ B}{N\ +}$	$\frac{N\ +}{\Longrightarrow}$
$\frac{+\ +}{}$	$\frac{+\ B}{+\ +}$	$\frac{+\ +}{\Longrightarrow}$

	Genotype	Phenotype	Ratio
(i)	$\frac{+\ B}{N\ +}$	Notch Bar female	$1\left(\frac{1}{4}\right)$
(ii)	$\frac{+\ B}{+\ +}$	Bar female	$\frac{1}{2}\left(\frac{1}{4}\right)$
(iii)	$\frac{N\ +}{\Longrightarrow}$	Notch male (lethal)	0
(iv)	$\frac{+\ +}{\Longrightarrow}$	Wild type male	$1\left(\frac{1}{4}\right)$

The final phenotypic ratio would be 1 : 1 : 1.

(i) If a gene exist in a duplicated state and if that gene's product is required for survival, then mutation in either (but not both) the original gene or its duplicate will not ordinarily threaten the survival of the organism.

(ii) Duplication of gene provides a buffer to mutation.

Problem 32

Female fruit flies homozygous for the X-linked white eye alleles are crossed to male with red eyes. On very rare ocasions, an offspring is a male with red eyes. Assuming these rare offspring are not due to new mutation in one of the mother's X-chromosomes that converted the white eye allele into a red eye allele. Explain how this red eyed male arose.

Solution:

White-eyed female × Red-eyed male

$\frac{X^{w}}{X^{w}}$ × $\frac{X^{w^+}}{\Longrightarrow}$

Gametes: (i) $\frac{X^{w}}{}$; (i) $\frac{X^{w^+}}{}$ (ii) $\longrightarrow$
(Normal)

Nondisjunction: (i) $\frac{X^{w}}{w}$; (ii) 0

♂ \ ♀	$\frac{X^{w}}{}$	$\frac{X^{w}}{X^{w}}$	0
$\frac{X^{w^+}}{}$	$\frac{X^{w}}{X^{w^+}}$	$\frac{X^{w^+}}{X^{w}}$ $\frac{w^+}{}$	$\frac{X^{w^+}}{0}$
$\longrightarrow$	$\frac{X^{w}}{\Longrightarrow}$	$\frac{X^{w}}{X^{w}}\!\!\rightarrow$	$\frac{0}{\longrightarrow}$

Male $\frac{X^{w^+}}{O}$ (XO); i.e., red eye

(i) As a result of non-disjunction, white-eyed female produces X^{w}, $\frac{X^{w}}{X^{w}}$, O (only autosome, no sex chromosomes).

(ii) The red-eyed male produces X-chromosome (X^{w^+}).

(iii) The gamete which is fertilized to the non-disjunction female gamete, which is lacking sex-chromosome. As a result XO male is produced, i.e., red-eyed male.

Problem 33

A *Drosophila* male carries an X-linked temperature sensitive recessive allele that is viable at 18°C but lethal at 29°C. What sex ratio would be expected among the progeny, if the progeny were received at 29°C and male were mated to (a) A normal XX female (b) An attached X female.

Solution:

(a) Male $\frac{X}{\Longrightarrow}$ × $\frac{X}{X}$ female

Gametes: $\frac{X}{}$, $\longrightarrow$; $\frac{X}{}$

Results: ♀ $\frac{X}{X}$, ♂ $\Longrightarrow$:

1 : 1 ratio, because the mutation is recessive and males get their X-chromosome from mother.

(b) Male $\left(\frac{X\ *}{\Longrightarrow}\right)$ × attached X female

Gametes: $\frac{X}{}$, $\longrightarrow$;

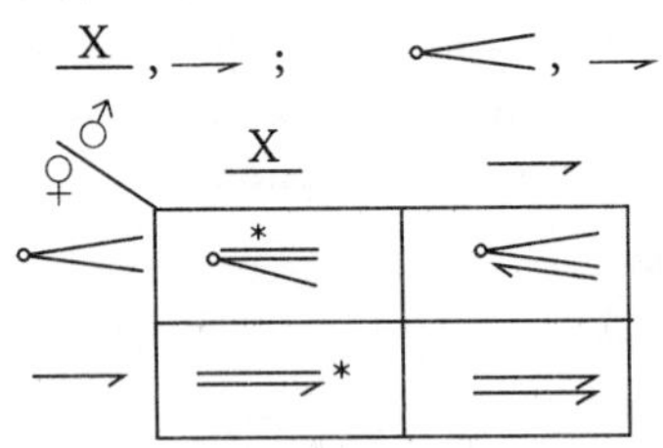

Observation:

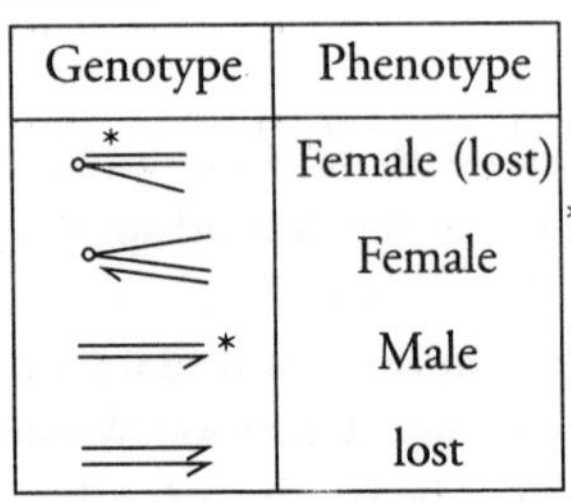

* X-chromosome from father

All females were found, because at 29°C males (who in this case get their X-chromosome from their father) dies.

Problem 34

A chromosome has the gene sequence A B C D E F and another chromosome has the sequence M N O P Q R. A reciprocal translocation between the chromosomes results – A B C P Q R in one chromosome and M N O D E F in other chromosome. Illustrate how these translocated chromosomes would pair with their normal counterparts in a heterozygous individuals during meiosis.

Solution:

Translocation heterozygous:

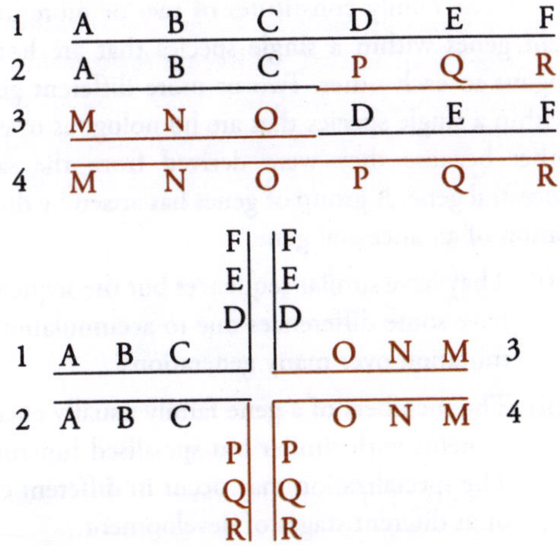

Problem 35

Compare and contrast the origins of Turner syndrome, Williams syndrome, Cri-du-chat syndrome and Down syndrome. Why are they called syndrome?

Solution:

(i) **Williams** syndrome is the result of a deletion of the 7q11.23 region of chromosome 7.

(ii) **Cri-du-chat** syndrome is the result of a deletion of a significant part of the short arm of the chromosome 5 (5p15.2 and 5p15.3).

(iii) Both **Turner** (XO) and **Down** syndrome (trisomy 21) result from meiotic **non-disjunction.**

Turner = 44 + XO

Down = Trisomy 21

The term syndrome is used to describe a set of phenotypes (often complex) that are generally present together.

Problem 36

A group of scientists conducted a field studies on a chain of islands and decided to karyotype two phenotypically identical groups of turtles which were found in different islands. The turtles on one island had 24 chromosomes while the turtles on another island have 48 chromosomes. How would you explain this observation? How do you think the turtles with 48 chromosomes came into being? If they mated the two types of turtles together, would you expect their offspring to be phenotypically normal? Would you expect them to be fertile? Explain.

Solution:

(i) The turtles are two distinct species with identical phenotypes.

(ii) The turtles with 48 chromosomes are polyploid relatives (i.e., tetraploid) of the species 24 chromosomes.

(iii) In higher animals generally self-fertilization does not ocur, so there were two animals (i.e., one male and one female) that became tetraploids.

(iv) It is easy to imagine that complete non-disjunction could occur during the first cell division of fertilized egg thereby creating tetraploid animal. This would have to happen independently (i.e., two individuals of opposite sex) to create tetraploid individual or species.

(v) If a tetraploid turtle is mated with diploid turtles, the offspring would be triploid and probably phenotypically normal, but they would be sterile as they always produce aneuploid gametes.

Problem 37

The karyotype of a young girl who is affected with familial Down syndrome revealed a total of 46 chromosomes. Her older brother, however, who is phenotypically unaffected, actually had 45 chromosomes. Explain how this could happen?

What would you expect to be the chromosomal number in parents of these two children?

Solution:

(i) The boy carries a translocation involving chromosome 21. Here all chromosomes of 21 is translocated to chromosome 14.

He would have one normal copy of chromosome 14, one normal copy of chromosome 21 and the translocated chromosome that contains both 14 and 21.

(ii) The boy is phenotypically normal because the total amount of genetic material is normal, although the total number of chromosomes is 45 (because chromosome 14 and chromosome 21 are fused to single chromosome).

(iii) His (Boy) sister has familial Down syndrome because she has inherited translocated chromosome, but she also must have one copy of 14 and two copies of 21. She has the equivalent of three copies of chromosome 21 (i.e., two normal copies and one copy fused with chromosome 14). This is why she has familial Down syndrome.

(iv) One of the parents of these two children is probably normal with regard to karyotype (i.e., the parent has 46 normal chromosome). The other parent would have a karyotype that would be like phenotypically normal boy.

Problem 38

A woman who is heterozygous Bb, has brown eyes. B (brown) is a dominant allele and b (blue) is recessive. In one of her eyes, however, there is a patch of blue colour. Give three different explanations for how this might have occured.

Solution:

(i) A mutation occurred during early embryogenesis to develop blue patch tissue.

(ii) (a) **1st explanation:**

A mitotic nondisjunction results two chromosomes carrying 'b' allele went to one cell and the two chromosomes carrying 'B' allele went to the other daughter cell.

(b) **2nd explanation:**

The chromosome carrying 'B' allele was lost or deleted.

(c) **3rd explanation:**

Here the 'B' allele was deleted which results the 'b' allele to exhibit **pseudodominance.**

Problem 39

What is a gene family? How are the gene families produced overtime? With regard to gene function, what is the biological significance of a gene family.

Solution:

Gene family constitutes of two or more different genes within a single species that are homologous to each other. Two or more different genes within a single species that are homologous to each other because they were derived from the same ancestral gene. A group of genes has arisen by duplication of an ancestral gene.

(i) They have similar sequences but the sequences have some differences due to accumulation of mutation over many generations.

(ii) The members of a gene family usually encode proteins with similar but speialised functions. The specialization may occur in different cells or at diferent stages of development.

Problem 40

State the common and unique features of the family members in the globin gene family.

Solution:

(i) The duplication may occur by an abnormal (misaligned) crossing over, which produces a chromosome with deletion and another chromosome with a gene duplication. Duplication may occur several times to produce many copies of a particular gene.

(ii) In addition, translocation may move the duplicated genes to other chromosome. So the members of the gene family may be dispersed among several different chromosomes.

(iii) Eventually, each member of a gene family will accumulate mutations, which may **subtly** alter their function.

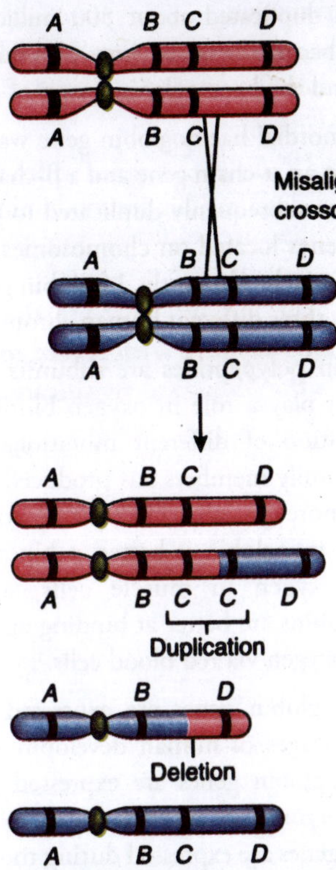

Fig. 7.15: Abnormal crossing over, leading to a duplication and a deletion.

(iv) All the members of the globin bind oxygen. Myoglobin tends to bind it more tightly. So it is a good at storing oxygen.

(v) Haemoglobin binds it more loosely, so it can transport oygen throughout the body (via red blood cells) and release it to the tissues that need oxygen. The polypeptides that form haemoglobins are predominantly expressed in red blood cells, whereas myoglobin genes are expressed in many different cell types.

(vi) The expression pattern of the globin genes changes during different stages of development. The ε- and ζ- globin genes are expressed in the early embryo. They are turned off near the end of the first trimester, and then the γ-globin genes exhibit their maximal expression during the second and third trimesters of gestation.

(vii) Following birth, the γ-globin genes are silenced, and the β-globin gene is expressed for the rest of a person's life. These differences in the expression of the globin genes reflect the differences in the oxygen transport needs of humans during the different stages of life.

(viii) The evolution of gene families has resulted in gene products that are better suited to a particular tissue or stage of development.

Problem 41

Evolution of the globin gene family in human.

(i) Chromosomal duplication provides raw material for the addition of more genes into a

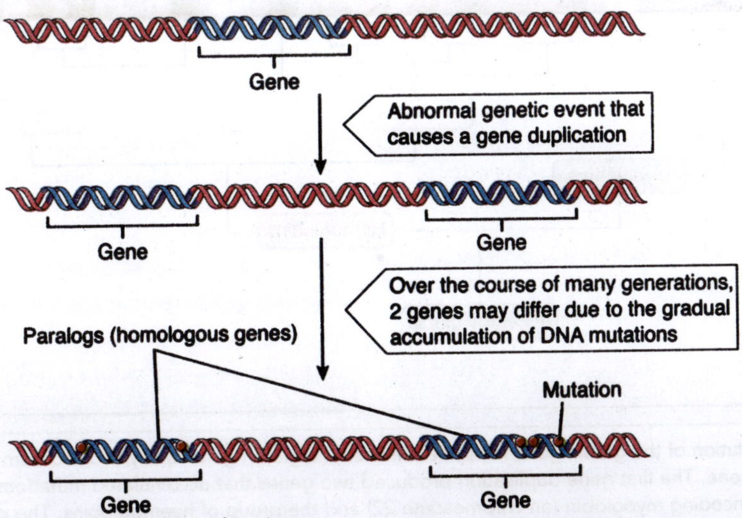

Fig. 7.16: Gene duplication and the evolution of paralogs.

species chromosome through many generations. This can lead to the formation of a gene family consisting of two or more genes that are similar to each other.

(ii) The members of a gene family are derived from the same ancestral gene. Over times two copies of an ancestral gene can accumulate different mutations. Therefore, after many generations, the two genes will be similar but not identical. During evolution, this type of event can occur several times, creating a family of many similar genes.

(iii) When two or more genes are derived from a single ancestral gene, the genes are said to be **homologous**. Homologous genes within a single species are called **paralogs** and constitute a gene family. For example, globin gene family in humans.

(iv) The globin genes encode polypeptides that are subunits of proteins that function in oxygen binding. For example, haemoglobin is a protein found in red blood cells; its function is to carry oxygen throughout the body.

(v) The globin gene family is composed of 14 paralogs that were originally derived from a single ancestral globin gene. According to an evolutionary analysis, the ancestral globin gene was first duplicated about 500 million years ago and became separate genes encoding myoglobin and the haemoglobin group of genes.

(vi) The primordial haemoglobin gene was duplicated into an α-chain gene and a β-chain gene, which was subsequenly duplicated to produce several genes located on chomosomes 16 and 11, respectively. Currently, 14 globin genes are found on three different human chromosomes.

(vii) All globin polypeptides are subunits of proteins that play a role in oxygen binding, the accumulation of different mutations in the various family members has produced globins that are more specialized in their function. For example, myoglobin is better at binding and storing oxygen in muscle cells, and the haemoglobins are better at binding and transporting oxygen via red blood cells.

(viii) Different globin genes are expressed during different stages of human development. The ε- and ζ-globin genes are expressed in very early embryonic life whereas the α-gobin and γ-globin genes are expressed during the second and third trimesters of gestation.

(ix) After birth the α-globin gene remains **turned on**, while the γ-globin genes are **turned off** and the β-globin gene is **turned on.**

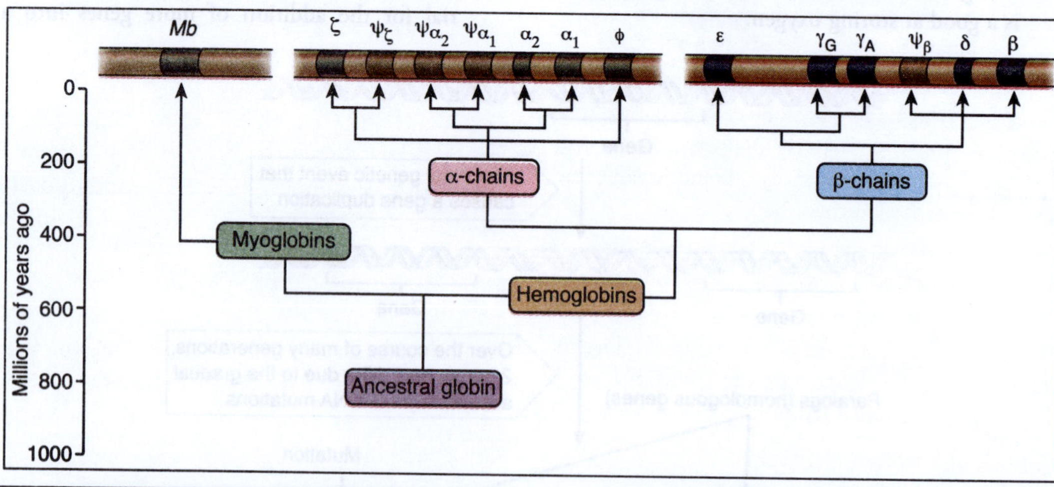

Fig. 7.17: The evolution of the globin gene family in humans. The globin gene family evolved from a single ancestral globin gene. The first gene duplication produced two genes that accumulated mutations and became the genes encoding myoglobin (on chromosome 22) and the group of haemoglobins. The primordial haemoglobin gene is then duplicated to produce several α-chain and β-chain genes, which are found on chromosomes 16 and 11, respectively. The four genes shown in grey colour are nonfunctional pseudogenes.

(x) These differences are the expression of the globin genes reflect the differences in the oxygen transport needs of humans during the embryonic, foetal and post-partum stages of life.

Problem 42

In the treatment of cancer, the basis for many types of chemotherapy and radiation therapy is that mutagens are more effective at killing dividing cells compared to non-dividing cells. Explain why? What are possible harmful side effects of chemotherapy and radiation therapy?

Solution:

(a) (i) The effects of mutations are cumulative. If one mutation occurs in a cell, this mutation will be passed to the daughter cell. If a mutation occurs in the daughter cells, now there will be two mutations. These two mutations will be passed to the next generation of daughter cells.

(ii) The accumulation of many mutations eventually kills the cells, that is why mutagens are more effective at killing dividing cells compared to nondividing cells. It is because the number of mutations accumulates to a lethal level.

(b) (i) There are two main side effects of chemotherapy and radiation therapy.

- Some normal cells (non cancereous) of the body, viz. skin cells and intestinal cells are actively dividing. These cells are also killed by chemotherapy and radiation therapy.
- The therapy may produce mutations that will cause non-cancerous cells to become cancerous.

For these reasons, maximal dose of chemotherapy or radiation therapy is recommended.

Problem 43

What type of mutation is the following (shown as mRNA)?

Wild type = …… 5′ AAUCCUUACGGA 3′……

Mutant = ……… 5′ AAUCCUACGGA 3′ ……

Solution:

The mutant has deletion of one base and this deletion will result in a frameshift (–1) mutation.

Problem 44

By base pair substitution, what are the synonymous changes that can be made starting with the codon CGG?

Solution:

With the assumption of single base pair substitutions CGG can be changed to the following: CGU, CGA, AGG or CGC. They would still code for arginine.

Problem 45

Which tautomer of thymine can form the most hydrogen bonds in pairing with other DNA bases?

Solution:

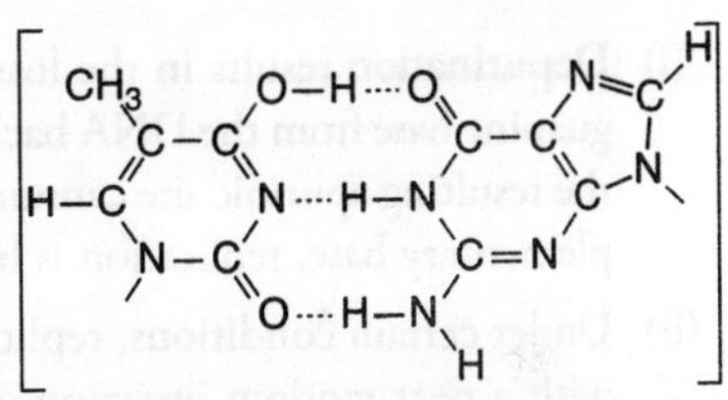

The enol form of thymine forms three hydrogen bonds with guanine.

Problem 46

A newly discovered SOS by Pass system is found to preferentially insert thymine opposite Apurinic sites. What type of mutations should be preferentially produced?

Solution:

(i) Apurinic sites have lost either an A or a G.

(ii) If thymine is preferentially inserted opposite these sites, GC : AT transitions should be produced.

Problem 47

Why are frameshift mutations more likely than missense mutations to result in proteins that lack normal function?

Solution:

(i) Frameshift mutations arise from addition or deletion of one or more bases in other than multiples of three.

(ii) When it is translated, this mutation will alter the reading frame and therefore, the amino acid sequence from the site of mutation to the end of the protein product.

(iii) Additionally, frameshift mutation often results in premature stop codons in the new reading frame, leading to shorten the protein products.

(iv) A missense mutation changes only a single amino acid in the protein product.

5′ AAA 3′ (Lysin) ⟶ 5′ GAA 3′ (Glutamic acid)

Problem 48

Describe two spontaneous lesions that can lead to mutations.

Solution:

(i) **Depurination** results in the loss of adenine or guanine base from the DNA backbone. Because the resulting apurinic site cannot specify a complementary base, replication is blocked.

(ii) Under certain conditions, replication proceeds with a near random insertion of a base opposite the apurinic site. At three-fourths of these insertions, mutations will result.

(iii) Deamination of cytosine yields uracil. If left unrepaired, the uracil will be paired with adenine in replication, ultimately resulting in a transition mutation.

(iv) **Deamination** of **5-methyl cytosine** yields thymine and thus frequently leads to C to T **transitions.**

(v) Oxidatively damaged bases such as 8-oxodG (8-oxo-7-hydrodeoxyguanosine) can pair with adenine, resulting in **transversion.**

(vi) Errors in the course of DNA replication leads to spontaneous indel mutation.

Problem 49

Do all nucleotide substitutions in the second position of a codon necessarily produce an amino acid replacement?

Solution:

All nucleotide substitutions in the second codon position result in either an amino acid replacement or terminator codon.

Problem 50

Do all nucleotide substitutions in the first position of a codon necessarily produce an amino acid replacement? Do all of them necessarily poduce a nonfunctional protein?

Solution:

(i) Both UUR and CUR code for leucine and both CGR and AGR code for arginine (R stands for any purine either A or G).

(ii) Hence, in each of these codons one of the possible substitutions in the first codon position does not result in amino acid replacement.

(iii) Even when an amino acid replacement does occur, the mutant protein may still functional, because more than one amino acid at a given position in a protein molecule may be compatible with normal or nearly normal function.

Problem 51

In a coding sequence made up of equal proportions of A, U, G and C, what is the probability that a random nucleotide substitution will result in a chain termination codon?

Solution:

(i) Any single nucleotide substitution creates a new codon with a random nucleotide sequence.

(ii) So the probability that it is a termination codon (UAA, UAG or UGA) is $\frac{3}{64}$ or approximately 5 percent.

Problem 52

The molecule-2-aminopurine (AP) is an analog of adenine that pairs with thymine. It also occasionally pairs with cytosine. What types of mutations will be induced by 2-aminopurine?

Solution:

(i) The base is replaced and then in subsequent rounds of DNA replication, the base analog pairs normally or abnormally.

(ii) Here the 2-aminopurine (AP) is incorporated in place of adenine opposite thymine, forming an AP-T base pair.

(iii) In the following round of DNA replication, the AP will usually pair with T, but when it occasionally pairs with C it forms AP-C pair.

(iv) In the next round of replication, the AP again pairs with T (forming an AP-T pair) but the C pairs with G (forming a mutant GC pair). Hence, the type of mutation induced by 2-amino purine is a change from an A-T pair to a G-C pair which is a **transition** mutation.

Problem 53

Twenty spontaneous white eye(w) mutants of *Drosophila* are examined individually to determine their reversion frequencies. Of these, 12 revert spontaneously at a frequency of 10^{-3} to 10^{-5}. Another 6 do not revert spontaneously at detectable frequency but can be induced to revert at frequencies ranging from 10^{-5} to 10^{-6} by treatment with alkylating agents. The remaining 2 cannot be made revert under any condition. Answer the following questions and explain your answers:

(a) Which class of mutants would you suspect to result from the insertion of transposable elements?

(b) Which classes of mutants are probably missense mutation?

(c) Which classes of mutants are probably deletions?

Solution:

(a) Insertions of transposable elements in eukaryotes are often characterized by genetic instability and high reversion rates. Therefore, twelve mutants with the highest reversion frequencies are probably transposable elements.

(b) The fact that alkylating agents can induce the second class of mutations to revert suggests that they are **missense mutation.** Although they do not revert spontaneously, this is probably because the exact reversal of a base substitution is an exceedingly rare spontaneous event.

(c) Deletion mutations cannot revert under any condition because there is too much missing genetic information to be restored. Therefore, the class of mutation that does not revert probably consists of deletion.

Problem 54

A mutation is isolated that cannot be induced to revert. What types of molecular changes might be responsible?

Solution:

Causes of non-reverting mutation include

(i) deletion of the gene

(ii) multiple nucleotide substitution in the gene

(iii) complex DNA rearrangement with more than one break point in the gene.

Deletions do not revert because the genetic information in the gene is **missing** and the others do not revert because the nucleotide substitution or DNA rearrangements are unlikely to be reversed precisely in one step.

Problem 55

There are 12 possible substitutions of one nucleotide pair for another (for example A-G). Which changes are transition and which are transversions?

Solution:

(i) Transitions are AT → GC, GC → AT, TA — CG and CG — TA (total four).

(ii) Transversions are AT → TA, AT — CG, GC — CG, GC — TA, TA — AT, TA — GC, CG — GC and CG — AT (total eight).

Problem 56

Explain why mutations induced by acridines in phage T4 or ICR-191 bacteria cannot be reverted by 5-bromouracil.

Solution:

(i) Acridines and ICR 191 induce mutations by **deletion** or **adding** one or more base pairs which results in a **frameshift.**

(ii) However, **5-bromouracil** induces mutations by causing the **substitution** of one base for another. This substitution cannot compensate for the frameshift resulting from ICR-191 and acridines.

Problem 57

(a) **Why is it impossible to induce nonsense mutations (represented at the mRNA level by triplets UAG, UAA) by treating wild type strains with mutagens that cause only A·T → G·C transitions in DNA.**

(b) **Hydroxylamine (HA) causes only G·C → A·T transitions in DNA. Will HA produce nonsense mutations in wild type strains?**

(c) **Will HA treatment revert nonsense mutations?**

Solution:

(a) (i) Because 5′ UAA 3′ does not contain G or C, a transition to G.C pair in the DNA cannot result in 5′ UAA 3′, 5′ UGA 3′ and 5′ UAG 3′, have the DNA antisense strand sequence of 3′ ACT 5′ and 3′ ATC 5′ respectively.

(ii) A transition to either of these stop codons occurs from the nonmutant 3′ ATT 5′ resulting in an RNA sequence of 5′ UAA 3′ itself a stop codon.

(b) Yes. An example is 5′ UGG 3′ which encodes Trp to 5′ UAG 3′.

(c) No. (i) In the three stop codons, the only base that can be acted on is G (in UAG for instance). (ii) Replacing G with an A would result in 5′ UAA 3′ a stop codon.

Problem 58

Contrast and compare the mutagenic effects of deaminating agents, alkylating agents and base analogs.

Solution:

All three agents are mutagenic because they cause **base substitutions.**

(a) (i) **Deaminating** agents oxidatively deaminate bases such that cytosine is converted into **uracil** and adenine is converted to **hypoxanthine.**

(ii) Uracil pairs with adenine instead of guanine and hypoxanthine pairs with cytosine rather than thymine.

(b) Alkylating agents donates an **alkyl group** to the **amino** or **keto** groups of nucleotides; thus altering base pairing affinities.

Example: 6-ethyl guanine acts like adenine, thus pairing with the thymine.

(c) Base analogs such as **5-bromouracil** and **2-amino purine** are incorporated as thymine and adenine respectively. Yet they base pair with guanine and cytosine.

Problem 59

What genetic defects result in the disorder xeroderma pigmentosum (XP) in humans? How do these defects create the phenotypes associated with the disorder?

Solution:

(i) Xeroderma pigmentosum is a form of human skin cancer caused by perhaps several rare autosomal genes that interfere with the repair of damaged DNA.

(ii) Studies with heterokaryons provided evidence for complementation, indicating that there may be as many as seven different genes are involved.

(iii) The photoreactivation repair enzyme appears to be involved.

Problem 60

Why X-rays are more potent mutagen than UV radiation.

Solution:

(i) X-rays are of higher energy and shorter wavelength than UV light.

(ii) They have greater penetrating ability and can create more disruption of DNA.

Problem 61

If a mutagen produces frameshift mutations, what kinds of mutations will it induce or revert?

Solution:

A frameshift mutagen will also revert a frameshift mutation.

Problem 62

If a deletion eliminates a single amino acid in a protein, how many base pairs are missing?

Solution:

(i) Exactly three.

(ii) Otherwise, there would be a frameshift, and all downstream amino acids would be different.

Problem 63

What are two ways in which pyrimidine dimers are removed from DNA in *E. coli*?

Solution:

Photoreactivation and excision repair.

Problem 64

We learned that UAG and UAA codons are two of the chain terminating nonsense triplets. On the basis of the specificity given in the test for aflatoxin B_1 and ethyl methane sulfonate (EMS), describe whether each mutagen would be able to revert these codons to wild type.

Solution:

(i) EMS induces primarily G·C → A·A **transitions.** UAG codons could not be reverted to wild type, because only the UAG → UAA change would be stimulated by EMS and that generates a **nonsense (ochre)** codon.

(ii) UAA codons would not be acted on by EMS, **Aflatoxin B_1**, induces primarily G·C → T·A transversions. Only third position of UAG codons would be acted on resulting in a UAG → UAU change (on the mRNA level) which produces tyrosine.

(iii) Therefore, if tyrosine were in acceptable amino acid at the corresponding site in the protein, **aflatoxin B_1**, could revert UAG codons.

(iv) Aflatoxin B_1, would not revert UAA codons because no G·C base pair appears in the corresponding position in the DNA.

Problem 65

A single mutation blocks the conversion of phenylalanine to tyrosine. (a) Is the mutant gene expected to be pleotropic? (b) Explain.

Solution:

(a) Yes.

(b) A block would result in the accumulation of phenylalanine and a decrease in the amount of tyrosine would be expected to result in several different phenotypic expressions.

Problem 66

How does ultraviolet light produce revertible mutation?

Solution:

(i) During replicating process, UV light produces mispairing alternations mostly in pyrimidines (Example : C → T transitions). Thymine may be altered to cytosine (modified pyrimidine base pairing potential of cytosine) which pairs with guanine.

(ii) A reverse mutation may occur when cytosine is changed to thymine (or a derivative of cytosine with the hydrogen bonding potential of thymine) which pairs with adenine.

(iii) The T-A base pair may thus be changed to a C-G and the reverse mutation may occur from C-G to T-A.

Problem 67

How does the mutagenic effect of 5-bromouracil differ from that of nitrous acid?

Solution:

(i) Nitrous acid acts as a mutagen on either replicating or non-replicating DNA and produces transitions from A to G or C to T, whereas 5-bromouracil does not affect non-replicating DNA but acts during the replication process causing GC $\rightleftharpoons$ AT transitions.

(ii) 5-Bromouracil must be incorporated into DNA during the replication process in order to induce mispairing of bases and thus mutation.

Problem 68

A mutant of *E. coli* is highly resistant to mutagenesis by a variety of agents, including ultraviolet light, aflatoxin B, and benzo-(a)-pyrene. Explain one possible cause of this mutant phenotype.

Solution:

(i) The mutant might lack the SOS system and perhaps carries a defect in the Umu C gene.

(ii) Such strains would not able to bypass replication blocking lesions of the type caused by the three mutagens listed.

(iii) Without the processing of premutational lesions, mutation would not be recovered in viable cells.

Problem 69

A certain compound that is an analog of the base cytosine can become incorporated into DNA. It normally hydrogen bonds just like cytosine does, but it is quite often isomerizes to a form that hydrogen bonds as thymine does. Do you expect this compound as mutagenic, and if so, what types of changes might it induce at the DNA level?

Solution:

Yes this compound is mutagenic.

It will cause CG to TA transitions.

Problem 70

5-bromouracil, 2-aminopurine, proflavin, ethyl methane sulphonate and nitrous acid are chemical mutagens. What does each do?

Solution:

(i) **5-bromouracil** (Pyrimidine analogue) and a **2-aminopurine** (Purine analogue) are incorporated into DNA as thymine and adenine respectively.

(ii) However, each undergoes **tautomeric shifts** more frequently than normal base. Both cause transitions.

Proflavin induces insertions and deletion by intercalating and buckling DNA.

Ethyl ethane sulphonate removes purine rings and thus promotes transitions and transversions.

Nitrous acid causes oxidative deamination of amino groups and promotes transitions by converting cytosine into uracil which acts like thymine and base pair with adenine. Adenine is deaminated into hypoxanthine just like guanine and base pair with cytosine.

Problem 71

In what way does 5-bromouracil (BU) function as a mutagen? What type of mutations would occur from its use?

Solution:

(i) BU is a base analogue mutagen. It is a thymine analogue and is incorporated into DNA during replication by base pairing with adenine.

(ii) Due to the bromine atom, a shift in the keto-enol equilibrium occurs, such that the enol (COH) form is more prevalent than it would be with thymine.

(iii) The enol form may base pair with guanine in future round of replication leading to A·T → G·C or T·A → C·G transition after one more round of replication.

Problem 72

What is the difference between mismatch repair and AP repair?

Solution:

(i) Both processes are similar. They entail removal of an incorrect base in a DNA double helix by an excision process followed by repair process.

(ii) Mismatch repair is triggered by a base pair that does not occupy the correct space in the double helix, i.e., by a non-Watson and Crick base pairing (not A·T or G·C).

(iii) AP repair is triggered by enzymes that recognize a missing base.

Problem 73

Contrast various types of DNA repair mechanisms known to counteract the effects of UV radiation. What is the role of visible light in repairing UV induced mutation?

Solution:

(i) The UV damages caused in cells are repaired after exposure of cells in visible light. This is called **protoreactivation**. Photoreactivation can lead to repair UV induced damages.

(ii) In this mechanism an enzyme DNA **photolyase** (encoded by the **phr** gene) activated by absorbing photon of light and cleaves or splits thymine dimers (T-T).

(a) **Excission repair:**

(i) Excision repair of damaged DNA involves the products of several genes, DNA polymerase, and DNA ligase. It also involves at least three steps.

(ii) Step 1: DNA repair endonuclease or endonuclease complex recognizes, binds to and excises the damaged base or bases.

Step 2: A DNA polymerase fills in the gap by using the undamaged complementary strand of DNA as template.

Step 3: The enzyme DNA ligase seals the break left by DNA polymerase to complete the repair process.

(iii) There are two types of excision repair—viz., **base excision repair** and **nucleotide excision** repair.

(iv) The excision repairs process can be activated by damage that distorts the DNA helix.

(b) **Recombinational repair:**

(i) When excision repair mechanism fails, this mechanism is required to repair error. To know the mechanism of recombinational repair, it is necessary to know the effect of a **thymine dimer** on replication.

(ii) When polymerase III reaches the thymine dimer, the replication fork fails to advance.

(iii) A thymine dimer is still capable of forming hydrogen bonds with two adenines because the chemical change in dimerization does not alter the groups that engage in hydrogen bonding.

(iv) However, the dimer introduces a distortion into the helix and when an adenine is added to the growing chain, polymerase III is added, and the editing function removes the adenine.

(v) The cycle begins again and polymerase is stalled at dimer.

(c) **SOS repair:**

If a gap is created on one of the newly synthesized strands, a rescue operation or SOS response allows the gap to be filled.

Problem 74

Name two enzymes which catalyze direct reversal of DNA damage.

Solution:

Photoreactivating enzyme (**DNA photolyase**) and **O^6 methylguanine methyl transferase.**

Problem 75

(a) What kind of DNA damage does UV light causes?

(b) Why is this damage harmful to a cell?

Solution:

(a) Pyrimidine (specifically thymine) dimers.

(b) If this damage goes unpaired, it can either stall DNA replication or cause error prone repair and therefore, constitute lasting damage.

Problem 76

(a) What kind of DNA damage does X-ray cause?

(b) Why is this damage harmful to a cell?

Solution:

(a) DNA strand breaks.

(b) Double strand breaks are difficult to repair properly and therefore, constitute lasting damage.

Problem 77

What is the difference between exonuclease and endonuclease?

Solution:

An exonuclease is a type of enzyme that degrades or cleaves an covalent bond between two nucleotides at one end of a strand. An endonuclease is a type of enzyme that cleaves the covalent bond between two adjacent nucleotides within (middle) a strand.

BIOCHEMICAL MUTATION

- **Why *Neurospora* has been chosen as the experimental organism by George W. Beadle and Edward L. Tatum?**

1. It is universally studied.
2. It requires very small space.
3. The four products in meiosis stay in the same ascus and segregation of eight ascospore of an ascus can be easily studied.
4. The organisms remain in haploid condition throughout all the stages of its life cycle except diploid zygote condition. So detection of any kind of mutation is very easy in this organism.
5. It is very easy to raise in the laboratory and can be easily propagated asexually.
6. It is capable of synthesizing all purines, pyrimidines, amino acid and all vitamins which need for its development.

- **Biochemical mutation detection in *Neurospora.***

Requirements:

(i) Wild type *Neurospora crassa.*

(ii) X-rays or UV rays.

(iii) Complete and minimal media.

Procedure:

(i) The conidia (spores) of wild type *Neurospora crassa* are irradiated with X-rays or U.V rays.

(ii) The irradiated spores are crossed with wild type *N. crassa* or opposite sex.

(iii) Hybrid is then allowed to form fruiting body.

(iv) Spores are then isolated from fruiting body.

(v) Isolated spores are then allowed to grow in both complete and minimal media.

Observation:

(i) All the spores grow in complete medium but some of the spores fail to grow in minimal media.

(ii) The spores which grow both in complete and minimal media are called **prototroph.** The spores which fail to grow in minimal media are called **auxotroph.**

(iii) Now the auxotrophs are shifted to the supplemented media containing nutrients (A-F).

Inference:

(i) In which supplemented medium the auxotroph is capable to grow or germinate, it will be presumed mutation would have been occurred for that nutrients.

Nuclear origin:

If the auxotroph is crossed with wild type of opposite sex and produces offspring like wild type: mutant (auxotroph) phenotype in 1 : 1 ratio, this observation suggest, it is nuclear origin.

Explanation:

Biochemical synthesis occur in as stepwise succession. Each step is controlled by an **enzyme.** That enzyme is regulated by a particular **gene.**

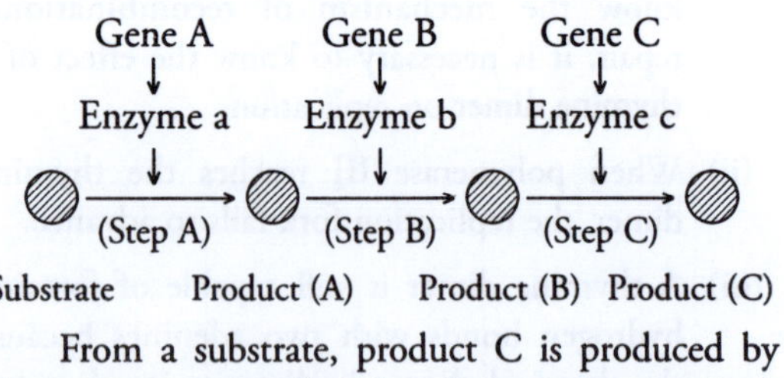

From a substrate, product C is produced by three steps controlled by three enzymes. Each enzyme is regulated by an individual gene.

In *Neurospora,* mutant is obtained through irradiation which was incapable of producing (synthesizing) a particular substance, because the gene which has been engaged, is no longer able to synthesize that particular enzyme, needed for that particular reaction.

From this analysis, this hypothesis arises that **one gene** controlled **one enzyme,** i.e. **one gene one enzyme.**

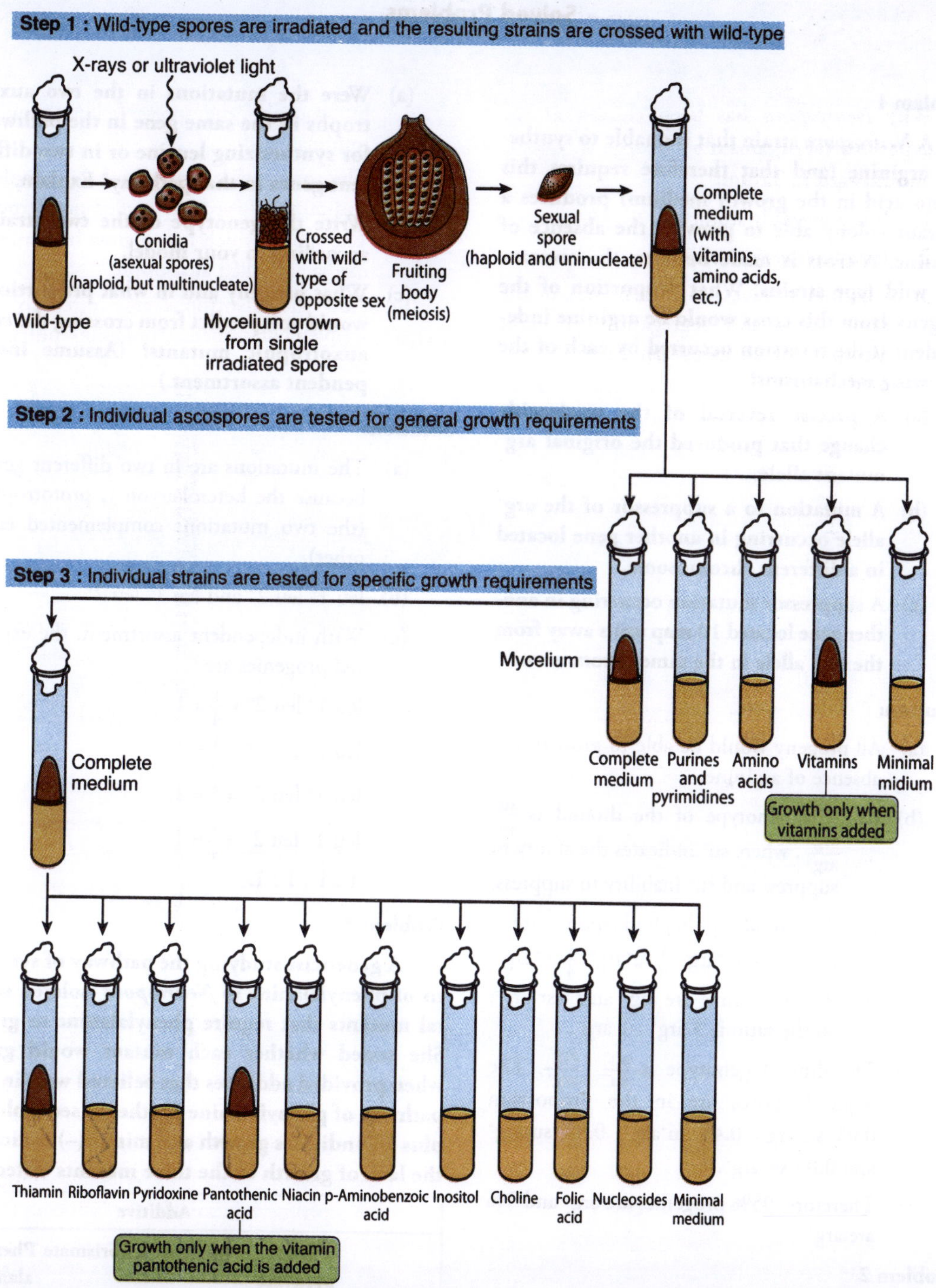

Fig. 7.18: Induction, isolation, and characterization of a nutritional auxotrophic mutation in *Neurospora*. In (a), most conidia are not affected, but one conidium (shown in red) contains such a mutation. In (b) and (c), the precise nature of the mutation is determined to involve the biosynthesis of tyrosine.

Solved Problems

Problem 1

A *Neurospora* strain that is unable to synthesize arginine (and that therefore requires this amino acid in the growth medium) produces a reverant colony able to grow in the absence of arginine. A cross is made between the reverant and wild type strains. What proportion of the progeny from this cross would be arginine independent if the reversion occurred by each of the following mechanisms?

(a) A precise reversal of the nucleotide change that produced the original arg^- mutant allele.

(b) A mutation to a suppressor of the arg^- allele occurring in another gene located in a different chromosome.

(c) A suppressor mutation occurring in another gene located 10 map units away from the arg^- allele in the same chromosome.

Solution:

(a) All progeny would be able to grow in the absence of arginine.

(b) (i) The genotype of the diploid is $\frac{su^+}{su^-}\ \frac{arg^+}{arg^-}$, where su^+ indicates the ability to suppress and su^- inability to suppress.

(ii) The resulting haploid spores are $\frac{1}{4}$ su^+arg^+, $\frac{1}{4}$ su^+arg^-, $\frac{1}{4}$ su^-arg^+, $\frac{1}{4}$ su^-arg^-.

(iii) The first three are arg^+ and last arg^-; so the ratio is 3 arg^+ : 1 arg^-.

(c) The diploid genotype is $\frac{su^+}{su^-}\ \frac{arg^-}{arg^+}$. The haploid spores are in the proportion 0.05 su^+arg^+, 0.45 su^+arg^-, 0.45 su^-arg^+ and 0.05 su^-arg^-.

Therefore, 95% progenies are arg^+ and 5% are arg^-.

Problem 2

After irradiating wild type cells of *Neurospora* (a haploid fungus), a geneticist finds two leucine requiring auxotrophic mutants. He combines the two mutants in a heterokaryon and discovers that heterokaryon is prototrophic.

(a) Were the mutations in the two auxotrophs in the same gene in the pathway for synthesizing leucine or in two different genes in that pathway? Explain.

(b) Write the genotype of the two strains according to your model.

(c) What progeny and in what proportions would you predict from crossing the two auxotrophic mutants? (Assume independent assortment.)

Solution:

(a) The mutations are in two different genes because the heterokaryon is prototrophic (the two mutations complemented each other).

(b) leu 1^+ leu 2^- and leu 1^- leu 2^+.

(c) With independent assortment, the expected progenies are

leu 1^+ leu 2^- = $\frac{1}{4}$ = 1

leu 1^- leu 2^+ = $\frac{1}{4}$ = 1

leu 1^+ leu 2^+ = $\frac{1}{4}$ = 1

leu 1^- leu 2^- = $\frac{1}{4}$ = 1

1 : 1 : 1 : 1.

Problem 3

A geneticist studying the pathway of synthesis of phenylalanine in *Neurospora* isolated several mutants that require phenylalanine to grow. She tested whether each mutant would grow when provided additives that believed were in the pathway of phenylalanine synthesis (see table), a plus (+) indicates growth and minus (–) indicates the lack of growth in the three mutants tested.

	Additive			
	Phenyl-pyruvate	**Prephenate**	**Chorismate**	**Phenyl-alanine**
Wild type	+	+	+	+
Mutant - 1	–	–	–	+
Mutant - 2	+	+	–	+
Mutant - 3	+	–	–	+

Where in the pathway to phenylalanine synthesis does each of the additives belong, if at all?

Solution:

(i) The wild type grows in the presence of all additives. This is not surprising since the wild type can grow by definition, in the absence of all the additives because it can synthesize phenylalanine de novo.

(ii) Mutant-1 cannot grow in the presence of any additive except **phenylalanine**, indicating that its mutation affects the step just before the end of the pathway at phenylalanine. In other words, each of the other additives occurs in the phenylalanine pathway before the point of the mutation in mutant-1.

(iii) Mutant-2 can grow if given any additive but chorismate, indicating that chorismate is at the beginning of the pathway and the mutation affects the pathway just after the step.

(iv) Finally mutant-3 can grow if given phenyl pyruvate or phenylalanine, indicating that its mutation affects the step before phenyl pyruvate and phenyl alanine but after the earlier part of the pathway.

Putting all of these information together indicates that the pathway to phenyl-alanine with mutants indicated is

Chorismate $\xrightarrow{2}$ prephenate $\xrightarrow{3}$ phenyl pyruvate $\xrightarrow{1}$ phenylalanine.

Problem 4

The following chart shows the growth (+) or lack of growth (–) of four mutant strains of *Neurospora* with various additives. The additives are in the pathway of niacin biosynthesis.

	Mutant			
Additives	**1**	**2**	**3**	**4**
Nothing	–	–	–	–
Niacin	+	+	+	+
Tryptophan	+	+	–	–
Kynurenine	+	+	–	–
3-Hydroxyanthranilic acid	+	+	+	–
Indole	–	+	–	–

Diagram the pathway and show which steps the various mutant block. Which compound would each mutant accumulate? When you complete this problem, compare your result. What effect on growth would you observe following mutation in the pathway of serine biosynthesis?

Solution:

? $\xrightarrow{2}$ indole $\xrightarrow{1}$ tryptophan $\xrightarrow{3}$ 3-hydroxyanthranilic acid $\xrightarrow{4}$ niacin.

Accumulation: 1 Indole; 2 ? 3 tryptophan or kynurenine; 4, 3 hydroxyanthranilic acid.

Without serine, the pathway would be blocked before the point of tryptophan production after indole.

Problem 5

The growth responses that follow were obtained by using four mutant strains of *Neurospora* and the related compounds A, B, C and D. None of the mutations grow on minimal medium. Draw all possible combinations:

	Growth Product			
Mutation	**C**	**D**	**B**	**A**
1	–	–	–	–
2	–	+	+	+
3	–	–	+	+
4	–	–	+	–

Solution:

(i) First nothing can be concluded about mutation 1, except that is lacking some essential growth factor, perhaps even unrelated to the biochemical pathway represented by mutation 2-4; nor can anything be concluded about compound C.

(ii) If it is involved in the pathway, it is a product synthesized prior to the synthesis of A, B and D.

(iii) We must now analyze these three compounds and the control of their synthesis by the enzymes encoded by genes 2, 3 and 4. Because product B allows growth in all these cases, it may be considered as the "end product".

(iv) It bypasses the block in all three instances. Using similar reasoning, product A precedes B in the pathway, since it allows a bypass in two or three steps. Product D precedes B, yielding the more complete solution.

C (?) ⟶ D ⟶ A ⟶ B

(v) Now, we have to decide which mutations control which steps. Since mutation 2 can be by-products D, B and A, it must control a step prior to all three products, perhaps the direct conversion to D, although we cannot be certain.

(vi) Mutation 3 is alleviated by B and A, so its effect must precede them in the pathway. Thus, we will assign it as controlling the conversion of D to A.

(vii) Likewise, we can assign mutation 4 to the conversion of A to B, leading to the more complete solution.

C (?) $\xrightarrow{2(?)}$ D $\xrightarrow{3}$ A $\xrightarrow{4}$ B

Problem 6

A series of mutations in the bacterium *Salmonella typhimurium* results in the requirement of either tryptophan or some related molecule in order for growth to occur. From the data shown here, suggest a biosynthetic pathway for tryptophan.

	Growth supplement			
Mutation	**Minimal medium**	**Pyrimidine**	**Thiazole**	**Thiamine**
thi-1	–	–	+	+
thi-2	–	+	–	+
thi-3	–	–	–	+

Solution:

The pathway, therefore, would be as follows:

AA ⟶ IGP ⟶ I ⟶ TRY

Problem 7

Four mutant strains of Neurospora require one or more of the supplemented metabolites A through D to grow. From the data below (where '+' = growth; '0' = no growth) draw the metabolic pathway for the synthesis of these metabolites and show, where the mutant strains (1, 2, 3 & 4) are blocked.

	Metabolite			
Mutant	**A**	**B**	**C**	**D**
1	+	0	0	0
2	+	+	+	+
3	+	0	+	0
4	+	+	+	0

Solution:

The pathway, therefore, would be as follows:

Precursor ⟶ D ⟶ B ⟶ C ⟶ A

- **Isochromosome:** A chromosome with identical arms and identical genes. The arms are mirror image of each other. It is usually produced by the horizontal division of centromere.
- **Compound chromosome:** A chromosome formed by the fusion or union of homologous chromosomes or from the fusion of the arms of homologous chromosomes is called compound chromosome. **Example:** Attached XY chromosome of *Drosophila* (Lillian Morgan discovered it).
- **Chromocentre:** Agregation of centromeres and adjacent heterochromatic elements of polytene chromosome is called chromocentre. **Example:** It is found in the polytene chromosome of *Drosophila*. It is absent in the polytene chromosome of *Anopheles* mosquitoes.
- **Chromosome banding:** Differential banding of the mitotic and meiotic chromosomes to produce a characteristic banding pattern (i.e., light and dark area occurs along the length of the chromosome) or selective staining of certain chromosomal region such as centromere, nucleolar organizer region and GC or AT rich region (not to be confused with the banding pattern present in polytene chromosome) is termed as chromosome banding.
- **Robertsonian translocation:** It result from fusion of nonhomologous chromosome.

Somatic Cell Genetics

8

Genetic study of cultured cells is called **Somatic Cell Genetics.**

I. Both undifferentiated and differentiated cell types can be cultured.

II. Cell hybridization or cell fusion produces binucleated cell (**heterokaryon**). It is followed by nuclear fusion to produce uninucleated hybrid cells (**Synkaryon**).

III. Several chemicals and virus (inactivated) are used in cell fusion are called **fusogen** (Polyethylene glycol = PEG, Sendai virus, Polyvinyl alcohol, etc.)

Somatic Hybridization

The process of in vitro fusion of protoplast (naked cell or cell without protoplast) derived either from somatic cell of same or two genetically different organisms is called somatic hybridization.

- **Main features of Somatic hybridization**

I. In this process somatic cells rather than sexual gametes are involved.

II. In this method tissue culture is essential for regeneration of somatic hybrid into whole organism.

III. Specific nutrient medium (differs from species to species) is required for regeneration of protoplast.

IV. It overcomes the barriers of cross incompatibility and permits hybridization between any two organisms whether related and unrelated.

V. In this method, there is equal contribution of cytoplasm from both parents, whereas in sexual hybridization cytoplasm is strictly contributed by maternal parent.

- **Types of Somatic hybrid**
- **Interspecific hybrids**

I. Hybrids betwen two different species of the same genus are referred to as **interspecific** hybrids.

II. Such hybrids have been obtained in sexually compatible as well sexually incompatible species.

(a) Sexually compatible: *Nicotiana tabacum* + *N. alata.*

(b) Sexually incompatible: *N. tabacum* + *N. nesophila.*

- **Intergenic hybrids**

I. Hybrids between two different genera of the same family are known as *intergenic* hybrids.

II. Such hybrids have been obtained between *Arabidopsis thaliana* + *Brassica campestris.*

- **Intertribal hybrids**

I. Hybrids between two different tribes are called **intertribal hybrids.**

II. It can only be produced by somatic hybridization.

- **Symmetrical hybrids**

I. Somatic hybrid cells that contain all chromosomes of both the species involved in the fusion are known to as **symmetrical hybrids.**

II. It has ben obtained in tobacco, carrot etc.

- **Asymmetrical hybrids**

I. Somatic hybrid cells that contain complete somatic complement of one species and only a part of somatic complement of another species are referred to as **asymmetrical hybrids.**

II. These hybrids have been obtained in tobacco, petunia plants.

- **Cybrids/Cytoplasmic hybrid/Heteroplast**

I. Somatic hybrid cells that contain normal protoplast of one species and enucleate (nucleus less) protoplast (cytoplast) of another species are known as **cybrids.**

II. In the experiment cybrid may also arise by

(a) Fusion between normal nucleate protoplast and a protoplast with nonviable nucleus.

(b) Elimination of one of the nuclei (cytochalasin treatment) after heterokaryon formation.

(c) Selective elimination of chromosome at later stage.

Somatic Hybridization

- **Basis of Somatic hybridization**

(a) Protoplast isolation.

(b) Fusion of protoplast to form viable heterokaryon.

(c) Fusion of nuclei and the division of hybrid cells and their subsequent growth.

(d) Identification, isolation or selection of hybrid cell.

- **Steps in Somatic hybridization**

(a) Isolation of protoplast.

(b) Fusion of protoplast.

(c) Hybrid identification.

(d) Selection of hybrid.

(e) Culture and regeneration of hybrid cells.

(f) Characterization of hybrid.

- **Steps of Somatic hybridization**

A. Protoplast isolation

I. Protoplasts are generally isolated by two methods, viz., mechanical and enzymatic.

II. Mechanical method is rarely used because a very small quantity of protoplasts are obtained by this method. Cells are often broken.

III. Enzymatic method is widely used because a large quantity of protoplast is obtained by this method. Cells are not broken.

IV. Sometimes both methods are used. In combined method, first cells are separated mechanically and then cell wall is dissolved enzymatically.

V. Removal of cell wall enhances the fusion of protoplast of diverse origin and uptakes of foreign DNA.

B. Protoplast fusion

I. It is a physical phenomenon.

II. Protoplasts must be isolated in large quantities and isolated protoplast should have regenerative property.

III. During fusion two or more protoplast come in contact and adhere with one another either spontaneously or in presence of inducing chemicals.

IV. After adhesion, membrane of protoplasts fuse in some localized areas and eventually the cytoplasm of two protoplast intermingle.

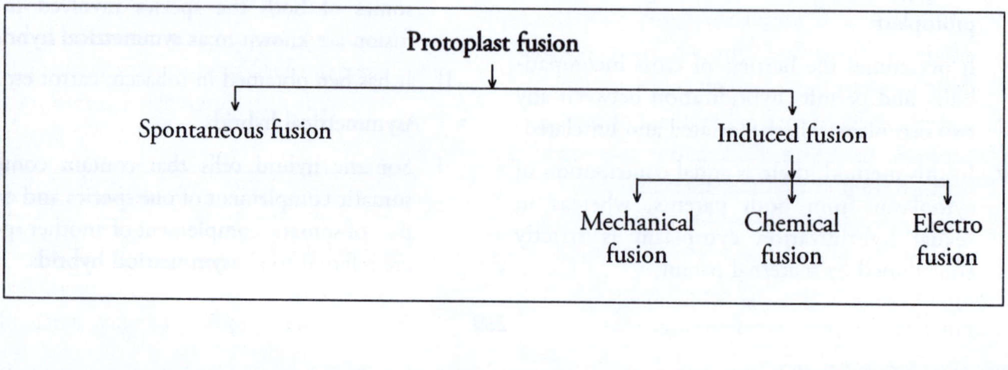

- **Homokaryon**: Union of the protoplast of the same species leads to the formation of homokaryon (fusion of nuclei)
- **Heterokaryon**: Union of the protoplast of the different species gives rise to heterokaryon.

  Nuclear fusion may occur and convert heterokaryon into a single nucleus, i.e., synkaryon.
- **Eukaryon**: Cell (Protoplast) contains nucleus of the same species.
- **Monospecific hybrid**: Somatic cell hybrids produced by fusion between two cells of the same species.
- **Interspecific hybrid**: Somatic cell hybrids produced by fusion between two cells or from different species.
- **Agent promoting cell fusion:**

  I. Inactivated virus: Sendai virus.

  II. Chemical agent: Polyethylene glycol (PEG).
- **HAT selection:** Hypoxanthine, Aminopterin and Thymidine.

  I. The human cells TK$^-$ (thymidine kinase is lacking) but have HGPRT$^+$ (Hypoxanthine-Guanine Phosphoribosyl Transferase) are crossed with mouse cell TK$^+$ but have HGPRT$^-$ (lac of HGPRT enzyme). Therefore, neither human or mouse cell can be able to grow in HAT medium.

  II. Nuclear fusion leads to form mouse-human hybrid cells (Synkaryon) which have HGPRT$^+$ because of presence of human chromosome with HGPRT gene and aso have TK$^+$ because of the mouse chromosome with TK gene.

  III. Thus only mouse-human hybrid cells survive in the presence of aminopterin of the HAT medium. Parental human TK$^-$ and mouse HGPRT$^-$ cells cannot survive in HAT medium and forms the basis of selection of hybrid.
- **G418 selection (Analogue of Neomycin):**

  I. Radiation hybrids can be selected for the presence of a particular human chromosome segment if it has been tagged by incorporation of a neomycin resistant (**Neo^R**) gene.

  II. **Neo^R** is a dominant selective marker.
- **Symmetric Hybrid:** Those somatic hybrid cells that contain all chromosomes of both the species involved in fusion are termed as symmetric hybrid. It is observed in tobacco and carrot. *Example: Datura innoxia + Atropa belladona.*
- **Asymmetric Hybrid:** Those somatic hybrid cells contain complete somatic complement of one species and only a part of somatic complement of another species is termed as symmetric hybrid. It is found in tobacco and petunia. *Example: Daucas carota + Oryza sativa.*

A. Spontaneous fusion:

During protoplast isolation, it fuses spontaneously and this phenomenon is called **spontaneous fusion**.

I. Simple physical contact is sufficient to bring about spontaneous fusion among the similar parental protoplasts.

II. During the enzyme treatment for isolation of protoplasts it is found that protoplasts from adjoining cells fuse through their plasmodesmata to form multinucleate protoplast.

III. Enlargement of plasmodesmatal connection between the adjacent cell allows the entry of organelles into neighbouring cells.

IV. Spontaneous fusion is strictly intraspecific and gives rise to **homokaryon.**

V. Spontaneous fusion has some practical importance in chromosome doubling.

B. Induced fusion:

Fusion of freely isolated protoplasts from different sources with the help of fusion inducing chemical agents is known as **induced fusion.**

I. Normally isolated protoplasts do not fuse with each other because the surface of the isolated protoplasts carries negative charge (–10 to –30 mV) outside of the plasma membrane.

II. Thus, there is strong tendency for protoplast to repel one another due to their same charges. So this type of fusion needs a fusion inducing chemical agents/systems which actually reduce the electronegativity of the isolated protoplasts and allow them to fuse with each other.

1. **Mechanical fusion**

I. In this process isolated protoplasts are brought into intimate physical contact mechanically under microscope using micromanipulator and perfusion micropipette.

II. This micropipette is partially blocked within 1 mm of the tip by a sealed glass rod.

2. **Chemofusion**

I. Several chemicals are used to induce protoplast fusion. Sodium nitrate ($NaNO_3$), polythylene glycol (PEG), calcium ions (Ca^{++}), polyvinyl alcohol, etc. are most commonly used protoplast fusion inducing agents which are commonly known as **Chemical fusogen.**

II. Generally chemofusion techniques are followed in most of induced fusion experiments.

III. Chemical fusogens cause the isolated protoplast to adhere to one another and leads to tight agglutination followed by fusion of protoplast.

3. **Electrofusion**

I. It is a modern technique of protoplast fusin which involves the use of mild electrical fields in protoplast suspension for inducing protoplast fusion.

II. This technique is very easy and simple.

III. It is applicable to those species whose protoplast exhibits a severe toxic response to polyethylene glycole used for chemofusion.

C. Selection of Hybrid:

I. After fusion, there is a mixture of parental types: homokaryon and heterokaryon.

II. For genetic analysis it is necessary to distinguish hybrid cells and unfused cells.

III. During selection procedure one must prevent growth of parent cells and permit growth of hybrid cells.

IV. Various methods are used for selection of heterokaryon or desired fusion product.

V. Non-toxic **flurochromes** (Fluorescein isothiocynate (FITC) or Rhodamine isothiocynate (RITC).

VI. For the recovery of hybrid cells, HAT (Hypoxanthine aminopterin thymidine) technique is generally used.

VII. Biochemical and molecular techniques such as isoenzyme pattern. RFLP, and PCR can also be used.

D. Culture and Characterization of Hybrid cells:

I. Hybrid cells are cultured in medium which is highly enriched with organic compound.

II. Auxin, cytokinins are always required in the culture medium.

III. In culture medium protoplasts regenerate a new cell within 2–4 days and cell division starts within 2–7 days depending upon the species.

Somatic Cell Hybridization through Cell fusion (Mouse and Human Cell)

I. G. Barski ands B Ephrussi (1960) developed the protocol of somatic cell hybridization.

II. At the present time hybrid cells are formed by mixing human cells with another animal cell, usually rodent (mouse or hamster). The cells are mixed in culture in the presence of polythylene glycol (Chemical induce fusion of cell membrane).

III. If a human cell fuses with rodent cell, the resulting cell will have the chromosome of both species. However, the human chromosomes are randomly eleminated during the cell division of hybrid. After many cycles of division the chromosome content of the cultured cells stabilizes.

IV. At this stage, each cell has a full complement of rodent chromosome and a random sample of human chromosome. Further division of a particular cell, that will produce a cell line carrying a characteristic set of chromosome 8, 13 and 17. Another cell line contains human chromosome 2, 6, 19 and 22.

V. To select interspecific hybrid cell, researchers grow the mixed cultures medium in which only such hybrid cells survive. In addition to nutritive materials, the preferred medium contains—hypoxanthine, aminopterin and thymidine. Here hybrid cells are capable of synthesizing DNA that undergoes cell division.

VI. The HAT medium prevents the growth of cells with some biochemical defect.

For example: Cells deficient in thymidine kinase (TK^-) or hypoxanthine phosphoribosyl transferase ($HPRT^-$) cannot grow in HAT medium.

VII. In particular cells deficient for either two enzymes thymidine kinase (TK) or hypoxanthine phosphorybosyl transferase (HPRT) cannot grow in HAT medium. Thus, if human cells that are TK^- (deficient in thymidine kinase) and mouse cells that are $HPRT^-$ (deficient in hypoxanthyne phosphoribosyl transferase) are placed in HAT medium, neither type will grow.

VIII. If two types of cells (TK^-, $HPRT^+$ × TK^+, $HPRT^-$) are fused, hybrid cell will grow, because the fused cell compensate for each other defects.
Here human cells deficient for TK enzyme (TK^-) but can produce HPRT (i.e., $HPRT^+$) but they can produce TK (i.e. TK^+).

IX. Here hybrid cells are formed by fusion of mouse and human cell. As a result they can produce both enzymes and are able to survive in HAT medium. Unfused cells (Human or mouse) or fused cells (mouse + mouse or human + human) are lost.

X. Hybrid cells have been selected by placing them in HAT medium. Individual cell lines can be established.

XI. By cytological analysis, human chromosomes are in this line can be established. Then each cell line can be tested for the presence of particular human gene. It is usually done by screening for the genes polypeptide product.

XII. In this way the presence of a gene can be corelated with the presence of a particular chromosome. Consequently, the location of gene to that chromosome can be identified.

- **Detection to locate Human Gene**

Experiment 1. Location of Human TK gene

Experiment was performed by Oscar Miller and his colleagues to detect the location of TK and HPRT genes.

Requirements:

I. Mouse cell—mutant for thymidine kinase (TK^-) and normal Hypoxanthine phosphoribosyl transferase ($HPRT^+$).

II. Human cell—normal for thymidine kinase (TK^+) and mutant for Hypoxanthine phosphoribosyl transferase ($HPRT^-$).

Procedure:

I. Mouse cell TK^- and $HPRT^+$ and human cell TK^+ and $HPRT^-$ were hybridized (fused) in culture.

II. Hybrid cells were formed by fusions. Hybrid cells were grown in HAT medium to select TK gene. During this period human chromosomes were randomly eliminated from the hybrid cell.

III. They were used to establish distinct cell lines which were then subjected to cytological analysis to determine which human chromosomes were present.

Mouse cell	×	**Human cell**
TK^+ $HPRT^-$	×	TK^- $HPRT^+$

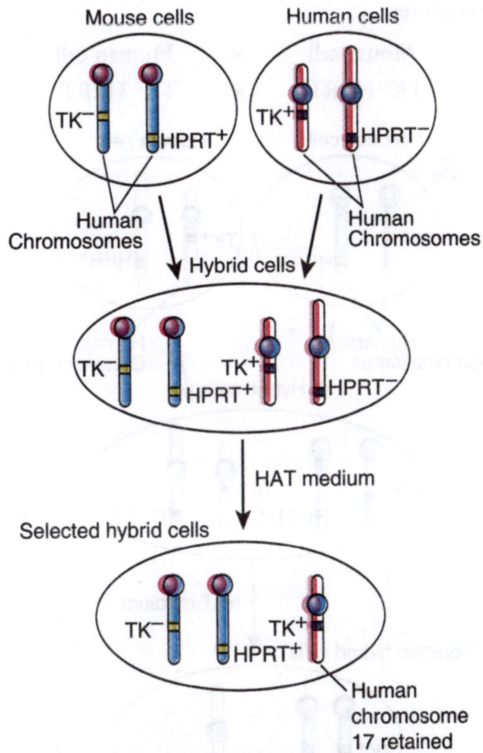

Fig. 8.1: Somatic cell hybridization to localize human TK^+ gene

Observation:

I. The selected cells carry the mouse $HPRT^+$ gene and human TK^+ gene.

II. In one of these selected cell lines human chromosome number 17 was found.

III Because this cell line survived in HAT medium and it must have been able to produce thymidine kinase (TK⁺) enzyme.

Inference:

Thus, the gene for thymidine kinase (TK⁺) enzyme must be located on chromosome number 17.

Experiment 2. Location of Human HPRT gene

Requirements:

I. Mouse cell—normal thymidine kinase (TK⁺) and mutant hypoxanthine phosphoribosyl transferase (HPRT⁻).

II. Human cell—mutant thymidine kinase (TK⁻) and normal hypoxanthine phosphoribosyl transferase (HPRT⁻).

Procedure:

Mouse cell	×	**Human cell**
TK⁺ HPRT⁻	×	TK⁻ HPRT⁺

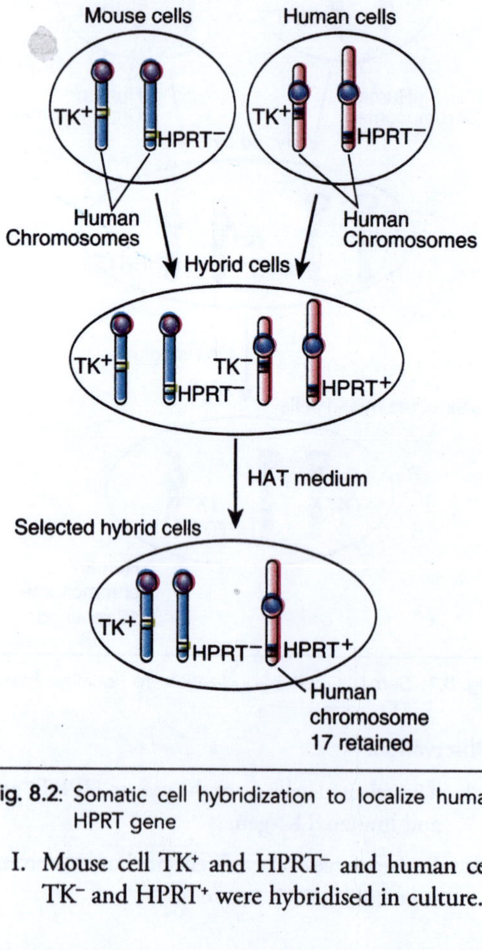

Fig. 8.2: Somatic cell hybridization to localize human HPRT gene

I. Mouse cell TK⁺ and HPRT⁻ and human cell TK⁻ and HPRT⁺ were hybridised in culture.

II. Hybrid cells were formed by fusions. Hybrid cells were grown on HAT medium to select HPRT⁺ genes. During that time human chromosomes were randomly eliminated from the hybrid cell.

III. They were used to establish distinct cell lines which were then subjected to cytological analysis to determine which human chromosomes were present.

Observation:

I. The selected cells carry the mouse TK⁺ gene and human HPRT⁺ gene.

II. In one of these selected cell lines human chromosome was found.

III. Because this cell line survived in HAT medium and it must have been able to synthesize.

Inference:

Thus, the gene for hypoxanthus phosphoribosyl transferase (HPRT) enzyme must be located on chromosome number 17.

Salvage pathway

I. Most animal cells can synthesize the purine and pyrimidine nucleotides *de novo* from deoxyribose or ribose sugar and amino acid rather than from already formed purine and pyrimidines.

II. When the *de novo* pathway is blocked, the salvage pathway converts purine and pyrimidine bases and nucleosides and nucleotides with the help of TK and HGPRT.

III. The folic acid antagonists **amethopterin and aminopterin** interfere with the donation of **methyl** ($-CH_3$) and **formyl** (–CHO) groups by **tetrahydrofolic acid** in the early stages of *de novo* synthesis of glycine, purines nucleoside triphosphates and **thymidine triphosphate**. These drugs are called **antifolates**.

IV. In this situation to the biosynthetic pathways, enzymes in most cells also can use purins and thymidine directly by **salvage pathways** that can bypass the metabolic blocks imposed by antifolates. The enzymes which are responsible to carry out salvage pathway are **Thymidine kinase** (TK) and **Hypoxanthine-guanine phosphoribosyl transferase** (HGPRT) and **Adenine phosphoribosyl transferase** (APRT).

V. A number of mutant cell lines have been discovered that are unable to carry out one of the salvage steps. For example, cell cultures lacking thymidine kinase (TK) have been selected and cultures have been established from human, who lacks APRT or HGPRT. These different types of salvage mutants become useful partners in cell fusion with one another or with the cells having salvage pathway enzymes, but that are differentiated and cannot grow the culture by themselves.

VI. The selective medium most often used to culture such fused cells is called HAT medium because it contains hypoxanthine (a purine) aminopterin and thymidine. Normal cell can grow in HAT medium. Salvage mutants cannot put their hybrid with normal cellscan (Fig. 8.3).

- **Limitation of Somatic Hybridization**

I. It does not always produce fertile plants.

II. Somatic hybridization can be used in species, where regeneration occurs, i.e., protoplast are **totipotent.**

III. Somatic hybrids are fertile (between closely related species), i.e., interspecific somatic hybrids are fertile and intergenic somatic hybrids are usually sterile.

IV. Viable somatic hybrid is not always possible. Generally, intergenic and intertribal hybrids exhibit high degree of sterility.

V. Somatic hybrids (interspecific) are mostly polypoid and often contain many unwanted traits. Several backcrosses are required to eliminate them.

VI. Limitation in selection method.

- **Features of Mouse-Human Somatic cell hybrids**
  - The mouse and human chromosomes are easily distinguished.
  - All the mouse chromosomes are usually retained in the hybrid cells but only few human chromosomes (1–20 chromosomes) are retained in hybrid clones.
  - Both (human and mouse) sets of chromosomes are expressed in hybrid cells.

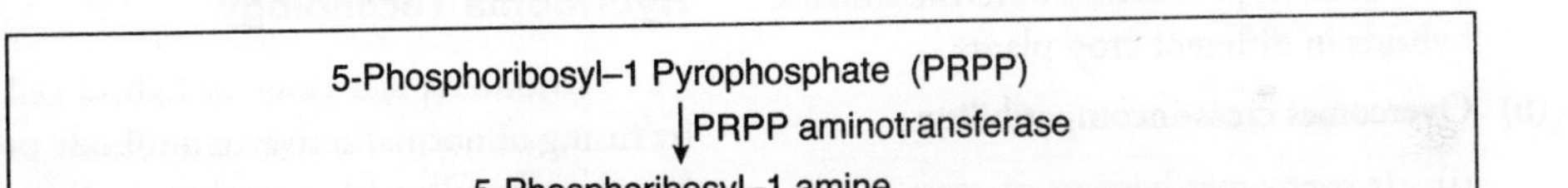

Fig. 8.3: Salvage pathways to obtain purine nucleotides and thymidylate

The hybrids known as 'pomato' or 'topato' is a small tomato like fruit, edible—either raw or cooked. It is developed by the fusion of tomato and potato protoplast. (Potato + tomato = Pomato)

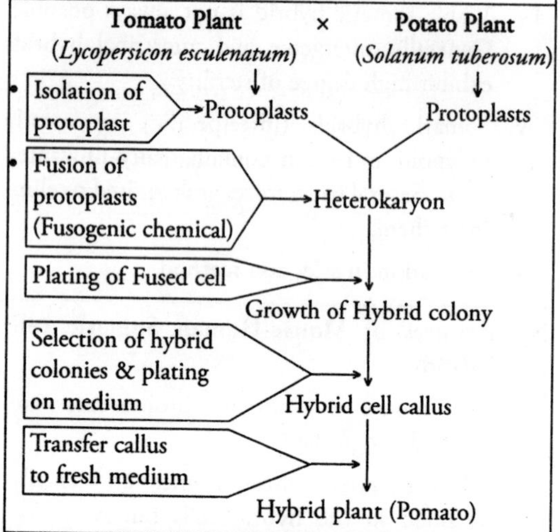

Development of Pomato plant I

- **Uses of Somatic Hybridization**

(a) Production of Somatic hybrids:

It is useful in production of fertile somatic hybrids in different crop plants.

(b) Overcomes cross-incompatibility:

(i) It overcomes barriers of cross incompatibility and makes possible hybridization between diverse organisms (plant).

(ii) It permits gene transfer between distantly related as well as unrelated species.

(iii) Thus, it is a powerful tool for modification and genetic improvement of polygenic character.

(c) Conservation of Heterosis:

(i) In somatic hybrids, segregation does not occur.

(ii) Therefore, heterosis can be easily conserved in such crossing.

(d) Modification of Cytoplasmic genes:

(i) In this method, cytoplasm is contributed equally by both the parents.

(ii) It (hybrid) contains a mixture of two partners.

(iii) Thus, modification of chloroplast and mitochondrial DNA is made in this method.

(e) Production of Allotetraploids:

(i) Somatic hybrids contain diploid chromosomes complements of both parent species.

(ii) This somatic hybridization is useful in the production of allotetraploid plants.

(f) Development of Transgenic hybrids:

(i) Through protoplast fusion, somatic hybrids are produced.

(ii) Genetic transformation of hybrid cells will lead to development of transgenic hybrids.

(g) Transfer of Cytoplasmic Male Sterility:

Somatic hybridization is useful on transfer of cytoplasmic male sterility from species to another.

Hybridoma Technology

Hybridoma is a clone of hybrid cell generated by fusing of normal activated antibody producing B lymphocyte cells with a myeloma cell (B-Lymphoid tumour cell) which retains the property of lymphocyte cell to produce antibody and exhibits immortal growth characteristics of myeloma cells.

For the production of a specific antibody, Kohler and Milstein fused the splenocytes (spleen cells) containing the B-lymphocyte and myeloma cells (lymphoid tumour cells) of the same strain and after fusion, it was raised in the HAT (Hypoxanthine-aminopterin-thymidine) medium. This technology is known as **Hybridoma technology**. This key technology has been devised by Kohlar & Milstein in 1975.

Objective:

I. To produce hybrid cell composed of an antibody producing lymphocyte and myeloma cell line of lymphocyte.

II. To produce specific antibody (monoclonal/ McAb) against predefined immunity from hybrid cell.

Requirement:

A. Mouse myeloma cells of B cells

(i) Cannot produce antibody

(ii) Contain genetic marker HGPRT⁻ trait.

B. B lymphocyte

Can produce specific antibody.

Procedure:

1. B lymphocyte cells of splenocytes are isolated from mouse spleen which are suitably immunised with antigen against which monoclonal antibodies are to be raised.
2. Mouse myeloma cells are selected. They are able to produce antibodies themselves. They lack genetic marker HGPRT⁻ (hypoxanthine guanine phosphoribosyl transferase) and cells are adapted to grow permanently in tissue culture medium.
3. In ordinary tissue culture medium splenocytes (spleen) cells quickly die whereas myeloma cells are adapted to grow permanently in the culture.
4. So far the production of specific antibodies against predetermined antigen with the immortalised myeloma cells and the B lymphocytes of splenocytes are fused with PEG or Sendai virus.
5. Fused cells are in incubated in the HAT medium for about 10-14 days.
6. In the HAT medium mutant form are taken. HGPRT⁻ myeloma cells are eliminated. In the HAT medium aminopterin usually blocks the *de novo* DNA synthesis. In the culture salvage pathway could be used to prepare cells having HGPRT cells component.
7. Mutant lack of HGPRT is unable to utilise salvage pathway (see Fig. 8.4).

Observation:

(i) After cell fusion three types of cells are observed (a) hybrid cell (hybridoma), (b) unfused myeloma cells, (c) unfused B lymphocyte cells.

(ii) As the cell population is cultured in HAT medium containing drug aminopterin, HGPRT⁻ myeloma cells are unable to divide. B lymphocyte cannot grow for a long period in the tissue culture medium.

(iii) In contrast only hybridoma cells are able to survive and proliferate in HAT medium since they have HGPRT gene (coming from B cell). These cells produce antibodies (a property of B cell) and are immortal (a property of myeloma cell).

(iv) Thus, the hybridoma (myeloma + B lymphocyte) are selected on HAT medium.

Inference:

The resulting clones of hybridoma cells, which secrete large quantities of monoclona antibody, can be cultured indefinitely.

This technology gave immunologists a powerful and versatile research tool. It is useful as diagnostic, imaging and therapeutic reagents in clinical medicine (e.g., Detecting pregnancy, diagnosing numerous pathogenic organisms, measuring blood levels of various drugs, histocompatibility antigens matching detecting antigens shed by certain tumours, etc.)

Somatic Hybridization	Sexual Hybridization
I. It involves fusion of protoplast (somatic cell).	I. It involves fusion of male and female gametes.
II. Cytoplasm is contributed by both parents.	II. Cytoplasm is exclusively contributed by female parents.
III. Tissue culture technique is required.	III. Tissue culture technique does not require.
IV. Hybrid contains somatic chromosomes of both parents.	IV. Hybrid contains haploid chromosomes of both parents.
V. Segregation does not occur.	V. Segregation occurs.

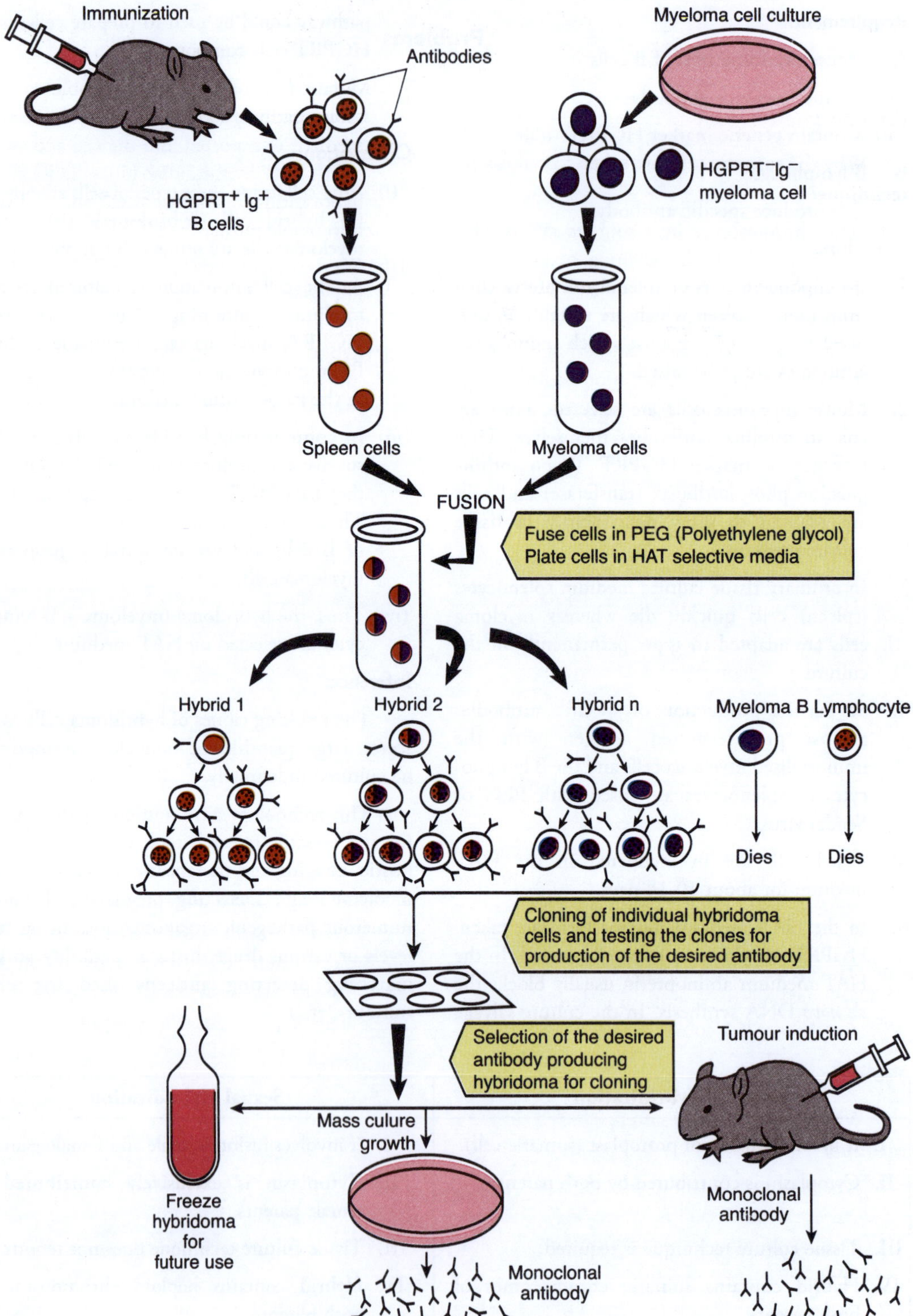

Fig. 8.4: Hybridoma Technology

Problems

Problem 1

Why the somatic cells are taken in cell fusion technique?

I. The chromosomes in a somatic cell is large and highly visible after staining, making it easy to identify them by type and to observe their distinctive arrangements.

II. Moreover, somatic cells can be fused to form heterokaryon (two nuclei in one cell).

- **Selection of Chromosome contents of Hybrids**

I. Hybrids can be selected for retention of a given human chromosome or chromosome fragments if it corrects and otherwise lethal abnormality will occur in rodent cells.

II. It occurs through HAT selection medium and G418 selection medium.

III. In HAT selection somatic cells hybrids can be forced to retain human chromosome 17 by using thyminidine kinase mutant (TK^-) rodent cells and growing the hybrids in HAT medium. TK mutant (TK^-) cells are killed in HAT medium but are rescued by human TK^+ gene on chromosome 17.

IV. In G418 selection, hybrids can be selected for the presence of a particular human chromosome segment if it has been tagged by incorporation of neomycin resistance (Neo^R) gene.

V. The neomycin analog G418 kills non-resistant cells. A typical example of a dominant selective marker is Neo^R.

Problem 2

Why hybrid cells are often selected on HAT medium?

I. The selective medium most often used to culture hybrid cell is called HAT medium because it contains **hypoxanthine** (a purine) **aminopterin** and **thymidine**.

II. Normal cells can grow in HAT medium because even though **aminopterin** blocks *de novo* synthesis of purines and thymidine monophsphate (TMP), the thymidine in the media is transported into the cell and converted to TMP by thimidine kinase (TK) and the hypoxanthine is transported and converted into usable purines by HGPRT.

III. On the other hand, TK^- or $HGPRT^-$ cells cannot grow in HAT medium because each lacks an enzyme of the salvage pathway. However, hybrids formed by the fusion of these two mutants will carry a normal TK^+ gene from the $HGPRT^-$ parent and normal $HGPRT^+$ gene from TK^- parent. Thus, the hybrid will produce both functional salvage pathway enzymes and grow on HAT medium.

IV. Likewise hybrids formed by fusion of mutant cells and normal cells can grow on HAT medium. Somatic cell hybrids can be forced to retain human chromosome 17 by using thymidine kinase deficient (TK^-) rodent cells and growing hybrids on HAT medium.

$$\begin{array}{ccc} \text{Mouse} & & \text{Human} \\ \dfrac{TK^-\ \ HGPRT^+}{TK^-\ \ HGPRT^+} & \times & \dfrac{TK^+\ \ HGPRT^-}{TK^+\ \ HGPRT^-} \\ & \downarrow & \\ & TK^+\ \ HGPRT^+ & \end{array}$$

Problem 3

Why the normal cells can grow in the HAT medium, mutant cells cannot grow in the HAT medium but the hybrid cells can grow in the HAT medium. Explain with example.

Solution

HAT medium is the medium provided with hypoxanthine, aminopterin and thymidine. The normal cells are provided with salvage pathway enzymes HGPRT (Hypoxanthine guanine phosphoribosyl transferase) TK (Thymidin kinase) and APRT (Adenine phosphoribosyl transferase).

I. HGPRT converts hypoxanthine (base) present in the medium to a **purine** and if medium is provided with thymidine, the enzyme TK converts it to the nucleotide **thymidylate.**

Similarly, the enzyme APRT converts adenine to **adenylate** so that the normal cells are able to grow in the HAT medium.

II. The mutant cells lack any salvage pathway enzyme producing gene. In mammalian cells, aminopterin of HAT medium act as an inhibitor of enzymes involved in DNA metabolism. Therefore, the mutant cells cannot follow *de novo* synthesis of purine primidine nucleotide. So they have to follow the secondary salvage pathway. As the mutant cells lack salvage pathway enzymes, they cannot follow this pathway. That is why the mutant cells cannot grow in the HAT medium.

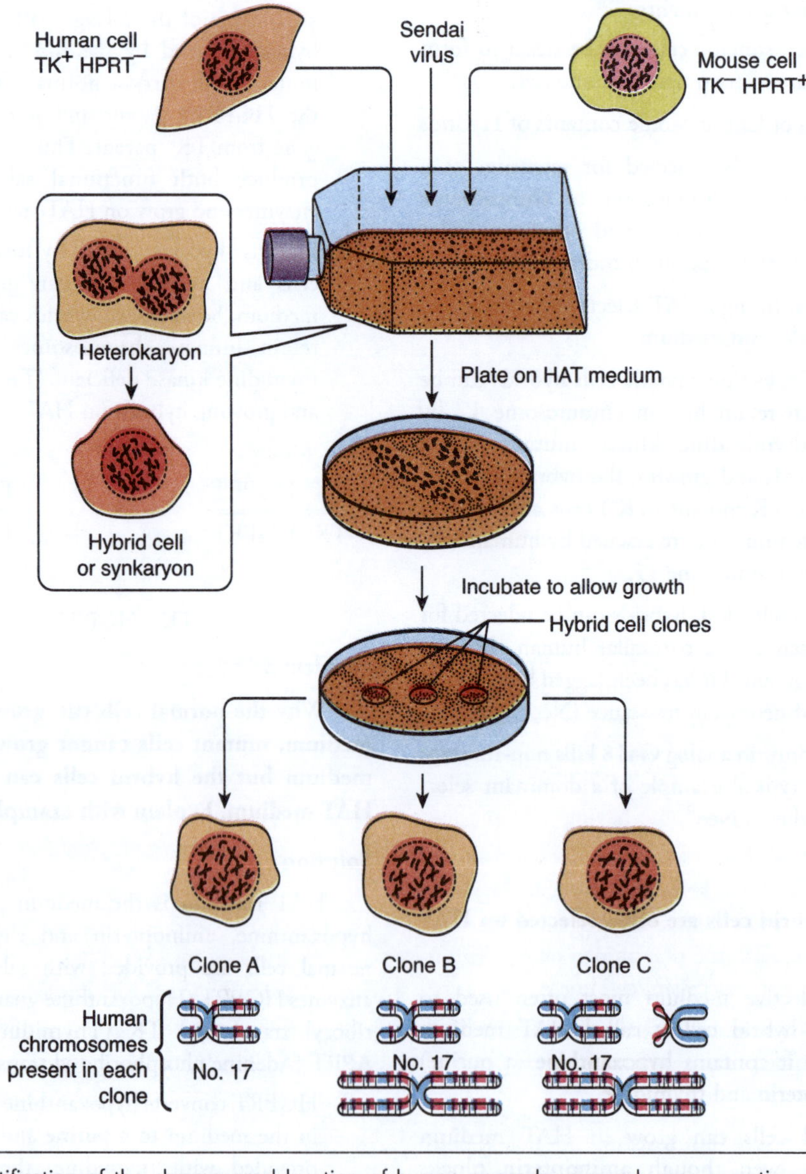

Fig. 8.5: Identification of the chromosomal location of human genes by mouse-human somatic-cell hybridization using the HAT selection procedure (after Gardner, Simmons and Snustad, 2001).

The selective medium most often used for culturing such fused cell is HAT medium which contains hypoxanthine (a purine), aminopterin and thymidine. Normal cells can grow in HAT medium but salvage mutants cannot grow in such medium. On the other hand, hybrid cells with normal cells can grow.

Explanation with Example:

I. Normally in mammalian cells aminopterin acts as an inhibitor of enzymes involved in DNA metabolism. HGPRT, TK enzymes can bypass aminopterin inhibition by making use of secondary or salvage pathway in the cell.

II. Normal cells in the absence of aminopterin can synthesize DNA even if they lack HGPRT activity ($HPRT^-$) or TK activity (TK^-).

III. In the presence of aminopterin $HGPRT^-$ and TK^- cell die. However, in the presence of aminopterin, $HGPRT^+$ and TK^+ cells can synthesize DNA, and survive in the HAT medium.

IV. Mouse cells that have the phenotype of $HGPRT^+$ and TK^- mixed with human cells that have the phenotype of $HGPRT^-$ and TK^+ in the presence of **Sendai virus** or **Polyethylene glycol**.

V. Fusion takes place in some of the cells and the mixture is grown in HAT (Hypoxanthine, Aminopterin and Thymidine) medium.

VI. In the presence of aminopterin unfused mouse cells (TK^-) and unfused human cells ($HGPRT^-$) cannot survive. However, the hybrid cells can survive, because they are $HGPRT^+$ and TK^+. The hybrid cell contains the human chromosome having TK^+ producing gene.

VII. This HAT method not only selects for hybrid clones but also localizes the TK gene to human chomosome 17, the one human chromosome found in every successful cell line.

Thus, the TK gene has been localized in the 17th chromosome in mouse-human hybrid cell.

Problem 4

In a research laboratory, it has been observed that cultured human cells produce an enzyme, called isomerase. Mouse cells do not produce this enzyme. When human cells are fused with mouse cells and human-mouse hybrid cell lines are established, isomerase 6 is detached in some of the hybrid cell lines, but not in others. The result screening six hybrid cell lines for isomerase 6 are shown below. From these results, which human chromosome carries the gene for isomerase 6?

Hybrid cell lines	Human chromosome present	Isomerase 6 present
A	1, 4, 7, 19, 22 X	No
B	3, 6, 7, 21, 22 X	Yes
C	3, 12, 21 X	No
D	6, 14, 15, 18 X	Yes
E	7, 14, 15, 19 X	No
F	5, 12, 15, 17, 21 X	No

Solution

The gene in chromosome 6 is present in cell lines B and D, but not any of the other cell lines.

Problem 5

A particular human enzyme is present only in clone B. The human chromosomes present in clones of B and C appear as pluses (+) in the following table. Determine the probable chromosomal location of the gene for the enzyme.

	Human chromosome							
Clone	1	2	3	4	5	6	7	8
A	+	+	+	+	–	–	–	–
B	+	+	–	–	+	+	–	–
C	+	–	+	–	+	–	+	–

Solution

I. If a gene is located on a chromosome, the gene must be present in the clones with the chromosome(+).

II. Chromosomes 1, 2, 5, 6 are present in clone B(+).

III. The gene in chromosome 1 have already been present in clone A, B, C. Similarly, gene in chromosome 2 is present as clones A and B.

IV. The chromosome 5 is also present in clones B and C.

V. The only chromosome that is unique to clone B is 6.

Therefore, gene is located on chromosome 6.

Problem 6

In a research laboratory in Kolkata, scientist have selected three mouse human hybrid clones and analyzed thus for the presence of human chromosomes. Analyze each clone for the presence or absence of particular human enzymes (+ = presence of human chromosomes or enzyme activity). Based on the following results indicate the probable chrmosomal location for each enzyme.

	Human chromosome						
Clone	**3**	**7**	**9**	**11**	**15**	**18**	**20**
X	–	+	–	+	+	–	+
Y	+	+	–	+	–	+	–
Z	–	+	+	–	–	+	+

	Enzyme				
Clone	**A**	**B**	**C**	**D**	**E**
X	+	+	–	–	+
Y	+	–	+	+	+
Z	–	–	+	–	+

Solution

I. Enzyme A is present in clones X and Y. The chromosome 11 is common to these two (X and Y) clones, therefore, enzyme is on chromosome 11.

II. Enzyme B is present only in X. Chromosome 15 is the unique to X. So the enzyme B is in present in Chromosome 15.

III. Enzyme C is present in clones Y and Z. The chromosome 18 is common to these two (Y and Z) clones. Therefore enzyme C is on chromosome 18.

IV. The enzyme D is present only in clone Y. The chromosome 3 is unique for D. Therefore, it is on chromosome 3.

V. The enzyme E is present in clones X, Y and Z. The chromosome 7 is common for all three clones (X, Y, Z). Therefore, enzyme E is on chromosome 7.

Problem 7

In a molecular laboratory at B.H.U., somatic hybridization results five-human hybrids for five enzymes are given in table along with the human chromosomal conduct of each clone. (+ = enzyme or chromosome present; – = absent) Deduce the location of gene on chromosome.

	Human chromosome																					
Clone	**1**	**2**	**3**	**4**	**5**	**6**	**7**	**8**	**9**	**10**	**11**	**12**	**13**	**14**	**15**	**16**	**17**	**18**	**19**	**20**	**21**	**22**
A	–	–	–	–	+	+	+	+	–	+	–		–	–	+	+	–	–	–	–	+	+
B	+	+	–	+	–	–	–	+	–	–	+		–	+	–	–	+	–	–	+	–	–
C	–	–	–	+	–	–	+	–	–	+	–		+	+	+	–	+	–	+	–	–	+
D	+	–	+	–	+	–	–	–	–	+	–		–	+	+	–	–	+	+	+	+	–
F	–	–	–	+	–	–	–	–	+	+	+		–	+	–	+	–	+	–	+	+	+

	Clone				
Human enzyme	**A**	**B**	**C**	**D**	**E**
P = Glutathione reductase	+	+	–	–	–
M = Malate dehydrogenase	–	+	–	–	–
Q = Adenosine deamanase	–	+	–	+	+
R = Galacto kinase	–	+	+	–	–
S = Hexo kinase	+	–	–	+	–

Solution

I. The enzyme P (glutathion reductase) is present in clones A and B. The chromosome 8 is common to both (A & B) clones. Therefore, enzyme P (glutathion reductase) is on chromosome 8.

II. The enzyme M (malate dehydrogenase) is present only in clone B. Chromosome 2 is unique for clone B.
Therefore, enzyme M (malate dehydrogenase) is on chromosome 2.

III. The enzyme Q (Adenosine deaminose) is present in clone B, D and E. The chromosome 20 is common to B, D and E. So, the enzyme Q adenosine deaminose) in chromosome 20.

IV. The enzyme R (Galactokinase) is present in clones B and C. Chromosome 17 is common to clones B and C. Therefore, enzyme R (galactokinase) is located on chromosome 17.

V. The enzyme S (Hexokinase) is available in clones A and D. Chromosome 5 is common to clones A and D.
Therefore, the enzyme S (Hexokinase) is on chromosome 5.

Enzyme	Chromosome
A = (Glutathione reductase)	8
M = (Malate dehydrogenase)	2
Q = (Adenosine deaminase)	20
R = (Galactokinase)	17
S = (Hexokinase)	5

Problem 8

Three mouse cell lines were scored for the presence (+) or absence (–) of human chromosomes with the results as follows:

	Human chromosome							
Clone	**1**	**2**	**3**	**4**	**5**	**14**	**15**	**18**
A	+	+	+	+	–	–	–	–
B	+	+	–	–	+	+	–	–
C	+	–	+	–	+	–	+	–

If a particular grene is located on chromosomes 3 which clone should be positive for the enzyme from the gene.

Solution

The clones A and C are positive on chromosomes are 3.

Problem 9

Data on the given table indicated the thymidine kinase activity of some mouse-human hybrid clones. Presence thymidine kinase activity is marked by '+' sign and lack of thymidine kinase activity indicated '–' sign.

		Human chromosome									
Clone	**Thymidine kinase activity**	**X**	**Y**	**1**	**2**	**4**	**7**	**10**	**15**	**17**	**21**
1	+	+	+	+	+	+	+	–	+	+	+
2	+	+	–	+	–	–	+	+	–	+	–
3	–	–	+	–	+	–	+	–	+	–	+
4	+	+	–	–	–	+	–	–	–	+	+
5	+	+	–	–	–	–	–	–	–	+	–
6	–	–	+	+	+	–	+	+	–	–	+
7	+	+	+	+	–	–	+	+	–	+	+

Identify thymidine kinase activity and chromosome.

Solution:

I. The clones 3 and 6 have lack of thymidine kinase activity.

II. In chromosome 17 has also lack of thymidine kinase activity. So the chromosome 17 has the direct correlation with thymidine kinase activity.

Problem 10

In a research laboratory of Delhi University South Campus, a number of human-mouse somatic cell clones (A–F) were examined for the expression of specific human genes and the presence of human chromosomes. The results are summerized below

Hybrid Cell Clone

Genes expressed	**A**	**B**	**C**	**D**	**E**	**F**
ENO (enolase-1)	–	+	–	+	+	–
MDH 1 (malate dehydrogenase-1)	+	+	–	+	–	+
PEPS (peptidase S)	+	–	+	–	–	–
PGM 1 (phospho glucomutase)	–	+	–	+	+	–
Chromosome (present or absent)						
1	–	+	–	+	+	–
2	+	+	–	+	–	+
3	+	+	–	–	+	–
4	+	–	+	–	–	–
5	–	+	+	+	+	+

Assign each gene to the chromosome upon which it is located.

Solution:

I. Enolase 1 (ENO-) is expressed in hybrid cell, i.e., coincides with chromosome 1, so it is located on chromosome 1. MDH-1 are located on chromosome 2. PEPS- are located on chromosome 4. PGM 1 also locatd on chromosome 1.

II. Eno 1 (enolase-1) is expressed in clone B, D and E. The chromosome 1 is common to these (B, D and E) clones. Therefore the enzyme Eno-1 is on chromosome 1.

III. The enzyme MDH 1 is present in clone A B D and F. The chromosome 2 is common to them (A B D & F). So it is on chromosome 2.

IV. The enzyme PEPs is expressed only in clone A and C. The chromosome 4 is common to them (A and B). So it is on chromosome 4.

V. The enzyme PGM 1 is present in clones B D and E. It is common to chromosome 1.

So it is also in chromosome 1.

Enzyme	Chromosome
ENO-1	1
MDH-1	2
PEPs	4
PGM 1	1

Problem 10

You fuse splenocytes (spleen cells) having a normal genotype for immunoglobin heavy chains (H) and right chains (L) with three myeloma cell preparations differing in their immunogobin genotype as follows: (a) H^+, L^+ (b) H^+, L^- and (c) H^-, L^-. For each hybridoma predict how many unique antigen binding sites, composed of one H and one L chain, theoretically could be produced and show the chain structure of the possible antibody molecules. For each possible antibody molecule indicate whether the chains would originate from the spleen (S) or from the myeloma (M) fussion partner (e.g., Hs Ls/Hm Lm).

Solution

(a) Four different antigen binding sites and 10 different antibody molecules: (i) Hs Ls/Hs Ls, (ii) Hm Lm/Hm Lm, (iii) Hs Lm/Hm Ls, (iv) Hm Ls/Hm Ls, (v) Hs Ls/Hm Lm, (vi) Hs Ls/Hs Lm, (vii) Hs Ls/Hm Ls, (viii) Hm Lm/Hs Lm, (ix) Hm Lm/ Hm Ls, (x) Hs Lm/Hm Ls.

(b) The different antigen-binding sites and 3 different antibody molecules: (i) Hs Ls/Hs Ls, (ii) Hm Ls/Hm Ls, (iii) Hs Ls/Hm Ls.

(c) One binding site one antibody molecule: Hs Ls/Hs Ls.

Quantitative Genetics

9

- **Quantitative Genetics**

A branch of genetics which deals with the study of the inheritance and genetic analysis of quantative or polygenic characteristics that are determined by multiple genes is called quantitative genetics.

The analysis of quantitative gonctics is based on means, variances and covariances.

- **Advantages of Quantitative Genetics**

It provides solution to the analysis of polygenic characters which is not possible through Mendelian genetics.

- **Limitation of Quantitative Genetics**

I. Polygenic characteristics are sensitive to environmental variation.

II. Therefore, quantitative estimates are also valid for a particular set of environmental condition.

III. These estimates may change with the change in the environment.

- **Continuous Trait**

I. This trait demonstrate true continuous variation, i.e., from one phenotypic extreme to the other (from minimum to maximum) with non-existance of clear-cut boundaries.

II. There is a continuous gradation from one phenotype to the next.

III. They are influenced by environmental factors.

Example: Weight of an organism, blood pressure of human.

- **Meristic Trait**

I. This type of trait does not demonstrate continuous variation.

II. The phenotypes are placed into discrete categories due to the intrinsic nature of the character.

III. The value of the phenotype of such trait is determind by counting. This is why they are known as **countable** character.

Example: Number of skin ridges forming fingerprint.
Number of bristles in the abdomen of fly.

- **Threshold (dichotomous) Trait**

I. These quantitative traits exhibit only two or a few phenotypic classes.

II. Their inheritance is determined by multiple genes and environmental factors.

- **Quasi Continuous Trait**

I. The polygenic trait having underlying continuous distribution but threshold effect makes it discontinuous appearances.

II. Such trait is said to be a **quasicontinuous trait.**

A rough estimate of the number of gene loci contributing to a quasi-quantitative trait can be obtained by determining the fraction of the F_2 (resulting from selfing the F_1 hybrid between two pure varieties) which is as extreme in its phenotype as that of the pure parental strains.

Number of gene loci	1	2	3	....	n
Fraction of F_2 as extreme as one parent	$\frac{1}{4}$	$\frac{1}{16}$	$\frac{1}{64}$	....	$\left(\frac{1}{4}\right)^n$

Qualitative trait	Quantitative trait
1. It is influenced by a single gene.	1. It is influenced by a multiple gene.
2. Phenotype of this trait falls into distinct categories. (Presence or absence of disease).	2. Phenotype of this trait is expressed on a continum.
3. Character's are of kind.	3. Character's are of degree.
4. Discontinuous variation, distinct phenotypic classes are present.	4. Continuous variation, phenotypic measurement from a spectrum.
5. Typically follows simple pattern of inheritance.	5. Does not follow simple pattern of inheritance.
6. Trait expression is typically unaffected by environmental factors.	6. Trait expression may be altered by environmental factors.

- **Glyph and metroglyph analysis:**

A small circle which depicts the position of a genotype on the graph is called glyph. The rays on the glyph indicate variation in various character.

Metroglyph is a semigraphic method of studying the pattern of morphological variability in larger number of germplasm lines or segregating population at a time.

- **Main features of metroglyph analysis:**

I. This is semigraphic approach.

II. The analysis is very simple.

III. The variability is depicted by glyph on the graph.

IV. Estimates are based on first order statistics.

- **D^2 statistics**

I. The statistical procedure which is used to measure the genetic diversity in a population is known as D^2 statistics.

II. It measures variation at two levels i.e. (a) Intracluster variaion, (b) Intercluster variation.

III. It also measures relative contribution of each character to the total divergence.

- **Main features of D^2 statistics**

I. It is a numerical approach.

II. Estimates are based on second order statistics.

III. The genetic diversity is depicted by cluster diagram.

IV. Analysis is possible from replicated data only.

- **Cluster diagram**

I. A line diagram which is drawn in D^2 statistics to depict the genetic diversity for various characters is called cluster diagram.

II. The greater the distance between the two cluster the greater is the genetic divergence and vice versa.

Metrograph Analysis	D^2 statistics
I. Estimates are based on first order statistics.	I. Estimates are based on second order statistics.
II. Analysis is very simple.	II. Analysis is complicated.
III. It is a semigraphic approach.	III. It is a numerical approach.
IV. Variability depicted by glyph on the graph.	IV. Genetic diversity is depicted by cluster diagram.

Polygenic Inheritance

- **Polygenic character and Polygenic inheritance:**

I. Characters which are governed by several genes and exhibit continuous variation are known as **polygenic** or **qantitative** character.

II. They show a normal or bell-shaped curve.

The mode of inheritance of polygenic or quantitative characters is called polygenic or quantitative or multifactor inheritance.

- **Oligogenic character and Oligogenic inheritance:**

Characters which are governed by one or few genes and exhibit discontinuous variations are called oligogenic or qualitative character.

The mode of inheritance of oligogenic or qualitative character is called oligogenic or qualitative inheritance.

- **Polygene and Oligogene:**

Several genes governing the same character are referred to as polygene.

One or few genes governing the same charcters are called Oligogene.

- **Main features of Oligogenic characters:**

I. They are governed by one or few genes.

II. They exhibit discontinuous variation.

III. Such characters are little influenced by environmental factors.

IV. The effect of each gene is easily detectable.

V. They do not exhibit transgression.

- **Main features of Polygenic characters:**

I. Such characters exhibit continuous variation.

II. They are governed by several genes.

III. Such characters are very sensitive to environment.

IV. The effect of individual gene is not detectable.

V. They exhibit transgressive segregation.

- **Transgressive segregation and transgressive segregants**

The phenomenon through which the appearance of F_2 individual with a lower or higher intensity of character than those present in both the parents involved in the cross is called transgressive segregation.

Segregrants which fall outside the range of both the parents for a character is known as **transgressive segregants.**

- **Features of Transgressive segregation**

I. Transgressive segregants are only possible from the crosses between two parents having mean value for a quantitative character.

II. Transgressive segregation results due to the fixation of dominant and recessive genes in separate individuals.

III. It occurs when parents are intermediate to the extreme values of the segregating population.

Parents: AAbb × aaBB

Gametes Ab aB

F_1 AaBb

F_2 = AaBb × AaBb

F_2 AABB and aabb (transgressive segregrants).

Example of Polygenic inheritance

Polygenic inheritance has been reported for various characters in plants and animals. It includes kernel colour in wheat and skin colour in man.

A. Kernel colour in wheat

I. A whole grain or seed of a cerel plant (corn, wheat, etc.) is called **kernel.**

II. Kernel colour in wheat is a quantitative traits and is controlled by two different genes.

H. Nilson Ehle (1908) studied the inheritance of kernel colour of wheat.

Requirement:

I. Dark red+ karnel variety (RRCC)

II. White karnel variety (rrcc)

Procedure:

I. H. N. Ehle crossed red and white kernel varieties and observed that F_1 was intermediate (light red) between the parental characters.

II. F_1 varieties were inbred & observed the F_2 result.

Parent	Dark red RRCC	×	White rrcc
Gametes	RC		rc
F_1	RrCc (Light red)		

$F_2 = F_1 \times F_1$

	Light red RrCc		Light red RrCc
Gametes	RC, Rc, rC, rc;		RC, Rc, rC, rc

♀ \ ♂	RC	Rc	rC	rc
RC	RRCC	RRCc	RrCC	RrCc
Rc	RRCc	RRcc	RrCc	Rrcc
rC	RrCC	RRCc	rrCC	rrCc
rc	RrCc	Rrcc	rrCc	rrcc

Observations

Genotype	Genotypic ratio	Number of contributing alleles for red	Phenotype	Phenotypic ratio	
RR CC	1	4	dark red	1	$\frac{1}{16}$
Rr CC Rr CC	2	3	medium dark red	4	$\frac{4}{16}$
RR Cc RR Cc	2				
Rr Cc Rr Cc Rr Cc Rr Cc	4	2	medium red	6	$\frac{6}{16}$
RR cc	1				
rr CC	1				
Rr cc Rr cc	2	1	light red	4	$\frac{4}{16}$
rr Cc rr Cc	2				
rr cc	1	0	white	1	$\frac{1}{16}$

Inference:

From the checkerboard it has been found that

I. Each of the contributing alleles R or C adds some red to the phenotype of kernel colour, so that genotypes of the white contain neither of these alleles.

II. Seed colour was **dark red** where **4 (four)** effective or contributing alleles were present. Similarly, where **3 contributing** alleles were present, the kernel colour was **medium dark red**. Presence of 2 effective alleles the colour of kerner was **medium red** and with **one** contributing allele, kernel colour was **light red**.

III. White kernel colour was developed in the absence of all the non-effective allele.

IV. The phenotypic ratio of the F_2 is 1 : 4 : 6 : 4 : 1. This ratio is a product of **binomial expansion**.

B. Skin colour in man: Skin pigmentation in human varies from light to dark.

I. Skin colour in human is under polygenic effect. It is controlled by two or more than two, possibly four-five genes.

II. The inheritance of skin colour in man was studied by C. V. Davenport (1913) in Jamaica.

III. He observed that two pairs of gene A-a and B-b cause the difference in skin pigmentation between Negro and Caucasian people.

IV. These genes were found to affect the character in additive fashion, thus a true Negro has four dominant genes (AABB) and Caucasians have four recessive genes (aabb).

Requirement

(a) Negro people (AABB)

(b) Caucasian people (aabb).

Procedure

I. A cross between Negro and Caucasians gives birth to a child with an intermediate skin colour termed **mulatoes.**

II. A mating of two such mulatoes produces wide variety of skin colour in the offspring ranging from skins as dark as Negro parent to as white as Caucasian parent.

Parent	Negro (black) × AABB	Caucasian (white) aabb
Gametes	AB	ab
F_1	AaBb (Mulato)	medium/intermediate (Skin colour)

$F_2 = F_1 \times F_1$

	Mulato × AaBb	Mulato AaBb
Gametes	AB, Ab, aB, ab;	AB, Ab, aB, ab

♀ \ ♂	AB	Ab	aB	ab
AB	AABB	AABb	AaBB	AaBb
Ab	AABb	AAbb	AaBb	Aabb
aB	AaBB	AaBb	aaBB	aaBb
ab	AaBb	Aabb	aaBb	aabb

Observations

Genotype	Genotypic ratio	Number of contributing or effective alleles for skin colour	Phenotype	Phenotypic frequency	
AA BB	1	4	Black	1	$\frac{1}{16}$
Aa BB, Aa BB	2	3	dark	4	$\frac{4}{16}$
AA Bb, AA Bb	2	3			
Aa Bb, Aa Bb, Aa Bb, Aa Bb	4	2	Mulato (medium)	6	$\frac{6}{16}$
aa BB	1	2			
AA bb	1	2			
Aa bb, Aa bb	2	1	light	4	$\frac{4}{16}$
aa Bb, aa Bb	2	1			
aa bb	1	0	white	1	$\frac{1}{16}$

From the checkerboard it is found that one out of sixteen has four black alleles result black skin. Four out of sixteen have three colour alleles that produce dark colouration. Six out of sixteen have two colour alleles develop intermediate, i.e. mulato skin colour. Four out of sixteen have one colour allele which causes light skin colour. One out of sixteen has no colour gene and resembles white skin colour.

Inference:

These result clearly indicated that A and B alleles produce same amount of pigment for dark colouration of skin. Therefore, the increase or decrease of the A and B genes result the variable skin colour in F_2 in the ratio of 1 (Negro) : 4 (dark) : 6 (Mulato) : 4 ((light) : 1 (white).

Problems

Problem 1

In an agriculture farm in Kalyani, certain stock of wheat grain colour is controlled by four loci according to an additive model. How many different gametes can a tetrahybrid produce? How many different genotypes will result if tetrahybrids are self-fertilized? What will be the phenotypic distribution of these genotypes?

Solution

I. Assuming P, Q, R and S loci with P and p, Q and q, R and r, and S and s respectively.

II. A gamete can get either allele at each four independently assorting loci, so there are 2^4 = 16 different gametes.

III. Three genotypes are possible for each locus, two homozygote and a heterozygote. Therefore, for four independent loci, there are $3^4 = 81$ different genotypes.

IV. Phenotypes are distributed according to the binomial distribution. This there will be a pattern of $(P+Q)^{2n} = (P+Q)^{2\times4} = (P+Q)^8$, a ratio of 1:8:28:56:28:8:1.

Problem 2

In certain plant height is determined by four independently assorting genes P, Q, R and S. Each segregating two alleles; with each gene one allele, denoted by the superscript zero, adds nothing to the stalk height of 10 centimeters whereas the other allele, denoted by the superscript one adds one centimeters, to the height. If all the alleles of these gene act additively to determine plant height. (a) What is the phenotype of the plant with the genotype $P^0P^1Q^0Q^1R^0R^1S^0S^1$ and (b) if this plant is selfed. What fraction of the offspring will be 10 centimeter tall?

Solution

I. The phenotype of the quadruple heterozygote should be the basic height 10 cm plus the contribution of each of the one superscript allele (4 cm), i.e., 14 cm (b).

II. Among the progeny of the selfed plant, only those that are homozygous for all the zero superscript alleles manifest the basic phenotype of 10 cm. These quadruple gene homozygous will have a frequency of $\left(\frac{1}{4}\right)^4 = \frac{1}{256}$.

Problem 3

A cross is made between two tomato plants. They have 20 centigram and 40 centigram weight of seed respectively. They produce an F_1 with seeds having 30 centigrams in weight. An $F_1 \times F_1$ cross produces 1000 plants. Among those four have seeds weighting 20 centigrams, four have seeds weighing 40 centigrams and other plants produce seeds with varying weights between these extremes. Determine the most probable gene pairs involved in the seed weights in the crosses.

Solution

I. The number of gene pairs determining a particular quantitative trait is approximately n where n is the exponent in the expression $\left(\frac{1}{4}\right)^n$.

II. The $\left(\frac{1}{4}\right)$ value derives from the proportion of F_2 individuals resembling one of the initial homozygous parents.

III. In the present problem probably four gene pairs are involved.

IV. Since the proportions of F_2 plants bearing one of the initial parental genotype is $\frac{4}{1000} = \frac{1}{250} = \left(\frac{1}{4}\right)^4$, i.e., $\frac{1}{256}$.

IV. So the probable gene pair will be 4.

Problem 4

Two homozygous varieties of tobacco plant have mean corolla length of 40.5 mm and 93.3 mm. The average of F_1 hybrids from these two varieties was of intermediate length. Among 444 F_2 plants, none was found to have flowers either as long or as short on the average of the parental varieties. Estimate the minimal number of pairs of alleles segregating from F_1.

Solution

I. If four pairs of alleles segregate from F_1, we expect $(\frac{1}{4})^4 = \frac{1}{256}$ of the F_2 to be as extreme as one of the other parental average.

II. Likewise if five pairs of alleles segregate from F_1, we expect $(\frac{1}{4})^5 = \frac{1}{1024}$ of the F_2 to be as extreme as one parent or the other.

III. Here none of 444 F_2 plants has flowers at the extreme; therefore, more than four loci (minimum free loci) are segregated from F_1.

Problem 5

A cross is made between two apple plants. They have 5 grams weight and 2 grams weight of apple respectively. They produce F_1 plants which have 3.5 gram weight apples. When two of these F_1 plants are crossed, the offsprings produce a range of apples weighing from 2 to 5 pounds. Of 200 offsprings, three produce apples weighing about 5 grams, and three produce apple about 2 grams in weight. Approximately how much does each effective gene contribute to the weight?

Solution

I. The number of gene pairs determining the particular quantitative trait is approximately n, where n is the exponent in the expression $(\frac{1}{4})^n$.

II. The $(\frac{1}{4})^n$ value derives from the proportion of F_2 individuals resembling one of the initial homozygous parents.

III. In the present problem, probably 3 allelic pairs are involved.

IV. Since the proportions of F_2 plants bearing one of the initial parental genotype is $\frac{3}{200} = \frac{1}{66}$, or, $\frac{1}{64}$(approx) = $(\frac{1}{4})^3$.

Therefore, 3 alleles pairs are involved.

V. Each effective allele contributes about $\frac{1}{2}$ gram over 2 gram base (3 gram difference divided by six effective alleles AABBCC = 5 gm aabbcc = 2 grams).

Problem 6

The mean internode length of a variety (X) of wheat was found to be 3.20 mm. The mean length of another variety (Y) was 2.10 mm. Crossing these two varieties produced as F_1 and F_2 with average internode lengths of 2.65 mm. About 6% of the F_2 had an internode length of 3.2 nm and another 6% had a length of 2.1 mm. Determine the most probable number of gene pairs involved in internode length and approximate contribution that each gene makes to the phenotype.

Solution

I. In one pair of genes, we expect about $\frac{1}{4}$ or 25% of the F_2 to be as extremes as one of the parents.

II. With two pairs of gene, we expect about $\frac{1}{16}$ or 6.25% as extreme as one of the parents.

III. Here woman postulates two pairs of genes.

Parent:	Variety (X) ×	Variety (Y)
	AABB	aa bb
	4 growth genes = 3.2 mm	No growth gene = 2.10 mm
F_1	AaBb 2 growth genes = 2.65 mm	

IV. The difference 2.65–2.10 = 0.55 mm is the result of growths. Therefore, each growth gene contributes $\frac{0.55}{2}$ = 0.275 mm to the phenotype.

F_2: No. of growth	Mean Internode length	Frequency	Genotypes
4	3.20	$\frac{1}{10}$	AABB
3	2.925 $\left(3.2+2.65 = \frac{5.85}{2}\right)$	$\frac{1}{4}$	AaBB AABb
2	2.65	$\frac{3}{8}$	AAbb, AaBb, aaBB, aaBb, Aabb
1	2.375 $\left(2.65+2.10 = \frac{4.75}{2}\right)$	$\frac{1}{4}$	aaBb Aabb
0	2.100	$\frac{1}{16}$	aa bb

Problem 7

Two varieties of wheat were scored for length of time (in days) to heading from which the following means were obtained: variety X = 13.0 days, variety Y = 27.6 days. From survey of 5,504,000 F_2 progeny, 86 were found to head out in 13 days or less. How many pairs of factors are probably contributing to early flowering?

Solution:

I. The number of pairs of factors contributing to early flowering is approximately n.

II. $\left(\frac{1}{4}\right)^n$ values are derive from the proportion of F_2.

III. From survey of F_2 population 5504000, 86 were found to head out.

$\left(\frac{1}{4}\right)^n = \left(\frac{86}{5504000}\right)$

or, $\left(\frac{1}{4}\right)^n = \frac{1}{6400}$

or, $\left(\frac{1}{4}\right)^n = \left(\frac{1}{4}\right)^7 = \frac{1}{16384}$

or, $\left(\frac{1}{4}\right)^8 = \frac{1}{65536}$

n = 7 less to satisfy 6400; or, n = 8, it satisfy 65536. So, 8 pairs of factors involved here.

Problem 8

Suppose the average skin colour of one population is 0.43 (reflect once of skin to red light of 685 nm wave length), the average skin colour of another population distinct from the first is 0.23 and hybrid population between the two population parents have skin colour as extreme as the average of the either population, estimate the numbers of segregating loci in hybrid parents that contribute the skin colour variability in their offspring.

Solution:

I. Let the number of segregating loci contribute the skin colour variability is approximately n.

II. $\left(\frac{1}{4}\right)^n = \frac{1}{150}$

or, $\left(\frac{1}{4}\right)^n = \left(\frac{1}{4}\right)^3 = \frac{1}{64}$. It is less than $\frac{1}{150}$

or, $\left(\frac{1}{4}\right)^n = \left(\frac{1}{4}\right)^4 = \frac{1}{256}$. It is more than $\frac{1}{150}$

Therefore, 3 or 4 loci are involved for skin colour variability.

Autozygosity Mapping

Autozygosity: It is a term used to mean homozygosity for markers identical by descent, i.e., two alleles are both copies of one specific allele that was present in recent ancestor.

I. Autozygosity mapping is a powerful tool if families can be found with multiple people affected by the same recessive condition in two or more sibships linked by consanguinity.

II. Small consanguineous families can generate significant **LOD scores**.

Features:

I. It occurs when individuals are homozygous at particular loci through identity of descent from a common ancestor.

II. In a consanguineous marriage, both partners share some of their ancestors' character and so many inherit a copy of the same ancestral allele at a locus.

III. Their child may then be homozygous or autozygous not only at disease locus but also at closely linked loci.

IV. In other words, all affected relatives in an inbred family (consanguineous marriage) will be homozgous for markers within the region surrounding the disease locus.

V. Thus, investigation can be made for shared areas of homozygosity in affected relatives.

In a pedigree with a relatively large number of affected individuals, only a small number of shared homozygous regions will be identified.

Significance:

I. Autozygosity mapping can be applied in both small inbred families and in genetic isolates with shared common genetic ancestor.

II. It is a powerful particular technique in large inbred families where more than one branch has affected individual.

III. Several of the genes that cause autosomal recessive trait, i.e., sensorineural hearing loss have been mapped.

QTL MAPPING

- **Locus**

  A unique chromosomal location of a gene or one of its allele or DNA sequence is termed as locus.

- **Quantitative Character**

  A heritable phenotypic character that is measured on a continuous scale and does not have a simple Mendelian basis is known as quantitative character.

- **Quantitative Trait Locus (QTL)**

  A locus (chromosomal site) at which two or more genes contribute in the expression of complex trait and its (quantitative trait) inheritance is called **quantitative trait locus (QTL)**.

- **Features of QTL mapping**

  I. Several molecular markers have been identified and mapped with the application of recombinant DNA technology. It may help to identify QTL by linkage analysis.

  II. The most frequently used molecular markers as **RFLP** (restrictions fragment length polymorphism), **SNP** (single nucleotide polymorphism), **copy number variants** such as **microsatellites** and **indel** polymorphisms serve as reference points along the chromosome.

  III. These markers are used to prepare detailed genetic map of several genomes (*Drosophila, Caenorhabditis, Arabidopsis,* etc.).

- **Protocol of QTL mapping**

  I. Two strains are inbred and produce F_1 offsprings or progenies.

  II. F_1 progenies are backcrossed to the parental stock. As a result, in the F_2 progenies different combination of parental chromosomes are observed.

  III. The F_2 progenies are first examined on the basis of their quantitative characters (fruit size). Afterwords, their sample cells are subjected to RFLP or SNP or microsatellite mapping. It will determine the molecular marker on their chromosome.

  IV. Finally, observation will be made to find out the association of molecular marker and the quantitative trait.

  In the way, the number of molecular marker and their association in the chromosome site along with quantitative characters can be determined.

- **Consegregate**
  - when both marker and a phenotypic trait of interest are expressed together, they are said to consegregate.
  - consistent consegregation establishes the presence of QTL at or near the RFLP marker.
  - QTL are responsible for a trait (under investigation) are closely linked along a single chromosome.

Mapping of QTLs in Tomato

Steven Tanksley and his colleagues (Paterson, DeVerna and Lanini) in 1990 conducted their research work to identify the loci responsible for variation in tomato fruit (***Lycopersicon***) weight by constructing detailed molecular maps for each of the tomatos 12 chromosomes.

I. The molecular genetic markers like RFLPs are available for many organisms for making systemic mapping of QTLs.

II. About 88 RFLP loci have been identified that are responsible for weight of tomato and are positioned the maps of the tomato chromosomes (12).

Requirement

I. Two tomato plants, viz., *L. pimpinellifolium* (LP) of one gram weight and *L. eseulentum* (LF) of 500 gram weight were taken.

II. RFLP markers.

Procedure:

I. *L. pimpinellifolium* (**LP**) plants were crossed to the *L. esculentum* (**LE**).

II. Fruit weight of LP plant was 1 gram and fruit weight of LE plant was 500 gram.

III. The F_1 plant was produced and its fruit weight was 10.5 gram (average).

IV. A single F_1 plant was self-fertilized to produce F_2 progeny. The fruit of the 188 F_2 plants was 11.1 gram (average).

V. To identify the location of QT loci on the genetic map, DNA was extracted from individual plants of F_2 followed by digestion with restriction enzyme (Eco RI).

VI. Southern blotting was done to determine RFLP marker.

Observation:

I. The average fruit weight of F_1 plants was obtained 10.5 gram.

II. The average fruit weight of F_2 plants (188) was obtained 11.1 gram.

III. The **LP** allele was derived from *L. pimpinellifollium* and **LE** allele was derived from *L. esculentum.*

IV. They observed that TG 167, RFLP locus was on chromosome 2. The homozygous for LP marker $\left(\frac{LP}{LP}\right)$ the fruit weight was 8.4 grams; the heterozygous plants, i.e., $\left(\frac{LP}{LE}\right)$ the fruit was 10.0 gram and the homozygous plants for LE marker $\frac{LE}{LE}$ was 17.5 gram.

Inference:

I. Thus at the RFLP(A) locus it seems that LE marker $\left(\frac{LE}{LE}\right)$ is associated with increased fruit weight which indicates that in *L. esculentum* there is an allele for increased fruit weight lying near TG 167 locus.

II. This analysis suggests that the existance of a QTL affecting fruit weight lies near TG 167 on chromosome 2.

They designate this QTL as fw 2.2.

III. Furthermore, Tanksley and Lippman concluded that there were five additional fruit weight loci, including one more on chromosome 2, two on chromosome 1, and one on each chromosome 3 and 11.

IV. They also pointed out that fw 2.2 QTL encoded a single gene ORF.

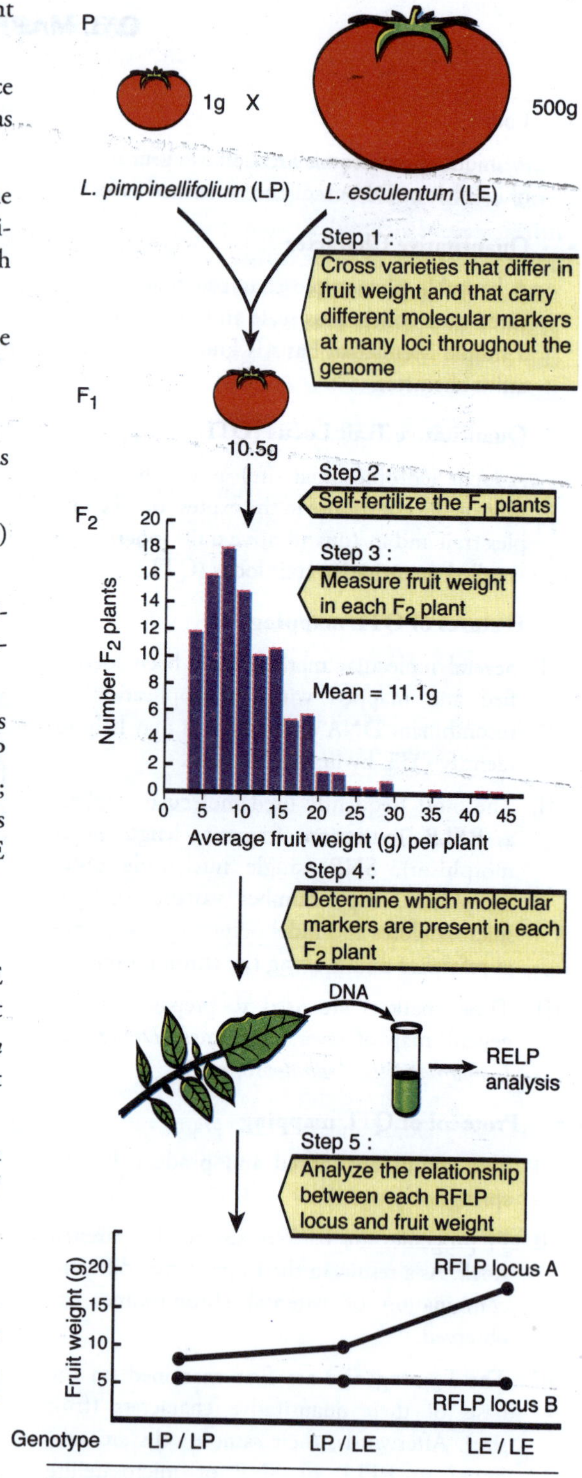

Fig. 9.1: Methods to identify QTL loci for tomatoe's weight.

Problems

Problem 1

A. Tanksley and others identified QT loci for fruit weight in tomatoes. The parents in the initial cross differed dramatically in the average weights of their fruits. *L. pimpinellifolium* 1 gram versus *L. esculentum* 500 gram. The F_1 fruits averaged 10.5 gram and the F_2 fruits averaged 11.1 gram. Why do these data indicate that dominance plays a role in determining fruit weight in tomatoes.

B. T. Lippman and Tanksley identified six QT loci affecting fruit weight. One locus fw 11.3 was located near the RFLP locus TG 36 on chromosome 11. Another locus fw 2.2 was located near the RFLP locus TG 167 on chromosomes 2. When F_2 plants were genotyped for these two loci, they found the following relationship between the genotypes and fruit weight.

		Genotype of F_2 planes		
QTL	RFLP locus	LP/LP	LP/LE	LE/LE
fw 11.3	TG 36	6.2	12.2	20.2
fw 2.2	TG 167	8.4	10.0	17.5

Which QTL shows dominance for the trait fruit weight? Which of the alleles LE or LP is dominant?

Solution

A. I. The mean fruit weights in P_1 $\left(\frac{1+500}{2} = 250.5\right)$, F_1 (10.5 gram) and F_2 (11.1 gram) indicate the dominance plays an important role in determing this quantitative trait.

II. The F_1 (10.5 gram) and F_2 (11. gram) averages are much more closer to the fruit weight of *L. pimpinellifolium* than *L. esculentum.*

III. This skew is clear evidence for dominance.

B. I. The fw 2.2 QTL shows dominance whereas fw 11.3 QTL does not.

II. For fw 2.2, the heterozygote's phenotype is close to the phenotype of the LP/LP homozygote, not midway between the two homozygote.

III. This observation indicates that the LP allele of the fw 2.2 is partially dominant over LE alleles.

IV. By contrast, the phenotype of the above mentioned heterozygote at fw 11.3. QTL is nearly midway between that of the two homozygotes. Thus, the alleles of this locus appear to act more or less additively to determine the fruit weight.

Problem 2

What is a quantitative trait locus (QTL)? Does a QTL contain one or multiple genes? What technique is commonly used to identify QTL?

Solution

Quantitative trait loci are sites within chromosomes that contain genes that affect quantitative trait.

It is possible for QTL to contain one gene or it may contain two or more closely linked genes.

QTL mapping involved linkage to known molecular markers is commonly used to determine the location of QTLs.

Problem 3

A restriction fragment polymorphism (RFLP) in *Drosophila* is linked to a QTL (quantitative trait locus) affecting the number of abdominal bristles. A probe for the RFLP hybridizes with a 6.5-kb restriction fragment, with a 4.0-kb restriction fragment, or with both (in heterozygotes). The QTL genotypes may be designated *QQ*, *Qq*, and *qq*. The average abdominal bristle numbers in these genotypes are 20, 18, and 16, respectively. The typical standard deviation in bristle number is about 2, so the effect of the favourable allele is equal to one phenotypic standard deviation. This would usually be regarded as a rather large effect. A mating is carried out as shown below, and each of the progenies is examined for bristle number and then subjected to electrophoresis and DNA hybridization to classify the RFLP genotype.

$$\frac{6.5\ Q}{4.0\ q} ♀♀ \times \frac{6.5\ Q}{4.0\ q} ♂♂$$

As expected, the progeny phenotypes with respect to the RFLP fall into three classes: those with the 4.0 band only (homozygous 4.0), those with both bands (4.0/6.5 heterozygotes), and those with the 6.5 band only (homozygous 6.5). In the boxes under the lanes in the gel shown below, write the average bristle number for each of the three RFLP phenotypes, assuming that the frequency of recombination between the RFLP locus and the QTL is 10%. Remember that there is no recombination in *Drosophila* males.

Phenotypes observed in population 6.5 4.0

Average bristle number

Solution

(i) Solving this kind of problem absolutely requires a Punnett square so that all the progeny genotypes and phenotypes can be displayed with their respective frequencies.

(ii) The appropriate format for this problem is shown in the table below, where the frequencies of the female gametes result from linkage with a recombination frequency of 0.1, and those in males result from random segregation with no recombination.

(iii) The average bristle number in each class of progeny is shown in bold type. From this tabulation it is clear that among flies with the 4.0 band only the ratio of Qq:qq genotypes is 1:9, so the average bristle number in this class is $\left(\frac{1}{10}\right) \times 18 + \left(\frac{9}{10}\right) \times 16 = 16.2$.

(iv) Among flies with the 4.0 and the 6.5 band, the ratio of QQ:Qq:qq genotypes is 1:18:1, so the average bristle number in this class is $\left(\frac{1}{20}\right) \times 20 + \left(\frac{18}{20}\right) \times 18 + \left(\frac{1}{20}\right) \times 16 = 18.0$.

(v) Finally, among flies with the 6.5 band only, the ratio of QQ:Qq genotypes is 9:1, so the average bristle number in this class is $\left(\frac{9}{10}\right) \times 20 + \left(\frac{1}{10}\right) \times 18 = 19.8$.

(vi) Note that the difference in mean phenotype between the RFLP genotypes is less than 1 phenotypic standard deviation, which means that even a difference be large, it would be very difficult to detect and that doing so would require large sample sizes.

Female gamete	Frequency	Male gamete (each with frequency 0.5) 6.5 *Q* 0.5			4.0 *q* 0.5		
6.5 *Q*	$\frac{(1-0.1)}{2} = 0.45$	0.225	$\frac{6.5\ Q}{6.5\ Q}$	20	0.225	$\frac{6.5\ Q}{4.0\ q}$	18
6.5 *q*	$\frac{0.1}{2} = 0.05$	0.025	$\frac{6.5\ q}{6.5\ Q}$	18	0.225	$\frac{6.5\ q}{4.0\ q}$	16
4.0 *Q*	$\frac{0.1}{2} = 0.05$	0.025	$\frac{4.0\ Q}{6.5\ Q}$	20	0.025	$\frac{4.0\ Q}{4.0\ q}$	18
4.0 *q*	$\frac{(1-0.1)}{2} = 0.45$	0.225	$\frac{4.0\ q}{6.5\ Q}$	18	0.225	$\frac{4.0\ q}{4.0\ q}$	16

Pedigree Analysis

10

- **PEDIGREE (FAMILY TREE):**

Pedigree is a family tree drawn with standard, genetic symbols (genealogical table, chart or diagram) representing the ancestry of an individual as well as inheritance patterns for specific phenotypic characters.

- **PEDIGREE ANALYSIS:**

It is a technique or scrutiny of established matings through which the exact mode of inheritance of a gene can be traced when sufficient family information is available.

- **STRUCTURAL COMPONENTS OF PEDIGREE:**

(i) **Proband/Propositus:** A member of a family (individual) whose phenotype first comes to the attention of the geneticist through whom the pedigree is ascertained, is called **proband** or **propositus** (female **proposita**). He or she is often marked in pedigree by an arrow (➻) or pointer (P).

(ii) A family pedigree chart conventionally has **circles** (○) for female and **squares** (□) for male.

(iii) The **unshaded** circles or squares designate that the people are **normal** for the characters being studied, while solid **black** squares (■) or circles (●) indicate **affected** individuals.

(iv) The direct horizontal lines connecting square and circle is called **marriage line** (○—□).

(v) A vertical line drawn from the marriage line is called **offspring line** (○┬□).

(vi) Horizontal line below the parents connecting offsprings (sibling) is called **sibship line** (○┬┴┬□).

(vii) The individuals on the same line from the same generation are denoted by **roman** letters I, II, III, IV, etc. on a side.

(viii) The offsprings are listed in order of birth from left to right and indicated by **Arabic** numbers (1, 2, 3, ...).

(ix) Twins that have same connection to the sibship line are identical, **monozygotic** and arise from the splitting of a single fertilized egg (○—○/□—□).

(x) Twins that have separate connection to the sibship line are **fraternal** or **dizygotic** and arise from two separately fertilized eggs (○—□).

(xi) Heterozygotes customarily are designated by colouring half of the symbols black [(◪,◒) = Heterozygous for autosomal recessive].

(xii) Carrier of a sex-linked recessive gene is designated by a black dot in the middle of the symbols (⊙).

(xiii) When the sex of the individual (child) is unknown, the symbol is **diamond-shaped** (◇).

(xiv) Mating between relatives is referred to as **consanguineous** and indicated by the symbol □=○.

(xv) Dead male is denoted by the symbol ⊠ and dead female is indicated by the symbol Ø

(xvi) Abortion or still-birth (sex unspecified) is indicated by the symbol ●.

(xvii) If one or two horizontal lines present below the offspring line indicate couples are childless (⊥⊥).

SYMBOLS USED IN PEDIGREE:

Description | **Symbols**

Individuals

- Normal male, female and unknown sex respectively
- Proband/Proposita
- Consultant
- Affected individual
- With two conditions
- Multiple individuals (number known)
- Multiple individuals (number unknown)
- Deceased individuals
- Still-birth (gestation)
- Single (bachelor)
- Bearer of specific genetic trait
- – do –
- Examined
- Autosomal recessive carrier
- Multiple traits
- Sex-linked female carrier

Relationship

- Marriage/Matings
- Marriage between relatives or consanguineous mating
- Divorce or Relationship no longer exists
- Extramarital relationship
- Biological parents known
- Biological parents unknown
- Siblings
- Twins
- Identical twins (Monozygotic)
- Unidentical twins (Dizygotic)

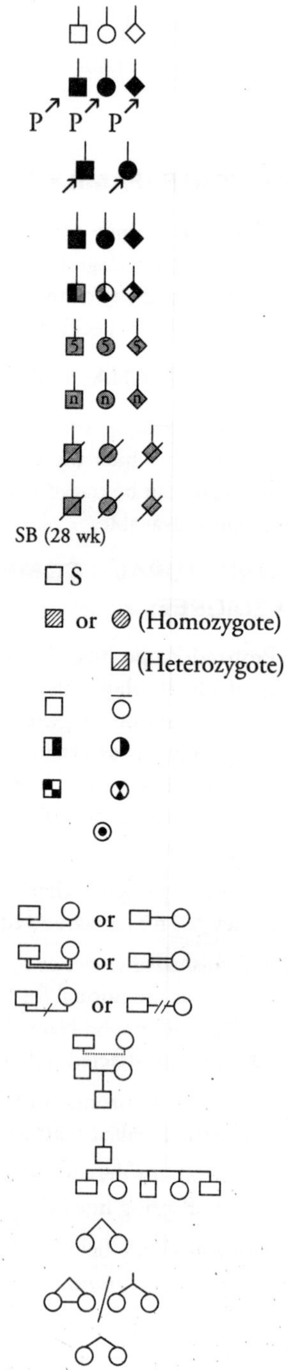

Description	Symbols
• Zygosity unknown	
• No children	
• Infertility (reason—azoospermia)	
• Adoptions	
• Adoption in	
• Adoption out	
• Illegitimate offspring	
• Surrogate mother	
• Surrogate ovum donation	
• Sperm donation	
• Ovum donation	
• Abortion	
• Spontaneous abortion	
• Affected spontaneous abortion	
• Termination of pregnancy	
• Identification as a medically examined individual	
• Record of more than one genetic trait	

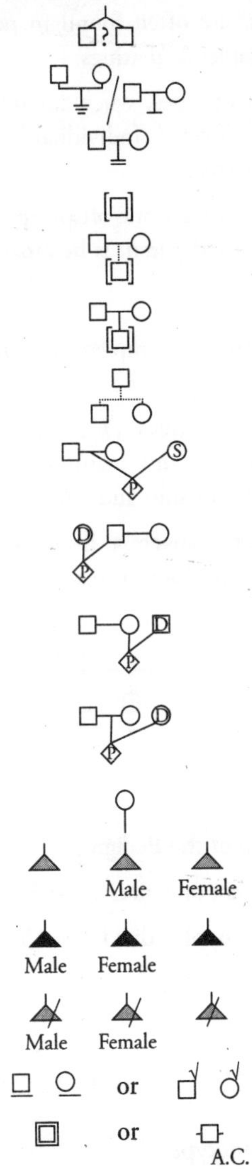

APPLICATION:

1. Pedigree is a very useful way to recognize different modes of inheritance, especially in humans.
2. Pedigree is used to identify dominant and recessive trait in human family.
3. Pedigree analysis can be used by the genetic counsellor to assess the risk that an individual will inherit a particular trait.

- **AUTOSOMAL RECESSIVE INHERITANCE:**

(i) Since the condition is autosomal, **both sexes** (males and females) are likely to be **affected equally**.

(ii) The pattern of transmission appears **horizontal** because unaffected parents have affected offsprings.

(iii) Trait often **skips** generation.

oth parents are affected, all children should be affected.

(v) Traits are often found in pedigree with consanguineous matings.

(vi) In most cases, when unaffected people mate with affected individuals, all offsprings are unaffected.

When at least one offspring is affected, it indicates unaffected parent is **heterozygous**.

Example:

1. **Albinism:** A complete depigmentation of skin due to deficiency of tyrosinase.
2. **Phenylketonuria:** Due to lack of phenylalanine hydroxilase and manifested by mental retardation, fair skin and epilepsy.
3. **Sickle-cell anaemia:** β-chain contains valine in the 6th position instead of glutamic acid.

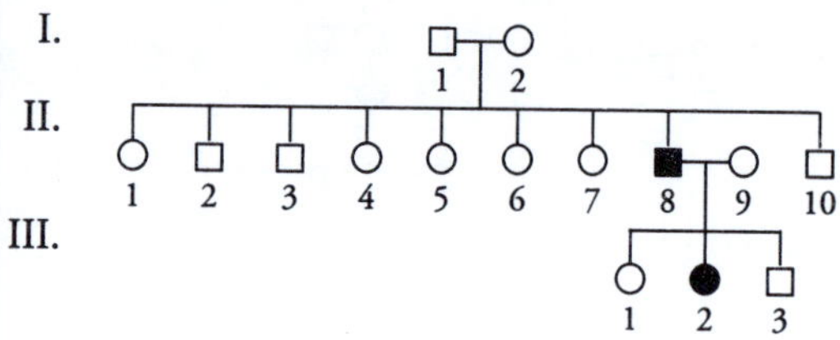

Description of the Pedigree:

(i) Total no. of generations: 3

(ii) Total no. of individuals: 15

(iii) Total no. of males: 6

(iv) Total no. of females: 9

1st generation: (i) Parents are normal

2nd generation: (i) Total no. of individuals: 10

(ii) Total no. of males: 4

(iii) Total no. of females: 6

(iv) No. of affected male: 1

(v) Total no. of affected female: none

3rd generation: (i) Total no. of individuals: 3

(ii) Total no. of male: 1

(iii) Total no. of females: 2

(iv) Total no. of affected male: none

(v) Total no. of affected female: 1.

Analysis of the Pedigree:

1. There is no sex bias in the pedigree as both males and females are affected. So it is not sex-linked inheritance.
2. In generation II-8 is affected, but the parents (I-1 and I-2) are normal but **heterozygous** for the affected trait. Here horizontal transmission occurs. Therefore, the trait may be due to **autosomal recessive gene.**
3. Mating between one affected (II-8) and one normal (II-9) produces both normal and affected progeny. It indicates normal individual is **heterozygous**. It is almost 1 : 1 ratio. Therefore, it is **autosomal recessive.**

Inference: The trait is autosomal recessive one.

Probable genotype: **— offspring —**

Generation	1	2	3	4	5	6	7	8	9	10
I.	$\frac{+}{a}$	$\frac{+}{a}$								
II.	$\frac{+}{+}$ / $\frac{+}{a}$	$\frac{+}{+}$ / $\frac{+}{a}$	$\frac{+}{+}$ / $\frac{+}{a}$	$\frac{+}{+}$ / $\frac{+}{a}$	$\frac{+}{+}$ / $\frac{+}{a}$	$\frac{+}{+}$ / $\frac{+}{a}$	$\frac{+}{+}$ / $\frac{+}{a}$	$\frac{a}{a}$	$\frac{+}{a}$	$\frac{+}{+}$ / $\frac{+}{a}$
III.	$\frac{+}{a}$	$\frac{a}{a}$	$\frac{+}{a}$							

- ## AUTOSOMAL DOMINANT INHERITANCE:

(i) Since the condition is autosomal, both sexes (males and females) are likely to be affected equally.

(ii) The pattern of transmission appears vertical because each affected person usually has an affected parent.

(iii) Phenotype (affected) tends to appear in every generation of the pedigree.

(iv) Affected parents transmit the phenotype almost equally among the sexes (male and female).

(v) Trait should not skip generation (unless trait lacks full penetration).

(vi) When one parent is affected, the risk of having affected child is 50%.

(vii) If the affected person marries a normal person, half of their children will be affected. It indicates affected individual is heterozygous.

(viii) Unaffected members of a family do not transmit the trait further.

(ix) All forms of transmission between sexes are observed, i.e., male to male, female to female, male to female and female to male.

Example:

1. **Huntington disease**: Neural degeneration leading to convulsions and premature death. However, it is late onset disease.
2. **Brachydactyly**: Shortening of the fingers.
   Polydactily: Occurrence of extra finger.
3. **Nail-patella syndrome**: Dystrophy of nails, absence of patellae and nephropathy.

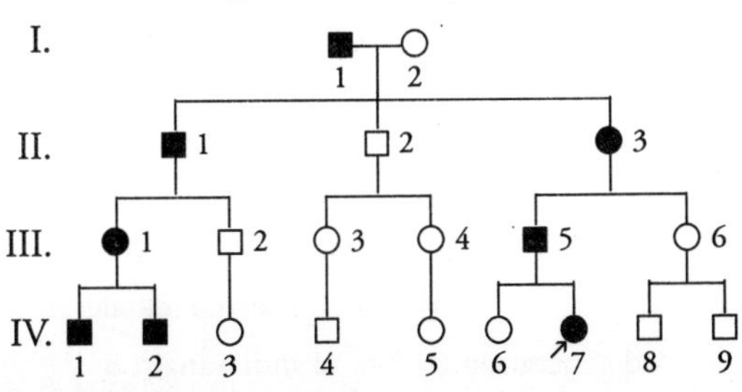

Description of the Pedigree:

(i) Total no. of generations: 4

(ii) Total no. of individuals: 20

(iii) Total no. of males: 10

(iv) Total no. of females: 10

1st generation: (i) Father: affected, mother: normal

2nd generation: (i) No. of individuals: 3
(ii) No. of males: 2
(iii) No. of female: 1
(iv) No. of affected male: 1
(v) No. of affected female: 1

3rd generation: (i) No. of individuals: 6
(ii) No. of male: 2
(iii) No. of female: 4
(iv) No. of affected male: 1
(v) No. of affected female: 1

4th generation: (i) No. of individuals: 9
(ii) No. of males: 5
(iii) No. of females: 4
(iv) No. of affected males: 2
(v) No. of affected female: 1

Analysis of the Pedigree:

1. Both males and females are affected. So there is no sex bias in the pedigree. Therefore, it is not sex-linked gene inheritance.
2. Here both males and females (I-1, II-1, II-3, III-1, III-5, IV-1, 2, 7) are affected. Males and females are almost equally affected, so it is **autosomal dominant inheritance.**
3. Traits tend to appear in every generation. So it is **autosomal dominant**.
4. The pattern of inheritance appears **vertical**, i.e., affected person (II-1, 3) usually has an affected parent. So it is **autosomal dominant** inheritance.
5. When one parent is affected, 50% children would be affected. Here II-1, II-3 has one affected offspring in each case, i.e. III-1, III-5.
6. In generation IV affected individuals are IV-1, IV-2 and IV-7. IV-7 is proposita.

Inference: Autosomal **dominant inheritance** is observed from the pedigree.

Probable genotype: — **offspring** —

Generation	1	2	3	4	5	6	7	8	9
I.	$\frac{d}{D}$	$\frac{d}{d}$							
II.	$\frac{D}{d}$	$\frac{d}{d}$	$\frac{D}{d}$						
II.	$\frac{D}{d}/\frac{D}{D}$	$\frac{d}{d}$	$\frac{d}{d}$	$\frac{d}{d}$	$\frac{D}{d}$	$\frac{d}{d}$			
IV.	$\frac{D}{d}/\frac{D}{D}$	$\frac{D}{d}/\frac{D}{D}$	$\frac{d}{d}$	$\frac{d}{d}$	$\frac{d}{d}$	$\frac{d}{d}$	$\frac{D}{d}/\frac{D}{D}$	$\frac{d}{d}$	$\frac{d}{d}$

- **SEX-LINKED RECESSIVE INHERITANCE:**

(i) Since the condition is sex-linked, the trait shows **sex bias.**

(ii) Males are more affected than female, because males are **hemizygous** (XY).

(iii) None of the offspring of unaffected male is affected but all his daughters are heterozygous "**carrier**".

(iv) Affected females come from affected fathers and affected or carrier mothers.

(v) The sons of affected females should be affected.

(vi) It follows **crisscross inheritance** (transmission of gene from male parent to female child to male grandchild) male to male through female or female to female through male.

(vii) Affected males result from mothers who are affected or who are known to be carriers (heterozygous).

(viii) When mother is carrier and father is healthy, 50% of the sons are affected.

(ix) When mother is healthy carrier and father is affected, 50% sons are affected and 50% sons are healthy.

(x) On rare occasions, a female may exhibit X-linked recessive trait.

  (i) She might be **Turner female** (XO).

  (ii) Physically female like appearance is due to **testicular feminisation** with XY-chromosome.

  (iii) Affected female may have carrier mother and affected father or carrier mother and normal father with fresh mutation affecting X-chromosome or occasionally as a result of non-random X inactivation.

Example:

1. **Colour blindness:** In ability to distinguish different (red and green) colours.
2. **Haemophilia:** Due to functionally defective antihaemophilic globulin.
3. **Testicular feminisation syndrome:** 44 + XY male, phenotypically as female.

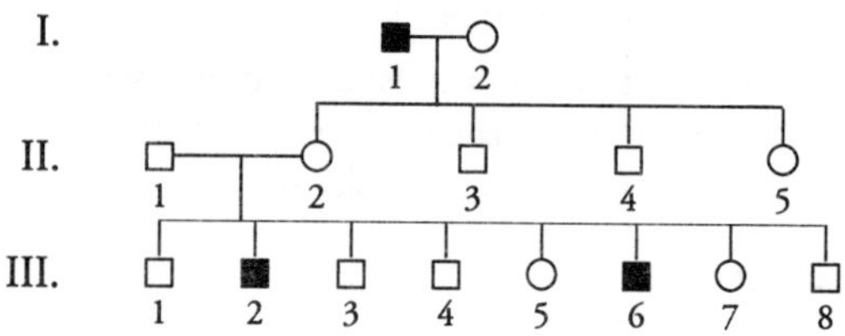

Description of the Pedigree:

(i) Total no. of generations: 3

(ii) Total no. of individuals: 15

(iii) Total no. of males: 10

(iv) Total no. of females: 5

1st generation: (i) Father: affected, mother: normal

2nd generation: (i) No. of individuals: 5
(ii) No. of males: 3
(iii) No. of females: 2
(iv) No. of affected male: none
(v) No. of affected female: none

3rd generation: (i) No. of individuals: 8
(ii) No. of males: 6
(iii) No. of females: 2
(iv) No. of affected males: 2
(v) No. of affected female: none

Pedigree Analysis:

1. Here the sex bias, i.e., males (I-1, III-2, III-6) are affected. Therefore, it may be sex-linked inheritance.
2. Males are more affected than females as males are hemizygous.
3. In generation I, male is affected and in generation II none is affected. So it is sex-linked recessive inheritance.
4. It follows **crisscross inheritance**, i.e., transmission of gene from male parent (I-1) through female to male grandchild (III-2).

Inference: The trait is inherited as sex-linked recessive one.

Probable genotype: — offspring —

Generation	1	2	3	4	5	6	7	8
I.	$\frac{xc}{y}$	$\frac{xc^+}{+}$						
II.	$\frac{xc^+}{y}$	$\frac{xc^+}{xc^+}$	$\frac{xc^+}{y}$	$\frac{xc^+}{y}$	$\frac{xc^+}{y}$			
III.	$\frac{xc^+}{y}$	$\frac{xc}{y}$	$\frac{xc^+}{y}$	$\frac{xc^+}{y}$	$\frac{xc^+}{xc^+}$ / $\frac{xc^+}{xc}$	$\frac{xc}{y}$	$\frac{xc^+}{xc^+}$ / $\frac{xc^+}{xc}$	$\frac{xc^+}{y}$

- **SEX-LINKED DOMINANT INHERITANCE:**

(i) Since the condition is sex-linked, the trait shows **sex bias.**

(ii) Affected males pass their conditions on to all daughters but none of their sons.

(iii) Affected females can transmit the disorder in both male and female offspring.

(iv) Females married to unaffected males pass the condition on to half of their sons and daughters.

(v) The trait does **not skip** generations.

(vi) Affected females come from affected mothers or fathers.

(vii) Both males and females are affected but disease is **less severe** in female.

(viii) Approximately half the children of an affected heterozygous female are affected.

Example:

1. **Hypophosphatemia:** A type of vitamin D resistant ricket.
2. **Xg blood group:** (i) **Histiocytosis X** is characterised by the presence of cytoplasmic rod structure called Langerhans cell granules or **X granules (Xg)**.

   (ii) Hypogammaglobulinaemia and measles pneumonitis.

   (iii) Faulty tooth enamel and dental discolouration.

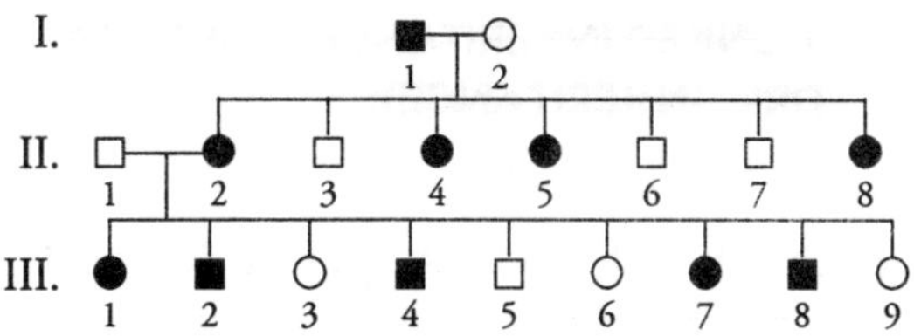

Description of the Pedigree:

I. Total no. of generations: 3

II. Total no. of individuals: 19

III. Total no. of males: 9

IV. Total no. of females: 10

1st generation: (i) Father: affected, Mother: normal

2nd generation: (i) No. of individuals: 8

(ii) No. of males: 4

(iii) No. of females: 4

(iv) No. of affected male: none

(v) No. of affected females: 4

3rd generation: (i) No. of individuals: 9

(ii) No. of males: 4

(iii) No. of females: 5

(iv) No. of affected males: 3

(v) No. of affected females: 2

Pedigree Analysis:

1. Sex bias is noticed, i.e., (II-2, II-4, II-5 and II-8) all females are affected.
2. Here the trait does not skip the generation.
3. Both males and females are affected.

Inference: The trait is sex-linked dominant.

Probable genotype: — **offspring** —

Generation	1	2	3	4	5	6	7	8	9
I.	$\frac{xc^{+}}{y}$	$\frac{xc}{xc}$							
II.	$\frac{xc}{y}$	$\frac{xc^{+}}{xc}$	$\frac{xc}{y}$	$\frac{xc^{+}}{xc}$	$\frac{xc^{+}}{xc}$	$\frac{xc}{y}$	$\frac{xc}{y}$	$\frac{xc^{+}}{xc}$	
III.	$\frac{xc}{xc^{+}}$	$\frac{xc^{+}}{y}$	$\frac{xc}{xc}$	$\frac{xc^{+}}{y}$	$\frac{xc}{y}$	$\frac{xc}{xc}$	$\frac{xc}{xc^{+}}$	$\frac{xc^{+}}{y}$	$\frac{xc}{xc}$

• Y-LINKED INHERITANCE (HOLANDANDRIC INHERITANCE):

(i) Only males are affected.

(ii) An affected male must transmit the disorder (Y-linked traits) to all of his sons but to none of his daughters.

(iii) It is also known as **holandric inheritance** or transmission.

Example:

1. **Hairy pinnae**
2. **Bilateral radio-ulnar synostosis** (radio-ulnae are fused)

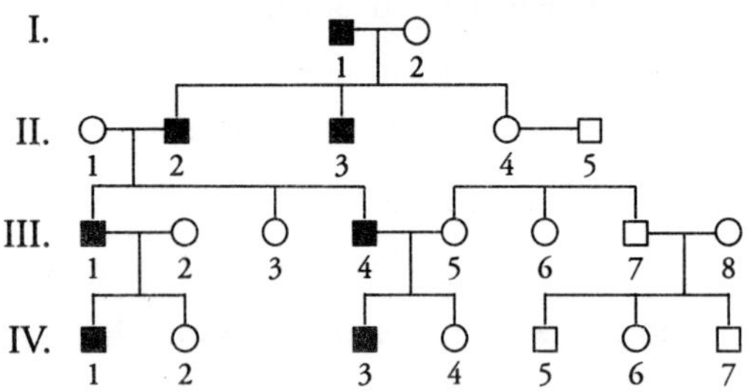

Description of the Pedigree:

(i) Total no. of generations: 4

(ii) Total no. of individuals: 22

(iii) Total no. of males: 11

(iv) Total no. of females: 11

1st generation: (i) Father: affected, Mother: normal

2nd generation: (i) No. of individuals: 5

(ii) No. of males: 3

(iii) No. of females: 2

(iv) No. of affected males: 2

3rd generation: (i) No. of individuals: 8

(ii) No. of males: 3

(iii) No. of females: 5

(iv) No. of affected males: 2

4th generation: (i) No. of individuals: 7

(ii) No. of males: 4

(iii) No. of females: 3

(iv) No. of affected males: 2

Pedigree Analysis:

(i) Traits appear in all the generations.

(ii) Only the males are affected.

(iii) In every generation males are affected (I-1; II-2, 3; III-1, 4; IV-1, 3).

(iv) Affected father transmits the trait to all his sons but not his daughter.

(v) Transmission occurs from father to son to grandson.

Inference: The trait appears to be Y-chromosome linked and the transmission is **holandric inheritance.**

Probable genotype: — **offspring** —

Generation	1	2	3	4	5	6	7	8
I.	$\frac{x}{yh}$	$\frac{x}{x}$						
II.	$\frac{x}{x}$	$\frac{x}{yh}$	$\frac{x}{yh}$	$\frac{x}{x}$	$\frac{x}{y}$			
III.	$\frac{x}{yh}$	$\frac{x}{x}$	$\frac{x}{x}$	$\frac{x}{yh}$	$\frac{x}{x}$	$\frac{x}{x}$	$\frac{x}{x}$	$\frac{x}{y}$
IV.	$\frac{x}{yh}$	$\frac{x}{x}$	$\frac{x}{yh}$	$\frac{x}{x}$	$\frac{x}{y}$	$\frac{x}{x}$	$\frac{x}{y}$	

A dominant condition may fail to manifest itself

(i) The penetrance of a character, for a given genotype, is the probability that person who has the genotype will manifest the character.

(ii) A dominant character is manifested in a heterozygous person and so should show 100% penetrance.

(iii) Nevertheless, many human characters although generally showing dominant inheritance; occasionally skip generation.

(iv) In the pedigree II-2 has an affected parent and an affected child and almost certainly carries the mutant gene, but is phenotypically normal. This would be described as non-penetrance.

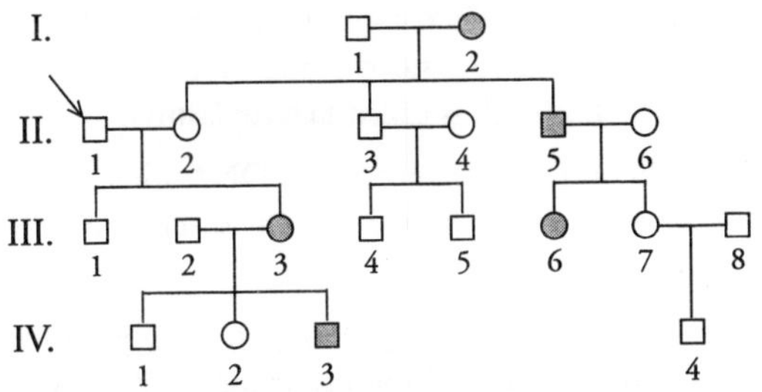

Complications to the basic Mendelian patterns: non-penetrance. Individual II2 carries the diseased gene but does not show symptoms. Other unaffected family members, such as II3, III1, IV1 or IV2, might also be non-penetrant gene carriers.

(v) Very often presence or absence of a character depends in the main and in normal circumstances on the genotype at one locus.

(vi) Non-penetrance is the pitfall in genetic counselling.

(vii) In the above pedigree III-7 was free of signs and she had no risk of affected children.

A common Recessive condition can mimic a dominant pedigree pattern

(i) A common recessive character such as blood group O may be seen in successive generation. It is due to the repeated matings of group O people with heterozygotes.

(ii) It produces a pattern, resembling dominant inheritance.

(iii) The classic Mendelian pedigree patterns are best seen with rare conditions, where there is little chance that somebody who marries into family might concidentally also carry the disease mutation that is segregating in the family.

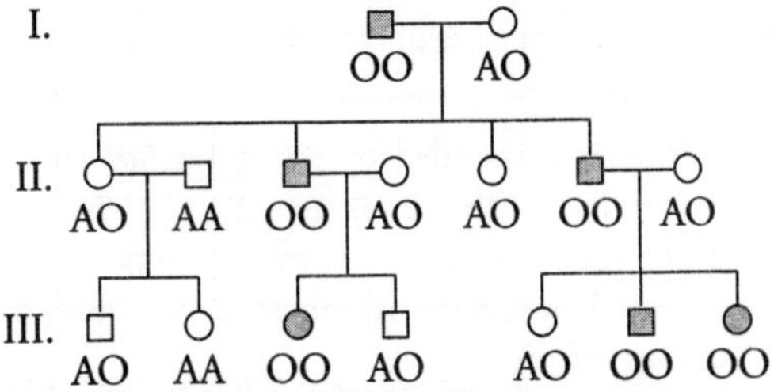

Complications to the basic Mendelian patterns: a common recessive condition giving an apparently dominant pedigree. If a recessive trait is sufficiently common that unrelatd people marrying into the family often carry it, the pedigree may misleadingly resemble that of a dominant trait. The condition in the figure is blood group O.

Male lethality may complicate X-linked pedigrees

(i) For some X-linked dominant conditions, absence of normal allele is lethal before birth.

(ii) Thus, affected males are not born.

(iii) These conditions affect only females who pass it onto half of their daughters but none of their sons.

(iv) If the family is large, 50% of males who inherit the mutant allele miscarry after birth.

Example: X-linked dominant incontentia pigments affected males abort spontaneously.

Rett syndrome in female:

(i) Developmental regression.

(ii) Develops normally for the first year or two.

(iii) Stops development, losing speech.

Rett syndrome in male:

(i) Usually lethal before birth

(ii) Rare survivors have severe encephalopathy.

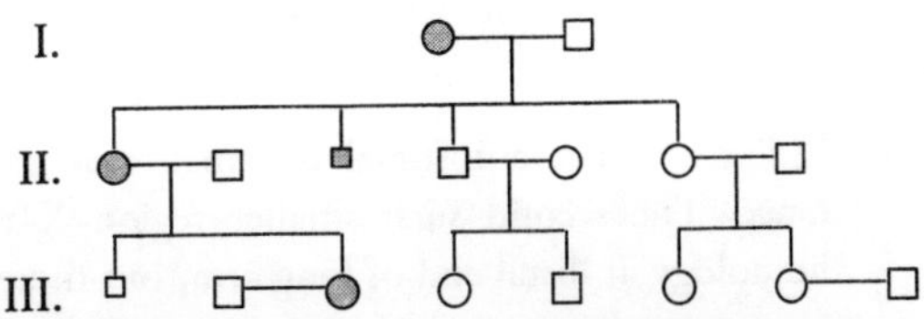

Complications to the basic Mendelian patterns: a male-lethal X-linked condition. In this family with X-linked dominant incontentia pigment (OMIM 208300), affected males abort spontaneously (small squares).

Genes in the Pseudoautosomal region

(i) Human X- and Y-chromosomes pair end to end rather than along the whole length.

(ii) Homology persists in between the short region of X- and Y-chromosomes.

(iii) Pairing is sustainded by an obligatory cross-over in a 2.6 Mb homology regions at the tips of the short arms, but crossover also occurs in a second homology region 0.32 Mb long at the tips of the long arm.

(iv) The terminal X-Y homology regions are known as **Pseudoautosomal** region.

(v) The distal 2.6 Mb of Xp and Yp contain homologous DNA sequences and in males they pair in prophase I of meiosis.

(vi) This ensures that at anaphase I each daughter cell receives one sex-chromosome, either X or Y. Pairing is sustained by an obligatory cross-over in this region. Thus, the few genes in these region segregate in a **pseudoautosomal** and not in sex-linked pattern.

Example: Leri-Weill dyschondrosteosis.

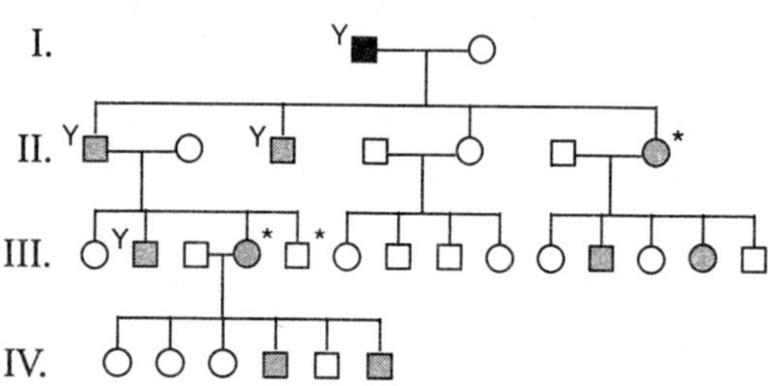

Pedigree of a pseudoautosomal condition: The allele causing this dominant condition is on the Y-chromosome in the affected males marked Y, but on the X chromosome in all other affected individuals. The three individuals marked with a red asterisk are the result of an X-Y crossover in their father. Note how the pedigree is indistinguishable from an autosomal dominant pedigree, although the causative gene is located on Xp or Yp.

(vii) Pedigree shows autosomal dominant inheritance. The second such smaller region X-Y homology at distal end of long arm, but there is no normally any crossing-over between those short sequences and so there is no special pedigree pattern associated with the genes in that second **pseudoautosomal region**.

Pedigree of Mitochondrially determined disease

(i) The mitochondrial genome is small (16.5 Kb) but is highly mutable in respect to nuclear genome.

(ii) Mitochondrial DNA replication is more error prone and the number of replications is much higher.

(iii) Mutation mtDNA are ten times higher than nuclear DNA. No mitochondrial DNA repair system exists and mtDNA prone to damage due to presence of free oxygen radicals.

(iv) Mitochondrially encoded diseases have two unusual features **matrillineal inheritance** and frequent **heteroplasmy** (the phenomenon of **heterogeneity** in composition of mtDNA, i.e., presence of both normal and mutant mtDNA in a single cell is called **heteroplasmy**).

(v) The number of mutant mtDNA molecule is directly proportional to the severity of the disease.

(vi) ATP production in tissue is the essential feature for mitochondria. Hence, tissues with high requirement of ATP are more sensitive to mutation in mtDNA.

(vii) Inheritance is matrilineal beause sperm do not contribute mitochondria to the zygote. Thus, mitochondrially inherited condition can affect both sexes but is passed on only to affected mothers.

(viii) In some patients with mitochondrial disease, every mitochondrial genome carries the causative mutation (**homoplasy**), but in other cases a mixed population of normal and mutant genomes is seen within each cell (**heteroplasmy**).

(ix) A single cell zygote cannot be mosaic for nuclear genes, but it contains many mitochondria. Therefore, heteroplasmic mothers have heteroplasmic child.

(x) In such cases, the proportion of abnormal mitochondrial genomes can vary widely between mother and child. This happens because during early development the germ line passes through a stage in which cells contain a very few mitochondria (**the mitochondrial bottleneck**).

Pedigree of Mitochondria inheritance

1.

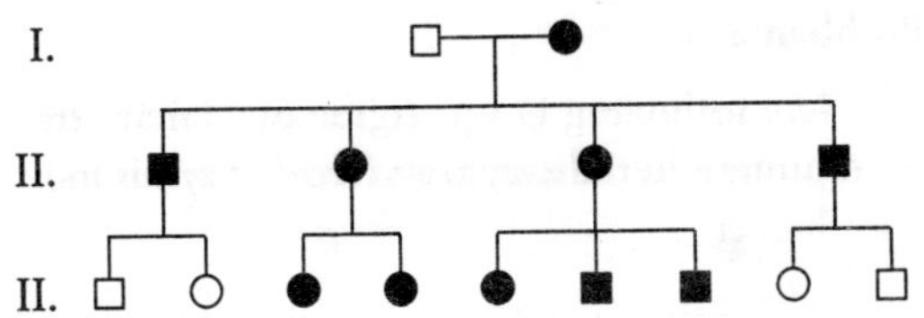

Typical pedigree chart of mitochondrial inheritance

2.

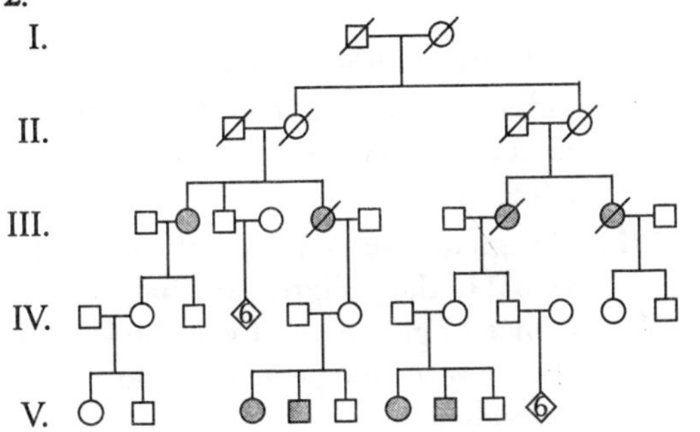

Pedigree of a mitochondrially determined disease: Note the very incomplete penetrance; although every child of an affected mother inherits mutant mitochindria through the ovum, only some are clinically affected. One cause of this can be variable heteroplasmy. [Family reported in Prezant TR, Agapian JV, Bohlman MC *et al.* (1993) *Nat. Genet.* 4, 289-294. With permission from Macmillan Publishers Ltd.]

3.

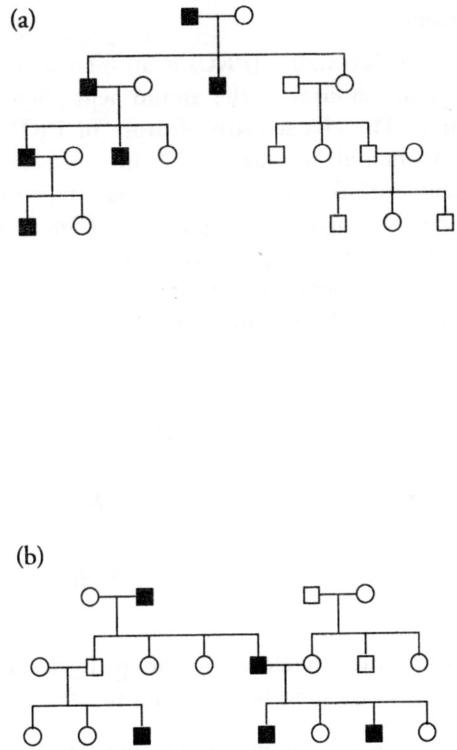

Pedigree demonstrating the inheritance of Y-linked trait (3a and b).

- **Half sib and Full sib progeny**

  The progeny having one parent common is called half sib progeny and such crosses are called half sib crosses.

  The progeny having two parent is called full sib progeny and mating between two individuals is called full sib crosses.

- **Dichotomous Character**

  A character which some people have and others do not have as compared to continuous character. *Example*: Polydactyle.

- **Anticipation**

  The tendency for the severity of a condition to increase in successive generation is termd as anticipation.

  (i) It is a hallmark of conditions caused by a very special genetic mechanism, i.e., dynamic mutation whereby a run of tandomly repeatd nucleotides such as CAG CAG CAG CAG CAG is meiotically unstable.

  (ii) A $(CAG)_{40}$ sequence in a parent may appear as a $(CAG)_{55}$ sequence in child.

  (iii) Certain of these unstable repeat sequences become pathogenic above some threshold size. *Example*: Fragile X syndrome.

Solved Problems

Problem 1

Phenylketonuria (PKU) is an inborn error in the metabolism of the amino acid phenylalanine. The characteristic feature of PKU is severe mental retardation. Many untreated patients with this trait have IQs below 20. The pedigree shown in the accompanying figure is of an affected family. The generations are labelled by Roman numerals, and the individuals in each generation by Arabic numerals.

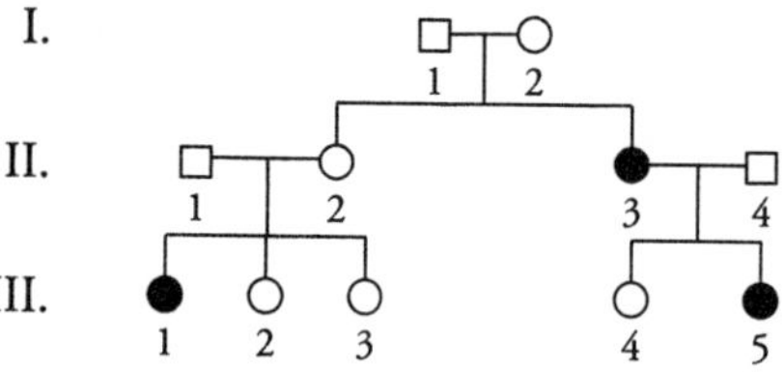

(a) What is the mode of inheritance of PKU?

(b) Which persons in the pedigree are known to be heterozygous for PKU?

(c) What is the probability that III-2 is a carrier (heterozygous)?

(d) If III-3 and 4 marry, what is the probability that their 1st child will have PKU?

Solution:

(a) The above pedigree indicates the autosomal recessive gene inheritance.

(b) The persons known to be heterozygous for PKU in this pedigree are I-1 and I-2, II-1 and II-2, II-4, III-2, 3 and 4.

(c) The genotype of III-2 may be AA/Aa. His parents are heterozygous. Therefore, probability of heterozygous is $\frac{2}{3}$.

(d) If III-3 and III-4 marry, the probability of PKU child will be as follows:

III-3 heterozygous probability is $\frac{2}{3}$ and III-4 heterozygous $\frac{1}{4}$; children of two heterozygous will express autosomal recessive trait.

First child of PKU will be $\frac{2}{3} \times \frac{1}{4} = \frac{1}{6}$.

Problem 2

The following is a pedigree of a fairly common human hereditary trait (shaded symbols):

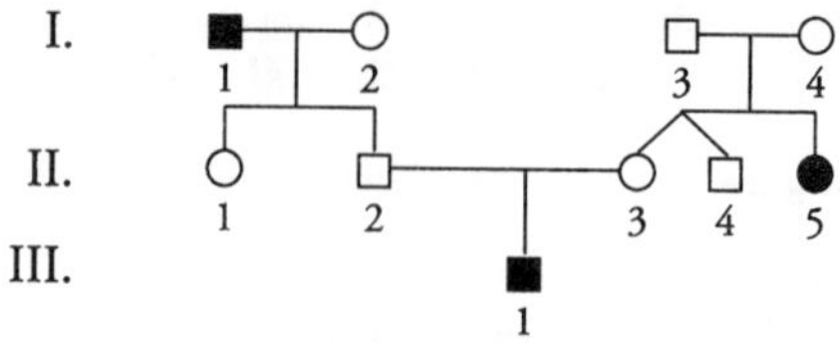

(a) Indicate whether you think the gene differences causing this effect are dominant or recessive, e.g., A (dominant) or a (recessive).

(b) Designate the genotype of each individual in the pedigree, and indicate the choice of genotypes if you think more than one genotype is possible for an individual.

Solution:

(a) The pedigree reveals that trait is autosomal recessive.

The affected offspring's parent do not show the trait.

(b) I-1, aa, I-2 AA/Aa, I-3 Aa, I-4 Aa, II-1 Aa, II-2 Aa, II-3 Aa, II-4 AA/Aa, II-5 aa, III-1 aa.

Problem 3

Indicate whether the trait in the following pedigree is caused by a dominant or a recessive gene:

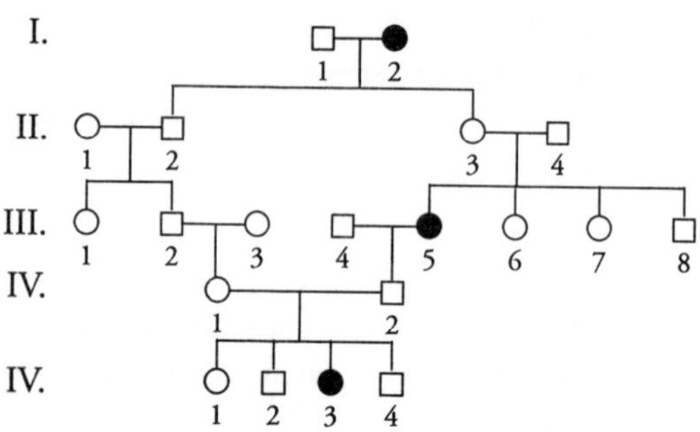

Solution:

(i) The above-mentioned pedigree reveals the inheritance of recessive gene.

(ii) Here the pattern of transmission is horizontal, i.e., unaffected parents (II-3, II-4) have affected (III-5) child and parents (IV-1 and IV-2) produce affected child (V-3). It indicates autosomal recessive inheritance.

(iii) Trait often skips generation.

That is why we can say it is autosomal recessive inheritance.

Problem 4

Assume the pedigree presented below to be straight forward, with no complications such as illegitimacy. Trait W, found in individuals represented by the shaded symbols, is rare in the population at large.

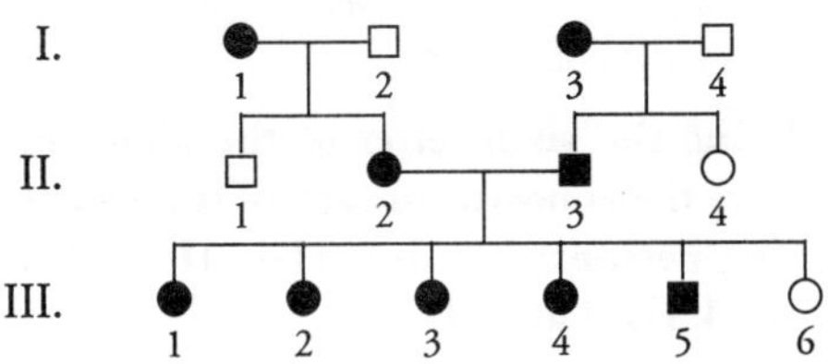

Tell which of the following patterns of transmission for W are consistent with or excluded by this pedigree:

(a) Autosomal recessive

(b) Autosomal dominant

(c) X-linked recessive

(d) X-linked dominant

(e) Y-linkage

Solution:

1. Transmission of 'W' trait appears to be caused by autosomal dominant gene.
2. Here vertical transmission occurs, i.e. affected person has affected parent. II-2 and II-3 have affected parent I-1 and I-3 respectively.
3. Trait should not skip generation, so it is autosomal dominant.
4. In case of autosomal dominant, if the affected person marries a normal individual, half of their children will be affected. Here I-1 and I-3 produce II-2 and II-3 offspring. It indicates affected persons are heterozygous.
5. II-2 and II-3 are heterozygous, i.e. Ww. After mating they produce WW, Ww, ww progeny. Here also affected progenies are found, i.e. III-1, III-2, III-3, III-4, III-5. The III-6 are non-affected, i.e. homozygous (ww).

Problem 5

The following is a pedigree of earlobe shape in human beings:

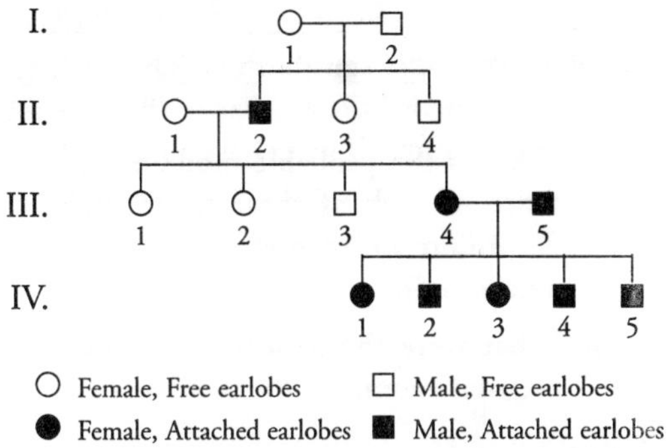

(a) Using the first letter of the alphabet, give the genotype of each of the following persons: I-1, II-1, III-4

(b) No ancestry information is given for II-1. How do you justify your designation of her genotypes?

Solution:

1. (a) I-1, Aa; II-1, Aa; III-4, aa. Attached earlobes appear to be autosomal recessive gene.

   (b) Attached earlobes of II-2, though his parents are normal. It indicates autosomal recessive inheritance. III-4 and III-5 affected, i.e., earlobed they have all earlobed children (IV-1 to 5). Therefore, this is autosomal recessive gene inheritance.

   (c) On the other hand, individuals in the pedigree with unshaded appearance have at least one dominant gene (A), i.e., they are heterozygous in nature.

2. II-1 has no ancestry information. She mates with affected (attached earlobes) males and produces attached earlobed child. Therefore, she is heterozygous. Her genotype will be Aa.

Problem 6

The following pedigree shows the incidence of two inherited traits in a human family.

Individuals with a progressive form of deafness are represented by solid black symbols and those with a rare congenital cataract condition by line symbols:

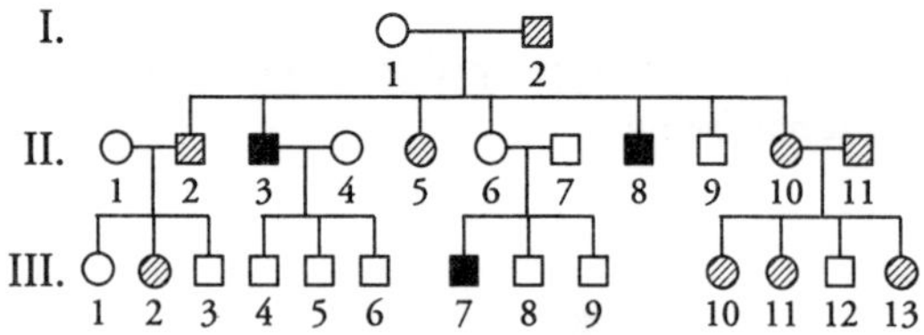

(a) What is the probable mode of inheritance of each of these traits (that is, dominant or recessive, X-linked or autosomal)?

(b) What were the genotypes of the two original parents in the pedigree?

Solution:

1. (a) From the pedigree it is seen that deafness is confined only to one sex (male) and is probably X-linked; congenital cataract is found in both sexes and is probably autosomal.

   (b) Cataract does not occur in males of generation III. But one male (III-7) is affected with deafness.

   (c) Specially the affected parents of generation II-2 and II-10, II-11 do not produce affected male offspring. Therefore, the genotype of II-2, II-10, II-11 must be heterozygous (Cc). So the trait is inherited as autosomal dominant. For cataract the genotype of I-1 is cc, and I-2 is Cc.

2. For deafness, it is sex-linked. Here affected offsprings II-3, II-8 have normal parents. Again, parents of III-7 is normal, so the trait is recessive. Therefore, deafness is sex-linked recessive.

   Genotype of I-1 is $X_D^+X_D$ and I-2 is X_D^+Y.

   In respect of two characters I-1 is $\frac{c}{c}\,\frac{X^{D^+}}{X_D}$ and I-2 is $\frac{c}{c}\,\frac{X^{D^+}}{Y}$.

Problem 7

The following is a pedigree of polydactyly, i.e., the occurrence of extra fingers in human beings:

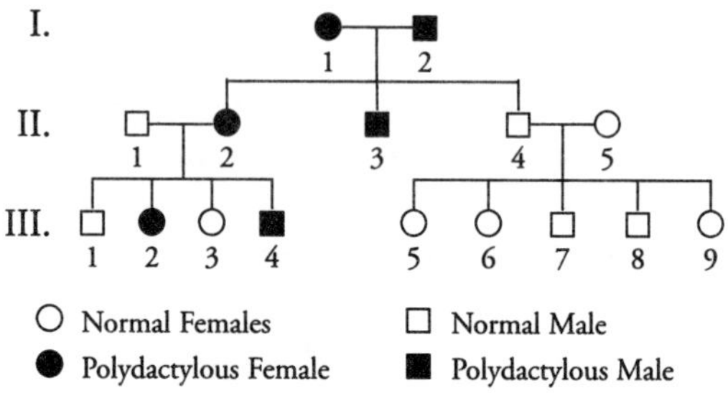

(a) Using the sixth letter of the alphabet, give the genotype of each of the following persons: I-1, I-2, II-1, II-2, II-3, II-4, III-1 and III-2.

(b) What sort of genetic cross does the marriage between II-1 & II-2 represent?

Solution:

1. (a) I-1 = Ff; I-2 = Ff; II-1 ff; II-2 Ff; II-3 FF/Ff; III-4 ff; III-5 ff; III-1 ff; III-2 Ff.

   (b) Polydactyly is caused by autosomal dominant gene. Therefore, polydactylous appear only when at least one member (parent) is affected. Polydactylous (I-1 and I-2) produce normal offspring (II-4). It indicates, normal is recessive and polydactylous is dominant.

   (c) All polydactylous have at least one F.

2. The marriage between II-1 & II-2, i.e., ff (homozygous) × Ff (heterozygous) is a kind of test-cross.

Human X-linked traits (Klug, Cummings and Spencer, 2006)

Traits	Description
1. Colour blindness (deutan type):	Insensitivity to green light
2. Colour blindness (protan type):	Insensitivity to red light
3. Fabry's disease:	Deficiency of galactosidase
4. G-6-P D deficiency:	Deficiency of glucose 6 phosphate dehydrogenase
5. Haemophilia A:	Classical form of blood clotting deficiency (factor VIII)
6. Haemophilia B:	Christmas disease, deficiency of clotting factor IX.

EXERCISE

1. Pedigree pattern of an autosomal recessive trait followed through three generations.

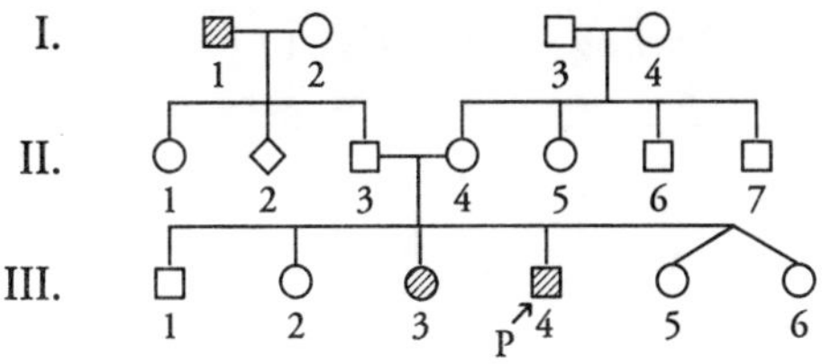

Hints: (i) Either I-3 or I-4 must be heterozygous.

(ii) II-3 and 4 must be heterozygous.

(iii) Recessive traits typically skip generations.

(iv) Recessive traits appear equally in both sexes.

2. Pedigree patterns of an autosomal recessive trait. Note the horizontal pattern of inheritance. Double line marriage figure indicates consanguineous mating:

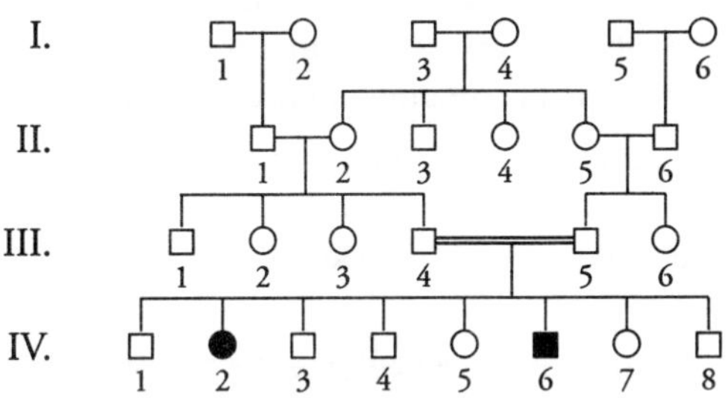

3. A representative pedigree for an autosomal recessive trait (PTC test) followed through IV generation.

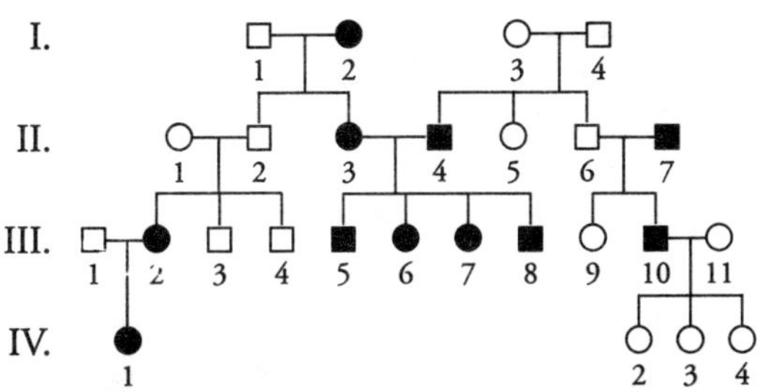

■ ● = Non-tasters(tt) □ ○ = Tasters (TT/Tt)

4. Predict the mode of inheritance and the most probable genotypes of each individual. Assume that alleles A and a control the expression of trait.

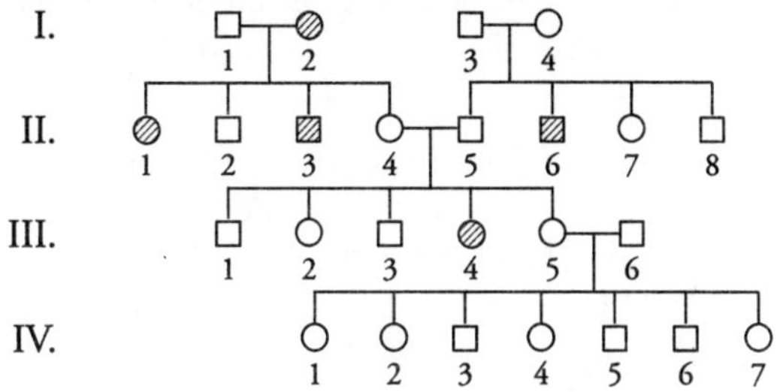

5. Phenylketonuria (PKU) is a rare autosomal recessive disease. A couple wants to have children but consults a genetic counsellor because the man has a sister with PKU and the woman has a brother with PKU. There are no other known cases in their families. They ask the genetic counsellor to determine one probability that their first children will have PKU. What is this probability?

Solution:

Let the allele causing the PKU phenotype be p and the respective normal allele be P. Then the man's sister and woman's brother must have been p/p. In order to produce these affected individuals, all four grand-parents must have been heterozygous normal (P/p). The pedigree can be summarized as follows:

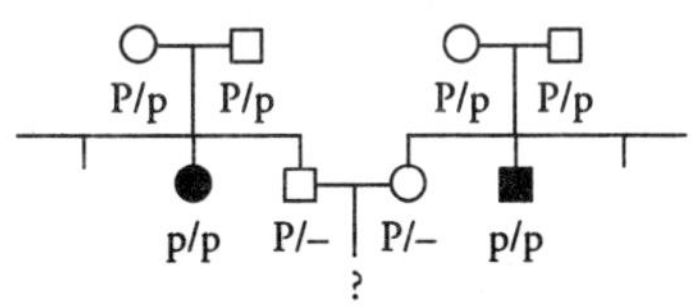

(i) The only way the couple can have a child with PKU, if both of them are heterozygous Pp (they are phenotypically normal).

(ii) Both the grand-parental matings are simple monohybrid crosses (P/p × P/p) expected to produce progeny in the following proportions:

$$\left.\begin{array}{l}\frac{1}{4}\text{PP}\\ \frac{1}{4}\text{Pp}\end{array}\right\}\frac{3}{4}\text{Normal}$$

$$\left.\frac{1}{4}\text{pp}\right\} = \frac{1}{4}\text{PKU}$$

(iii) We know that the man and woman are normal so far each of them the probability of being heterozygous is $\frac{2}{3}\left(1:2 = \frac{2}{3}\right)$ because with the normal P/– class $\left(P/p \text{ or } \frac{P}{P}\right)$, ⅔ are P/p and ⅓ are P/P. Therefore, using the product rule, the probability of both the man and the woman being heterozygous is $\frac{2}{3} \times \frac{2}{3} = \frac{4}{9}$.

(iv) If they are both heterozygous, then one quarter (¼) of their children would have PKU. So the probability that their first child will have PKU is ¼ and the probability of their being heterozygous and of their first child having PKU is $\frac{4}{9} \times \frac{1}{4} = \frac{4}{36} = \frac{1}{9}$.

6. The following pedigree was obtained for a rare kidney disease:

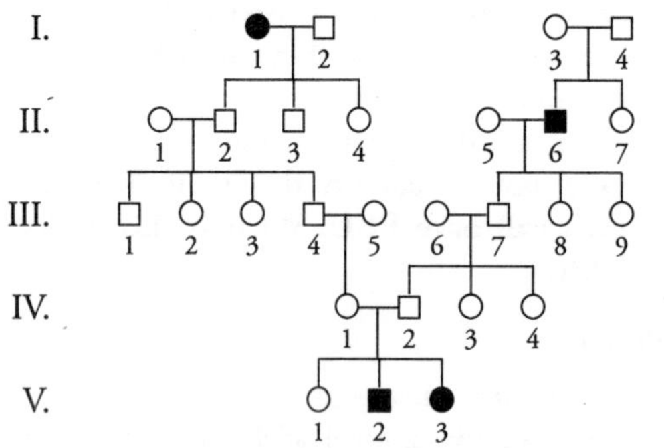

(a) Deduce the inheritance of this condition, stating your reasons.

(b) If individuals III(1) and III(8) marry, what is the probability that their first child will have the kidney disease?

Solution:

(a) (i) Here both sexes are **affected equally**. So it is autosomal inheritance.

(ii) The pattern of inheritance is **horizontal**, because unaffected parents have affected offspring.

(iii) Here traits skip generations.

So it is a **autosomal recessive inheritance.**

(b) If III-1 and III-8 marry, the result of kidney disease will be as follows:

The parent must be heterozygous and the one quarter $\left(\frac{1}{4}\right)$of their children would have kidney disease.

7. Describe and explain the mode of inheritance of the pedigree.

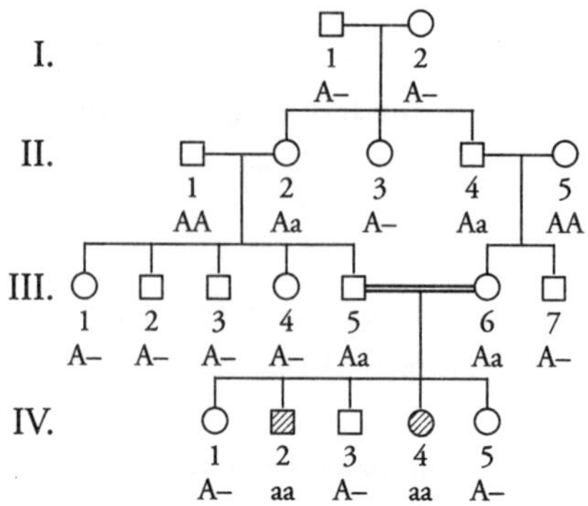

Solution:

(i) Pedigree of a rare recessive phenotype determined by a recessive allele a.

Gene symbols normally are not included in pedigree charts, but genotypes are inserted here for reference.

(ii) Note that individuals II-1 and II-5 marry into family, they are assumed to be normal because the heritable condition under scrutiny is rare.

(iii) Note also that it is not possible to be certain of the genotype in some individuals with normal phenotype; such individuals are indicated by A–.

8. A representative pedigree for an autosomal dominant trait that followed through three generations:

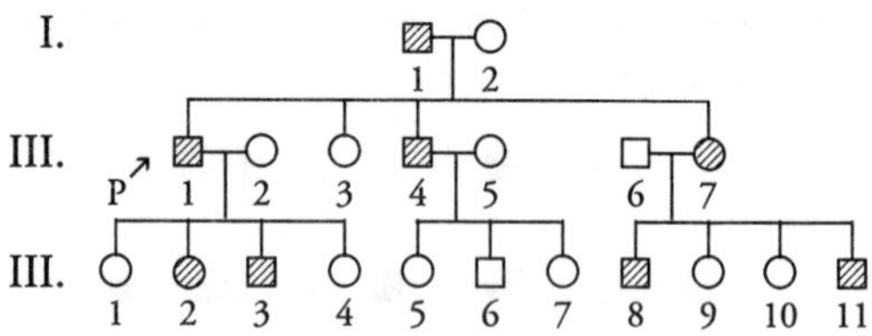

Hints: (i) I-1 is heterozygous for a dominant allele.

(ii) Dominant traits seldom skip generations.

(iii) Affected individuals all have an affected parent.

(iv) Dominant autosomal traits appear equally in both sexes.

9. (i) Pedigree of a dominant phenotype determined by a dominant allele A.

(ii) In this pedigree all the genotypes have been deduced.

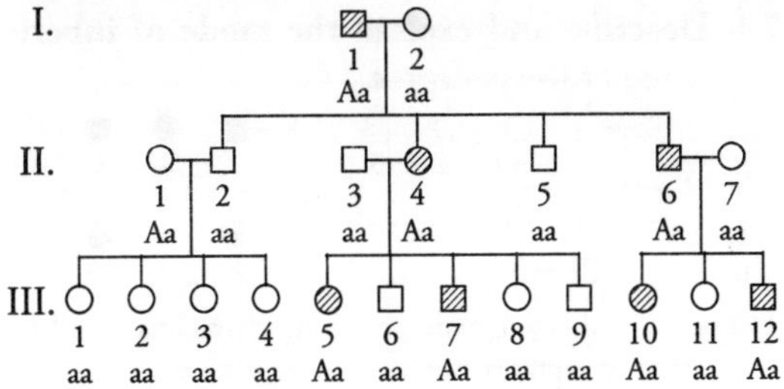

10. (i) Pedigree of an autosomal dominant trait.

(ii) Vertical pattern of inheritance is indicated.

(iii) Solid squares and circles indicating the affected males and females while hollow squares and circles are unaffected males and females.

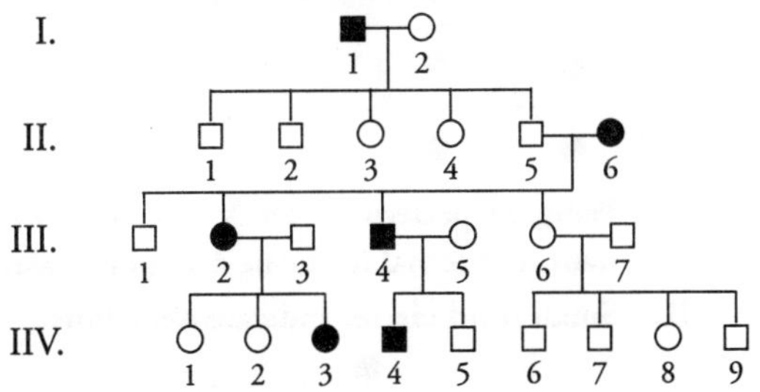

11. Three generations' pedigree showing a segregation of an autosomal dominant disorder and alleles (A and B) at a locus that may or may not be linked to the disease locus.

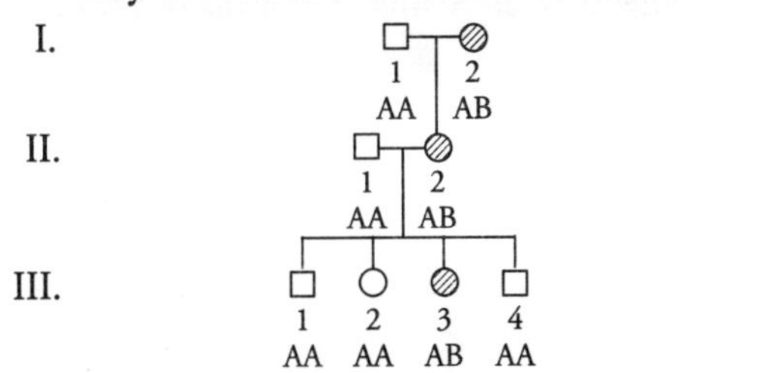

12. The following is a family pedigree for polydactyle (a autosomal dominant trait) in man:

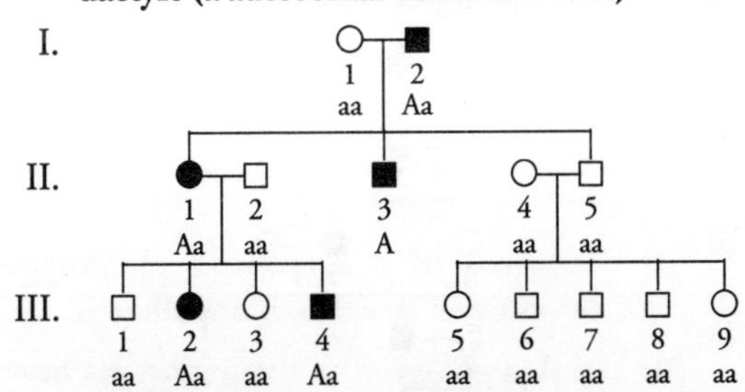

○ = Normal female, □ = Normal male

■ = Polydactyle male, ● = Polydactyle female

13. Pedigree pattern of X-linked recessive trait.

(i) Note the oblique pattern of inheritance.

(ii) Hollow circle with middle dot indicates carrier females.

A.

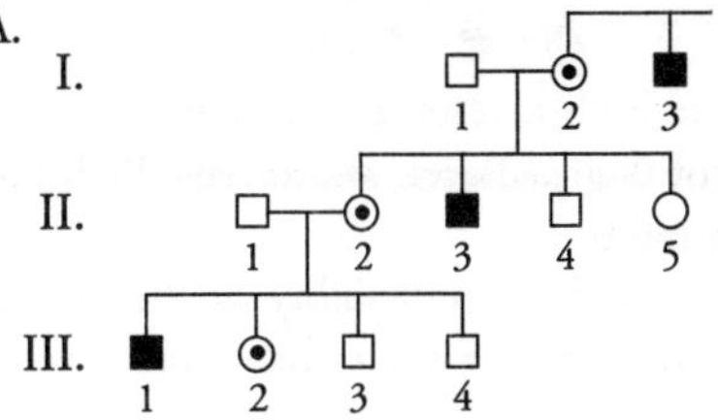

B. In this diagram it is observed that an affected female can result from the mating of an affected male and a carrier female.

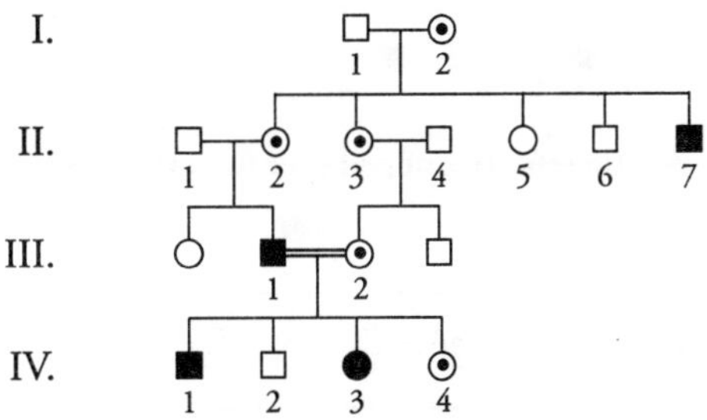

C. In this diagram it is observed that an affected male mating with a normal non-carrier female has all normal sons and all carrier daughters.

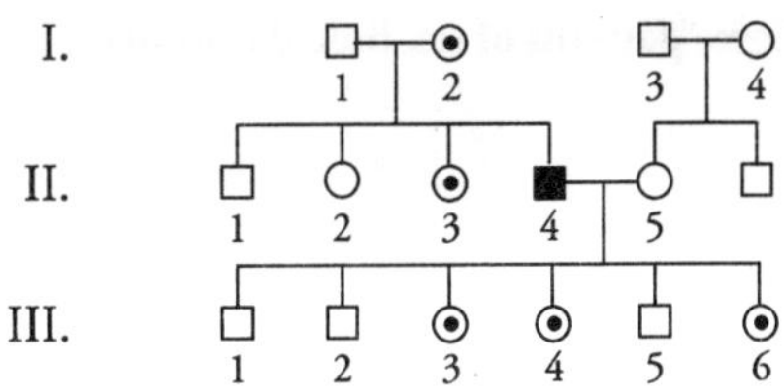

14. A human pedigree of the X-linked (recessive) colour blindness trait.

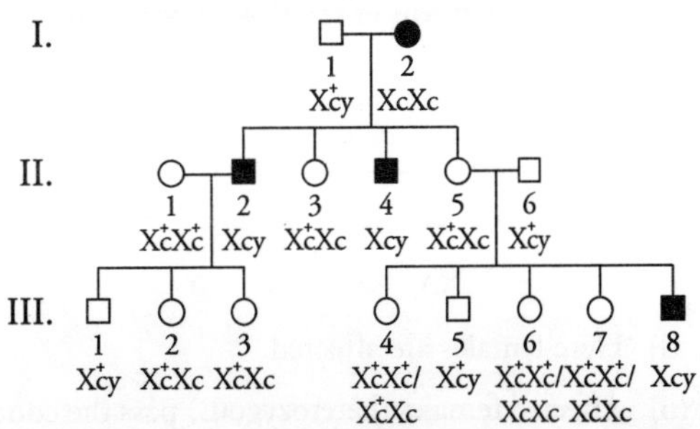

15. A representative pedigree for colour blindness in humans is inherited as an X-linked recessive trait.

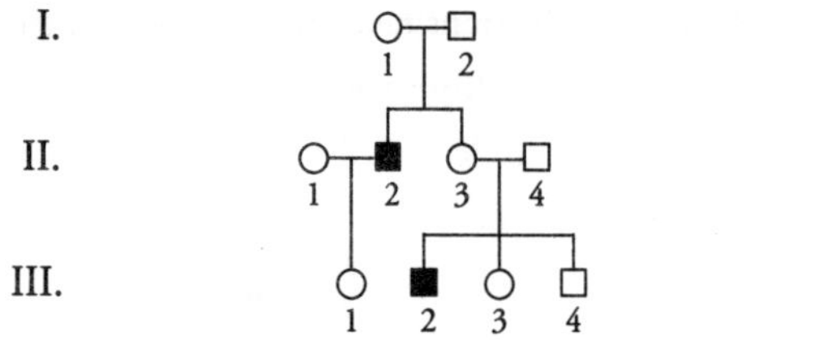

16. The following pedigree shows the X-linked recessive trait:

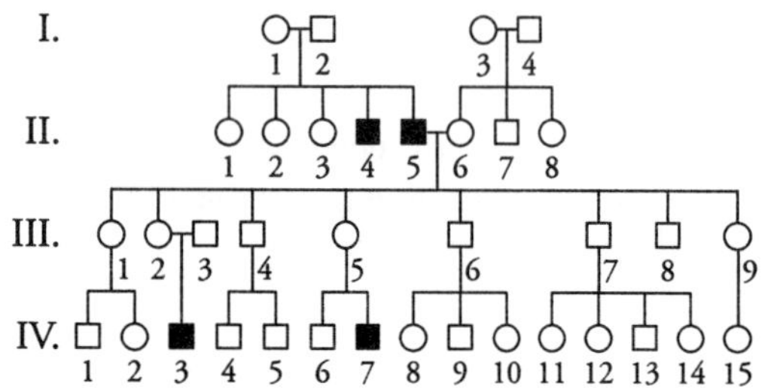

17. A representative of pedigree for a recessive trait:

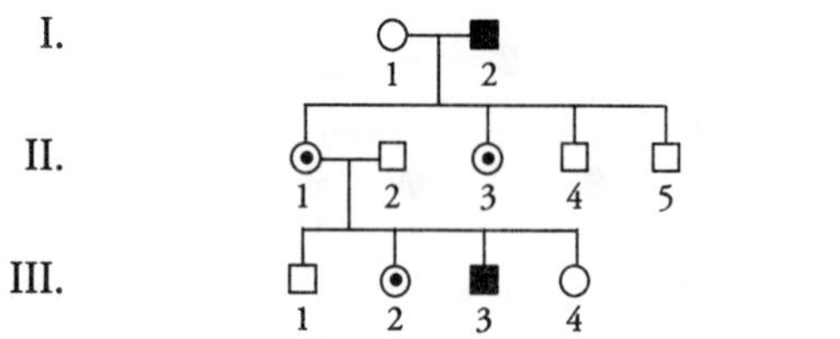

Hints: ■ = affected, ○ = carrier

18. Pedigree patterns of sex-linked dominant trait:

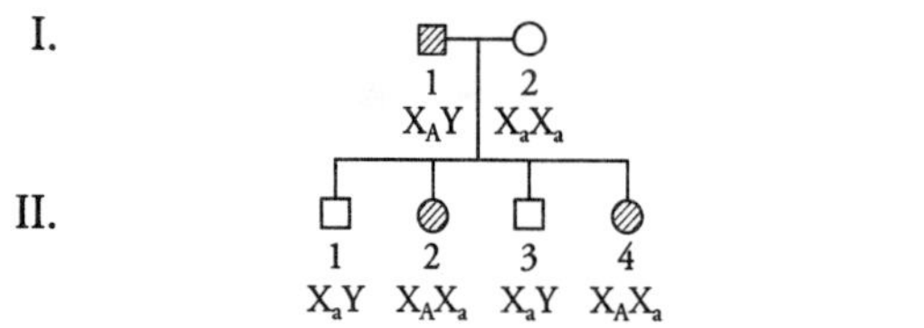

Here all the daughters are affected form.

19. Pedigree patterns of sex-linked dominant trait:

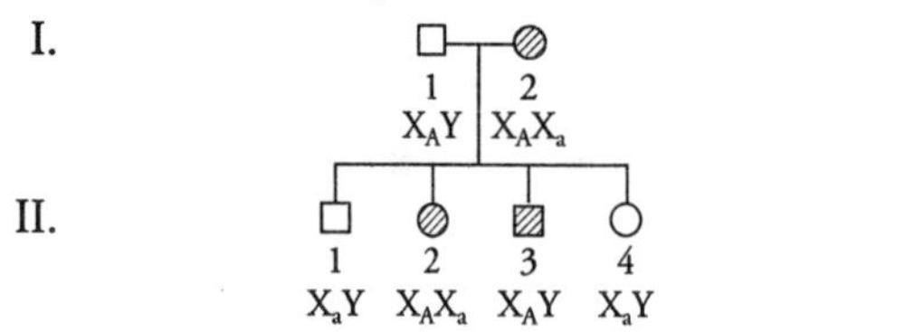

(i) Here females are affected.

(ii) Affected females (heterozygous) pass the condition to one half $(\frac{1}{2})$ of their progeny.

20. A representative of pedigree for sex-linked dominant inheritance:

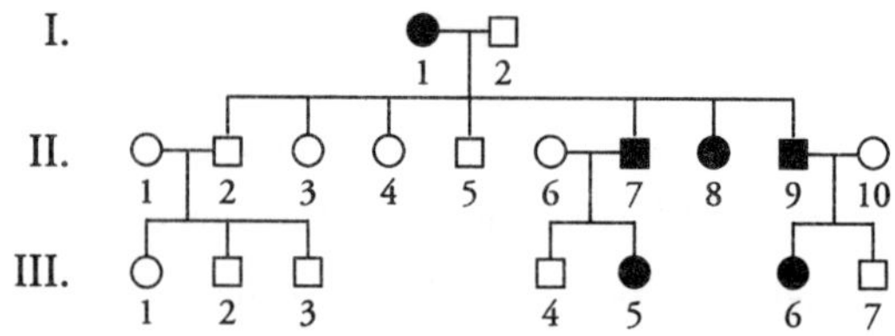

Hints: (i) Affected male passes its characters to all his daughters but to none of his sons.

(ii) Affected males come from affected mother.

21. A representative of pedigree for an X-linked dominant trait:

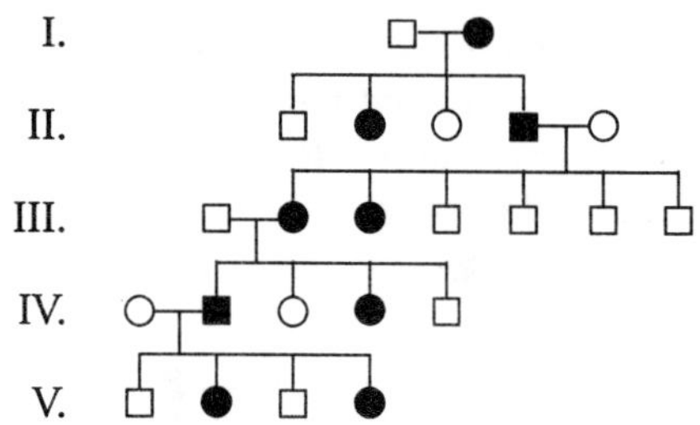

22. I. Pedigree pattern of an X-linked dominant trait lethal in the hemizygous male.

II. Small solid circles indicate abortions.

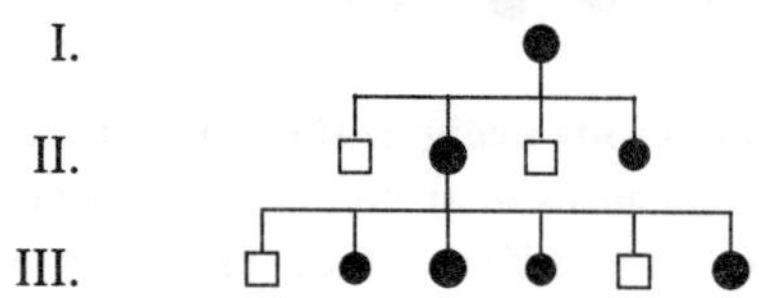

23. Pedigree of an X-linked dominant disorder:

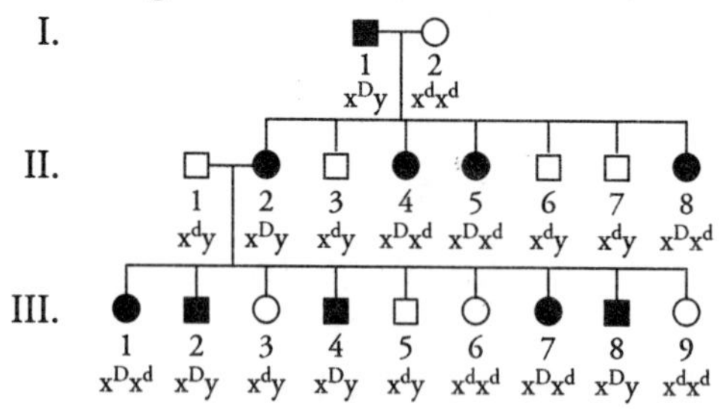

24. Study the pedigree below:

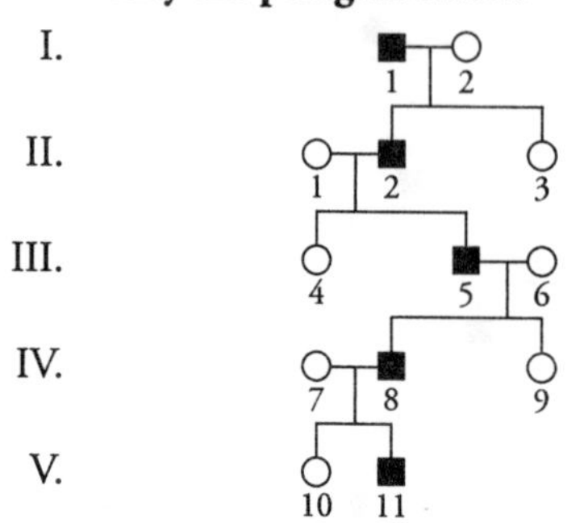

(a) What is the probable mode of inheritance?

(b) Are you positive about this conclusion?

(c) With regard to (b) why or why not?

Ans. (a) Probable mode of inheritance—Y-linked gene inheritance (holandric transmission).

(b) No.

(c) Could also be sex-limited, that is, the allele only expresses in the males.

25. A representative pedigree for Y-linked trait:

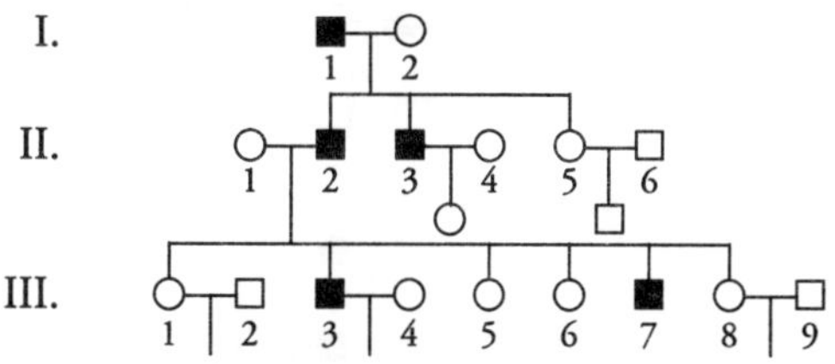

26. Assume that all of the pedigrees illustrated below show one of the following modes of inheritance.

(a) Holandric, (b) Sex-influenced, (c) Sex-limited, (d) Sex-linked recessive.

Assign the most probable mode of inheritance to each of the pedigrees, using each choice only once.

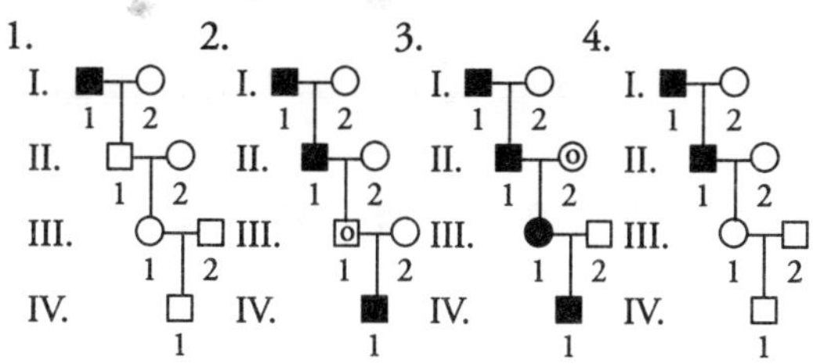

Key: □ & ○ = as normal phenotypes, ■ & ● = as affected phenotypes, ◎ & ⊡ = as normal phenotype but carriers—(heterozygous)

Solution:

1. It is sex-linked recessive inheritance. Here X-linked trait (I-1) denotes Y-chromosome to his son.
2. It is a kind of sex-limited transmission, the trait is transmitted through the daughter III-1 to the male (IV-1).
3. It is a sex influenced transmission. Both male and female can express the trait, if it shows dominance in the males and recessiveness in the females.
4. It is a holandric transmission. It occurs through male to male. But the expression of the trait in consecutive generation is broken by the female III-1.

27.

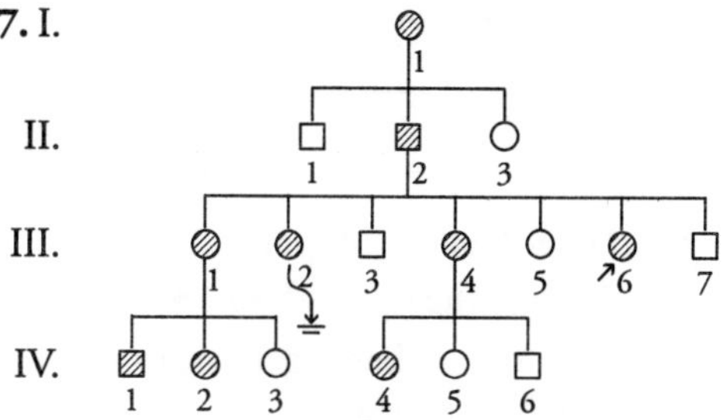

(a) Analyse the mode of inheritance.

(b) Give genotypes of all the members of the pedigree.

28.

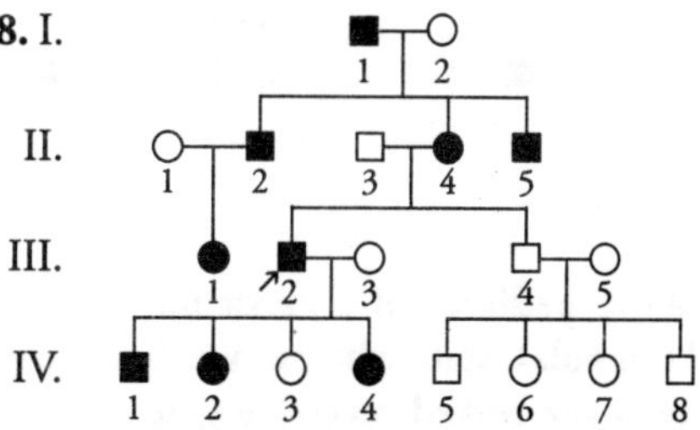

(a) Describe the pedigree.

(b) Explain the mode of inheritance and genotypes of all the members of the pedigree.

29.

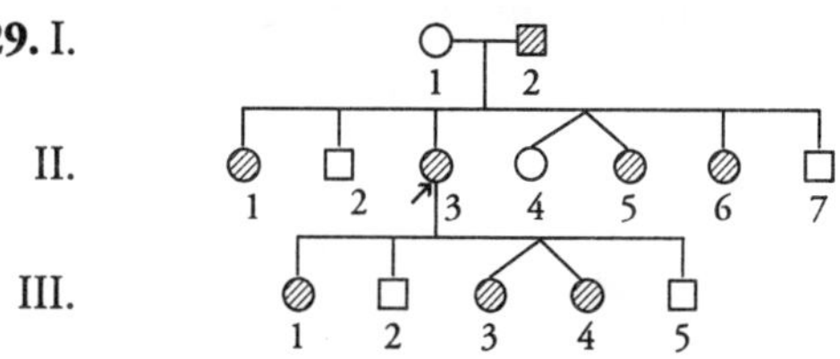

(a) Describe the pedigree.

(b) Explain the mode of inheritance and genotypes of all the members of the pedigree.

30.

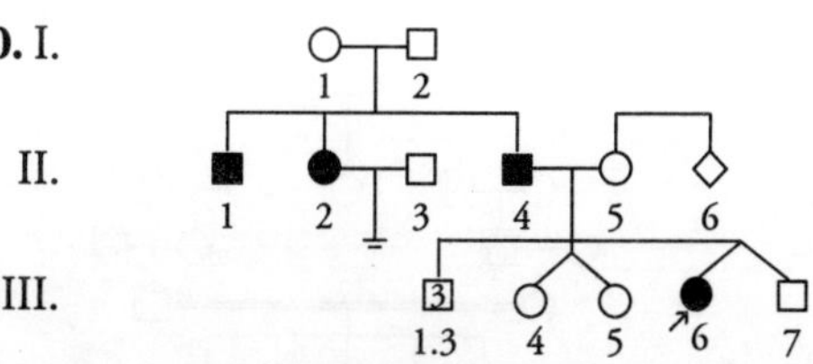

(a) Describe the pedigree.

(b) Explain the mode of inheritance and genotypes of all members of pedigree.

31.

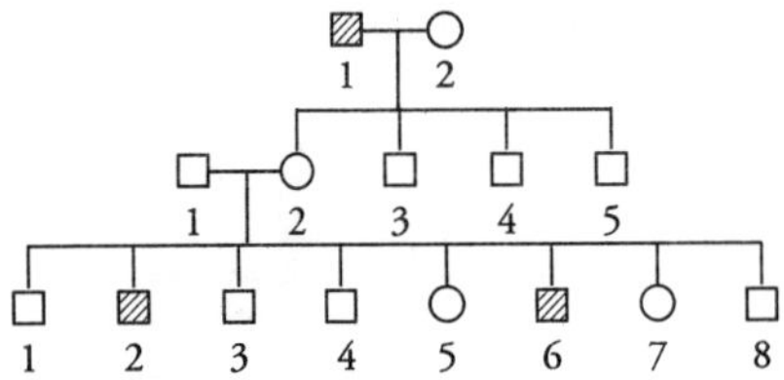

(a) Describe the pedigree.

(b) Analyse the mode of inheritance.

(c) Give the probable genotypes of all the members of pedigree.

32.

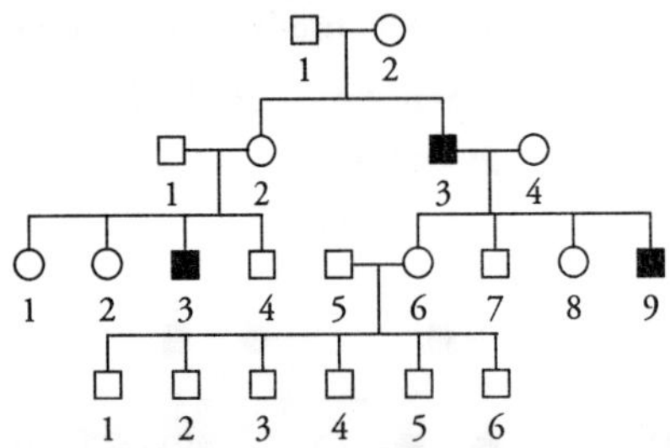

In the above pedigree of a particular rare trait (shaded symbols), state whether you believe it is caused by a sex-linked recessive gene or by an autosomal dominant gene that is expressed only in males.

33. What are the possible modes of inheritance in pedigrees a–c? What modes of inheritance are not possible for a given pedigree?

(a)

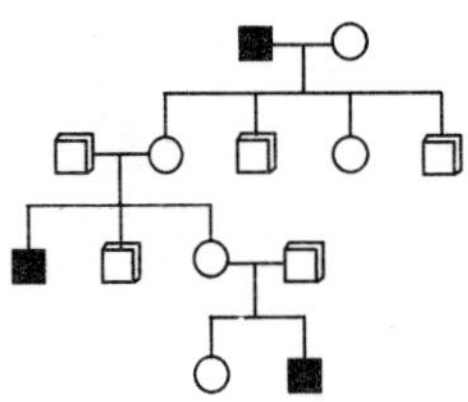

(b)

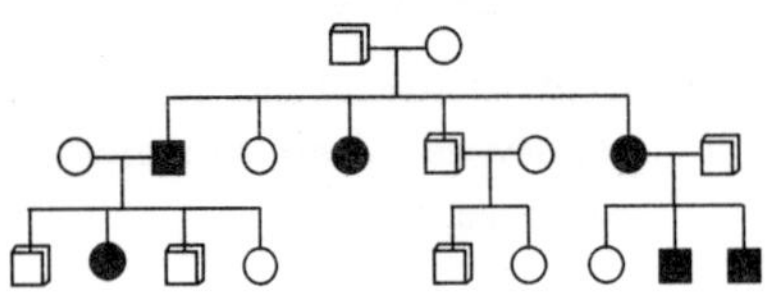

(c)

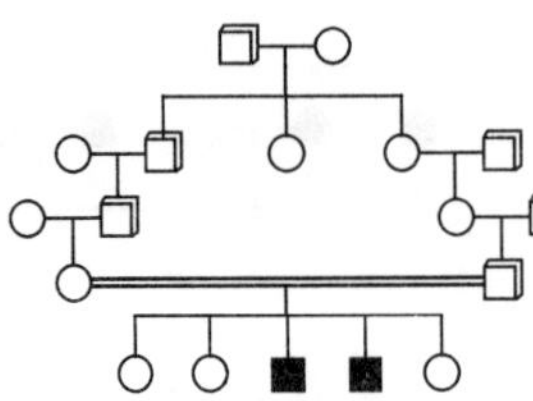

34. In pedigrees a–d in figure below, which show the inheritance of rare human traits, including twin production, determine which modes of inheritance are most probable, possible, or impossible:

(a)

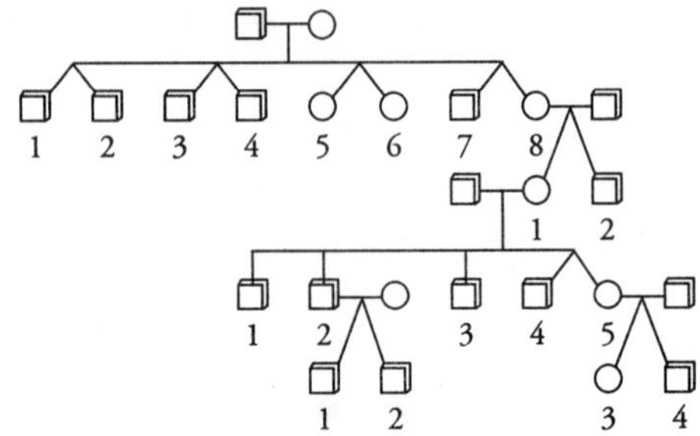

(b)

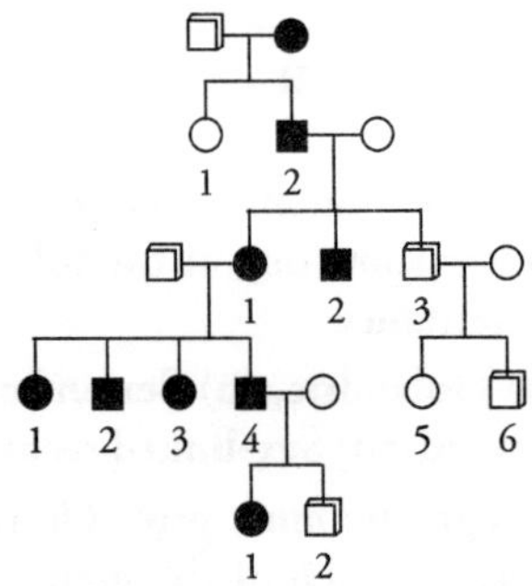

(c)

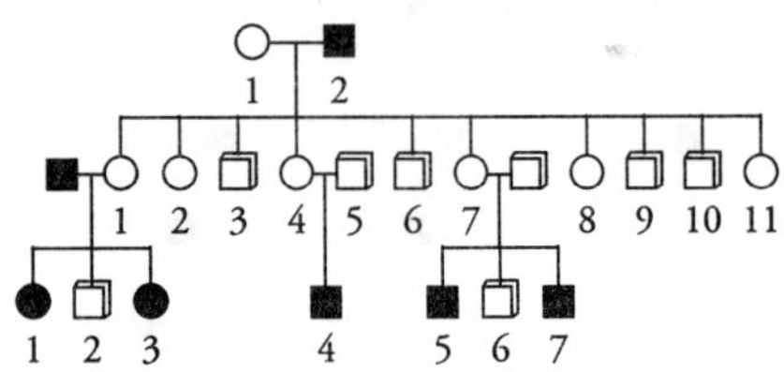

(d)

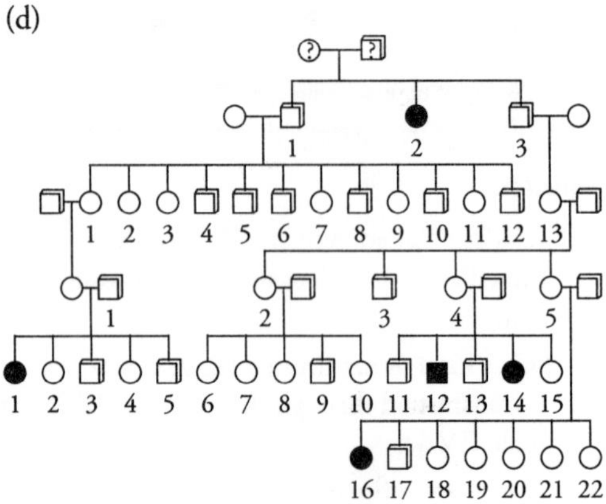

35. Determine the possible modes of inheritance for each trait in pedigrees a–c:

(a)

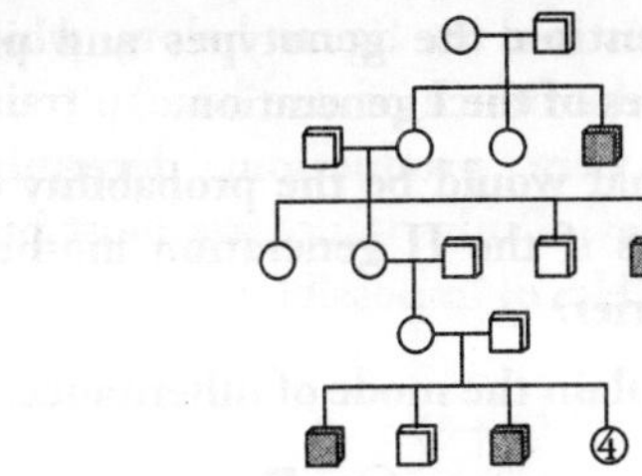

(b)

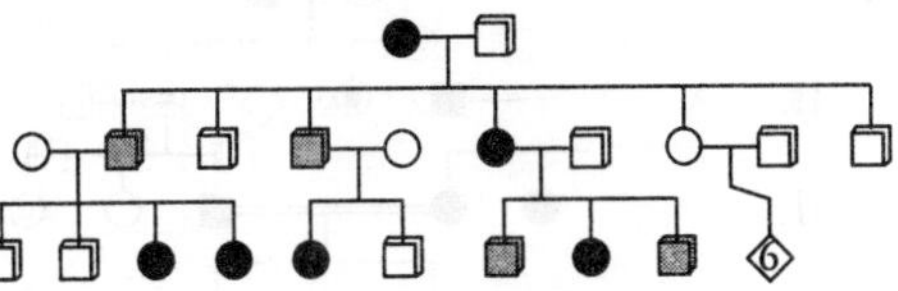

(c)

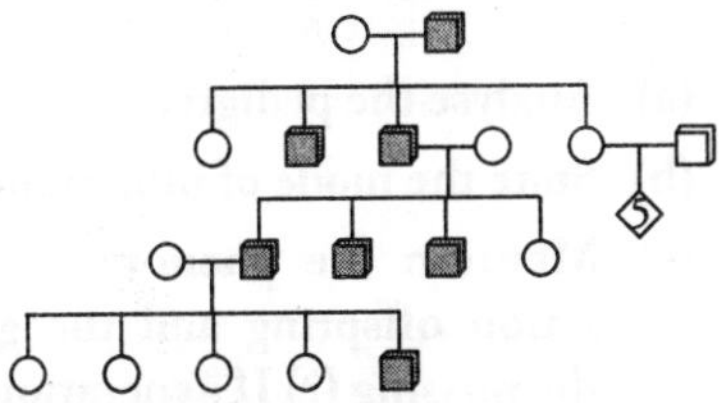

36. I.

II.

III.

IV.

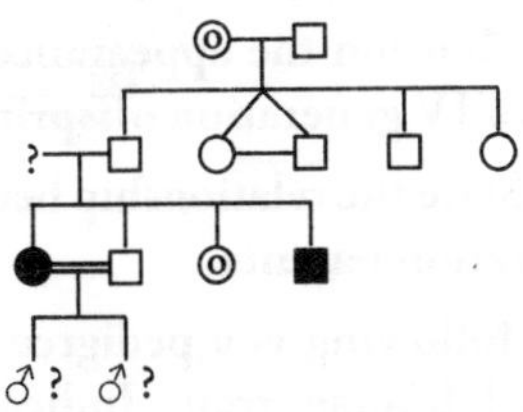

(a) Explain the pedigree with genotypes.

(b) State the mode of inheritance.

37. I.

II.

III.

IV.

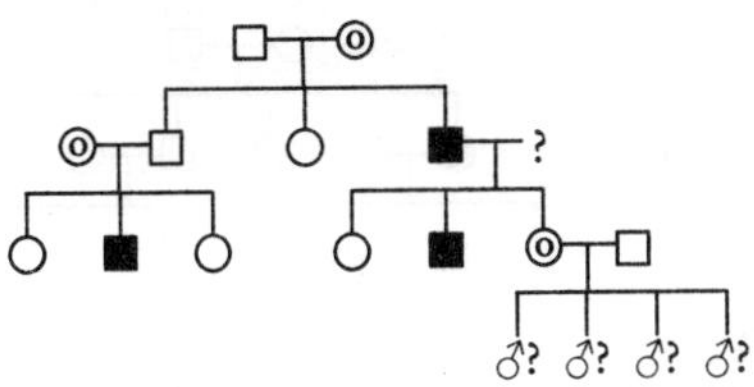

(a) Explain the pedigree.

(b) Write the genotypes and phenotypes of the offspring of IV generation.

(c) State the mode of inheritance.

38. I.

II.

III.

IV.

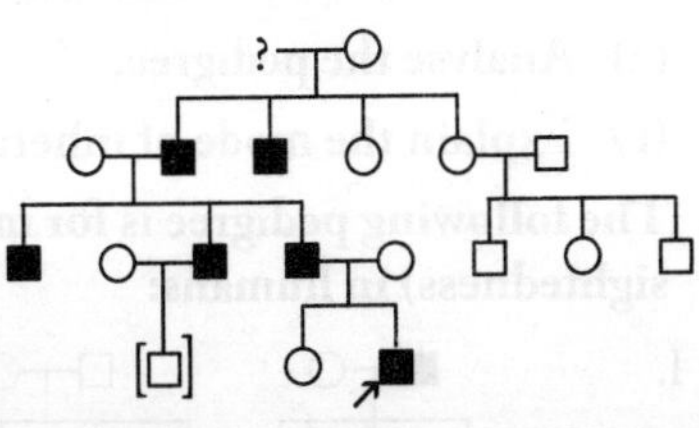

(a) Explain all the symbols in the pedigree.

(b) Give the genotype of I(1), II(3) and IV(1).

[P.G., M.G., E.G. & I.G. 22]

39. I.

II.

III.

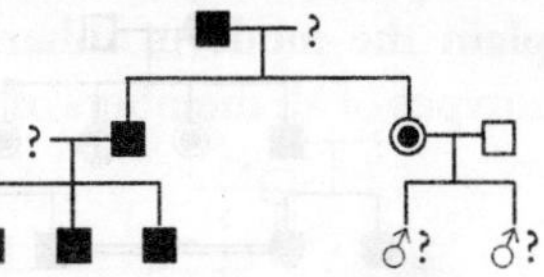

(a) Analyse the pedigree with genotypes.

(b) State the mode of inheritance.

40. I.

II.

III.

IV.

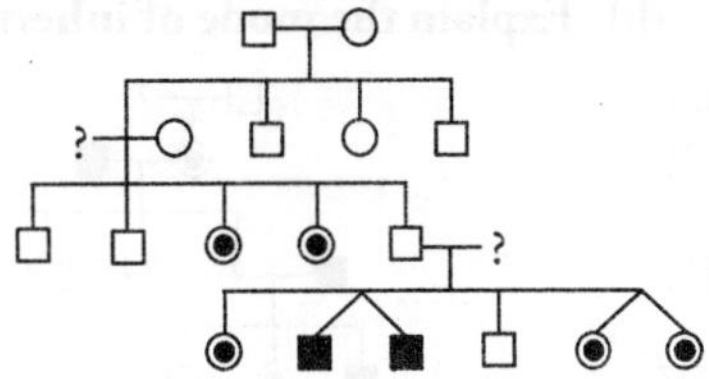

(a) Find out the genotype of the missing parents?

(b) State the pattern of inheritance and explain the IV generation offspring.

41. I.

II.

III.

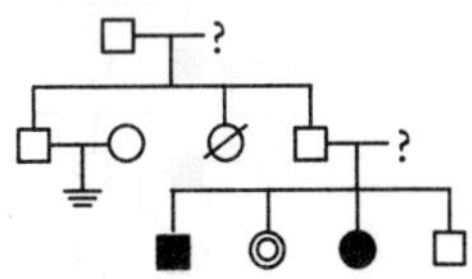

(a) Explain the pedigree symbols used in the chart.

(b) State the mode of inheritance.

42. I.

II.

III.

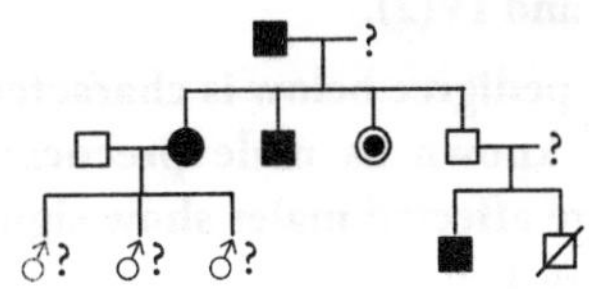

(a) Complete the pedigree with pedigree symbols.

(b) Find the genotype of the missing parents.

(c) State the mode of inheritance.

43. I.

II.

III.

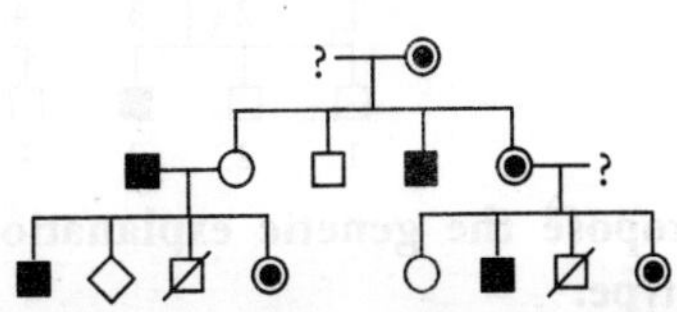

(a) Find out the genotype of the missing parents (?) in generations I and II.

(b) Explain the offsprings of generation III and state the mode of inheritance.

44.

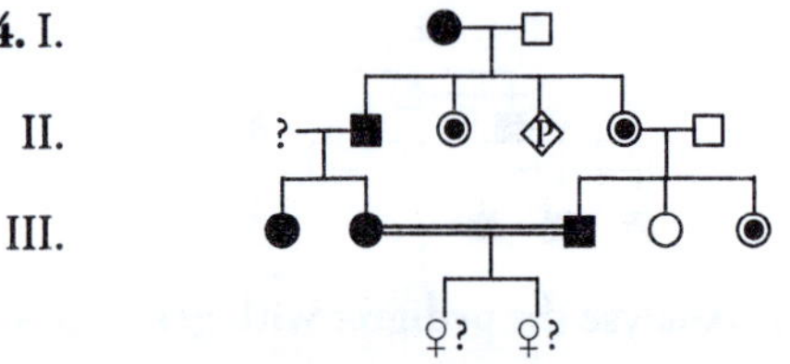

(a) Analyse the pedigree.

(b) Explain the mode of inheritance.

45.

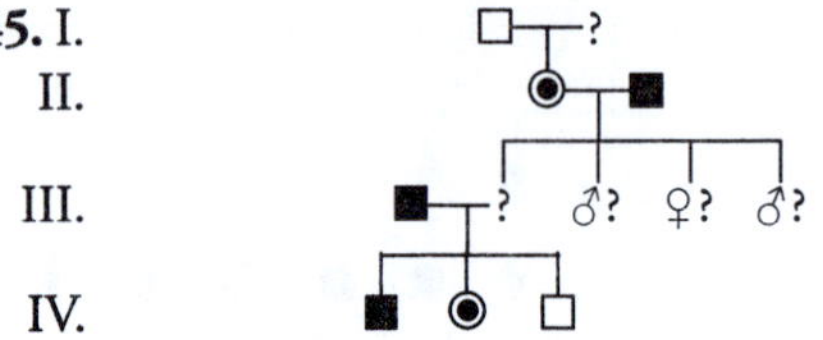

(a) Find out the genotype of the missing parent and III generation offspring.

(b) State the mode of inheritance.

46.

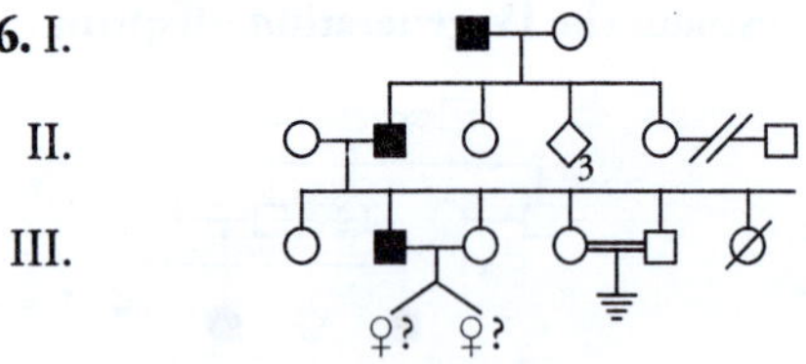

(a) Analyse the pedigree.

(b) State the mode of inheritance.

(c) Put the appropriate symbols in IV(1) and IV(2).

47. The pedigree below is characteristic condition known as male precocious puberty, where affected males show signs of puberty by age 4:

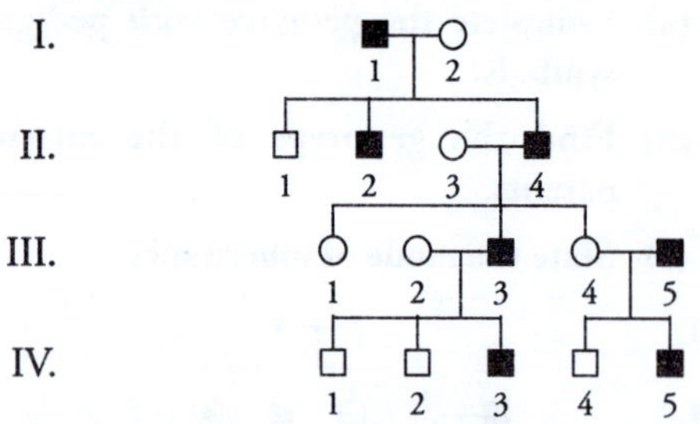

Propose the genetic explanation of this phenotype.

48.

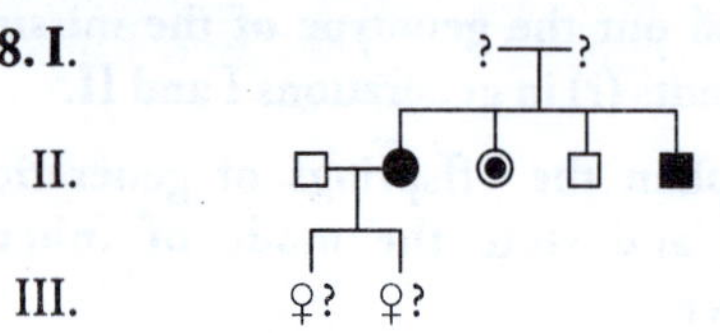

(a) Mention the genotypes and phenotypes of the I generation.

(b) What would be the probability of the sons if the II generation mother is a carrier?

(c) Explain the mode of inheritance.

49.

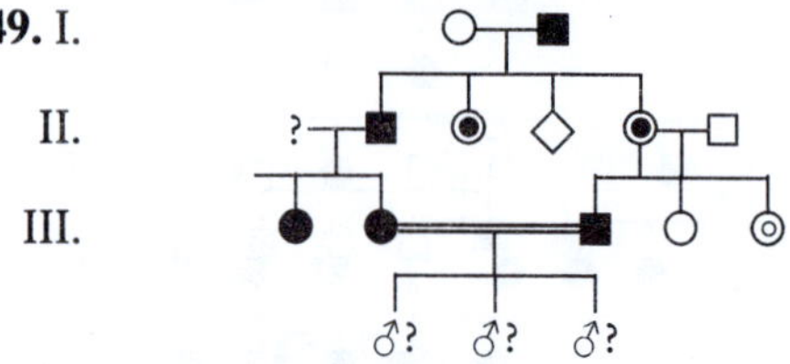

(a) Analyse the pedigree.

(b) State the mode of inheritance.

(c) Mention the phenotype of II generation offspring and the genotype of the missing (?) II generation parent.

(d) Mention the appearance and genotype of IV generation offspring.

(e) State the relationship between III generation parents.

50. The following is a pedigree for a rare sex-linked human trait. Indicate with a dot those females who are carriers and with a P those females who are potential carriers:

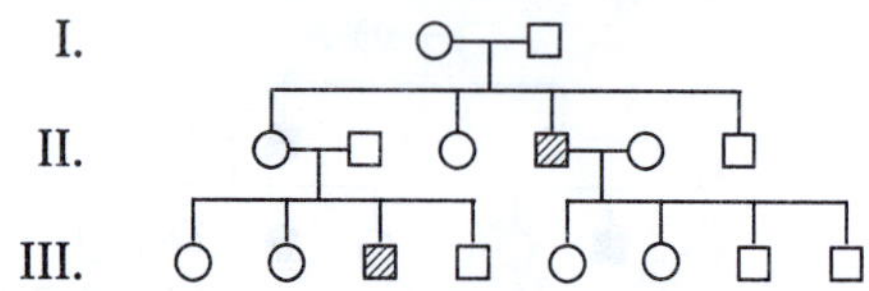

51.

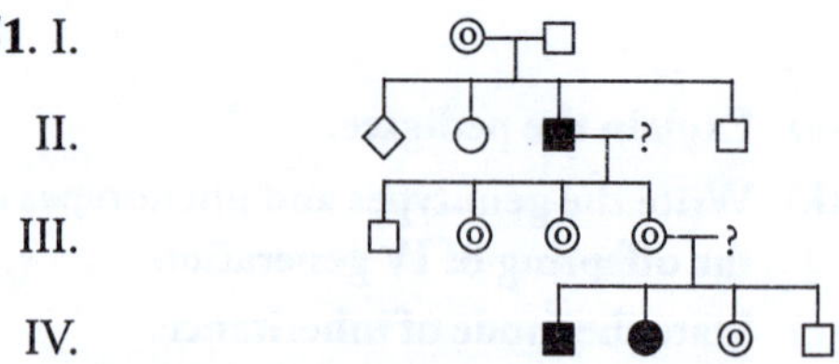

(a) Analyse the pedigree.

(e) Explain the mode of inheritance.

52. The following pedigree is for myopia (near-sightedness) in humans:

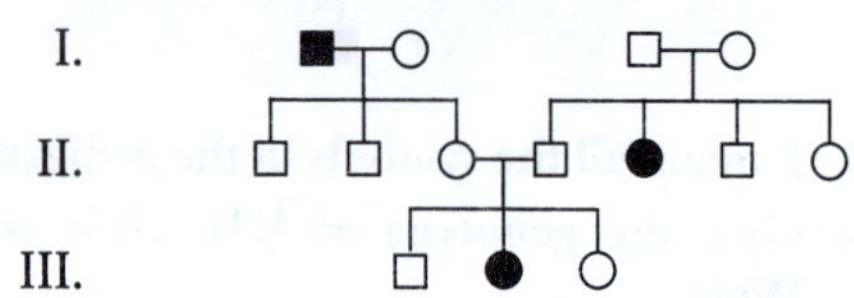

(a) **Predict whether the disorder is inherited as the result of dominant or recessive trait.**

(b) **Determine the most probable genotype for each individual based on your prediction.**

53.

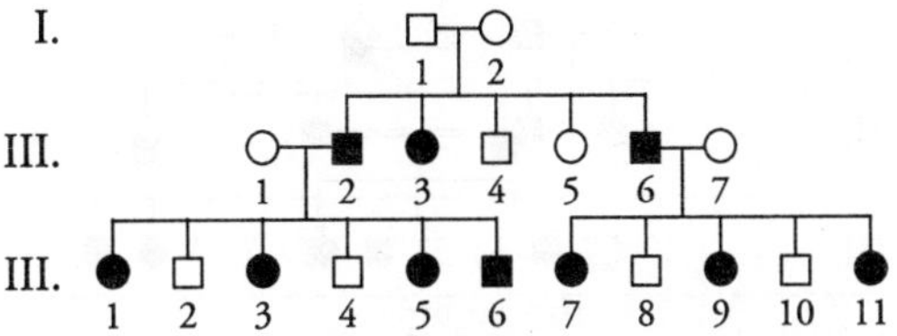

Determine the probable inheritance mode for the trait shown in the affected individuals by answering the following questions:

(a) **Y-linked inheritance can be excluded at a glance. What two other mechanisms of inheritance can be definitely excluded?**

(b) **Of the remaining mechanisms of inheritance, which is most likely to be excluded. Why?**

54.

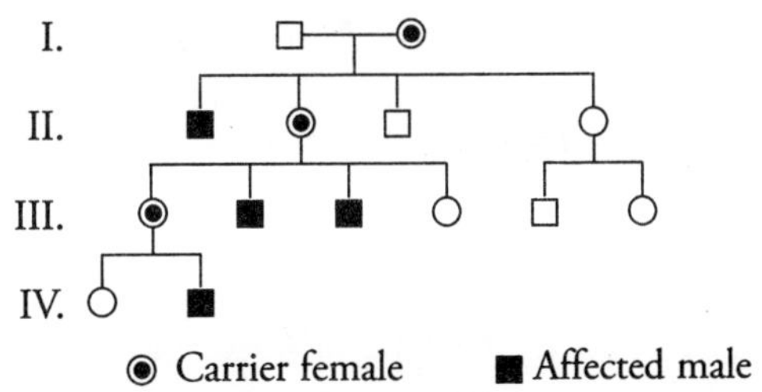

Fig. 10.1: Typical pedigree of X-linked inheritance in which affected males do not reproduce, for example, Duchenne muscular dystrophy.

55.

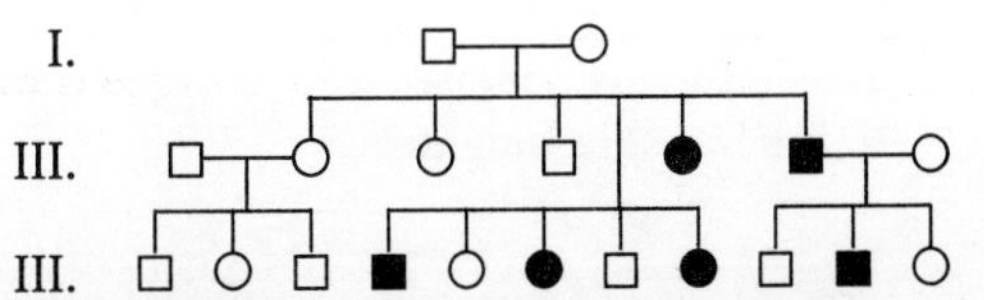

Fig. 10.2: Typical pedigree chart of autosomal dominant inheritance

56.

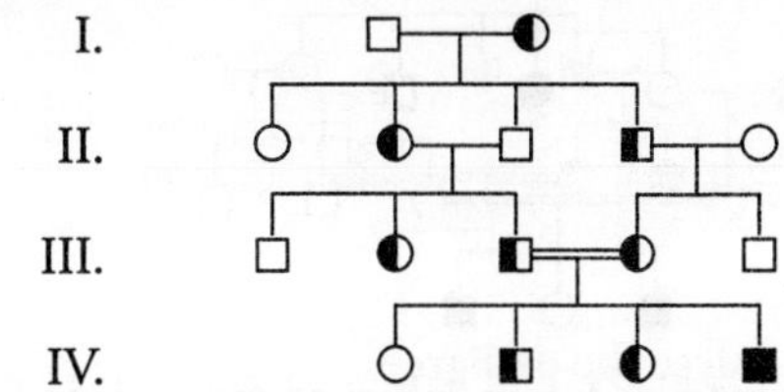

Fig. 10.3: Typical pedigree chart of autosomal recessive inheritance. Marriage between cousins in the third generation and its result in the fourth are shown. The recessive gene was inherited from a common female ancestor (first generation).

57.

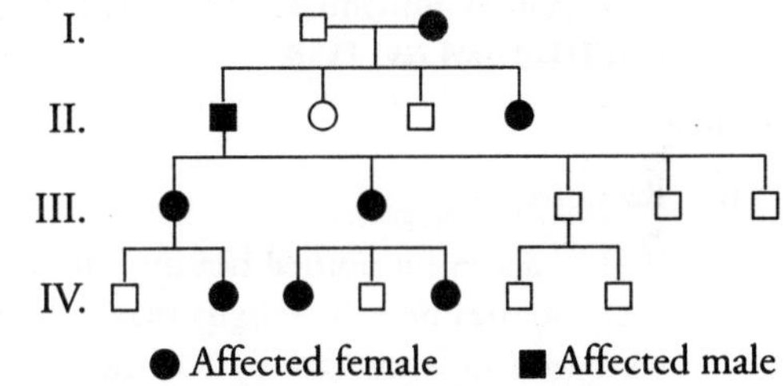

Fig. 10.4: Typical pedigree of X-linked dominant inheritance

58.

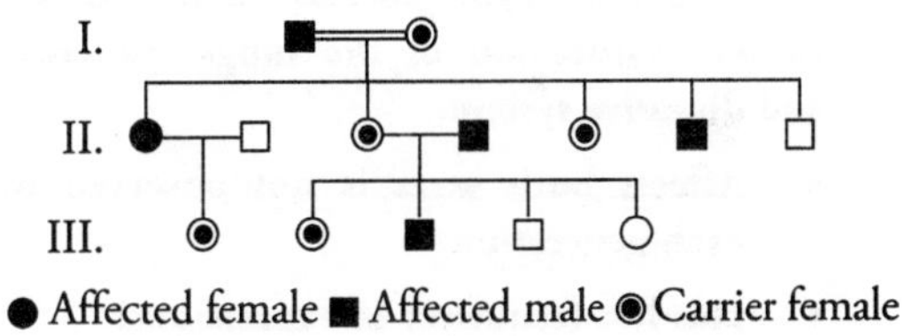

Fig. 10.5: X-linked recessive inheritance. A pedigree of hemophilia showing consanguinity between cousins (generation I) and an affected female (generation II).

59.

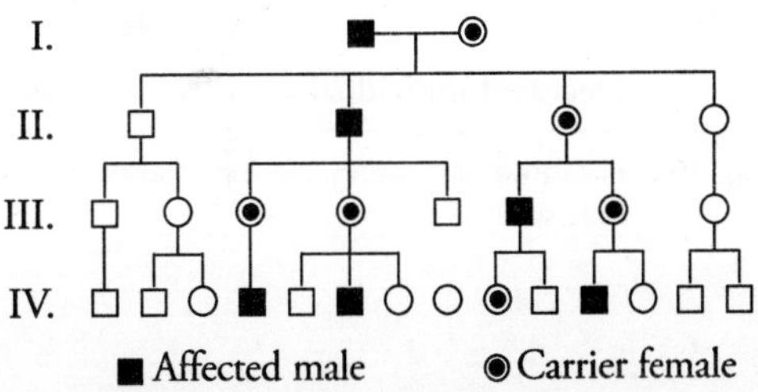

Fig. 10.6: Typical pedigree of X-linked recessive inheritance in which affected males reproduce, for example, red-green colour blindness.

60. I.

II.

III.

IV.

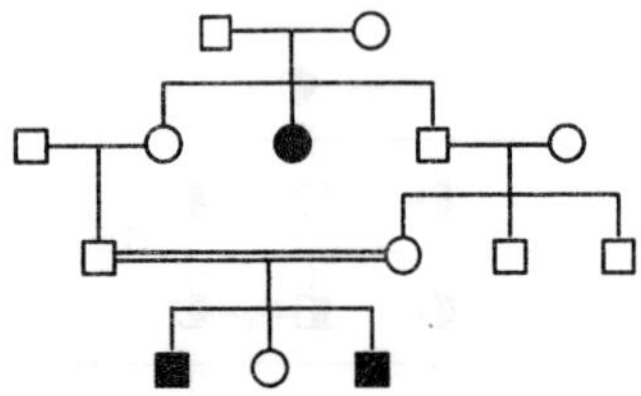

It is a pedigree for an inherited form of deafness (affected individuals are represented by a solid symbols).

(a) **Is the mode of inheritance of this trait recessive or dominant?**

(b) **Choose suitable systems, and give the genotypes of individuals (i) II=1, (ii) II-3, (iii) III-2 and (iv) IV-3.**

Solution:

(a) Recessive

(b) If D = alleles for normal hearing and d = alleles for deafness; (i) Dd/DD (has a child with woman whose family has the gene, unlikely to be carrying the deaf allele). (ii) dd, (iii) Dd, (iv) dd.

61. A typical pedigree segregation pattern for a recessive trait—*cystic fibrosis*—a disease that causes dysfunction of the lungs, pancreas and digestive system.

- **Affects both sexes is not observed in each generation.**
- **Has low frequency of expressions?**

I.

II.

III.

IV.

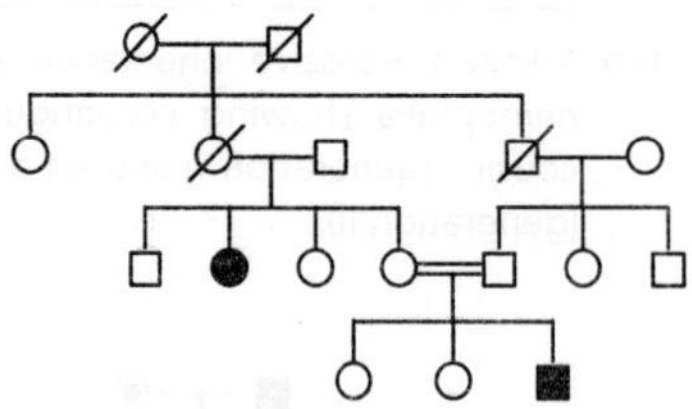

Deceased individual = Ø/☒

Fig. 10.7: Pedigree illustrating the inheritance of cystic fibrosis.

The probability of heterozygous parents having children expressing cystic fibrosis.

Parents: Normal (heterozygous) $\frac{CF}{cf}$ × Normal (heterozygous) $\frac{CF}{cf}$

Gametes: CF cf CF cf

♂ \ ♀	CF	cf
CF	$\frac{CF}{CF}$	$\frac{cf}{CF}$
cf	$\frac{CF}{cf}$	$\frac{cf}{cf}$

Results:

$\frac{CF}{CF}, \frac{CF}{cf}, \frac{CF}{cf} = 3$, normal; $\frac{cf}{cF}$ cystic fibrosis.

62. A pedigree for a human family in which the neurodegenerative Huntington disease is found.

- **Affects both sexes**
- **Appears in each generation within an affected family**
- **Affects about half of all the children in an affected family.**

I.

II.

III.

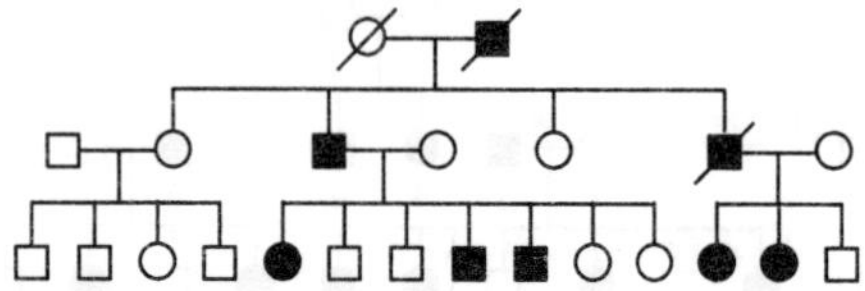

Fig. 10.8: Pedigree illustrating the inheritance of Huntington's disease.

Biostatics (A+B)

11

INTRODUCION

- **STATISTICS:**

It refers to the subject of scientific activity which deals with the theories and methods of collection, analysis and interpretation of such data.

- **BIOSTATISTICS:**

This term is used when tools of statistics are applied to the data that is derived from biological sciences.

- **BIOMETRICS OR BIOMETRY:**

The science which deals with the application of various statistical concepts and procedures to the study of biological problems is termed as Biometrics or Biometry.

- **BIOMETRICAL GENETICS:**

A branch of genetics which utilizes statistical and mathematical concepts and procedures for the study of genetical problems is called Biometrical Genetics.

It is also known as Mathematical Genetics or Statistical Genetics.

- **CONTRIBUTION OF R. A. FISHER IN BIOSTATISTICS:**

R. A. Fisher was a British statistician and geneticist. He extended and refined the field of statistics. His contributions are in experimental design, analysis of variance, and genetic variance.

- **CONTRIBUTION OF FRANK YATES IN BIOSTATISTICS:**

Frank Yates was an English statistician and biometrician. He worked with Fisher and published different books, viz., *Design* and *Analysis, Statistical Tables for Biological Agricultural* and *Medical Research, Sampling Methods for Census* and *Survey*, etc.

- **CONTRIBUTION OF KARL PEARSON IN BIOSTATISTICS:**

Karl Pearson was an English statistician and one of the founders of modern statistics. He applied statistics to biological problems of heredity and evolution. He developed concept on Chi-squares and coefficient of variation.

- **STANDARD DEVIATION:**

It is the square root of the arithmetic mean squares of deviation. In short SD may be defined as "Root-Mean-Square Deviation from Mean". It is usually denoted by the Greek small letter sigma (σ).

$$\sigma = \sqrt{\frac{\sum(x-\bar{x})^2}{N}}$$

- **COEFFICIENT OF STANDARD DEVIATION:**

It is the ratio of the standard deviation to its arithmetic mean.

Coefficient of standard deviation = $\frac{\sigma}{\bar{x}}$.

- **STANDARD ERROR OF MEAN:**

The difference or deviation between the value of statistic of a particular sample and the corresponding population parameter is known as Sampling Error or Standard Error. Thus, standard error is a measure of variation of the mean.

Calculation:

It is the standard deviation (SD) of the sample divided by the square root of the number of observation of the sample.

(i) SE of $(\bar{x})$ = $\frac{SD}{\sqrt{n}}$ (SD = standard deviation, n = size of the sample).

(ii) If the population has same SD, SE of $(\bar{x}_1 - \bar{x}_2)$ = SD $\sqrt{\frac{1}{n_1}+\frac{1}{n_2}}$ ($\bar{x}_1$ and $\bar{x}_2$ = population, and n_2 = size of the sample).

(iii) If two random samples with $\bar{x}_1$, σ_1, n_1 and $\bar{x}_2$, σ_2, n_2 are drawn from different population, $SE(\bar{x}_1 - \bar{x}_2) = \sqrt{\frac{\sigma_1^2}{n_1}+\frac{\sigma_2^2}{n_2}}$.

- **VARIANCE:**

The square of standard deviation is called Variance. It is denoted by σ^2.

Variance = $(SD)^2 = \sigma^2$.

- **COEFFICIENT OF VARIATION:**

It is the product of the coefficient of standard deviation × 100.

Coefficient of variation = $\frac{SD}{AM} \times 100$,

i.e., $\frac{\sigma}{\text{Arithmetic mean}} \times 100$.

- **ARITHMETIC MEAN:**

It is obtained by summing up all the observations in a sample and dividing the total by the number of observations.

- **VARIABILITY:**

The presence of differences among the individuals of a population due to differences in their genetic constitution or the environment in which they are grown is called Variability.

- **PHENOTYPIC VARIATION:**

The total variability, which is observable, is called Phenotypic Variability. It is measured in terms of phenotypic variance.

- **GENOTYPIC VARIATION:**

The inherent or genetic variability, which remains unaltered by environmental factors, is called Genotypic Variations. It is measured in terms of genotypic variance.

A—CHI-SQUARE TEST

It is a statistical test involve in the calculation of a quantity which is used to compare an "observed" ratio with an "expected" or "theoretical" ratio and to determine how closely the former fits with the latter.

The measure of chi-square enables us to find out the degree of discrepancy between observed frequencies and theoretical or expected frequencies and thus to determine whether the discrepancy between observed and theoretical frequencies is due to the error of sampling or due to the chance.

The chi-square is computed on the basis of frequencies in a sample and thus the value of chi-square so obtained is a "statistic". Chi-square is not a parametric test as its value is not derived from the observation in a population. Hence, chi-square test is a *Non-parametric Test.*

Chi-square (χ^2) test:

A statistical test to determine if the "observed" numbers deviate from those "expected" or "theoretical" number under a particular hypothesis.

$$\chi^2 = \frac{\sum_{i=1}^{n}\left[|(O-E)| - \frac{1}{2}\right]^2}{E}$$

where O = observed frequency, E = expected frequency, $\frac{1}{2}$ = 0.5 = Yates correction*.

Chi-square (χ^2) test was first used in testing statistical hypothesis by *Karl Pearson* in the year 1900.

- **WORKING PROCEDURE:**

1. Calculate all the expected frequencies, i.e., E for all values of i = 1, 2, 3, ..., n.
2. Take the difference between each observed frequency O and the corresponding expected frequency E for each value of i, i.e., (O – E).

- Yates correction = $\frac{1}{2}$ or 0.5, which is used to find out the formula of probability with accuracy and used only where degrees of freedom = 1, and expected classes are small.

3. Square the difference for each value of i, i.e., $(O - E)^2$ for all values of i = 1, 2, 3, ..., n.
4. Divide each square difference by the corresponding expected frequency, i.e., calculate $\frac{(O-E)^2}{E}$ for all values of i = 1, 2, 3, ..., n.
5. Add all these quotients obtained in the step 4, i.e.,

$$\chi^2 = \frac{\sum_{i=1}^{n}\left[|(O-E)|-\frac{1}{2}\right]^2}{E}$$

$\frac{1}{2}$ = Yates correction

It is the required value of chi-square.

- **IMPORTANT CHARACTERISTICS OF CHI-SQUARE:**

I. The value of chi-square is always positive as each pair is squared up.

II. χ^2 (chi-square) will be zero if each pair is zero and it may assume any value extending to infinity, when the differences between the observed frequency and expected frequency in each pair are unequal. Thus, chi-square lies between 0 and ∞.

III. Chi-square is a statistic, not a parameter.

USEFUL TERMS

- **STATISTICAL HYPOTHESIS:**

I. Any statement or assertion about a statistical population or the values of its parameters is called Statistical Hypothesis.

II. There are two types of hypothesis, viz., simple and composite.

(a) Simple Hypothesis: A statistical hypothesis which specifies the population completely (i.e., the probability distribution and all parameters are known) is called **Simple Hypothesis.**

(b) Composite Hypothesis: A statistical hypothesis which does not specify the population completely (i.e., either the form of probability distribution or some parameters remain unknown) is called **Composite Hypothesis.**

- **TEST OF HYPOTHESIS:**

A test of hypothesis is a procedure which specifies a set of "rules for decision" whether to "accept" or "reject" the hypothesis under consideration (i.e., null hypothesis).

- **NULL HYPOTHESIS:**

I. A statistical hypothesis which is set up (i.e., assumed) and whose validity is tested for possible rejection on the basis of sample observations is called a "Null Hypothesis".

II. It is denoted by H_0 and tested against alternatives.

III. Tests of hypothesis deal with rejection or acceptance of null hypothesis only.

Prof. R. A. Fisher remarked, "Null hypothesis is the hypothesis which is to be tested for possible rejection under the assumption it is true."

- **ALTERNATIVE HYPOTHESIS:**

I. The negation of null hypothesis is called the Alternative Hypothesis.

II. It is denoted by H_1 and H_∞.

III. It is not tested, but its acceptance (rejection) depends on the rejection (acceptance) of the null hypothesis.

IV. Alternative hypothesis contradicts the null hypothesis.

- **CRITICAL REGION OR REJECTION REGION:**

I. The set of values of the test statistic which leads to rejection of the null hypothesis is called Critical Region of the test.

II. The probability with which a true null hypothesis is rejected by the test is often referred to as **Size of the Critical Region.**

III. On the other hand, which lead to the acceptance of null hypothesis gives us a region called a **Acceptance Region.**

IV. The critical value separates the rejection region from the acceptance region.

- **TEST STATISTIC:**

I. After setting up the null hypothesis and alternative hypothesis, the test statistic is computed.

II. It is a statistic based on appropriate probability distribution.

III. It is used to test whether the null hypothesis set-up should be accepted or rejected.

- **LEVEL OF SIGNIFICANCE:**

I. The maximum probability with which a true null hypothesis is rejected is known as Level of Significance of the test.

II. It is denoted by α.

III. In framing decision rules, the level of significance is arbitrarily chosen in advance depending on the consequences of statistical decision.

IV. Generally, 5% or 1% level of significance is taken although other levels such as 2% or $\frac{1}{2}$% are also used.

- **DEGREES OF FREEDOM:**

It is the value of a sample which is freely variable without affecting the mean or it is an integer used to determine whether a chi-square value is statistically significant.

The number of data that are given in the form of a series of variables in a row or column or the number of frequencies that are put in cells in a contingency table which can be calculated independently is called the Degrees of Freedom and is denoted by df.

- **CALCULATION OF DEGREES OF FREEDOM:**

(i) If the data is given in the form of a series of variables in a row or column, then the degrees of freedom (df) = (the number of items in the series) – 1, i.e., df = n – 1, where 'n' is the number of observations.

(ii) When the number of frequencies are put in cells in a contingency table, the degrees of freedom will be the product of (number of rows less one) and the (number of columns less one), i.e., df = (R – 1) (C – 1), where 'R' is the number of rows and 'C' is the number of columns.

- **CONDITIONS FOR USING THE CHI-SQUARE TEST:**

I. Each of the observations making up the sample for this test should be independent of each other.

II. The χ^2-test applied in a four-fold table, will not give a reliable result with one degree of freedom if the expected value in any cell is less than 5. In such cases, to apply χ^2-test Yates correction is necessary.

III. The total number of observations used in this test must be large, i.e., $n \geq 50$.

IV. This test is used only for drawing inferences by testing hypothesis. It cannot be used for estimation of parameter or any other value.

V. It is wholly dependent on the degrees of freedom.

VI. The frequencies used in χ^2-test should be absolute and not relative in terms.

VII. The observation collected for χ^2-test should be on random basis of sampling.

TYPES

- **CHI-SQUARE:**
  1. Goodness of fit.
  2. Contingency chi-square.
  3. Homogeneity chi-square.

1. Test for Goodness of Fit (*Pearsonian* χ^2)

Chi-square test is also applied as a test of "goodness of fit" to determine whether the actual or observed numbers or frequencies are similar or in "good agreement" with the expected or theoretical number of frequencies.

Thus, the test is called "goodness of fit".

$$\chi^2 = \sum \frac{\left[|(O-E)| - \frac{1}{2}\right]^2}{E}$$

where O = observed frequency, E = expected frequency, $\frac{1}{2}$ = 0.5 = Yates correction.

Working Procedure:

In order to test "the goodness of fit" of the observed results:

I. It is necessary to find deviation between the observed and the expected results.

II. If the expected values or frequencies in any observation are less than 5 in one degrees of freedom, Yates correction is necessary, i.e., reduction of $\left[|(O-E)|-\frac{1}{2}\right]$. In other words, subtract 0.5 from absolute difference between observed and expected frequencies (generally in monohybrid cross).

III. Here the degrees of freedom (df) = the number of items in the series –1 = n – 1.

Null Hypothesis (assume) for Goodness of Fit:

I. 1 : 1 Monohybrid test cross ratio.

II. 3 : 1 Simple monohybrid Mendelian ratio (F_2).

III. 9 : 3 : 3 : 1 Simple Mendelian dihybrid ratio (F_2).

IV. 1 : 1 : 1 : 1 Dihybrid testcross ratio.

Example:

In a cross between black and white coat coloured mice, the F_2 individual segregated into 787 black and 277 white coat coloured individuals. If you have to test that these results agree with the expected ratio 3 : 1, then apply chi-square P = 5%.

Solution:

I. *Null Hypothesis:* 3 : 1

II. *Alternative Hypothesis:* 1 : 1

III. *Calculations:* $\chi^2 = \dfrac{(|O-E|-0.5)^2}{E}$.

Observed (O)	*Expected* (E)	$|(O-E)|$	$\{(|O-E|)-0.5\}$	$\chi^2 = \frac{(|O-E|-0.5)^2}{E} = \frac{D^2}{E}$
Black = 787	798	787 – 798 = –11	$|11| - 0.5 = 10.5$	$\frac{(10.5)^2}{798} = \frac{110.25}{798} = 0.138$
White = 277	266	277 – 266 = 11	$|11| - 0.5 = 10.5$	$\frac{(10.5)^2}{266} = \frac{110.25}{266} = 0.414$
Total = 1064				$\chi^2 = 0.552$

IV. Critical Value: The control value of χ^2 at 0.05 for df = 2 – 1 = 1 is 3.84.

V. Decision: Since calculated value of χ^2 = 0.55, so the null hypothesis is accepted.

Therefore, the data has good fit to the ratio 3 : 1.

Example:

Following are tested some of Mendel's reported results with the garden pea. Test each for goodness of fit:

Cross	*Progeny*	*Hypothesis*
(a) Green × Yellow pods	(F_2) 428 : 152	3 : 1
(b) Violet red × White flower	(F_1) 47 : 40	1 : 1
(c) Round yellow × Wrinkled green	(F_1) 31 : 26 : 27 : 26	1 : 1 : 1 : 1

Solution (a):

I. *Null Hypothesis:* 3 : 1

II. *Alternative Hypothesis:* 1 : 1

III. *Calculation*:

Observed (O)	*Expected* (E)	$\|(O-E)\|$	$(O-E)^2$	$\frac{(O-E)^2}{E}$
428	$\frac{3}{4} \times 580 = 435$	$\|428-435\| = \|-7\|$	49	$\frac{49}{435} = 0.113$
152	$\frac{1}{4} \times 580 = 145$	$\|152-145\| = \|7\|$	49	$\frac{49}{435} = 0.113$
Total = 580				$\chi^2 = 0.451$

IV. *Critical Value*: The control value of χ^2 at 0.05 and for $2 - 1 = 1$ degrees of freedom is 3.84.

V. *Decision*: Since the calculated value of chi-square (χ^2) = 0.451 < critical value of χ^2 for df = 3.84, so the null hypothesis is accepted, i.e., there is no significant variation with data. So it is the result of F_2-monohybrid cross.

Solution (b):

I. *Null Hypothesis*: 1: 1

II. *Alternative Hypothesis*: 3: 1

III. *Calculation*:

Observed (O)	*Expected* (E)	$\|(O-E)\|$	$(O-E)^2$	$\frac{(O-E)^2}{E}$
47	$\frac{87}{2} = 43.5$	$47 - 43.5 = \|3.5\|$	12.25	$\frac{12.25}{43.5} = 0.281$
47	$\frac{87}{2} = 43.5$	$40 - 43.5 = \|-3.5\|$	12.25	$\frac{12.25}{43.5} = 0.281$
Total = 87				$\chi^2 = 0.562$

IV. *Critical Value*: The control value of χ^2 at 0.05 and for $2 - 1 = 1$ degrees of freedom is 3.84.

V. *Decision*: Since the calculated value of chi-square χ^2 = 0.562 < critical value of χ^2 for 1 df = 3.84, so the null hypothesis is accepted, i.e. the variation is non-significant.

Solution (c):

I. *Null Hypothesis*: 1 : 1 : 1 : 1

II. *Alternative Hypothesis*: 9 : 3 : 3 : 1

III. *Calculation*:

Observed (O)	*Expected* (E)	$\|(O-E)\|$	$(O-E)^2$	$\frac{(O-E)^2}{E}$
31	$\frac{1}{4} \times 110 = 27.5$	$31 - 27.5 = \|3.5\|$	12.25	$\frac{12.25}{27.5} = 0.445$
26	$\frac{1}{4} \times 110 = 27.5$	$26 - 27.5 = \|-1.5\|$	2.25	$\frac{2.25}{27.5} = 0.082$
27	$\frac{1}{4} \times 110 = 27.5$	$27 - 27.5 = \|-0.5\|$	0.25	$\frac{0.25}{27.5} = 0.009$
26	$\frac{1}{4} \times 110 = 27.5$	$26 - 27.5 = \|-1.5\|$	2.25	$\frac{2.25}{27.5} = 0.082$
Total = 110				$\chi^2 = 0.618$

IV. *Critical Value*: The control value of chi-square at 0.05 and for 4 – 1 = 3 degrees of freedom is 7.82.

V. *Decision*: Since the calculated value of chi-square (χ^2) = 0.618 < critical value of χ^2 for 3 df = 7.82, so the null hypothesis is accepted, i.e., the variation is non-significant.

2. Test for Independence of Attributes or Contingency Chi-square:

I. It is applied to test the association between the attributes when the sample data is presented in the form of contingency table with any number of rows or columns.

II. It is occasionally desirable to compare one set of observations taken under particular conditions to those of a similar nature taken under different conditions.

III. In this case there are no definite expected values; the question is, whether the results are dependent (contingent upon) or independent of conditions under which they are observed. This test is, therefore, called a test for "independence of attributes or contingency test".

Working Procedure:

I. Set up (a) Null Hypothesis (H_0): No association exists between the attributes.

(b) Alternative Hypothesis (H_1): An association exists between the attributes.

II. Calculate the expected frequency 'E' corresponding to each cell by the formula

$$E_{ij} = \frac{R_i \times C_j}{N},$$

where R_i = sum total of the row in which E_{ij} is lying,

C_j = sum total of the column in which E_{ij} is lying,

N = total sample size.

III. Calculate the chi-square by the formula

$$\chi^2 = \sum \frac{(O-E)^2}{E}.$$

[E = expected value and O = observed value.]

Here degrees of freedom df = (R – 1)(C – 1)

[R = number of rows, C = number of columns in the contingency table].

IV. Find from the table the value of chi-square for the given value of the level of significance (α) and for the degrees of freedom (df).

If no value for α is mentioned, then the table $\alpha = 0.05$.

V. (i) Compare the computed value of chi-square with the tabled value of χ^2.

(ii) If the calculated value of χ^2 < tabulated value, then accept the hypothesis (H_0).

If χ^2 > tabulated value, reject the null hypothesis and accept the alternative hypothesis.

N.B.: It is used in (2 × 3) contingency tests.

Or,

Working Procedure:

I. Set up (a) Null Hypothesis (H_0): No association exists between the attributes.

(b) Alternative Hypothesis (H_1): An association exists between the attributes.

II. Here contingency table has only 2 rows and 2 columns with four cell frequencies, viz., a, b, c, d as shown below:

	a	b	R_1
	c	d	R_2
Total =	C_1	C_2	N

III. The frequency 'a' is placed in the upper left cell and 'd' is placed in the lower right cell. The frequencies 'b' and 'c' are placed in other diagonal position.

IV. For simplicity, R_1 and R_2 denote row totals, and C_1 and C_2 denote the column totals, i.e.,

$R_1 = a + b$, $R_2 = c + d$

$C_1 = a + c$, $C_2 = b + d$

$N = a + b + c + d = R_1 + R_2 = C_1 + C_2$

V. Therefore,

$$\chi^2 = \frac{N\{(|ad-bc|)-\frac{N}{2}\}}{R_1 \times R_2 \times C_1 \times C_2} \quad \left[\frac{N}{2} = \text{Yates correction}\right]$$

VI. Degrees of freedom = (R – 1)(C – 1) = 2 – 1 × 2 – 1 = 1.

[For 1 df, Yates correction is to be done.]

N.B.: It is used in (2 × 2) contingency test.

Example:

The following table gives the classification of 100 workers according to sex and the nature of work. Justify whether the nature of work is independent of the sex of the worker:

Sex	*Skilled*	*Unskilled*
Male	40	20
Female	10	30

Solution:

I. *Null Hypothesis*: Nature of work independent of the sex of worker (assume).

II. *Alternative Hypothesis*: Dependent of the sex of the worker (assume).

III. *Calculation*:

Sex	*Skilled*	*Unskilled*	*Total*	
Male	40 a	20 b	60	R_1
Female	10 c	30 d	40	R_2
	50	50	100	
	C_1	C_2		

$$\chi^2 = \frac{N\{(|ad-bc|)-\frac{N}{2}\}^2}{R_1 \cdot R_2 \cdot C_1 \cdot C_2}$$

$ad = 40 \times 30 = 1200$
$bc = 20 \times 10 = 200$
$R_1 = a + b = 60$
$R_2 = c + d = 40$
$C_1 = 50(a + c)$ $\quad C_2 = 50(b + d)$
$N = R_1 + R_2 = C_1 + C_2 = 100$

IV. *Degrees of Freedom*: $(R-1)(C-1) = (2-1)(2-1) = 1$.

V. *Critical Value*: The table value of chi-square at 1 df at 0.05 is 3.84.

VI. *Decision*: Now calculated $\chi^2 = 15.04 > \chi^2_{.05,1} > 3.84$. So the null hypothesis is rejected.

Example:

Test whether the prevalence of carriers of filaria is associated with sex:

Sex	*No. of Carriers*	*No. of Non-carriers*	*Total Studied*
Male	78	412	490
Female	57	553	610

Solution:

I. *Null Hypothesis*: Carrier of filaria is not associated with sex (assume).

II. *Alternative Hypothesis*: Carrier of filaria is associated with sex (assume).

III. *Calculation*:

$$\chi^2 = \frac{N\{(|ad-bc|)-\frac{N}{2}\}^2}{R_1 \cdot R_2 \cdot C_1 \cdot C_2} \qquad \begin{array}{l} R_1 = a + b,\ a + c = C_1 \\ R_2 = c + d,\ b + d = C_2 \end{array}$$

$N = a + b + c + d = R_1 + R_2 = C_1 + C_2$

$a = 78,\ b = 412,\ c = 57,\ d = 553$

$R_1 = 490, R_2 = 610, C_1 = 78 + 57 = 135, C_2 = 965.$

$$\chi^2 = \frac{1100\{|78 \times 553 - 412 \times 57| - \frac{1100}{2}\}^2}{490 \times 610 \times 135 \times 965}$$

$$= \frac{1100\{|43134 - 23484| - 550\}^2}{490 \times 610 \times 135 \times 965}$$

$$= \frac{1100\{19650 - 550\}^2}{490 \times 610 \times 135 \times 965}$$

$$= \frac{1100 \times 19100 \times 19100}{490 \times 610 \times 135 \times 965}$$

$$= \frac{11 \times 19100 \times 19100}{2989 \times 130275}$$

$= 11 \times 6.39 \times 0.1467$

$= 1.6137 \times 6.39 = 10.31$

IV. *Degrees of Freedom*: $(R - 1)(C - 1) = (2 - 1) \times (2 - 1) = 1 \times 1 = 1$.

V. *Critical Value*: The table value of χ^2 = at 0.05 for 1 df is 3.84.

VI. *Decision*: Now calculated χ^2 is 10.31, it is greater than tabulated value, i.e., $10.31 > \chi^2_{.0.05} = 3.84$.

Null hypothesis is rejected. Alternative hypothesis is accepted.

Example:

A random sample of 500 students were classified according to economic condition of their family and also according to merit:

Merit	*Economic Condition*			*Total*	
	Rich	*Middle Class*	*Poor*	*Total*	
Meritorious	42	137	61	240	R_1
Not Meritorious	58	113	89	260	R_2
Total	100	250	150	500	N
	C_1	C_2	C_3		

Test whether the two attributes Merit and Economic Condition are associated or not.

Solution:

I. *Null Hypothesis*: Attributes are independent.

II. *Alternative Hypothesis*: Attributes are associated (dependent).

III. *Calculation*: $E = \frac{R \times C}{N}$

Merit	*Rich*	*Middle Class*	*Poor*	*Total*
Meritorious	$\frac{240 \times 100}{500} = 48$	$\frac{240 \times 250}{500} = 120$	$\frac{240 \times 150}{500} = 72$	240
Not Meritorious	$\frac{260 \times 100}{500} = 52$	$\frac{260 \times 250}{500} = 130$	$\frac{260 \times 150}{500} = 78$	260
Total	100	250	150	500

$$\chi^2 = \frac{(42-48)^2}{48} + \frac{(137-120)^2}{120} + \frac{(61-72)^2}{72} + \frac{(58-52)^2}{52} + \frac{(113-130)^2}{130} + \frac{(89-78)^2}{78}$$

$$= \frac{36}{48} + \frac{289}{120} + \frac{121}{72} + \frac{36}{52} + \frac{289}{130} + \frac{121}{78}$$

$$= 9.75 + 2.41 + 1.68 + 0.69 + 2.22 + 1.55 = 9.30$$

IV. *Degrees of Freedom*: 3 – 1 = 2.

V. *Critical Value*: The table value of χ^2 at 0.05 for 2 df = 5.99.

VI. *Decision*: Now calculated χ^2 is 9.30, it is greater than tabulated value, i.e., $9.30 > \chi^2_{.0.05,\ 2} = 5.99$; therefore, null hypothesis is rejected. Alternative hypothesis is accepted. Therefore, merit and economic conditions are associated.

3. Homogeneity chi-square:

A test of "homogeneity" must be performed to decide whether the separate samples are sufficiently uniform to be added together.

Working Procedure:

I. The chi-square of each individual sample should be calculated based on expected ratio. Since these chi-squares are to be added, the Yates correction factor should not be used, even though only 1 degree of freedom may be involved in each calculation.

II. The individual chi-square should be summed to give a total chi-square. In this process the total chi-square accumulates a number of degrees of freedom equal to the sum of the degree of freedom in the individual chi-square.

The total chi-square value has two components:

(a) The chi-square contributed by the departure of the pooled data from the expected ratio.

(b) The chi-square contributed by the differences between individual samples.

To calculate (b) and (a) and subtract the value from total chi-square,

III. Calculate chi-square for the summed data of all samples.

IV. Subtract that chi-square of the summed data from the summed chi-square to obtain the homogeneity chi-square.

V. Accompany this, by subtracting the number of degrees of freedom between these respective values to obtain degrees of freedom for homogeneity chi-square.

Example:

Four of the self-fertilized F_1 plants that Mendel observed for segregating of yellow and green seeds colour showed the following results among their seeds:

PLANT	1	2	3	4
Yellow Seed	25	32	14	70
Green Seed	11	7	5	27

Test the homogeneity of the four plants for the 3 : 1 ratio, and determine whether the data can be summed to calculate chi-square.

Solution:

I. *Null Hypothesis*: 3: 1 (assume)

II. *Calculation*:

	Observed (O)	*Expected* (E)	|(O – E)|	$(O-E)^2$	$\frac{(O-E)^2}{E}$	χ^2	df
1.	25	27	–2	4	$\frac{4}{27} = 0.148$	0.59	2 – 1 = 1
	11	9	–2	4	$\frac{4}{9} = 0.44$		
2.	32	29.25	2.75	7.56	$\frac{7.56}{29.25} = 0.148$	1.03	2 – 1 = 1
	07	9.75	–2.75	7.56	$\frac{7.56}{9.75} = 0.775$		
3.	14	14.25	–0.25	0.0625	$\frac{0.0625}{14.25} = 0.004$	0.017	2 – 1 = 1
	05	4.75	0.25	0.0625	$\frac{0.0625}{4.75} = 0.013$		
4.	70	72.75	–2.75	7.56	$\frac{7.56}{72.75} = 0.103$	0.415	2 – 1 = 1
	27	24.25	2.75	7.56	$\frac{7.56}{24.25} = 0.312$		

$\sum\chi^2 = 2.06$ $\sum df = 4$

Summed

141	143.25	2.25	5.06	$\frac{5.06}{143.25} = 0.035$	0.141	2 – 1 = 1
50	47.75	2.25	5.06	$\frac{5.06}{47.75} = 0.106$		

	Chi-square	*Degrees of Freedom*	*Probabilities*
Total	2.06	4	
Summed data	0.14	1	
Homogeneity	1.92	3	0.70 – 0.50

III. *Critical Value*: The control value of chi-square at 0.05 for 3 df is 7.82.

IV. *Decision*: Since calculated χ^2 value at 3 df is 1.92 < critical value, so the null hypothesis is accepted, i.e., homogeneity persists among the samples.

Example:

The following results were obtained by four experimenters in a genetics laboratory class who back-crossed a stock of *D. melanogaster* heterogeneous for the recessive genes black body colour and sepia eye colour (b/+; se/+) to a stock homogeneous for the recessives (b/b; se/se):

	Phenotypes			
Experimenter	*Wild type*	*Black*	*Sepia*	*Black sepia*
1	42	38	38	36
2	28	30	25	25
3	102	94	100	93
4	75	80	84	72

Test these data to see whether they are homogeneous and can be pulled.

Solution:

I. *Hypothesis*: Since this is a back cross to a homozygous recessive stock, the expected ratio will be 1 wild type : 1 black : 1 sepia : 1 black sepia, i.e., 1 : 1 : 1 : 1 (assume).

II. *Calculation*:

Experiment (O)	*Observed (E)*	*Expected*	$\|(O-E)\|$	$(O-E)^2$	$\frac{(O-E)^2}{E}$	χ^2	df
1.	42	38.5	3.5	12.25	$\frac{12.25}{38.5} = 0.318$		
	38	38.5	–0.5	0.25	$\frac{0.25}{38.5} = 0.006$	0.492	4 – 1 = 3
	38	38.5	–0.5	0.25	$\frac{0.25}{38.5} = 0.006$	0.49	
	36	38.5	–2.5	6.25	$\frac{6.25}{38.5} = 0.162$		
2.	28	27	1	1	$\frac{1}{27} = 0.37$		
	30	27	3	9	$\frac{9}{27} = 0.33$	0.667	
	25	27	–2	4	$\frac{4}{27} = 0.15$	0.67	4 – 1 = 3
	25	27	–2	4	$\frac{4}{27} = 0.15$		
3.	102	97.25	4.75	22.56	$\frac{22.56}{97.25} = 0.23$	0.608	4 – 1 = 3
	94	97.25	–3.25	10.56	$\frac{10.56}{97.25} = 0.108$		
	100	97.25	2.75	7.56	$\frac{7.56}{97.25} = 0.08$	0.61	
	93	97.25	–4.25	18.06	$\frac{18.06}{97.25} = 0.19$		
4.	75	77.75	–2.75	7.56	$\frac{7.56}{77.75} = 0.097$		
	80	77.75	2.25	5.06	$\frac{5.06}{77.75} = 0.06$	1.077	4 – 1 = 3
	84	77.75	6.25	39.06	$\frac{6.25}{77.75} = 0.50$	1.08	
	72	77.75	–5.75	33.06	$\frac{33.06}{77.75} = 0.42$		

$\sum\chi^2 = 2.85$ $\sum df = 12$

Summed

Observed (*O*)	*Expected* (*E*)	$\lvert(O-E)\rvert$	$(O-E)^2$	$\frac{(O-E)^2}{E} = \chi^2$	df
247	240.5	6.5	42.25	$\frac{42.25}{240.5} = 0.175$	
242	240.5	1.5	2.25	$\frac{2.25}{240.5} = 0.009$	$4-1=3$
247	240.5	6.5	42.25	$\frac{42.25}{240.5} = 0.175$	
226	240.5	14.5	210.25	$\frac{210.25}{240.5} = 0.874$	

$\sum\chi^2 = 1.233$

III. *Critical value*: The control value of chi-square at 0.05 for 3 df is 7.82.

IV. *Decision*: Since calculated χ^2 value at 9 df is 1.62 < critical value, so the null hypothesis is accepted.

USES OF CHI-SQUARE:

The chi-square test is very powerful test for testing the hypothesis of a number of statistical problems.

(a) Test of Goodness of Fit:

With the help of this test, probabilities of association between two attributes are measured.

(b) Test of Independence of Attributes:

I. In this test attributes are classified into a two-way table or contingency table.

II. This test discloses whether there is any association or relationship between two or more variables.

(c) Test of Homogeneity:

I. This test may be used to test the homogeneity of the attributes in respect of a particular characteristics.

II. It may also be used to test the population variance.

Solved Problems

1. In *Drosophila* wild type allele (+), responsible for red colouration of the eye, is dominant over its mutant allele white (w). From a cross the following result was obtained:

Red-eyed female 126; White-eyed male 112.

Test the result and comment on segregation.

- It is a result from a cross between homozygous white-eyed female and hemizygous red-eyed male.

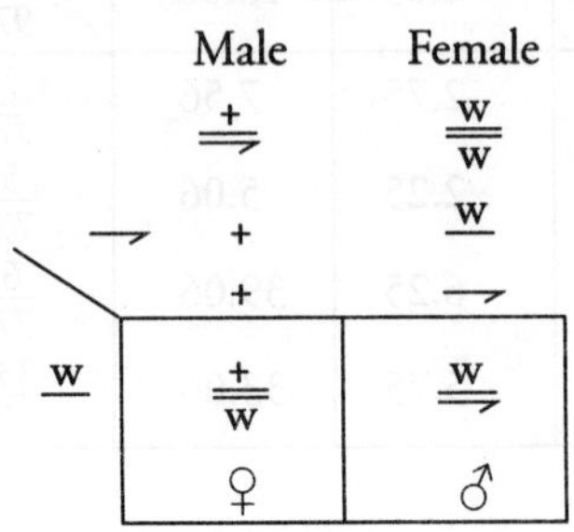

Result: Red-eyed female $\left(\frac{+}{w}\right)$ and white-eyed male $\left(\frac{w}{\rightarrow}\right)$.

- *Null Hypothesis*: It is a result of cross between homozygous white-eyed female and hemizygous red-eyed male. Such cross naturally produces offspring in 1: 1 ratio.
- *Alternative Hypothesis*: 3 : 1.
- *Calculation*:

Observed (O)	*Expected (E)*	$(\lvert O-E\rvert-0.05)$	$(\lvert O-E\rvert-0.05)^2$	$\frac{(\lvert O-E\rvert-0.05)^2}{E}$	χ^2
Red-eyed female = 126	119	$\lvert 126-119\rvert-\frac{1}{2}$ $=7-0.5=\lvert 6.5\rvert$	42.25	$\frac{42.25}{119}=0.36$	0.72
White-eyed male = 112	119	$\lvert 112-119\rvert-\frac{1}{2}$ $=7-0.5=\lvert 6.5\rvert$	42.25		
Total = 238					

- *Degrees of freedom*: 2 – 1 = 1.
- *Critical value*: The table value of chi-square for 2 – 1 = 1 df is 3.84.
- *Decision*: Since the calculated value of chi-square (χ^2) is 0.72 < critical value of χ^2 for 1 df is 3.84, so the null hypothesis is accepted, i.e., there is no significant variation with the result.

2. In the evening primrose pure red-flowered plants are crossed with white-flowered plants, F_1 are all pinked-coloured. Inbred F_1 plants produced red 113, pink 242, white 129. This phenotypic ratio also seems to be genotypic ratio of the F_2 of a Mendelian monohybrid cross involving a gene responsible for flower pigmentation.

Analyse the result with suitable statistical test.

Solution:

- *Null Hypothesis*: 1 : 2 : 1.
- *Alternative Hypothesis*: 3 : 1.
- *Calculation*:

Observed (O)	*Expected (E)*	$(\lvert O-E\rvert-0.05)$	$(\lvert O-E\rvert-0.05)^2$	$\frac{(\lvert O-E\rvert-0.05)^2}{E}$	χ^2
Red = 113	121	$\lvert 113-121\rvert-\frac{1}{2}$ $=\lvert -8\rvert-0.5=7.5$	56.25	$\frac{56.25}{121}=0.46$	
Pink = 242	242	$\lvert 242-242\rvert-\frac{1}{2}$ $=0-0.5=0.5$	0.25	$\frac{0.25}{242}=0.001$	0.921
White = 129	121	$\lvert 129-121\rvert-\frac{1}{2}$ $=\lvert -8\rvert-0.5=7.5$	56.25	$\frac{56.25}{121}=0.46$	
Total = 238					

- *Degrees of freedom*: 3 – 1 = 2.
- *Critical value*: The central value at 0.05 for 2 df is 5.99.

[P.G., M.G., E.G. & I.G. 23]

- *Decision*: Since the calculated value of chi-square (χ^2) is 0.921 < critical value of χ^2 for 2df is 5.99, so the null hypothesis is accepted, i.e. the variation is non-significant.

3. Crossing of purple-eyed straight-winged *Drosophila* with red-eyed curved-winged one produced dihybrid red-eyed straight-winged females in F_1. On crossing such F_1 females with double recessive purple-eyed curved-winged males gives following phenotypes:

Red-eyed straight-winged 339; Purpled-eyed straight-winged 612; Red-eyed curved-winged 725; Purple-eyed curved-winged 348.

Find out whether or not F_1 generation obeys the testcross ratio.

Solution:

- *Null Hypothesis*: 1 : 1 : 1 : 1
- *Alternative Hypothesis*: 9 : 3 : 3 : 1
- *Calculation*:

Observed (*O*)	*Expected* (*E*)	$\lvert(O-E)\rvert$	$(O-E)^2$	$\frac{(O-E)^2}{E}$	χ^2
Red-eyed and straight wing = 339	506	339 – 506 = –167	27889	$\frac{27889}{506} = 55.116$	
Purpled-eyed and straight wing = 612	506	612 – 506 = 106	11236	$\frac{11236}{506} = 22.205$	221.44
Red-eyed and curved wing = 725	506	725 – 506 = 219	47961	$\frac{47961}{506} = 94.784$	
Purpled-eyed and curved wing = 348	506	348 – 506 = –158	24964	$\frac{24964}{506} = 49.335$	
Total = 2024					

- *Degrees of freedom*: 4 – 1 = 3.
- *Critical value*: The control value of chi-square for 3 df at 0.05 is 7.82.
- *Decision*: Since the calculated value of chi-square (χ^2) for 3 df 221.4 < critical value 7.82, so the hypothesis is rejected, i.e., there is significant difference between the observed and expected result.

4. Mendel self-fertilized pea plants with round and yellow peas. In the next generation he recovered the following number of peas:

315 round and yellow peas,

108 round and green peas,

101 wrinkled and yellow peas,

32 wrinkled and green peas.

What is your hypothesis about the genetic control of the phenotype? Do the data support this hypothesis?

Solution:

- Hypothesis about the genetic control, i.e., *null hypothesis*: 9 : 3 : 3 : 1.

- *Calculation*:

Observed (*O*)	*Expected* (*E*)	$\lvert(O-E)\rvert$	$(O-E)^2$	$\frac{(O-E)^2}{E}$	χ^2	df
315	$\frac{9}{16} \times 556 = 312.75$	$315 - 312.75 = 2.25$	5.06	$\frac{5.06}{312.75} = 0.016$		
108	$\frac{3}{16} \times 556 = 104.25$	$108 - 104.25 = 3.75$	14.06	$\frac{14.06}{104.25} = 0.135$		
101	$\frac{3}{16} \times 556 = 104.25$	$101 - 104.25 = -3.25$	10.56	$\frac{10.56}{104.25} = 0.101$	0.470	3
32	$\frac{1}{16} \times 556 = 34.75$	$32 - 34.75 = -2.75$	7.56	$\frac{7.56}{34.75} = 0.218$		
Total = 556						

- *Critical value*: The control value of chi-square at 0.05 for 3 df is 7.82.
- *Decision*: Since the calculated χ^2 value at 3 df is 0.470, it is less than the critical chi-square value, i.e., $0.470 < \chi^2_{0.05}, 3 = 7.82$; so the null hypothesis is accepted.

5. Two curly-winged flies when mated, produce 61 curly- and 35 straight-winged progeny. Use a chi-square test to determine whether these numbers fit a 3 : 1 ratio.

Solution:

- *Null Hypothesis*: 3 : 1.
- *Alternative Hypothesis*: 2 : 1.
- *Calculation*:

Observed (*O*)	*Expected* (*E*)	$(\lvert O-E\rvert-0.05)$	$(\lvert O-E\rvert-0.05)^2$	$\frac{(\lvert O-E\rvert-0.05)^2}{E}$	χ^2	*df*
Curly-winged = 61	$\frac{3}{4} \times 96 = 72$	$\lvert 61 - 72\rvert - 0.5 = \lvert -11\rvert - 0.5$	110.25	$\frac{110.25}{119} = 1.531$		1
Straight-winged = 35	$\frac{1}{4} \times 96 = 24$	$\lvert 35 - 24\rvert - 0.5 = \lvert 11\rvert - 0.5 = 10.5$	110.25	$\frac{110.25}{24} = 4.593$	6.124	
Total = 238						

- *Critical value*: The control value at 0.05 for 1 df is 3.84.
- *Decision*: Since the calculated χ^2 value at 1 df is 6.723, it is greater than the critical value, i.e., $6.124 > \chi^2_{0.05} = 3.84$; so the null hypothesis is rejected.

6. Find whether or not the following phenotype distribution in a sample of 96 flies from a particular fruit fly population has a goodness of fit with the Mendelian 9 : 3 : 3 : 1 distribution:

Phenotypes:	*Grey body red eye*	*Black body red eye*	*Grey body scarlet eye*	*Black body scarlet eye*
Frequencies:	60	16	15	5

Solution:

- *Null Hypothesis*: 9 : 3 : 3 : 1.
- *Alternative Hypothesis*: 1 : 1 : 1 : 1.

- *Calculation*:

Observed (*O*)	*Expected* (*E*)	$\lvert(O-E)\rvert$	$(O-E)^2$	$\frac{(O-E)^2}{E}$	χ^2
60	54	+6	36	$\frac{36}{54} = 0.67$	
16	18	–2	04	$\frac{4}{18} = 0.22$	
15	18	–3	09	$\frac{9}{18} = 0.50$	1.56
5	6	–1	01	$\frac{1}{6} = 0.17$	
Total = 96					

- *Critical value*: The control value of chi-square at 0.05 for 3 df is 7.82.
- *Decision*: Since the calculated value of chi-square (χ^2) 1.56 < critical value of χ^2 for 3 df is 7.82, so the null hypothesis is accepted, i.e., the result is good fit for 9 : 3 : 3 : 1 hypothesis.

7. Random testing of ABO blood group in the offspring of only AB couples in an Europian population obtained the following distribution of blood groups:

A 312; AB 575; B 313.

Test whether the data is consistent with the normal segregation of alleles in the population.

Solution:

- *Null Hypothesis*: There is no significant difference between the observed and expected number of progeny blood group as AB × AB produces AA : AB : BB, i.e. 1 : 2 : 1 ratio.
- *Calculation*:

	AA	AB	BB	Total
Observed (O)	312	575	313	1200
Expected (E)	300	600	300	1200
O – E = D	312 – 300 = 12	575 – 600 = –25	313 – 300 = 13	
D^2	144	625	169	
$\frac{D^2}{E}$	$\frac{144}{300} = 0.48$	$\frac{625}{600} = 1.04$	$\frac{169}{300} = 0.56$	

$\chi^2 = 0.48 + 1.04 + 0.56 = 2.08$

- *Critical value*: The control value of χ^2 at 0.05 for 3 – 1 = 2 df is 5.99.
- *Decision*: Since the calculated χ^2 value at 2 df is 2.08 < critical value, so the null hypothesis is accepted, i.e., the variation between observed and expected data is non-significant.

8. A sample of South Indian population has found 40 individuals having homozygous curly hair, 40 individuals having heterozygous curly hair and 20 individuals having straight hair. Test statistically whether this sample is in equilibrium or not.

Solution:

To test the sample for genetic equilibrium we have to follow the Hardy-Weinberg law of population genetics for the calculation of gene frequency or allelic frequency for this generation. Then we can calculate the genotypic frequency for the next generation.

- *Calculation of gene or allele frequency*:
  (i) A gene 'A' for curly hair, and gene 'a' for straight hair.
  (ii) p the gene frequency for allele 'A' and q the gene frequency for allele 'a'.
  (iii) p^2 represent the frequency of homozygous dominant genotype, i.e., AA. 2pq represents the frequency heterozygous genotype Aa. q^2 represents the frequency of homozygous aa.
  (iv) In the present problem gene frequency of allele A, i.e., $p_A = p^2 + \frac{1}{2}pq$ and gene frequency of allele a, i.e., $q_a = q^2 + \frac{1}{2}pq$.

	AA	Aa	aa	Total
	40	40	20	100
Genotypic frequency	0.4	0.4	0.2	

Gene frequency of A, i.e., $p_A = 0.4 + \frac{1}{2} \times 0.4 = 0.4 + 0.2 = 0.6$

Gene frequency of a, i.e., $q_A = 0.4 + \frac{1}{2} \times 0.4 = 0.4 + 0.2 = 0.6$

  (v) Therefore, gene frequency in the present generation is $p_A = 0.6$ and $q_a = 0.4$.
  (vi) Calculation of genotypic frequency for the next generation:

$$(p_A + q_a)^2 = p_A^2 + 2p_Aq_a + q_{aa}^2$$
$$(0.6 + 0.4)^2 = 0.36 + 0.48 + 0.16$$

- *Null Hypothesis*:
  (i) If the genotypic frequency of present generation is exactly or more or less close to the genotypic frequency of next generation, then the population is called in equilibrium.
  (ii) Here in present generation genotypic frequencies are 0.40, 0.40 and 0.20 and the next generations will be 0.36, 0.48 and 0.16.

  Therefore, there is no significant deviation between observed and expected genotypic frequency of this population.

- *Calculation*:

	AA	Aa	aa
Observed (O)	0.4	0.4	0.2
Expected (E)	0.36	0.48	0.16
$\lvert O - E \rvert = D$	$0.4 - 0.36 = 0.04$	$0.4 - 0.48 = 0.08$	$0.2 - 0.16 = 0.04$
D^2	0.0016	0.0064	0.0016
$\frac{D^2}{E}$	$\frac{0.0016}{0.36} = 0.0044$	$\frac{0.0064}{0.48} = 0.0133$	$\frac{0.0016}{0.16} = 0.01$

$\chi^2 = 0.0044 + 0.0133 + 0.012 = 0.0277$

- *Critical value*: The control value of χ^2 at 0.05 for df = 2 – 1 = 1 is 3.84.
- *Decision*: Since the calculated value of chi-square $\chi^2 = 0.0277 <$ critical value of χ^2 for 1 df = 3.84, so the null hypothesis is accepted, i.e., deviation is non-significant.

9. Find whether or not the following observed distribution of phenotypes in a sample of 384 *Drosophila* flies have a significant goodness of fit with proposed Mendelian 9 : 3 : 3 : 1 distribution ($\alpha = 0.05$):

Phenotypes:	**AB**	**Ab**	**aB**	**ab**	**Total**
Number of animals:	**232**	**76**	**58**	**18**	**384**

Solution:

- *Null Hypothesis*: 9 : 3 : 3 : 1.
- *Alternative Hypothesis*: 1 : 1 : 1 : 1.
- *Calculation*:

Observed (O)	Expected (E)	|(O – E)| = |D|	D^2	$\frac{D^2}{E}$
232	$\frac{384}{16} \times 9 = 216$	232 – 216 = 16	256	$\frac{256}{216} = 1.185$
76	$\frac{384}{16} \times 3 = 72$	76 – 72 = 04	16	$\frac{16}{72} = 0.223$
58	$\frac{384}{16} \times 3 = 72$	58 – 72 = –14	196	$\frac{256}{72} = 2.723$
18	$\frac{384}{16} \times 1 = 24$	18 – 24 = –6	36	$\frac{36}{16} = 2.25$
Total = 384				$\sum \chi^2 = 6.381$

- *Critical value*: The control value of chi-square at 0.05 level and for 4 – 1 = 3 degrees of freedom is 7.82.
- *Decision*: Since the calculated value of chi-square (χ^2) = 6.381 < critical value of chi-square for 3 df is 7.82, so the null hypothesis is accepted, i.e,. the variation is non-significant.

10. In *Drosophila*, the wild type allele (+) responsible for normal bristle is dominant over mutant allele forked (*f*). From a cross the following progenies are obtained:

Normal bristle female 238; Normal bristle male 108; Forked male 122.

Test the result.

Solution:

In the present cross, the uneven distribution of dominant character between the two sexes tempted as to speculate that the gene under consideration resides on the X-chromosome.

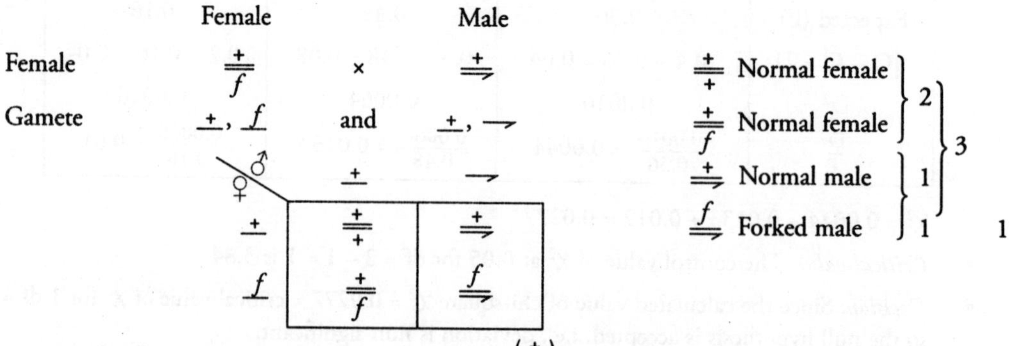

As a result of cross between heterozygote female $\left(\frac{+}{f}\right)$ and wild type male, three types of attributes appeared (3 : 1) or (1 : 2 : 1).

- *Null Hypothesis*: 3 : 1, i.e., 1 : 2 : 1 (male : female : male)
- *Alternative Hypothesis*: 1 : 1.

- *Calculation:*

Observed (O)	*Expected (E)*	$\lvert(O-E)\rvert$	$(O-E)^2$	$\frac{(O-E)^2}{E}$	χ^2
Wild type male = 108	117	108 – 117 = –9	81	$\frac{81}{117} = 0.69$	
Wild type female = 238	234	238 – 234 = +4	16	$\frac{16}{234} = 0.06$	0.96
Forked male = 122	117	122 – 117 = +5	25	$\frac{25}{117} = 0.21$	

- *Degrees of freedom*: 3 – 1 = 2.
- *Critical value*: The table value of χ^2 at 0.05 level for 2 df is 5.99.
- *Decision*: Since the calculated chi-square (χ^2) value 0.96 < critical value of χ^2 for 2 df = 5.99, so the null hypothesis is accepted, i.e. deviation is non-significant.

11. Evaluate the association between the abilities to test PTC and sex from the following scores: Taster male 80; Taster female 60; Non-taster male 40 and Non-taster female 20.

Solution:

- *Null Hypothesis*: No close relationship between PTC taster and non-taster with sex (male and female).
- *Alternative Hypothesis*: Relationship between PTC taster and non-taster with sex (male and female).
- *Calculation*:

PTC	*Male*	*Female*	*Total*	
Tester	80 (a)	60 (b)	140	(a + b) R_1
Non-taster	40 (c)	20 (d)	60	(c + d) R_2
	a + c = 120	b + d = 80	a + b + c + d = 200	
	C_1	C_2		

$$\chi^2 = N\frac{\left\{|ad-bc|-\frac{N}{2}\right\}^2}{R_1 \cdot R_2 \cdot C_1 \cdot C_2}$$

$$= \frac{200 \times \left\{|1600-2400|-\frac{200}{2}\right\}^2}{140 \times 60 \times 120 \times 80}$$

$$= \frac{200\{|-800|-100\}^2}{140 \times 60 \times 120 \times 80} = \frac{200 \times 700 \times 700}{140 \times 60 \times 120 \times 80}$$

ad = 80 × 20 = 1600
bc = 60 × 40 = 2400
R_1 = a + b = 140, C_1 = a + c = 120
R_2 = c + d = 60, C_2 = b + d = 80
N = $R_1 + R_2 = C_1 + C_2$ = 200

- *Degrees of freedom*: (R – 1) (C – 1) = (2 – 1) (2 – 1) = 1.
- *Critical value*: The table value of chi-square (χ^2) at 1 df at 0.05 is 3.84.
- *Decision*: Since calculated value of chi-square (χ^2) at 1 df is 1.215 < critical value = 3.84, so the null hypothesis is accepted, i.e., deviation is non-significant.

12. In a laboratory among 60 *Drosophila*, 30 males and 30 females are found. Further segregated by the presence of mutation 21 mutated males, 25 mutated females, 9 normal males and 5 normal females are found. Find out the relationship between mutation and sex.

Solution:

In this case there is no definite expected value, the question is whether the result is dependent or independent of the condition under which they are observed.

- *Null Hypothesis*: No significant relationship between mutation and sex.
- *Alternative Hypothesis*: Presence of significant relationship between mutation and sex.
- *Calculation*:

Sex	*Mutated*	*Normal*	*Total*
Male	21 (a)	9 (b)	30
Female	25 (c)	5 (d)	30
	a + c = 46	b + d = 14	60
	C_1	C_2	

$$\chi^2 = \frac{N\{|ad-bc|-\frac{N}{2}\}^2}{R_1 \cdot R_2 \cdot C_1 \cdot C_2}$$

ad = 21 × 5 = 105

bc = 9 × 25 = 225

$R_1 = a + b = 30$, $C_1 = a + c = 46$

$R_2 = c + d = 30$, $C_2 = b + d = 14$

$N = R_1 + R_2 = C_1 + C_2 = 60$

$$= \frac{60\{|105-225|-\frac{60}{2}\}^2}{30\times30\times46\times14}$$

$$= \frac{60\{|-120|-30\}^2}{30\times30\times46\times14} = \frac{60\times90\times90}{30\times30\times46\times14} = \frac{540}{644} = 0.838 = 0.84$$

$$= \frac{540}{644} = 0.838 = 0.84$$

- *Degrees of freedom*: (R – 1) (C – 1) = (2 – 1) (2 – 1) = 1.
- *Critical value*: The table value of chi-square (χ^2) at 0.05, for 1 df is 3.84.
- *Decision*: Since the calculated value of chi-square (χ^2) is 0.84 < critical value 3.84, so the null hypothesis is accepted, i.e., variation is non-significant.

13. In a sample of owls, it is found that red male 35, red female 70, grey male 50, grey female 45. Colouration is due to the plumage. Is the colouration independent of sex of the sample?

Solution:

- *Null Hypothesis*: The experimental result may show that colouration of the individuals are independent or not contingent of the sex.
- *Alternative Hypothesis*: This result may show that the colouration of the individuals is not independent of sex, i.e., dependents on contingent of sex.
- *Calculation*:

Sex	*Mutated*	*Normal*	*Total*
Male	35 (a)	50 (b)	85
Female	70 (c)	45 (d)	115
	a + c = 105	b + d = 95	200
	C_1	C_2	

$$\chi^2 = \frac{N\{|ad-bc|-\frac{N}{2}\}^2}{R_1 \cdot R_2 \cdot C_1 \cdot C_2}$$

$$= \frac{200\{|1575-3500|-\frac{200}{2}\}^2}{85\times115\times105\times95}$$

$$= \frac{200\{|-1925|-100\}^2}{85\times115\times105\times95} = \frac{200\times1825\times1825}{85\times115\times105\times95} = \frac{200\times3330625}{97505625}$$

$$= \frac{200\times3330625}{97505625}$$

$$= 200\times0.0341582 = 6.831$$

ad = 35 × 45 = 1575
bc = 50 × 70 = 3500
$R_1 = a + b = 85$, $C_1 = a + c = 105$
$R_2 = c + d = 115$, $C_2 = b + d = 95$
$N = R_1 + R_2 = C_1 + C_2 = 200$

- *Degrees of freedom*: (R – 1) (C – 1) = (2 – 1) (2 – 1) = 1.
- *Critical value*: The table of chi-square at 1 df at 0.05 is 3.84.
- *Decision*: Now calculated $\chi^2 = 6.831 > \chi^2_{0.005,1} > 3.84$

  So, the null hypothesis is rejected.

14. In *Drosophila* sp. a wild type gene (+) responsible for normal wing shape is recessive to mutant allele curly (cy). This dominant mutation like any other dominant mutation of *Drosophila* is homozygous lethal. From a cross following results are recovered:

Wild type male 56; Wild type female 52; Curly male 108; Curly female 112.

From above observation, determine (a) whether gene under consideration is X-linked, Y-linked or autosomal, (b) genotype of parents and (c) genotype of offsprings. (d) Substantiate your claim statistically.

Solution:

(a) (i) In this problem wing shape is under consideration. Hence, it is monohybrid cross.

(ii) Here sex ratio is 1 : 1 (164 : 164). It indicates absence of any sex-linked recessive lethal gene or sex-specific lethal or sex-transforming mutants are involved in this cross.

(iii) Here dominant and recessive trait are present (2 : 1 ratio). Therefore, both characters are equally expressed in both sexes.

The above-mentioned observations indicates that gene under consideration is "autosomal".

Since gene for wild type is recessive, then individuals with wild type phenotype will be homozygous recessive.

Parents have at least one strand of wild type allele in their genetic configuration (since dominant mutation is homozygous lethal).

Curly: Wild type = 2 : 1 indicates that both parents have gene for curly shape wing in their genetic configuration. Hence, parents are heterozygous for this particular gene.

(b) Genotype of parents: $\frac{cy}{+}$

(c) Genotype of offsprings: $\frac{cy}{+}$, $\frac{cy}{+}$, $\frac{+}{+}$, $\frac{cy}{cy}$(Lethal)

Phenotypic ratio, (curly): (wild type) = 2 : 1.

(d) Statistical Analysis:

- *Null Hypothesis*: The result shows dominant phenotype: recessive phenotype = 2 : 1.

- *Calculation:*

Observed(O)	*Expected (E)*	$\|(O-E)\| = D$	D^2	$E—D^2$	χ^2	df
Curly male = 108	$328 \times \frac{2}{6} = 109.33$	$108 - 109.33 = -1.33$	1.7689	$\frac{1.7689}{109.33}$	0.016	
Curly female = 112	$328 \times \frac{2}{6} = 109.33$	$112 - 109.33 = 2.67$	7.1289	$\frac{7.1289}{109.33}$	0.065	$4 - 1 = 3$
Wild type male = 56	$328 \times \frac{1}{6} = 54.66$	$56 - 54.66 = 1.34$	1.795	$\frac{7.075}{54.66}$	0.033	
Wild type female = 52	$328 \times \frac{1}{6} = 54.66$	$52 - 54.66 = -2.66$	7.075	$\frac{70.075}{54.66}$	0.129	

Total = 328 Total = 0.243

- *Critical value*: The control value of chi-square (χ^2) for 3 df at 0.05 is 7.82.
- *Decision*: Since the calculated value of chi-square for 3 df is 0.243 < critical value 7.82, so the null hypothesis is accepted, i.e., there is no significant variation between observed and expected results.

15. In *Drosophila* sp. a wild type gene (+) responsible for grey colouration of body is dominant over mutant allele yellow. From a cross following progenies are recovered:

Yellow females 88; Wild females 78.

Determine (a) whether gene under consideration is autosomal or sex-linked, (b) genotype of parents, (c) genotype of offsprings, (d) Substantiate your claim statistically.

Solution:

(a) (i) Here only colour of the body is under consideration. Hence, it is a monohybrid cross.

(ii) Since among progenies males are absent, i.e., sex ratio ≠ 1 : 1. Moreover, distribution of dominant and recessive characters among sexes is not equal. Therefore, the gene under consideration is sex-linked.

(b) (i) Since both yellow and wild type females are present, we can conclude that none of the alleles are lethal.

(ii) Therefore, the genetic constitution of the offspring and that of the parents must have two different loci for lethal genes which complement each other and produce healthy offspring in **trans condition** (females) but become lethal when present in **cis condition.**

(iii) Now, female parents are healthy; they have lethal genes in trans condition $\frac{l_1 +}{+ l_2}$ and in males no lethal genes are present $\frac{+\,+}{\rightharpoonup}$.

(iv) Since both recessive (yellow) and dominant (grey) phenotypes are present among offsprings, the parent female is heterozygous for colour of the body as well as trans heterozygous for lethal genes. The male parent must carry recessive gene for colour of the body and lethal genes are absent.

Genotype of parent: $\frac{y\,l_1\,+}{+\,+\,l_2}$ ♀ and $\frac{y\,l_1\,+}{\rightharpoonup}$ ♂

(c) Genotype of offspring: $\frac{y\,l_1\,+}{\rightharpoonup}$, $\frac{y\,+\,l_2}{\rightharpoonup}$, $\frac{y\,l_1\,+}{y\,+\,+}$, $\frac{y\,+\,l_2}{y\,+\,+}$.

♂ \ ♀	$y\,l_1\,+$	$+\,+\,l_2$
$y\,+\,+$	$\frac{y\,l_1\,+}{y\,+\,+}$ yellow ♀	$\frac{+\,+\,l_2}{y\,+\,+}$ wild ♀
$\rightharpoonup$	$\frac{y\,l_1\,+}{\rightharpoonup}$ lethal	$\frac{+\,+\,l_2}{\rightharpoonup}$ lethal

(d) Statistical Analysis:

- *Null Hypothesis*: Ratio of wild and yellow females is 1 : 1.
- *Calculation*:

Observed(O)	*Expected (E)*	$\|(O-E)\| = D$	D^2	$\frac{D^2}{E}$	χ^2	df
Yellow females = 88	83	$(83 - 88) - 0.5 = 4.5$	20.25	$\frac{20.25}{83}$	0.243	$2 - 1 = 1$
Wild females = 78	83	$(78 - 83) - 0.5 = 4.5$	20.25	$\frac{20.25}{83}$	0.243	$2 - 1 = 1$

Total = 166 Total = 0.486

- *Critical value*: The control value of chi-square for 1 df at 0.05 is 3.84.
- *Decision*: Since the calculated value of χ^2 for 1 df is 3.84, so the null hypothesis is accepted, i.e., variation is non-significant.

16. Map distance between A – B is 20 m.u. The following progenies are obtained:

AB = 85; ab = 75; Ab = 25; aB = 15.

Find out whether the actual map distance represents C.O. distance.

Solution:

20 m.u.

A ———————— B

We know 1% crossing over represents 1 map unit. Map distance between A – B = 20 m.u. Crossing-over classes are Ab = 25, aB = 15.

Null Hypothesis: The actual map distance represents the crossing over distance.

Alternative Hypothesis: The actual map distance does not represent the crossing over distance.

Calculation:

Observed (O)	*Expected (E)*	$\|(O-E)\| = D$	D^2	$\frac{D^2}{E}$	χ^2	df
AB = 85	80	$(85 - 80) = 5$	25	$\frac{25}{80} = 0.3125$	3.125	$4 - 1 = 3$
ab = 75	80	$(75 - 80) = -5$	25	$\frac{25}{80} = 0.3125$		
Ab = 25	20	$(25 - 20) = 5$	25	$\frac{25}{20} = 1.25$		
aB = 15	20	$(15 - 20) = -5$	25	$\frac{25}{20} = 1.25$		

Total = 200 Total = 0.486

- *Critical value*: The control value of χ^2 for 3 df at 0.05 is 7.82.
- *Decision*: Since the calculated chi-square (χ^2) for 3 df is 3.125 < critical value of χ^2 = 7.82, so the null hypothesis is accepted. Therefore, actual map distance represents crossing over distance.

17. Map distance between b – vg is 17.5 m.u. The following progenies are observed: ++ = 415; bvg = 400; b+ = 95; +vg = 90.

Find out whether actual map distance represents C.O. (crossing over) distance.

Solution:

17.5 m.u.

b ———————— vg

We know, 1% crossing over represents 1 map unit. 17.5% C.O. is present between b and vg. Crossing over classes are b+ = 95; +vg = 90.

- *Null Hypothesis*: The actual map distance represents the crossing over distance.
- *Alternative Hypothesis*: The actual map distance does not represent the crossing over distance.
- *Calculation*:

Observed (O)	*Expected (E)*	$\|(O-E)\| = \|D\|$	D^2	$\frac{D^2}{E}$	χ^2	df
+ + = 415	412.5	(415 – 412.5) = 2.5	6.25	$\frac{6.25}{412.5} = 0.015$		
bvg = 400	412.5	(400 – 412.5) = –12.5	156.25	$\frac{156.25}{12.5} = 0.378$	1.106	4 – 1 = 3
b + = 95	87.5	(95 – 87.5) = 7.5	56.25	$\frac{56.25}{87.5} = 0.642$		
+ vg = 90	87.5	(90 – 87.5) = 2.5	6.25	$\frac{6.25}{87.5} = 0.071$		

Total = 1000

- *Critical value*: The control value of chi-square (χ^2) for 3 df at 0.05 is 7.82.
- *Decision*: Since the calculated chi-square for a 3 df is 1.106 < critical value of chi-square 7.82, so the null hypothesis is accepted. We can conclude that actual map distance represents C.O. distance.

18. In pea plant, the wild type gene responsible for normal length is dominant over recessive allele short, responsible for shortness of plant. From a cross between two tall plants the following results are obtained:

Experiment	**Tall**	**Short**
1	**102**	**33**
2	**129**	**42**
3	**148**	**51**
4	**70**	**24**
5	**108**	**38**

Recheck the results of the data whether they can be pulled together or not.

Solution:

- *Null Hypothesis*: Since this is a monohybrid cross (Tall × Tall), the expected ratio will be 3 : 1.
- *Calculation*:

Sl. No.	*Observed (O)*	*Expected (E)*	$\|(O-E)\| = \|D\|$	D^2	$\frac{D^2}{E}$	χ^2	df
1	*Tall* = 102	101.25	102 – 101.25 = 0.75	0.5625	$\frac{0.5625}{101.25} = 0.005$	0.021	1
	Short = 33	33.75	33 – 33.75 = 0.75	0.5625	$\frac{0.5625}{33.75} = 0.016$		
2	*Tall* = 129	128.25	129 – 128.25 = 0.75	0.5625	$\frac{0.5625}{128.25} = 0.004$	0.017	1
	Short = 42	42.75	42 – 42.75 = 0.75	0.5625	$\frac{0.5625}{42.75} = 0.013$		
3	*Tall* = 148	149.25	148 – 149.25 = 1.25	1.56	$\frac{1.56}{149.25} = 0.01$	0.041	1
	Short = 51	49.75	51 – 49.75 = 1.25	1.56	$\frac{1.56}{49.75} = 0.031$		

Contd.

Sl. No.	Observed (O)	Expected (E)	$\lvert(O-E)\rvert = \lvert D\rvert$	D^2	$\frac{D^2}{E}$	χ^2	df
4	*Tall* = 70	70.5	70 – 70.5 = 0.5	0.25	$\frac{0.25}{70.5} = 0.003$	0.013	1
	Short = 24	23.5	24 – 23.5 = 0.5	0.25	$\frac{0.25}{23.5} = 0.070$		
5	*Tall* = 108	109.5	108 – 109.5 = 1.5	2.25	$\frac{2.25}{109.5} = 0.02$	0.081	1
	Short = 38	36.5	38 – 36.5 = 1.5	2.25	$\frac{2.25}{36.5} = 0.061$		

$\chi^2 = 0.173$ df = 5

Summed:

Observed (O)	Expected (E)	$\lvert(O-E)\rvert = \lvert D\rvert$	D^2	$\frac{D^2}{E}$	χ^2	df
Tall = 557	558.75	(557 – 558.75) = 1.75	3.06	$\frac{3.06}{558.75} = 0.005$	0.021	2 – 1 = 1
Short = 188	186.25	(188 – 186.25) = 1.75	3.06	$\frac{3.06}{186.25} = 0.016$		

	Chi-square	*Degrees of freedom*
Total	0.173	5
Summed	0.021	1
Homogeneity	0.152	4

- *Critical value*: The control value of chi-square (χ^2) for 4 df at 0.05 is 9.49.
- *Decision*: Since the calculated chi-square (χ^2) value is 0.152 < critical value = 9.49, so the null hypothesis is accepted, i.e., deviation is non-significant. We can conclude that homogeneity persists among the sample.

19. According to the census report (2004) the number of women to every thousand men in the following five states/union territories are as follows:

In Andaman 846, in Delhi 821, in Dadra and Nagar Haveli 812, in Chandigarh 777 and in Daman and Diu 710.

By appropriate statistical method determine whether 1 : 1 sex ratio among human population could be attributed to above five regions.

Solution:

- *Null Hypothesis*: Sex ratio is 1 : 1 among human population of five regions.
- *Calculation*:

Sl. No.	Observed (O)	Expected (E)	$\lvert(O-E)\rvert = \lvert D\rvert$	D^2	$\frac{D^2}{E}$	χ^2	df
A	Men = 1000	923	1000 – 923 = 77	5929	$\frac{5929}{77} = 6.42$	12.84	1
	Women = 846	923	846 – 923 = –77	5929	$\frac{5929}{77} = 6.42$		
D	Men = 1000	910.5	1000 – 910.5 = 89.5	8010.25	$\frac{8010.25}{910.5} = 8.79$	17.58	1
	Women = 821	910.5	821 – 910.5 = –89.5	8010.25	$\frac{8010.25}{910.5} = 8.79$		

Sl. No.	Observed (O)	Expected (E)	$\lvert(O-E)\rvert = \lvert D\rvert$	D^2	$\frac{D^2}{E}$	χ^2	df
DN	Men = 1000	906	1000 – 906 = 94	8836	$\frac{8836}{906} = 9.75$	19.5	1
	Women = 812	906	812 – 906 = –94	8836	$\frac{8836}{906} = 9.75$		
C	Men = 1000	888.5	1000 – 888.5 = 111.5	12432.25	$\frac{12432.25}{888.5} = 13.99$	27.98	1
	Women = 777	888.5	777 – 888.5 = –111.5	12432.25	$\frac{12432.25}{888.5} = 13.99$		
DD	Men = 1000	855	1000 – 855 = 145	21025	$\frac{21025}{855} = 24.59$	49.18	1
	Women = 710	855	710 – 855 = –145	21025	$\frac{21025}{855} = 24.59$		

Male = 5000, Female = 3966 $\chi^2 = 127.08$ df = 4

Summed:

Observed (O)	Expected (E)	$\lvert(O-E)\rvert = \lvert D\rvert$	D^2	$\frac{D^2}{E}$	χ^2	df
Male = 5000	4483	(5000 – 4483) = 517	267289	$\frac{267289}{4483}$	59.62	2 – 1 = 1
Female = 3966	4483	(3966 – 4483) = –517	267289	$\frac{267289}{4483}$	59.62	

$\chi^2 = 119.24$

	Chi-square	Degrees of freedom
Total	127.08	5
Summed	119.2	1
Homogeneity	7.88	4

- *Critical value*: The control value of chi-square (χ^2) for 4 df at 0.05 is 9.49.
- *Decision*: Since the calculated chi-square (χ^2) value is 4 df at 0.05 is 7.88 < critical value of χ^2 for 4 df 9.49. So the null hypothesis is accepted.

20. The gene producing sickle-shaped red blood cells in humans is called Hb^S and normal allele Hb^A. Various investigations have tested incidence of heavy infections of malarial parasite *Plasmodium falciparum* in African children who are heterozygous for sickle-cell gene (Hb^S/Hb^A) and in children who are homozygous normal (Hb^A/Hb^A). In an investigation by Allison and Clyde following observations are procured:

	Heavy infections	*Non/Highly infections*
Hb^S/Hb^A	36	100
Hb^A/Hb^A	152	255

Test whether heterozygotes in the sample are better protected against heavy malarial infections.

Solution:

According to the given data infection incidence are as follows:

Genotype	*Heavy infections*	*Non/Light infections*
Hb^S/Hb^A	36	100
Hb^A/Hb^A	152	255

- *Null Hypothesis*: Results of the investigation are independent of genetic configuration.
- *Alternative Hypothesis*: Results of investigations are dependent on genetic configuration.

Genotype	*Heavy infections*	*Non/Light infections*	Total	
Hb^S/Hb^A	36 (a)	100 (b)	136 (a + b)	R_1
Hb^A/Hb^A	152 (c)	255 (d)	407 (c + d)	R_2
	C_1 = 188 (a + c)	C_2 = 355 (b + d)		

$$\chi^2 = \frac{N\{|ad-bc|-\frac{N}{2}\}^2}{R_1 \times R_2 \times C_1 \times C_2}$$

ad = 36 × 255 = 9180
bc = 100 × 152 = 15200
R_1 = a + b = 136, C_1 = a + c = 188
R_2 = c + d = 407, C_2 = b + d = 355
N = R_1 + R_2 = C_1 + C_2
= 136 + 407 = 543

$$= \frac{543\{|9180-15200|\frac{543}{2}\}^2}{136 \times 407 \times 188 \times 355}$$

$$= \frac{543\{6020-271.5\}^2}{55352 \times 66740} = \frac{543 \times 5748.5 \times 5748.5}{55352 \times 66740}$$

$$= \frac{543 \times 33045252}{55352 \times 66740} = 597.0 \times 0.0081 = 4.83 = 4.8$$

- *Degrees of freedom*: (R – 1)(C – 1) = (2 – 1) × (2 – 1) = 1
- *Critical value*: The control value of chi-square for 1 df at 0.05 is 3.84.
- *Decision*: Since the calculated value of chi-square (χ^2) at 1 df is 4.8 > critical value of chi-square 3.84, so the null hypothesis is rejected, i.e., deviation is significant. So we conclude that heterozygous are better protected.

21. In the Botany garden of Midnapur College garden pea, orange pods (orp) are recessive to the green pods (Orp) and sensitivity to pea mosaic virus (mo) is recessive to resistance to the virus (Mo). A plant with orange pods and sensitivity to the virus was crossed to a true-breeding plant with green pods and resistance to the virus. The F_1 plants were then test-crossed to plants with orange pods and sensitivity to the virus. The following results were obtained.

160 orange pods, virus sensitive.
165 green pods, virus resistant.
36 orange pods, virus resistant.
39 green pods, virus sensitive.

Total sample is 400

(i) Conduct a chi-square analysis to determine if these genes are linked.
(ii) If they are linked calculate the map distane between the two genes.

Solution:

A. Chi-square analysis

- *Hypothesis*: Genes are not linked, i.e., test-cross ratio is 1 : 1 : 1 : 1.
- *Alternative Hypothesis*: Genes are linked.
- *Calculation*: $x^2 = \frac{(|0-E|)^2}{E}$

Sl. No.	Observed	Expected	$(\|0-E\|)$	$(\|0-E\|)^2$	$\frac{(\|0-E\|)^2}{E}$	x^2	ds
1	160	100	$(\|160-100\|)$	$(60)^2$	$\frac{3600}{100} = 36.00$	36.0	1 } 4
2	165	100	$(\|165-100\|)$	$(65)^2$	$\frac{4225}{100} = 42.25$	42.3	1
3	36	100	$(\|36-100\|)$	$(64)^2$	$\frac{4095}{100} = 40.95$	41.0	1
4	39	100	$(\|39-100\|)$	$(61)^2$	$\frac{3721}{100} = 37.21$	37.2	1

- *Critical value*: The central value of chi-square at 0.05 for 3 df is 7.82.
- *Decision*: Some calculated chi-square value at 3 df is 156.5 < critical value, so the null hypothesis is rezected. So, aternate hypothesis, i.e., genes are linked is accepted.

B. Calculate map distance

- *Map distance*: $\frac{\text{Number of non-parental offspring}}{\text{Total number of offspring}} \times 100 = \frac{36+39}{36+39+160+165} \times 100 = \frac{75}{400} \times 100$

  $= 18.8$ mu.

The genes are approximately 18.8 mu apart.

22. In the Botany garden of Serampore College, a trait garden peas have curling leaves. A dihybrid cross was made between a pea plant with yellow pods curling leves to wild type plant with green pods and normal leave. All F_1 plants had green pod with normal leaves, the results were 117 green pods normal leave, 115 yellow pods, curling leaves, 78 green pods curling leaves. 80 yellow pods, normal leaves.

(a) Conduct a chi-square analysis to determine if these two genes are linked.

(b) If they are linked calculate the map distane between the two genes.

How accurate do you think this distances?

Solution:

A. Chi-square analysis

- *Hypothesis*: Genes are not linked, i.e., test-cross ratio is 1 : 1 : 1 : 1.
- *Calculation*: $x^2 = \frac{(|0-E|)^2}{E}$ = Total 390. (117 + 115 + 78 + 80 = 390)

$$x^2 = \left(\frac{117-98}{98}\right) + \left(\frac{115-98}{98}\right) + \left(\frac{78-98}{98}\right) + \left(\frac{80-98}{98}\right)$$

$$x^2 = 3.68 + 2.93 + 4.08 + 3.31 = 14.02.$$

- *Critical value*: The central value of chi-square (x^2) at 0.05 for 3 df is 7.82.
- *Decision*: Some calculated chi-square value at 3 df is 14.02 > critical value, so the null hypothesis is rejected. Therefore, genes are linked.

B. Calculate map distance

- *Map distance*: $\frac{\text{Number of non-parental offspring}}{\text{Total number of offspring}} \times 100 = \frac{78+80}{117+115+78+80} \times 100 = \frac{158}{390} \times 100$

  $= 40.5$ mu.

The value is close to 50 mu. It is probably a significant underestimate of the true distance between the two genes.

B—STUDENT 't'-DISTRIBUTION

The 't'-test was used with small sample (n < 30) and was worked out by W. S. Gossett whose pen-name was "Student". Hence, this test is also called "Student's 't'-test".

The 't' may be defined as quantity representing the difference between the sample mean and true mean or population mean expressed in terms of the standard error.

$$t = \frac{\text{Difference between sample mean}}{\text{Standard error of the difference between means}}$$

$$t = \frac{|\overline{X}_1 - \overline{X}_2|}{SE} \quad [\overline{X}_1 \text{ and } \overline{X}_2 = \text{Mean, SE = Standard Error}]$$

Types:

(a) Unpaired t-test,

(b) Paired t-test.

Criteria for applying t-test:

I. Random samples are drawn from normal population.

II. For testing the equality of two population means, the population variances are regarded as equal.

III. The samples are less than 30.

IV. In case of two samples some adjustments in degrees of freedom for 't' are made.

- **Properties of 't'-Distribution:**

I. The shape of the 't'-distribution curve varies with the degrees of freedom.
The degrees of freedom is defined as size of sample minus one.

II. 't'-distribution is symmetrical distribution with mean zero.

III. 't'-distribution is asymptotic to x-axis, i.e., it extends to infinity on either side.

IV. Its graph is similar to that of normal distribution. 't'-distribution has a greater spread than normal distribution.

V. The larger the number of degrees of freedom, the more closely 't'-distribution resembles standard normal distribution.

- **Application of 't'-Distribution:**

1. To test the significance of a single mean when the population variance is unknown.
2. To test the significance of difference between two sample means the population variance being equal and unknown.
3. To test the significance of an observed sample correlation coefficient or difference between means of two samples (*dependent sample or paired observation*).

(a) Unpaired 't'-test:

This test is applied to unpaired data of independent observations made on individuals of two different groups or samples drawn from two populations.

Working procedure:

I. Calculate means of the two samples and find the observed difference between means of two samples $|\overline{X}_1 - \overline{X}_2|$.

II. Calculate the standard error of difference between the two means.

III. Calculate the 't'-value, i.e., the ratio between the observed difference and the standard error by substituting the above values in the formula

$$t = \frac{|\overline{X}_1 - \overline{X}_2|}{SE} \quad \overline{X}_1 \text{ and } \overline{X}_2 = \text{mean, } S\sigma_1 \text{ and } S\sigma_2 = \text{SD}, \; n_1 \text{ and } n_2 = \text{size of sample}$$

$$SE \text{ of } (\overline{X}_1 - \overline{X}_2) = SD\sqrt{\frac{1}{n_1} + \frac{1}{n_2}}$$

$$SD = \sqrt{\frac{\sum(X_1 - \overline{X}_1)^2 + \sum(X_2 - \overline{X}_2)^2}{n_1 + n_2 - 2}}$$

IV. Determine the pooled degrees of freedom from the formula $df = (n-1) + (n_2 - 1) = n_1 + n_2 - 2$.

V. Compare the calculated value with the table value at particular degrees of freedom.

Example:

In a nutritional study, 13 children were given a usual diet plus vitamins A and D tablets, while the 2nd comparable group of 12 children was taking the usual diet. After 12 months, the gain in weight in pounds was noted as given in the table. Can you say that vitamins A and D were responsible for this difference?

A	5	3	4	3	2	6	3	2	3	6	7	5	3
B	1	3	2	4	2	1	3	4	3	2	2	3	–

Solution:

I. *Null Hypothesis*: Vitamins (A and D) are responsible for the gain weight difference.

II. *Alternative Hypothesis*: Vitamins (A and D) are not responsible for the gain weight difference.

III. *Calculation*:

Sl. No.	Gr A(X)	$X-\overline{X}=D$	$(X-\overline{X})^2=D_1^2$	Gr B(Y)	$Y-\overline{Y}=D_2$	$(Y-\overline{Y})^2=D_2^2$
1	5	$5-4=-1$	1	1	$1-2.5=-1.5$	2.25
2	3	$3-4=-1$	1	3	$3-2.5=-0.5$	0.25
3	4	$4-4=-0$	0	2	$2-2.5=-0.5$	0.25
4	3	$3-4=-1$	1	4	$4-2.5=-1.5$	2.25
5	2	$2-4=-2$	4	2	$2-2.5=-0.5$	0.25
6	6	$6-4=-2$	4	1	$1-2.5=-1.5$	2.25
7	3	$3-4=-1$	1	3	$3-2.5=-0.5$	0.25
8	2	$2-4=-2$	4	4	$4-2.5=-1.5$	2.25
9	3	$3-4=-1$	1	3	$3-2.5=-0.5$	0.25
10	6	$6-4=-2$	4	2	$2-2.5=-0.5$	0.25
11	7	$7-4=-3$	9	2	$2-2.5=-0.5$	0.25
12	5	$5-4=-1$	1	3	$3-2.5=-0.5$	0.25
13	3	$3-4=-1$	1	–		
	$\sum 52$		$\sum D_1^2=32$	$\sum 30$		$\sum D_2^2=11$

Student 't'-distribution

Gr A	Gr B
$n=13$	$n=12$
$\overline{X}=4$	$\overline{Y}=2.5$
$D_1^2=32$	$D_2^2=11$

$$SD(A-B)=\sqrt{\frac{\sum(X_1-\overline{X}_1)^2+\sum(X_2-\overline{X}_2)^2}{n_1+n_2-2}}$$

$$=\sqrt{\frac{D_1^2+D_2^2}{13+12-2}}=\sqrt{\frac{32+11}{23}}=\sqrt{\frac{43}{23}}$$

$$SE=SD\sqrt{\frac{1}{n_1}+\frac{1}{n_2}}=\sqrt{\frac{43}{23}}\times\sqrt{\frac{1}{13}+\frac{1}{12}}=\sqrt{\frac{43}{23}}\times\sqrt{\frac{12+13}{156}}$$

$$=\sqrt{\frac{43}{23}}\times\sqrt{\frac{25}{156}}=\sqrt{1.87}\times\sqrt{0.16}=1.37\times 0.4=0.548=0.55$$

$$t=\frac{\overline{X}-\overline{Y}}{SE}=\frac{4-2.5}{0.55}=\frac{1.5}{0.55}=2.72$$

IV. *Level of significance*: 5% level, i.e., 0.05.

V. *Critical value*: Tabulated value at 0.05 for df 23 is 1.71.

VI. *Decision*: The calculated value of $|t|=2.72$

$|t|=2.72>t_{0.05,\,23}=2.07$

So the null hypothesis is rejected.

(b) Paired 't'-test:

It is applied to paired data of independent observations from one sample only when each individual gives a pair observations.

I. To study the role of a factor or cause when the observations are made before and after its play.
II. To compare the effect of two drugs, given to the same person.
III. To compare the results of two different laboratory techniques.
IV. To compare observations made at two different sides in the same body.

Working procedure:

I. Find the difference in each set of paired observations before and after ($X_1 - X_2 = D$).
II. Calculate the mean of the difference ($\overline{D}$).
III. Work out the SD of differences and then the SE of mean from the same $\frac{SD}{\sqrt{n}}$.
IV. Calculate 't'-value by substituting the above values in the formula $t = \frac{\overline{D}-0}{\frac{SD}{\sqrt{n}}} = \frac{\overline{D}}{\frac{SD}{\sqrt{n}}} = \frac{\overline{D}}{SE}$.

$$SD = \sqrt{\frac{\sum D^2 - \frac{(\sum D)^2}{n}}{n-1}}.$$

V. Determine the degrees of freedom. Being one and the same sample df should be n – 1.
VI. Compare the calculated value with the table value at a particular degrees of freedom.

Example:

Ten students were given intensive coaching in Statistics. The scores obtained in 1st and 5th tests are given below:

Sl. No.	1	2	3	4	5	6	7	8	9	10
Marks in 1st	50	52	53	60	65	67	48	69	72	80
Marks in 5th	65	55	65	65	60	67	49	82	74	86

Does the score from test 1st to test 5th show an improvement? Test at 5% level of significance.

Solution:

I. *Null Hypothesis*: No improvement has been occurred (assume).
II. *Alternative Hypothesis*: Improvement (assume).
III. *Calculation*:

Sl. No.	*Marks in 1st Test*	*Marks in 5th Test*	*Difference* $X_1 - X_2 = D$	D^2
1	50	65	–15	225
2	52	55	–3	09
3	53	65	–12	144
4	60	65	–5	25
5	65	60	5	25
6	67	67	0	0
7	48	49	–1	1
8	69	82	–13	169
9	72	74	–2	04
10	80	86	–6	36
			$\sum D = -52$	$\sum D^2 = 638$

$$\overline{D} = \frac{\sum D}{n} = \frac{-52}{10} = 5.2$$

$$SD = \sqrt{\frac{\sum D^2 - \frac{(\sum D)^2}{n}}{n-1}} = \sqrt{\frac{638 - \frac{(-52 \times -52)}{10}}{10-1}} = \sqrt{\frac{638-270.4}{9}} = \sqrt{\frac{367.6}{9}} = \frac{19.17}{3} = 6.391.$$

SE of difference

$$\frac{SD}{\sqrt{n}} = \frac{6.391}{\sqrt{10}} = \frac{6.391}{3.16} = 2.022 \qquad t = \frac{\overline{D}}{SE} = \frac{-5.2}{2.022} = 2.57.$$

IV. *Level of significance*: 0.05.

V. *Critical value*: The critical value of 't' at 0.05 for 10 – 1 = 9 df =, i.e. t0.05,9 = 1.833.

VI. *Decision*: Since calculated 't' = 2.57 > tabulated $t_{0.05,\,9}$ = 2.26, so the null hypothesis is rejected, i.e., alternative hypothesis is accepted.

Example:

An IQ test was administered to 5 persons before and after they are trained. The results are given below:

Candidates	I	II	III	IV	V
IQ before training	110	120	123	132	125
IQ after training	120	118	125	136	121

Test whether there is any change in IQ after training programme. It is given that $t_{0.01}$ = 4.6 for df 4.

Solution:

I. *Null Hypothesis*: No change in IQ after the training (assume).

II. *Alternative Hypothesis*: Change in IQ after the training (assume).

III. *Calculation*:

Candidates	*IQ before training*	*IQ after training*	*Difference*	D^2
I	110	120	110 – 120 = –10	100
II	120	118	120 – 118 = +2	04
III	123	125	123 – 125 = –2	04
IV	132	136	132 – 136 = –4	16
V	125	121	125 – 121 = +4	16
			$\sum D = -10$	$\sum D^2 = 140$

Mean Difference $(\overline{D}) = \frac{-10}{5} = -2$

Estimated SD (S) $= \sqrt{\frac{D^2 - \frac{(\sum D)^2}{n}}{n-1}} = \sqrt{\frac{140 - \frac{(-10 \times -10)}{5}}{5-1}} = \sqrt{\frac{140 - \frac{100}{5}}{4}} = \sqrt{\frac{120}{9}} = \sqrt{30}$

SE, i.e., Standard Error of Difference $= \frac{S}{\sqrt{n}} = \frac{\sqrt{30}}{\sqrt{5}} = \sqrt{6}$

$$t = \frac{\overline{D}}{SE} = \frac{-2}{2.45} = -0.816, \text{ i.e., } |t| = 0.816$$

IV. *Level of significance* = 0.01.

V. *Critical value*: Critical value of 't' at 0.05 for df 4, i.e. 5 – 1 is 4.6.

VI. *Decision*: The calculated value of $|t|$ = 0.816 < $t_{0.01}$ = 4.6.

So the null hypothesis is accepted, i.e., no significant difference after training programme.

Solved Problems

1. A research-minded student started fish-breeding and wanted to see whether there is any growth difference at 100th day of hybrid fishes at reciprocal parental crosses. The data (wt.) of fishes are given below:

Male × Female	1	2	3	4	5	6	7	8	9	10
X: Rohu × Mrigal:	15	12	8	14	16	16	9	5	11	14
Y: Mrigal × Rohu:	12	9	13	10	8	12	13	14	9	10

Find out whether sex of parents of given crosses have any effect on hybrid fishes growth or not by applying suitable statistical method.

Solution:

- *Null Hypothesis*: Sex of the parents have no effect on the growth of the hybrid fishes.
- *Alternative Hypothesis*: Sex of the parents have some effect on the growth of the hybrid fishes.
- *Calculation*:

Sl. No.	R × M (X)	$X-\overline{X} = D_1$ ($\overline{X} = 12$)	$(X-\overline{X})^2 = D_1^2$	M × R (Y)	$Y-\overline{Y} = D_2$	$(Y-\overline{Y})^2 = D_2^2$
1	15	15 – 12 = +3	09	12	12 – 11 = +1	01
2	12	12 – 12 = –0	00	9	9 – 11 = –2	04
3	8	8 – 12 = –4	16	13	13 – 11 = +2	04
4	14	14 – 12 = +2	04	10	10 – 11 = –1	01
5	16	16 – 12 = +4	16	8	8 – 11 = –3	09
6	16	16 – 12 = +4	16	12	12 – 11 = +1	01
7	9	9 – 12 = –3	09	13	13 – 11 = –2	04
8	5	5 – 12 = –7	49	14	14 – 11 = +3	09
9	11	11 – 12 = –1	01	9	9 – 11 = –2	04
10	14	14 – 12 = +2	04	10	10 – 11 = –1	01

$\sum X = 120$ $\qquad \sum D_1^2 = 124$ $\qquad \sum Y = 110$ $\qquad \sum D_2^2 = 38$

$n = 10$ $\qquad n = 10$

$\overline{X} = \frac{120}{10} = 12$ $\qquad \overline{Y} = \frac{110}{10} = 11$

$\sum(X-\overline{X})^2 = D_1^2 = 124$ $\qquad \sum(Y-\overline{Y})^2 = D_2^2 = 38$

$$SD = \sqrt{\frac{D_1^2 + D_2^2}{n_1 + n_2 - 2}} = \sqrt{\frac{124+38}{10+10-2}} = \sqrt{\frac{162}{18}}$$

$$SE(X-Y) = SD\sqrt{\frac{1}{n_1}+\frac{1}{n_2}} = 3\sqrt{\frac{1}{10}+\frac{1}{10}} = 3\sqrt{\frac{2}{10}} = 3\sqrt{0.2} = 3 \times 0.447 = 1.341$$

$$|t| = \frac{\overline{X}-\overline{Y}}{SE} = \frac{12-11}{1.34} = \frac{1}{1.34} = 0.75$$

- *Level of significance* = 5% level, i.e. 0.05.
- *Critical value*: Tabulated value of |t| at 0.05 for degrees of freedom 20 – 2 = 18 is 2.10.

- *Decision*: The calculated value of $|t| = 0.75\ |t| = 0.75 \leq t_{0.05,18}$.

So the null hypothesis is accepted, i.e., there is no significant deviation among the hybrid fishes obtained through reciprocal crosses.

2. The body weights (kg) of 8 adult males and of eight (8) adult females are presented respectively in the first and second columns of table. Find out whether or not the mean weight of males is significantly higher than that of females:

Table: Weight (kg)

Males **(X_1)**	***Females*** **(X_2)**
50	**49**
58	**52**
60	**51**
55	**56**
59	**55**
56	**53**
54	**52**
64	**48**
456	**416**

[C.U. (*Zoology Hons.*) 2004]

Solution:

- *Null Hypothesis*: Mean weight of males is not significantly higher than females.
- *Alternative Hypothesis*: Mean weight of males is significantly higher.
- *Calculation*:

Sl. No.	Male X_1	$X_1 - \overline{X_1} = D_1$	D_1^2	Female X_2	$X_2 - \overline{X_2} = D_2$	D_2^2
1	50	50 – 57 = –7	49	49	49 – 52 = –3	09
2	58	58 – 57 = +1	01	52	52 – 52 = 0	0
3	60	60 – 57 = +3	09	51	51 – 52 = –1	01
4	55	55 – 57 = –2	04	56	56 – 52 = +4	16
5	59	59 – 57 = +2	04	55	55 – 52 = +3	09
6	56	56 – 57 = –1	01	53	53 – 52 = +1	01
7	54	54 – 57 = –3	09	52	52 – 52 = 0	00
8	64	64 – 57 = +7	49	48	48 – 52 = –4	16
	456		$\sum D_1^2 = 126$	416		$\sum D_2^2 = 52$

$\overline{X_1} = \frac{456}{8} = 57$ $\qquad\qquad$ $\overline{X_1} = \frac{416}{8} = 52$

$n = 8$

$\sum D_1^2 = 126$ $\qquad\qquad$ $\sum D_2^2 = 52$

$$SD = \sqrt{\frac{\sum(X_1 - \overline{X_1})^2 + \sum(X_2 - \overline{X_2})^2}{n_1 + n_2 - 2}} = \sqrt{\frac{126+52}{8+8-2}} = \sqrt{\frac{178}{14}} = \sqrt{12.714} = 3.57$$

$$SE(X_1 - X_2) = 3.57\sqrt{\frac{1}{8}+\frac{1}{8}} = 3.57\sqrt{\frac{2}{8}} = 3.57\sqrt{0.25} = 3.57 \times 0.5 = 1.785$$

$$|t| = \frac{57-52}{1.785} = \frac{5}{1.785} = 2.80$$

- *Level of significance*: 5%, i.e., 0.05.
- *Critical value*: Tabulated value at 0.05 for df 16 – 2 = 14 is 2.14.
- *Decision*: The calculated value of $|t| = 2.80 = 2.80 > t_{0.05,\ 2.14}$

So the null hypothesis is rejected, i.e., mean weight of males is not significantly higher than females.

3. Application of fertilizer was tested for the yield of rice grown in 10 plots. Another seeds of 10 plots of similar size and condition were taken as control. Test the effect of fertilizer:

Plot No.	1	2	3	4	5	6	7	8	9	10
Fertilizer applied	16	14	18	15	13	17	16	15	14	13
Fertilizer not applied	10	12	11	9	13	13	12	14	13	11

Solution:

- *Null Hypothesis*: No significant effect of fertilizer on yield of rice grown.
- *Alternative Hypothesis*: Significant effect of fertilizer in yield of rice grown.
- *Calculation*:

Plot No.	*Fertilizer applied* (X)	$X_1-\overline{X}_1 = D_1$	D_1^2	*Fertilizer not applied* (X_2)	$X_2-\overline{X}_2 = D_2$	D_2^2
1	16	16 – 15.1 = 0.90	0.81	10	10 – 11.8 = –1.8	3.24
2	14	14 – 15.1 = –1.1	1.21	12	12 – 11.8 = +0.2	0.04
3	18	18 – 15.1 = +2.9	8.41	11	11 – 11.8 = –0.8	0.64
4	15	15 – 15.1 = –0.1	0.01	09	09 – 11.8 = –2.8	7.84
5	13	13 – 15.1 = –2.1	4.41	13	13 – 11.8 = +1.2	1.44
6	17	17 – 15.1 = +1.9	3.61	13	13 – 11.8 = +1.2	1.44
7	16	16 – 15.1 = +0.9	0.81	12	12 – 11.8 = +0.2	0.04
8	15	15 – 15.1 = –0.1	0.01	14	14 – 11.8 = 2.2	4.84
9	14	14 – 15.1 = –1.1	1.21	13	13 – 11.8 = +1.2	1.44
10	13	13 – 15.1 = –2.1	4.41	11	11 – 11.8 = –0.8	0.64

$\sum X = 151$ $\quad\sum D_1^2 = 24.90 \sum X_2 = 118$ $\quad\sum D_2^2 = 21.60$

$\overline{X}_1 = \frac{151}{10} = 15.1$ $\quad\overline{X}_2 = \frac{118}{10} = 11.8$

n = 10 $\quad$ n = 10

$\overline{X}_1 = 15.1$ $\quad\overline{X}_2 = 11.8$

$\sum D_1^2 = 24.9$ $\quad\sum D_2^2 = 21.6$

$$SD(X_1-X_2) = \sqrt{\frac{\sum(X_1-\overline{X}_2)^2 + \sum(X_2-\overline{X}_2)^2}{n_1+n_2-2}} = \sqrt{\frac{\sum D_1^2 + \sum D_2^2}{10+10-2}} = \sqrt{\frac{24.9+21.6}{18}} = \sqrt{\frac{46.5}{18}} = \sqrt{2.584} = 1.607$$

$$SE = 1.607\sqrt{\frac{1}{10}+\frac{1}{10}} = 1.607\sqrt{0.2} = 1.607 \times 0.447 = 0.718$$

$$|t| = \frac{\overline{X}-\overline{Y}}{SE} = \frac{15.1-11.8}{0.718} = \frac{3.3}{0.718} = 4.596$$

- *Level of significance*: 5% level, i.e., 0.05.
- *Critical value*: Tabulated value at 0.05 for df 10 + 10 – 2 = 18 is 2.10.
- *Decision*: The calculated value of $|t|$ is 4.596 or, $|t| = 4.59 > t_{0.05,\,18} = 2.10$.

So the null hypothesis is rejected, i.e., there is significant difference between the control and fertilizer used in rice plant growth. Therefore, fertilizer has some impact on rice plant growth.

4. Body length of fishes of a species was obtained from two ponds. They were measured as follows (in cm):

Pond A	20	24	20	28	22	20	24	32	24	26
Pond B	12	10	8	10	6	4	14	20	10	6

Calculate whether the mean difference in total body length between the two ponds of fish is significant or not.

Solution:

- *Null Hypothesis*: There is no significant difference in total body length between the fishes of the two ponds.
- *Alternative Hypothesis*: Presence of significant difference in body length between the fishes of the two ponds.
- *Calculation*:

Sl. No.	Pond A (X)	$X-\overline{X}=D_1$	D_1^2	Pond B (Y)	$Y-\overline{Y}=D_2$	D_2^2
1	20	20 – 24 = –4	16	12	12 – 10 = +2	04
2	24	24 – 24 = 0	0	10	10 – 10 = 0	0
3	20	20 – 24 = –4	16	08	08 – 10 = –2	04
4	28	28 – 24 = +4	16	10	10 – 10 = 0	0
5	22	22 – 24 = –2	04	06	06 – 10 = –4	16
6	20	20 – 24 = –4	16	04	04 – 10 = –6	36
7	24	24 – 24 = 0	0	14	14 – 10 = +4	16
8	32	32 – 24 = +8	64	20	20 – 10 = +10	100
9	24	24 – 24 = 0	0	10	10 – 10 = 0	0
10	26	26 – 24 = +2	04	06	06 – 10 = –4	16
	$\Sigma X = 240$		136	$\Sigma Y = 100$		192

Pond A (X)

$$\overline{X} = \frac{240}{10} = 24$$

n = 10

$$\Sigma D_1^2 = 136$$

Pond B (Y)

$$\overline{Y} = \frac{100}{10} = 10$$

n = 10

$$\Sigma D_2^2 = 192$$

$$SD(X-Y) = \sqrt{\frac{\Sigma(X-\overline{X})^2 + \Sigma(Y-\overline{Y})^2}{n_1+n_2-2}} = \sqrt{\frac{\Sigma D_1^2 + \Sigma D_2^2}{10+10-2}} = \sqrt{\frac{136+192}{18}} = \sqrt{\frac{328}{18}} = \sqrt{18.22} = 4.27$$

$$SE = SD\sqrt{\frac{1}{n_1}+\frac{1}{n_2}} = 4.27\sqrt{\frac{1}{10}+\frac{1}{10}} = 4.27\sqrt{0.2} = 4.27 \times 0.45 = 1.92$$

$$|t| = \frac{\overline{X}-\overline{Y}}{SE} = \frac{24-10}{1.92} = \frac{14}{1.92} = 7.29$$

- *Level of significance*: 5% level, i.e. 0.05.
- *Critical value*: Tabulated value at 0.05 for df 20 – 2 = 18 is 2.10.
- *Decision*: The calculated value of $|t| = 7.29$

$$|t| = 7.29 > t_{0.05,\,18} = 2.10$$

So the null hypothesis is rejected.

5. In order to find the effect of *Azolla* growth on the rice-field and experimentally grow *Azolla* in 10 similar field plots before rice-planting and other 10 similar plots were taken as control without *Azolla* growth. Rice was grown in all these plots and yields were noted:

Plot No.	1	2	3	4	5	6	7	8	9	10
With Azolla	15.3	15.8	16.1	17.0	15.5	16.5	16.2	15.4	17.1	16.3
Without Azolla	14.5	13.8	15.9	13.9	14.8	14.9	15.2	15.0	14.1	13.7

Verify whether there is any significant effect of Azolla growth on the gain yield of rice.

Solution:

- *Null Hypothesis*: There is no significant effect of Azolla growth in gain yield of rice.
- *Alternative Hypothesis*: Azolla growth has some effect on the gain yield of rice.
- *Calculation*:

Plot No.	*With Azolla* (X)	$X-\overline{X} = D_1$	$(X-X^2) = D_1^2$	*Without Azolla* (Y)	$Y-\overline{Y} = D_2$	D_2^2
1	15.3	15.3 – 16.12 =–0.82	0.6724	14.5	14.5 – 14.58 = –0.08	0.0064
2	15.8	15.8 – 16.12 =–0.32	0.1024	13.8	13.5 – 14.58 = –1.08	1.1664
3	16.1	16.1 – 16.12 =–0.02	0.0004	15.9	15.9 – 14.58 = +1.32	1.7424
4	17.0	17.0 – 16.12 = +0.88	0.7744	13.9	13.9 – 14.58 = –0.68	0.4624
5	15.5	15.5 – 16.12 =–0.62	0.3844	14.8	14.8 – 14.58 = +0.22	0.0484
6	16.5	16.5 – 16.12 = +0.38	0.1444	14.9	14.9 – 14.58 = +0.32	0.1024
7	16.2	16.2 – 16.12 = +0.08	0.0064	15.2	15.2 – 14.58 = +0.62	0.3844
8	15.4	15.4 – 16.12 =–0.72	0.5184	15.0	15.0 – 14.58 = +0.42	0.1764
9	17.1	17.1 – 16.12 = 0.98	0.9604	14.1	14.1 – 14.58 = –0.48	0.2304
10	16.3	16.3 – 16.12 = +0.18	0.0324	13.7	13.7 – 14.58 = –0.88	0.7744

$\sum X = 161.2$ $\qquad \sum D_1^2 = 3.596 \quad Y = \sum 145.8 \qquad \sum D_2^2 = 5.0940$

$N = 10$

$\overline{X} = \frac{161.2}{10} = 16.12$ $\qquad \overline{Y} = \frac{145.8}{10} = 14.58$

$N_1 = 10$ $\qquad N_2 = 10$

$\sum(X-\overline{X})^2 = \sum D_1^2 = 3.596$ $\qquad \sum(Y-\overline{Y})^2 = \sum D_2^2 = 5.0940$

$$SD = \sqrt{\frac{D_1^2+D_2^2}{N_1+N_2-2}} = \sqrt{\frac{3.596+5.0940}{10+10-2}} = \sqrt{\frac{8.690}{18}} = \sqrt{0.482} = 0.694$$

$$SE_{(X-Y)} = SD\sqrt{\frac{1}{N_1}+\frac{1}{N_2}} = 0.694\sqrt{\frac{1}{10}+\frac{1}{10}} = 0.694\sqrt{\frac{2}{10}} = 0.694\sqrt{0.2} = 0.694 \times 0.447 = 0.310$$

$$|t| = \frac{\overline{X}-\overline{Y}}{SE} = \frac{16.12-14.58}{0.310} = \frac{1.54}{0.310} = 4.967.$$

- *Level of significance*: 5% level, i.e., 0.05.
- *Critical value*: Tabulated value at 0.05 for df 20 – 2 = 18 is 2.10.
- *Decision*: The calculated value of $|t| = 5.09$

$$|t| = 4.99 > t_{0.05,18} = 2.10.$$

So the null hypothesis is rejected.

6. The students of two schools were measured for their heights, one school was in east coast and another in west coast where there is slight difference in weather. The sampling results are as follows:

East coast	**43**	**45**	**48**	**49**	**51**	**52**			
West coast	**47**	**49**	**51**	**53**	**54**	**55**	**55**	**56**	**57**

Find whether there is any impact of weather on height taking other variable constant. Apply suitable statistical method.

Solution:

- *Null Hypothesis*: Weather has no impact on the height of the students.
- *Alternative Hypothesis*: Weather has some impact on the heights of students.
- *Calculation*:

Sl. No.	*East Coast* (X)	$X-\overline{X} = D_1$	$(X-\overline{X})^2 = D_1^2$	*West Coast* (Y)	$Y-\overline{Y} = D_2$	$(Y-\overline{Y})^2 = D_2^2$
1	43	43 – 48 = –5	25	47	47 – 53 = –6	36
2	45	45 – 48 = –3	09	49	49 – 53 = –4	16
3	48	48 – 48 = 0	00	51	51 – 53 = –2	04
4	49	49 – 48 = +1	01	53	53 – 53 = 0	00
5	51	51 – 48 = +3	09	54	54 – 53 = +1	01
6	52	52 – 48 = +4	16	55	55 – 53 = +2	04
7				55	55 – 53 = +2	04
8				56	56 – 53 = +3	09
9				57	57 – 53 = +4	16
	288		$\sum D_1^2 = 60$	477		$\sum D_2^2 = 90$

$\overline{X} = \frac{288}{10} = 48$ $\qquad$ $\overline{Y} = \frac{477}{9} = 53$

X $\qquad$ Y

$n_1 = 6$ $\qquad$ $n_2 = 9$

$\overline{X} = 48$ $\qquad$ $\overline{Y} = 53$

$\sum D_1^2 = 60$ $\qquad$ $\sum D_2^2 = 90$

$$SD(X-Y) = \sqrt{\frac{\sum(X-\overline{X})^2 + \sum(Y-\overline{Y})^2}{n_1+n_2-2}} = \sqrt{\frac{\sum D_1^2 + \sum D_2^2}{n_1+n_2-2}} = \sqrt{\frac{60+90}{6+9-2}} = \sqrt{\frac{150}{13}} = \sqrt{11.5} = 3.39$$

$$SE = SD\sqrt{\frac{1}{n_1}+\frac{1}{n_2}} = 3.39 \times \sqrt{\frac{1}{6}+\frac{1}{9}} = 3.39 \times \sqrt{\frac{5}{18}} = 3.39 \times \sqrt{0.28} = 3.39 \times 0.52 = 1.76$$

$$|t| = \frac{\overline{X}-\overline{Y}}{SE} = \frac{48-53}{1.76} = \frac{-5}{1.67} = \frac{500}{176} = 2.840$$

- *Level of significance*: 5% level, i.e., 0.05.
- *Critical value*: Tabulated of t at 0.05 for df 9 + 6 – 2 = 13 is 2.16.
- *Decision*: The calculated value of |t| is 2.840

$$|t|\ 2.840 > t_{0.05,\ 13} = 2.16.$$

So the null hypothesis is rejected.

7. Find whether or not there is a significant difference between the mean inter-orbital widths (mm) of 2 groups of pigeons, consisting of 10 and 8 animals respectively:

Group I (X_1):	**11.9**	**11.4**	**11.9**	**11.4**	**11.2**	**12.2**	**12.6**	**12.2**	**12.7**	**12.5**
Group II (X_2):	**10.8**	**10.9**	**10.5**	**11.0**	**10.6**	**10.5**	**10.8**	**10.5**		

Solution:

- *Null Hypothesis*: No significant difference between the mean inter-orbital width of two groups of pigeons.
- *Alternative Hypothesis*: Presence of significant difference between inter-orbital widths of two groups of pigeons.
- *Calculation*:

Sl. No.	Group I (X_1)	$X_1-\overline{X}_1 = D_1$	$(X_1-\overline{X}_1)^2 = D_1^2$	Group II (X_2)	$X_2-\overline{X}_2 = D_2$	$(X_2-\overline{X}_2)^2 = D_2^2$
1	11.9	11.9 – 12 = –0.1	0.01	10.8	10.8 – 10.7 = +0.1	0.01
2	11.4	11.4 – 12 = –0.6	0.36	10.9	10.9 – 10.7 = +0.2	0.04
3	11.9	11.9 – 12 = –0.1	0.01	10.5	10.5 – 10.7 = –0.2	0.04
4	11.4	11.4 – 12 = –0.6	0.36	11.0	11.0 – 10.7 = +0.3	0.09
5	11.2	11.2 – 12 = –0.8	0.64	10.6	10.6 – 10.7 = –0.1	0.01
6	12.2	12.2 – 12 = +0.2	0.04	10.5	10.5 – 10.7 = –0.2	0.04
7	12.6	12.6 – 12 = +0.6	0.36	10.8	10.8 – 10.7 = +0.1	0.01
8	12.2	12.2 – 12 = +0.2	0.04	10.5	10.5 – 10.7 = –0.2	0.04
9	12.7	12.7 – 12 = +0.7	0.49			
10	12.5	12.5 – 12 = +0.5	0.25			
	120.0		$\sum X_1^2 = 0.256$	85.6		$\sum X_2^2 = 0.28$

Gr I

$n_1 = 10$

$\overline{X}_1 = \frac{120}{10} = 12$ mm

Gr II

$n_2 = 8$

$\overline{X}_2 = \frac{85.6}{8} = 10.7$ mm

$$SD(X_1-X_2) = \sqrt{\frac{\sum(X_1-\overline{X}_1)^2+\sum(X_2-\overline{X}_2)^2}{n_1+n_2-2}} = \sqrt{\frac{\sum D_1^2+\sum D_2^2}{n_1+n_2-2}} = \sqrt{\frac{2.56+0.28}{10+8-2}} = \sqrt{\frac{2.84}{16}} = \sqrt{0.1775} = 0.42 \text{ (App.)}$$

$$SE = SD\sqrt{\frac{1}{n_1}+\frac{1}{n_2}} = 0.42\sqrt{\frac{1}{10}+\frac{1}{8}} = 0.42\sqrt{\frac{4+5}{40}} = 0.42\sqrt{\frac{9}{40}} = 0.199$$

$$|t| = \frac{\overline{X}_1-\overline{X}_2}{SE} = \frac{12.0-10.7}{0.199} = \frac{1.3}{0.199} = 0.6533$$

- *Level of significance*: 5% level, i.e., 0.05.
- *Critical value*: Tabulated of |t| at 0.05 for df = 10 + 8 – 2 = 16 is 2.12.
- *Decision*: The calculated value of |t| is 6.533

$$|t|\ 6.533 > t_{0.05,\ 16} = 2.12.$$

So the null hypothesis is rejected, i.e., there is significant difference between the mean inter-orbital width of two groups of pigeons.

8. Ten rats were fed with rice in first month and body weights of the rats were recorded. In the next month they were fed with grams and their weights were measured again. The respective weights of ten rats in two months are as follows:

Weight in* 1st *month	**50**	**60**	**58**	**52**	**51**	**62**	**58**	**55**	**50**	**65**
Weight in* 2nd *month	**56**	**58**	**68**	**61**	**56**	**59**	**64**	**60**	**50**	**62**

Test the given data to find the impacts of grams in rats nutrition.

Solution:

- *Null Hypothesis*: Grams have no impact in rats nutrition.
- *Alternative Hypothesis*: Grams have impact on rat nutrition.
- *Calculation*:

Sl. No.	*Weight in* 1st *Month* (X_1)	*Weight in* 2nd *Month* (X_2)	*Difference* $X_1 - X_2 = D$	D^2
1	50	56	50 – 56 = –6	36
2	60	58	60 – 58 = +2	04
3	58	68	58 – 68 = –10	100
4	52	61	52 – 61 = –9	81
5	51	56	51 – 56 = –5	25
6	62	59	62 – 59 = +3	09
7	58	64	58 – 64 = –6	36
8	55	60	55 – 60 = –5	25
9	50	50	50 – 50 = 0	00
10	65	62	65 – 62 = +3	09
			$\Sigma D = -41 + 8 = -33$	325

n = 10 $\qquad \overline{D} = \frac{\Sigma D}{n} = \frac{-33}{10} = -3.3.$

$$SD = \sqrt{\frac{\Sigma D^2 - \frac{(\Sigma D)^2}{n}}{n-1}} = \sqrt{\frac{325-\frac{(-33)^2}{10}}{10-1}} = \sqrt{\frac{325-\frac{1089}{10}}{9}} = \sqrt{\frac{325-108.9}{9}} = \sqrt{\frac{216.1}{9}} = \sqrt{24.1} = 4.9.$$

$$SE = \frac{SD}{\sqrt{n}} = \frac{4.9}{\sqrt{10}} = \frac{4.9}{3.16} = 1.55.$$

$$|t| = \frac{\overline{D}}{SE} = \frac{3.3}{1.55} = 2.13.$$

- *Level of significance*: 0.05 level.
- *Critical value*: Tabulated value at 0.05 level at df 10 – 1 = 9 is 2.26.
- *Decision*: The calculated value of $|t|$ is 2.13 < $|t|_{0.05,\,9}$ = 2.26.

  So the null hypothesis is accepted, i.e. grams have no impact on rat nutrition.

9. Albino rats were administered with an aurvedic medicine at the rate of 10 mg/10 kg body weight per day for 7 days. Initial and final body weights of the rats were recorded as shown in the table:

Rat No.	**1**	**2**	**3**	**4**	**5**	**6**	**7**	**8**	**9**	**10**
Initial body weight	**110**	**115**	**102**	**98**	**112**	**110**	**97**	**120**	**102**	**110**
Final body weight	**109**	**116**	**100**	**95**	**108**	**112**	**98**	**115**	**98**	**111**

Determine whether the drug has any significant effect on the gain or loss of body weight of the rat.

Solution:

- *Null Hypothesis*: Drug has no significant effect in gain or loss of body weight of rat.
- *Alternative Hypothesis*: Drug has some effect on gain or loss of body weight of rat.
- *Calculation*:

Sl. No.	*Initial weight* (X_1)	*Final weight* (after application of medicine) (X_2)	*Difference* $X_1 - X_2 = D$	D^2
1	110	109	110 – 109 = +1	01
2	115	116	115 – 116 = –1	01
3	102	100	102 – 100 = +2	04
4	98	95	98 – 95 = +3	09
5	112	108	112 – 108 = +4	16
6	110	112	110 – 112 = –2	04
7	97	98	97 – 98 = –1	01
8	120	115	120 – 115 = +5	25
9	102	98	102 – 98 = +4	16
10	110	111	110 – 111 = –1	01
			14	$\sum D^2 = 78$

$$\bar{D} = \frac{\sum D}{n} = \frac{14}{10} = 1.4$$

$$SD = \sqrt{\frac{\sum D^2 - \frac{(\sum D)^2}{n}}{n-1}} = \sqrt{\frac{78 - \frac{(14)^2}{10}}{10-1}} = \sqrt{\frac{78 - \frac{196}{10}}{9}} = \sqrt{\frac{78 - 19.6}{9}} = \sqrt{\frac{58.4}{9}} = \sqrt{6.49} = 2.55$$

$$SE = \frac{SD}{\sqrt{n}} = \frac{2.55}{\sqrt{10}} = \frac{2.55}{3.16} = 0.806$$

$$|t| = \frac{\bar{D}}{SE} = \frac{1.4}{0.806} = 1.736 = 1.74$$

- *Level of significance*: 5%, i.e., 0.05.
- *Critical value*: The critical value of 't' at 0.05 level for 10–1=9 df, i.e., $|t|_{0.05,\,9}$ = 1.833.
- *Decision*: Since calculated $|t|$ = 1.74 < tabulated $|t|_{0.05,\,9}$ = 1.833, so the null hypothesis is accepted, i.e., drug has no significant effect in gain/loss of body weight of rat.

10. Ten individuals each two strains (A and B) of a species were measured for a particular trait in the environment with the following results:

Strain A	10	12	8	14	12	9	7	7	13	16
Strain B	13	9	14	12	10	13	15	11	14	12

Is there significant difference in trait between 2 strains?

Solution:

- *Null Hypothesis*: There is no significant difference in trait.
- *Alternative Hypothesis*: There is significant difference in trait.
- *Calculation*:

Sl. No.	A (X)	$X-\overline{X} = D_1$	$(X-\overline{X})^2 = D_1^2$	B (Y)	$Y-\overline{Y} = D_2$	$(Y-\overline{Y})^2 = D_2^2$
1	10	10 – 10.8 = 0.8	0.64	13	13 – 12.3 = 0.7	0.49
2	12	12 – 10.8 = 1.2	1.44	09	09 – 12.3 = –3.3	10.89
3	08	08 – 10.8 = –2.8	7.84	14	14 – 12.3 = 1.7	2.89
4	14	14 – 10.8 = 3.2	10.24	12	12 – 12.3 = –0.3	0.09
5	12	12 – 10.8 = 1.2	1.44	10	10 – 12.3 = –2.3	5.29
6	09	09 – 10.8 = –1.8	3.24	13	13 – 12.3 = 0.7	0.49
7	07	07 – 10.8 = –3.8	14.44	15	15 – 12.3 = 2.7	7.29
8	07	07 – 10.8 = –3.8	14.44	11	11 – 12.3 = –1.3	1.69
9	13	13 – 10.8 = 2.2	4.84	14	14 – 12.3 1.7	2.89
10	16	16 – 10.8 = 5.2	27.04	12	12 – 12.3 = –0.3	0.09

$\sum X = 108$, $n_1 = 10$ $\sum D_1^2 = 85.6$ $\sum Y = 123$, $n_2 = 10$ $\sum D_2^2 = 32.1$

$\overline{X} = \frac{108}{10} = 10.8$ $\overline{Y} = \frac{123}{10} = 12.3$

$$SD(X-Y) = \sqrt{\frac{(X-\overline{X})^2 + (Y-\overline{Y})^2}{n_1 + n_2 - 2}} = \sqrt{\frac{85.6+32.1}{20-2}} = \sqrt{\frac{117.7}{18}} = \sqrt{6.539} = 2.557$$

$$SE = SD\sqrt{\frac{1}{n_1}+\frac{1}{n_2}} = 2.557\sqrt{\frac{2}{10}} = 2.557\sqrt{0.2} = 2.557 \times 0.447 = 1.14$$

$$|t| = \frac{\overline{X}-\overline{Y}}{SE} = \frac{10.8-12.3}{1.14} = \frac{1.5}{1.14} = 1.31$$

- *Level of significance*: 0.05.
- *Critical value*: Tabulated value of $|t|$ for 18 df at 0.05 is 2.10.
- *Decision*: The calculated value of 't' = 1.31 < $|t|_{0.05,\,18}$ = 2.10.

  So the null hypothesis is accepted, i.e., there is no significant difference in trait.

11. Ten soldiers visit a rifle range for two consecutive weeks. For the first week their scores are 67, 24, 57, 55, 63, 54, 56, 68, 33, 43 and during second week they score in same order 70, 38, 58, 58, 56, 67, 68, 72, 42, 38. Examine if there is any significant difference in the performance.

Solution:

- *Null Hypothesis*: There is no significant difference in the performance of soldiers.
- *Alternative Hypothesis*: There is significant difference in the performance of soldiers.
- *Calculation*:

Sl. No.	X	$X-\overline{X} = D_1$	$(X-\overline{X})^2 = D_1^2$	Y	$Y-\overline{Y} = D_2$	$(Y-\overline{Y})^2 = D_2^2$
1	67	67 – 52 = 15	225	70	70 – 56.7 = 13.3	176.89
2	24	24 – 52 = –28	784	38	38 – 56.7 = –18.7	349.69
3	57	57 – 52 = 05	25	58	58 – 56.7 = 1.3	1.69
4	55	55 – 52 = 03	09	58	58 – 56.7 = 1.3	1.69
5	63	63 – 52 = 11	121	56	56 – 56.7 = –0.7	0.49
6	54	54 – 52 = 02	04	67	67 – 56.7 = 10.3	106.09
7	56	56 – 52 = 04	16	68	68 – 56.7 = 11.3	127.69
8	68	68 – 52 = 16	256	72	72 – 56.7 = 15.3	234.09
9	33	33 – 52 = –19	361	42	42 – 56.7 = –14.7	216.09
10	43	43 – 52 = –09	81	38	38 – 56.7 = –18.7	349.69

$\sum X = 520$ $\quad \sum D_1^2 = 1882$ $\quad \sum Y = 567$ $\quad \sum D_2^2 = 1564.1$

$\overline{X} = \frac{520}{10} = 52$ $\quad \overline{Y} = \frac{567}{10} = 56.7$

$D_1^2 = 1882, n_1 = 10$ $\quad D_2^2 = 1564.1, n_2 = 10$

$$SD(X-Y) = \sqrt{\frac{\sum(X-\overline{X})^2 + \sum(Y-\overline{Y})^2}{n_1+n_2-2}} = \sqrt{\frac{1882+1564.1}{20-2}} = \sqrt{\frac{3446.1}{18}} = \sqrt{191.45} = 13.84$$

$$SE = SD\sqrt{\frac{1}{n_1}+\frac{1}{n_2}} = 13.84\sqrt{\frac{2}{10}} = 13.84\sqrt{0.2} = 13.84 \times 0.447 = 6.186$$

$$|t| = \frac{\overline{X}-\overline{Y}}{SE} = \frac{52-56.7}{6.186} = \frac{-4.7}{6.186} = 0.759 = 0.76$$

- *Level of significance*: 0.05.
- *Critical value*: Tabulated value of $|t|$ for df 18 at 0.05 is 2.10.
- *Decision*: The calculated value of 't' is 0.76, $|t| < t_{0.05, 18} = 2.10$; so the null hypothesis is accepted, i.e., there is no significant difference in the performance of soliders.

12. Two strains of mice were grown under the same environment and fed with same ratios. Twenty adult mice in each strain of the same age and sex were then measured for body weight (in gm) and results were as follows:

Strain	*Mean body weight*	*Variance*
A	**16**	**16**
B	**28**	**25**

Is there any significant differences in body weight between two strains?

Solution:

Since data is qualitative, we shall verify it by using 't-test'.

- *Null Hypothesis*: There is no significant difference in the mean body weight.
- *Alternative Hypothesis*: There is significant difference in the mean body weight.
- *Calculation*:

Strain	*Mean*	N	*Variance*	*Variance of mean*	*Variance of mean* $S^2_{\bar{x}-\bar{y}}$	SE = $\sqrt{S^2_{\bar{x}-\bar{y}}}$
A	16	20	16	$\frac{S^2_x}{n} = \frac{16}{20} = 0.8$	0.8 + 1.25 = 2.05	$\sqrt{2.05}$ = 1.43
B	28	20	25	$\frac{S^2_y}{n} = \frac{25}{20} = 1.25$		

$$|t| = \frac{\text{Difference between sample means}}{\text{Standard Error of the difference between means}} = \frac{16-28}{1.43} = \frac{12}{1.43} = 8.39.$$

- *Degrees of freedom*: 20 + 20 – 2 = 38.
- *Level of significance*: 0.05.
- *Critical value*: Tabulated value $|t|$ for 38 df at 0.05 is 2.02 (approximate), i.e., 40 df.
- *Decision*: Since the calculated value of $|t|$ is 8.39 > $t_{0.05,\,38}$ = 2.02, so the null hypothesis is rejected. The differences between mean body weights persist.

13. Hourly oxygen consumptions (in ml/100 g of body weight) were found to be as follows in a sample of 10 birds respectively before and after the exposure to a drug.

Birds:	**1**	**2**	**3**	**4**	**5**	**6**	**7**	**8**	**9**	**10**
Before:	**160**	**176**	**162**	**179**	**184**	**158**	**180**	**167**	**182**	**176**
After:	**130**	**148**	**131**	**145**	**158**	**130**	**143**	**142**	**150**	**147**

Find whether or not oxygen consumption is significantly higher before exposure to drug than that after exposure.

Solution:

- *Null Hypothesis*: Oxygen consumption is not significantly higher before exposure than after.
- *Alternative Hypothesis*: Oxygen consumption is higher before exposure than after.
- *Calculation*:

Sl. No.	Before (X_1)	After (X_1)	D = ($X_1 - X_2$)	D^2
1	160	130	160 – 130 = 30	900
2	176	148	176 – 148 = 28	784
3	162	131	162 – 131 = 31	961
4	179	145	179 – 145 = 34	1156
5	184	158	184 – 158 = 26	676
6	158	130	158 – 130 = 28	784
7	180	143	180 – 143 = 37	1369
8	167	142	167 – 142 = 25	625
9	182	150	182 – 150 = 32	1024
10	176	147	176 – 147 = 29	841

$\sum D = 300$ $\sum D^2 = 9120$

$n_1 = 10$

$$\bar{D} = \frac{\Sigma D}{n} = \frac{300}{100} = 30$$

$$SD = \sqrt{\frac{\Sigma D^2 + \frac{\Sigma D}{n}}{n-1}} = \sqrt{\frac{9120 + \frac{(300)^2}{10}}{9}} = \sqrt{\frac{9120 + \frac{90000}{10}}{9}} = \sqrt{\frac{9120 - 9000}{9}} = \sqrt{\frac{120}{9}} = \sqrt{13.33} = 3.65$$

$$SE = \frac{SD}{\sqrt{n}} + \frac{3.65}{\sqrt{10}} = \frac{3.65}{3.10} = 1.155$$

$$|t| = \frac{\bar{D}}{SE} = \frac{30}{1.16} = 25.86$$

- *Level of significance*: 5% level, i.e., 0.05.
- *Critical value*: Tabulated value at 0.05 level at 10 – 1 = 9 df is 2.26.
- *Decision*: The calculated value of $|t|$ is 25.86 > $|t|$ = 2.26[0.05(9)].

So the null hypothesis is rejected.

Table 11.1: The probabilities of exceeding different chi-square values for degrees of freedom from 1 to 50 when the expected hypothesis is true

	Probabilities									
df	0.95	0.90	0.70	0.50	0.30	0.20	0.10	0.05	0.01	0.001
1	0.004	0.016	0.15	0.46	1.07	1.64	2.71	3.84	6.64	10.83
2	0.10	0.21	0.71	1.39	2.41	3.22	4.61	5.99	9.21	13.82
3	0.35	0.58	1.42	2.37	3.67	4.64	6.25	7.82	11.35	16.27
4	0.71	1.06	2.20	3.36	4.88	5.99	7.78	9.49	13.28	18.47
5	1.15	1.61	3.00	4.35	6.06	7.29	9.24	11.07	15.09	20.52
6	1.64	2.20	3.83	5.35	7.23	8.56	10.65	12.59	16.81	22.46
7	2.17	2.83	4.67	6.35	8.38	9.80	12.02	14.07	18.48	24.32
8	2.73	3.49	5.53	7.34	9.52	11.03	13.36	15.51	20.09	26.13
9	3.33	4.17	6.39	8.34	10.66	12.24	14.68	16.92	21.67	27.88
10	3.94	4.87	7.27	9.34	11.78	13.44	15.99	18.31	23.21	29.59
11	4.58	5.58	8.15	10.34	12.90	14.63	17.28	19.68	24.73	31.26
12	5.23	6.30	9.03	11.34	14.01	15.81	18.55	21.03	26.22	32.91
13	5.89	7.04	9.93	12.34	15.12	16.99	19.81	22.36	27.69	34.53
14	6.57	7.79	10.82	13.34	16.22	18.15	21.06	23.69	29.14	36.12
15	7.26	8.55	11.72	14.34	17.32	19.31	22.31	25.00	30.58	37.70
20	10.85	12.44	16.27	19.34	22.78	25.04	28.41	31.41	37.57	45.32
25	14.61	16.47	20.87	24.34	28.17	30.68	34.38	37.65	44.31	52.62
30	18.49	20.60	25.51	29.34	33.53	36.25	40.26	43.77	50.89	59.70
50	34.76	37.69	44.31	49.34	54.72	58.16	63.17	67.51	76.15	86.66

Abridged from Table IV of Fisher and Yates, Statistical Tables for Biological, Agricultural and Medical Research, Oliver and Boyd Ltd. Edinburgh, by permission of the authors and publisher.

Table 11.2: Distribution of *t* values ("Student's" distribution)*

	Probabilities							
df	0.90	0.70	0.50	0.30	0.10	0.05	0.01	0.001
1	0.16	0.51	1.00	2.00	6.31	12.70	63.66	636.62
2	0.14	0.44	0.82	1.39	2.92	4.30	9.92	31.60
3	0.14	0.42	0.76	1.25	2.35	3.18	5.84	12.92
4	0.13	0.41	0.74	1.19	2.13	2.78	4.60	8.61
5	0.13	0.41	0.73	1.16	2.02	2.57	4.03	6.87
6	0.13	0.40	0.72	1.13	1.94	2.45	3.71	5.96
7	0.13	0.40	0.71	1.12	1.90	2.36	3.50	5.41
8	0.13	0.40	0.71	1.11	1.86	2.30	3.36	5.04
9	0.13	0.40	0.70	1.10	1.83	2.26	3.25	4.78
10	0.13	0.40	0.70	1.09	1.81	2.23	3.17	4.59
11	0.13	0.40	0.70	1.09	1.80	2.20	3.10	4.44
12	0.13	0.40	0.70	1.08	1.78	2.18	3.05	4.32
13	0.13	0.39	0.69	1.08	1.77	2.16	3.01	4.22
14	0.13	0.39	0.69	1.08	1.76	2.14	2.98	4.14
15	0.13	0.39	0.69	1.07	1.75	2.13	2.95	4.07
16	0.13	0.39	0.69	1.07	1.75	2.12	2.92	4.02
17	0.13	0.39	0.69	1.07	1.74	2.11	2.90	3.96
18	0.13	0.39	0.69	1.07	1.73	2.10	2.88	3.92
19	0.13	0.39	0.69	1.07	1.73	2.09	2.86	3.88
20	0.13	0.39	0.69	1.06	1.72	2.09	2.84	3.85
21	0.13	0.39	0.69	1.06	1.72	2.08	2.83	3.82
22	0.13	0.39	0.69	1.06	1.72	2.07	2.82	3.79
23	0.13	0.39	0.68	1.06	1.71	2.07	2.81	3.77
24	0.13	0.39	0.68	1.06	1.71	2.06	2.80	3.74
25	0.13	0.39	0.68	1.06	1.71	2.06	2.79	3.72
26	0.13	0.39	0.68	1.06	1.71	2.05	2.78	3.71
27	0.13	0.39	0.68	1.06	1.70	2.05	2.77	3.69
28	0.13	0.39	0.68	1.06	1.70	2.05	2.76	3.67
29	0.13	0.39	0.68	1.06	1.70	2.04	2.76	3.66
30	0.13	0.39	0.68	1.06	1.70	2.04	2.75	3.65
40	0.13	0.39	0.68	1.05	1.68	2.02	2.70	3.55
60	0.13	0.39	0.68	1.05	1.67	2.00	2.66	3.46
120	0.13	0.39	0.68	1.04	1.66	1.98	2.62	3.37
∞	0.13	0.38	0.67	1.04	1.64	1.96	2.58	3.29

Abridged from Table III of Fisher and Yates, Statistical Tables for Biological, Agricultural and Medical Research, published by Oliver and Boyd Ltd. Edinburgh, by permission of the authors and publisher.

SECTION-II

MOLECULAR GENETICS

12 DNA and Topology of DNA

DNA

1. List the forces that hold the DNA double helix together as a stable unit.

The DNA double helix is held together by two types of bonds: viz. covalent and hydrogen.

Solution:

Covalent bonds occur within each linear strand and strongly bond the bases, sugars and phosphate groups (both within each component and between components).

Covalent bonds:

(a) In bases and sugars:

C—C
C—N
C—H
C—O
O—H
N—H

C:C
↑ ↑
Shared electron

(b) Phosphodiester linkage:

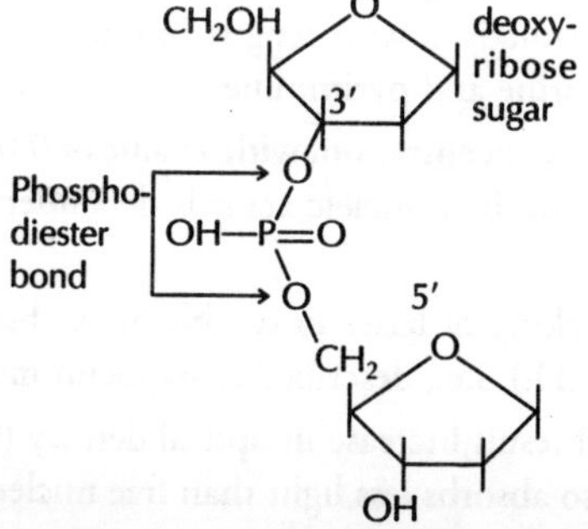

(ii) Hydrogen bonds occur between the two strands, between a base from one strand and a base from the other strand in complementary pairing. These hydrogen bonds are individually weak but collectively strong.

A = T (double hydrogen bond)
G ≡ C (triple hydrogen bond)

2. Define base complementarity.

Solution:

The chemical affinity provided by the hydrogen bond between the nitrogen bases of nucleic acids is termed as **base complementarity.**

3. Denaturation and Renaturation

Solution:

1. The hydrogen bonds between the nitrogenous bases of two strands of DNA are weak and break easily when heated or when pH of the medium is changed.
2. The separation of DNA strand is called **denaturation** or **melting**:
3. When denatured DNA is cooled slowly hydrogen bonds are established again and the separated polynucleotide strands pair again restoring the double helical structure. This is known as **renaturation** or **annealing.**

Denaturation:

(i) It means the formation of single strand from double stranded on by breaking the H-bonds between the nitrogenous bases (A = T ad G= C) of double helix by heating the solution or by the change of pH of the medium, i.e., action of alkali (pH-12).

(ii) Thus, the separation of DNA strand is called denaturation or melting.

Renaturation:

(i) The ability of separated (denatured) complementary strands of nucleic acids to reunite (pair again) and restoring the stable double stranded structure is known as **renaturation or annealing.**

(ii) It is used to measure the relatedness of nucleic acids from different organs, different types of cells or from different species.

- ***Steps of Renaturation:***

(a) A rate of determining step in renaturation is the collisions or nucleation of complementary sequences.

(b) A rapid pairing of complementary bases or zippering of the strands to form duplex molecule or DNA.

(c) Renaturtion rate obeys the law of **mass action** and increase with **DNA concentration.**

- ***Characteristics of Renaturation:***

(a) It is a slow process compared to denaturation.

(b) It is not the rewinding of the helix, but the precise collisions between complimentary strands such that base pairs are formed at the correct positions.

(c) Since this is a result only of random motion, it is a concentration dependent process and takes several hours in laboratory.

- ***Mechanism of Renaturation:***

(i) The molecular details of renaturation can be understood through a hypothetical molecular sequences.

IA	IB	II	IC
...ATGA...	...ATGA......	...CCCC......	...ATGA...
...TACT...	...TACT......	...GGGG......	...TACT...
IA[1]	IB[1]	II[1]	IC[1]

(ii) Assume that each single strand contains 50,000 bases and that base sequences are complementary.

(iii) However, any short sequences of bases (say, 4-6 bases long) certainly appear many times in such a molecule and can provide sites for base pairing.

(iv) Random collisions between noncomplementary sequences such as IA and II[1] will be ineffective, but a collision between IA and IC[1] will result base pairing $\left(\frac{\text{ATGA}}{\text{TACT}}\right)$. However, this will be short lived because the bases surrounding these short complement tracts are not able to pair and stacking stabilization will not occur.

(v) Thus, at elevated temperature that has brought about strand separation, these paired regions rapidly become disrupted.

(vi) However, as soon as two sequences such as IB and IB[1] pair, the adjacent bases will also rapidly pair and the entire double stranded

(i) Nucleic acid absorbs ultraviolet (UV) light most strongly at wave length 260 nm due to the interaction between UV light and the ring systems of the purine and pyrimidine.

(ii) The absorbance at 260 nm, A_{260} is proportional to concentration, with a value of 0.02 units per microgram DNA per millilitre. The amount of light absorbed nucleic acids is dependent on the structure molecule.

(iii) The more ordered structure, i.e., the regular stacking of bases in double helix absorbs less UV light and this will result decrease in optical density (O.D.), i.e., described as **hypochromicity** (A_{260} = 1.00).

The free nucleotide absorb, more light which will result increase in optical density (O.D.), i.e., **hyperchromacity** A_{260} (1.60). The single stranded DNA also absorbs less light than free nucleotide (A_{260} = 1.37) but more than ds DNA, i.e., it is hyperchromicity (decreases O.D.).

	UV	O.D.
D. S. DNA	A260	1.00
S. S. DNA	A260	1.37
Free nucleotide	A260	1.50

DNA molecule will "**zip up**" in a few seconds.

Renaturation Kinetics:

- ***DNA complexity:***

DNA complexity is the total length of different sequences of DNA present in a given preparation.

- ***Kinetic complexity:***

It is the complexity of the DNA component which is measured through the association of DNA in the solution followed by random collisions between the complementary strands which obeys the **second order reaction of kinetics.**

If DNA is sheared into uniform fragments (300–10,000 bp) denatured and kept at low concentration, so that effects of mechanical entanglement are small.

The renaturation is both a biomolecular and **second order reaction** and described by rate law.

Therefore, the rate of renaturation of denatured DNA is expressed as follows:

$\frac{d[A]}{dt} \propto [A]$, [A] concentration of SS DNA

$\frac{d[A]}{dt} \propto [A][B], [A] = [B]$ for duplex DNA

$-\frac{d[A]}{dt} = K[A]^2$, K = Second order rate constant.

Rearranging and integrating it.

Let, $[A]_0$ = the initial concentration of A

$[A]$ = the concentration of A at time 't'

$$\int_{[A]_0}^{[A]} -\frac{d[A]}{[A]^2} = K\int_0^t dt$$

$$\text{or,} \quad \frac{1}{[A]} = \frac{1}{[A]_0} + Kt = \frac{1+[A]_0Kt}{[A]_0}$$

$$\text{or,} \quad \frac{[A]}{[A]_0} = \frac{1}{1+[A]_0Kt}$$

$$\text{or,} \quad f = \frac{1}{1+[A]_0Kt} \qquad \text{... (I)}$$

The above-mentioned equation (1) tells us

1. After a very long time, the fraction that is dissociated approaches zero—that is ultimately, renaturation is complete.
2. The rate of renaturation increases as the initial DNA concentration is increased.

In ordinary chemical kinetics, concentration is expressed as the number of molecules per unit volume, but in renaturation kinetics the best measure of concentration is actually the number of copies of each sequence per unit volume.

Let Co be the initial concentration of single stranded DNA which is expressed in base pair/cm^3 and x is the number of base pairs in each unique sequence.

$$\text{or,} \quad [A]_0 = \frac{Co}{x} \qquad \text{... (II)}$$

Substituting the value of $\frac{Co}{x}$ for $[A]_0$ in the equation (1).

$$\text{or,} \quad f = \frac{1}{1+\frac{Cot}{x}K} = \frac{x}{x+CotK} \qquad \text{... (III)}$$

[f = fraction of single stranded molecule; t = time in second]

When half of the sample of molecules (SS DNA) renatured, then f = 0.5 or half.

$$\frac{1}{2} = \frac{1 \times x}{x+Co\frac{t}{2}K}$$

$$\text{or,} \quad 2x = x + Cot\frac{1}{2}K$$

$$\text{or,} \quad Co\frac{t}{2}K = 1x$$

$$\text{or,} \quad Co\frac{t}{2} = \frac{x}{K}$$

If $x = 1$ [for poly(dA) Poly(dU)]

$Cot\frac{1}{2} = \frac{1}{K}$ [∵ K = rate constant]

The value of 'K' can be evaluated by multiplying the value of unit at concentration, i.e., Co and the time required for 50% of renaturation.

Cot and Cot curve analysis:

- ***Cot:***

The parameter controlling the ressociation reaction is the product of initial cocentration (moles/lit) of single stranded DNA (Co) and the time (second) of incubation for reassociation (t) is simply described as the Cot.

(i) The reassociation is both a biomolecular reaction and second order reaction and described as rate law below:

$\frac{d[C]}{dt} \propto C$ C concentration of SS DNA

$\frac{d[C]}{dt} \propto C^2$ C^2 = ds DNA

$$\frac{dC}{dt} = -K\,C^2 \qquad \text{... (1)}$$

(ii) By integrating the equation, between the limits of the initial concentration of DNA (Co) at time t = 0 and the concentration C that remains single stranded after t.

$$\frac{C}{Co} = \frac{1}{1+KCot}$$

(iii) Cot value also varies depending on the temperature of reassociation and the concentration of monovalent cations. Generally reassociation temperate is 65°C and a concentration of 0.3 M NaCl.

- ***Cot curve:***

(i) It is prepared by plotting the fraction of DNA that has reassociated $\left[1-\frac{C}{Co}\right]$ against the log of Cot values.

(ii) The initial shape of the curve reflects the fact that in a mixture of unique sequence fragments, each with one complement, initial matches take more time to make.

(iii) Then as many as single strands are converted to duplexes, matches are made more quickly, reflecting an increase in the slope of the curve.

(iv) Near the end of the reaction, a few remaining single strands require relatively greater time to make final maches.

(v) The Cot curves for Poly-U: Poly A from two bacterio phages (MS2 and T4) and one bacteria (*E. Coli*) source each with different genome size (MS-2 = 35 × 10^2 bp, T4 = 1.7 × 10^5 bp, *E. coli* 4.2 × 10^6 bp) follow similar pattern, i.e., sigmoidal and the ratio of Cot values between 10% and 90% reactions is almost 100% in each (Fig. 12.1).

(vi) The curves obtained are shifted further and further to the right indicating extended reassociation time (Fig. 12.2).

(vii) Renaturation curve, i.e., Cot curve of eukaryotic DNA is quite complex as compared to *E. coli* DNA.

(viii) The curve shows three distinct regions, viz., highly repetitive, moderately repetitive and non-repetitive or unique sequences.

(ix) Highly repetitive sequences are the earliest to renature, i.e., **fast component.** Moderately repetitive sequences are the next to reassociate, i.e., **intermediate component** and the unique sequences renature towards the end, i.e., **slow component** (Fig. 12.3).

The rate constant (K) is the characteristic of rate at which single strand collision in the solution under the physical process of renaturation. So it is independent of the complexity of DNA or the molecular weight of DNA and the value of x.

Since x is also constant, Cot$\frac{1}{2}$ is a constant although neither Co nor t$\frac{1}{2}$ is a constant. Thus, for any particular DNA concentration, Co that is chosen t$\frac{1}{2} = \frac{x}{KCo}$.

Cot$\frac{1}{2}$ is the product of DNA concentration and time (t) required to proceed to half completion of reaction and is directly proportional to the unique length of reassociation DNA that is the point in the reaction when one half of the DNA is present as double stranded fragments is

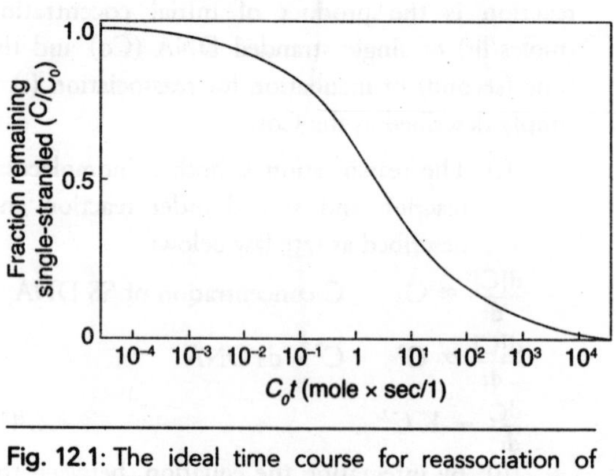

Fig. 12.1: The ideal time course for reassociation of DNA (C/C_0) when, at time zero, all DNA consists of unique frgments of single-stranded complements. Note that the abscissa (C_0t) is plotted logarithmically.

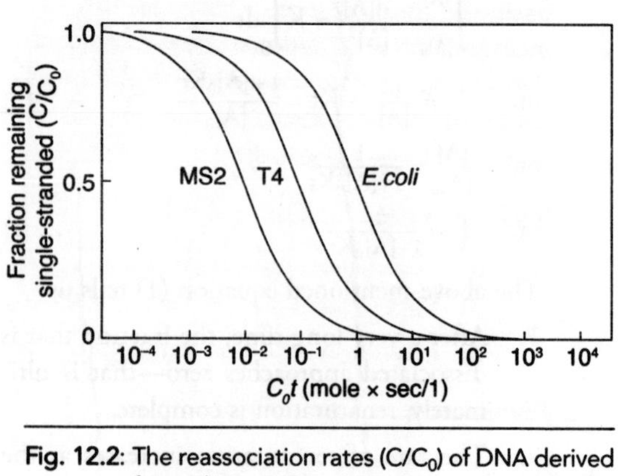

Fig. 12.2: The reassociation rates (C/C_0) of DNA derived from phage MS2, phage T4, and *E. coli*. The genome of T4 is larger than MS2 and that of *E. coli* is larger than T4.

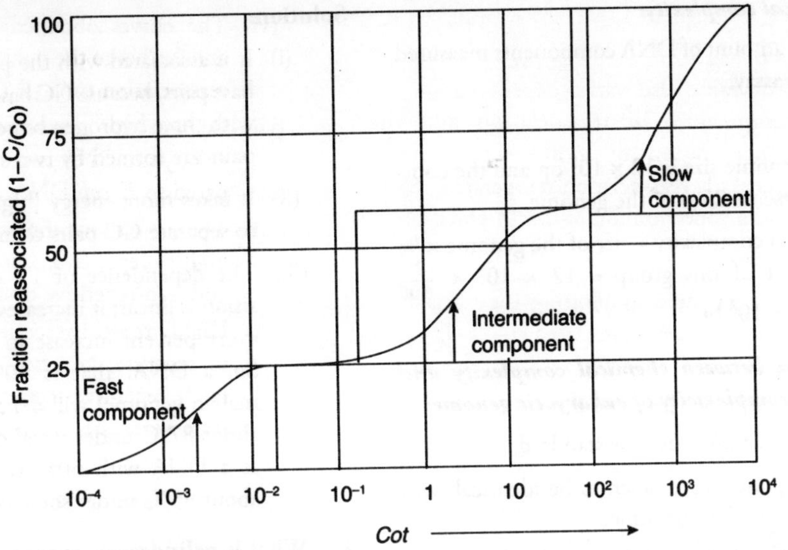

Fig. 12.3: The reassociation kinetics of eukaryotic DNA showing three types of components. Arrow indicates $C_0t\frac{1}{2}$ values of each component.

termed as Cot$\frac{1}{2}$ or half reaction time, provided that all pairs of single-stranded DNA complements consists of unique nucleotide sequences and all are about the same size.

Cot$\frac{1}{2}$ varies directly with the complexity of DNA, designated as *x*. Complexity represents the length in unique DNA fragments laid end to end.

The measurement of renaturation rate to obtain the value of Cot$\frac{1}{2}$ is called Cot analysis.

The Cot$\frac{1}{2}$ is proportional to the number of base-pair per molecule. This means that the measured value of Cot$\frac{1}{2}$ can be used to obtain the size of a particular sequence and the number of copies of the sequence in the organism from which DNA was isolated.

Technique used for Cot analysis

(i) Absorbance measurements.

(ii) Membrane filtration.

(iii) Hydroxyapatite chromatograph.

Significance of Cot analysis

(i) Cot analysis is used as a means of detecting repetitive sequences in DNA.

(ii) A higher Cot$\frac{1}{2}$ means slower reaction and lower Cot$\frac{1}{2}$ means faster reaction.

(iii) If there is no repetitive DNA (because repetitive DNA reassociates faster). The Cot$\frac{1}{2}$ reaction will be directly proportional to the DNA content. Therefore, Cot$\frac{1}{2}$ will indicate the length of all the different sequences in a genome, which will be less than the length of the total DNA in a genome, when there is repetition. This is called **Kinetic complexity**.

- ***Kinetic complexity*:**

It is the complexity of the DNA component which is measured through the reassociation of DNA solution followed by random collisions between the complementary strands which follow second order reaction of kinetics. This can be calculated by Cot$\frac{1}{2}$.

Example:

E. coli genome has 0.004 pg (picogram) of DNA, i.e., 4.2×10^6 bp with Cot$\frac{1}{2}$ = 9.

Kinetic complexity of genome can be calculated as follows (If Kinetic complexity is X):

$$X = \frac{\text{Cot}\frac{1}{2} \text{ of DNA of any genome} \times \text{base pair}}{\text{Cot}\frac{1}{2} \text{ of } E.\ Coli \text{ genome}}$$

$$= \frac{\text{Cot}\frac{1}{2} \text{ of any genome} \times 4.2 \times 10^6}{9}$$

- ***Chemical complexity:***

It is the amount of DNA components measured by chemical assay.

Example:

If the genome size is 12×10^8 bp and the component represents 25% of the genome.

Chemical complexity = size of the genome × % of component of this group = $12 \times 10^8 \times \frac{25}{100}$ $= \frac{12 \times 10^8}{4} = 3 \times 10^8$ bp.

- ***Relation between chemical complexity and kinetic complexicity of eukaryotic genome:***

I. Usually *E. Coli* is used as standard.

II. Its components are taken to be identical with the length of the genome.

III. Thus, complexity of DNA can be determined by comparing its Cot$\frac{1}{2}$ with that of standard DNA of known DNA complexity.

4. What is melting profile?

Solution:

(i) On heating hydrogen bonds of double helical DNA break, i.e. **denaturation** of DNA into single strands at a specific 'melting' temperature and their presence is marked by their increased absorbtion of UV light.

(ii) When absorbtion at 260 nm ($0.D_{260}$) is monitored and plotted against temperature during heating, a melting profile of DNA is obtained. This profile (outline) is termed as "melting profile" or thermal denaturation curve or "**helix coil transition**" of DNA.

5. What is T_m?

Solution:

(i) The mid-point of the thermal denaturation curve or profile that represents the point at which 50% of the double-stranded nucleic acid molecule is dissociated into single strand is called melting temperature or T_m.

(ii) The melting temperature of DNA ranges between 82-90°C.

6. Why T_m is related to base composition?

Solution:

(i) It is associated with the proportion of GC base pairs, because GC base pairs are formed with three hydrogen bonds, while AT base pairs are formed by two such bonds.

(ii) It takes more energy (higher temperature) to separate GC pairs than an A:T pairs.

(iii) The dependence of T_m on base composition is linear; it increases about 0.4°C for every percent increase in (G≡C) content. So, a DNA, that is 40% G≡C (mammalian genome) will denature with T_m of about 87°C under usual condition, where as a DNA with 60% will have a T_m of about 95°C under some condition.

7. What is palindromic sequence?

Solution:

(i) A sequence of words, phrases, nucleotides that reads the same either backward or forward, regardless of directions from which it starts is called palindromic sequence.

(ii) It often occurs as sites for restriction, endonuclease recognition and cutting.

(iii) Ex: rotator, ⟵ ⟶
malayalam, AND MADAM DNA

8. What are the features of DNA breathing?

Solution:

(i) It involves transient changes in the axis of the helix, relative dispositions of base pairs and back-bone and base unpairing either individually or coordinately followed by proton exchange.

(ii) The transient opening of the double helix and chemical reactions occur in atoms which remain protected completely when the base pairing is undisrupted.

9. What happens to DNA in acidic solution?

Solution:

Strong hot acid can break the phosphodiester backbone of DNA and leads to release the free bases and breakdown products of deoxyribose.

10. Why RNA not DNA is sensitive to alkaline hydrolysis?

Solution:

(i) Since RNA has ribose sugar which possess 2′ OH^- group, it is more sensitive to alkaline solution than DNA.

(ii) Alkaline hydrolysis of RNA (0.3 M NaOH at 37°C for 18 minutes) yields both the nucleosides 2′ phosphates and nucleosides 3′ phosphate.

(iii) These isomers are readily interconverted under acidic conditions via the intermediate formation of nucleotide 2′–3′ cyclic phosphates.

11. Write down the differences between DNA and RNA.

Solution:

	DNA	RNA
Location	(i) Primarily in the nucleus, also in mitochondria and chloroplast.	(i) Primarily in the cytoplasm and also in the nucleus.
Composition	(i) Pentose sugar is deoxyribose ($C_5H_{10}O_4$).	(i) Pentose sugar is ribose ($C_5H_{10}O_5$).
	(ii) Pyrimidines are cytosine and thymine.	(ii) Pyrimidines are cytosine and uracil.
Structure	(i) Double-stranded and helical.	(i) Single-stranded and non-helical.
Function	(i) Store genetic information and transmit it from cell to cell.	(i) Perform protein synthesis.
Reaction (stain)	(i) Feulgen positive.	(i) Feulgen negative.

12. Write down the differences between Purine and Pyrimidine.

Solution:

Character	Purine	Pyrimidine
1. Ring	(i) Double ringed compound (ii) Formed of a pyrimidine ring and an imidazole ring (iii) Ring formed of nine atoms: 4N + 5C	(i) Single ringed compound. (ii) Formed of pyrimidine ring (iii) Rings formed of six atoms 2N + 4C
2. Bonding	Glycosidic bond is formed between nitrogen atom at 9th position in pyrimidine and carbon at 1st position in pentose sugar.	Glycoside bond is formd between nitrogen at 1st postion in pyrimidine and carbon at 1st position in pentose sugar.

13. Mention resemblances between Z-DNA and B-DNA.

Solution:

1. Both are double helical/stranded.
2. Two strands of double helix are antiparallel in both DNAs.
3. Both forms exhibit G ≡ C pairing.

14. What is the difference between major and minor grooves of double stranded DNA?

Solution:

The deep, wide groove made by turns in double stranded DNA constitutes the major groove.

The shallow, narrow groove between the antiparallel single strand of DNA makes the minor groove.

15. Write down the differences between B-DNA and Z-DNA.

Solution:

B-DNA	Z-DNA
1. It is right-handed helical DNA.	1. It is left-handed helical DNA.
2. One complete helix requires 10 bp in B-DNA.	2. One complete helix requires 12 bp in Z-DNA.
3. B-DNA follows cyclical way.	3. Z-DNA follows zig-zag course.

16. Write down the differences between Mitochondrial DNA and Nuclear DNA.

Solution:

Mitochondrial DNA	Nuclear DNA
1. It does not contain histone and has no nucleosome.	1. It contains histones and is package into nucleosome.
2. Reassociation of mitochondrial DNA is rapid.	2. Reassociation of nuclear DNA is not so rapid.
3. It is generally circular except in *Paramoecium* and *Tetrahymena*.	3. It is linear.
4. The genome size is small, containing smaller nubber of genes.	4. The genome size is large, containing many genes.
5. The mitochondrial DNA consists of a mixture of opening circles and twisted or supercoils.	5. It does not have any such structure.
6. Repetitive sequence of G and C are found only in a small proportion.	6. It is common in nuclear DNA.

17. State briefly molecular architecture of centromere.

Solution:

The chromosomal site for attachment of spindle fibre identified by constricted region is called **centromere**. The centromere is associated with a plate-like protoplasmic body known as **kinetochore**. The location of centromere and hence the kinetochore is directly controlled by a chromosomal DNA segment called **centromeric DNA**.

A. **In Yeast:**

(i) The DNA sequences of centromeres have been analyzed extensively in the yeast *Saccharomyces cerevisiae*. These sequences are called CEN sequences. Sequence analysis of cloned centromere of yeast chromosome (each) consists of 112-125 bp and they are organized into three sequence domains called centromere DNA elements (CDEs), viz., CDE I, CDE II, CDE III.

(ii) CDE I and CDE III are relatively short and highly conserved. CDE I has 8 bp sequence [RTCACRTG where R is purine (i.e., either A or G)]. CDE III consists of 26 bp sequence domain and is also AT rich.

(iii) CDE III is the largest domain and contains 78-86 bp. More than 90 percent or up to 95 percent is composed of AT base pairs.

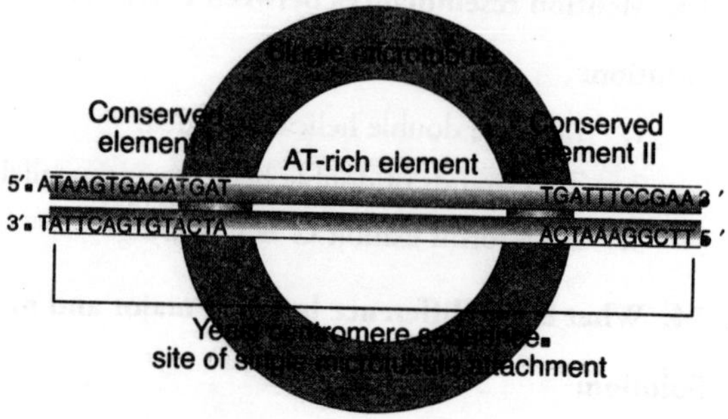

Fig. 12.4: Structure and DNA sequence organization of yeast centromeres.

(iv) CDE II and CDE III appeared to mediate interaction with microtubules through CBF_2 and CBF_3 (Centromeric binding factor 2 and 3) respectively. Kinesin-related protein Kar^3p, these proteins are important link between kinetochore microtubules and centromere.

B. **In *Drosophila*:**

(i) Centromeric DNA of *Drosophila* is about 420 kb with highly repetitive satellite DNA.

(ii) It contains 10 bp sequence AATAACATAG, which is tandemly repeated in many times the centromere of all chromosomes of *Drosophila*.

C. **In Mouse:**

(i) Highly repetitive "satellite" DNA is localized in the centromeric regions of mice GAAAAATGA
CTTTTTACT

(ii) This DNA occupies specific sites in all the mouse chromosome except Y-chromosome.

D. **In Human:**

(i) The centromeric DNA of human is called α satellite DNA or **alphoid sequence.** It is about 170 bp in length and is repeated many times in a tandem array.

(ii) **Alphoid sequence** contains a conserved sequence called **CENP box** with the consensus sequence.

PyTTCGTTGGAAPUCGGGA which serves as binding sites for centromeric protein B (CENP-B) with the centromere.

17. Highlight the molecular architecture of telomere.

Solution:

The eukaryotic chromosome terminates at either end in a structure called **telomere.** Telomere forms a cap at each end of the chromosome. It provides stability to the chromosome. Most telomeric sequences may be divided into two types.

A. **Simple telomeric sequences:**

(i) These sequences are species-specific and consist of a series of tandemly repeated DNA sequences.

(ii) In most vertebrates, the repeating sequences is the hexa nucleotide TTAGGG. These are generally GC rich with guanine residues clustered on one strand and cytidine residues on the other.

Organism	Telomeric repeat
1. *Tetrahymena thermophilia*	5′ TTGGGG 3′
2. *Trypanosoma*	5′ TAGGGG 3′
3. *Paramoecium*	5′ TTGGGG 3′
4. Round-worm	5′ TTAGGG 3′
5. Mouse and Human	5′ TTAGGG 3′

This DNA is not double stranded all the way out to the end. Although these sequences are variable in different species. The basic repeat unit in all species have the pattern $5' \dots T_{1\text{-}4}\ A_{0\text{-}1}\ G_{1\text{-}8} \dots 3'$

(iii) The 'G' rich strands run in the 5′-3′ direction towards each end of the chromosome and end 12 to 15 nucleotide past end of the 'C' rich strand. As a result overhang of 12-15 nucleotides in length of G rich strand.

(iv) A square planar structure (G–tetard/G–quartet) is formed by unorthodox interactions between guanosine residues.

(v) The $(TTAGGG)_n$ array of a human telomere often extend about 10-15 kb. A protein complex **telosome (shelterin)** recognise and bind to telomeric DNA.

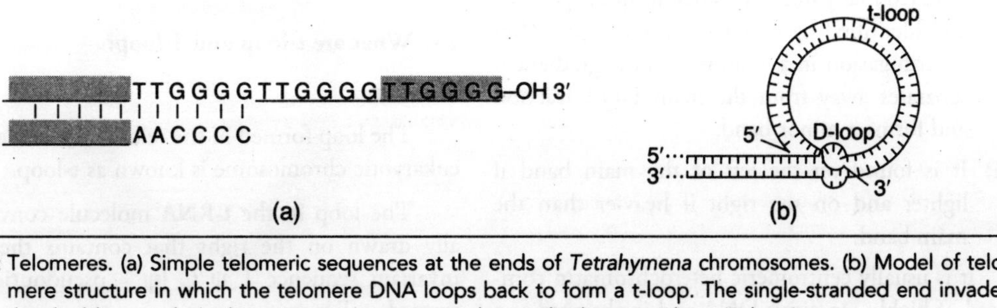

Fig. 12.5: Telomeres. (a) Simple telomeric sequences at the ends of *Tetrahymena* chromosomes. (b) Model of telomere structure in which the telomere DNA loops back to form a t-loop. The single-stranded end invades the double-stranded telomeric sequences to produce a displacement loop (D-loop).

(vii) As a result of natural difficulty in replicating lagging strand at the end of telomere, the G rich strand has single stranded over hang at its 3′ end that is typically 150-200 bp loop. This can fold back and form base pairs with each other (G–quartet), C rich strand to form a telomeric loop known as the **T loop** Ex *Oxytricha fallax* (ciliate); *Trypanosoma bruce* (protozoa) *Pisum sativum* (pea). It protects chromosme ends.

(viii) The single-stranded end invades the double-stranded telomeric sequences, causing a displacement loop or **D loop** to form.

B. **Telomere-associated sequences:**

(i) These regions are internal to the simple telomeric sequences.

(ii) It also consists of repetitive sequences.

(iii) **Telosome (Shelterin)** is a very large protein complex. It contains several components that recognise and bind to telomeric DNA.

(iv) The protein complex components are TRF1, TRF2 (telomeric repeat binding factors), POT-1 (Protection of Telomeres-1).

(v) These sequences vary among the organisms.

(vi) The significance of such sequences remains unknown.

19. What is telomere silencing?

Solution:

Telomeric region associatd with protein **RAP-1** (Repressor Activato Protein) that inhibits gene expression in this region, i.e., known as telomere silencing.

20. What is satellite DNA? What does it signify?

Solution:

(i) This DNA usually consists of short sequences repeated many times (in tandem) in the genome and identified by its rapid reassociation. After centrifugation in a cesium chloride gradient it separates away from the main DNA fraction and forms a minor band.

(ii) It is found on the left of the main band if lighter and on the right if heavier than the main band.

It is usually centromeric heterochromatin composed of highly repetitive DNA and is observed as a **secondary** or **satellite** bands formed in the density gradient.

21. What is cryptic satellite DNA?

Solution:

This DNA has a buoyant density which is comparable to bulk genomic DNA and does not form satellite band. It is identified by restriction mapping.

Example:

Most of the centromeric DNA of human chromosomes comprise **cryptic satellite** DNA, called **alphoid DNA (α satellite DNA).**

22. What are mini- and micro-satellite DNA?

Solution:

Minisatellite

(i) A short DNA sequence usually 12 bp in length (12–100 bp) is found in cluster with repeated over and over.

(ii) Minisatellites tend to be unstable. The number of copies of a particular sequence often increases or decreases from one generation to the next.

(iii) They are used to identify individual in criminal or paternity cases.

Microsatellite

(i) A short DNA sequence usually 2–4 bp repeats many times in tandem ($\leq$ 200 bp).

(ii) They are highly polymorphic and distributed throughout the genome; so they make ideal genetic marker.

(iii) It is used to analyze the relationship between different ethnic human population.

23. What are t-loop and T-loop?

Solution:

The loop formed in the telomere at the end of a eukaryotic chromosome is known as t-loop.

The loop in the t-RNA molecule conventionally drawn on the right that contains the nearly invariant sequence T ψ C [ψ = pseudouridine] is termed as T-loop.

24. What are SINEs and LINEs?

Solution:

SINEs:

(i) These are short sequences of repetitive DNA interspersed in eukaryotic chromosome in many copies.

(ii) It consists of sequences 100–400 bp in length.

(iii) It can move in the genome by retrotransposon. They do not encode the enzymes. They need for movement.

Ex: Alu1(**"Al you one"**). In humans Alu family is the most abundant SINE family in the genome.

LINEs:

(i) These are long sequences of repetitive DNA interspersed in eukaryotic chromosome in many copies.

(ii) They have sequences of 1000 to 7000 bp or longer.

(iii) In human, the most prominent example is a family designated L1. Members of this sequence family are about 6400 bp long and are present up to 100,000 times.

25. Why LINEs are referred to as retrotransposon?

Solution:

LINEs are often referred to as retrotransposon because their mechanism of transposition resembles that used by retroviruses.

26. What is selfish DNA?

Solution:

Selfish DNA is the large quantities of **repetitive nontranscribed, noncoding** segment of genome and contributes nothing to the well being of the organism. In most cases such DNA is selectively neutral except for the energetic burden of replicating it.

27. What is C-value? Why is it regarded as a paradox?

Solution:

The amount of DNA in picograms in a genome (haploid set of chromosome) of a given species is called **C-value.**

(i) There is a rough correlation between the DNA content and the number of genes in a cell.

(ii) However, this correlation breaks down in several cases of closely related organisms where the DNA content per haploid cell (C-value) varies widely. This discripancy between the size of the genome and the structural or organizational complexity of an organism is called **C-value paradox.**

(iii) There are two aspects of paradox:

(a) Huge variations in C-values between certain species whose apparent complexity does not vary correspondingly.

(b) The apparent absolute excess of DNA compared with the amount of that could be expected to code for proteins.

(iv) The C-value paradox is probably explained not by extra genes but by extra noncoding DNA in some organism.

28. State briefly about unique sequence and repetitive sequence.

Solution:

Unique sequence:

(i) A class of DNA sequences each of which is present in one to a few copies per genome, called unique sequence or single copy DNA.

(ii) Most of the known genes that code for protein in the cell are in the unique sequence class of DNA. Conversely, not all the unique sequence DNA contain protein-coding sequences.

Repetitive sequence:

(i) A class of DNA sequences each of which is repeated or reiterated many times in a genome, called repetitive sequence.

(ii) They are tandemly arranged (arranged one after the other) in the genome.

(iii) Some tandemly repeated sequences are not associated with genes. In fact, the greatest amount of tandemly repeated DNA are associated with centromere and telomere.

(iv) It may be slightly repetitive (1-10 copies), middle repetitive (10 several hundred) and highly repetitive (several hundred to several million copies).

- **NOTES ON TRIPLE-STRANDED DNA**

I. DNA may contain some repeated segments of pyrimidines and purines such as $(TC)_n$ paired with $(AG)_n$. These segments can form a three-stranded helix or H DNA.

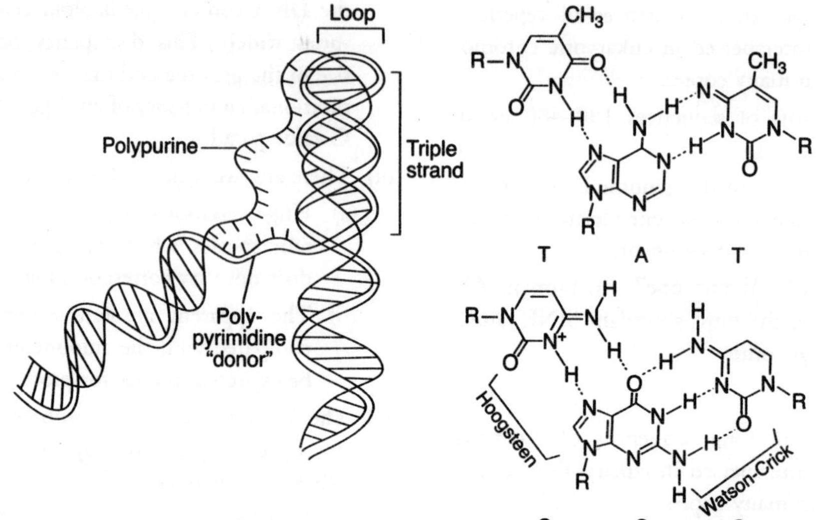

Fig. 12.6: Showing triple stranded DNA (left) by Hoogsteen base pairs (right)

II. The triplex is formed when half of the normal duplex is disrupted and the polypyrimidine/polypurine in strands folds back and inserts itself into the major groove of the remaining duplex. The polypyrimidine strand involved in triplex structure is known as **donors.**

III. (i) Triple-stranded DNA is formed by laying a **third strand** into the major groove of the DNA.

(ii) Complementary hydrogen-bonding interaction is responsible for the specificity of the third-strand interaction.

(iii) The third strand pairs in a **Hoogsteen base pairing** scheme. The base pairing scheme involved in triple-stranded DNA (TAT, CGC, GGC and AAT).

(iv) The central strand of the triplex must be **purine rich** since a pyrimidine does not have hydrogen-bonding surfaces with more than one hydrogen bond.

(v) Thus, triple-stranded DNA requires a **homopurine homopyrimidine** region of DNA.

If third strand is **purine rich** it forms **reverse Hoogsteen hydrogen bonds** in an antiparallel with purine strand of Watson-Crick helix.

If third strand is **pyrimidine rich** it forms **Hoogsteen hydrogen bonds** in a parallel orientation with Watson-Crick paired purine strand.

29. What functions exist in unique, repetitive and highly repetitive DNAs?

Solution:

Unique DNA, make up the bulk of structural genes. It has a large component that is transcribed.

Repetitive DNA is composed of dispersed DNA (short and long interspersed elements—SINEs and LINEs), multiple copies of transcribed DNA (e.g., ribosomal RNA histones) and diverged copies of ancestral genes (e.g., globin family genes).

Highly repetitive DNA usually makes up the centromeric and telomeric regions of the chromosome.

30. Would you expect to find most protein-coding genes in unique-sequence DNA in moderately repetitive DNA or in highly repetitive DNA?

Solution:

The most protein-coding genes are expected to find in unique-sequence DNA.

31. What is junk DNA?

Solution:

The DNA which is not useful to the organism and is made up of untranscribed and parasitic (**selfish**) sequences is called **junk DNA**. Junk DNA is mainly short and long interspersed elements (SINEs and LINEs).

32. An important feature of triplex DNA formation is that it is sequence specific. The synthetic third strand incorporates itself into a triple helix, so a thymine in the synthetic DNA will bind near an AT pair in the biological DNA and a cytosine in the synthetic DNA will bind near a GC pair. From a practical point of view, this opens the possibility of synthesizing a short strand of DNA that will form a triple helix at a particular target site. For example, if the sequence of a particular gene is known, researchers can make synthetic piece of DNA that will form triple helix somewhere within that gene according to the T to AT and C to GC rule. Triplex DNA formation known to inhibit gene transcription. In other words, when synthetic DNA binds within the DNA of a gene, the formation of triplex DNA prevents that gene from being transcribed into RNA. Discuss how this observation might be used to combat disease.

Solution:

(i) Triplex DNA formation opens the exciting possibility of designing synthetic pieces of DNA to inhibit the expression of particular genes. Theoretically such a tool could be used to combat viral disease or to inhibit growth of cancer cells.

(ii) To combat viral disease, a synthetic DNA could be made that specifically binds to an essential viral gene and thereby prevents viral proliferation.

(iii) To inhibit cancer, a synthetic DNA could be made to bind to oncogene. Inhibition of oncogene could prevent cancer.

At this point, a primary obstacle in applying this approach is devising a method of getting the synthetic DNA into living cells.

33. (a) What parts of a nucleotide (namely, phosphate sugar and/or bases) occupy the major and minor grooves of double stranded DNA and what parts are found in the DNA back bone?

(b) If a DNA binding protein does not recognize a specific nucleotide sequence. Do you expect that it recognizes the major groove, the minor groove or the DNA back bone? Explain.

Solution:

(a) The nucleotide bases occupy the major and minor grooves.

Phosphate and sugar are found in the backbone.

(b) (i) If a DNA binding proteins does not recognize a nucleotide sequence, it probably is not binding in the grooves, but instead is binding to the DNA back bone (i.e., sugar phosphate sequence).

(ii) DNA binding proteins that recognize a base sequence must bind into major or minor groove of the DNA, which is where the bases would be accessible to a DNA binding protein.

(iii) Most DNA binding proteins that recognize a base sequence fit into the major groove.

(iv) By comparison, other DNA binding proteins such as histones which do not recognize a base sequence bound to the DNA backbone.

34. Suppose you have recently identified an organism that was scraped from an asteroid that hit the earth. (Fortunately no one was

injured). After analyzing you discover that triplex DNA in this alien organism is composed of a double helix, with the third helix wound within the major groove. How would you propose that this DNA is able to replicates itself? In your answer, be specific about the base pairing rules within the double helix and which part of the triple helix DNA would be replicates first.

Solution:

(i) First, the double helix could unwind and replicate itself. This would produce two double helices.

(ii) The third strand (bound in the major groove) could replicate itself via a semiconservative mechanism.

(iii) This new strand could be copied to make a copy that is identical to the strand that lies in the major groove.

At this point you would have two double helices and two strands that could lie in the major groove. These could assemble to make two triple helices.

- **Packaging of DNA in Prokaryotes**

(i) Prokaryotes like *E. coli*, though do not have a defined nucleus, DNA is scattered in the cell. DNA being negatively charged is held together with some proteins that have positive charge in a region called **nucleoid.**

(ii) In *E. coli* the length of the DNA is 1.36 nm.

(iii) DNA is organised in the form of large loops held by proteins in the nucleoid region.

- **Packaging of DNA in Eukaryotes**

In eukaryotes DNA is packed in the form of a number of chromosomes. The packaging of DNA in eukaryotes is much more complex than prokaryotes. It involves a number of molecules.

(a) **Histones:**

(i) These are basic proteins and are positively charged.

(ii) The charge of a protein depends on the presence of the type of amino acids with charged side chain.

(iii) Histones are made rich with basic amino acids like **lysines** and **arginines.** They both carry positive charges in their side chain.

(b) **Histone octamer:**

Histones are organised to form a group of eight molecules (Two copies of each histone H_{2A}, H_{2B}, H_3 and H_4) called histone octamer.

(c) **Nucleosome:**

(i) The negatively charged DNA is wrapped around the positively charged histone octamer to form a structure, known as **nucleosome.**

(ii) It is the subunit of chromatin. A typical nucleosome contains about 200 bp of DNA helix.

(iii) Human cell has 6.6×10^9 bp it means 3.3×10^7 nucleosome.

(d) **Chromatin:**

(i) A number of nucleosomes together form a structure called chromatin.

(ii) It is invisible as thread-like stained or coloured bodies in nucleus.

(iii) In chromatin nucleosomes as visible as beads on strings structure when seen in electron microscope.

(e) **Chromosome:**

(i) During metaphase stage the chromatin gets visible as chromosome in the cell.

(ii) The chromatin gets further coiled and condensed at metaphase of cell division.

This packaging of chromatin requires additional set of proteins that are collectively called **non-histone chromosomal poroteins.**

In human cell DNA is packed in 46 chromosome.

Human DNA

(i) Diploid chromosome 46. Haploid content i.e., 23 chromosomes of human DNA is made up of 3.3×10^9 bp. Hence, diploid content, i.e., 46 chromosomes of human DNA is made up of $3.3 \times 10^9 \times 2 = 6.6 \times 10^9$ bp.

(ii) The distance between two consecutive base pair (bp) is 0.34 nm = 0.34×10^{-9} m (34Å). (1 nm = 10^{-9} m).

(iii) Total length of a DNA double helix in human cell is $6.6 \times 10^9 \times 0.34 \times 10^{-9}$ m.

Total number of bp = 6.6×0.34 m = 2.2 m.

(iv) The dimension of nucleus is 10^{-6} m (approx).

It is really amazing to know how this long polymer 2.2 metres of DNA is packed in a nucleus which is just 10^{-6} m.

It is packed in the form of 46 chromosomes.

35. Difference of Euchromatin and Heterochromatin.

Solution:

(i) In euchromatin, nucleosomes are loosely packed with nucleosome-free regions that can bind regulatory proteins.

(ii) In heterochromatin, the nucleosomes are densely packed and associated with heterochromatin protein 1 (HP-1).

(iii) In comparison with euchromatin, heterochromatin is gene poor, contains much repetitions of DNA sequence and replicates late during S phase of the cell cycle.

36. State briefly about Histone code:

Solution:

The idea that the pattern of covalent modification of histones in nucleosomes determines the activity of the DNA in the vicinity.

(i) The various histone modifications influence chromatin structure and function by acting as binding sites for a wide range of histone and non-histone effector protein (equal weight of histone and nonhistone protein).

(ii) These proteins contain domains that recognize specific histone modification. Bromodomains recognize acetylated lysins and chromosodomains recognize methylated lysine.

- **Modification of histones in nucleosomes may comprise a histone code:**

(i) Histones in nucleosome are subject to many different modifications, that affect specific amino acid residues in the tails.

(ii) Common modification includes acetylation, monomethylation, dimethylation or trimethylation of lysines and phosphorylation of serines. These are affected by large families; **histone acetyl transferases** (HATs), **histone methyl transferases** (HMTs) and **histone kinases.**

(iii) Histone deacetylases (HDACs), histone demethylases and histone phosphates reverse these effects.

(iv) Less frequent modifications include monomethylation, dimethylation or trimethylation of arginines, phosphorylation of threonines H3T3 and H3T11 and ubiquitylation of lysines H2AK119 and H2BK120.

37. What is nucleosome?

Solution:

Nucleosome is the fundamental ultrastructural subunit of chromatin consisting of 200 bp of DNA and octamer of histone protein where DNA wrapts around its surface like thread around a spool and one molecule of histone H is also bound outside the core histone octamer.

38. What is nucleosome core particle?

Solution:

After a prolonged treatment of micrococcal nuclease on chromatin, the DNA is removed from both the free ends, yielding particles containing DNA of 140-146 nucleotide pairs. This reduced form of nucleosome is called core particles.

The core is an octamer histone containing two copies of each histones H_2A, H_2B, H_3 and H_4, called histone octamer.

39. What is linker DNA?

Solution:

The DNA linking the nucleosome core particle is termed as linker. Its length varies from 15-100 bp (8-114 bp). The linker DNA is folded or coiled in the normal state of chromatin.

40. What do you mean by nucleosome phasing?

Solution:

Nucleosome does not form randomly on DNA, but occupies preferential sites. This property of non-random distribution of nucleosome is known as nucleosome phasing.

41. What is chromatosome?

Solution:

It is a unit consisting of a core nucleosome plus H_1 protein and approximately 165-168 bp of DNA.

42. What do you mean by chromatin remodelling?

Solution:

The chromatin fibres when complexed with histones and folded into various levels of compaction, make the DNA inaccessible to interaction with important nonhistone proteins. To accommodate these non-histone proteins DNA interaction chromatin must be induced by the process of ATP-dependent alternation in the structure of nucleosome that either moves the nucleosome or enables them to be moved by other protein is termed as chromatin remodelling.

43. State briefly the different process of nucleosome remodelling.

Solution:

A. **Acetylation** and **Deacetylation**

(i) It is linked to the gene activation.

(ii) It increases the region of active gene and decreases the region of inactive gene.

(iii) Here Histone **acetyl transferase** (HAT) enzyme is involved.

(iv) Acetylated nucleosomes are associated with the regions of chromosome that are **transcriptionally active** and deacylated nucleosomes are associated with **transcriptionally repressed** chromatin.

(v) The optimum balance of acetylation and deacetylation of histone regulates gene expression. A number of histone deacetylases (HDAC) have been characterized. These are three types — HDAC I, HDAC II and HDAC III.

(vi) Both the classes I and II HDAC directly interact with repressor which facilitates the targeted **deacetylation** of **core histone** in a gene programmed to be repressed. Classes I and II deacetylase function as **corepressor.**

B. **Methylation of Histones**

Methylation of different N terminal tail is associated with both repressed and active chromatin depending on the particularly amino acid, i.e., modified in the histone tail.

(i) Lysine **methyl transferases** methylate $H_3 - H_4$. Methylation of lysine initiates chromatin condensation by deacetylation and thereby function as **silencer.**

(ii) Arginine **methyl transferase (AHMT)** also plays as critical role in transcriptional regulation.

While methylation of histones in nucleosomes is positively correlated in gene activity in eukaryotes, methylation of nitrogenous base cytosine within polynucleotide chains of DNA, forming 5-**methyle cytosine**, is negatively correlated with gene activity.

C. **Phosphorylation**

(i) Phosphorylation results from the action of enzyme **kinase** which adds pO_3^{-2} group to the OH group of serine and histidine.

(ii) Phosphorylation of H_3 and H_4 has been implicated in **chromatin condensation** during mitosis.

H_3-phosphorylation (serine) is directly linked to **induction of gene expression.**

44. What are the chromatin or nucleosome remodelling complexes?

Solution:

(i) Two classes of protein complexes constitute the chromatin or nucleosome remodelling complex.

(ii) One classes involves enzymes that modify nucleosomes by **acetylating** or **deacetylating** core histones.

(iii) The other class of chromatin remodellers is ATP-dependent nucleosome remodelling complex.

Ex.: SWI/SNF (Switch, Sniff), the ISWI (initiation switch) family, the NuRD family and the INO 80 family.

45. What happens during chromatin remodelling?

Solution:

(i) Chromatin remodelling involves a change in the interaction between DNA and histones in nucleosome.

(ii) When remodelling is completed, promoter sequence is freed from the histones and accessible to protein that initiates transcription.

46. In what circumstances does a double stranded DNA molecule form an A helix? a B helix and a Z helix?

Solution:

B helix is considered to be the common form of DNA.

(i) DNA forms an A helix under condition of dehydration.

(ii) A Z helix is favoured by segments with alternating Gs and Cs.

47. What non-covalent interactions are involved in maintaining the double helical conformation of DNA?

Solution:

(i) Hydrophobic and hydrophilic properties of the bases and phosphate groups respectively.

(ii) Hydrogen bonding between the bases of a base pair.

48. What would happen if double-stranded DNA is placed in distilled water?

Solution:

(i) There are no cations in distilled water, so there is no complex formation with phosphate.

(ii) As a result the repulsion among them would be great enough to denature DNA.

49. Why are salt concentration and temperature important to the renaturation of DNA?

Solution:

(i) The salt concentration must be high enough so that electrostatic repulsion between the phosphates in the two strands is eliminated; usually 0.15 to 0.50 M NaCl is used.

(ii) The temperature must be high enough to disrupt the random, intra-strand hydrogen bonds. However, temperature cannot be too high, or stable inter-strand base pairing will not occur. The optimal temperature for renaturation is 20-25°C below the value of Tm.

50. All DNA molecules more in the same direction when electrophoresed. Why? Is this also generally true of proteins?

Solution:

(i) All DNA molecules have negative charge and therefore, move in the same direction when electrophoresed.

(ii) This is not true for protein molecules, because they may have a net positive or negative charge.

Unusual bases of tRNA: methylguanine (GMe), dimethylguanine (GMe_2), methyl cytosine (Me), ribothymine (T), pseudo uridine (ψ), dihydrouridine (DHU, H_2U, U_2H), Inosine (I) and methylinosine (IMe, MeI).
Unusual bases of DNA: PBSI and PBS-2 (virus): uracil occurs in place of thymine in DNA, bacteriophages: cytosine is replaced by 5-hydroxymethyl cytosine (HMC).

- **Characteristics of histone modification in different types of histones: (Nomenclature amino acid of histones — blank nuclear pattern; a = promoter, b = enhancer)**

Amino acid	Constitutive heterochromatin	Facultative heterochromatin	Euchromatin
H_3K_4	—		trimethylated[a] monomethylated[b]
H_3K_9	trimethylated	dimethylated	acetylated
H_3K_{27}	—	trimethylated	—
H_4R_3	—	—	methylated
H_4K_5			acetylated
H_4K_{12}	acetylated	acetylated	—
H_4K_{20}	trimethylated	—	monomethylated

51. Whether DNA is placed in distilled water the two strands come apart. Explain why?

Solution:

There are no counter ions to shield the phosphate from one another. The positive ions of the salt stabilizes the DNA and thus raises the value of T_m.

52. Why does salt concentration affect T_m?

Solution:

By shielding the negatively charged phosphates from one another the positive ions of the salt stabilizes the DNA and thus raises the value of T_m.

53. In preparing hybrid DNA by renaturation of denatured DNA molecules of different types, some unrenatured DNA usually remains. Mention several procedures that could be used to eliminate this DNA, if it was unwanted.

Solution:

Hydroxyapatite chromatograpy, filtration through nitrocellulose filters, digestion with nucleases specific for single stranded DNA.

54. A covalently closed circular DNA molecules that is relaxed (neither positively nor negatively supercoiled) is partially denatured and 50 base pairs are broken. How does the degree of supercoiling change?

Solution:

Each turn of the helix contains ten base pairs, so the DNA acquires about 5 positive turns of supercoiling.

55. Mention the several ways in which linear DNA molecule can form a circle.

Solution:

A linear DNA molecule can form circle

(i) by cohesion of single-stranded termini.

(ii) by recombination between redundant termini.

(iii) by exonucleolytic conversion of double-stranded termini to single-stranded termini followed by hydrogen bonding between single strand.

Numerical Problems

1. In the following table are base ratios of certain nucleic acid fractions extracted from seven different species. For each one write the type of nucleic acid that is involved (i.e., DNA or RNA) and whether you think it is double strand or single strand:

Species	A	G	T	C	U	$\frac{A+T}{G+C}$	$\frac{A+G}{T+C}$
(a)	23	27	23	27	—	—	—
(b)	27	23	27	23	—	—	—
(c)	23	23	27	27			
(d)	23	27	—	27	23	—	—
(e)	23	27	—	23	27		
(f)	—	—	—	—	—	1.00	0.67
(g)	—	—	—	—	—	0.67	1.00

Solution:

(a) $\frac{A+G}{T+C}$, i.e., double-stranded DNA. $\frac{23+27}{23+27} = \frac{50}{50} = 1$, i.e., Absence of uracil = not RNA.

(b) $\frac{27+23}{27+23} = \frac{50}{50} = 1$, i.e., double-stranded DNA.

(c) $\frac{23+23}{27+27} = \frac{46}{54} = 0.852$ (approx.) $= 0.85 \neq 1$, i.e., single-stranded DNA.

(d) $\frac{23+27}{23+27} = \frac{50}{50} = 1$, i.e., double-stranded RNA (Presence of uracil).

(e) $\frac{23+27}{27+23} = \frac{50}{50} = 1$, i.e., double-stranded RNA (Presence of uracil).

(f) $\frac{A+G}{T+C} = 0.67$, i.e., $\neq 1$. It is single-stranded DNA.

(g) $\frac{A+G}{T+C} = 1$, i.e., double-stranded DNA.

2. State whether *true* or *false*:

(a) A + T = G + C; (b) A = G, C = T; (c) $\frac{A}{T} = \frac{C}{G}$; (d) $\frac{T}{A} = \frac{C}{G}$; (e) A + G = C + T; (f) $\frac{G}{C} = 1$; (g) $\frac{A}{T} = 1$.

(h) Hydrogen bonding provides stability to the double helix in aqueous solution.

(i) Hydrophobic bonding provides stability to double helix in aqueous cytoplasm.

(j) When separated two strands of double helix are identical.

(k) Once base sequence of one strand of a DNA double helix is known. The base sequence of 2nd strand can be deduced.

(l) The structure of DNA double helix is fully conserved during its replication.

(m) Each nucleotide pair contains two phosphate groups, two deoxyribose molecules and two bases.

Solution:

(a) A + T = G + C (False); (b) A = G, C = T (False); (c) True; (d) True; (e) A + G = C + T (True); (f) True; (g) True; (h) True; (i) True; (j) False; (k) True; (l) False; (m) True.

3. A single-stranded DNA phage has $\frac{A}{T}$ base ratio of 0.33, $\frac{G}{C}$ ratio of 2.0 and $\frac{A+T}{G+C}$ ratio of 1.33. What is the $\frac{A+G}{T+C}$ ratio in this molecule?

Solution:

$\frac{A}{T} = 0.33, \frac{G}{C} = 2.0, \frac{A+T}{G+C} = 1.33, \frac{A+G}{T+C} = ?$

A = 0.33T, G = 2C, A+T = 1.33 (G+C)

$\frac{A+G}{T+C} = \frac{0.33T+2C}{3C+C} = \frac{0.33\times3C+2C}{4C} = \frac{0.99C+2C}{4C} = \frac{2.99C}{4C} = 0.7475 = 0.75.$

0.33T+T = 1.33(G+C)

T(0.33+1) = 1.33(2C+C)

1.33T = 1.33(3C)

T = 3C

4. DNA has $\frac{A}{T}$ base ratio 0.50 and $\frac{G}{C} = 0.33$, $\frac{A+T}{G+C} = 0.75$; write the $\frac{A+G}{T+C}$ ratio.

Solution:

$\frac{A}{T} = 0.5, \frac{G}{C} = 0.33,$

$A = 0.5T,\ G = 0.33C,$

$$\frac{A+G}{T+C} = \frac{0.5T+0.33C}{0.665C^{*}+C} = \frac{0.5\times 0.665C+0.33C}{1.665C} =$$

$$\frac{0.3325C+0.33C}{1.665C} = \frac{(0.3325+0.33)C}{1.665C} = \frac{0.6625}{1.665}$$

$= 0.3978$

$\frac{A+T}{G+C} = 0.75$

$A+T = 0.75\ (G+C)$

$0.5T+T = 0.75(0.33C+C)$

$T(1.5) = 0.75(1.33C)$

$T = \frac{0.75}{1.5}(1.33C) = \frac{1}{2}\times 1.33C = 0.5\times 1.33C$

$T = 0.665C^{*}$

5. Which of the following statements are true about double-stranded DNA?

(a) A + C = T + G; (b) A + G = C + T; (c) A + T = G + C; (d) $\frac{A}{G} = \frac{C}{T}$; (e) $\frac{A}{G} = \frac{T}{C}$; (f) $\frac{A+G}{T} = 1$.

Solution:

We know, A + G = T + C, i.e., $\frac{A+G}{T+C} = 1$, i.e., ds DNA.

(a) $\frac{A+C}{T+G}$ = F; (b) True for ds DNA; (c) A + T = G + C = F; (d) $\frac{A}{G} = \frac{C}{T}$ = F; (e) F; (f) F.

6. If the G + C content of a DNA sample 48 percent, what will be the proportion of the four different nucleotides?

Solution:

As G + C = 48%, A + T = 100 – 48 = 52%, G ≡ C, G and C are 24% each, on the other hand A + T = 52%, A and T are 26% each.

7. When the base composition of a DNA sample from *Mycobacterium tuberculosis* was determined, 18% of the bases are found to be adenine.

What is the percentage of cytosine?

Solution:

18% is adenine, 18% is all thymine, for a total of 36% for A + T. Therefore, G + C will be 64% and C will be half of that, i.e. = 32%.

8. Suppose that a non-supercoiled DNA molecule is composed of 4800 base pairs. How many helical turns are present? How long is the linear, double helical form of this molecule?

Solution:

$\frac{4800}{10}$ = 480 turns are present (∵ one turn = 10 bp)

one turn = 34Å, therefore 480 × 34 = 16320Å. So, it will be 1.632 μm. [1Å = 1 × 10^{-4} micron].

9. The human DNA contains 3,000,000 kb and is estimated to contain 10,000 genes. If we make the simple assumption that the average size of the coding region of a gene (the part that codes for protein) is 2 kb from this data.

(a) What can be concluded about the average spacing of the central point of genes on chromosomal DNA?

(b) What can be concluded about the proportion of the genome that codes for protein?

(c) What occupies the non-coding regions?

Solution:

(a) The average spacing between the central points of genes must be 3,000,000 divided by 100,000, i.e.,

$$\frac{3{,}000{,}000}{100{,}000} = 30 \text{ kb}.$$

(b) Since the protein coding segment is only 2 kb, it is clear that only $\frac{1}{15}$ or about 6% of the genome codes for protein.

(c) The remaining sequences must be composed of regulatory regions, adjacent to genes, as well as repetitive DNA and non-repetitive spacer DNA between the protein-coding regions and introns within the protein-coding regions.

10. The genome of the bacterium *Hemophilus influenzae* is 1830 kb in size, and sequencing has shown that it has 1703 genes.

(a) What is the average spacing between the central points of genes?

(b) If the average size of a coding region of a gene is assumed to be 1 kb, what is the average distance between genes?

(c) **What is in this region?**

(d) **What proportions of the genome is expected to be introns?**

(e) **What is the average space between coding regions?**

Solution:

(a) The average spacing between the central points of genes must be $\frac{1830}{1703} = 1.0748$ kb $= 1.074 \times 1000 = 1074 = 1074$ bp.

(b) Coding region of a gene is 1 kb, i.e., 1000 bp. Average spacing between the central points of gene is 1074 bp.

Therefore, average distance between the genes is 1074 – 1000 = 74 bp.

(c) Regulatory regions.

(d) 0% $\left[\frac{74}{1{,}830{,}000} \times 100 = 0.04\%\right]$

(e) 74 base pairs.

11. If DNA of an *E. coli* has 4.2×10^6 nucleotide pairs in its DNA and if an average cistron contains 1500 nucleotide pairs, how many cistrons does it possess?

Solution:

$$\text{Number of cistrons} = \frac{4.2\times 10^6}{1500} = \frac{42\times 10^6}{15\times 10^2\times 10}$$

$$= \frac{42\times 10^3}{15} = 2.8 \times 10^3 = 2800.$$

12. A double-stranded DNA molecule is 100,000 bp (100 kb) long.

(a) **How many nucleotides does it contain?**

(b) **How many complete turns are there in the molecule?**

(c) **How long is the DNA molecule?**

Solution:

(a) $2 \times 100{,}000 = 200{,}000$ nucleotides.

(b) One turn requires 10 bp.

So, 10 bp = 1 turn

$1 = \frac{1}{10}$

$100{,}000 = \frac{1}{10} \times 100{,}000 = 10^4$ turns.

(c) DNA helix has single turn 3.4 nm in length. 10^4 turns have 3.4×10^4 nm length.

13. A virus particle contains double-stranded DNA with 200,000 base pairs.

(a) **How many nucleotides would be present?**

(b) **How many complete spirals would occur in each strand?**

(c) **How many atoms of phosphorus would be present?**

(d) **What would be length of the DNA configuration in the virus?**

Solution:

(a) $2 \times 200{,}000 = 400{,}000$ nucleotides.

(b) One turn requires 10 bp

So, 10 bp = 1 turn

$1 = \frac{1}{10}$

$200{,}000 = \frac{1}{10} \times 200{,}000 = 20{,}000$ complete spirals.

(c) 400,000 phosphorus atoms.

(d) DNA helix has one spiral = 3.4 nm in length or 34 Å.

Therefore, 20,000 spirals $= 3.4 \times 20{,}000 = 68{,}000$ nm is length

or, $34 \times 20{,}000 = 680{,}000$ Å.

14. A DNA molecule has 1000 bp. Calculate the length of the molecule and the number of complete spirals in the molecule. The molecule has 350 guanine bases. How many adenine bases should be present there?

Solution:

1 spiral = 10 base pairs

So, 10 bp = 1 turn; or, 1 bp = $\frac{1}{10}$ spiral

1000 bp = $\frac{1}{10} \times 1{,}000 = 100$ spirals.

1000 bp length, i.e. $2 \times 1{,}000 = 2{,}000$ nucleotide long

Guanine 350, cytosine is also 350.

Total $350 \times 2 = 700$

2000 – 700 = 1300 nucleotide remain

Therefore, A + G = 1300

Adenine = $\frac{1300}{2} = 650$.

15. A DNA segment contains 100 nucleotide pairs.

(i) What is the length of the DNA segment?

(ii) Calculate the number of spirals in the molecule.

(iii) There are in total 70 adenine bases. Calculate the number of guanine present in the segment.

Solution:

(i) The length between nucleotide pair is 3.4 Å.

Therefore, the length of DNA is 3.4 × 100 = 3400 = 340 Å.

(ii) One spiral requires 10 bp.

So, 10 bp = 1 spiral

$1 \text{ bp} = \frac{1}{10}\text{spiral}$

$1000 \text{ bp} = \frac{1}{10} \times 1000 = 100$ spirals.

(iii) Total number of nucleotide is 100 × 2 = 200.

Adenine base is 70, so, thymine is also 70.

Total (A + T) = 140, Rest G + C = 200 – 140 = 60

The number of guanine $\frac{60}{2} = 30$.

16. Phage P-1 has a double-stranded DNA with 91,500 bp (91.5 kb).

(a) How many full double helical turns does this DNA contain?

(b) How long is the DNA in microns (1 μm; 1 micron = 10^4Å)?

(c) What is the molecular mass of this DNA?

(d) How many phosphorus atoms does the DNA contain?

Solution:

(a) One turn requires 10 bp.

So, 10 bp = 1 turn

$1 \text{ bp} = \frac{1}{10} \text{turn}$

$91{,}500 \text{ bp} = \frac{1}{10} \times 91{,}500 = 9150$ complete helical turns.

(b) The spacing between the base pair is about 3.4Å. One spiral has 3.4 × 10 = 34Å.

$10^4 \text{Å} = \frac{1}{10}$ micron

$1 \text{Å} = \frac{1}{10^4}$ micron $= 1 \times 10^{-4}$ micron

$3.4\text{Å} = 3.4 \times 10^{-4}$ μm (micron).

The length of the DNA = $3.4 \times 91{,}500 \times 10^{-4} = 3111 \times 10^2 \times 10^{-4} = \frac{3111}{10^2} = \frac{3111}{100} =$ 31.11 μm.

The length is 31.11 μm (micron).

(c) One base pair has a molecular mass of about 660 daltons (D).

Therefore, 91,500 b = $915 \times 66 \times 10^3 = 6039 \times 10^4$ dalton, or, $\frac{6039}{1000} \times 10^4 = 6.039 \times 10^4$ kD $= 6.04 \times 10^4$ kD.

(d) Each base pair has two phosphorus atoms, one on each strand.

Therefore, 91,500 bp has 91500 × 2 = 183,000 phosphorus atoms.

17. The DNAs of bacteriophages T_2 and T_4 contain 2×10^5 bp. How many genes of average size (encoding proteins of about 40,000 molecular weight) can these phages contain?

Solution:

The average molecular mass of an amino acid is about 110 D (dalton), a protein with molecular mass of 40,000 D contains about 364 amino acids (40,000 ÷ 110 = 363.636 = 364 (approx.).

Each amino acid requires 3 bp of DNA to code for it. Therefore, a protein containing 364 amino acids needs a gene of about 364 × 3 = 1092 bp. Thus, the number of genes of average size in a DNA having 2×10^5 bp will be $\frac{2 \times 10^5}{1092} = 183.150 = 183$.

18. The double-stranded DNA genome of human herpes simplex virus 1 has a molecular mass of about 1.0×10^5 kD.

(a) How many base pairs does this virus contain?

(b) How many full double helical turns does this DNA contain?

(c) How long is this DNA in microns?

Solution:

(a) Herpes simplex virus 1 has ds DNA genome of 1.0×10^5 kD, i.e., $1.0 \times 10^5 \times 10^3$ daltons (1 kD = 1000 daltons).

One bp has a molecular mass of about 660 daltons (D).

$$660 \text{ daltons} = 1 \text{ bp}$$

$$1 \text{ dalton} = \frac{1}{660} \text{bp}$$

$$1.0 \times 10^5 \times 10^3 \text{ dalton} = \frac{1}{660} 1.0 \times 10^5 \times 10^3$$

$$= 151.51515 \times 10^3 = 15515.15.$$

This virus has 151,515 bp.

(b) One turn requires 10 bp.

So, 10 bp = 1 turn

$$1 \text{ bp} = \frac{1}{10} \text{turn}$$

$$151{,}515 \text{ bp} = \frac{1}{10} \times 151{,}515 = 15{,}511.5 \text{ turns}$$

This DNA contains 15,151.6 turns.

(c) The spacing between the base pair is 3.4Å or one spiral has 3.4 × 10 = 34Å.

It (DNA) contains 15,151.5 turns (spiral); then the length is 15,151.5 × 34 = 515,151Å

1 micron (μm) = 104Å

$$\therefore 515{,}151\text{Å} = \frac{515151}{10^4} = 51.5151 \text{ microns}$$

$$= 51.52 \ \mu\text{m}.$$

19. How many proteins of average size could be encoded in a virus with a DNA genome having 12,000 bp, assuming no overlap genes?

Solution:

Average size protein contains 364 amino acid. [i.e., average molecular mass of protein is 40,000 dalton and the molecular mass of amino acid (average) is 110 D. Average protein contains $\frac{40{,}000}{110} = 364$ amino acid.]

Each amino acid requires 3 bp of DNA to code for it. A protein containing 364 amino acid needs a gene of about 1092 bp (364 × 3 = 1092).

Thus, the number of protein of average size in a DNA having 12,000 bp will be $\frac{12{,}000}{1092} = 10.98901$ = 10.99 = 11 (approx.).

20. If nucleotides were arranged at random in a piece of single-stranded RNA 10^6 nucleotides long and if these base composition of this RNA was 20 percent A, 25 percent C, 25 percent U and 30 percent G, how many times would you expect the specific sequence 5′-GUUA-3′ to occur?

Solution:

The chance of finding 5′-GUUA-3′ is (0.30 × 0.25 × 0.25 × 0.20) = 0.00375.

(i) In a molecule, i.e., single-stranded RNA 10^6 nucleotides long, there are nearly 10^6 groups of four bases.

(ii) The first group of four is bases 1, 2, 3 and 4, the second group is bases 2, 3, 4, 5 and so on. Thus, the number of times this sequence is expected to appear is 0.00375 × 10^6 = 3750.

21. Two double-stranded DNA molecules from a population of T_2 phages were denatured to single strands by heat treatment. The result was the following four single-stranded DNA:

1. TAGCTCC → and 3. GCTCCTA →

2. ATCGAGG ← 4. CGAGGAT ←

These separated strands were then allowed to renature. Diagram the structures of the renatured molecules most likely to appear when (a) strand 2 renatures with strand 3 and (b) strand 3 renatures with strand 4. Label the strands and indicate sequences and polarity.

Solution:

(a) The sequence CGAGG in molecule 2 is complementary to the sequence GCTCC in molecule 3. These can pair to give

3. GCTCCTA →

2. ATCGAGG ←

Each strand has two unpaired bases sticking out. These bases are complementary to each other, so that two strands pair up.

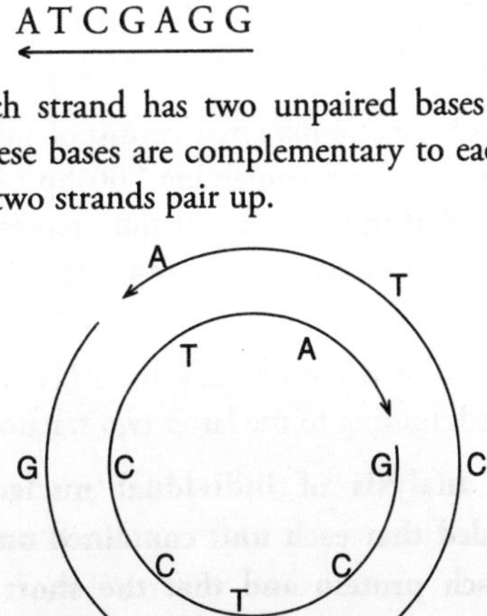

(b) The sequence in molecule 3 is complementary to the sequence in molecule 4. It also has opposite polarity, so that the two strands pair up.

3. G C T C C T A →

4. C G A G G A T ←

22. If a DNA molecule is 7 × 10⁷ bp cm long, how many nucleosome could be present in its most highly compacted state? (Assume linker DNA has 60 base pairs.)

Solution:

In this case there are 200 bp per nucleosome. A DNA molecule is 2.4 cm long contains about 7×10^7 bp (0.34 nm per base pair).

So, 200 bp for 1 nucleosome

1 bp for $\frac{1}{200}$ nucleosome

7×10^7 bp for $\frac{1}{200} \times 7 \times 10^7 = \frac{7000}{200} \times 10^4 = 3.5 \times 10^4$ nucleosomes.

$\therefore$ 3.5×10^4 nucleosomes would be present.

23. The nucleus of a single-celled organism contained a single linear chromosome having 7×10^6 nucleotide pairs of DNA coalesced with three types of histone like proteins.

A short micrococcal nuclease digestion yielded DNA fraction consisting of 700, 1400 and 2100 base pairs.

Predict what these fractions represent. What conclusion can be drawn?

Solution:

(i) The chromatin fibre may consist of a variation of nucleosomes containing 700 bp of DNA. The 1400 and 2100 bp fractions represent two and three nucleosomes respectively, linked together.

(ii) Enzymatic digestion may have been incompleted, leading to the latter two fractions.

24. The analysis of individual nucleosomes revealed that each unit contained one copy of each protein and that the short linker DNA contained no protein bound to it. If the entire chromosome consists of nucleosomes (discounting any linker DNA), how many are there, and how many total proteins are needed to form them? The linear chromosome contains 7×10^6 bp DNA.

Solution:

The chromosome contains 7×10^6 bp of DNA. Nucleosome containing bp is 7×10^2.

The number of nucleosome $= \frac{7 \times 10^6}{7 \times 10^2} = 10^4$ nucleosomes.

The chromosome thus contains 10^4 copies of each of the 3 proteins, for a total 3×10^4 molecules.

25. Analysis then revealed the organism's DNA to be double helix similar to Watson-Crick model, but containing 20 base pairs per complete turn of the right-handed helix. The physical size of the nucleosome was exactly double the volume occupied by that found in all other eukaryotes, by virtue of increasing the distance along the fibre axis by a factor of two. Compare the degree of compaction (the number of turns per nucleosome) of this organism's nucleosome with that found in eukaryotes.

Solution:

(i) The unique organism compacts a length of DNA consisting of 35 complete turns of the helix, i.e., 700 bp per nucleosome/ 20 bp per turn $= \left(\frac{700}{20}\right)$ into each nucleosome.

(ii) The normal eukaryote compacts a length of DNA consisting of 20 complete turns of the helix (200 bp per nucleosome/10 bp per turn $= \frac{200}{10} = 20$ into a nucleosome half the volume of that in the unique organism.

The degree of compaction is less in the unique organism.

26. The A-T and C-G base pair complementary relationship within a DNA molecule allows one to determine the composition of each polynucleotide strand and various base pair relationships within the entire molecule from a minimal amount of information.

Given that one polynucleotide strand is composed of adenine (A) = 0.20, guanine = 0.16, thymine = 0.36 and cytosine (C) = 0.28.

(a) What is the $\frac{A+G}{T+C}$ ratio in a single polypeptide strand?

(b) What is the $\frac{A+T}{C+G}$ ratio in this single polypeptide strand?

(c) What is the $\frac{A+G}{T+C}$ ratio in the complementary polynucleotide strand?

(d) What is the $\frac{A+T}{C+G}$ ratio in the complementary polynucleotide strand?

(e) What is the $\frac{A+G}{T+C}$ ratio in the entire DNA molecule?

(f) What is the $\frac{A+T}{C+G}$ ratio in the entire DNA molecule?

Solution:

(a) $\frac{0.20+0.16}{0.36+0.28} = \frac{0.36}{0.64} = 0.5625 = 0.56.$

(b) $\frac{0.20+0.36}{0.28+0.16} = \frac{0.56}{0.44} = 0.1.2727 = 1.27.$

(c) $\frac{0.20+0.16}{0.36+0.28} = \frac{0.36}{0.64} = 0.56.$

The complementary strand would be $\frac{1}{0.56}$ = 1.7857 = 1.79.

(d) $\frac{A+T}{C+G}$ = 1.27. The complementary strand would be $\frac{1}{1.27}$.

(e) This ratio must be 1.00, because base pair complementary relationships

$A+G = 0.20+0.16 = 0.36$

$T+C = 0.36+0.28 = 0.64$

(f) $A+T = 0.20+0.36 = 0.56+5.6+1.12$

$C+G = 0.28+0.16 = 0.44+0.44 = 0.88$

$\therefore \frac{A+T}{C+G} = \frac{1.12}{0.88} = 1.27.$

27. The following table lists hypothetical data relative to base ratios of nucleic acid fractions extracted from four different sources. For each one, assess whether the nucleic acid was DNA or RNA and secondly, whether it would be single-stranded or double-stranded.

Source	Thymine (T)	Guanine (G)	Cytosine (C)	Adenine (A)	Uracil (U)
1	31	19	19	31	0
2	19	31	31	19	0
3	0	31	31	19	19
4	31	31	19	19	0

Solution:

1. $\frac{A+G}{T+C} = \frac{31+19}{31+19} = \frac{50}{50} = 1$. It is double-stranded, absence of uracil, it is DNA.
2. $\frac{A+G}{T+C} = \frac{19+31}{19+31} = \frac{50}{50} = 1$. Double-stranded DNA, due to absence of uracil.
3. $\frac{A+G}{U+C} = \frac{19+31}{19+31} = \frac{50}{50} = 1$. Double-stranded RNA, due to presence of uracil.
4. $\frac{A+G}{T+C} = \frac{19+31}{31+19} = \frac{50}{50} = 1$. Double-stranded DNA, due to absence of uracil.

28. Consider a segment of single-stranded DNA that is six nucleotides long.

(a) How many different sequences are possible? Only use four common nucleotides in determining your answer.

(b) What would your answer be if the DNA was double-stranded and six nucleotide base pairs long?

Solution:

(a) Four nucleotides and DNA is six nucleotides long, $4^6 = 4096$. Sequences are 4096.

(b) $4^6 = 4096$. Being double-stranded no changes in the sequences.

29. Given that the ratio of $\frac{A+G}{T+C}$ in one polynucleotide strand of a DNA molecule is 0.6.

(a) What is the $\frac{A+G}{T+C}$ ratio in the complementary polynucleotide strand?

(b) What is the $\frac{A+T}{C+G}$ ratio in the entire DNA molecule?

Solution:

(a) (i) Consider A + G = 0.6 and T + C = 1.0, and their ratio is 0.6.

(ii) Then the complementary strand would be A + G = 1.0 and T + C = 0.6, and their ratio is $\frac{1}{0.6} = 1.6666 = 1.67$.

(b) This has to be 1.0 because of base complementary relationship: $\frac{0.6+1.0}{1.0+0.6} = \frac{1.6}{1.6} = 1$.

30. The DNA in each somatic cell of a particular mammalian species contains 3.255×10^{-15} gm of phosphorus.

(a) How many phosphorus atoms are contained in the DNA of each of these cells?

(b) How many nucleotide pairs make up the DNA of each of these cells?

(c) How long is the total DNA in each of these cells?

Solution:

(a) $\frac{3.255 \times 10^{-15}}{30.97}$ (molecular weight of P) $= 1.05 \times 10^{-14}$

$1.05 \times 10^{-14} \times 6.023 \times 10^{23}$ (Avogadro number) $= 6.32 \times 10^{7}$ molecules.

(b) $\frac{6.32 \times 10^{9}}{2} = 3.16 \times 10^{9}$ base pairs.

(c) Assuming 3.4Å per bp

$3.16 \times 109 \times 3.4 = 1.074 \times 10^{10}$Å

1.074×10^{10}Å $= 1.074 \times 10^{3}$ mm $= 1074$ mm.

31. In gel electrophoresis, do smaller double-stranded molecules move more slowly or more rapidly than larger molecules?

Solution:

Smaller molecules move faster because they can penetrate the pores of the gel more easily.

32. The genome size of *Drosophila melanogaster* is 165,000 kilobases (kb), approximately $\frac{2}{3}$ of the genome is euchromatic and $\frac{1}{3}$ is heterochromatic. In salivary gland chromosomes, the euchromatic part of the genome, becomes polytene and the giant chromosome exhibits approximately 5000 transverse bands used as land-marks for the location of genes, chromosome break-points and other cytogenetic features. What is the approximate DNA content of an average band in the salivary gland chromosomes?

Solution:

(i) The euchromatic part of the *Drosophila* genome contains approximately 165,000 kb $\times \frac{2}{3} = 110{,}000$ kb of DNA.

(ii) This amount of DNA is distributed over 5000 bands for an average of $\frac{110{,}000\text{kb}}{5000} =$ 22 kb.

33. The polytene chromosome of *Drosophila* contains about 5000 bands. The size of the eukaryotic genome is 120×10^{6} bp and there are about 1,3600 genes. Assume that amount of DNA between the bands is negligible.

(a) What is the average amount of DNA per band.

(b) What is the average number of gene per band?

Solution:

(a) $\frac{120 \times 10^{6}}{5000} = \frac{120 \times 10^{6}}{5 \times 10^{3}} = 24 \times 10^{3} = 24$ kb/band.

(b) $\frac{13{,}600}{500} = 2.7$ gene per band.

34. The *E. coli* chromosome is 4700 kb in length and haploid human genome contains 3×10^{7} bp of length. Mention their (*E.coli* and Human) length in millimetres?

Solution:

E. coli = 4,700 kb = 4,700,000 bp

$4{,}700{,}000 \times 3.4$Å/bp $\times \frac{10^{-7}}{\text{Å}} = 1.6$ mm

Human beings $= 3 \times 10^{9}$ bp $\times$ 3.4Å/bp $\times \frac{10^{-7}\text{mm}}{\text{Å}}$

= 1,020 mm.

In other words, there is approximately 1 metre of DNA in human genome.

35. No further coiling or compaction of this unique chromosome occurs in the unique organism. Compare this situation with that of a eukaryotic chromosome. Do you think an interphase human chromosome 7×10^{6} bp in length would be shorter or longer chromatin fibre?

Solution:

(i) The eukaryotic chromosome contains still another level of condensation in the form of solenoids, which are dependent on the H_1 histone molecule associated with linker DNA. Solenoids condense the eukaryotic fibre by still another factor of five.

(ii) The length of the unique chromosome is compacted into 10^4 nucleosomes; each containing an axis length twice that of the eukaryotic fibre. The eukaryotic fibre consists of $\frac{7\times10^6}{2\times10^2} = 3.5 \times 10^4$ nucleosomes, 3.5 times more than the unique organism. However, they are compacted by the factor of five in each solenoid.

Therefore, the chromosome of the unique organism is longer chromatin fibre.

36. If the length of *E. coli* DNA is 1.36 mm, how many base pairs will be present in *E. coli*?

Solution:

(i) The distance between two consecutive base pairs (bp) is 0.34×10^{-9} m.

(ii) Hence the number of base pairs in 1.36 mm long DNA = $\frac{1.36}{0.34\times10^{-9}} = \frac{136}{34\times10^{-9}}$ mm.

Since, 1 m = 1000 mm.

$\therefore \frac{4}{10^{-9}} \times 10^3 = \frac{4\times10^3}{10^{-9}} = \frac{4}{10^{-6}} = 4 \times 10^6$

= 4000000 bp.

37. The length of DNA in a typical mammalian cell is approximately 2.2 metres. This DNA is wrapped around a number of nucleosomes. Each nucleosome contains about 200 base pairs of DNA. How many nucleosomes are present in a mammalian cells.

Solution:

We know that length of DNA = number of base pairs distance between consecutive base pairs.

Distance between two consecutive base pairs (bp) is 0.34×10^{-9} mm.

Human DNA is made of 6.6×10^9 bp (The total length of a DNA double helix in a typical human cell = $6.6 \times 10^9 \times .34 \times 10^{-9}$ mm).

200 bp of DNA are present in a nucleosome.

So, 6.6×10^9 bp of DNA are present in

$\frac{1\times6.6\times10^9}{200} = 3.3 \times 10^7 = 33000000$ nucleosome.

38. Let's assume the linker DNA average 54 bp in length. How many molecules of H_2A would you expect to find in a DNA sample that is 46,000 bp in length?

Solution:

(i) There are 146 bp around the core histones. Here the linker region is 54 bp, we expect 200 bp of DNA (146 + 54) for each nucleosome and linker region.

(ii) The template DNA is about 46000 bp. Therefore, number of nucleosome = $\frac{46000}{10^{-9}}$ = 230.

(iii) There are 2 molecules of H_2A in a nucleosome. So, 230 × 2 = 460 molecules of H_2A in 46×10^3 bp sample of DNA.

39. 1 bp of DNA is approximately 0.34 nm in length. A bacterial chromosome is about 4 million bp in length and is organized into about 100 loops that are about 40,000 bp in length.

A. If it was stretched out linearly, how long (unmicrometer) would one loop be?

B. If a bacterial chromosomal loop is circular what would be its diameter? (Note circumference-πD, where D is the diameter of the circle).

C. Is the diameter of the circular loop calculated in part B small enough to fit inside a bacterium? The dimensions of the bacterial cytoplasm, such as *E. coli* are roughly 0.5 μm wide and 1.0 μm long.

Solution:

A. One loop is 40,000 bp, one bp is 0.34 nm which is equal to 0.34×10^{-3} μm. It will be $40{,}000 \times 0.34 \times 10^{-3} = 13600 \times 10^{-3} = 13.6$ μm.

B. Circumference = πD.

13.6 μm = πD.

D = $\frac{13.6\times7}{22}$ = 4.32 = 4.3 μm.

C. No, it is too big to fit inside of *E. coli.* Super coiling is needed to make the loops more compact.

40. The linear chromosome of Phage T_2 is 52 μm long. The chromosome consists of double stranded DNA with 0.34 nm between each base pair. How many base pair does a chromosome of T_2 contain?

Solution:

(i) The first step is to put the lengths in the same units. 52 μm is 52 millionths a meter or 52000×10^9 m = 52,000 nm.

(ii) One base pair occupy 0.34 nm in double helix.

So the number of bp = $\frac{52000}{0.34} = \frac{52 \times 10^5}{34} = 1.52941 \times 10^5$ = 152941 base pair.

41. A double-stranded DNA molecule is 1 cm long, and the percentage of adenine is 15%. How many cytosines would be found in this DNA molecule?

Solution:

The linear length of 1 base pair is 0.34 nm, which equals to 0.34×10^9 m.

One centimetre equals to 10^{-2} meters.

$$0.34 \times 10^{-9}\,\text{m} = 1\text{ bp}$$

$$1 = \frac{1}{0.34 \times 10^{-9}}$$

$$10^{-2}\,\text{m} = \frac{1 \times 10^{-2}}{0.34 \times 10^{-9}} = \frac{1}{34} \times 10^9 \times 10^{-2} \times 10^2$$

$$1 = \frac{1}{34} \times 10^9\text{ bp} = 2.94 \times 10^7\text{ bp}$$

There are approximately 2.9×10^7 bp in this DNA molecule which equals $2.9 \times 10^7 \times 2 = 5.8 \times 10^7$ nucleotides.

If there is 15% adenine, then 155 must also be thymine. Therefore, 100 – (15+15) = 70% are guanine and cytosine. 35% are guanine and 35% are cytosine.

Cytosine will be $(5.8 \times 10^7) \times 0.35 = 2.03 \times 10^7$ cytosin about 2 billion.

42. Consider a length of duplex DNA packaged into nucleosomes with linkers of 55 base pair (bp) between nucleosomes. How many nucleosomes would be expected in a region of 1 kb? In a region of 50 kb?

Solution:

(i) Each nucleosome includes 145 bp or (200–55) and with a linker size of 55 base pair (the total DNA length per nucleosome is 200 bp).

(ii) This value corresponds to 5 nucleosomes per kb and as 1 kb = 1000 bp, i.e., $\frac{1000}{200} = 5$ nucleosomes/1 kb. In a region of 50 kb, i.e., 50000 bp, i.e., $\frac{5000}{200}$ = 250 nucleosome/50 kb.

43. What are the roles of core histone proteins compared to the role of histone H_1 in the compaction of eukaryotic DNA.

Solution:

(i) The role of core histone is to form the nucleosomes.

(ii) In a nucleosome, the DNA is wrapped 1.65 times around the core histone.

(iii) Histone H_1 binds to the linker region. It may play a role in compacting the DNA into a 30 nm fibre.

44. Consider how histone proteins bind to DNA and then explain why a high salt concentration can remove transform DNA.

Solution:

(i) Histones are positively charged and DNA is negatively charged.

(ii) They bind to each other by these ionic interaction.

(iii) Salt is composed of positively charged and negatively charged ions. For example, NaCl when dissoved in water, it is dissociated into Na^+ and Cl^-.

(iv) Chromatin is exposed to a salt such as NaCl, the positively charged Na^+ could bind to the DNA and the negatively charged Cl^- could bind to positively charged histone.

This would prevent the histones and DNA from binding to each other.

45. An RNA molecule folds back upon itself to form a hairpin structure held together by a

region of base pairing. One segment of the molecule in the paired region has the base sequence 5′-UAUCGUAU-3′. What is the base sequence with which this segment is paired.

Solution:

5′-UAUCGUAU-3′

3′-AUAGCAUA-5′

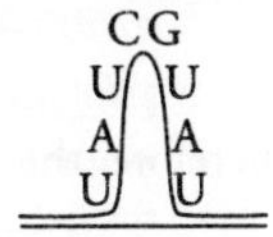

46. Which of the following sequences are palindromes and which are not?
Explain: (a) 5′-CCGG-3′, (b) 5′-TTTT-3′, (c) 5′-GCTAGC-3′, (d) 5′-CCGCTC-3′, (e) 5′-AAGGTT-3′.

Solution:

A DNA palindrome reads the same forward and backward, but along opposite stand.

(a) $\frac{5'\text{-CCGG-}3'}{3'\text{-GGCC-}5'}$ It is palindrome because the complementary strand shows the same.

(b) $\frac{5'\text{-TTTT-}3'}{3'\text{-AAAA-}5'}$ Complementary strand is 5′-AAAA-3′. It is not palindrome.

(c) $\frac{5'\text{-GCTAGC-}3'}{3'\text{-CGATCG-}5'}$ It is palindrome because the complementary strand shows the same.

(d) $\frac{5'\text{-CCGCTC-}3'}{3'\text{-GGCGAG-}5'}$ It is not palindrome. Partner strand show the other sequence

(e) $\frac{5'\text{-AAGGTT-}3'}{3'\text{-TTCCAA-}5'}$ It is palindrome because the palindrome strands show the other sequence.

47. Which of the following types of RNA would likely be more stable (i.e., resistance to nuclease digestion) in the living cell ribosomal mRNA or tRNA?

Solution:

Ribosomal RNA.

It is complexed with several dozen proteins to form a compact structure that is not easily attacked by RNase.

48. The DNA molecules below are denatured and then allowed to reanneal. Which of the two is least likely to reform the original structure? What would prevent that molecule from doing so?

(a) ATATGTATATATAGAT
TATACATATATATCTA

(b) GCCTATACGTGCACCA
CGGATATGCACGTGGT

Solution:

The DNA molecules of (a) would be least likely to reform the original structure.

It could have extensive intrastrand base pairing.

49. (a) GCCTATACGTGCACCA
CGGATATGCACGTGGT

(b) ATATGTATATATAGAT
TATACATATATATCTA

Which of the two molecules would have the highest Tm? Why? Which would require a higher temperature for renaturation?

Solution:

(i) The DNA molecules of (a) would have highest Tm. (ii) It has highest GC content.

The DNA molecules of (b) would require higher temperature for reannealing (relative to its Tm) in order to overcome infrastrand base pairing.

50. The ds DNA molecule extracted from a newly discovered virus were found to be 102 μm in length. (a) How many complete turns of the two chains are present? (b) How many nucleotides are present in one such molecule?

Solution:

Hints: $1\ \mu m = 10^4 Å$

$102\ \mu m = 102 \times 10^4 Å$

$1\ \text{spiral} = 34Å$

$34Å = 1\ \text{spiral}$

$102 \times 10^4 = \frac{1}{34} = 102 \times 10^4 = 30{,}000\ \text{spiral}$

$30{,}000 \times 10 = 300 \times 10^3\ \text{bp}.$

51. The ds DNA molecule of Epstein Barr virus contains 172, 280 bp.

(a) How many nucleotides are present?

(b) What would be the length of the molecule?

Solution:

Hints:

(a) $172280 \times 2 = 344560$

(b) $\frac{172280}{10} = 17228$ turn;

$34 \times 17228 = 58572Å$

or, $\frac{58572}{10^4} = 58.572 = 58.58$ μm.

The relationship betwen the Tm and GC content can be expressed: Tm = 69 + 0.41(%GC).

52. Calculate melting temperature of *E. coli* DNA that has about 40% of GC.

Solution:

Tm = 69 + 0.41 × 40 = 69 + 16.4 = 85.4°C.

53. Estimate the % GC of DNA from human liver cell where Tm = 75 °C.

Solution:

75 = 69 × 0.41 × (% GC),

or, 0.41 × (GC%) = 75 – 69

or, GC(%) = $\frac{6 \times 100}{41}$ = 14.634 = 15%.

54. Relationship between the melting Tm and GC content can expressed in this such simplified form, by the formula Tm = 69 + 0.41 (% GC).

(a) Calculate the melting temperature of *E. coli* DNA that has about 40% GC.

(b) Estimate the % GC of DNA from a human liver cell where Tm = 75 °C.

Solution:

(a) Tm = 69 + 0.41 (% GC)

Tm = 69 + 0.41 × 41 = 69 + 16.4 = 85.4°C.

(b) Tm = 75°C

75 = 69 + 0.41 (GC%)

0.41 (GC) = 75 – 69 = 6

GC (%) $\frac{6}{0.41} = \frac{600}{41}$ = 14.634 = 15%.

55. Calculate kinetic complexity of the following data taking Cot$\frac{1}{2}$ of *E. coli* = 4.0

	% of genome	Cot$\frac{1}{2}$
(i)	25 = 0.25	0.0013
(ii)	30 = 0.30	1.9
(iii)	15 = 0.45	630

Solution:

The Kinetic complexity =

$$\frac{\text{Cot}\frac{1}{2} \text{ of DNA of any genome} \times 4.2 \times 10^6}{\text{Cot}\frac{1}{2} \text{ of DNA of } E. coli}$$

Therefore, the Kinetic complexity

(i) $= \frac{0.0013 \times 0.25 \times 4.2 \times 10^6}{4}$

$= \frac{13 \times 25 \times 402 \times 10^6}{10^4 \times 10^2 \times 4} = \frac{1365}{4} = 341.25$

(ii) $= \frac{0.30 \times 1.9 \times 4.2 \times 10^6}{4} = \frac{2.394}{4} \times 10^6$

$= .5985 \times 10^6 = 5.985 \times 10^5 = 6 \times 10^5$

(iii) $= \frac{0.45 \times 630 \times 4.2 \times 10^6}{4} = \frac{1190.7 \times 10^6}{4}$

$= 297.675 \times 10^6 = 2.976 \times 10^8 = 3 \times 10^8$

56. The DNA of the bacterial virus T_4 produces a Cot$\frac{1}{2}$ of about 0.5 and contains 10^5 nucleotide pairs are present in the genome of the virus MS2 and the bacterium *E. coli* whose respective DNAs produce Cot$\frac{1}{2}$ values of 0.001 and 10.0?

Solution:

T_4 = Cot$\frac{1}{2}$ = 0.5 bp = 10^5

MS2 has Cot$\frac{1}{2}$ = 0.001; base pair = X

$\frac{0.5}{10^5} = \frac{0.001}{X}$; or, $X = \frac{10^5 \times 0.001}{0.5} = \frac{10^5 \times 10^3}{0.5}$

$= 10^5 \times 10^{-3} = \frac{10^3}{5} = 200$ bp,

when Cot$\frac{1}{2}$ = 10.0.

$\frac{0.5}{10^5} = \frac{10}{X}$; or, $X = \frac{10 \times 10^5}{0.5} = \frac{10 \times 10^6}{5} = 2 \times 10^6$ bp.

57. The DNA of the phage T_4 contains 10^5 nucleotide pairs and exhibits Cot$\frac{1}{2}$ 0.5. The unknown organism produces Cot$\frac{1}{2}$ of 20. How many bp is present in the genome of this organism?

Solution:

$T_4 = 10^5$ bp; Cot$\frac{1}{2}$ = 0.5

$$\frac{\text{Cot}\frac{1}{2}}{\text{bp}} = \frac{\text{Cot}\frac{1}{2}}{\text{unknown bp (X)}};$$

$\frac{0.5}{10^5} = \frac{20}{X}$;

or, $X = \frac{20 \times 10^5}{0.5} = \frac{20 \times 10^6}{5} = 4 \times 10^6$.

58. The size of the lambda (λ) DNA is 59 kb and its Cot$\frac{1}{2}$ is 0.08. The Cot$\frac{1}{2}$ of baculovirus is 0.22. What is the size of the baculovirus? DNA assuming no repetitive sequence in either DNA or in an equivalent GC content or in both?

Solution:

The two DNAs have no repetitive sequences and do not differ in their GC contents.

λ DNA = 50 kb; Cot$\frac{1}{2}$ = 0.08

baculovirus DNA = ? Cot$\frac{1}{2}$ = 0.22

$$\frac{\text{Cot}\frac{1}{2}\text{ of }\lambda}{\text{DNA size in bp}} = \frac{\text{Cot}\frac{1}{2}\text{ of baculovirus}}{\text{DNA size (in bp) baculovirus}}$$

or, $\frac{0.08}{50\text{ Kb}} = \frac{0.22}{\text{DNA size of baculovirus}}$

or, DNA of baculovirus $= \frac{0.22 \times 50\text{Kb}}{0.08} = \frac{22 \times 50\text{Kb}}{8}$

$= 2.75 \times 50 = 137.5$ kb.

59. *E. Coli* DNA has a Cot$\frac{1}{2}$ of about 8 and its DNA contain 4.2 milion base pairs. A bacteriophage has a Cot$\frac{1}{2}$ of 0.04. Assuming no difference in GC content and no repetitive sequences, how many base pairs does the phage contain?

Solution:

E. Coli Cot$\frac{1}{2}$ = 8 DNA = 4.2 milion bp

Phage Cot$\frac{1}{2}$ = 0.04 DNA = ?

$$\frac{8}{4.2\text{ million bp }(E.\ Coli)} = \frac{0.04}{\text{DNA (phage)}}$$

or, DNA of phage $= \frac{0.04 \times 4.2}{8} = \frac{0.168}{8}$

$= 0.021 \times 10{,}0000 = 21000$ bp.

60. A hypothetical bacterium has a DNA with 10^7 bp. The phage DNA 5×10^5 base pairs long. If the Cot$\frac{1}{2}$ of DNA is 40, what is the half Cot of the phage DNA assuming no difference in GC contain and no repetitive DNA.

Solution:

Bacterial DNA = 10^7 bp Cot$\frac{1}{2}$ = 40

Phage DNA = 5×10^5 bp

$$\frac{\text{Cot}\frac{1}{2}\text{ of bacterial DNA}}{\text{Length of DNA (bp)}} = \frac{\text{Cot}\frac{1}{2}\text{ of phage DNA}}{\text{Length of phage DNA (bp)}}$$

or, $\frac{40}{10^7} = \frac{\text{Cot}\frac{1}{2}\text{ of phage DNA}}{5 \times 10^5}$

or, Cot$\frac{1}{2}$ of phage DNA $= \frac{40 \times 5 \times 10^5}{10^7} = \frac{200}{10^2}$

$= 2.$

In non-repetitive DNA sequences Cot$\frac{1}{2}$ is directly proportional to the size of the genome. Bacterium has genome size 10^7 bp and phage DNA is 5×10^5 bp, i.e., 5.

Bacterium genome : Phage genome

10^7 5×10^5

i.e., $\frac{100 \times 10^5}{5 \times 10^5} = 20$ times greater than phage.

or, Phage genome $\frac{1}{20}$th of bacterium.

Therefore, Cot$\frac{1}{2}$ of phage of gene is also $\frac{1}{20}$, i.e., $\frac{1}{20} \times 40 = 2$.

- **Fuelgen staining**

  In this cytochemical staining, free deoxyribose aldehyde, produced from mild acid hydrolysis of DNA, binds with **Schiff's reagent** and develops magenta colour or deep pink purple colour. Detection and localization of DNA can be made with this technique.

- **Sonication:**

  A procedure that uses ultrasound to cause random breaks in DNA molecule.

- **Ancient DNA:**

  Preserved DNA from an archaeological or fossil specimen.

B. TOPOLOGY OF DNA

- **Supercoiling**

The term Supercoiling literally means—coiling of the coil. Supercoiling or super-twisting or super-helicity is generally used to describe a special feature of DNA. It means the helical coiling is over and above that of the twisting of the two strands of the DNA double helix. These are interchangeable terms to define the same feature.

- **Supercoiled DNA**

A form of DNA structure in which the helix is coiled upon itself. Such structure can exist in stable forms only when the ends of the DNA are not free, as in covalently closed circular DNA.

- **Positive supercoiling**

(i) When the circular duplex winds about itself in the same direction as the helix twists (right-handed, i.e., called positive supercoiling).

(ii) If a linear DNA is twisted more tightly than normal before the free ends are joined together, an overwound or positively supercoiled DNA is formed.

(iii) Here **linking number** increases.

- **Negative supercoiling**

(i) When the DNA duplex winds about itself in the opposite direction as the helix twists (i.e. called negatively supercoiled).

(ii) Here linking number decreases.

- **Linking number**

The total number of times that two strands of a closed, circular DNA duplex cross over each other.

(i) It is a topological property because it does not vary when DNA duplex is twisted or deformed in any way as long as both DNA strands remain intact.

(ii) If the interwound of the polynucleotide strand is in right-handed direction, then the linking number is **positive.**

If the interwound is in left-handed direction, then the linking number is **negative.**

(iii) It is measured by the number of base pair of the DNA divided by the base pairs occupying each complete turn of the helix.

Example:

1. **The virus SV 40 contains 5200 base pairs. If the DNA is relaxed, 10.4 base pairs occupy each complete turn of the helix, what is the linking number of that DNA?**

$L = \frac{5200}{10.4} = \frac{52000}{104} = 500$, $\therefore$ linking number is 500.

2. **The length of a relaxed DNA molecule is 210 base pairs and has turn of 10.5 base pairs. Calculate the linking number of that DNA.**

$L = \frac{210}{10.5} = \frac{2100}{105} = 20$, $\therefore$ linking number is 20.

Linking number(Lk^0)

The linking number of a fully relaxed DNA is called Lk^0.

The linking number has two geometric components, viz., **Twist** and **Writhe.**

- **Twist (Tw)**

(i) It is the measure of coiling of the axis.

(ii) The number of helical turns of one strand about the other that is the number of times one strand completely wraps around the other.

(iii) It can be easily determind by counting the number of times the two strands cross each other.

- **Writhe (Wr)**

It is the average number of times the helix axis crosses over itself when viewed from all direction, i.e. number of superhelical turns.

(i) It determines the local twisting or spatial relationship of neighbouring base pair.

(ii) In torsionally stressed DNA, the long axis of double helix crosses over itself repeatedly in three-dimensional space, it is called **Writhe.**

(iii) It has two forms:

(a) **Plectonemic writhe:** Here long axis is twisted around itself.

(b) **Toroid** or **Spatial writhe:** Here long axis is wound in a cylindrical manner as often DNA wraps around protein.

(i) Twist (**Tw**) and Writhe (**Wr**) are interconvertible. CCC DNA can readily undergo distortions that convert some of its twist to writhe and some of its writhe (Wr) to twist (Tw) without breakage of any covalent bond.

(ii) Only the constraint is that twist number (Tw) and writhing number (Wr) must remain equal to the linking number.

- **Constraint:** Lk = Tw + Wr.

- **Linking difference**

The extent of supercoiling is measured by the difference between Lk and Lk^0 which is called **linking difference.**

Δ Lk = Lk – Lk^0

Lk = Linking number at twisted condition
Lk^0 = Linking number at relaxed condition.

(i) If DNA is underwound and Lk < Lk^0 and ΔLk < 0, then the DNA is said to be **negatively supercoiled.**

(ii) If DNA is overwound and Lk > Lk^0 and ΔLk > 0, then the DNA is said to be **positively supercoiled.**

(iii) ΔLk quantities the degree of torsional strain of a given molecule and its propensity to undergo supercoiling.

(iv) ΔLk and Lk^0 are dependent upon the length of the DNA molecule.

- **Superhelical density**

It is the measure of the turns removed relative to those present in the relaxed DNA. It is also called **specific linking** difference.

It is assigned by symbol σ or λ.

$$\sigma/\lambda = \frac{Lk - Lk^0}{Lk^0} = \frac{\Delta Lk}{Lk^0}.$$

- **Topological isomer**

Molecules of DNA that are the same for their linking numbers are called **topological isomers.**

Molecules identical in sequence and lengths that differ only in topology are called **topoisomers.**

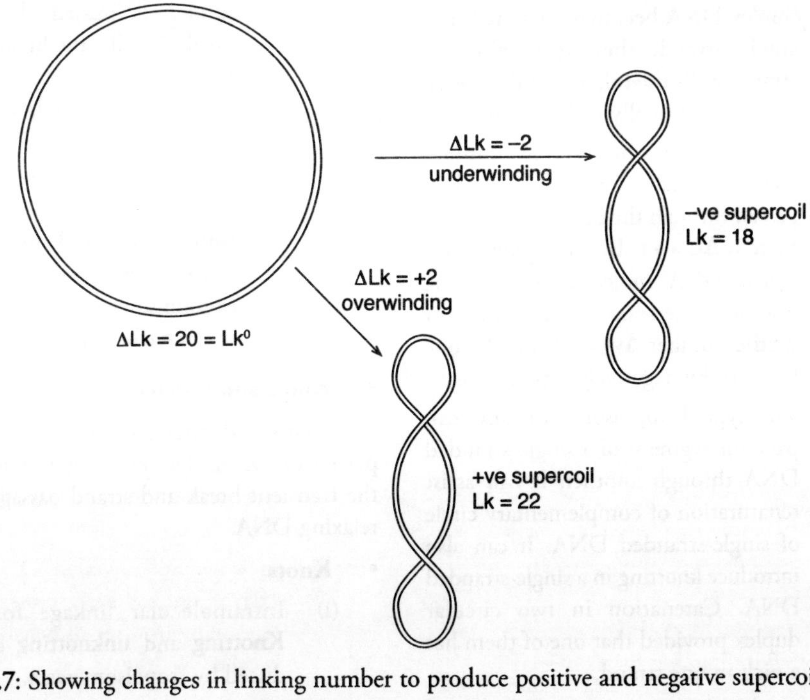

Fig. 12.7: Showing changes in linking number to produce positive and negative supercoil.

- **DNA topoisomerase**

A class of enzymes that can catalyze interconversion of DNA from one topological form to another and are required during replication, transcription and recombination to relax the torsional strain generated when helicase unwinds the duplex.

They are also called relaxing enzyme, swivelase, untwisting enzyme, nicking closing enzyme, etc.

- **Types:**
  - **Topoisomerase-I:**

(i) This enzyme relaxes highly negatively supercoiled DNA with high efficiency, the rate of relaxation becomes slower as the linking number approaches completely relaxed DNA.

(ii) It is coded by the gene top. The mutant of the gene increases the supercoiling in the nucleoid.

(iii) In *E. coli* topoisomerase-I binds to DNA and forms a stable complex in which one strand of the DNA has been nicked. Its 5′ phosphate end is covalently linked to a tyrosine residue of the enzyme.

(iv) The enzyme binds to a region in which duplex DNA becomes separated into single strand, then it breaks one strand, pulls the other strand through the gap and finally seals the break.

(v) In step one, the linking number changes. Each time one strand is passed through the break in the other, there is **ΔL = +1**. But in highly supercoiled DNA molecule, the linking number taken up by a change of **writhe** number **Δw** = **+1**, i.e., by one less turn by negatively supercoiling.

(vi) The type I topoisomerase also can pass one segment of a single-stranded DNA through another. It can assist renaturation of complementary circle of single-stranded DNA. It can also introduce knotting in a single-stranded DNA. Catenation in two circular duplex provided that one of them has a nick in one strand.

- **Topoisomerase-II:**

(i) Topoisomerases-II have been characterized from both prokaryotes and eukaryotes. They relax both positive and negative supercoils.

(ii) The reaction is ATP-dependent. Probably one unit of ATP is hydrolyzed for each catalytic reaction. The reaction is mediated by making a double-stranded break in one DNA duplex and passing another duplex region through it.

(iii) A formal consequence of two strands transfer is that the linking number is always changed in multiple of two. It can be used also to introduce or resolve catenated duplex circles and knotted molecules.

(iv) The hydrolysis of ATP may be used to drive the enzyme through conformational changes that provide the force needed to push one DNA duplex through the break made in other duplex.

(v) The enzyme makes double-stranded break in one duplex, and unbroken duplex is passed through end of break. Finally, the break is sealed and the enzyme releases DNA. In this way topoisomerase-II relaxes both positive and negative supercoils.

(vi) There are at least two types of topoisomerases-II, viz., Eukaryotic topoisomerase-II and topoisomerases-II from bacteriophage T_4. A third type of topoisomerase-II has been reported in **Archaebacteria**.

- **Knots and Catenanes**

Knots and catenanes are the topological properties of DNA. The enzymes topoisomerases use the transient break and strand passage reaction for relaxing DNA.

- **Knots**

(i) Intramolecular linkage forms a knot. **Knotting** and **unknotting** can be introduced by tropoisomerases.

(ii) Some site-specific reactions give rise knotted DNA duplex.

(iii) The enzyme topoisomerase-I first binds to a region in which duplex DNA becomes separated into single strand.

(iv) Then it breaks one strand and the single strand passage reaction is performed with two parts of the same circular DNA molecule. As a result, a knot is produced.

- **Catenane**

(i) Interlocking of two DNA molecules form **catenane.**

(ii) The enzyme topoisomerase-I can also **catenate** and **decatenate** the two circular duplex molecule of DNA, provided that one of them has a nick in one strand.

(iii) When one circle is passed through the another, they become interlocked. This structure is called catenate and the activity is known as **catenating the DNA.**

(iv) Topoisomerase-II can catenate or decatenate covalently closed circular DNA molecule by introducing a double-stranded break in one DNA and passing the other DNA molecule through the break.

(v) It should be noted that some site-specific reactions give rise knotted DNA duplex.

- **State briefly the importance of supercoiling in transcription.**

(i) It has an important influence in transcription at both initiation and elongation.

(ii) When DNA is negatively supercoiled, the two strands are less tightly wound around one another.

(iii) Negative supercoiling may assist initiation by making it easier for the strand of DNA to be separated.

(iv) Supercoiling also has a continuing involvement with transcription. As RNA polymerase transcribes DNA, unwinding and rewinding occurs.

(v) The consequence of the rotation of the DNA is often referred to as **twin** model for transcription. As RNA polymerase pushes forward along the double helix, it generates positive supercoils (more tightly wound DNA) ahead and leaves negative supercoils (partially unwound DNA) behind.

(vi) For each helical turn traversed by RNA polymerase, + 1 turn is generated a head and – 1 turn behind.

Too much positive supercoiling can actually prevent the continued forward mutation of the polymerase.

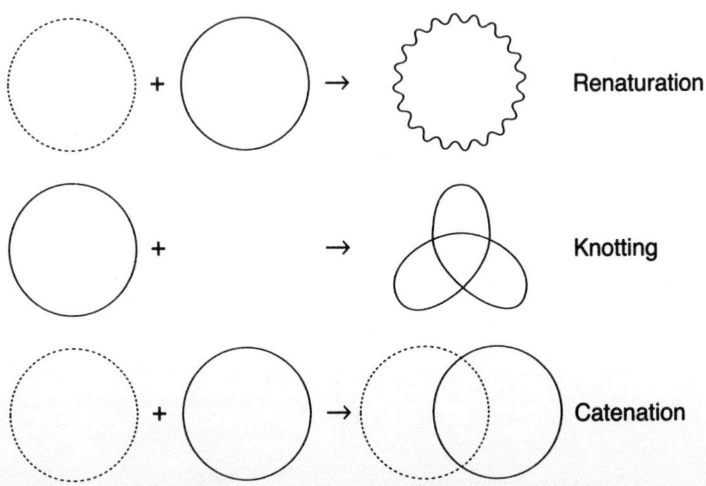

Fig. 12.8: The single-strand passage reaction of type I topoisomerases can assist renaturation of complementary circles of single-stranded DNA, can knot single-stranded DNA, and can catenate circles of (nicked) duplex DNA.

- **"Some promoters are influenced by the extent of supercoiling while others are not". Explain.**

(i) The dependence of supercoiling is determined by its surrounding sequence or sequence context.

(ii) This would predict that some promoters have sequences that are easier to melt and are therefore, less dependent on supercoiling, whereas others have more difficult sequences and have a greater need to be supercoiled.

(iii) An alternative is that the location of the promoter might be important if different regions of the bacterial chromosome have different degrees of supercoiling.

- **Why is positive supercoiling called overwinding and negative supercoiling called underwinding?**

(i) The double helix (DNA) that can undergone both underwinding and overwinding, can induce supercoiling of the helix.

(ii) In positive supercoiling the right-handed double helical conformation of DNA is twisted about itself in the **clockwise** direction to form a tighty coiled structure, that is why supercoiling is called overwinding.

(iii) In negative supercoiling left-handed double helical conformation of DNA undergoes uncoiling in its own axis in left-handed twisting motion, i.e., **anticlockwise** direction that tends to unwind the helix. So negative supercoiling is called under-winding.

- **State briefly the importance of supercoiling in transcription.**

(i) It has an important influence on transcription at both initiation and elongation.

(ii) When DNA is negatively supercoiled, the two strands are less tightly wound around one another.

(iii) Negative supercoiling may assist initiation by making it easier for the stands of DNA to be separated.

(iv) Supercoiling also has a continuing involvement with transcription. As RNA polymerase transcribes DNA and negative supercoils (partially unwound DNA) behind.

(v) For each helical turns traversed by RNA polymerase, + turn is generated ahead and −1 turn behind.

Too much positive supercoiling can actually prevent the continued forward motion of the polymerase.

Solved Problems

Problem 1

The linking number of relaxed DNA molecule is 20. If two turns are removed from this DNA molecule, what will be the linking difference of this DNA molecule? Calculate the superhelical density of this molecule and comment on it.

Solution:

The linking number of relaxed DNA is 20.

$Lk^0 = 20$

Two turns are removed, i.e., 20 – 2 = 18. Therefore, linking number, i.e., Lk = 18.

∴ Linking difference (ΔL) = $Lk - Lk^0 = 18 - 20 = -2$.

As we know superhelical density (d) = $\frac{\Delta L}{Lk^0}$.

$\sigma = \frac{-2}{20} = -0.1.$

The negative sign denotes the changes of linking number and comes about as a result of underwinding of DNA and induces negative supercoil.

Problem 2

A covalently closed circular (CCC) DNA is 360 bp in length.

(a) If its linking number (Lk) is 36 and twisting number (Tw) is 36, what will be its writhing number (Wr)?

(b) If its twisting number (Tw) is 36 and writhing number is –4, what will be its linking number (Lk)?

(c) If its linking number (Lk) is 32 and writhing number (Wr) is 0, what will be its twisting number (Tw)?

Solution:

(a) Here Lk = 36 Tw = 36 Wr = ?

We know, Lk = Tw + Wr = 36 + Wr

Wr = 36 – 36 = 0, ∴ Wr = 0.

(b) Here Tw = 36 Wr = -4 Lk = ?

We know, Lk = Tw + Wr = 36 – 4 = 32

∴ Lk = 32.

(b) Here Lk = 32 Wr = 0, Tw = ?

We know, Lk = Tw + Wr

32 = Tw + 0, ∴ Tw = 32.

Problem 3

In a DNA molecule Lk and Lk^0 is 20. After overwinding and underwinding, ΔLk of this molecule becomes +2 and –2 respectively. Comment on this result with diagramatic representation.

Solution:

(i) After underwinding, $\Delta Lk = -2$.

We know, when $\Delta Lk < 0$, then the DNA is said to be negatively supercoiled and Lk will be less than Lk^0, i.e., Lk = 18.

$\Delta Lk = Lk - Lk^0 = 18 - 20 = -2.$

(ii) After overwinding, $\Delta Lk = 2$.

We know, when $\Delta Lk > 0$, then the ΔNA is said to be positively supercoiled and Lk will be greater than Lk^0, i.e., Lk = 20 + 2 = 22.

For diagrammatic representation, see Fig. 12.7.

Problem 4

What is the superhelica density (σ) of a close circular DNA with length of 4200 bp, and a linking number (Lk) of 374? What is the superhelical density of the same density, when Lk = 412? Are these molecules negatively or positively supercoiled?

Solution:

(i) First calculate Lk^0 by dividing the length of the closed circular DNA (bp) by 10.5 bp/turned: 4200 bp/10.5 bp = 400.

(ii) We can now calculate $\Delta Lk = Lk - Lk^0 = 374 - 400 = -26$. Substituting the values of Lk and Lk^0 into the above equation, superhelical density (σ) = $\Delta Lk/Lk^0 = -\frac{26}{400} = -0.065$. Since the superhelical density is negative, these DNA molecules are negatively supercoiled.

(iii) When the same DNA molecule has Lk of 412, $\Delta Lk = 412 - 400 = 12$ and $\sigma = \frac{12}{400} = 0.03$. The superhelical density is positive and the molecule is positively supercoiled.

C. CODON

Essential features of Genetic code

1. **The codon is triplet:**
   (i) Each set of three nitrogenous bases is called codon and specifies a particular amino acid.
   Example: UUU = phenyl alanine.
   (ii) There are 61 codons that code for amino acid and 3 codons do not code for amino acid.
2. **Codons are commaless:**
   (i) There is no punctuation or comma between the adjacent codons.
   (ii) It means that after one amino acid is coded the second amino acid will automatically coded by the next three letters and that no letters are wasted as the punctuation marks.
   C A U G A C
3. **Code is non-overlapping:**
   (i) In translating mRNA molecules, the sequence of bases read in blocks of three corresponding the sequence of amino acids, without any overlapping bases.
   (ii) Thus a non-overlapping code means that a base in mRNA is not used for different codons.
   Example: the bases from the starting are AUG/CCA/AUC. The sequences of codon AUG CCA AUC non-overlapping and not the following AUG, GCC, CAA, AUC.

   AUG CCA AUC

   Codon-1. AUG
   Codon-2. GCC
   Codon-3. CAA
   Codon-4. AUC

   (iii) However, it has been shown that in the bacteriophage $\phi x174$ there is a possibility of overlapping genes and codons.
4. **Code is degenerate:**
   (i) More than one codon may specify the same amino acid, this is called degeneracy of the code.
   (ii) Except methionine (AUG) and tryptophan (UGG) which have a single codon each.
   (iii) It is of two types—partial and complete.
   In partial degeneracy, the first two nucleotides are identical but the third (i.e., 3′-base) nucleotide of the degenerate codons differs.
   Example: CUU & CUC codes for leucine.
   In complete degeneracy, any of the four bases can take third position and still codes for the same amino acid.
   Example: UCU, UCC, UCA and UCG codes for serine.
   (iv) Degeneracy also provides a mechanism of minimizing mutational lethality.
5. **Code of Universal:**
   (i) Almost all micro- and macro-organisms, i.e., virus to human use the same genetic code or same genetic language.
   (ii) A few exceptions, different code exists in mitochondria in some eukaryotes.
6. **Code is nonambiguous/unambiguous:**
   (i) A particular codon will always code for the same coded amino acid, wherever it is found.
   (ii) However, there are some exceptions AUG and GUG both may code for methionine as initiation codon. GUG for valine, when it is not an initiation codon.
7. **Code has initiator (start) and terminator (stop) codon:**
   (i) In most organism AUG (methionine), GUG (valine) bacteria in act as start codon.
   (ii) In most cases, UAA (ochre), UAG (amber), UGA (opal) act as terminator or nonsense codon.

8. **Principle of collinearity:**

(i) It explains the linear relationsip among DNA, RNA and polypeptide.

(ii) The linear order of nucleotides in DNA determines the linear order of codons in mRNA.

(iii) In turn, linear order of codons in mRNA determines the linear order of amino acid in polypeptide chain.

- **Confirmation of genetic code by studying phage MS2:**

The code is triplet in nature, degenerate unambiguous and commaless, but it contains punctuation with respect to start and stop signals. Experiments have been made by Walter Fiers and his coworkers in 1970 end on 1972.

Requirement:

(i) MS2 is a RNA containing bacteriophage.

(ii) It contains 3500 ribonucleotides making up only three genes.

(iii) The amino acid sequence of the coat protein was reportd on 1970 and the nucleotide sequence of the genes and a number of nucleotides on each end were reported in 1972.

(iv) The coat protein has 1970 amino acid and other two genes have 1972 amino acid.

Observation:

(i) When chemical natures of the gene and protein are compared, a colinear relationship is evident.

(ii) It is based on the coding dictionary. The linear sequence of nucleotides (the sequence of triplet codons) corresponds precisely with linear sequence of amino acid in the protein.

(iii) The codon for the first amino acid is AUG, the common initiator codon and the codon for the last amino acid are followed by two consecutive terminator codons UAA and UAG.

(iv) The other two genes and their protein products were sequenced providing similar confirmation.

Inference:

(i) The analysis clearly shows that the genetic code as establishd in bacterial system is identical in virus.

(ii) The code is also identical in eukaryotes.

Exception to the Universal Code

Codon	General code word	Altered code word	Source
UGA	Stop	trp	Human, *Mycoplasma* Yeast mitochondria
		cys	Euplotes (protozoan ciliate)
UAG	Stop	gln	*Paramoecium*
		Ala/leu	Mitochondria of some higher plants
UAA	Stop	Gln	*Paramoecium* Tetrahymena Stylonychia
AUA	ile	met	Human mitochondria *Drosophila* mitochondria
AGA	arg	stop	Human mitochondria
CUA	leu	thr	Yeast, mitochondria

- **The genetic code is degenerate and not quit universal.**

(i) The genetic code is a 3 letter code an there are 4 possible bases to choose from a each of the three base positions in a codon

(ii) There are, therefore, $4^3 = 64$ possibl codons, which are more than sufficient t encode the 20 major types of amino acid

(iii) The genetic code is degenerate because o average each amino acid is specified b about three different codons. Some amin acids (such as leucine, serine, and arg nine) are specified by as many as si codons. Others are much more poor represented. The degeneracy of geneti code most often involves the third base the codon.

(iv) Sixtyone codons (64−3) specify an amin acid. Three stop codons (UAA, UG UAG) do not encode any amino acid.

(v) The genetic code for mitochondrial mRNA (mtDNA) conforms to the universal code except for a few variants. For example, in the mitochondrial genetic code in humans and many other species **four codons** are used differently. UGA (stop codon) encodes **tryptophan**, AUA encodes methionine, instead of encoding isoleucine and AGA and AGG are **stop codons** instead of encoding arginine.

(vi) Although more than one codons can specify an amino acid, the number of different cytoplasmic tRNA molecules is quite a bit smaller and only 22 types of mitochondrial tRNA are made.

(vii) The interpretation of more than 60 sense codons with a much smaller number of different tRNAs is possible because base pairing in RNA is more flexible than in DNA. Pairing of codon anticodon follows the normal A-U and G-C rules for the first two base positions in a codon. However, at third position there is some flexibility (**base wobble**) and GU base pairs are tolerated here.

(viii) The genetic code is same throughout nearly in all life forms. However, **mitochondria** and **chloroplast** have a limited capacity for protein synthesis. During evolution their genetic codes have diverged slightly from that used in a cytoplasmic ribosomes.

(ix) Translation of nuclear encoded mRNA contains until one of three stop codons is encountered (**UAA, UAG** or **UGA**) but in mammalian mitochondria there are four possibilities (**UAA, UAG, AGA, AGG**).

(x) Depending on the surrounding sequence some codons in a few of nuclear encoded mRNA can be interpreted differently from normal. For example, in a wide variety of cells, the stop codon **UGA** is alternatively interpreted as selenocystein with some nuclear encoded mRNAs, and UAG can sometimes be interpreted to encode **glutamine.**

- DNAs, RNAs and protein of various masses can be separated by gel electrophoresis.
- The most common gel used in nucleic acid electrophoresis is agarose, but polyacrylamide is usually used in protein electrophoresis.
- SDS-PAGE is used to separae polypeptides according to their masses.

Measurement of DNA molecule

1 kilobase	(1 kb)	1000 bp	10^3 bp (ds)	10^3 Nucleotide (ss)
1 megabase	(1 Mb)	1000 kb	10^6 bp (ds)	10^6 Nucleotide (ss)
1 gigabase	(1 Gb)	1000 Mb	10^9 bp (ds)	10^9 Nucleotide (ss)

Solved Problems

Problem 1

What does it mean when we say that genetic code is degenerate? Discuss the universality of the genetic code.

Solution:

(i) Genetic code is degenerate—it means that more than one codon can specify the same amino acid.

(ii) For example, GGG, GGA and GGU all specify glycine.

In general, the genetic code is nearly universal because it is used in the same way by viruses, prokaryotes, fungi, plants and animals.

Problem 2

Mention the salient features of genetic code by studying the following nucleotide sequnce of mRNA strand and the polypeptide translated from it.

AUG UUU UCU UUU UUU UCU UAG
Met Phe Ser Phe Phe Ser

Solution:

(a) **Triplet nature:** Genetic code is triplet code where three adjacent nitrogen bases specify the amino acid.

(b) **Nonoverlapping:** Adjacent codons do not show overlapping.

(c) **Commaless:** Reading of genetic codons is continuous without any comma or punctuation.

(d) **Initiation codon:** AUG function as start signal or initiation codon that also codes for methionine.

(e) **Termination codon:** There is termination or nonsense codon like UAG (amber) which terminates polypeptide.

(f) **Colinearity:** There is colinearity between the sequence of codons and amino acids in the polypeptide chain synthesized.

Problem 3

A single base mutation in a codon may not always result in a loss or gain function. Is this statement is correct or not? Justify.

Solution:

(i) In many codons the change of third base is insignificant. This change does not alter the amino acid of that codon.

As for example: GCU, GCC, GCA and GCG, they code for amino acid alanine.

(ii) They differ in this last base. It means single base mutation in the last base of codon may not produce change in the amino acid seqeunce in the protein or its gene function is normal.

Problem 4

Poly A codes for polylysine. If a C is added to the 5′ end of the molecule, the polylysine has a different amino acid at the amino terminus and if a C is added to the 3′ end, there is a different amino acid at the carboxyl terminus. What are the amino acids?

Solution:

(i) Codons are read from the 5′ end of the mRNA. The first amino acid in the growing polypeptide is at amino end.

(ii) Therefore, if C is added to the 5′ end, it would be CAA which specify glycine (gln) at the amino end.

(iii) If C is added to 3′ end it would be AAC which specifies aspergine (Asn) at the carboxyl end.

Problem 5

The synthetic polymer, poly-U, is used as an mRNA molecule in an in vitro protein-synthesizing system that does not a special start codon. A simple cytosine nucleotide is added to one end of the poly-U.

Solution:

(i) Codons are read from the 5′ end of the mRNA. The first amino acid in the growing polypeptide is at amino end.

(ii) Therefore, if the C were added to the 3′ end, the last codon would be UUC, which also codes for phenyl alanine owing to the degeneracy of the code.

(iii) On the other hand, adding C to the 5′ end would make the first codon CUU which specifies valine (val) at the amino end of the polyphenyl alanine chain.

Problem 6

In a heteropolymer experiment using $\frac{1}{2C}:\frac{1}{4A}:\frac{1}{4G}$, how many different triplets will occur in the synthetic RNA molecule. How often will be most frequent triplet occur?

Solution:

There will be $(3)^3$ or 27 triplets produced. The most frequent will be CCC, present $\left(\frac{1}{2}\right)^3$ or, $\frac{1}{8}$ of the time.

Problem 7

Calculate how many triplet codons would be possible had evolution serized on six bases (three complementary base pairs) rather than four bases within the structure of DNA. Would six bases be sufficient using a two-letter code. Assuming 20 amino acids and start-and-stop codons.

Solution:

(i) Six things taken three at a time will produce $(6)^3$ or 216 triplet codes.

(ii) If the code was a doublet, there would be $(6)^2$ or 36 two letters codes, more than enough to accommodate 20 amino acids and start-and-stop codon.

Problem 8

In a regular copolymer experiment, in which UAC is represented over and over, how many different triplets will occur in the synthetic RNA, and how many amino acids will occur in the polypeptide when this RNA is translated?

Solution:

(i) The synthetic RNA will repeats four triplets – UUA, UAC, ACU, CUU – ever and ever.

(ii) Because both UUA and CUU encode leucine while ACU and UAC encode threonine and tyrosine respectively.

(iii) The polypeptides synthesizes under the directory of such an RNA containing three amino acids in the repeating sequence leu-leu-thr-tyr.

Problem 9

What polypeptide products are made when alternating polymer AAGAAGAAG ...... is used in an in vitro protein synthesizing system that does not need a start codon?

Solution:

(i) Translation could be in one of the three reading frames. AAGAAGAAG .... AGAAGAAGA ...... or GAAGAA ......

(ii) The first codes for polylysine, the second codes for polyarginine and the third for polyglutamic acid.

Problem 10

What polypeptide products are made when the alternating polymer CAC GCA CGC ACG ... is used in an in vitro protein synthesizing system that does not need a start codon?

Solution:

(i) The RNA would be translated with alternating sequence of codons CAC codes for Hist (His), GCA for Alanine (Ala), CGC for Arginine (Arg) and ACG for theronine (thr). His–Ala–Arg–Thr ......

Problem 11

What polypeptide products are made when the alternating polymer AUA UAU ... is used in an in vitro protein synthesizing system that does not need a start codon?

Solution:

The RNA would be translated as an alternating

sequence of the codons AUA-UAU-AUA-UAU- which corresponds to a polypeptide with repeatory sequence. Illeu-Tyr-Illeu-Tyr-.

Problem 12

Some codons in the genetic code were determined experimentally by the translation of random polymer. If a riboneucleotide polymer is synthesized that contains $\frac{1}{4}$ G and $\frac{3}{4}$ U in random order, which amino acids would the resulting polypeptide chain contain, and in what frequencies?

Solution:

In random sequence there are $2^3 = 8$ possible codons. Their identities, coding capacities and frequencies are

GGG (Gly) = $\left(\frac{1}{4}\right)^3 = \frac{1}{64}$

GGU (Gly) = $\left(\frac{1}{4}\right)^2\left(\frac{3}{4}\right) = \frac{3}{64}$

GUG (Val) = $\left(\frac{1}{4}\right)^2\left(\frac{3}{4}\right) = \frac{3}{64}$

UGG (Trp) = $\left(\frac{1}{4}\right)^2\left(\frac{3}{4}\right) = \frac{3}{64}$

GUU (Val) = $\left(\frac{1}{4}\right)\left(\frac{3}{4}\right)^2 = \frac{9}{64}$

UUG (Leu) = $\left(\frac{1}{4}\right)\left(\frac{3}{4}\right)^2 = \frac{9}{64}$

UGU (Cys) = $\left(\frac{1}{4}\right)\left(\frac{3}{4}\right)^2 = \frac{9}{64}$

UUU (Phe) = $\left(\frac{1}{4}\right)^3 = \frac{27}{64}$

Therefore, the amino acids expected in a random polymer and their frequencies are glycine $\left(\frac{3}{64}+\frac{3}{64}\right) = \frac{4}{64} = 6.25\%$.

Valine $\frac{3}{64} + \frac{9}{64} = \frac{12}{64} = 18.75\%$

Trp = $\frac{3}{64} = 4.69\%$

Leu = $\frac{9}{64} = 14.06\%$

Cysteine = $\frac{9}{64} = 14.06\%$

Phenyl alenine $\frac{27}{64} = 42.19\%$.

Problem 13

How many different codons would exist in a genetic code that had codons that were four bases long.

Solution:

$4^4 = 256$

Problem 14

(a) How many of the 64 codons can be made from the three nucleotides A, U and G?

(b) How many of the 64 codons can be made from the four nucleotides A, U, G and C with one or more Cs in each codon?

Solution:

(a) (i) The problem is related to probability. There are four bases, so the probability of cytosine at the first position in a codon is $\frac{1}{4}$.

(ii) Convensely the probability of a base other than cytosine is $\left(1-\frac{1}{4}\right) = \left(\frac{3}{4}\right)$.

(iii) These same probabilities apply to the other two position in the codon. Therefore, the probability of a codon without a cytosine is $\left(\frac{3}{4}\right)^3 = \frac{27}{64}$.

(b) (i) This question involves the relative frequencies of codons that have one or more cytosines.

(ii) We have already calculated that the probability of a codon not having a cytosine, so all the remaining codons have one or more cytosin. At least one codon is $1-\frac{27}{64} = \frac{37}{64}$.

Problem 15

Random copolymers were used in some of the experiments directed toward deciphering the genetic code. For each of the following ribonucleotide mixtures, give the expected codons and their frequencies, and give the expected proportions of amino acids that would be found in a polypeptide directed by the copolymer in a cell-free protein-synthesizing system.

(a) 2U : 1C

(b) 1U : 1C : 2G

Solution:

(a) (i) The probability of a U at any position in a codon is $\frac{2}{3}$ and the probability of C at any position in a codon is $\frac{1}{3}$.

(ii) Thus, the codons, their relative frequencies and the amino acids for which they code are as follows:

Codons	Amino acid	Probabilities	Percentage
1. UUU	Phenyl alanine (Phe)	$\left(\frac{2}{3}\right)^3 = \frac{8}{27} = 0.296$	29.6 %
2. UUC	Phenyl alanine (Phe)	$\left(\frac{2}{3}\right)^2 . \frac{1}{3} = \frac{4}{27} = 0.148$	14.8 %
3. UCC	Serine (Ser)	$\frac{2}{3} \cdot \left(\frac{1}{3}\right)^2 = \frac{2}{27} = 0.741$	7.41%
4. CCC	Proline (Pro)	$\left(\frac{1}{3}\right)^3 = \frac{1}{27} = 0.037$	3.7 %
5. CUC	Leucine (Leu)	$\left(\frac{1}{3}\right)^2 . \frac{2}{3} = \frac{2}{27} = 0.741$	7.41%
6. CUU	Leucine (Leu)	$\frac{1}{3} \cdot \left(\frac{2}{3}\right)^2 = \frac{4}{27} = 0.148$	14.8 %
7. UCU	Serine (Ser)	$\left(\frac{2}{3}\right)^2 . \frac{1}{3} = \frac{4}{27} = 0.148$	14.8 %
8. CCU	Proline (Pro)	$\frac{2}{3} \cdot \left(\frac{1}{3}\right)^2 = \frac{2}{27} = 0.74$	7.41%

(iii) 2U : 1C gives $2^3 = 8$ codons.

(b) (i) The probability of U at any position in a codon is $\frac{1}{4}$, the probability of C at any position in a codon is $\frac{1}{4}$ and the probability of G at any position in a codon is $\frac{1}{2}$.

(ii) Thus, the codons, their relative frequencies and the amino acids for which they code are as follows:

Codons	Amino acid	Probabilities	Percentage
1. UUU	Phe	$\left(\frac{1}{4}\right)^3 = \frac{1}{64} = 0.0156$	1.56
2. UUC	Phe	$\left(\frac{1}{4}\right)^2 . \frac{1}{4} = \frac{1}{64} = 0.0156$	1.56
3. UCC	Serine	$\frac{1}{4} \cdot \left(\frac{1}{4}\right)^2 = \frac{1}{64} = 0.0156$	1.56
4. UCU	Serine	$\left(\frac{1}{4}\right)^2 . \frac{1}{4} = \frac{1}{64} = 0.0156$	1.56
5. CUU	Leucine	$\frac{1}{4} \cdot \left(\frac{1}{4}\right)^2 = \frac{1}{64} = 0.0156$	1.56
6. UUG	Leucine	$\left(\frac{1}{4}\right)^2 . \frac{1}{2} = \frac{1}{32} = 0.03125$	3.125
7. CUC	Leucine	$\left(\frac{1}{4}\right)^2 . \frac{1}{4} = \frac{1}{64} = 0.0156$	1.56
8. UGG	Trp	$\frac{1}{4} \cdot \left(\frac{1}{2}\right)^2 = \frac{1}{16} = 0.0625$	6.25
9. GUU	Val	$\left(\frac{1}{4}\right)^2 . \frac{1}{2} = \frac{1}{32} = 0.03125$	3.125
10. UGU	Cys	$\left(\frac{1}{4}\right)^2 . \frac{1}{2} = \frac{1}{32} = 0.03125$	3.125
11. CCC	Proline	$\left(\frac{1}{4}\right)^3 = \frac{1}{64} = 0.0156$	1.56
12. CGG	Arg	$\frac{1}{4} \cdot \left(\frac{1}{2}\right)^2 = \frac{1}{16} = 0.0625$	6.25
13. CGC	Arg	$\left(\frac{1}{4}\right)^2 . \frac{1}{2} = \frac{1}{32} = 0.03125$	3.125
14. GCG	Ala	$\left(\frac{1}{2}\right)^2 . \frac{1}{4} = \frac{1}{16} = 0.0625$	6.25
15. CUG	Leucine	$\left(\frac{1}{4}\right)^2 . \frac{1}{2} = \frac{1}{32} = 0.3125$	3.125
16. CGU	Arg	$\frac{1}{4} . \frac{1}{2} . \frac{1}{4} = \frac{1}{32} = 0.3125$	3.125
17. UGC	Cys	$\frac{1}{4} . \frac{1}{2} . \frac{1}{4} = \frac{1}{32} = 0.3125$	3.125
18. GGC	Gly	$\left(\frac{1}{2}\right)^2 . \frac{1}{4} = \frac{1}{16} = 0.0625$	6.25
19. UCG	Serine	$\frac{1}{4} . \frac{1}{4} . \frac{1}{2} = \frac{1}{32} = 0.3125$	3.125
20. GCU	Ala	$\frac{1}{2} . \frac{1}{4} . \frac{1}{4} = \frac{1}{32} = 0.3125$	3.125
21. GGU	Gly	$\left(\frac{1}{2}\right)^2 . \frac{1}{4} = \frac{1}{16} = 0.0625$	6.25
22. GGG	Gly	$\left(\frac{1}{2}\right)^3 = \frac{1}{8} = 0.125$	12.5
23. CCG	Proline	$\left(\frac{1}{4}\right)^2 . \frac{1}{2} = \frac{1}{32} = 0.3125$	3.125
24. GUC	Val	$\frac{1}{2} . \frac{1}{4} . \frac{1}{4} = \frac{1}{32} = 0.3125$	3.125
25. CCU	Proline	$\left(\frac{1}{4}\right)^2 . \frac{1}{4} = \frac{1}{64} = 0.0156$	1.56
26. GUG	Val	$\left(\frac{1}{2}\right)^2 . \frac{1}{4} = \frac{1}{16} = 0.0625$	6.25
27. GCC	Ala	$\frac{1}{2} \cdot \left(\frac{1}{4}\right)^2 = \frac{1}{32} = 0.3125$	3.125

(iii) 1U : 1C : 2G gives $3^3 = 27$ codons.

In sum:

Phe = 1.56 + 1.56	3.12
Ser = 1.56 + 1.56 + 3.125	6.245 = 6.25
Leu = 1.56 + 3.125 + 1.56 + 3.125	9.37
Pro = 1.56 + 3.125 + 1.56	6.245 = 1.25
Cys = 3.125 + 3.125	6.25
Trp = 6.25	6.25
Val = 3.125 + 3.125 + 6.25	12.5
Gly = 6.25 + 6.25 + 12.5	25%
Arg = 6.25 + 3.125 + 3.125	12.5%
Ala = 3.125 + 3.125 + 6.25	12.5%

Problem 16

Random copolymers were used in some of the experiments that revealed the characteristics of the genetic code. For each of the following ribonucleotide mixtures, give the expected proportions of the amino acid that would be found in a polypeptide directed by the copolymer in a cell-free protein-synthesizing system.

(a) 4A : 6C, (b) 5G : 1C, (c) 1A = 3U : 1C, (d) 1A : 1U = 1G : 1C.

Solution:

(a) (i) The probability of A at any position of codon is $\frac{4}{10}$ and C at any position of codon is $\frac{6}{10}$.

(ii) 4A : 6C gives 2^3 = 8 codons.

(iii) The probable codons are AAA, AAC, ACC, CAC, CCC, CAA, CCA, ACA.

Their relative frequencies and the amino acids for which they code are as follows:

Codons	Amino acid	Probabilities	Percentage
AAA	Lys	$\left(\frac{4}{10}\right)^3 = (0.4)^3 = 0.64$	6.4%
AAC	Asn	$(0.4)^2 \times (0.6) = 0.096$	9.6%
ACC	Thr	$0.4 \times (0.6)^2 = 0.144$	14.4%
CAC	His	$(0.6)^2 \times (0.4) = 0.144$	14.4%
CAA	Gln	$(0.4)^2 \times (0.6) = 0.096$	9.6%
CCA	Pro	$(0.6)^2 \times (0.4) = 0.144$	14.4%
ACA	Thr	$(0.4)^2 \times (0.6) = 0.096$	9.6%
CCC	Pro	$(0.6)^3 = 0.216$	21.6%

(b) 4G : 1C, the probability of G at any position of codon is $\frac{4}{5}$; C is $\frac{1}{5}$. 4G:1C gives 2^3 = 8 codons.

The codons, their relative frequencies and the amino acids for which they code are as follows:

Codons	Amino acid	Probabilities	Percentage
GGG	Gly	$\left(\frac{4}{5}\right)^3 = (0.8)^3 = 0.512$	1.2%
GGC	Gly	$(0.8)^2 \times (0.2) = 0.128$	12.8%
GCC	Ala	$0.8 \times (0.2)^2 = 0.032$	3.2%
CGC	Arg	$(0.2)^2 \times 0.8 = 0.032$	3.2%
CCG	Pro	$(0.2)^2 \times 0.8 = 0.032$	3.2%
CGG	Arg	$0.2 \times (0.8)^2 = 0.128$	12.8%
CCC	Pro	$(0.2)^3 = 0.008$	0.8%
GCG	Ala	$0.2 \times (0.8)^2 = 0.032$	3.2%

(c) (i) 1A : 3U : 1C gives 3^3 = 27 different possible codons.

(ii) Their probaility, i.e., 1A : 3U : 1C : $\frac{1}{5}$ = 0.20, $\frac{3}{5}$ = 0.60 and $\frac{1}{5}$ = 0.20.

(iii) UAA chain terminater and remaining 100 – 2.4 = 97.6%, or, 0.976.

Codons	Amino acid	Probabilities
AAA	Lys	$0.2 \times 0.2 \times 0.2 = 0.008$
AAU	Asn	$0.2 \times 0.2 \times 0.6 = 0.024$
AUU	Ileu	$0.2 \times 0.6 \times 0.6 = 0.072$
UAU	Tyr	$0.6 \times 0.6 \times 0.2 = 0.072$
UAA	Stop	$0.6 \times 0.2 \times 0.2 = 0.024$
AUA	Ille	$(0.2)^2 \times 0.06 = 0.024$
AUA	Leu	$0.6 \times 0.6 \times 0.2 = 0.072$
CUA	Leu	$0.2 \times 0.6 \times 0.2 = 0.024$
CCA	Pro	$0.2 \times 0.2 \times 0.2 = 0.008$
CAA	Gln	$0.2 \times 0.2 \times 0.2 = 0.008$
ACA	Thr	$0.2 \times 0.2 \times 0.2 = 0.008$
ACC	Thr	$0.2 \times 0.2 \times 0.2 = 0.008$
CAC	Hist	$0.2 \times 0.2 \times 0.2 = 0.008$
AAC	Asn	$0.2 \times 0.2 \times 0.2 = 0.008$
AUC	Ille	$0.2 \times 0.6 \times 0.2 = 0.024$
CCC	Pro	$0.2 \times 0.2 \times 0.2 = 0.008$
UCC	Ser	$0.6 \times 0.2 \times 0.2 = 0.024$
UUC	Phe	$0.6 \times 0.6 \times 0.2 = 0.072$
CUU	Leu	$0.2 \times 0.6 \times 0.6 = 0.072$
CUC	Leu	$0.2 \times 0.6 \times 0.2 = 0.024$
UUU	Phe	$0.6 \times 0.6 \times 0.6 = 0.216$
CCU	Pro	$0.2 \times 0.2 \times 0.6 = 0.024$
UCU	Ser	$0.6 \times 0.2 \times 0.6 = 0.072$
UCA	Ser	$0.6 \times 0.2 \times 0.2 = 0.024$
UAC	Tyr	$0.6 \times 0.2 \times 0.2 = 0.024$
CAU	His	$0.2 \times 0.2 \times 0.6 = 0.024$
ACU	Thr	$0.2 \times 0.2 \times 0.6 = 0.024$

$$\text{Leu} = 0.072+0.072+0.024+0.024 = \frac{0.192}{0.976} = 0.1967 = 9.67\%$$

$$\text{Ser} = 0.024+0.024+0.072 = \frac{0.12}{0.976} = 0.1229 = 12.29 = 12.3\%$$

$$\text{Pro} = 0.008+0.008+0.024 = \frac{0.04}{0.976} = 0.0409 = 4.09 = 4.1\%$$

$$\text{His} = 0.024+0.008 = \frac{0.032}{0.976} = 0.0327 = 3.278 = 3.28\%$$

$$\text{Tyr} = 0.024+0.072 = \frac{0.096}{0.976} = 0.09836 = 9.836 = 9.84\%$$

$$\text{Gln} = 0.008 = \frac{0.008}{0.976} = 0.00819 = 0.8196 = 0.82\%$$

$$\text{Thr} = 0.008+0.008+0.024 = \frac{0.04}{0.976} = 0.0409 = 4.09 = 4.1\%$$

$$\text{Phe} = 0.216+0.072 = \frac{0.288}{0.976} = 0.295 = 29.5\%$$

$\text{Ile} = 0.072 + 0.024 + 0.024 = \frac{0.12}{0.976} = 12.295 = 12.3\%$

$\text{Asn} = 0.024 + 0.008 = \frac{0.032}{0.976} = 0.0327 = 3.278 = 3.28\%$

$\text{Lys} = 0.008 = \frac{0.008}{0.976} = 0.00819 = 0.819 = 0.82\%.$

Problem 17

Random copolymers were used in some of the experiments directed toward deciphering the genetic code. For the following ribonucleotide mixture, give the expected codons and their frequencies and give the expected proportions of amino acids that would be found in a polypeptide directed by the copolymer in a cell-free protein-synthesizing system.

5U : 1G

Solution:

(i) Here the probability of U at any position of codon is $\frac{5}{6}$ and G is $\frac{1}{6}$.

(ii) It gives $2^3 = 8$ codons.

Codons	Amino acid	Probabilities	Percentage
UUU	Phe	$\left(\frac{5}{6}\right)^3 = \frac{125}{216} = 0.5787$	58%
UUG	Leu	$\frac{5}{6}\times\frac{5}{6}\times\frac{1}{6} = \frac{25}{216} = 0.1157$	12%
UGG	Trp	$\frac{5}{6}\times\frac{1}{6}\times\frac{1}{6} = \frac{25}{216} = 0.1157$	12%
GUG	Val	$\frac{1}{6}\times\frac{5}{6}\times\frac{1}{6} = \frac{5}{216} = 0.02314$	2.13%
GGG	Gly	$\frac{1}{6}\times\frac{1}{6}\times\frac{1}{6} = \frac{1}{216} = 0.00462$	0.5%
GGL	Gly	$\frac{1}{6}\times\frac{1}{6}\times\frac{5}{6} = \frac{5}{216} = 0.02314$	2.13%
UGU	Cys	$\frac{5}{6}\times\frac{1}{6}\times\frac{5}{6} = \frac{25}{216} = 0.1157$	12%
GUU	Val	$\frac{1}{6}\times\frac{5}{6}\times\frac{5}{6} = \frac{25}{216} = 0.1157$	12%

Problem 18

Two populations of RNAs are made by the random combination of nucleotides. In population A, the RNAs contain only A and G nucleotides (3A : 1G), whereas, populations B, the RNAs contain only A and U nucleotides (3A : 1U). In what ways other than amino acid content will the proteins produced by translating the population A RNAs differ from those produced by translating the population B RNAs?

Solution:

(i) In population A, the codons that can be produced encode

$$\left.\begin{matrix}\text{AAA}\\ \text{AAG}\end{matrix}\right(\text{Lys}),\ \left.\begin{matrix}\text{AGG}\\ \text{AGA}\end{matrix}\right(\text{Arg}),\ \left.\begin{matrix}\text{GAG}\\ \text{GAA}\end{matrix}\right(\text{Glu}),\ \left.\begin{matrix}\text{GGA}\\ \text{GGG}\end{matrix}\right(\text{Gly})$$

(ii) All of these are sense codons, so long polypeptide chains containing these amino acids will be synthesized.

(iii) In population B, the codons that can be produced encode AAA(Lys), AAU(Agn), AUA, AUA(Ileu), UAU(Tyr), UUA(Leu), UUU(Phe), UAA(Stop).

The frequency of the stop codons will be $\text{UAA} = \frac{1}{4} \times \frac{3}{4} \times \frac{3}{4} = \frac{9}{64} = 0.14$ or 14%. Therefore, remaining sense codons are $100 - 14 = 86\%$, i.e., the polypeptide produced in the population B will be shorter than population A.

(iv) If stop codon appears about 14% of the time, there will be an average $\frac{1}{0.14} = 7 : 14$ (7 sense, other nonsense codon).

Problem 19

If an artificial mRNA contains two parts of uracil and to one part of cytosine, name the amino acids and the proportions in which they should be incorporated into protein.

Solution:

(i) 2U : 1C

The probability of U at any position in a codon is $\frac{2}{3}$ and C is $\frac{1}{3}$.

(ii) 2U and 1C gives $2^3 = 8$ codons.

Codons	Amino acid	Probabilities	Percentage
1. UUU	Phe	$\left(\frac{2}{3}\right)^3 = \frac{8}{27} = 0.296$	29.6%
2. UUC		$\frac{2}{3}\times\frac{2}{3}\times\frac{1}{3} = \frac{4}{27} = 0.148$	14.8%
3. UCC	Ser	$\frac{2}{3}\times\frac{1}{3}\times\frac{1}{3} = \frac{2}{27} = 0.074$	7.4%
4. UCU		$\frac{2}{3}\times\frac{1}{3}\times\frac{2}{3} = \frac{4}{27} = 0.148$	14.8%
5. UUU	Pro	$\frac{1}{3}\times\frac{1}{3}\times\frac{1}{3} = \frac{1}{27} = 0.037$	3.7%
6. CCU		$\frac{1}{3}\times\frac{1}{3}\times\frac{2}{3} = \frac{2}{27} = 0.074$	7.4%
7. CUC	Leu	$\frac{1}{3}\times\frac{2}{3}\times\frac{1}{3} = \frac{2}{27} = 0.074$	7.4%
8. CUU		$\frac{1}{3}\times\frac{2}{3}\times\frac{2}{3} = \frac{4}{27} = 0.148$	14.8%

A. REPLICATION

Q. 1 What is primase? Mention its role in replication.

Solution:

- **Primase** is a specialized RNA polymerase, a product of dna G gene dedicated to make a short RNA **primer** (5-10 nucleotides long) on ss DNA template which does not require a free 3′ OH end to initiate DNA synthesis.

(i) These primers (5-10 nucleotides) are subsequently extended by DNA polymerase III. Although DNA polymerases incorporates only deoxyribonucleotides into DNA, they can initiate synthesis using either an RNA primer or a DNA primer annealed to the DNA template.

(ii) **Primase** does not require specific DNA sequences to initiate synthesis of a new RNA primer. Before priming by primase, preprimer intermediate is formed with the help of six prepriming proteins (dna B, dna C, n, n′, n″ and i proteins). For the synthesis of primer, primase needs several accessory proteins which combine with primase.

(iii) The **continuous replication** of leading strand is initiated by a **single RNA primer** by primase. In contrast, **the discontinuous replication** of lagging strand requires **repeated priming** by primase, i.e., one RNA primer for each **Okazaki fragment.** Because a single replication fork can replicate millions of base pairs, synthesis of lagging strand requires hundreds to thousand of **Okazaki fragments** and their associated RNA primers.

(iv) Once it is activated, it (primase) synthesizes RNA primer using the most recently exposed lagging strand template, regardless of the sequence (Watson, Baker *et. al.*, Molecular Biology of Gene 2004).

Q. 2 "The replication process is generally referred to as semidiscontinuous process." Why?

Solution:

After initiating of replication at ori C, two replication forks move in opposite direction.

(i) The 3′-5′ template strand is polymerized by DNA polymerase, continuously in 5′-3′ direction, called **leading strand.**

(ii) The other strand of 5′-3′ template is polymerized in a frequently interrupted manner and produces a short fragments of DNA, i.e., (**Okazaki fragments**) as helix unwind, in 3′-5′ direction of the template. Thus, this strand is replicated discontinuously in 5′-3′ direction called **lagging strand.**

(iii) So the lagging strand is synthesized discontinuously with a **Okazaki** fragments in an opposite direction.

(iv) The leading strand grows in 5′-3′ direction continuously, towards the fork. The template strand (3′-5′) is called **forward template.**

On the other hand, lagging strand grows in 5′-3′ direction **discontinuously,** away from the fork. The template strand is called **retrograde template.** The overall growth of lagging strand is 3′-5′.

Therefore, the whole DNA replication process is generally referred to as **semidiscontinuous process** as one strand (leading strand) replicates continuously and other strand (lagging strand) replicates discontinuously.

Q. 3 DNA polymerase at the replication fork synthesizes thousands to millions of base pairs without releasing from the template but in the absence of other proteins the DNA polymerases that act at the replication fork are only able to synthesize 20-100 base pairs before releasing the template.

How is the processivity of these enzymes increased so dramatically at the replication fork?

Solution:

The key to high processivity of the DNA polymerases that act at replication forks in their association with proteins are called **sliding DNA clamps.**

(i) These proteins are composed of multiple identical subunits that assemble in the shape of "**doughnut**". The hole in the centre of the clamp is large enough to encircle the DNA double helix and leave a room for one or two water molecules between the DNA and protein.

(ii) **Sliding DNA clamps** also bind tightly with DNA polymerase at replication fork. Thus, the clamp encircles the newly synthesized double-stranded DNA and the polymerase associates with the primer: template junction.

(iii) In the absence of the **sliding clamp**, a DNA polymerase dissociates and diffuses away from the template DNA, on average, once every 20-100 base pairs are synthesized.

(iv) In the presence of **sliding clamp**, the DNA polymerase still disengages its active site from

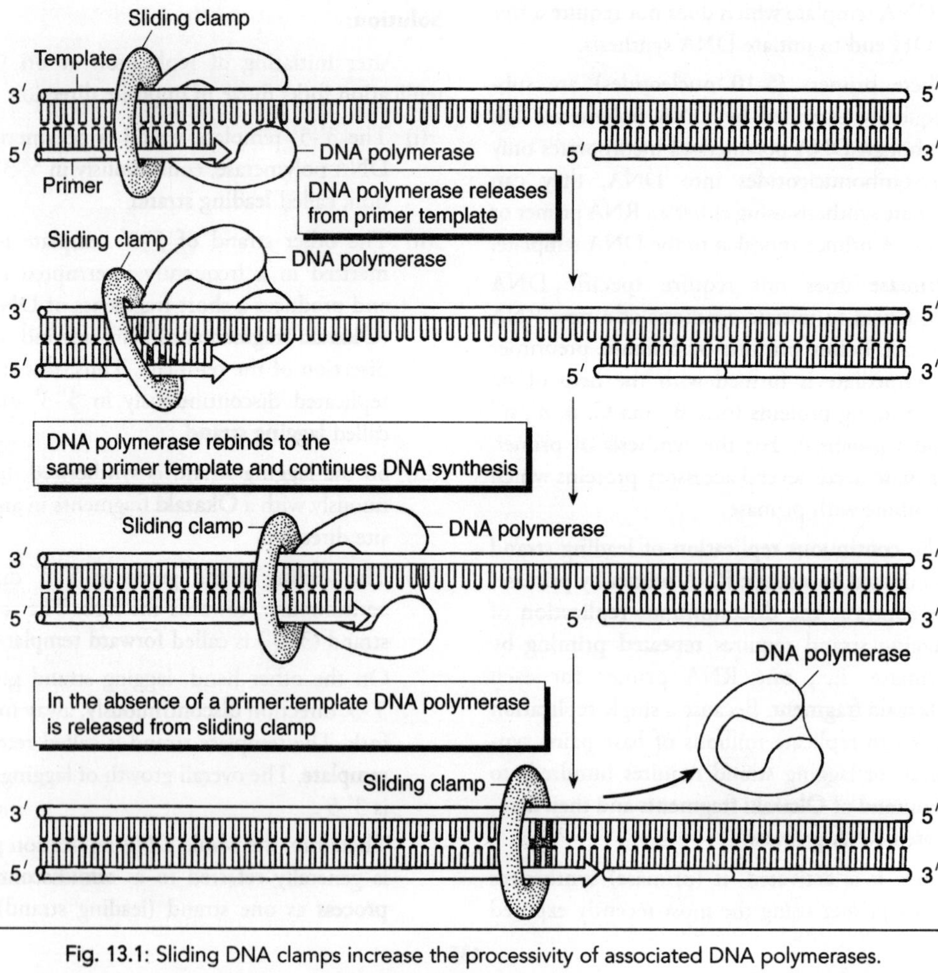

Fig. 13.1: Sliding DNA clamps increase the processivity of associated DNA polymerases.

3′ OH end of the DNA frequently, but the association with the **sliding clamp** prevents the polymerase from diffusing away from the DNA.

(v) A special class of protein complexes called **sliding clamp loaders** catalyzes the opening and placement of sliding clamp on the DNA. These enzymes couple ATP binding and hydrolyze the placement of the sliding clamp around primer template junction on the DNA.

(vi) By keeping the DNA polymerase in close proximity to the DNA, the sliding clamp ensures that the DNA polymerase rapidly rebinds the same primer template junction, vastly increasing the processivity of the DNA polymerases.

Q. 4 What role does a helicase play in DNA replication? Why is an ATPase a necessary part of this role?

Solution:

(a) (i) **Helicase** is a **hexameric ring-shaped** protein, the product of dna B gene, which encircles one of the two single strands of DNA at the replication fork near the single-stranded/double-stranded junction.

(ii) Helicases are **ATP-dependent unwinding enzymes** which promote the separation of two parental strands at **ori C** and creates two **'Y'-shaped** areas, called **replication bubble.** Each Y is called **replication fork.**

(iii) Each DNA helicase moves along ssDNA in a defined direction. This characteristic of each DNA helicase is called **polarity** (5′-3′ or 3′-5′).

- Helicase I—Stimulate strand separation of leading strand that has 3′-5′ polarity.
- Helicase II—Stimulate strand separation of lagging strand that has 5′-3′ polarity.

(b) (i) DNA helicases hydrolyse ATP when they are bound to single strand of DNA. Hydrolysis of ATP can change the shape of the protein molecule in a cyclical manner that allows the protein to perform mechanical work.

(ii) ATP is the source of energy in all cells. **Helicase** contains an ATPase activity that breaks down two molecules of ATP to break hydrogen bond for every base pair separation.

Q. 5 How might continuous and discontinuous modes of DNA replication be distinguished experimentally?

Solution:

If nascent DNA is labelled by exposure to ^3H thymidine for very short periods of time:

(i) Continuous replication predicts that the label would be incorporated into chromosome size DNA molecules.

(ii) On the other hand, discontinuous replication predicts that label would first appear in small pieces of nascent DNA (prior to covalent joining, catalyzed by polynucleotide ligase).

Q. 6 *E. coli* cells are grown for many generations in a medium in which the only available nitrogen is the heavy nitrogen ^{15}N. They are then transferred to a medium containing ^{14}N as the only source of nitrogen.

(a) What distribution of ^{15}N and ^{14}N would be expected in the DNA molecules of cells that had grown for one generation in the ^{14}N containing medium assuming that DNA replication was (i) conservative, (ii) semiconservative and (iii) dispersive?

(b) What distribution would be expected after two generations of growth in the ^{14}N containing medium assuming (i) conservative, (ii) semiconservative and (iii) dispersive replication?

Solution:

(a) (i) One half of the DNA molecule with ^{15}N in both strands and $\frac{1}{2}$ with ^{14}N in both strands.

(ii) All the DNA molecules with one strand containing ^{15}N and the complementary strand containing ^{14}N.

(iii) All DNA molecules with both strands containing roughly equal amount of ^{15}N and ^{14}N.

(b) (i) $\frac{1}{4}$ of the DNA molecules with ^{15}N in both strands and $\frac{3}{4}$ with ^{14}N in both strands.

(ii) $\frac{1}{2}$ of the DNA molecules with one strand containing ^{15}N and complementary strand containing ^{14}N and other $\frac{1}{2}$ ^{14}N with ^{14}N in both strands.

(iii) All DNA molecules with both strands containing $\frac{1}{2}$ ^{15}N and $\frac{3}{4}$ ^{14}N.

Q. 7 What do you mean by proof-reading of DNA polymerase?

Solution:

A molecular mechanism for detecting errors (mismatch base paired nucleotide) for later corrections during replication, transcription and translation. It is also known as **editing.**

A DNA polymerase I (**Kornberg enzyme**) has both exonuclease activity (3′→5′ and 5′→3′) and polymerase activity (5′→3′).

(i) DNA Pol I removes RNA primer from **Okazaki fragments** (exonuclease activity) and fill the resultant gap (polymerase activity 5′-3′).

(ii) It also removes thymine dimers produced by U.V. radiation and fills the gap.

(iii) During leading strand synthesis, a non-complementary nucleotide is occasionally inserted erroneously. To compensate such accuracies, the mismatched nucleotide is removed by exonuclease activity (3′-5′) and put the proper (complementary) nucleotide and continuing the synthesis.

This is the proof-reading function of DNA polymerase.

Q. 8 Compare semiconservative, conservative and dispersive modes of DNA replication.

Solution:

Semiconservative	Conservative	Dispersive
(i) In this mode of replication each daughter duplex or DNA is hybrid, i.e., one old and one new strand.	(i) Here no hybrid is formed. Each duplex is pure either with two old or two new strands.	(i) Hybrid duplex, i.e., new and old DNA coexist in the same strand after replication.
(ii) Hybridization is 50 : 50, i.e., one strand in a duplex is pure either old or new.	(ii) No hybridization but each strand is either old or new, i.e., pure.	(ii) Different degree of hybridization. Each strand is hybrid, i.e., contain old and new DNA.

Q. 9 Describe the structure of DNA polymerase III in prokaryotic cell. Show with proper diagrams how the same DNA polymerase III enzyme carries out the polymerization of nucleotide on both strands of DNA in *E. coli.*

[M.Sc. (Zoo.) C.U. 2003]

Solution:

(a) (i) DNA polymerase III is considered to be the enzyme responsible for the polymerization essential for replication. It is a **multisubunit complex** or holo-enzyme having **ten subunits.**

(ii) It is the product of the gene polymerase C (also known as Dna ε gene).

(iii) Its molecular weight is 900,000 Da.

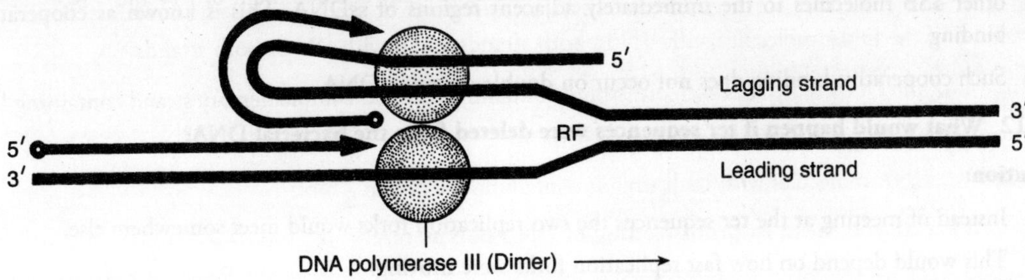

Fig. 13.2: Concurrent DNA synthesis both leading and lagging strands by DNA polymerase III.

(iv) Polymerase III — Core enzyme : (i) α, (ii) ε, (iii) θ
— Second subgroup : γ, δ, δ[1], χ and ψ.

Subunit	*Gene synthesis*	*Molecular wt* (Da)	*Function*	*Grouping*
α (alpha)	dna E	130,000	5′–3′ polymerization } Elongated polynucleotide chain	Core enzyme
ε (epsilon)	dna Q	27,500	3′–5′ Exonuclease } Elongated polynucleotide chain	Core enzyme
θ (theta)	hol E	10,000		Core enzyme
γ (gamma)	dna X	45,000	Serve as clamp loader	
δ (delta)	hol A	35,000	Serve as clamp loader	
δ[1] (delta prime)	hol B	33,000	Serve as clamp loader	
χ (chi)	hol C	15,000	Serve as clamp loader	
ψ (psi)	hol D	12,000	Serve as clamp loader	
β (beta)	dna N	40,000	Sliding clamp structure	
τ (tau)	dna X	11,000	dimerization of core	

Q. 10 What are the enzymes involved in DNA replication in *E. coli*?

Solution:

(i) The combination of all the proteins that function at the replication fork is called **replisome**, consists of a primosome (a primase and helicase) and two polymerase III holoenzymes form at Y-junction on DNA.

(ii) One polymerase acts processively synthesizing the **leading strand**, while the other forms **Okazaki fragments** initiated by primers created by primase.

(iii) **DNA polymerase I** completes Okazaki fragments eliminating the RNA primer of the previous Okazaki fragments and replacing it with DNA.

(iv) Finally, **DNA ligase** connects the fragments.

(v) **SSB proteins** and **topoisomerase** are involved. Initiation proteins bind at initiation site and termination proteins bind to terminator sites.

Q. 11 Why does the SSB proteins of *E. coli* bind so much better to single-stranded DNA (ssDNA) than to double-stranded DNA (dsDNA)?

Solution:

(i) SSB protein binds cooperatively to single-stranded DNA but not double-stranded. This means once one molecule of SSB binds to single-stranded DNA, it greatly promotes or facilitates the binding of other SSB molecules to the immediately adjacent regions of ssDNA. This is known as cooperative binding.

(ii) Such cooperative binding does not occur on double-stranded DNA.

Q. 12 What would happen if ter sequences were deleted from the bacterial DNA?

Solution:

(i) Instead of meeting at the **ter** sequences the two replication forks would meet somewhere else.

(ii) This would depend on how fast replication forks were moving.

(iii) If one fork was advancing faster than the other fork, they would meet closer to where the slower moving fork started.

- **Replisome:**

  It is a dynamic complex of enzymes and all the proteins that function at the replication fork during elongation.

Replisome component	Function during replication
• Helicase	• Unwinds DNA ahead of replication fork.
• Ligase	• Joins fragments of repaired lagging strand.
• DNA polymerase	• DNA synthesis, repair of gaps in lagging strand.
• Primase	• Synthesis of primer.
• RNase H	• Removes RNA primer from lagging strand.
• SSB protein	• Stabilizes ssDNA.

- **Replicon:**

  (i) A replicating genetic unit including a length of DNA and its site for initiation of replication, or it may be defined as all the DNA replicated from a particular origin as **replicon.**

  It has two controlling components, viz., **replicator** and **initiator.**

  (ii) The number of replicons may vary in a genome from one in bacteria (*E. coli*) to 500 in yeast to several thousand in plants and animals.

  (iii) *E. coli* has only one origin of replication, the entire chromosome is a single **replicon**. The genetic locus is **ori** C (245 bp).

  In contrast the presence of multiple origins of replication divides each eukaryotic chromosome into multiple replicons—one for each origin of replication.

- **Replicator:**

  (i) It is defined as an entire set of cis-acting DNA sequences that is sufficient to direct the initiation of DNA replication.

  (ii) This is in contrast to the origin of replication which is the site on DNA where the DNA is unwound and DNA synthesis initiates. Although the origin of replication is always a part of replicator.

- **Klenow fragment:** A fragment of DNA polymerase I. Created by cleaving with protease that lacks 5′–3′ exonuclease activity of parent strand.

- **Topoisomerase:**

A class of enzymes that convert DNA from one topological form to another. During replication these enzymes facilitates the unwinding of the double helical structure as well as relax the torsional strain generated by helicase during unwinding.

(i) It converts supercoiled to uncoiled and vice versa; positive to negative coil.

(ii) It is of two types: 1. **Topoisomerase I** cuts only one strand of DNA duplex by hydrolyzing a 3′-5′ phosphodiester bond and it converts negative supercoil to relaxed state. 2. **Topoisomerase II** cuts both strands of DNA molecule and converts positive to negative supercoil.

- **SSBP (Single Strand Binding Protein) HDP (Helix Destabilising Protein):**

(i) The newly generated single-stranded DNA by helicases, another kind of protein binds to this newly exposed single-stranded DNA to stabilize the separated strands and thereby prevents their reassociation. This protein is called **SSB protein**.

(ii) The SSBP is a tetramer and can wrap around 30-36 base length of single-stranded DNA in **co-operative manner.**

(iii) Binding of one SSBP promoting the binding of another SSBP to the immediately adjacent regions of ssDNA is called **cooperative binding**. SSBP molecules bound to immediately adjacent regions of ssDNA can also bind to each other. This strongly stabilizes the interaction of SSBP with ssDNA.

(iv) Cooperative binding ensures that ssDNA is rapidly coated by SSBP as it emerges from the DNA helicases.

(v) Once converted SSBP, ssDNA is held in an elongated state that facilitates its use as a **template** for DNA or **RNA primer** synthesis.

Its activity: (1) It stabilizes single-stranded DNA (ssDNA). (2) Inhibition of nuclease activity. (3) Low affinity to dsDNA.

Q. 13 Contrast the function of a helicase with that of a topoisomerase.

Helicase	Topoisomerase
(i) Helicase unwinds DNA molecule.	(i) Topoisomerase changes the topology of the DNA.
(ii) Here hydrogen bonds are broken.	(ii) Here the number of supercoils are increased or decreased.
(iii) Helicases do not break DNA strand.	(iii) Transient strand breakage is necessary part of topoisomerase.
(iv) It is ATP-dependent protein.	(iv) It is not so.

Q. 14 Compare lagging strand and leading strand.

Solution:

Points	Lagging strand	Leading strand
1. DNA synthesis	Replication occurs discontinuously	Replication occurs continuously
2. Movement of DNA polymerase III	Away from the fork in 5′→3′ direction	Moves towards the fork in 5′→3′ direction
3. Okazaki fragments grows	5′→3′ direction	No Okazaki fragments
4. Role of polymerase I & ligase	Play vital role	No role
5. Initiation of DNA synthesis	Many times	Once
6. RNA primer	Many times	Once

Q. 15 Differentiate between Unidirectional and Bidirectional replication.

Solution:

Unidirectional replication	Bidirectional replication
1. DNA replication proceeds in one direction from the point of origin.	1. DNA replication proceeds in two directions from the point of origin.
2. One of the two ends of replication eye or bubble will be stationary and other end will move with replication.	2. None of the two ends will be stationary and both will move with replication.
3. Autoradiographic studies show that labelling is seen only at one fork.	3. Labelling on both the replication forks.
4. Replication of DNA of coliphage p_2 and mitochondria are unidirectional.	4. Replication of DNA of both prokaryotes and eukaryotes are bidirectional.

Q. 16 Compare the properties of DNA polymerase I, II and III of *E. coli*.

Solution:

Character	Polymerase I	Polymerase II	Polymerase III
1. 5′-3′ polymerase activity	+	+	+
2. 3′-5′ exonuclease activity	+	+	+
3. 5′-3′ exonuclease activity	+	–	–
4. Structure	Polypeptide	Polypeptide	Multimeric complex
5. *E. coli* gene	Polymerase A	Polymerase B	dna E, dna N, dna θ, dna α and other subunit
6. Function	Principle repair, polymerase, primer excision	Error prone repair, polymerase (sos inducible)	Unassigned principle replication polymerase
7. Polymerase molecules/cell	400	?	15

Q. 17 How can DNA replication be both rapid and accurate?

Solution:

(i) Replication is rapid because DNA pol III remains attached to its template substrate (processive).

(ii) It is accurate because of the proof-reading activity of DNA pol III and multiple steps of leading- and lagging-strand synthesis, including the removal of misincorporated nucleotides, coordinated by **replisome**.

Q. 18 What is a processive enzyme? Explain why this is an important feature for DNA polymerase.

Solution:

A processive enzyme is one that remains clamped to one of its substrates.

(i) In the case of DNA polymerase, it remains clamped to the template strand as it makes new daughter strand.

(ii) It is important to ensure a faster rate of DNA synthesis.

Q. 19 If a eukaryotic chromosome has 25 origins of replication, how many replication forks does it have at the beginning of DNA replication?

Solution:

Two replication forks emanate from each origin of replication. Therefore, 50 replication forks will be formed.

Q. 20 Differentiate between Prokaryotic and Eukaryotic replications.

Solution:

Prokaryotic replication	Eukaryotic replication
1. Occurs continuously and rapidly.	1. Restricted to 's' phase of cell cycle.
2. Number of replication origin is one.	2. Number of replication origin is more than one, even thousands and developmentally regulated. Mammals have 50-100,000 replicons, each origin of which replicate 40–200 kb DNA.
3. When circular DNA is copied, replication begins at a single point and replication forks move around the circle, a theta(θ)-shaped structure is produced.	3. Replication fork proceeds in opposite direction from each origin until they reach the terminal point at which they meet the neighbouring replicon.
4. Primase generates a primer extended by polymerase III.	4. Primase generates a primer extended by pol α and completed by pol σ.
5. The length of RNA primer is more or less fifty nucleotides.	5. The length of RNA primer is nine nucleotides.
6. The length of Okazaki fragments of lagging strands is more or less 1000–2000 bases.	6. The length of Okazaki fragments of lagging strand is 100–200 base pairs.
7. Replication fork progresses rapidly, i.e., at the speed of 600 nucleotides per second (in *E. coli* 2.2 Mb/hr).	7. Each replication fork progresses slowly at a speed of 50 nucleotides per second (in *Drosophila* 18 Kb/hr).
8. Replicative polymerase-polymerase III holoenzyme.	8. Replicative polymerase pol δ/pol α.
9. Primer removal made by RNase H and polymerase I.	9. Primer removal occurs by RNase H_1 and MF-1 nuclease.
10. Lagging strand repaired by Pol I and ligase.	10. Lagging strand repaired by pol δ/pol ε and DNA ligase I.
11. Component of topoisomerase is DNA gyrase.	11. Component of topoisomerase is topoisomerase II.
12. Most prokaryotic chromosomes are circular and thus have no end.	12. It is linear and has unique terminal called telomere and replicates by telomerase.

Q. 21 Compare Continuous and Discontinuous replication.

Solution:

Continuous replication	Discontinuous replication
1. It occurs on leading strand of replication fork.	1. It occurs on lagging strand of replication fork.
2. It starts at the beginning of replication.	2. It starts somewhat later than that of leading strand.
3. Progress in one direction, i.e., towards the replication fork.	3. Progress in opposite direction, i.e., away from the replication fork.
4. Synthesis of DNA takes place continuously.	4. Short fragments (Okazaki fragments) are synthesized and subsequently they are joined, i.e., synthesis becomes continuous.
5. In vivo needs no RNA primer.	5. Needs RNA primer.
6. Only DNA polymerase III is involved.	6. Primase, DNA polymerase III, DNA polymerase I and DNA ligase are required.

Q. 22 Highlight unidirectional DNA replication.

Solution:

In unidirectional DNA replication, there is one replication origin and one growing fork but both the strands are being replicated. *Example*: Plasmid-ColEI.

(i) A **plasmid** is a circular piece of DNA that replicates independently of the cell's chromosome.

(ii) M. Lovett, L. Katz and D. R. Helsinki used some kinds of marker in DNA to locate the two forks at different times during replication.

(iii) If both forks move relative to the marker they will know that both forks are active and DNA replication is **bidirectional** and on the other hand, if only one moves, replication must be **unidirectional**.

(iv) The **marker** Lovett and his colleagues chose the site at which ColEI DNA is cut by the **restriction endonuclease** EcoRI. EcoRI recognizes the sequence 5′ GAATTC-3′. EcoRI has only one such recognition site in the whole ColEI molecule, so cutting ColEI DNA with EcoRI creates linear molecules whose ends are always at the same site (the EcoRI restriction site).

(v) Clearly, only one fork moves relative to the ends of the linearized DNA molecule. Therefore, the replication of ColEI DNA is **unidirectional**.

Q. 23 Notes on DNA polymerases in eukaryotic cells

Solution:

DNA polymerase	5′ → 3′ polymerase activity	3′ → 5′ exonuclease activity	Cellular function
α (alpha)	Yes	No	Initiation of nuclear DNA synthesis and DNA repair; has primase activity.
β (beta)	Yes	No	DNA repair and recombination of nuclear DNA.
γ (gamma)	Yes	Yes	Replication and repair of mitochondrial DNA.
δ (delta)	Yes	Yes	Leading- and lagging-strand synthesis of nuclear DNA, DNA repair, and translation DNA synthesis.
ε (epsilon)	Yes	Yes	Unknown; probably repair and replication of nuclear DNA.
ζ (zeta)	Yes	No	Translation DNA synthesis.
η (eta)	Yes	No	Translation DNA synthesis.
θ (theta)	Yes	No	DNA repair.
ι (iota)	Yes	No	Translation DNA synthesis.
κ (kappa)	Yes	No	Translation DNA synthesis.
λ (lambda)	Yes	No	DNA repair.
μ (mu)	Yes	No	DNA repair.
σ (sigma)	Yes	No	Nuclear DNA replication (possibly), DNA repair, and sister-chromatid cohesion.

Q. 24 State briefly about Rolling circle model of DNA replication:

Solution:

A model of DNA replication in which the growing point or replication fork rolls around a circular template strand and the newly synthesized strand displaces the strand from the previous replication, producing a series of contiguous copies of the template strand.

Example: (i) Replication of single-strand DNA virus $\phi \times 174$

(ii) Transfer of *E. coli* sex factor (F-factor).

- **Steps of replication:**

(a) Conversion of positive (+) strand into double helical form (replicative form or RF).

(b) Replication of RF.

(c) Synthesis (production) of positive (+) strand from RF.

(a) **Conversion of positive (+) strand into double helical form (RF):**

(i) Single-stranded parental DNA (positive strand) of $\phi \times 174$ is converted into duplex circle by the synthesis of complementary negative (–) strand. This duplex circle becomes **supercoiled.**

(ii) The double-stranded [one parental positive (+) and one daughter negative (–) strand] circular form of DNA is known as **replicative form (RF).**

(iii) The negative strand is synthesized with the help of **primosome** (primase + six priming proteins) and **DNA Pol. III.**

(b) **Replication of RF:**

(i) An endonuclease (nickase) encoded by $\phi \times 174$ gene A protein creates a nick in the positive strand of the double-stranded replicative form.

(ii) One strand becomes a polynucleotide with two ends, one end having 3′ OH group, the other having 5′ PO_4 group where the gene A protein covalently linked.

(iii) The free 3′ end created by the nick serves as the primer for the positive (+) strand elongation, as the other end of the positive

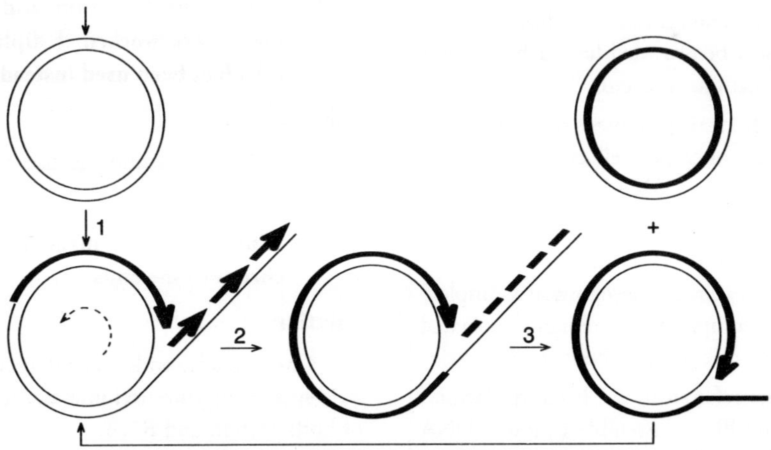

Fig. 13.3: Rolling circle replication

A sequence-specific cut (↓) generates a nicked molecule.

1. The polymerase elongates the chain from the 3′ -end of the nick. The 5′ -end of the parental strand unrolls. This is also copied and Okazaki fragments grow.
2. As the circle rolls one full turn, a full length complementary strand (—) forms and is liberated by a sequence-specific cut (↑) in 3.
3. This molecule has a nick and can re-enter the pathway by providing a 3′ -terminus for further extension by polymerase.

The lagging strand following ligation yields a daughter double-stranded molecule.

strand is displaced. The negative strand is the template.

(iv) Replication starts by adding bases at 3′ OH end by displacing the 5′ end which rolls out as free tail and is then copied in 5′-3′ direction in small segment.

(v) When a complete positive strand has been synthesized, a new origin is produced between the old and new strands. The Gene A protein cleaves the new origin and ligates the 3′ and 5′ ends of the old positive strand to form close circular structure.

(c) **Production of positive (+) strand from the RF:**

(i) When the positive (+) strand reaches double length, the circle can be considered to be rolling counter-clockwise and trailing out the progeny single-stranded DNA.

(ii) The single-stranded positive (+) strand is packaged into the phage again.

Notes on Licensing factor:

(i) A licensing factor is something in the nucleus, that is necessary for replication and is inactivated or destroyed after one round of replication. New licensing factors must be provided for further rounds of replications to occur.

(ii) The eukaryotic genome is replicated once and only once per cell cycle. Therefore, one segment of DNA has been replicated in **S-phase**.

(iii) The origin of replication is bound by an **ORC** (Origin Recognition Complex) protein complex which remains associated with the origin throughout the cell cycle.

(iv) Proteins referred to as "**licensing factors**" bind to ORC to assemble a protein-DNA complex called **prereplication complex** (Pr-RC) that is licensed to initiate replication.

(v) A set of six related proteins **Mcm proteins** (Mcm 2 – Mcm 7) are loaded on the origin of replication complex (**ORC**) at the late stage of mitosis or soon after mitosis. The Mcm protein acts as licensing factors that allow replication to initiate.

Q. 25 Why the proteins that replicate DNA are called biological machine?

Solution:

Many of the proteins taking part in DNA replication, interact through protein-protein contacts and are assembled only where (at the origins) and when needed to coordinate a complex biological process.

Q. 26 What would happen if during replication the topoisomerases were unable to reattach the DNA fragments of each strand after unwinding (relaxing) the DNA molecule?

Solution:

The chromosomes would become hopelessly fragmented.

Q. 27 If thymine makes up 15 percent of the bases in a specific DNA molecule, what percentage of the bases is cytosine?

Solution:

If the DNA is double-stranded, A = T, G ≡ C and A + T + G + C = 100. If T = 15%, A also 15%, then G + C = [100 – (15 × 2)] = 70% or C = 35%.

Q. 28 Would the Meselson and Stahl experiment have worked if diploid eukaryotic cells had been used instead?

Solution:

Yes, DNA replication is also semiconservative in diploid eukaryotes.

Q. 29 What special mechanism replicates chromosome ends?

Solution:

Chromosome ends, i.e., telomeres are extended by a special enzymes (**telomerase**) that is composed of both protein and RNA.

Q. 30 State briefly about telomerase.

Solution:

(i) Telomeric DNA consists of **TTAGGG** repeats. The 3′ end of the parental DNA is extended by DNA synthesis using RNA component (**TERC**) of telomerase as a template.

(ii) In vertebrate, TERC has a tandem repeat (**AAUCCCAU**) near its 5′ end that contains a complementary template sequence for guiding the synthesis of additional telomeric hexa-nucleotide sequences at the terminal end.

(iii) Therefore, the leading strand is extended by the repeats **(TTAGGG)n**. By extending the leading strand, the telomerase provides a template for the lagging strand to extend (by a standard DNA-dependent RNA polymerase).

(iv) After finishing the addition of hexanucleotides repeats by the telomerase, the lagging strand cannot be extended to the extreme 5′ end. As a result the G-rich strand is left with an overhanging 3′ end close to 200 nucleotides long.

Q. 31 State briefly the role of telomerase in telomeric DNA replication.

or

How does telomerase resolve the terminal primer problem?

Solution:

(i) Telomerase is a non-typical reverse transcriptase with two components, viz., Protein and RNA.

(ii) Its RNA components **TERC** (Telomerase RNA component) and a protein component **TERT** (Telomerase Reverse Transcriptase).

(iii) Telomerase uses the 3′OH of the G + T telomeric strand as a primer for synthesis of tandem **TTGGGG** repeats. Ony d GTP and d TTP are needed for the activity.

(iv) At the 5′ end TERC RNA is a hexanucleotide sequence which consists of a template RNA (coded by TLC-1) and a catalytic protein (EST-2).

(v) The short RNA component includes a sequence of 15-22 bases pair with C+A rich repeating sequence. The telomerase acts discontinuously. The template RNA is positioned on the DNA primer and extends the telomeric DNA synthesis by adding several nuceotides on the leading strand. Then enzyme translocates to begin again.

(vi) One of the protein subunits, i.e., **reverse transcriptase** acts as RNA template to synthesize G+T rich repeating sequences. For further extension of leading strand the necessary template provided from DNA polymerase to complete the DNA synthesis of the lagging strand.

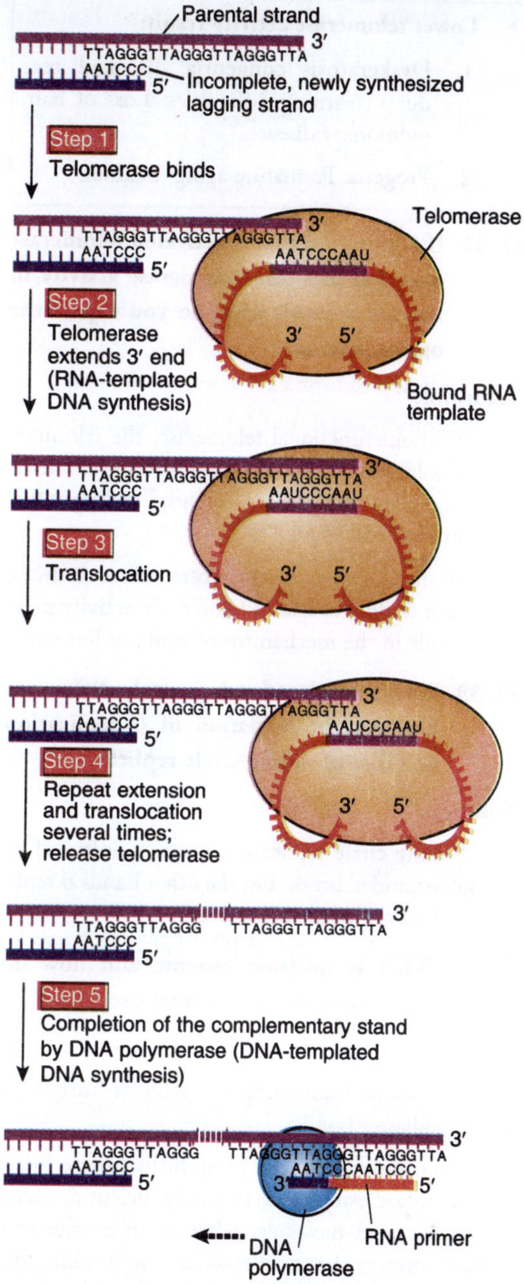

Fig. 13.4: Telomere DNA replication by telomerase

- **Lower telomerase activity result:**
  1. **Dyskeratosis congentia:** Blocked tear ducts, tearning difficulties. Loss of hair, pulmonary disease.
  2. **Progeria:** Premature aging: baldness.

Q. 32 If a mutation that inactivated telomerase occurred in a cell (telomerase activity in the cell = zero), what do you expect the outcome to be?

Solution:

(i) Without functional telomerase, the telomeres would shorten at each replication cycle, leading to eventual loss of essential coding information and death.

(ii) In fact, some current observations indicate that decline or loss of telomerase activity plays a role in the mechanism of aging in humans.

Q. 33 What is the fundamental difference between the initiation of θ replication and that of rolling circle replication?

Solution:

Rolling circle replication must be initiated by a single-stranded break. On the other hands θ replication does not need such a break.

Q. 34 What is nuclease enzyme and how do endonucleases and exonucleases differ?

Solution:

A nuclease is an enzyme capable of breaking a phosphodiester bond.

An endonuclease can break or hydrolyze internal phosphodiester bonds in a polynucleotide chain or nucleic acid molecule, whereas an exonuclease breaks down nucleic acid molecule by breaking the phosphodiester bonds at 3′ or 5′ terminal nucleotides.

Q. 35 What is the chemical difference between the groups joined by DNA polymerase and DNA ligase?

Solution:

(i) DNA polymerase joins a 5′ triphosphate with a 3′ OH group and the outermost two phosphates are released.

(ii) The resulting phosphodiester bond contains one phosphate. Thus, it catalyzes the synthesis of DNA from deoxyribonucleotide and a template DNA.

(iii) DNA ligase forms a covalent bond between the 5′ end of one polynucleotide chain and the 3′ end of the another polynucleotide chain.

(iv) It is also called polynucleotide joining enzyme.

- Origin of replication tend to be rich in AT base pairs.
- A bacterial chromosome contains a single origin which is responsible for initiating replication once in every division cycle.
- The ori C origin in *E. coli* is a sequence of 245 bp. Any DNA molecule with this sequence can replicate in *E. coli*.
- Replication of the circular bacterial chromosome produces θ structure, in which the replicated DNA starts out as a small replicating eye. Replication proceeds until the eye occupies the whole chromosome.
- The bacterial origin contains sequences that are methylated on both strands. Replication produces hemimethylated DNA, which can not function as an origin.
- There is delay before hemimethylated origins are remethylated to convert them to the functional state, and this is responsible for preventing initiation from occuring too frequently.
- In eukaryotic chromosome, each replicon is activated once and only once in a single division cycle.
- Origin of replications are isolated as **ARS** sequence in yeast by virtue of their ability to support replication.
- The core of **ARS** is an 11 bp AT rich sequence and is bounded by the **ORC** protein complex, which remains bounded throughout the cell cycle.
- It is controlled by **MCM** licensing factors that are associated with **ORC**.

Q. 36 Why can RNA polymerase and primase initiate DNA replication, whereas DNA polymerase cannot?

Solution:

Neither RNA polymerase nor primase requires a primer which provides the free 3′ OH group that DNA polymerase needs in order to initiate replication.

Q. 37 A diploid organism has 4.5×10^8 bp in its DNA. The DNA is replicated in 3 minutes. Assuming that all replication forks move at a rate of 10^4 bp per minute, how many replicons (replication unit) are present in the organism's genome?

Solution:

(i) Since a replication fork moves at a rate of 10^4 bp per minute and each replication has two replication forks moving in opposite directions.

(ii) In one replicon, replication occurs at a rate of 2×10^4 bp/minute.

(iii) Assume for the purposes of this calculation only that DNA replication is distributed among similarly sized replicons initiating replication at the same time.

The diploid organism has 4.5×10^8 bp in its DNA. Since all the DNA replicates in 3 minutes therefore in one minute it replicates $\frac{4.5\times10^8}{3}$ bp.

2×10^4 bp = 1 replicon

1 bp = $\frac{1}{2\times10^4}$ replicon

$\frac{4.5\times10^8}{3} = \frac{4.5\times10^8}{3} \times \frac{1}{2\times10^4} = \frac{4.5\times10^8}{6}$ replicon

= 7500 replicon.

So the number of replicon in diploid genome is 7500 replicon.

Q. 38 A space probe returns from Jupiter and brings with it a new microorganism for study. It has double-stranded DNA as its genetic material. However, studies of the replication of the alien DNA reveal that, although the process is semiconservative, DNA synthesis is continuous on both the leading strand and lagging strand templates. What conclusions can you draw from this result?

Solution:

(i) DNA replication in Jovian bug (microorganism) does not occur as it does in *E. coli*. Assume that the double-stranded DNA is antiparallel as it is in *E. coli*.

(ii) The Jovian DNA polymerase must be able to synthesize DNA in the 5′-3′ direction (on the leading strand) as well as in the 3′–5′ direction (on the lagging strand).

(iii) This is unlike as DNA polymerase on earth.

Q. 39 Why DNA synthesis expected to be more complex in eukaryotes than bacteria?

Solution:

(i) There is much greater amount of DNA to be replicated and replication is slower.

(ii) There are multiple initiation sites for replication in eukaryotes in contrast to the single replication origin in prokaryotes.

(iii) Replication occurs at different sites during different intervals of the S phase.

Q. 40 Suppose that *E. coli* synthesizes DNA at a rate of 100,000 nucleotides per minute and takes 40 minutes to replicate its chromosome.

(a) How many base pairs are present in the entire *E. coli* chromosome?

(b) What is the physical length of the chromosome in its helical configuration—that is, what is the circumference of the chromosome if it were opened into a circle?

Solution:

(a) In *E. coli* 100,000 bp or 100 kb are added to each growing chain per minute. In 40 minutes the chain should be about 40 × 100 kb or 40 × 100,000 bp = 4,000,000 bp or, 4×10^6 bp..

(b) Distance between the nucleotide pair (bp) is 3.4Å or 0.34 nm. Therefore, the length of the chromosome will be 4×10^6 bp × 0.34 nm/bp = 1.36×10^6 nm or 1.3 mm.

Q. 41 What must be done to two precursor fragments before they can be joined together?

Solution:

The RNA primer must be removed from the precursor fragment that was made first, because DNA ligase cannot join DNA to RNA.

Q. 42 Consider a hypothetical phage whose DNA replicates exclusively by rolling circle replication. A phage with radioactive DNA in both strands infects a bacterium and is allowed to replicate in a nonradioactive medium. Assume that only a daughter DNA from the elongating branch ever gets packaged into progeny phage particles.

(a) What fraction of parental radioactivity will appear in progeny phage?

(b) How many progeny phages will contain radioactive DNA and how will the number be affected by recombination?

Solution:

(a) (i) In rolling circle replication, one parental strand remains in the circular part and the other is at the terminus of the branch.

(ii) Therefore, only half of the parental radioactivity will appear in the progeny.

(b) (i) If we assume no recombination between progeny DNA molecules, one progeny phage will be radioactive.

(ii) If a single recombination occurs, two particles will be radioactive.

(iii) If recombination is frequent and occurs at random, the radioactivity will be distributed among virtually all the progeny.

Q. 43 What experimental techniques can be used to separate DNA molecules of mass 3 × 10^7 daltons isolated from Lamda bacteriophage and DNA molecules of mass 1.3 × 10^8 daltons isolated from T_2 bacteriophage?

Solution:

(i) Sucrose velocity density gradient centrifugation is the standard technique for separating DNA molecules in this size range.

(ii) Pulsefield gel electrophoresis could also be used.

Q. 44 Why must each of the giant DNA molecules in eukaryotic chromosomes contain multiple origin of replication?

Solution:

(i) In eukaryotes, the rate of DNA synthesis at each replication fork is about 2500 to 3000 nucleotide pairs per minute.

(ii) Large eukaryotic chromosome often contains 10^7 to 10^8 nucleotide pairs.

(iii) A single replication fork could not replicate the giant DNA in one of these large chromosomes fast enough to permit the observed cell generation time.

Q. 45 Pol A mutant of *E. coli* lack the 3′–5′ exonuclease activity of DNA polymerase I. Will the rate of DNA synthesis be altered in these mutants? What effect(s) will these Pol A mutants have on the phenotype of the organism?

Solution:

No, *E. coli* strain carrying Pol A mutation that eliminate the 3′–5′ exonuclease activity of DNA; polymerase I will exhibit usually high mutant rate.

Q. 46 (a) Why isn't DNA primase activity required to initiate rolling circle replication?

(b) DNA primase is required for the discontinuous synthesis of the lagging strand, which occurs on the single-stranded tail of the rolling circle. Why?

Solution:

(a) (i) Rolling circle replication begins when an endonuclease cleaves one strand of a circular DNA double helix.

(ii) This cleavage produces a free 3′-OH on one end of the cut strand, allowing it to function as primer.

(b) The discontinuous synthesis of the lagging strand requires the de novo initiation of each Okazaki fragment, which requires DNA primase activity.

Q. 47 Consider a long linear DNA molecule, one end of which rotated four times with respect to the other end in the unwinding direction.

(a) If the two ends are joined to keep the molecule in the unwound state, how many base pairs will be broken?

(b) If the underwound molecule is allowed to form a supercoil, how many twists will be present?

Solution:

(a) There are ten base pairs for every turn of the double helix. With four turns of unwinding $4 \times 10 = 40$ base pairs will be broken.

(b) Each twist compensates for the under-winding of full turn of the helix; hence there will be four twists.

Q. 48 Endonuclease SI can break single-stranded DNA but does not break double-stranded linear DNA. However, SI can cleave supercoiled DNA, usually making a single break. Why does this occur?

Solution:

(i) The supercoiled form is in equilibrium with an underwound form having many unpaired bases.

(ii) At the instant that a segment is single-stranded, SI can attack it. Because a nicked circle lacks the strain of supercoiling, there is no tendency to have single-stranded regions, and so a nicked circle is resistant to SI cleavage.

Q. 49 Circular DNA molecules of the same size migrate at different rates in electrophoretic gels, depending on whether they are supercoiled or relaxed circles. Why?

Solution:

(i) The rate of movement in the gel is determined by the overall charge and by the ability of the molecule to penetrate the pores of the gel.

(ii) Supercoiled and relaxed circles migrate at different rates because they have different conformations.

Q. 50 The dna A gene product of *E. coli* is required for the initiation of DNA synthesis at ori C. What is its function? How do we know that Dna A protein is essential to the initiation process?

Solution:

(i) Dna A protein initiates the formation of the replication bubble by binding to the 9-bp repeats of ori C.

(ii) Dna A protein is known to be required for the initiation process because bacteria with temperature-sensitive mutations in the dna A gene cannot initiate DNA replication at restrictive temperature.

Q. 51 In the yeast *S. cerevisiae*, haploid cells carrying a mutation called est I (for ever-shorter telomere) lose distal telomere sequences during each cell division. Predict the ultimate phenotypic effect of this mutation on the progeny of these cells.

Solution:

(i) The chromosome of haploid yeast cells that carry the est I mutation become shorter during each cell division.

(ii) Eventually, chromosome instability results from the complete loss of telomeres and cell death occurs because of deletion of essential genes near the ends of the chromosomes.

Q. 52 Why DNA polymerases are only able to synthesize DNA in the presence of both a template strand and a primer strand? What are the functions of these two strands?

Solution:

(i) DNA polymerase can only extend DNA chains with a free 3′-OH, because the mechanism of

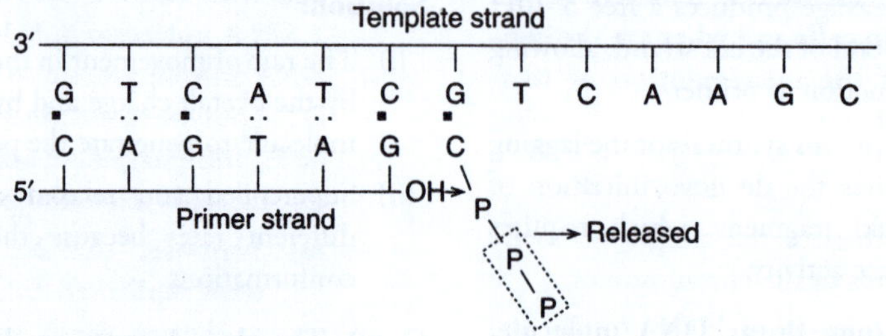

extensions involves a nucleophilic attack by the 3′-OH on the interior phosphorus of the deoxyribonucleoside triphosphate precursor with the elimination of pyrophosphate.

(ii) The strand with 3′-OH is the primer strand, it is extended during synthesis.

The **template strand** specifies the nucleotide sequence of the strand being synthesized; the new strand will be complementary to the template strand.

Q. 53 DNA from species A labelled with ^{14}N and randomly fragmented, is renatured with an equal concentration of DNA from species B, labelled with ^{15}N and randomly fragmented and then centrifuged to equilibrium in CsCl. Five percent of the total renatured DNA has a hybrid density. What fraction of the base sequences is common between the two species?

Solution:

(i) The hybrid molecules that can be produced from single-stranded fragments from species A and B are AA, AB, BA and BB, and these are expected equal amounts.

(ii) The molecules with hybrid ($^{14}N/^{15}N$) density are AB and BA, and these account for 5 percent of the total.

(iii) The same sequences can also renature as AA or BB which together must make up another 5 percent, so a total of 10 percent of the base sequences are common between the species.

Q. 54 How can autoradiography be used to distinguish between uni- and bidirectional replication of DNA?

Solution:

(i) If cells are grown in medium containing ^3H thymidine for a short period of time and are then transferred to nonradioactive medium for further growth (a pulse-chase experiment).

(ii) Uni- and bidirectional replication predict different labelling patterns and these patterns can be distinguished by autoradiography.

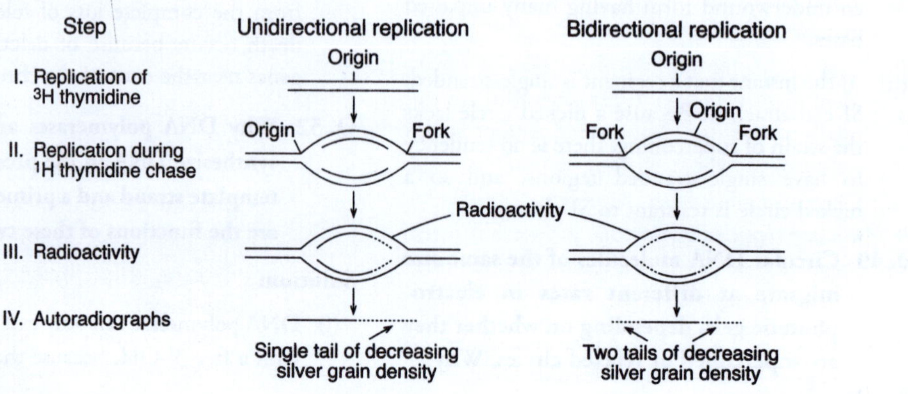

Q. 55 What is the evidence that telomere length is regulated in cells and what are the consequences of the misregulation of telomeric length?

Solution:

(i) Telomerase synthesizes the simple telomeric repeats at the ends of the chromosomes.

(ii) The enzyme made up of both protein and RNA and the RNA component has a base sequence that is complementary to the telomere repeat unit.

(iii) The RNA component is used as template for the telomere repeat, so that if the RNA components were altered, the telomere repeat would be as well. Thus, the mutant in this question is likely to have an altered RNA component.

Q. 56 Autoradiography is a technique that allows radioactive areas of chromosomes to be observed under microscope. The slide is covered with a photographic emulsion, which is exposed by radioactive decay. In the regions of exposure, the emulsion forms silver grains on being developed. The tiny silver grains can be seen on top of the (much larger) chromosomes. Devise a method to find out which regions in the human karyotype replicate during the last 30 minutes of the S phase. (Assume a cell cycle in which the cell spends 10 hours in G_1, 9 hours in S and 4 hours in G_2 and 1 hour in M.)

Solution:

(i) Assuming cells spend 4 hours in G_2, there are 4.5 hours from the last 30 minutes of S to metaphase in M.

(ii) Late-replicating chromosomal regions can be identified by adding ^{3}H thymidine to the medium waiting 4.5 hours and then preparing a slide of metaphase chromosomes.

(iii) Chromosomal regions displaying silver grains are late replicating, as cells that were at earlier stages of S when ^{3}H was added will be unable to reach metaphase in 4.5 hours.

Q. 57 M-13 of an *E. coli* bacteriophage whose capsid holds a closed circular DNA molecule with 2.221 T, 1,296 C, 1,315 G, and 1,575 A nucleotides. M13 lacks a gene for DNA polymerase and so must use bacterial DNA polymerases for replication. Unlike λ or T_4, this phage does not form concatamer during replication and packaging.

(a) Suppose, the M13 chromosomes were replicated in a manner similar to the way the *E. coli* chromosome is replicated, using semidiscontinuous replication mechanism.

(b) Suppose, the M13 chromosomes were replicated in a manner similar to the way the λ chromosome is replicated, using rolling circle replication. How would the rolling circle replication mechanism need to be modified?

Solution:

(a) (i) Since M13 has a closed circular genome with (A) ≠ (T) and (G) ≠ (C), it must have a single-stranded DNA genome. Bidirectional replication would require the initial synthesis of a complementary strand.

(ii) To produce many phages, many rounds bidirectional replication would be necessary. However, upon completion of replication and prior to packaging, the nongenomic strand of the resulting double-stranded molecules would need to be selectively degraded.

(b) (i) To produce single-stranded molecules with the same sequence and base composition as the packaged M13 genome, rolling circle replication must use a complementary template.

(ii) Therefore, DNA polymerase must initially synthesize the genome complementary strand to make a double-stranded molecule; then, a nuclease could nick the genomic strand to create a displacement fork.

(iii) Continuous rolling circle replication using the intact complementary strand as the leading strand template and without discontinuous replication on the displaced genomic strand will generate single-stranded M13 genome.

(iv) To prevent octamer formation; the newly replicated DNA must be cleaved by an endonuclease after exactly one genome has been replicated. To form a closed circle, the molecule's ends would need to be ligated.

Q. 58 In 1994, telomerase activity was discovered in human cancer cells in culture (in vitro). Although telomerase is not active in human somatic tissue, this discovery indicated that humans do contain the genes for telomerase protein and telomerase RNA. Since inappropriate activation of telomerase can cause cancer, why do you think the genes coding for this enzyme have been maintained in the human genome throughout evolution? Are there any types of body cells where telomerase activation would be advantageous or even necessary? Explain.

Solution:

(i) Telomerase activity is present in germ-line tissue to maintain telomere length from one generation to the next.

(ii) In other words, telomeres cannot shorten indefinitely without eventually eroding genetic information.

Q. 59 A researcher completely labels the chromosomes of cells grown in tissue culture with tritiated thymidine that is incorporated into newly synthesized DNA. The cells are then placed into a tritiated thymidine-free colchicine solution and are allowed to continue cell division. The colchicine prevents chromosome segregation by interfering with spindle fibre formation.

(a) Discounting sister chromatid exchanges, what proportion of the total number of chromatids would be labelled in prophase of the first cell cycle following placement into the tritium-free colchicines solution?

(b) Prophase of the second cell cycle?

(c) Prophase of the fourth cell cycle?

Solution:

(a) All of the chromatids would be labelled.

(b) Only $\frac{1}{2}$ of the chromatids because semi-conservative replication.

(c) $\frac{1}{8}$; the proportion of chromatids labelled will decrease by $\frac{1}{2}$ for each replication.

Q. 60 What is the chemical difference between the groups joined by DNA polymerase and DNA ligase?

Solution:

(i) DNA polymerase joins a 5′ triphosphate with a 3′OH group and the outermost two phosphates are released, the resulting phosphodiester bond contains one phosphate.

(ii) DNA ligase joins a single 5′ phosphate group with a 3′OH group.

Q. 61 State briefly the role of DNA ligase.

Solution:

After elimination of RNA primer a nick remains between the two DNA fragments. This gap is joined by DNA ligase.

(i) DNA ligase catalyzes the covalent closure of nicks in DNA molecule by using energy form NAD (Nicotinamide adenine dinucleotide) or ATP (Adenosine triphosphate).

(ii) The AMP of ligase-AMP intermediate forms a phosphodiester linkage with 5′ phosphate at the nicks.

(iii) The 3′OH group then attacks this phosphate and displaces AMP leading to the formation of phosphodiester bond and nick is then sealed.

(iv) The DNA ligase alone has no activity at breaks in DNA where one or more nucleotides are missing—so called gaps. The gaps can be filled by the combined action of DNA polymerase and ligase.

Q. 62 Mention the difference in the function of ligase in *E. coli* and T_4.

Solution:

The *E. coli* and T_4 ligases share the property of

sealing nicks that have 3′OH and 5′ phosphate termini.

(i) Both enzymes undertake a two-step reaction involving an enzyme-AMP complex.

(ii) The *E. coli* enzymes uses NAD as cofactor while T_4 enzyme uses ATP.

(iii) The AMP of the enzyme complex becomes attached to the 5′ phosphate of the nick and then phosphodiester bond is formed with the 3′OH terminus of the nick, releasing the enzyme and the AMP.

Q. 63 Discuss the similarities and differences in the synthesis of DNA in the lagging and leading strands. What is the advantage of a primosome and a replisome as opposed to having all replication enzymes functioning independently of each other?

Solution:

The leading strand is primed once at the origin and then DNA polymerase III synthesizes DNA continuously in the direction of the replication fork.

In the lagging strand short piece of DNA (**Okazaki fragments**) are made. This requires many RNA primers. The primers are removed by DNA polymerase I, which then fills the gaps with DNA. DNA ligase then covalently connect the okazoki fragments.

Having the enzymes within a complex such as **primosome** or **replisome** provides coordination among the different steps in the replication process and thereby allows it to proceed faster and more efficiently.

Q. 64 If a strain of *E. coli* over produced the Dam (DNA adenine methyl transferase) enzyme, how would affect such a strain to have more or less chromosome per cells compared to a normal strain of *E. coli*? Explain why?

Solution:

(i) If a strain of *E. coli* overproduced the Dam enzyme, DNA would replicate more rapidly.

(ii) The GATC methylation sites in the origin of replication (ori C) have to be fully methylated for DNA replication to occur.

(iii) The Dam enzyme recognizes the 5′-GATC-3′ sequence, binds there and covalently attaches the methyl group on to the adenine base, forming **methyl adenine.**

(iv) Immediately after DNA replication, a delay occurs before the next round of DNA replication, because the two copies of newly replicated DNA are **hemimethylated.**

(v) A strain that overproduced Dam would rapidly convert the hemimethyated DNA into fully methylated DNA and more quickly allow the next round of DNA replication to occur.

(vi) For this reason, the overproducing strain might have more copies of the *E. coli* chromosome because it would not have a long delay in DNA replication.

Q. 65 What would happen to DNA replication in a mutant strain of *E. coli* that had a mutation in the gene for DNA ligase, such that the enzyme was functional at 37 °C and non-functional at 42 °C.

Solution:

(i) DNA replication could occur normally at 37 °C but not at 42 °C.

(ii) At the higher temperature; the mutant would not be able to grow.
It would be lethal.

Q. 66 How long would it take *E. coli* to replicate 10^5 base pair of DNA?

Solution:

E. coli DNA elongates at the rate of about 1000 nucleotides per second per fork. Since the DNA replication is bidirectional and two forks are used, the effective rate of elongation is 1000 × 2 = 2000 nucleotides per second. If there were no pause, 10^6 base pairs could replicate in $\frac{10^6}{2000} = 50$ seconds.

Q. 67 How long would it take *E. coli* to replicate its entire genomes (4.2 × 10^6 bp)?

Solution:

(i) *E. coli* DNA elongates at the rate of about 1000 bp per second per fork. Since the DNA replication is bidirection and two forks are used, the effective rate of elongation is 2000 bp per second.

(ii) If there is no pause 4.2×10^6 bp could replicate in $\frac{4.2 \times 10^6}{2000}$ = 210 seconds = 35 minutes.

Q. 68 Chromosome 4 of the chicken (*Gallus gallus*) contains 2 × 10⁸ bp. How long would it take to replicate this molecule if replication starts at one end and proceeds to the other, assuming a typical eukaryotic rate of DNA synthesis is 50 bp per second?

Solution:

It will take $\frac{2 \times 10^8}{50} = 4 \times 10^6 = \frac{4 \times 10^6}{60 \times 6\ 0 \times 24}$ = 46.29 days = 46.3 days.

Q. 69 A strain of *E. coli* has circular genome of 4.3 × 10⁶ bp length. Replication starts at one location in the genome and proceeds in both directions. Under one particular set of growth conditions, 800 bp of DNA are formed per second at each replication.

(a) At how many revolutions per minute (rpm) is each replication fork rotating?

(b) How long does take to replicate the entire *E. coli* genome under these conditions.

Solution:

(a) (i) In each complete 360 degree turn of the double helix there are 10 bp of DNA.

(ii) Hence, 800 bp of DNA correspond, to $\frac{800}{10}$ = 80 complete turns. So each replication fork makes 80 revolutions per second and hence, it will be 80 × 60 = 4800 revolutions per minute.

(b) (i) *E. coli* DNA elongates at the rate of 800 bp per second per fork. As replication is bidirectional and two forks are used, the effective rate of replication would be 2 × 800 = 1600 bp per second.

(ii) So it will take $\frac{4.3 \times 10^6}{1600} = \frac{4300000}{1600}$ = 2687.5 seconds = $\frac{26875}{60}$ = 44.79 minutes.

Processive enzyme and processivity

An enzyme which glides along DNA and does not dissociate from the DNA template stand as it catalyzes the covalent attachment of nucleotide.

The tendency of an enzyme to remain bound to one or more of its substrate and repetitively continue its catalytic function without dissociation from its substrate.

PCNA (Proliferating cell nuclear antigen) and mention its function

(i) PCNA is a DNA clamp that act as a processivity factor for DNA polymerase δ in eukaryotic cells and is essential for replication.

(ii) Rf-C (Replicate factor C) is required for PCNA to load on to DNA.

(iii) RFC induces conformational change of PCNA that allows it to encircle.

DNA polymerase switch

It refers to a replication factor C dependent process which provides transition from initiation to elongation of DNA replication by substitution of Pol α primase by DNA polymerase δ.

B. TRANSCRIPTION

- **LEADER AND TRAILER SEQUENCE:**

(a) (i) The portion of mRNA molecule from the 5′ end to the initiation codon AUG is called **leader sequence.**

(ii) It is about 100 bp and has three regions, middle region is 20–25 bp and highly conserved and known **shine dalgarno sequence** (SDS), i.e., ribosome binding sites.

(b) (i) The length of messenger RNA (mRNA) from the nonsense codon (UAA, UGA, UAG) to the 3′ end is called **trailer sequence.**

- **UPSTREAM AND DOWNSTREAM:**

It is a convention on DNA related to the position and direction of transcription by RNA polymerase (5′ → 3′). A particular base sequence from where transcription begins is fixed and is mostly purine (A-adenine) base, i.e., AUG or GUG. It is called **zero point.**

- **Upstream** is the direction (5′) from which RNA polymerase has come, i.e., sequence preceding the start site (left side of the zero point) is designated by negative value (–1).
- **Downstream** is the direction (3′) of transcription, i.e., right side of the zero point and is designated as (+1).

- **SHINE DALGARNO SEQUENCE:**

(i) The nucleotides **AGGAGGU** present in the leader sequence of prokaryotic genes that serve as a ribosome binding site.

(ii) The sequence (**AGGAGGU**) of complementary between the messenger RNA and the 16S ribosomal RNA is referred to as **shine Dalgarno sequence** (SDS).

- **SENSE STRAND/ANTITEMPLATE STRAND/CODING STRAND:**

In DNA duplex, the strand that has the same polarity or sense as mRNA transcript, i.e., it bears the same sequence as the mRNA (except for possessing T instead of U) is called **coding or sense strand or antitemplate strand.**

- **ANTISENSE STRAND/TEMPLATE STRAND/NONCODING STRAND:**

(i) In DNA duplex, the strand which directs the synthesis of the mRNA via complementary base pairing is called the **template strand** or **antisense** or **noncoding strand.**

(ii) During transcription it produces sense mRNA.

- **PRIBNOW BOX AND GOLDBERG HOGNESS BOX (TATA BOX):**

A DNA region where RNA polymerase binds to initiate transcription, i.e., the regulatory site of DNA, which have two conserved or, almost unchanged sequences –35 region (bases) and –10 bases upstream from initiating site with 5′ **TATAAT** 3′ in prokaryote is called **Pribnow box** (Pribnow, 1975) for fixing RNA polymerase to select:

(i) Precise nucleotide at which transcription will begin,

(ii) Orient polymerase for transcription,

(iii) Unwinding of DNA at TATA box,

(iv) Antisense strand.

In most eukaryotes a consensus sequence comparable to –10 region in prokaryotes which is rich in adenine and thymine residues about seven base sequences 5′ TATAATA 3′ located –25 on the promoter DNA is called **Hogness box** or **TATA box** (D. Hogness).

- **RNA SPLICING:**

This is a phenomenon of nuclease cleavage of introns and ligation of adjacent exons to generate a contiguous molecule, i.e., a mature mRNA consisting only of exons. Since the first and last exons of primary transcript become the 5′ and 3′ ends of the mRNA, all intervening Introns are spliced out.

Characteristics:

(i) The splicings of pre-mRNA molecules occur exclusively in the nucleus.

(ii) The exon-intron junctions of mRNAs show specific sequences that are highly conserved. These sequences are functional in the splicing reactions.

(iii) Introns typically begin with GU dinucleotide sequence at the 5′ splice site and an AG dinucleotide sequence at the 3′ splice site (GU-AG rule).

(iv) Other less well-conserved sequences are found flanking these. These common configurations of the pre-mRNA are recognized by small nuclear ribonucleo-protein particles or **SnRNPs** which catalyze the cutting and splicing reactions.

(v) The splicing events occur in **splicing complexes** called **spliceosomes,** i.e., a complicated intranuclear machine which ensures that all of the splicing reactions occur in concert to remove the introns from the pre-mRNAs. It is very large, being 40S in yeast and 60S in mammal.

It consists of pre-mRNA bound to small nuclear ribonucleoprotein particles (**SnRNPs** or **Snurps**). These are found only in the nucleus. Each SnRNP is abundant in nucleus and at least 10^5 copies per cell. They are rich in **uridine residues**—named as $U_1, U_2, \ldots, U_6$.

Splicing sites:

I. Three types of short sequences within the primary transcript dictate the sites of splicing.

II. Splicing sites → Splice donor; → Splice acceptor; → Branch sites

III. (a) **Splice donor:**

(i) 3′ ends of an exon abuts the 5′ end of the intron.

(ii) **GU** dinucleotide sequence at the beginning of introns is flanked on either side by a few purine bases (**As** or **Gs**).

(b) **Splice aceptor:**

(i) 3′ ends of the intron where it joins with succeeding exon.

(ii) The final nucleotides are always **AG** preceded by a run of 12-14 pyrimidine.

(c) **Branch site sequence:**

(i) It is located upstream (–30 nucleotide) of splice acceptor.

(ii) The key A nucleotide that is a part of a sequence called **branch point sequence** (18 to 38 nucleotide upstream) within it.

Mammal = YNCURAY

Yeast = UACUAAC

Y = Pyrimidine
N = Nucleotide
R = Purine
A = Adenine

Spliceosome complex:

(i) RNA splicing is catalyzed by a spliceosome formed from the assembly of U_1 U_2 U_5 and U_4/U_6 SnRNPs plus other components.

(ii) The SnRNA of U_1 bears a nucleotide sequence that is homologous to the 5′ end of the intron. Base pairing from this homology promotes its binding. It represents the initial step of splicing.

(iii) A preassembled $U_4/U_5/U_6$ particles join the complex. This occurs through association of the particle with bound U_1 and U_2 SnRNPs.

Splicing process:

After assembly of the spliceosome, the reaction occurs in two steps.

(a) **Step I:**

(i) Intron removal begins by a cleavage event at the splice donor site, i.e., first exon-intron junction (GU).

(ii) The G at the 5′ of the intron folds back and forms unusual 2′-5′ phosphodiester bond with the Adenine nucleotide of the branch point sequence (2′ OH-Adenine-5′ OH G).

(iii) The 'A' itself remains in normal 3′-5′ linkage with its adjacent nucleotides of intron.

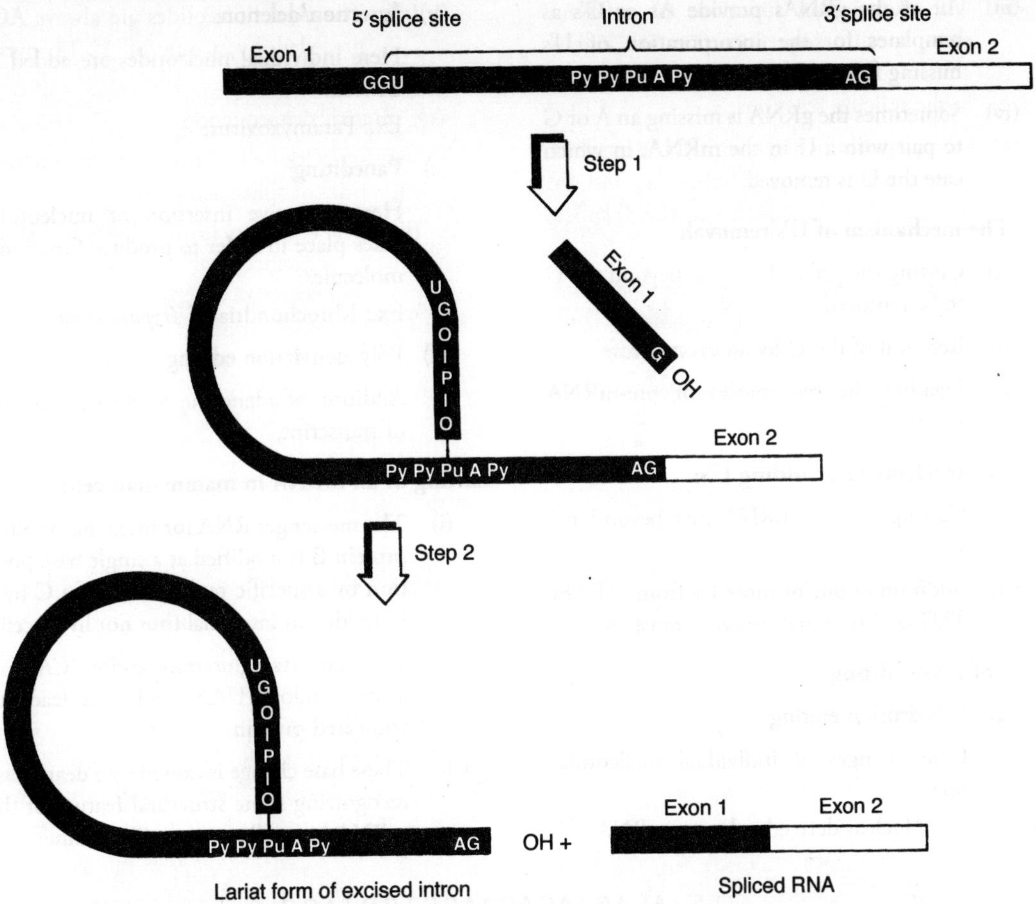

Fig. 13.5: Diagram of the two-step pathway of excision of introns from nuclear pre-mRNAs (hnRNAs)

This reaction produces **lariat**-shaped intermediate.

(iv) Here U_4 SnRNA released.

(b) **Step II:**

(i) The 3′OH of the first exon which is created at the first step.

(ii) The second cut is at splice acceptor site. Cleaving the RNA molecule at 3′ splice site and is accompanied by precise splicing of adjacent exons and introns sequence is released as a lariat.

- **RNA EDITING:**

It is the unusual type of RNA processing where certain nucleotide sequence of pre-mRNA is changed or modified by the post-transcriptional insertion or deletion or substitution of uridine nucleotides to yield the final mRNA which is translated into protein.

- **Features:**

(i) A special RNA molecule called **guide RNA** (gRNA) is involved in these process. Editing occurs 3′–5′ direction by the action of one or more guide RNAs.

(ii) The 5′ end of the first gRNA hybridizes to an unedited region at the 3′ border of editing in the pre-mRNA; the 5′ ends of the rest of the gRNAs hybridize to edited region progressively closer to the 5′ end of the region to be edited in the pre-mRNA.

(iii) All of the gRNAs provide A's or G's as templates for the incorporation of U's missing from mRNA.

(iv) Sometimes the gRNA is missing an A or G to pair with a U in the mRNA, in which case the U is removed.

- **The mechanism of U's removal:**
  - (i) Cutting the pre-mRNA just beyond the U to be removed.
  - (ii) Removal of the U by an exonuclease.
  - (iii) Ligating the two pieces of pre-mRNA together.

- **The mechanism of adding U's:**
  - (i) Cutting the pre-mRNA just beyond the U.
  - (ii) Addition of one or more Us from UTP by **TUTase** (Terminal uridylyl transferase).

Types of RNA Editing:

(a) **Substitution editing:**

Here changes of individual nucleotides occur.

Ex.: Nuclear derived eukaryotic RNA.

(b) **Insertion/deletion:**

Here individual nucleotides are added or deleted.

Ex.: Paramyxovirus.

(c) **Panediting:**

Here extensive insertion of nucleotides takes place in order to produce functional molecules.

Ex.: Mitochondria of *Trypanosoma.*

(d) **Polyadenylation editing:**

Addition of adenosine residues at the end of transcript.

Editing of an mRNA in mammalian cells

(i) The messenger RNA for intestinal apolipo protein B is modified at a single base position by a specific replacement of a C by a U residue in intestinal (but not liver) cells.

(ii) This converts a glutamine codon (CAA) to a stop codon (UAA) and thus leads to truncated protein.

(iii) These base change is caused by a deaminase recognizing some structural feature of the mRNA in a cell-type specific manner.

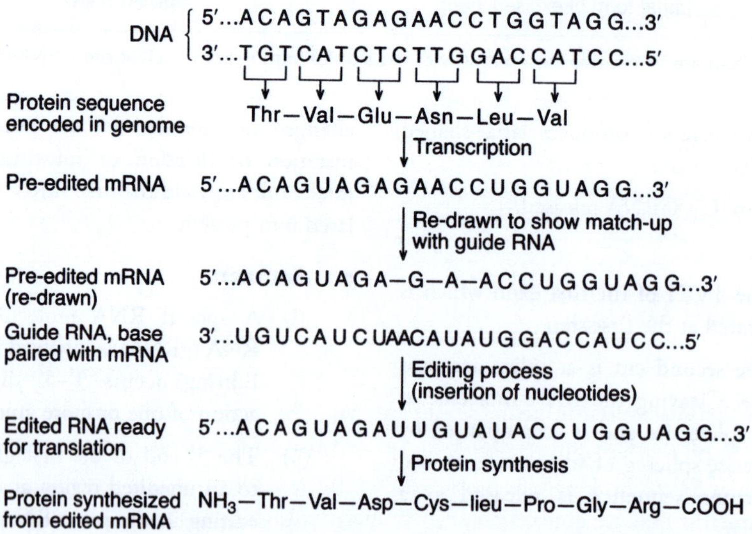

Fig. 13.6: Editing of an mRNA. The guide RNA base pairs with pre-mRNA, with some gaps in the base complementarity. Gaps are filled by insertion of bases (here U) which generates the shift in the reading frame in the mRNA. The protein that is translated from the edited mRNA has a new amino acid sequence as shown in bottom.

- **Cap:**

Cap is the structure at the 5′ end of eukaryotic mRNA introduced after transcription by linking the terminal phosphate of 5′ GTP (5′ end of guanine nucleotide) to the terminal base of the mRNA. The reaction is catalyzed by nuclear enzyme **guanylyl transferase.** This G forms a 5′-5′ bond with terminal A or G of the RNA.

The added G (GTP) is methylated giving the structure of the former ^{7}Me G5′ PPP 5′NP…

This process is known as **capping.**

Cap 0 (i) A Cap that posseses single methyl group (CH_3) is known as Cap 0.
(ii) Found in unicellular eukaryotes.
(iii) Enzyme involves **7th guanine methyl transferase.**

Cap 1 (i) The addition of another methyl group (CH_3) to the 2′-0 position of the penultimate base.
(ii) The reaction is catalyzed by another enzyme called **2′-0 methyl transferase.**
(iii) A cap which has two methyl groups (CH_3) is known as Cap 1. It occurs in higher eukaryotes.

Cap 2 (i) It has three methyl groups (CH_3).
(ii) Methylation can occur in the same second base (if and only if it is an adenine).
(iii) This event happens only when the position is occupied by adenine.
(iv) The enzyme **2′-0 methyl adenosine N^6 methyl transferase.**

- **Short lived mRNA:**
  (i) Bacterial mRNAs are short-lived.
  (ii) Average functional half-life 2 minutes.
- **Moderately long-lived mRNA:**
  (i) In eukaryotes, mRNAs have functional half-life about 6 hours. In some cases 24 hours.
  (ii) The mRNA surviving 24 hr to 4 days are called moderately long-lived **RBC-mRNA.**
- **Long lived mRNA:**
  (i) mRNA stored in seeds of plenty long-lived lotus seeds 1700 years.

- **RNA i (RNA INTERFERENCE):**

A phenomenon of silencing of the expression of a specific gene by a short double-stranded RNA (20 nucleotides long) whose sequence matches a portion of the mature mRNA encoded by the gene.

(i) It has been first discovered in nematode ***C. elegans***. This type of RNA has also been reported in ***Trypanosona, Drosophila*** and ***Arabidopsis***. It also exists in mammals, including human beings.

(ii) The molecules 21-28 bp long, are produced from larger, double-stranded RNA molecules by the action of proteins that are double-stranded RNA specific **endonucleases.**

These endonucleases dice large RNA into small pieces, they are called **Dicer** enzymes.

(iii) RNA pathways appear to be involved in the formation of heterochromatin around the centromeres of chromosomes.

- **Ribozyme:**

RNA molecule that has catalytic activity and is able to cleave and form specific phosphodiester bond in a manner analogous to protein is called **ribozyme.**

In ***Tetrahymena thermophila*** a 6-4 kb precursor of ribosomal RNA can catalyze the cutting and splicing which leads to the removal of part of its own length.

Features of Ribozyme

(i) Ribozymes are found in the intron of an RNA transcript which removes itself from the transcript and in the RNA component of the RNase P complex.

(ii) Many natural ribozymes catalyze either their own cleavage or cleavage of other RNAs.

(iii) They have also been found to catalyze the amino transferase activity of the ribosome.

(iv) They are large or small. The large ones range in size from several hundred to 3000 nucleotides long.

Types: There are eight classes of ribozymes based on unique characters in the sequence.

1. *Tetrahymena* Group I intron; 2. Group II intron; 3. RNase P; 4. Hammerhead ribnozyme; 5. Hairpin ribozyme; 6. Hepatitis delta virus ribozyme; 7. Varkud satellite ribozyme.

- **Micro RNA (miRNA):**

These are short (18-25 nt) RNAs produced from cellular RNA with stem loop structure and can control the expression of cellular genes by causing destruction of specific RNA by bocking their translation.

- **Short interfering RNA (siRNA):**

The short (21-nt base pair) double stranded trigger RNA created by Dicer during RNAi process and involved in triggering RNA interference in eukaryotes.

Important distinctions between the actions of siRNA and miRNA

(i) The siRNAs silence the genes by inducing degradation of target mRNA while miRNAs tend to silence the genes by interfering with accumulation of the protein product of the target mRNAs.

(ii) The siRNAs are formd by Dicer action on double-stranded RNAs that usually contain at least one strand that is foreign to the cell or derive from transposon. On the other hand, miRNAs are formed by Dicer action on the double-stranded part of a stem loop RNA that is a normal product.

(iii) The siRNAs base pair perfectly with the target mRNAs whereas the miRNAs usually base pair imperfectly with their target mRNAs.

Antisense RNA:

It is a single-stranded RNA synthesized from sense strand (coding strand) DNA, which is complementary in sequence to the bonafide mRNA, that normally hybridizes with cognate mRNA. It can form duplex with mRNA and inhibits gene expression as well as protein synthesis.

- **Features**

(i) Antisense refers to the orientation of complementary strands during transcription. The two complementary strands of DNA are referred to as **sense** (= coding or plus) and **antisense** (= noncoding or minus).

(ii) Transcrption occurs using the antisense strand as template, resulting in an mRNA that is identical in sequence to the sense strand (except for replacement of uracil for thymine).

(iii) Antisense RNA is synthesized using the sense strand as template and therefore, it has a sequence complimentary to mRNA.

(iv) When a cell has both the mRNA (i.e., sense strand of RNA) plus a complementary antisense copy, the two single strands anneal to form double-stranded RNA.

The duplex can either inhibit protein synthesis (translation) by blocking the ribosome binding site, or inhibit the mRNA splicing by blocking splice site.

(v) When antisense sequences are made in the laboratory they are usually synthesized as DNA as this is more stable than RNA.

(vi) In this case, DNA : RNA duplex is digested by RNase H. RNase H is a cellular enzyme that normally functions during replication. It recognises and cleaves the RNA backbone of an DNA : RNA duplex, targeting the antisense DNA : mRNA duplex for further degradation. RNase H recognizes a 7 base pair heteroduplex, so the region of homology between the antisense DNA and target mRNA need not be very long.

Strategies used to create antisense effect:

Strategy 1

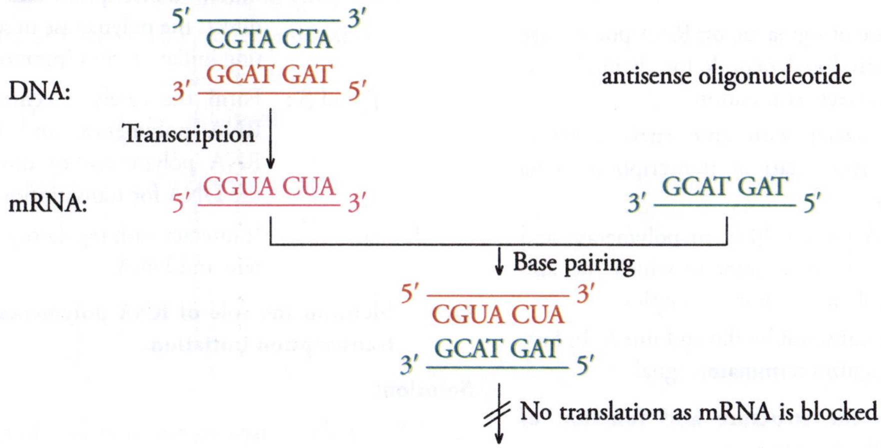

Strategy 2

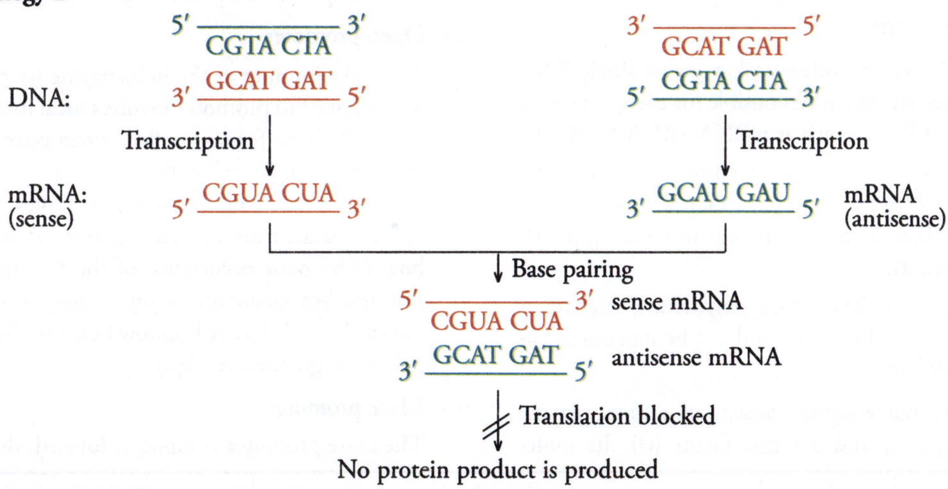

- **Significance**
  - Control of X chromosome inactivation. Antisense RNA binds to the Xist gene to prevent its expression and thus prevents X chromosome inactivates.
  - Control circadian Rhythm in Neurospora.
  - Control of HIV gene expression.
  - Control RNA editing.

Solved Problems

Q. 1 Mention different steps in Transcription.

Solution:

Steps of Transcription are Initiation, Elongation and Termination.

Q. 2 Mention the role of nus A protein in Transcription initiation.

Solution:

(i) After release of sigma factor, RNA polymerase binds to **nus A** polypeptide for chain elongation and correct termination.

(ii) Sigma associates with core enzyme before initiation. After start of transcription sigma dissociates.

(iii) The nus A protein bind to polymerase and proceed up to terminator to which rho also binds to polymerase nus A complex.

(iv) Release of transcript by rho and nus A. In fact, nus A recognizes terminator signal.

Q. 3 Discuss the structure and function of *E. coli* RNA polymerase.

Solution:

In *E. coli* and other prokaryotes a single RNA polymerase enzyme is responsible for the synthesis of all kinds of RNAs (such as mRNA, tRNA & rRNA).

(i) It consists of six polypeptides (six subunits) two identical alpha (α) subunits, one beta (β), one beta prime (β'), omega (ω) and sigma (σ) subunits.

(ii) The complete DNA polymerase enzyme is termed holoenzyme and can be represented as $\alpha_2\beta\beta'\omega\sigma$.

This holoenzyme consists of a core enzyme ($\alpha_2\beta\beta'\omega$) and a sigma factor ($\sigma$). Its molecular weight is 490,000 da.

(iii) The core enzyme is the component of holoenzyme and carries out polymerization and sigma factor is involved in recognition of transcription start signals on the DNA.

Functions:

1. Sigma (σ): It gives signal to polymerase to initiate transcription and also directs the polymerase in selecting initiation sites (promoter).
2. β and β': Form the catalytic centre of RNA polymerase and help RNA polymerase to unwind the DNA for transcription.
3. σ: It interact with regulatory protein and DNA.

Q. 4 Mention the role of RNA polymerase in transcription initiation.

Solution:

RNA polymerase—sigma complex burrows into DNA duplex and sigma recognize proper site.

(a) **Open promoter:**

The tight binding of the holoenzyme ($\alpha_2\beta\beta'\omega$ + σ (sigma)) to promoter involves local melting of the DNA to form so-called "**open promoter complex**". It is "open" in the sense that approximately ten base pairs of the DNA duplex open up or separate from the left end of the **Pribnow box.** (The poor occurrence of the GC in the Pribnow box facilitates the unwinding or opening of the ds DNA at Pribnow box. GC bp are held strongly than AT bp.)

(b) **Close promoter:**

The close promoter complex is formed, due to

Subunit	*Gene*	*Molecular weight (datons)*	*Location*	*Function*
1. α_2	rpo A	41,000	Core enzyme	Promoter binding
2. β	rpo B	155,000	Core enzyme	Nucleotide binding
3. β'	rpo C	165,000	Core enzyme	DNA template binding
4. ω	rpo Z	12,000	Core enzyme	Assembly of subunits
5. σ	rpo D	95,000	Sigma factor	Initiation

relatively loose binding of RNA polymerase and promoter. It is "close" in the sense that the DNA duplex remains intact with no "opening" or melting base pairs.

Q. 5 What is Sigma cycle?

Solution:

(i) The sigma factor required for promoter recognition by RNA polymerase and is dispensible once initiation has taken place.

(ii) In *E. coli* sigma subunit is synthesized at approximately 30% of the corresponding rate of synthesis of other subunits of RNA polymerase.

(iii) The sigma factor helps in recognition of start signal on DNA molecule and directs RNA polymerase in selecting the initiation sites.

(iv) Once RNA synthesis is initiated and attains elongation up to 8-9 bases, sigma factor dissociates (Russel, 1998) from holoenzyme and passes to cytoplasm for repetion of same function, called **sigma cycle** (Fig. 13.7).

Q. 6 State briefly about chain elongation during transcription.

Solution:

(i) During elongation phase of transcription, RNA polymerase (core enzyme) directs the sequential binding of ribonucleotides to the growing chain in the 5′-3′ direction.

(ii) The region of DNA unwound by RNA polymerase is called **transcription bubble.** Within the bubble the nascent RNA chain remains base-paired with DNA template forming a DNA-RNA hybrid.

(iii) However, in those parts of the gene behind the bubble, which have already been transcribed, the DNA double helix reforms displacing the RNA which hangs out of the transcription complex as a single strand with free 5′ end.

(iv) It proceeds at uniform rate from 40 to 50 bases per second at 37°C in *E. coli.*

(v) RNA polymerase makes only one strand during transcription which means the copies of only one strand. Therefore, transcription is **asymmetric.**

(vi) When both strands of DNA are used as template **alternately** that is called **symmetrical transcription.**

Q. 7 Mention the role of 'nus. A' in Chain Elongation during Transcription.

Solution:

(i) In λ phage and *E. coli* there is a specific sequence, called 'nut' site, which lies between promoter and terminater site.

(ii) As the nut site is transcribed in RNA it binds with specific protein called **N-protein.** The function of N-protein requires several protein, called **nus proteins**, viz., nus A, B, E and G (nut-N utilization and nus-N utilization substance).

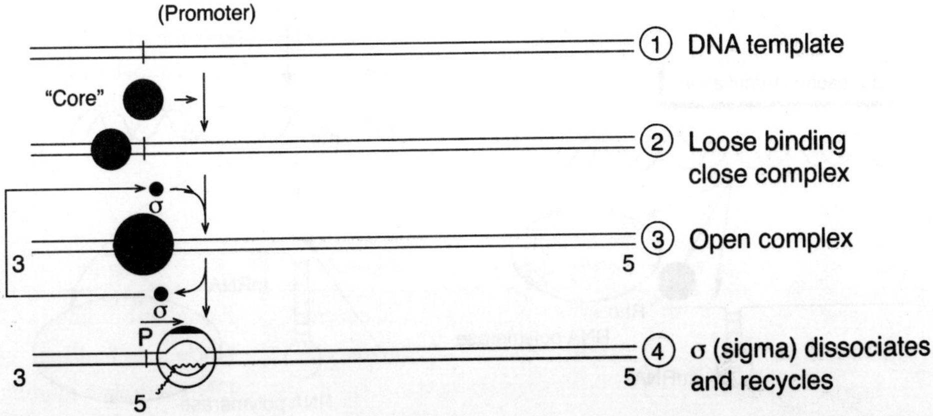

Fig. 13.7: Sigma cycle

(iii) N-protein binds to "nut site" of newly synthesized RNA, nut has two protein binding sites, i.e., box A and box B.

(iv) Box-B interacts with N-protein. Box-A interacts with nus B, E.

(v) The whole complex remains bound to nut site of RNA and move along the DNA template by preventing termination.

Q. 8 How does termination of transcription occur?

Solution:

Termination of transcription occurs at a specific base sequence in the DNA molecule. It has three important regions.

1. There is an inverted repeat base sequence **(Pallindromic base sequence)** containing central non-repeating sequences (segment). In DNA ABCDEF XYZ F′E′D′C′B′A′ pallindromic sequences are invariably found in terminator site and capable of **intrastrand base pairing**, i.e., forming **"stem and loop"** or **"hairpin configuration"** in transcript RNA. When hairpin forms in transcript RNA, it slows down polymerase movement and such pausing is associated with termination.

2. The second region is near the loop end of the presumed and sequence having high G + C content.

3. Most of the prokaryotic mRNAs terminate with sequence of 5′ UUUUUU A 3′ suggesting that 3′ AAAAAA T 5′ sequence is present in DNA.

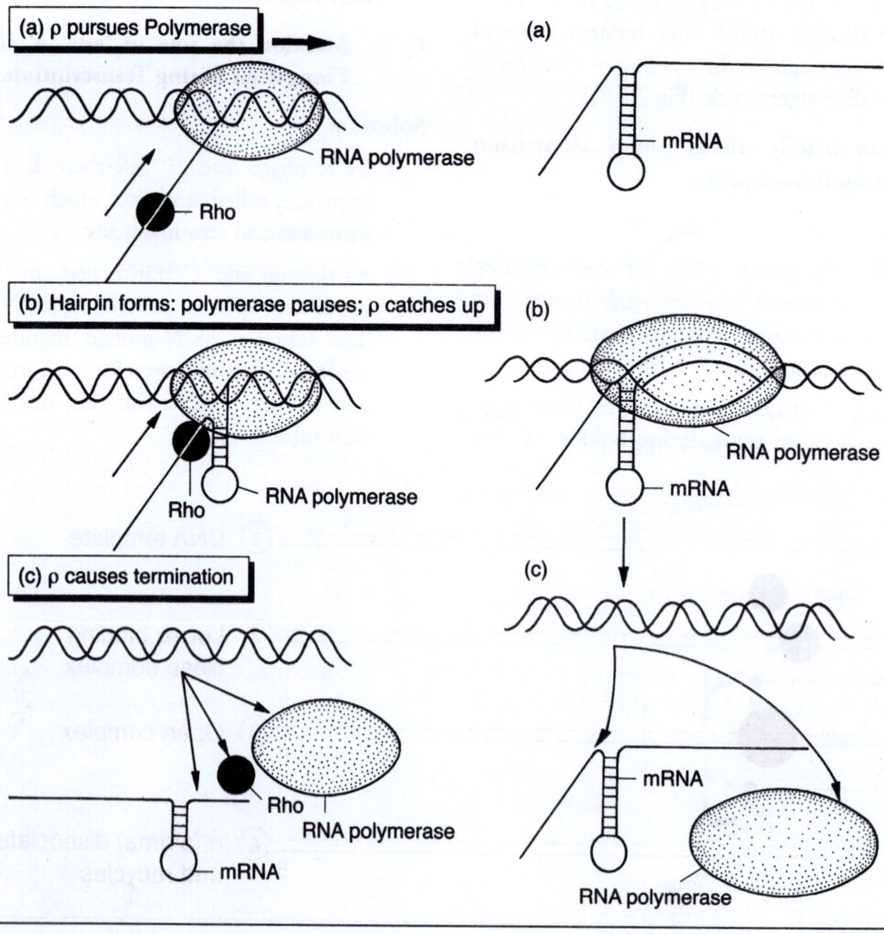

Fig. 13.8: (A) A model of ρ-dependent transcription termination. (B) The termination factor ρ-independent.

(a) Rho-dependent transcription termination:

At the end of transcription, both new and the RNA polymerase are released from DNA template. Some prokaryotic gene requires a protein called **Rho (ρ)** factor (Hexamer Mw 55000) for dissociating the components of transcription complex. These are said to be rho-dependent termination.

(i) Rho (ρ)-dependent terminations have only one part, the inverted repeat.

(ii) There is no string of T's or other second sequence motif.

Therefore, polymerase presumably pauses at the hairpin but cannot release the transcript itself. Rho performs that function.

(iii) Rho acts as an ATP-dependent unwinding enzyme and binds at the 5′ end of nascent mRNA on nut site.

(iv) When RNA polymerases having a pause due to hairpin structure, rho catches up and mediates termination with nascent transcript and RNA polymerases by dynamic process driven by hydrolysis of ATP molecule to ADP and AMP (ip).

In this way, it destabilizes RNA-DNA bonds and releases the mRNA molecule from template accompanied with RNA polymerase and rho factor.

(b) Rho-independent transcription termination:

(i) In this case strings 'T' are the key features of the transcription termination signal.

(ii) RNA polymerase transcribes inverted repeats, providing a transcript that tends to form 'hairpin'.

(iii) When hairpin forms, RNA polymerase pauses and slows down.

(iv) Due to the presence of a long run of 'U' bases in mRNA and 'A' bases in the DNA template, there will be a very weak rU–dA (RNA-DNA hybrid) base pairs needing very little energy to break. [They have melting temperature 20° lower than rU–dA or rA–dT pair.]

(v) Since RNA polymerase slows down due to the hairpin structure, rU–dA bonds break at any point causing release of mRNA.

Q. 9 Why is the discovery of self-splicing introns considered by some to be important as the discovery of the DNA double helix?

Solution:

(i) The discovery that the primary transcript of *Tetrahymena* rRNA excised its own intron was the first demonstration that RNA could function as biological catalyst.

(ii) Many more RNA catalysts, called ribozymes, have been identified, providing support for the view that RNA was the genetic material in the first cell.

Q. 10 In both prokaryotes and eukaryotes, describe what else is happening to the RNA while RNA polymerase is synthesizing a transcript from the DNA template.

Solution:

(i) In prokaryotes, translation is beginning at the 5′ end while 3′ end is still being synthesized.

(ii) In eukaryotes, processing (capping and splicing) is taking place at the 5′ end while 3′ end is still being synthesized.

Q. 11 Are there similarities between the DNA replication bubbles and transcription bubbles found in eukaryotes? Explain.

Solution:

(i) Yes, both replication and transcription are performed by large, multisubunit molecular machines (the replisome and RNA polymerase II respectively) and both require helicase activity at the fork of the bubble.

(ii) However, transcription proceeds in only one direction and only one DNA strand is copied.

Q. 12 What is the primary function of sigma factor? Is there a protein in eukaryotes analogous to the sigma factor?

Solution:

(i) Sigma factor, as part of the RNA polymerase holoenzyme, recognizes and binds to the –35 and –10 regions of bacterial promoters.

(ii) It positions the holoenzyme to correctly initiate transcription at the start site.

(iii) In eukaryotes, TBP (TATA binding protein) and other GTF (general transcription factors) have an analogous function.

Q. 13 A researcher was mutating prokaryotic cells by inserting segments of DNA. In this way she made this following mutation:

Original T T G A C A T 15 to 17 bp T A T A A T

Mutant T A T A A T 15 to 17 bp T T G A C A T

(a) What does this sequence represent?

(b) What do you predict on the effect of such a mutation? Explain.

Solution:

(a) The original sequence represents the –35 and –10 consensus sequences (with the correct number of intervening spaces) of a bacterial promoter. Sigma factor, as a part of the RNA polymerase holoenzyme, recognizes and binds to these sequences.

(b) (i) The mutated (transposed) sequences would not be a binding site for sigma factor.

(ii) The two regions are not in the correct orientation to each other and therefore would not be recognized as a promoter.

Q. 14 A certain *Drosophila* protein coding gene has one intron. If a large sample of null alleles of this gene is examined, will any of the mutant sites be expected

(a) in the exon?

(b) in the intron?

(c) in the promoter?

(d) in the intron-exon boundary?

Solution:

(a) Yes, the exons encode the protein, and so null mutations would be expected to map within exons.

(b) Possibly, sequences near the boundaries of introns and within them are necessary for correct splicing. If these sequences are altered by mutation, correct splicing will be disrupted. Although they may be transcribed, their translation is unlikely.

(c) Yes, if the promoter is deleted or altered such that GTFs cannot bind, transcription will be disrupted.

(d) Yes, sequences near the boundaries of introns and within them are necessary for correct splicing.

Q. 15 What is an Enhancer and Enhansosome? State briefly the functional features of Enhancer.

Solution:

Enhancer is a set of cis-acting DNA sequence that increases the utilization of some eukaryotic promoters and can function in either direction (**bidirectional**) in any location (upstream or downstream) relative to the promoter. It increases the rate of transcription and its enhancing acivity is independent of orientation with respect to the gene. Its consensus sequence is 'AGAQCAGQ'. **Enhansosome** is a macromolecular assembly responsible for interaction between enhancer elements and the promoter regions of genes.

(i) Promoter sequences are located close to the start site of transcription while enhancer located at a distance. Enhancer can stimulate any promoter that is placed nearest vicinity.

(ii) However sequences within enhancer are often composed of the same sequences found adjacent to the promoter.

Ex.: Immunoglobin gene heavy chain enhancers contain octamer **ATGCAAAT**, i.e., also present in immunoglobin promoter.

(iii) Enhancer action involves binding to a regulatory protein factor followed by direct interaction with proteins of the transcriptions apparatus.

Q. 16 What is a Silencer?

Solution:

It is a cis acting transcription regulatory sequence, that acts upon eukaryotic gene to reduce the rate of transcription of a gene or genes located some distance away in either direction.

Ex.: In ***Drosophila*** silencing complexes are P_CG (Polycomb growth) proteins that silence same developmental genes.

Q. 17 What components of the introns of nuclear genes that encode proteins in higher eukaryotes are conserved and required for the correct excision of intron sequences from primary transcripts by spliceosome?

Solution:

"Self-splicing" of RNA precursors demonstrates that RNA molecules can also contain catalytic sites. This property is not restricted to proteins.

Q. 18 Which two elements are almost always present in the promoters of eukaryotic genes that are transcribed by RNA polymerases II? Where are these elements located relative to the transcription start site? What are their functions?

Solution:

(i) The two elements are TATA and CAAT boxes.

(ii) The TATA and CAAT boxes are usually centered at position –30 and –80 respectively, relative to the start point (+1) of transcription.

(iii) The TATA box is responsible for positioning the transcription start-point, it is the binding site for the first basal transcription factor that interacts with promoter.

The CAAT box enhances the efficiency of transcriptional initiation.

Q. 19 How does RNA editing contribute to protein diversity in eukaryotes? What role(s) do guide RNAs play in RNA editing?

Solution:

(i) RNA editing sometime leads to the synthesis of two or more distinct polypeptides from a single mRNA.

(ii) Guide RNAs are partially complementary to the pre-mRNAs that are edited, and serve as templates for the editing process.

(iii) During editing, uracils are inserted in pre-mRNAs opposite adenines in the guide mRNAs.

Q. 20 What role did human disease systemic lupus erythematosus play in characterization of human snRNPs?

Solution:

(i) Some individuals with systemic lupus erythematosus produce antibodies that react with proteins in snRNPs.

(ii) These antibodies have been used to immunoprecipitate snRNPs, facilitating their purification and subsequent characterization.

Q. 21 A scientist is studying the transcription of a gene in mouse cells growing in culture. How can she determine whether the gene is transcribed by RNA polymerase I, RNA polymerase II or RNA polymerase III?

Solution:

She should determine what effect treatment of the cells with α amantin has on the transcription of gene.

(i) If α amanitin has no effect on transcription, the gene is transcribed by RNA polymerase I.

(ii) If α amanitin blocks transcription, the gene is transcribed by RNA polymerase II.

(iii) If α amanitin decreases, but does not completely block transcription, the gene is transcribed by RNA polymerase III.

Q. 22 The primary transcript or pre-mRNA of a nuclear gene in a chimpanzee has the sequence: 5′-G-exon-1-AGGUAAGC-intron-CAUC-exon 2–A–3′.
After the introns has been excised, what is the most likely sequence of the mRNA?

Solution:

(i) Introns contain highly conserved dinucleotide termini: 5′-GT-AG-3′ in the DNA nontemplate strand or 5′-GU-AG-3′ in the RNA transcript.

(ii) Thus, the intron sequence is almost certain to be 5′-GUAAGC-intron-CAG-3′. With the precise excison of the intron, the sequence of the mRNA will be 5′-G-exon-1-AGUC-exon 2-A-3′.

Q. 23 What factors influence the concentration and longevity of mRNA molecules?

Solution:

(i) The concentration of mRNA is determined by the rate of transcription relative to the rate of degradation.

(ii) The longevity of an mRNA is influenced by the length of its poly (A) tail, the structure of its 3′ untranslated region and the metabolic state of the cell.

Q. 24 What enzymes are involved in the maturation of short interfering RNA's and micro-RNAs?

Solution:

The **Dicer enzymes** cleave double-stranded RNA to produce siRNA and miRNA molecules.

Q. 25 Arrange the following events in chronological order, beginning with earliest— (a) splicing of an RNA molecule, (b) migration of mRNA molecule into cytoplasm, (c) transcription of a gene, (d) degradation of an mRNA molecule, (e) polypeptide synthesis.

Solution:

c → a → b — e — d.

Q. 26 In what ways are antisense RNA and RNAi gene silencing procedures the same, different?

Solution:

(i) Both antisense RNA and RNAi make use of RNA strands that are complementary to the mRNA target strand to block its translation.

(ii) Antisense RNA uses **single-stranded RNA** that is complementary to the mRNA.

RNAi involves the use of **double-stranded RNA** where one strand is complementary to the mRNA and the other strand is equivalent to the mRNA to silence the expression of the target gene. RNAi makes use of the RNA-induced silencing complex (**RISC**) to block the gene expression.

Q. 27 Write briefly about the Eukaryotic RNA polymerase.

Solution:

Eukaryotic cells have three distinct transcribing enzymes, each responsible for synthesizing different groups of RNA.

- RNA polymerase I
- RNA polymerase II
- RNA polymerase III

- **RNA polymerase I (A)**
  (i) It is located in the nucleolus.
  (ii) It synthesizes precursors of most rRNA.
  (iii) It is sensitive to α amanitin.
- **RNA polymerase II (B)**
  (i) It is located in the nucleoplasm.
  (ii) It synthesizes mRNA precursors.
  (iii) It is very sensitive to α amanitin.
- **RNA polymerase III (C)**
  (i) It is located in the nucleoplasm.
  (ii) It synthesizes the precursors of tRNA and rRNA, and other small nuclear (snRNA) and cytoplasmic RNAs (scRNA).
  (iii) It is moderately sensitive to α amanitin.

Enzyme	*Function*	*Sensitive*
Polymerase I (A)	(i) Transcription of the 45S rRNA precursor. (ii) That is a single polycistronic unit containing 5-8S, 18S and 28S rRNA genes.	(i) Insensitive to α amanitin. (ii) Sensitive to actinomycine D.
Polymerase II (B)	Transcription of all protein encoding genes and most genes for small nuclear RNAs.	Very sensitive to α amanitin.
Polymerase III (C)	Transcription of tRNA genes 5S rRNA genes and genes U6 snRNA and the various scRNAs.	Moderately sensitive to α amanitin.

The promoters for eukaryotic RNA polymerases I, II and III.

Features	Promoters for RNA polymerase		
	I	II	III
1. Location in relation start point	Upstream	Upstream	both upstream and downstream
2. Number of accessory factor	Small (two)	Large (generally upstream and regulatory	Small (three)
3. Number of different promoters recognized	One (rRNA genes)	Many (all mRNA coding genes)	A small set (tRNA and other small RNAs)
4. Organization	Bipartite (core promoter and upstream control element)	Modular (some of many elements)	Bipartite or tripartite (internal promoter) or (upstream promoter)
5. Variation in promoter sequence	little	The greatest for the three polymerase	Greater than that for RNA polymerase

Compare the features of Promoters and Enhancers

Features	Promoters	Enhancers
1. Organization	Modular	Modular
2. Elements	Some are common in enhancers	Some are common in promoter
3. Location	(i) Relatively fixed with regard to start point (ii) Upstream start point (iii) In single orientation	(i) At greatly variable distance from the start point (ii) Upstream or downstream start point (iii) In either direction
4. Densist of elements	Relatively lower than that in enhancer	Much greater than promoter

Q. 28 Difference between Replication and Transcription:

Solution:

Replication	Transcription
1. The enzyme DNA polymerase is involved in this process.	1. The enzyme RNA polymerase is involved in this process.
2. Polymerization of deoxyribonucleotides d ATP, d GTP, d TTP, d CTP.	2. Polymerization of ribonucleotides—ATP, GTP, UTP, CTP.
3. Here both strands act as template; therefore, it is **symmetric**.	3. Here one strand (antisense) acts as templates; therefore, it is **asymmetric**.
4. DNA remains bound to template and forms double helix.	4. RNA after transcription "peels off" or detaches from DNA and passes to cytoplasm.
5. Synthesis occurs from whole DNA or chromosome.	5. Synthesis occurs from small segment of DNA, called cistron or gene.
6. No modification of DNA.	6. Nascent RNA is modified to form functional or mature RNA.
7. Replication occurs in dividing cell in 's' period.	7. Transcription occurs in any cell and in any stage of cell cycle.

Q. 29 Similarities between Replication and Transcription:

Solution:

1. In transcription and replication, polymerization of complementary bases (nucleotides) on DNA template occurs according to Watson Crick base pairing rules.
2. In both cases synthesis occurs in 5′-3′ direction.
3. Synthesis occurs in chromosome and nucleus (eukaryotes).

Q. 30 Degradasome and Exosome

Solution:

Degradasome: It is a multienzyme complex in bacteria (*E. coli*) involved in mRNA degradation including ribonuclease R (RNase E), Polynucleotide phosphorylase (PNPase) and an ATP dependent helicase (Rh1B), a glycolytic enzyme (enolase), RNase III and RNase II. Degradation takes place in 5′-3′ direction.

Exosome: It is a complex of several exonucleases involved in degradation of eukaryotic mRNA. It occurs through **deadenylation** and **decapping**.

Q. 31 State briefly the processing events that occurs during the production of tRNA in *E. Coli*.

Solution:

1. In bacteria, the 5′ end of the tRNA is cleaved by RNase P.
2. The 3′ end is cleaved by a different endonuclease and then a few nucleotides are digested away by an exonuclease that removes nucleotides until it reaches a CCA sequence.

Q. 32 What is the differene between a structural gene and a non-structural gene?

Solution:

1. Structural gene encodes mRNA that is translated into a polypeptide sequence.

   Non-structural gene encodes RNAs that are never translated.
2. Products of structural gene include tRNA and rRNA which function during translations, microRNA which is involved in gene regulation, 7sRNA which is part of a complex known as SRP; scRNA, small cytoplasmic RNA found in bacteria, the RNA of RNase P, telomerase RNA, which is involved in telomere replication, sno RNA which is involved in rRNA trimning and snRNA which is a component of spliceosome. In many cases, the RNA from non-structural gene becomes part of a complex composed of RNA molecular and protein subunits.

Q. 33 A single codon in a double-stranded DNA molecule undergoes an inversion.

(a) If the original codon in the mRNA is 5′–AUG–3′ what is the codon in the transcript from the inversion.

(b) Would inversion of any codon conserve the amino acid that is specified?

(c) What codons would upon inversion, yield chain termination codons?

(d) Which amino acids would inversion of each of the chain termination codons specify?

Solution:

(a) (i) The original codon must have the sequence of bases $\frac{5'\text{–ATG–}3'}{3'\text{–TAC–}5'}$.

(ii) The bottom strand is transcribed to form 5′–AUG–3′, the inversion has the sequence 5′–CAT–3′/3′–GTA–5′. Again, the bottom strand is transcribed the mRNA therefore contain. 5′–AUG–3′ ("The reverse complement" of the original codon) which specifies His.

(b) None of the reverse complement codons are synonymous. No synonymous codon with complementary base is in the second position.

(c) The chain termination codons are 5′–UAA–3′, 5′–UAG–3′, 5′–UGA–3′ which have reverse complement.

5′–UUA–3′ (Leu), 5′–CUA–3′ (Leu) and 5′– UCA–3′ (Ser) respectively.

(d) Same answer as in part (c).

Q. 34 A mutation within a gene sequence changes the start codon to a stop codon. How will this affect the transcription of the gene.

Solution:

This will not affect transcription. However, it will affect translation by preventing the initiation of polypeptide synthesis.

General Transcription factors in Eukaryotic structural genes

Accurate transcription of protein coding gene involves the assembly of RNA polymerase II and a number of other protein, called general transcription factors (GTFs), on the core promoter.

General Transcription Factors	Function
1. TFIID: Composed of TATA binding protein(TBP)	1. Recognizes TATA box of eukaryotic structural gene promoter.
2. TFIIB:	2. (i) Binds to TFIID and then enables RNA polymerase II to bind to the core promoter. (ii) It also promotes TFIIF binding.
3. TFIIE:	3. (i) Binds to RNA polymerase II and plays role in its ability to bind to TFIIB and the core promoter. (ii) It also plays a role in the ability of TFIIE and TFIIH to bind to RNA polymerase II.
4. TFIIF:	4. (i) Plays a role in the formation and or the maintenance of the open complex. (ii) It may exert its effects by facilitating the binding of TFIIH to RNA polymerase II and regulating the activity of TFIIH.
5. TFIIH:	5. (i) A multisubunit protein that has multiple roles. (ii) Certain subunits act as helicases and promote the formation of open complex. (iii) Other subunits phosphorylate the carboxyl terminal domain (CTD) of RNA polymerase II, which releases interaction with TFIIB and thereby allows RNA polymerase II to proceed to the elongation phase.

Q. 35 In bacteria, what event marks the end of initiation stage of transcription?

Solution:

The release of sigma factor indicates the end of initiation and beginning of chain elongation stage of transcription.

Q. 36 When RNA polymerase transcribes DNA, only one of the two DNA strands is used as template. Explain how RNA polymerase determines which DNA strand is the template strand.

Solution:

(i) The binding of sigma factor RNA polymerase depends on the sequence of promoter.

(ii) RNA polymerase binds to the promoter in such a way that the –35 sequence TTGACA and the –10 sequence TATAAT are within the coding strand, while the –35 sequence

AACTGT and the –10 sequence ATATTA are within the template strand.

(iii)

Coding strand	–35TTGACA	TATAAT
Template strand	–35AACTGT	ATATTA

Q. 37 At molecular level, describe how sigma (σ) factor recognizes bacterial promoters. Be specific about the structure of σ-factor, the type of chemical bonding.

Solution:

(i) Sigma factor can slide along the major groove of the DNA. In this way, it is able to recognize base sequences that are exposed in the groove.

(ii) When it encounters the promoter sequence, hydrogen bonding between the bases and the sigma factor protein can promote tight and specific interaction.

Q. 38 What sequence elements are found within the core promoter of structural genes in eukaryotes? Describe their locations and specific function.

Solution:

(i) Eukaryotic promoters are somewhat variable with regard to the pattern of sequence elements.

(ii) In the case of structural genes that are transcribed by RNA polymerase II, it is common to save a TATA box which is about 25 upstream from a transcriptional start site.

(iii) The TATA box is important in the identification of the transcriptions start site and the assembly of the RNA polymerase and various transcription factors.

(iv) The transcriptional start site defines where transcription actually begins.

Q. 39 What two elements are almost always present in the promoters of eukaryotic genes that are transcribed by RNA polymerase II, where these elements located relative to the transcription start sites? What are their functions?

Solution:

(i) TATA and CAAT boxes. They are usually centred at position –30 and –80 respectively relative to the start point (+1) of transcription.

(ii) TATA box is responsible for positioning the transcription start point. It is the binding site for the first basal transcription factor that interacts with the promoter.

(iii) CAAT box enhances the efficiency of transcription initiation.

Q. 39 Study the mRNA segment given below and answer the questions:

```
       'a'                                   'b'
     ⏞    G A G      U U U U C C C C A     ⏞
5' ─┴─┴─┴─┴─┴─┴........┴─┴─┴─┴─┴─┴─┴─┴─┴─┴─┴─┴─ 3'
```

(i) What is the name of a given codon 'a'?

(ii) Which amino acid does it code for?

(iii) Which molecule will decipher this codon and how?

(iv) Write codon for 'b'.

(v) Write the bases of template strand of the structural gene from which this mRNA is formed.

Solution:

(i) Initiator codon (AUG)

(ii) Methionine

(iii) tRNA with anticodon UAC

(iv) Terminator codon or stop codon UAG/ UGA/UAA/

(v) Template strand.

```
3' ─┬─┬─┬─┬─┬─┬─........─┬─┬─┬──┬─┬─┬──┬─┬─┬─ 3'
    T A C C T C          A A A  A G G  G G T
```

Q. 41 If the sequence of nucleotides in mRNA is 5′ AUG UUU UUC UUC UAC 3′, then

(a) What will be the sequence present on the anticodon site of tRNA?

(b) What will be the sequence in the template and coding strand of DNA?

Solution:

(a) Given the coding strand (mRNA) is

mRNA 5′ AUG UUU UUC UUC UAC 3′

Anticodon in tRNA	5′ UAC AAA AAG AAG AUG 5′
Amino acid	Methionine Phenylalanine Phe Phe Tyr

(b)

Coding strand 5′ A T G T T T T T C T T C T A C 3′

template 3′ T A C A A A A A G A A G A T G 5′

↓

mRNA 5′ A U G U U U U U C U U C U A C 3′

Q. 42 Study the coding strand in a transcription unit as follows:
5′ ATG CCT AGG TCC AGG CAT 3′
Write down the sequence in mRNA. Write down in template strand and anticodon for each code and their corresponding amino acids.

Solution:

Given the coding strand (mRNA) as under:

Coding strand 5′ A T G C C T A G G T C C A G G C A T 3′

template 3′ T A C G G A T C C A G G T C C G T A 5′

mRNA will be formed for template in 5′–3′ direction which is as follows:

mRNA: 5′ A U G C C U A G G U C C A G G C A U 3′

Anti-codon	3′UAC	GGA	UCC	AGG	UCC	GUA 5′
Amino acid	Methionine	Proline	Arginin	Serin	Arginin	Histine

Q. 42 Study the mRNA segment given above which is complete to be translated into polypeptide chain.

5′ ‘a’ UAC GAG .... AGA UUU ‘b’ 3′

(a) Write the codons ‘a’ and ‘b’.

(b) What do they code for?

(c) How is peptide bond formed between two amino acids?

Solution:

(a) a = AUG; b = UAA/UGA/UAG

(b) AUG code for methionine, UAG (ambar), UGA (opal), UAA (ochre) nonsense codon or stop codon. They do not code for any amino acid.

(c) Peptide bond is formed between –COOH group of ‘P’ site amino acid and NH_2-group of A site amino acid with the help of ribozyme peptidyl transferase provided by ribosome.

Q. 44 What is an R loop? In an R loop experimentally to which strand of DNA does the mRNA bind, coding strand or the template strands.

Solution:

(i) R loop is a loop of DNA that occurs when RNA is hybridized to double-stranded DNA.

(ii) While RNA is hydrogen bonding to one of the DNA strands, the other strand does not have a partner to hydrogen bond with soil bubbles and a loop.

(iii) RNA is complementary to the template strand, so that is the strand it binds to.

Regulation of transcription in prokaryote

- **Signal factor (δ):** Responsible for recognition of consensus promoter sequence and transcription.
- **Promoter recognition:** *E. coli* δ^{70} is responsible for the recognition of –10 to 35 consensus sequence.
- **Heat shock protein:** About 17 proteins specification expressed in *E. coli,* when temperature is increase about 37°C.

C. TRANSLATION

Q. 1 What is Polysome? How does it form?

Solution:

Most mRNAs are simultaneously being translated by several ribosomes, spaced about 90 nucleotides along the mRNA molecules. The size or configuration of these translation complex, i.e., several ribosomes, each with a polypeptide, and one mRNA is called **polyribosome** or **polysome**.

(i) When one ribosome moves 90 nucleotides or 30 codons towards 3′ end of mRNA, a second ribosome forms initiation complex and starts translating the same mRNA and forms the second polypeptide.

(ii) A third, fourth and so on also follow the above steps.

(iii) Thus, a single mRNA molecule is able to direct the simultaneous synthesis of multiple polypeptides using an array of ribosome.

(iv) In silk moth "fibroin" mRNA is translated simultaneously by 80 ribosomes, and 80 polypeptides are simultaneously formed from one mRNA.

Major steps in translation (Prokaryotes)

Initiation:

(i) Dissociation of the 70s ribosome.

(ii) Formation of the mRNA-30S subunit initiation complex.

(iii) Formation of the 30s initiation complex.

(iv) Formation of the 70s initiation complex.

Elongation:

(i) Binding of aminoacyl-tRNA to the ribosomal A site.

(ii) GTP hydrolysis dissociation of EF-TU. GTP complex.

(iii) Formation of peptide bond (Transpeptidation)

(iv) Peptidyl-tRNA movement from A to P site.

Termination:

(i) Recognition of terminal signal.

(ii) Release of the completed polypeptide chain.

(iii) Processing of the polypeptide chain.

Prokariotic translation control

(i) The bacterial mRNA is very short-lived (only 1-3 minutes).

(ii) It allows shifts in mRNA secondary structure. Messenger RNA secondary structure can govern translation initiation (as in the replicase gene of two MS2 classes of phages).

(iii) Riboswitch can also be used to control translation initiation via mRNA secondary structure.

Q. 2 State briefly about the charging of tRNA in *E. coli*.

Solution:

(i) Attachment of amino acid to tRNA is called **charging** or **activation of tRNA**.

(ii) tRNA plus amino acid is called **charged tRNA** or **amino acyl tRNA**.

(iii) 20 kinds of tRNA, each is specific for one amino acid.

(iv) Amino acyl tRNA **synthetases**.

(v) 20 different amino acids.

(vi) ATP as energy source.

(a) Activation of amino acid

(i) Each of the 20 amino acids occurs in an inactive stage.

(ii) The activation of amino acids takes place through their carboxyl groups.

(iii) Each amino acid is catalysed by its own specific activating enzyme, called aminoacyl tRNA synthetase, and ATP to forms aminoacyl adenylate or activated amino acid or aminoacyl adenylic acid (Klug and Cummings, 2006).

(iv) A covalent linkage is formed between 5′ phosphate group of ATP and carboxyl end of amino acid, and this molecule remains attached with the enzyme.

$$\text{AA} + \text{ATP} \xrightarrow{\text{Aminoacyl tRNA synthetase}} \text{AA}{\sim}\text{AMP}{\sim}\text{E} + \text{PP}$$

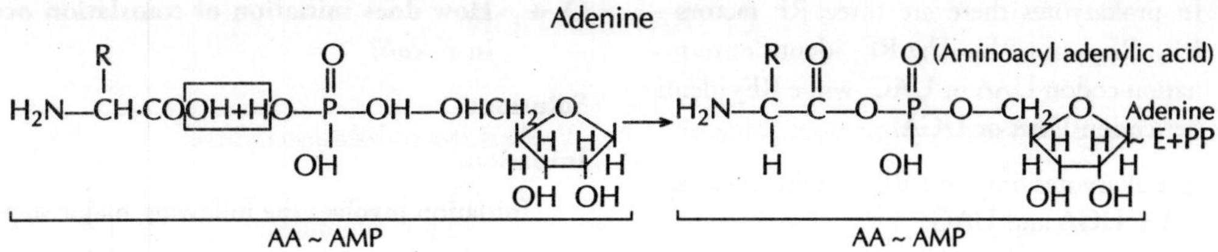

(b) Attachment of activated amino acid to tRNA

(i) Charging or attachment of activated amino acid to tRNA is also specific.

(ii) **Aminoacyl synthetase** and ATP make the amino acid active and then help in forming AA ~ tRNA complex.

(iii) The synthetase has two active sites—one for recognizing the specific amino acid and the other for recognizing the specific tRNA molecule.

(iv) tRNA molecule has two selection sites—one for recognizing its **specific aminoacyl tRNA synthetase** and the anticodon for recognizing the **codon in mRNA.**

(v) Aminoacyl tRNA synthetase catalyzes the formation of **acyl bond** between aminoacyl adenylate (AA ~ AMP) to 3′ terminal adenine nucleotide of tRNA. This ester bond is formed between 3′ OH tRNA and carboxyl group of amino acid.

(vi) A tRNA covalently coupled to its amino acid is called a **charged tRNA.**

I. Amino acid + ATP (adenosine – Ⓟ + P + P ⟶ AA ~ Ⓟ + P + P

RCHNH$_2$COOH + Adenosine ~ Ⓟ + P + P ⟶ NH$_2$RCH C(=O)OH + OHⓅ + P + P

II. AA ~ Ⓟ – adenosine + tRNA ⟶ AA ~ tRNA + adenosine + P

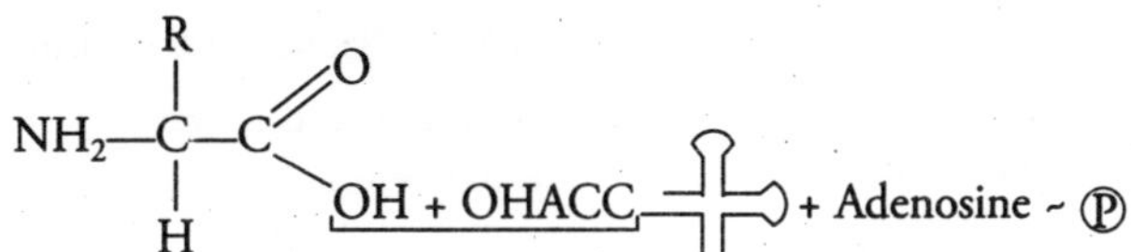

Q. 3 How does termination of translation occur in *E. coli*.?

Solution:

Recognition of termination signal

(i) The polypeptide chain continuously elongates until a termination codon on mRNA reaches to ribosome.

(ii) The termination codon (**UAA–ochre, UAG– amber UGA–opal**) is also called **nonsense codon**, because no tRNA anti-codon pairs or matches with them.

(iii) It is not necessary that termination codon is the last codon of mRNA, e.g., in bacteria and bacteriophages polygenic mRNAs are common and they consist of a number of initiation and termination codons.

AGGAGGU AUG UGA AGGAGGU AUG UGA AGGAGGU AUG UGA

(iv) After translocation brings one of the termination codons into the 'A' site the ribosome does not bind with aminoacyl tRNA-EF-Tu-GTP complex. Then it receives the signal of termination.

Release of polypeptide chain

(i) When a termination codon is translocated into the 'A' site, the ribosome instead of binding with a complex containing amino acids binds a peptide release factor (RF).

(ii) In prokaryotes there are three RF factors—RF_1, RF_2 and RF_3. The **RF_1** identifies termination codon **UAA** or **UAG** while **RF_2** identifies codon **UAA** or **UGA**.

In Eukaryotes only one RF protein recognizes **UAA, UGA** and **UAG.**

(iii) The resultant RF termination codon. Ribosome complex effectively blocks further chain elongation.

(iv) The function of RF_3 binds GTP and seems to stimulate the action of RF_1 and RF_2. It is also known as stimulating factor or s-factor.

(v) With the aminoacyl site, so clogged, the completed polypeptide remains esterified to the final tRNA occupying the 'peptidyl site'.

(vi) The release factors interact with peptidyl transferase causing hydrolysis of the polypeptide chain at the 'P' site. This result cleavage and release of chain from the tRNA molecule at the 'P' site of ribosome.

The ribosome then dissociates into 50s and 30s subunits.

Notes on EF–Ts EF–Tu cycle

EF–Ts–EF–Tu cycle:

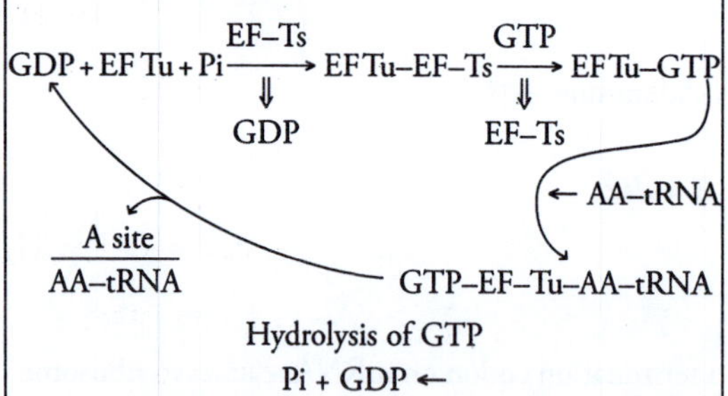

EF–Ts and EF–Tu are required for a transfer RNA to attach to the A site of ribosome. EF–Tu attached to a GDP. GDP is displaced by EF–Ts, which in turn is displaced by GTP. A transfer RNA attaches and is brought to the ribosome. If the codon-anticodon fit is correct, the transfer RNA attches at the A site with the help of the hydrolysis of GTP to GDP + Pi allowing the **EF–Tu to release.** Since EF–Tu has a strong affinity for GDP, the role of EF–Ts is to displace to GDP and later to be replaced by GTP.

Q. 4 How does initiation of translation occur in *E. coli*?

Solution:

Initiation:

Initiation involves the following major steps:

(i) Dissociation and association of ribosomal subunits (30S and 50S).

(ii) Binding of mRNA (mediated by Shine–Dalgarno sequence 5′ AGGAGGU 3′) to ribosome (30S subunit).

(iii) Selection of the right AUG start signal.

(iv) Positioning of N-formyl met $tRNA_f$met on the ribosome at the 'P' site.

(v) Three auxillary proteins, known as "initiation factors", IF_1, IF_2 and IF_3 are required for translation.

(a) Dissociation Ribosomal subunit

(i) Association and dissociation of ribosomal subunits take place during protein synthesis.

(ii) It is also dependent on the cellular concentration of Mg^{2+}.

(iii) Low concentration of Mg^{2+} and the influence of IF_1 and IF_3 lead to dissociation of 70S to 30S and 50S subunits.

(iv) Within the cell the Mg^{2+} concentration is high enough and ribosomes exist as 70s.

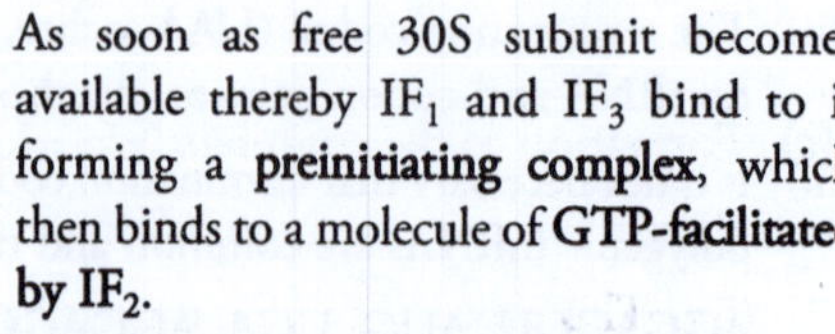

(b) Formation of 30S initiation complex

(i) As soon as free 30S subunit becomes available thereby IF_1 and IF_3 bind to it forming a **preinitiating complex**, which then binds to a molecule of **GTP-facilitated by IF_2.**

(ii) An initiating factor **IF_3** influences the binding of mRNA to 30S subunit of ribosome.

(iii) Binding of mRNA and 30S subunit is very precise and involves complementary base pairing between mRNA and 16S rRNA of 30S ribosome subunit.

(iv) Before the initiator codon 'AUG', there is a stretch of purine-riched nucleotide sequence in the leader, i.e., 5′ AGGAGGU 3′ in mRNA, called **Shine Dalgarno Sequence.**

(v) This sequence binds with the pyrimidine-riched complementary sequence present on 16s rRNA of 30s ribosome subunit, i.e., 3′ UCCUCCA 5′.

SDS		fmet	arg	ala	phe	ser		
5′ AGGAGGU	......	AUG	CGA	GCU	UUU	AGU	......	3′ mRNA
3′ UCCUCCA	......							

3′ end of ribosomal RNA

In this way mRNA and 30S ribosome unite to form 30S initiation complex.

(vi) The **first codon** translated in all mRNA is **AUG** which codes for methionine. This AUG is called **initiation codon** or **start codon.**

(vii) Naturally, other AUG codons also occur internally in an mRNA where they encode methionine residues, internal to the protein.

(viii) Two different tRNAs are used to these two types of AUG codon. **$tRNA_f^{met}$** is used for the initiation codon and is called **initiator tRNA,** whereas $tRNA_m^{met}$ is used for **internal AUG codon.**

(ix) The initiator or start codon is AUG, but sometimes GUG (val) or UUG (leu) is recognized by initiator tRNA, i.e., **$tRNA_f^{met}$**.

(x) In prokaryotes, the first normal methionine is attached with tRNA by its COOH. Then the free $—NH_2$ group is blocked by formyl group (—CHO) by the enzyme transformylase. Hence, it is called N formyl methionine. Thus, only —COOH group of $_f^{met}$ remains free to form peptide bond with $—NH_2$ group of second amino acid.

(xi) A special initiator tRNA whose anticodon is 5′ CAU 3′, anticodon is complementary to AUG.

$$\text{Formyl tetrahydrofolate} + NH_2\text{—Methionyl } tRNA_f^{met} \xrightarrow{\text{Transformylase}} \text{N formylmethionyl } tRNA_f^{met}$$

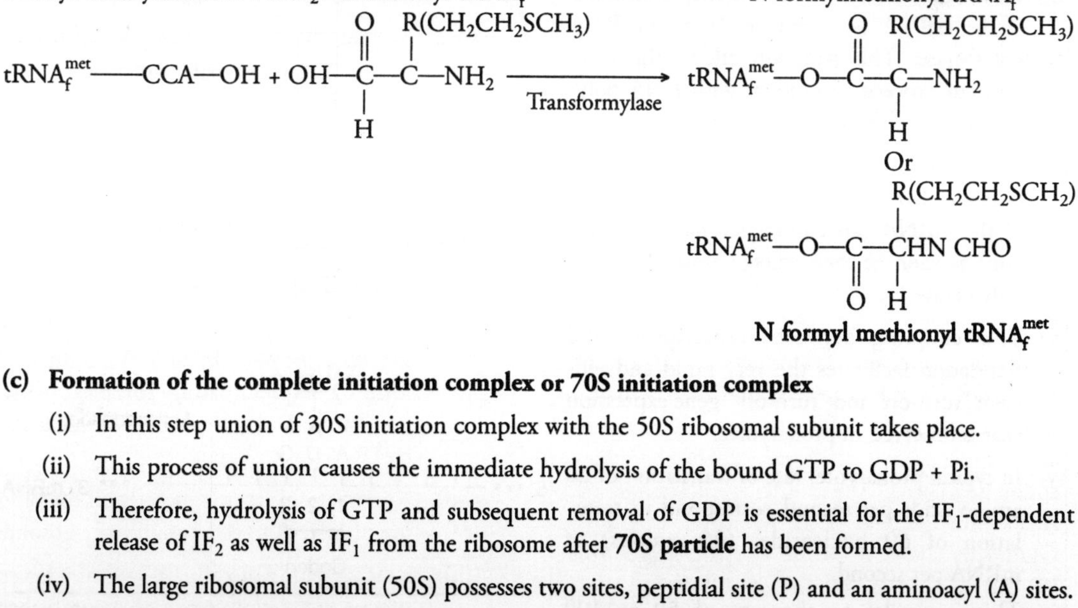

N formyl methionyl $tRNA_f^{met}$

(c) Formation of the complete initiation complex or 70S initiation complex

(i) In this step union of 30S initiation complex with the 50S ribosomal subunit takes place.

(ii) This process of union causes the immediate hydrolysis of the bound GTP to GDP + Pi.

(iii) Therefore, hydrolysis of GTP and subsequent removal of GDP is essential for the IF_1-dependent release of IF_2 as well as IF_1 from the ribosome after **70S particle** has been formed.

(iv) The large ribosomal subunit (50S) possesses two sites, peptidial site (P) and an aminoacyl (A) sites.

(v) Recently 3rd site in 50S has been proposed, called E site (exist site).

Role of Antibiotics in translation

Antibiotic	Role/Action
1. Streptomycin	(i) Inhibits prokaryotic translation initiation. (ii) Reduces fidelity of translation in prokaryotes by causing mis-reading of codon.
2. Erythromycin	Inhibits prokaryotic translocation.
3. Tetracyclin	Blocks binding of tRNAs to small subunits of prokaryotic ribosome.
4. Puromycin	Forms covalent bond with growing polypeptide and releases it from ribosome causing premature termination.
5. Chloramphenicol	Inhibits prokaryotic peptidyl transferase.

Q. 5 State why transcription and translation are coupled phenomena in prokaryotes.

Solution:

(i) In prokaryotes translation begins before transcription is completed. The mRNA molecules are both synthesized and translated in the 5′-3′ direction and because there is no nuclear membrane separating transcription and translation as in eukaryotes, i.e., transcription and translation machinery are located in the same compartment.

(ii) Thus, the ribosome can commence translation of the mRNA as it emerges from the RNA polymerase. This situation allows the ribosome to proceed in tandem with RNA polymerase as it elongates the transcript.

(iii) The 5′ end of an RNA is synthesized first and thus translation which also starts at the 5′ end of the mRNA can commence nascent transcript as soon as they emerge from the RNA polymerase.

(iv) This coupling between transcription and translation facilitates the very rapid and efficient "turn-on" and "turn-off" gene expression that is observed in prokaryotes.

(v) In typical prokaryote, rate of translation of 20 amino acids per second correspond to translation of 60 nucleotides (20 codons) of mRNA per second.

This is similar to the rate of 50 to 100 nucleotides per second synthesized by RNA polymerase.

(vi) In contrast to the situation in prokaryotes, translation in eukaryotes is completely separate from transcription. Indeed, these events occur in separate compartment of the cell; transcription occurs in the nucleus, whereas translation occurs in the cytoplasm.

Perhaps, due to the lack of coupling to transcription, eukaryotic translation proceeds at the more leisurely speed of 2-4 amino acids per second.

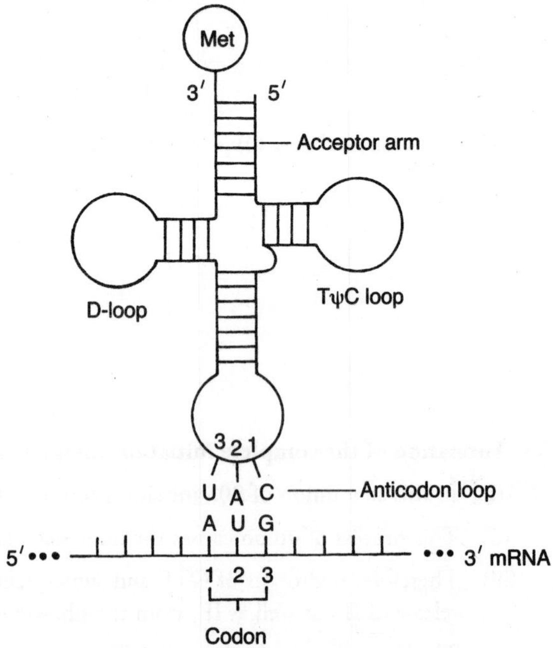

Fig. 13.9: tRNA: Codon and anticodon base positions are numbered from the 5′ end. The 3′ position in the codon (5′ in anticodon) is the wobble base.

Q. 6 Contrast the roles of tRNA and mRNA during translation and list all enzymes that participate in the transcription and translation processes.

Solution:

I. Transfer RNAs are 'adaptor' molecules, i.e., they provide a way for amino acids to interact with sequences of bases in nucleic acids.

II. Amino acids are specifically and individually attached to the 3′ end of tRNAs which posse-sses a three-base sequence (the anticodon) to base pair with three bases of mRNA.

(i) Messenger RNA, on the other hand, contains a copy of the triplet codes which are stored in DNA.

(ii) The sequences of bases in mRNA interact thrice at a time, with anticodons of tRNAs.

Enzymes involved in transcription include:

(i) RNA polymerase (in *E. coli*) [$\alpha^2\beta\beta'\omega\sigma$]

(ii) RNA polymerases I, II, III (in Eukaryote).

Enzymes involved in translation include:

(i) Aminoacyl tRNA synthetase

(ii) Peptidyl transferase

(iii) GTP-dependent release factors.

Q. 7 State briefly the initiation of eukaryotic translation.

Solution:

The eukaryotic mRNA contains a short recognition sequence surrounding the initiating AUG codon, 5′-ACCAUGG-3′ (**Kozak sequence**).

(a) Dissociation of 80S ribosome:

(i) The 80S ribosomes are dissociated with the help of initiating factor eIF-3.

(ii) The initiation factor eIFA and eIF binds to 40S subunit and shifts towards dissociation of 80S ribosome subunit into 60S and 40S.

(b) Formation of 43S pre-initiation complex:

(i) The binary complex of eIF-2 with GTP binds methionine charged tRNA (met tRNAimet) to form a ternary complex (met tRNAimet—eIF-2-GTP). It binds to 40S subunit (carrying eIF-3) to form **43S pre-initiation complex.**

(ii) The initiation factor eIF-2B activates eIF-2 by replacing its GDP with GTP, eIF-1 and eIFIA aid in scanning to initiation codon.

(c) Formation of 48S pre-initiation complex:

(i) The 43S pre-initiation complex with eIF-4F (a cap binding protein) allows the 40S subunit to bind 5′ end of mRNA which is aided by eIF-4F, eI4A, eIF-4B and high energy bond of ATP. The 43 pre-initiation complex with met tRNA, eIF-2, eIF-3 and eIF-1A binds to mRNA to form 48S pre-initiation complex.

(ii) The eIF1 and eIF-1A act synergistically to promote the formation of stable **48S pre-initiation complex.**

(d) Formation of 80S subunit:

(i) 60S ribosomal subunit binds to the 40S subunit carrying mRNA and tRNA positioned at the AUG initiation codon to form 80S initiation complex.

(ii) The eIF4A (RNA helicase) can unwind hairpins of 5′ leaders of eukaryotic mRNA: aided by eIF4B and ATP.

(iii) The eIF4G is an adaptor protein, capable of binding to a variety of other protein including eIF4E (cap binding protein). eIF3 (40S-ribosomal subunit binding protein) and Pab1p (a poly-(A) binding protein). By interacting with these protein, eIF4G can recruit 40S ribosomal subunit to the mRNA and stimulate translation.

(iv) The eIF-5B stimulates the association of 2 ribosomal subunit with eIF-5 and GTP.

Q. 8 How is the correct amino acid paired with each codon?

Solution:

(i) The tRNA molecule acts as an "adapter" that pairs each codon with correct amino acid. Think of an adapter plug that makes a monitor compatible with a computer. The card from the monitor may not fit directly into the computer but will fit if it is attached to the correct adapter, which translates the shape of the monitor card into a shape that fits into the computer.

(ii) In a similar way, one end of the tRNA molecule binds to an amino acid. The anticodon–containing end binds (by base complementarity) to a codon in mRNA. In this way, the tRNA translates the codon of the mRNA into the amino acid of the protein.

Q. 9 Difference between Prokaryotic and Eukaryotic translation:

Solution:

Prokaryotic translation	Eukaryotic translation
1. In prokaryotes most messenger RNA contain the information for several cistron or genes, i.e., **poly-cistronic** message which forms polyprotein.	1. Here mRNA carries information for one polypeptide chain, i.e., **mono-cistronic**-mRNA form single protein.
2. Here, both, transcription and translation occur simultaneously.	2. RNA synthesis (transcription) and translation occur in different places.
3. IF_1, IF_2, IF_3 and SDS, AUG of mRNA and 16S rRNA participate in initiation of translation.	3. Initiation is made by many (10) eIF factors and AUG of mRNA.
4. Ribosome enter the mRNA at AUG codon or at nearby **shine Dalgarno sequence** (5′-AGGAGGU-3′).	4. Ribosome enters the mRNA at the capped 5′ end of mRNA and then advances to AUG codon by linear scanning, i.e., at **kozak sequence** (5′-ACCAUGG-3′).
5. mRNA lacking cap translated poorly.	5. 5′ end of mRNA is capped with 7 methyl guanosine residue. This 'cap' is essential for efficient translation.
6. Initiating amino acid (methionine) needs to be formylated. [Due to this reason two tRNAs for methionine, i.e., $tRNA_f^{met}$ and $tRNA_A^{met}$ are found.]	6. Initiating amino acid (methionine) is not formylated [only one tRNA for methionine, i.e., $tRNA_i^{met}$ is found].
7. No initiation factors or cofactors are required for initial contact between ribosomes and mRNA.	7. ATP and a number of protein factors are needed for ribosomes to engage the mRNA.
8. Small ribosomal subunit (30S) can engage mRNA before binding of initiator met $tRNA_f^{met}$.	8. Small ribosomal subunit (40S) binds stably to mRNA only after initiator met tRNA has bound.

Q. 10 What is the evidence that the ribosomal RNA, not the ribosomal proteins, carries out the key steps in translation.

Solution:

The crystal structure of ribosomes clearly shows that the decoding centre in the 30S subunit and the peptidyl transferase centre in the 50S subunit are composed entirely of rRNA.

Q. 11 Why is the attachment of amino acid to the correct tRNA considered to be such an important step in protein synthesis?

Solution:

The results of experiments have shown that anticodon of a tRNA charged with an incorrect amino acid still interacts with the complementary codon in mRNA, thus causing the wrong amino acid to be incorporated into the growing polypeptide chain. No proof-reading or editing mechanism exits in the ribosome to remove these incorrect amino acids.

Q. 12 How does initiation of translation differ in prokaryotes and eukaryotes?

Solution:

(i) In prokaryotes, translation starts when 16S rRNA in the 30S ribosomal subunit binds to the Shine-Dalgarno sequence near the 5′ end of an RNA that is still being transcribed.

(ii) The initiation complex then assembles at that sites, which is upstream of an initiator AUG codon.

(iii) In contrast eukaryotic mRNAs are transported out of the nucleus to the cytoplasm where an elongation factor binds to the 5′ cap.

(iv) The initiation complex can assemble and scan the mRNA by moving in a 3′ direction, unwinding the RNA in search of an AUG codon that is in the proper sequence context for translation initiation.

Q. 13 (a) In how many cases in the genetic code would you fail to know the amino acid specified by a codon if you knew only the first two nucleotides of the codon?

(b) In how many cases would you fail to know the first two nucleotides of the codon if you knew which amino acid is specified by it?

Solution:

(a) By studying the table listing the genetic code, we will discover that there are eight cases in which knowing the first two nucleotides does not tell us the specific amino acid.

(b) If we know the amino acid, we would not know the first two nucleotides in the cases Arg, Ser and Leu.

Q. 14 What is post-translational processing and why is it important for protein function?

Solution:

(i) Some proteins cannot function until they are modified by the addition of certain molecules, such as phosphate or carbohydrates.

(ii) In addition, some proteins do not function until they have been transported to a particular cellular compartment.

(iii) Proteins destined for the nucleus contain nuclear localization sequences, whereas those destined for secretion, insertion into the membranes or transport into organelles contain signal sequences at their amino ends. These signal sequences are cleaved during targeting and are not part of the mature protein.

Q. 15 In protein synthesizing system in vitro, the addition of specific human mRNA to the *E. coli* translational apparatus (ribosome, tRNA and so forth) stimulates the synthesis of a protein very much like that specified by the mRNA. What does this result show?

Solution:

(i) It suggests very little evolutionary change between *E. coli* and humans with regard to the translational apparatus.

(ii) The code is universal, the ribosomes are interchangeable and the participating enzymes are interchangeable.

Q. 16 How might various antibiotics be used to study the process of translation?

Solution:

(i) Most of the antibiotics (**Streptomycin** and **neomycin, tetracycline, chloram-phenicol, erythromycin**) are inhibitor of protein synthesis.

(ii) Streptomycin and Neomycin bind to a particular protein in the 30S subunit and thereby prevent binding of $tRNA_f^{met}$ to the P site.

(iii) Tetracyclins inhibit binding of charged tRNA, chloramphenicol inhibit the peptidyl transferase. Erythromycin binds to free 50S subunit and prevent formation of 70S ribosome.

(iv) Puromycin causes premature chain termination.

Q. 17 In a haploid, a nonsense suppressor su 1 acts on mutation 1 but not on mutation 2 or 3 of gene P. An unlinked nonsense suppressor su 2 works on P mutation 2 but not on 1 or 3. Explain this pattern of suppression in terms of the nature of the mutation and the suppressors.

Solution:

(i) With the assumption that the three mutations of gene P are all nonsense mutations. Three different possible stop codons might be the cause (**amber, opal** and **ochre**).

(ii) A suppressor mutation would be specific to one type of nonsense codon. For example, amber suppressor would suppress **amber** mutants but not **opal** or **ochre.**

Q. 18 The first amino acid in bacterial polypeptide chain is methionine. The start codon in the mRNA is GUG, which codes for valine. Why is not the first amino acid formylmethionine or valine?

Solution:

(i) The first amino acid on a polypeptide chain is carried by the initiator tRNA, which carries formyl methionine.

(ii) This occurs even when the start codon is GUG (valine) or UUG (leucine).

(iii) The formyl group can be later removed to yield methionine as the first amino acid.

Q. 19 What is adaptor? Mention its utility.

C.U. 2015

Solution:

I. A transfer RNA is an adaptor molecule composed of RNA, typically 76 to 90 nucleotide length.

II. It serves as the physical link between the mRNA and the amino acid sequence of protein.

(i) Each amino acid is first attached to its own specific "adaptor" piece of nucleic acid (in an enzyme-catalyzed reaction).

(ii) The order of assembly of amino acids is then determined by a specific recognition between the adaptor and the nucleic acid which is serving as the information plate.

(iii) In this way, amino acids could be lind up by the template in a specific order.

Q. 20 Suppressors of frameshift mutations are now known. Propose a mechanism for their action.

Solution:

If the anticodon on a tRNA molecule also were altered by mutation to be four bases long with the fourth base on the 5′ side of the anticodon.

It would suppress the insertion. Alterations in the ribosome can also induce frameshifting.

Q. 21 What do the process of DNA synthesis, RNA synthesis and polypeptide synthesis have in common.

Solution:

(i) DNA, RNA and protein synthesis all involve the synthesis of long chains of repeating subunits.

(ii) All three processes can be divided into three stages—initiation, elongation and termination.

Q. 22 Which of the following is the usual cause of chain termination? (i) mRNA synthesis stops at a chain termination codon, (ii) the tRNA corresponding to a chain termination codon cannot be charged with an amino acid, (iii) there are no tRNA molecules that have anticodons corresponding to chain termination codons.

Solution:

(iii) There are no tRNA molecules that have anticodons corresponding to chain termination codons

Q. 23 Poly-U codes for polyphenylalanine. If a G is added to the 5′ end of the molecule, the polyphenylalanine has a different amino acid at the amino terminus and if a G is added to the 3′ end, there is a different amino acid at the carboxyl terminus. What are the amino acids?

Solution:

With a G at the 5′ end, the first codon is GUU, which codes for valine.

With a G at the 3′ end, the last codon is UUG, which codes for leucine.

Q. 24 If the DNA molecule shown here is transcribed from the left to right, what is the sequence of the mRNA and the amino

acid sequence? What is the sequence of the mRNA and the amino acid sequence if the segment is inverted?

5′AGACTTCAGGCTCAACGTGGT3′

3′TCTGAAGTCCGAGTTGCACCA5′

Solution:

If transcription occurs from left to right and the template is 3′–5′ and transcription occurs from 5′ to 3′ end, the sequence of mRNA is as follows:

5′AGACUUCAGGCUCAACGUGGU3′

which codes for Arg-Leu-Gln-Ala-Gln-Arg-Gly.

To invert the molecule, the strands must be interchanged in order to preserve correct polarity and so the sequence of the inverted molecule is

5′AGAACGTTGAGCCTGAAGGGT–3′

3′ TCTTGCAACTCGGACTTCCCA–5′

These codes for an mRNA with sequence AGA ACG UUG AGC CUGAAG GGT, codes for Arg-Thr-Leu-Ser-Leu-Lys-Gly.

Q. 25 The following DNA sequence represents part of a transcribed gene. Assume that all of the nucleotide triplets that contain a C constitutes intron DNA and all others exon DNA. (a) Show the RNA transcript and (b) show the processed mRNA:

DNA: TAC | CCC | CAC | GAG | TTA | TAT | ATA | CGG | GGG | GTT | AAA | CTC | CAT | CAT | CAT

Solution:

(a) RNA:AUG | GGG | GUG | CUC | AAU | AUA | UAU | GCC | CCC | CAA | UUU | GAG | GUA | GUA | GUA

(b) Simply splice out the triplets of RNA that contains a C.

mRNA:AUG | GGG | GUG | AAU | AUA | UAU | UUU | GAG | GUA | GUA | GUA

Q. 26 Assume that a genetic situation exists in which a particular gene has three alleles. Call them A_1 A_2 and A_3. Also, assume that three different polypeptides resulting from these alleles can randomly combine with each other in all combinations to form functional protein molecules, each consisting of two of the polypeptides (dimer).

(a) How many different protein molecules could possibly be formed a very large population of this diploid organism?

(b) What protein molecules are possible, using the symbols given in the above question?

(c) What is the maximum number of different protein molecules that could be found in any organism?

Solution:

(a) Protein molecules formed in this case is $\underline{|3}$. Since three alleles exist, that is $3 = 3 \times 2 \times 1 = 6$.

(b) A_1A_1, A_1A_2, A_1A_3, A_2A_2, A_2A_3, A_3A_3.

(c) Since the organism is a diploid, only three different protein molecules could be found in any organism. For example, A_1A_1, A_1A_2, A_2A_2 or other comparable combinations.

Q. 27 The following interrupted length of DNA constitutes a gene in eukaryotic organism. Which side of the DNA duplex (top or bottom) is the side transcribed? Give reason for your decision.

3′–TACCGACCC...... TGCAAT–5′

5′–ATGGCTGGG...... ACGTTA–3′

Solution:

(i) The top side of the duplex is the transcribed DNA strand. The TAC triplet corresponds to AUG, which is the start codon in most eukaryotes.

(ii) Also the sequence ends with ATT, which relates to the UUA stop codon.

Q. 28 Biochemical analysis shows that a particular protein molecule has four identical NH^+ terminal amino acid residues and four identical COO^- terminal residues. What can be suggested relative to the structure of this protein?

Solution:

(i) Four polypeptides make up this protein molecule because of the four NH^+ and COO^- terminal residues.

(ii) Also the four are probably identical polypeptides that is a homotetramer.

(iii) Even considering that all four might begin with Met, like most polypeptides, the chances that the other terminal residues would be the same by random is $\left(\frac{1}{20}\right)^4 = \frac{1}{160,000}$.

Q. 29 Analysis of certain enzyme showed it to be a single polypeptide chain of 480 amino acids.

(a) Disregarding leader regions, caps, starts and stops, how many ribonucleotides would you expect to find in the mRNA molecule translated to produce this enzyme?

(b) If all of the amino acid residues were equally represented in this enzyme, how many of residues would be leucine?

Solution:

(a) One amino acid is represented by three codons (ribonucleotides). Therefore, 480 amino acids would represent by 480 × 3 = 1440 ribonucleotides.

(b) Normally, the ratio of leucine is $\frac{1}{20}$; therefore, in 480 amino acids, it will be $\frac{1}{20} \times 480 = 24$.

Q. 30 A segment of DNA has the following deoxyribonucleotide sequence:

T T A T C T T C G G G A G A G A A A

If transcription begins on the left and proceeds towards the right, what amino acid sequence would correspond to this DNA?

Solution:

DNA:	TTA	TCT	TCG	GGA	GAG	AAA
RNA:	AAU	AGA	AGC	CCU	CUC	UUU
Amino acids:	Asn	Arg	ser	pro	leu	phe

Q. 31 The following strand of mRNA has been processed and is ready for translation:

5′–AUG CUA UAC CUC CUU UAU CUG UGA–3′

(a) How many amino acid residues will make up this polypeptide corresponding to this mRNA?

(b) How many different kinds of amino acid will make up this polypeptide?

(c) How many different tRNA molecules would be necessary to translate this mRNA into the polypeptide?

Solution:

(a) 7, UGA is stop codon.

(b) AUG = Meth

$\underbrace{\text{CUA CUC CUU CUG}}_{\text{Leu}} \quad \underbrace{\text{UAC UAU}}_{\text{Tyr}}$

Three different kinds of amino acids, viz., Meth, Leu and Tyr make up this polypeptide. Some degeneracy is involved.

(c) Could be seven because different codons will involve different anticodons.

Q. 32 At one point in the translation of an mRNA, a sequence of four transfer RNA molecules have the anticodons UUU, AAA, UGA and CCU respectively. From this information, construct the segment of the duplex DNA that coded for this sequence.

Solution:

Anticodons: U U U A A A U G A C C U
Codons: A A A U U U A C U G G A
Transcribe
DNA strand: T T T A A A T G A C C T
Complementary
DNA strand: A A A T T T A C T G G A
Duplex DNA: T T T A A A T G A C C T
A A A T T T A C T G G A

Q. 33 An alanine tRNA molecule charged with alanine is isolated and chemically treated such that the alanine is replaced with glycine. The treated amino acid tRNA complex is then introduced into a cell-free peptide synthesizing system. This treated tRNA will become bound to which mRNA codon?

Solution:

(i) Treated tRNA alanine becomes tRNA glycine. GCU is one codon for alanine.

(ii) This codon will not change, because of a change in amino acid, specificity is determined between codon and anticodon.

Q. 34 A synthetic mRNA molecule is produced from a mixture of guanine and uracil in the relative frequencies of 4 :1 respectively. Assuming that the sequence will form in a random linear manner, what triplets would be formed, and what would be the relative frequencies of each?

Solution:

G = 0.80 and U = 0.20

GGG = $(0.80)^3$ = 0.512

GGU, UGG, GUG = 0.80 × 0.80 × 0.20 = 0.128 (each of them)

GUU, UUG, UGU = 0.80 × 0.20 × 0.20 = 0.032 (each of them)

UUU = $(0.20)^3$ = 0.008

Q. 35 The codon UCA specifies the amino acid serine.

(a) How many single ribonucleotide substitutions in this codon could be result in chain termination?

(b) How many single ribonucleotide substitutions in this codon would fail to change the amino acid?

Solution:

(a) Two (2): UCA ⟶ UGA (opal)
UCA ⟶ UAA (ochre)

(b) Three (3): UCA ⟶ UCC, UCG, UCU } Serine

Q. 36 The following sequence of bases is present along one chain of a DNA duplex that has opened up at a replication fork, and synthesis of an RNA primer on this template begins by copying the base (backet).

3′ T C T G A T A [T] C A G T A C G...5′

(a) If the RNA primer consists of eight nucleotides, what is its base sequence?

(b) In this intact RNA primer, which nucleotide has a free hydroxyl (–OH) terminus and what is the chemical group on the nucleotide at the end of the primer?

(c) If replication of the other strand of the original duplex proceeds continuously (with a few or no intervening RNA primers), does the replication most likely move from left to right or right to left direction?

Solution:

(a) (i) The first base in the DNA that is copied is a T, and so the first base in the RNA is an A.

(ii) RNA grows by addition to the 3′ OH group and so the 5′ P end remains free. Therefore, the sequence of the eight nucleotide primer is

5′ AGUCAUGC–3′

(b) The C at the 3′ end has a free –OH. The A at the 5′ end has a free triphosphate.

(c) Right to left, because this direction would require discontinuous replication of the strand shown.

Q. 37 Determine the transcribed DNA strand, the complementary DNA strand and the mRNA codons that correspond to the following sequence of anticodons:

UAC—CGU—AAC—UCC

Solution:

The anticodons exist as triplets of ribonucleotides on the tRNA molecules. Also, recall that RNA molecules contain uracil rather than thymine. The corresponding complementary relationships, therefore, are as follows:

tRNA anticodons:	UAC—CGU—AAC—UCC
mRNA codons:	AUG—GCA—UUG—AGG
DNA transcribed strand:	TAC—CGT—AAC—TCC
DNA complementary strand:	ATG—GCA—TTG—AGG

Q. 38 The following polyribonucleotides were used in vitro system to synthesize polypeptides. Which amino acids would be expected to be incorporated into the polypeptide in each case:

(a) Poly G

(b) Poly GU

(c) Poly UA

Solution:

(a) Polyribonucleotide strand: GGG | GGG | GGG | GGG, etc.

Polypeptide: Gly—Gly—Gly—Gly, etc.

(b) Polyribonucleotide strand: GUG UGU GUG UGU, etc. = GUG | UGU | GUG | UGU

Polypeptide = Val—Cys—Val—Cys—, etc.

(c) Polyribonucleotide strand: UA UA UA UA UA UA, etc. = UAU | AUA | UAU | AUA |, etc.

Polypeptide = Tyr—Ile—Tyr—Ile, etc.

Q. 39 The following is the nucleotide sequence of DNA:

TAC GTC TCC AGC GGA GAC CTT TTC CGG TCG CAA CTG AGG TTG ATC

The strand is transcribed from left to right and codes for a small peptide.

(a) Which end is 3′ and which the 5′ end?

(b) What is the sequence of the complementary DNA strand?

(c) What is the sequence of the transcript?

(d) What is the amino acid sequence of the peptide?

(e) In which direction along the transcript does translation occur?

(f) Which is the amino ($-NH_2$) end of the peptide, the carboxyl (–COOH) end?

Solution:

The DNA strand is transcribed from left to right, and the first nucleotides incorporated into an RNA transcript from its 5′ end, the 3′ end of the DNA strand must be at the left.

The original strand

(a) 3′–TAC GTC TCC AGC GGA GAC CTT TTC CGG TCG CAA CTG AGG TTG ATC–5′

(b) The complementary strand is as follows: 5′–ATG CAG AGG TCG CCT CTG GAA AAG GCC AGC GTT GAC TCC AAC TAG–3′

(c) The sequence of transcript is 5′–AUG CAG AGG UCG CCU CUG GAA AAG GCC AGC GUU GAC UCC AAC UAG–3′

(d) 5′– Met–Gln–Arg–Ser–Pro–Leu–Glu–lys–Ala–Ser–Val–Asp–Ser–Asn–ter–3′

(e) Translation is form 5′–3′ along the mRNA and so in this example, translation also is from left to right.

(f) The amino end of the peptide is synthesized first (Met) and the carboxyl end last (Asn).

Q. 40 The average mass of the 20 amino acids is about 137 daltons. Estimate the approximate length of an mRNA molecule that encodes a polypeptide with a mass of 65,760 daltons. Assume that the polypeptide contains equal amount of all 20 amino acids.

Solution:

(i) Based on this assumption, the polypeptide would contain about 480 amino acids $\left(\frac{65,760}{137}\right.$ daltons per amino acid).

(ii) Since each codon contains three nucleotides, the coding region of the mRNA would have to be 480 × 3 = 1440 nucleotides long (3 nucleotide per amino acid).

Q. 41 Is the DNA that serves as the template for the RNA polymerase transcribed in the 5′ → 3′ or 3′ → 5′ direction? Which end of an mRNA molecule is translated first? Which end of a polypeptide is synthesized first?

Solution:

The template strand of DNA remain in the 3′ → 5′ direction and the 5′ end of an mRNA molecule is transcribed first. Therefore, the 5′ end of the mRNA is translated first.

The amino ($-NH_2$) end of the polypeptide is synthesized first.

Q. 42 What are the translation initiation and stop codons in the genetic code? In a random sequence of the four ribonucleotides, what is the probability that any 3 adjacent nucleotides will be a start codon? A stop codon: In an mRNA molecule of random sequence, what is the average distance between stop codons?

Solution:

The translation initiation codon is AUG or GUG which codes for methionine. Stop codons are UAA. UAC and UGA.

The probability of a start codon is $\frac{1}{64}$ and the probability of stop codon is $\frac{3}{64}$.

Average distance between stop codons is $\frac{64}{3}$ = 21.3 (codons).

Q. 43 What anticodon sequence would pair with the codon 5′–AUG–3′, assuming no modified bases are present in the anticodon?

Solution:

In the absence of modified bases, the base pairing is the usual A with U and G with C. However, the orientation of the codon and anticodon are antiparallel, and so the anticodon sequence is 3′–UAC–5′.

Q. 44 In human β-globin, polypeptide is 146 amino acid long. How long is the coding portion of the human β-globin mRNA?

Solution:

(i) Each amino acid is specified by a codon containing three nucleotides. Therefore, the 146 amino acids in β-globin will be specified by 438 (146 × 3) nucleotides.

(ii) However, a termination codon must be present at the end of the coding sequence, bringing the length to 438 + 3 = 441 nucleotides.

(iii) In the case of β-globin and many other proteins, the amino terminal methionine (specified by the initiation codon AUG) is removed from the globin during synthesis.

(iv) Adding the initiation codon increase the coding sequence of the β-globin mRNA to 444 nucleotides (441 + 3).

Q. 45 If the coding segment of an mRNA with the sequence 5′ AUG UUU CCC AAA GGG–3′ is translated, what amino acid sequence will be produced?

Solution:

(i) (Amino terminus)-methionine, phenylalanine-proline-lysine-glycine-(carboxyl terminus).

(ii) AUG is the methionine initiation codon followed by phenyl alanine, codon UUU, proline codon CCC, the lysine codon AAA and the glycine codon GGG.

Q. 46 If the average molecular mass of an amino acid is assumed to be 100 daltons, how many nucleotides approximately will be present in an mRNA coding sequence specifying a single polypeptide with molecular mass of 27,000 daltons?

Solution:

(i) $\frac{27,000}{100}$ = 270 amino acid is present.

(ii) 3 nucleotide per amino acid. Therefore, 270 × 3 = 810 nucleotides will be present.

Q. 47 An *E. coli* gene has been isolated and shown to be 68 nm long. What is the maximum number of amino acids that these gene could encode?

Solution:

(i) Assuming 0.34 nm per nucleotide pair in B DNA, a gene 68 nm long would contain 200 nucleotide pairs.

$$\left(\begin{array}{l} 0.34 \text{ nm} = 1 \text{ bp, or, } 1 \text{ nm} = \frac{1}{0.34} \text{ bp} \\ \therefore\ 18 \text{ nm} = \frac{1}{0.34} \times 68 = 200 \text{ bp} \end{array}\right)$$

(ii) Given the triplet code, this gene would contain $\frac{200}{3}$ = 66.7 triplets, one of which must specify chain termination.

(iii) Disregarding the partial triplet this gene could encode a maximum of 65 amino acids.

Q. 48 The human α-globin gene is 141 amino acids long. How many nucleotides in

mRNA required to encode human α-globin?

Solution:

(i) One amino acid = 3 nucleotides. Therefore, 141 amino acids: 141 × 3 = 423 nucleotides.

(ii) 423 + 3 (one codon) for chain termination.

∴ 426 nucleotides.

Q. 49 What is Wobble?

Solution:

Wobble refers to the possibility of forming base pairs other than AU and GC in the third position of a codon-anticodon pair.

Q. 50 What are the three types of hydrogen bonds in tRNA?

Solution:

Base-base, base-ribose and ribose-ribose bonds. The second and third types are used only for tertiary folding.

Q. 51 What is meant by suppressor tRNA and how does such a molecule arise?

Solution:

A tRNA molecule that either can read a chain termination codon or with low frequency, either misread a codon or mischarged.

Such molecule arises from mutations in genes for tRNA molecules.

Q. 52 Why are Ochre (UAA) suppressors usually weak suppressors?

Solution:

Binding to a UUA anticodon must be weak. This is necessary because UAA often occurs singly as a natural stop codon.

Q. 53 A certain amber suppressor inserts tryptophan (Trp) in response to amber codon. What change in the anticodon of tRNA^{TrP} probably created this suppressor?

Solution:

(i) Tryptophan codon is 5′ UGG 3′ which is recognized by the anticodon 3′ ACC 5′ of the wild type tRNA.

(ii) The amber codon is 5′ UAG 3′ which is most likely recognized by the anticodon 3′ AUC 5′ of the amber suppressor tRNA.

(iii) Thus, comparing this anticodon (3′ AUC 5′) with the wild type one (3′ ACC 5′) shows that change in the tRNA^{Trp} anticodon that created the amber suppressor was 3′ ACC 5′ → 3′ AUC 5′. In other words, C was changed to a U.

Q. 54 A certain ochre suppressor inserts glutamine (Gln) in response to ochre codon. What is the likest change in the anticodon at RNA^{Gln} that created this suppressor strain?

Solution:

(i) The glutamine codon is 5′ CAA 3′, which is recognised by anticodon 3′ GUU 5′ of the wild type tRNAGln.

(ii) The ochre codon is 5′ UAA 3′, which is almost recognised by anticodon 3′ AUU 5′ of the ochre suppressor tRNA.

(iii) Thus, comparing this anticodon (3′ AUU 5′) with the wild type one (3′ GGU 5′) shows that the change in tRNA^{Gln} anticodon that created ochre suppressor was 3′ GUU 5′ → 3′ AUU 5′.

In other words, G was changed to A.

Q. 55 What is nick and nick translation?

Solution:

A single stranded break, involving the absence of one or more nucleotides in a double-stranded DNA molecule is known as **nick**.

(i) The repair of a nick with DNA polymerase I, usually introduces labelled nucleotides into DNA molecule.

(ii) Here DNA pol I simultaneously degrades the DNA (5′–3′ exonuclease action) of a nick and elongates the DNA (5′–3′ polymerase action) behind the nick.

(iii) The result is movement (translation) of nick in the 3′ direction in a DNA strand.

Q. 56 Why is it said that genetic code is non-overlapping and degenerate?

Solution:

The genetic code is non-overlapping because consecutive amino acids are specified by consecutive codons. The code is degenerate because more than one can specify the same amino acid.

For example: GGC, GGA and GGU all specify glycine.

Q. 57 What is Kouzak rule?

Solution:

(i) In eukaryotes, translation frequently begins at the **AUG** closest to the 5′ terminus of the mRNA molecule, although the efficiency with which a given '**AUG**' is used to initiate translation depends on the contiguous nucleotide sequence.

(ii) The sequences of bases around AUG play an important role in determining whether or not, it will be selected as the start codon by scanning ribosome.

(iii) The optimum initiation sequence is
5′-GCC(A/G)CCAUGG-3′
-6-5-4-3-2-1+1+2+3+4

The purine(A/G) three bases upstream from the AUG initiator codon and the G immediately following it are the most important influencing initiation efficiency by ten-fold or more.

(iv) The changes of other bases in the sequence cause smaller decrease in initiation efficience. These sequence requirements for optimal translation initiation in eukaryotes are called **Kozak's rule.** (After Marilyn Kozak who 1st proposed it).

Q. 58 What is the role of amino acyl-tRNA synthetase? The ability of the aminoacyl tRNA synthetases to recognize tRNAs has sometimes been called "the second genetic code". Why has the function of this type of enzyme been described this way?

Solution:

(i) The role of **aminoacyl tRNA synthetase** is to specifically recognize tRNA molecules and attach correct amino acid to them.

(ii) This ability is sometimes described as the second genetic code because the specificity of the attachment is a critical step in deciphering the genetic code.

(iii) For example: If a tRNA has a 3′ GGG 5′ anticodon it will recognise 5′-CCC-3′ codon which should specify proline.

It is essential that the aminoacyl tRNA synthetase known as proline-tRNA-synthetase recognise this tRNA and attaches proline to the 3′ end.

(iv) The other aminoacyl tRNA synthetase should not recognize this tRNA.

Problem 59

The first amino acid in certain bacterial polypeptide chain is methionine. The start codon in the mRNA is GUG, which codes for valine. Why isn't the first amino acid formyl-methionine or valine?

Solution:

(i) The first amino acid in a polypeptide chain is carried by the initiator tRNA which always carries **formyl methionine.**

(ii) This occurs even when the start codon is GUG (valine) or UUG (leucine).

(iii) The formyl group can be latter removed to yield methionine as the first amino acid.

Problem 60

A tRNA has the anticodon sequence 3′-CAG-5′. What amino acid does it carry?

Solution:

(i) The anticodon 3′-CAG-5′ would be complementary to a codon with sequence 5′-GUG-3′.

(ii) According to the genetic code, this codon specifies the amino acid valine. Therefore, this tRNA must carry valine at its acceptor stem.

Problem 61

What anticodon could recognize all three codons for isoleucine assuming wobble? Write the anticodon in the 3′ → 5′ direction.

Solution:

UAI.

Problem 62

If a tRNA molecules carries a glutamic, what are the two possible anticodon sequences that it could contain? Be specific about the 5′ and 3′ ends.

Solution:

The codons for glutamic acid are GAA and GAG. The possible anticodons are 3′-CUU-5′ and 3′-CUC-5′.

Problem 63

A tRNA has an anticodon sequence 3′-GUU-5′. What amino acid does it carry?

Solution:

The anticodon is 3′-GUU-5′. The mRNA sequence is 5′-CAA-3′. According to genetic code the codon specifies the amino acid proline. Therefore, the tRNA must carry proline at its acceptor arm. CAA is proline amino acid.

Problem 64

A tRNA has an anticodon sequence 3′-GGU-5′. What amino acid does it carry?

Solution:

The anticodon is 3′-GGU-5′. The 5′-CCA-3′. According to genetic code, the codon specifies the amino acid proline. Therefore, tRNA must carry proline at its acceptor arm, CCA is proline.

Problem 65

If a tRNA anticodon sequence 3′-CCI-5′. What codon(s) can it recognise?

Solution:

The anticodon is 3′-CCI-5′. It can recognise 5′-GGU-3′, 5′-GGC-3′ and 5′-GGA-3′.

Problem 66

Highlight the significance of modified bases within tRNA molecules.

Solution:

(i) Bases that have been chemically modified can occur various locations through out the tRNA molecule.

(ii) The significance of all of these modification is not fully known.

(iii) However, within the anticodon region, base modification alters base pairing to allow the anticodon to recognize two or more different bases within the codon.

Problem 67

What amino acids could replace methionine if one base mutation occured.

Solution:

(i) The codon for methionine (internal as well as initiator) is AUG.

(ii) If A is replaced =
GUG(val) CUG(Leu) UUG(Leu)
If U is replaced =
AGG(arg) ACG(thr) AAG(Lys)
If G is replaced =
AUC (Ileu) AUU (Ileu) AUA (Ileu)

(iii) Therefore, a one base change in codon result, six different amino acids viz. valine, leucine, arginine, isoleucine, theronine and lysine.

Problem 68

As it necessary for a cell to make 6′ different tRNA molecules corresponding to th 6′ codons for amino acids? Explain your result?

Solution:

(i) No, it is not.

(ii) On the basis of wobble rules, the 5′ base in the anticodon of tRNA can recognize two or more bases in the third (3′) position in the mRNA.

(iii) Therefore, any given cell type synthesizes fewer than 61 types of tRNA.

Problem 69

Select the events of bacterial translation which occurs primarily within 30s subunit or within 50s subunit or at the interface betwen these two ribosomal subunits.

(a) mRNA-tRNA recognition (b) peptidyl transfer reaction (c) exist of the polypeptide chain from the ribosome (d) binding of initiation factors IF_1m IF_2 and IF_3.

Solution:

(a) On the surface of the 30s subunit and at the interface between the two subunits.

(b) Within the 50s subunit

(c) From the 50s subunit

(d) To the 30s subunit.

Problem 70

Mutant tRNAs may act as nonsense and missense suppressor. Explain it at molecular level.

Solution:

(i) A suppressor is a second-site mutation that suppress the phenotypic effects of a first mutation.

(ii) Intergenic suppressor mutations in tRNA genes can act as nonsense or missense suppressor.

For example: Lets suppose a first mutation puts a stop codon into a structural gene.

(iii) A second mutation in a tRNA gene can alter the anticodon region of a tRNA, so that the anticodon recognizes a stop codon, but inserts an amino acid at this site.

(iv) A missense suppressor is a mutation in a tRNA gene, that changes the anticodon, so that it puts in the wrong amino acid at a normal codon that is not a stop codon. These mutant tRNAs are termed missense tRNA.

For example: a tRNA that normally recognizes glutamic acid (**GAA/GAG**) may incur a mutation that changes its anticodon sequences, so that it recognizes glycine codon (**GGA, GGG, GGC, GGU**) instead of

	codon anticodon		codon anticodon
glutaric acid	GAA = C[U]U CCU }**		GGA = CCU }**
	GAG = C[U]C CCC }**		UGG = CCC }**
			GGC = CCG
** glycin			GGU = CCA

(v) Like nonsense suppressor, missense suppressors can be produced by mutation in the anticodon region of tRNAs, so that the tRNA recognizes an incorrect codon.

(vi) Alternatively, missense suppressor can also be produced by mutation in aminoacyl-tRNA synthetases that cause them to attach the incorrect amino acid to tRNA.

Problem 71

Describe the sequence in bacterial mRNA that promotes recognition by the 30s subunit.

Solution:

(i) Most bacterial mRNAs contain a Shine-Dalgarno sequence which is necessary for the binding of the mRNA to the small ribosomal RNA.

(ii) This sequence 'AGGAGGU' is complementary to sequence (UCCUCCA) in the 16S rRNA.

(iii) Due to this complementarity, these sequences will hydrogen bond to each other during the initiation stage of translation.

S. D. sequence

mRNA-5′……A G G A G G U……3′

16S rRNA-3′……U C C U C C A……5′

Region of base pairs

Problem 72

How does a eukaryotic ribosome select its start codon? Describe the sequences in eukariotic mRNA that provides optimal context for a start codon.

Solution:

(i) The ribosome binds at the 5′ end of the mRNA and then scans in 3′ direction in search of an AUG start codon.

(ii) If it finds one that reasonably obeys Kozak's rules, it will begin translation at that site.

(iii) Aside from an AUG start codon, two other important features are guanosine at the +4 position and a purine at the –3 position.

Problem 73

Eukaryotic mRNAs are post transcriptionally modified at their 5′ and 3′ ends. What role does each of the modifications play in translation.

Solution:

(i) A eukaryotic mRNA is modified to contain a 5′ 0 7 **methyl G** cap and 3′ poly (A) tail.

(ii) The 5′ cap is required early in translation initiation. It binds to the eIF–4F complex just prior to binding of a complex 40s ribosomal subunit, the initiator Met tRNA and other eIF proteins.

(iii) Transcription initiation is stimulated by the looping of the poly (A) tail close to the 5′ end. This occurs when poly (A) binding protein (PAB) binds to eIF-4G, which is part of the eIF-4F complex.

Problem 74

In recombinant DNA, researchers typically prefer ligating restriction fragments that have sticky ends (single-stranded overhangs) rather than those that have blunt ends. Can you propose a reason why?

Solution:

The sticky ends are preferred because a sticky end will join only with other sticky ends that are complementary in sequence, whereas blunt ended fragments will join with any other blunt ended fragments.

Problem 74

What is the average distance between restriction sites for each of the following restriction enzymes. Assume that the DNA substrate has random sequence with equal amounts of each nucleotide. The symbol N stands for any nucleotide, R for any prime (A and G) and Y for any pyrimidine (T or C).

(a) TCG A Taq I, (b) GGT ACC Kpn I, (c) GTN AC Mac III, (d) GGNNCC Nla IV, (e) GRCCYC Acy I.

Solution:

(i) The average distance between restriction sites equals the reciprocal of the probabiity at occurrence of the restriction sites.

(ii) Therefore, we must calculate the probability of occurrence of each restriction sites on a random DNA sequence.

(a) 4 nucleotide is present, i.e., $\frac{1}{4} \times \frac{1}{4} \times \frac{1}{4} \times \frac{1}{4}$ or $\left(\frac{1}{4}\right)^4 = \frac{1}{256}$, the reciprocal of this 256.

(b) GGTACC = 6 nucleotide i.e., $\left(\frac{1}{4}\right)^6 = \frac{1}{4096}$ = 4096.

(c) GTNAC = 4 (because N = any nucleotide site, is 1) = $\left(\frac{1}{4}\right)^4 = \frac{1}{256}$; 256.

(d) GG NNCC = 4 i.e. $\left(\frac{1}{4}\right)^4 = 256$.

(e) GR CC YC = $\frac{1}{4} \times \frac{1}{2} \times \frac{1}{4} \times \frac{1}{4} \times \frac{1}{2} \times \frac{1}{4}$ (because the probabilities for R = $\frac{1}{2}$ and Y = $\frac{1}{2}$) = $\frac{1}{1024}$ or, 1024.

Problem 75

The restriction enzyme A as I has the restriction site 5′-GACNNNN/NNGTC-3′. Where N means any nucleotide (matched by complementary nucleotide in the other strand) and cleavage at the site of vertical bar. What is the probability that two random A as I sites have compatible sticky Y ends?

Solution:

(i) 5′ G A C NNNN 3
3′ C T G NN’

The two N’s in the overhangs would have to match perfectly.

(ii) The probability of this event is $\left(\frac{1}{4}\right)^2 = \frac{1}{16}$.

Problem 76

Which of the following DNA sequences are pallindromes? Which are mirror image.

**(a) TAAGCTTA
ATT CGAAT**
**(b) GC GAA GCG
CG CTT CGC**
**(c) GATCAGATC
CTA GTC TAG**
**(d) GTTG
CAAC**

Solution:

(a) T AA GC TTA / A TT CG AAT } Pallindrome

(c) GATC[A]GATC / CTAG[T]CTAG } Pallindrome (central symmetry)

(b) G C G A A G C G / C G C T T C G C } Mirror image

(d) G T T G / C A A C } Mirror image

Problem 77

Will the sequences 5′-GGCC-3′ and 3′-GGCC-5′ in a double-stranded DNA molecule be cut by the same restriction enzyme.

Solution:

No, because the 5′–3′ polarity of the sequence is different.

Problem 78

The restriction enzyme Fae I has the restriction site CATGI, where the 5′ end is written at the left and the cleavage site is indicated by the vertical bar. If an mRNA has the sequence CAUG, where AUG is the start codon for translation, would Fae I be a good enzyme translation, would Fae I be a good enzyme to use to cleave the DNA immediately after the initiation codon?

Solution:

(i) The mRNA has the sequence 5′-CAUG-3′ the transcribed DNA strand has the sequence 3′-GTAC-5′, and the non-transcribed strand has the sequence 5′-CATG–3′.

(ii) Hence, Fae I cleaves the coding sequence immediately after the initiation codon.

Problem 79

The restriction enzyme Aoc II has the restriction site GDGCH/C where D denotes A or G or C (with a complementary base in the other strand) and H denotes A or C or T (with a complementary base in the other strand). If two random Aoc II sites are cleaved, what is the probability that their sticky ends are compatible.

Solution:

There are 3 × 3 = 9 possible. Aoc II sites and so the probabilities that any two random sites have compatible error is $\frac{1}{9}$ or $\left(\frac{1}{3} \times \frac{1}{3}\right)$.

Problem 80

The restriction enzyme 81 has restriction site GC NNAC, where N denotes any nucleotide and the restriction enzyme Acc I was restriction sites GT MK AC, where M denotes A or C (with complementary base on the other strand). What is the probabioity that two random HPY 81 sites are identical? What is the probabilities that a random Hpy 81 site is also Acc I site?

Solution:

GC NNAC = 4 two random Hpy 81 = $\left(\frac{1}{4}\right)^2 = \frac{1}{16}$

Hpy site is also Acc I = $\left(\frac{1}{2}\right)^2 = \frac{1}{4}$.

Problem 81

What would be the effect of mutation in the initiating codon of an mRNA?

Solution:

(i) It would eliminate initiation at that point.

(ii) A subsequent AUG in the middle of the message might then function as start codon.

(iii) But the resulting protein product would be incomplete truncated at the amino terminus.

Problem 82

For eukaryotic proteins, what is the function of sorting signals? Explain.

Solution:

(i) An amino acid sequence of post-translation modification that directs a protein to its correct location (i.e., compartment) within cells.

(ii) Protein destined for the E. R. Golgi complex, lysosomes, plasma membrane or secretory vesicles have an SRP sorting signal at their terminal end.

(iii) Nuclear protein has a nuclear localization sequence and so on.

(iv) These sorting signals are recognized by specific cellular components that sort the protein to their correct destination.

WOBBLE HYPOTHESIS

1. The codon of mRNA and anticodon of tRNA are complementary to each other. On that basis, each codon welcomes one tRNA having the complementary base sequence in their anticodon part. Here codon-anticodon recognition ratio is 1 : 1.
2. Certain amino acid can be brought to the ribosome by several alternative tRNA types having different anticodons whereas certain other amino acid brought to the ribosome by only one tRNA.
3. Certain tRNA species can bring their specific amino acids in response to several codons, not just one, through loose kind of base pairing at the 3′ end of the codon and 5′ end of the anticodon. This loose pair is called *wobble*.

   This hypothesis states that the two bases of the mRNA codon and anticodon have some play (or wobble) that permit it to pair with more than one base.

   Therefore, wobble is a situation in which third nucleotide of an anticodon (at the 5′ end) can form two alignments. This nucleotide can form hydrogen bond not only with its normal complementary nucleotide in the third position of the codon but also with different nucleotide in that position. These are 'wobble rules' that dictate which nucleotide can and cannot form new hydrogen-bonded associates through wobble.

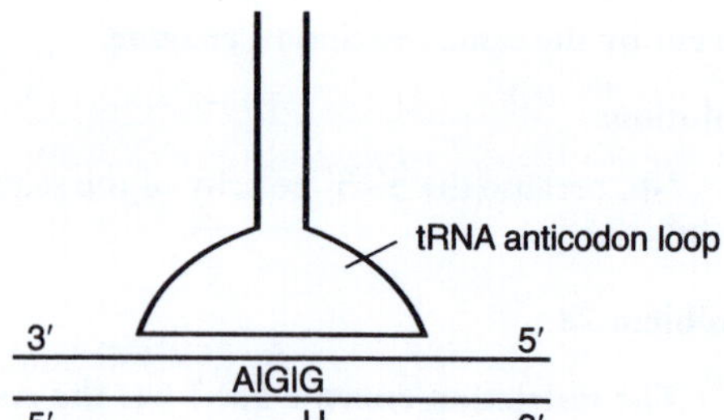

Example: I stands for Inosine, one of the rare bases found in tRNA often in the anticodon.

4. All the codons for serine can pair with the three different tRNAs ($tRNA^{Ser}{}_1$, $tRNA^{Ser}{}_2$, $tRNA^{Ser}{}_3$). In some species there can be additional tRNA species which could represent as $tRNA^{Ser}{}_4$, that has an anticodons identical with one of the three anticodon but differs in nucleotide sequence elsewhere. These four tRNAs are called **Iso accepting tRNAs**, because they accept the same amino acid but they are transcribed from different tRNAs.

tRNA	*Anticodon*	*Codon*
$tRNA^{Ser}{}_1$	AGG + Wobble	UCU UCC
$tRNA^{Ser}{}_2$	AGU + Wobble	UCA UCG
$tRNA^{Ser}{}_3$	UCG + Wobble	AGU AGC

Q. What is t_mRNA? Mention its function.

Solution:

A small RNA, called transfer-messenger RNA (t_mRNA), i.e., both a tRNA and mRNA.

(i) It releases ribosome from mRNA that lacks stop codons (possibly because of partial degradation of the mRNA) and also targets the truncated protein products for degradation.

(ii) It prevents accumulation of stalled ribosome and remove truncated protein that could be toxic.

Microbial Genetics and Complementation Mapping

14

A. MICROBIAL (BACTERIAL) GENETICS

- **Prototroph:**

  Wild type strains of organisms (usually microorganism) that can survive and proliferate in a defined minimal medium are regarded as **prototroph.**

- **Auxotroph:**

  A mutant microorganism that will not grow on a minimal medium but that requires some additional nutritional substance (amino acid, vitamin) for growth.

- **Minimal Medium:**

  (i) A culture medium for microorganism that contains the minimal necessities (glucose, lactose, inorganic ions, Na^+, K^+, Mg^{++}, Ca^{++} and NH_4^+ present in organic salts) for growth of the wild type bacteria.

  (ii) Here wild type bacteria can grow but mutant bacteria cannot grow.

- **Complete Medium:**

  A culture medium contains all the nutrients which bacteria synthesize for growth.

  Here both wild type and mutant bacteria can grow.

- **Supplemented Medium:**

  (i) Minimal medium plus one or more nutrients for which bacteria are deficient is called **supplemented medium.**

  (ii) In this medium, mutant bacteria can be detected.

- **Selective Medium:**

  (i) Complete medium that lacks one or more nutrients or contains antibiotics can be utilized to detect mutant bacteria is called selective medium.

  (ii) Supplemented medium is a kind of selective medium.

- **The characters of an Hfr strain.**

Solution:

(a) *Higher recombination frequency*:

  (i) Hfr is the abbreviated form of high frequency rate. This rate is in respect to the formation of recombinants at conjugation.

  (ii) In case of F^+ strain the frequency of the formation of recombinants is about 10^{-7} but in case of Hfr the frequency is 10^{-4} which is about 1000 times higher.

(b) *Transfer of Hfr factor*:

  (i) When the F factor becomes integrated with the main bacterial chromosome, then the bacterial strain is regarded as the **Hfr strain.**

  (ii) In case of transfer of the chromosome from donor to the recipient the integrated F factor (Hfr) goes last, i.e., after entering the whole of the bacterial chromosome.

(c) *Conversion to other strain*:

  (i) Conversion of F^- strain (recipient) into the donor strain (F^+) does not occur at the time of conjugation as in case of cross between F^+ and F^-.

(ii) The entrance of the integrated F factor is very rare because it transfers at the very last.

(d) *Formation of F′ factor:*

(i) In case of Hfr strain there is the chance of transformation of F′ (F prime) factor. By some external shock the integrated F factor of Hfr strain may exist out from the main chromosome.

(ii) At the time of exist an illegitimate crossing-over may occur between the main chromosome and the F factor. Due to this illegitimate crossing-over some markers get attached with the F factor. Then the F factor is regarded as the **F′ factor**.

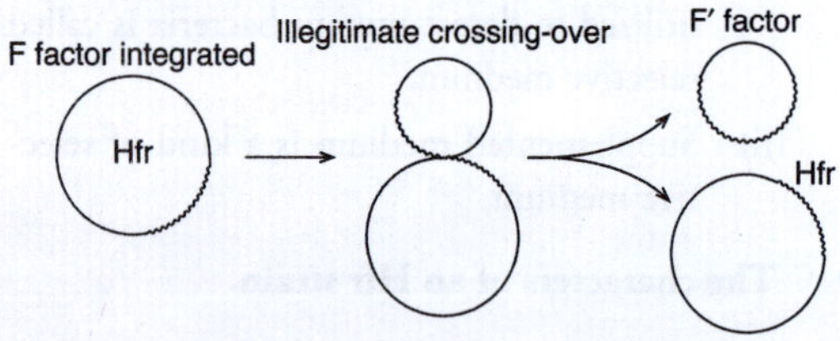

Fig. 14.1: Formation of F′ factor.

Characters of the F factor:

It is a small molecule of DNA, about (2.5×10^5) nucleotides long and about 2% of the length of the *E. coli* chromosome.

E. coli Chromosome:

(i) It is circular and is attached with the plasma membrane by a chromatin body. It is attached to the invaginated part of the cell membrane.

(ii) The DNA molecule of *E. coli* is double-stranded, molecular weight 2×10^9, length 1000 μm and the number of nucleotides is 3×10^6 pairs.

Nature of existence of F factors

(i) It can exist as an autonomous self-replicating circular molecule, rather like a very small second chromosome, which normally replicates whenever the cell divides. Bacteria with an autonomous F are referred to as F^+.

(ii) In an F^+ cell, a single cross-over can occur between the circular F and the circular bacterial chromosome, and so recombine F into the continuity of the bacterial chromosome, where it behaves exactly as a member of the bacterial genes. Bacteria with an integrated F is known as Hfr.

(iii) When F+ bacteria is incubated overnight in medium containing **acridine orange**. As a result it is converted into F^-.

(iv) Acridine orange has no effect on Hfr strand. Acridine orange inhibit replication of autonomous F factor.

The F factor cycle:

The sex or fertility factor F, can exist as an autonomous entity in an F^+ bacterium, or by a single recombination event it can be inserted into the continuity of bacterial chromosome producing an Hfr cell. F can be spontaneously lost from F^+ cell and so restore the F^- state.

3. The most obvious property of the F factors is to enable donar bacteria to conjugate with recipient bacteria by the synthesis of a new surface component, this component is an appendage identified by C. Brington in 1964 is called **F-pillus** or **sex pillus**.

4. The F factor contains at least seven genes concerned with the formation of the F-pillus and the transfer of genetic material from donor to recipient bacterium.

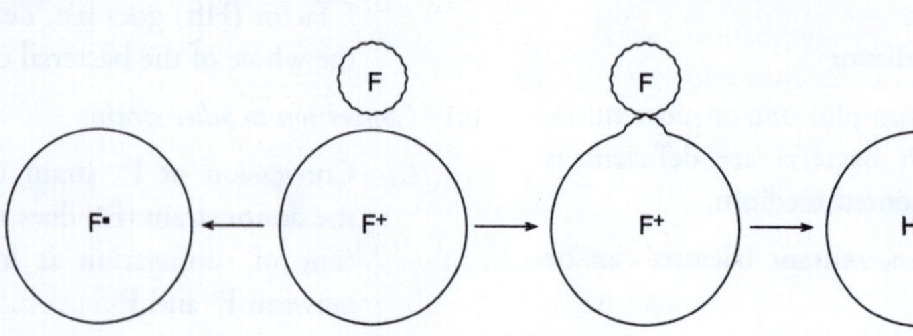

Fig. 14.2: Formation of Hfr

Solved Probolems

Q. 1 What are the genotypic differences among F⁺ cells, F⁻ cells and Hfr cells ?

Solution:

F^+ cells have autonomous F factor, i.e., fertility factor and F^- cells have no F factor or fertility factor. In Hfr, F factor is integrated within bacterial chromosome.

Q. 2 What are the phenotypic differences among F^-, F^+ and Hfr ?

Solution:

F^+ and Hfr cells have F-pilli and F^- cells have no pilli.

Q. 3 By what mechanism are F^- cells converted to F^+ cells and F^+ cells to Hfr cells ?

Solution:

(i) F^- cells are converted into F^+ cells by the conjugative transfer of F factors from F^+ cells.

(ii) Hfr cells are formed when F factors in F^+ cells become integrated into chromosome of these cells.

Q. 4 What is F′ factor? Of what use are F′ factors in genetic analysis ?

Solution:

An F factors that also carries genes derived from the bacterial chromosome. F′ factors are useful for genetic mapping particularly for its trans complementation test in bacteria.

Q. 5 How are F′ factors formed ?

Solution:

F′ factors are formed by anomalous/aberrant excision of F factor from Hfr chromosome.

Q. 6 Mention the funcion of the F-pillus

Solution:

Only due to F-pillus the donor can conjugate with the recipient in culture medium. F-pillus has the structure of an axial hole or longitudinal groove.

Q. 7 Compare Hfr and F^+

Solution:

F^+	Hfr
1. F factor remains free in cytoplasm.	1. F factor remains integrated within chromosome of bacteria.
2. F factor replicates independently.	2. F factor replicates with bacterial chromosome.
3. Low recombination, i.e., 10^{-7}.	3. High recombination, i.e., 10^{-4}.
4. F factor transferred: F^+ to F^-, $F^+ \times F^- = F^+$.	4. F factor rarely transferred: Hfr × F^- = Hfr + F^-.

Q. 8 List all differences between (a) $F^+ \times F^-$, (b) Hfr × F^- and (c) $F' \times F^-$.

Solution:

(a) $F^+ \times F^- \longrightarrow F^+ + F^+$

(i) During conjugation, a copy of F factor is transferred from donor (F^+) to recipient and convert the recipient to F^+ state. Thus, both donor and recipient become F^+.

(ii) During conjugation, F factor replicate independently of donor chromosome, one copy is passed to recipient cell and other remains in F^+ cell.

(iii) During conjugation F factor goes first. Here recombination is very low.

(b) Hfr × $F^- \longrightarrow$ Hfr + F^-

(i) During conjugation, usually a part of the chromosome is transferred.

(ii) In Hfr strain F factor goes last. Therefore, in this conjugation chance of transfer of F factor is very very low.

(iii) Here frequency of recombination (bacterial chromosome) is very high.

(c) $F' \times F^-$.

The frequency of recombination is high as well as frequency of transfer of whole F factor is also high.

Q. 9 State briefly the genetic recombination (Sexual reproduction) in bacteria.

Solution:

(i) **Conjugation**: Unidirectional transfer of genetic material by direct physical contact between donor (male) and recipient (female) bacteria is termed as conjugation.

(ii) **Transformation**: It is a change in the genotype of a cell or organism brought about by the incorporation of naked foreign DNA (exogenous) containing limited amount of genetic information into the host DNA (endogenous).

(iii) **Transduction**: Bateriophage mediated gene (DNA) transfer from one baterium to another is known as transduction.

Q. 10 State differences between generalized transduction & specialized transduction?

Solution:

Generalized transduction	Specialized transduction
1. Transducing particles contain only host (bacterial) DNA.	1. Transducing particles carry recombinant DNA(phage & host).
2. Generalized transducing phages are formed due to packing errors in lytic cycle.	2. Specialized transducing phages are formed due to defective excission of prophage during integrated excision cycle.
3. Recombinant progenies are haploid. *Ex*: Phages P-1 & P-22	3. Recombinant may be partial diploid. Ex: λ Phage.

Q. 11 What does the term co-transduction ?

Solution:

(i) Co-transduction refers to the simultaneous transduction of two different genetic markers to a single recipient cell.

(ii) Here relatively closely linked markers are co-transducted.

Q. 12 How can co-transduction frequencies be used to map genetic markers ?

Solution:

(i) The frequency of co-transduction of any two markers will be inverse function of the distance between them on the chromosome.

(ii) As such this frequency can be used as estimate of the linkage distance. Specific co-transduction linkage function must be prepared for each phage host system studied.

Q. 13 What is abortive transduction ?

Solution:

When the exogenote (donor bacterial DNA) is not integrated into the endogenote and remains free, it is called **abortive transduction**. The recipient bacteria that contain this non-integrated DNA and are partially diploid called **abortive transductant**.

Q. 14 What is complete transduction?

Solution:

Here exogenote integrated into the endogenote and replicate as a part of the chromosome and passed to all daughter cells. This process is called **complete transduction**.

Q. 15 What do you mean by sexduction ?

Solution:

(i) Gene transfer mediated by F′ factor (may carry from a single bacterial gene to $\frac{1}{2}$ bacterial chromosome) is called **sexduction** or **F duction**. Incorporation of bacterial chromosomal genes into F factor and their subsequent transfer by conjugation to recipient cell occurs here.

(ii) If an F′ plasmid is transferred upon conjugation with an F^- strain, the recipient generated is stable merozygotes, carrying a complete bacterial genome plus a donor fragment on autonomously replicated plasmid.

(iii) The process of creating a merozygote by a F′ element is called **sexduction** or **F duction**.

Q. 16 Use of F′ factors in genetic analysis.

Solution:

F′-factors are useful for genetic analyses where two copies of a gene must be present in the same cell. *Example*: Determining dominance relationship.

Q. 17 Compare in table form similarities and differences of the mechanism through which (a) transduction, (b) transformation and (c) sexduction occur.

Solution:

	Transformation	Transduction	Sexduction
1. Agent-mediating DNA transfer	Active uptake of free DNA	Bacteriophage	F-factor
2. Size of DNA units transferred	Small (about $\frac{1}{200}$ to $\frac{1}{100}$ of a chromosome)	Small (about $\frac{1}{100}$ to $\frac{1}{50}$ of a chromosome)	Variable
3. State of donor DNA in recombinant cell	Single strand integrated	Integrated into host chromosome (except autonomous in abortive transduction)	Initially added to the host cell as separate plasmid may undergo recombination with host chromosome to yield stable sexductant.

Q. 18 With respect to F^+ and F^- bacterial matings, answer the following questions:

(a) How was it established that physical contact was necessary ?

(b) How was it established that chromosome transfer was unidirectional ?

(c) What is the physical and chemical basis of a bacterium being F^+ ?

Solution:

(a) The requirement of physical contact between bacterial cells during conjugation was established by placing a sintered glass filter in a U-tube such that the medium can be exchanged but the bacteria cannot come in contact, but genetic material (DNA) can pass. Under this condition, conjugation does not occur.

(b) By treating cells with **Streptomycin**, an antibiotic, it was shown that recombination would not occur, if one of the two bacterial strains was inactivated. However, if the other was similarly treated, recombination would occur. Thus unidirectionality was suggested, with one strain being the donor strain and other being the recipient.

(c) An F^+ bacterium contains a circular, double-stranded structurally, independent, DNA molecule which can direct recombination.

Q. 19 What roles do IS elements play in the integration of F factors?

Solution:

(i) IS elements are short (800–1400 nucleotide pairs) DNA sequences that are transposable—that is capable of moving from one position in a chromosome to another position or from one chromosome to another chromosome.

(ii) IS elements mediate recombination between non-homologous DNA molecules, *viz.,* F factors and bacterial chromosome.

Q. 20 How is a cross performed with bacteriophage?

Solution:

(i) A phage cross is performed by simultaneously infecting the same host cells with two phage of interest (usually mutants) and screening for recombinants (usually wild type) among the progeny.

(ii) The ratio of phage to bacteria is kept high enough (about 5–10 phages of each genotype infected with both parental phages.

(iii) A major advantage of bacteriophage for such genetic studies is that large number of progeny (10^8 to 10^9) can easily be screened for rare recombinants.

Q. 21 What are the major difference between crossing over in bacteria and eukaryotes?

Solution:

(i) In bacteria, crossing-over usually occurs between a fragment of chromosome from a donor cell and an intact circular chromosome in a recipient cell.

(ii) As a result, crossover must occur in pairs that insert segments of the donor cell's chromosome into the recipient cell's chromosome.

(iii) Single crossover or any odd number of crossovers, will destroy the integrity of the circular chromosome and yield a linear DNA molecule in its place.

(iv) In eukaryotes there is no disintegration of chromosome.

Q. 22 What are F Prime Type I and Type II?

Solution:

F Prime (F′) Type I:

(i) Here separation of F factor from the integrated chromosome occurs aberrantly at a low frequency.

(ii) It yields plasmid containing F factor and small segment of chromosome. Here F′ has lost some sequences but carries some host DNA at either side of the integrated F.

F Prime (F′) Type II:

It contains all F′ plus some host DNA from both sides of the point where F integrates.

Q. 23 Differences between $F^+ \times F^-$ and *Hfr* × F^- matings

$F^+ \times F^-$ conjugation	Hfr × F^- conjugation
1. (a) F plasmid is free in the cell. (b) The cell acts as donor (male).	1. (a) F. plasmid is integrated into the chromosome. (b) Integration may be at different sites of the chromosome containing insertion sequences (is element) which can recognize F-DNA. (c) Integration at different sites gives rise to different Hfr strain.
2. (a) Replication of F plasmid is independent of chromosomal replication. (b) Replication of F plasmid starts at Ori T and proceeds by rolling circle mechanism.	2. (a) Replication of integrated F-DNA occurs with the replication of chromosomal DNA. (b) Replication of integrated F-DNA begins at Ori T of the F-DNA and proceeds by rolling circle mechanism.
3. Transfer of F-plasmid from donor to recipient takes about 2 minutes and the whole F plasmid is transferred.	3. F integrated chromosome transfer takes about 100 minutes. The whole chromosome transfer occurs very rarely. Generally, very small part of the F-DNm is transferred.
4. The F^- recipient is transformed into F^+ donor.	4. Generally, F^- recipient remains F^-, because very rarely the complete DNA is transformed.
5. Only the F plasmid-borne genes are transferred. No chromosomal genes are transferred.	5. Chromosomal genes are transferred along with a few F-plasmid genes; the mating cells break apart before the entire chromosome is transferred. Only several hundred chromosomal genes can be transferred.
6. Recombination of chromosomal genes does not occur.	6. Transferred chromosomal genes may be incorporated into the recipient chromosome by homologous crossing-over resulting in genetic recombination.

Q. 24 Physical contact is essential for Bacterial recombination.

Solution:

Requirement

1. Bernard Davis' U tube:
   (i) Two separate pieces of curved glass tube were prepared and fused at base to form a U shaped culture tube.
   (ii) The two arms of the tube are separated by sintered glass filter with a pore too small to allow the passage of genetic material (i.e. DNA molecules) but too small to permit the passage of bacterial cells.
2. Two strains of Bacteria: (A) F^+ met^- bio^- thr^+ leu^+ thi^+, (B) F^- met^+ bio^+ thr^- leu^- thi^-.
3. Others: Fluid (nutrition medium) in the tube, cotton ball, stopped and suctioner pressure.
4. Selective medium.

Procedure

1. The U tube was filled with nutritional medium and each side was inoculated with a different auxotropic strain of *E. coli*.
2. They (*E. coli*) were F^+ met^- bio^- thr^+ leu^+ thi^+ in one arm of the tube and F^- met^+ bio^+ thr^- thi^- are placed in other arm of the tube.
3. During incubation by application of pressure and suction the medium was pumped back and forth through the filter, so that bacterial cells share a common medium.
4. However, any genetic material that was released from a bacterium could pass through filter.
5. After four hours of incubation in a 'U' tube, bacteria from the either side of the tube are placed on a selective media/minimal media i.e. lack of metionine, biotin, threonine, leucine, thimine but contain all other nutrients essential for growth.

Observation

1. Taking the bacteria from the two arms of the tube was plotted on the selective medium separately.
2. No prototrophs were formed.

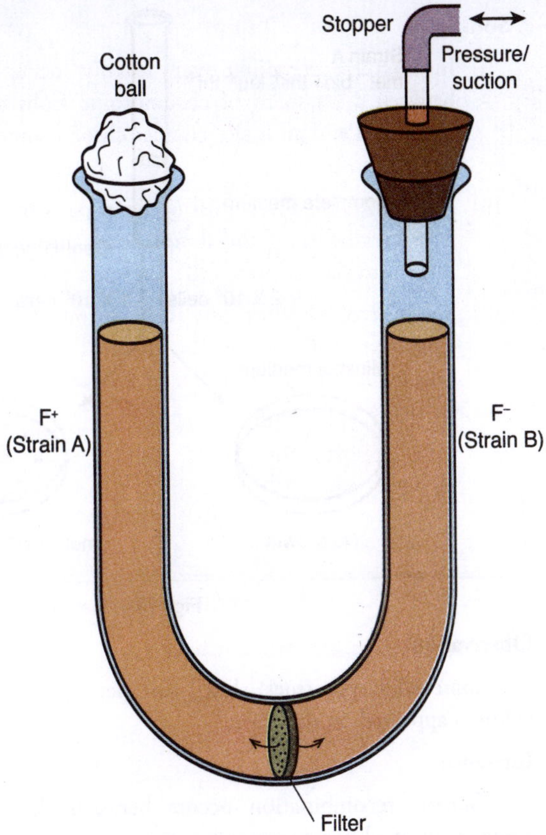

Fig. 14.3: Davis' U-tube experiment

Inference

Davis logically concluded that physical contact i.e., cell to cell contact was essential for genetic recombination and appearance of prototrophic strain of bacteria.

Q. 25 Transfer of genetic material through conjugation.

Solution:

Joshua Lederberg and Edward L. Tatum discovered that *E. coli* cells can transfer genetic material (DNA) through conjugation.

Requirement

Two auxotrophic strain

(i) They crossed the two auxotrophic strain (by mixing the two strain and grown together).
(ii) After that they (the mixed strain) were placed on minimal medium.

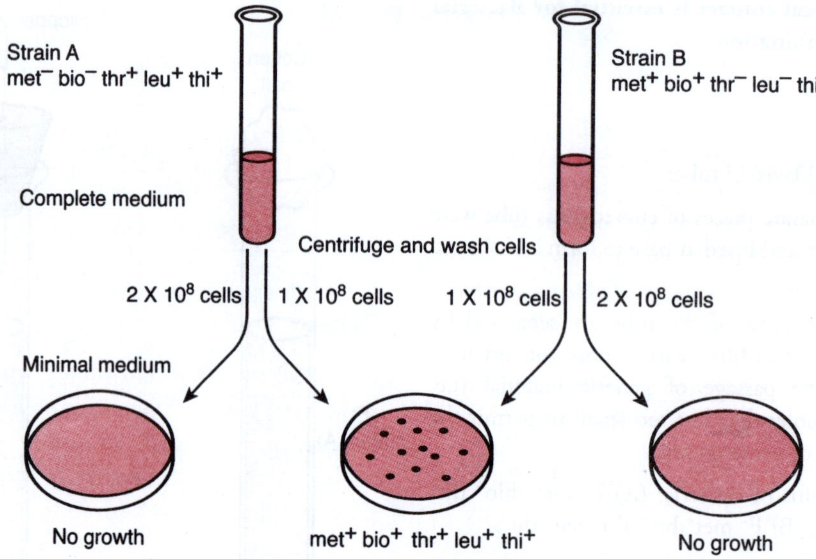

Fig. 14.4: Genetic recombination in bacteria

Observation

Some wild type (met$^+$, bio$^+$, thr$^+$, leu$^+$, thi$^+$) colonies appeared on the plate.

Inference

Genetic recombination occurs between the above mentioned two auxotrophic strain.

Q. 26 Bacteriophage T_4 has a linear chromosome but a circular genetic map. How can the two structures reconciled?

Solution:

Linkage maps are based on population averages with circularly permuted chromosomes, genes that are close together on some chromosomes are at opposite ends of other chromosomes. At the population level, circularly permuted chromosomes yield a circular genetic map.

Q. 27 Mutant bacteria resistant to drugs and antibiotics.

Solution:

(i) Wild type of *E. coli* cells are killed by antibiotics, such as amphiciline used tetracycline. Phenotypically they are Amp^S and Tet^S.

(ii) The mutant alleles that make *E. coli* resistant to these antibiotics are designated as amp^S and tet^r respectively.

(iii) The bacteria that contain these alleles can grow on medium containing antibiotics where wild type strain of *E. coli* cannot.
Thus, antibiotics can be used to select bacteria that carry gene for resistance.

(iv) The resistance gene functions as dominant selective maker.

Q. 28 Recombination in Bacteria.

Solution:

(i) In this event gene transfer occurs from one bacterium to another.

(ii) Here gene transfer occurs unidirectional rather than bidirectional.

(iii) Recombination in bacteria usually occurs between a fragment of donar chromosome and a complete chromosome of eukaryote.

(iv) The recipient cells become partial diploid (merodiploid) containing a linear piece of donor chromosome and complete circular recipient chromosome. As a result crossover must occur in pairs and insert a segment of donor chromosome into recipient chromosome.

(v) If a single crossover occurs (or any odd number of crossing-over) occurs, it will destroy the integrity of the recipient chromosome producing nonviable linear DNA molecule.

Numerical Probolems

Problem 1

You want to determine the titre plaque forming units [pfu] per millilitre of phage particles in a phage suspension. You dilute the suspension by a factor of 10^7 and plate 0.2 ml. You observe 120 plaques. What is titre of the original phage suspension.

Solution:

$$\text{Titre} = \frac{\text{Number of plaque} \times \text{dilution factor}}{\text{Volume in millilitre}}$$

$$= \frac{120 \times 10^7}{0.2} = \frac{120 \times 10^7 \times 10}{2} = 6 \times 10^9 \text{ pfu/ml.}$$

Problem 2

In the microbiology laboratory of R. K. M. Vidyamandira, a scholar has a suspension of phages and he wants to know its titre (plaque-forming units per millilitre). He has diluted the suspension by a factor 10^6 and plate 0.05 ml. He observed 250 plaques. What was the titer of the original suspension?

Solution:

$$\text{Titre} = \frac{\text{Number of plaque} \times \text{dilution factor}}{\text{Volume in millilitre}}$$

$$= \frac{250 \times 10^6}{0.05} = \frac{25 \times 10^7 \times 100}{5} = 5 \times 10^9 \text{ pfu/ml.}$$

Problem 3

DNA was extracted from a wild type strain of bacteria and used to transform a mutant strain unable to synthesize the amino acids alanine (ala), proline (pro) and arginine (arg). The transformant classes are as follows:

8400 ala^+ pro^+ arg^+; 840 ala^+ pro^- arg^-; 2100 ala^+ pro^- arg^+; 1400 ala^+ pro^+ arg^-; 420 ala^- pro^+ arg^+; 840 ala^- pro^+ arg^-; 840 ala^- pro^- arg^+.

(a) What are the linkage distances between the genes?

(b) What is the linkage order ?

Solution:

Distance	*Transformant*		*Number of colonies*	*Frequency*
ala – pro	ala^+	pro^-	840 + 2100 = 2940	$\frac{4200}{14000} = 0.3$
	ala^-	pro^+	420 + 840 = 1260	
	ala^+	pro^+	8400 + 1400 = 9800	
			Total 14000	
ala – arg	ala^+	arg^-	840 + 1400 = 2240	$\frac{3500}{14000} = 0.25$
	ala^-	arg^+	420 + 840 = 1260	
	ala^+	arg^+	8400 + 2100 = 10500	
			Total 14000	
pro – arg	pro^+	arg^-	1400 + 840 = 2240	$\frac{5180}{14000} = 0.37$
	pro^-	arg^+	2100 + 840 = 2940	
	pro^+	arg^+	8400 + 420 = 8820	
			Total 14000	

(a) Linkage distance: ala – pro = 0.03 = 30%; ala – arg = 0.25 = 25%; pro – arg = 0.37 = 37%

(b) Order of gene: pro —— ala —— arg

pro – ala: 0.3; ala – arg: 0.25; pro – arg: 0.37

Problem 4

Following are the number of colonies of different transformant classes from a cross of trp^+, his^+, tyr^+ as the donor with trp^-, his^-, tyr^- as the recipient:

1. **trp^- his^- tyr^+ = 685,**
2. **trp^- his^+ tyr^- = 418,**
3. **trp^- his^+ tyr^+ = 3660**
4. **trp^+ his^- tyr^- = 2600**
5. **trp^+ his^- tyr^+ = 107**
6. **trp^+ his^+ tyr^- = 1180**
7. **trp^+ his^+ tyr^+ = 11940**

Calculate the linkage distances and linkage map between the three loci trp, his and tyr.

Solution:

Distance	*Transformant*		*Number of colonies*	*Frequency*
ala – pro	trp^+	his^-	2600 + 107 = 2707	$\frac{6785}{19905} = 0.34$
	trp^-	his^+	418 + 3660 = 4078	
	trp^+	his^+	1180 + 11940 = 13120	
			Total 19905	
trp – tyr	trp^+	tyr^-	2600 + 1180 = 3780	$\frac{8125}{20172} = 0.40$
	trp^-	tyr^+	685 + 3660 = 4345	
	trp^+	tyr^+	107 + 11940 = 12047	
			Total 20172	
his – tyr	his^+	tyr^-	418 + 1180 = 1598	$\frac{2390}{17990} = 0.13$
	his^-	tyr^+	685 + 107 = 792	
	his^+	tyr^+	3660 + 11940 = 15600	
			Total 17990	

(a) Linkage distance:

trp – his = 0.34 = 34%; trp – tyr = 0.40 = 40%; his – tyr = 0.13 = 13%

(b) Order of gene:

tryp —— his —— tyr

0.34 0.13

0.40

Problem 5

In a cross of *E. coli* F^+ $tryp^+$ his^+ try^+ with F^- $tryp^-$ his^- tyr^- the following results were obtained:

A	B	C	D	E	F	G
$tryp^+$	**$tryp^+$**	**$tryp^+$**	**$tryp^-$**	**$tryp^-$**	**$tryp^-$**	**$tryp^+$**
his^+	**his^-**	**his^+**	**his^+**	**his^+**	**his^-**	**his^-**
tyr^+	**tyr^-**	**tyr^-**	**tyr^-**	**tyr^+**	**tyr^+**	**tyr^+**
11,945	**2,700**	**1,180**	**400**	**3,600**	**600**	**100**

Calculate the recombination frequencies and draw a genetic map of the region.

[M.Sc. (Zoo) C.U. 2001]

Solution:

Distance	*Transformant*		*Number of colonies*	*Frequency*
trp – his	tryp$^+$	his$^-$	2700 + 100 = 2800	$\frac{6800}{19925} = 0.34127$
	tryp$^-$	his$^+$	400 + 3600 = 4000	
	tryp$^+$	his$^+$	11945 + 1180 = 13125	
			Total 19925	
his – tyr	his$^+$	tyr$^-$	1180 + 400 = 1580	$\frac{2280}{17825} = 0.1279 = 0.13$
	his$^-$	tyr$^+$	600 + 100 = 700	
	his$^+$	tyr$^+$	11945 + 3600 = 15545	
			Total 17825	
tryp – tyr	tryp$^+$	tyr$^-$	2700 + 1180 = 3880	$\frac{80800}{20125} = 0.40$
	tryp$^-$	tyr$^+$	3600 + 600 = 4200	
	tryp$^+$	tyr$^+$	11945 + 100 = 12045	
			Total 20145	

(a) Linkage distance:

tryp – his = 0.34 = 34%; his – tyr = 0.13 = 13%; tryp – tyr = 0.40 = 40%

(b) Order of gene: tryp —— his —— tyr

0.34 0.13

0.40

Problem 6

In PI transduction experiment, the donor bacteria were synP$^+$, supM$^+$, trpZ$^+$, the recipients were synP$^-$, supM$^-$, trpZ$^-$. Initial selection was for supM$^-$ transductants.

1. M$^+$P$^+$Z$^+$ = 48, 3. M$^+$P$^+$Z$^-$ = 500,

2. M$^+$P$^+$Z$^-$ = 120, 4. M$^+$P$^-$Z$^+$ = 0.

(a) What is the marker order? (b) What are the transduction frequencies between M and P, M and Z?

Solution:

(a) The rarest class is M$^+$P$^-$Z$^+$ = 0, i.e., quadruple crossing-over product. The initial selection was supM$^+$. Therefore, the order of the genes is

supM —— synP —— trpZ.

(b) Total number of progeny = 48 + 120 + 500 + 0 = 668, supM$^+$ and synP$^+$ are co-transduced in classes (1) and (2) but not in classes (3) and (4).

Therefore, co-transduction frequency between M and P = $\frac{48+120}{668} = 0.25$

supM$^+$ and trpZ$^+$ are co-transducted in classes (1) and (4) but not in (2) and (3).

Therefore, co-transduction frequency between M and Z = $\frac{48+0}{668} = \frac{48}{668} = 0.072$.

Problem 7

In a generalized transduction experiment, donor *E. coli* cells have the genotype trpA$^+$ supC$^+$ pyrF$^+$ and recipient cells trpA$^-$ supC$^-$ pyrF$^-$. PI mediated transduction supC$^+$ were initially selected. Results are as follows:

1. **supC$^+$ trpA$^+$ pyrF$^+$ = 36**
2. **supC$^+$ trpA$^+$ pyrF$^-$ = 114**
3. **supC$^+$ trpA$^-$ pyrF$^+$ = 0**
4. **supC$^+$ trpA$^-$ pyrF$^+$ = 453.**

(a) What is order of gene ?

(b) Determine co-transduction frequency for supC and trpA; supC and pyrF.

Solution:

(a) $supC^+$ initially selected. In 3-point mediated quadruple crossing-over product will be rarest frequency, i.e., here 0.

Donor has the genotype $trpA^+$ $supC^+$ $pyrF^+$. First gene will be $supC^+$ and terminal $pyrF^+$. So the order of gene is supC — trpA — pyrF.

(b) Total progeny = 36 + 114 + 0 + 453 = 603.

$supC^+$ and $trpA^+$ are both transductants in classes 1 and 2 but not in 3 and 4.

Therefore, transduction frequency between supC and trpA $= \frac{36+114}{603} = \frac{150}{603} = 0.25$.

$supC^+$ and $pyrF^+$ are both transductants in classes 1 and 3, but not in 1 and 2.

Therefore, the co-transduction frequency between $supC^+$ and $pyrF^+$ $= \frac{36+0}{603} = \frac{36}{603} = 0.06$.

Problem 8

Co-transduction of genes leu^+ and trp^+ to recipient leu trp cells produced the following transductants:

leu^+ trp^+ 369, leu^- trp^+ 31, leu^+ trp^- 46.

What is the map distance between leu and trp?

Solution:

Total progeny (transductants) = 369 + 31 + 46 = 446.

One crossing-over between leu and trp produces two classes of transductants, viz., leu^- trp^+ and leu^+ trp^- = 31 + 46 = 77.

Map distance will be $\frac{77}{446} \times 100 = 17.264 = 17.3\%$.

The map distance between leu and trp is 17.3%.

Problem 9

Using bacteriophage p 22 you performed a three-factor cross in *Salmonella typhimurium*. The cross was between as Arg^- leu^- His^- recipient bacterium and bacteriophage p 22 which was grown on an Arg^+ leu^+ and His^+ strain. You selected for 1000 Arg^+ transductants and tested them on several selective media. By replicaplating you obtained the following results:

Arg^+ leu^- His^- = 585

Arg^+ leu^- His^+ = 300

Arg^+ leu^+ His^+ = 114

Arg^+ leu^+ His^- = 1

(a) What is the order of the three markers?

(b) What are the co-transduction frequencies?

Solution:

(a) Arg^+ leu^- His^- is the largest class with only Arg^+ gene transferred. The next largest class is Arg^+ and His^+. Therefore, Arg and His are closer to each other than Arg and leu. The order of the genes is Arg, His, leu.

(b) Co-transduction frequency is the percentage of cells that received two markers. For Arg^+ and His^+ this includes the Arg^+ leu^- His^+ = 300 and Arg^+ leu^+ His^+ = 114.

The co-transduction frequency (Arg and His) = 300 + 114 $= \frac{414}{1000} = 41.4\%$, co-transduction frequency of Arg and leu $= \frac{114+1}{1000} = 11.5\%$.

Problem 10

In *E. coli* four Hfr strains donate the following genetic markers shown in the order donated:

Strain 1:	Q	W	D	M	T
Strain 2:	A	X	P	T	M
Strain 3:	B	N	C	A	X
Strain 4:	B	Q	W	D	M

All these Hfr strains are derived from the same F^+ strain. What is the order of these markers on the chromosome of the original F^+?

[M.Sc. (Zoo) V.U. 2003]

Solution:

I. Each Hfr strain donates genetic markers from the fixed point on the circular chromosome, and the earliest markers are donated with the highest frequency, because not all markers are donated by each Hfr.

II. Only the early marker must be donated by each Hfr.

Alignment of gene transfer for the Hfr strain:

Strain 1: Q W D M T
Strain 2: M T P X A
Strain 3: X A C N B
Strain 4: Q W D M B

The circular linkage map of the order

Q W D M T P X A C N B

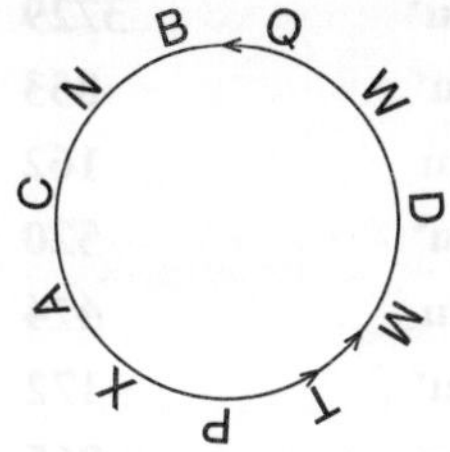

Problem 11

In *E. coli* four Hfr strains donate the following genetic markers shown in the order donated:

Strain 1:	**M**	**Z**	**X**	**W**	**C**
Strain 2:	**L**	**A**	**N**	**C**	**W**
Strain 3:	**A**	**L**	**B**	**R**	**U**
Strain 4:	**Z**	**M**	**U**	**R**	**B**

All these Hfr strains are derived from the same F^+ strain. What is the order of these markers on the circular chromosome of the original F^+?

Solution:

Alignment of gene transfer for the Hfr strains:

Strain 1: M Z X W C
Strain 2: W C N A L
Strain 3: A L B R U
Strain 4: B R U M Z

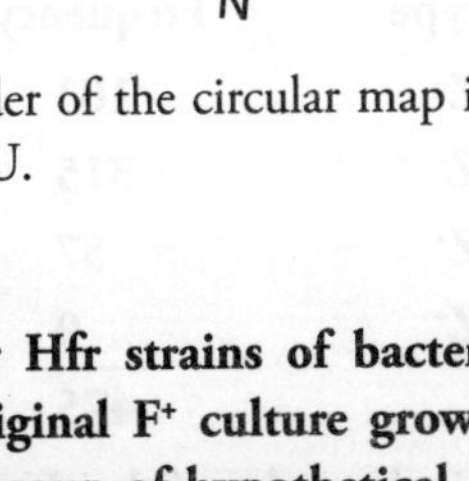

The order of the circular map is M Z X W C N A L B R U.

Problem 12

In four Hfr strains of bacteria all derived from an original F^+ culture grown over several months, a group of hypothetical genes was studied and shown to be transferred in the following orders:

Hfr strain	**Order of transfer**
1	**E R I U M B**
2	**U M B A C T**
3	**C T E R I U**
4	**R E T C A B**

Assuming that B is the first gene along the chromosome, determine the sequence of all genes shown. One strain creates an apparent dilemma. Which one is it? Explain why the dilemma is only apparent and not real.

Solution:

Alignment of gene transfer for the Hfr strains:

Strain 1: E R I U M B
Strain 2: U M B A C T
Strain 3: C T E R I U
Strain 4: B A C T E R

Starting with B, the sequence of gene is B A C T E R I U M.

Strain 4 creates an apparent dilemma which is resolved by realizing that the F factor is integrated opposite orientation, thus the gene enters in opposite sequence starting with gene R: M U I R E T C A B.

Problem 13

When interrupted mating technique was used with five different strains of Hfr bacteria, the following order of gene entry and recombination was observed:

Hfr strain	**Order of transfer**
1	**T C H R O**
2	**H R O M B**
3	**M O R H C**
4	**M B A K T**
5	**C T K A B**

On the basis of these data, draw a map of bacterial chromosome. Do the data support the concept of circularity ?

Solution:

Alignment of gene transfer for the Hfr strain:

Strain 1: T C H R O
Strain 2: H R O M B
Strain 3: C H R O M

Strain 4: M B A K T

Strain 5: B A K T C

(i) The order of gene is T C H R O M B A K T

(ii) Yes, it supports the concept of circulation.

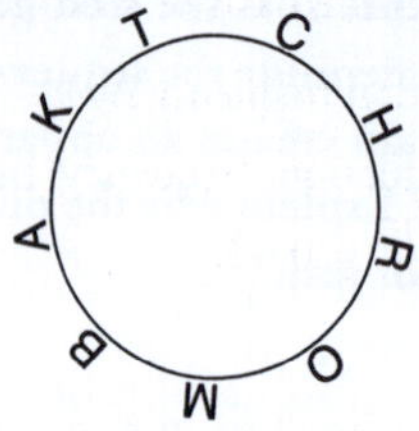

Problem 14

Generalized transduction experiment uses a met E^+ pyr D^+ strain as donor and met E^- pyr D^- as recipient. Met E^+ transductants are selected and then tested for pyr D^+ allele. The following numbers were obtained:

met E^+ pyr D^- = 857; met E^+ pyr D^+ = 1

Do these results suggest that these loci are closely linked? What other explanations are there for the lone "double"?

Solution:

(i) No, closely linked loci would be expected to be co-transducted; the greater the co-transduction frequency, the closer the loci are.

(ii) Because only 1 of 858 (1 + 857) met E^+ was also pyr D^+, the genes are not closely linked.

(iii) The lone met E^+ pyr D^+ could be the result of co-transduction or it could be a spontaneous mutation of pyr D to pyr D^+ or it could be the result of co-infection by two separate transducing phages.

Problem 15

In an experiment, *E. coli* was infected with two strains of T-4 bacteriophages, one was mutant for rapid lysis (r) minute (m) and turbid (tu) and the other was wild type for all three genes. The lytic products of this mixed infection were plated and classified as follows:

Genotype	Number of plaques
r m tu	3467
$r^+m^+tu^+$	3729
r m tu^+	853
r^+m tu	162
r^+m tu^+	520
r m^+tu	474
r m^+tu^+	172
r^+m^+tu	965

(a) Is the r gene closer to the m gene or the tu gene?

(b) As a matter of linkage, which gene is probably located in between the other two?

Solution:

Total progeny 10,342

r+m+tu+ + r m tu =	3467 + 3729 =	7196
r m tu+ + r+m+tu =	853 + 965 =	1,818
r+m tu+ + r m+tu =	520 + 474 =	994
r+m tu + r m+tu+ =	162 + 172 =	334
		10,342

(i) Distance between r – m = 994 + 334 = 1,328

Distance between m – tu = 1,818 + 334 = 2,152

(ii) r appears to be middle gene. The fewest number would require two cross-overs.

Problem 16

In an experiment in Microbiology laboratory of A. P. C. College, Kolkata, a research student conducted a transduction experiment. He used a donor $X^+Y^+Z^+$ and an $X^-Y^-Z^-$ recipient. X^+ transductants were selected and found the following results:

Phenotype	Frequency
$X^+Y^-Z^-$	103
$X^+Y^+Z^+$	315
$X^+Y^+Z^+$	57
$X^+Y^-Z^+$	0
	475

(a) What is the order of the three marker?

(b) What is the co-transduction frequency of X and Y and of X and Z?

(c) What are the genetic distances between X and Y and between Y and Z? Given a length of markers 91.5 kb

Solution:

(a) The rarest combination is $X^+Y^-Z^+ = 0$. Observing rarest and greatest frequency, we can calculate that Y is the middle gene, i.e., XYZ.

(b) Co-transduction frequency of X and Y

$$= \frac{\text{The number of } X^+ Y^+ \text{ transductants}}{\text{Total number of } X^+ \text{ transductants}} \times 100$$

$$= \frac{315+57}{(103+315+57+0)} \times 100 = \frac{372}{475} \times 100$$

$= 0.78315 \times 100 = 78.3\%$.

Co-transduction frequency of X^+ and Z^+

$$= \frac{\text{The number of } X^+ Z^+ \text{ transductants}}{\text{Total number of } X^+ \text{ transductants}} \times 100$$

$$= \frac{57+0}{475} \times 100 = 0.12 \times 100 = 12\%.$$

(c) If two genes are transduced, they are closed linked and lie with 91.5 kb of each other.

$x = \left(1 - \frac{d}{L}\right)^3$, or $d = L\left(1 - \sqrt[3]{x}\right)$

$\because L = 91.5$; so, $d = 91.5\left(1 - \sqrt[3]{.78}\right)$

$= 91.5(1 - 0.92) = 91.5 + 0.08 = 7.32$.

[x = co-transduction frequency, L = size of chromosome, d = distance between genes]

Problem 17

In an experiment in Microbiology laboratory of Kalyani University, a research fellow conducted a transduction experiment. He used Q2 phase to transduce DNA with three markers Awk$^+$ kat$^+$ and Nrd$^+$ into an Awk$^-$ kat$^-$ Nrd$^-$ recipient. He selected for Awk$^+$ transductants and observed the following:

Phenotype	Frequency
Awk$^+$ Kat$^-$ Nrd$^-$	638
Awk$^+$ Kat$^-$ Nrd$^+$	309
Awk$^+$ Kat$^+$ Nrd$^+$	115
Awk$^+$ Kat$^+$ Nrd$^-$	1

What is the order of these three markers? What are the co-transduction frequency between Awk and kat and between Awk and Nrd?

Solution:

The Awk$^+$ Nrd$^-$ Kat$^+$ is the least frequent transductant class (1 in 1063). Therefore, the order is Awk–Nrd–Kat$^+$, on the basis of 4 crossover events.

A co-transduction frequency Awk – Kat $= \frac{1+115}{1063}$ $= 0.11$. Co-transduction frequency between Awk and Nrd is $\frac{309+115}{1063} = 0.40$.

Problem 18

Bacteriophage lambda has genome size of 50 Kb pairs. What is the approximate genetic length of the prophage in minutes, when inserted into an *E. coli* strain with genome size of 5 Mb? (There are 100 minutes in the entire *E. coli* genetic map)

Solution:

The *E. coli* strain contains 5 Mb, i.e., $5 \times 100 =$ 500 Kb pairs, which implies approximately 50 Kb/ minute $\left(\frac{5000}{100} = 50 \text{ minutes}\right)$. Because the 1 genome is 50 Kb, the genetic length of prophage is one minute.

Problem 19

P2 and P4 are bacteriophages of *E. Coli.* They have the following properties (1) when P2 phage infects a bacterium, the bacterium eventually bursts, yielding about 100 P2 progeny. (2) When P4 infects a bacterium the bacterium survives because P4 is detective phage. (3) When P2 phage and P4 phage co-infect the same bacterium lysis of the bacterium yields 100 P4 progeny and no P2 progeny, because the presence of P4 inhibits the growth of P2. Suppose, 3×10^8 P2 and 2×10^8 P4 are added to 10^8 bacteria cells.

(a) How many bacteria will not be infected with either phage?

(b) How many bacteria will survive?

(c) How many bacteria will produce P2 progeny?

(c) How many bacteria will produce P4 progeny?

(Hint: If the average number of phage per cell in a mixed suspension is *x*, then probabilities that a cell remains uninfected is e^{-x}).

Solution:

(i) The mean number of P2 phage per cell is 3×10^{-8} and P4 phage per cell is 2×10^{-8}. It may be that P2 phage is 3 and P4 phage is 2.

(ii) The cell remaining uninfected by P2 is $e^{-x} = e^{-3} = 0.04978 = 0.050$. In the same way, cell remaining uninfected by P4 is $e^{-2} = 0.135$.

(a) The number of bacteria uninfected by either phage will be $0.050 \times 0.135 \times 10^8 = 6.75 \times 10^5$.

(b) Survivors will be $6.75 \times 10^5 + 0.050 \times (1 - 0.135) \times 10^8 = 6.75 \times 10^5 + 0.050 \times 0.865 \times 10^8 = 10^5(6.75 + 0.04325 \times 10^3) = (6.75 + 43.25) \times 10^5 = 50 \times 10^5 = 5 \times 10^6$.

(c) Producers of P2 progeny = $(1 - 0.050) \times 0.135 \times 10^8 = 0.95 \times 0.135 \times 10^8 = 0.12825 \times 10^8 = 1.28 \times 10^7$.

(d) Producer of P4 progeny = $(1 - 0.050) \times (1 - 0.135) \times 10^8 = 8.22 \times 10^7 = 0.95 \times 0.865 \times 10^8 = 0.8217 \times 10^8 = 8.22 \times 10^7$.

Problem 20

The experimental result from Microbiology, Department of R. K. Mission Vidyamandira three independent integrations of the F factor into the chromosome of a bacterial species related to *E. coli* yielded four different Hfr derivatives (Hfr P, Hfr Q and Hfr R) each with a different origin and possibly a different direction of transfer of markers. These were examined in interrupted mating experiments and were found to transfer chromosomal genes at the times shown in the following table.

Genetic Marker

Hfr	ala	arg	his	ser	thr	met
P	–	24	12	–	–	4
Q	9	25	–	–	3	–
R	–	–	–	11	–	29

(a) Draw a diagram of the origin and direction of transfer of each Hfr showing the relative positions of the chromosomal markers and the distances between them in minutes.

(b) Draw a composite genetic map using the above mentioned data.

Solution:

(a) (i) Consider each Hfr in turn and starting with the earliest entering gene, write down the name of each gene as it enters.

(ii) Difference in time of entry between adjacent genes.

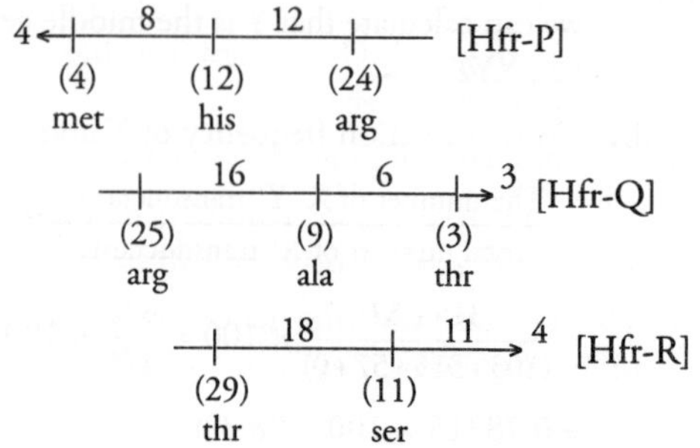

(b) (i) Arrange the maps of the Hfr strains (P, Q & R), so that many shared genes between two or more Hfr strains coincide.

(ii) A composite map is drawn based on the overlaps between the individual Hfr strains.

(iii) The composite maps are shown below with the help of individual maps.

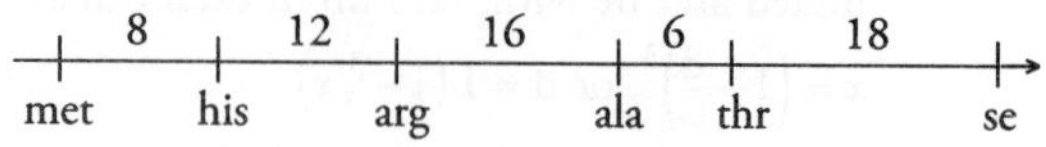

Composite map

Problem 21

It has been reported from the Microbiology Department of Calcutta University that four independent integrations of the F factor into the chromosome of an unusual strain of *E. coli* yielded four different Hfr derivatives of the strain (A, B, C & D) each with a different origin and possibly a different of transfer of markers. These were examind in interrupted-mating experiments and were found to transfer chromosomal genes at the times shown in the following table:

Genetic Marker

Hfr	his	lac	leu	lip	pheS	pro	pyrD	terC	tonA	trp
A	–	20		11	–	–	3	–		–
B	–	–	18	–	–	–	31	–	20	25
C	–	13	–	22	–	6	–	2	–	–
D	19	–	–	–	12	4	–	8	–	–

(a) Draw a circular genetic map, with position 0 (also 100) minute(s) at the top, showing the order of the chromosomal genes and the distance (in minutes) between the adjacent genes (the marker leu is 2 minutes on the standard map). Annotate the genetic map with arrows indicating the origin and direction of transfer of each Hfr and the distance (in minutes) from the origin of transfer to the first marker transferred.

(b) How does the genetic map of this particular strain compare with that of standard *E. coli* strain?

Solution:

(a) (i) Considering each Hfr in turn, starting with the earliest entering genes, the name of the gene in each Hfr strain is written.

(ii) The difference in time of entry between the adjacent gene is the distance in minutes between the genes.

(iii) When a partial genetic map of gene transfer from each Hfr has been made, arrange the maps, so that any shared markers between two or more Hfr strains coincide.

(iv) This process yields the genetic map where arrows indicate direction of transfer and the numbers are times of transfer in minutes.

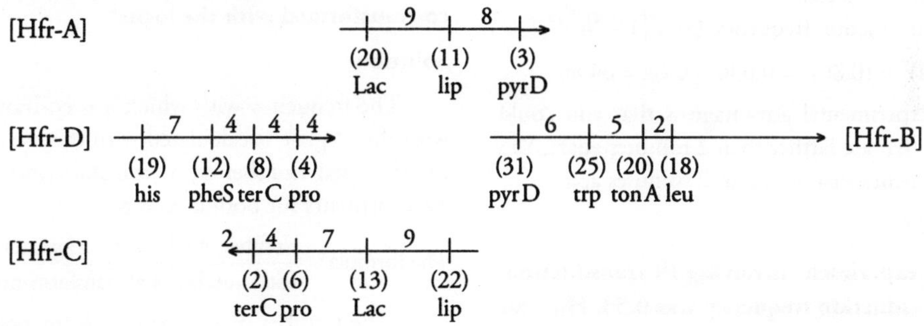

(v) Arrange the composite genetic map in the form of a circle 100 minutes with 0 minute at the top and leu is 2 minutes in standard map.

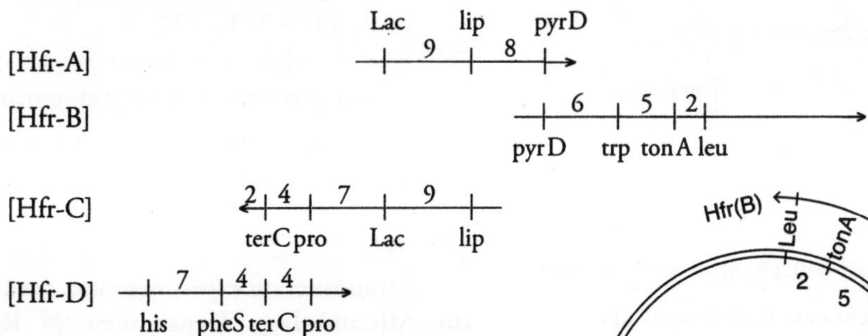

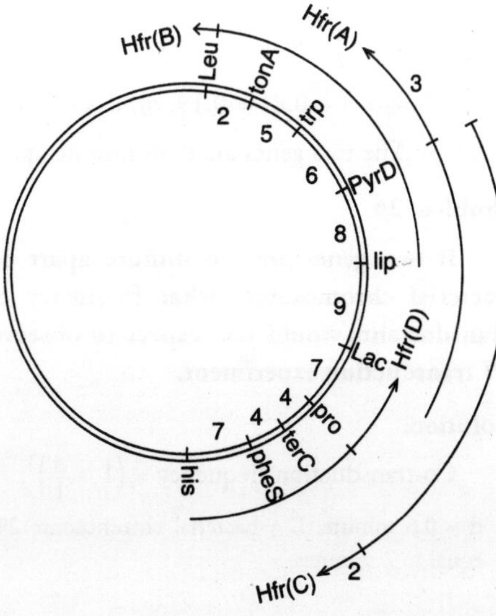

(b) As compared with the standard map a group of markers are present in antiguous segment of the chromosome in opposite order.

It is due to the conversion of the chromosome with breakpoints, ton A – pro and trp – ter C interval in the standard map.

Problem 22

A new strain of P1 has been identified that packages larger pieces of the *E. coli* chromosome. This new P1 strains packages pieces of the *E. coli* chromosome is

5 minutes long. If two genes are 0.7 minute apart along the *E. coli* chromosome, what would be the co-transduction frequency using a normal strain of P1 and using this new strain of P1 that packages large pieces? What would be the experimental advantage of using this new P1 strain?

Solution:

Co-transduction frequency = $\left(1 - \frac{d}{L}\right)^3$.

* d = distance between gene in minutes; L = length of the size of chromosome.

For the normal strain,

Co-transduction frequency (x) = $\left(1 - \frac{0.7}{2}\right)^3$

$= (1 - 0.35)^3 = (0.65)^3 = 0.274625 = 0.275 = 27.5\%$.

For the new strain,

Co-transduction frequency (x) = $\left(1 - \frac{0.7}{5}\right)^3$

$= (1 - 0.14)^3 = (0.86)^3 = 0.636 = 0.64 = 64\%$.

The experimental advantage is that you could map genes that are farther than 2 minutes apart. You could map genes that are up to 2 minutes apart.

Problem 23

In an experiment involving PI transduction, the co-transduction frequency was 0.53. How far apart are the two genes?

Solution:

Cotransduction frequency = $\left(1 - \frac{d}{L}\right)^3$.

$$0.53 = \left(1 - \frac{d}{L}\right)^3 = \left(1 - \frac{d}{L}\text{minutes}\right)^3.$$

or, $\left(1 - \frac{d}{2}\right) = \sqrt[3]{.53} = 0.80926 = 0.81$

$\left(1 - \frac{d}{2}\text{minutes}\right) = 0.81,$

$\frac{d}{2} = 1 - 0.81 = 0.19$, or, d = 0.3 minutes.

The two genes are 0.38 minute apart.

Problem 24

If two genes are 0.6 minute apart on the bacterial chromosome, what frequency of co-transductants would you expect to observe in a PI transduction experiment.

Solution:

Co-transduction frequency = $\left(1 - \frac{d}{L}\right)^3$.

* d = 0.6 minute; L = bacterial chromosome 2%, i.e., equal to 2 minutes.

c.t.f. = $\left(1 - \frac{0.6}{2}\right)^3 = (1 - 0.3)^3 = (0.7)^3$

$= 0.343 \times 1000 = 34300 = 34.3\%$.

Problem 25

In a transformation experiment in the Microbiology Department of Bijaygarh College, donor DNA from an a^+b^+ strain was used to transfer a recipient strain of genotype a b. The transformed classes were isolated and their frequencies determined to be

$a^+ b^+ = 307$

$a^+ b = 215$

$a b^+ = 278$

The total number of transformants was 800. What is the frequency with which the b locus is co-transformed with the locus?

Solution:

The frequency with which it is co-transformed with the a^+ gene is calculated with the use of values for the total number of a^+ transformants and the transformants for both a^+ and b^+.

The formula is $\frac{\text{Number of } a^+b^+ \text{ co-transformants}}{\text{Total number of } a^+ \text{ transformants}} \times 100\%$

(i) The number of a^+b^+ co-transformants is 307.

(ii) The a^+ transformants are represented by two classes a^+b^+ = 307 and a^+b = 215, the total being 307 + 215 = 522.

(iii) The ab^+ class is irrelevent to the question, because its members are not transformants for a^+.

(iv) Thus, the co-transformation frequency for a^+ and b^+ is $\frac{307}{522} \times 100 = 58.8\%$.

Problem 26

A transduction experiment was conducted in the Microbiology Department of Rabindra Mahavidyalaya, Champadanga, Hooghly. The donor was c^+d^+e and recipient was cde. Selection was for c^+. The four classes of transductant from this experiments are as follows:

Class	Genetic composition	Number of individuals
1	$c^+d^+e^+$	57
2	c^+d^+e	76
3	c^+de	365
4	c^+de^+	2

(a) Determine the co-transduction frequency for c^+ and d^+.

(b) Determine the co-transduction frequency for c^+ and d^+.

(c) Which of the co-transduction frequencies calculated in (a) & (b) represents the greater actual distance between genes? why?

Solution:

(a) The co-transduction frequency for c^+ & d^+

$$= \frac{\text{The number of } c^+d^+ \text{ transductants}}{\text{Total number of } c^+ \text{ transductants}} \times 100$$

$$= \frac{57+76}{500} \times 100 = \frac{133}{500} \times 100 = 26.6\%$$

(b) The co-transductants frequency for c^+ & e^+

$$= \frac{\text{The number of } c^+e^+ \text{ transductants}}{\text{Total number of } c^+ \text{ transductants}} \times 100$$

$$= \frac{57+2}{500} \times 100 = \frac{59}{500} \times 100 = 11.8\%.$$

(c) (i) The greater actual distance is between c^+ and e^+ genes.

(ii) The principle involved is that the closer two genes are on chromosome, the greater is the likelihood that they will be co-transducted.

(iii) Thus as the distance between the genes increases the co-transduction frequency decreases.

(iv) Since the co-transduction frequency of c^+ and e^+ is 11.8% and co-transduction frequency of c^+ and d^+ is 26.6%.

The genes c^+ and e^+ are farther apart than genes c^+ and d^+.

Problem 27

Consider the following data pertaining to P1 transduction:

Donor	Recipient	Selected marker	Unselected marker	%
aroA PyrD$^+$	aroA$^+$ PyrD	PyrD$^+$	aroA	5
aroA$^+$ cmlB	aroA cmlB$^+$	aroA$^+$	cmlB	26
cmlB PyrD$^+$	cmlB$^+$ PyrD	PyrD$^+$	cmlB	5

Choose the correct order:

(a)	**aroA**	**cmlB**	**PyrD**
(b)	**aroA**	**PyrD**	**cmlB**
(c)	**cmlB**	**aroA**	**PyrD**

Solution

(i) Closer genes have higher co-transduction frequencies. The PyrD cmlB genes show highest co-transduction frequency. So they are closest together and (c) is eliminated.

(ii) The genes aroA PyrD show the lowest co-transduction frequency, so they are the farthest apart. So the (b) is eliminated.

(iii) The aroA and cmlB genes show an intermediate co-transduction frequency as would be expected if cmlB is between aroA and PyrD. Thus, the correct order is (a) aroA cmlB PyrD.

Problem 28

Order cheA cheB cda and supD from the following data.

Marker	Cotransduction or %
CheA-eda	15
cheA-supD	5
cheB-eda	28
cheB-supD	2.7
eda-supD	0

Solution

(i) Since relatively closer genes show a relatively higher co-transduction frequency, pairs of genes (loci) can be ordered in terms of their proximity.

(ii) The order (closest together to farthest apart) is cheB-eda, cheA-eda, cheA-supD, cheB-supD, eda-supD.

The gene order consistent with the relationship eda-cheB-cheA-supD.

EXERCISE

1. Lederberg and his colleagues (Nester, Schafer, and Lederberg, 1963, Genetics 48(4) : 529-551) determined gene order and relative distance between gene using three markers in the bacterium, *Bacillius subtilis*. DNA from prokaryotic strains (trp^+ his^+ tyr^+) was used to transform the autotroph. The seven classes of transformants with their numbers, are tabulated as follows:

trp^+ his^- tyr^-	trp^- his^+ tyr^-	trp^- his^- tyr^+	trp^+ his^+ tyr^-	trp^+ his^- tyr^+	trp^- his^+ tyr^+	trp^+ his^+ tyr^+
2,600	418	685	1,180	107	3,660	11,940

Outline the techniques used to obtain these data taking the loci in pairs. Calculate co-occurrence. Construct the most consistent linkage map of these loci.

2. DNA from a bacterial strain that is a^+ b^+ c^+ is used to transform a strain that is a^- b^- c^-. Following are the numbers of each transformed genotype appeared. What can we say about the relative positions of the genes:

Genotype	Number
a^+ b^- c^-	214
a^- b^+ c^-	231
a^- b^- c^+	206
a^+ b^+ c^-	11
a^+ b^+ c^+	6
a^- b^- c^+	93
a^- b^+ c^+	14

3. In *E. coli* the three loci ara, leu and ilvH are within $\frac{1}{2}$ minute, map distance apart. To determine the exact order and relative distance, the photograph (ara^+ leu^+ $ilvH^+$) was infected with transducing phage P1. The lysate was used to infect the auxotroph (ara^- leu^- $ilvH^-$). The ara^+ classes of transductants were selected to produce the following data:

ara^+ leu^- $ilvH^-$	ara^+ leu^+ $ilvH^-$	ara^+ leu^- $ilvH^+$	ara^+ leu^+ $ilvH^+$
32	9	0	340

What is the gene order and what are the relative co-transduction frequencies between gene? Why do some classes occur so infrequently?

4. A cross is made between two *E. coli* strains: Hfr arg^+ bio^+ leu^+ X F^- arg^- leu^-. Interrupted mating studies show that arg^+ enters the recipient last, so arg^+ recombinants are selected on a medium containing bio and leu only. These recombinants are tested for the presence of bio^+ and leu^+.

The following numbers of individuals are found for each gonotype:

arg^+	bio^+	leu^+	320
arg^+	bio^+	leu^-	8
arg^+	bio^-	leu^-	0
arg^+	bio^-	leu^-	48

(a) What is the gene order ?

(b) What are the map distances in recombination units ?

5. A cross is made among Hfr met^+ thi^+ pur^+ X F^- met^- thi^- pur^-. Interrupted mating studies show that met^+ enters the recipient last. So met^+ exconjugants are selected on medium containing thi and pur only. These exconjugants are selected for the presence of the thi^+ and pur^+. The following number of individuals are found with each genotype:

met+	thi+	pur+	280
met+	thi+	pur–	0
met+	thi–	pur+	6
met+	thi–	pur–	52

(a) Why was met left out of the selection medium ?

(b) What is the gene order ?

(c) What are the map distances in recombination units ?

(d) Why is there no individual of the met^+ thi^+ pur^- genotype ?

[M.Sc. (Zoo) V.U. 2003]

6. An F^+ strain, marked at 10 loci, gives rise spontaneously to Hfr progeny, whenever the F factor becomes incorporated into the chromosomes of the F^+ strain. The F factor can integrate into the circular chromosomes at many points, so that the various Hfr segregates transfer the genetic markers in different sequences. For any Hfr strain the order of markers entering early can be determined by interrupted mating experiments using a waring blender. From the following data for several Hfr strains derived from the same F^+ determine the order of markers in the F^+ strain:

Hfr strain	Markers donated in order
1	— Z–H–E–R →
2	— O–K–S–R →
3	— K–O–W–I →
4	— Z–T–I–W →
5	— H–Z–T–I →

7. The order of markers transferred by four Hfr strains to F^- cell is as follows:

Hfr strain	Order of markers donated
1	G E B D N A
2	P Y L G E B
3	X T J F P Y
4	B E G L Y P

Draw a map showing the order of markers on the chromosome of original F^+ chromosome.

8. In *E. coli* four Hfr strains transfer their markers in the order shown:

Strain 1	A B C D E
Strain 2	G H I E D
Strain 3	J K L G H
Strain 4	J A B C D

All the Hfr strains were derived from the same F^+ strain. Arrange the markers on the circular chromosome showing the origin and direction of transfer with an arrow in each case.

[M.Sc. (Zoo) C.U. 2002]

9. Consider the following transduction data:

Donor	Recipient	Selected Marker	Unselected Marker	%
$aceF^+$ dhl	aceF dhl^+	$aceF^+$	dhl	88
$aceF^+$ leu	aceF leu^+	$aceF^+$	leu	34

Is dhl or leu closer to aceF?

10. Consider the following data pertaining to P1 transduction:

Donor	Recipient	Selected Marker	Unselected Marker	%
aroA $pyrD^+$	$aroA^+$ pyr D	$pyrD^+$	aroA	5
$aroA^+$ cmlB	aroA $cmlB^+$	$aroA^+$	cmlB	26
cmlB $pyrD^+$	$cmlB^+$ pyrD	$pyrD^+$	cmlB	54

Choose the correct order:

(a) aroA cmlB pyrD
(b) aroA pyrD cmlB
(c) cmlB aroA pyrD

Competence

The ability of recipient bacterial cells to interact and integrate DNA to give rise to genetically transferred cell (bacterium) is called competence.

Several genes and proteins, such as DNA receptor protein, the PSTC (pilus, secretion, twitching, competence) protein, a competence nuclease and competence proteins are required for competence.

Problem:

Two strain of *E. coli* ab^+ and a^+b are inown to exchange genetic material leading to the production of a^+b^+ recombinants. However, when these two above mentioned strains are grown is opposite arms of a U tube (like Davies) no a^+b^+ recombinants are produced. Name the parasexual process responsible for the formation of a^+b^+ recombinants when both strains are grown together.

Solution:

(i) The parasexual process is conjugation, i.e., (cell to cell contact). The two strains of *E. coli* a^+b and ab^+ exchange their genetic material through conjugation and form a^+b^+ recombinants.

(iii) The glass filter separating the two arms of U tube prevents contact between cells in these arm.

B. COMPLEMENTATION MAPPING

- **COMPLEMENTATION:**

The ability of two recessive mutation to restore the wild type phenotype either partially or completely in trans (heterozygote) position is known as **complementation.**

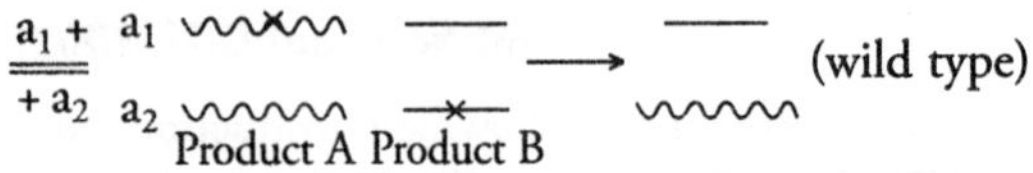

Here,

a_1—produces two products (A and B) and A is defective.

a_2—produces also two products (A and B) and B is defective.

- **COMPLEMENTATION TEST (CIS TRANS TEST):**

Introduction of two recessive mutations simultaneously into the same cell but on different chromosomes to determine whether they are alleles of the same gene, that is, whether they affect the same genetic function is known as **complementation test.**

If the mutations are allelic, the genotype $\frac{m_1 +}{+ m_2}$ will exhibit mutant phenotype (i.e. they are non-complementary) whereas if they are non-allelic it will exhibit the wild type phenotype (i.e. they complement each other).

- **CIS TEST:**

(i) The two-mutation presence in the cis or coupling configuration (on the same chromosome) is called **cis heterozygote.**

(ii) The **cis heterozygote** must have the wild type phenotype. If the cis heterozygote has mutant phenotype, the trans test cannot be used to determine whether the two mutations in the same gene.

Thus, trans test can never be used with dominant mutation. The cis test is an important control which establishes the validity of **trans** or **complementation test.**

- **TRANS TEST:**

It is commonly known as **complementation test.** Here two mutations are in repulsion configuration.

(i) If the trans heterozygote has the mutant phenotype, then the two mutations are in the same unit of function or the same gene.

(ii) If the trans heterozygote has wild type phenotype, then two mutations are in two different units of function or two different genes.

In this case, the two mutations are said to complement each other (Fig. 14.5).

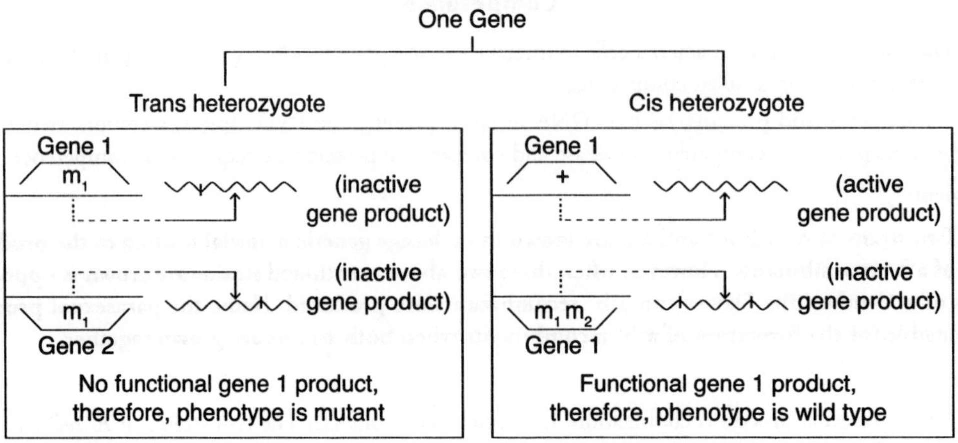

Fig. 14.5: cis trans test, showing two mutations m_1 and m_2 are both in same gene.

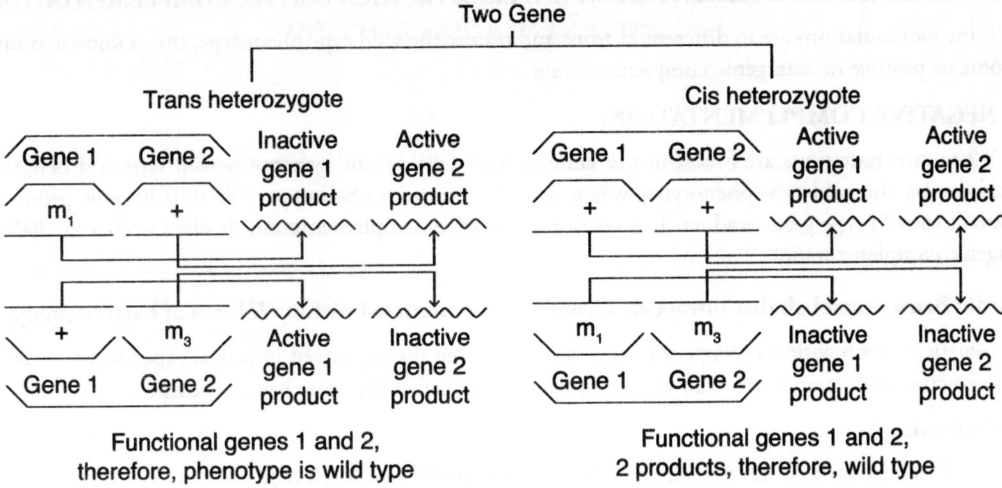

Fig. 14.6: cis trans test, showing mutation m_3 in second chromosome.

Difference between Complementation and Recombination

Complementation	Recombination
1. It involves the interaction of gene products (usually protein), i.e., normal phenotype produced by chromosomes carrying two different mutations when they are present in common protoplasm.	Recombination, involves direct interaction between the chromosomes carrying the actual breakage, and reunion of chromosome takes place.
2. It does not depend upon recombination of the two chromosomes.	It occurs through breakage and reunion of chromosome which results wild type and mutant phenotype.
3. It occurs in every trans-heterozygote containing mutation at different genes.	It is normally detected by examining progeny of the heterozygotes & trans-heterozygotes will never produce more than 25% gamete with wild type chromosome.
4. It can occur only one gene and another.	It can occur within a gene.

- **LIMITATION OF COMPLEMENTATION TEST:**

There are four instances where complementation cannot be applied.

1. Dominant or co-dominant mutation.
2. Genes in which mutation occurs that exhibits intragenic complementation.
3. Polar mutation (mutation that affects the expression of adjacent genes) within operon.
4. Cis acting genes (operator or promoter of operon).

- **COMPLEMENTATION GROUP:**

A collection of mutants mutually unable to complement each other when tested in pairwise combination in trans is termed complementation group and corresponds to a cistron. The inference is that both affect the same function.

- **COMPLEMENTATION MAP:**

It represents a complementation group as discrete non-overlapping lines which can be constructed by performing pairwise complementation analysis of several mutants.

- **INTERGENIC COMPLEMENTATION (INTERCISTRONIC/POSITIVE COMPLEMENTATION):**

If the two mutations are in different cistrons and restore the wild type phenotype, this is known as **inter-cistronic** or **positive** or **intergenic** complementation.

- **NEGATIVE COMPLEMENTATION:**

When two mutations are found in the same genome, the second genome is wild type and hence cell should display the wild type phenotype. When it exhibits mutant phenotype, i.e., mutant gene product is dominant to wild type gene product. It is termed as **negative complementation.** It often occurs in **allelic** or **intragenic complementation.**

How did Benzer establish that intragenic recombination occurred within *rII* locus of bacteriophage T_4?

Genetic recombination in bacteriophages was discovered during mixed infection experiments in which two distinct mutant strains were allowed to simultaneously infect the same bacterial culture.

Requirements:

(a) *E. coli* bacteria: It has two strains, viz., *E. coli* B and *E. coli* K-12 (λ).

(b) T_4 bacteriophage.

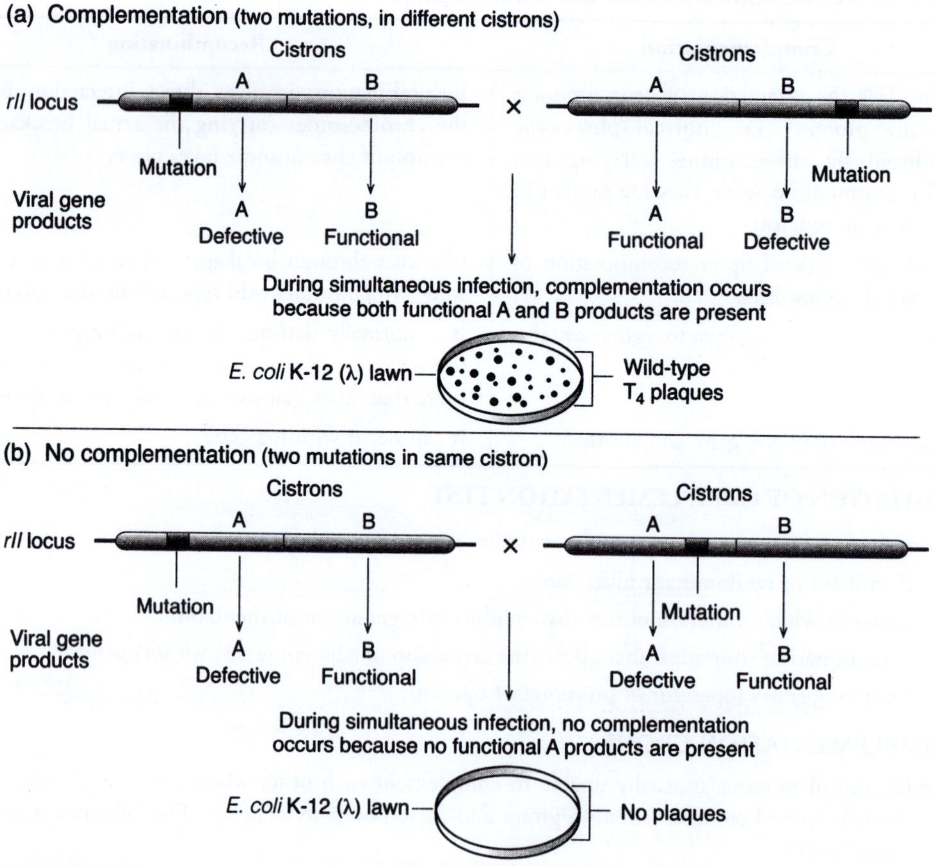

Fig. 14.7: Comparison of two rII mutations that either (a) complement one another or (b) do not complement one another. Complementation occurs when each mutation is in a separate cistron. Failure to complement occurs when the two mutations are in the same cistron.

(i) A single locus rII is present in T_4 bacteriophage infecting *E. coli*. The mutant rII locus of T_4 bacteriophage is responsible for the formation of rough plaque.

(ii) It had largest number of rapid lysis (r) mutant.

It can be distinguished from other r-loci by the inability of rII mutants to produce plaques in lysogenic K strain of *E. coli* and only the ability to produce plaques in B strain of *E. coli.*

(iii) Wild type T_4 is able to grow in and lyses cells of either *E. coli* B and *E. coli* K strain. In contrast, rII mutants of T_4 can grow and lyse the *E. coli* B, producing large clear plaques but are unable to grow in cells K-12 (λ).

(iv) Strain B is the permissive host for rII mutants and strain K is non-permissive host for the same.

Experimental process:

(i) S. Benzer worked with rII mutants on non-permissive *E. coli* strain K-12 (λ).

(ii) K-12 bacteria were simultaneously infected with a pair of different rII mutant strains.

(iii) After allowing a sufficient time for single lytic cycle to occur.

Observation:

(i) Benzer sometimes found that certain pairs of rII mutant strains lysed the K-12 (λ) bacteria.

(ii) This was initially quite puzzling, since only the wild type rII was supposed to be capable of lysing K-12 bacteria.

Inference:

(i) The result indicated that mutant strains of rII, each of which was thought to contain a defect in the same gene which showed a wild type character.

(ii) Benzer reasoned that during simultaneous infection each mutant strain provided something that the other lacked, thus restoring wild type function.

(iii) Benzer called this phenomenon as **complementation**, those mutants fail to complement one another were placed in the **same complementation group**, while those that did complement one another were assigned to a **different complementation group**.

(iv) Benzer coined the term **cistron**, which he defined as the smallest functional genetic unit, to describe a complementation group.

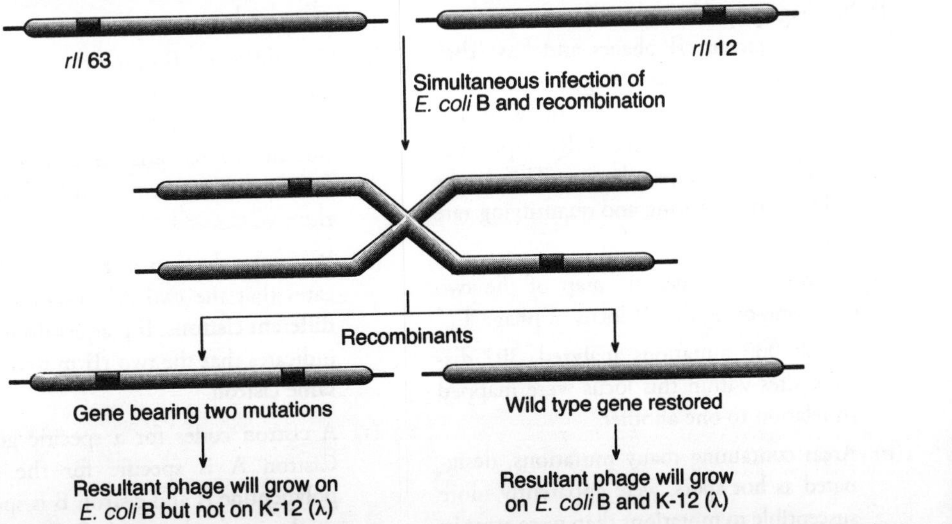

Fig. 14.8: Illustration of intragenic recombination between two mutations in the rII locus of phage T_4. The result is the production of a wild-type phage that will grow on both *E. coli* B and K-12 and a phage that has incorporated both mutations into the rII locus. It will grow on *E. coli* B, but not on *E. coli* K-12.

In modern terminology—**cistron** represents a **gene.**

- **Notes on rII locus:**

(i) A single locus rII in T_4 bacteriophage infects *E. coli* bacteria. This locus is known as **rII locus** and mutant at this locus is responsible for the formation of **rough plaques** (colonies).

(ii) Mutants at the rII locus produce distinctive plaques when plated in *E. coli* strain B, allowing their easy identification.

(iii) Benzer isolated many independent rII mutants. He assumed that most mutations (randomly isolated) would represent different locations within rII locus.

(iv) The key to Benzer analysis was that rII mutant phages, capable of infecting and lysis *E. coli.* But could not successfully lyse a second related strain *E. coli* K-12 (λ).

(v) Wild type phage by contrast could lyse both B and K-12 strain.

(vi) If phages from any two different mutant strains were allowed to simultaneously infect *E. coli* B; exchange between the two mutant sites within the locus would produce rare wild type recombinants.

(vii) If the phage population, which contains more than 99.9 percent rII phages and less than 0.1 percent wild type phages, were then allowed to infect strain K-12, the wild type recombinants would successfully reproduce and produce wild type plaques. This is the critical step in recovering and quantifying rare recombinants.

(viii) Benzer produced a genetic map of the two cistrons composing the rII locus of phage T_4.

(a) Of, 20,000 mutations analyzed, 307 distinct sites within this locus were mapped in relation to one another.

(b) Areas containing many mutations, designated as **hot spots,** were apparently more susceptible to mutations than were areas in which only one or a few mutations were found.

(c) He discovered areas within the cistrons in which no mutations were localized. He estimated that as many as 200 recombinational units that are arranged in specific order.

(d) He finally demonstrated that gene is not indivisible particle, instead consists of mutational and recombinational units that are arranged in a specific order.

- **STATE BRIEFLY ABOUT COMPLEMENTATION TESTING IN RESPECT TO rII LOCUS**

Complementation test:

A genetic test to determine whether two mutations occur within the same gene. The production of wild type phenotype when two recessive mutations with similar phenotype are combined in a diploid or heterokaryon. **Complementation** means that **mutations** are in **different genes.**

(i) It was discovered that T-4 rII mutants could lyse and form plaques on *E. coli* strain B but could not lyse and form plaques on *E. coli* strain K-12 (λ).

(ii) The wild type phage (r+) could form plaques on both strain B and strain K-12 (λ); therefore, *E. coli* strain K-12 (λ) can be used as selective host.

(iii) Two different rII mutants mixed in a high titre can effect a mixed infection that is a large number of viral particles relative to the number of bacteria. It means two or more viruses will infect one individual host bacterium.

(iv) If plaques develop in K-12 (λ) strain, it indicates that the two rII mutations must be **in different cistrons.** If plaques do not develop, it indicates that the two rII mutations are **in the same cistron.**

(v) A cistron codes for a specific gene product. Cistron A is specific for the synthesis of **polypeptide A** and cistron B is specific for the synthesis of **polypeptide B.** Both polypeptides are necessary for the formation of a **functional protein.**

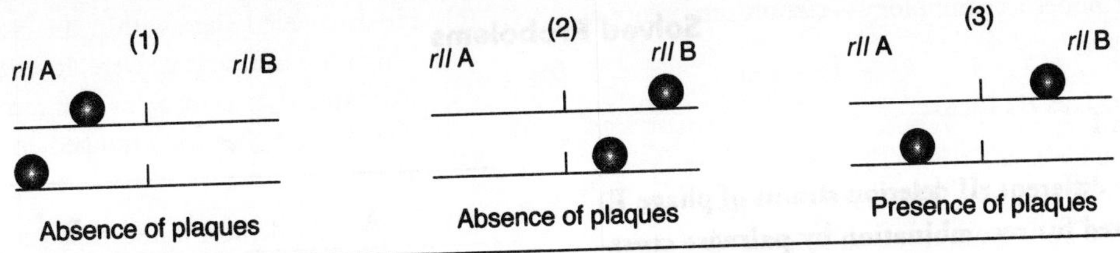

Fig. 14.9: Schematic diagram showing trans arrangement of rII gene

(vi) From the Figure 14.9:

(a) **In the situation 1:**

(i) Here the 2 mutations are on separate DNA strands (trans arrangement).

(ii) No plaque formation will occur on K-12 (λ) in this case because there is no way to generate a normal polypeptide A for the protein to be functional.

(b) **In the situation 2:**

(i) Here it is also a trans arrangement.

(ii) No plaque formation will occur on K-12 (λ) again because there is no way to generate a normal polypeptide B for the protein to be functional.

(c) **In the situation 3:**

(i) It is also a trans arrangement.

(ii) However, plaque formation will occur.

(iii) Phage multiplication can occur on K-12 (λ) because normal polypeptide A is specified from one viral DNA strand while polypeptide B is specified by the other viral DNA strand.

(iv) A functional protein will then result because of complementation and the viral mutants will lyse K-12 (λ).

(vii) In cis arrangement, along with mixed infection, lysis occurs on K-12 (λ). The cis serves as control. All combination in cis should yield plaques.

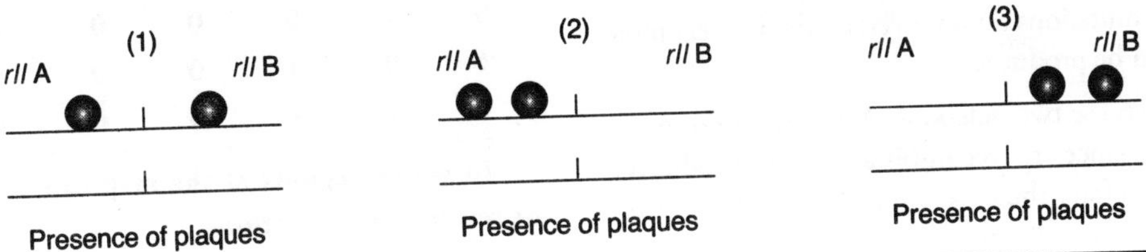

Fig. 14.10: Schematic diagram of cis arrangement of rII gene

- **Deletion mapping:** The shortcut mapping procedure by using of strains carrying deletion within a defined region to map a point mutation of unknown location in a gene or on a chromosome.
- **Process of deletion mapping:**

  (i) The principle deletion mapping involves crossing a strain which has a point mutation in a gene and other carrying deletion in a different region of the same gene.

  (ii) If recombination occurs between two strains to produce wild type (+) gene, the point mutation cannot lie within the region of the deletion

  If recombination cannot produce any wild type (+) genes, then it is reasonable to conclude that point mutation and deletion are found within the same stretch of DNA or gene.

Solved Probolems

Problem 1

Five different rII deletion strains of phage T_4 were tested for recombination by pairwise crossing in *E. coli* B. The following results were obtained, where + = r^+ recombinants produced and 0 = no r^+ recombinants produced:

	A	B	C	D	E
E	0	+	0	+	0
D	0	0	0	0	
C	0	0			
B	+	0			
A	0				

Draw a deletion map compatible with these data.

Solution:

If two deletion mutations overlap, then no r^+ recombinants can be produced. Conversely, if two deletion mutations do not overlap, then r^+ recombinants can be produced.

(i) From the two deletions, A and B do not overlap since r^+ recombinants are produced. Therefore, these two mutations can be represented as follows:

(ii) The next deletion C does not give r^+ recombinants with any of the other four deletions. Therefore, C is an extensive deletion that overlaps all of the four. It is represented as

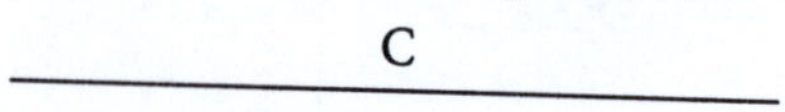

(iii) The deletion D does not give r^+ recombinant with A, C or E. In turn, E gives r^+ recombinants with B and D, but not with A and C. Thus, D must overlap A, B and C, but not with E. On the other hand, E must overlap A and C, but not with B and D. Therefore, map can be drawn as follows:

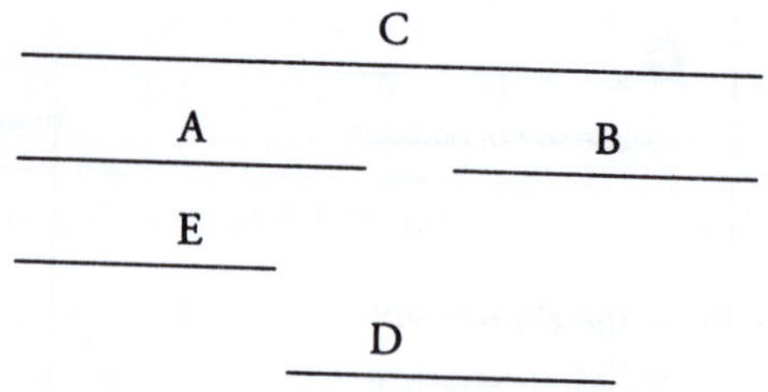

Problem 2

Seven different rII point mutants (1 to 7) of phage T_4 were tested for recombination crosses in *E. coli* B with five deletion strains A B C D E. The following results were obtained, where + = r^+ recombinants produced and 0 = no r^+ recombinants produced:

	A	B	C	D	E
1	0	+	0	+	+
2	+	0	0	+	+
3	0	+	0	+	0
4	+	+	0	+	0
5	+	0	0	0	+
6	0	+	0	0	+
7	+	+	0	0	+

In which regions of the map can you place the seven point mutations?

Solution:

(i) Point mutation seven (7) gives r^+ recombinant with A, B and E, but not with C and D. Therefore, 7 is located in the region defined by the part of deletion D. That is not involved in the overlap with A and B.

(ii) Similarly, point mutation four (4) gives r^+ recombinant with A, B and D, but not with E. Thus, 4 must be in a region defined by a segment of deletion E that does not overlap deletion A. Since 4 does not give r^+ recombinant with C deletion, C must overlap the site defined by point mutation 4.

(iii) Strain C has extensive deletion and covers all the point mutations (1 – 7).

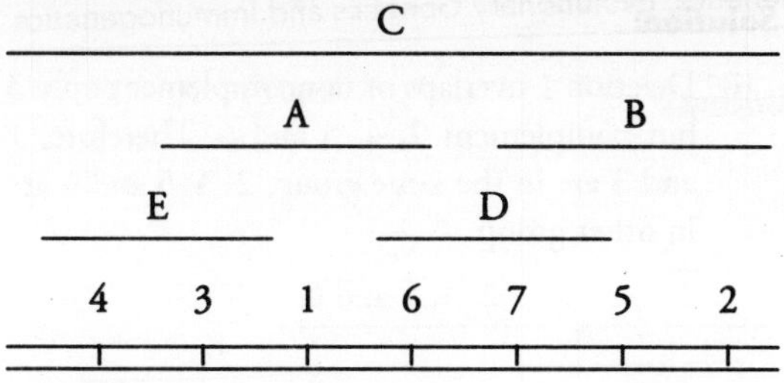

Problem 3

In *Drosophila* mutant A, B, C, D, E, F and G, all have the same phenotype; the absence of red pigment in the eyes. In pairwise combinations in complementation test, the following results were produced, where + = complementation and – = no complementation:

	A	B	C	D	E	F	G
G	+	–	+	+	+	+	–
F	–	+	+	–	+	–	
E	+	+	–	+	–		
D	–	+	+	–			
C	+	+	–				
B	+	–					
A	–						

(a) How many genes are present? (b) Which mutants have defects in the same gene?

Solution:

(a) (i) Here A is complement with B, C, E and G. Therefore, B, C, E and G are in separate cistron. On the other hand, F, D are non-complement with A. So they are on the A cistronic group, i.e., A, F, D are in the same cistron.

(ii) B is non-complement with G. Therefore, B and G are in the another group, i.e., they are in the one cistronic group.

(iii) C is non-complement with E. So it is in the other group, i.e., C and E are in one cistronic group. So, there are three genes present.

(b) A, D, F are in one gene; B and G in other gene, and C and E in an other gene.

Problem 4

Seven deletion mutants within the A cistron of the rII region of phage T_4 were tested in all pairwise combinations for wild type recombinants. In the adjacent table of results + = recombination, 0 = no recombination. Construct a topological map for these deletion:

	1	2	3	4	5	6	7
1	0	+	0	0	+	0	0
2		0	0	0	+	+	0
3			0	0	+	+	0
4				0	+	0	0
5					0	0	0
6						0	0
7							0

Solution:

If two deletions overlap to any extent, no wild type recombinants can be formed.

(i) Deletion 1 overlaps with 3, 4, 6, 7 but not with 2 and 5. Therefore, 3, 4, 6, 7 are in one group and 2, 5 are in other group.

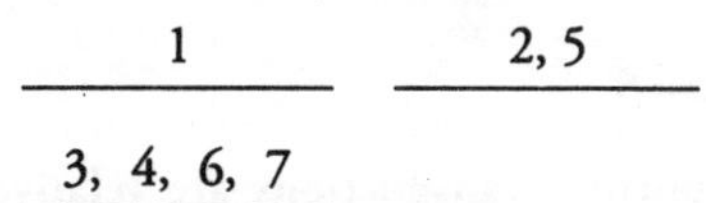

(ii) Deletion 2 overlaps with 3, 4 and 7, but not with 5 and 6.

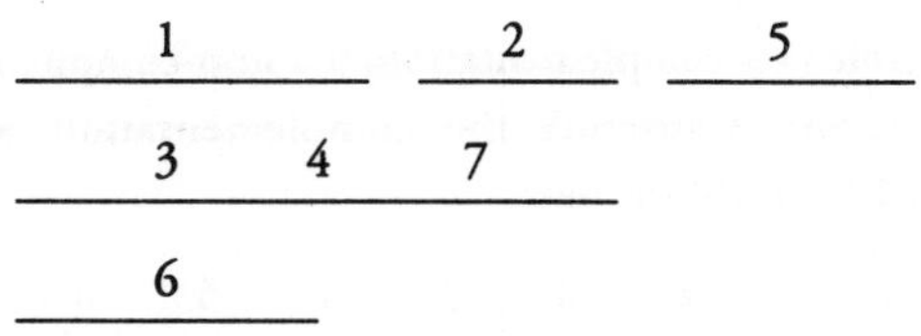

(iii) Deletion 3 overlaps with 1, 2, 4 and 7 but not with 5 and 6.

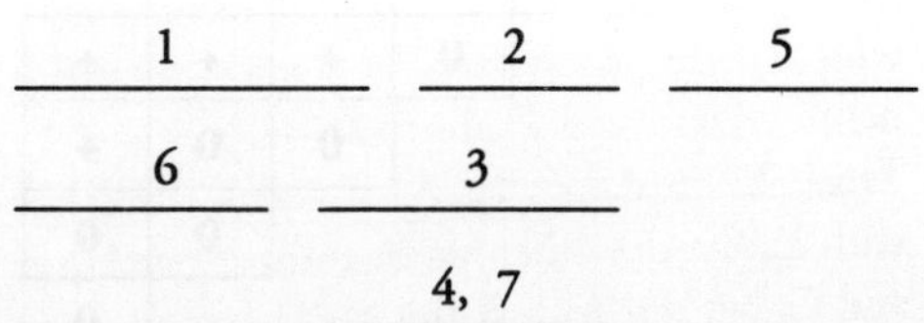

(iv) Deletion 4 overlaps with 1, 2, 3, 6 and 7 but not with 5.

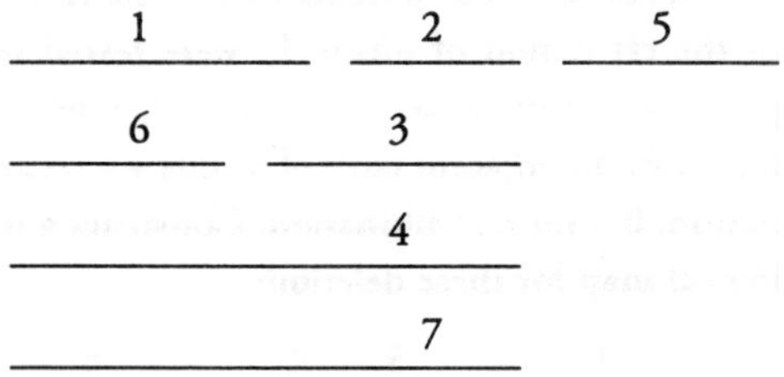

(v) Deletion 5 overlaps with 6, 7, but not with 1, 2, 3 or 4. To satisfy this condition, 5 will be shifted to the left side of 1 and extend 7 to the part region 5. Now segment 5 can overlap with 1 and 7 without overlapping 1, 2, 3 and 4.

(vi) Deletion 6 overlaps with 1, 4, 5 and 7, but not with 2 and 3.

(vii) Deletion 7 overlaps with all other regions just as it was diagrammed in step (v).

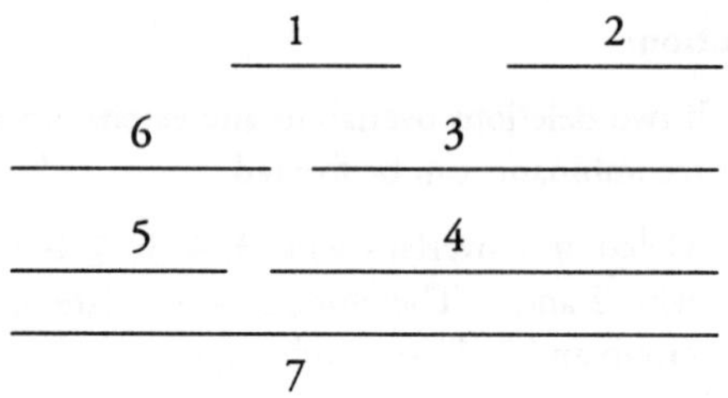

Problem 5

Abortive transductions are relatively stable merozygotes which can be used for complementation tests. Six mutants are tested in all pairwise combinations, yielding the result shown in the table (+ = complementation, 0 = non-complementation). Construct the complementation map with consistent data.

1	2	3	4	5	6	
0	+	0	+	+	+	1
	0	0	+	+	+	2
		0	+	+	+	3
			0	0	+	4
				0	0	5
					0	6

Solution:

(i) Deletion 1 overlaps or noncomplement only 3 but complement 2, 4, 5 and 6. Therefore, 1 and 3 are in the same group; 2, 4, 5 and 6 are in other group.

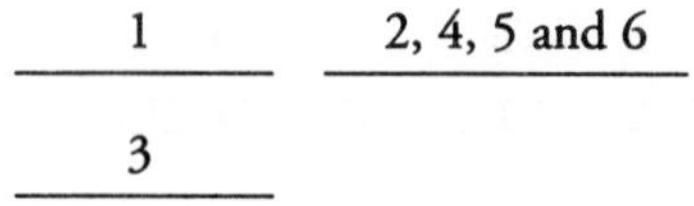

(ii) Deletion 2 overlaps with 3, not with 4, 5 and 6. Therefore, 2 remains in separate group and 4, 5 and 6 in other group.

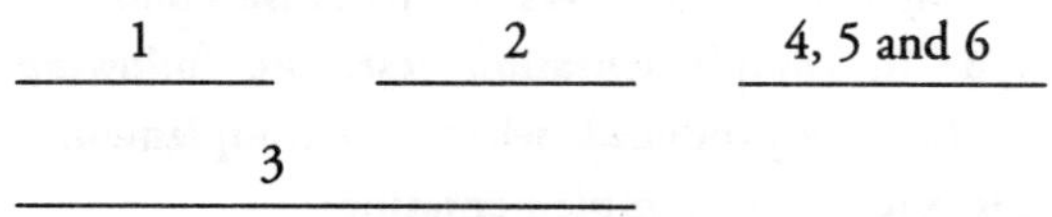

(iii) Deletion 3 overlaps with 1, 2, but not with 4, 5 and 6.

(iv) Deletion 4 overlaps with 5, but not with 6. On the other hand, 5 overlaps with 4 and 6, but not with 1, 2 and 3.

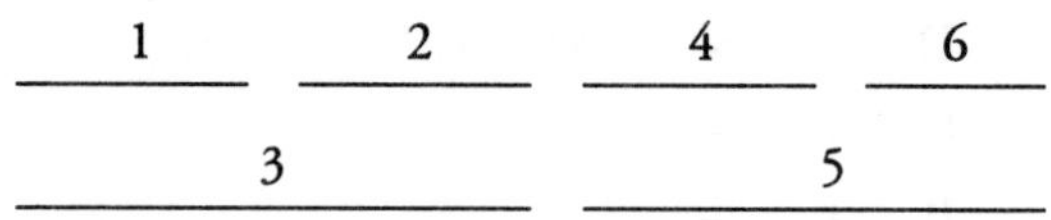

Complementation map is given below:

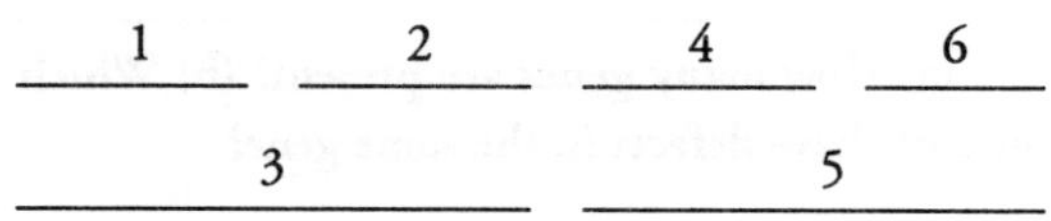

Problem 6

Five deletion mutations within the B cistron of the rII region of phage T_4 were tested in all pairwise combinations for wild type recombinants. In the following table of results, + = recombination, 0 = no recombination. Construct topological map for these deletions:

	1	2	3	4	5
1	0	+	+	0	0
2		0	+	0	+
3			0	+	0
4				0	0
5					0

Solution:

(i) Mutant 1 overlaps with 4 and 5, i.e., no recombination, but recombination occurs in 2 and 3. Therefore, 1, 4, 5 are in the same group, but 2 and 3 on the other group.

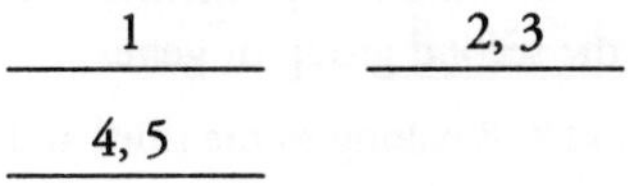

(ii) Mutant 2 overlaps with 4, but not with 1, 3 and 5; on the other hand, mutant 3 overlaps with five (5).

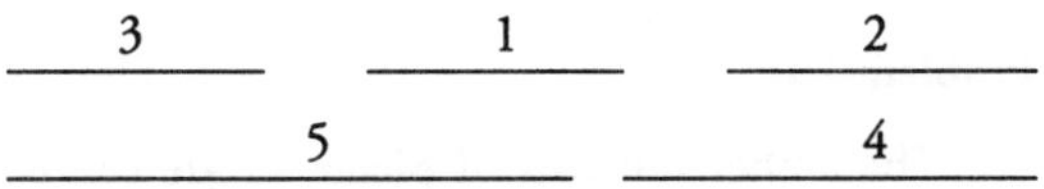

Mutant 4 overlaps with 1, 2 and 5, but not with 3.

(iii) Therefore, the complementation map is as follows:

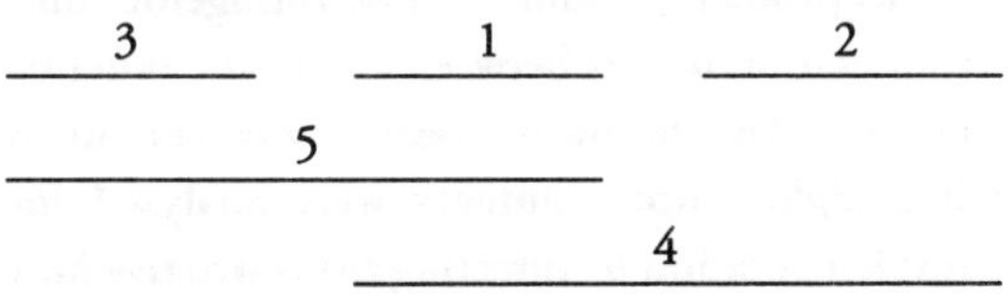

Problem 7

Six deletion mutants within the A cistron of the rII region of phage T_4 were tested in all pairwise combinations for wild type recombinants. On the following table, + = recombination, 0 = no recombination. Construct a topological map for these deletion:

1	2	3	4	5	6	
0	0	0	0	0	0	1
	0	0	0	0	+	2
		0	+	0	0	3
			0	+	+	4
				0	+	5
					0	6

Solution:

(i) Mutant 1 overlaps with all, so it is top mutant. Mutant 2 overlaps with all except 6. So 6 belongs to separate group.

(ii) Mutant 3 overlaps with 1, 2, 5 and 6, but not with 4. Mutant 4 overlaps only with 1 and 2.

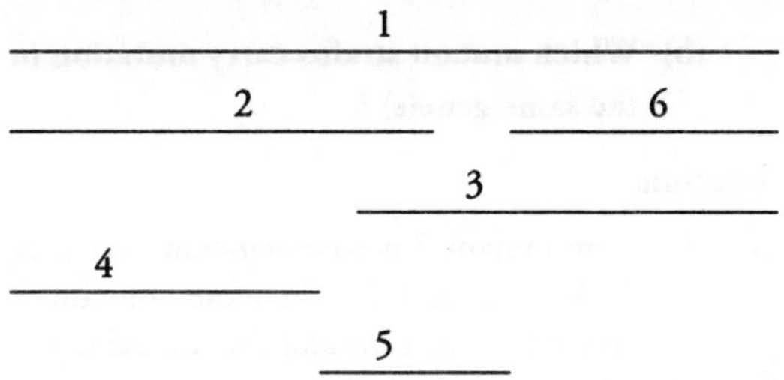

(iii) Mutant 5 overlaps with all 1, 2, 3 and not with 4, 6. On the other hand, mutant 6 overlaps with 1, 3, but not with 2, 4 and 5. Therefore, the complementation map is as follows:

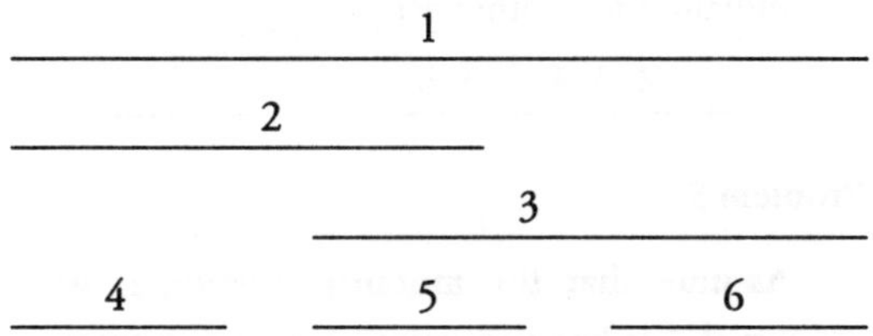

Problem 8

Eight independently isolated mutants of *E. coli*, all of which are unable to grow in the absence of histidine (his$^-$), were examined in all possible cis and trans heterozygotes (partial diploids). All the cis heterozygotes were able to grow in the absence of histidine. The trans heterozygotes yielded two different responses. Some of them grew in the absence of histidine and some did not. The experimental results using + to indicate growth and 0 to indicate no growth, are given in the following table:

Growth of Trans Heterozygotes (Histidine)

Mutant	1	2	3	4	5	6	7	8
8	0	0	0	0	0	0	+	0
7	+	+	+	+	+	+	0	
6	0	0	0	0	0	0		
5	0	0	0	0	0			
4	0	0	0	0				
3	0	0	0					
2	0	0						
1	0							

(a) How many genes are defined by these eight mutations ?

(b) Which mutant strains carry mutation in the same gene(s) ?

Solution:

(a) (i) Here mutant 7 is complementary to 1, 2, 3, 4, 5, 6 and 8. Therefore, mutant 7 belongs to one gene and the rest belong to other gene.

(ii) So there are two genes present in these eight mutants.

(b) 1, 2, 3, 4, 5, 6 and 8 are in one gene and mutant 7 is in other gene.

1, 2, 3, 4, 5, 6, 8	7

Problem 9

Assume that the mutants (histidine) were examined. Cis heterozygotes were able to grow in the absence of histidine. The trans heterozygotes yielded two different responses; some of them grow in absence of histidine, some did not. Results are given below:

Growth of Trans Heterozygotes (Histidine)

Mutant	1	2	3	4	5	6	7	8
8	+	+	+	+	+	+	0	0
7	+	+	+	+	+	+	0	
6	+	+	+	+	0	0		
5	+	+	+	+	0			
4	+	+	0	0				
3	+	+	0					
2	0	0						
1	0							

(a) How many genes would they have defined ?

(b) Which mutations would have been in the same gene(s) ?

+ indicates growth, 0 indicates no growth.

Solution:

(i) Mutant 1, 2 are non-complementary, so they belong to the same group. The rest (3, 4, 5, 6, 7, 8) in the other group or gene.

(ii) The mutant 3, 4 are non-complementary, so they belong to the second group or gene.

(iii) Similarly, 5, 6 and 7, 8 belong to the third and fourth group or gene respectively.

(a) Four genes are present.

(b)

1, 2	3, 4	5, 6	7, 8

Problem 10

The amber mutants of phage T_4 are conditional lethal mutants. They grow on *E. coli* strain CR63, but are lethal on *E. coli* strain B. An amber mutant almost never exhibits intragenic complementation with any other amber mutant; for this problem, assume that no intragenic complementation occurs between any of the mutants involved. The following results were obtained when eight amber mutants were analysed for complementation by infecting the restrictive host (*E. coli* strain B) with each possible pair of mutants. The results of mixed infection by pairs of mutants are shown as 0, if no progeny is produced, and as +, if progeny phage results from the infection with that particular pair of mutants:

Mutant	1	2	3	4	5	6	7	8
8	+	+	+	+	+	+	0	0
7	+	+	+	+	+	+	0	
6	+	+	+	+	+	0		
5	0	+	0	+	0			
4	+	+	+	0				
3	0	+	0					
2	+	0						
1	0							

(a) Indicate how many different genes are located by these eight amber mutation.

(b) Which mutations are located in the same gene or genes ?

Solution:

(i) Mutant 1, 3, 5 are non-complementary, so they are in one gene, the mutant 4 is complementary to all. So it belongs to separate group.

(ii) Mutant 2 is also complementary to all, it belongs to a separate group. Mutant 6 is also complementary to all, so it also belongs to separate group or gene.

(iii) Mutant 7, 8 also belong to a separate gene as they are non-complementary.

1, 3, 5	7, 8	6	4	2

(a) Five genes are present.

(b) Mutation 1, 3, 5 are in one gene; 7, 8 are in second gene; mutation 2, 4 and 6 identify genes 3, 4 and 5 respectively.

Problem 11

Six point mutants are known to be resided in three cistrons. Complete the following table where + = complementation and 0 = non-complementation:

	1	2	3	4	5	6
1	0	+		+	+	+
2		0	+			+
3			0	0		
4				0		
5					0	+
6						0

Solution:

(i) Mutants 3 and 4 fail to maintain complement. So they are in the same cistron. Suppose, they belong to A, i.e., $\frac{3,\,4}{A}$.

(ii) Mutant 2 complements with 3 and 6; so mutant presents 2 in a separate cistron B, i.e., $\frac{2}{B}$.

(iii) Mutant 5 complements with 6 and 1, so it belongs to the cistron where 6 and I mutants do not present. Therefore, it presents either in cistron A(3, 4) or cistron B(2). On the other hand, mutant 6 complements with 1, 2 and 5. Therefore, it must be present in cistron A(3, 4). From this situation it is clear that mutant 5 must be present in cistron B(2).

(iv) Mutant 1 is complement with 2, 4, 5, 6, so it remains in a separate cistron, i.e., C.

3, 4, 6	2, 5	1
A	B	C

Problem 12

Given the topological map of 6 deletion mutants shown below, predict the results of recombination experiments involving the 5 point mutants (a to e) with each of the 6 deletion mutants (1 to 6). Complete the accompanying table of results by using + for recombination and 0 for no recombination:

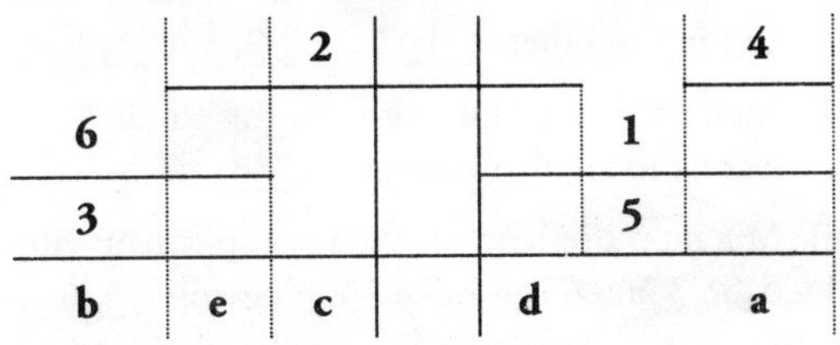

Solution:

(i) From the above cistron mutant a overlaps with 1, 4 and 5. So a × (1, 4, 5) = 0.

(ii) Mutant b overlaps with 3 and 6. So b × (3, 6) = 0. In this way, e overlaps with 2, 3 and 6; c overlaps with 2 and 3; d overlaps with 1, 2 and 5.

Deletion

Point mutant	1	2	3	4	5	6
a	0	+	+	0	0	+
b	+	+	0	+	+	0
c	+	0	0	+	+	+
d	0	0	+	+	0	+
e	+	0	0	+	+	0

Problem 13

The complementation data shown in the table are observed. The number refers to particular mutation. The symbols + and – referred to particular mutations do and do not complement

respectively. How many genes are represented? Assign the mutation to the genes.

[V.U. M.Sc.(Zool) 2002]

Mutant:	1	2	3	4	5	6	7
1	−	+	+	+	+	+	−
2		−	+	−	+	+	+
3			−	+	+	−	+
4				−	+	+	+
5					−	+	+
6						−	+
7							+

Solution:

(i) Mutant 1 overlaps with 7, i.e., they are non-complement. So they are in the same cistron and rest in other <u>1, 7</u> <u>2, 3, 4, 5, 6</u>

(ii) Mutant 2 overlaps with 4, therefore, 2, 4 belong to another cistron <u>2, 4</u> .

(iii) Mutant 3 overlaps, i.e., non-complement with 6. So 3 and 6 belong to other cistron <u>3, 6</u> .

(iv) Now the rest 5 is complement with 6 and 7. So it remains in a separate cistron.

(v) Therefore, cistrons are four, i.e., <u>1, 7</u> <u>2, 4</u> <u>3, 6</u> <u>5</u> .

The map can be drawn in this way:

<u>1, 7</u> <u>2, 4</u> <u>3, 6</u> <u>5</u> .

Problem 14

In Benzer's fine structure analysis of the rII locus in phage T_4, he was able to perform complementation testing of any pair of mutation once it was clear that the locus contained two cistrons. Complementation was assayed by simultaneously infecting *E. coli* K-12 with two phage strains, each with an independent mutation, neither of which could alone lyse K-12. From the data that follow, determine which mutations are in which cistron, assuming that mutation 1 (M – 1) is in the A cistron and mutation 2 (M – 2) is in the B cistron and also mention if there be any cases where the mutation cannot be properly assigned:

Test pair	Results*
1, 2	+
1, 3	−
1, 4	−
1, 5	+
2, 3	−
2, 4	+
2, 5	−

* **= + or – indicates complementation or the failure of complementation respectively.**

Solution:

(i) Mutant 1 and mutant 5 complement one another; therefore, they are not in the same cistron. Thus, mutant 5 must be in the B cistron.

(ii) Mutant 2 and mutant 4 complement one another; therefore, mutant 4 is in the A cistron.

(iii) Mutant 3 fails to complement either mutant 1 and mutant 2, and so it would seem to be in both (A and B) cistrons. This is why mutant 3 overlaps both A and B cistrons.

(iv) It might be a double mutation with one in each cistron. It might be a deletion that overlaps both the A and B cistrons.

Problem 15

Seven mutant strains of T_4 can grow individually only in suppressor strains of *E. coli.* All possible pairwise mixed infections of non-suppressor *E. coli* cells are performed with seven strains to test for the ability to carry out a complete infection. In the result given, 0 = no phage growth; + = phage growth and lysis:

	1	2	3	4	5	6	7
1	0						
2	+	0					
3	0	+	0				
4	0	0	0	0			
5	+	+	+	0	0		
6	0	+	0	0	+	0	
7	+	0	+	+	+	+	0

What is the minimum number of genes that are represented in this collection of seven mutant strains ?

Solution:

(i) From this data, strain 4 clearly cannot complement any other. It must represent along deletion or less likely a multipoint mutation or a dominant mutation inhibiting growth under all conditions of mixed infection.

(ii) Strain 1 appears to be mutant in the same gene (no complementation) as strain 3 and strain 6. Suppose, it is gene (cistron) A $\frac{1\ 3\ 6}{A}$. Therefore, 2, 5, 7 remain in other cistron.

(iii) Strain 2 cannot complement strain 7. So they belong to cistron B $\frac{2\ \ 7}{B}$.

(iv) Strain 5 complements all strain except itself and deletion strain 4. It belongs to cistron C $\frac{5}{\quad}$.

These genes or cistrons, viz., A, B and C are represented in the collection.

Problem 16

Five point mutants (a through e) were tested for wild recombinants with seven deletion mutants drawn below. What is the order of point mutants of this deletion map (+ = recombination, 0 = no recombination)?

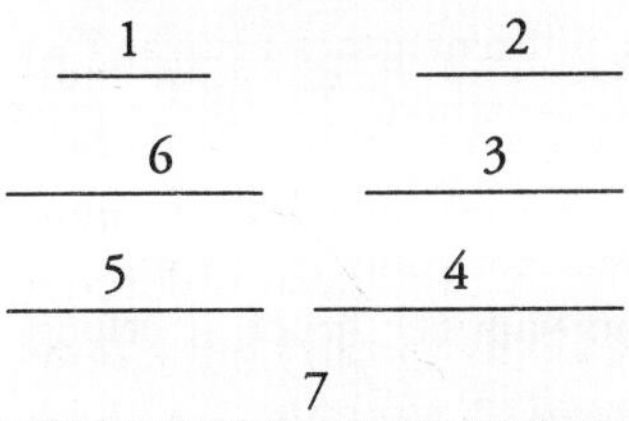

deletion

	1	2	3	4	5	6	7
a	0	+	0	0	+	+	0
b	+	+	+	0	+	0	0
c	+	+	+	+	0	0	0
d	0	+	+	0	+	0	0
e	+	0	0	0	+	+	0

Solution:

(i) From this data it is clear that deletion 7 cannot complement any other. It represents an extensive long deletion $\underline{\quad 7 \quad}$.

(ii) Mutant a does not recombine with 1, 3, 4 and 7. Therefore, it must be in a region common to all of these deletions.

(iii) 5 small deletion and only 'c' is unable to complement it. c lies in a region common to 5 to 7.

(iv) Similarly, deletion 2 is very small and only 'e' is unable to complement it. Mutant e lies in a region common to 2, 3, 4 and 7.

(v) Deletion 6 covers b, c and d, whereas deletion 1 covers d. Therefore, map order in this region is c b d ......

(vi) Mutant 6 does not recombine with 4, 6 and 7.

(vii) Deletion 3 covers both a and e, while deletion 1 covers a and d. Therefore, map order in this region is d a e.

(viii)

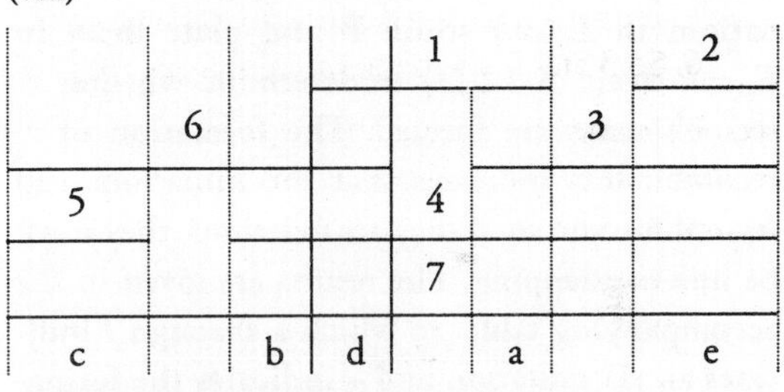

(ix) The complete order is c b d a e.

Problem 17

Pairs of rII mutants of T_4 phages have tested in both cis and trans by mixed infection of *E. coli* K-12 (λ). Comparisons were made as to the number of phage particles that were reproduced per host bacterium, called burst size. Results of six different rII mutants r-1, r-2, r-3, r-4, r-5 and r-6 were as follows:

Mixed infection	cis burst size	trans burst size
r-1 × r-3	260	220
r-5 × r-6	211	0
r-3 × r-6	235	236
r-4 × 4-5	241	233
r-1 × r-2	244	0
r-2 × r-4	229	240

If we assign mutation r-4 to the A cistron, what are the locations of the other rII mutations relative to the A and B cistrons.

Solution:

All of the cis results are expected to show a normal viral burst because a wild type chromosome is always available in such arrangements to effect a lysis. The trans burst results yield the following information:

r = A (given)

then r-5 = B (r-5 and r-4 complement)

r-6 = B (r-6 and r-5 do not complement)

r-3 = A (r-3 and r-6 complement)

r-2 = B (r-2 and r-4 complement)

r-1 = B (r-1 and r-2 do not complement)

Problem 18

You have reason to believe that a number of rII mutants of bacteriophage T_4 are deletion mutations. You cross them in all possible combinations in *E. coli* strain B and plate them in *E. coli* strain K-12 (λ) to determine whether r^+ recombinants are formed. The formation of r^+ recombinants indicates that the mutations can recombine and so, if they are deletions, they must be non-overlapping. The results are given in the accompanying table, in which a through *f* indicates an rII mutation and + indicates the formation of r^+ recombinant progeny in the cross:

	a	b	c	d	e	f
a	–	–	–	–	–	–
b		–	–	+	+	–
c			–	+	–	+
d				–	–	+
e					–	+
f						–

Assemble a deletion map for these mutations, using a line to indicate the DNA segment that is deleted in each mutant.

Solution:

(i) Mutant a is non-complementary to cc def.

(ii) Mutant b is non-complement to a, c and non-complementary to f.

(iii) Mutant d is non-complementary to a and e.

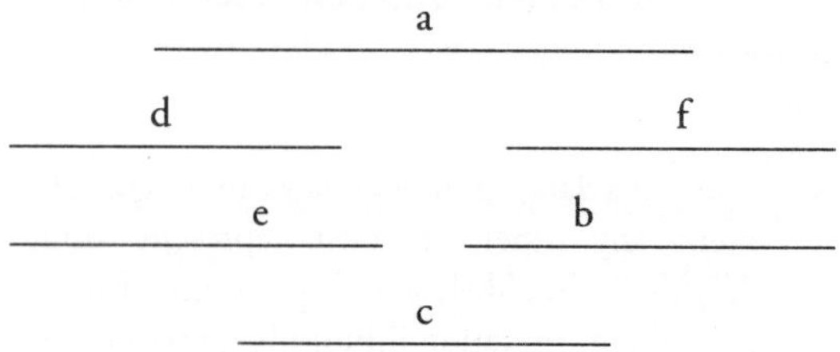

Problem 19

Six different mutations are known to reside in three nonallelic genes. From the following results of functional allelism tests whereby the mutants were crossed with each other, determine which mutations are located in the same genes:

Key: + = wild type; 0 = mutant

	1	2	3	4	5	6
1	0	+	+	0	+	+
2		0	+	+	+	0
3			0	+	0	+
4				0	+	+
5					0	+
6						0

Solution:

(i) 1 and 2 are in different genes because they undergo complementation. Therefore, gene 1 and gene 2 .

(ii) Mutation (6) does not complement with (2) but complements with (1); hence, it belongs to (2) gene 1 , gene 2, 6 .

(iii) Mutation (5) complements with (1) and (2), so it belongs to a separate gene.

gene 1 , gene 2, 6 , gene 5 .

(iv) Mutation (4) complements with (5) and (6), but not with (1); therefore, it belongs to (1).

genes 1, 4 , genes 2, 6 , gene 5 .

(v) Mutation (3) complements with (4) and (6), but does not complement with (5). Therefore, it goes with gene (5).

1, 4 , 2, 6 , 3, 5 .

Problem 20

Recall that two cistrons in the bacteriophage T_4, rIIA and rIIB, are located adjacent to each other. Assume that we doubly infect *E. coli* strain B in the combinations indicated below, where m designates the exact position of a mutation. What types of bacteriophage progeny would be expected in this burst.

	A cistron	B cistron
(a)	m	
	m	
(b)	m	
		m
(c)		m m
		m
(d)		m

Solution:

(a) Only mutant progeny, the two mutants are at exactly the same point.

(b) Mostly mutant progeny, but some wild type by recombination between the two mutation sites.

(c) Both mutant and wild type progeny in a 1 : 1 ratio.

(d) Mostly mutant progeny, but a few wild type by recombination between the two mutation sites.

Problem 21

Wild type phage T_4 grows on both *E. coli* B and *E. coli* K-12 (λ), producing turbid plaques. The rII mutants of T_4 grow on *E. coli* B, producing clear plaques, but do not grow on *E. coli* K-12 (λ). This host range property permits the detection of a very low number of r^+ phages among a large number of rII phages. With such a sensitive system, it is possible to determine the genetic distance between two mutations within the same gene—in this case, the rII locus. Suppose *E. coli* B is mixedly infected with rIIx and rIIy, the two separate mutants in the rII locus. Suitable dilutions of progeny phages are plated on *E. coli* B and *E. coli* K-12 (λ). A 0.1 ml sample of a thousand fold dilution plated on *E. coli* B produces 672 plaques. A 0.2 ml sample of undiluted phage plated on *E. coli* K-12 (λ) produces 470 plaques. What is the genetic distance between the two rII mutations?

Solution:

(i) The plaque produced in *E. coli* K-12 (λ) are r^+, while those on *E. coli* B may be either r^+ or r^-. Thus, the total number of progeny in 1 ml = (dilution factor) × (progeny phage per ml) = $1000 \times \left(\frac{6.72}{0.1}\right) = 6.72 \times 10^6$ ml.

(ii) Since *E. coli* B is co-infected with rx and ry, the only way to obtain an r^+ progeny phage is to have a crossover within rII locus.

The progeny resulting from a crossover would be $\frac{1}{2}r^+$ and $\frac{1}{2}$rIIxy recombinants.

(iii) The number of recombinant phage is twice the number of r^+ phage which can be assayed for by growth on *E. coli* K-12 (λ).

(iv) The number of recombinant progeny in 1 ml = 2 × (number of r^+ phage/ml) = $\left(2 \times \frac{470}{0.2}\right)$ = 4,700/ml. The map distance between rIIx and rIIy is $\left(\frac{4700}{6.72} \times 10^6\right) \times 100\% = 0.07$ mμ.

The genetic distance is 0.07 mμ.

Problem 22

In recombination studies of rII locus in phage T_4, what is the significance of the value determined by calculating phage growth on the K-12 versus the B strains of *E. coli* following simultaneous infection in *E. coli* B? Which value is always greater?

Solution:

(i) By performing plaque analysis on *E. coli* B, where wild type and mutant phages are both lytic, the total number of phages per millimetre can be determined.

(ii) Because almost all cells are rII mutants of one type or another, the value is much larger.

(iii) To avoid total lysis of the plate, extensive dilution is necessary.

(iv) On K-12, rII mutation will not grow but wild type phages will. Because wild type phages are the rare recombinants, there are relatively few of them and extensive dilution is not required.

Problem 23

Each of the two mutations p_1 and p_2 can produce albino rats. The genes are closed together in one chromosome and may be allelic. The cross $p_1p_1 \times p_2p_2$ produces progeny with wild type pigmentation. Are the mutation in the same gene?

Solution:

(i) Here the progenies are wild type, so the mutations are complement with each other and both p_1 and p_2 must be heterozygous.

(ii) This indicates that the genotype of the progeny is $\frac{p_1 +}{+ p_2}$, which implies that p_1 and p_2 are mutations in different genes.

(iii) Therefore, we can say that mutations (p_1 and p_2) are not in the same gene.

Problem 24

Construct a map from the following two factor phage crosses, data indicating map distances.

Cross	% recombination
(i) $r_1 \times r_2$	0.10
(ii) $r_1 \times r_3$	0.05
(iii) $r_1 \times r_4$	0.17
(iv) $r_2 \times r_3$	0.15
(i) $r_2 \times r_4$	0.10
(i) $r_3 \times r_4$	0.32

Solution:

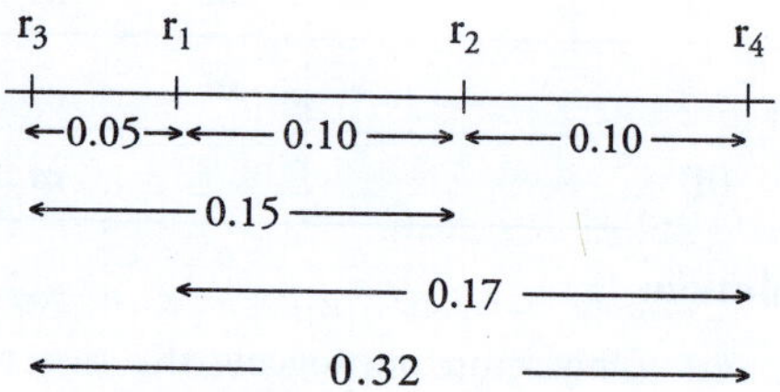

EXERCISE

1. **Four T_4 phage deletion strains are tested for recombination by pairwise crossing in *E. coli*. We get these results (+ = recombination, 0 = no recombination):**

	1	2	3	4
1	0	+	+	+
2	+	0	0	+
3	+	0	0	0
4	+	+	0	0

Explain the results in terms of a genetic map.

2. **A mutant phage m known to contain no deletion, recombines with mutant 1 to 4 as follows:**

	1	2	3	4
m	0	+	0	+

Explain the nature of mutation in phage m and show its position on the map.

3. **Four mutant T_4 strains are tested for complementation in *E. coli*. We get the following results (+ = complementation, 0 = no complementation):**

	1	2	3	4
1	0	0	+	+
2	0	0	0	+
3	+	0	0	+
4	+	+	+	0

(a) Explain the above results in terms of functional relationship between mutations 1 to 4.

(b) Draw a map showing the relative location of these mutations. What ambiguity remains?

15 Gene Regulation

- **What is operon ?**

A group of genes making up a regulatory or control unit. The unit includes operator, promoter and structural genes.

- **What do you mean by inducible and repressible system ?**

1. The enzyme whose full level of transcription does not occur unless their effector molecules are present is called the **inducible enzyme**. Such effector is called **inducer** and such regulation is called **inducible system.**

Example: Synthesis of β-galactosidase enzyme is induced by lactose, lactose-inducer.

2. The enzyme whose full level of transcription occurs until such time as their effector reaches a critical concentration at which transcription is inhibited. Such effector is called **corepressor** and such regulation is called **repressible system.**

Example: Tryptophan synthesis. In absence of tryptophan the enzymes are synthesized at full level.

Remarks:

(i) Inducible systems are off until the effector is present.

(ii) Repressible systems are on until the effector is present.

- **What is regulator and repressor in gene regulation ?**

(a) *Regulator*

(i) It is a DNA segment responsible for production of **repressor** molecule.

(ii) Mutation of such gene produces nonfunctional repressor and fails to regulate enzyme synthesis, i.e., called constitutive synthesis and mutation called constitutive mutation.

(b) *Repressor*

(i) A special group of protein molecules.

(ii) Each repressor blocks the synthesis of one or more proteins.

(iii) Repressor for lac operon composed of four (4) polypeptide (tetramer) and blocks synthesis of three enzymes—β-galactosidase, permease (M protein) and trans-acetylase.

- **How does repressor act ?**

The function of repressor depends upon the effector (inducer and corepressor).

(a) *In inducible system*:

(i) Inducer binds with repressor and makes it inactive, results transcription of structural gene to produce mRNA.

Active repressor + inducer → Inactive repressor

(ii) For example, lactose (inducer) binds with lac repressor and repressor becomes inactive. Synthesis of enzyme occurs.

(b) *In repressible system*:

(i) Normally, the repressor is inactive, addition of corepressor results active repressor.

Apo-repressor + corepressor → Active repressor.

(ii) Active repressor stops transcription, and synthesis of enzyme does not occur.

(iii) Tryptophan **(a.a)** acts as corepressor in *E. coli*. It produces active repressor and stops tryptophan biosynthesis.

Remarks:

(i) Inducer and corepressor act in opposite direction.

(ii) Active or inactive condition of repressor quality altered on the basis of presence or absence of inducer and corepressor.

- **State briefly about operator and its role in gene regulation in prokaryotes.**

(i) It is a specific segment of DNA where active repressor binds. Operator repressor interaction stops transcription of corresponding mRNA molecules by preventing the movement of RNA polymerase.

(ii) Gilbert (1976) showed that the lac operon repressor binding on operator covers some part of the promoter to which RNA polymerase binds, i.e., overlaps existence between polymerase binding site and repressor binding site, so that when repressor binds on operator, polymerase cannot bind.

(iii) Three operator site (O_1, O_2 & O_3) are present: O_1 = located slightly upstream of the promoter. O_2 = downstream from O_1 within Lac Z gene. O_3 = slightly upstream from the promoter.

(iv) Operator mutations:

Operator (O) → O^+ wild type
Operator (O) → O^c constitutive (Mutation of O_1)
Operator (O) → O^0 permanently off

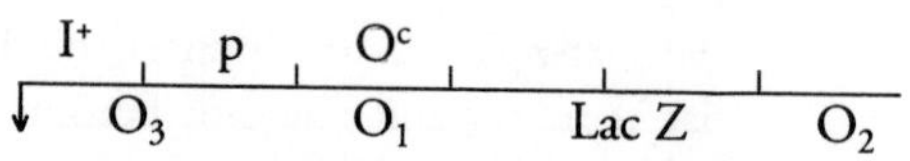

Fig. 15.1: Constitutive synthesis of lac operon (operator constitutive mutant).

- **State briefly about promoter and its role in gene regulation in prokaryotes.**

(i) Promoter (P) lies before operator (O). It has two compartment: (a) RNA polymerase binding site and (b) Catabolic activator protein (CAP) binding site.

(ii) Interaction of promoter and polymerase initiates transcription of neighbouring structural genes.

(iii) (a) Promoter mutation may block synthesis or even in absence of active repressor or inactive repressor.

(b) "**Up promoter mutation**" may increase the binding efficiency of polymerase.

"**Down promoter mutation**" may decrease the binding efficiency of polymerase (Watson, 1970).

- **Lactose is not the real inducer! Discuss.**

(i) Since in the repressed state of lac operon, *E. coli* cells will not synthesize β-galactoside permease.

(ii) Question arises, how lactose enters the cell? It is believed that a minimal amount (basal level) of enzyme is always present in the cell to start the metabolic pathway.

(iii) It has been shown that lactose is not the real inducer of lac operon, rather it is an anti-inducer. Since it binds to repressor to increase its affinity to operator. On the other hand, the bound protein in inactive repressor is **allolactose.**

(iv) It has also been shown that while β-galactosidase breaks lactose into glucose and galactose, a side reaction changes galactose into **allolactose** and **galactobiose.**

(v) Therefore, in the regulation of lac operon, lactose is taken up and a part of it through galactose, is converted into allolactose, which then induces the operon a outweights the anti-inducing effect of lactose.

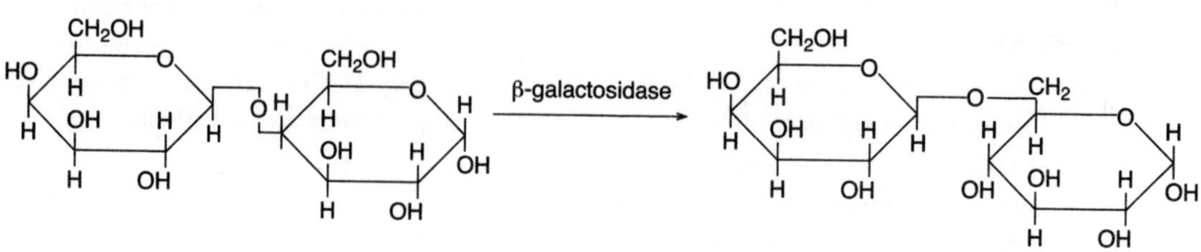

(a) Lactose (1, 4 linkage)

(b) Allolactose (1, 6 linkage)

- **Difference between Inducible and Repressible systems.**

INDUCIBLE SYSTEM	*REPRESSIBLE SYSTEM*
1. Regulator gene can produce active repressible molecule, called **repressor.**	1. Regulator gene produces inactive repressor molecules called **apo-repressor.**
2. **Repressor** molecule can bind with the operator, so the transcription of the structural genes are **switched off.**	2. Apo-repressor cannot bind with the operator so the transcription of the structural genes are permanently **switched on.**
3. **Inducers** can bind with **repressor** rendering it **inactive.**	3. **Corepressor** can bind with **apo-repressor** rendering it **active.**
4. **Effector (inducer)** induces the production of mRNA of structural genes; hence, it is called **inducible system.**	4. **Effector (corepressor)** inhibits the formation of mRNA of structural genes; hence it is called **repressible system.**
Active repressor + Inducer ↓ Inactive repressor ↓ Transcription of mRNA	Inactive repressor (apo) + Corepressor ↓ Active repressor ↓ No transcription of mRNA
Example: Lac operon	*Example*: Tryptophan operon

Inducible system — Lac operon

(a) **No Inducer**

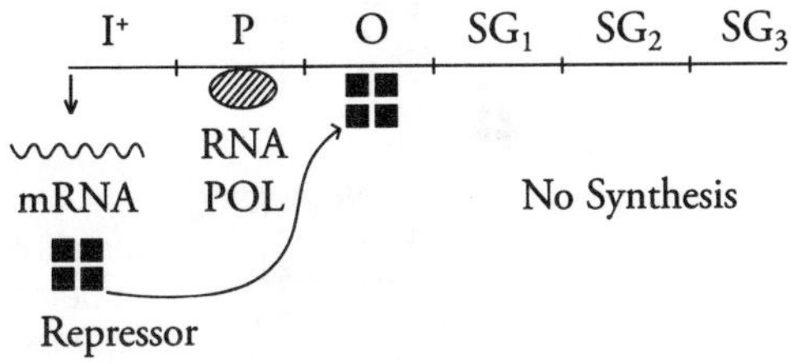

(b) **Presence of Inducer**

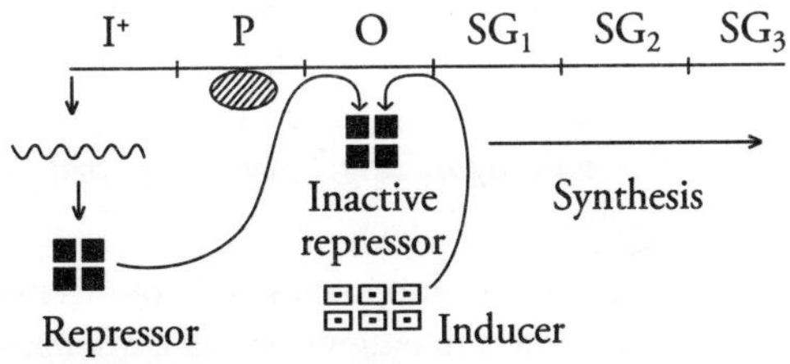

Repressible system — Tryptophan operon

(a) **No Corepressor**

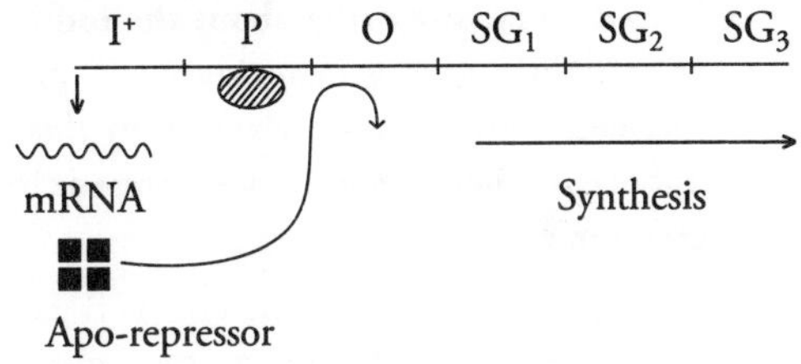

(b) **Presence of Corepressor**

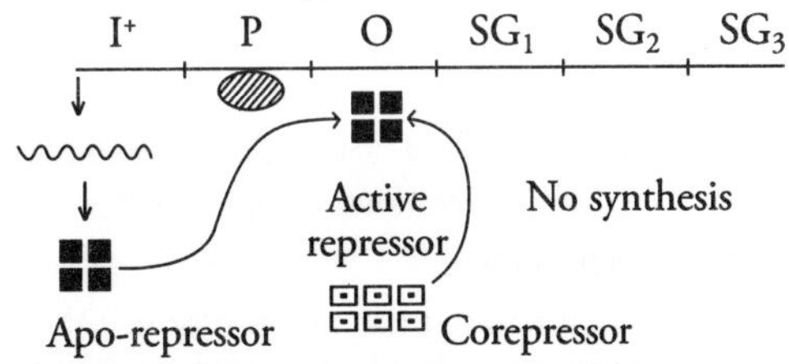

- **What is allolactose?**

It is actually an alternate form of lactose called **allolactose** (Greek—allos other) when β-galactosidase cleaves lactose to galactose and glucose, it rearranges small fraction of the lactose to **allolactose.**

- **Gratuitous inducer:**

Isopropyl thiogalactoside (IPTG) is a chemical analogue of lactose. It behaves as inducer but does not serve as substrates for the enzymes that are subsequently synthesized. Therefore, IPTG is known as **gratuitous inducer.**

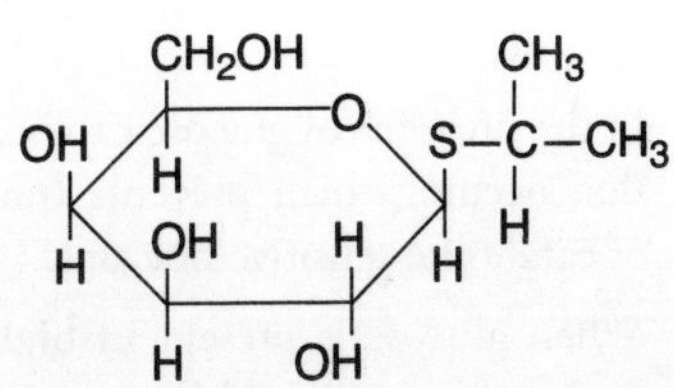

IPTG (Isopropyl thiogalactoside)

- **Distinguish the effects you would expect from (a) missense mutation, (b) nonsense mutation of Z gene (β-galactosidase) of lac gene operon.**

(a) (i) A **missense** mutation results in partial or complete loss of β-galactosidase activity, but there would be no loss of permease or transacetylase activity.

(ii) Here a change of single base pair in the DNA leads to a different codon in mRNA that may code for different amino acids in the polypeptide.

(b) (i) A **nonsense** mutation has a polar effect, unless the mutation is very close to the normal chain terminating codon for β-galactosidase.

(ii) Permease and transacetylase activity would be lost, in addition to β-galactosidase activity.

Closer the mutation to the operator, the more severe polar effect.

- **How does lactose bring about the induction of synthesis of β-galactosidase, M protein (permease) and transacetylase ? Why this does not occur when glucose is also present in the medium ?**

(a) (i) Lactose induces synthesis of β-galactosidase, β-galactoside permease and transacetylase.

(ii) Lactose is converted into allolactose by β-galactosidase. Then it binds to repressor, one molecule for each polypeptide and causes a conformational change in the repressor.

(iii) As a result it has no affinity to DNA and repressor falls off the operator.

(iv) Now the RNA polymerase can bind to the promoter and initiate transcription of the operon.

(b) (i) In the presence of glucose catabolic repression occurs, which prevents transcription of catabolite sensitive operons.

(ii) When glucose is present in high concentration, the cells' cAMP concentration is low, as the glucose concentration decreases cellular concentration of cAMP increases correspondingly.

Notes on:

- **Mutant regulator gene (I^{-d}).**

(i) It produces a repressor that cannot bind to the operator and is, therefore, constitutive like i^- allele.

(ii) It produces a "bad subunit" which is not only unable to bind itself to operator DNA but also acts as part of tetramer and spoils the whole tetramer, thereby prevents any good subunits from binding.

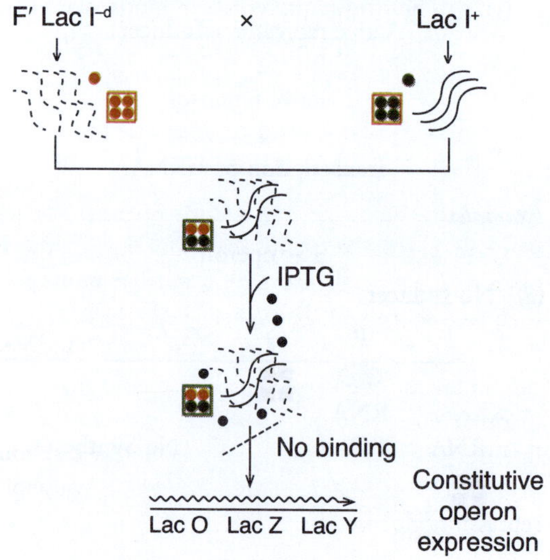

Fig. 15.2: The lac I^{-d} mutation affecting binding to lac DNA.

(iii) The poisoning effect can be produced *in vitro* by mixing "good" and "bad" subunit.

(iv) Hence, the dominant nature of this mutation, these mutations are not cis-dominant. Since this spoiled repressor cannot bind to either operator (merodiploid), this kind of spoiler mutation is widespread in nature and it is called by the generic name **dominant negative.**

- **Mutant regulator (I^S).**

(i) It produces unaltered repressor molecule.

(ii) It binds to the operator normally, but is not capable of recognizing the inducer molecule **lactose**, so that transcription cannot be initiated since the defective repressor remains suck on the operator.

(iii) In some case, I^S strains are induced to some degree by using extremely high concentration of inducer (Gardner, Simmons and Snustad).

(iv) In partial diploid, it is transdominant to i^+ allele.

(v) I^S map central position of the i gene.

(vi) *In vitro* they can bind to operator but exhibit low affinity to IPTG.

(vii) These mutant repressor binds irreversibly to both operators and remains bound there even in the presence of inducer (Fig. 15.3).

- **What is catabolic repression or glucose effect? How does it occur?**

Lactose alone can activate the operon. The presence of glucose prevents the induction of the **lac operon.** *E. coli* cells metabolize **glucose** more easily than **lactose.** In fact, *E. coli* cells keep the lac operon turned down as long as glucose is present. This selection in favour of glucose metabolism and against use of other energy sources has long attributed to the influence of some breakdown product or **catabolite** of glucose. It is, therefore, known as **catabolite repression/glucose** effect (Fig. 15.4).

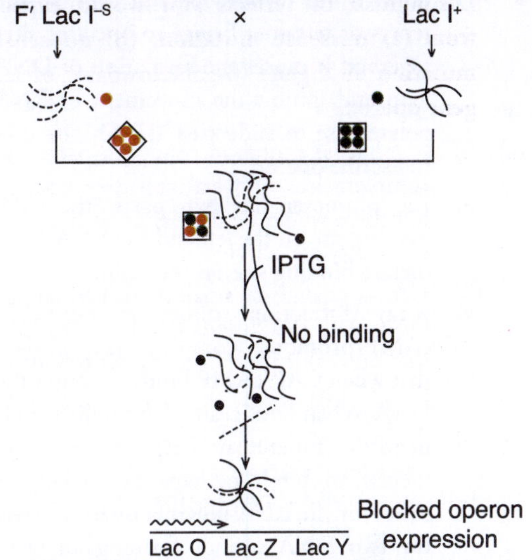

Fig. 15.3: The lac I^{-S} mutation affecting binding to inducer (IPTG).

(i) The positive controller of the operon is a complex composed of two parts cAMP and a binding protein called CAP for catabolite activator protein.

(ii) The protein binds cAMP and the resulting complex binds to the lac control region and helps RNA polymerase bind there.

(iii) Lac repressor must bind to O_1 and O_2 or O_1 and O_3 not O_2 and O_3. Each repressor

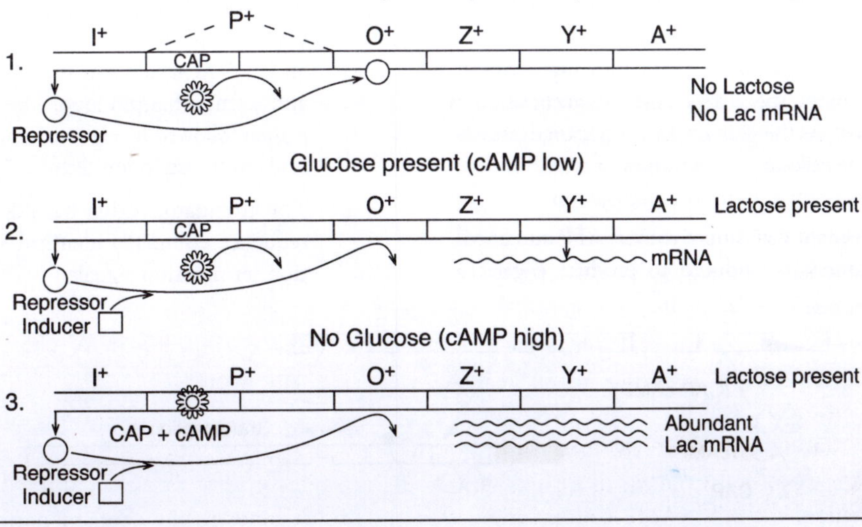

Fig. 15.4: Control of lac operon by presence or absence of glucose.

dimer binds to one operator sites, so the repressor tetramer brings to operator sites together. It causes the formation of DNA loop and inhibit the movement of RNA polymerase to slide past the O_1 site and transcribe operon.

(iv) Lac promoter has two parts, the CAP binding site on the left and the RNA polymerase binding site on the right.

(v) X-ray diffraction studies of the CAP/cAMP/promoter complex have shown that when CAP/cAMP binds, it causes the DNA to bend noticeably (about 90°). This bending apparently helps RNA polymerase to bind; perhaps by making it easier for the RNA polymerase to separate the two DNA strands forming an open promoter complex.

(vi) CAP/cAMP may also stimulate transcription by direct protein-potein contact between CAP and RNA polymerases. cAMP/CAP binds to the upstream part of the promoter and facilitates binding of RNA polymerase to the downstream part. This greatly enhances transcription of the operon. The physiological significance of this positive control mechanism is that it can operate only when cAMP concentration is elevated, this occurs when glucose concentration is low and there is a correspondence need to metabolize an alternate energy source.

(vii) When glucose is present in high concentration, the cell's c-AMP concentration is low. As the glucose concentration decreases, the cellular concentration of cAMP is necessary for activation of lac operon.

(viii) Mutant that cannot convert ATP into cAMP cannot be induced to produce β-galactosidase, because the concentration of cAMP is not great enough to activate lac operon.

CAP protein is made by crp gene.

(a) High glucose ⟶ inactive adenylate cyclase

ATP —×→ No cAMP

(b) Low glucose —×→ Active adenylate cyclase

ATP ⟶ cAMP

(c) cAMP + CAP ⟶ [cAMP CAP]

- **Positive control of the operon**

(a) Without cAMP, CAP cannot bind to its target site in the lac control region. Cyclic AMP binds to CAP and render it capable of binding to this DNA site.

(b) The CAP-cAMP complex binds to its DNA target site.

(c) Binding the CAP-cAMP complex to DNA allows RNA polymerase to bind to the lac promoter. Transcription now occurs assuming that lactose is present and the operator is, therefore, unoccupied by repressor.

- **Write briefly about the attenuator control of tryptophan operon.**

(i) Attenuation occurs by premature termination of transcription at an attenuator site located within mRNA leader sequence (162 nucleotide long). Attenuator is a DNA sequence that when transcribed gives rise to an RNA molecule that folds back on itself and forms a hairpin loop. The loop region is then followed by eight uridine residues. It leads to **transcription termination**.

(ii) The attenuator region has nucleotide pair sequence essentially identical to transcription termination signals.

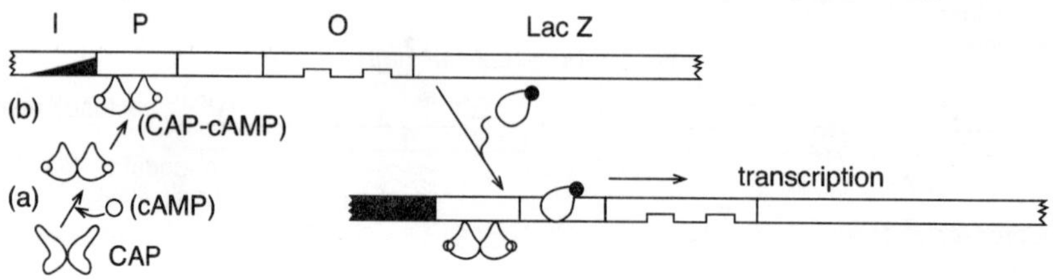

Fig. 15.5: Catabolite control of lac operon.

(iii) The leader sequence has **SDS** (8-12) [**ribosome-binding site**], AUG initiation codon and small open reading frame and two contiguous trp codons (UGG, UGG).

(iv) The mRNA transcript of leader region (14 amino acids/codons) contains two successive (tandem) trp codon for tryptophan (UGG) preceded upstream by an AUG sequence that promotes initiation of translation by ribosome.

(v) There are four regions of leader peptide mRNA that can fold and form secondary structure by complementary base pairing. The first region 1 (nucleotide 60-68) pairing with region 2 (nucleotiode 75-83) results in transcription **pause signal**. The region 3 (nucleotide 110-121) pairing with region 4 (nucleotide 126-134) generates hairpin that precedes the U8 sequence, i.e., essential signal for intrinsic termination.

(vi) A different structure is formed if region 1 is prevented from pairing with region 2, region 2 pairing with region 3. Then region 4 has no available pairing partner, so it is compelled to remain single-stranded. So the terminator hairpin cannot be formed. Position of ribosome determines which structure will be formed.

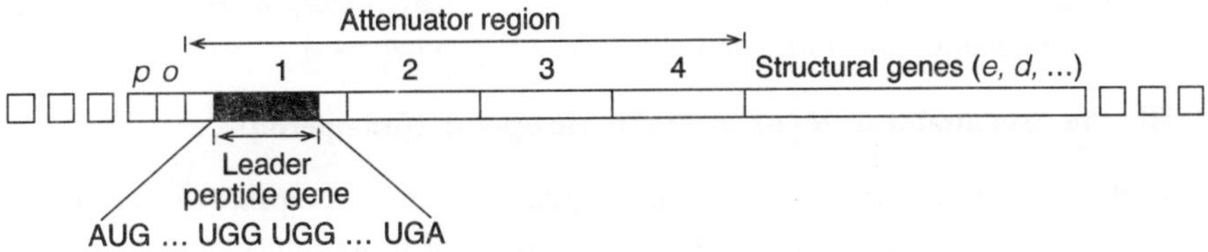

Fig. 15.6: Attenuator region of the trp operon, which contains the leader peptide gene. This region is transcribed into the leader transcript.

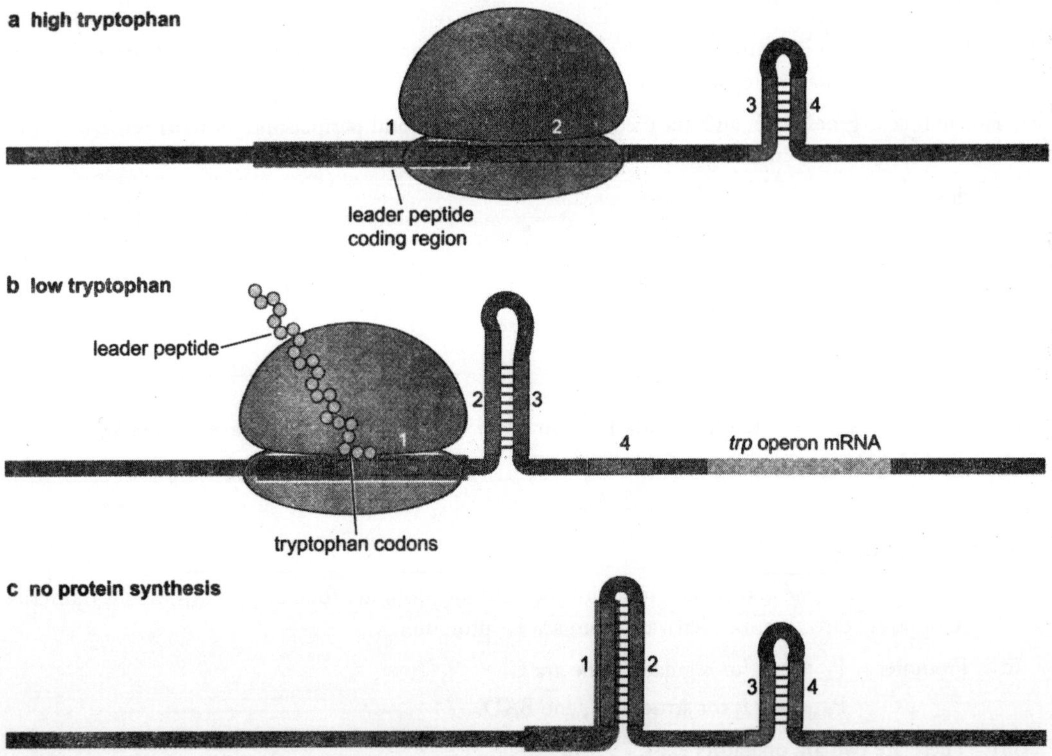

Fig. 15.7: Transcription termination at the trp attenuator which is controlled by the availability of tryptophan.

A. **In Presence of tryptophan:**

(i) Transcription and translation are coupled in prokaryotes. When **tryptophan levels are high**, RNA polymerase that has initiated transcription pauses at a specific site and then terminates before getting to trp E. When adequate tryptophan is present in cells, tryptophan charged $tRNA^{trp}$ is produced. This allows translation of trp leader sequence through the 2 trp codons UGG to the UGA, which termination codon lies between regions 1 and 2.

(ii) This translation of the trp leader region prevents base pairing between 2 and 3 complementary mRNA leader sequences, which in turn permits the formation of the transcription termination, hairpin involving leader sequences the region 3 and 4.

B. **Absence of tryptophan:**

(i) With the low level of tryptophan, charged $tRNA^{trp}$ is unavailable and the ribosome stalls at region 1 during the translation of triplets calling for charged $tRNA^{trp}$.

(ii) This stalling allows region 2 and 3 to pair which prevents region 3 from pairing 4 to form transcription termination hairpin.

(iii) This event influences the secondary structure of the transcript such that the termination hairpin does not form, attenuation is bypassed.

- **State briefly the structural features of an Arabinose operon (Ara operon)**

This inducible operon is unique because the same regulatory protein is capable of exerting both positive and negative control and as a result, either induces or represses gene expression.

(i) The metabolism of sugar L-arabinose is under the direction of the enzymatic (arabinose isomerase, ribulose Kinase and ribulose $5PO_4$ epimerase) products of the three structural genes, viz. ara B, ara A and ara D.

(ii) Their transcription is controlled by the regu-latory protein ara C, encoded by the ara C gene that interacts with two regulatory regions ara O_1 and ara O_2.

(iii) Two unlinked genes ara E and ara F coding for membrane and periplasmic proteins respectively, that are involved in L-arabinose uptake.

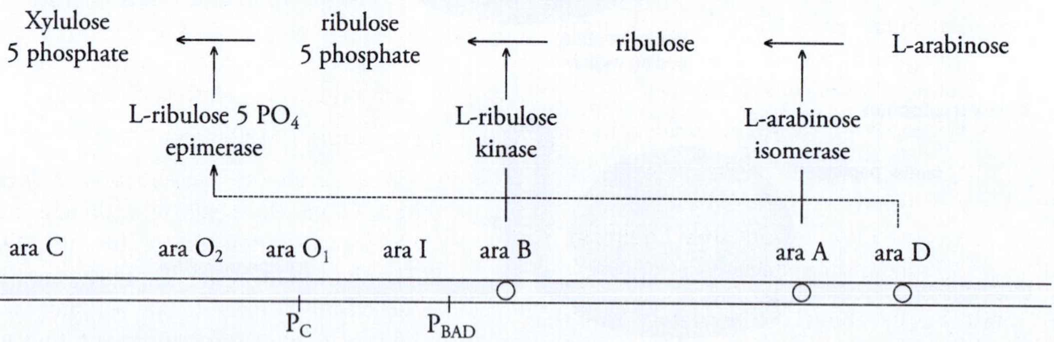

Components of ara operon:

(i) The **regulator gene** is **ara C** and it lies just beyond the promoter for ara_{BAD} (P_{BAD}) and is transcribed in opposite direction, i.e., leftward from ara P_C promoter.

(ii) Promoter = P_C = It is for regulator gene ara C.

P_{BAD} = It is for structural gene BAD.

(iii) There are **two ara operators** exist.

- ara O_1: It regulates transcription of control gene called ara C.

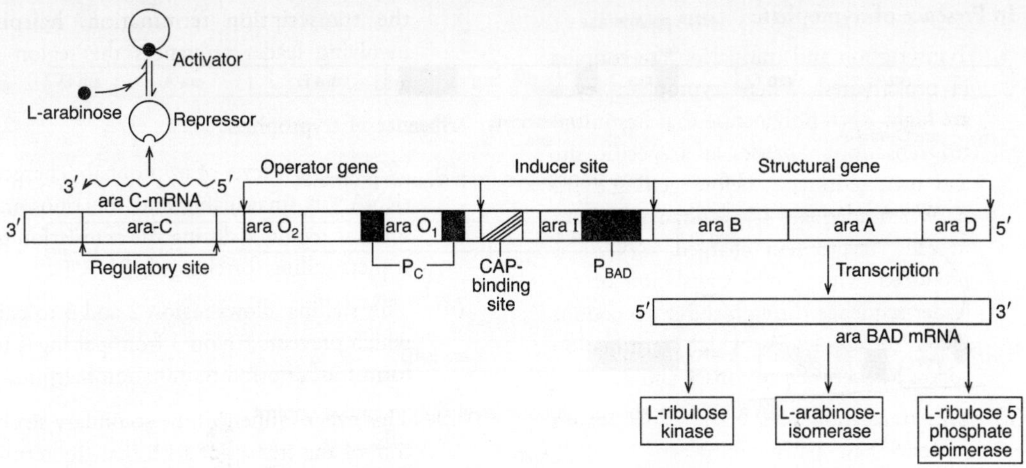

Fig. 15.8: Structure of ara operon in *E. coli*.

- ara O_2: Located far upstream of the promoter P_{BAD} between –265 & –294.

(iv) CAP-binding site = It lies adjacent to ara operon promoter (about 200 bp upstream).

(v) **ara I:**

(i) Initiator involved in the initiation of transcription. It has two sites I_1 and I_2.

(ii) I_1 = located –56 to –78 bp; I_2 = located –35 to –51 bp.

(iii) Each of which can bind one monomer of ara C.

(vi) **Structural gene:** It has three structural genes ara B, ara A and ara D, which are, collectively known as ara BAD.

- ara B = It synthesizes ribulose kinase.
- ara A = It synthesizes L-arabinose isomerase.
- ara D = It synthesizes L-ribulose 5 phosphate epimerase.

- **State briefly about the regulatory mechanism of ara operon.**

One regulatory protein, i.e., ara C mediates both positive and negative control.

A. **In the absence of arabinose:**

(i) In the absence or low concentration of arabinose and cAMP (but glucose level high), when no ara BAD products are needed, the ara C exerts negative control.

(ii) Ara C protein binds coordinately to both ara I_1 site and ara O_2 site (second operator site). A conformational change occurs in DNA which brings these sites together to form a DNA loop of about 210 bp. In this way arabinose operon is repressed.

(iii) Binding at either regulatory region involves a dimer of ara C. When both regions are bound, the dimer interact leading to the loop that causes repression.

(iv) Presumably this loop inhibits the access of RNA polymerase to the promoter region. The region of DNA located between ara I and ara O_2, which loops out, is critical to the formation of repression complex.

B. **In the presence of arabinose:**

(i) When arabinose concentration is high, ara C protein binds to arabinose and looses its attachment to ara O_2. Conformation of ara C protein (acts as repressor) changes to an activator and prefers to bind as dimer to ara I_1 and ara I_2. It breaks up the repression loop and allows the transcription of the operon, i.e., the operon is derepressed.

(ii) If glucose is absent and CAP-cAMP is high concentration, it occupies the CAP-binding site. CAP-cAMP further stimulates transcription by binding to a site upstream of ara I.

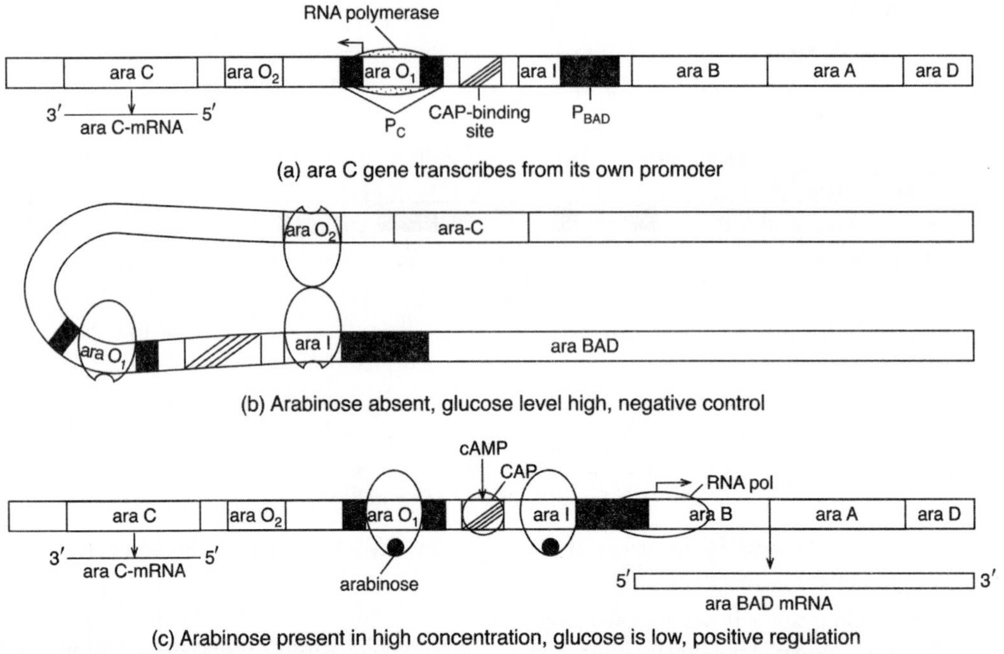

Fig. 15.9: Regulation of the ara operon

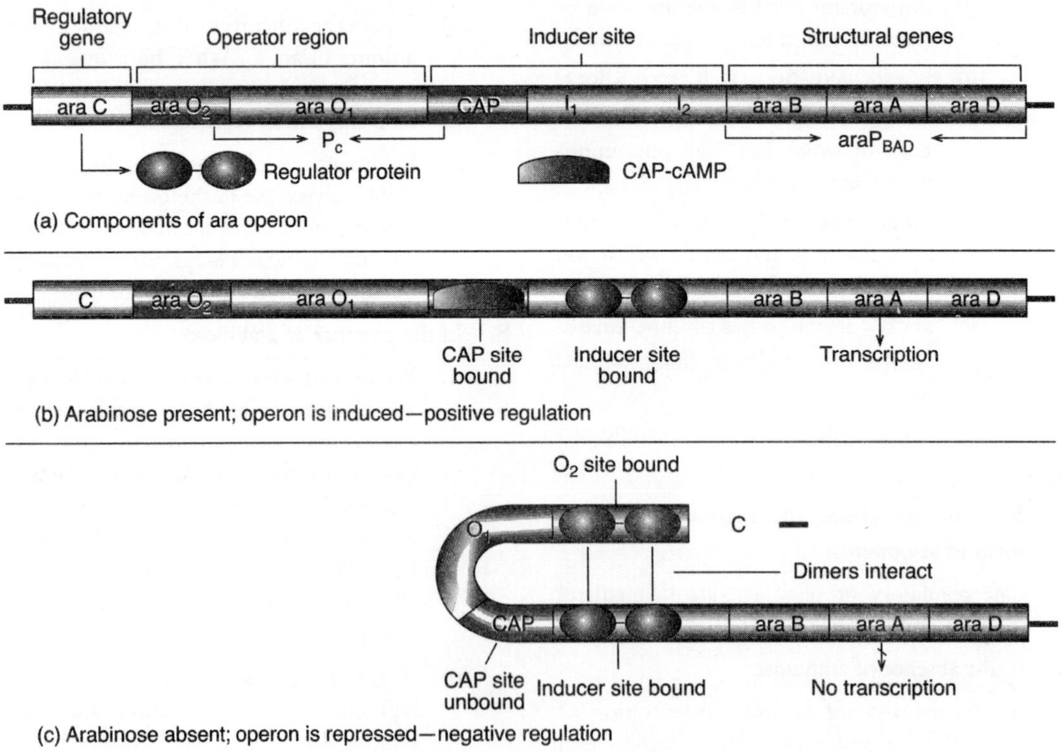

Fig. 15.10: Genetic regulation of the ara operon. The regulatory protein of the ara C gene acts as either an inducer (in the presence of arabinose) or a repressor (in the absence of arabinose).

Autoregulation of ara C:

The ara C protein regulates its own synthesis by repressing transcription of its gene. This phenomenon is called **autoregulation.**

(i) Ara C gene is transcribed from Pc in the leftward direction, which puts O_1 in a position to control this transcription.

(ii) As the level of ara C protein rises, it binds to ara O_1 and inhibits leftward transcription, thus preventing an accumulation of too much repressor.

This kind of regulation, where a protein controls its own synthesis, is called **autoregulation**.

Repression loop:

(i) The operator of lac gene is located just upstream from the beginning of actual coding sequence of lac Z gene. Crystallographic studies show that the actual region of repressor binding of the **primary operator** (O_1) consists of 21 b.p. (Fig. 15.11).

(ii) Two other **auxiliary operators** are O_2 and O_3.

(iii) The operators (O_2) consists of 401 base pairs downstream from the primary operator within lac Z gene and the another, operator (O_3) is 93 base pairs upstream from the primary operator (O_1) just above CAP site.

(iv) All three operators must be bound for maximum repression.

(v) Binding by the repressor at two operator sites distorts the conformation of DNA, causing it to bend away from the repressor.

(vi) When the model is created involving dual binding of operators O_1 and O_3, the 93 base pairs of DNA that intervene must just out by forming a repression loop.

Importance:

(i) This model positions the promoter region that binds RNA polymerase on the inside of the loop which prevents access during repression.

(ii) The repression loop positions the CAP-binding site in a way to facilitate CAP interaction with RNA polymerase upon subsequent induction.

(iii) The finding of DNA looping during repression is similar to transitions that are predicted to occur in eukaryotic system.

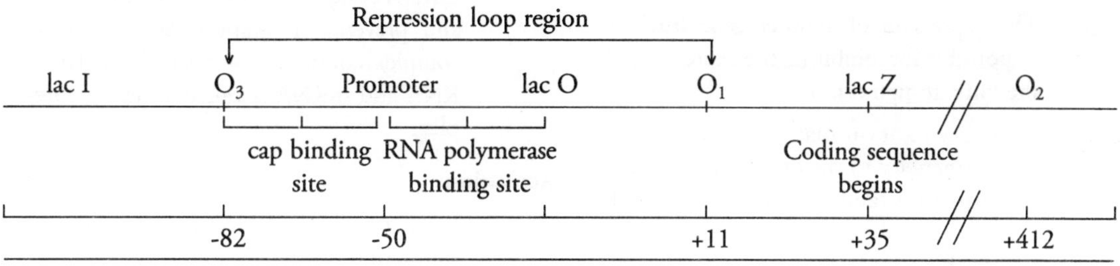

Fig. 15.11

- **Regulion:**

(i) A system that includes dispersed genes under common control is referred to as a regulon.

(ii) CAP and cAMP are involved in the co-ordinated regulation of many operon (such as galactose, arabinose, etc.). Such a network of operon with a common regulator is termed as regulon.

- **Ribo switch:**

A region in the UTR (untranslated region) of an mRNA that can bind to a small molecule that alters gene expression by influencing transcription or translation.

(i) It contains two modules, viz., aptamer and expression platform.

(ii) An aptamer can bind a ligand and an expression platform whose change in conformation can cause a change in expression of a gene.

- **Role of Antisense RNA in gene regulation**

An RNA sequence complementary to part of mRNA is called an antisense RNA.

The antisense regulatory RNAs act by pairing with mRNA to either inhibit or activate translation.

Inhibition

Prokaryote:

(i) Osmoregulation refers to the ability to control the amount of water inside the cell, because the solute concentration in the external environment is rapidly changed between hypotonic and hypertonic conditions, bacteria must have osmoregulation mechanism to maintain their internal cell volume.

(ii) Lets consider a trait known osmoregulation in *E. coli.* In *E. coli* an outer membrane protein encoded by **omF** gene is important for osmoregulation.

(iii) At low molarity the **omF** protein is preferentially produced whereas at high osmolarity its synthesis is decreased.

(iv) The expression of another gene **micF** is responsible for inhibiting the expression of the **omF** at high osmolarity.

(v) It has been known that the micF RNA is complementary to the **ompF** mRNA, it is an antisense strand of RNA. When micF gene is transcribed its RNA product binds to **ompF** mRNA via hydrogen bonding between their complementory regions. The binding of the **micF RNA** to the **ompF** mRNA prevents the **ompF** mRNA from being translated.

(vi) The RNA transcribed from **micF** gene is called **antisense RNA** because it is complementary to the **ompF** mRNA which is sense strand of mRNA that encodes a protein. The **micF** RNA does not encode a protein.

(vii) The double stranded RNA structure formed between the **micF** antisense RNA and **ompF** mRNA.

(viii) They have region complements to each of other. The **micF** antisense RNA structure to **ompF** mRNA for double stranded in which **ompF** mRNA cannot be translated.

Eukaryote:

(i) A small regulatory RNA relieves oxidative stress in *E. coli.* One of the genes depressed in the presence of hydrogen peroxide is **oxyS** which encodes a regulatory RNA called **OxyS**. This RNA binds to several mRNAs.

(ii) For any given interaction, only short stretches of **OxyS** are complementary to the target RNA and able to anneal with it.

(iii) Two separate regions of **OxyS** bind to the mRNA of gene **flhA** which encodes a transcriptional activator protein.

The complementary regions are very short (7 nucleotide). One of the complementary regions is near the AUG translation start and other is more than fortinucleotides upstream.

(iv) Base pairing between OxyS and **flhA** mRNA conceals the ribosome binding site and prevents translation. Such bipartite complex composed of a small regulatory RNA and mRNA is called a **kissing complex**.

Activation:

(i) Small regulatory RNAs can also activate translation.

(ii) **DsrA** regulatory RNA from *E. coli.* Its translation is controlled from a gene **rpoS** which encodes sigma factor for RNA polymerase.

(iii) The 5′ end of **rpoS** mRNA is self-complementary and can form a hairpin that hides the ribosome binding site and the translation start site.

(iv) Two virtually contiguous regions of **DsrA** bind to the rpoS mRNA and when binding occurs, the ribosome binding site is freed and translation can occur.

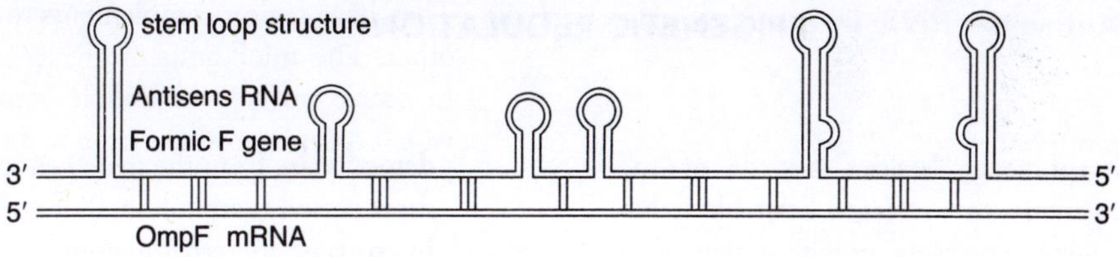

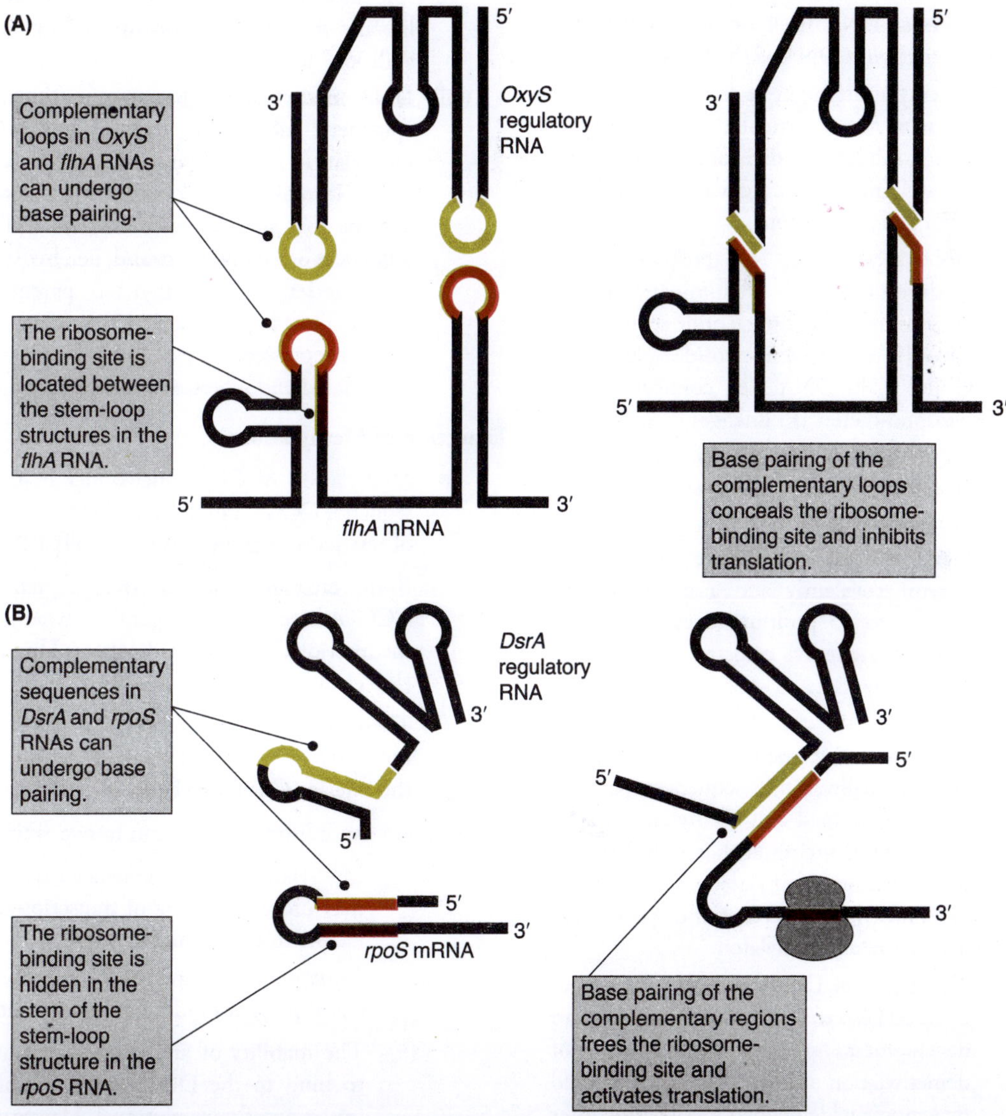

Fig. 15.12: Regulation of translation of target mRNAs by (A) *OxyS* and (B) DsrA regulatory RNAs. (After S. Altuvia and E., G., H. Wagnerm 2000. *Proc. Nati. Acad. Scl. USA* 97, 9824.)

EPIGENETIC REGULATION

The prefix epi means "besides"/outsides or "in" addition to epigenetic, that refers to heritable stable alternation in gene expression potential that arise during development and cell proliferation. The study of "epigenesis" means the interpretation of the genotype during development to give the phenotype. This process identifies post replicative modification of either DNA itself or of proteins that ultimtly associate with DNA as the key mediator.

DNA methylation might be responsible for the stable maintenance of a particular gene expression pattern through mitotic cell division. DNA methylation is now recognised to be a chief contributor to the stability of gene expression.

(i) DNA methylation is a post replicative modification that is predominately found in cytosine of the dinucleotide sequence of **CpG**. [It refers to dinucleotide sequence of C and G in DNA. i.e., connected by a phospho-diester (P) linkage].

In vertebrates and plants CpG island (1000-2000 bp) occur near the many promoters of gene.

(ii) Here **methyl transferase** (methylase) enzyme covalently add a methyl group ($-CH_3$) to -5 position of cytosine base, forming 5 methyl cytosine. The methylated sequence is $5'–C^m\ G–3'$ $m = CH_3$
$3'–G\ C^m–5'$

(iii) Usually cytosine residues on both strands of short palindromic sequences (C^m G/ G C^m) is termed **full methylated.** Replicating of this site result that each daughter duplex has one methylated CG strand and one unmethylated GC strand. Such a site is called **hemi methylated.**

(iv) The extent of DNA methylation changes in an orchestrated way during mammalian development, starting with a wave of demethylation during cleavage, followed by genome wide denovo methylation after implantation.

(v) Demethylation is an active process to remove the methyl group with the help of **demethylase** from the male genome after a short period (within hours) of methylation. In contrast maternal genome is only passively demethylated during subsequent cleavage division.

(vi) The addition of methyl group to the cytosine base does not change the primary literally means to act "on top of" or "in addition" to genetics.

(vii) **DNA methylation** is best mechanism of **epigenetic inheritance.** Pattern of DNA methylation are stably maintained following DNA replication by **methyl transferase enzyme** that specifically methylates **CpG sequences** of a daughter strand, i.e., hydrogen bonded to a methylated parental strand. Thus a gene that has been methylated or repressed in a parental cell remains methylated and repressed in progeny cells.

Detection of Methylation:

(i) Methylation of CG dinucleotides in the sequence CCGG can be detected by the use of restriction enzyme **Msp I** and **Hpa II.**

(ii) Both enzymes cleaves the sequence CCGG. **Msp I** cleaves regardless, whether the interior C is methylatd, wheras **Hpa II** cleaves only unmethylted DNA.

(iii) Therefore **Msp I** restriction site that are not cleaved by **Hpa II** are sites at which the interior C is methylated.

Methylation can inhibit transcription in two ways:

1. (i) Methylation of CG sequence (island) may prevent binding of transcription factors to the promoter region.

   Example: Methylated CG sequence could prevent binding of an element.

   (ii) The inability of an activator protein to bind to the DNA would inhibit initiation of transcriptional. However, CG methylation does not slow down the movement of RNA polymerase along a gene.

2. Methylation inhibits transcription by converting chromatin from open to close conformation of protein known as **methyl –CpG– binding protein** that bind methylated sequences. Once bound to the DNA, the methyl CpG binding protein recruit other protein to the region that causes the chromatin to become more compact.

   Example: Methyl CpG binding proteins may recruit histone se acetylase to a methylated CpG island near a promoter.

DNA methylation is heritable:

(i) Methylated DNAs sequences are inherited during cell division.

(ii) The DNA in a particular cell may become methylated or hemi-methylated.

(iii) This hemi-methylated DNA is efficiently recognized by DNA methyl transferase which makes it fully methylated. This process is called **maintenance methylation** because it preserves methylated condition in future cell.

Epigenetic regulation during development:

(i) DNA methylation is generally considered as a repressive modification and associated with gene **silencing.**

(ii) The paternal genome is actively demethylated prior to the first cell division. On the other hand, maternally derived DNA (genome) is demethylated after several cleavage divisions.

Epigenetic regulation in Mammalian X chromosome inactivation:

(i) Females have to silence one of their two X chromosomes in order to maintain 1 : 1 ratio of X chromosome with male.

(ii) X inactivation occurs shortly after implantation. The key element appears to be non-coding subsequent coating of the X chromosome is believed to be sufficient for the X inactivation.

(iii) The of Xist leads to chromosome wide transcriptional silencing and is accompanied by DNA methylation of histone H_3 lysine 9 leading to chromatin condensation and conversion of most of the inactive X heterochromatin.

(iv) Recently it has been known that X inactivation involves the formation dsRNAs between Xist and its complementary strand. These ds Xist RNAs are then processed to si RNA mediate X inactivation.

mRNA stability

The short lived mRNA produces fewer protein molecules than long lived mRNA and it affects the stability of mRNA as well as gene expression.

It is of two routes:

1. Deadenylation depending pathway:
   (i) It begins with the enzymatic trimming of the length of poly A tail on the mRNA.
   (ii) When a poly A tail is trimmed to a length of 25-60 nucleotides, the mRNA become susceptible to a decapping enzyme that removes the 5′ cap and renders the molecule unable to initiate translation.
   (iii) From this state the mRNA is rapidly degraded by exonuclease.
2. Deadenylation independent pathway:
   (i) It is initiated either with decapping or with endonuclease cleavage of the mRNA and then digestion goes to completion by exonuclease activity.
   (ii) Deadenylate independent pathway is particularly active for mRNAs that contain early chain terminator codon or unspliced introns as it prevents the accumulation of truncated polypeptide in the cell.

Lac I^Q and Lac I^{SQ} mutants (Q stands for quantity and S stands for super quantity)

(i) It is produced by the mutation of promoter region of regulator gene. Base pair mutation results in the increase in the rate of transcription of the repressor gene, with greater increase of the Lac I^{SQ}.

(ii) Since Lac I^Q and Lac I^{SQ} mutants produce more lac repressor molecules than wild type producing these mutants reduce the efficiency of induction of the lac operon.

Solved Problems

1. Production of β-galactosidase in different *E. coli* genotypes, some of which are heterogenous bearing F′ lac factors.

	Genotypes	β-galactosidase Production: Induced Bacteria	β-galactosidase Production: Non-induced Bacteria
1.	$i^- z^-$	–	–
2.	$i^- z^+$	+	+
3.	$i^+ z^+$	+	–
4.	$i^+ z^-$	–	–
5.	$i^- z^+/F' i^- z^+$	+	+
6.	$i^+ z^+/F' i^- z^-$	+	–
7.	$i^+ z^-/F' i^- z^+$	+	–
8.	$i^- z^-/F' i^+ z^+$	+	–
9.	$i^- z^+/F' i^+ z^-$	+	–

2. Production of β-galactosidase and cross reacting material (CRM) in different *E. coli* genotypes.

	Genotypes	β-galactosidase Production: Induced Bacteria	β-galactosidase Production: Non-induced Bacteria
1.	$i^+o^+z^+/F'i^-o^+z^+$	+	–
2.	$i^+o^+z^+/F'i^-o^+z^-$	+	–
3.	$i^+o^cz^+/F'i^-o^+z^-$	+	+
4.	$i^-o^cz^+/F'i^+o^+z^-$	+	+
5.	$i^+o^+z^-/F'i^-o^cz^+$	+	+
6.	$i^-o^+z^-/F'i^-o^cz^+$	+	+
7.	$i^+o^+z^+/F'i^-o^cz^-$	+	–
8.	$i^+o^+z^-/F'i^+o^cz^+$	+	+
9.	$i^+o^cz^-/F'i^+o^+z^+$	+	–
10.	$i^+o^cz^+/F'i^+o^+c^z$	+CRM	+
11.	$i^+o^+z^+/F'i^+o^cc^z$	+CRM	+CRM

- **Notes on CRM (Cross-reacting Material)**

(i) Antibodies sensitive to normal enzyme (tryptophan synthetase/β-galactosidase) can be prepared in rabbit serum and shown to react with extract from many of the mutant tryptophan synthetase/β-galactosidase stocks. The reacting substance is called **cross-reacting material (CRM)** and tryptophan synthetase mutant/β-galactosidase mutants producing CRM are designating as **CRM⁺**.

CRM produced by tryptophan synthetase/β-galactosidase is called CRM⁺.

CRM produced by mutant typtophan synthetase/β-galactosidase is called CRM⁻.

(ii) Because of its sensitivity, the immunological reaction indicates that a protein similar to tryptophan synthetase or β-galactosidase is produced in normal or almost normal amounts by these mutants but nevertheless is unable to perform the necessary enzymatic role in tryptophan or lactose metabolism. This means that such mutation affects a critical portion of the tryptophan synthetase or β-galactosidase molecule.

(iii) In contrast to CRM⁺ mutants more drastic effects on the tryptophan synthetase or β-galactosidase molecule seem to occur in CRM⁻ stocks, bacteria whose mutant product cannot be detected by the usual test for CRM.

Cz-mutations: CRM is produced constitutively when cz is on the o^c. CRM is produced by induction when cz is on o^+.

3. Using (+) for productions, (0) for non-production of the enzymes permease (p) and β-galactosidase (β-gal), complete the following table:

	Genotypes	Inducer absent (Permease)	Inducer absent (β-gal)	Inducer present (Permease)	Inducer present (β-gal)
1.	$i^+ o^+ y^+ z^+$	0	0	+	+
2.	$i^- o^+ y^+ z^+$	+	+	+	+
3.	$i^s o^+ y^+ z^+$	0	0	0	0
4.	$i^+ o^c y^+ z^+$	+	+	+	+
5.	$i^+ o^o y^+ z^+$	0	0	0	0
6.	$i^- o^c y^+ z^+$	+	+	+	+
7.	$i^- o^o y^+ z^+$	0	0	0	0
8.	$i^s o^c y^+ z^+$	+	+	+	+
9.	$i^s o^o y^+ z^+$	0	0	0	0

4. **Using (+) for production and (–) for non-production of enzymes, complete the following table:**

Genotypes	No lactose		Lactose	
	β-gal(z)	β-per(y)	β-gal(z)	β-gal(y)
1. $i^+p^+o^+z^+y^+$	–	–	+	+
2. $i^+p^+o^cz^+y^+$	+	+	+	+
3. $\frac{i^+p^+o^cz^+y^-}{i^+p^+o^+z^-y^+}$	+	–	+	+
4. $\frac{i^+p^+o^cz^-y^+}{i^+p^+o^+z^+y^-}$	–	+	+	+
5. $i^-p^+o^+z^+y^+$	+	+	+	+
6. $\frac{i^+p^+o^+z^+y^-}{i^-p^+o^+z^-y^+}$	–	–	+	+
7. $\frac{i^-p^+o^+z^+y^-}{i^+p^+o^+z^-y^+}$	–	–	+	+
8. $i^sp^+o^+z^+y^+$	–	–	–	–
9. $\frac{i^sp^+o^+z^+y^+}{i^+p^+o^+z^+y^+}$	–	–	–	–
10. $i^{-d}p^+o^+z^+y^+$	+	+	+	+
11. $\frac{i^{-d}p^+o^+z^+y^-}{i^+p^+o^+z^-y^+}$	+	+	+	+
12. $\frac{i^{-d}p^+o^+z^-y^+}{i^+p^+o^+z^+y^-}$	+	+	+	+
13. $\frac{i^-p^+o^+z^+y^-}{i^+p^-o^+z^-y^+}$	–	–	+	–
14. $\frac{i^-p^-o^+z^+y^+}{i^+p^+o^+z^-y^-}$	–	–	–	–
15. $\frac{i^+p^-o^+z^+y^-}{i^+p^+o^cz^-y^+}$	–	+	–	+
16. $\frac{i^+p^-o^cz^+y^-}{i^+p^+o^+z^-y^+}$	–	–	–	+
17. $\frac{i^-p^+o^+z^+y^+}{i^+p^+o^+z^-y^-}$	–	–	+	+
18. $\frac{i^sp^+o^+z^+y^-}{i^+p^+o^+z^-y^+}$	–	–	–	–
19. $\frac{i^sp^+o^+z^+y^-}{i^+p^+o^+z^-y^+}$	–	–	–	–
20. $\frac{i^+p^+o^cz^-y^+}{i^+p^+o^+z^+y^-}$	–	+	+	+
21. $\frac{i^-p^+o^cz^+y^-}{i^+p^+o^+z^-y^+}$	+	–	+	+
22. $\frac{i^sp^+o^cz^+y^+}{i^+p^+o^cz^+y^+}$	+	+	+	+
23. $\frac{i^dp^+o^+z^+y^-}{i^+p^+o^+z^-y^+}$	+	+	+	+

5. **Mutants were instrumental in the elaboration of the model for the regulation of lactose operon.**

   (a) Discuss why o^c mutants are cis-dominant but not trans-dominant.

   (b) Explain why i^s mutants are trans-dominant to the wild type i^+ allele but i^- mutants are recessive.

   (c) Discuss the consequences of mutation in the repressor gene promoter as compared with mutation in the structural gene promoter.

Solution:

(a) (i) The o^c mutants have an altered base pair in the operator, that prevents the repressor from recognizing operator. As a result structural genes cis (adjacent) to the o^c are transcribed constitutively.

(ii) o^c has no effect on the other lactose operon in the same cell because o^c codes for no product that could diffuse through the cell and affect other DNA sequence.

(b) (i) The i^s mutants produce super-repressor molecules that have lost their ability to recognize the inducer. As a result super-repressor molecules bind to normal lactose operator in the cell; no matter where they are and prevent transcription of structural gene.

(ii) In this case, trans-dominance is seen because the i^s mutants produce diffusible product. Even with the normal repressor (from an i^+) molecule in the cell, the super-repressor molecules are "stuck" on the operator.

(iii) i^- mutants are recessive.

- It causes a change in the repressor molecule such that it cannot bind to the operator.
- However, if an i^+ gene is present in partial diploid with i^-, a diffusible, functional repressor will be made that can bind to the operator and block transcription.

(c) (i) Mutation in the repressor gene promoter results either in an increase or decrease in the level of expression of repressed gene. Such mutations are likely to have little effect on the control of expression of lac operon structural genes since the repressor molecule itself is unaltered.

(ii) Promoter mutation for structural genes can also increase or decrease the level of expression of the operon.

(iii) Most known promoter mutation causes an almost complete loss of expression of three genes.

6. In a series of experiment performed by Jacob and Monad, the following results are obtained: [M.Sc. (C.U.) 1981]

Genotype	Units of enzyme produced			
	β-galactosidase		Galactoside transacetylase	
	Non-induced	Induced	Non-induced	Induced
(i) $i^sz^+a^+$	0.1	0.1	1	1
(ii) $i^sz^+a^+/F'i^+z^+a^+$	0.1	0.2	1	3
(iii) $i^+o^cz^+a^+$	25	95	15	100
(iv) $o^+z^+a^+/F'o^cz^+a^-$	180	440	1	220
(v) $i^so^+z^+a^+/F'i^+o^cz^+a^+$	190	219	150	220

Explain these operating lac operons on the basis of these results.

Solution:

(i) $i^sz^+a^+$ → In spite of the presence of normal **z** and **a** genes, they cannot produce their normal product both in induced and non-induced condition. This is due to the mutation of i gene, i.e., no inducer can bind. Therefore, it may be concluded that the i gene has the regulatory effect on these structural genes.

(ii) $i^sz^+a^+/F'i^+z^+a^+$ → Even in the presence of i^+ gene the product of the structural genes are almost nil, because the mutation of regulator gene (as super-repressor), i.e., i^s. We can conclude that it is dominant over i^+ (normal), i.e., dominant negative.

(iii) $i^+o^cz^+a^+$ → Here the production of enzyme is constitutive, i.e., production occurs in both induced and non-induced condition.

This is due to constitutive mutation of operator, i.e., o^c. The configuration of this region has changed and no repressor molecule produced by regulator cannot bind with the operator. Hence, the mechanism is permanently switched on.

(iv) $o^+z^+a^+/F'o^cz^+a^-$ → The production of z is constitutive. Here o^+ is dominant over o^c in cis-arrangement. oc is cis-dominant and expressed constitutively even in the presence of o^+ operator.

(v) $i^so^+z^+a^+/F'i^+o^cz^+a^+$ → The production is constitutive.

7. The synthesis of enzyme 'E' is inducible and dependent on three loci; a regulator gene (Rg), a structural gene (sg) and an operator (o). Haploid bacteria strains are found with the following properties:

(a) Enzyme synthesis occurs only when inducer is present.

(b) Enzyme synthesis never occurs.

(c) Enzyme synthesis occurs in the presence or absence of inducer.

1. List all possible genotypes producing each of the phenotypes listed above.

2. When F′ element F $Rg^+o^+sg^-$ is introduced into these haploid cells, the resultant partial diploids which can synthesize 'E' and which cannot.

Solution:

1. The genotypes are as follows:

(a) $Rg^+\ o^+\ sg^+$ [Rg^+ o^+ sg^+]

(b) $Rg^+\ o^+\ sg^-$; $Rg^s\ o^+\ sg^+$ [Rgs o^+ sg^+]

(c) $Rg^-\ o^+\ sg^+$; $Rg^+\ o^c\ sg^+$

(i) $Rg^-\ o^+\ sg^+$; (ii) $Rg^+\ o^c\ sg^+$.

2. After introducing F $Rg^+\ o^+\ sg^-$ the resultant partial diploid will be

	Non-induced	Induced
(a) F $Rg^+ o^+ sg^-/Rg^+ o^+ sg^+$	—	+
(b) (i) F $Rg^+ o^+ sg^-/Rg^+ o^+ sg^-$	—	—
(ii) F $Rg^+ o^+ sg^-/Rg^s o^+ sg^+$	—	—
(c) (i) F $Rg^+ o^+ sg^-/Rg^- o^+ sg^+$	—	+
(ii) F $Rg^+ o^+ sg^-/Rg^+ o^c sg^+$	+	+

- Rg^s mutations are dominant to Rg^+; Rg^s produces an altered repressor which cannot be activated by the inducer.
- Alarmone—It is an intra cellular signal molecule that is produced due to harash environmental factors or nutritional stress condition and regulate gene expression at transcriptional lelvel. Example: cAMP.

  cis—when the mutations are located on the same homologous chromosome.

  trans—when the two mutations are located opposite homologous chromosome.

  o^c mutation is dominant in cis-position.

8. Consider mutations involving the promoter of the lac operon. These mutations prevent RNA polymerase binding and are designated p^-. Predict the phenotypes of the following partial diploids assuming no glucose in the growth medium. lac z = the gene for β-galactosidase:

Genotype	**Phenotype for β-galactosidase production**	
	Non-inducer	**Inducer**
(a) $i^+p^-o^+z^+/i^+p^+o^+z^+$ =		
(b) $i^+p^-o^+z^+/i^+p^-o^+z^+$ =		
(c) $i^+p^-o^+z^+/i^+p^+o^+z^-$ =		
(d) $i^-p^-o^+z^+/i^-p^-o^+z^+$ =		

Solution:

Genotype	Phenotype for β-galactosidase production		
	Non-inducer	Inducer	Remark
(a) $i^+p^-o^+z^+/i^+p^+o^+z^+$	–	+	p^+-dominant. The lac operon with wild type promoter will still function even if the other does not. Phenotype is wild type.
(b) $i^+p^-o^+z^+/i^+p^-o^+z^+$	–	–	With neither promoter able to bind RNA polymerase. Neither lac operon can function.
(c) $i^+p^-o^+z^+/i^+p^+o^+z^-$	–	–	The operon with wild type z gene has non-functional promoter and the operon with wild type promoter has non-functional z gene. No production (mutant).
(d) $i^-p^-o^+z^+/i^-p^-o^+z^+$	–	–	Even in the absence of active repressor, neither lac operon can be productive without functional promoter.

9. The table below presents the genotypes and phenotypes of several lac mutants, but with one missing notation (represented by an x) in each case. From the information given, fill in the missing notation for each genotype. Give your reasoning (sometimes x can represent wild type) z = lac z the gene for β-galactosidase:

Genotype	Phenotype for β-galactosidase production: Non-inducer	Inducer	
(a) $i^xo^+z^+$	+	+	
(b) $i^xo^+z^+/i^+o^+z^+$	–	–	
(c) $i^xo^+z^+/i^+o^+z^+$	–	+	(2 possible answers)
(e) $i^+o^xz^+/i^+o^+z^+$	+	+	
(f) $i^+o^xz^+/i^+o^+z^+$	–	+	
(g) $i^+o^+z^x/i^+o^+z^+$	–	–	

Solution:

(a) Since the mutants, phenotype is constitutive, its genotype must carry a mutation. Therefore, i^x must be i^-.

(b) Since the mutation in i prevents lac function and is dominant, i^x must be i^s.

(c) Since the partial diploid behaves as wild type, i^x is either i^+ or i^- (i^- is recessive).

(d) Since the mutation is constitutive, i^x must be i^-.

(e) Since the mutant is constitutive, o^x must be o^c.

(f) Since the phenotype of the partial diploid is wild type, o^x must be o^+.

(g) Since the partial diploid fails to make β-galactosidase, even with wild type forms of the control loci (i^+ o^+) z^x must be z^-.

10. The table below gives the genotypes (with respect to the lac operon) of several partial diploid *E. coli* strain. Fill the phenotypes using a '+' for β-galactosidase sythesis and a '–' for no β-galactosidase synthesis. Glucose is absent in all cases. Give a brief explanation of your reasoning:

Genotype	Phenotype for β-galactosidase production: Non-inducer	Inducer
(a) $i^+o^+z^+/i^+o^+z^+$		
(b) $i^+o^+z^-/i^+o^+z^+$		
(c) $i^-o^+z^+/i^+o^+z^+$		
(d) $i^so^+z^+/i^+o^+z^+$		
(e) $i^+o^cz^+/i^+o^+z^+$		
(f) $i^+o^cz^-/i^+o^+z^+$		
(g) $i^so^cz^+/i^+o^+z^+$		

Solution:

Genotype	Phenotype for β-galactosidase production: Non-inducer	Inducer	Reasoning
(a) $i^+o^+z^+/i^+o^+z^+$	–	+	wild type
(b) $i^+o^+z^-/i^+o^+z^+$	–	+	z^+ is dominant
(c) $i^-o^+z^+/i^+o^+z^+$	–	+	i^+ is dominant
(d) $i^so^+z^+/i^+o^+z^+$	–	–	i^s is dominant
(e) $i^+o^cz^+/i^+o^+z^+$	+	+	o^c is dominant
(f) $i^+o^cz^-/i^+o^+z^+$	–	+	o^c is only cis-dominant
(g) $i^so^cz^+/i^+o^+z^+$	+	+	o^c is dominant even in the presence of repressor because the repressor cannot bind to the o^c operator.

11. The i^- allele allows constitutive enzyme production by the y^+ and z^+ factors. The action of i^- might be explained by one of two alternatives. (1) i^- produces an internal inducer, thus eliminating the need for lactose in the medium to induce enzyme synthesis, i^+ produces no repressor, or (2) i^+ produces a repressor substance which in the absence of lactose inducer blocks enzyme formation, but in the presence of lactose inducer becomes inactivated and allows enzyme synthesis. i^- produces no repressor.

(a) **Assuming dominance of the i^- allele under first alternative in an *E. coli* partial heterozygote of the constitution i^+/i^-, would internal inducer be produced?**

(b) **Would enzymes be produced inductively or constitutively ?**

(c) **Assuming dominance of the i^+ allele under the second alternative in partial heterozygotes of the constitution i^+/i^-, would repressor be produced ?**

(d) **Would enzymes be produced constitutively or inductively ?**

(e) **From the pattern of reactions exhibited by different heterozygotes in the following table, determine which of the two assumptions is correct:**

Genotype	Lactose absent		Lactose present	
	P	β gal	P	β gal
1. $y^-z^-i^-/y^+z^+i^-$	+	+	+	+
2. $y^+z^+i^-/y^-z^-i^-$	+	+	+	+
3. $y^-z^-i^-/y^+z^+i^+$	0	0	+	+
4. $y^-z^-i^+/y^+z^+i^-$	0	0	+	+

(f) **Is the repressor substances diffusible, or can it only act on loci in cis position with the i locus ? How can this be determined from the information in the table of part (e) ?**

Solution:

(a) Yes, it produces internal inducer.

(b) Enzymes are produced constitutively.

(c) Yes, repressor will be produced.

(d) Enzymes will be produced inductively.

(e) From the table (3) and (4) the i^- allele fails to produce enzymes in the absence of the external inducer (lactose). It fails to exhibit dominance, this is why the first assumption is incorrect.

Under the second assumption, it is dominant and produces repressor which, in the absence of lactose inducer, blocks enzyme synthesis as seen in (3) and (4).

(f) Heterozygote (4) as $y^+ z^+$ on one DNA segment and the i^+ allele on the other. Yet in the absence of lactose inducer, the repressor made by i^+ still prevents the production of enzymes by y^+ and z^+. Therefore, repressor must be able to act at a distance (diffusible).

12. By what mechanism does the presence of tryptophan in the medium in which *E. coli* cells are growing results in premature termination or attenuation of transcription of the trp operon?

Solution:

(i) Transcription and translation are coupled in prokaryotes. When tryptophan present in cells, tryptophan-charged $tRNA^{trp}$ is produced. This allows translation of trp leader sequence through the two UGG trp codon to the trp leader sequence UGA termination codon.

(ii) This translation of the trp leader region prevents base pairing between the partially complementary mRNA leader sequences 74-85 and 108-119 which in turn permits the formation of the transcription termination "hairpin" involving leader sequences 108-119 and 126-134.

13. Is the CAP–cAMP effect on transcription of the lac operon an example of positive or negative regulation? Why?

Solution:

Positive regulation.

(i) The CAP–cAMP complex has a positive effect on the expression of the lac operon.

(ii) It functions in **turning on** the transcription of the structural gene in the operon.

14. What consequences would mutation in the catabolite activator protein (CAP) gene of *E. coli* have for the expression of a wild type lac operon?

Solution:

(i) The CAP in a complex with cAMP is required to facilitate RNA polymerase binding occurs only in the absence of glucose and only if the operator is not occupied by repressor.

(ii) A mutation in the CAP gene, then would render the lac operon incapable of expression because RNA polymerase would not be able to recognize promoter.

15. In the presence of high intracellular concentration of tryptophan, only short transcripts of trp operon are synthesized because of attenuation of transcription to the structural genes. This is mediated by the recognition of two Trp codons in the leader sequence. If these codons were mutated to be amber (UAG) nonsense mutations, what effect would this have on the regulation of the operon in the presence or absence of tryptophan ? Explain.

Solution:

(i) For a wild type trp operon, the absence of tryptophan results in antitermination, that is, the structural genes are transcribed and the tryptophan biosynthetic enzymes are made.

(ii) This occurs because the absence of tryptophan results in non-availability or at least very low level of trp-tRNA-trp which in turn causes the ribosome translating the leader sequence to stall at the **Trp codon.**

(iii) With the ribosome stalled at the location, the RNA being synthesized just ahead of the ribosome by RNA polymerase assumes a secondary structure that favours continued transcription of the structural genes by RNA polymerase.

(iv) If two trp codons are mutated to stop codons, then the mutant operon would function in the same way as the wild type operons do in the absence of tryptophan.

(v) In case wild type trp operon, in the presence of tryptophan, turns the transcription of structural genes off, the presence of tryptophan leads to the accumulation of trp-tRNA-trp, which allows the ribosome to read the two trp codons and to stall at the normal stop codon of the leader sequence.

(vi) When stalled in that position, the antitermination is synthesized. Instead a termination signal is formed resulting termination of transcription.

(vii) In a mutant trp operon with two stop codons instead of trp codons, the ribosome will still stall at the premature stop codons even in the presence of tryptophan. It results antitermination signal so that the structural genes will be synthesized.

So both in presence or absence of tryptophan the mutant trp operon will show no attenuation, i.e., structural genes will be synthesized and tryptophan biosynthetic enzymes will be produced.

16. The SOS repair genes in *E. coli* are negatively regulated by the Lex A gene product, called Lex A repressor. When a cell sustains extensive damage to its DNA, the Lex A repressor is inactivated by the rec A gene product (Rec A), and transcription of the SOS genes is increased dramatically.

One of the SOS genes is in uvr A gene. You are studying the function of the uvr A gene product in DNA repair. You isolate a mutant strain that shows constitutive expression of the uvr A protein. You name this mutant strain uvr A^C. The following simple diagram shows the lexA and uvrA operons:

p^{lexA} O^{lexA} lexA p^{uvrA} O^{uvrA} uvrA

(a) Describe two different mutations that would result in a uvrA constitutive phenotype. Indicate the actual genotypes involved.

(b) Outline a series of genetic experiments, using partial diploid strains, that would

allow you to determine which of the two possible mutations you have isolated.

Solution:

(a) Cell one constitutive mutation $lexA^-$ (mutation in the repressor gene product) and the other O^{urvA-} (mutation in the operator).

(b) One can make partial diploid strains using F′. O^{urvA-} will be dominant to O^{urvA+} and $lexA^-$ will be recessive to $lexA^+$. O^{urv-} will act in cis.

17. The rate of transcription of the trp operon in *E. coli* is controlled by both (1) repression/derepression and (2) attenuation. By what mechanisms do these two regulatory processes modulate trp operon transcripts levels?

Solution:

(i) Repression/depression of the trp operon occurs at the level of transcription initiation, modulating the frequency at which RNA polymerase initiates transcription from the trp operon promoters.

(ii) Attenuation modulates trp transcript levels by altering the frequency of termination of transcription within the trp operon leader region (trpL).

18. Would attenuation of the type that regulates the level of trp transcripts in *E. coli* be likely to occur in eukaryotic organisms?

Solution:

No, since transcription (nucleus) and translation (cytoplasm) are not coupled in eukaryotes, attenuation of the type occurring in prokaryotes would not possible.

19. What is the major structural difference between the operator region in the lac operon of *E. coli* and phage λ operators O_L and O_R?

Solution:

(i) The operator region of the lac operon in *E. coli* contains a single repressor-binding site, whereas the lambda operators O_L and O_R each contains three repressor binding sites that are differentially occupied under different conditions.

(ii) The differential affinity of lambda repressor for the three binding sites in these operators plays an important role in the maintenance of the lysogenic state of a lambda prophage *E. coli*.

20. The N protein of phage λ is a transcriptional antiterminator. What role does it play on the lambda regulatory cascade?

Solution:

(i) The N protein, the product of the first gene transcribed from promoter P_L, prevents the termination of transcription at two sites (t_{R_1} and t_{L_1}), which results in the transcription of the delayed early genes located downstream from these sites.

(ii) N protein binds to **nut** sites just upstream from the termination sites, and together with the **nus** protein and ribosomal protein S-10, modifies the specificity of RNA polymerase so that termination does not occur at sites t_{R_1} and t_{L_1}.

21. A gene R codes for a protein that is a negative regulator of transcription of a gene S. Is the gene S transcribed in an R^- mutant? How does the situation differ if the product of R is a positive regulator of S transcription?

Solution:

(i) A negative regulator of transcription is needed to turn transcription off; hence in the R^- mutant, transcription of S is constitutive.

(ii) The opposite happens in the positive regulation. A positive regulator of transcription is needed to turn transcription on.

So if the product of R is a positive regulator, transcription does not occur in R^- cells.

22. For each of the following partial diploid genotypes of *E. coli*, state whether β-galactosidase is made and whether its synthesis is inducible or constitutive. The convention for writing partial diploid genotypes is to put

the plasmid genes at the left of the slash and the chromosomal genes at the right:

(a) lac Z^+ lac Y^-/lac Z^- lac Y^+

(b) lac O^c lac Z^- lac Y^+/lac Z^+ lac Y^-

(c) lac P^- lac Z^+/lac O^c lac Z^-

(d) lac I^+ lac P^- lac Z^+/lac I^- lac Z^+

(e) lac I^+ lac P^- lac Y^+/lac I^- lac Y^-

Solution:

The location of structural gene does not matter because they function in the same manner whether present in the plasmid or in a chromosome.

(a) (i) In this partial diploid, there are no cis-dominant mutations. The plasmid operon is lac Z^+ and can make the enzyme, whereas the chromosomal operon is lac Z^- and cannot.

(ii) The cell will produce the enzyme from the plasmid gene as long as the operon can be turned on. Turning it on requires the presence of functional regulatory elements (recall that lac if a gene is not tested—for example, lac I—it is assumed to be wild type). It is also necessary that the inducer be able to enter the cell which it can in this example because the chromosomal genotype can supply the lac Y permease.

(b) (i) This partial diploid has the cis-dominant mutation lac O^c in the plasmid and so the genes in the plasmid operon are always expressed.

(ii) However, the lac Z gene in the plasmid makes a defective enzyme. The chromosome has a functional lac Z gene from which active enzyme can be made, but its synthesis is under the control of a normal operator (because the operator genotype is not indicated, it is wild type). Hence, enzyme synthesis must be inducible.

(iii) Thus, the partial diploid makes a defective enzyme constitutively and a normal enzyme inducibly, and so the overall phenotype is that the cell can be induced to make β-galactosidase.

(c) (i) A promoter mutation which is cis-dominant, is in the plasmid which means that no lac mRNA can be made from this operon.

(ii) The chromosome contains a cis-dominant lac O^c mutation that drives constitutive synthesis of an mRNA, but the enzyme is defective.

(iii) Thus, there is no way for the cell to make active β-galatosidase.

(d) (i) The plasmid genotype contains a promoter mutation, so no mRNA can be produced from the lac Z structural gene. However, in the lac I gene in the plasmid has its own promoter, and so lac repressor molecules are present in the cell.

(ii) The chromosomal genotype alone would make enzyme constitutively because of the lac mutation, but the presence of the functional lac product may be made by the plasmid means that any synthesis that occurs must be induced.

(iii) The chromosomal operon can provide both β-galactosidase and permease, and so β-galactosidase is inducible in this genotype.

(e) (i) This genotype differs from that in part (d) by the presence of a lac Y mutation in the chromosome.

(ii) Again, the plasmid contributes only lac I repressor to the cell, so any synthesis of the enzyme must be inducible.

(iii) However, the chromosomal genotype is lac Y^-. Because the lac Y^+ allele in the plasmid cannot be expressed, no inducer can enter the cell. So, the cell is unable to make any enzyme.

23. Is it necessary for the gene that codes for the repressor of a bacterial operon to be near the structural genes? Why or why not?

Solution:

(i) The repressor gene need not be near because the repressor is diffusible.

(ii) For example, in trp repressor, the trp repressor gene is located and is quite far from the structural gene.

24. Consider a eukaryotic transcriptional activator protein that binds to an enhancer sequence and promotes transcription. What change in regulation would you expect from a duplication in which several copies of the enhancer were present instead of just one?

Solution:

(i) The mutant gene should bind more of the activator protein or bind it more efficiently.

(ii) Therefore, the mutant gene should be induced with lower levels of the activator protein or expressed at higher levels than the wild type gene or both.

25. The permease of *E. coli* that transports the α-galactoside melibiose can also transport lactose, but it is temperature-sensitive. Lactose can be transported into the cell at 30 °C but not at 37 °C. In a strain that produces the melibiose permease constitutively, what are the phenotypes of lac Z^- and lac Y^- mutants at 30 °C and 37 °C?

Solution:

(i) At 37 °C, both lac Z^- and lac Y^- are phenotypically lac$^-$.

(ii) At 30 °C, the lac Y^- strain is lac$^+$ and the lac Z^- strain is lac$^-$.

26. Is an attenuator a region of DNA, that likes an operator bound with a protein? Is RNA synthesis ever initiated at an attenuator?

Solution:

(i) The attenuator is not a binding site for any protein and RNA synthesis does not begin at an attenuator.

(ii) An attenuator is strictly a potential termination site for transcription.

27. If a wild type *E. coli* strain is grown in a medium without lactose or glucose, how many proteins are bound to the lac operon? How many are bound if the cells are grown in glucose?

Solution:

(i) In the absence of glucose or lactose, two proteins, viz., the lac repressor and CRP-cAMP are bound.

(ii) In the presence of glucose, only the repressor is bound.

28. How does CAP-cAMP recruit polymerase to the promoter?

Solution:

(a) It occurs in two-step process.

(i) Formation of closed promoter complex

(ii) Conversion of closed promoter complex to open promoter complex.

$$R + P \underset{K_\beta}{\rightleftharpoons} RP_c \underset{K_2}{\longrightarrow} RP_0$$

R = RNA polymerase; P = Promoter;
RP_c = Closed promoter complex;
RP_0 = Open promoter complex.

(b) CAP-cAMP directly stimulates the first step by increasing K_β.

(c) If γ CAP-cAMP has little effect on K_2, the second step is not accelerated.

(d) Thus, the net effect of CAP-cAMP is to increase the rate of open promoter complex formation.

29. (a) State briefly about the three operators of the lac operon. (b) What are the relative effects of removing each or both of the auxiliary operators?

Solution:

(a) (i) There are three operators—viz. one major operator (O_1) near the transcription start site and two auxiliary operators (one upstream and one downstream).

(ii) The classical (major) operator (O_1) centered at position +11, the down-stream auxiliary operator (O_2) centered at position +412 and the upstream auxiliary operator (O_3) centered at position –82.

(iii)

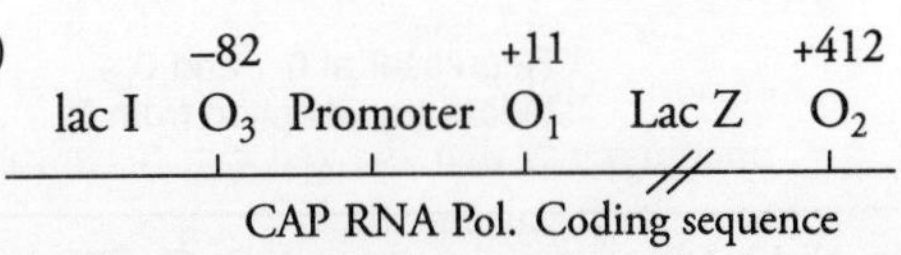

The map of three lac operators

(b) (i) Benno Müller-Hill and others have investigated the auxiliary operators and have discovered that they are not just trivial copies of the major operator. Instead, they play a significant role in repression.

(ii) Removal of either of the auxiliary operators decreased repression only slightly, but removal of both auxiliary operators decreased repression about 50 folds.

(iii) All three operators together repress transcription 1300 folds; two operators together repress 440 to 770 folds, but the classical operator itself repress only 18 folds.

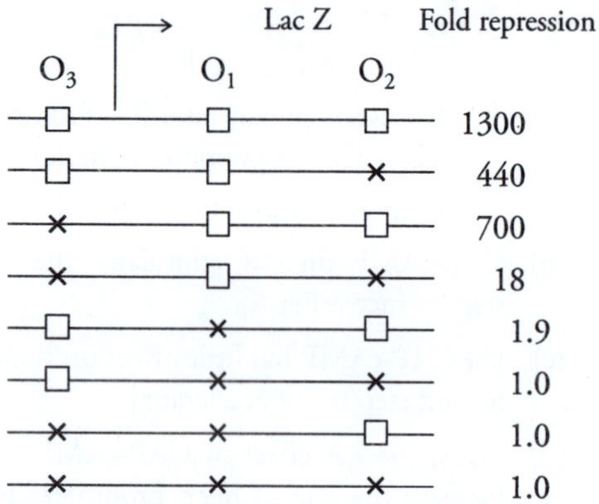

Effects of mutation in the three lac repressors.

30. What determines whether phage lambda (λ) will enter the lytic or lysogenic state when it infects an *E. coli*?

Solution:

(i) The decision between lysogeny and lytic growth hinges on a delicate balance between the races of two regulatory gene (C_I and Cro) products, viz. lambda repressor protein and Cro proteins.

(ii) If C_I gene manages to produce enough repressor which binds or blocks O_R1, O_R2, O_L1 and O_L2 and turning off all early transcription including transcription Cro genes. This leads to **lysogeny**.

(iii) On the other hand, Cro gene codes for Cro protein which blocks O_R3 and O_L3 and turning off C_I genes transcription. This leads to **lytic infection.**

(iv) The winner of the race (between C_I and Cro) is determined by the concentration of the gene product C_{II} which is determined by cellular protease concentration. The C_{II} concentration leads to **lysogeny**.

(v) C_{II} activates P_{RE}, i.e., facilitates the binding of RNA polymerase to P_{RE}.

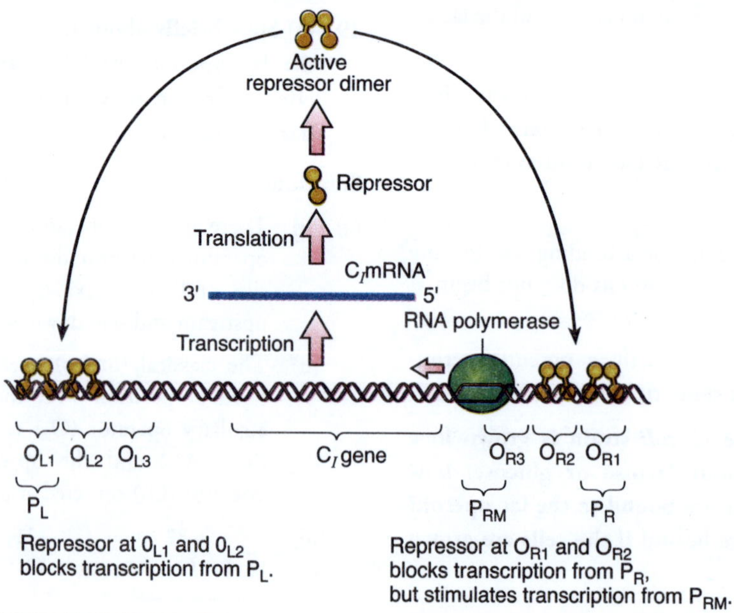

Fig. 15.13: Diagram of the phage λ C_{II}-C_{III}-PRE regulatory circuit involved in determining whether lytic or lysogenic development will occur.

31. What are the differences in action of the λ promoters PRE and PRM?

Solution:

(i) The promoters P_{RE} and P_{RM} are both promoters of the repressor operon of phage λ. Transcription from these promoters allows production of the C_I repressor protein, which favours lysogeny.

(ii) Initially, the promoter P_{RE} is activated. For it to be a transcription site, it must be activated by the product of the C_{II} gene which lies in the right side.

(iii) This promoter P_{RE} produces a messenger with a long leader that is translated very efficiently. Once the repressor binds at the operators of the left and right operons, the C_{II} gene no longer transcribed and therefore, P_{RE} is no longer a site for transcription.

(iv) However, the repressor gene can still be transcribed from the P_{RM} promoter, which does not need the product of the C_{II} gene. This promoter produces a transcript with no leader and thus is translated very inefficiently. At that point, however, only a very small quantity of repressor is needed to maintain lysogeny.

(v) Thus, the two promoters are the sites for initiation of the repressor operon under different circumstances, one early in the infection stage and one after lysogeny stage.

32. When lambda phage infects an *E. coli*, it can either undergo lytic cell growth and lyse the host cell or enter lysogenic pathway.

(a) Will lytic or lysogenic development occur in a cell infected with a λ phage with a deletion of the C_I gene?

(b) Which will occur in a cell infected with a lambda phage with a deletion of the Cro gene?

Solution:

(a) (i) The C_I gene encodes the λ repressor. Thus, lytic development occurs in cells infected with lambda phage that carries a deletion of the C_I gene.

(ii) Cro protein will occupy the lambda operator sites because no lambda repressor will be present if the C_I gene has been deleted.

(b) (i) In contrast lysogeny will occur in bacteria infected with λ phage that harbour a deletion of the Cro gene.

(ii) Repressor will be synthesized by expression of the C_I gene, and in the absence of Cro protein, it will bind to the λ operator sites and block the expression of the λ lytic genes.

33. What is the major structural difference between the operator region in the lac operon of *E. coli* and the phage lambda operators O_L and O_R?

Solution:

(i) The operator region of the lac operon in *E. coli* contains single repressor binding site whereas the lambda operators O_L and O_R each contains three repressor binding sites (O_{L_1}, O_{L_2}, O_{L_3} and O_{R_1}, O_{R_2}, O_{R_3}) that are differentially occupied under different conditions.

(ii) The differential affinity of lambda repressor for the three binding sites in these operators plays an important role in the maintenance of the lysogenic state of lambda prophage in *E. coli*.

34. Bacteriophage λ, after infecting *E. coli*, will either produce many progenies that are released by lysis of the cell (lytic growth) or phage DNA integrates into the chromosome because transcription from the major promoters of the phage has been shut down. The repressor protein cI encoded by phage λ, bound to two phages is produced. Mutations in the cI gene that destroy the binding ability of the repressor lead to lytic life-cycle exclusively. That is, all cells infected by the phage will burst and release progeny phages. Another type of mutation gives the same phenotype—lytic growth only. Such mutations called λ vir, arise at a much lower frequency than the cI mutations (about I in

10^{12} compared to 1 on 10^6 for cI mutants). What do you think these mutations are and why are they less frequent than cI mutations?

Solution:

(i) This problem requires understanding the types of regulatory mutations that can affect negative regulation.

(ii) The lack of negative regulation (by Cl) in the λ life-cycle leads to the lytic cycle of growth only.

(iii) Such mutations could be either in the encoding the negative regulator or the site to which the repressor binds.
It is told that Cl mutations are defects in the gene encoding the repressor.

(iv) The λ vir mutations could be mutations in the site to which the repressor binds but since repressor has to bind with two sites, there must be two mutations in a λ vir mutant. Therefore, these would arise less frequently (at a frequency predicted for two independent mutational events combined: 1 in $10^6 \times 1$ in 10^6 or 1 in 10^{12}).

35. The ara I site is required for induction of ara BAD. I⁻ mutants do not express ara BAD. Starting with an I⁻ mutant, a second mutation arose that resulted in constitutive arabinose synthesis. A southern blot using a probe from the regulatory region and early part of the ara B gene showed a very different set of restriction fragments than were seen in the starting strain. Based on the altered restriction pattern and constitutive expression, propose a hypothesis about the nature of the mutation.

Solution:

(i) To solve this problem, it is needed to think how changes in restriction pattern could arise, what effects they could have and what is necessary to get expression.

(ii) The fact is that the experiment began with a strain that lacked the inducing site, I, and that there is constitutive synthesis which means that the normal regulation is lacking.

(iii) Constitutive synthesis could result from a deletion that fused the ara BAD genes to another promoter (one that is under the growth conditions used).

(iv) A deletion would lead to a different pattern of restriction fragments that could observed in the southern hybridization analysis.

36. A mutation in the phage lambda Cl gene results in a nonfunctional Cl gene product. What phenotype would you expect the phage to exhibit?

Solution:

(i) The Cl gene product is a repressor protein that acts to keep the lytic functions of the phage repressed when λ (lambda) is in the lysogenic stage.

(ii) A Cl mutant strain would lack the repressor and be unable to repress lysis, so the phage would always follow a lytic pathway.

37. Five lambda (λ) mutants have the molecular phenotypes shown in the left column of the following table:

Mutant	Molecular phenotype	Lytic growth	Lysogenic growth	Inducible by uv light
1.	**The Cro protein is unable to bind DNA**			
2.	**The N protein does not function**			
3.	**The CII protein does not function**			
4.	**The Q protein does not function**			
5.	**PRM is unable to bind RNA polymerase**			

Fill the table to indicate whether each mutant will be able to undergo lytic or lysogenic growth. For mutants able to follow a lysogenic pathway, state whether they can be induced by UV light.

Solution:

Mutant	Molecular phenotype	Lytic growth	Lysogenic growth	Inducible by UV light
1.	The Cro protein is unable to bind DNA	no	yes	no
2.	The N protein does not function	no	no	no
3.	The CII protein does not function	yes	no	yes
4.	The Q protein does not function	no	yes	no
5.	PRM is unable to bind RNA polymerase	yes	no	yes

38. DNase protection experiments were helpful to elucidate the functions of different DNA sequences in the lac promoter.

(a) What is DNase protection experiment and how does it provide this information?

(b) How are the binding sites for the lac repressor, RNA polymerase, and CAP-cAMP arranged at the 5′ end of the lac operon?

(c) What effects would you expect each of the following mutations to have on the coordinate induction of the lac operon by lactose (in the absence of glucose)? Explain your reasoning.

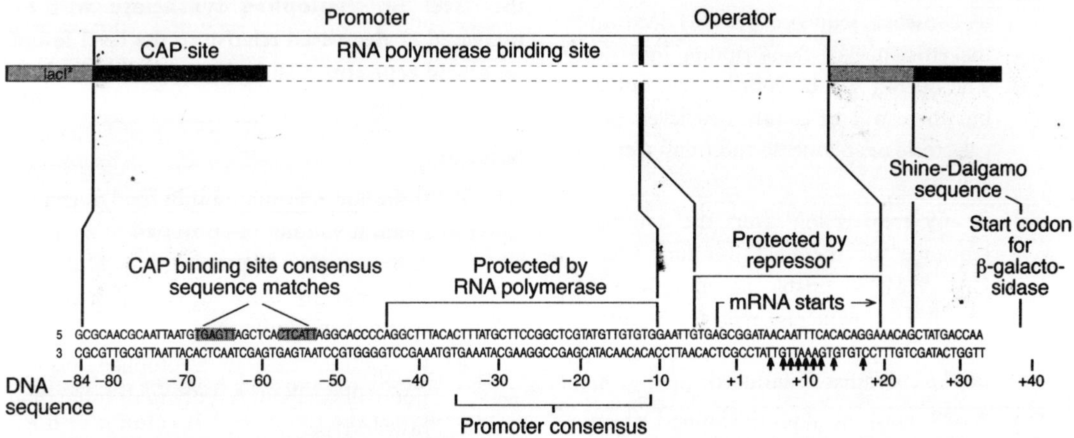

Fig. 15.14: Base-pair sequence of the promoter and operator for the lactose operon of *E. coli*. Also shown are locations of some known lac O^c mutations (indicated by arrows).

(i) a deletion of base pairs from +3 to +18

(ii) a TA– to GC transversion at –12

(iii) a TA– to GC transversion at –69

(iv) a GC– to AT transition at +28

(v) a GC– to AT transition at +9

(d) Would any of the mutations listed in (c) affect catabolite repression of the lac operon?

Solution:

(a) (i) A DNase protection experiment is in vitro method to identify DNA sites that are bound by a protein.

(ii) After a purified protein is allowed to bind a DNA segment, the complex is treated with DNase, and sequences unprotected by protein binding are digested.

(iii) DNase protection experiment defines the location of the operator, promoter and CAP-cAMP binding site.

(b)

Site	Nucleotide pair
Lac repressor:	–3 to +21
RNA polymerase:	–44 to –8 including –10 to –35 consensus sequence
CAM–CAP complex	–54 to –58 and –65 to –69

(c) (i) This deletion disrupts the operator; so the operon would be expressed constitutively.

(ii) A –12 transversion alters the –10 promoter consensus sequence, possibly decreasing the efficiency of transcription initiation. The operon still be coordinately induced but there will be diminished levels of β-galactosidase, permease and trans-acetylase activity.

(iii) A –69 transversion alters the consensus sequence for the CAP binding site. If CAP-cAMP is unable to bind the CAP site, RNA polymerase will not be recruited to the promoter and the operon will not be coordinately induced.

(iv) A +28 transition alters the Shine-Dalgarno sequence and by affecting translation initiation, could result in diminished or absent β-galactosidase and due to polar effects, diminished or absent expression of permease and trans-acetylase.

(v) A +9 transition alters the operator. It could either have no effect, cause the repressor to have more affinity for the operator (preventing coordinate induction in a cis-dominant manner) or cause the repressor to have less affinity for operator (leading to constitutive expression).

(d) None of the mutants will prevent catabolite expression.

39. The mutant *E. coli* strains described in the following table are individually inoculated into two different media, one with supplement tryptophan and one without

Mutant	Phenotype
(a)	**Apo-repressor is unable to bind to tryptophan.**
(b)	**A point mutation in the trp operator prevents binding by an active Trp repressor.**
(c)	**The trp E gene has a nonsense mutation.**
(d)	**The levels of Trp-tRNA Trp are decreased due to a mutation in a gene for Trp-amino-acyl synthetase.**
(e)	**Three adjacent G-C base pairs in region 4 are mutated to A-T base pair.**

For each mutant and medium state whether the level of tryptophan synthetase will be increased or decreased relative to the level found on a wild type strain and where possibly by how much and why?

Solution:

(a) (i) If the apo-repressor cannot bind to tryptophan, it will not be converted to an active repressor when tryptophan is present. This will lead to constitutive expression of tryptophan synthetase.

(ii) In medium without tryptophan, expression will be about seventy folds more than in wild type.

In medium with tryptophan, expression will be reduced only through attenuation, and so it will be about seventy folds more than in wild type.

(b) The trp operon will exhibit constitutive expression, so mutant b will show the same expression pattern as mutant I.

(c) (i) The trp E gene is the first gene transcribed in the trp operon.

(ii) A nonsense mutation could have a polar effect, leading to diminished or absent translation of trpB and trpA, which encode tryptophan synthetase.

(iii) Therefore, in **medium without tryptophan** where the operon is not repressed, mutant c would produce diminished level of tryptophan to synthetase compared to wild type cells.

In **medium with tryptophan**, the level will be same as in wild type cells (very low).

(d) (i) Trp-tRNA Trp molecules are needed to attenuate transcription at the trp operon. Therefore, if the levels of Trp-tRNA Trp are always low, transcription of the trp operon will not be attenuated even when tryptophan levels are high.

(ii) Therefore, in **medium with tryptophan**, mutant d will have eight-to-ten folds higher levels of tryptophan synthetase than wild type cells.

In medium without tryptophan, attenuation does not occur, so mutant (d) will have levels of tryptophan synthetase that are similar to the wild type.

(e) In mutant (e), the 3 : 4 attenuator structure shown will not form. So attenuation will not occur when tryptophan levels are high.

Tryptophan synthetase levels will be the same as in mutant (d).

40. Bacteriophage λ after infection of a cell can integrate into the chromosome of the cell if the repressor protein, C_I, binds to and shuts down phage transcription immediately. (A strain containing a bacteriophage integrated into the chromosome is called a lysogen.) The alternative fate is the production of many more viruses and lysis of the cell. In an Hfr cross, a donor strain that is a lysogen was crossed with a lysogenic F^- recipient cell and no phages were produced. However, when Hfr lysogen donor strain transferred its DNA to a nonlysogenic F^- recipient cell, the recipient cell burst, releasing new generation of phages. Why did crossing to a nonlysogenic cell result in phage growth and release but crossing to a lysogenic recipient did not?

Solution:

Non-lysogenic strain did not have C_I repressor protein, so incoming infecting phage could not go into the lytic cycle.

41. Given the data below, explain which strains and growth conditions are important for reaching the following conclusions:

(a) Arabinose induces coordinate expression of the ara BAD genes (encoding kinase, isomerase and epimerase).

(b) The ara C genes encode a positive regulator of ara BAD expression.

Genotypes	Arabinose	Kinase	Isomerase	Epimerase
1. $C^+ B^+ A^+ D^+$	no	–	–	–
2. $C^+ B^+ A^+ D^+$	yes	+	+	+
3. $C^- B^+ A^+ D^+$	no	–	–	–
4. $C^- B^+ A^+ D^+$	yes	–	–	–

Solution:

(a) 1 ($C^+ B^+ A^+ D^+$) arabinose absent.

2 ($C^+ B^+ A^+ D$) arabinose present.

(b) 2 ($C^+ B^+ A^+ D^+$) arabinose present

4 ($C^- B^+ A^+ D^+$) arabinose present.

42. What do you mean by strong promoter and week promoter?

Solution:

(a) (i) An efficient promoter can direct synthesis of RNA transcript at a relatively fast rate.

(ii) Strong promoter usually controls genes whose translation products are needed in large amounts by the cells.

(b) (i) An inefficient promoter directs synthesis of RNA transcripts at a relatively low rate.

(ii) It controls genes whose translation products are needed in only small amounts by the cell.

43. What is hybrid promoter? Give example.

[C.U. Zoo (Cytogenetics) M.Sc. 2009]

Solution:

(i) Lac promoter is not very strong, so many vectors have been designed with a hybrid tac promoter, (trp and lac) which combines the strength of the trp (tryptophan operon) promoter with the inducibility of the lac promoter.

(ii) The trp promoter is much stronger than the lac promoter because of its –35 box.

(iii) Molecular biologists have combined –35 box of the trp promoter with the –10 box of the lac promoter plus the lac operator.

(iv) The –35 box of the trp promoter makes the hybrid promoter string and the lac operator makes it inducible by IPTG.

44. How much enhancement of trp transcription is possible in the following *E. coli* cells as the tryptophan concentration falls from abundant to very low? Why?

(a) Wild type

(b) Mutants in which the two tryptophan codons in the leader are changed from UGGUGG to UGAUGG [Hint: UGA is stop codon]

(c) Mutants in which two tryptoplan codons in the leader are changed from UGG UGG to UGGUGC.

(d) Mutants in which the string of Ts in the attenuator is replaced by a string of G's.

Solution:

(a) 700 folds; 70-fold from derepression and 10-fold from relief of alternation.

(b) 70-fold; the 10-fold enhancement from relief of attenuation is lost because a ribosome traversing the leader will fall off the ribosome at the new UGA (stop) codon.

(c) 70-fold; the 10-fold enhancement from relief of attenuation is lost because the tryptophan codon is missing, so the ribosome will not stall in the absence of the amino acid.

(d) 70-fold; attenuation should not work because the string of G's would not be recognized as part of a proper terminator.

45. Researchers have identified mutations in the promoter region of the lac I gene that makes it more difficult for the lac operon to be induced. These are called lac I^Q mutants, because a greater quantity of lac repressor protein is made. Explain why an increased transcription of the lac I gene make it more difficult to induce the lac operon.

Solution:

(i) An increasse in the amount of lac repressor protein makes it easier to repress the lac operon.

(ii) When the cells become exposed to lactose, allolactose levels slowly rise. Some of the allolactose binds to the lac repressor protein and causes it to be released from the operator site.

(iii) If many lac repressor proteins are found within the cell, more allolactose is needed to ensure that no unoccupied repressor proteins can repress the operon.

46. With regard to key proteins that affect the choice between the lytic and lysogenic cycles which may be degraded by cellular proteases.

Solution

(i) After infection, the key protein affected by cellular proteases is cII. If protease level are high, as under good growth conditions, cII is degraded. This promotes lytic cycle.

(ii) Under starvation condition, the protease levels are low This prevents degradation of cII and thereby, promotes lysogenic cycle.

(iii) After lysogeny has been established, the key protein affected by cellular proteases is the λ repressor.

Agents such as UV light activate cellular proteases that digest the λ repressor. This permits the induction of the lytic cycle.

47. How do the λ repressor and the cro protein affect transcription from P_{R1} and P_{RM}?

Explain where these proteins are binding to cause their effects?

Solution

(i) The O_R regions contain three operators sites, and two promoters. P_{RM} and P_R transcribe in opposite directions.

(ii) The λ repressor will first bind to O_{R_1} and then O_{R_2}. The binding of the λ repressor to O_{R_1} and O_{R_2} inhibits tyranscription from PR and thereby switches off the lytic cycle.

(iii) Early in the lysogenic cycle, the λ repressor protein concentration may become so high that will occupy OR_3. Later, when the λ repressor concentration begins to drop, it will first be removed from OR_3. This allows the transcription from P_{RM} and maintain **lysogenic cycle**.

(iv) The cro protein has its highest affinity for O_{R_3} and so it binds there first. This blocks transcription from P_{RM} thereby **switches off** lysogenic cycle.

(v) The cro protein has similar affinity for O_{R_2} and O_{R_1} and so it may occupy either off these sites next. It will bind to both O_{R_2} and O_{R_1}. This turns down the expression from P_R which is not needed in later stages of lytic cycle.

48. Would you expect the regulation of a gene to be affected by an inversion of its promoter sequence? By an inversion of an enhancer sequence? Explain your answer.

Solution

(i) An inversion not only inverts the sequence, but also exchange the strands themselves because the DNA strands in duplex are antiparallel.

(ii) The inversion of a promoter sequence would, therefore, have a profound effect, because the inverted promoter is in the opposite strand pointing to the wrong direction.

(iii) On the other hand, one would expect little or no effect of the inversion of an enhancer element because enhancers are targets for DNA binding proteins irrespective of their strand or orientation in the DNA.

49. When glucose is present in an *E. coli* cell, is the concentration of cyclic AMP high or low? Can a mutant with either an inactive adenyl cyclase gene or an inactive Crp gene synthesize beta galactosidase? Does the binding of cAMP-CRP to DNA affect the binding of a repressor?

Solution

(i) The cAMP concentration is low in the presence of glycose.

(ii) Both types of mutation are phenotypically lac^-.

(iii) The cAMP-CRP binding does not interfere with repressor binding because the DNA binding sites of cAMP-CRP and lac are distinct.

50. Is an attenuator a region of DNA that like an operator binds with a protein? Is RNA synthesis ever initiated at an atenuator?

Solution

(i) The attenuator is not a binding site for any protein and RNA synthesis does not begin at attenuator.

(ii) An attenuator is strictly potential termination site for transcription.

51. Among amino acid biosynthetic operons that have attenuation, some leader mRNA sequences contain very low codons for amino acid that is synthesized (for example trp operon) whereas others contain many more codons for the amino acid (for example in the phe operon). Explain why this might be expected.

Solution

(i) Amino acid that normally exists among the lowest levels of charged tRNA in the cell should have few codons in the leader mRNA, otherwise stalling would take place even at normal low levels.

(ii) On the other hand, amino acid that exists among the highest levels of charged tRNA in the cell should include multiple codons in the leader mRNA, otherwise stalling might not take place often enough.

Problem 52

Trp repressor is a positive alloasteric protein. Explain. **[C.U. 2015]**

Solution

(i) In this repressible system, the product of the trpR gene is an **inactive** (apo) **repressor.** TRP protein does not recognise the operator sequence of trp operon. When tryptophan level in cell is very low.

(ii) The effector molecules, i.e., tryptophan (co-repressor) bind to the apo repressor and cause conformational changes in the three-dimensional structure of these proteins. These changes are called allosteric change of the protein.

(iii) It changes the activity of the protein, i.e., trp repressor is able to bind to the trp operator site. This inhibit the ability of RNA polymerase to transcribe the operon.

Problem 53

The repressor protein of lactose and tryptophan operons are functionally dissimilar. **[C.U. 2012]**

Solution

(i) The regulator gene products are called **activator** in positive control system and **repressor** in negative control system.

(ii) The regulator protein can bind to the regulator protein binding sites (RPBs) near to promoter of the operon and it depnds on the presence or absence of **effector** molecules (inducer or co-repressor).

(iii) Inducer or co-repressor can bind to the regulator gene product (repressor or aporepressor) and causes conformational changes of the regulator protein product.

(iv) In lac operon, regulator protein product is **repressor** molecule, i.e., **360 amino acid long.** It is tetramer of four identical polypeptide. It binds to the lac operator, i.e., RPBs and prevents RNA polymerase from catalyzing the transcription of structural genes (Lac Z, Lac Y and Lac A).

On the other hand, in trp operon regulator gene product is aporepressor (inactive).

It normally cannot able to bind operator as it is inactive, i.e., apo repressor. In presence of co-repressor it becomes active and binds to operator and switch off, i.e., prevent the initiation of transcription of trp operon-protein coding genes by RNA polymerase.

Problem 54

In what ways the action of the lac repressor and trp repressor similar and different? In other word, discuss the similarities and differences with regard to their binding to operator sites and their effects on transcription.

Solution

The two proteins are similar in a sense that both bind to a segment of DNA and repress transcription.

They are different in three ways:

(i) They recognize different effector molecules (i.e., lac repressor recognizes allolactose and the trp repressor recognizes **tryptophan**).

(ii) Allolactose causes the lac repressor to release from the operator while tryptophan causes trp repressor to bind to its operator.

(iii) The sequences of the operator sites that these two proteins recognize are different from each other.

Problem 55

In a molecular genetics laboratory scientists have identified mutations in the promoter region of lac I gene that makes it more difficult for the lac operon to be induced. These are called IQ mutants because a greater quantity of lac repressor protein is made. Explain why an increased transcription of lac 1 gene makes it more difficult to induce the lac operon.

Solution

(i) An increase in the lac repressor protein makes it easier to repress the lac operon.

(ii) When the cells become exposed to lactose, allolactose levels slowly rise. Some of the allolactose bind to the lac repressor protein and cause it to be released from the operator site.

(iii) If many more lac repressor proteins are found within the cell, more allolactose is needed to ensure that no unoccupied repressor proteins can repress the operon.

Problem 56

Some mutations have a cis-effect whereas others have a trans effect. Explain the molecular difference between cis- and trans-mutation. Which type of mutator (cis or trans) can be complemented on a mesozygote experiment?

Solution

(i) A mutation that has a cis-effect is within a genetic regulatory sequence. As for example, an operator site that affects the binding of a genetic regulations protein.

(ii) A cis-effect mutation affects only the adjacent genes that the genetic regulations sequence controls.

(iii) A mutation having trans-effect is usually in a gene that encodes a genetic regulatory protein.

(iv) A trans-effect mutator can be complemented in a mesozygote experiment by the introduction of a normal gene that encodes a regulatory protein.

Problem 57

What is diauxic growth? Explain the roles of AMP and the catabolic activator protein in this process.

Solution

(i) The phenomenon in which a cell first uses up one type of sugar (such as glucose) before it begins to metabolize a second sugar (such as lactose) is termed as **diauxic growth.**

(ii) In this case, this phenomenon is regulated by gene, when bacterial cell is exposed to both sugars the uptake of glucose causes the cAMP levels in the cell to fall.

(iii) When this occurs the catabolite activator protein (CAP) is removed from the operon, so it is not able to be activated as CAP.

Problem 58

The sequential uses of two sugars by a bacterium is known as diauxic growth. It is a common phenomenon among many bacterial species (a) Less glucose of the two such sugars as available, it is typical that the bacterium metabolizes glucose first, and then a second sugar after glucose has been used. Among *E. coli* and related species, diauxic growth is regulated by intracellular cAMP levels and the catabolic activator protein (CAP). Summarize the effects of glucose and lactose on the ability of the lac represents cAMP-CAP complex to regulate lac operon.

Solution

(i) In the absence of lactose, the lac repressor has the dominating effect of shutting off the lac operon.

(ii) Even when glucose is also absent and the cAMP-CAP complex is formed, the presence of bound lac repressor prevents the expression of lactose.

(iii) The effects of cAMP-CAP complex are exerted only in the presence of lactose. When lactose is present, glucose is absent, the cAMP-CAP complex acts to enhance the rule of transcription.

(iv) However, when both lactose and glucose are present the inability of CAP to bind to the lac operon decreases the rate of transcription.

Summarise the effect

Sugar present	Transcription of the lac operon
1. None	1. The operon is turned off due to the dominating effect of the lac repressor protein.
2. Lactose	2. (i) The operon is maximally turned on. (ii) The repressor is removed from the operator site and the cAMP-CAP complex is bend to the CAP site.
3. Glucose	3. The operon is turned off due to the dominating effect of the lac repressor protein.
4. Lactose	4. (i) The expression of the lac operon is greatly decreased.

(ii) The lac repressor is removed from the operator site and the catabolic activator protein is not bound to the CAP site.

(iii) The absent of CAP at the CAP site makes difficult for RNA polymerase to begin transcription.

(iv) However, a little more transcription occurs under three conditions than in the absence of lactose, when the repressor is bound.

Problem 59

Explain how a mutation could affect the regulation of ara operon of the mutation prevented ara C protein from binding to the following sites: (A) ara O_2, (B) ara O, (C) ara I, (D) ara O_2 and (E) ara I.

Solution

A. (i) Without ara O_2, the repression of the ara operon could not occur. The operon would be constitutively expressed at high levels, because ara C protein could still activate transcription of the ara operon by binding to ara I.

(ii) The presence of arabinose would have no effect. The binding of arabinose to ara C is not needed to form an ara C dimer at ara I.

(iii) The dimer is able to form because the loop has been broken. This point may be figured out if an ara C dimer is bound to ara O_I in the presence or absence of arabinose.

B. (i) Without ara O_I the ara C protein would be overexpressed. It would probably require more arabinose to alleviate repression.

(ii) In addition, activation might be higher because there would be more ara C protein available.

C. (i) Without ara I, transcription of the ara operon cannot be activated.

(ii) You might get a very low level of constitutive transcription.

Problem 60

If a wild type *E. Coli* strain is grown in a medium without lactose or glucose, how many proteins are bound to lac operon? How many are bound if the cells are grown in glucose?

Solution

(i) In absence of lactose or glucose, two proteins are bound—lac repressor and CRP-cAMP.

(ii) In the presence of glucose, only the repressor is bound.

Trp Operon

Regulator: (i) It is not closely linked to the trp operon.

(ii) It encodes trp repressor (inactive, i.e., aporepressor).

Promoter: (i) It lies close to the promoter and known as Primary promoter (P_1).

(ii) There is also a week promoter lies operator end distal of the trp D structural gene.

Operator: It lies within the primary promoter (P_1) region.

Structural gene: Five contiguous structural genes (trp E, D, C, B & A) encoding enzymes involved in the biosynthesis of amino acid tryptophan.

Leader sequence: (i) It consists of 162 bp prior to the first structural gene E.

(ii) It has four regions, region 1 has the codon for tryptophan, region 2, 3 & 4 regulate the mRNA synthesis.

Transcription terminator: Transcription terminator sequence (t_1 and t^1) are located down stream from trp A.

Genetic Engineering and Biotechnology 16

- **RECOMBINANT DNA TECHNOLOGY:**

It is a method which involves in splicing of DNA fragment from one organism and to insert this spliced DNA into another unrelated organism, so as to obtain the expression of a new character (recombinant DNA) and it is then cloned.

This term now refers to methodologists that are used to cut, join, recombine and graft DNA.

- **Steps:**
  - (i) Selection and isolation of target DNA commonly referred to as **foreign DNA** or **passenger DNA**. At the same time appropriate vector would be selected to carry the donor DNA.
  - (ii) Appropriate **restriction endonuclease** would be selected by which donor DNA and vector DNA would be spliced and the spliced DNA be incorporated into the vector's DNA.
  - (iii) The donor DNA bearing vector is known as **recombinant** or **chimeric vector** (DNA).
  - (iv) Transformation of host cell, i.e., transfer and maintenance of this recombinant DNA into the host cell (bacteria).
  - (v) Identification of transformant cell and their selection forms non-transformant.
  - (vi) Amplification or cloning of recombinant DNA to get multiple copies in a cell.
  - (vii) Cloned DNA is isolated from the cloning vector. DNA is purified and then used for specific purposes.

- **TRANSPOSABLE ELEMENT — MOBILE ELEMENT OR JUMPING GENES:**

The DNA sequences capable of moving from one place to another on the same or different chromosome of a genotype, without any RNA intermediate and sometime causing a change in DNA function when they insert themselves around the genome into all uninfected chromosome, are termed as **mobile element, IS element** or **jumping genes.**

- **CHARACTERISTICS OF TRANSPOSABLE ELEMENT:**
  - (i) They are DNA sequences which code for enzymes that bring about insertion of an identical copy of themselves into new DNA site.
  - (ii) **Transposable elements** are recognized by the presence of **inverted repeat** DNA sequences at their ends.
  - (iii) **Transposition** events involve both **recombination** and **replication** processes which produce two daughter copies of the original transposable elements. One copy remains at original site and another appears at the target site.
  - (iv) The insertion of transposable element always **disrupt** the integrety of the target gene.
  - (v) Transposable elements sometimes **activate** dormant gene because they carry the signal of initiation of transcription.

- **MECHANISM OF TRANSPOSITION:**

Transposons mobilize directly as DNA whereas retroelements employ an RNA intermediate.

- **Steps of Transposition**
  - → Proliferation
  - → Stages of transposition
  - → Target site duplication
  - → Target preference

A. Proliferation:

Prior to transposition, transposable elements have the potential power to increase their number or copies actively or passively.

B. Stages of transposition:

(a) **Donar cut stage**: Enzyme **transposase** catalyze endolytic cleavage reaction that separates the transposable elements from the host DNA.

(b) **Strand exchange stage: Transposase** catalyze a pair of trans esterification reactions which join the 3′ ends of transposable element to host DNA at the target site.

(c) **Repair stage:**

(i) DNA synthesis undertaken by the cell which fills any remaining gaps.

(ii) **Retro transposons** precede these three stages by transcription and reverse transcription to generate a **cDNA** for processing and integration.

(iii) **Nonviral retroelements** integrate into host DNA through repair stage of **transposition.**

C. Target site duplication:

(a) The sites into which **transposable elements** insert are called target sites. First a staggered cut is made in the target site and transposable element is then inserted and joined to the single-stranded ends.

(b) The gaps are repaired by DNA polymerase and DNA ligase with two direct repeats of the sequence at the target site. Direct in this case means that the two sequences are repeated in the same orientation. The direct repeats are called **target site duplication.**

D. Target preference:

Transposable elements vary in their preference for target sites. Some integrate at specific sequences and others appear to integrate at random.

- **Ac/Ds ELEMENT:**

1. Barbara McClintock discovered two controlling elements in the kernel colouration of maize. The deep purple colour is due to a factor encoded by the allele C which can synthesise purple pigment anthocyanin.

2. If C gene mutates no pigment synthesis and kernel appears white. Insertion of Ds element into C results in the mutations that prevent pigment synthesis. Mutation in C locus is unstable and can revert to wild type. This reversion is caused by Ds trasposon out of C with the help of Ac.

3. **Ac** (activator) is the regulatory component as well as autonomous element which is capable of transposition by itself since it specifics enzyme transposase located on chromosome 9 in maize.

   On the other hand **Ds** (dissociator) is receptor component and non-autonomous element which cannot transpose by itself. Since it is an aberrant version of Ac element and cannot specify transposase.

4. Transposition of Ds is dependent on the presence of active Ac element. The term dissociation refers to the ability of particular Ds elements to induce chromosomal breaks at the insertion site only in the presence of Ac element.

5. Ac and Ds belong to the family transposons. Multiple copies of Ac and Ds elements are often found in maize.

6. Ac elements, 4563 bp long encode a single mRNA of 807 codons (encode transposase) and are bonded by 11 bp inverted repeats at both ends. Ac elements generate 8 bp direct repeats at their site of insertion. Ds elements are formed with a varying degree of deletion (194 bp to 2.5 kb) of transpose gene of Ac elements.

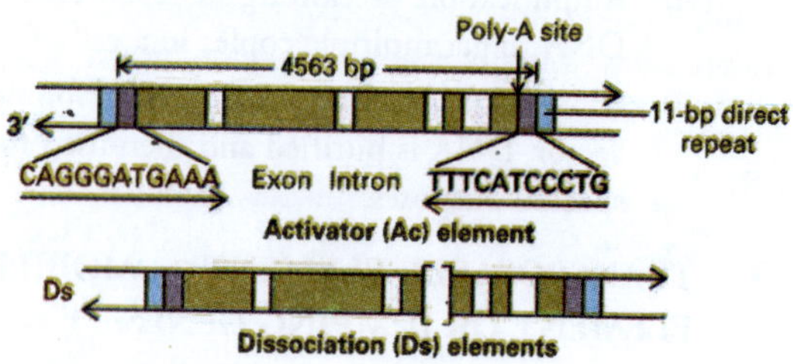

Structure of Ac/Ds Elements

- ***COPIA*-LIKE ELEMENTS:**

1. *Drosophila melanogaster* contains middle DNA sequences with transposon-like structure similar to that of yeast Ty element. They are called **copia**, because they are present in abundant (**copius**) quantities.
2. The ***Copia*-like elements** constitute at least seven families, ranging in size from 5 to 8.5 kb. Members of each family appear at 10–100 positions in the *Drosophila* genome.
3. Each member carries a long, direct terminal repeat (DTR) of 267 bp and a short imperfect inverted repeat (IR) of 17 bp and has a long 4227 bp ORF and shows homologies to the gag and pol genes of retroviruses.
4. *Copia*-like elements transcribe abundantly poly $(A)^+$ mRNAs.

For example, the white-apricot (w^{ap}) mutation for eye colour is caused by the insertion of an element for the *copia* family into the white locus.

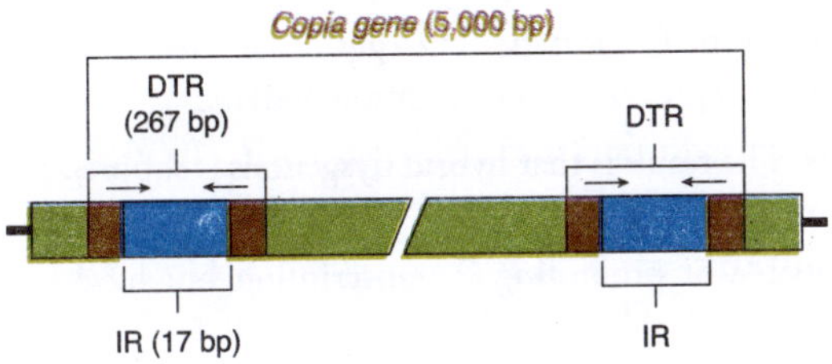

Fig. 16.1: Structure of *Copia* elements

- **P-ELEMENTS:**

It is a type of transposable genetic element in *Drosophila* used as the basis of cloning vector as tool for the insertional mutagenesis for germ line transformation and other types of genetic engineering, which is responsible for **hybrid dysgenesis.**

Features

(i) The P-element is an insertion form of classic transposable element. Individual elements vary in length but are homologous in sequence.

(ii) The largest elements are 2907 nucleotide pairs ~2.9 kb long including terminal repeats of 31 nucleotide pairs.

(iii) The complete **P-elements** have four ORF (open reading frame) or exons interrupted by three intorns. It posseses terminal inverted repeats (IRs) of 31 bp and are bounded by host protein. The subterminal inverted repeats are bounded by the **P-element transposase.**

(iv) On transposition it generates 8 bp direct repeats of target DNA on the outer side of the inverted repeats.

(v) It encodes a **transposase** and a repressor of transposition by alternate splicing.

(vi) The primary transcripts are of 2.5 kb (short RNA) or 3.0 kb (long RNA) depending on the termination site.

- **In somatic tissue:**

(i) Intron 1 and 2 are excised but intron 3 remains.

(ii) This creates a coding region of ORF 0 – ORF 1 – ORF 2 which terminates at intron 3.

(iii) Trnaslation of this RNA yield a protein of 66 kD or kDa.

(iv) This protein is a **repressor** of transposon activity.

- **In germ line tissue:**

(i) Additional splicing event occurs to remove intron 3.

(ii) This connect all open reading frames (ORF) into an mRNA.

(iii) The transcribe mRNA is translated into an 87 kD or kDa protein.

(iv) This protein is the **transposase.**

(v) Presence only in germ line.

Genetic cross: Hybrid dysgenesis

In ***Drosophila melanogaster***, a syndrome of abnormal germ line traits, including high mutation rate, high frequency of chromosomal aberrations, distorted segregation at meiosis or non-disjunction and sterility which results from transposable element activity. The phenomenon by which such abnormal hybrids are produced is called **hybrid dysgenesis.**

- **P. male × P. female cross**

(i) In this cross both male and female parents chromosomes contain P elements.

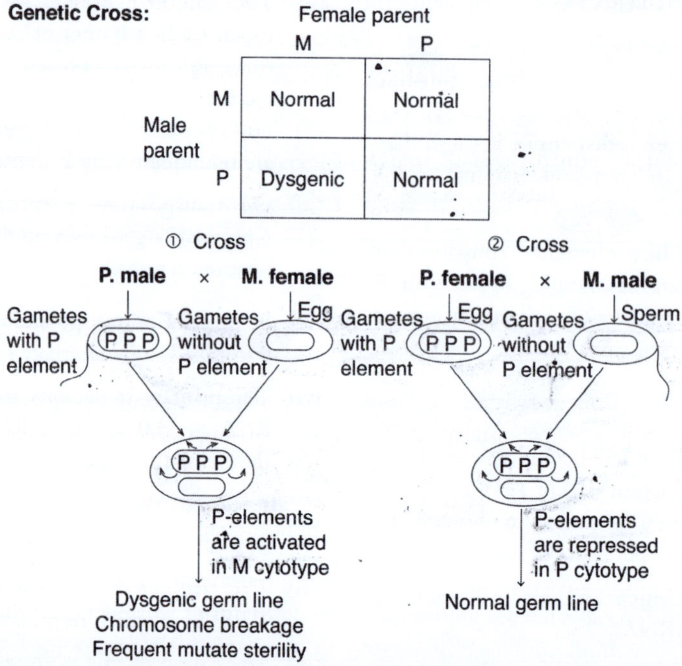

Fig. 16.2: *P* element-mediated hybrid dysgenesis in *Drosophila*, Cytotype is inherited maternally. The P cytotype represses *P* element movement, whereas the M cytotype permits it. Hybrid dysgenesis is asymmetrical; it is induced by P male × M female crosses but not by M male × P female crosses.

(ii) The presence of P elements in the female parent results in the **P. cytotype.** The eggs of such female contain a protein maternal factor in the form of the 66 kDa repressor of transposition activity.

(iii) This repressor prevents transposition of all P elements. The result is that **hybrid dysgenesis** is suppressed for several generations until the repressor is diluted out.

(iv) Thus, P × P crosses produces offspring that are capable of promoting P transcription but have no P-elements to transpose.

P male × P female cross — Repressor prevents transposition of all *P* elements

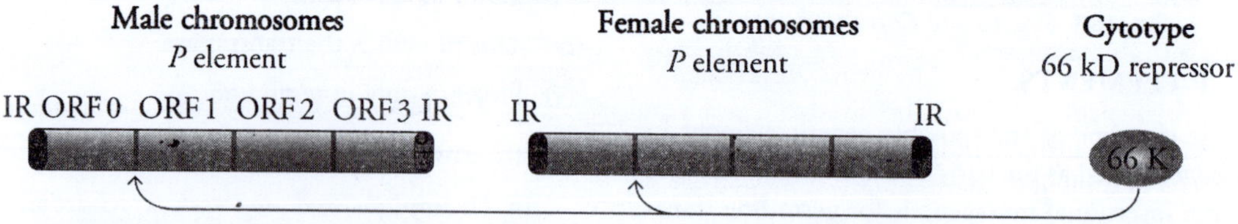

- **P. male × M. female cross:**

(i) In this type of cross the P-element is present in the male parent but absent in the female parent.

(ii) The absence of P-element in female parent gives in the M cytotype in which no repressor protein is found in the egg.

(iii) This permits the P-element in the male parent to encode 87 kDa transpose proteins. As a result transposition of the P-elements occurs, and the hybrid offsprings show pronounced **hybrid dysgenesis.**

(iv) Thus, only when the P-element from the male parent and the M cytotype from the female are brought together, transposition can occur and results **hybrid dysgenesis.**

P male × M female cross — *P* element synthesizes transposase → hybrid dysgenesis →

Male chromosomes	**Female chromosomes**	**Cytotype**
P-element	No P-elements	Nothing

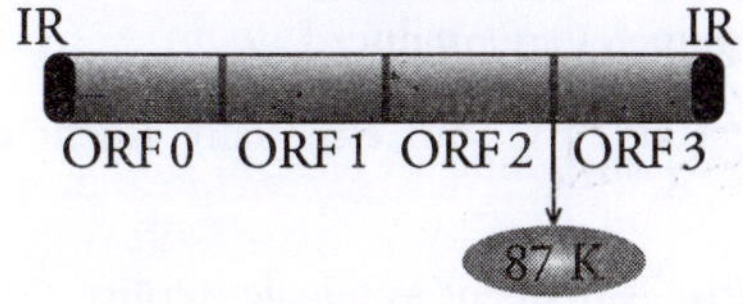

- **M. male × P. female cross:**

(i) In this reciprocal or reverse cross, no P-elements are present in the male parent, but the female parent has P-element.

(ii) The presence of P-elements gives the female parent *P* cytotype. The eggs, therefore, contain repressor proteins (66 kD) which prevent transposition of the P-element and therefore, no transposase is produced.

(iii) Thus no transposition of the P-element takes place and the hybrid dysgenesis is greatly reduced.

M male × P female cross—Repressor prevents transposition of all *P* elements

Male chromosomes	**Female chromosomes**	**Cytotype**
No P-elements	P-element	66 kD repressor

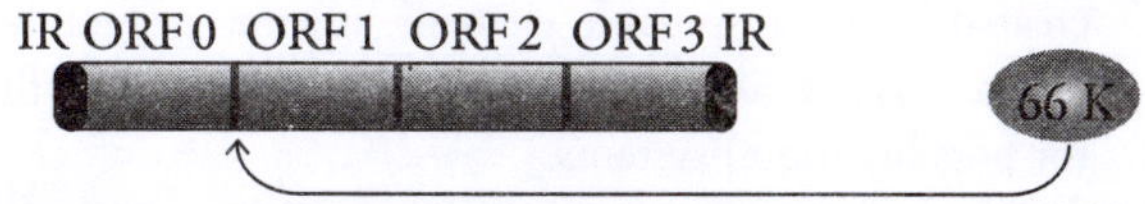

P element mediated transformation of *Drosophila*

1. P element mediated transformation is one of the most efficient methods for introducing cloned genes into higher eukaryotes. *Drosophila* geneticists have harnessed P element as vector to introduce cloned gene into *Drosophila* genome.
2. The gene of interest is cloned into the transformation plasmid, which also contains a dominant gene for eye colour (e.g., wild type red eye). The genes are flanked by P element ends (31 bp inverted repeats). The transformation plasmid and the helper plasmid containing the tranposase gene are co-injected into fertilized eggs from a homozygous recessive mutant white-eyed female.
3. The P element sequences along with the two trans genes insert into the DNA of the embryo's germ cells and the gene is transcribed and translated. Progeny of this fly will express the eye colour marker (red), if the transformation plasmid inserted into the parent's germ cells.

1. Clone gene into transformation plasmid — Red eye gene + Transposase gene (Helper plasmid) + Gene of interest + P element 31 bp repeat

↓

2. Inject plasmids into *Drosophila* embryo

↓

Embryo develops into white-eyed adult

↓

3. Mate progeny with white eyed flies

↓

White eyed ♂ × White eyed ♀

↓

Red eyed progeny contain transgene

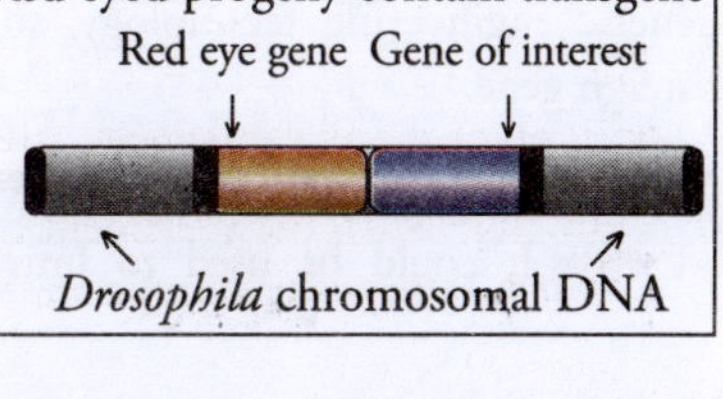

Features of Hybrid Dysgenesis:

(i) In males and females affected by **GD (gonadial dysgenic) sterility**, the organs derived from somatic tissues are normal, but the **gonads** are **rudimentary** and no gametes are produced.

(ii) **GD sterility** is highly **temperature** dependent. It is highest at 29°C, but essentially absent at 24°C.

(iii) In a particular strain, **GD sterility** in males is only one fourth as frequent as female sterility. GD sterility is probably caused by loss of germ line cells.

(iv) **High mutability** is associated with the gonadal dysgenesis. The mutation includes those caused by insertion or excision of transposable genetic elements and chromosomal rearrangements.

(v) **Chromosomal rearrangements** may be as high as 10% of the X chromosomes in gametes of dysgenic hybrids. The most common types are **insertions** and **reciprocal translocation** involving two breakpoints in transposable elements.

- **Significance:**

(i) **Hybrid dysgenesis** reduces interbreeding and it is a step on the path to speciation; consequently they will no longer be able to interpret and will become separate species. If dysgenic system is created by a transposable element in some geographic location, another element may create a different system in some other location. As far as no flies in the two areas will be dysgenic for two (or possibly more) systems.

(ii) P and M strains of *Drosophiula* frequently produce sterile offspring.

(iii) Due to the prolonged lack of mixing of their (P & M strain genes, they will tend to become genetically isolated.

(iv) If this tenders them intersterile and the populations become genetically isolated, further separation may occur.

(v) Multiple dysgenic systems, therefore, lead to inability to mate and to speciation.

- **TRANSGENE:**

A gene (exogenous) introduced in the genome of an organism by genetic manipulation in order to alter its genotype is known as transgene.

- **Transgenic organism (Transgenic means transfer of exogenous DNA in the genome):**

An organism (Animal/Plant) whose genome has been changed by the introduction of different genes of the same species or different species with genetic engineering technology, so as to contain foreign gene.

- **Practical application of transgenic technology:**

(i) It could be used to introduce new and sometimes novel genes into various organism.

(ii) To inactivate genes in order to learn more about their normal function in embryonic development, tissue differentiation and development of cancer.

- **Potential disadvantages of this technique:**

(i) If the engineered gene is essential for normal embryonic development, homozygous offspring may die before birth.

(ii) Due to host specific differences, the modified cells may fail to be incorporated into developing embryo.

- **Methods of transgenesis:**

Transgenesis

→ **Microinjection** of DNA into the fertilized egg.
Example: cow, sheep, goat, fish.

→ **Infection** of preimplantation of embryos with **retroviral** vectors.
Example: mice.

→ **Transfection** of totipotent embryonic stem cells.
Example: mice.

- **Transfection:**

1. The uptake of DNA by eukaryotic cell followed by the incorporation of genetic markers present in the DNA into the cell's genome.
2. It may be
   (a) **Transient**, i.e. the introduced gene(s) is (are) gradually lost from the daughter cells of transfected cells.
   (b) **Stable**, i.e. the introduced gene(s) is (are) retained and expressed in all the cells derived from the transfected cells.

- **CHROMOSOME PAINTING:**

This is a technique for visualizing each of the chromosome in distinct bright colours by the use of differentially labelled chromosome specific DNA strands for hybridization with chromosome to label each chromosome with a different colour.

- **Characteristics:**

(i) It simplifies greatly the distinction between the chromosomes of similar shape, size and karyotyping of the chromosome.
(ii) Chromosome painting is done at the stage of cell cycle when chromosomes are specially compact and easy to visualize.
(iii) Sometime selective chromosome painting is done in the interphase.
(iv) The probes which are used for chromosome painting are specific for sites scattered along the length of each chromosome.
(v) The probes are labelled with one of the two dyes that fluorescence at different wave length.
(vi) The labelled probe can hybridize only the chromosome from which it was derived, each chromosome is differently labelled.

- **Application:**

1. It is useful in karyotyping the chromosome.
2. It can be done in interphase stage to the specific chromosome and its arrangements.
3. The multicolour FISH (combination of chromosome banding with FISH) can detect chromosomal translocation which are associated with genetic disorder by **CML, Philadelphia chromosome.**

- **DNA FOOT PRINTING:**

It means finding the target DNA sequence or binding site of DNA binding protein.

It is a technique to determine the length of the DNA sequence that binds to a particular protein, based on the idea that phosphodiester bonds in the region covered by the protein are protected from degradation by **DNase**. The resulting (remaining or protected) DNA fragments are electrophorised, the protein-binding site shows up a **gap** or **foot print** in the pattern where the protein protected the DNA from degradation.

- **DMS FOOT PRINTING:**

In this method, the DNA methylating agent **DMS** (Dimethyl sulphate) instead of **DNase** is used to attack the DNA-protein complex. The DNA is then broken at the methylated sites. Unmethylated or hypermethylated sites show up on electrophoresis of the labelled DNA fragments and demonstrate where the protein bound to the DNA.

- **SELEX (SYSTEMIC EVOLUTION OF LIGANDS BY EXPONENTIAL ENRICHMENT):**

It is a method for enriching nucleic acid, usually RNAs containing a aptamer (functional region). Functional molecules are selected by a method like affinity chromatography, then amplified by PCR, then selected and amplified again and again effecting the exponential enrichment.

- **FLUORESCENT IN SITU HYBRIDIZATION (FISH):**

A technique of in situ hybridization in which a probe labelled with fluorescent dye [**Fluorescein isothiocyanate** (FITC)] binds to a specific site on a chromosome and makes itself visible by its fluorescence.

- **Characteristics:**
  - (i) A sample of dividing cells is dried on a microscope slide and treated with formaldehyde so that the chromosomes become denatured but do not lose their characteristic metaphase morphologies.
  - (ii) A probe labelled with fluorescent tag is added to the denatured chromosome, which will be hybridized with the complementary region of the chromosome.
  - (iii) The position at which the probe hybridizes to the chromosomal DNA is visualized by detecting the fluorescent signal emitted by the labelled DNA.
- **Advantages:**
  - (i) All clones can be mapped by FISH protocol.
  - (ii) FISH is great enough to detect a single gene within an entire set of chromosome.
  - (iii) It requires only a single sample on a microscope slide.
  - (iv) FISH protocol enables direct comparisons between the map positions of cloned loci and aberrations.
- **Disadvantages:**
  - (i) Its resolving power is less.
  - (ii) The metaphase bands and interbands used as guide posts in FISH protocols containing the equivalent of 4 to 8 Mb or 5 to 10 cM of chromosomal material.
  - (iii) FISH provides a rough idea of its map position.
- **Polymerase Chain Reaction (PCR):**

It is a rapid versatile in vitro method for amplifying single molecule of DNA, involving repeated cycle of denaturation, hybridization to oligonucleotide primers and synthesis of polynucleotide by the action of **taq polymerase** at high temperature.

- **Requirements of PCR:**
  1. DNA template
  2. DNA polymerase: The widely used heat stable DNA polymerase is **Taq polymerase** (obtained from *Thermus aquaticus*).
  3. Primers: Two oligonucleotides.
  4. Deoxy nucleotide triphosphate: d ATP, d TTP, d CTP and d GTP.
  5. Thermal cycler: Reaction is done in a thermal cycler and the process is automated by computer device.

Thermostable DNA polymerase	
Polymerase	***Source***
• Plaque DNA polymerase	*Pyrococcus furiosus*
• PWO DNA polymerase	*Pyrococcus woeesi*
• Vent polymerase	*Thermococcus litoralis*
• Tma polymerase	*Thermologa maritima*

- **Steps involved in PCR:**
  1. Isolation and purification of DNA segment to be amplified.
  2. Cyclical amplification of DNA
     - (a) Denaturation
     - (b) Annealing
     - (c) Chain extension
  3. Screening of amplified DNA.

Types of PCR:

(a) **Reverse transcriptase PCR (RT PCR)**

Here an mRNA transcript of a gene is used and from this a number of copies of DNA can be produced.

(b) **Inverse PCR**

- (i) It is an approach that uses incomplete sequences information to amplify a target gene.
- (ii) In this case, a sequence of a part of a long DNA molecule is to be known.

(c) **Quantitative Real time PCR (Q RT PCR)**

It is used to amplify and simultaneous quantify a targated DNA molecule.

(d) **Arbiratory Primed PCR (AP-PCR)**

It is a type of RAPD where single primer of 10 to 15 bases are used to amplify genomic DNA in PCR.

- **RACE (Rapid Amplification of cDNA ends):**

A modified RT-PCR based technique for mapping and identifying the 5′ and 3′ end of an RNA molecule.

- **Multiplex PCR:**

A PCR carried out with more than one pair of primers and hence targeting two or more sites in the DNA being studied.

- **Repetitive DNA PCR:**

A clone finger printing technique that uses PCR to detect the relative positions of repeated sequences in cloned fragments.

- **Reverse Transcriptase PCR (RT-PCR):**

A two-step method for detecting and quantifying a particular RNA in an RNA mixture by first converting the RNAs to cDNA by reverse transcriptase and then performing polymerase chain reaction using primers specific for the RNA of interest.

(i) In the first step, cDNA is synthesized from RNA using a reverse primer [oligo (dT) for example mRNA] and enzyme reverse transcriptase (RT). Then a forward primer is used to make ss DNA to ds DNA.

(ii) In the second step, the specific cDNA is amplified by PCR (Tth DNA polymerase is used).

- **Properties of RT-PCR enzyme:**

Tth DNA polymerase is isolated from *Thermus thermophilus* (Tth) bacterium.

(i) It can reversely transcribe DNA from RNA.

(ii) It can multiply DNA into DNAs by its polymerase activity.

(iii) It can withstand a temperature of 98°C like **Taq polymerase.**

- **RT–PCR (Real time PCR) or Real time RT–PCR:**

It is a technique of quantifying the amplification of a specific region of a DNA molecule.

(i) In the first step reverse primer and reverse transcriptase are used. Then a forward primer is used to make it double-stranded.

(ii) In the next step, DNA amplification is done in presence of highly sensitive fluorescent dye **SYBR green** which binds to double-stranded DNA which emits 1000 fold greater fluorescence.

(iii) As the DNA is amplified by PCR, the staining is measured in real time (hence, the name of the technique) by means of special thermal cycler that uses laser detection of the fluorescence.

(iv) The amount of experimental mRNA is quantified and is measured by comparing the rate of production of stained amplified DNA with the control of known amount of starting mRNA.

Traditional PCR	Real time PCR
1. Post PCR processing is needed.	1. DNA is quantified during reaction.
2. Low sensitivity.	2. High sensitivity.
3. The range of detection is very low (about 2 log).	3. The range of detection is high (multiple log).
4. Used for semi-quantification.	4. Can be used for quantification.
5. Size of amplified DNA ranges from about 100 bp to 3 kb.	5. Amplification size generally 150-300 bp, small size is best for normal assay.

- **The advantages of quantitative (real time) PCR over traditional PCR?**

(i) Quantitative (real time) PCR is more sensitive than traditional PCR.

(ii) It is quicker; no post PCR analysis of DNA is needed as DNA is detected as it is synthesized.

(iii) DNA produced by real time PCR can be quantified.

(iv) As all reactions occur in a single tube during real time PCR, there is less chance of contamination.

Application of Real time PCR:

1. Powerful tool for basic research and for medical diagnostics.
2. In biological research, the most important use of realtime PCR is the relative and absolute quantification of gene expression under various conditions.
3. It has been effectively used to check drug resistance of HIV.
4. It is extremely useful in detection and identification of bacterial strain.
5. It is also very effectively useful in identification of specific DNA sequence in clinical oncology.

- **Construction of DNA Library:**

(i) To construct a DNA library, target DNA (genomic DNA) and vector DNA are digested with same restriction enzyme and are incubated together with ligase.

(ii) The resulting pool or library of hybrid vectors are then introduced into *E. coli* and then cells are placed onto a filter placed over agar medium.

(iii) Once colonies have grown, the filter is removed from the plate and prepared for hybridization.

(iv) Cells are lysed, the DNA is denatured and the filter is incubated with labelled probe. The clone is identified by autoradiography.

- **Difference between Genomic library and cDNA library**

Genomic library	cDNA library
1. It is made by cloning.	1. It contains DNA fragments derived from cellular mRNA by reverse transcriptase.
2. A genomic clone could contain promoter and regulatory elements as well as introns.	2. A cDNA clone would have no introns or control elements, but if supplied with a bacterial promoter, it can still be expressed.

- **Difference between Chromosome walking and Chromosome jumping**

Chromosome walking	Chromosome jumping
1. A technique for analyzing long stretches of DNA (10–20 kb) by sequentially identifying adjacent, overlapping clones in a genomic library.	1. A technique for isolating clones from a genomic library that are not contiguous but a skip region (many kbp apart, 100 kb or more) between known points in the chromosome.
2. In this technique relatively long segment of a chromosome can be available for studying overlapping clones.	2. This is usually done to bypass regions that are difficult to walk through.

- **Recombinant DNA molecule:**

A DNA molecule constructed in vitro by joining all or parts of two different DNA molecules.

- **Gene cartridges:**

These are purified DNA fragments obtained from plasmid. They contain a central region and a gene for particular protein. *Example*: Ampicillin resistance gene cartridges consists of 2296 bp EcoRI/PvUII fragments of pBR322.

- **Molecular markers:**

(i) The markers (DNA sequence) revealing variations at DNA level and can be used for evaluation of phenotypic variability. (ii) It possesses unique genetic properties.

- **Characteristics of Cloning vectors**

Essential components:

(a) **Origin of replication (ori)**

(i) It must have an **origin of DNA replication** that allows it to replicate in the host.

(ii) Control the copy number of the linked or foreign DNA.

(b) **Selectable marker**

(i) Two dominant selectable marker genes must be present — (a) one conferring antibiotic resistance (tet^R, amp^R, kan^R) or complementing a genetic mutation in the host such that only transformed cells live on a selecting medium (b) Another that distinguishes recombinant molecules from nonrecombinant ones by being inactivated when a fragment is cloned into it.

(c) **Recognition site (cloning sites)**

(i) A unique **restriction endonuclease cleavage site** must be present in one of the marker genes.

(ii) However, many commonly used vectors contain several recognition and cleavage site for several restriction enzyme in a small region. These are called **multiple cloning sites** (MCS) which makes the vector more **versatile.**

General characteristics

(i) Small size, making them easy to manipulate once they are isolated.

(ii) Easy to transfer from cell to cell (usually by transformation).

(iii) Easy to isolate from the host organism.

(iv) Easy to detect and select.

(v) Multiple copies help in obtaining large amounts of DNA.

(vi) Presence of clusters of unique restriction sites called polylinker or polycloning sites or multiple cloning sites (**MCS**).

(vii) Method to detect the presence of inserted DNA (e.g., alpha complementation).

(viii) Two types of vectors are commonly used in cloning.

(ix) One type is derived from bacterial viruses, i.e., phages, other type of vector is derived from bacterial plasmids.

Alpha complementation

(i) It refers to the ability of β-galactosidase to be expressed as two protein fragments which assemble to form a functional protein.

(ii) In a cells without an insert in the plasmid β-galactosidase is active and splits x-gal to form a blue dye.

(iii) In cells with an insert, the alpha fragment is not made and β-galactosidase is inactive. These cells remain white on media with x-gal.

Characteristics of Plasmid—Cloning vector

(i) It can be really isolated from the cells.

(ii) It posseses an *Ori sequence* which controls the replication of the plasmid within the bacteria (*E. coli*).

(iii) It possesses a single restriction site for one or more restriction enzyme(s).

(iv) Presence of at least two selectable marker genes. The common selectable marker genes are the antibiotic resistance genes [ampicillin (amp^R), tetracyclin (Tet^R), kanamycin (kan^R)].

Due to the presence of such marker genes, the bacteria that contain plasmids, can resist the presence of these antibiotics and grow there.

(v) It can be introduced into bacterial cell and cells carrying the plasmid with or without the insert, and can be selected or identified.

(vi) They will not transfer the inserted DNA to the host genome.

(vii) They have replication and amplification capabilities in the host.

(viii) They do not occur free in nature but are found in bacterial cell.

Plasmid Vectors

- **Characteristics of p^{BR32}**

It is the first described artificial cloning vector published in 1977 by F. Bolivar and R. Rodriguez and hence pBR.

(i) It is 4.362 kb long and most commonly used vector and regarded as **workhouse of gene cloning laboratory.**

(ii) It contains ampicillin (amp^R) and tetracyclin (tet^R) resistance genes and various (7) restriction sites.

(iii) Three of such restriction sites, Pst-1, Pvu-1 and Sca-1 in the amp^R region, other three Hind III Bam HI and sal-1 in the tet^R region and the last one ECO RI which is present outside the selectable marker region. Thus, six of the sites which lie in the antibiotic resistant regions (If any foreign DNA fragment is ligated any of these sites, then the antibiotic resistance property becomes lost, it is known as **insertional inactivation**).

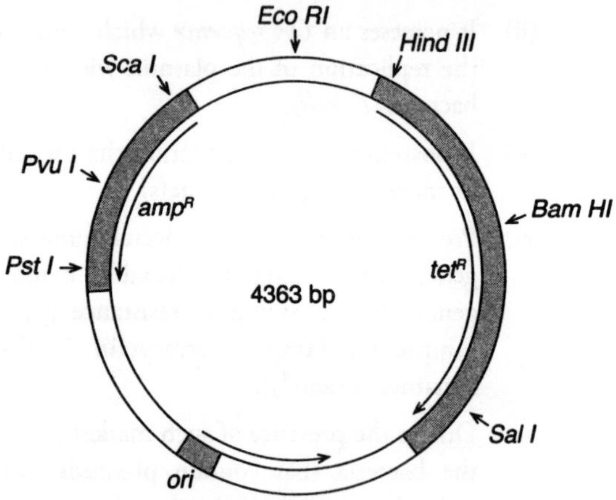

Fig. 16.3: The map shows the positions of the amp^R, tet^R and origin of replication and recognition sites in P^{BR322}.

Characteristics of $p^{UC\ 19}$ plasmid:

This is a phagemids vectors which contain component of both phage chromosomes and plasmids.

p^{UC} for plasmid and University of California where they were constructed.

(i) It is 2.69 kb long with high copy number nearly 100 copies.

(ii) It consists of ColE1 ori (origin of replication) a selectable marker amp^R gene (ampicillin resistance), lac Z gene (for β galactosidase) and lac I (for production of repressor molecule).

(iii) It has a number of unique restriction sites clustered in one region called a poly-linker or multiple cloning site (MCS).

(iv) In pUC 19, MCS is incorporated into lac Z′ gene without interfering the function of other gene.

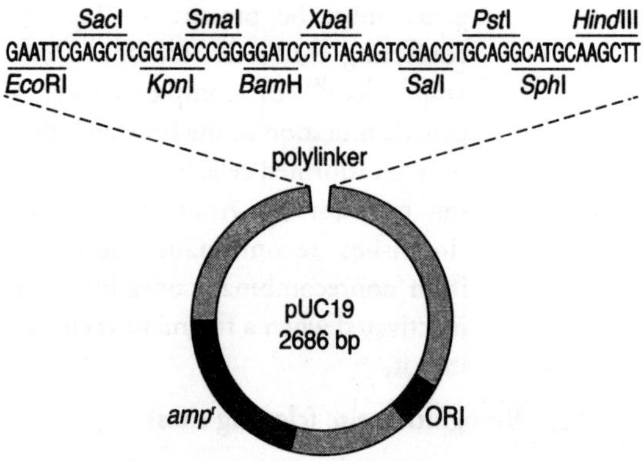

Fig. 16.4: Genetic map of the plasmid cloning vector pUC 19. It contains a polylinker which can recognize several different restriction enzymes, an ampicillin-resistance gene (*amp*^R^) for selective amplification, and a replication origin (*ori*) for proliferation in the host cell.

(iv) The MCS has the unique-sites for several restriction endonucleases such as Eco RI, Sac I, Kpn I, Xma I; Sma I, Bam HI, Xba I, Sal I, Hind II, Acc II, Bsp MI, Pst I, Hind III.

(v) When the cells carrying unchanged pUC19 are grown in the presence of IPTG, (isopropylthiogalactoside) the repressor of lac I gene cannot bind to promoter operator of lac Z gene. Therefore, lac Z gene encoded in the plasmid is transcribed.

(vi) The lac Z gene protein combines with chromosomal DNA transcribed protein and forms hybrid β-galactosidase.

(vii) The MCS tagged in the lac Z gene of pUC19 plasmid which does not interfere the production of hybrid β-galactosidase.

(viii) On the other hand, if a substrate 5-bromo-4 chlorionolye-β galactosidase (x-gal) is present in the medium the hybrid β galactosidase hydrolyses it to produce a blue product. Therefore, colonies contain unmodified pUC19 appear blue in colour.

(ix) When the DNA insert is cloned in lac Z gene, it will inactivated and will not express β galactosidase.

(x) Consequently the colonies formed by *E. coli* cells containing recombinant plasmid will be white in colour.

This plasmid vector has been designed for blue white screening.

- **Characteristics of Cosmid vector:**

Cosmids are hybrid or recombinant plasmid vectors derived from plasmids which contain cohesive cos site of phage λ (for λ cos site + plasmid). It was developed by Collins and Hohn (1978).

(i) It is a small (30–47 kb) multicopy plasmid that carry λ cos site.

(ii) It contains origin of replication (Ori-c).

(iii) It contains ampicillin resistance (amp^R) gene.

(iv) It accepts foreign DNA inserts of 35–45 kb. The DNA of interest is ligated between the two cos sites using restriction enzymes and ligase.

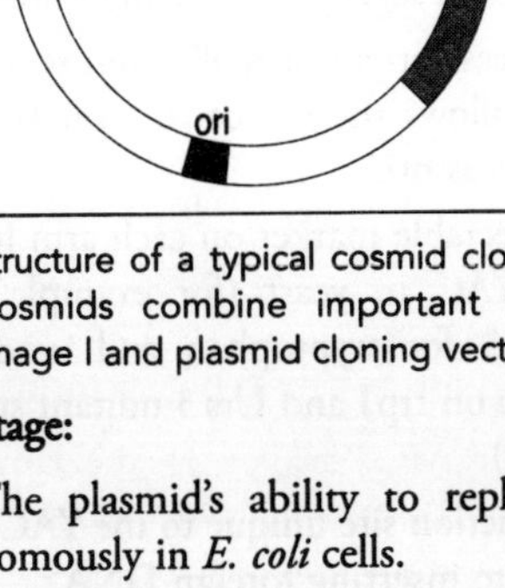

Fig. 16.5: Structure of a typical cosmid cloning vector. Cosmids combine important features of phage l and plasmid cloning vectors.

Key advantage:

1. The plasmid's ability to replicate autonomously in *E. coli* cells.
2. The in vitro packaging capacity of the λ chromosome, which facilitates efficient transformation of *E. coli* cells.
3. The cosmid vector carries the origin of replication, an antibiotic resistance gene of the plasmid and the λ cos site which is needed to package DNA in λ heads.

- **Characteristics of Shuttle vectors:**

Some cloning vectors receive a eukaryotic transcriptional promoter and can be expressed in both *E. coli* and eukaryotic organisms (such as yeast). Since they can shuttle between bacterial and eukaryotic cells, they are called shuttle vectors.

(i) The vector must replicate in many organisms (bacteria, yeast and plants).

(ii) Presence of two Ori sequences (ori E and Ori Euk) so that it can replicate in *E. coli*, i.e., ori^E and in eukaryotic cells (yeast), i.e., Ori^{Euk}.

(iii) Presence of tetracyclin resistance gene (Tet^R) to use as selectable marker in *E. coli* cells.

(iv) Presence of URA 3 gene to act as a selectable marker in yeast.

(v) Presence of Autonomous replication sequence, so that it can replicate in yeast.

(vi) The vector must be stable, non-pathogenic and non-stress-inducing.

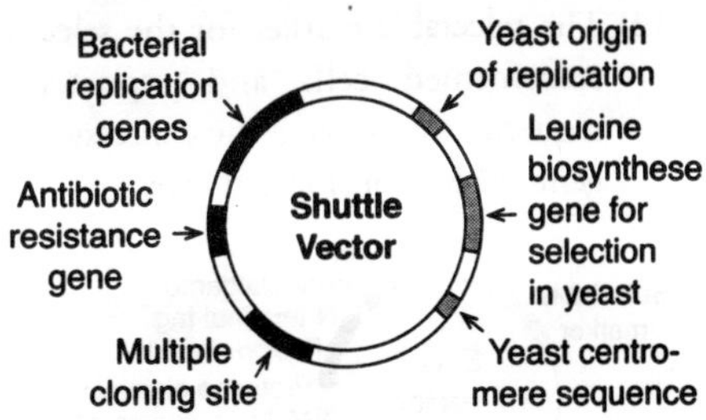

Fig. 16.6: A shuttle vector for yeast.

Expression vector

(i) An expression vector is a cloning vector containing the regulatory sequences (Promoter, i.e., for initiation, structural gene and terminator sequences, i.e., for termination) that allow the transcription and expression signal (initiation site-ribosome binding site and termination signal) for translation of a cloned gene or genes.

(ii) A foreign gene carried by an expression vector into the recipient cell will have complete expression, i.e., the gene will be transcribed into mRNA and then translated into the protein (the gene produce).

(iii) They are responsible for both gene expression and gene cloning.

- **Characteristics of Expression vectors:**

(i) They have unique restriction sites for cloning near its well characterised promoters and its regulatory elements.

(ii) They may have ribosome binding sites and transcription termination signals.

(iii) The desired gene expressed is to be inserted into the restriction sites in the polylinker near the promoter and the gene must be inserted with the end encoding the amino terminus proximal to the promoter.

(iv) (a) The **promoter** would provide efficient transcription of the gene. The **terminator** can improve the amount and stability of the mRNA produced.

(b) The **operator** permits regulation by means of repressor that binds to it.

(c) **Ribosome binding** site provides signal sequences for efficient translation of mRNA.

(d) The **selectable** marker for the selection of transformed cells and reporter gene sequence for assaying the expression of the gene in the transformed cells.

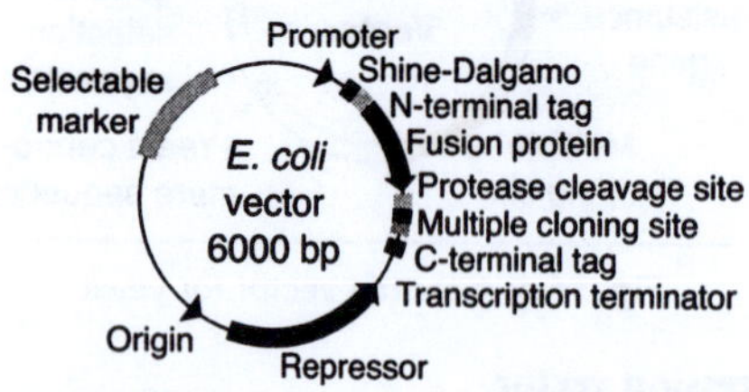

Fig. 16.7: Restriction map of *E. coli* expression vector.

Fosmid Vector:

(i) These are similar to cosmid but are based on bacterial F plasmid.

(ii) The cloning vector is limited, as a host (*E. coli*) can only contain one fosmid molecule.

(iii) Its copy number is very low compares to high copy number in cosmids. Low copy number offers its higher stability.

(iv) It can hold DNA inserts of up to 40 kb in size.

(v) Fosmid clones are used to help assess the accuracy of the Public Human Genome sequence.

Example: The pEpi FOS™-5 is a 7, 518 bp long fosmid cloning vector.

(i) Presence of chloramphenicol-resistance antibiotic resistance marker.

(ii) Presence of bacteriophage λ cos site.

(iii) Presence of bacteriophage loxP site.

(iv) Bacteriophage T-7 RNA polymerase promoter flanking the cloning site.

(v) Eco-72 I (blunt) restriction enzyme recognition site.

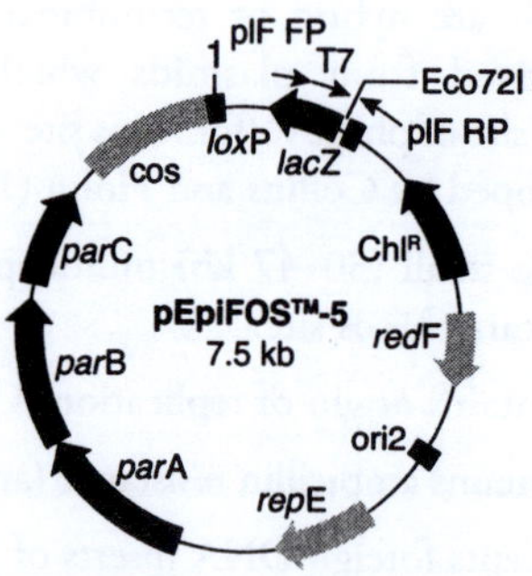

Fig. 16.8: EpiFOS™-5 fosmid vector.

YAC (Yeast artificial chromosomes)

These vectors are propagated in *S. cerevisiae* rather than *E. coli*. It can accommodate foreign DNA inserts of 200–500 kb.

Features

(i) Two yeast telomeres (TEL) one at each end.

(ii) An yeast centromere sequence (CEN) allowing regulated segregation during mitosis.

(iii) An yeast origin of replication sequence (ARS) that allows the vector to replicate in a yeast. It acts as **ori**.

(iv) A selectable marker on each arm for detecting the YAC in yeast (for example TRPI and URA3 for tryptophan and uracil independence on trpI and Ura3 mutant strain respectively).

(v) Restriction site unique to the YAC that can be used to inserting foreign DNA.

It is especially valuable in investigation such as Human Genome protect.

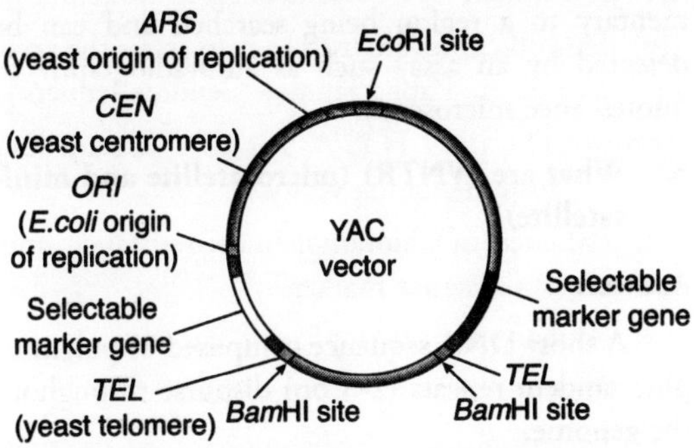

Fig. 16.9: Yeast artificial chromosome vector.

BAC (Bacterial artificial chromosome)

They are based on the naturally occurring F plasmid of *E. coli*. It can accommodate foreign DNA inserts of 150–300 kb.

Features

(i) It contains ori gene, i.e., origin of replication and rep E gene for maintenance of F factor.

(ii) It contains genes (par A, B & C) for plasmid copy number.

(iii) It also accommodates a selective maker, i.e. antibiotic resistance gene, e.g., chloramphenicol acetyle transferase (chl^R) for selection of the plasmid.

(iv) It has multiple cloning site (restriction site) for the insertion of foreign DNA.

Advantage:

F. factor origin of replication keeps the copy number of plasmid one per cell.

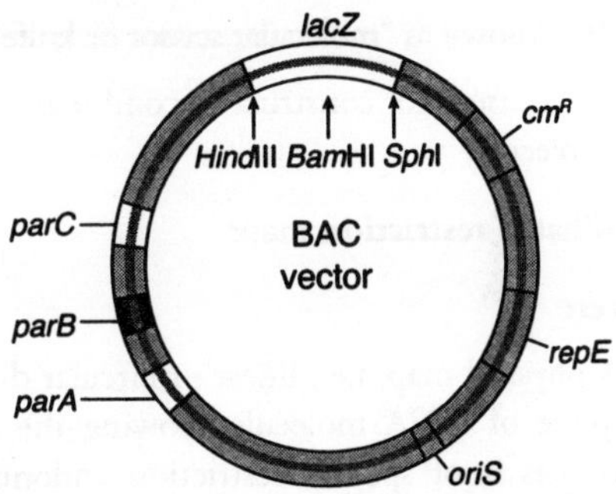

Fig. 16.10: Bacterial artificial chromosome vector.

Mammalian Artificial Chromosome (MAC) Vectors

Features:

(i) Mammalian artificial chromosome (MAC) and human artificial chromosome (HAC): these vectors are linear vectors that contain centromeres, telomeres, origin of replication recognition sequences and transcriptionally active chromosomal domains.

(ii) These vectors resemble microchromosomes, mitotically and cytogenetically stable in the absence of any selection, bind centromere proteins specific for active centromeres, *viz.*, CENP-B, CENP-C and CENP-E, and are about 6-10 megabase in size (~10–20% the size of human chromosomes).

(iii) MACs and HACs can be constructed in cultured cells by artificially changing the structure of natural chromosomes.

(iv) Alternatively, cloned chromosome elements may be transferred into cultured cells leading to the *de novo* formation of MACs and HACs.

Strategies for constructing MAC/HAC

1. Long synthetic arrays of alpha satellite DNA is combined with telomeric DNA and genomic DNA.
2. Telomere repeats and selectable markers are integrated into a YAC that has human centromeric DNA, including centromere protein B (CENP-B) boxes.
3. PAC vector is used to obtain high concentrations of centromere, telomere origins of replication and structural gene sequences.

Importance:

MACs and HACs are expected to provide a strategy for an efficient and predictable production of transgenic animals, and to a safe and efficient somatic gene therapy in humans.

Question Answers

1. What is restriction endonuclease?

Answer:

(i) An endonuclease that recognizes a specific short sequence of DNA, i.e., restriction sites and cleaves the DNA molecule.

It protects or prevents or restricts cell from the viral infection by degrading the invading nucleic acid of virus.

(ii) It is known as "**molecular scissor** or **knife**".

(iii) It is used to construct recombinant DNA molecule.

2. What is restriction map?

Answer:

A physical map, i.e., linear or circular diagram of a piece of DNA molecule showing the recognition sites is of specific restriction endonuclease separated by length marked in numbers of bases.

3. What is RFLP (Restriction Fragment Length Polymorphism)?

Answer:

A genetic polymorphism or variation among individuals that is detected by comparing the length and number of DNA fragments released by digestion with one or more restriction enzymes.

4. Mention the utility of RFLP.

Answer:

1. It is used in forensic study to detect the criminals.
2. The disease caused by gene mutation which creates new restriction site, such as Hbs, β-thalassaemia can be detected by the RFLP map by comparing a diseased person with a healthy person.

5. What is probe?

Answer:

A purified fragment or chemically synthesized DNA or RNA, that has been labelled and complementary to a region being searched and can be detected by an assay such as autoradiography or fluorescence microscopy.

6. What are (VNTR) (microsatellite and minisatellite)?

Answer:

A short DNA sequence composed of polymorphic tandem repeats (2-5 bp) dispurse throughout the genome.

The short repeated DNA sequences (10-100 bp) present as tandem repeats. The numer varies between individual genome.

7. What is the difference between a VNTR and a microsatellite? Why would it be in appropriate to call most VNTRs microsatellites?

Answer:

VNTRs and microsatellites are both polymorphic tandem repeats of relatively short nucleotide sequences. It includes both micro and mini satellite. Microsatellites, however, are repeats of sequences only 2-5 nucleotide pairs in length.

8. Why are Flavr Savr tomatoes preferred over natural tomato?

Answer:

(i) Flavr Savr variety tomatoes are preferred because expression of a native tomato gene has been blocked.

(ii) This gene produces enzyme **polygalacturonase** which acts on pectin and makes the tomatoes mere soft and soggy which also develops bruises.

(iii) The production of this enzyme was reduced in the Flavr Savr transgenic tomato.

(iv) The non-availability of this enzyme prevents over ripening because the enzyme is essential for degradation of cell walls. Thus, fruits remain fresh for a longer period than the fruits of normal variety of tomato.

9. Why is Poly dT an effective primer for reverse transcriptase?

Answer:

(i) Reverse transcriptase is often used to promote the formation of cDNA (complementary DNA) from mRNA molecule.

(ii) Eukaryotic mRNAs typically have a 3′ poly A tail. The poly dT segment provides a double stranded section which serves to prime the production of the complementary strand.

10. What is DNA cloning?

Answer:

The ability to construct recombinant DNA molecules and maintain them in cells is called DNA cloning.

This process typically involves a **vector** that provides the information necessary to propagate the cloned DNA and an insert DNA that is inserted within the vector and includes DNA of interest.

- **Positional cloning:**

  The strategy of gene cloning based on the location or position of a cloned gene or other DNA sequence associated with genetic disease, in the genetic map, is called positional cloning.

- **Directional cloning:**

  The strategy of insertion of foreign DNA into two different restriction sites of a vector such that the orientation of the insert is only one direction.

11. What is cloning vectors?

Answer:

Cloning vectors or vehicles which are small self-replicating genetic elements (DNA), that can carry a desired foreign DNA fragment when inserted into it and deliver it into a host cell.

12. Define replacemnt vector. **[C.U. 2018]**

Answer:

A cloning vector (λ) so designed in which part of the normal genome is replaced by foreign DNA.

Example: λ (lambda) phage.

13. What is STS (Sequence tagged site)?

Answer:

It is a short stretche (100-500 bp) of DNA sequence, easily recognizable, unique position in the human genome and is usually detected by PCR with defined primer.

It is used to orient and identify mapping data for the construction of physical genome maps.

14. When cloning is better to use restriction enzymes that produce blunt ends or sticky ends? Mention significance of sticky.

Answer:

(i) It is better to use restriction enzymes that produce sticky ends. It creates single stranded ends to both the cut cloning vector and DNA insert.

(ii) This single strands hybridize to each complementary base pair and so fix the DNA insert into the vector.

(iii) DNA molecules cut by the same restriction endonuclease have the same sticky ends.

(iv) These can be joined with the same complementary end of DNA fragments.

15. State briefly the characteristics of vector's DNA.

Answer:

1. They contain an origin of replication (Ori-c) that allows them to replicate independently of the chromosome of the host.
2. They contain selectable marker that allows cells that contain the vector (and any attached DNA) to be readily identified.
3. They have single sites for one or more restriction enzymes (multicloning sites or polylinker). This allows the DNA fragments to be inserted at a defined point.
4. Can be recovered easily from host cell.

16. (a) What is plasmid? (b) Why it is considered as a suitable cloning vector?

Answer:

(a) Plasmids are extra chromosomal, self-replicating and double stranded closed and circular DNA molecule present in the bacterial cell.

They contain several informations for their own replication.

(b) (i) They can be isolated from cells but not found naturally.

(ii) They must possess at least one cleavage site for one or more restriction enzymes.

(iii) Insertion of linear molecule at one of these sites does not alter its replication.

(iv) They can be introduced into bacterial cells and cells carrying plasmid with or without insertion can be selected or identified.

17. What are DNA Library, cDNA Library and Genomic Library?

Solution:

- **DNA Library:**

A DNA library is a population of identical vectors that each contains a different DNA insert.

- **cDNA Library:**
  - (i) DNA copy of mRNA molecule is known as copy DNA or cDNA (complementary DNA). Most of the clones of cDNA is prepared by the use of reverse transcriptase. Therefore, the library made from cDNA is called cDNA library.
  - (ii) cDNA library represents the DNA of eukaryotic organism, not prokaryote.
  - (iii) It is intron free representative of DNA having only coding sequence.
- **Genomic Library:**
  - (i) It is a complete collection of cloned DNA fragments, which comprise the entire genome of an organism.
  - (ii) It is constructed by a shotgun experiment where whole genome of a cell is cloned in the form of random and unidentified clones.

18. What is Blotting (Blots)?

Answer:

The process of transferring DNA, RNA or proteins to a solid support (usually a sheet of nitro-cellulose paper) for hybridization after it has been separated by electrophoresis. **Blots** are named according to the material that is analyzed.

- **Southern blot:** A technique for transferring DNA from an agarose gel to nitro-cellulose filter on which the DNA can be detected by a suitable radiolabelled probe (DNA).
- **Northern blot:** A technique for transferring RNA from an agarose gel to a nitro-cellulose filter, on which the RNA can be detected by suitable radiolabelled probe (DNA).
- **Western blot:** A technique for transferring protein from an electrophoretic gel to a nitro-cellulose filter, on which the protein is detected by a suitable radiolabelled probe (antibody). It is also called Immunoblot.
- **Zoo blot:** Hybridization of a cloned DNA from one species to DNA from the other organism to determine the extent to which the cloned DNA is evolutionary conserved.
- **Noah's ark blot:** A blot with DNA from males and females of a variety of organism is called Noha's ark blot.

19. What is Genetic Engineering?

Answer:

It is the most fundamental mechanism of biotechnologies and deals with identification, isolation of desired gene and then inserting this gene of interest into another organism. The desired gene expresses itself in the organism by the gene product.

20. What is recognition sequence?

Answer:

A sequence of DNA which is 4 to 8 base pair long and **pallindromic** in nature is called recognition sequence. It acts as recognition site for restriction enzymes.

21. Mention the steps of Genetic Engineering.

Answer:

(i) Gene isolation.

(ii) Selection of vector.

(iii) Cloning of desired gene.

(iv) Specific gene transfer.

(v) Expression of desired gene.

22. What is taq polymerase?

Answer:

A thermostable enzyme, isolated from a thermostable bacterium called *Thermus aquaticus* and commonly used in polymerase chain reaction (PCR) is called taq polymerase.

23. Why heat stable DNA polymer as c from thermophilic bacteria required for the PCR?

Answer:

(i) The temperature of the PCR reaction mix is raisd to 95 °C for DNA renaturation.

(ii) Normal enzymes are also denatured by this high temperature.

(iii) Thermophilic bacteria that permanently live in environments with high temperature have evolved enzymes that are heat stable.

24. What is DNA profiling or fingerprinting or typing?

Answer:

The nucleotide forms the code, therefore, the coded genetic information can be profiled to produce most authentic identity card of any organism. The molecular technology that facilitates the identification of an individual based on the unique characteristic of their DNA, observed by the banding pattern of DNA fragments obtained by restriction enzyme digestion and followed by southern blot hybridization using **minisatellite probe** is known as DNA profiling or DNA fingerprinting or DNA typing.

25. On which technique DNA fingerprinting is based?

Answer:

Restriction Fragment Length Polymorphism (RFLP) technique.

26. What is molecular archaeology?

Answer:

Study of DNA fragments from mummified remains of humans and other extinct mammals such as **wholly** mammoth is called molecular archaeology.

27. What is Reporter System?

Answer:

Genetically manipulated system that allows an investigator to determine that a specific locus is active by measuring the phenotypic expression of an associated locus.

Example: In the firefly **luciferase** reported system, the desired result (transcription of a particular promoter) causes transgenic tobacco plant to **glow**.

28. (a) What features at a minimum, must a transposon contain?

(b) What are the minimum element that a transposable must have in order to be transposed in a cell that harbours a separate complete transposon?

Answer:

(a) Genes for transposase and terminal inverted repeat.

(b) Terminal inverted repeats that the other transposon's transposase recognizes.

29. Mutations in the transposase genes of IS 1 render the transposon immobile. What can we conclude from this finding?

Answer:

Transposase is necessary for transposition.

30. Mention the areas of application of PCR?

Answer:

- *Diagnosis of Pathogens*: To detect the presence of certain specific sequences of pathogens present in the infected cells or tissues.
- *Prenatal Diagnosis*: Useful for several genetic diseases, like β-thalassemia, Hemophilia.
- *DNA Fingerprinting*: Use in forensic medicine for identification of criminal and disputed parentage.

- *Gene Tagging*: Utilized for developing molecular markers closely linked to specific genes for economic importance.
- *In Research*: Isolation of new micro-organism and identification.
- *In Molecular Archaeology*: Used to clone the DNA fragments from the mummified animals.

31. Why is the term "jumping gene" is misleading nickname for some transposons?

Answer:

DNA does not always jump. In the replicative transposon, the transposon does not even leave its original location. Instead it replicates and one of the copies goes to the new location.

32. A certain transposon's transposase creates staggered cuts in the host DNA five base pairs apart. What consequence does this have for the host DNA surrounding the inserted transposon?

Answer:

The inserted transposon will be surrounded by 5 bp direct repeats of host DNA.

33. Which of the following pairs of DNA sequences could qualify as the terminal repeats of bacterial IS element?

(a) 5′ GAATCCGCA 3′ and 5′ ACGCTAAG 3′

(b) 5′ GAATCCGCA 3′ and 5′ CTTAGGCGT 3′

(c) 5′ GAATCCGCA 3′ and 5′ GAATCCGCA 3′

(d) 5′ GAATCCGCA 3′ and 5′ TGCGGATTC 3′

Answer:

The pair in (d) have inverted repeats and could therefore qualify.

34. What are Cointegrate and Cointegration?

Answer:

Cointegrate: The product of fusion of two circular elements (DNA) to form an intermediate single structure during transposition is termed as cointegrate.

Cointegration: The process of fusion of two DNA elements or replicons, one originally containing a transposable elements and the other lacking it, is called cointegration.

35. What do you mean by autonomous and non-autonomous transposon?

Answer:

These mobile elements carry a pair of terminal repeats, and transposase genes can promote their own transposition. These elements are called **autonomous transposon.**

These mobile DNA segments carry only the terminal inverted repeats, those are the cis acting sequences needed for transposition. In a cell that also carries an **autonomous transposon** encoding a **transposase** that will recognize these terminal inver-ted repeats, the nonautonomous element will be able to transpose. However, in the absence of "helper transposon" (donate transposase) non-autonomous transposon remains frozen, unable to move.

36. Mention the characteristics of Poly-A-Retrotransposons.

Answer:

(i) They do not have the terminal inverted repeats instead the two ends of the element have distinct sequences.

(ii) One end is called 5′ UTR 3′ (untranslated region) whereas the other end has 3′ UTR 5′ followed by a stretch of A-T base pairs called poly A sequence. These elements are also flanked by short target site duplications.

(iii) It contains two genes ORF_1 and ORF_2. ORF_1 encodes an RNA binding protein and ORF_2 encodes a protein with both **reverse transcriptase** activity and an **endonuclease** activity. It plays an essential role in recombination.

(iv) Like their DNA and viral like transposon counterparts, poly-A-retrotransposon exists commonly in both autonomous or non-autonomous forms.

37. Transposable elements have been referred to as "jumping genes" because they appear to jump from one position to another, leaving the old locus and appearing at a new locus. In light of what we now know concerning the mechanism of transposition, how appropriate is the term "jumping genes" for bacterial transposable elements?

Answer:

In bacteria transposition occurs by two different modes.

(i) The conservative mode results in true jumping genes, because in this case, the transposable element excises from its original position and inserts at a new position.

(ii) In replicative mode, the transposable elements move to a new location by replicating into the target DNA, leaving behind a copy of the transposable element at the original site.

Therefore, in **replicative mode**, transposable elements are not really jumping genes, because a copy does remain at the original site.

38. What do you mean by conservative, replicative and Retroelement transposition?

Answer:

Conservative transposition involves the excision of the element from one site and in its integration at another. DNA replication does not occur and the element is lost from the site of original chromosome. *Example*: Tn 10.

Replicative transposition involves duplication of the element, one copy remaining at the donor site and one copy integrating at the target site.

Retroelement transposition is also replicative because the element yielding the transcript which is converted into a cDNA for integration into the genome.

39. What is BAC vector?

Answer:

It is a stable bacterial artificial chromosome (BAC) which introduces foreign DNA of 80-100 kb in bacteria. Hence, BAC is used in construction of genomic library.

40. What is bioinformatics?

Answer:

It is the study of information content and information flow in biological system and process and it is conceptualizing biology in terms of molecules (in the sense of physical chemistry) and applying "**informatics techniques**" (derived from disciplines such as applied mathematics, computer science and statistics) to understand and organize the information associated with these molecules on a large scale.

41. What is DNA microarray technology?

Answer:

An array is an arrangement of point in rows and columns. DNA microarray refers to DNA chips which are high density miniatured microarrays of large number of DNA sequences which are spotted or anchored in a fixed location in a systemic order on a solid support (glass or nylon).

42. What is knockout mice?

Answer:

Transgenic mice that carry a desired gene or gene of interest replaced by the other gene is called knockout mice.

43. What is electroporation?

Answer:

Introduction of naked DNA into cells by exposing them for very brief periods to high voltage electrical pulses which are thought to induce transient pores in the plasma membrane, is called **electroporation.**

44. What is lipofection?

Answer:

Introduction of DNA enclosed with liposome (vesicle encircled by layered phospholipid) into the cell, is known as lipofection. Liposome protect the DNA from the degradation of enzymes.

45. Why is sodium hydroxide used for making cDNA?

Solution:

(i) Reverse transcriptase polymerizes DNA with the use of RNA as a template.

(ii) This RNA : DNA hybrid is then treated with sodium hydroxide to degrade the RNA (NaOH hydrolyzes RNA by catalyzing the formation of a 2′, 3′-cyclic phosphate).

46. What is retrotransposon?

Answer:

Transposable element that utilizes reverse transcriptase to transpose through an RNA intermediate, is termed retrotransposon.

Example: Copia elements in *Drosophila*.

47. What is IS element?

Answer:

It is a smallest transposable element. It has inverted repeat sequence at both ends. It has a gene code for transposase which helps it in transposition.

48. Why is ligase needed to make recombinant DNA? What would be the immediate consequence if the cloning process of someone forgot to add it?

Answer:

(i) Ligase is an essential enzyme within all cells that seal breaks in the phosphate sugar backbone of DNA.

(ii) In DNA replication, it joins okajaki fragments to create a continuous strand and in cloning it is used to join the various DNA fragments with vector.

49. In the PCR process, if we assume that each cycle takes 5 minutes, how many fold amplification would be accomplished in 1 hour?

Answer:

(i) Each cycle takes 5 minutes and doubles the DNA.

(ii) In one hour there would be $\frac{60}{5}$= 12 cycles.

(iii) So the DNA would be amplified 2^{12} = 4096 folds.

50. Arrange the following cloning vectors in order of the size of the DNA inserts that they can carry. (a) BAC (b) Cosmid (c) YAC (d) Plasmid pBR322.

Answer:

Plasmid (15–20 kb) < Cosmid (35 – 45 kb) < BAC (150–300 kb) < YAC (300 kb and larger).

Plasmid < Cosmid < BAC < YAC.

51. How is a gene is isolated and amplified by cloning?

Answer:

(i) Genomic DNA is cut up with restriction enzymes and spliced into a vector chromosome.

(ii) It is then replicated in a bacterial cell.

52. What are ectopic expression and ectopic integration?

Answer:

- **Ectopic expression:**

(i) The occurrence of gene expression in a tissue in which it is normally not expressed.

(ii) It can be caused by the juxtaposition of novel enhancer elements with a gene.

- **Ectopic integration:**

In a transgenic organism, the insertion of an introduced gene at a site other than its usual locus.

53. How is DNA amplified without cloning?

Answer:

(i) The polymerase chain reaction is used.

(ii) Two specific primers that flank the region in question are hybridized to denatured DNA. Then DNA polymerase shuttles back and forth between the primers, amplifying the flanked sequence exponentially.

54. How is amplified DNA used in genetics?

Answer:

(i) The sequence of a region or an entire genome is obtained as probe.

(ii) The sequence to be inserted as a transgene to modify a recipient genome.

55. What transposable element is more abundant in the human genome?

Answer:

The Line known as L1 is the most abundant human transposon. It accounts for 17% of all human DNA.

56. In the diagnosis of sickle cell anaemia, assume that the β-globin intron is cut out, cloned and used as a probe in a Southern analysis of Mst II-cut genomic DNA from a certain foetus. The blot shows two bands, one 1 : 1 and another 1.3 kb in size. What is the genotype of the foetus?

Answer:

(i) The foetus is heterozygous for sickle cell anaemia.

(ii) The sickle cell mutation which affects an Mst II restriction fragment is larger (1.3 kb) than normal (1 : 1 kb). Because both bands are present, therefore, the person is heterozygous.

57. Why was cDNA and not genomic DNA used in the commercial cloning of the human insulin gene?

Answer:

The commercial cloning of human insulin gene was introduced in bacteria are not capable of processing introns. Genomic DNA would include the introns, whereas cDNA is a copy of processed (and thus intron-free) mRNA.

58. Is bacterial operator a Docking element?

Answer:

Yes, the operator is the location at which repressor functionally binds through interactions between the DNA sequence and the repressor protein.

59. The sequence fragment of DNA in *Drosophila* was used in a BLAST search. The best (closest) match was to a kinase gene from *Neurospora*. Does this match that the *Drosophila* sequence contains a kinase gene? (BLAST = Basic Local Alignment Search Tool)

Answer:

(i) A level of 35% or more amino acid identities at comparable positions in two polypeptides is indicative of a common three dimensional structure.

(ii) It also suggests that the two polypeptides are likely to have at least some aspect of their function in common.

(iii) However, it does not prove that your particular sequence in fact encodes a kinase.

60. How are sequence maps of genomes produced?

Answer:

From sequencing large number of clones either are mapped first by using molecular markers and then sequenced or clones are sequenced at random and then aligned by overlaps.

61. How can genomic sequence maps help find a gene of interest?

Answer:

If a mutant gene has been mapped on a linkage map, its relative position on the linkage map can be compared with its genome sequence and the candidate genes are tested — for example, for the presence of mutation.

62. What is the nature of the information in the genome?

Answer:

It can be viewed as the sum of the functional products (Proteins and RNA) or as an array of docking sites for molecules such as proteins to bind to DNA or RNA.

63. What is contig?

Answer:

A contig is a collection of clones that contain overlapping segments of DNA that spans a particular region of a chromosome.

64. The word 'contig' is derived from the word contiguous. Explain the derivations.

Answer:

Contig describes a set of adjacent DNA

sequences or clones assembled by using overlapping sequences or restriction fragments. Because the pieces are assembled into a continuous whole, it makes sense that contig is derived from the word contiguous.

65. A FISH analysis with a probe of unknown sequence (the label on the tube fell off) reveals a fluorescent spot at one end of every chromosome. Could this be

(a) a unique gene probe?

(b) a centromere probe?

(c) a telomere probe?

Answer:

(i) Detection of hybridization at one end of every chromosome would not indicate that the probe used is from a unique gene (because FISH indicates more than one region of homology) or from the telomere (because there is a telomere at each end of a chromosome and not just at one end).

(ii) However, it is possible that the probe is from the centromere because it hybridizes one location on each chromosome.

66. Explain how would you determine that two clones in a contig are overlapping.

Answer:

(i) If two clones are overlapping, one could conduct a Southern blotting experiment.

(ii) In this approach, one of the clones is used as probe.

(iii) If the band corresponding to the second clone is labelled, this means that two clones are overlapping.

67. In sequencing a 1-gigabase genome with five-fold coverage by using BACs, how many clones would you be handling?

Answer:

(i) Assume that the average BAC contains 200 kb. It would take 5000 BAC clones to hold a one gigabase genome.

(ii) For five-fold coverage it would be handling 5000 × 5 = 25000 clones.

68. What is the difference between a contig and a scaffold?

Answer:

(i) A scaffold is also called a super coiling.

(ii) Contigs are sequences of overlapping reads assembled into units. On the other hand, scaffold is a collection of contigs joined together.

69. In an in situ hybridization experiment, a certain cloned bound to only the X-chromosome in a boy with no disease symptoms. However, in a boy with Duchenne muscular dystrophy (X-linked recessive disease), it bound to the X-chromosome and to an autosome. Explain that could this clone be useful in isolating DMD gene.

Answer:

(i) Yes, the clone hybridizes to and spans a translocation breakpoint that entails the X-chromosome and an autosome.

(ii) Because the DMD gene normally maps to the X-chromosome, it is of interest that translocation of the X is found in a patient with DMD.

(iii) If the translocation is the cause of DMD mutation, the clone identifies a least part of the gene or its location or both.

70. Transgenic mice are now routinely produced and studied in research laboratories throughout world.

(a) How are transgenic mice produced?

(b) What kinds of information can be obtained from studies performed on transgenic mice?

(c) Does this information have any importance of the practice of medicine? If so, what?

Answer:

(a) Transgenic mice are usually produced by microinjecting the genes of interest into the pronuclei of fertilized eggs or by infecting preimplantating embryos with retroviral vectors containing the gene of interest.

(b) Transgenic mice provide valuable tools for studying gene expression, mammalian development and the immune system of mammals.

(c) (i) Transgenic mice are of major importance in medicine. They provide the model system most closely related to humans.

(ii) They have great value in developing the tools and technology that will be used for human gene therapy.

71. What is antisense technology?

Answer:

The use of antisense mRNA to prevent or inhibit the translation of a natural mRNA is called antisense technology.

72. In what ways are antisense RNA and RNAi gene silencing procedures the same or different?

Answer:

(i) Both antisense RNA and RNAi make use of RNA strands that are complementary to the mRNA target strand to block its translation.

(ii) Antisense RNA uses single stranded RNA that is complementary to the mRNA. RNAi involves the use of double-stranded RNA where one strand is complementary to the mRNA and the other strand is equivalent to the mRNA, to silence this expression of the target gene.

RNAi makes lose of the RNA induced silencing complex (RISC) to block gene expression.

73. What is the Colony hybridization?

Answer:

The process by which a labelled DNA probe is used to screen DNA genomic library is called colony hybridization.

74. What is gene therapy and how does it differ from supportive (replacement) therapy?

Answer:

(a) In gene therapy, a normal copy of a gene is introduced into the tissues of a patient who is suffering from a genetic desease as a result of the presence of a specific defective gene, thus, correcting the genetic defect in those tissues and allowing the patient to lead a normal life.

(b) Supportive therapy involves the replacement of malfunctioning cells with normal cells, such as in bone marrow transplants.

75. (a) Explain the difference between a microsatellite repeat and a minisatellite repeat.

(b) Which of the two is more useful in DNA fingerprinting? Why?

Answer:

(a) Microsatellite repeats consist of 2–5 nucleotide long repeats occurring at many locations throughout the entire genome. It is also called STR (short tandem repeat). Minisatellite repeats consist of longer repeating unit (30–35 bp) containing a shorter, common core sequence and tend to occur at fewer sites within the genome.

(b) Minisatellite repeats are more useful in DNA fingerprinting because their base sequences are more unique and do not occur so frequently within the genome that they are difficult to separate and distinguish electrophoretically.

76. What are CP^G islands? Of what value are CP^G islands in positional cloning of human genes?

Answer:

CP^G islands are clusters of cytosines and guanines that are often located just upstream (5′) (near promoter region) from the coding regions of human genes. The methylation of cytosine bases usually inhibit transcription.

Their presence in nucleotide sequences can provide hints as to locate the gene in human chromosomes.

77. What is functional genomics? How does it differ from comparative genomics?

Answer:

- Functional genomics seek to understand functional similarities of genomes across phylogenetic and evolutionary distances.
- Comparative genomics analyzes the arrangement and organization of families of genes within and among genomes.

78. What is gel electrophoresis?

Answer:

(i) Electrophoresis is the movement of charged molecules in an electric field, negatively charged molecules migrate towards positive electrode (anode) and positively charged molecules migrate towards negative electrode (cathode).

(ii) It is a standard method for separating DNA molecules of different lengths. It is applied in size analysis of DNA fragment and separation of RNA molecules.

79. What are agarose gel electrophoresis and polyacrylamide gel electrophoresis (PAGE)?

Answer:

(i) Agarose is a polysaccharide that forms a gel with pores (100-300 nm in diameter), depending upon the concentration of agarose in the gel 0.3% can be used for molecules between 5-50 kb length of DNA and a 5% gel for 50-100 bp DNA molecule.

(ii) Polyacrylamide gels have smaller pores size than agarose gel, therefore, allows precise separation of molecules of 10-150 bp.

(iii) This is used in DNA sequencing, PCR, etc.

80. What is the chemical nature of polyacrylamide gel?

Answer:

It consists of a chain of acrylamide monomers ($CH_2 = CH—CO—NH_2$), cross-links with N—N methylenebisacrylamide unit ($CH_2 = CH—CO—NH_2—CH_2—NH—CO$).

81. What do you mean by PFGE?

Answer:

New technique pulse field gel electrophoresis is used for the separation of large size DNA 2×10^4 to 107 bp in length. Here short pulses of electricities are used in different directions; electric field is turned on and off.

82. What are RFLPs and how did they arise?

Answer:

(i) RFLPs are restriction fragment length polymorphisms, that is the variations in the length of DNA base-sequences from one individual to another generated by restriction endonuclease sites.

(ii) They can arise if deletion, insertion, point mutation alter a restriction site within or near gene, or if such a change occurs in a noncoding region between genes.

- **Genetic selection:**

The simplest procedure for identifying a clone of interest is known as genetic selection.

- **Green fluorescent protein (GFP):**

A naturally occurring fluorescent protein is synthesized by the jelly fish (*Aequorea victoria*). It can be used to study gene expression at protein level.

- **Single nucleotide polymorhism (SNPs or Snips)**

It is single base pair change—a point mutation—at a site (the SNP locus). SNPs are the most common type of DNA polymorphism and occurs at a frequency of about 1 per 350 bp. It may be found as noncoding SNPs or coding SNPs (eSNPs).

83. Explain why it is necessary to use the technique of Southern blotting for RFLP mapping.

Answer:

(i) When chromosomal DNA is isolated and digested with a restriction endonuclease, this produces thousand of DNA fragments of different sizes.

(ii) This makes it impossible to see any particular band on a gel, if you simply stain the gel for DNA.

(iii) Southern blotting allows us to detect one or more RFLPs that are complementary to the radioactive probe that is used.

84. What is gene targeting?

Answer:

It is a form of in vivo site directed mutagenesis involving homologous recombination between a targeting vector containing one allele and an endogenous gene represented by different allele. Two types of vectors, viz., integration vector and transplacement vector are used.

85. Mention the application of Gene targeting?

Answer:

(a) Gene knockout (targeted disruption)
(b) Allele replacement
(c) Gene knock in
(d) Gene therapy.

86. The hypothetical bacterium *Xenobacterium giganticus* produces restriction endonuclease. According to the convention for naming these enzymes, what woud you call it?

Answer:

Xgil (*Xenobacterium giganticus*).

87. Some restriction endonucleases recognize 5-base sites in target DNA. Approximately how frequent are 5-base sites?

Answer:

The probabilities of a restriction site is 4^n 5 base sites, i.e., $4^5 = 1024$.

88. A small piece of DNA that contains a 6 base palindromic recognition site for a restriction endonuclease GACGATATCAACT. Find the sequence of the recognition site.

Answer:

GACGATATCAACT

Recognition site is GATATC, because in opposite strand the sequence is same but in opposite direction CTATAG.

89. How does gene cloning produce many copies of a gene?

Answer:

(i) In conventional gene cloning, many copies are made because the vector replicates with a high copy number within the cell and the cell divides to produce many more cells.

(ii) In PCR, the replication of the DNA to produce many copies is facilitated by primers, deoxyribonucleoside triphosphates (dNTP) and Taq polymerase.

90. Explain the basis for using an antibody as a probe in western blotting experiment.

Answer:

(i) The products of structural genes are proteins with particular amino acid sequence.

(ii) Antibodies can specifically recognise protein due to their amino acid sequence.

(iii) Therefore, an antibody can detect whether or not a cell making a particular type protein.

91. While doing a PCR, 'denaturation' step is missed. What will be the effect on the process.

Answer:

In the absence of denaturation of DNA, the primer will fail to anneal to the template and polymerase chain reaction will not occur.

92. What modification is done on the Ti plasmid of Agrobacterium tumefaciens to convert it into a cloning vector?

Answer:

The part of Ti plasmid (*Agrobacterium tumefaciens*) responsible for giving virulance is deleted.

93. Restriction enzyme should not have more than one site of action in the cloning site of a vector. Comment

Answer:

In case a restriction enzymes has more than one recognition sites in a vector, the vector will get fragmented on treatment with the restriction enzyme.

94. A geneticist plans to use the polymerase chain reaction (PCR) to amplify part of the DNA sequence.

5′-GCGAAACGATCCTCATCCTGTCTCT TGATCAGAGCTTGATCCCCTG-3′
3′-CGCTTTGCTAGGAGTAGGACAGAGA ACTAGTCTCGAACTAGGGGAC-5′

Specify the sequence of the primers that should be used including the polarity and deduce the sequence amplified DNA fragment.

Answer:

(i) The primer should be able to pair with the chosen primer sites which must be oriented with their 3′ ends facing each other.

(ii) Thus, the "forward" primer (the one that is elongated in left to right direction) should have the sequence 5′-ACGAT-3′ and the 'reverse' primer (the one that is elongated in right to left direction) should have the sequence 3′-ACTAG-5′. Because the convention for writing nucleic acid sequences is to put 5′ end on the left. The reverse primer is 5′-GATCA-3′.

95. A plasmid of 6200 base pair (bp) includes at least three complete copies of a viral gene arranged in tandem. Cleavage of the plasmid with the restriction enzyme HindIII results in fragments of 900, 1300 and 4000 bp. The tandem copies of the gene are contained within only one of the HindIII fragments. The gene encodes a protein of 405 amino acid. Which HindIII fragment is likely to contain the gene copies?

Answer:

(i) To encode a protein of 405 amino acids requires at least (405 × 3 = 1215) 1215 bp. Three tandem copies of the gene would therefore require 1215 × 3 = 3645 bp and four copies 4860 bp.

(ii) The smallest HindIII fragment is not long enough to contain even one copy of the gene.

(iii) The 1300 bp fragment could contain one copy but not two or more. Only the largest fragment could include three copies of the gene and it could include more than three copies.

96. Assuming a random arragement of nucleotide on a piece of DNA, what are the probabilities that a restriction endonuclease, whose recognition sites consist of four bases (a four cutter) will cut DNA? What are the probabilities for a six-cutter and eight-cutter?

Answer:

In a DNA with a random resequence, a four cutter will find sites approximately once in $\frac{1}{4^n}$, where n is the number of nucleotide.

(i) Once in 44 bases, i.e., $\frac{1}{4^4} = \frac{1}{256} = 0.0039$

(ii) Once in for 6-cutter, i.e., $\frac{1}{4^6} = \frac{1}{4096} = 0.00024$

(iii) Once in for 8-cutter, i.e., $\frac{1}{4^8} = \frac{1}{65336} = 0.00152.$

97. In a random sequence of a piece of DNA consisting of equal properties of all four nucleotides, what are the probabilities that a particular short sequence of nucleotides matches restriction on sites for

(a) a restriction enzyme with a 4-base cleavage site?

(b) a restriction enzyme with a 6-base cleavage site?

(c) a restriction enzyme with a 8-base cleavage site.

Answer:

The probability of a restriction site is $\frac{1}{4^n}$, where n is the number of nucleotides in the site.

(a) 4-base, i.e., $\frac{1}{4^4} = \frac{1}{256} = 0.0039$

(b) 6-base, i.e., $\frac{1}{4^6} = \frac{1}{4096} = 0.000244$

(c) 8-base, i.e., $\frac{1}{4^8} = \frac{1}{65536} = 0.00015258.$

98. If a *Haemophilus influenzae* DNA were essentially a random sequence of 4.6×10^6 bp with equal proportion of all four nucleotides, approximately how many restriction fragments would be expected from cleavage with

(a) a "4 cutter" restriction enzyme.

(b) a "6 cutter" restriction enzyme.

(c) a "8 cutter" restriction enzyme.

Answer:

Sequence is 4.6×10^6 bp

(a) 4 cutter restriction enzyme = $\frac{4.6 \times 10^6}{256}$

$= \frac{460 \times 10^4}{256} = 1.796 \times 10^4 = 1.8 \times 10^4$.

(b) 6 cutter restriction enzyme = $\frac{4.6 \times 10^6}{4096}$

$= \frac{4600 \times 10^3}{4090} = 1.12 \times 10^3$.

(c) 8 cutter restriction enzyme = $\frac{4.6 \times 10^6}{65536}$

$= 0.000070 \times 10^6 = 70$.

99. The human genome (haploid) contains about 3×10^7 nucleotide pairs of DNA. If you digest a preparation of human DNA with Not I, a restriction endonuclease that recognizes and cleaves the octamaric sequence 5′-GCGGC-CGC-3′, how many different restriction fragments would you expect to produce? Assume that the four bases (G, C, A and T) are equally prevalent and randomly distributed in the human genome.

Answer:

(i) Assuming that the four bases are present in equal amounts and are randomly distributed, the chance of a specific nucleotide occuring at a given site is $\frac{1}{4}$.

(ii) The chance of a specific dinucleotide sequence (e.g., AG) occuring is $\frac{1}{4} \times \frac{1}{4} = \left(\frac{1}{4}\right)^2 = \frac{1}{16}$ and the probability of specific octanes nucleotide sequence $\left(\frac{1}{4}\right)^8 = \frac{1}{65,536}$.

(iii) Therefore, Not I will cleave such DNA molecules an average of once in every 65,536 nucleotide pairs.

(iv) If a DNA molecule is cleaved at n sites, n + 1 fragments will result. A genome of 3×10^9 nucleotide pairs should contain about 45,776, i.e., $\left(\frac{3 \times 10^9}{65536}\right)$ or $\left(\frac{3000000}{65,536}\right) \times 10^3 = 45.77636 \times 10^3 = 45776$.

(v) If the entire human genome consisted of a single molecule of DNA. Not I would cleave it into 45,776 + 1 fragments.

(vi) Given that these cleavage sites are distributed on 24 different chromosomes, complete digestion of the human genome with Not I should yield about 45,776 + 24 restriction fragments.

100. The restriction enzyme Hind III cuts DNA at the sequence AAGCTT, and the restriction enzyme Hpa II cuts DNA at the sequence CCGG. On average how frequently will each enzyme cut double-stranded DNA? (In other words, what is the average spacing between restriction sites?)

Answer:

(i) Here we consider only one strand of DNA, because both sequences will be present on the opposite strand at the same site owing to the symmetry of the sequences.

(ii) The frequency of the six base long Hind III sequence is $\left(\frac{1}{4}\right)^6$ or $\frac{1}{4096}$, because there are four possibilities at each of the six positions. Therefore, the average spacing between Hind III sites is approximately 4 kb.

(iii) For Hpa II the frequency of the four base long sequence is $\left(\frac{1}{4}\right)^6$ or $\frac{1}{256}$. The average spacing between Hpa II sites is approximately 0.25 kb.

101. Calculate the average distances (in nucleotides pair) between the restriction sites in organism X for the following restriction enzymes, assuming an AT:GC ratio of 50:50.

Alu I	**5′ AGCT 3′**
	3′ TCGA 5′
ECO-RI	**5′ GAATTC 3′**
	3′ CTTAAG 5′
Acyl	**5′ GP_uCGP_yC 3′**
	3′ CP_yGCP_uG 5′

Answer:

This problem assumes a random equal distribution of nucleotides.

Alu 1: $\left(\frac{1}{4}\right)^4$ = on average once in every 256 bp.

Eco-RI: $\left(\frac{1}{4}\right)^6$ = on average once in every 4096 bp.

Acyl: $\left(\frac{1}{4}\right)^4 \times \left(\frac{1}{4}\right)^2$ = on average once in every 1024 bp.

102. For the following restriction endonucleases calculate the average distance between restriction sites in an organism whose DNA has a random sequence and equal proportion of all four nucleotides. The symbol R means any purine (A or G) and Y means any pyrimidine (T or C), but an R-Y pair must be either A-T or G-C.

Taq I: **5′–T C G A–3′**
3′–A G C T–5′

Bam HI: **5′–G G A T C C–3′**
3′–C C T A G G–5′

Hae H II: **5′–R G C G C Y–3′**
3′–Y C G C G R–5′

Answer:

(i) For each site the probability equals the product of the probability of each of the nucleotide pairs in turn, $\frac{1}{4}$ for any of the specified pairs and $\frac{1}{4}$ for R-Y.

(ii) The average spacing is the reciprocal of the probability. Hence, for Taq I, the average distance between sites is $4^4 = 256$ base pairs, for Bam HI it is $4^6 = 4096$ and for Hae III it is $4^4 \times 2^2 = 256 \times 4 = 1024$ base pairs.

103. How frequently would the restriction enzymes Taq1 (restriction site TCGA) and Mac III (restriction site GTNAC in which N is any nucleotide) cleave ds DNA molecules containing random sequences of

(a) $\frac{1}{6}$ A, $\frac{1}{6}$ T, $\frac{1}{3}$ G and $\frac{1}{3}$ C?

(b) $\frac{1}{3}$ A, $\frac{1}{3}$ T, $\frac{1}{6}$ G and $\frac{1}{6}$ C?

Answer:

(a) (i) The probability of Taq I site is $\frac{1}{6} \times \frac{1}{3} \times \frac{1}{3} \times \frac{1}{6} = \frac{1}{324}$. So Taq cleavage is expected in every 324 base pair.

(ii) The probability of Mac III site is $\frac{1}{3} \times \frac{1}{6} \times 1 \times \frac{1}{6} \times \frac{1}{3} = \frac{1}{324}$. So Mac III cleavage is expected in every 324 base pairs.

(b) (i) For Taq: Probability is $\frac{1}{3} \times \frac{1}{6} \times \frac{1}{6} \times \frac{1}{3} = \frac{1}{324}$. i.e., Taq cleavage is expected in every 324 bp.

(ii) For Mac III: Probability is $\frac{1}{6} \times \frac{1}{3} \times 1 \times \frac{1}{3} \times \frac{1}{6} = \frac{1}{324}$, i.e., Mac III cleavage is expected in every 324 bp.

104. What are the main stages in producing transgenic mouse?

Answer:

(i) Recombinant DNA technology is first used to build the DNA vector to insert into the mouse.

(ii) Along with the gene of interest and vector DNA are the promoter sequences that enable expression of gene in mouse.

(iii) The vector is injected into the nucleus of newly fertilized eggs that are grown for few generations.

(iv) Viable embryos then implanted are pregnant females.

105. To produce transgenic plants, plant tissue is exposed to a *A. tumefaciens* and then grown in media containing kanamycin, carbenicillin and plant growth hormones. Explain the purpose behind each of these three agents. What would happen if you left out the kanamycin.

Answer:

(i) A kanamycin-resistance gene is contained within the T DNA. Exposure to kanamycin selects for growth of plant cells that have incorporated the T DNA into their genome.

(ii) The carbenicillin kills the *A. tumefaciens.*

(iii) The plant growth hormones promote the regeneration of an entire plant from somatic cells.

(iv) If kanamycin were left out, it would not be possible to select for growth of cells that had taken up T DNA.

106. What is a gene knockout? Is an animal or plant with a gene knockout a heterozygote or hamozygote? What might you conclude if a gene knockout does not have a phenotypic effect.

Answer:

(i) The term knockout refers to an organism in which the function of a particular gene has been eliminated.

(ii) For an autosomal genes in animals and plants, a gene knockout is a homozygote for a defect in both copies of the gene.

(iii) If a gene knockout has no phenotypic effect the gene may be redundant. In other words, there may be multiple genes within the genome that can carryout the same function. Another reason, a gene knockout may not have a phenotypic effect, i.e., due to environment.

(iv) For an example mouse gene is required for a synthesis of a vitamin, the knockout mouse that has lacking this gene would have a normal phenotype, it would survive just fine.

Sometimes scientist have trouble knowing the effects of a gene knockout until they modify the environmental conditions in which the animal are raised.

107. What is molecular farming? Compared to the production of proteins by bacteria—why might it be adventageous?

Answer:

The term molecular farming refers to the practice of making transgenic animals and plants, where genes have been transferred with a view to obtain large scale production of proteins encoded by these genes in milk, urine or blood of such animals and also to the manufacture of medical products by agricultural plants. Such biological (plant and animals) organisms are often called bioreactors and the approach is referred to as **molecular farming** or **gene farmering.**

It can be advantageous when bacterial cells are unable to make a functional protein product from a human gene.

(a) *Example*: Some proteins are post-transcriptionally modified by the attachment of carbohydrate molecules. This type of modification does not occur in bacteria, but it may occur correctly in transgenic animals and plants.

(b) Dairy cows produce large amounts of milk, which may improve the yield of the human product.

(c) Likewise plants can produce a large amount of recombinant protein.

108. (a) Describe some of the possible applications of the technology used to create transgenic animals.

(b) State briefly some of the potential disadvantages and problems associated with the use of such technology.

Answer:

(a) The technology used to create transgenic animals could be used to introduce new and sometimes novel genes in various organisms to inactivate genes in order to learn more about their normal functions in embryonic development, tissue differentiation, development of cancer functioning of immune system and variation on this technique can be employed in gene therapy.

(b) (i) If the engineered gene is essential for normal embryonic development, homozygous offspring may die before birth.

(ii) Due to host specific differences, the modified cells may fail to be incorporated into the developing embryo or to be transmitted to subsequent generations of offpring.

(iii) Nonhomologous recombination occurs more often than homologous rcombination.

109. Describe the targeting methods in cystic fibrosis gene therapy. Provided the CFTR gene gets to the patient lung cells, would you expect this to be a permanent cure for the patient or would it be necessary to perform this gene therapy on a regular basis (say, monthly).

Answer:

(i) In cystic fibrosis gene therapy, an aerosol spray, containing the normal CF gene in a retrovirus or liposome is used to get normal CF gene into the lung.

(ii) The epithelial cell on the surface of the lung will take up the gene.

(iii) However, these surface epithelial cells have a finite life span, so it is necessary to have repeated applications of the aerosol spray.

(iv) Scientists hope to devise methods, whereby the normal CF gene will be able to penetrate more deeply into the lung tissue.

110. Describe the strategy to produces human protein in the milk of livestock.

Answer:

(i) Milk proteins are encoded by genes with promoter and regulatory sequences that direct the expression of these genes within the cells of the mammary gland.

(ii) To get other proteins expressed in the mammary gland, the strategy is to link the promoter and regulatory sequences from a milk-specific gene to the coding sequence of the gene that encodes the human protein from a cell.

(iii) If the target protein does not already have a signal sequence, it is possible to use a signal sequence from a milk-specific gene to promote the secretion of the target protein from the mammary gland cells and into the milk.

(iv) During this process, the signal sequence is cleaved from the secreted protein.

111. Which of the following would appropriately be described as a transgenic organism?

(a) The sheep "Dolly" which was produced by cloning.

(b) A sheep that produces human α-1-antitrypsin in its milk.

(c) The Flavr Savr strain of tomato.

(d) A hybrid strain of corn produced from crossing two inbred strains of corn (the inbred strains were not transgenic).

Answer:

(a) (i) No, Dolly was not produced using recombinant techniques.

(ii) Pieces of DNA were not cut and combined in a new way.

(b) Yes.

(c) Yes.

(d) No, the hybrids simply contain chromosomal genes from two different parental strain.

112. With regard to genetically modified organism, name two that have not been successful and two that have.

Answer:

1. Two unsuccessful examples are
   (i) Oil eating bacteria
   (ii) Flavr Savr tomato.
2. Two very sucessful examples are
   (i) Bacteria that make human insulin
   (ii) Bt varieties of agricultural crop.

113. Name the first transgenic cow? Which gene was introduced in this cow?

Answer:

(i) The first transgenic cow is Rosie.

(ii) Human lactoalbumin gene was introduced in it.

114. Transgenic animals are the animals in which foreign gene is expressed. Such animals can be used to study the fundamental biological process, phenomenon as well as for producing products used for mankind. Give one example for each type.

Answer:

(a) Study of biological aspect of growth by insulin like growth factors (IGFI and IGFII).

(b) Transgenic cow Rosie whose milk is riched in lactoalbumin.

115. What is recombinant DNA vaccine? Give two examples.

Answer:

Recombinant DNA vaccines are immunogenic proteins or other biochemicals, which initiates development of antibodies against some infectious diseases and are developed with the help of recombinant DNA.

Example: Vaccines against Hepatitis B, influenza B and meningitis.

116. Why does Bt toxin does not kill bacteria Bacillus?

Answer:

(i) *B. thuringiensis* bacteria form protein crystal during a particular phase of their growth. These crystals contain a endotoxin, known as **insecticidal crystalline protein (Icp)** called **Bt toxin** or **cry** toxin, the **cry** gene encodes **Cry protein** (70 kg-130 kD).

(ii) This Bt toxin protein exists in an inactive **protoxin** form in bacteria.

(iii) Only when inactive Bt toxin is ingested by insect, it is converted into an active form of toxin due to the alkaline pH of the gut which solubilise the crystal.

(iv) Hence, Bt toxin protein is produced by bacteria which does not kill it. It kills insects that ingest it.

117. Mention the action of Bt toxin protein in insect.

Answer:

(i) As the crystal of inactive Bt toxin solubilise in the gut (alkaline pH) of insect, the Bt toxin gets converted into active toxin.

(ii) Activated toxin binds to the surface of the epithelial cells of midgut and creates pores in them.

(iii) This causes swelling and ultimately lysis of cells eventually causing the death of the insect.

118. (a) What are cry proteins? (b) Name an organism that produce it. (c) How has man exploited this protein to his benefit?

Answer:

(a) Cry proteins are toxic proteins encoded by genes, named **cry**.

(b) They are produced by bacteria ***Bacillus thuringiensis***.

(c) (i) These cry proteins are toxic to insect.

(ii) The genes encoding them have been introduced from bacteria into crops like cotton and corn, called Bt cotton and Bt corn.

(iii) These crops have become resistant to the insects/pests and thereby crop production increases.

119. What is genetically modified (GM) crops?

Answer:

(i) An organism carrying foreign genes (foreign DNA) is termed **genetically modified** or **transgenic organism.**

(ii) The transgenic crop contains genes that has been artificially inserted, instead of the plant acquiring it through hybridization.

(iii) The foreign gene which is insereted is known as **transgene**, it may come from another unrelated plant or different organism.

(iv) Transgenic crop contains or expresses a transgene.

120. What are transgenic bacteria? Illustrate with an example.

Answer:

Transgenic bacteria are genetically modified bacteria that contain foreign DNA which has been introduced from any other sources (DNA).

Example:

(i) In transgenic *E. coli* bacteria, the genes of DNA sequence A and B chains of human insulin are introduced in the plasmid.

(ii) This transgenic *E. coli* can produce human insulin chains.

121. A plasmid DNA and a linear DNA (both are same size) have one site for a restriction endonuclease. When cut and separated an agarose gel electrophoresis plasmid shows one DNA band while linear DNA shows two fragments. Explain.

Answer:

1. Plasmid DNA is circular, when cut by **endonuclease**, it becomes linear but is not divided into segments. Therefore, it forms a single band on agarose gel.

2. A linear DNA will cut into two segments by endonuclease and will form two bands on agarose gel.

122. How does one visualize DNA on agarose gel?

Answer:

On agarose gel, DNA segments appear as fluorescent bands of orange colour because they are strained wth **Ethidium bromide** stain and are irradiated with UV radiation.

123. A copy of the wild-type white gene (w^+) from *Drosophila* was insertd in the middle of an incomplete P element contained within a plasmid. The plasmid was mixed with another plasmid that contained a complete P element and the mixture was carefully injected into *Drosophila* embryos homozygous for a null mutation (w^-) of the white gene. The adult that developed from these injected embryos, all had white eyes, but when they were mated to uninjected white flies, some of their progeny had red eyes. Explain the origin of these red-eyed progeny.

Solution:

(i) The complete P element in one of the plasmids would produce the P transposase, the enzyme that catalyzes P element transposition, in the germ lines of injected embryos.

(ii) The incomplete P element in other plasmid would be a target for this transposase.

(iii) If this incomplete P element were mobilized by the transposase to jump from its plasmid into the chromosomes of the injected embryo, the fly that developed from this embryo would carry a copy of the wild type white gene in its germline (P element movement is limited to the germ line, therefore, the incomplete P-element would not jump into the chromosome of the somatic cells, such as those that eventually form the eye).

(iv) Such a genetically transformed fly would, in effect, have the germ line genotype $\frac{w^-}{w^-}$; $P(w^+)$ or w^-/y; $P(w^+)$, where $P(w^+)$ denotes the incomplete P-element that contains the w^+ gene. This element could be inserted on any of the chromosome.

(v) If the transformed fly were mated to an uninjected white fly, some of its offspring would inherit the $P(w^+)$ insertion, because it carries a wild-type white gene, and would cause red eyes to develop.

The red eyed progeny are, therefore, the result of genetic transformation of a mutant white fly by the w^+ gene within the incomplete P-element.

124. A geneticist has two strains of *Drosophila* one, a long-standing laboratory stock with white eyes, is devoid of P-elements; the other recently derived from wild-type flies collected in a fruit market, has P elements in its genome. Which of the following crosses would be expected to produce dysgenic hybrid offspring:

(a) White female × wild-type males

(b) White males × wild-type females

(c) White females × white males

(d) Wild type females × wild-type males

Answer:

White females × wild-type males

(i) The white females lack P-elements in their genome and they have M cytotype which is the cellular condition that would permit P-elements to become active if they were present.

(ii) This condition is transmitted to the offspring through the egg.

(iii) The wild type males have P-elements in their genomes, and they also have the P cytotype, which repress the activity of these elements. However, the P cytotype cannot be transmitted through the sperm.

(iv) Thus, when the wild-type males are crossed to the white females, the offspring inherit P-elements from their fathers and the M-cytotype from their mothers.

(v) This combination of factors allows the paternally inherited P-elements to become active in the germline tissues of the offspring and hybrid dysgenesis ensues.

EST (Expressed Sequence Tag):

A partial sequence of a cDNA (300-5000 bp) clone that can be used to identify sites in a gene.

Genomics:

Genomics is a term coined by Thomas Roderick in 1986. It refers to mapping sequencing, analysis and function of genome, and the study of one or more complete genome sequences.

Proteome:

The sequence and expression patterns of all proteins, i.e., the total set of proteins expressed from the transcriptome of a cell. It is a wide term which includes the entire collection of proteins that an organism can produce.

Proteomics:

The term 'proteomics' was coined by Marc Wilkins in 1994. The study of protein expression structure and function and the interaction of all proteins of a specific organism.

The transcriptome:

The sequence and expression patterns of all transcripts.

The interactome:

The complete set of physical interactions between proteins and DNA segments, between proteins and RNA segments and between proteins.

The Phenome:

The description of the complete set of phenotypes produced by the inactivation of gene function for each gene in the genome.

Anchor markers:

Markers that are mapped both genetically and physically are called anchor markers.

Structural genomics and Functional genomics:

Study of the genome structure, and the study of the genome function.

Comparative genomics:

Study of the genome evolution.

Meta genomics:

A new field combining molecular biology and genetics in an attempt to identify and characterize the entire genetic material isolated from the environmental sample.

Databases:

(i) A repository of sequences (DNA or amino acid) which provide a centralized and homogenous view of its contents.

(ii) Here data are collected in machine readable form which can be manipulated by software to appear in varying arrangements and subsets.

(iii) Genetic information is stored in different databases.

GDB:

It is the official central repository for genome mapping data created by Human Genome project. Full form of GDB is Genome Database.

Entrez:

The integrated information database retrieval system of NCBI (National Centre for Bioinformatics) is called entrez.

Application of Bioinformatics:

A. The manipulation and analysis of individual nucleic acid and protein sequences called as gene therapy.

B. 1. The maintenance of sequence data in database is for easy manipulation and inspection searched from ncbi site. The best known sequence data base include EMBL & GenBank (which contains all known DNA sequence data), Swiss—prot and TrEMBL (which contain protein sequences derived from translation).

2. Analytical methods particularly for the comparison of DNA and protein sequences (sequence similarity searching, sequence alignment) used to identify unknown sequences, assign functions, identification of structural and functional motifs, protein domains called as physiochemical modelling.

C. Analysis of physiogenetic relationships are done multiple sequence alignments by PHYLIP.

D. Statistical analysis including cluster of genes whose transcription pattern varies are also designated.

E. Development of large database systems (Entrez, SRS) to encompass many types of biological informations.

F. Manipulation and development of different drug.

RESTRICTION MAP

A restriction map is a compilation of the number, order and distance between restriction enzyme cutting sites along a cloned segment of DNA. The **restriction map** is analogous to the arrangement of genes on a **linkage map**.

Map units are expressed in **base pairs (bp)** or for longer distances **kilobase pairs (kb)**.

(i) Restriction map provides information about the length of a cloned insert and the location of restriction enzyme sites within the DNA.

(ii) Fragments generated by cutting DNA with restriction enzymes can be separated by gel electrophoresis, a method that separates fragments by size, with smallest pieces moving farthest. The fragments appear as a series of bands that can be visualized by staining the DNA with **ethidium bromide** and viewing under ultraviolet illumination.

EXAMPLE OF RESTRICTION MAPPING

Example 1:

Let us consider the construction of a relatively simple restriction map. A cloned DNA segment of 70 kb is taken.

(i) Three samples of the cloned DNA are digested with restriction enzymes. One sample with of enzyme Eco-RI, a second sample is digested with Bam-HI and third sample with both Eco-RI and Bam-HI.

(ii) The DNA restriction fragments of each reaction are separated according to their molecular size by agarose gel electrophoresis, stained with **ethidium bromide** and **photographed.**

(iii) They are then measured by comparing them with a set of standard run in adjacent lanes. The map is constructed by analyzing the fragments.

Results:

(i) When the 5.0 kb DNA is cut with Eco-RI, 4.5 kb and 0.5 kb DNA fragments result. This indicates that there is only one restriction site for Eco-RI in the DNA and that this site is 0.5 kb from one end of the molecule.

(ii) Similarly, when DNA is cut with Bam-HI, two fragments 2.0 kb and 3.0 kb result indicating that this enzyme has one cutting site located 2 kb from one end of the molecule.

(iii) At this point, we know there is one restriction site for each enzyme, but we do not know the relationship between the two.

From the information available two different maps are also possible.

(a) **In model 1:** The Eco-RI site is 0.5 kb from one end and the Bam-HI site is 2.0 kb from that same end.

(b) **In model 2:** The Eco-RI site is 0.5 kb from one end and the Bam-HI site is 2.0 kb from the other end or 3.0 kb from that end.

- **Model 1** predicts that cutting with Eco-RI and Bam-HI will produce three fragments of 0.5 kb, 1.5 kb and 3.0 kb (going from left to right along the DNA).
- **Model 2** predicts that cutting with both enzymes will produce three fragments of 0.5 kb, 2.5 kb and 2.0 kb.

The actual data show three fragments with sizes 2.5 kb, 2.0 kb and 0.5 kb, thereby validating model 2.

Results:

Uncut	*Eco-RI*	*Bam-HI*	*Eco-RI + Bam-HI*
5.0 kb	4.5 kb	3.0 kb	2.5 kb
	0.5 kb	2.0 kb	2.0 kb
			0.5 kb

Interpretation:

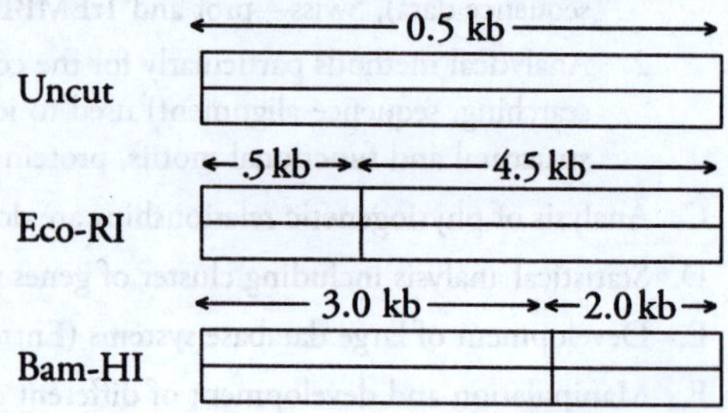

Construct models:

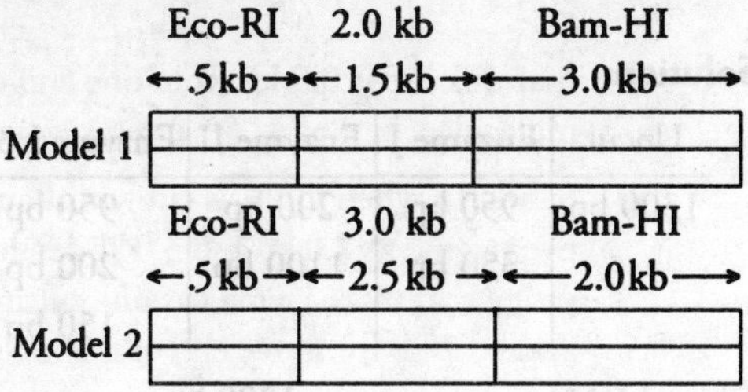

Conclusion: Eco-RI + Bam-HI data indicate model 2 is correct.

Example 2:

(i) **A population of Cloned DNA fragments prepared:** Cloned 7.0 kb DNA fragments.

(ii) **Results:** Uncut: 7.0 kb

Cut: (i) Hind-III: 6.2 kb and 0.8 kb

(ii) Sal-I: 5.8 kb and 1.2 kb

(iii) Hind-III and Sal-I: 5.8 kb, 0.8 kb, 0.4 kb

(iii) **Interpretation:**

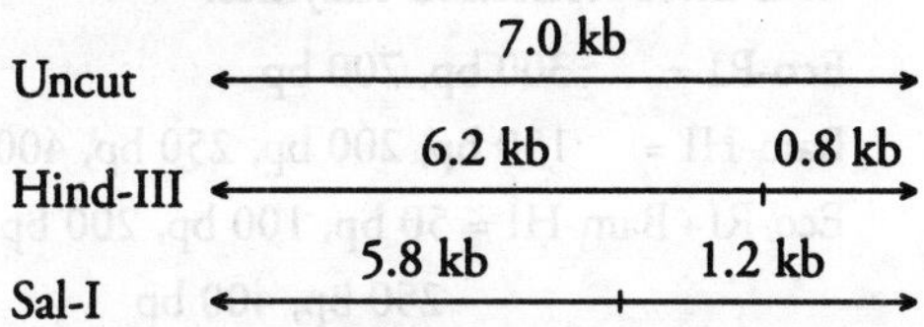

(iv) **Construct models:**

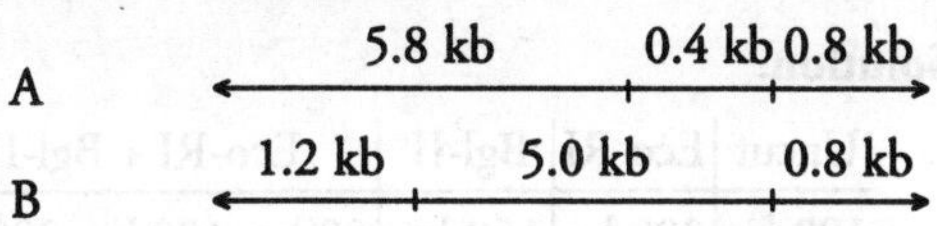

(v) **Conclusion:** Hind-III and Sal-I: data indicate model A is correct.

Example 3:

(i) **A population of cloned DNA fragments:** Cloned 10 kb DNA molecules.

(ii) **Results:**
- Uncut: 10 kb DNA
- Cut with restriction endonuclease A: 2 kb and 8 kb DNA
- Cut with restriction endonuclease B: 3 kb and 7 kb
- Cut with both A and B: 2 kb, 3 kb and 5 kb.

(iii) **Interpretation:**

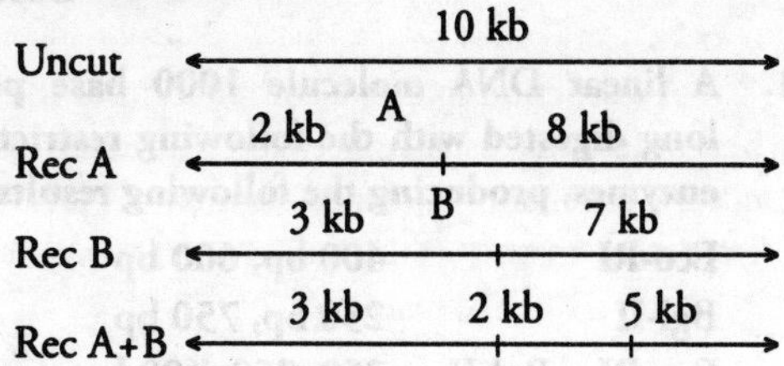

(iv) **Construction of map:**

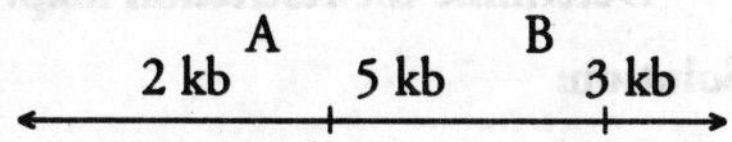

(v) **Conclusion:** Enzyme A cuts near one end of the molecule. Enzyme B must cut either near the same end or near the other end. The size of the fragments produced by both enzymes acting together rules out the first alternative and leads to the unambiguous order of restriction nuclease cutting sites shown above.

Example 4:

You are mapping the gene responsible for a human genetic disease and it is found that the gene is linked to a RFLP detected with probe B 101. You hybridize labelled B 101 DNA to DNAs from a panel of mouse human hybrid cells. The table shows the human chromosome present in each hybrid cell line whether or not the probe is hybridized to DNA from each. Which human chromosome carries the gene?

Cell line	Human chromosome content	Hybridization probe
A	1, 2, 17	–
B	3, 5, 9, 12	+
C	5, 18, 12	–
D	1, 3, 9	+
E	2, 9, 8	–
F	2, 3, 7	+

Results:

(i) Positive cell lines are B, D, F, i.e., B(3, 5, 9, 12) D(1, 3, 9) F(2, 3, 7).

(ii) Chromosome 3 is common and the RFLP maps to chromosome 3.

(iii) The diseased gene must be linked to RFLP.

Solved Problems

1. A linear DNA molecule 1000 base pairs long digested with the following restriction enzymes, producing the following results:

Eco-RI 400 bp, 600 bp
Bgl-II 250 bp, 750 bp
Eco-RI + Bgl-II 250, 350, 400 bp

Determine the restriction map.

Solution:

(i) Each enzyme alone produces two fragments, so the molecule has one site for each enzyme.

(ii) Since we get different sized fragments with each enzyme, the sites must be located asymmetrically along the DNA.

Uncut	Eco-RI	Bgl-II	Eco-RI + Bgl-II
1000	400 bp	250 bp	250 bp
	600 bp	750 bp	350 bp
			400 bp

Uncut: ← 1000 bp →

Eco-RI: ← 400 bp | 600 bp →

Bgl-II: ← 250 bp | 750 bp →

Restriction map: ← 400 bp | 600 bp →
350 | 250
↑ Eco-RI ↑ Bgl-II

2. You have recovered a cloned DNA segment of interest and determined that the insert is 1300 bp in length. To characterise this cloned segment, you have isolated the insert and decided to construct a restriction map. Using this enzyme I and enzyme II followed by gel electrophoresis, you determine the number and size of this fragments produced by enzymes I & II alone and in combination as shown here:

Enzymes	**Restriction fragments**
I	**350 bp, 950 bp**
II	**200 bp, 1100 bp**
I & II	**150 bp, 200 bp, 950 bp**

Construct a restriction map from these data, showing the position of the restriction sites relative to one another and distance between base pairs.

Solution:

Uncut	Enzyme I	Enzyme II	Enzyme I & II
1300 bp	950 bp	200 bp	950 bp
	350 bp	1100 bp	200 bp
			150 bp

Cloned DNA (uncut): ← 1300 bp →

Cut with enzyme I: ← 950 | 350 →

Cut with enzyme II: ← 1100 | 200 →

Cut with enzyme I & II: ← 950 | 150 | 200 →
Enzyme II

Restriction map: ← 950 bp | 150 bp | 200 bp →
Enzyme I   Enzyme II

3. A linear DNA molecule 1000 bp long gives the following size fragments when treated with these restriction enzymes:

Eco-RI = 300 bp, 700 bp

Bam-HI = 150 bp, 200 bp, 250 bp, 400 bp

Eco-RI + Bam-HI = 50 bp, 100 bp, 200 bp, 250 bp, 400 bp

Derive restriction map.

Solution:

Uncut	Eco-RI	Bgl-II	Eco-RI + Bgl-II
100 bp	300 bp	150 bp	50 bp, 100 bp, 200 bp
	700 bp	200 bp	250 bp, 400 bp
		250 bp	
		400 bp	

Uncut: ← 1000 bp →

Eco-RI: ← 300 bp | 700 bp →

Bam-HI: ← 200 bp | 150 bp | 250 bp | 400 bp →

Eco-RI + Bam-HI: ← 200 bp | 100 | 50 | 250 bp | 400 bp →

Restriction map:
Eco-RI
300 bp ↓ 700 bp
200 bp ↑ 150 bp ↑ 250 bp ↑ 400 bp
Bam-HI Bam-HI Bam-HI

4. A piece of DNA 900 bp long is cloned and then cut out of the vector for analysis. Digestion of this linear piece of DNA with three different restriction enzymes singly and in all possible combinations of pairs gave the following restriction fragment size data:

Enzymes (s)	***Restriction fragments***
Eco-RI	**200 bp, 700 bp**
Hind-III	**300 bp, 600 bp**
Bam-HI	**50 bp, 350 bp, 500 bp**
Eco-RI + Hind-III	**100 bp, 200 bp, 600 bp**
Eco-RI + Bam-HI	**50 bp, 150 bp, 200 bp, 500 bp**
Hind-III + Bam-HI	**50 bp, 100 bp, 250 bp, 500 bp**

Construct a restriction map from these data.

Solution:

Uncut	Eco-RI	Hind-III	Bam-HI	Eco-RI + Hind-III	Eco-RI + Bam-HI	Hind-III + Bam-HI
900 bp	200, 700 bp	300, 600 bp	50, 350, 500 bp	100, 200, 600 bp	50, 150, 200, 500 bp	50, 100, 250, 500 bp

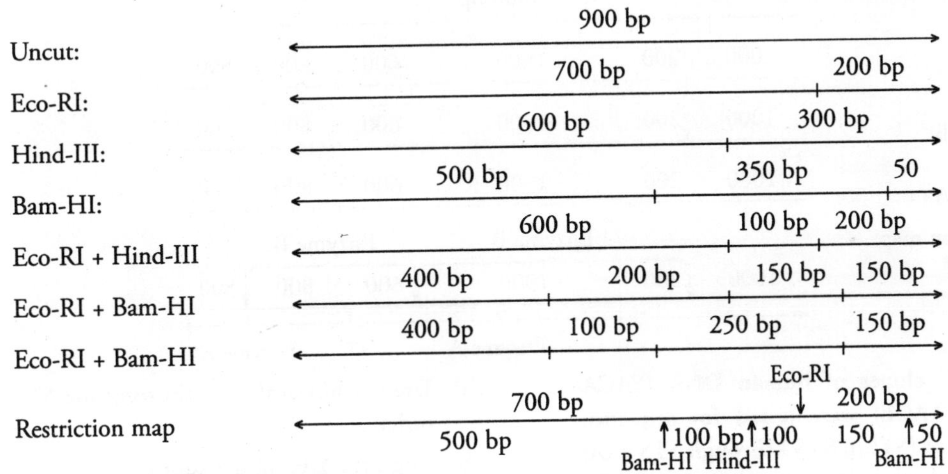

5. A piece of DNA 5000 bp long is digested with restriction enzymes A and B singly and together. The DNA fragments produced were separated by DNA electrophoresis and their sizes were calculated, with following results:

	Digestion with	
A	**B**	**A + B**
2,100 bp	**2,500 bp**	**1,900 bp**
1,400 bp	**1,300 bp**	**1,000 bp**
1,000 bp	**1,200 bp**	**800 bp**
500 bp		**600 bp**
		500 bp
		200 bp

[P.G., M.G., E.G. & I.G. 39]

Each A fragment was extracted from the gel and digested with enzymes B and each B fragment was extracted from the gel and digested with enzyme A. The sizes of the resulting DNA fragments were determined by gel electrophoresis with the following results:

A fragment		Fragments produced by digestion with B	B fragment		Fragments produced by digestion with A
2100 bp	→	**1900, 200 bp**	**2500 bp**	→	**1900, 600 bp**
1400 bp	→	**800, 600 bp**	**1300 bp**	→	**800, 500 bp**
1000 bp	→	**1000 bp**	**1200 bp**	→	**1000, 200 bp**
500 bp	→	**500 bp**			

Construct restriction map of the 5000 bp DNA fragment.

Solution:

Uncut	Enzyme B (fragment A)	Enzyme A (fragment B)	Enzyme A + B
5000 bp	1900, 200 bp 800, 600, 1000, 500 bp	1900, 600, 800, 500 bp 1000, 200 bp	1900, 1000, 800, 600, 500 bp 200 bp

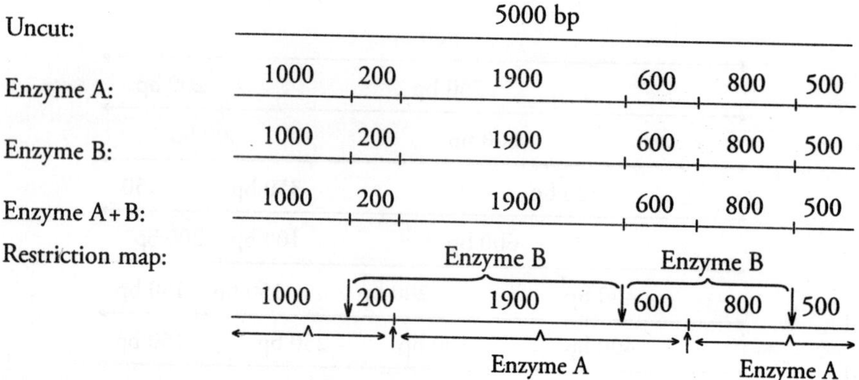

6. **Five YAC clones of human DNA (YACA- through YACE) were tested for sequence-tagged sites (STSs), STS I through STS 7 (An STS is a short unique sequence that is PCR amplified by specific primer pair). The results are shown in the following table, in which a plus sign shows that the YAC contain that STS.**

	STS						
YAC	**1**	**2**	**3**	**4**	**5**	**6**	**7**
A	+	–	+	+	–	–	–
B	+	–	–	–	+	–	–
C	–	–	+	+	–	–	+
D	–	+	–	–	+	+	–
E	–	–	+	–	–	–	+

(a) Draw a physical map showing the STS order.

(b) Align the YAC into a contig.

Solution:

(a) To determine the physical map showing the STS order, simply list the STS that are positive using parentheses, if the order is unknown and align them with one another to form a consistent order.

YAC–A			**1**	**4**	**3**	
YAC–B		**5**	**1**			
YAC–C				**4**	**3**	**7**
YAC–D	**(6, 2)**	**5**				
YAC–E					**3**	**7**

(b) When the sequence of STS is known, the YAC can be aligned as follows, although precise details of overlapping and the locations of ends are as follows:

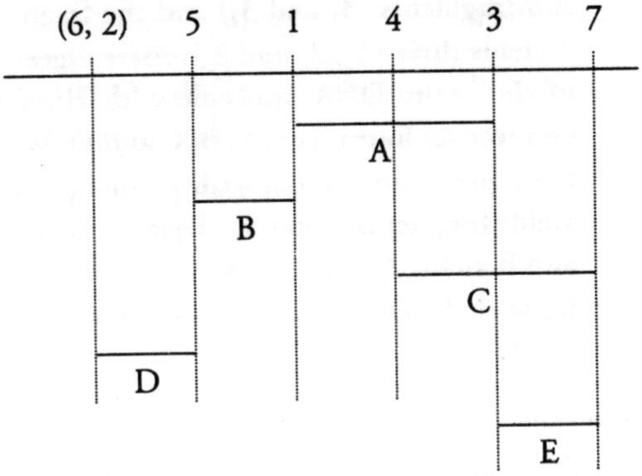

7. A *Neurospora* geneticist wanted to clone the gene CYS-1 which was believed to be near the centromere on chromosome five. Two RFLP markers (RFLP-1 and RFLP-2) were available in that vicinity and so he made the following cross.

Oak Ridge CYS-1 × Maurice villa CYS-1

Then 100 ascospores were tested for RFLP and CYS-1 genotypes and the following results were obtained.

RFLP–1	**0**	**M**	**0**	**M**	**0**	**M**
RFLP–2	**0**	**M**	**M**	**0**	**M**	**0**
CYS locus	**CYS**	**+**	**+**	**CYS**	**CYS**	**+**
Total genotypes	**40**	**43**	**2**	**3**	**7**	**5**

(a) Is CYS-1 in this region of the chromosome?

(b) If so, draw a map of the loci in this region labelled with map units.

What would be a suitable next step in cloning the CYS-1 gene?

Solution:

(i) The cross is CYS-1 RFLP-1° RFLP-2° × CYS-1$^+$ RFLP-1^M RFLP-2^M

The parental type will have the genotype of either strain and be in majority. The recombinant type will have a mixed genotype and be in minority of the ascospores clearly the first two ascospores are parental type and remaining types are recombinants.

(ii) (a) The CYS-1 locus is in this region of chromosome 5. If it was not in this region, linkage to either of the RFLP loci would not be observed.

(b) CYS-1 to RFLP-1 = 2+3 = 5;

$\frac{5}{100} \times 100 = 5\% = 5$ map unit (mu)

CYS-1 to RFLP-2 = 7+5 = 12;

$\frac{12}{100} \times 100 = 12\% = 12$ map unit (mu)

RFLP-1 to RFLP-2 = 2+3+5+7 = 17;

$\frac{17}{100} \times 100 = 17\% = 17$ map unit (mu)

|——5 mu——|——12 mu——|

RFLP-1 CYS-1 RFLP-2

↑…………………… 17 mu ……………………↑

(c) A number of strategies could be tried. Since this mutant is auxotrophic, functional complementation can be attempted. Positional cloning or chromosome walking from the RFLPs also is a very common strategy.

8. A *Neurospora* geneticist has just isolated a new mutation that causes aluminum insensitivity (al) in a wild-type strain, called Oak Ridge. She wishes to isolate the gene from the genome sequence and therefore, needs to map it by using RFLP. It so happens that there are RFLPs between Oak Ridge and another wild-type Mauriceville. For reasons that we do not need to go into, she suspects that it is located near the tip of the right arm of Chromosome 4. Luckily, there are three RFLP markers (1, 2 and 3) available in the vicinity and so the following cross is made.

al (Oak Ridge wild type) X al$^+$ (Mauriceville wild-type) one hundred progenies are isolated and tested for al and the six RFLP alleles 1°, 2°, 3°, 1^M, 2^M and 3^M. The results were as follows, where O and M represent the RFLP alleles and al and + represent al and al$^+$ respectively.

RFLP–1	**O**	**M**	**O**	**M**	**O**	**M**
RFLP–2	**O**	**M**	**M**	**O**	**O**	**M**
RFLP–3	**O**	**M**	**M**	**O**	**M**	**O**
al locus	**al**	**+**	**al**	**+**	**al**	**+**
Total genotypes	**34**	**36**	**6**	**4**	**12**	**8**

(a) Is the al locus in fact in this vicinity?

(b) If so, to which RFLP is it closest?

(c) How many map units separate the three RFLP loci?

Solution:

(i) This is a mapping problem but with the twist that some of the markers are traditional types and other are molecular markers (in this case RFLPs).

(ii) Nevertheless, the principle of mapping is the same as that which we used before. In other words, it is based on recombination frequency.

(iii) In any recombination analysis, we must be clear about the genotype of the parents before we classify progeny into recombinant classes. In this case, we know that the Oak Ridge parent must contain all O alleles and Mauriceville all M alleles, therefore, the parents are

$$al\ 1^O\ 2^O\ 3^O \times al^+\ 1^M\ 2^M\ 3^M$$

and knowing this makes determining recombinant classes easy.

(iv) We see from the data, that in the parental classes are the two most common (34 and 36). We first of all notice that the al alleles are **tightly linked to RFLP 1 (all progeny are al I^O or $+I^M$)**. Therefore, the **al locus is definitely on chromosome 4.**

(v) There 6 + 4 = 10 recombinants between RFLPs 1 and 2. So the loci must be $\left(\frac{10}{100} \times 100 = 10\%\right)$ 10 map units apart.

There 12 + 8 = 20 recombinants between RFLP 2 and RFLP 3; that is they are $\left(\frac{20}{100} \times 100\right)$ = 20 map units apart.

There are 6 + 4 + 12 + 8 = 30 recombinants between RFLPs 1 and 3 showing that these loci must flank RFLP 2. Therefore, the map is

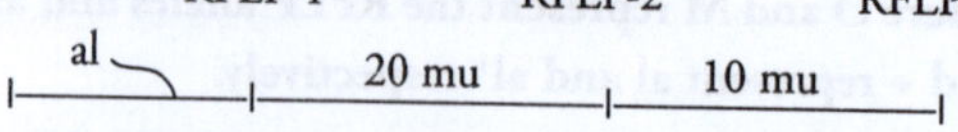

(vi) There are evidently no double recombinants which would have been of the types MOM and OMO.

9. You have purified DNA molecules and you wish to map restriction enzyme sites along its length. After digestion with Eco-RI, you obtain four fragments 1, 2, 3 and 4. After digestion of each of these fragments with Hind II, you find that fragment 3 yields two sub-fragments (3_1 and 3_2) and the fragment 2 yields three (2_1, 2_2 and 2_3). After digestion of the entire DNA molecule with Hind II, you recover four pieces A, B, C and D. When these pieces are treated with Eco-RI piece D yields fragments 1 and 3_1, A yields 3_2 and 2_1 and B yields 2_3 and 2_4. The C piece is identical with 2_2. Draw the restriction map of this DNA.

Solution:

1. Digestion with Eco-RI = 1, 2, 3 and 4 fragments
2. Digestion with Hind II = 2_1, 2_2 and 2_3, 3_1 and 3_2 (each fragment)
3. Digestion with Hind II entire DNA
4. Digestion with Eco-RI with (A, B, C & D) pieces

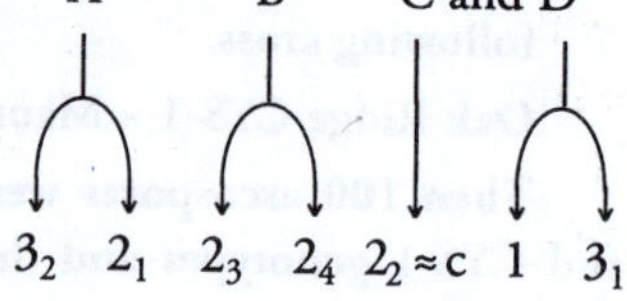

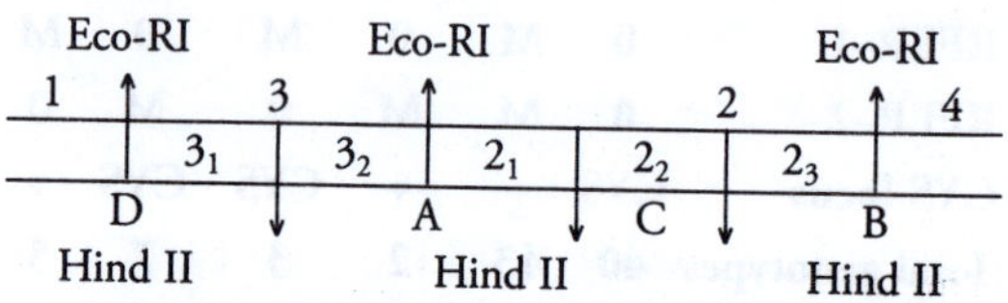

10. A linear fragment of DNA is cleaved with individual restriction enzymes Hind III and Sma I and then with combination of the two enzymes. The fragments obtained are

Hind III = 2.5 kb, 5.0 kb

Sma I = 2.0 kb, 5.5 kb

Hind III and Sma I = 2.5 kb, 3.0 kb and 2.0 kb

(a) Draw the restriction map.

(b) The mixture of fragments produced by the combined enzymes is cleaved with the enzyme Eco-RI resulting the loss of the 3 kb fragment (band stained with the Ethidium bromide on an agarose gel) and the appearance of a band

stained with ethidium bromide representing 1.5 kb fragments. Mark the Eco-RI cleavage site on the restriction map.

Solution:

(a)

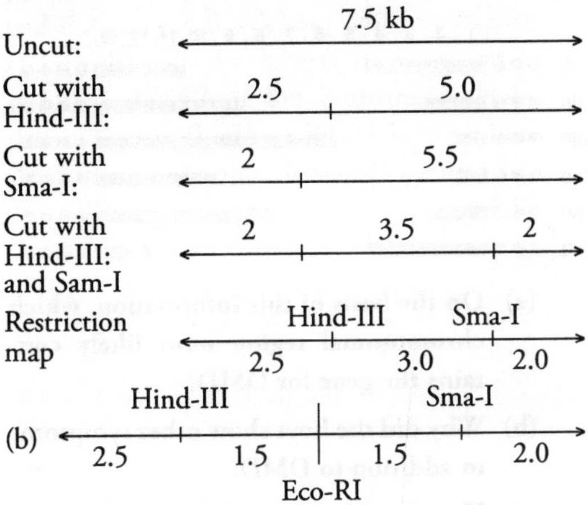

11. A DNA fragment produced by the restriction enzyme Sal I is inserted to a unique Sal I cloning site in a vector molecule. Digestion with restriction enzyme produces the following fragment sizes that originate from the inserted DNA.

(a) Sal I = 20 kb

(b) Sal I + Eco-RI = 7 kb, 13 kb

(c) Sal I + Hind III = 4 kb, 5 kb, 11 kb

(d) Sal I + Eco-RI + Hind = III 3 kb, 4 kb, 5 kb, 8 kb

What restriction map of the insert is consistent with these fragment sizes?

Solution:

(a) The cloned Sal I fragment is 20 kb in length and there are no Sal I sites within it.

(b) There is a single Eco-RI site 7 kb from one end.

(c) There are two Hind III sites within the fragment.

(d) The Hind III sites must flank the Eco-RI site dividing the 7 kb Eco-RI fragment into 3 kb and 4 kb subfragments and the 13 kb Eco-RI fragment into 5 kb and 8 kb subfragments. The 3 kb and 8 kb fragments must be adjacent because digestion with Sal I and Hind III alone produces a 11 kb fragments. Except for the left to right orientation, the inferred restriction map must be as follows:

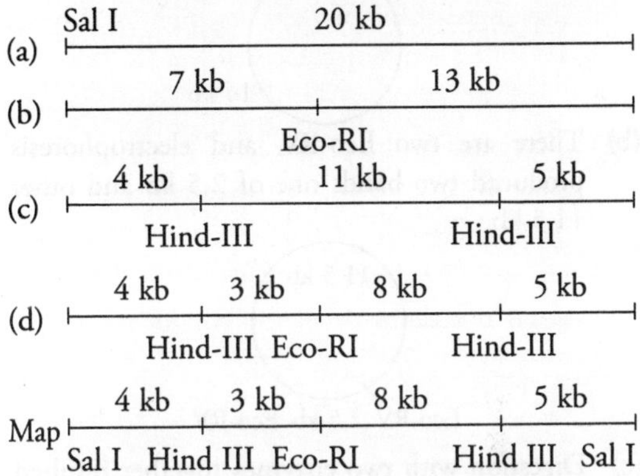

12. A circular bacterial plasmid containing a gene for tetracycline resistance was cut with restriction enzyme Bgl II. Electrophoresis showed one band of 14 kb.

(a) What can be deduced from this result?

(b) This plasmid was cut with Eco-RV, and electrophoresis produced two bands, one of 2.5 kb and other 11.5 kb. What can be deduced from this result?

(c) Digestion with both enzymes together resulted in three bands of 2.5, 5.5 and 6 kb. What can be deduced from this result?

(d) Plasmid DNA cut with Bgl II was mixed and ligated with donor DNA fragments, also cut with Bgl II to make recombinant DNA molecules. All recombinant clones were proved to be tetracycline sensitive. What can be deduced from this result?

(e) One recombinant clone was cut with Bgl II and fragments of 4 and 14 kb were observed. Explain the result?

(f) The same clone was treated with Eco-RV and fragments 2.5, 7 and 8.5 were observed.

Explain these results by showing a restriction map of the recombinant DNA.

Solution:

(a) There is one Bgl II site and the plasmid is 14 kb.

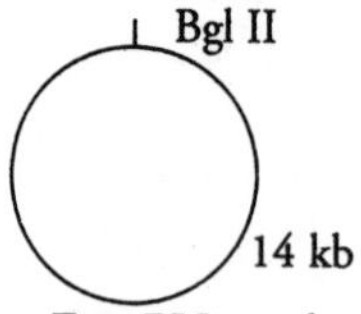

(b) There are two Eco-RV, and electrophoresis produced two bands one of 2.5 kb and other 11.5 kb.

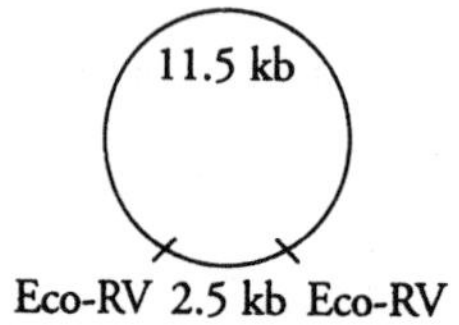

(c) Digestion with two enzymes together resulted in three bands of 2.5, 5.5 and 6 kb.

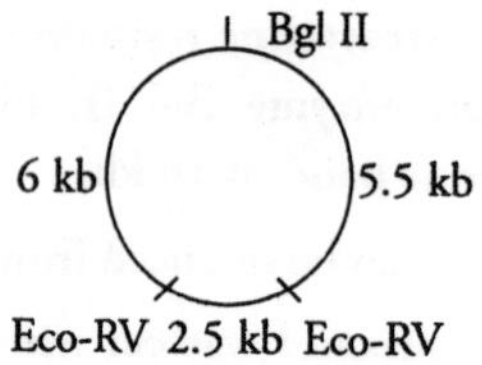

(d) The Bgl II site must be within the tet gene.

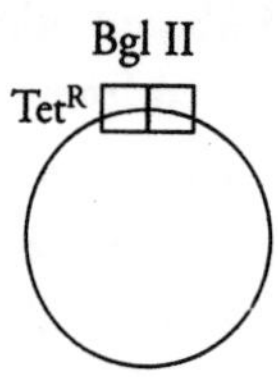

(e) There was an insert of 4 kb.

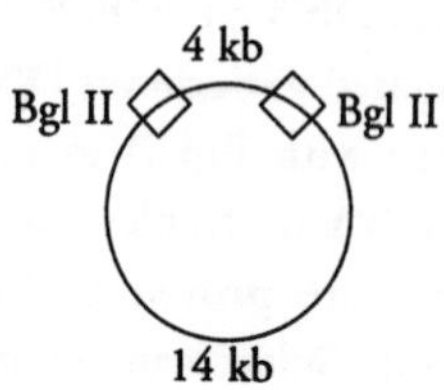

(f) There was an Eco-RV site within the insert.

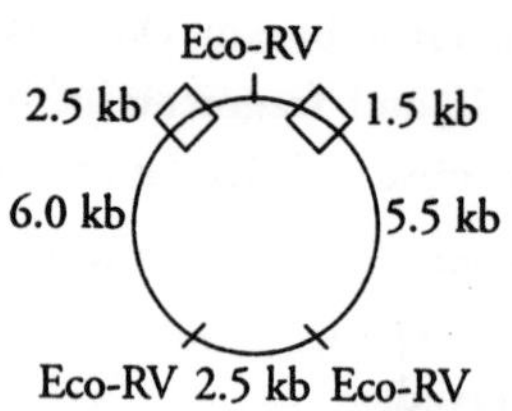

13. Duchenne muscular dystrophy (DMD) is an X-linked recessive human disease affecting muscles. Six small boys had DMD, together with various other disorders, and they were found to have small deletions of the X chromosome, as shown here:

(a) On the basis of this information, which chromosomal region most likely contains the gene for DMD?

(b) Why did the boys show other symptoms in addition to DMD?

(c) How would you use DNA samples from these six boys and DNA from unaffected boys to obtain an enriched sample of DNA containing the gene for DMD, as a prelude to cloning the gene?

Solution:

(a) The only region that all the deletions are lacking is the chromosomal region labelled 5, so this region presumably contains the gene for DMD.

(b) The other symbols probably result from the deletion of the other regions surrounding the DMD region.

(c) (i) If the DNA from all the DMD deletions is denatured (that is, its strands separated) and bound to some kind of filter, the normal DNA can be cut by shearing or by restriction enzyme treatment, denatured and passed through the filter containing the deleted DNA. Most DNA will bind to the filter, but the region-5 DNA will pass through this.

(ii) This process can be repeated several times. The filtrate DNA can be cloned and then used in FISH analysis to see if it binds to the DMD X-chromosome. If not, it becomes a candidate for the DMD containing sequence.

14. The restriction enzyme Eco-RI was used to cut (1) a linear DNA molecule containing 10 Eco-RI sites and (2) a circular DNA molecule also containing 10 Eco-RI sites.

(a) How many fragments are generated from the linear molecule?

(b) Can all the fragments of the linear molecule and circular molecule circularize? If not, how many can? Which ones cannot and why?

Solution:

(a) The Eco-RI digest will generate 11 fragments from the linear molecule (n + 1 fragments, where n is the number of sites) and 10 fragments from the circular molecule.

(b) (i) Only nine of the fragments (n – 1) from the linear molecule will circularize, since the two end fragments will only have one Eco-RI cohesive end.

(ii) All of the fragments from the circular molecule can circularize. This assumes that all fragments are long enough to circularize.

15. The maps of the sites for restriction enzyme R in the wild-type and the mutated cystic fibrosis genes are shown schematically in the following figure

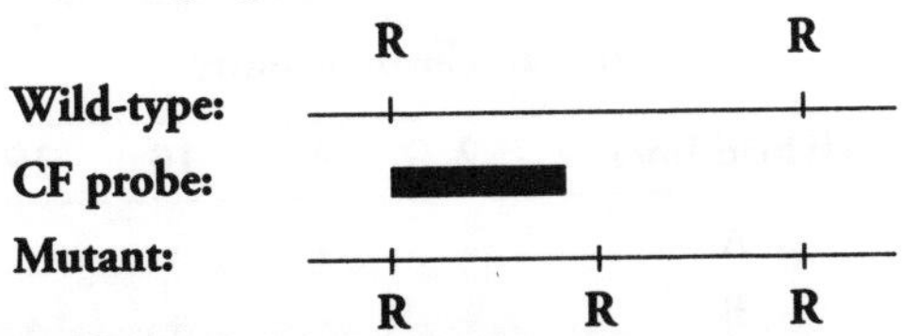

Samples of DNA obtained from a foetus (F) and her parents (M and P) were analyzed by gel electrophoresis followed by the Southern blot technique and hybridization with the radioactively labelled probe designated "CF probe" in the previous figure. The autoradiographic results are shown in the following figure

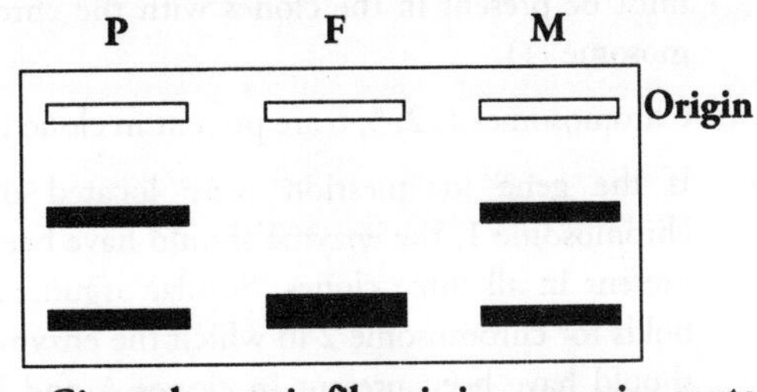

Given the cystic fibrosis is a recessive mutation, will the foetus be affected? Explain.

Solution:

(i) Chromosomes bearing CF mutations have a shorter restriction fragment than chromosome bearing wild-type alleles.

(ii) Both parent lanes (M and P) have two bands, indicating that each parent has a normal and a mutant chromosome. The parents are, therefore, heterozygous for the CF trait.

(iii) The foetus lane (F) shows only one (lower molecular weight) band. The size of the band indicates that the foetus has only mutant chromosome.

The intensity of the band is about twice that of the same sized band in the parent lanes. This is because the diploid genome of each parent only has one.

(iv) Since the foetus is homozygous for the CF trait, it will have CF (open box represent origin where DNA was loaded before electrophoresis).

16. A particular human enzyme is present only in clone B. The human chromosomes present in clones A, B and C appear as pluses (+) in the following table. Determine the probable chromosomal location of the gene for the enzyme.

	Human Chromosome							
Clone	1	2	3	4	5	6	7	8
A	+	+	+	+	–	–	–	–
B	+	+	–	–	+	+	–	–
C	+	–	+	–	+	–	+	–

Solution:

(i) If a gene is located on a chromosome, the gene must be present in the clones with the chromosome (+).

(ii) Chromosomes 1, 2, 5, 6 are present in clone B.

(iii) If the gene in question were located on chromosome 1, the enzyme should have been present in all three clones. Similar argument holds for chromosome 2 in which the enzyme should have been present in clones A and B and so on for the rest of the chromosomes.

(iv) The only chromosome that is unique to clone B is 6. Therefore, the gene is located on chromosome 6.

17. You have selected three mouse-human hybrid clones and analyzed them for the presence of human chromosomes. You then analyze each clone for the presence or absence of particular human enzymes (+ = presence of human chromosome or enzyme activity). Based on the following results indicate the probable chromosomal location for each enzyme.

Human Chromosome

Clone	3	7	9	11	15	18	20
X	–	+	–	+	+	–	+
Y	+	+	–	+	–	+	–
Z	–	+	+	–	–	+	+

Enzyme

Clone	A	B	C	D	E
X	+	+	–	–	+
Y	+	–	+	+	+
Z	–	–	+	–	+

Solution:

Enzyme	Clone	Chromosome
A	X & Y	11
B	X	15
C	Y & Z	18
D	Y	3
E	X, Y and Z	7

18. The results of an analysis of five human-mouse hybrids for five enzymes are given in table along with the human chromosomal content of each clone (+ = enzyme or chromosome present; – = absent). Deduce which chromosome carries which gene.

Human Enzyme	Clone A	B	C	D	E
1. Glutathione reductase	+	+	–	–	–
2. Malate dehydrogenase	–	+	–	–	–
3. Adenosine deaminase	–	+	–	+	+
4. Galactokinase	–	+	+	–	–
5. Hexoaminidase	+	–	–	+	–

Human chromosome

Clone	1	2	3	4	5	6	7	8	9	10	11	12	13	14	15	16	17	18	19	20	21	22
A	–	–	–	–	+	+	+	+	–	+	–	–	–	–	+	+	–	–	–	–	+	+
B	+	+	–	+	–	–	–	+	–	–	+	–	–	+	–	–	+	–	–	+	–	–
C	–	–	–	+	–	–	+	–	–	+	–	+	+	+	+	–	+	–	+	–	–	+
D	+	–	+	–	+	–	–	–	–	+	–	–	–	+	+	–	–	+	+	+	+	–
E	–	–	–	–	–	–	–	–	+	+	+	+	–	+	–	+	–	+	–	+	+	+

Solution:

Enzymes	Clones	Chromosome
1. Glutathione reductase =	A and B	8
2. Malate dehydrogenase =	B	13
3. Adenosine deaminase =	B, D & E	20
4. Galactokinase =	B and C	4
5. Hexoaminidase =	A and D	5

19. The accompanying table shows the only human chromosomes present in stable human-mouse cell hybrid lines

Human Chromosomes

Hybrid lines	2	4	10	19
A	–	+	+	–
B	+	–	+	+
C	–	+	+	+
D	+	+	–	–

The presence of four enzymes I, II, III and IV was investigated: I was present in A, B and C, but absent in D; II was present in B, and D but

absent in A and C; III was present in A, C and D but not in B; and IV was present in B and C but not in A and D. On what chromosomes are the genes for the four enzymes?

Solution:

(i) Enzyme I was present in A, B and C but absent in D. Here the presence of enzyme is indicated by + sign and absent by – sign. From the above table chromosome 10 satisfies this condition. Therefore, the gene for enzyme 1 is located on chromosome 10.

(ii) Enzyme II was present in B and D but not in A and C. Chromosome 2 satisfies this. Therefore, the gene for enzyme II is located on chromosome 2.

(iii) Enzyme III was present in A, C and D, but not in B; Chromosome 4 satisfies this. Therefore, the gene for enzyme III is located on chromosome 4.

(iv) Enzyme IV was present in B and C, but not in A and D. Chromosome 19 satisfies this. Therefore, the gene for enzyme IV is located on chromosome 19.

Enzyme	Chromosome
I	10
II	2
III	4
IV	19

20. The accompanying table shows the presence or absence of seven sequence-tagged sites (STSs), numbered 1 through 7, in each of six BAC clones, designated A through F. A + in the table means that PCR primers specific to the STS are able to amplify the sequence from the clone, means that the STS cannot be amlified from the clone. Use these dat a to make a map of the order of the BAC clones and of the STS markers located within the clones. In orienting the map, put BAC clone A at the top.

	STS marker						
BAC clone	**1**	**2**	**3**	**4**	**5**	**6**	**7**
A	–	–	+	+	–	–	–
B	–	+	–	–	–	–	+
C	+	–	–	–	+	–	–
D	+	–	–	–	–	+	–
E	–	–	–	+	–	+	–
F	–	+	–	–	+	–	–

Solution:

(i) The principle in ordering the BAC clones is that the DNA fragments present in any two (or more) clones the contain the same STS marker must overlap.

(ii) The table contains no information bearing on the relative sizes of the BAC clones or on the distance between the STS markers.

(iii) Thus, each clone is drawn as below though it had the same size, and the STS markers are spaced equally. Considering the STS markers shared between clones and starting with clone A, we obtain the map shown below:

STS Marker

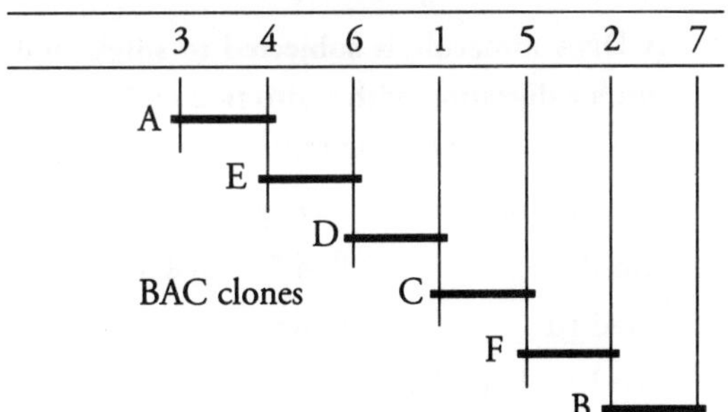

(iv) The mirror-image map with STS 3 on the right and STS 7 on the left is also a valid solution.

EXERCISE

1. A plasmid 3 kb in length contains a gene for ampicillin resistance and gene for tetracycline resistance. The plasmid has a single site for each of the following enzymes: Eco-RI, Bgl-II, Hind-III, Pst-I and Sal-I. If DNA is cloned into Eco-RI site, resistance to either antibiotic is not affected. DNA cloned into the Bgl-II, Hind-III or Sal-I sites abolishes tetracycline resistance and DNA inserted into the Pst-I site eliminates ampicillin resistance. If the plasmid is digested completely with enzyme mixes, the following fragments result:

Mixture	Fragment size (kb)
Eco-RI + Pst-I	0.7, 2.3
Eco-RI + Bgl-II	0.3, 2.7
Eco-RI + Hind-III	0.08, 2.92
Eco-RI + Sal-I	0.85, 2.15
Eco-RI + Bgl-II + Pst-I	0.3, 0.7, 2.00

Draw a restriction map of the plasmid and indicate the locations of the resistance genes and the sites of enzyme cleavage.

2. A DNA molecule is subjected to single and double digestion with restriction endonucleases and the following results are obtained.

Enzymes	Fragments (kb)
Eco-RI	2.9, 4.5, 7.4, 8.0
Hind-III	3.9, 6.0, 12.9
Eco-RI + Hind-III	1.0, 2.0, 2.9, 3.5, 6.0, 7.4

Draw a restriction map of the plasmid and indicate the locations of the resistance genes and the sites of enzyme cleavage.

[M.Sc. (Zoo) C.U. 2005]

3. Three mouse-human cell lines were scored for the presence (+) or absence (–) of human chromosomes with the results as follows:

Human Chromosomes

Clone	1	2	3	4	5	14	15	18
A	+	+	+	+	–	–	–	–
B	+	+	–	–	+	+	–	–
C	+	–	+	–	+	–	+	–

If a particular gene is located on chromosome 3 which clones should be positive for the enzyme from the gene?

4. You are mapping the gene responsible for a human genetic desease. You find that the gene is linked to a RFLP detected with a probe called X-21. You hybridize labelled X-21 DNA to DNAs from a panel of mouse human hybrid cells. The table below shows the human chromosomes present in each hybrid cell line and whether or not the probe is hybridized to DNA from each. Which human chromosome carries the disease?

Cell line	Human chromosome centre	Hybridization to X-21
A	1, 5, 21	+
B	6, 7	–
C	1, 22, Y	–
D	4, 5, 18, 21	+
E	8, 21, Y	–
F	2, 5, 6	+

Hints: Positive cell lines are A, D & F.

$$\begin{matrix} A = \\ D = \\ F = \end{matrix} \begin{pmatrix} 1, 5, 21 \\ 4, 5, 18, 21 \\ 2, 5, 6 \end{pmatrix} \text{ 5 is common}$$

Only chromosome 5 is present in all cell lines, whose DNA-hybridized to disease-linked probe.

DNA SEQUENCING

Determination of nucleotide or base sequence of a DNA molecule is known as DNA sequencing.

Types:

1. First generation DNA sequencing:
   (i) Maxam and Gilbert procedure.
   (ii) Sanger-Coulson method.
2. Second generation sequencing method:
   (i) Much faster and cheaper.
   (ii) Do not require DNA cloning.
3. Third generation sequencing method:
   (i) It can use single DNA molecule.
   (ii) It is rapidly refined.

Sanger-Coulson method of DNA sequencing:

It was developed by F. Sanger and coworkers. It is also known as '**dideoxy sequencing**'. It involves the extension of a short primer by DNA polymerase. Both linear and circular DNA can be sequenced with the dideoxy DNA sequencing method.

(i) In this sequencing method in which target DNA is denatured and annealed to an oligonucleotide primer which is then extended by DNA polymerase using a mixture of deoxynucleotide triphosphate (dNTPs) and chain terminating dideoxynucleotide triphosphates (ddNTPs).

[The chain terminating nucleotides are formed of sugars lacking of OH groups at both 2′ & 3′ carbon (hence the term dideoxy). Without 3′ OH group, a dideoxynucleotide cannot be used for further DNA polymerization. The chain terminating nucleotides permit synthesis to be stopped at a known base.

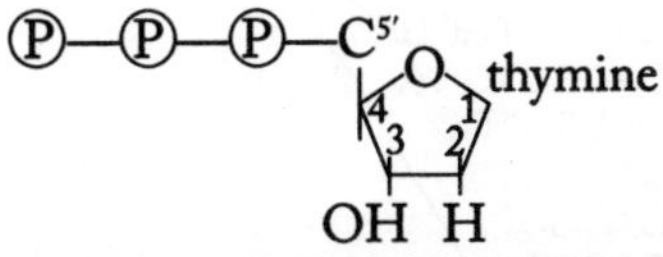

deoxythymidine thriphosphate (dTTP)

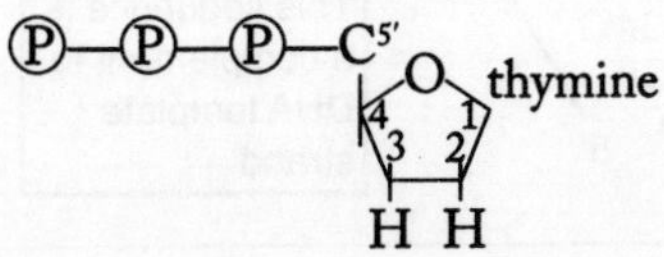

Dideoxythydine thriphosphate (ddTTP)

(ii) This primer is specifically constructed, so that 3′ end located next to the sequence of interest. This primer or any nucleotide should be labelled, so that final product can be detected on gel.

(iii) ddNTPs lack 3′ OH group to which the next dNTP of the growng DNA is added. Without the 3′ OH no more nucleotides can be added and DNA polymerase falls off.

- **Requirements:**

(i) A radio labelled primer with 3′ OH.

(ii) Single stranded DNA (obtained by heating or treated with NaOH or cloned in M13 phage or phagemid).

(iii) DNA polymerase (**Klenow fragment**). It is obtained by removing the first 323 amino acids of the DNA polymerase I polypeptide in view of its exonuclease activity. As a result Klenow fragment has only DNA polymerase activity.

(iv) Four reaction tubes are labelled with 'G', 'A', 'T' and 'C', each containing single-stranded DNA.

- **Procedure:**

(i) Target DNA is denatured to form a single-stranded DNA.

(ii) Four reaction tubes are labelled with 'A', 'T', 'G' and 'C'. Each contains single-stranded DNA which will be sequenced.

(iii) Four nucleoside triphosphates (dNTPs) is added in each tube. At least one of the nucleoside triphosphates is radioactively labelled, generally with ^{32}P (^{32}PDCTP).

(iv) A small amount of ddGTP is added to tube 'G', ddATP to tube 'A' ddTTP to tube 'T' and ddCTP to tube 'C'. The concentration of ddNTPs should be maintained to about 1% of the concentration of dNTPs, i.e., dXTP : ddXTP (where X can be any one of the four bases) is kept at approximately 100 : 1, so that probability of termination of a given X in the nascent chain is about $^1/_{100}$. This yields population of fragments terminating at all potential (X) termination sites, when a distance of a few nucleotides from the original primer terminus.

(v) DNA polymerase (Knelow fragment) is added to each tube. DNA synthesis starts and chain elongates. In each tube ddNTP is randomly

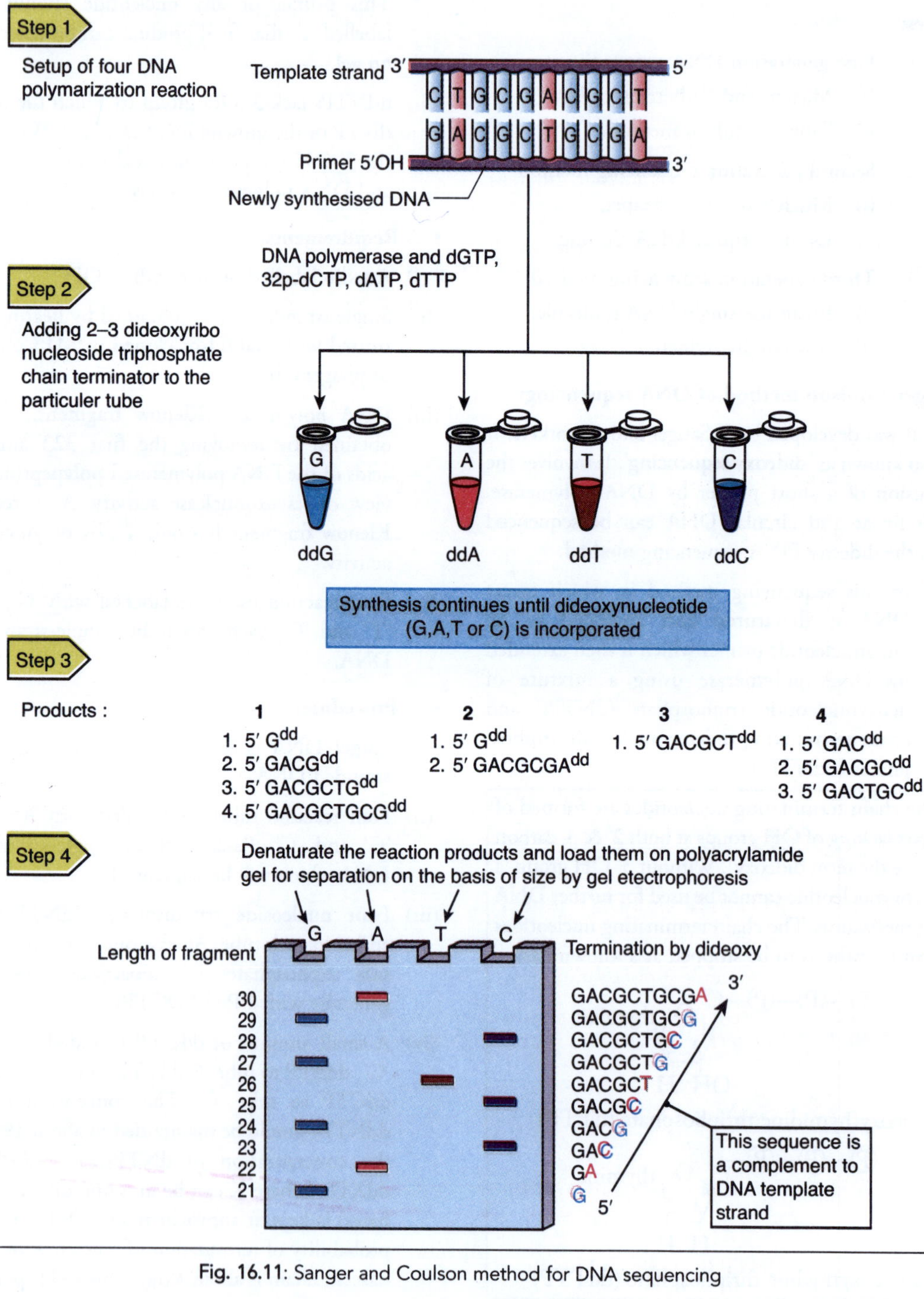

Fig. 16.11: Sanger and Coulson method for DNA sequencing

incorporated and chain terminates because ddNTPs lack 3′OH to which the next dNTP of the growing chain is added. Without 3′OH no more nucleotide can be added.

- **Observation:**

(i) After completion of the reaction, the reaction product of each tube is loaded in the well of high resolution polyacrylamide gel in such a way that the DNA fragment of each tube are separated by electrophoresis in four different lanes with bands.

(ii) The location of different bands are revealed by autoradiogaphy.

(iii) The DNA sequence of newly synthesized strand is visualized from autoradiogram by reading up on the sequencing ladder from bottom to the top to give the sequence in 5′ to 3′ orientation.

- **Interference:**

(i) The shortest fragment migrates greatest distance and gives rise to the band nearest to anode and each subsequent band will be one nucleotide longer than the previous one.

(ii) The DNA fragment near the bottom represents the nucleotide position closest to the primer end (5′ end).

(iii) The sequence is read from the bottom to the top. The sequence is the complementary to that of template strand.

- **Pyrosequencing:**

In 1998 Pal Nyren Mathias Uhlen and Mostafa Ronaghi developed this method.

(i) In this method a sequencing primer is hybridized to a single-stranded template, and sequence is based on detection of the pyrophosphate (PPi) released during DNA synthesis.

(ii) The prosequencing is simple and robust. It contains the DNA template, sequencing primer, APS (adenosine-5′ phosphosulphate) chemiluminescent enzyme, **luciferin** and the enzymes DNA polymerase (Klenow fragments of *E. coli* DNA Polymerase I), ATP sulfurylase, luciferase and apyrase.

(iii) To this reaction mixture, the four dNTPI are added subsequently one after another, e.g., in the sequence G, C, T, A.

Steps along with reaction in Pyrosequencing

Step-1 **Release PPi by DNA polymerase:**

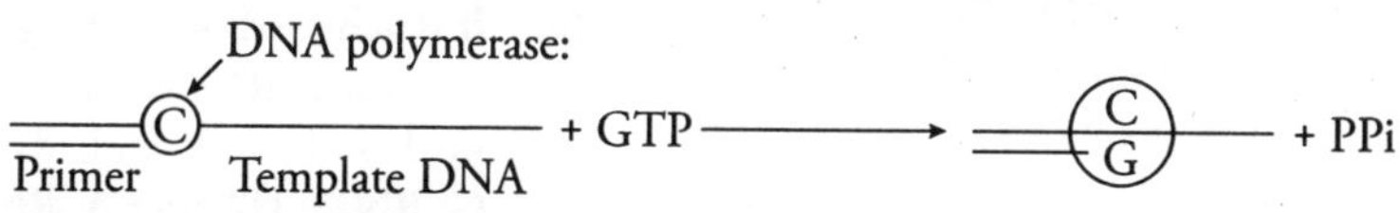

Step-2 **ATP synthesis by ATP sulfurylase:**

$$\text{Adenosine 5-phosphosulphate (APS)} + \text{PPi} \xrightarrow{\text{ATP sulfurylase}} \text{ATP} + SO_4^-$$

Step-3 **Luciferase uses ATP to generate visible light:**

$$\text{Luciferin} + \text{ATP} \xrightarrow{\text{Luciferase}} \text{Oxyluciferin} + \text{light}$$

Step-4 **Pyrase action degrades NTPs:**

$$\text{dNTP (including ATP)} \xrightarrow{\text{Apyrase}} \text{dNDP} + \text{dNMP} + \text{Phosphate}$$

- **Application:**
  (i) Confirmatory sequencing of disease gene.
  (ii) Genotyping of individual.
  (iii) Identification of bacterial species and strain.
  (iv) Comparative sequencing for variations.

- **Automated sequencing:**
  (i) This procedures use different fluorescent labels in the place of radioactive labelling of four dideoxynucleotides used for chain termination.
  (ii) Using this technique a DNA sequence containig thousand of nucleotides can be determined in a few hours.
  (iii) Each dideoxynucleotide is linked with a **fluorescent dye** that imparts different colours to all the fragments terminting in that nucleotide.
  (iv) In the four-track system of automated DNA sequencing, each of the four dideoxynucleotides is used in a separate reaction and the products are run in four adjacent lanes of the gel. But in single-track system, a different **fluorochrome** is attached to each of the four dideoxynucleotides and all of them are used in same reaction.
  (v) DNA fragments of different colours are separated by their respective size in a single electrophoretic gel. Reaction products are subjected to polyacrylamide gel electrophoresis. The negatively charged DNA strands migrate through the pores of the gel and forwards the positive poles. The small size DNA fragments migrate faster.
  (vi) The bands produced in the polyacrylamide gel are identified with the help of a **fluorescence detector**, which identifies the fluorescent signal emited by each band.
  (vii) The **fluorochromes** are excited by a laser beam and the resulting fluoroscence signal is sensed by a photovoltaic cell. The resulting data are fed directly into a computer which determines the sequence.
  (viii) A tracing electrogram of emitted light of the four dyes is generated by a computer. Colour of each dye represents the different nucleotides. Computer converts the data of emitted light in the nucleotide sequences.

- **Advantages:**
  (i) Radioactivity is not used.
  (ii) Gel processing and autoradiography are not needed.
  (iii) The sequence data are directly fed into a store in computer.

- **RNA sequencing:**
  (i) RNA sequencing sometime becomes necessary when one has to determine the position of modified nucleotide present in RNAs.
  (ii) RNA sequencing involves the production of cDNA by RT-PCR and then sequencing the cDNA product. If the primers for RT-PCR are correctly designed, only the desired mRNA would be copied.
  (iii) It is achieved by cleaving 5′ end labelled RNA by using RNases that cleaves a particular nucleotide at 3′ end. RNase T1 cleaves after G, RNase U2 after A, RNase P by M1 after A and U, and *Bacillus cereus* RNase after U and C. The fragments are analyzed by PAGE.

Solved Problems

Problem 1

How was the dideoxy technique improved with fluorescent dye?

Solution:

Use of fluorescent dyes has allowed for the automation process and the elimination of radioactive tag.

Problem 2

How are 2′ – 3′ dideoxy ribnoucleotide triphosphates used in DNA sequencing protocol?

Solution:

(i) The 2′ – 3′ dideoxyriboucleotide triphosphates function as specific terminators of DNA synthesis.

(ii) When 2′ – 3′ dideoxyriboucleotide monophosphate is added to the end of nascent DNA chain, that chain no longer be extended by DNA polymerase, because the absence of 3′ OH is required for chain extension.

(iii) By using approximate ratios of 2′ – 3′ dideoxyribonucleosides triphosphate and DNA chains are produced that are terminated at all possible nucleotide positions.

(iv) Separation of these nascent DNA chains by gel electrophoresis and detection of their position in the gel with radioactive nucleotide or fluorescent dyes are then used to determine their nucleotide sequences.

Problem 3

Highlight about the Sanger DNA sequencing method.

Solution:

(i) This method uses dideoxynucleotides to terminate DNA synthesis by yielding a series of DNA fragments whose sizes can be measured by electrophoresis.

(ii) The last base of each of the fragments is known because we know which dideoxynucleotide was used to terminate each reaction.

(iii) Therefore, ordering three fragments by size, each fragment one (known) base longer than the next tells us the base sequence of the DNA.

Problem 4

A piece of DNA has the sequence 3′ GGCG-TATTC 5′. It is sequenced using dideoxy method. How many bands are found on the ladder gel? How many bands and of what size are found for each reaction mixture?

Solution:

(i) Since the piece of DNA is nine bases long, the total number of bands in all four lanes of a sequencing gel add up to nine.

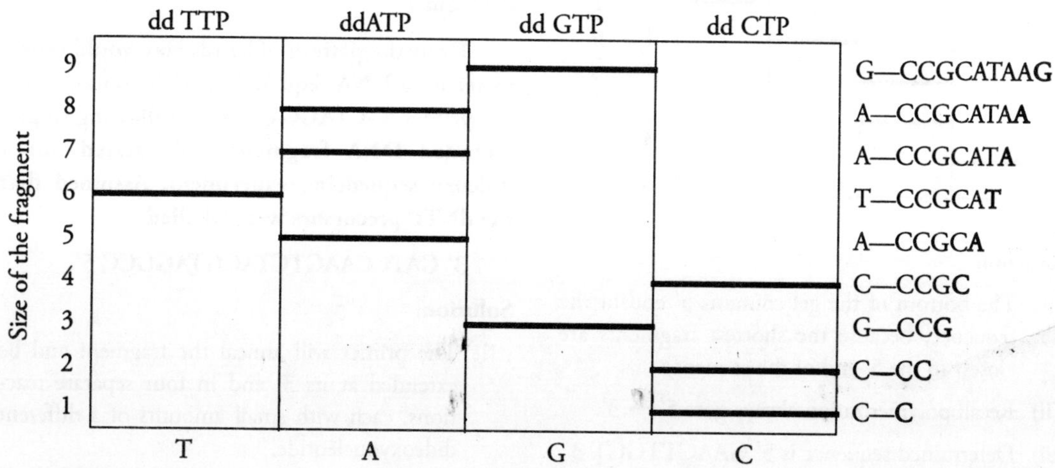

(ii) Four reaction mixtures occur, each with one of the dideoxynucleotides. In the reaction mixture with ddTTP, chain termination occurs at the adenine in the piece of DNA, i.e., a DNA segment is synthesized which is six bases long.

(iii) In the reaction mixture with ddATP, chain termination occurs opposite to each of the thymines, producing five DNA segments and eight nucleotides.

(iv) In the reaction mixture with ddATP, chain termination occurs opposite to the cytosine in positions three and nine.

(v) Add in the reaction mixture with ddCTP, chain termination occurs after the synthesis of segments, opposite to guanine in positions one, two and four bases long.

(vi) It is noted that the gel gives us the sequence of the complement strand of the original piece of single-stranded DNA.

Problem 5

The following gel pattern was obtained in an attempt to sequence a DNA molecule using Sanger method. What is the sequence of the DNA molecule in question?

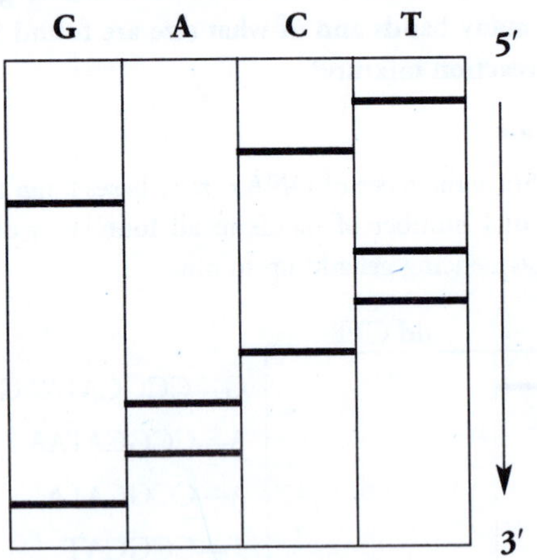

Solution:

(i) The bottom of the gel contains 5′ end of the sequence, because the shortest fragments are closest to the 5′ end of the sequence.

(ii) Recall polymerization always goes 5′ → 3′.

(iii) Determined sequence is 5′ GAACTTGCT 3′.

This means that the template sequence is 5′ AGCAAGTTC 3′.

Problem 6

The following gel pattern was obtained in an attempt to sequence DNA molecule using sanger method. What is the sequence of the DNA molecule in question?

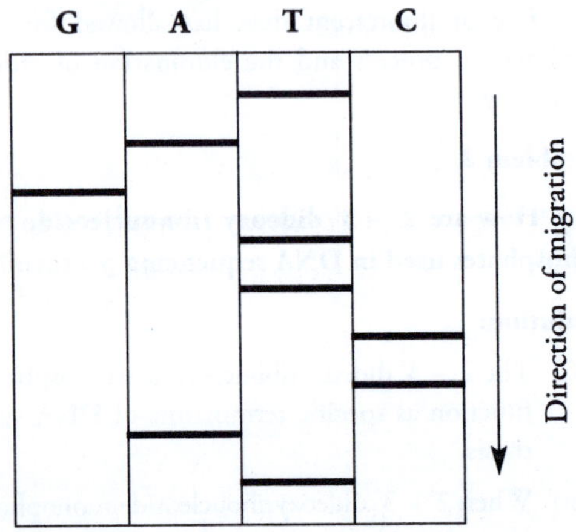

Solution:

(i) The bottom of the gel contain 5′ end of the synthesized sequence.

(ii) That is because the shortest fragments are closet to the 5′ end of the sequence.

(iii) Remember polymerization always goes 5′ → 3′. The determined sequence is 5′ TACCTTGAT 3′. This means, the template sequence is 5′ ATCAAGGTA 3′.

Problem 7

Draw the pattern of bands you would expect to see on a DNA sequencing gel if you annealed the primer 5′-CTAGG-3′ to the following single-stranded DNA fragment and carried out a dideoxy sequencing experiment. Assumed that the dNTP precursors were labelled.

3′ GATCCAAGTCTACGTAGGCC 5′.

Solution:

(i) The primer will anneal the fragment and be extended at its 3′ end in four separate reactions, each with small amounts of a different dideoxynucleotide.

(ii) **In each reaction some chains will be prematurely terminated when the dideoxynuceotide is incorporated.**

(iii) **By using dNTP precursors, all extension products will be labelled and be observed as distinct labelled bands after separation on a denaturing polyacrylamide gel and signal detection.**

(iv) Gel analysis 3′-GATCCAAGTCTACGTATAG GCC-5′

Primer: 5′-CTAGG: 3′ GATCCAAGTCTACGTATAGGCC-5′

5′-**CTAGG**TTCAGATGCATATCCGG-3′

(v) Extension product

ddNTP added

5′ CTAGGTTCAGATGCATATCCGG

5′ CTAGGTTCAGATGCATATCCG

5′ CTAGGTTCAGATGCATATCC

5′ CTAGGTTCAGATGCATATC

5′ CTAGGTTCAGATGCATAT

5′ CTAGGTTCAGATGCATT

5′ CTAGGTTCAGATGCAT

5′ CTAGGTTCAGATGCA

5′ CTAGGTTCAGATGC

5′ CTAGGTTCAGATG

5′ CTAGGTTCAGAT

5′ CTAGGTTCAGA

5′ CTAGGTTCAG

5′ CTAGGTTCA

5′ CTAGGTTC

5′ CTAGGTT

5′ CTAGGT

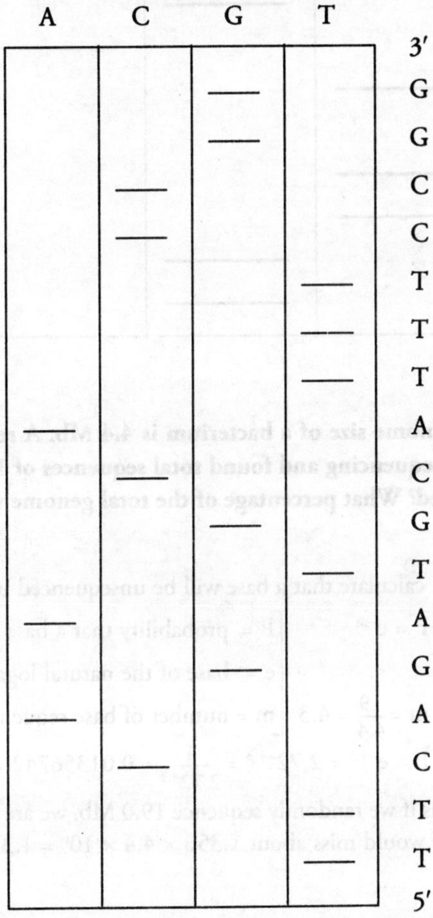

Problem 8

Draw the expected gel pattern derived from the dideoxy sequencing method for a template strand with the following sequence.

5′-CAGCGAATGCGGAA-3′.

Solution:

(i) Since DNA synthesis proceeds 5′ – 3′ direction, the base will be 5′-TTCCGCATTCGCTG-3′.

5′-TTCCGCATTCGCTG-3′.

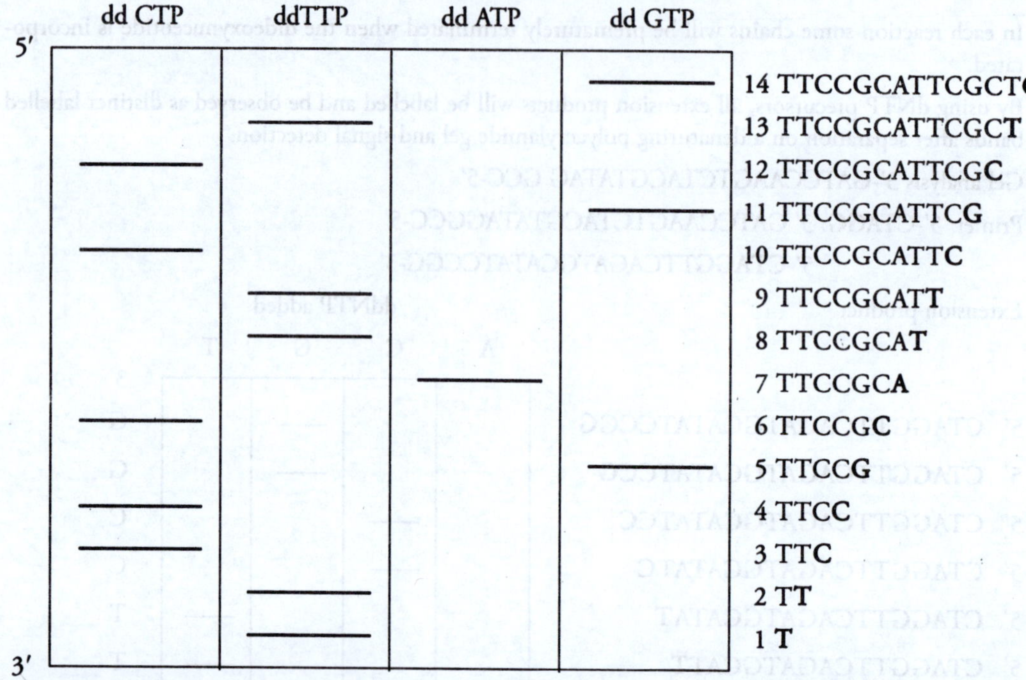

Problem 9

The genome size of a bacterium is 4.4 Mb. A researcher of R. K. M. Vidyamandira carried short gun DNA sequencing and found total sequences of 19 Mb. What is the probability that a base will be unsequenced? What percentage of the total genome will be left unsequenced?

Solution:

We can calculate that a base will be unsequenced using the following formula.

$P = e^{-m}$

P = probability that a base will be unsequenced.

e = base of the natural logarithm, e = 2.72

m = number of base sequenced divided by the total genome size.

Here, $m = \frac{19}{4.4} = 4.3$

$P = e^{-4.3} = 2.72^{-4.3} = \frac{1}{2.72^{4.3}} = 0.01356742 = 0.01356$ or, 1.356%.

It means if we randomly sequence 19.0 Mb, we are likely to miss 1.356 of the genome with genome size 4.4 Mb. We would miss about $1.356 \times 4.4 \times 10^6 = 1.356 \times 4.4 \times 10^6 = 5.9664 \times 10^6 = 5966400$ bp.

SECTION-III

EVOLUTIONARY GENETICS AND IMMUNOGENETICS

Population Genetics

17

HARDY-WEINBERG EQILIBRIUM

- **POPULATION:**

A population is a local group belonging to a single species within which mating actually or potentially occurs.

- **GENE POOL:**

Sum total of all different alleles present in the breeding members of a population at a given time. Here the set of genetic information is carried by all interbreeding members of a population.

- **GENE FREQUENCY:**

The proportion of a particular allele or gene in relation to all alleles at a given locus in a particular population is termed as gene frequency.

- **GENOTYPIC FREQUENCY:**

The total number of a kind of individuals from a population all of which exhibit similar characters with respect to the locus is known as genotypic frequency.

- **ASSUMPTIONS OF HARDY-WEINBERG LAW:**

In a large panmictic population, allelic frequency or gene frequency remains constant generation after generation if the population remains free from evolutionary forces like mutation, migration, selection, genetic drift, and the genotypic frequency maintains the equilibrium.

- **LARGE POPULATION SIZE:**

The population is infinitely large and mates at random. In a large population sampling errors are negligible.

- **NO SELECTION:**

No selection is operative, i.e., each genotype under consideration can survive just like any other (no differential mortalities) and all genotypes, produced by random mating are equally viable and fertile, i.e., efficient in production of progeny.

- **CLOSED POPULATION:**

Population is closed, i.e., neither immigration of individuals from another population nor emigration from the population under consideration is allowed.

- **NO MUTATION:**

There is no mutation from one allelic state to another. Mutation may be allowed if the forward and backward mutation rates are equivalent.

- **MEIOSIS IS NORMAL:**

Meiosis is normal, so that chance is the only factor operative in gametogenesis.

Example:

A locus has two alleles, i.e., A and a; the frequency of allele A in both egg and sperm is represented by p and the frequency of a is represented by q.

The sum of p and q represents 100 percent of alleles that gene is population, then $p + q = 1$.

In a random mating state the probability that sperm and egg both contain A alleles is $p \times p = p^2$ and q both contain a alleles, i.e., $q \times q = q^2$, similarly, the chance the gamets will carry unlike alleles, i.e., $p \times q$, $q \times p$, i.e., 2pq.

The equation is $(p + q)^2 = p^2 + 2pq + q^2$

$$A + a = AA \quad 2Aa \quad aa$$

Explanation and Application of H-W Law:

(i) The Hardy-Weinberg law states that when a population is in equilibrium, the genotypic frequency will be in the proportion p^2, 2pq, q^2.

(ii) Take two groups of population differing in a single gene, i.e. one is 'AA' and other is 'aa'.

(iii) Consider the two genes or alleles 'A' and 'a' in a stable population, i.e. population is large, random mating (Panmictic), any individual has equal chance to mate with any of individual of opposite sex and there is no migration or mutation.

(iv) The frequency of each allele can be expressed as percentage or fraction of I. Thus if we consider 'p' is the frequency of 'A' and 'q' is the frequency of a, the p + q must be 1.

(v) The genotypes of individual resulting from the cross between A and a are as below:

♂ \ ♀	A(p)	a(q)
A(p)	$AA(p^2)$	Aa(pq)
a(q)	Aa(pq)	$aa(q^2)$

Thus, genotypes of resultant cross will be AA, 2Aa and aa and their frequencies will be p^2, 2pq, q^2 respectively. Therefore sum of the genotypic frequencies in the next generation will be

$$F_1 \rightarrow \frac{p^2}{AA} + \frac{2pq}{2Aa} + \frac{q^2}{aa} = 1.00$$

or, $\quad (p + q)^2 = (1)^2 = 1$

In the new generation (F_1) we can calculate the frequency of two alleles as follows:

- Frequency of A $= p^2 + \frac{1}{2} \times 2pq = p^2 + pq$
  $= p(p + q) = p.1 = p.$
- Frequency of a $= q^2 + 12 \times 2pq = q^2 + pq$
  $= q(p + q) = q.1 = q.$

(vi) If random mating occurs in the new generation (F_2), i.e. $F_1 \times F_1$

♂ \ ♀	$AA(p^2)$	Aa(2pq)	$aa(q^2)$
AA (p^2)	AA × AA $p^2 \times p^2 = p^4$	AA × Aa $p^2 \times 2pq = 2p^3q$	AA × aa $p^2 \times q^2 = p^2q^2$
Aa (2pq)	AA × Aa $p^2 \times 2pq = 2p^3q$	Aa × Aa $2pq \times 2pq = 4p^2q^2$	Aa × aa $2pq \times q^2 = 2pq^3$
aa (q^2)	AA × aa $p^2 \times q^2 = p^2q^2$	Aa × aa $2pq \times q^2 = 2pq^3$	aa × aa $q^2 \times q^2 = q^4$

	Types of mating	Sum of mating frequency	Offspring genotypes and frequency		
			AA	Aa	aa
1	AA × AA	p^4	p^4	x	x
2	AA × Aa	$4p^3q$	$2p^3q$	$2p^3q$	x
3	AA × aa	$2p^2q^2$	x	$2p^2q^2$	x
4	Aa × Aa	$4p^2q^2$	p^2q^2	$2p^2q^2$	p^2q^2
5	Aa × aa	$4pq^3$	x	$2pq^3$	$2pq^3$
6	aa × aa	q^4	x	x	q^4
	Total		$p^4 + 2p^3q + p^2q^2$	$2p^3q + 4p^2q^2 + 2pq^3$	$p^2q^2 + 2pq^3 + q^4$

Sums: (AA) $= p^4 + 2p^3q + p^2q^2 = p^2(p^2 + 2pq + q^2) = p2.1 = p^2$

(Aa) $= 2p^3q + 4p^2q^2 + 2pq^3 = 2pq(p^2 + 2pq + q^2) = 2pq.1 = 2pq$

(aa) $= p^2q^2 + 2pq^3 + q^4 = q^2(p^2 + 2pq + q^2) = q2.1 = q^2$

- Genotypic frequency of AA:

$$= p^4 + 2p^3q + p^2q^2$$

$$= p^2(p^2 + 2pq + q^2) = p^2(p + q)^2 = p^2.1 = p^2$$

- Genotypic frequency of Aa:

$= 2p^3q + 4p^2q^2 + 2pq^3$

$= 2pq(p^2 + 2pq + q^2) = 2pq(p + q)^2 = 2pq.1 = 2pq$

- Genotypic frequency of aa:

$= p^2q^2 + 2pq^3 + q^4 = q^2(p^2 + 2pq + q^2)$

$= q^2(p + q)^2 = q^2.1 = q^2$

Now in this F_2 generation, allele frequency of A and a.

Frequency of $A = p^2 + \frac{1}{2}pq = p^2 + pq = p^2(p + q) = 1$

Frequency of $a = q^2 + \frac{1}{2}pq = q(q + p) = 1$.

In the F_2 generation the frequency of A(p) and a(q) remain same as parent generation. Therefore, we can say that allele frequency remain constant from generation after generation.

- **Deviation from/Exception to Hardy Weinberg Equilibrium:**

A. Non-random mating

(i) Random mating is the key assumption of Hardy-Weinberg equilibrium.

(ii) If mating is not random, the simple relationsip between allele frequencies and genotype frequencies breakdown.

There are two ways in which members of a population might mate non-randomly, viz., assortive mating, consangineous mating.

B. Selection

(i) Natural selection occurs when there is differential survival or reproduction among individuals due to their phenotype.

(ii) If zygote produced by random mating have different survival rate.

(iii) We would not expect genotypic frequencies of the individuals that develop from the zygote to conform to the Hardy-Weinberg's predictions.

C. Population structure subdivision

(i) When a population is single interbreeding unit, it is said to be panmictic and members of the population is able to exchange genes in population.

(ii) In real population, population is subdivided and is genetically heterogenous. Gene frequencies deviation from HWE.

D. Inbreeding

It results decrease of heterozygosity in population beyond that expected on HWE.

E. Migration

(i) When individual move from one territory to another, they carry genes with them.

(ii) The introduction of genes by recent migration within a population can alter allele and genotypic frequencies within a population and disrupt Hardy-Weinberg equilibrium.

- **Assortive mating:**

Individuals may be more likely to mate with other individuals of the same or other genotypes. This is called assortive mating. It results decrease of heterozgous individuals in the next generation predicted by HWE.

- **Disassortive/Negative assortive mating:**

Here individuals are not to mate with individuals of their own genotype. It results the number of heterozygous individuals in excess of those expected under HWE.

Solved Problems

Problem 1

From a sample of 278 American Indians, the following MN blood types were obtained: M 78, N 61 and MN 139. Calculate the gene frequency of M and N.

Solution:

Genotypes/Phenotypes	M	MN	N	Total
Number of individuals	78	139	61	278
Genotypic frequency	$\frac{78}{278}$ = 0.28	$\frac{139}{278}$ = 0.5	$\frac{61}{278}$ = 0.219	$\frac{278}{278}$ = 1

Gene frequency of M in population

$= 0.28 + \frac{1}{2} \times 0.5 = 0.28 + 0.25 = 0.53.$

Gene frequency of N in population

$= 0.219 + \frac{1}{2} \times 0.5 = 0.219 + 0.25 = 0.469 = 0.47.$

Therefore, gene frequency of M = 0.53 and N = 0.47.

Problem 2

The MN blood type is determined by a pair of codominant alleles, L^M and L^N. A population of 600 persons had blood type with the following results: 300 were type M, 240 were type MN and 60 were type N. Calculate the gene frequency of L^M and L^N.

Solution:

Genotypes/Phenotypes	M	MN	N	Total
Number of individuals	300	240	60	600
Genotypic frequency	$\frac{300}{600}$ = 0.5	$\frac{240}{600}$ = 0.4	$\frac{60}{600}$ = 0.1	$\frac{600}{600}$ = 1

Gene frequency of L^M in population

$= 0.5 + \frac{1}{2} \times 0.4 = 0.5 + 0.2 = 0.7.$

Gene frequency of L^N in population

$= 0.1 + \frac{1}{2} \times 0.4 = 0.1 + 0.2 = 0.3.$

Problem 3

A sample of 1000 persons were tested for MN blood antigens and was found to be distributed: M 380; MN 460; N 160. What is the frequency of genes M and N?

Solution:

Genotypes/Phenotypes	M	MN	N	Total
Number of individuals	380	460	160	1000
Genotypic frequency	$\frac{380}{1000}$ = 0.38	$\frac{460}{1000}$ = 0.46	$\frac{160}{1000}$ = 0.16	$\frac{1000}{1000}$ = 1

Gene frequency of M in population

$= 0.38 + \frac{1}{2} \times 0.46 = 0.38 + 0.23 = 0.61.$

Gene frequency of N in population

$= 0.16 + \frac{1}{2} \times 0.46 = 0.16 + 0.23 = 0.39.$

Problem 4

Three genotypes were observed at the P_{gm-1} locus in a human population. In a sample of 1110 individuals, the three genotypes occurred in the following numbers (1 and 2 represent two different alleles):

Genotypes:	**1/1**	**1/2**	**2/2**	
Numbers:	**634**	**391**	**85**	

Calculate genotypic and allelic frequencies.

Solution:

Genotypes	$\frac{1}{1}$	$\frac{1}{2}$	$\frac{2}{2}$	Total
Number of individuals	634	391	85	1110
Genotypic frequency	$\frac{634}{1110}$ = 0.571	$\frac{391}{1110}$ = 0.352	$\frac{85}{1110}$ = 0.765	

Allelic frequency of '1' is $0.571 + \frac{1}{2} \times 0.352 = 0.571 + 0.176 = 0.747.$

Allelic frequency of '2' is $0.077 + \frac{1}{2} \times 0.352 = 0.077 + 0.176 = 0.253.$

Therefore, genotypic frequencies are $\frac{1}{1} = 0.571$, $\frac{1}{2} = 0.352$ and $\frac{2}{2} = 0.077$.

The frequency of allele '1' is = 0.747 and allele '2' is = 0.253.

Problem 5

Two human serum haptoglobins are determined by two alleles at a single locus. In a sample of 219 Egyptians, three genotypes occurred in the following numbers (1 and 3 represent the two alleles):

Genotypes:	**1/1**	**1/3**	**3/3**
Numbers:	**9**	**135**	**75**

Calculate genotypic and allelic frequencies.

Solution:

Genotypes	$\frac{1}{1}$	$\frac{1}{3}$	$\frac{3}{3}$	Total
Number of individuals	9	135	75	219
Genotypic frequency	$\frac{9}{219}$ =0.0410	$\frac{135}{219}$ =0.6164	$\frac{75}{219}$ =0.3424	

Allelic frequency of '1' is $0.0410 + \frac{1}{2} \times 0.6164 = 0.0410 + 0.3082 = 0.3492 = 0.349$.

Allelic frequency of '3' is $0.3424 + \frac{1}{2} \times 0.6164 = 0.3424 + 0.3082 = 0.6506 = 0.651$.

Problem 6

One hundred fruit flies (*Drosophila melanogaster*) from Kolkata were tested for their genotype at the alcohol dehydrogenase locus using starch-gel electrophoresis. Two alleles were present, S and F, for slow and fast migration, respectively. The following results were obtained: SS 66; SF 20; FF 14. What are the allelic and genotypic frequency in this population? Is this population in Hardy-Weinberg equilibrium?

Solution:

Genotypes/Phenotypes	SS	SF	FF	Total
Number of individuals	66	20	14	100
Genotypic frequency	$\frac{66}{100}$ = 0.66,	$\frac{20}{100}$ = 0.2,	$\frac{14}{100}$ = 0.14,	$\frac{100}{100}$ = 1

Gene frequency of S in population

$= 0.66 + \frac{1}{2} \times 0.2 = 0.66 + 0.1 = 0.76$

Gene frequency of F in population

$= 0.14 + \frac{1}{2} \times 0.2 = 0.14 + 0.1 = 0.24$

Gene frequency of S is 0.76 and F is 0.24 and genotypic frequency of S = 0.66 and genotypic frequency of F = 0.14.

Using the gene frequency of S and F, i.e. 0.76 and 0.24.

The expected number of genotypes will be $(0.76 + 0.24)^2 = 0.5776 + 0.3648 + 0.0576$

	SS	SF	FF
Observed Number	0.66	0.20	0.14
Expected Number	0.5776	0.364	0.0576

The genotypic frequency of observed and expected numbers are not exactly same, so this population is not in Hardy-Weinberg equilibrium.

Notes: Although it is statistically done through chi-square test.

Problem 7

One hundred persons from a small town in Gujrat were tested for their MN blood types. The genotypic data are: MM 41; MN 38 and NN 21. Calculate gene frequency of M and N. Is the population in Hardy-Weinberg equilibrium?

Solution:

Genotypes	MM	MN	NN	Total
Number of individuals	41	38	21	100
Genotypic frequency	$\frac{41}{100}$ = 0.41,	$\frac{38}{100}$ = 0.38,	$\frac{21}{100}$ = 0.21,	$\frac{100}{100}$ = 1

Gene frequency of M is $0.41 + \frac{1}{2} \times 0.38 = 0.41 + 0.19 = 0.6$

Gene frequency of N is $0.21 + \frac{1}{2} \times 0.38 = 0.21 + 0.19 = 0.4$

Therefore, gene frequency of M is 0.6 and N is 0.4.

Expected number of genotypes will be $(0.6 + 0.4)^2 = 0.36 + 0.48 + 0.16$

	MM	MN	NN
Observed Number	0.41	0.38	0.21
Expected Number	0.36	0.48	0.16

The genotypic frequency of observed and expected numbers are not exactly same. So this population is not in Hardy-Weinberg equilibrium.

Problem 8

How many A and a alleles are present in a population consisting of 10AA, 15Aa and 4aa individuals? What are the allelic frequencies?

Solution:

Genotypes	AA	Aa	aa	Total
Number of individuals	10	15	4	29
Number of alleles	20	30	8	58

Total number of 'A' alleles is $20 + \frac{1}{2} \times 30 = 20 + 15 = 35$

Total number of 'a' alleles is $8 + \frac{1}{2} \times 30 = 8 + 15 = 23$

Gene or allelic frequency 'A' is $\frac{35}{58} = 0.603$ and allelic frequency of 'a' is $\frac{23}{58} = 0.3965 = 0.397$.

Allelic frequencies A = 0.603 and a = 0.397.

Problem 9

Among 205 Central Americans, the frequencies of the L^M and L^N alleles were 0.78 and 0.22 respectively. Calculate the percentage of individuals with M, MN and N type blood.

Solution:

According to Hardy-Weinberg law, here

$p = 0.78, q = 0.22.$

$(p + q)^2 = (0.78 + 0.22)^2$

$= (0.78)^2 + 2 \times 0.78 \times 0.22 + (0.22)^2$

$= 0.6084 + 0.3432 + 0.0484.$

Percentage of individuals = 60.84% + 34.32% + 4.84%.

Therefore, M type 60.84%, MN type 34.32% and N type 4.84%.

Problem 10

A population of soyabeans is segregating for the colours golden, light green and dark green produced by the codominant genotypes C^GC^G, C^GC^D and C^DC^D respectively. A sample from this population contains 2 golden, 36 light green and 162 dark green. Determine the frequencies of the alleles C^G and C^D.

Solution:

Genotypes	C^GC^G	C^GC^D	C^DC^D	Total
Number of individuals	2	36	162	200
Genotypic frequency	$\frac{2}{200} = 0.01$	$\frac{36}{200} = 0.18$	$\frac{162}{200} = 0.81$	$\frac{200}{200} = 1$

Gene (allelic) frequency of $C^G = 0.01 + \frac{1}{2} \times 0.18 = 0.01 + 0.09 = 0.1$.

Gene (allelic) frequency of CD $= 0.81 + \frac{1}{2} \times 0.18 = 0.81 + 0.09 = 0.9$.

Therefore, the gene frequency of C^G is 0.1 and C^D is 0.9.

Problem 11

The M-N blood system in man is governed by a pair of codominant alleles (L^M and L^N). A sample of 208 people was tested for the presence of the M and N antigens and found to contain 119 group of M(L^ML^M), 76 group MN(L^ML^N) and 13 group N(L^NL^N). (a) Calculate the gene frequencies of L^M and L^N. (b) If the frequency of L^M = 0.3, how many individuals in a sample of size 500 would be expected to belong to group MN?

Solution:

(a)

Genotypes	L^ML^M	L^ML^N	L^NL^N	Total
Number of individuals	119	76	13	208
Genotypic frequency	$\frac{119}{208} = 0.5721$,	$\frac{76}{208} = 0.3653$,	$\frac{13}{208} = 0.0625$	$\frac{208}{208} = 1$

Gene frequency of $L^M = 0.5721 + \frac{1}{2} \times 0.3653 = 0.5721 + 0.1826 = 0.7547 = 0.755$

Gene frequency of $L^N = 0.0625 + \frac{1}{2} \times 0.3653 = 0.0625 + 0.1826 = 0.2451 = 0.245$

(b) Here, $L^M = 0.3$, L^N will be $1 - 0.3 = 0.7$ (as $p + q = 1$)

$(L^M + L^N)^2 = (0.3 + 0.7)^2 = 0.9 + 0.42 + 0.49.$

$MN = 2L^ML^N = 0.42.$

In a sample of size 500, the MN individuals would be 500 × 0.42 = 210.00 = 210.

Problem 12

What is the frequency of heterozygotes Aa in a random mating population in which frequency of all dominant allele is 0.19?

Solution:

Let p is the gene frequency of 'A' and q is the gene frequency of 'a'.

$p^2 + 2pq = 0.19, q^2 = 1 - 0.19 = 0.81.$

Therefore, $q = \sqrt{0.81} = 0.9.$

As $p + q = 1$, therefore, $p = 1 - 0.9 = 0.1$

Therefore, heterozygotes Aa, i.e., $2pq = 2 \times 0.9 \times 0.1 = 0.18$ or 18%.

Problem 13

What is the frequency of heterozygotes Aa in a random mating population if the frequency of recessive phenotypes (aa) is 0.09?

Solution:

aa = 0.09, i.e. $q^2 = 0.09$, $q^2 = \sqrt{0.09} = 0.3.$

As $p + q = 1$, $p = 1 - 0.3 = 0.7.$

Therefore, the frequency of heterozygotes Aa, i.e., $2pq = 2 \times 0.7 \times 0.3 = 0.42.$

Problem 14

In a population of mice, there are two alleles of the A locus (A_1 and A_2). Tests showed that in this population there are mice of genotype 384 of A_1/A_1, 210 of A_1/A_2 and 260 of A_2/A_2. What are the frequencies of the two alleles in the population?

Solution:

Genotypes	A_1/A_1	A_1/A_2	A_2/A_2	Total
Number of individuals	384	210	260	854
Genotypic frequency	$\frac{384}{854}$	$\frac{210}{854}$	$\frac{260}{854}$	
	= 0.449,	= 0.246,	= 0.304	

Gene frequency of A_1 allele $= 0.449 + \frac{1}{2} \times 0.246 = 0.449 + 0.123 = 0.572 = 0.57.$

Gene frequency of A_2 allele $= 0.304 + \frac{1}{2} \times 0.246 = 0.304 + 0.123 = 0.427 = 0.43.$

The frequency of A_1 is 0.57 and A_2 is 0.43.

Problem 15

Two hundred students of R. K. Mission Vidyamandira, were tested for their MN blood types and the following blood types were obtained MM 114, MN 76 and NN 10. Calculate the gene frequencies of M and N.

Solution:

Genotypes	MM	MN	NN	Total
Number of individuals	114	76	10	200
Genotypic frequency	$\frac{114}{200}$	$\frac{76}{200}$	$\frac{10}{200}$	$\frac{200}{200}$
	= 0.57,	= 0.38,	= 0.05	= 1

Gene frequency of M is $0.57 + \frac{1}{2} \times 0.38 = 0.57 + 0.19 = 0.76.$

Gene frequency of N is $0.05 + \frac{1}{2} \times 0.38 = 0.05 + 0.19 = 0.24.$

Therefore, gene frequency of M is 0.76 and N is 0.43.

Problem 16

In shorthorn cattle, the genotype C^RC^R is phenotypically red, C^RC^W is roan (a mixture of red and white) and C^WC^W is white. (a) If 108 red, 48 white and 144 roan animals were found in a sample of shorthorns from western valley of Himachal Pradesh, calculate the estimated frequencies of the C^R allele and the C^W allele in the gene pool of the population. (b) If this population is completely panmictic, what zygotic frequencies would be expected in the next generation? (c) Is the population in equilibrium?

Solution:

(a)

Genotypes	C^RC^R	C^RC^W	C^WC^W	Total
Number of individuals	108	144	48	300
Genotypic frequency	$\frac{108}{300}$	$\frac{144}{300}$	$\frac{48}{300}$	
	= 0.36	= 0.48	= 0.16	

Gene frequency of C^R allele $= 0.36 + \frac{1}{2} \times 0.48 = 0.36 + 0.24 = 0.6.$

Gene frequency of CW allele $= 0.16 + \frac{1}{2} \times 0.48 = 0.16 + 0.24 = 0.4.$

(b) Here $p = C^R = 0.6$ and $q = C^W = 0.4$

$$(0.6 + 0.4)^2 = (0.6)^2 + 2 \times 0.6 \times 0.4 + (0.4)^2$$

Expected zygotic frequency in the next generation will be 0.36 + 0.48 + 0.16.

(c) The genotypic frequency of observed and expected data are exactly similar. So this population is in equilibrium.

Problem 17

In a large interbreeding population 81% individuals are homozygous for a recessive character. In the absence of mutation or selection what percentage of the next generation would be homozygous recessives, homozygous dominants and heterozygous?

Solution:

Suppose q is the gene frequency for recessive character. Therefore, genotypic frequency, i.e., $q^2 = \frac{81}{100} = 0.81$, so $q = \sqrt{0.81} = 0.9$.

Therefore, gene frequency of dominant character, i.e., $p = 1 - 0.9 = 0.1$.

Genotypic frequency of the next generation will be $(p + q)^2 = (0.1 + 0.9)^2 = (0.1)^2 + 2 \times 0.1 \times 0.9 + (0.9)^2 = 0.01 + 0.18 + 0.81 = 1\% + 18\% + 81\%$.

Therefore, in the next generation 1% would be homozygous dominant, 81% would be homozygous recessive and 18% would be heterozygous.

Problem 18

A group of 228 students of R. K. Mission Vidyamandira were invited to taste the chemicals. The following results were obtained. 160 tasters and 68 non-tasters. Calculate gene frequency of taster (T) and non-taster (t).

Solution:

Since the taster (T) is dominant over non-taster, 160 out of 228 tasters includes both p^2 (TT) and 2pq (Tt). The genotypic frequency of non-taster q^2 (tt) is, thus,

$$q^2 = \frac{68}{288} = 0.298$$

$q = \sqrt{0.298} = 0.5454$, i.e., non-taster (q) = 0.545.

The gene frequency of taster, i.e., $p = 1 - 0.545 = 0.455$.

Therefore, gene frequency of taster is 0.455 and non-taster is 0.545.

Problem 19

White wool is dependent upon a dominant allele B and black wool upon its recessive allele b. Suppose that a sample of 900 sheep of Ladhak gave the following data: 891 white and 9 black. Estimate the allelic frequencies.

Solution:

White wool is dominant, i.e., BB or Bb.

Black wool is recessive, i.e., bb.

$p^2(BB) + 2pq(Bb) + q^2(bb) = 1$.

As p is the gene frequency for B and q is the gene frequency of b,

∴ Genotypic frequency of black wool, i.e.,

$q^2(bb) = \frac{9}{900}$ (∵ 891 + 9 = 900).

$q^2 = 0.01$, ∴ $q = \sqrt{0.01} = 0.1$.

As $p + q = 1$, $p = 1 - 0.1 = 0.9$.

So the gene frequency of B(p) = 0.9 and b(q) = 0.1.

Problem 20

Black stripe colour in zebra is dependent upon a dominant allele B and lack of such stripe upon its recessive allele b. In a large random mating population 800 zebra gave the following data: 738 black stripe and 62 lack of such stripe. Calculate gene frequency.

Solution:

$p^2(BB) + 2pq(Bb) + q^2(bb) = 1.0$

As the population are random mating, therefore, genotypic frequency of recessive allele (b), i.e., $q^2(bb) = \frac{62}{800} = 0.0775$, $b = \sqrt{0.0775} = 0.278 = 0.28$.

As $p + q = 1$, the frequency of allele $B(p) = 1 - 0.28 = 0.72$.

Problem 21

In a randomly breeding human population 36% of the persons cannot taste PTC. Non-taster allele (a) and taster allele (A). What is the gene frequency of a and A?

Solution:

The gene frequency of non-taster (a) is q.

Therefore, the genotypic frequency of non-taster (aa), i.e., $q^2 = \frac{36}{100} = 0.36$, $q = \sqrt{0.36} = 0.6$.

As $p + q = 1$, $p = 1 - 0.6 = 0.4$.

The gene frequency of a = 0.6 and A = 0.4.

Problem 22

The ability of certain people to taste a chemical called PTC is governed by a dominant allele T, and the inability to taste PTC by its recessive allele t. If 24% of a population is homozygous taster and 40% is heterozygous taster, what is the gene frequency of t?

Solution:

Suppose, the gene frequency of taster is p and non-taster is q.

Therefore, p^2 (homozygous taster) = 24% = 0.24.

2pq (heterozygous tasters) = 40% = 0.40 and non-tasters are 100 – (24 + 40) = 100 – 64 = 36, i.e., 36%, i.e., $q^2 = 0.36$, $q = \sqrt{0.36} = 0.6$.

The gene frequency of t, i.e., q = 0.6.

Problem 23

About 70 percent of all non-resident Indians can taste the chemical phenyl-thiocarbamide and the remainder cannot. The ability to taste is determined by the dominant allele T and the inability to taste is determined by the recessive allele t. If the population is assumed to be Hardy-Weinberg equilibrium, what are the genotypic and phenotypic frequencies in this population?

Solution:

As 70 percent are tasters, 30 percent must be non-tasters (tt).

Let q is the gene frequency of non-taster allele. The genotypic frequency of non-tasters is 0.30, i.e., $q^2 = 0.30$, $q = \sqrt{0.30} = 0.55$.

As the population is in Hardy-Weinberg equilibrium, i.e., $p + q = 1$, we can write, $p = 1 - q$.

$\therefore p = 1 - 0.55 = 0.45$.

$$(p + q)^2 = (0.45 + 0.55)^2$$
$$= (0.45)^2 + 2 \times 0.45 \times 0.55 + (0.55)^2$$
$$= 0.2025 + 0.495 + 0.3025.$$

Therefore, TT = 0.2025 = 0.20 = 20%

Tt = 0.495 = 0.50 = 50%

tt = 0.3025 = 0.30 = 30%.

Taster will be 20% + 50% = 70% and non-tasters will be 30%.

Problem 24

Albinism is an autosomal recessive trait. They are easily recognised by lack of pigment in skin, hair and iris. One source estimates the frequency of albinos as 1 in 10,000. What percentage of the population is heterozygous for this gene?

Solution:

The gene frequency of recessive allele is q. The genotypic frequency is q^2.

$q^2 = \frac{1}{10,000} = 0.0001$, $q = \sqrt{0.0001} = 0.01$.

Since, $p + q = 1$, $p = 1 - q = 1 - 0.01 = 0.99$

Heterozygous population, i.e., $2pq = 2 \times 0.99 \times 0.01 = 0.0198 = 0.02$ or 2%.

Therefore, 2% are heterozygous.

Problem 25

Albinism is the phenotypic expression of a homozygous recessive genotype. One source estimates the frequency of albinos as in 1 in 18,000. What percentage of population is heterozygous for this gene?

[C.U. 2001]

Solution:

Let the gene frequency of recessive allele is q.

Therefore, genotypic frequency is q^2.

$q^2 = \frac{1}{18,000} = 0.000055$, $q = \sqrt{0.000055} = 0.0074$

$p = 1 - q = 1 - 0.0074 = 0.9926$.

Heterozygous population, i.e.,

$2pq = 2 \times 0.9926 \times 0.0074 = 0.01469$

$= 0.1469 \times 100 = 1.46900 = 1.469\%$.

Problem 26

The ability vs. inability to taste Phenyl thiocarbamide (PTC) is determined by a single pair of autosomal alleles of which T for taster is dominant over t for non-taster. Of the 4,000 persons tested. 3,000 were tasters and 1000 persons non-tasters. Assuming that the population satisfies the conditions for Hardy-Weinberg equilibrium,

(i) Calculate the frequencies of alleles T and t and the frequencies of the genotypes TT, Tt and tt.

(ii) State whether TT, Tt and tt are in equilibrium. Justify your answer.

[C.U.(Zoo. Hons.) 1993]

Solution:

Let the gene frequency of non-taster, i.e., t allele is q. Therefore, genotypic frequency is q^2.

1,000 persons are non-tasters among 4,000 persons.

$q^2 = \frac{1,000}{4,000} = 0.25, q = \sqrt{0.25} = 0.5.$

As population is in equilibrium, i.e., p + q = 1,

p = 1 − q = 1 − 0.5 = 0.5.

(i) Gene frequencies of T and t alleles are 0.5 and 0.5.

Genotypic frequencies are

$(p + q)^2 = (0.5 + 0.5)^2 = 0.25 + 0.50 + 0.25$

TT = 0.25, Tt = 0.50 and tt = 0.25.

(ii) From this population the gene frequency of non-taster, i.e.,

$q = \sqrt{0.25} = 0.5.$

Therefore, p = 1 − 0.5 = 0.5.

In the subsequent generation,

$(p + q)^2 = (0.5 + 0.5)^2 = 0.25 + 0.50 + 0.25.$

It is same as previous. Therefore, it is in equilibrium.

Problem 27

M and N blood groups of man are due to segregating genes L^M and L^N. Neither of them is dominant over the other. A sample of 426 individuals was typed for MN blood group with the following results:

M	**MN**	**NN**
238	**152**	**36**

(a) Calculate the frequencies of L^M and L^N alleles.

(b) If the frequency of L^M = 0.3, how many individuals in a population of 1,000 would be expected to belong to blood group MN?

[C.U.(Zoo. Hons.) 1998]

Solution:

(a)

Genotypes/Phenotypes	MM	MN	NN	Total
Number of individuals	238	152	36	426
Genotypic frequency	$\frac{238}{426}$	$\frac{152}{426}$	$\frac{36}{426}$	$\frac{426}{426}$
	=0.558,	=0.356,	=0.084	=1

Gene frequency of M in population = $0.558 + \frac{1}{2} \times 0.356 = 0.558 + 0.178 = 0.736 = 0.74$

Gene frequency of N in population = $0.084 + \frac{1}{2} \times 0.356 = 0.084 + 0.178 = 0.26$

Frequencies of L^M alleles = 0.74 and L^N alleles = 0.26.

(b) The frequency of L^M = 0.3 as p + q = 1.
i.e. $L^M + L^N = 1$, $L^N = 1 - 0.3 = 0.7$.

Therefore, 2pq = MN = 2 × 0.7 × 0.3 = 0.42.

Among 1000 population, it would be 0.42 × 1000 = 420.00 = 420.

Problem 28

If the population of an Island contains individuals in the proportion of 0.4225 AA, 0.4550 Aa, 0.1225 aa, would you consider the population to be in Hardy-Weinberg equilibrium?

[C.U.(Zoo. Hons.) 1997]

Solution:

Genotypes/Phenotypes	AA	Aa	aa
Genotypic frequency	0.4225	0.4550	0.1225

Gene frequency of A = $0.4225 + \frac{1}{2} \times 0.4550 = 0.4225 + 0.2275 = 0.65$

Gene frequency of a = $0.1225 + \frac{1}{2} \times 0.4550 = 0.1225 + 0.2275 = 0.35$

$(A + a)^2 = (0.65 + 0.35)^2 = 0.4225 + 0.4550 + 0.1225.$

The genotypic frequency of P_1 and F_1 are exactly similar, so it is in equilibrium.

Problem 29

The distribution of MN blood types in a population of 1000 individuals is as follows:

(i) M blood group 350.

(ii) MN group 450 and N group 200 individuals. In this population work out: (a) The frequency of M and N alleles, (b) whether the population is in equilibrium and (c) what is the expected distribution of these three genotypes when the population is in equilibrium?

[C.U.(Zoo. Hons.) 2000]

Solution:

(a)

Genotypes/Phenotypes	MM	MN	NN	Total
Number of individuals	350	450	200	1000
Genotypic frequency	$\frac{350}{1000}$ =0.35	$\frac{450}{1000}$ =0.45	$\frac{200}{1000}$ =0.2	$\frac{1000}{1000}$ = 1

Gene or allele frequency of M $= 0.35 + \frac{1}{2} \times 0.45 = 0.35 + 0.225 = 0.575.$

Gene or allele frequency of N $= 0.2 + \frac{1}{2} \times 0.45 = 0.2 + 0.225 = 0.425.$

(b) In the next generation genotypic frequency would be $= (0.575 + 0.425)^2$ $= 0.330 + 0.488 + 0.180 = 0.33 + 0.48 + 0.18.$

These are not exactly similar to genotypic frequency of the former generation.

So, it is not in equilibrium.

(c) When the population is in equilibrium, the expected distribution of genotypes would be 0.35 + 0.45 + 0.2 or 35% (NM), 45% (MN) and 2% (NN).

Problem 30

Black colour in rats is dependent upon a dominant allele B and brown colour upon its recessive allele b. Suppose that a sample of 800 rats gave the data as 720 black and 80 brown. Estimate the allelic frequencies.

[C.U.(Zoo. Hons.) 2001]

Solution:

Let the gene frequency of allele b is q.

The genotypic frequency is q^2, i.e., $\frac{80}{800} = 0.1.$

$q^2 = 0.1, \therefore q = \sqrt{0.1} = 0.316.$

As $p + q = 1$. Therefore, $p = 1 - 0.316 = 0.684.$

Therefore, allelic frequencies are 0.684 and 0.316.

Problem 31

Among a Jewish Community of Central European ancestry, about 1 in 3,600 children is born with Tay-sachs disease (an autosomal recessive condition). Derive what fraction of the individuals in this population are carrier.

[C.U.(Zoo. Hons.) 2002]

Solution:

Let the gene frequency of Tay-sachs disease is q. So the genotypic frequency is q^2, i.e., $\frac{1}{3600}$.

$q^2 = 0.0002777 = 0.00028$

$q = \sqrt{0.00028} = 0.0167 = 0.017.$

As $p + q = 1$, $p = 1 - 0.017 = 0.983.$

The carriers are heterozygous, i.e.,

$2pq = 2 \times 0.983 \times 0.017 = 0.0334$

$= 0.0334 \times 100 = 3.3400 = 3.34\%.$

Problem 32

Consider the populations that have the genotypes shown in the following table:

Population	A/A	A/a	a/a
1	1.0	0.0	0.0
2	0.0	1.0	0.0
3	0.0	0.0	1.0
4	0.50	0.25	0.25
5	0.25	0.25	0.50
6	0.25	0.50	0.25
7	0.33	0.33	0.33
8	0.04	0.32	0.64
9	0.64	0.32	0.04
10	0.986049	0.013902	0.000049

(a) What are p and q in each population?

(b) Which of the populations are in Hardy-Weinberg equilibrium and why?

Solution:

(a) 1. $p = AA + \frac{1}{2} Aa = 1 + \frac{1}{2} \times 0 = 1 + 0 = 1$

$q = aa + \frac{1}{2} Aa = 0 + \frac{1}{2} \times 0 = 0 + 0 = 0$

$(p + q)^2 = (1 + 0)^2 = 1 + 0 + 0.$

Yes, it is in Hardy-Weinberg equilibrium as genotypic frequency is exactly similar to ancestral population.

2. $p = 0 + \frac{1}{2} \times 1 = 0 + 0.5 = 0.5$

$q = 0 + \frac{1}{2} \times 1 = 0 + 0.5 = 0.5$

$(0.5 + 0.5)^2 = 0.25 + 0.50 + 0.25.$

No, it is not in H-W equilibrium as genotypic frequency is not exactly similar to parental one.

3. $p = 0 + 0 = 0$

$q = 1 + 0 = 1$

$(0 + 1)^2 = 0 + 0 + 1.$

Genotypic frequency is similar to ancestral one. So it is in equilibrium.

4. $p = 0.50 + \frac{1}{2} \times 0.25 = 0.50 + 0.125 = 0.625$

$q = 0.25 + \frac{1}{2} \times 0.25 = 0.25 + 0.125 = 0.375$

$(0.625 + 0.375)^2 = 0.3906 + 0.4687 + 0.1406.$

No, it is not in H-W equilibrium as it is not exactly similar to the ancestral one.

5. $p = 0.25 + \frac{1}{2} \times 0.25 = 0.25 + 0.125 = 0.375$

$q = 0.50 + \frac{1}{2} \times 0.25 = 0.50 + 0.125 = 0.625$

$(0.375 + 0.625)^2 = 0.1406 + 0.4687 + 0.3906.$

No, it is not same as 4.

6. $p = 0.25 + \frac{1}{2} \times 0.50 = 0.25 + 0.25 = 0.5$

$q = 0.25 + \frac{1}{2} \times 0.50 = 0.25 + 0.25 = 0.5$

$(0.5 + 0.5)^2 = 0.25 + 0.50 + 0.25.$

Yes, it is in H-W equilibrium as genotypic frequency is exactly similar to that of ancestral population.

7. $p = 0.33 + \frac{1}{2} \times 0.33 = 0.33 + 0.165 = 0.495$

$q = 0.33 + \frac{1}{2} \times 0.33 = 0.33 + 0.165 = 0.495$

$(0.495 + 0.495)^2 = 0.2450 + 0.490 + 0.2450.$

No, it is not in H-W equilibrium as genotypic frequency is not similar to ancestral one.

8. $p = 0.04 + \frac{1}{2} \times 0.32 = 0.04 + 0.16 = 0.20$

$q = 0.64 + \frac{1}{2} \times 0.32 = 0.64 + 0.16 = 0.80$

$(0.2 + 0.8)^2 = 0.04 + 0.32 + 0.64.$

Yes, it is in H-W equilibrium as genotypic frequency is similar to parental generation.

9. $p = 0.64 + \frac{1}{2} \times 0.32 = 0.64 + 0.16 = 0.80$

$q = 0.04 + \frac{1}{2} \times 0.32 = 0.04 + 0.16 = 0.20$

$(0.8 + 10.2)^2 = 0.64 + 0.32 + 0.04.$

Yes, it is in H-W equilibrium as genotypic frequency is similar to the parental one.

10. $p = 0.986 + \frac{1}{2} \times 0.0139 = 0.986 + 0.069 = 0.168.$

$q = 0.004 + \frac{1}{2} \times 0.0139 = 0.004 + 0.069 = 0.069.$

$(0.168 + 0.069)^2 = 0.02826 + 0.02338 + 0.00483.$

No, it is not in H-W equilibrium.

Problem 33

Which of the following populations are in Hardy-Weinberg equilibrium?

Population	AA	Aa	aa
a	0.25	0.54	0.25
b	0.10	0.74	0.16
c	0.64	0.27	0.09
d	0.46	0.50	0.04
e	0.81	0.18	0.01

Solution:

(a) Gene frequency of A = $0.25 + \frac{1}{2} \times 0.54 = 0.25 + 0.27 = 0.52$.

Gene frequency of a = $0.25 + \frac{1}{2} \times 0.54 = 0.25 + 0.27 = 0.52$.

Genotypic frequency $(0.52 + 0.52)^2 = 0.2704$ (AA) + 0.5408(Aa) + 0.2704(aa).

These are not similar to the parental genotypic frequency. So, it does not seem to be in equilibrium.

(b) Gene frequency of A = $0.10 + \frac{1}{2} \times 0.74 = 0.1 + 0.37 = 0.47$.

Gene frequency of a = $0.16 + \frac{1}{2} \times 0.74 = 0.1 + 0.37 = 0.47$.

Genotypic frequency $(0.47 + 0.47)^2 = 0.2209$ (AA) + 0.4418(Aa) + 0.2209(aa).

These are not similar to that of parental genotypic frequency. So it does not seem to be in equilibrium.

(c) Gene frequency of A = $0.64 + \frac{1}{2} \times 0.27 = 0.64 + 0.135 = 0.775$.

Gene frequency of a = $0.09 + \frac{1}{2} \times 0.27 = 0.09 + 0.135 = 0.225$.

Genotypic frequency = $(0.775 + 0.225)^2 = 0.600625 + 0.34875 + 0.050625$.

Genotypic frequency are not similar to parental one. So it is not in equilibrium.

(d) Gene frequency of A = $0.46 + \frac{1}{2} \times 0.50 = 0.46 + 0.25 = 0.71$.

Gene frequency of a = $0.04 + \frac{1}{2} \times 0.50 = 0.04 + 0.25 = 0.29$.

Genotypic frequency of A = $(0.71 + 0.29)^2 = 0.5041 + 0.4118 + 0.0841$.

It is not in equilibrium as genotypic frequency is not similar to parental one.

(e) Gene frequency of A = $0.81 + \frac{1}{2} \times 0.18 = 0.81 + 0.09 = 0.9$.

Gene frequency of a = $0.01 + \frac{1}{2} \times 0.18 = 0.01 + 0.09 = 0.1$.

Genotypic frequency $(0.9 + 0.1)^2 = 0.81 + 0.18 + 0.01$.

Genotypic frequency is similar to parental one. So it seems to be in equilibrium.

Problem 34

The following gives the genotypic frequencies for five populations. Which is in Hardy-Weinberg equilibrium? For those are not, suggest some hypotheses to explain why they are not in equilibrium?

Genotype

Population	AA	Aa	aa	Total
1	10	80	10	100
2	40	20	40	100
3	0	150	100	250
4	2	16	32	50

Solution:

1. Genotypes	AA	Aa	aa
Genotypic frequency	$\frac{10}{100}$ = 0.1	$\frac{0.80}{100}$ = 0.8	$\frac{0.10}{100}$ = 0.1

Gene frequency of A = $0.1 + \frac{1}{2} \times 0.8 = 0.1 + 0.4 = 0.5$.

Gene frequency of a = $0.1 + \frac{1}{2} \times 0.8 = 0.1 + 0.4 = 0.5$.

Genotypic frequency = $(0.5 + 0.5)^2 = 0.25 + 0.50 + 0.25$.

These are not similar to the parental one. So it is not in equilibrium. It may be due to heterozygote advantage.

2. Genotypes	AA	Aa	aa
Genotypic frequency	$\frac{40}{100}$ = 0.4	$\frac{20}{100}$ = 0.2	$\frac{40}{100}$ = 0.4

Gene frequency of A = $0.4 + \frac{1}{2} \times 0.2 = 0.4 + 0.1 = 0.5$.

Gene frequency of a = $0.4 + \frac{1}{2} \times 0.2 = 0.4 + 0.1 = 0.5$.

Genotypic frequency = $(0.5 + 0.5)^2 = 0.25 + 0.50 + 0.25$.

These are not similar to the parental one. So it does not seem to be in equilibrium. It may be due to heterozygote disadvantage, there could be **Wahlund effect. [The effect of subdivision on a population causing it to contain fewer heterozygotes than predicted despite the fact that all subdivisions are in Hardy-Weinberg equilibrium.]**

3. Genotypes	AA	Aa	aa
Genotypic frequency	0	$\frac{150}{250}$ = 0.6	$\frac{100}{250}$ = 0.4

Gene frequency of A = $0 + \frac{1}{2} \times 0.6 = 0 + 0.3 = 0.3$.

Gene frequency of a = $0.4 + \frac{1}{2} \times 0.6 = 0.4 + 0.3 = 0.7$.

Genotypic frequency = $(0.3 + 0.7)^2 = 0.09 + 0.42 + 0.49$.

These are not similar to the parental population. So it does not seem to be in equilibrium. It looks like AA is lethal.

4. Genotypes	AA	Aa	aa
Genotypic frequency	$\frac{2}{50}$ = 0.04	$\frac{16}{50}$ = 0.32	$\frac{32}{50}$ = 0.64

Gene frequency of A = $0.04 + \frac{1}{2} \times 0.32 = 0.04 + 0.16 = 0.20 = 0.2$.

Gene frequency of a = $0.64 + \frac{1}{2} \times 0.32 = 0.64 + 0.16 = 0.80 = 0.8$.

Genotypic frequency = $(0.2 + 0.8)^2 = 0.04 + 0.32 + 0.64$ in equilibrium.

Problem 35

A population called the "founder generation" consisting of 2,000 AA individuals, 2,000 Aa individuals and 6,000 aa individuals is established on a remote island. Mating within this population is random, the three genotypes are selectively neutral and mutations occur at a negligible rate.

(a) What are the frequencies of alleles A and a in the founder generation? (b) Is the founder generation at Hardy-Weinberg equilibrium? (c) What is the frequency of the A allele in the second generation (subsequent to the founder generation)? (d) What are the frequencies of AA, Aa and aa genotypes in the second generation? (e) Is the second generation in Hardy-Weinberg equilibrium? (f) What are the frequencies for the AA, Aa and aa genotypes in the third generation?

Solution:

(a)

Phenotypes/Genotypes	AA	Aa	aa	Total
Number of individuals	2,000	2,000	6,000	10,000
Genotypic frequency	$\frac{2,000}{10,000}$ =0.2	$\frac{2,000}{10,000}$ =0.2	$\frac{6,000}{10,000}$ =0.2	$\frac{10,000}{10,000}$ =1

Gene frequency of A in population = $0.2 + \frac{1}{2} \times 0.2 = 0.2 + 0.1 = 0.3$.

Gene frequency of a in population = $0.6 + \frac{1}{2} \times 0.2 = 0.6 + 0.1 = 0.7$.

(b) Gene frequency, i.e. p = 0.3 and q = 0.7
$(0.3 + 0.7)^2 = 0.09 + 0.42 + 0.49$.

It is not as founder. As the population is 10,000, the number of individuals will be = 900(AA) + 4200 (Aa) + 4900(aa).

The founder members are not equal to next generation. So it is not in equilibrium.

(c) Conditions are random mating, selective neutral alleles and no new mutation.

The gene frequency of 'A' is 0.3.

(d) The genotypic frequencies for the second generation would be
$(0.3 + 0.7)^2 = 0.09(AA) + 0.42(Aa) + 0.49(aa)$.

(e) The genotypic frequency of 2nd generation is 0.09 + 0.42 + 0.49.

The gene frequency will be $0.09 + \frac{1}{2} \times 0.42 = 0.09 + 0.21 = 0.30$
$0.49 + \frac{1}{2} \times 0.42 = 0.49 + 0.21 = 0.30$.

(f) In third generation, it would be $(0.3 + 0.7)^2 = 0.09 + 0.42 + 0.49$. It is same, i.e., 0.09 + 0.42 + 0.49.

Problem 36

Two hypothetical populations found on opposite sides of a mountain in the Arizona

desert have two alleles (A^F, A^S) of a single gene A with the following three genotypes.

Genotypes	A^FA^F	A^FA^S	A^SA^S
Population 1	38	44	18
Population 2	0	80	20

(a) What is the gene frequency of A^F and A^S in the two populations? (b) Do the two populations appear to be in Hardy-Weinberg equilibrium? (c) A huge flood opened a canyon in the mountain range separating populations 1 and 2. They were then able to migrate such that the two populations which are equal size mixed completely and mated at random. What are the frequencies of the three genotypes in the next (A^FA^F, A^FA^S, A^SA^S) generation of the single new population of lizards?

Solution:

(a)

Population 1/Genotypes	A^FA^F	A^FA^S	A^SA^S	Total
Number of individuals	38	44	18	100
Genotypic frequency	$\frac{38}{100}$ = 0.38,	$\frac{44}{100}$ = 0.44,	$\frac{18}{100}$ = 0.18	$\frac{100}{100}$ = 1

Gene frequency of AF in population = $0.38 + \frac{1}{2} \times 0.44 = 0.38 + 0.22 = 0.60 = 0.6$.

Gene frequency of AS in population = $0.18 + \frac{1}{2} \times 0.44 = 0.18 + 0.22 = 0.40 = 0.4$.

Population 2/Genotypes	A^FA^F	A^FA^S	A^FA^S	Total
Number of individuals	0	80	20	100
Genotypic frequency	0	$\frac{80}{100}$ = 0.8,	$\frac{20}{100}$ = 0.2	

Gene frequency of A^F in population = $0 + \frac{1}{2} \times 0.8 = 0 + 0.4 = 0.4$.

Gene frequency of A^S in population = $0.2 + \frac{1}{2} \times 0.8 = 0.2 + 0.4 = 0.6$.

(b) (i) $(0.6 + 0.4)^2 = (0.06)^2 + 2 \times 0.6 \times 0.4 + (0.4)^2$

Genotypic frequency = $0.36(A^FA^F) + 0.48 (A^FA^S) + 0.16(A^SA^S)$.

This is not similar to parent population. So it does not seem to be in equilibrium.

(ii) $(0.4 + 0.6)^2 = (0.4)^2 + 2 \times 0.4 \times 0.6 + (0.6)^2 = 0.16(A^FA^F) + 0.48(A^FA^S) + 0.36(A^SA^S)$

This is not similar to that of parent population. So it does not seem to be in equilibrium.

(c)

	A^FA^F	A^FA^S	A^SA^S	Total
Population 1:	0.38	0.44	0.18	0.1
Population 2:	0	0.80	0.20	0.1
Av. Genotypic Frequency	$\frac{0.38}{2}$ = 0.19	$\frac{1.24}{2}$ = 0.62	$\frac{0.38}{2}$ = 0.19	

Gene frequency of A^F in population = $0.19 + \frac{1}{2} \times 0.62 = 0.19 + 0.31 = 0.50 = 0.5$.

Gene frequency of A^S in population = $0.19 + \frac{1}{2} \times 0.62 = 0.19 + 0.31 = 0.50 = 0.5$.

Average Gene frequency of

$A^F = 0{\cdot}6 + 0{\cdot}42 = 1{\cdot}02 = 0.5$.

$A^S = 0{\cdot}4 + 0{\cdot}62 = 1{\cdot}02 = 0.5$.

Problem 37

Two populations begin with the following genotypic frequencies:

Population 1 0.33 AA 0.14 Aa 0.53 aa

Population II: 0.24 AA 0.32 Aa 0.44 aa

Are the two populations in equilibrium? If random mating occurs between the two populations, determine their genotypic frequencies in the next generation.

[C.U.(Zoo. Hons.) 1990]

Solution:

Population I:

Gene frequency of (A) = $0.33 + \frac{1}{2} \times 0.14 = 0.33 + 0.07 = 0.40 = 0.4$.

Gene frequency of (a) = $0.53 + \frac{1}{2} \times 0.14 = 0.53 + 0.07 = 0.60 = 0.6$.

Population I:

$(0.4 + 0.6)^2 = 0.16(AA) + 0.48(Aa) + 0.36(aa)$.

Genotypic frequency is not similar to the parent population, so population (I) does not seem to be in equilibrium.

Population II:

Gene frequency of (A) = $0.24 + \frac{1}{2} \times 0.32 = 0.24 + 0.16 = 0.40 = 0.4$.

Gene frequency of (a) = $0.14 + \frac{1}{2} \times 0.32 = 0.44 + 0.16 = 0.60 = 0.6$.

Population II:

$(0.4 + 0.6)^2 = 0.16(AA) + 0.48(Aa) + 0.36(aa)$.

Genotypic frequency is not similar to the parent population, so population (II) does not seem to be in equilibrium.

After random mating of the two populations, gene frequency of A = $\frac{0{\cdot}4+0{\cdot}6}{2} = \frac{1{\cdot}0}{2} = 0.5$;

a = $\frac{0{\cdot}6+0{\cdot}4}{2} = \frac{1{\cdot}0}{2} = 0.5$.

Genotypic frequency = $(0.5 + 0.5)^2 = 0.25(AA) + 0.50\ (Aa) + 0.25\ (aa)$.

Problem 38

A sample of 300 plants from a population is examined for the electrophoretic mobility of an enzyme that varies according to the genotype determined by two alleles, F and S, of a single gene. The results are 7 plants with genotype FF. 106 with genotype FS and 187 with genotype SS. What are the allele frequencies of F and S? What are the expected number of genotypes assuming random mating?

Solution:

Genotypes	FF	FS	SS	Total
Number of individuals	7	106	187	300
Genotypic frequency	$\frac{7}{300}$ =.0233,	$\frac{106}{300}$ =0.353,	$\frac{187}{300}$ =0.623	

Gene frequency of F = $0.0233 + \frac{1}{2} \times 0.353 = 0.0233 + 0.1998 = 0.2$.

Gene frequency of S = $0.623 + \frac{1}{2} \times 0.35 = 0.623 + 0.1765 = 0.7995 = 0.8$.

The expected number of genotypes = $(0.2 + .8)^2 = 0.04 + 0.32 + 0.64$.

Among 300 plants: $0.04 \times 300 =$ 12.00 (FF)

$0.32 \times 300 =$ 96.00 (FS)

$0.64 \times 300 =$ 192.00 (SS)

Total 300.00

Problem 39

Electrophoretic mobility of glucose phosphate isomerase of 16 mice shows the following data (FF = fast, SS = moves slow, FS = moves intermediate). Calculate allele frequency.

FF = 5; FS = 7 and SS = 4.

Solution:

Genotypes	FF	FS	SS	Total
Number of individuals	5	7	4	16
Genotypic frequency	$\frac{5}{16}$ =0.312	$\frac{7}{16}$ =0.437	$\frac{4}{16}$ =0.25	

Gene frequency of F = $0.312 + \frac{1}{2} \times 0.437 = 0.312 + 0.218 = 0.530 = 0.53$.

Gene frequency of S = $0.25 + \frac{1}{2} \times 0.437 = 0.25 + 0.219 = 0.469 = 0.47$.

Problem 40

In Caucasians, the M-shaped hairline that receeds with age (sometimes to nearly complete baldness) is due to an allele that is dominant in males but recessive in females. The frequency of baldness allele is approximately 0.3. Assuming random mating, what frequencies of the bald and nonbald phenotypes are expected in males and females?

Solution:

The frequencies of bald and nonbald alleles are 0.3 and (1 – 0.3) or 0.7 respectively.

(i) Bald females are homozygous and occur in frequency $(0.3)^2 = 0.09$. Nonbald females occur with frequency $1 - 0.09 = 0.091$.

(ii) Bold males are homozygous or heterozygous and occur with frequency = $(0.3)^2 + 2 \times 0.3 \times 0.7 = 0.09 + 0.42 = 0.51$.

Nonbald males occur with frequency 01 – 0.51 = 0.49.

Problem 41

In a randomly mating population that contains n equally frequent alleles, what is the expected frequency of heterozygotes?

Solution:

The frequencies of each allele is $\frac{1}{n}$ and there are n possible homozygotes. Therefore, the total frequency of homozygotes is $n \times (\frac{1}{n})^2 = \frac{1}{n}$.

It implies that the total frequency of heterozygote is $(1 - \frac{1}{n})$.

Problem 42

For an autosomal gene with two alleles in a randomly mating population, what is the probability that a heterozygous female will have a heterozygous offspring?

Solution:

(i) An Aa females produces $\frac{1}{2}$A and $\frac{1}{2}$a gametes and these yields heterozygous offspring if they are fertilized by a and A respectively.

(ii) The overall probability of a heterozygous offspring is $\left(\frac{1}{2}\right)q + \left(\frac{1}{2}\right)p = \left(\frac{1}{2}\right)(p + q) = \frac{1}{2}$.

Problem 43

A population is on Hardy-Weinberg equilibrium for the allele F and f, where F is dominant. A study reveals that 18% of the population is phenotypically F. What is the expected frequency of heterozygotes in this population?

Solution:

FF(p^2) and Ff (2 pq) make up 18% of the population, i.e. 0.18.

The remaining is ff (q^2) = 1 – 0.18 = 0.82

$q^2 = 0.82, \therefore q = \sqrt{0.82} = 0.9055$

$p = 1 - 0.9055 = 0.0945.$

The expected frequency of heterozygotes is 2pq $= 2 \times 0.0945 \times 0.9055 = 0.1711$.

0.1711 = 17.11% of the population is expected to be heterozygotes.

Problem 44

It is suspected that the excretion of the strongly odorous substance methanethiol is controlled by a recessive allele m in humans. Non-excretion is governed by the dominant allele M. The frequency of m on Iceland is 0.40.

(a) What is the probability of being an excretor in Iceland?

(b) What is the probability of a nonexcretor individuals being heterozygous?

(c) What is the probability of both parents being heterozygous nonexcretor?

Solution:

(a) $q = m = 0.40 \therefore q^2 = (0.40)^2 = 0.16 = 16\%$.

(b) $2pq = Mm = 2 \times 0.60 \times 0.40 = 0.48 = 48\%$ [1 – 0.4 = 0.6]

$p^2 = MM = (0.6)^2 = 0.36.$

And $\frac{2pq}{(p^2+2pq)} = \frac{0.48}{0.36+0.48} = \frac{0.48}{0.84} = 0.571$

$= 57.1\%.$

(c) $2pq \times 2pq = 0.48 \times 0.48 = 0.2304 = 23.04\%$.

Problem 45

An anaemic condition in humans called thalassaemia is known to be governed by a pair of codominant alleles. The homozygous genotype T^mT^m results in severe anaemia (thalassaemia major) and the heterozygous genotype T^mT^n results in mild anaemia (Thalassaemia minor). Normal individuals are T^nT^n. The distribution of this disease among tribal population at Chattisgarh was found to be approximately 4 major, 400 minor and 9596 normal. Does this population is in equilibrium?

Solution:

Genotype/Phenotype	T^mT^m	T^mT^n	T^nT^n	Total
Number of individuals	4	400	9596	1000
Genotypic frequency	$\frac{4}{10000}$	$\frac{400}{10000}$	$\frac{9596}{10000}$	
	=0.0004,	=0.04,	=0.9596	

Gene frequency of $T^m = 0.0004 + \frac{1}{2} \times 0.04 = 0.0004 + 0.02 = 0.0204$.

Gene frequency of $T^n = 0.0506 + \frac{1}{2} \times 0.04 = 0.9596 + 0.02 = 0.9796$.

Expected genotypic frequency: $(p+q)^2 = (0.0204)^2 + 2 \times 0.0204 \times 0.9796 + (0.9796)^2$

	T^mT^m	T^mT^n	T^nT^n
Observed frequency =	0.0004	0.04	0.9596
Expected frequency =	0.0004	0.0399	0.9596

These data are obviously a good fit and there is no deviation between observed and expected genotypic frequency.

Problem 46

Consider a population in which random mating is occurring with an insignificant level of mutation, migration, and selection. What will be the zygotic composition of the next generation of a very large population if the initial population has the following genotypic frequencies?

0.20 HH, 0.40 Hh, 0.40 hh.

Solution:

Offspring genotypes

Mating	Frequency	HH	Hh	hh
1. HH×HH:	$(0.20\times0.20)=0.04$	0.04	–	–
2. HH×Hh:	$(0.20\times0.40)\times2=0.16$	0.08	0.08	–
3. HH×hh:	$(0.20\times0.40)\times2=0.16$	–	0.16	–
4. Hh×Hh:	$(0.40\times0.40)=0.16$	0.04	0.08	0.04
5. Hh×hh:	$(0.40\times0.40)\times2=0.32$	–	0.16	0.16
6. hh×hh:	$(0.40\times0.40)\times0.16$	–	–	0.16
		0.16,	0.48,	0.36

The zygotic composition in the next generation is 16% (HH), 48% (Hh) and 36% (hh).

$$\left[\begin{aligned} H &= 0.20+\tfrac{1}{2}\times0.40=0.20+0.2=0.4\\ h &= 0.40+\tfrac{1}{2}\times0.40=0.40+0.2=0.6\\ (0.4+0.6)^2 &= 0.16+0.48+0.36\end{aligned}\right]$$

Problem 47

The following values are the observed numbers of individuals in a population with each of these genotypes. Is this population in Hardy-Weinberg equilibrium?

KK = 640, Kk = 960, kk = 400.

Solution:

Genotype/Phenotype	KK	Kk	192	Total
Number of individuals	640	960	400	2000
Genotypic frequency	$\frac{640}{2000}=0.32$	$\frac{960}{2000}=0.48$	$\frac{400}{2000}=0.2$	

Gene frequency of K $= 0.32+\frac{1}{2}\times0.48=0.32+0.24=0.56$.

Gene frequency of k $= 0.2+\frac{1}{2}\times0.48=0.2+0.24=0.44$.

Expected genotypic frequency $=(0.56+0.44)^2=0.3136+0.4928+0.1936$.

Genotypic frequency	KK	Kk	kk
Observed frequency =	0.32	0.48	0.2
Expected frequency =	0.3136	0.4928	0.1936

The data, i.e., observed and expected is more or less similar, so it appears to be in H-W equilibrium.

(Required Chi square test to confirm it).

Problem 48

The following gives the (genotypic) frequencies for three population. Which one is in Hardy-Weinberg equilibrium?

Population I: 10 AA 80 Aa 10 aa Total: 100
Population II: 0 AA 150 Aa 100 aa Total: 250
Population III: 2 AA 16 Aa 32 aa Total: 50

C.U. Zoology (Hons.) Part II-2009

Solution:

Population	Genotypic frequency			Gene frequency		Equilibrium/ not
	AA	Aa	aa	A	a	
I	0.1	0.8	0.1	0.5	0.5	not
II	0	0.6	0.4	0.3	0.7	not
III	0.04	0.32	0.64	0.2	0.8	yes

$(0.2+0.8)^2=0.04+0.32+0.64$

Genotyping frequency is similar to parental one. So it is in equilibrium.

In case of I and II, population genotypic frequencies are not similar to parental one. So they are not in equilibrium.

Problem 49

A survey of 50 *Drosophila* by Dobzhansky, and his colleagues in New Mexico found the following numbers of the various inversion genotypes. ARAR, 18; ARPP, 19; ARST, 5; PPPP, 5: PPST, 3: and STST, 0.

(a) Treating the inversions ARPP and ST as if they are in three different alleles, calculate the frequency of the three inversion types.

(b) Using these frequencies calculate expected numbers of six genotypes if they were in Hardy-Weinberg proportions.

Solution:

	Type	Proportion of inversion
ARAR = 18	AR	$= 18 + \frac{1}{2} \times 19 + \frac{1}{2} \times 5$
ARPP = 19		$= 18 + 9.5 + 2.5 = 30.0 = 30$
ARST = 5	PP	$= 5 + \frac{1}{2} \times 19 + \frac{1}{2} \times 3$
PPPP = 5		$= 5 + 9.5 + 1.5 = 16.0 = 16$
PPST = 3	ST	$= 0 + \frac{1}{2} \times 3 + \frac{1}{2} \times 5$
STST = 0		$= 0 + 1.5 + 2.5 = 4.0 = 4$

	AR	PP	ST	
Proportion	30	16	4	= 50
Frequency	$\frac{30}{50} = 0.6$	$\frac{16}{50} = 0.32$	$\frac{4}{50} = 0.08$	

Takes AR = p

PP = q,

ST = r,

AR + PP $\quad \frac{AR}{AR} \quad \frac{AR}{PP} \quad \frac{PP}{PP}$

$(p + q)^2 = (.6 \times .32)^2 = 0.36 + 0.384 + 0.1024$

1. Among 50 inversion =

$0.36 \times 50 + 0.384 \times 50 + 0.1024 \times 50$

$= 18 \quad = 19.2 \quad = 5.12$

$(AR + ST)^2$ = AR/AR AR AR/ST ST/ST

$(p + r)^2 = (0.6 \times 0.08)^2 = 0.36 + 0.096 + 0.0064$

Amg 50 = $0.36 \times 50 = 0.96 \times 50 + 0.0064 \times 50$

$= 18 \quad = 4.8 = 0.32$

PPST = $2 \times 0.32 \times 0.08 = 0.0512$

$50 \times 0.0512 = 2.56$

Among 50 inversions

ARAR = $(AR)^2$	$6 \times .6 = .36 = .36$	$.36 \times 50$	18
ARPP = 2ARPP	$2 \times .6 \times .32 = .384$	$.384 \times 50$	19.2
ARST = 2ARST	$2 \times .6 \times .08 = .096$	$.096 \times 50$	4.8
PPPP = PP × PP	$.32 \times .32 = .1024$	$.1024 \times 50$	5.12
PPST = 2(PPST)	$2 \times .32 \times .08 = .0512$	$.0512 \times 50$	2.56
STST = ST × ST	$.08 \times .08 = .0064$	$.0064 \times 50$	0.32

Problem 50

From the following genotypic counts, calculate genotype frequencies and allele frequencies CC:42, CT:16, TT:32.

Solution:

The three genotype frequencies are:

CC	CT	TT	Total
42	16	32	90
$\frac{42}{90} = 0.467$	$\frac{16}{90} = 0.178$	$\frac{32}{90} = 0.356$	

Alleles frequencies are

$f_C = 0.467 + \frac{1}{2} \times 0.178 = 0.467 + 0.087 = 0.55$

$f_T = 0.350 + \frac{1}{2} \times 0.178 = 0.356 + 0.089 = 0.45$

Exercise

1. In a garden of Serampore College Botany department, a certain plant population has 289 individuals, some bearing white flowers, others red flowers. The red allele is dominant over the white allele of the 289 plants, 246 red flowers. Assuming that the population is in equilibrium, (a) What is the frequency of the white allele? (b) How many of the red plants are expected to be heterozygous?

Ans. (a) 0.386, (b) 0.474 = Heterozygous flower 0.474 × 289 = 136.986 = 137

2. Calculate the frequency of alleles A_1 & A_2 for the following population.

Population:

	AA	Aa	aa	Total
(a)	68	42	24	= 134
(b)	72	122	91	= 285
(c)	20	40	40	= 100
(d)	30	8	12	= 50

Ans.

	A	a
(a)	0.664	0.336
(b)	0.467	0.533
(c)	0.4	0.6
(d)	0.68	0.32

3. Which of the following genotype frequencies AA, Aa and aa respectively satisfy the Hardy-Weinberg principle?

	AA	Aa	aa
(a)	0.36	0.55	0.09
(b)	0.49	0.42	0.09
(c)	0.29	0.42	0.29
(d)	0.64	0.27	0.09
(e)	0.12	0.87	0.01

Ans.

	p	q	AA	Aa	aa
(a)	0.635	0.365	0.403	0.46	4
(b)	0.7	0.3	0.49	0.42	0.09
(c)	0.5	0.5	0.25	0.5	0.25
(d)	0.775	0.225	0.60	0.349	0.051
(e)	0.555	0.325	0.455	0.614	0.105

4. The incidence of recessive albinism is 0.008 in a human population. If mating for trait is random in the population, what is the frequency of the recessive allele?

Ans. (0.089)

5. The following data for M-N blood types were obtained from a village of Malda district of West Bengal.

Place	Number of individual	M	MN	N
(a) Bulbulchandi:	240	68	55	117
(b) Gour	98	55	35	8

Calculate the gene frequencies of M and N alleles.

Ans. (a) p = 0.4 & q = 0.6; (b) p = 0.74 & q = 0.26

6. Hartnup disease is an autosomal-rcessive disorder of intestinal and renal transport of amino acids. The frequency of affected newborn infants is about 1 in 14,000. Assuming random mating, what is the frequency of heterozygotes:

Ans. $[q = \sqrt{\frac{1}{14000}} = 0.0085; p = 1 - 0.0085 = 0.9915$

Frequency of heterozygote = 2pq = $2 \times .9915 \times .0085 = .0168 = .017$ **of at 1 in 80]**

7. In human dominant allele T can taste the substance phenylthiocarbamide (PTC). In a population in which the frequency of this allele is 0.4, what is the probability that a particular taster is homozygous?

Ans. [Taster (TT + Tt) $= (.04)^2 + 2 \times .04 \times .06 = .16 + .48 = .64$

Among all taster $\cdot\frac{16}{64} = 0.25]$

8. If A_2 is recessive and 4 out of 400 individuals are A_2A_2 (rest are the dominant phenotype), what is the estimated frequency of A_2? What proportion of the population is expected to be heterozygous?

Ans. $\left[\frac{4}{400} = \frac{1}{100} = 0.01; \sqrt{0.01} = 0.1\right.$
$\left. 2 \times 0.1 \times 0.9 = 0.18\right]$

9. If the initial genotypic frequencies are 0.04, 0.32 and 0.64 for A_1A_1, A_1A_2 and A_2A_2, what are the frequencies after one generation of complete self-fertilization?

Ans. [0.04, 0.32, 0.64]

10. The frequency of new born infants homozygous for a recessive lethal allele is about 1 in 2500. What is the expected frequency of carriers of these alleles in the population.

Ans. $[q^2 = \frac{1}{25000} = 0.00004 = 4 \times 10^{-5}$

$q = \sqrt{4 \times 10^{-5}} = 6.3 \times 10^{-4} = .006324$

$p = 1 - 0.006324 = 0.993676 = 0.012568 = 0.0126]$

$2pq = 0.012568 = 0.0126]$

11. If q = 0.3 and there are Hardy-Weinberg proportions. What is the most common genotype and what is its frequency? What is the least frequent genotype and its frequency?

Ans. $\left[a = 0.3;\ A = 1 - 0.3 = 0.7;\right.$
Most common $(0.7)^2 = 0.49$;
Least $(0.3)^2 = 0.09]$

12. In a large random mating population the frequencies of the genotypes GG, GG and gg are 0.4, 0.32 and 0.64 respectively. Assume that the change of environment induces the population to reproduce by self-fertilization. Predict the frequencies of the genotype in this population after many generations of self-fertilization.

Ans. Ultimate frequency of GG = 0.2 & 79 = 0.8

13. In a survey of moths collected from Gorumara forest, a researcher found 51 dark variety and 49 light variety. The dark moths carry dominant allele (D) and light moths are homozygous for recessive (dd) allele. If the population is in Hardy-Weinberg equilibrium, what is the estimated frequency of the recessive allele in the population? How many of the dark moths in the sample are likely to be homozygous for the dominant allele?

Ans. $[51 + 49 = 100;\ dd = \frac{49}{100} = 0.49;$

$q = \sqrt{.49} = 0.7;\ D(p^2) = (.3)^2 = 0.09$ in 100 species, it will be $0.09 \times 100 = 9]$

14. In a population of 2000 poisonous snake, a genetic difference with respect to venom exists at a single locus. The alleles are incompletely dominant. The population shows 100 individuals homozygous for the t allele (genotype tt, non-poisonous) 800 heterozygous (genotype Tt mildly poisonous) and 1100 homozygous for the T allele (genotype TT, lethally poisonous).

(a) What is the frequency of the t allele in the population.

(b) Are the genotypes in Hardy-Weinberg eqiulibrium?

Ans. (a) t = 0.25

	TT	Tt	tt	
(b) $(0.75+0.25)^2 =$	(0.5625	+0.375	+0.0625)	
in 2000 =	1125	+ 750	+ 125	Expected
	1100	+ 800	+ 100	Observed

Close to equilibrium please perform chi square.

Multiple Alleles

In case of Multiple allele:

(i) More than one allele at same locus is called multiple alleles.

(ii) If there are three alleles at a single locus, i.e., A_1, A_2 and A_3, their respective frequencies are p, q and r, then $p + q + r = 1$.

(iii) If there is random mating between all gametes, their genotypes are as below:

♀ \ ♂	$A_1(p)$	$A_2(q)$	$A_3(r)$
$A_1(p)$	$A_1A_1(pp = p^2)$	$A_1A_2(pq)$	$A_1A_3(pr)$
$A_2(q)$	$A_1A_2(pq)$	$A_2A_2(q^2)$	$A_2A_3(qr)$
$A_3(r)$	$A_1A_3(pr)$	$A_2A_3(qr)$	$A_3A_3(r^2)$

Thus, genotypic frequency in the next generation will be

$$p^2(A_1A_1) + 2pq(A_1A_2) + 2qr(A_2A_3) + 2pr(A_1A_3) + r^2(A_1A_3) =$$

$$p^2 + q^2 + r^2 + 2pq + 2qr + 2pr = (p + q + r)^2 = (1)^2 = 1.$$

- The frequency of $A_1 = p^2 + \frac{1}{2} \times 2pq + \frac{1}{2} \times 2pr$
  $= p^2 + pq + pr$
  $= p(p + q + r) = p.1 = p.$
- The frequency of $A_2 = q^2 + \frac{1}{2} \times 2pq + \frac{1}{2} \times 2qr$
  $= q^2 + pq + qr$
  $= q(q + p + r) = q.1 = q.$
- The frequency of $A_3 = r^2 + \frac{1}{2} \times 2pr + \frac{1}{2} \times 2qr$
  $= r^2 + pr + qr$
  $= r(r + p + q) = r.1 = r.$

Therefore, this law is applicable in ABO blood group.

Solved Problems

Problem 1

Among 192 students of Serampore College, the following phenotypes were observed: A 63, B 31, AB 6 and O 92. Calculate the gene frequencies of genes I^A, I^B and i.

Solution:

Assuming random mating the gene frequencies of I^A, I^B and i represented by p, q and r respectively.

In the equilibrium population, the frequency of the three alleles can be described as follows:

$p(A) + q(B) + r(O) = 1$ and the distribution of genotypes will be given by

$$(p + q + r)^2 = p^2(AA) + q^2(BB) + r^2(O) + 2pq(AB) + 2qr(BO) + 2pr(AO).$$

As 'O' allele is recessive, the frequency of type 'O' blood in the population is equal to the frequency of the recessive genotype r^2. Thus,

$$r^2 = \frac{92}{192} = 0.479.$$

$$\therefore r = \sqrt{0.479} = 0.692 = 0.69.$$

A allele is present in two genotypes AA and AO.

The frequency of AA genotype is represented by p^2 and the AO genotype by 2pr. Therefore, the frequency of $A + O = p^2 + 2pr + r^2$.

Genotypic frequency of $A = \frac{63}{192} = 0.328.$

$\therefore A + O = p^2 + 2pr + r^2$

$(p + r)^2 = p^2 + 2pr + r^2 = 0.328 + 0.479 = 0.807$

$(p + r)^2 = A + O, \therefore p + r = \sqrt{A+O}$

$p + 0.69 = \sqrt{0.328+0.479} = \sqrt{0.807} = 0.898.$

As $p + q + r = 1 \therefore p = 0.898 - 0.69 = 0.208$

$q = 1 - (p + r) = 1 - (0.2 + 0.69) = 1 - 0.89 = 0.11.$

$p = 0.2,\ q = 0.1,\ r = 0.69 = 0.7.$

Problem 2

Among 600 students of R. K. Mission Vidyamandira the following ABO blood types are obtained:

O 288, A 280, B 19, AB 13

Solution:

Let the gene frequency of A, B, O be p, q and r respectively.

As the population are in equilibrium, the frequency of three alleles can be described as follows:

$$p(A) + q(B) + r(O) = 1$$

As 'O' allele is recessive, the frequency of type 'O' blood in the population is equal to the frequency of the recessive genotype r^2.

Thus, $r^2 = \frac{288}{600} = 0.48, \quad r = \sqrt{0.48} = 0.69.$

A allele is present in two genotypes AA and AO.

The frequency of AA genotype is represented by p^2 and AO genotype by 2pr. Therefore, the frequency of

$$A + O = p^2 + 2pr + r^2 = (p + r)^2$$

Genotypic frequency of A = $\frac{280}{600} = 0.466 = 0.47.$

$\therefore\ p + r = \sqrt{A+O}$

$p + 0.69 = \sqrt{0.47+0.48} = \sqrt{0.95} = 0.97$

$p = 0.97 - 0.69 = 0.28$

As $p + q + r = 1;$

$q = 1-(p+r) = 1-(0.28+0.69) = 1-0.97 = 0.03$

$p = 0.28,\ q = 0.03,\ r = 0.69.$

Problem 3

In a human population of 100 people, 17 have type A blood, 17 have type B blood, 2 have type AB and 64 have type O. If this population is in equilibrium, what are the allelic frequencies?

Solution:

Let the gene frequency of A, B, O are p, q and r respectively. As the population are in equilibrium the frequency of three alleles can be described as follows:

$$p(A) + q(B) + r(O) = 1$$

As 'O' allele is recessive, the frequency of type 'O' blood in the population is equal to the frequency of the recessive genotype r^2. Thus,

$r^2 = \frac{64}{100} = 0.64. \qquad \therefore\ r^2 = \sqrt{0.64} = 0.8.$

A allele is present in two genotypes. AA and AO. The frequency of AA genotype is represented by p^2 and the AO genotype by 2pr. Therefore, the frequency of $A + O = p^2 + 2pr + r^2 = (p + r)^2$.

Genotypic frequency of A = $\frac{17}{100} = 0.17.$

$A + O = (p + r)^2$

or, $p + r = \sqrt{A+O} = \sqrt{0.17+0.64} = \sqrt{0.81}$

$p + r = 0.9 \ \therefore\ p = 0.9 - 0.8 = 0.1$

As $p + q + r = 1;\ q = 1-(p+r) = 1-0.9 = 0.1$

$\therefore\ p = 0.1,\ q = 0.1,\ r = 0.8.$

Problem 4

A sample 1000 students of Raja Pearymohan college showed the following distribution of blood groups A 430, B 140, AB 50, O 380. Calculate the frequencies of I^A, I^B and i.

Solution:

Let the gene frequency A, B, O are p, q and r respectively. As O allele is recessive, the frequency of type 'O' blood in the population is equal to the frequency of the recessive genotype r^2. Thus,

$r^2 = \frac{380}{1000} = 0.38; \quad \therefore\ r = \sqrt{0.38} = 0.616 = 0.62.$

A allele is present in two genotypes AA and AO. The frequency of AA genotype is represented by A^2 and the AO genotype by 2pr. Therefore, genotypic frequency of A = $\frac{430}{100} = 0.43.$

$(p + r)^2 = p^2 + 2pr + r^2 = AA + AO + 0.38 = 0.43 + 0.38 = 0.81.$

$(p + r)^2 = (A + O)$

$\therefore\ p + r = \sqrt{A+O} = \sqrt{0.81} = 0.9$

$p = 0.9 - 0.62 = 0.28$

$p + q + r = 1.$

$\therefore\ q = 1 - (p + r) = 1 - 0.9 = 0.1.$

$\therefore\ p = 0.28,\ q = 0.1,\ r = 0.62.$

Problem 5

The ABO blood group system is governed by a multiple allelic system in which some codominant relationships exist. Three alleles I^A, I^B and i form the dominance hierarchy (I^A, I^B) > i.

(a) A sample of human population was blood grouped and found to contain 23 group AB, 441 group O, 371 group B and 65 group A. Calculate allelic frequency of I^A, I^B and i.

(b) Given the gene frequency I^A = 0.36, I^B = 0.20 and i = 0.44. Calculate the percentage of population expected to be of groups A, B, AB and O.

Solution:

Let the gene frequency of A, B, O are p, q and r respectively. As the populations are in equilibrium, the frequency of three alleles can be described as follows:

$$p^2(A) + q^2(B) + r^2(O) = 1$$

As 'O' allele is recessive, the frequency of type 'O' blood in the population is equal to the frequency of the recessive genotype r^2. Thus,

$r^2 = \frac{441}{900} = 0.49, \quad \therefore\ r = \sqrt{0.48} = 0.7.$

A allele is present in two genotypes AA and AO. The frequency of Aa genotype is represented by p^2, the AO genotype by 2pr. Therefore, the frequency of

$$A + O = p^2 + 2pr + r^2 = (p + r)^2.$$

Genotypic frequency of A $= \frac{65}{900} = 0.072.$

$(p + r)^2 = A + O$

$p + r = \sqrt{A+O} = \sqrt{0.072+0.49} = \sqrt{0.562} = 0.75,$

$p + r = 0.75$

$p + 0.7 = 0.75$

$p = 0.75 - 0.7 = 0.05.$

(a) $p + q + r = 1$

$q = 1 - (p + r) = 1 - (0.05 + 0.7)$

$= 1 - 0.75 = 0.25$

$p = 0.5, q = 0.25, r = 0.7.$

(b) $(p + q + r)^2 = (0.36 + 0.2 + 0.44)^2$

$= 0.1296(p^2) + 0.3168(2pr) + 0.04(q^2) + 0.176(2qr) + 0.144(2pq) + 0.1936(r^2)$

$= 0.4464(A) + 0.216(B) + 0.144(AB) + 0.1936(O)$

$= 44.64\% + 21.6\% + 14.4\% + 19.36\%$

(A) (B) (AB) (O)

Problem 6

In a large random mating human population the frequencies of the I^A, I^B and i alleles are 0.7, 0.2 and 0.1 respectively. Calculate the expected frequencies for each blood type.

Solution:

Let p, q, r be the gene frequencies of the A, B and O blood group. The genotypic frequencies will be $(p + q + r)^2$.

$(0.7 + 0.2 + 0.1)^2 = (0.7)^2 + (0.2)^2 + (0.1)^2 + 2 \times 0.7 \times 0.2 + 2 \times 0.7 \times 0.1 + 2 \times 0.2 \times 0.1 = 0.49(p) + 0.04(q) + 0.01(r) + 0.28(2pq) + 0.14(2pr) + 0.04(2qr)$

$A(p^2 + 2pr)$ $B(q^2 + 2qr)$ $AB\ (2pq)$ $O(r^2)$

$= 0.49 + 0.14 = 0.04 + 0.04 = 0.28 = 0.01$

$= 0.63(63\%) = 0.08(8\%) = 0.28(28\%) = 0.01(1\%)$

Problem 7

173 people of Uttarpara showed the following blood types O 78; A 71; B 17 and AB 7. Calculate gene frequencies of A, B and O.

Solution:

Let p, q and r be the gene frequencies of A, B and O.

As 'O' allele is recessive, the frequency of type 'O' blood in the population is equal to the frequency of the recessive genotype r^2.

Thus, $r^2 = \frac{78}{173} = 0.45, \therefore\ r = \sqrt{0.45} = 0.67.$

A allele is present in two genotypes AA and AO. The frequency of AA genotype is represented by p^2 and AO genotype by 2pr. The genotypic frequency of A $= \frac{71}{173} = 0.41.$

$(p + r)^2 = p^2 + 2pr + r^2 = A + O$

$(p + r)^2 = A + O$

$p + r = \sqrt{A+O} = \sqrt{0.41+0.45} = \sqrt{0.86}$

$= 0.927 = 0.93;\ p + 0.67 = 0.93;$

$\therefore\ p = 0.93 - 0.67 = 0.26.$

As $p + q + r = 1,$

$q = 1 - (q + r) = 1 - 0.93 = 0.07.$

So, $p = 0.26, q = 0.07, r = 0.67.$

Problem 8

Consider a population of Bandel that consists of persons having phenotypes of the ABO blood types in the frequencies given below:

Blood type

O	A	B	AB
460	420	90	30

(a) What are the gene frequencies of A, B and O in the population?

(b) How many of the persons with the A blood type would be homozygous for that allele?

Solution:

Let p, q and r be the gene frequencies of A, B and O blood group.

Thus, $r^2 = \frac{460}{1000} = 0.46$, $\therefore$ $r = \sqrt{0.46} = 0.678$.

$(p + r) = \sqrt{A+O} = \sqrt{\frac{460}{1000}+\frac{460}{1000}} = \sqrt{0.42+0.46}$

$p + r = \sqrt{0.88} = 0.938$.

$\therefore$ $p = 0.938 - 0.678 = 0.26$

$p + q + r = 1$

or, $q = 1 - (p + r) = 1 - 0.938 = 0.062$.

Hence, $p = A = 0.26$

$q = B = 0.062$

$r = O = 0.678$.

(b) $p = A = 0.26$

$p^2 = (0.26)^2 = 0.0676 = 0.068$

$2pr = 2 \times 0.26 \times 0.678 = 0.35256 = 0.353$

Total A group $= 0.068 + 0.353 = 0.421$.

Proportion of homozygous A $= \frac{0.068}{0.421}$

$= 0.16152 = 0.162$.

Problem 9

In a tribal group in Purulia, the frequencies of alleles determining the ABO blood groups were estimated as 0.74 for I^O, 0.16 for I^A and 0.10 for I^B. Assuming random mating, what are expected frequencies of ABO genotypes and phenotypes?

Solution:

$A = I^A = 0.16$

$B = I^B = 0.10$

$O = I^O = 0.74$

$(0.16 + 0.10 + 0.74)^2$

$= (0.16)^2 + 2 \times 0.16 \times 0.74 + (0.10)^2 + 2 \times 0.10 \times 0.74 + (0.74)^2 + 2 \times 0.16 \times 0.10$

$= 0.0256 + 0.2368 + 0.01 + 0.148 + 0.5476 + 0.032$

$= 0.2624 + 0.158 + 0.5476 + 0.032$.

The phenotype frequencies of A $= (I^A)^2 + 2I^AI^O = 0.256 + 0.2368 = 0.2624$.

The phenotype frequencies of B $= (I^B)^2 + 2I^BI^O = 0.01 + 0.148 = 0.158$.

The phenotype frequencies of O $= (I^O)^2 = (0.74)^2 = 0.5476 = 0.548$.

The phenotype frequencies of AB $= 2 \times I^A \times I^B = 2 \times 0.16 \times 0.10 = 0.032$.

A = 0.262, B = 0.158, AB = 0.032, O = 0.548.

Problem 10

The following table lists the frequencies for the blood group phenotypes observed in 190177 persons in North Bengal:

Phenotype	Observed data
A	79,334
B	16,280
AB	5,781
O	88,782
	1,90,177

Apply Hardy-Weinberg relationship to these data.

Solution:

Let the gene frequencies are p, q and r for A, B and O blood group respectively.

Phenotype	Genotypes	Algebraic frequency	Genotypic frequency	
A	I^AI^A and I^AI^0	$p^2 + 2pr$	$\frac{79,334}{1,90,177} = 0.41715$	= 0.417
B	I^BI^B and I^BI^0	$q^2 + 2qr$	$\frac{16,280}{1,90,177} = 0.08656$	= 0.086
AB	$I^A I^B$	$2pq$	$\frac{5,781}{1,90,177} = 0.030039$	= 0.030
O	I^0	r^2	$\frac{88,782}{1,90,172} = 0.4668$	= 0.467

The r^2(O) is homozygous recessive =

$r^2(O) = 0.467, \quad \therefore r = \sqrt{0.467} = 0.683.$

$(p + r)^2 = p^2 + 2pr + r^2$

$p + 0.683 = \sqrt{p^2 + 2pr + r^2} = \sqrt{0.417 + 0.467}$

$= \sqrt{0.884} = 0.940$

$p + 0.683 = 0.94$

$p = 0.940 - 0.683 = 0.257$

$p + q + r = 1.$

$\therefore q = 1 - (p + r) = 1 - 0.940 = 0.060.$

With the frequencies of p, q and r all of the genotypic frequencies can be calculated and tabulated.

Genotype	Algebraic frequency	Calculation	Genotypic frequency
I^AI^A	p^2	$(0.257)^2$	= 0.0660 = 0.066
I^BI^B	q^2	$(0.06)^2$	= 0.0036 = 0.004
I^i	r^2	$(0.683)^2$	= 0.4664 = 0.467
I^Ai	2pr	$2 \times 0.257 \times 0.683$	= 0.3510 = 0.351
I^Bi	2qr	$2 \times 0.06 \times 0.683$	= 0.08196 = 0.082
I^AI^B	2pq	$2 \times 0.257 \times 0.06$	= 0.03084 = 0.031
			1.001

Problem 11

Consider a population of Sonarpur that consists of persons having phenotypes of the ABO blood types in the frequencies given in the next page:

Blood type			
O	**A**	**B**	**A B**
250	**42**	**672**	**36**

What are the gene frequencies in the population?

Solution:

Let the gene frequencies of A, B & O are p, q and r respectively.

As 'O' blood group is recessive, the frequency of the type O blood in the population is equal to the frequency of the recessive genotype r^2.

$r^2 = \frac{250}{1000} = 0.25 \qquad \therefore r = \sqrt{0.25} = 0.5$

$(p + r)^2 = p^2 + 2pq + q^2$

$(p + r)^2 = A + O$

$p + r = \sqrt{A+O} = \sqrt{0.042+0.25} = \sqrt{0.292} = 0.54$

$= 0.54 - 0.5 = 0.04$

$p = 0.04, \ r = 0.5.$

$\therefore q = 1 - (0.04 + 0.5) \pm 1 - 0.54 = 0.46.$

Problem 12

In a large random mating population of Andaman, the frequencies of the I^A, I^B, i alleles are 0.15, 0.25 and 0.60 respectively. Calculate the expected frequencies for each blood type.

Solution:

Let p, q and r be the gene frequencies of I^A, I^B and i alleles respectively.

$(0.15 + 0.25 + 0.60)^2$

$= (0.15)^2 + (0.25)^2 + (0.60)^2 + 2 \times 0.15 \times 0.25 + 2 \times 0.15 \times 0.60 + 2 \times 0.25 \times 0.60$

$= 0.0225 + 0.0625 + 0.36 + 0.075 + 0.18 + 0.30$

$= 0.0225 + 0.18 + 0.0625 + 0.30 + 0.36 + 0.75.$

$A(p^2 + 2pr) \quad B(q^2 + 2qr) \quad O(q^2)\ AB(2pq)$

$= \quad 0.2025 \quad + \quad 0.3625 \quad + \quad 0.36 + 0.75.$

Problem 13

The gene for coat colour in rabbits can exist in four alleles termed C (full coat colour), C^{Ch} (chinchilla) C^h (Himalayan), and C (albino). In a population of rabbits in Hardy-Weinberg-equilibrium, the allele frequencies are C = 0.34, C^{Ch} = 0.17, C^h = 0.44, c = 0.05.

Assume that C is dominant to the other three alleles C^{Ch} is dominant to C^h and c and C^h is dominant to c.

(a) What is the frequencies of albino rabbits?

(b) Among 1000 rabbits, how many would you expect to have Himalayan coat colour?

(c) Among 1000 rabbits, how many would be heterozygous with a chinchilla coat colour?

Solution:

In three or more alleles $(p + q + r + s)^2 = p^2 + q^2 + r^2 + s^2 + 2pq + 2qr + 2pr + 2sr + 2sp + 2qs = 1$.

Let p = C, q = C^{Ch}, r = C^h and s = c.

(a) The frequency of albino rabbits is $(s^2 = (0.05)^2 = 0.0025 = 0.25\%$.

(b) Himalayan is dominant to albino and recessive to full coat colour and chinchilla. Himalayan rabbit is represented by $r^2 + 2rs = (.44)^2 + 2 \times 0.44 \times 0.05 = 0.1936 + 0.044 = 0.2376 = 0.24 = 24\%$.

Among 1000 rabbits it will be $0.24 \times 1000 = 240.00 = 240$.

(c) Heterozygous chinchilla is represented by $2qr + 2qs = 2q(r + s) = 2 \times 0.17(0.44 + 0.05) = 0.34 \times 0.49 = 0.1666 = 0.17 = 12\%$.

Among 1000 rabbits $0.17 \times 1000 = 170.00 = 170$ would be heterozygous chinchilla.

Problem 14

In human MN blood group is determined by two codominant alleles, M and N. The following data were obtained from various populations from different states of India.

	Place	**Percentge**		
		MM	**MN**	**NN**
1.	**Hariyana**	**83.5**	**15.6**	**0.9**
2.	**Tamil Nadu**	**2.4**	**30.4**	**67.2**
3.	**Delhi**	**84.5**	**14.4**	**1.1**
4.	**West Bengal**	**45.7**	**43.1**	**11.2**
5.	**Orissa**	**39.9**	**44.0**	**16.1**

(a) Calculate the allele frequencies in these five population.

(b) Which populations appear to be H.W. equilibrium?

(c) Which populations do you think have had significant intermixing due to migration?

Solution:

(a)

1. **Hariyana**

M = 83.5% = 0.835

MN = 15.6% = 0.156

N = 00.9% = 0.009

$$p^M = 0.835 + \frac{1}{2} \times 0.156 = 0.835 + 0.078 = 0.913$$

$$q^N = 0.009 + \frac{1}{2} \times 0.156 = 0.009 = 0.078 = 0.087$$

2. **Tamil Nadu**

$$p^M = 0.024 + \frac{1}{2} \times 0.304 = 0.024 + 0.152 = 0.176$$

$$q^N = 0.672 + \frac{1}{2} \times 0.304 = 0.672 + 0.152 = 0.824.$$

3. **Delhi**

$$p^M = 0.845 + \frac{1}{2} \times 0.144 = 0.845 + 0.072 = 0.917$$

$$q^N = 0.672 + \frac{1}{2} \times 0.144 = 0.011 + 0.072 = 0.083.$$

4. **West Bengal**

$$p^M = 0.457 + \frac{1}{2} \times 0.431 = 0.457 + 0.2155 = 0.6725 = 0.673.$$

$$q^N = 0.112 + \frac{1}{2} \times 0.431 = 0.112 + 0.2155 = 0.3275 = 0.327.$$

5. **Orissa**

$$p^M = 0.399 + \frac{1}{2} \times 0.44 = 0.399 + 0.22 = 0.619.$$

$$q^N = 0.161 + \frac{1}{2} \times 0.44 = 0.161 + 0.22 = 0.381.$$

(b) In F_1

		MM	MN	NN
1.	$(0.913 + 0.087)^2 =$	0.8335 +	0.1588 +	0.00756
	Expected	83.3%	15.9%	0.76%
	Observed	83.5	15.6	0.9

Expected numbers are almost equal and so it is in equilibrium.

2.	$(0.176 + 0.824)^2 =$	0.03 +	0.290 +	0.678
	E =	0.30%	0.29%	67.8%
	O =	2.4	30.4	67.2
3.	$(0.917 + 0.083)^2 =$	0.84 +	0.152 +	0.0068
	E =	84%	15.2%	0.68%
	O =	84.5	14.4	1.1
4.	$(0.673 + 0.327)^2 =$	0.452 +	0.440 +	0.1069
	E =	45%	44%	10.69%
	O =	45.7	43.1	11.1
5.	$(0.619 + 0.381)^2 =$	0.383 +	0.471 +	0.145
	E =	38.3%	47%	14.5%
	O =	39.9	44	16.1

In general the expected values are almost similar, i.e., agree pretty well with an equilibrium.

(c) In Hariyana, allele frequency = 0.913 and 0.087.

In Delhi, allele frequency = 0.917 and 0.083.

Based on allele frequency population seems to have interbred.

In West Bengal, allele frequencies are 0.673 and 0.327

In Orissa, allele frequencies are 0.619 & 0.381

Here also two populations seem to have interbred.

Problem 15

In a large random mating isolated population of Sagar island, the frequencies of the I^A, I^B and i alleles of the A, B, O blood type genes are 0.15, 0.28 and 0.58 respectively. If the genotypes of the A, B, O blood type genes are in Hardy-Weinberg proportion, what fractions of the people who have type A blood in this population are expected to be homozygous for the I^A allele.

Solution:

[$p + q + r = 1$

$r^2 = 0.58$, $r = \sqrt{0.58}$ = $r = 0.76$

$(p + r)^2 = p^2 + 2pr + r^2$

or, $(p + 0.76)^2 = 0.14 + 0.58 = 0.72$

or, $p + 0.76 = \sqrt{0.72} = 0.848 = 0.85$

$\therefore$ $p = 0.85 - 0.76 = 0.09$

$p^2 = I^AI^A = (0.09)^2 = 0.0081 = 0.81\%$

$\therefore$ $I^AI^A = 0.81\%$]

Problem 16

In a tribal group of Jhargram (Midnapur) the frequencies of allele determining the ABO blood groups were estimated as 0.74 for I^O, 0.16 for I^A and 0.10 for I^B. Assuming random mating what are the expected frequencies of ABO genotypes and phenotypes.

Solution:

[$I^O = r^2 = 0.74$, $\therefore$ $r = \sqrt{0.74}$ = 0.86

$p + q + r = 1$

$q = 1 - (p + r) = 1 - 0.9 = 0.1$

$(p + r)^2 = p^2 + 2pr + r^2$

$(p + 0.86)^2 = 0.16 + 0.74 = 0.9$

$p + 0.86 = \sqrt{0.9} = 0.948 = 0.95$

$p = 0.95 - 0.86 = 0.09$

p^2 2pr 2pq 2qr

$(.09 + 0.1 + .86)^2$

$= .0081 + 2 \times .09 \times 0.86 + 2 \times .09 \times 0.1 + 2 \times 0.1 \times 0.86$

$= (0.86)^2 + (.01)^2$

A = .0081 + .1548 B = 0.01 + 0.172 a = .74 AB = 2.009

= .1629 = 0.182]

Problem 17

In land snail, multiple allele at a single locus determine the shell colour. The allele for brown (C^B) is dominant to allele for pink (C^p) and to the allele for yellow (C^Y). The dominance hierarchy among these allele is $C^B > C^p > C^Y$. In one population sample of snail, the following colour phenotypes are obseerved.

Brown	Pink	Yellow	Total
236	**231**	**33**	**= 500**

Assuming the population is in Hardy-Weinberg equilibrium. Calculate the frequencies of the C^B, C^p and C^Y alleles.

Hints:

[$p + q + r = 1$

$r^2 = \frac{33}{500} = 0.066$, $\therefore$ $r = \sqrt{0.066} = 0.2569 = 0.257$

$(p^2 + 2pr) = \frac{236}{500} = 0.472$;

$p^2 + 2pr + r^2 = (p + r)^2$, or, $.472 + .066 = (p + 0.257)^2$,

or, $0.538 = (p + .257)^2$, or, $p + .257 = \sqrt{.538} = 0.733$

$\therefore$ $p = 0.733 - 0.257 = 0.476$

$q^2 + 2qr = \frac{231}{500} = 0.462$

$(q + r)^2 = q^2 + 2qr + r^2$

$q + r = \sqrt{.462 + .066} = \sqrt{.528} = 0.7266 = 0.723$

$q = 0.73 - 0.257 = 0.466$

$\therefore$ $p = 0.476$; $q = 0.466$; $r = 0.257$].

Sex Linkage

- **Basic features:**

(i) Females are chromosomally XX and males are XY. Therefore, females are homogametic and males are heterogametic.

(ii) Five genotypes can occur for sex-linked pair of all alleles A and a. Three in females, i.e., AA, Aa and aa and two in males A and a.

(iii) If we assign the gene frequencies p and q to A and a respectively, the equilibrium genotype values in females are same as for an autosomal gene p^2(AA), 2pq(Aa) and q^2(aa).

(iv) In males (hemizygotes), the gene frequencies are expressed directly, and at equilibrium, there will be p(A) and q(a) genotypes.

(v) If alleles are X-linked, females may be homozygous or heterozygous.

(vi) Males carry only a single g allele for each X-linked locus.

(vii) For X-linked alleles in female, the Hardy-Weinberg frequncies are the same as those for autosomal loci $p(X^AX^A)$, $2pq(X^AX^B)$, $q^2(X^BX^B)$. In males, however, the frequencies of the genotypes are $p(X^AY)$ and $q(X^BY)$.

(viii) For this reason X-linked traits are more frequent among males than females.

Explanation:

The frequency of affected males A_2Y is q (A_1 = p, A_2 = q), while the frequencies of affected females, A_2A_2 is q^2. If q is low (say 0.005), the frequencies of affected males and females are 0.005 and 0.000025 respectively. In other words, the ratio of affected males to females is $q^2:q$ or 1 : q.

Since there is only one X-chromosome in males and two in females, the average frequency of sex-linked gene in a breeding population with *equal numbers of males and females* will be sum of one-third of its frequency in males plus two-thirds of its frequency in females.

$$p = \frac{1}{3}\text{ (p males)} + \frac{1}{3}\text{ (p females)} = \frac{\text{p males + 2p females}}{3}$$

If the number of males and females are *unequal*, the gene frequencies can be calculated as follows:

$$p = \frac{2naa + nAa + nA(y)}{2Nf + Nm}$$

Nf = Total no. of females, Nm = Total no. of males, nAA = No. of female AA genotype, nAa = No. of female Aa genotype, nA(y) = No. of male A genotype.

Example:

A population consists of 0.20A, 0.80a in males and 0.20AA, 0.60Aa, 0.20aa in females; the frequency of A is = 0.2 in males and 0.5 in females.

According to the formula, the equilibrium values expected with p is equal to 0.4 and q to 0.6, i.e., 4A : 6a males.

16AA, 48Aa, 36aa females.

$$\text{Frequency of A} = \left[\frac{0.2+2\times0.5}{3}\right] = \left[\frac{0.2+1}{3}\right] = \frac{1.2}{3} = 0.4$$

$$a = 1 - 0.4 = 0.6.$$

Solved Problems

Problem 1

If haemophilia (type-A sex-linked recessive) has a frequency of 1 per 5000 males in certain equilibrium population. What is the frequency of female haemophilia?

Solution:

Let the frequency of colour blind recessive allele be q for male (XY) and q^2 for female (XX).

Therefore, (♂) (♀)

q : q^2

Frequency is 1 per 5000, i.e., $\frac{1}{15000} = 0.0002$

q = 0.0002

$q^2 = (0.0002)^2 = 0.0002 \times 0.0002 = \frac{2}{10^4} \times \frac{2}{10^4}$

$= 4 \times 10^{-8}$.

Problem 2

One percent male exhibits sex-linked recessive gene. What is the frequency of (a) heterozygous women? (b) homozygous recessive woman?

Solution:

1 out of 100 has recessive allele 'a'.

q is the frequency for male (XY).

$q = \frac{1}{100} = 0.01, \quad \therefore p = 1 - 0.01 = 0.99.$

(a) The frequency of heterozygous woman is $2 \times 0.01 \times 0.99 = 0.0198 = 198.00 = 1.98\%$.

(b) The frequency of homozygous recessive woman is $0.01 \times 0.01 = 0.0001 = 0.0100 = 0.01\%$.

Problem 3

About 8% of men in a population are red and green colour-blind (due to sex-linked recessive gene). Assuming random mating in the population, with respect to colour- blindness (a) what percentage of women would be expected to be colour-blind? (b) what percentage of woman would be expected to be heterozygous? (c) what percentage of men would be expected to have normal vision two generations later?

Solution:

Gene frequency of colour-blind recessive gene in male is q(XY) and in female q^2(XX).

$q = \frac{8}{100} = 0.08$

Normal vision (p) = 1 – 0.08 = 0.92.

(a) Female, i.e., $q^2 = 0.08 \times 0.08 = 0.0064$

$0.0064 \times 100 = 0.6400$

0.64% would be expected to colour-blind.

(b) Heterozygous = $2 \times 0.08 \times 0.92 = 0.1472$ = 14.72%.

Women heterozygous would be expected 14.72%.

(c) Normal vision 0.92 = 92%.

Frequencies of alleles will not change in two generations as random mating occurs in H-W equilibrium.

Problem 4

Colour-blindness is due to a sex-linked recessive gene. A survey of 500 men from a local population revealed that 20 were colour-blind. (a) What is the gene frequency of the normal allele in the population? (b) What percentage of females in this population would be expected to be normal?

Solution:

Gene frequency of colour-blind recessive gene for male is q and female is q^2.

$q = \frac{20}{500} = 0.04.$

Normal vision = 1 – 0.04 = 0.96.

(a) Frequency of normal allele is 0.96 = 96%.

(b) $q^2 = (0.04)^2 = 0.0016$ or 0.16% colour-blind in female.

Women with normal vision = 1 – 0.0016 = 0.9984 = 99.84%.

Problem 5

A genetic disease of man called haemophilia is governed by a sex-linked recessive gene which

constitutes 1% of the gamets of the gene pool of a certain population. (a) What is the expected frequency of haemophilia among men of this population? (b) What is the expected frequency of haemophilia among women?

Solution:

Let the frequency of haemophilic gene (sex-linked recessive) be q for male (XY) and q^2 for female (XX).

It is 1% in population in man

$q = \frac{1}{100} = 0.01.$

(a) Expected frequency of haemophilia for men is $0.01 \times 100 = 1\%$.

(b) Expected frequency of haemophilia among women is q^2, i.e.,

$(0.01)^2 = 0.01 \times 0.01 = \frac{1}{100} \times \frac{1}{100} = 1 \times 10^{-4}$

$= 0.0001$ i.e. $00.0001 \times 100 = 0.0100 = 0.01\%$.

Problem 6

The phenotypic frequency of colour-blind males in a certain population is 0.08%. This form of colour-blindness is caused by a sex-linked recessive allele. What are the expected frequencies of the three genotypes among women?

Solution:

Let the frequency of colour-blind allele in male be q and in female be q^2.

$q = 0.08$, Normal $= p = 1 - 0.08 = 0.92$.

The expected frequencies of three genotypes =

$(0.92)^2 + 2 \times 0.92 \times 0.08 + (0.08)^2$

$= 0.8464 + 0.1472 + 0.0064.$

84.64%, 14.72%, 0.64%.

Problem 7

Two alleles have been found at the X-linked phosphoglucomutase gene in *Drosophila persimilies* population in California. The frequency of $P_{gm}A$ allele is 0.25 while the frequency of $P_{gm}B$ allele is 0.75. Assuming that the population is in H-W equilibrium, what are the expected genotypic frequencies in males and females?

Solution:

For sex-linked genes males (XY) have only one copy of the X-chromosome. So the genotypic frequency is equal to allele frequency.

$p^A = 0.25$ and $q^B = 0.75$.

The frequency of male flies with genotype $X^P{}_{gmA}Y = 0.25$.

The frequency of male with genotype $X^P{}_{gmB}Y$ is $= 0.75$.

Three genotypes exist for females:

$X^P{}_{gmA}\,X^P{}_{gmA}$, $X^P{}_{gmA}\,X^P{}_{gmB}$ and $X^P{}_{gmB} \times X^P{}_{gmB}$.

$\therefore$ Genotypic frequencies in females are $(0.25)^2$, $2 \times 0.25 \times 0.75$, $(0.75)^2 = 0.0625, 0.375, 0.5625$.

Problem 8

The most common form of haemophilia is due to sex-linked recessive allele with a frequency of 0.0001. What are the expected frequency of the two male genotypes and three female genotypes in the population?

Solution:

Let the frequency of haemophilia allele is q for male and q^2 for female.

$q = 0.0001$, $p = 1 - 0.0001 = 0.9999$.

The frequency of two male genotypes are 0.0001 and 0.9999.

The frequency in females are

p^2,	$2pq$,	q^2
$(0.0001)^2$,	$2 \times 0.0001 \times 0.9999$,	$(0.9999)^2$
$= 10^{-8}$	$0.00019 = 0.0002$	0.9998

Problem 9

The white eyes of *Drosophila* is due to a sex-linked recessive gene w and wild type (red) eye colour to its dominant allele w^+. A laboratory population of *Drosophila* was found to contain 170 red-eyed males and 30 white-eyed males. (a) Estimate the frequency of the w^+ allele and the w allele in the gene pool. (b) What percentage of the females in this population would be expected to be white-eyed?

Solution:

(a)

Observed no. of males	Genotypes of males	Phenotypes of males
170	w^+Y	Red-eyed (wild type)
30	wY	White-eyed
200		

The gene frequency of white eye gene in male is q, i.e., $\frac{30}{200} = 0.15$, i.e., 15%.

Red eye, i.e., frequency p = 1 – 0.15 = 0.85, i.e., 85%.

So w^+ allele is 85%, i.e., 0.85.

w allele is 15%, i.e., 0.15.

(b) Since, the female possesses two X-chromosomes, their expectation may be as likely as autosomal gene.

$p^2(w^+w^+) + 2pq(w^+w) + q^2(ww)$

$= (0.85)^2 + 2 \times 0.85 \times 0.15 + (0.15)^2$

$= 0.7225 + 0.255 + 0.0225.$

0.2225, i.e., 22.25% of all females, in this population is expected to be white-eyed.

Problem 10

In human about one man in 20 is colour-blind, colour blindness being due to a sex-linked recessive gene. What proportion of woman would be expected to be colour-blind?

Solution:

Let the gene frequency of colour-blind gene in male be q, i.e., $q = \frac{1}{20} = 0.05$.

Normal gene, i.e., p = 1 – 0.05 = 0.95.

In female, the gene frequency of colour-blind gene is q^2, i.e., $(0.05)^2 = 0.0025$.

The proportion of colour-blind woman would be 0.0025 × 100 = 0.2500 = 0.25%.

Problem 11

Assume that an X-linked recessive trait is observed in 2.5% of all of the males. What are the genotype frequencies among the females of this population if the population is in Hardy-Weinberg equilibrium?

Solution:

q = 2.5% = 0.025

p = 1 – 0.025 = 0.975

Genotypes of the females are

Homozygous recessive $(q^2) = (0.025)^2 = 0.000625$

Homozygous dominant $(p^2) = (0.975)^2 = 0.9506$

Heterozygous (2 pq) $= 2 \times 0.025 \times 0.975 = 0.04875.$

Problem 12

Haemophilia is caused by an X-linked recessive allele. In a particular population of Salt Lake the frequency of males with haemophilia is $\frac{1}{4000}$. What is the expected frequency of females with haemophilia?

Solution:

Let the frequency of haemophilia is q for male (XY) and q^2 for female (XX).

Therefore, (♂) (♀)

q : q^2

$q = \frac{1}{4000} = 0.00025$

$\therefore q^2 = (0.00025)^2 = \left(\frac{2.5}{10^5}\right)^2 \times \left(\frac{2.5}{10^4}\right)^2 = 6.25 \times 10^{-8}.$

Problem 13

Excessive secretion of male sex hormones results in premature sexual maturation in males and masculization of the sex characters in females. This disorder is called androgenital syndrome and in Switzerland there is an autosomal recessive form of the disease that affects about one in 5000 new-borns.

(a) Assuming random mating, what is the allele frequency of the recessive?

(b) What is the frequency of heterozygous carriers?

(c) What is the expected frequency of the condition among the offsprings of the first cousins?

Solution:

(a) Let the frequency of androgenital syndrome is q for male (XY) and q^2 for female (XX).

Therefore, ♂ ♀

q q^2

$$q^2 = \frac{1}{5000} = 0.0002$$

$$\therefore\ q = \sqrt{.0002} = 0.014.$$

The allele frequency of recessive is 0.014.

(b) The frequency of heterozygote is 2pq.

Here q = 0.014 and p = 1 – 0.014 = 0.986.

Thus, 2 × 0.014 × 0.986 = 0.0276 = 0.028

or, $\frac{28}{1000} = \frac{1}{36}$,

i.e., three percent of the population are carrier, even though only about 0.02 percent of the population are affected.

(c) With first cousin mating, the inbreeding coefficient $F = \frac{1}{6} = 0.062$.

The expected proportion of the affected people:

$$q^2(1 - F) + qF = 0.002(1 - 0.062) + 0.014 \times 0.062$$
$$= 0.002 \times 0.938 - 0.014 \times 0.062$$
$$= 0.001876 - 0.000868$$
$$= 0.001008 = 0.001.$$

In other words, there is about a five-fold greater risk of homozygosities for this allele among the offsprings of first cousin.

Problem 14

In some population approximately 8% of the males have a particular type of colour blind (X-linked recessive gene). What proportion of the females would you expect to have the trait of what proportion of females wound you expect to become for the alleles.

Solution:

Let the frequency of colour-blind gene in male is q, i.e. q = 8% = 0.08.

Normal gene, i.e., p = 1 – 0.08 = 0.92.

In female, frequency of colour-blind is $q^2 = (0.08)^2 = 0.0064$.

Carrier would be 2 × 0.92 × 0.08 = 0.1472.

Problem 15

Colour-blindness in human is due to sex-linked recessive gene. A survey on 500 men from a local population revealed 20 were colour-blind. What is the frequency of the normal allele in the population? What percentage of the female in this population would be expected to be carrier?

Solution:

Let the frequency of colour-blind gene in male is q, i.e., $\frac{20}{500} = \frac{1}{25} = 0.04$.

Frequency of normal allele = 1 – 0.04 = 0.96.

2 × p × q = 2 × 0.96 × 0.04.

8%, i.e., $q = \frac{8}{100} = 0.08$.

1 – 0.08 = 0.92.

2 × 0.92 × 0.08 = 0.1472.

Problem 16

If an X-linked recessive trait is present in 2 percent of the males in a population with random mating, what is the frequency of the trait in females? What is the frequency of carrier females?

Solution:

(i) The frequency in males implies that q = 0.2.

(ii) The expected frequency of the trait in females is $q^2 = 0.0004$.

(iii) The expected frequency of heterozygous females is 2pq = 2(0.98)(0.02) = 00392.

Exercise

1. Red green colour blindness is caused by a sex-linked recessive gene. About 64 women out of 10,000 are colour blind. What proportion of men would be expected to show the trait of mating random.

[Ans: $\frac{64}{15000} = 0.0064$, $\therefore q^2 = 0.0064$.
$q = 0.08$, probabilities of male = 0.08]

2. In some populations, approximately 8% of the males have a particular type of colour blindness (an X-linked recessive trait). What proportion of females would you expect to have this trait? What proportion of females would you expect to be carriers for this allele?

[Ans: 8% = 0.08, $\therefore q^2 = (0.08)^2 = 0.0064$
$q = 0.08$, $1 - 0.08 = 0.92$
Carriers: $2 \times 0.08 \times 0.92 = 0.147$]

3. If an X-linked recessive trait is present in 2 percent of the males in a population with random mating. What is the frequency of the trait in females? What is the frequency of the carrier females?

[Ans: 2% males bear the character, i.e., $\frac{2}{100} = 0.02$
$q^2 = 0.004$, Female $1 - 0.02 = 0.98$
Carriers: $2pq = 2 \times 0.02 \times 0.98 = 0.0392$

4. A selectively neutral, recessive character appears in 40 percent of the males and n 16 percent of the females in a large randomly interbreeding population. What is the gene frequency? What proportion of females are heterozygous for it? What proportion of females are heterozygous trait?

[Ans: Since the frequency of the trait is different in males than females, the character might be caused by sex-linked recessive gene. In female gene frequency q^2 and in male it is q. In male $(q) = \frac{40}{100} = 0.4$. Then in female $q^2 = (0.4)^2 = 0.16$. If it is sex-linked, proportion is 16%.

In female heterozgous = $2pq = 2 \times 0.4 \times 0.6$.

In male no existence of heterozygous = 0.48]

5. In a population of *Drosophila*, an X-linked recessive allele causing yellow body colour is present in genotypes at frequencies typical of random mating the frequency of recessive allele is 0.20. Among 1000 females and 1000 males what are the expected number of yellow and wild type phenotypes in each sex?

[Ans: The frequency of the yellow allele y is 0.2. In males the expected frequencies of wild type $\left(\frac{+}{+}\text{ and }\frac{y}{+}\right)$ and yellow $\left(\frac{y}{y}\right)$ are $(0.8 + 0.2) = 0.64 + 2 \times 0.8 \times 0.2 = 0.64 + 0.32 = 0.96$ and yellow = 0.04 respectively. Among 1000 females $0.96 \times 1000 = 96000$ wild type, $0.04 \times 1000 = 40.00$ yellow.

In males, the expected frequencies of wild type $\left(\frac{+}{y}\right)$ and yellow $\left(\frac{y}{\equiv}\right)$ are 0.8 and 0.2. Among 1000 flies $0.8 \times 1000 = 800$ and yellow $0.2 \times 1000 = 200$.]

Evolutionary Steps 18

- **MICRO EVOLUTION:**

Evolutionary changes on the small scale such as changes in gene frequency, interaction of variations, natural selection, genetic drift which operates up to species level are termed as micro evolution.

- **MACRO EVOLUTION:**

Evolution on grand scale, i.e., above the species level (genus, family, order and class) which operates through fragmentation of population into more than one adaptive type.

- **MEGA EVOLUTION:**

(i) Evolutionary changes which results in appearance of new and major organizational changes is called mega evolution.

(ii) Example: Evolution of vertebrate classes such as fish to amphibia from amphibia to reptilia, etc.

- **QUANTUM EVOLUTION:**

A rapid increase in the rate of evolution through genetic drift and selection over a relatively short period of time is termed as quantum evolution.

- **CO-EVOLUTION:**

Evolution in two or more species in which the evolutionary changes of each species influence the evolution of the other species.

- **MOLECULAR EVOLUTION:**

(i) Evolution at the molecular level of DNA and protein sequences.

(ii) The study of molecular evolution is primarily based on PCR, DNA sequencing and technique for localizing and isolating gene and gene products.

- **CONVERGENT EVOLUTION:**

The process of evolution by which a similar characters, in genetically unrelated species evolves independently, mostly because they have been subjected to similar environmental selective pressure.

Insect wing (membranous)
Bird's wing (forelimb with feather) → flying
Bat's wing (forelimb with skin)

- **DIVERGENT EVOLUTION:**

The process of evolutionary changes away from the ancestral type and the same basic structures become adapted by specialization to different function in response to occupying different ecological niches is termed as divergent evolution.

Forelimbs in vertebrates	Humerus Radius & ulna Carpals, meta Carpals & Phalanges	Bird → Flying (Primary) Bat → Flying (Secondary) Horse → Running on land Whale → Swimming

- **PARALLEL EVOLUTION:**

The process of evolution by which similar characters in related lineages whose common ancestor was phenotypically different and it leads to the production of analogous structures is termed as parallel evolution.

Antelope (artiodactyle) = (two-toed) → running
Horse (perissodactyle) = (one-toed) → running

- **BRADYTELIC EVOLUTION:**

A relatively slow rate of evolution for a long time is called bradytellic evolution.

- **HOROTELIC EVOLUTION:**

A comparatively average rate of evolution following rapid rate is called horotelic evolution.

- **TACHYTELIC EVOLUTION:**

A relatively rapid rate of evolution through which a group proliferates into many lines of evolution is called tachytelic evolution.

- **CONCERTED EVOLUTION:**

The process of evolution by which a series of nucleotide sequences or different members (genes) of a gene family evolves in concert i.e. each gene locus in the family comes to have the same genetic variant is termed as concerted evolution.

- **MOSAIC EVOLUTION:**

Evolution of different characters at different rates within a lineage or clade, hence more or less independently one another, is called mosaic evolution.

- **SOFT SELECTION:**

A selection in which number of survivors is determined by competition for a limiting factor such as food and space. On the other hand, it may be defined as the selection where selection coefficient depends on frequency and density of the genotype.

- **HARD SELECTION:**

A selection in which survival of an individual depends solely on how well its genotype equips it for the microenvironment where the individual settles.

- **CLINAL SELECTION:**

Selection that changes gradually along a geographic gradient.

- **WILLISTON'S RULE:**

During the evolution of lineage, serially homologous parts tend to reduce in number, but get more and more differentiated.

Example: In fishes, the maximum number of bones in the skull is counted to be 150. Whereas in an mammals only about 28 skull bones are present.

- **COPES RULE:**

Organisms have a tendency towards increase in body size over geological time in a lineage of population.

Example: Evolutions of horse, camel and others exhibit increase in size.

- **COACERVATES:**

Droplets of proteins or other substances that form spontaneously in colloidal suspensions surrounding themselves with a membranous shell.

(i) They possess simple but persistent organization.

(ii) Although most are unstable, some can maintain themselves for extended periods.

(iii) They can increase in size reaching a critical size, they divide spontaneously.

Selective Sweep:

It is the reduction or elimination of variation among nucleotides near a mutation in DNA. It results from beneficial alleles having recently reached fixation due to strong positive natural selection.

MOLECULAR TINKERING IN EVOLUTION

(i) Natural selection directs and favours the formation of new and useful structures in organisms to bring about their evolutionary transformation.

(ii) According to Francois Jacob (1977) the operation of natural selection in the process of organic evolution is not at all comparable to that of an engineer but it is as crude as that of **tinkerer**.

(iii) An engineer works according to a preconceived plan and with the help of machine and material that are solely designed for his task. On the other hand, tinkerers do their job without any preconceived plan but they perform their job with the pre-existing materials and tools around themselves.

(iv) Natural selection often acted like a tinkerer on the pre-existing structure, either transforming an organ to give it a **new function** or combining different organs to produce **more complex** one.

Example:

(a) Wings of birds from reptilian forelimbs. Ear ossicles of mammals from reptilian jawbone.

(b) Molluscan eyes and vertebrates eyes are functionally similar but in molluscan photoreceptor cells of the retina point towards light and in vertebrates away from light.

(c) **Lysozyme** (hydrolytic enzyme) is found in blood neutrophils of all vertebrates. A very similar protein with some modifications **lactoalbumin** is found in milk secreted from mammalian mammary gland.

- **PUNCTUATED EQUILIBRIUM:**

A pattern of evolution seen in many but not all lineages in the fossil record in which a **long interval** or **period of stasis** or **equilibrium** is punctuated or broken by a relatively short period of **rapid change** or **bursts of speciation** and macroevolutionary events during which new taxa arise is called punctuated equilibrium.

It is seen during cladogenesis.

Example: Fossil record of Bryozoa.

- **PHYLECTIC GRADUALISM:**

The process of persistent accumulations and gradual progressive evolutionary changes over time is known as phylectic gradualism.

(i) The transformation of a lineage over time is termed **anagenesis.**

(ii) It is very slow and is unable to produce major events in evolution.

Example: Cyanobacteria.

- **CAMBRIAN EXPLOSION:**

An explosive radiation of multicellular eukaryotes took place within a short geological period (Cambrian period, about 545 million years ago) largest number of animal groups with unique body plans emergd (found from fossil record). This phenomenon is popularly called the Cambrian explosion.

- **K-T BOUNDARY AND K-T EVENT:**

The period between the very end of Mesozoic era (actually Cretaceous period) and the beginning of Cenozoic era (actually tertiary period).

Dinosaurs are surely a fascinating group of animals. Despite their diversity and successful adaptation all kinds of dinosaurs wiped out from the earth at the end of Cretaceous period. Not only dinosaurs, ichthyosaurs, plesiosaurs, and pterosaurs were also vanished. The occurence of mass extinction of animals at the boundary of Cretaceous of tertiary period is popularly known as **K-T event.**

A catastrophic crash of large meteorite at the end of Cretaceous period and the beginning of the tertiary period is thought to be the cause of extinction.

Difference between Punctuated Equilibirum and Phylectic Gradualism

Punctuated Equilibirum	Phylectic Gradualism
1. It occurs during cladogenesis.	1. It occurs during anagenesis.
2. Morphological change is sudden, rapid and occurs only rarely.	2. Morphological change is gradual and continuous.
3. New species originate abruptly.	3. New species originate gradually.
4. The subpopulations of the ancestral species transform into new species.	4. An ancestral species can be transformed into new species.
5. Phyletic replacement involves splitting off of a new lineage/species, which then replaces its ancestor abruptly.	5. Phyletic replacement does not involve splitting of the lineage.

SELECTION

- **SELECTION:**

Selection or natural selection is a process whereby one phenotype and therefore one genotype leaves relatively more offsprings than another genotype measured by both reproduction and survival. Selection is thus a matter of reproductive success. It includes three parameters—survival rate, relative fitness and selection coefficient.

- **RELATIVE FITNESS:**

It is a measure of reproductive success of a genotype. Fitness of an organism is a measure of its ability to pass its genes to the next generation. It is expressed as relative parameter, usually defined as **relative fitness** and abbreviated 'w'.

$$w = \frac{\text{Reproductive success of individual genotype}}{\text{Reproductive success of best genotype}}$$

- **SELECTION COEFFICIENT(s):**

It is a force or factor acting on each genotype to reduce its relative fitness. This force or factor is called selection coefficient. It is usually represented by 's'.

$$w = 1 - s$$

Thus, as the selection coefficient increases, fitness decreases.

- **STABILIZING SELECTION (CENTRIPETAL SELECTION):**

It is the action of natural selection in keeping a population constant. Here phenotypic features coincide with optimal environmental condition and competition is not severe.

Features:

(i) It operates in constant environment.

(ii) It favours average or intermediate values.

(iii) It checks accumulation of mutation in the gene pool.

(iv) It tends to arrest variance and evolutionary changes, but maintains adaptiveness.

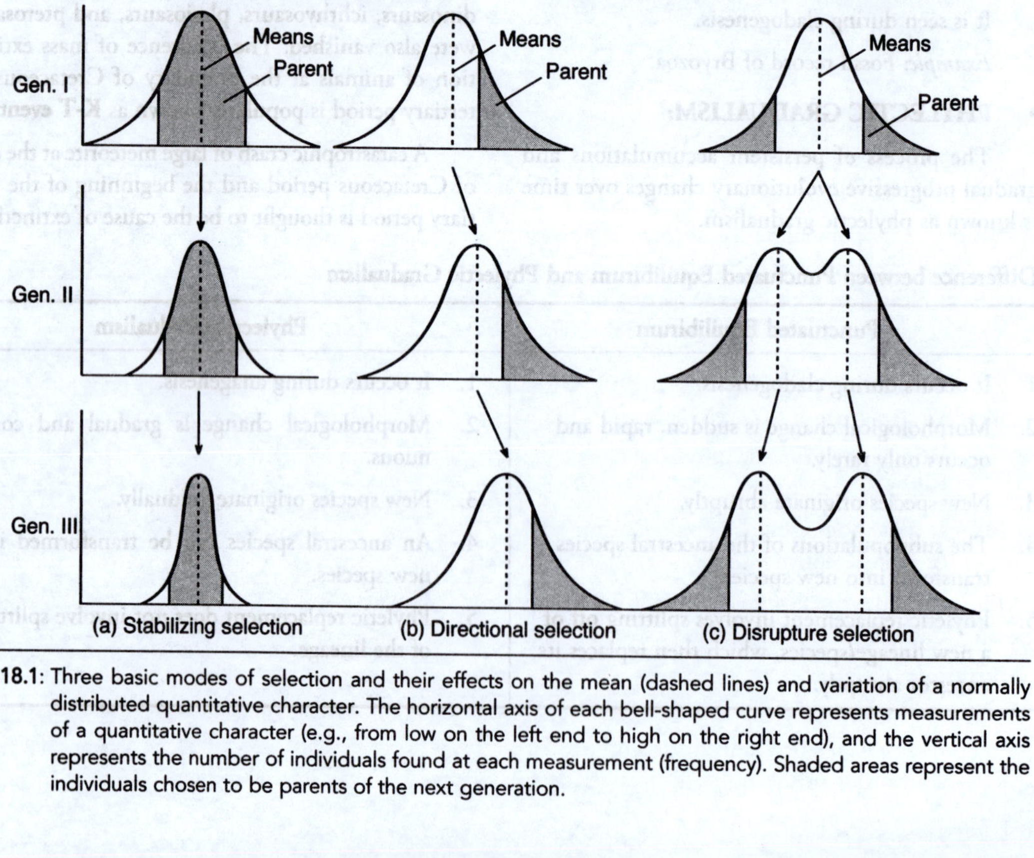

Fig. 18.1: Three basic modes of selection and their effects on the mean (dashed lines) and variation of a normally distributed quantitative character. The horizontal axis of each bell-shaped curve represents measurements of a quantitative character (e.g., from low on the left end to high on the right end), and the vertical axis represents the number of individuals found at each measurement (frequency). Shaded areas represent the individuals chosen to be parents of the next generation.

Example: Birth weight of newborns provide a good example of stabilizing selection. The optimum birth weight is 7.3 pounds and newborn infants less than 5.5 pounds have highest probability of mortality. The optimum is associated with lowest mortality.

- **DIRECTIONAL SELECTION:**

This form of selection operates in response to gradual changes in environmental condition. Here the selective force changes the frequency of alleles in a given direction either towards fixation or towards elimination.

Features:

(i) Directional selection operates when environment is changing in one direction.

(ii) It favours non-average or specialised phenotypes and eliminates the normal or average individuals.

(iii) It favours accumulation of those mutations that increase fitness of the population in the changing environment.

(iv) It brings about progressive evolution.

Example: (a) Industrial melanism, (b) Insecticides (DDT or other) resistance in mosquito.

- **DISRUPTIVE SELECTION (CENTRIFUGAL SELECTION):**

Selection that tends to favour the survival of organisms in a population that are at opposite phenotype extremes for a particular character and eliminates individual with intermediate values.

Features:

(i) It operates in heterogenous environment.

(ii) It pushes the phenotype within a population away from the mean and favour the values of the extreme.

(iii) It splits homogenous population into two or more adaptive forms.

(iv) It occurs when population is subjected to divergent selection pressure.

Example:

(a) On the basis of bristle number of *Drosophila*. When individuals with highest and lowest bristle number were selected, the population showed a non-overlapping divergence in only 12 generation.

(b) British lands nail (*Cepaea nemoralis*)

(i) They are found in grass field and forest. Predator bird (song bird) in low vegetation dark shell without light coloured bad snail are preyed by bird (thrushes) and in forest dark shelled light coloured snails are eaten up.

(ii) The populations as is divided into light coloured light bands in grassland and dark snails without band in forest.

- **CYCLIC SELECTION:**

Stabilizing and directional selection is constant, when selective environment is stable.

When environment is not stable between seasons or generations the optimum phenotype and also optimum genotype may show fluctuation because of the selection operating in one direction and in opposite direction for the next. This type of selection is called cyclic selection.

Features:

(i) It occurs when selective environment is unstable.

(ii) It helps in maintaining genetic differences in a population and fixes all the alleles of the gene pool.

- **COMPONENTS OF SELECTION:**

(1) Gametic selection, (2) Zygotic selection, (3) Sexual selection, (4) Fecundity selection.

- **GENERAL METHOD OF DETERMINING CHANGE IN ALLELIC FREQUENCY DUE TO NATURAL SELECTION:**

Solution:

Let there are two alleles viz. (A) and (a) for a gene in population. (A) is dominant over (a); p is the initial gene frequency of dominant allele (A) and q is the initial gene frequency of recessive allele (a).

If the population is at equilibrium, the gene frequencies will remain unaltered, but if the population exhibit differential reproduction due to selection against recessive allele (a), the selection coefficient(s) represent(s) the force of natural selection operating

against recessive alleles. It means that genotypes AA and Aa produce cent percent (1) offsprings and aa produces 1–s.

The genetic contribution from them will be

$AA = p^2 \times 1 \quad Aa = 2pq \times 1 \quad aa = q^2(1-s)$.

Allele frequency changes after one generation of selection against recessive homozygotes.

	Genotypes			
	A_1A_1	A_1A_2	A_2A_2	Total
1. Initial genotypic (zygote) frequency	p^2	$2pq$	q^2	1
2. Fitness (w)	1	1	$1-s$	
3. Frequency after selection	$p^2 \cdot 1$	$2pq \cdot 1$	$(1-s)q^2$	$1-sq^2$
4. Relative genotypic frequency after selection	$\frac{p^2}{1-sq^2}$	$\frac{2pq}{1-sq^2}$	$\frac{(1-s)q^2}{1-sq^2}$	

Allelic frequency after selection:

$$p^1 = p + \frac{1}{2}2pq = \frac{p^2}{1-sq^2} + \frac{1}{2} \times \frac{2pq}{1-sq^2} = \frac{p^2+pq}{1-sq^2} = \frac{p(p+q)}{1-sq^2}$$

$$q^1 = \frac{(1-s)q^2}{1-sq^2} + \frac{1}{2}\frac{2pq}{1-sq^2} = \frac{(1-s)q^2+pq}{1-sq^2} = \frac{q^2-sq^2+pq}{1-sq^2} = \frac{q(q-sq+p)}{1-sq^2} = \frac{q\{(p+q)-sq\}}{1-sq^2} = \frac{q(1-sq)}{1-sq^2}$$

Change in allelic frequency due to selection, i.e.

$$\Delta q = q^1 - q = \frac{q(1-sq)}{1-sq^2} - q = \frac{q(1-sq)-q(1-sq^2)}{1-sq^2} = \frac{q-sq^2-q+sq^3}{1-sq^2} = \frac{sq^2(q-1)}{1-sq^2} = \frac{-sq^2(1-q)}{1-sq^2}$$

As $p = 1-q$, so $\Delta q = -\frac{sq^2 \cdot p}{1-sq^2}$

(when $\Delta q = 0$, no further change occurs in allelic frequencies).

There is negative sign in the equation to the left side of $\frac{sq^2 \cdot p}{1-sq^2}$, because the values of s, p and q are always positive or zero, Δq is negative or zero. Thus, the values of q will decrease with selection.

- **Genetic load:**
  - (i) A reduction in the average fitness of individuals in population is due to the deleterious genes or gene combinations in the population.
  - (ii) It has many forms such as "mutational load", "segregational load" and "recombinational load".
- **Genetic polymorphism:**
  - (i) The stable coexistence of two or more discontinuous genotypes in a population over a succession of generation.
  - (ii) When the frequencies of two alleles are carried to an equilibrium, the condition is called balanced polymorphism.
- **Genetic identity (I)** = $\frac{\text{Arithmetic mean of the products of allele frequency}}{\text{Reproductive success of best genotype}}$
- **Genetic distance:** It is a quantitative measure of divergence between the phenetic appearance of two populations (or species) based on their difference in gene frequencies.

  Nei's D index = –ln I = –ln 0.9899 = 0.0101.

  (Nei's D index is that the rate of evolutionary change is equal for all loci being evaluated.)

- **Selection against Recessive lethal:**

The limiting case of selection against recessive alleles occur when the recessive homozygotes have zero fitness, i.e., when they die before attaining reproductive age or are sterile.

Example: **Phenylketonuria**. The fitness of the homozygote (aa) if left untreated, w = o, therefore, selection coefficient s = 1.

We know the frequency of allele 'a' after one generation of selection becomes

$$q^1 = \frac{q-sq^2}{1-sq^2}.$$

$$\text{As } s = 1, \text{ then } q^1 = \frac{q-q^2}{1-q^2} = \frac{(1-q)q}{1-q^2} = \frac{(1-q)q}{(1+q)(1-q)}$$

$$q^1 = \frac{q}{1+q}$$

The change in allelic frequency, i.e.,

$$\Delta q = \frac{-sq^2p}{1-sq^2} = \frac{-pq^2}{1-q^2} = \frac{-(1-q)q^2}{(1+q)(1-q)} = \frac{-q^2}{1+q}.$$

- **Calculation of allelic frequency after a given number of generations:**

Let the frequency of 'a' in the initial and following generations be $q_0, q_1, q_2, \ldots, q_t$

$$q_1 = \frac{q_0}{1+q_0} \text{ and } q_2 = \frac{q_1}{1+q_1}$$

Substituting the value of q_1 in the equation for q_2, we get

$$q_2 = \frac{\frac{q_0}{1+q_0}}{1+\frac{q_0}{1+q_0}} = \frac{\frac{q_0}{1+q_0}}{\frac{1+q_0+q_0}{1+q_0}} = \frac{\frac{q_0}{1+q_0}}{\frac{1+2q_0}{1+q_0}} = \frac{q_0}{1+2q_0}$$

Similarly, after 't' generations of selection,

$$q_t = \frac{q_0}{1+tq_0}.$$

The number of generations, t required to change the allelic form of a certain value q_0 to another q_t

$$\therefore q_0 - q_t(1+t_{q_0}) = q_1 + t_{q_0 q_1}; \quad t_{q_0 q_1} = q_0 + q_t$$

$$\therefore t = \frac{q_0 - q_t}{q_0 q_t}$$

- **Selection against Dominants and Selection without Dominance:**

Selection is more effective against dominant alleles than against recessive alleles because a dominant allele is expressed in the heterozygotes as well as in the homozygotes. Assume that dominance with respect to fitness is complete, so that dominant homozygotes and heterozygotes have equal fitness.

Genotype:	AA	Aa	aa
Fitness (w):	$1-s$	$1-s$	1

Allele frequency changes after one generation of selection against carrying the dominant allele

	Genotypes			
	AA	Aa	Aa	Total
1. Initial genotypic (zygote) frequency	p^2	$2pq$	q^2	1
2. Fitness (w)	$1-s$	$1-s$	1	
3. Frequency after selection	$p^2(1-s)$	$2pq(1-s)$	$q^2 \cdot 1$	$1-s+sq^2$
4. Relative genotypic frequency after selection	$\frac{p^2(1-s)}{1-s+sq^2}$	$\frac{2pq(1-s)}{1-s+sq^2}$	$\frac{(1-s)q^2}{1-s+sq^2}$	

Allelic frequency after selection:

$$p^1 = \frac{p^2(1-s)}{1-s+sq^2} + \frac{1}{2} \times \frac{2pq(1-s)}{1-s+sq^2} = \frac{(1-s)p^2+pq(1-s)}{1-s+sq^2} = \frac{(1-s)(p^2+pq)}{1-s+sq^2} = \frac{(1-s)(p+q)p}{1-s+sq^2} = \frac{(1-s)\cdot 1\cdot p}{1-s+sq^2} = \frac{(1-s)p}{1-s+sq^2}$$

Change in allelic frequency due to selection, i.e.,

$$\Delta p = p^1 - p$$

$$= \frac{(1-s)p}{1-s+sq^2} - p = \frac{(1-s)p - p(1-s+sq^2)}{1-s+sq^2}$$

$$= \frac{p-sp-p+sp-spq^2}{1-s+sq^2} = \frac{-spq^2}{1-s+sq^2}. \therefore \mathbf{\Delta p = \frac{-spq^2}{1-s+sq^2}}.$$

As long as both alleles are in population (p and q are positive) and there is selection (s is positive) the values of spq^2 and $1 - s + sq^2$ will be positive. Therefore, Δp will have a negative value; the frequency of A will gradually decrease to zero. If a dominant allele is sterile or lethal, that (s = 1) change in allele frequency is

$$\Delta p = \frac{-spq^2}{1-s+sq^2} = \frac{-pq^2}{1-1+q^2} = \frac{-pq^2}{q^2} = -p.$$

The frequency of the dominant allele then becomes zero in one single generation of selection, i.e., neither the homozygotes nor heterozygotes for the dominant allele leave any progeny.

- **Allele frequency changes after one generation when there is no dominance:**

Selection just considered always goes in one direction, toward elimination of deleterious alleles and establishment or fixation of the favoured allele.

Consider some instances where fitness of the heterozygotes is intermediate between the fitness of two homozygotes.

Genotype:	A_1A_1	A_1A_2	A_2A_2
Fitness (w):	1	$1-\frac{1}{2}s$	$1-s$

Allele frequency changes after one generation of selection when there is no dominance

	Genotypes			
	A_1A_1	A_1A_2	A_2A_2	Total
1. Initial genotypic (zygotic) frequency	p^2	$2pq$	q^2	1
2. Fitness (w)	1	$1-\frac{1}{2}s$	1	
3. Frequency after selection	$p^2 \cdot 1$	$2pq\left(1-\frac{1}{2}s\right)$	$q^2 \cdot (1-s)$	$1-sq$
4. Relative genotypic frequency after selection	$\frac{p^2}{1-sq}$	$\frac{2pq\left(1-\frac{s}{2}\right)}{1-sq}$	$\frac{q^2 \cdot (1-s)}{1-sq}$	

Allelic frequency after selection:

$$q^1 = \frac{q^2(1-s)}{1-sq} + \frac{1}{2}\,\frac{2pq\left(1-\frac{s}{2}\right)}{1-sq} = \frac{q^2-q^2s+pq-\frac{pqs}{2}}{1-sq}$$

$$= \frac{q^2-q^2s+(1-q)q-\frac{pqs}{2}}{1-sq} = \frac{q^2-q^2s+q-q^2-\frac{pqs}{2}}{1-sq} = \frac{+q-q^2s-\frac{pqs}{2}}{1-sq}.$$

Change in allelic frequency due to selection, i.e.,

$$\Delta q = q^1 - q$$

$$= \frac{q-q^2s-\frac{pqs}{2}}{1-sq} - q = \frac{q-q^2s-\frac{pqs}{2}-q+q^2s}{1-sq} = \frac{-\frac{pqs}{2}}{1-sq}.$$

It will be negative as long as both alleles persist in the population and there is selection. The equilibrium condition $\Delta q = 0$ will be reached when q = 0, i.e., allele selected against A_2 is completely eliminated.

- **Heterozygote Advantage:**

Selection just in favour of the heterozygote over both homozygotes is known as **overdominance** or **heterosis.**

Genotypes:	AA	Aa	aa
Fitness (w):	$1-s_1$	1	$1-s_2$

Allele frequency changes after one generation of selection when there is overdominance

	Genotypes			
	AA	Aa	Aa	Total
1. Initial genotypic (zygotic) frequency	p^2	$2pq$	q^2	
2. Fitness (w)	$1-s_1$	1	$1-s_2$	
3. Frequency after selection	$p^2(1-s_1)$	$2pq\cdot 1$	$(1-s_2)q^2$	$1-s_1p^2-s_2q^2$
4. Relative genotypic frequency after selection	$\frac{p^2(1-s_1)}{1-s_1p^2-s_2q^2}$	$\frac{2pq}{1-s_1p^2-s_2q^2}$	$\frac{(1-s_2)q^2}{1-s_1p^2-s_2q^2}$	

Allelic frequency after selection:

$$q^1 = \frac{(1-s_2)q^2+\frac{1}{2}2pq}{1-s_1p^2-s_2q^2} = \frac{(1-s_2)q^2+pq}{1-s_1p^2-s_2q^2}$$

$$= \frac{q^2-s_2q^2+pq}{1-s_1p^2-s_2q^2} = \frac{q^2+pq-s_2q^2}{1-s_1p^2-s_2q^2} = \frac{q(q+p)-s_2q^2}{1-s_1p^2-s_2q^2}$$

$$= \frac{q-s_2q^2}{1-s_1p^2-s_2q^2} = \frac{q(1-s_2q)}{1-s_1p^2-s_2q^2}.$$

Change in allelic frequency due to selection, i.e., $\Delta q = q^1 - q$

$$= \frac{q(1-s_2q)}{1-s_1p^2-s_2q^2} - q = \frac{q(1-s_2q)-q(1-s_1p^2-s_2q^2)}{1-s_1p^2-s_2q^2}$$

$$= \frac{q-s_2q^2-q+s_1p^2q+s_2q^3}{1-s_1p^2-s_2q^2} = \frac{s_1p^2q-s_2q^2+s_2q^3}{1-s_1p^2-s_2q^2}$$

$$= \frac{s_1p^2q-s_2q^2(1-q)}{1-s_1p^2-s_2q^2} = \frac{s_1p^2q-s_2pq^2}{1-s_1p^2-s_2q^2} = \frac{pq(s_1p-s_2q)}{1-s_1p^2-s_2q^2}$$

$$\Delta q = \frac{pq(s_1p-s_2q)}{1-s_1p^2-s_2q^2}.$$

The equilibrium condition $\Delta q = 0$ will be satisfied only when the numerator is zero. If both alleles are present in a population (p and q are greater than zero) this will occur only when

$s_1p = s_2q$

or, $s_1(1-q) = s_2q$

or, $s_1 - s_1q = s_2q$

or, $s_2q + s_1q = s_1$

or, $q(s_1+s_2) = s_1$

$\therefore q = \frac{s_1}{s_1+s_2}$

Correspondingly the equilibrium frequency of A allele is

$s_1p = s_2q$

$s_1p = s_2(1-p)$

or, $s_1p + s_2p = s_2$

or, $p(s_1+s_2) = s_2$

$\therefore p = \frac{s_2}{s_1+s_2}$

These two equilibrium frequencies are stable because selection will change the two alleles frequencies until the equilibrium values are reached.

(i) If p is greater than its equilibrium value, i.e., $p > s_2(s_1+s_2)$, then $s_1p > s_2q$ and Δq will be positive.

Consequently q will increase at the expense of p until $s_1p = s_2q$.

(ii) On the other hand, if $p < s_2(s_1+s_2)$, then $s_1p < s_2q$ and Δq will be negative, leading to a decrease in the value of q until the equili-brium frequencies are reached.

The equilibrium frequencies in overdominance depend on the relative magnitudes of the two selection coefficients, not on their actual values.

- **MUTATION AND SELECTION:**

The ultimate outcome of selection is the elimination of deleterious allele. Such allele nevertheless remains in population owing to mutation. Selection and mutations have opposite effects with respect to deleterious alleles. The net effect of the two processes will be nil when the number of deleterious alleles eliminated by selection is the same as that produced by mutation.

The frequency of allele a is q, owing to selection $\Delta q = \frac{-spq^2}{1-sq^2}$.

Because a allele will be at low frequency, the denominator will be nearly 1 and the approximate change of allele frequency due to selection will be $\Delta q = -spq^2$.

The a allele will however, increase in frequency by the amount up_1, owing to mutation from **A** to **a** (we may ignore back mutation from **a** to **A** because the frequency of a is low). There will be equilibrium between mutation and selection when

$$spq^2 = up$$

$$\therefore sq^2 = u, \text{ or, } q^2 = \frac{u}{s};$$

$$\therefore q = \sqrt{\frac{u}{s}},$$

when $s = 1$, $q = \sqrt{u}$.

- **PHYLECTIC SPECIATION:**

Gradual transformation of one species to another species over a long period by accumulation of slow change in the gene pool is called **phylectic** (Autogenous speciation).

(i) Neither there is splitting of the parental gene pool nor evolution of reproductive isolation.

(ii) Here the gene pool of parental species modifies due to changes in gene freqencies, chromosomal inversions and other structural changes in the chromosome.

- **QUANTUM (SUDDEN) SPECIATION:**

In this type of speciation chromosomal rearrangement (inversion, translocations and ploidy) leads to reproductive isolation and results in speciation.

It occurs suddenly and intermittently.

- **GRADUAL SPECIATION:**

Here the new species originate from daughter population of the same species by gradual accumulation of many minute gene over a long period of time under the influence of natural selection.

- **SYMPATRIC SPECIATION:**

Sympatric speciation is that which occurs between populations occupying the same geographic area. New species develop by the appearance of some biological barriers in the individuals of an initially random mating population

(i) Sympatric species occur in the same geographic area.

(ii) Populations are not separated by an physical barrier.

(iii) Sympatric population may be separated by preferences for different habitats or niches resources.

Examplew: Tree hopper (*Enchenopa binotata*) has six sibling species.

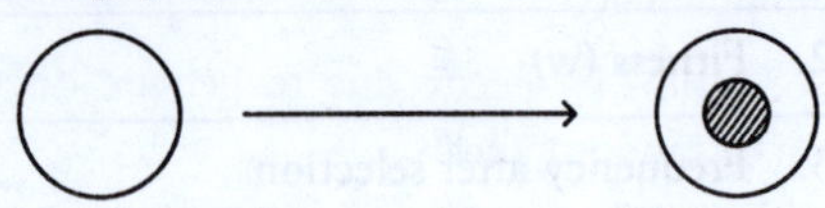

Ancestral (Parental) New species

- **PARAPATRIC SPECIATION:**

Speciation in which the new species is formed through reproductive isolation among members of contiguous (or in contact) but nonoverlapping or spatialy distinct population within same geographic range.

(i) Here moderate gene flow occurs.

(ii) New species evolve from contaguous population.

Example:

1. (a) Carrion crow: black-coloured (*Corvus corone*) – West Europe

   (b) Hooded crow: grey body with black-headed tail – East Europe, Hybrid zone (interbreed) and new species developed.

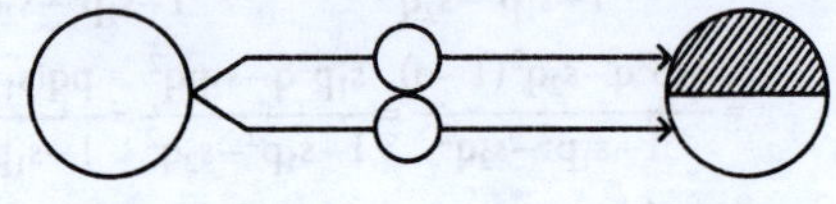

Ancestral (Parental) New species developed

2. Greenbul (*Andropadus virens*)

   (a) Rain forest (smaller bird)

   (b) Grassland (larger bird – longer bills, leggs and wing), Hybrid forms in ecotone.

- **ALLOPATRIC SPECIATION:**

Allopatric speciation is the divergence of physically or geographically isolated population of a species occupying separate geographical areas into new species.

Here gene flow is more than negligible rate.

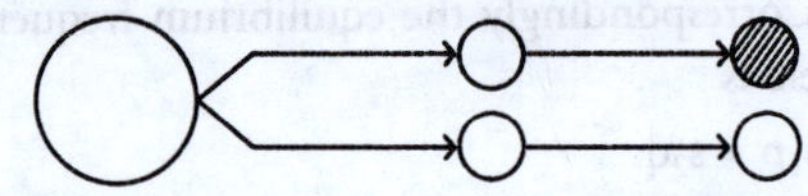

Ancestral (Parental) New species

(a) Peripatric or Transilience Speciation

A founder effect model of allopatric speciation in which the new species is formed from a small population isolated (bud off) at the edge or periphery of the ancestral (parental) population of geographic range.

Speciation is speedy due to genetic drift and fixation of random alleles.

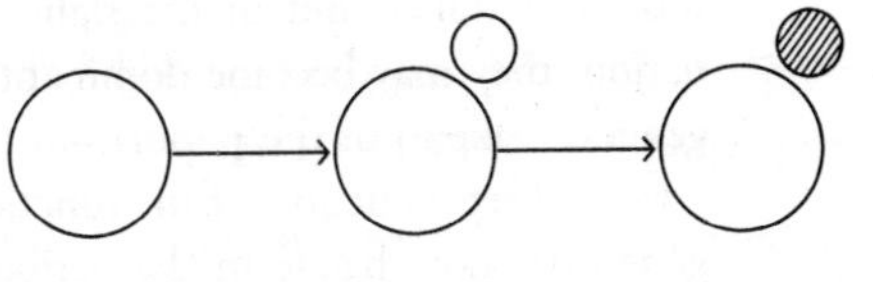

Example: Plumage variation of kingfisher (*Tanyisptera hydrocharis galeata*) of New Guinea.

(b) Vicariant Speciation

In this type of speciation, a large widespread populations are separated by the emergence of an extrinsic or topographic or ecological barrier and one or both populations diverge from ancestral population (state).

Example: Emergence of isthmus Panama in the Pliocene divided many marine organism into pacific and caribean population.

Neutral theory and Synthetic theory at the molecular level

All regions of a gene do not mutate equally in all lineages at all times. Some regions in a structure of a gene (DNA) are constraint for change while others are susceptible to incidental change. The constraint region of within the structure of a gene directly affects the gene product.

The rate of mutation at the constraint sites in the gene is slow, but is faster in the other regions that do not affect function of the gene product.

(i) The **Shine-Dalgarno Sequence** (SDS) in the leader region, TATA sequence in the promoter region and the base at the second position of the codons are constraint sequences of a functional gene.

(ii) If random base replacements occur in the SDS, the resulting mRNA would fail to bind to the smaller subunit of ribosome and affects translation.

(iii) If mutation occurs in **TATA** sequence, the RNA polymerase would fail to bind with promoter and affects transcription.

(iv) If mutation occurs in the second position of the codon, it would often result in an altered amino acid of the polypeptide chain.

(v) If mutation occurs in coding region due to loss or gain of a base, it would often change the frame-shift, resulting in the formation of biologically inactive protein.

(vi) Under such conditions, the mutations in the constraint region of a gene are often harmful while those in non-constraint regions are neither harmful nor useful but neutral in activity.

Neutral theory of molecular evolution describes the fate of purely neutral mutation, i.e., those that neither enhance nor lower fitness. One possible fate is that a mutation will become fixed, i.e., attains a frequency of 1.0 — entirely by chance.

(vii) The probability that this will occur equals 'U', the rate at which neutral mutation arises. In each generation, therefore, probability is 'U' that a mutation that occured as sometime in the past will become fixed.

After the passage of 't' generation, the fraction of mutation that will have become fixed is, therefore, $\lambda = Ut$.

(viii) If two species diverged from a common ancestor 't' generation ago, the expected fraction of fixed mutation in both species is D = 2 Ut, since various mutations have become fixed in both lineages. If the mutation in question are base pair changes, a fraction D = 2 Ut of the base pairs of gene should differ between the species assuming that all base pairs are equally likely to mutate. Thus, the average mutation per base pair per generation is $U = D/2t$.

Thus, we can estimate U if we can measure the fraction of base pairs in a gene that differ between the two species (D) and if we can estimate the number of generation since they diverged from their common ancestor (t).

If U_0 is the rate of selectively neutral mutation per gene per generation, the substitution rate (the rate of replacement in a population) is independent of population size and is $k = U_0$ per generation (the substitution per unit time is k/g where 'g' is the length of generation). Thus, two population of a species that have been isolated for 't' generation are expected to differ by $2U_0t$ substitution.

(ix) Variation within a population at equilibrium measured the frequency of hetrozygotes per nucleotide site is $4NeU_0$, where Ne = effective populaion size. The quantity of $4NeU_0$ is often denotd by θ.

(x) The neutral mutation θ can be estimated by 'S', the number of segregating (variable) nucleotide sites in a sample of gene copies (sequence) or by 'd' the average number of nucleotide differ as between pairs of sequences in a sample.

(xi) Natural selection can modify these pattern in three ways:

(a) **Purifying selection**: Eleminates or reduces the frequency of deleterious mutation in a population.

(b) **Positive selection** (directional): It fixes a sequence genes that includes advantageus mutation.

(c) **Balancing selection**: Owing to a factor such as heterozygote advantage, frequency dependent selection or variable selection maintains variant sequences in a population.

- **Reconciliation between Neutral theory and Synthetic theory**

(i) During a given interval, the number of nucleotide or amino acid replacement is found to be rare in a locality in a particular season. This may be the result of **natural selection.**

(ii) During the adjustment of an evolutionary lineage to a new environment, simultaneous changes are required in many functions. Then chance itself can cause an abrupt increase in the rate of genetic changes.

(iii) In a particular season of the annual climatic cycle, some species become rare in a locality and abundant in other seasons. The rarity is the **bottleneck** of the species. When species passes through a population bottleneck the neutral alleles can gain momentous importance and in the right combination, they may become dominant in the genetic makeup of the population for survival and reproduction. If the function of a gene does not change in the period of its evolution, there is no reason to expect fluctuation in rate at which it evolves.

(iv) The magnitude of the variation makes it possible to assert that most differences among alleles in a population are adaptively neutral. They are unselectd variation. In contrasts neo-Darwinians are accustomed to thinking of evolution resulting from accumulation of favourable or beneficial mutation.

Here a conflict arises between Darwinian and neutralists.

The reconciliation of the two points of view lies in that even though majority of the mutations are neutral, a few may have adaptive effect and they would be able to supply enough raw materials for the novelties by natural selection.

- **State briefly the Neutral theory of Molecular evolution**

At the molecular level most evolutionary changes and most variation within and between the species is not caused by natural selection, but by genetic drift of mutant alleles that are neutral.

(i) A neutral mutation is one that does not affect an organisms ability to survive and reproduce. In population genetics, mutations in which natural selection does not affect the spread of the mutation in a species are termed **neutral mutation.**

(ii) Neutral mutations that are inheritable and not linked to any genes under selection will either be lost or will replace all other alleles of the gene. This loss or fixation of the genes proceeds on the basis of random sampling, i.e., known as a **genetic drift.**

(iii) A neutral mutation that is in linkage disequilibrium with other alleles that are under selection may proceed to loss or fixation via genetic hitchhiking and or back ground selection. The rate of fixation is the rate at which gene mutate to selectively neutral alleles.

The rate of fixation is equal to the neutral mutation rate. Suppose, the population size is N and each generation, a fraction of μ of the 2N copies of gene, mutate to selectively neutral alleles. Let also assume each new mutant is unique. In each generation selectively neutral mutation appear in the population will be $2N\mu$. As each mutant is unique and its frequency will be $\frac{1}{2}N$. This probability of fixation of new mutation will be $\frac{1}{2}N$. Number of each mutant appearing in each generation is $2n\mu$. This rate of fixation of selectively neutral mutant is $2N\mu \times \frac{1}{2}N = \mu$. Therefore, for selectively neutral mutation the rate of molecular evolution is equal to rate at which the mutation occurs in population.

(iv) Many mutations in genome may decrease organisms ability to reproduce (fitness). These mutations are selected against and not passed on to 'future generations'. The most commonly observed mutations defectable as variations in organism and populations appear to have no visible effect in fitness of individuals and are therefore neutral.

(v) The rate of evolution does not depend on population size, or the efficiency of selection peculiarities of mating system. Furthermore, if neutral mutation rate is constant, the nucleotide and amino acid substitution should occur in clockwise fashion in all evolving lineages. The neutral theory has had an enormous impact on the study of evolute at molecular level.

(vi) Neutral mutations are also the basis for using molecular clocks to identify speciation and adaptive radiation.

MOLECULAR PHYLOGENY

The graphic representation of evolutionary relationship among biological entities, i.e., phylogenic tree (individual population, species or higher taxa) based on the divergence observed within a single homologous gene by using a combination of molecular data (such as DNA and protein sequences, presence or absence of transposable element and gene order data) and statistical technique is called **molecular phylogeny** or **phylogenies.**

Most phylogenetic trees generated by molecular data analyses and are prone to uncertainties which may differ in some respect from the true tree is sometimes called **inferred tree.**

- **Features:**

  (i) All living things (both extant and extinct) on the earth share a single common ancestor that lived 4 billion years (approx) ago.

  (ii) All phylogenetic trees include some portion of their common ancestor with branches which connect two/more adjancent nodes.

  (iii) Terminal nodes indicate the latest taxa and internal node represent the common ancestors before branching.

  (iv) The length of the branches indicate the degree of divergence between the **genes** representd by the **nodes.** Internal nodes represent a common ancestor before branching.

  (v) As internal node representing a common ancestor to all other nodes is called **rooted tree.**

  In **unrooted** trees, there is no common ancestors, the tree specifies only the relationship between nodes.

  (vi) Rooted trees provides information on the past evolutionary event, that led to the organisms or DNA sequences being studied. Unrooted trees does not tell us there evolutionary pathway.

- **Methods of Reconstruction of Phylogenic relationship**

  A. Distance-Matrix approach
  B. Parsimony based approach
  C. Maximum likelyhood approach.

- **Distance-Matrix Approach**
  - (i) It is the oldest and simplest method for phylogenetic tree construction.
  - (ii) This approach is statistically based and classifies organism on the basis of overall similarity.
  - (iii) The unweighted pair group method with arithmatic average (**UPGMA**) is a distance-matrix approaches.
  - (iv) The distance-matrix method approaches work equally well with **morphological** as well as **molecular** data and also with combination of both.
  - (v) The weakness of this approach (UPGMA) is that it presumes a constant rate of evolution across all lineages which is not always true.

Steps:

(i) Aligning the DNA sequences to obtain the comparative data.

(ii) Mathematical approaches to sequence alignment have also been devised. Distance measures can be calculated by comparing character states of several taxa or gene to determine the overall similarity or difference between each pair of possible taxa or genes.

(iii) The differences are coded and assembled into a data matrix called **distance-matrix.**

(iv) The coefficient of similarity between each possible pair of taxa is calculated by comparing all characters state of each taxon or gene with those of every other taxon or OTCs (Operational Taxonomic Unit). The coefficients are assembled into a **similarity matrix.**

- **Conversion of alignment data into a phylogenetic tree:**
  - (i) The multiple sequence alignment is converted into numerical data that can be analyzed mathematically in order to construct the tree.
  - (ii) The simplest approach is to convert the sequence information into a distance matrix which is simply a table showing the evolutionary distances between all pairs of nucleotide in a data set.
  - (iii) The evolutionary distance is expressed as the number of nucleotide differences per nucleotide site for each sequence pair.
  - (iv) The distance matrix is used to establish the lengths of the branches connecting these two sequences in the reconstructed tree.

UPGMA

(Unweighted Pair Group Method with Arithmatic mean)

(i) The method is based on the assumption that after two species diverged from each other, the rate of accumulation of mutation in homologous region of DNA is almost same in the two species.

(ii) We can expect that two closely related species, which have diverged more recently have greater sequence similarity compared to distantly related species.

Preparation of distance matrix

(i) For constructing phylogenetic tree, we should take most closely related species first. Here humans and Chimpanzees, because they have fewest number of nucleotide differences. The number

Distance matrix: Total 10,000 bp sequences.

	Human	Chimpanzee	Gorilla	Orangutan	Monkey
Human	0	145	151	298	751
Chimpanzee	145	0	157	294	755
Gorilla	151	57	0	304	739
Orangutan	298	294	304	0	710
Monkey	751	755	739	710	0

change 10,000 is 145. i.e., average number of base pair change in each species is $\frac{145}{2}$ = 72.5. That is about 72.5 sequences changed in human and 72.5 sequences changed in chimpanzees, since they were separated from each other.

(ii) As a result, we have a total 145 sequence changes between humans and primates. So, percentage of nucleotide differences in human versus chimpanzee $\frac{72.5 \times 100}{10{,}000}$ = 0.725.

(iii) Humans and chimpanzees are closely related to gorillas. Sequence differences between humans and gorillas are 151 and between chimpanzees and gorillas are 157. So the average difference is $\frac{151+157}{2} = \frac{308}{2}$ = 154. Then average number of changes that occur in each species is $\frac{154}{5}$ = 77.

(iv) About 77 changes take place in human or chimpanzee and also 77 changes occur in gorilla, since they have diverged from a common ancestor. So the percentage of nucleotide difference in gorilla versus human or chimpanzee $\frac{77 \times 100}{10{,}000}$ = 0.77.

(v) Next step, orangutan is closely related to human, chimpanzee and gorilla. Sequence differences between human and orangutan are 298 and between chimpanzee and orangutan are 294 and between gorilla and orangutan are 304. So the average number of differences between orangutan and human/chimpanzee/gorilla is

$$\frac{(298+294+304)}{3} = 298.6 = 299.$$

(vi) Then average number of changes occur in each species is $\frac{299}{2}$ = 149.5. Since they have diverged from a common ancestor, the percentage of nucleotide differences in orangutan versus human/chimpanzee/gerilla is $\frac{149.5 \times 100}{10{,}000}$ = 1.495.

(vii) Monkeys are related to the rest four species: The differences are 751 between human and monkey 755 between chimpanzee and monkey. 739 between gorilla and monkey, 710 between orangutan and monkey. So the average number of differences between monkey and human/chimpanzee/gorilla/orangutan is 739, i.e.

$$\frac{(751+755+739+710)}{3} = \frac{2955}{4} = 738.75$$

(viii) Then average number of changes occurred in each species is $\frac{739}{2}$ = 369.5. Since they have diverged from a common ancestor. The percentage of nucleotide differences in monkeys versus human/chimpanzee/gorilla/orangutan is $\frac{369.5 \times 100}{10{,}000}$ = 3.695.

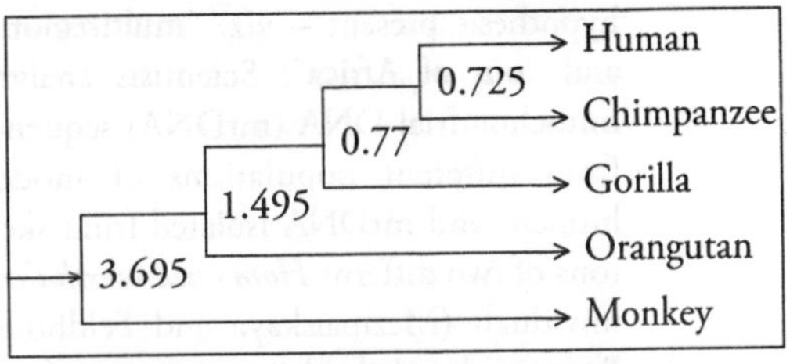

Phylogenetic tree showing evolutionary relationship among 5 species using UPGMA method.

Q. Highlight the origin and evolution of man through molecular genetic studies.

Ans.

Populations that are closely related share genetic properties, but distant related populations do not. Thus, by analyzing variations in genes, gene product and DNA sequences, it is possible to determine the relatedness of different species or organisms.

(i) Both nuclear and mitochondrial DNA sequence (circular DNA and lacks of intron) variation has been investigatd to know the origin and evolution of man. Genetic studies of modern human (not DNA, sequence and chromosome DNA) and fossil record indicated. The nuclear genome contains the preponderance of human polymorphism, but the mitochondrial genome has the unique property of being transmitted exclusively through female. Therefore, variation in mitochondrial DNA provides a way of tracing maternal lineages in human ovolutionary history. Mitochondrial DNA accumulates mutation approximately 10-fold faster than nuclear DNA. Its molecular clock is much faster. This increased rate of accumulating mutations allows scientists to examine evolution over a shorter period of time.

(ii) By counting the number of mutation that occurs between the ancestral mtDNA sequence and the modern human mt-DNA sequences and divide the number by known mutation rate. Then it is possible to calculate the time of existence of common ancestor of human.

(iii) Regarding human evolution, there are two hypothesis present – viz. "**multiregional**" and "**out of Africa**". Scientists analyzed mitochondrial DNA (mtDNA) sequences from different populations of modern humans and mtDNA isolated from skeletons of two differnt *Homo neanderthalensis* inividuals (Mezmaiskaya and Feldhofer). Because the skeletal remains were found over 1000 miles from each other, they likely represent different Neanderthal populations.

(iv) By comparing mtDNA sequences from different human populations it was determined that the last common ancester between *Homo sapiens* and *Homo neanderthalensis* lived almost 6×10^5 years ago. This analysis also revealed that *Homo neanderthalensis* did not interbreed with the direct ancestor of *Homo sapiens.* This data revealed that *Homo neanderthalensis* was evolutionary more closely related than chimpanzees. African population exhibits a greater number of mtDNA sequence changes than other modern human population, which is consistent with their mt-DNA being the oldest and having the greatest amount of time to accumulate differences.

(v) After analyzing the DNA sequences of Y-chromosome in males from different populations throughout the world, the data supported the model (hypothesis) that the *Homo sapeins* male also originated from Africa.

(vi) Along with the mitochondrial DNA, Y-linked DNA and autosomal DNA sequences have been investigated. The result indicated that modern human population originated in Africa. Scientist determined that a group of early *Homo sapiens* migrated out of Africa across the southern end of the Red sea and among the southern coast of Asia to Australia ($6 \times 10^4 - 75 \times 10^3$ years ago). A second wave of *Homo sapiens* migrated from the Nile River to Eurasia and then fanned out across Europe and Asia (45×10^3 to 50×10^3 years ago).

(vii) To clarify human origin and evolution a study had been made among six indigenous tribal populations inhabiting Andaman Nicobar Islands in the Bay of Bengal.

Scientist analyzed mtDNA of five Onge (native Andamanese), five great Andamanese, five Nicobarese individuals. The mtDNA of all five Onge and all five great Andamanese individuals were more similar to Indian and East Asian population than West Eurasians. In contrast, all five Nicobarese individual, possessed mtDNA sequence that were similar to the West Eurasian mtDNA sequences.

Thus, Onge and Great Andamanese likely coionized the island upon the first migration out of Africa and Nicobarese likely came to island after the second migration from Africa.

Q. Chromosomal similarity between Apes and Man.

Ans.

(i) A combination of molecular data and fossil evidences suggest that the great apes and man evolved from a common ancestor.

(ii) The studies on DNA, Chromosome numbers and banding pattern of chromosomes have further strengthened these fact.

(iii) The banding pattern of individual human chromosome is very similar with that of apes.

(iv) The banding pattern of human chromosome number 3 and 6 are very similar to the pattern with those of some chromosome in the chimpanzee (Fig. 18.2).

(v) When genomes of chimpanzees and humans are compared, they are more than 99 percent identical. This high degree of identity implies that chimpanzee and humans are quite closely related.

(vi) Total amount of DNA in human diploid cells and that of the apes is more or less similar DNA. DNA hybridization indicates 2.5% difference between chimpanzee and man.

Handy Man	: *Homo habilis*
Java Man	: *Homo ecretus* or *Pithecanthropus erectus*
Peking Man	: *Homo erectus pekinensis*
Heidelberg Man	: *Homo sepiens neidelbergensis*
Neanderthal Man	: *Homo sapiens neanderthalensis*
Cro magnan Man	: *Homo sapiens fusilis*

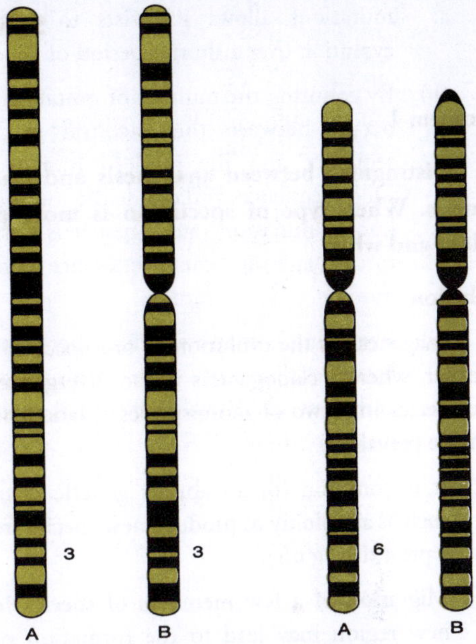

Fig. 18.2: Diagrammatic representation of chromosomes 3 and 6 of man and chimpanzee.

Differences between Apes and Man

Characteristics	Apes	Man
1. Posture:	(i) Semi-erect. (ii) Forelimbs are longer than hind limbs.	(i) Erect. (ii) Hind limbs are more powerful.
2. Eye Brow ridges:	(i) Bony eye brow ridges are prominent.	(i) Brow ridges are not so prominent.
3. Cranium	(i) Small in size. (ii) Cranial capacity 100-500 cc.	(i) Large in size. (ii) Cranial capacity 1450-1600 cc.
4. Locomotion	(i) Quardipedal. (ii) Sole of the foot does not lie on the ground while walking. (iii) Only outer edges of hand and forelimbs touch the ground.	(i) Bipedal. (ii) Sole of the foot lies flat on the ground while walking.

Gene tree:

A diagram showing the real or estimated ancestral relationship among a set of protein or nucleic acid sequences.

Bootstrapping:

Analysis of the multiple data sets formed by random sampling with replacement from an actual data set in order to estimate a degree of confidence in a particular branch or branching pattern in a gene tree.

Solved Problems

Problem 1

Distinguish between anagenesis and cladogenesis. When type of speciation is more prevalent and why?

Solution

Anagenesis is the evolution of one species into another, whereas cladogenesis is the divergence of one species into two or more species. Cladogenesis is more prevalent.

(i) It is common for an abrupt genetic change such as alloploidy to produce new species from a pre-existing one.

(ii) Migration of a few members of species into new region may lead to the formation of a new species in the region (i.e., allopatric speciation).

Problem 2

Discuss the evidence in favour of the punctuated equilibrium model of evolution. What mechanisms could account for this pattern of evolution? In contrast, what type of genetic changes are consistent with gradualism?

Solution

(i) The fossil record is the main evidence in favour of punctuated equilibrium.

(ii) Palentologists rarely find a gradual transition of fossil form. The transition period in which environmental pressure and genetic changes cause a previous species to evlolve into a new species is thought to be so short that few, if any, of the transitional members would be preserved as fossil. Therefore, the fossil record primarily contains representatives from the long equilibrium periods.

Rapid evolutionary change is consistent with known genetic phenomena, including single gene mutations that have dramatic effect on phenotypic characteristics, the founder effect, genetic events such as changes in chromosome (inversion, translocation) or chromosome number, which may abruptly create individual with new phenotypic trait.

In some cases, however, gradual changes are observed in certain species for a long period of time.

In addition, the gradual accumulation of mutation is known to occur from the molecular analysis of DNA.

Problem 3

What is meant by molecular clock? How is the concept related to the neutral theory of evolution?

Solution

The steady rate of change of molecular variant (nucleotide/amino acid) overtime leads to evolutionary divergence of organism is termred as molecular clock.

(i) Neutral mutations occur at a rate proportional to the rate of mutation per generation.

(ii) Therefore, the genetic divergence between species, i.e., due to neutral mutations reflects the time elapsed since their common ancestor.

(iii) The concept is related to the theory of evolution because it assumes that most genetic variation is due to the accumulation of neutral or nearly neutral mutation.

Problem 4

Mention the assumption of molecular clock.

Solution

(i) The mutation at molecular level are incorporated at fixed or regular rates over a time.

(ii) The proportional rate of fixation of mutation in one gene relative to the rates of fixation of mutation in other genes remains the same throughout any line of descent.

(iii) The fixation of molecular mutation does not occur on their adaptive or selective value but on a fixed rate.

(iv) The lines of descent leading from common

ancestor to all its descendants have similar rates of fixed mutation.

Problem 5

What is the rate of amino acid substitutions? Would you expect the rate of nucleotide substitution to be faster or slower than the rate of amino acid substitution? Why?

Solution

The rate of amino acid substitution is the change in amino acid sequence per unit time.

Nucleotide substitution would be faster because there are fewer constraints and the third position changes fast.

Problem 6

State briefly the problems of molecular clock?

Solution

(i) All changes are not neutral.

(ii) Changes are influenced by natural selection. Certain changes are favoured while some are eliminated.

(iii) The structure of certain important genes is conserved over a long period. Their mutation rate is very slow.

Problem 7

Mention the significance of molecular clock.

Solution

(i) Molecular information about changes in the sequence of nucleotides in DNA and RNA or in the sequence of amino acids in different proteins are obtained.

(ii) These information help in the preparation of phylogenetic trees and estimation of approximate time of separation of each group from other.

Problem 8

Compare the predictions of the neutrality theory and the selection theory in influencing the amount of genetic variation.

Solution

(i) Neutrality considers genetic variation a function of mutation and genetic drift predicting alleles to transient and fewer in small population.

(ii) Selection predicts constant alleles and more variation in small population.

Problem 9

A species of antelope contains 20 chromosomes per set. The species is divided by a mountain range into two separate population, which we will call the eastern and western populations. When comparing the karyotypes between these two populations, it was discovered that the members of the eastern population are homozygous for a large inversion within the chromosome 14. How would this inversion affect the interbreeding between the two populations? Could such an inversion play an important role in speciations.

Solution

(i) Inversion do not affect the total amount of genetic material as well as the phenotype of the organism.

(ii) Therefore, if members of the two populations were to interbreed, the offspring would probably be viable, because they would have inherited a normal amount of genetic material from each parent. However, such offspring would be inversion heterozygotes.

(iii) During meiosis (gamete formation) crossing over may create chromosome that have too much or too little genetic material. If this unbalanced chromosomes are passed to the next generations of offspring, the offspring may survive.

(iv) For this reason, inversion heterozygotes (phenotypially normal) may not be very fertile because many of their offsprings will die. Since inversion heterozygotes are less fertile, this would tend to keep eastern and western populations reproductively isolated.

(v) Over this time, this would aid in the independent evolution of the two populations and would ultimately promote the evolution of the two populations into separate species.

Problem 10

A codon for leucine is UUA. A mutation causing a single base substitution in a gene can change this codon in the transcribed mRNA into GUA (valine), AUA (isoleucine), CUA (leucine), UGA (opal stop codon), UAA (ochre stop codon), UCA (serine), UUG (leucine), UUC (phenyl alenine) or UUU (phenyl alanine). According to neutral theory which of these mutations would you expect to see within the genetic variation of a natural populations? Explain.

Solution

According to neutral theory, the neutral mutation will accumulate to the greater extent in a population.

(i) Leucine (CUA) is a nonpolar amino acid. For a UUA codon a single base changes create CUA (leucine); UUG (leucine); they are **silent mutation**. So they would be most likely to occur in a natural population.

(ii) Likewise, conservative substitutions to other nonpolar amino acid such as isoleucine (AUA), valine (GUA) and phenylalanine (UUC or UUU) may bot affect the protein structure and functions. So they may also occur and not be eliminated rapidly by natural selection.

(iii) The polar amino acid serine (UCA) is a nonconservative substitution, one would predict that it is more likely disrupt protein function. Therefore, it may be less likely to be found.

(iv) The stop codon UGA (opal) and UAA (ochre) would be expected to diminish or eliminate protein function. Particularly, if they occur earlier in the coding sequence. These types of mutations are selected again and therefore are not usually found in natural population.

Problem 11

Explain how homologous genes have sequences that are similar but not identical.

Solution

(i) Homologous genes are derived from the same ancestral gene. Therefore, at the beginning they had identical sequences.

(ii) Over time, each gene accumulates random mutations that the other homologous genes did not acquire. These random mutations change the gene from its original sequence.

(iii) Therefore, much of the sequence between homologous genes remains identical, but some of the sequences will be altered due to the accumulation of independent random mutations.

Problem 12

Highlight the evolution of cytochrome.

Solution

Cytochrome is an evolutionary conservative molecule. It is formed of slightly over 100 amino acids. It has 104 amino acids in vertebrates and a few more in certain lower forms.

(i) Some amino acids in **cytochrome c** have not changed at all; example amino acids at 20 specific positions are unchanged from mould to man.

(ii) One region of cytochrome 'c' formed of eleven amino acids (in position from 70-80) has remained unchanged in all organisms.

(iii) The cytochrome c of closely related vertebrates either do not differ or differ in only a few amino acid residues.

Example: **Cytochromes c** of man and chimpanzee have identical composition.

(iv) The greater are the phylogenetic differences, the more is the differnce in cytochrome composition.

Example: Human **cytochrome c** differs from **cytochrome c** of *Drosophila* in 24 amino acids and from *Neurospora* in 40 amino acids.

Solved Problems (Numerical)

Problem 1

In a population there are three genotypes A_1A_1, A_1A_2 and A_2A_2 numbering 40, 50 and 10 respectively. Suppose that each genotype produces 80, 90, 10 zygotes in F_1 generation. Calculate the fitness and selection coefficient values of each genotype.

Solution:

	Genotypes			
	A_1A_1	A_1A_2	A_2A_2	Total
1. Number of zygote in one generation	40	50	10	100
2. Number of zygote produced by each genotype in next generation	80	90	10	180

Computation:

1.	Average number of progeny per individual in next generation	$\frac{80}{40} = 2$	$\frac{90}{50} = 1.8$	$\frac{10}{10} = 1$
2.	Fitness (Relative reproductive efficiency, i.e. w)	$\frac{2}{2} = 1$	$\frac{1.8}{2} = 0.9$	$\frac{1}{2} = 0.5$
3.	Selection coefficient [(s) = 1 – w]	1–1 = 0	1–0.9 = 0.1	1–0.5 = 0.5

Problem 2

In a population the dark peppered moth (*Biston carbonaria*) released and recaptured was 154 & 90 whereas for light varieties (*B. betularia*) released and recaptured was 64 and 10 respectively. Calculate relative fitness, selection coefficient.

Solution:

	Population	*Dark*	*Light*
(a)	Released	154	64
(b)	Recaptured	90	10
(c)	Survival rate	$\frac{90}{154} = 0.58$	$\frac{10}{64} = 0.15$
(d)	Relative fitness (w)	$\frac{0.58}{0.58} = 1$	$\frac{0.15}{0.58} = 0.26$
(e)	Selection coefficient [(s) = 1–w]	1–1 = 0	1–0.26 = 0.74

Problem 3

In a population the dark peppered moth (*Biston carbonaria*) released and recaptured was 154 and 82 whereas for light varieties (*B. betularia*) released and recaptured was 64 and 16 respectively. Calculate relative fitness, selection coefficient and delta q (change of gene frequency) of the population. **[C.U.(M.Sc. Zoo) 2005]**

Solution:

	Population	*Dark*	*Light*
(a)	Released	154	64
(b)	Recaptured	82	16
(c)	Survival rate	$\frac{82}{154} = 0.53$	$\frac{16}{64} = 0.25$
(d)	Relative fitness (w)	$\frac{0.53}{0.53} = 1$	$\frac{0.25}{0.53} = 0.47$
(e)	Selection coefficient [(s) = (1–w)]	1–1 = 0	1–0.47 = 0.53

Since we do not know the frequency of heterozygotes we will have to calculate $q^2(64)$ as the percentage of the total population.

$$q^2 = \frac{64}{154 + 64} = \frac{64}{218} = 0.2935.$$

$\therefore q^2 = 0.2935, \therefore q = \sqrt{0.2935} = 0.541.$

$p = 1-q = 1-0.541 = 0.459.$

Change of gene frequency, i.e.,

$$q^1 = \frac{q-sq^2}{1-sq^2} = \frac{0.541-0.53 \times 0.541 \times 0.541}{1-0.53 \times 0.541 \times 0.541}$$

$$= \frac{0.541(1-0.28673)}{1-0.53 \times 0.292681} = \frac{0.71327 \times 0.541}{1-0.1551209}$$

$$= \frac{0.385879}{0.8448791} = 0.4567 = 0.457.$$

$$\Delta q = q^1 - q = 0.457-0.541 = -0.084$$

$$\text{or, } q = \frac{-sq^2(1-q)}{1-sq^2} = \frac{-sq^2 p}{1-sq^2}$$

$$= \frac{-0.53 \times 0.541 \times 0.541 \times 0.459}{1-0.53 \times 0.541 \times 0.541} = \frac{-0.0712}{1-0.15512}$$

$$= \frac{-0.0712}{0.8448} = \frac{-0.0712}{0.845} = -0.0842 = -0.084.$$

Problem 4

Consider a population in which the frequency of allele A is p = 0.7 and the frequency of allele a is q = 0.3 and where the alleles are codominant. What will be the allele frequencies after one generation if the following occur ?

(a) $w_{AA} = 1$ $w_{Aa} = 0.9$ $w_{aa} = 0.8$

(b) $w_{AA} = 1$ $w_{Aa} = 0.95$ $w_{aa} = 0.9$

(c) $w_{AA} = 1$ $w_{Aa} = 0.99$ $w_{aa} = 0.98$

(d) $w_{AA} = 0.8$ $w_{Aa} = 1$ $w_{aa} = 0.8$

Solution:

(a) Here $p = 0.7$ and $q = 0.3$

$w_{AA} = 1, \quad w_{Aa} = 0.9, \quad w_{aa} = 0.8.$

$$p_1 = \frac{p^2 w_{AA} + \frac{1}{2} 2pq w_{Aa}}{p^2 w_{AA} + 2pq w_{Aa} + q^2 w_{aa}}$$

$$= \frac{0.7 \times 0.7 \times 1 + \frac{1}{2} \times 2 \times 0.7 \times 0.3 \times 0.9}{0.7 \times 0.7 \times 1 + 2 \times 0.7 \times 0.3 \times 0.9 + 0.3 \times 0.3 \times 0.8}$$

$$= \frac{0.49 + 0.189}{0.49 + 0.378 + 0.072} = \frac{0.679}{0.94} = 0.722$$

$$= \frac{0.3 \times 0.3 \times 0.8 + 0.7 \times 0.3 \times 0.9}{0.94}$$

$$= \frac{0.072 + 0.189}{0.94} = \frac{0.261}{0.94} = 0.2776 = 0.278.$$

$\therefore p_1 = 0.722, \quad q_1 = 0.278.$

(b) $p_1 = 0.7, \quad q = 0.3$

$w_{AA} = 1, \quad w_{Aa} = 0.95, \quad w_{aa} = 0.9.$

$$p_1 = \frac{0.7 \times 0.7 \times 1 + 0.7 \times 0.3 \times 0.95}{0.7 \times 0.7 \times 1 + 2 \times 0.7 \times 0.3 \times 0.95 \times 0.3 \times 0.3 \times 0.9}$$

$$= \frac{0.49 + 0.1995}{0.49 + 0.399 + 0.081} = \frac{0.6895}{0.97} = 0.7108$$

$= 0.711.$

$$q_1 = \frac{0.3 \times 0.3 \times 0.9 + 0.7 \times 0.3 \times 0.95}{0.97}$$

$$= \frac{0.1995 + 0.081}{0.97} = \frac{0.2805}{0.97} = 0.289.$$

$\therefore p_1 = 0.711, \quad q_1 = 0.289.$

(c) $p = 0.7, \quad q = 0.3$

$w_{AA} = 1, \quad w_{Aa} = 0.99, \quad w_{aa} = 0.98.$

$$p_1 = \frac{0.7 \times 0.7 \times 1 + 0.7 \times 0.3 \times 0.99}{0.7 \times 0.7 \times 1 + 2 \times 0.7 \times 0.3 \times 0.99 + 0.3 \times 0.98}$$

$$= \frac{0.49 + 0.2079}{0.49 + 0.4158 + 0.0882} = \frac{0.6979}{0.994} = 0.702.$$

$$q_1 = \frac{0.3 \times 0.3 \times 0.98 + 0.7 \times 0.3 \times 0.98}{0.994}$$

$$= \frac{0.0882 + 0.2079}{0.994} = \frac{0.2761}{0.994} = 0.2978 = 0.298$$

$\therefore p_1 = 0.702, \quad q_1 = 0.298.$

(d) $p = 0.7, \quad q = 0.3$

$w_{AA} = 0.8, \quad w_{Aa} = 1, \quad w_{aa} = 0.8.$

$$p_1 = \frac{0.7 \times 0.7 \times 0.8 + 0.7 \times 0.3 \times 1}{0.7 \times 0.7 \times 0.8 + 2 \times 0.7 \times 0.3 \times 1 + 0.3 \times 0.3 \times 0.8}$$

$$= \frac{0.392 + 0.21}{0.392 + 0.42 + 0.072} = \frac{0.602}{0.884} = 0.6809 = 0.681$$

$$q_1 = \frac{0.3 \times 0.3 \times 0.8 + 1 \times 0.7 \times 0.3}{0.884}$$

$$= \frac{0.072 + 0.21}{0.884} = \frac{0.282}{0.884} = 0.319.$$

$\therefore p_1 = 0.681, \quad q_1 = 0.319.$

Problem 5

Consider the allele frequency p = 0.7 and q = 0.3, where selection occurs against homozygous recessive genotype. What will be the allele frequency after one generation if s = 1.0, s = 0.5, s = 0.1 and s = 0.01.

Solution:

(i) $p = 0.7$, $q = 0.3$, $s = 1$ after selection the frequency of recessive allele is q^1.

$$q^1 = \frac{q - sq^2}{1 - sq^2} = \frac{0.3 - 1\times0.3\times0.3}{1 - 1\times0.3\times0.3} = \frac{0.3 - 0.09}{1 - 0.09} = \frac{0.21}{0.91}$$

$= 0.23.$

$q^1 = 0.23$ and $p^1 = 1 - 0.23 = 0.77.$

(ii) Here $s = 0.5$, $p = 0.7$, $q = 0.3$.

$$q^1 = \frac{0.3 - 0.5\times0.3\times0.3}{1 - 0.5\times0.3\times0.3} = \frac{0.3 - 0.045}{1 - 0.045} = \frac{0.255}{0.955}$$

$= 0.267$

$\therefore q^1 = 0.267$, $p^1 = 1 - 0.267 = 0.733.$

(iii) Here $s = 0.1$, $p = 0.7$, $q = 0.3$.

$$q^1 = \frac{0.3 - 1\times0.3\times0.3}{1 - 1\times0.3\times0.3} = \frac{0.3 - 0.009}{1 - 0.009} = \frac{0.291}{0.991}$$

$= 0.293.$

$q^1 = 0.293$, $p^1 = 1 - 0.293 = 0.707.$

(iv) Here $s = 0.1$, $p = 0.7$, $q = 0.3$.

$$q^1 = \frac{0.3 - 0.01\times0.3\times0.3}{1 - 1\times0.3\times0.3} = \frac{0.3 - 0.0009}{1 - 0.0009} = \frac{0.2991}{0.9991}$$

$= 0.299.$

$q^1 = 0.299$, $p^1 = 1 - 0.299 = 0.701.$

Problem 6

In a population there are three genotypes G^1G^1, G^1G^2 and G^2G^2 numbering 16, 10 and 20 respectively. Suppose that each genotype produces 128, 40 and 40 zygotes in F_1 generation. Calculate the fitness and selection coefficient values of each genotype.

Solution:

	Genotypes			
	G^1G^1	G^1G^2	G^2G^2	Total
1. Number of breeding adults	16	10	20	46
2. Number if offspring produced by all the adult genotypes in the next generation	128	40	40	208

3. Computation:

(i) Average number of progeny per individual in next generation $\left(\frac{b}{a}\right)$	$\frac{128}{16} = 8$	$\frac{40}{10} = 4$	$\frac{128}{16} = 8$
(ii) Fitness (Relative reproductive) efficiency (w)	$\frac{8}{8} = 1$	$\frac{4}{8} = 0.5$	$\frac{2}{8} = 0.25$
(iii) Selection coefficient [(s) = 1–w]	$1 - 1 = 0$	$1 - 0.5 = 0.5$	$1 - 0.25 = 0.75$

Problem 7

In a laboratory colony of fruit flies, the fitness of the genotypes of an electrophoretic locus (malate dehydrogenase) are determined. Three genotypes FF, FS and SS have fitness 0.85, 1.0 and 0.6 respectively. What is the equilibrium frequency of slow (SS) allele?

Solution:

Here the fitness of three genotypes FF, FS and SS are given; the locus exhibiting heterozygote advantages with selection coefficient of the two homozygotes.

$s_1 = 1 - 0.85 = 0.15$, $s_2 = 1 - 0.6 = 0.4$.

Let q is the frequency of slow alleles, therefore equilibrium frequency of

$$q = \frac{s_1}{s_1 + s_2} = \frac{0.15}{0.15 + 0.4} = \frac{0.15}{0.55} = 0.272.$$

Problem 8

For a pair of alleles, completely dominant A has an initial frequency of 0.6 and adaptive value of 1. Its recessive allele has a frequency of 0.4 and an adaptive value of 0.5. What is the frequency of a in the next generation?

Solution:

Let the frequency of recessive allele 'a' is q, i.e., q = 0.4.

Adaptive value (w) = 0.5.

The selection coefficient (s) = 1 – w = 1 – 0.5 = 0.5

The frequency of 'a' allele in the next generation will be q^1.

$$q^1 = \frac{q - sq^2}{1 - sq^2} = \frac{0.4 - 0.5 \times (0.4)^2}{1 - 0.5 \times (0.4)^2}$$

$$= \frac{0.4 - 0.5 \times 0.16}{1 - 0.5 - 0.16} = \frac{0.4 - 0.08}{1 - 0.08} = \frac{0.32}{0.92}$$

$$= 0.3478 = 0.348.$$

Problem 9

The fitness of three genotypes are w_{AA} = 0.9, w_{Aa} = 1.0, w_{aa} = 0.7.

(a) If the population starts at allele frequency p = 0.5, what is the value of p in the next generation?

(b) What is the predicted equilibrium frequency?

Solution:

Here the fitness of three genotypes are given $w_{AA} = 0.9$, $w_{Aa} = 1.0$, $w_{aa} = 0.7$.

Here the locus exhibiting heterozygote advantage with selection coefficient

$s_1 = 1 - 0.9 = 0.1$, $s_2 = 1 - 0.7 = 0.3$.

(a) p = 0.5. In the next generation frequency will be p^1 and q = 1 – p = 0.5.

$$p^1 = \frac{p(1 - s_1 p)}{1 - s_1 p^2 - s_2 q^2}$$

$$= \frac{(1 - 0.1 \times 0.5) \times 0.5}{1 - 0.1 \times 0.5 \times 0.5 - 0.3 \times 0.5 \times 0.5}$$

$$= \frac{(1 - 0.5) \times 0.5}{1 - 0.025 - 0.075} = \frac{0.5 \times 0.095}{1 - 0.1}$$

$$= \frac{0.475}{0.9} = 0.528.$$

(b) Equilibrium frequency

$$\hat{p} = \frac{s_2}{s_1 + s_2} = \frac{0.3}{0.1 + 0.3} = \frac{0.3}{0.4} = 0.75.$$

Problem 10

In an experiment 120 melanic moths were exposed and 96 survived while only 74 of 120 typicals survived. Estimate the relative fitness of the two morphs. What is the expected frequency of the melanic allele after one generation, given that its initial frequency is 0.4.

Solution:

Population:	melanic	typical
Exposed	120	120
Survived	96	74
Survival rate	$\frac{96}{120} = 0.8$	$\frac{74}{120} = 0.6167$
Relative fitness	$\frac{0.8}{0.8} = 1$	$\frac{0.6168}{0.8} = 0.771$

Melanic varieties frequency 0.4, therefore, typical has frequency 0.6 i.e., (1 – 0.4)

$P\ (0.6 + 0.4)^2 =$

$F_1 = 0.36 + 0.48 + 0.16$ (Genotypic frequency)

Gene frequency will be $0.16 + 0.48 \times \frac{1}{2}$

= 0.16 + 0.24 = 0.40.

$\therefore$ q = 0.4.

Problem 11

In an industrialized region, the fitness of *B. betularia* moths is 1 for the dark form (DD, Dd) and 0.47 for the light form (dd). The allelic frequencies at a certain time are p = 0.40 (D-allele) and q = 0.60 (d-allele). Calculate genotypic frequency and relative frequency of DD, Dd and dd after selection. Assume that the allele frequencies are (1) p = 0.10 q = 0.90 and (2) p = 0.90, q = 0.10. Calculate corresponding values of Δq and compare them with each other and with the one obtained in the first part of this problem.

Solution:

	Genotypes				
	DD	Dd	dd	Total	Frequency of d
1. Initial frequency	0.16	0.48	0.36	1.00	0.60
2. Fitness (w)	1	1	0.47		
3. Frequency after selection	$1\times0.16 = 0.16$	$1\times0.48 = 0.48$	$0.36\times0.47 = 0.17$	0.81	
4. Relative genotypic frequency	$\frac{0.16}{0.81} = 0.1972 = 0.20$	$\frac{0.48}{0.81} = 0.592 = 0.59$	$\frac{0.17}{0.81} = 0.2098 = 0.21$	1.00	

$q^1 = 0.21 + \frac{1}{2} \times 0.59 = 0.21 + 0.295 = 0.505 = 0.51$

$\Delta q = q^1 - q = 0.51 - 0.60 = -0.09.$

1. $\Delta q = \frac{-spq^2}{1-sq^2}$ $s = 1 - 0.47 = 0.53,\ q = 0.90,\ p = 0.10$

$$\therefore \Delta q = -\frac{0.53\times0.1\times0.9\times0.9}{1-0.53\times0.9\times0.9} = -\frac{0.53\times0.81\times0.1}{1-0.53\times0.81} = \frac{-0.04293}{1-0.4293} = \frac{-0.04293}{0.5707} = -0.075.$$

(As q = 0.9 and p = 0.1)

2. $$\Delta q = \frac{-0.53\times0.1\times0.1\times0.9}{1-0.53\times0.1\times0.1} = -\frac{0.00477}{1-0.0053} = \frac{0.00477}{0.9947} = -0.004795 = -0.0048.$$

The rate of change is largest when q = 0.6, smaller when q = 0.9 and smallest when q = 0.1.

Problem 12

At a locus (Est – 6) coding for an estarase, there are two alleles, Est – 6^F and Est – 6^s, in an experimental population of *Drosophila*. First instar larvae of three genotypes are placed in cultures and the numbers of emerging adults are recorded. The results of two experiments are as follows:

Experiment:	Number of larvae			Number of adults		
	FF	FS	SS	FF	FS	SS
1	160	480	360	80	240	90
2	360	480	160	90	240	80

Assuming the relative fitness depends only on larval viability, what are the fitness of three genotypes in each experiment? Do you think that a stable polymorphic equilibrium might occur in the population and if so at what frequency?

Solution:

	Experiment 1			Experiment 2		
	FF	FS	SS	FF	FS	SS
1. Number of larvae (a)	160	480	360	360	480	160
2. Number of Adults (b)	80	240	90	90	240	80
3. Survival rate $\left(\frac{b}{a}\right)$	$\frac{80}{160} = 0.5$	$\frac{240}{480} = 0.5$	$\frac{90}{360} = 0.25$	$\frac{90}{360} = 0.25$	$\frac{240}{480} = 0.5$	$\frac{80}{160} = 0.5$
4. Relative fitness (w)	$\frac{0.5}{0.5} = 1$	$\frac{0.5}{0.5} = 1$	$\frac{0.25}{0.5} = 0.5$	$\frac{0.25}{0.5} = 0.5$	$\frac{0.5}{0.5} = 1$	$\frac{0.5}{0.5} = 1$

Fitness	FF	FS	SS
Exp. (1)	1.0	1.0	0.5
Exp. (2)	0.5	1.0	1.0

This is a case of frequency dependent selection because the fitness of the homozygotes are inversely related to their frequency. We would expect a stable polymorphic equilibrium. Stable equilibrium will occur when allelic frequency is 0.50.

Problem 13

In a region where industrial pollution has been under control for a number of years, the fitness of *Biston betularia* moths is 0.47 for the dark form and 1 for the light form. Calculate the change in allelic frequency Δp after one generation of selection when (1) p = 0.40 (2) p = 0.10 and (3) p = 0.90.

Solution:

In case of selection against dominant $\Delta p = \frac{-spq^2}{1-s+sq^2}$.

$w = 0.47, \; s = 1-0.47 = 0.53.$

(i) $p = 0.40, \; q = 1-p = 0.60$

$$\Delta p = \frac{-0.53\times0.40\times0.60\times0.60}{1-0.53+0.53\times0.60\times0.60} = \frac{-0.40\times0.1908}{1-0.53+0.1908} = \frac{-0.07632}{1.1908-0.53} = \frac{-0.07632}{0.4753} = -0.11549 = -0.12$$

(ii) $p = 0.10, \; q = 0.90, \; s = 0.53$

$$\Delta p = \frac{-spq^2}{1-s+sq^2} = \frac{-0.53\times0.1\times0.9\times0.9}{1-0.53+0.53\times0.9\times0.9} = \frac{0.1\times0.4293}{1-0.53+0.4293} = \frac{0.04293}{1.4293-0.53} = \frac{0.04293}{0.8993} = -0.04773$$

$= 0.048$

(iii) $p = 0.90, \; q = 0.10, \; s = 0.53$

$$\Delta p = \frac{-spq^2}{1-s+sq^2} = \frac{-0.53\times0.9\times0.1\times0.1}{1-0.53+0.53\times0.1\times0.1} = \frac{-0.9\times0.0053}{1-0.53+0.0053} = \frac{-0.00477}{1.0053-0.53} = \frac{-0.00477}{0.4753} = -0.0100.$$

Problem 14

In a large experimental *Drosophila* population, the fitness of a recessive phenotype is calculated to be 0.90 and the mutation rate to the recessive allele is 5 × 10⁻⁵. If the population is allowed to come to equilibrium, what allelic frequency can be predicted?

Solution:

Here mutation and selection are working in opposite direction. The change in allele frequency, i.e., $\hat{q} = \sqrt{\frac{m}{s}}$ (μ = mutation rate, s = selection coefficient).

$w = 0.9, \; s = 1-w = 1-0.9 = 0.1, \; \mu = 5\times10^{-5}$

$q = \sqrt{\frac{5\times10^{-5}}{0.1}} = \sqrt{5\times10^{-4}} = 2.2\times10^{-2} = 0.022$, and $\hat{p} = 1-0.022 = 0.978.$

Problem 15

Consider changes in the gene frequencies of a population due to zygotic selection or during early development.

(a) Calculate the change in frequency after one generation of an allele that is recessive, lethal in the homozygous condition and has a frequency q of 1%.

(b) Calculate the change in frequency after one generation that is recessive and has a selection coefficient 0.5 and a frequency q of 1%.

Solution:

(a) The change in gene frequency is calculated

$$\Delta q = \frac{q^2}{(1+q)}$$

In this example, $\Delta q = \frac{-(0.01)^2}{(1+0.01)}$

$[\therefore q = 1\% = \frac{1}{100} = 0.01] = \frac{0.0001}{1.01} = 0.000099$

$= 0.00009.$

Note that change in gene frequency over one generation when the initial gene frequency is low.

The change is a loss of only 0.00009 or, 0.0001 or, one allele in $\frac{1}{10{,}000}$.

(b) This is a situation in which homozygous condition might be deleterious but not necessarily lethal. The selection coefficient is less than one and the equation used in this case is

$$\Delta q = \frac{-sq^2(1-q)^2}{(1-sq^2)}$$

$$\Delta q = \frac{0.5\times(0.01)^2(1-0.01)}{1-0.5\times(0.01)^2} = \frac{0.5\times0.0001\times0.99}{1-0.0000495}$$

$$= \frac{0.0000495}{1-0.0000495} = \frac{0.00005}{1-0.00005} = \frac{0.00005}{0.99995}$$

$= 0.00005.$

Note the difference between (a) and (b) even though both populations began with q equal to 1%. The change in gene frequency is much slower in the latter case, dependent upon the intensity of the selection coefficient. A two-fold difference exists between their selection coefficient.

Population	Selection coefficient	Δq
(a)	1.0	−0.0001
(b)	0.5	−0.00005

$$\frac{1.0}{0.5} = 2X \;\&\; \frac{-0.0001}{-0.00005} = 2X$$

Problem 16

Genotype Aa yields 80 offspring per generation per mating pair. Overdominance exists since the AA genotype yields 60 offspring and the aa genotype yields 48 offspring per generation per mating pair.

(a) What is the adaptive value (w) of AA?

(b) What is the selection coefficient (s) of aa?

Solution:

Genotype	Offsprings	Adaptive value (w)	Selection coefficient(s)
Aa	80	$\frac{80}{80} = 1.0$	$1.0 - 1.0 = 0$
AA	60	$\frac{60}{80} = 0.75$	$1 - 0.75 = 0.25$
aa	48	$\frac{48}{80} = 0.6$	$1 - 0.6 = 0.4$

(a) The adaptive value (w) of A is 0.75.

(b) The selection coefficient(s) of aa is 0.4.

Problem 17

What are the adaptive values of the three genotypes – HH, Hh and hh under the different conditions listed below?

Condition	Average number of progeny produced by each genotype		
	HH	**Hh**	**hh**
(a)	**312**	**354**	**0**
(b)	**248**	**246**	**30**
(c)	**420**	**367**	**0**

Solution:

Condition	Genotypes	Adaptive value
(a)	HH	$\frac{312}{354} = 0.88$
	Hh	$\frac{354}{354} = 1.00$
	hh	$\frac{0}{354} = 0.00$
(b)	HH	$\frac{248}{248} = 1.00$
	Hh	$\frac{246}{248} = 0.99$
	hh	$\frac{30}{248} = 0.12$
(c)	HH	$\frac{420}{420} = 1.00$
	Hh	$\frac{367}{420} = 0.87$
	hh	$\frac{0}{420} = 0.00$

Problem 18

Calculate the change in frequency of allele (Δq) after one generation, the allele is recessive, lethal in the homozygous condition, and has a frequency q of 3.2%.

Solution:

In this condition

$$\Delta q = \frac{-q^2}{(1+q)} \quad [\because q = 3.2\% = \frac{32}{1000} = 0.032]$$

$$\Delta q = \frac{-(0.032)^2}{(1+0.032)} = \frac{0.001024}{1.032} = 0.0009922$$

$$= 0.00099.$$

Problem 19

In a particular population, the genotype frequency of AA is 0.25, Aa is 0.50 and aa is 0.25. If the fitness (adaptive value) of these genotypes are 1.0, 0.8 & 0.5 respectively, calculate the following:

(a) Selection coefficient of AA, Aa and aa.

(b) Genotype frequency of aa after one generation of selection.

Solution:

(a)

Genotype	Adaptive value (w)	Selection coefficient(s)
AA	1.0	1 – 1 = 0.0
Aa	0.8	1 – 0.8 = 0.2
aa	0.5	1 – 0.5 = 0.5

(b) AA = 1 × 0.25 = 0.25

Aa = 0.8 × 0.50 = 0.40

aa = 0.5 × 0.25 = 0.125

Sum 0.775

$$aa = \frac{0.125}{0.775} = 0.161$$

The genotype frequency of aa after one generation of a selection coefficient would be 0.161.

Problem 20

A rare recessive allele is completely lethal when homozygous. If the frequency of this allele in a population is $\frac{1}{10,000}$, what will be the change in gene frequency Δq after one generation?

Solution:

$$\Delta q = \frac{-q^2}{(1+q)} \quad [\because q = \frac{1}{10,000} = 0.0001]$$

$$\Delta q = \frac{-(0.0001)^2}{1+0.0001} = \frac{0.0001 \times 0.0001}{1.0001} = \frac{0.00000001}{1.0001}$$

$$= \frac{0.00000001}{1.0001} \times 10^4 = \frac{1 \times 10^{-8} \times 10^4}{1.0001}$$

$$= \frac{1}{1.0001} \times 10^{-4} = \frac{999}{10^7} \times 10^{-4} = 9.99 \times 10^{-9}.$$

Problem 21

Warfarin is a rat-killer that acts by hindering blood coagulation. Continued use of the poison has resulted in the evolution of resistance in most rat populations because of selection favouring a resistance mutation R. The normal sensitive allele may be denoted by S. In the absence of warfarin, the relative fitness of SS, RS and RR are in the ratio 1.00:9.77:0.46 and in the presence of warfarin, the ratio of fitness is 0.68:1.00:0.37.

(a) With regard to persistence or loss of the R and S alleles, what is the expected result of the continued use of the chemical?

(b) What would happen it warfarin were no longer used?

(c) Under what circumstances might R allele become fixed?

Solution:

(a)

Rodenticide	SS	RS	RR
(i) Absence of warfarin:	1.00	0.77	0.46
(ii) Presence of warfarin:	0.68	1.00	0.37

In the presence of warfarin, there is heterozygote superiority. So, continued use would be expected to result in a stable equilibrium in which both R and S persisted in the population.

(b) In the absence of warfarin, the genotype SS is favoured and RS is intermediate in fitness, so allele S would become fixed and R would become lost.

(c) (i) The allele R could become fixed in a small population. Since in small population there are chances of fluctuation in allele frequency, the S allele might become lost.

(ii) However in small population if the chance increases, the R allele might be lost.

Problem 22

By DNA sequencing we have found twenty nucleotide sites that differ among the species. Prepare distance matrix.

Species		1	2	3	4	5	6	7	8	9	10	11	12	13	14	15	16	17	18	19	20
P	=	A	G	G	C	C	A	A	G	C	C	A	T	A	G	C	T	G	T	C	C
Q	=	A	G	G	C	A	A	A	G	A	C	A	T	A	C	C	T	G	A	C	C
R	=	A	G	G	C	C	A	A	G	A	C	A	T	A	G	C	T	G	T	C	C
S	=	A	G	G	C	A	A	A	G	A	C	A	T	A	C	C	T	G	T	C	C

Solution:

1. Evolutionary difference between P and Q: (5C → A, 9C → A, 14G → C, 18T → A) i.e. 4 $= \frac{4}{20} = 0.2$
2. Evolutionary difference between Q and R: (5A → C, 14C → G, 18A → T), i.e., $3 = \frac{3}{20} = 0.15$.
3. Evolutionary difference between R and S: (5C → A, 14G → C), i.e., $2 = \frac{2}{20} = 0.10$.
4. Evolutionary distance between P and R: (9C → A), i.e., $1 = \frac{1}{20} = 0.05$.
5. Evolutionary distance between P and S: (5C → A, 9C → A, 14G → C) i.e. $3 = \frac{3}{20} = 0.15$.
6. Evolutionary distance between Q and S: (18A → T), i.e., $1 = \frac{1}{20} = 0.05$.

Distance Matrix

	P	Q	R	S
P	—	0.2	0.5	0.15
Q	0.2	—	0.15	0.05
R	0.05	0.15	—	0.10
S	0.15	0.05	0.10	—

Problem 23

The following DNA sequences are observed among four closely related species; prepare distance-matrix:

	1	2	3	4	5	6
W	G	C	G	A	T	G
X	G	T	G	T	T	G
Y	G	T	T	G	C	A
Z	G	T	C	C	C	A

Solution:

1. Evolutionary distance between W and X: (2C → T, 4A → T) i.e. $2 = \frac{2}{6} = 0.333 = 0.34$.
2. Evolutionary distance between X and Y: (3C → T, 4T → G, 5T → C, 6G → A), i.e., 4 $= \frac{4}{6} = \frac{2}{3} = 0.666 = 0.67$.

3. Evolutionary distance between Y and Z: (3T → C, 4G → C), i.e., 2 = $\frac{2}{6} = \frac{1}{3}$ = 0.34.
4. Evolutionary distance between W and Z: (2C → T, 3G → C, 4A → C, 5T → C, 6G → A), i.e., 5 = $\frac{5}{6}$ = 0.84.
5. Evolutionary distance between X and Z: (3G → C, 4T → C, 5T → C, 6G → A), i.e., 4 = $\frac{4}{6}$ = 0.67.
6. Evolutionary distance between W and Y: (2C → T, 3G → T, 4A → G, 5T → C, 6G → A), i.e., 5 = $\frac{5}{6}$ = 0.84.

Distance-matrix

	W	X	Y	Z
W	—	0.34	0.84	0.84
X	0.34	—	0.67	0.67
Y	0.84	0.67	—	0.34
Z	0.84	0.67	0.34	—

Problem 24

By DNA sequencing the following sequences are observed from four closely related species; prepare distance-matrix.

Species	**1**	**2**	**3**	**4**	**5**	**6**	**7**	**8**	**9**	**10**
A =	G	C	T	G	A	T	G	A	G	T
B =	A	T	C	A	A	T	G	A	G	T
C =	G	T	T	G	C	A	A	C	G	T
D =	G	T	C	A	A	A	T	G	C	A

Solution:

1. Evolutionary distance between A and B: (1G → A, 2C → T, 3T → C, 4G → A), i.e., 4 = $\frac{4}{10}$ = 0.4
2. Evolutionary distance between B and C: (1A → G, 3C → T, 4A → G, 5A → C, 6T → A, 7G → A, 8A → C), i.e., 7 = $\frac{7}{10}$ = 0.7
3. Evolutionary distance between C and D: (3T → C, 4G → A, 5C → A, 7A → T, 8C → G, 9G → C, 10T → A), i.e., 7 = $\frac{7}{10}$ = 0.7
4. Evolutionary distance between A and D: (2C → T, 3T → C, 4G → A, 6T → A, 7G → T, 8A → G, 9G → C, 10T → A), i.e., 8 = $\frac{8}{10}$ = 0.8
5. Evolutionary distance between A and C: (2C → T, 5A → C, 6T → A, 7G → A, 8A → C), i.e., 5 = $\frac{5}{10}$ = 0.5
6. Evolutionary distance between B and D: (1A → G, 6T → A, 7G → T, 8A → G, 9G → C, 10T → A), i.e., 6 = $\frac{6}{10}$ = 0.6

	A	B	C	D
A	—	0.4	0.5	0.8
B	0.4	—	0.7	0.6
C	0.5	0.7	—	0.7
D	0.8	0.6	0.7	—

Problem 25

The following DNA sequences are found from the four closely related species, prepare distance-matrix.

	1	2	3	4	5
M =	G	T	A	C	A
N =	G	A	C	A	G
O =	G	T	C	A	A
P =	G	A	C	C	G

Solution:

1. **Evolutionary distance between M and N:** (2T → A, 3A → C, 4C → A, 5A →), i.e., 4 = $\frac{4}{5}$ = 0.8.
2. **Evolutionary distance between N and O:** (2A → T, 5G → A), i.e., 2 = $\frac{2}{5}$ = 0.4.
3. **Evolutionary distance between O and P:** (2T → A, 4A → C, 5A → G, 5A → G), i.e., 3 = $\frac{3}{5}$ = 0.6.
4. **Evolutionary distance between M and P:** (2T → A, 3A → C, 5A → G), i.e., 3 = $\frac{3}{5}$ = 0.6.
5. **Evolutionary distance between N and P:** (4A → C), i.e., 1 = $\frac{1}{5}$ = 0.2.
6. **Evolutionary distance between O and M:** (3A → C, 4G → A), i.e., 2 = $\frac{2}{5}$ = 0.4.

	M	N	O	P
M	—	0.8	0.4	0.6
N	0.8	—	0.4	0.2
O	0.4	0.4	—	0.6
P	0.6	0.2	0.6	—

Problem 26

With regard to allele frequency, what does it mean to say that an allele is fixed? That an allele is lost?

Solution:

Fixation means that the allele frequency equals to 1; loss means that it is 0.

MUTATION

- **MUTATION:**

The process by which a gene or chromosome changes structurally or changes in the nucleotide sequence of genetic material whether by substitution, duplication, insertion, deletion or inversion is called mutation.

Calculation:

(i) Let us consider a population with two alleles A_1 and A_2. A_1 is dominant over A_2. The mutation from A_1 to A_2 occurs at a rate of μ per gamete per generation. Assume also that at a given time the frequency of A_1 is p_0. In the next generation fraction of μ of all A_1 alleles become A_2 by mutation.

(ii) The frequency of p_1 will, therefore, be the initial frequency (p_0) minus (–) the frequency of mutated alleles (μp_0) or,

$p_1 = p_0 - \mu p_0 = p_0(1 - \mu)$.

(iii) In the next generation a fraction μ of the remaining A_1 alleles will mutate to A_2 and the frequency of A_1 will become

$p_2 = p_1 - \mu p_1 = p_1(1 - \mu)$.

Substituting the value of p_1, we get

$p_2 = p_1(1 - \mu) = p_0(1 - \mu)(1 - \mu) = p_0(1 - \mu)^2$.

After 't' generation it will be $p_t = p_0(1 - \mu)^t$.

Assume that A_1 mutate to A_2 at the rate of u as before and that A_2 mutates to A_1 at a rate v. If at a given time the frequencies of A_1 and A_2 are p_0 and q_0 respectively, then after one generation, the frequency of A_1 will be

$p_1 = p_0 - up_0 + vq_0$.

This is because the fraction up_0 of the A_1 allele is changes to A_2 but the fraction of vq_0 of the alleles changes to A_1. If the change in frequency of A_1 is represented by Δp,

$\Delta p = p_1 - p_0$.

Substituting the value of p_1,

$\Delta p = p_0 - up_0 + vq_0 - p_0 = vq_0 - up_0$.

An equilibrium between forward and backward mutations will exist when there is no net change in gene frequencies, i.e., when Δp = 0. If we represent the equilibrium allele frequencies by $\hat{p}$ and $\hat{q}$, the requirement that Δp = 0 implies $u\hat{p} = v\hat{q}$.

This result says that the allelic frequencies will be at equilibrium when the number of A_1 alleles changes to A_2 alleles is the same as the number of A_2 changes to A_1. Since p + q = 1 and therefore, q = 1 – p, there will be equilibrium, when

$u\hat{p} = v(1-\hat{p})$, or, $u\hat{p} + v\hat{p} = v$, $\hat{p}(u + v) = v$

or, $\hat{p} = \frac{v}{u+v}$ and because $\hat{p} + \hat{q} = 1$. $\therefore \hat{q} = \frac{u}{u+v}$.

Solved Problems

Problem 1

At a particular locus, there are two alleles B and b. The mutation rate of B to b is 3.5 × 10⁻⁴, whereas the mutation rate of b to B is 6 × 10⁻⁸. What is the equilibrium frequency of the b allele assuming no other factor is operating in this population to disturb the Hardy-Weinberg equilibrium?

Solution:

Let frequency of B = p and b = q

B mutate to b at the rate of $u = 3.5 \times 10^{-4}$

b mutates to B at the rate of $v = 6 \times 10^{-8}$.

Equilibrium frequency of b = $\hat{q}$.

$$\hat{q} = \frac{u}{u+v} = \frac{3.5\times10^{-4}}{3.5\times10^{-4}+6\times10^{-8}} = \frac{10^{-4}\times3.5}{10^{-4}(3.5+6\times10^{-8})}$$

$$\hat{q} = \frac{3.5}{3.5+\frac{6}{10^4}} = \frac{3.5}{3.5+0.0006} = \frac{3.5}{3.5006}$$

$\hat{q} = 0.9998$.

Problem 2

Consider a population in which p = 0.9 and q = 0.1. If the forward mutation rate A → a is 5 × 10⁻⁵ and the reverse mutation rate a → A is

2×10^{-8}, calculate the equilibrium frequency $\hat{q}$ of the a allele.

Solution:

$$u = 5 \times 10^{-5},\ v = 2 \times 10^{-5},\ p = 0.9,\ q = 0.1$$

$$\hat{q} = \frac{v}{u+v} = \frac{5 \times 10^{-5}}{5\times10^{-5}+2\times10^{-5}} = \frac{10^{-5}\times 5}{10^{-5}(5+2)} = \frac{10^{-5}\times 5}{10^{-5}\times 7}$$

$$\hat{q} = \frac{5}{7} = 0.714.$$

Problem 3

The initial frequency of the A_1 allele is p_0 and that of the A_2 allele is q_0; the mutation rate of A_1 to A_2 is $u = 5 \times 10^{-5}$ and the back mutation rate of A_2 to A_1 is 1×10^{-5}. Calculate the change in frequency of A_2 allele (Δq) due to one generation of mutation when (a) $p = 0.9$, $q = 0.1$; (b) $p = 0.5$, $q = 0.5$; (c) $p = 0.1$, $q = 0.9$.

Solution:

$$\Delta q = up_0 - uv_0$$

(a) $u = 5 \times 10^{-5},\ p = 0.9,\ q = 0.1,\ v = 1 \times 10^{-5}$

$$\Delta q = 5 \times 10^{-5} \times 0.9 - 1 \times 10^{-5} \times 0.1$$
$$= 10^{-5}(4.5 - 0.1) = 10^{-5} \times 4.4 = 4.4 \times 10^{-5}$$

(b) $\Delta q = 5 \times 10^{-5} \times 0.5 - 1 \times 10^{-5} \times 0.5$

$$= 10^{-5}(2.5 - 0.5) = 10^{-5} \times 2 = 2 \times 10^{-5}$$

(c) $\Delta q = 5 \times 10^{-5} \times 0.1 - 1 \times 10^{-5} \times 0.9$

$$= 10^{-5}(0.5 - 0.9) = 10^{-5} \times -0.4$$
$$= -0.4 \times 10^{-5} = \frac{-4}{10} \times 10^{-5} = -4 \times 10^{-6}$$

Problem 4

In *E. coli* the rate of mutation from histidine independence (his^+) to histidine requirement (his^-) and the rate of reverse mutation have been estimated as:

$his^+ \longrightarrow his^-$ 2×10^{-6}

$his^- \longrightarrow his^+$ 4×10^{-8}

Assuming that no other processes are involved, what are the equilibrium frequency of the two alleles?

Solution:

Here $u = 2 \times 10^{-6}$, $v = 4 \times 10^{-8}$.

Equilibrium frequency for his^+ p and his^- q

$$\hat{p} = \frac{v}{u+v} \text{ and } \hat{q} = \frac{u}{u+v}.$$

$$\hat{p} = \frac{4 \times 10^{-8}}{2 \times 10^{-6} + 4 \times 10^{-8}} = \frac{4 \times 10^{-8}}{10^{-6}(2 + 4 \times 10^{-2})}$$

$$= \frac{4 \times 10^{-8}}{10^{-6}\left(2 + \frac{4}{100}\right)} = \frac{4 \times 10^{-2}}{2 + 0.04} = \frac{0.04}{2.04} = 0.0196.$$

$$= 0.02.$$

$$\hat{q} = 1 - 0.02 = 0.98.$$

Problem 5

If two alleles of loci A and a, can be inter-converted by mutation $A \underset{v}{\overset{u}{\rightleftharpoons}} a$ and mutation rate u is of 6×10^{-7} and v is a mutation rate of 6×10^{-8} and what will be the frequency of A and a at mutation equilibrium, assuming no selective difference, no migration and no random fluctuation caused by genetic drift?

Solution:

Let frequency of A is $\hat{p}$ and a is $\hat{q}$. $u = 6 \times 10^{-7}$, $v = 6 \times 10^{-8}$.

$$\hat{p} = \frac{v}{u+v} = \frac{6 \times 10^{-8}}{6 \times 10^{7} + 6 \times 10^{-8}} = \frac{6 \times 10^{-8}}{10^{7}(6 + 6 \times 10^{-1})}$$

$$= \frac{6 \times 10^{-8}}{10^{7}(6 + 0.6)} = \frac{6 \times 10^{-1}}{6.6} = \frac{0.6}{6.6} = 0.09.$$

$$\hat{p} = 1 - p = 1 - 0.09 = 0.01.$$

Problem 6

Assume that the forward and backward mutation rates of a certain locus of *Drosophila* are

$A \longrightarrow a$ 2×10^{-5}

$a \longrightarrow A$ 6×10^{-7}

What are the expectd allelic equilibrium frequencies if no other process are involved? Is the calculation of mutation equilibrium frequencies of alleles different for diploid and haploid organisms.

Solution:

Let frequency of A be $\hat{p}$ and a is $\hat{q}$.

$U = 2 \times 10^{-5}$ and $V = 6 \times 10^{-7}$

$$\hat{p} = \frac{V}{U+V} = \frac{6 \times 10^{-7}}{2\times10^{-5}+6\times10^{-7}}$$

$$= \frac{6}{2 \times 10^{7} \times 10^{-5}(1 + 3 \times 10^{-2})} = \frac{3}{100\left(1 + \frac{3}{100}\right)}$$

$$= \frac{3}{100 \times \frac{103}{100}} = \frac{3}{100} = 0.029 = 0.03.$$

$\hat{q} = 1 - 0.03 = 0.97.$

The allelic frequencies are calculated in the same way for haploid and diploids.

Problem 7

Consider a locus with alleles A and a in a large, randomly mating population under the influence of mutation.

(a) If the mutation rate of A to a is 6×10^{-5} and the back mutation rate is 7×10^{-7}, what is the equilibrium frequency of a?

(b) If q = 0.9 in generation n, what would it be one generation later, under only the influence of mutation?

Solution:

(a) Equilibrium frequency of $\hat{q} = \frac{u}{u+v}$, $q = \hat{q}$;

$u = 6 \times 10^{-5}$, $v = 7 \times 10^{-7}$

$$\hat{q} = \frac{u}{u+v} = \frac{6 \times 10^{-5}}{6 \times 10^{-5} + 7 \times 10^{-7}} = \frac{6 \times 10^{-5}}{10^{-5}(6 + 7 \times 10^{-2})}$$

$$= \frac{6}{6 + 7 \times 10^{-2}} = \frac{6}{6 + 0.07} = \frac{6}{6.07} = 0.9884 = 0.988.$$

(b) If q = 0.9 at generation n, then q_{n+1}.

$[u = 6 \times 10^{-5},\ v = 7 \times 10^{-7}]$

$$q_{n+1} = q_n + up_n - vq_n$$

$$= 0.9 + 6 \times 10^{-5} \times 0.1 - 7 \times 10^{-7} \times 0.9$$

$$[\because p_n = 1 - 0.9 = 0.1]$$

$$= 0.9 + 10^{-5}(6 \times 0.1 - 7 \times 0.9 \times 10^{-2})$$

$$= 0.9 + 10^{-5}(0.6 - 6.3 \times 10^{-2})$$

$$= 0.9 + 10^{-5}(0.6 - 0.063)$$

$$= 0.9 + 10^{-5}(0.537) = \frac{9}{10} + \frac{0.537}{10^5}$$

$$= \frac{90000 + 0.537}{10^5} = \frac{90000.537}{10^5}$$

$$= 0.90000537 = 0.9000054.$$

Problem 8

Assume that at a certain locus the mutation rate of A ⟶ a is 10^{-6} and that there is no back mutation. What will be the frequency of A after 10, 1000 and 100,00 generation of mutation?

Solution:

Initial frequency p_0, after 't' generations it will be $p_t = p_0(1-\mu)^t$. If we assume that initial frequency is $p_0 = 1.00$, we obtain

$$p_{10} = (1 - 10^{6})^{10} = 0.99999,$$

$$p_{1000} = (1 - 10^{-6})^{1000} = 0.999$$

$$p_{100,000} = (1 - 10^{-6})^{1000} = 0.905.$$

Problem 9

Assume that a large population of insects reared under laboratory conditions have the following genotype frequencies, well ascertained due to distinct incomplete dominance.

Genotype	Number
DD	22,400
Dd	67,200
dd	50,400
	1,40,000

The population is in perfect Hardy-Weinberg equilibrium, and because of carefully controlled conditions, the researchers ruled out any gene frequency changes. Additional experiments revealed a reverse mutation rate of $\frac{1}{14,925}$. Reverse mutations are often more easily detected in a population than forward mutations. Based upon these data, what is the probable rate of forward mutation?

Solution:

	DD	Dd	dd	Total
	22,400	67,200	50,400	= 1,40,000
Genotypic frequency:	0.16	0.48	0.36	

Gene frequency of D = $p = 0.16 + \frac{1}{2} \times 0.48$

$= 0.16 + 0.24 = 0.40$

Gene frequency of d = $q = 0.36 + \frac{1}{2} \times 0.48$

$= 0.36 + 0.24 = 0.60$

Mutation rate $v = \frac{1}{14,925} = 0.000067.$

Since $u \times p = v \times q$, or, $u \times 0.4 = 0.000067 \times 0.6$

$$u = \frac{0.000067 \times 0.6}{0.4} = \frac{0.000067 \times 0.3}{0.2} = \frac{0.000201}{2}$$

$= 0.000105 = 0.0001.$

Forward mutation rate is 0.0001 or, $\frac{1}{10,000}$.

Problem 10

Given the mutational equilibrium where m is the forward mutation rate, an expression relating to q, m and g at equilibrium can be derived as $q = \frac{m}{(g+m)}$

If m is equal to 10^{-6} and g is equal to 10^{-7}, predict the equilibrium distribution of the three genotypes BB, Bb and bb in the population.

Solution:

Here m = 10^{-6} and g = 10^{-7}

$$q = \frac{m}{g+m} = \frac{10^{-6}}{10^{-7}+10^{-6}} = \frac{6\times10^{-5}}{0.000001+0.0000001}$$

$$= \frac{0.000001}{0.0000011} = \frac{1}{1.1} = \frac{10}{11} = 0.909 = 0.91$$

$p = 1 - 0.91 = 0.09$

$\therefore \quad p^2 = BB = (0.09)^2 = 0.0081$

$2pq = Bb = 2 \times 0.09 \times 0.92 = 0.2638 - 0.164$

$q^2 = bb = (0.91)^2 = 0.8281 - 0.828.$

Problem 11

Achondroplasia is a dominant trait that causes a characteristic form of dwarfism. In a survey of 50,000 births, five infants with achondroplasia were identified. Three of the affected infants had affected parents, while two had normal parents. Calculate the mutation rate for achondroplasia and express the rate as the number of mutant genes per given number of gametes.

Solution:

(i) Three of the affected infants had affected parents. Only two new genes from mutation enter into the problem.

(ii) The gene is dominant, therefore, each new case of achondroplasia arose from a single new mutation.

(iii) There are 50,000 births, therefore, 50,000 × 2 = 1,00,000 genes (alleles) are involved.

Therefore, the frequency of mutation is $\frac{2}{1,00,000}$ = 2×10^{-5}.

Problem 12

Purifying selection eliminates deleterious alleles from a population, but recurrent mutation replenishes them. Suppose the recessive lethal alleles of the B gene are created at the rate of 2×10^{-6} per generation. What is the expected frequency of the lethal alleles in a population in mutation-selection equilibrium?

Solution:

The frequency of lethal alleles is given by the equation $q = \sqrt{\frac{u}{s}}$

[u = mutation rate, s = selection coefficient]

Here, $u = 2\times10^{-6}$, $s = 1$.

The expected frequency of lethal allele is

$$q = \sqrt{\frac{2\times10^{-6}}{1}} = (2\times10^{-6})^{\frac{1}{2}} = 0.00141 = 0.0014.$$

Problem 13

In some regions of Nalanda, the frequency of the Hb^S allele is 0.2. If this frequency is the result of a dynamic equilibrium due to the superior fitness of Hb^SHb^A heterozygotes and if Hb^SHb^S homozygotes are essentially lethal, what is the intensity of selection against Hb^AHb^A homozygotes?

Solution:

$p = \frac{s_2}{s_1+s_2}$

$p = 0.2$

or, $0.2 = \frac{s_2}{s_1+s_2}$

or, $0.2 + 2s_2 = s_2$

[p = gene frequency
s = 1, because $\frac{Hb_s}{Hb_s}$ homozygodes die
s_1 = selection coefficient of $\frac{Hb_s}{Hb_s}$
s_2 = selection coefficient of $\frac{Hb_A}{Hb_A}$]

or, $s_2 - 0.2s_2 = 0.2$

or, $s_2(1-0.2) = 0.2$, or, $0.8s_2 = 0.2$

or, $s_2 = \frac{0.2}{0.8} = \frac{1}{4} = 0.25.$

Problem 14

In some regions of West Midnapore the frequency of Hb^S allele is 0.1. If this frequency is the result of a dynamic equilibrium due to the superior fitness of Hb^SHb^A heterozygotes and if Hb^SHb^S homozygotes are essentially lethal, what is the intensity of selection against Hb^AHb^A homozygotes?

Solution:

$p = \frac{s_2}{s_1+s_2}$

$p = 0.1$

or, $0.1 = \frac{s_2}{(1+s_2)}$

or, $0.1(1+2s_2) = s_2$

or, $0.1 + 0.1s_2 = s_2$ or, $s_2 - 0.1s_2 = 0.1$

or, $s_2(1-0.1) = 0.1$, or, $0.9s_2 = 0.1$

or, $s_2 = \frac{0.1}{0.9} = 0.11$.

[p = gene frequency
s = 1, because $\frac{Hb^s}{Hb^s}$ homozygodes die
s_1 = selection coefficient of $\frac{Hb^s}{Hb^s}$
s_2 = selection coefficient of $\frac{Hb^s}{Hb^A}$]

Problem 15

A completely recessive allele g is lethal in homozygous condition. If the dominant allele G mutates to g @ 10^{-6} per generation, what is the expected frequency of the lethal allele when the population reaches mutation-selection equilibrium?

Solution:

The frequency of lethal allele is given by the equation $q = \sqrt{\frac{u}{s}}$

[u = mutation rate, s = selection coefficient]

Here, $u = 10 \times 10^{-6}$, $s = 1$

$$q = \sqrt{\frac{10\times10^{-6}}{1}} = \sqrt{10\times10^{-6}} = \sqrt{\frac{10}{10^{-6}}}$$

$$= \sqrt{0.000001} = 0.001.$$

Problem 16

Neurofibromatosis type 1 is a common Mendelian condition. In a population survey of 50,000 people in one town. 12 cases are identified, of which eight cases belong to one large affected family.

(a) Based on these figures, what is the mutation rate in the neurofibromin gene?

(b) Name some limitations to the validity of calculating the mutation rate from a survey like this.

Solution:

(a) (i) 12 cases are identified. Eight cases belong to one large group. Rest (12 – 8) = 4 cases in the town appear to be new mutations.

(ii) Number of people 50,000, i.e., 1,00,000 genes are involved. Therefore, 4 new mutations per 1,00,000 genes are inherited.

Then $\mu = \frac{4}{1,00,000} = \frac{1}{25,000}$ gametes = 0.04×10^{-3} gamete.

(b) (i) The population is small and may not, therefore, be representative of the larger wider population.

(ii) If there is a bias towards older, retired population, the reproducing sub-population may be smaller.

(iii) The figure may be distorted by the migration of younger people away from the town.

(iv) In addition, four 'new mutation' cases should be verified by proper examination of the parent.

Problem 17

If $u = 10^{-6}$ and $v = 10^{-7}$, what is Δq for q = 0.0, 0.25, 0.75 and 0.17, when Δq is the largest? Explain why?

Solution:

$\Delta q = up - vq$

$q = 0.0$ $p = 1$, $\therefore \Delta q = 10^{-6}\times 1 - 10^{-7}\times 0.$

$q = 0.25$, $p = 0.75$, $\Delta q = 10^{-6}$.

$\Delta q = 10^{-6}\times 0.75 - 10^{-7}\times 0.25$

$= 10^{-6}\left(0.75 - \frac{0.25}{10}\right) = 10^{-6}(0.75 - 0.025)$

$= 0.725\times 10^{-6} = 7.25\times 10^{-7}$.

$q = 0.5$, $p = 0.5$

$\Delta q = 10^{-6}\times 0.5 - 10^{-7}\times 0.5 = 10^{-7}\times .5 - 10^{-7}\times .5$

$= 10^{-7}(5 - 0.5) = 4.5\times 10^{-7}$

$q = 0.75$ $p = 0.25$

$\Delta = 10^{-6}\times 0.25 - 10^{-7}\times 0.75 = 10^{-7}(2.5 - 0.75)$

$= 1.75\times 10^{-7}$.

$\Delta q = 1$, $p = 0$

$\therefore \Delta q = 0\times 10^{-6} - 1\times 10^{-7} = 10^{-7}$.

Δq is the largest, when p = 1 and q = 0, because all alleles are wild-type.

MIGRATION

- **MIGRATION:**

When an individual migrates from one population to another, it carries genes that are representative of its own ancestral population. It results in the exchange of genes with one another to a greater or lesser extent. This phenomenon is called **gene flow.**

- **Measurements of Migration:**

(i) Assume that individuals from a large source or continent migrate at a certain rate into an island and there interbreed with residents. Although reciprocal gene flow may occur, we will assume it so low that it has a negligible effect on the allelic frequencies in the source of population.

Native (Island) Migrants (continents)

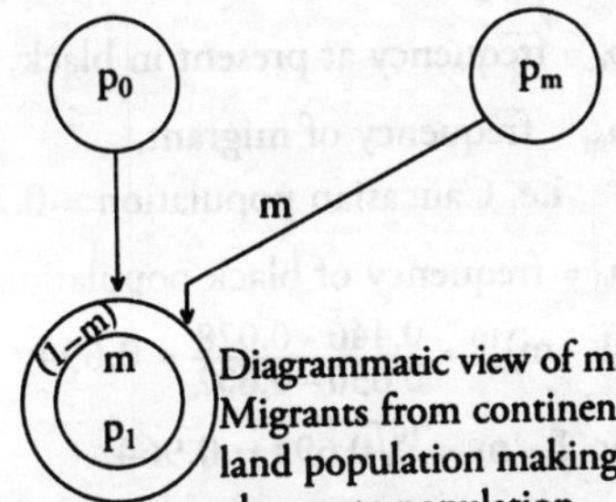

Diagrammatic view of migration. Migrants from continent to island population making a conglomerate population.

Conglomerate

(ii) Let be the population migrants moving into island population in each generation m. So that proportion of non-migrant (residual) is $1-m$.

(iii) Assume that the frequency of migrant is p_m and the frequency of island population before migration is p_0, the allelic frequency of population after migration is p_1.

$$p_1 = (1-m)p_0 + mp_m = p_0 - mp_0 + mp_m$$
$$= p_0 - m(p_0 - p_m).$$

Therefore, new allelic frequency will be original frequency (p_0) multiplied by the proportion of reproducing individuals that are residents $(1-m)$, plus the proportion of reproducing migrant individuals (m) multiplied by their gene frequency (p_m).

Reorganising the statements: The new allelic frequency (p_1) will be the original allelic frequency (p_0) minus the proportion of migrant individuals (m) multiplied by the difference in allelic frequency between residents and migrants (p_0-p_m).

(iv) The change in allelic frequencies after one generation of gene flow is the difference between the frequencies after gene flow or migration and before migration or gene flow.

$$\Delta p = p_1 - p_0 = p_0 - m(p_0 - p_m) - p_0$$
$$[\text{as } p_1 = p_0 - m(p_0 - p_m)]$$
$$= -m(p_0 - p_m) = m(p_m - p_0).$$

- **Interpretation:**

(i) No change of allelic frequency will occur ($\Delta p = 0$), if $m = 0$ or, $p_0 = p_m$.

(ii) If p_0 is smaller than pm, the frequency of second generation increases because of migration or gene flow, i.e., Δp is positive.

(iii) If p_0 is greater than p_m, the frequency of second population decreases, i.e., Δp is negative.

As a result of two effects, there is a stable equilibrium frequency of population in the island, i.e., $p_m = p_0$.

Calculation:

The difference in allele frequency between the island and continent population:

- **After first generation:**

$$p_1 - p_m = (1-m)p_0 + mp_m - p_m$$
$$= p_0 - mp_0 + mp_m - p_m$$
$$= p_0(1-m) - p_m(1-m) = (1-m)(p_0 - p_m)$$

- **After second generation**: Difference in allelic frequency will be $p_2 - p_m = (1-m)^2(p_0 - p_m)$.

- **After 't' generation**: Difference in allelic frequency will be $p_1 - p_m = (1-m)^t(p_0 - p_m)$.

Rearranging the equation, $(1-m)^t = \frac{p_t - p_m}{p_0 - p_m}$.

Migration coefficient:

An expression of the proportion of migrant genes entering the population per generation is known as migration coefficient.

Solved Problems

Problem 1

Determine the frequency of allele A in an island population after one generation of migration from the main land under the following conditions.

(a) $p_1 = 0.6, p_m = 0.1, m = 0.2$

(b) $p_1 = 0.2, p_m = 0.7, m = 0.3$

(c) $p_1 = 0.1, p_m = 0.2, m = 0.1$

Solution:

(a) $p^1 = p_1 - m(p_1 - p_m) = 0.6 - 0.2(0.6 - 0.1)$
$= 0.6 - 0.2 \times 0.5 = 0.6 - 0.10 = 0.5.$

$\therefore p^1 = 0.5.$

(b) $p^1 = 0.2 - 0.3(0.2 - 0.7) = 0.2 - 0.06 + 0.21$
$= 0.2 + 0.21 - 0.06 = 0.41 - 0.06 = 0.35$

$\therefore p^1 = 0.35.$

(c) $p^1 = 0.1 - 0.1(0.1 - 0.2) = 0.1 - 0.01 + 0.02$
$= 0.1 + 0.02 - 0.01 = 0.12 - 0.01 = 0.11$

$\therefore p^1 = 0.11.$

Problem 2

In a particular population, the frequency of allele t was 0.25 in a migrant population and 0.45 in the conglomerate population. If the migration rate was 0.1, calculate the frequency of 't' in the original native population.

Solution:

$p_1 = 0.45, \quad p_m = 0.25, \quad m = 0.1$

$p_1 = (1 - m)p_0 + mp_m$

or, $0.45 = (1 - 0.1)p_0 + 0.1 \times 0.25$

or, $0.45 = 0.9p_0 + 0.025$

or, $0.9p_0 = 0.45 - 0.025 = 0.425$

or, $p_0 = \frac{0.425}{0.9} = 0.472.$

Problem 3

In the USA, people of mixed Caucasian and black descent are considered members of black population. Racial admixture can hence be seen as a process of gene flow from the Caucasian to black population. The frequency of R^0 allele at the locus determining the Rhesus blood group p = 0.028 in Caucasian population. Among the African population from which Black come, the frequency of R^0 is 0.630, which can be considered as the initial allele frequency of US black population. The African ancestors of US blacks came to United States about 10 generation ago and the frequency of R^0 among US blacks is at present $p_t = 0.446$.

Calculate the rate of migration in this population. [C.U. (M.Sc. Zoo) 2005]

Solution:

According to the formula, $(1 - m)^t = \frac{p_t - p_m}{p_0 - p_m}$.

m = migration rate, t = years = 10,

p_t = frequency at present in black, i.e., 0.446.

p_m = frequency of migrant,
i.e. Caucasian population = 0.28.

p_0 = frequency of black population = 0.630.

$(1 - m)^{10} = \frac{0.446 - 0.028}{0.630 - 0.602} = 0.694$

or, $1 - m = \sqrt[10]{0.694} = 0.964,$

or, $m = 1 - 0.964 = 0.036.$

Therefore, the gene flow from US Caucasian into US black population has occurred at a rate equivalent to an average 0.036 × 100 = 3.6% per generation.

Problem 4

The following data refer to the R^0 allele in the Rh blood system:

Frequency in Western European = 0.62

Frequency in Eastern European = 0.45

Frequency in Mongols = 0.03.

What is the total proportion of alleles that have entered the Eastern European populations?

Solution:

We know, $p_1 = (1 - m)p_0 + mp_m$
$= mp_m + p_0 - mp_0$

or, $p_1 - p_0 = m(p_m - p_0)$

or, $m = \frac{p_1 - p_0}{p_m - p_0}$.

p_1 = frequency of conglomerate population

p_0 = frequency of native population

p_m = frequency of migrant population

m = migration rate

Here, $p_m = 0.03$, $p_0 = 0.62$, $p_1 = 0.45$.

$$m = \frac{p_1 - p_0}{p_m - p_0} = \frac{0.45 - 0.62}{0.03 - 0.62} = \frac{-0.17}{-0.59} = 0.288.$$

Therefore, 0.288 entered into Eastern European population.

Problem 5

In a population of nine hundred butterflies, the frequency (p) of the fast allele of enzyme phosphoenol pyruvate is 0.6 and the frequency of the slow form (q) is 0.4. Ninety butterflies migrate to this population and the migrants have a slow allelic frequency of 0.8. Calculate the allelic frequencies of the new population.

Solution:

Frequency of fast allele 'p' is 0.6 and slow allele 'q' is 0.4

Among 900 butterflies p has $0.6 \times 900 = 540.0$

Total alleles are $540 \times 2 = 1080$.

Slow alleles are $0.4 \times 900 = 360.0 = 360 \times 2 = 720$

Migrants have slow allelic frequency 0.8. Total number of migrants 90.

Total $90 \times 0.8 = 72.0$. Total alleles are $72 \times 2 = 144$

Fast allelic frequencies 0.2, $90 \times 0.2 = 18.0$. Alleles $18 \times 2 = 36$.

Allelic frequency of new population will be

$$p^1 = \frac{0.1080 + 36}{1800 + 180} = \frac{1116}{1980} = 0.56$$

$$q^1 = \frac{720 + 144}{1800 + 180} = \frac{864}{1980} = 0.436 = 0.44.$$

Problem 6

In a particular population, the frequency of allele t was 0.25 in a migrant population and 0.45 in the conglomerate population. If the migration rate was 0.1, calculate the frequency of t in the original native population.

Solution:

$$m = \frac{p_1 - p_0}{p_m - p_0}$$

p_1 = frequency of conglomerate population

p_0 = frequency of native population

p_m = frequency of migrant population

m = migration rate

$m = 0.1$, $p_m = 0.25$, $p_1 = 0.45$

$$0.1 = \frac{0.45 - p_0}{0.25 - p_0} = 0.1(0.25 - p_0) = 0.45 - p_0,$$

or, $0.025 - 0.1p_0 = 0.45 - 0.025$.

$p_0(1 - 0.1) = 0.425$, or, $0.9p_0 = 0.425$

$$p_0 = \frac{0.425}{0.9} = 0.472.$$

Problem 7

A population of 80 adult squirrels resides on campus and the frequency of the Est′ alleles among these squirrels is 0.70. Another population of squirrels is found in a nearby woods and the frequency of Est′ allele is 0.5. During a severe winter, 20 of the squirrels from the woods population migrate to campus in search of food and join the campus population. What will be the allelic frequency of Est′ in the campus population after migration?

Solution:

Frequency of Est′ allele in campus population is 0.7.

Frequency of Est′ allele in nearly wood's population is 0.5.

Among 80 adults squirrel population Est′ (p) has $80 \times 0.7 = 56.0 = 56$.

Total number of alleles are $56 \times 2 = 112$.

Number of migrant from wood's population is 20.

Est′ allele frequency in wood's population is 0.50.

Among 20 migrant Est′ has $0.5 \times 20 = 10.0 = 10$.

Total number of allele are $10 \times 2 = 20$.

After migration total number of alleles are 112 + 20 = 132.

Total population 80 + 20 = 100.

Total number of alleles are $100 \times 2 = 200$.

Allelic frequency of Est′ after migrations is $\frac{132}{200}$ = 0.66. Here total population (native + migrant) is 80 + 20 = 100.

Therefore, proportion of migrant is 0.20 and native is 0.80.

The frequency native population, i.e., $p_0 = 0.70$ and migrant population is $p_m = 0.50$.

The frequency of conglomerate population is $p'' = mp_m + (1 - m)p_0$.

[Here $p_0 = 0.70$, $p_m = 0.5$, $m = 0.20$, $1 - m = 0.80$.]

$$p'' = 0.20 \times 0.5 + (1 - 0.20) \times 0.7$$
$$= 0.100 + 0.80 \times 0.7 = 0.1 + 0.560 = 0.66.$$

Problem 8

In a donor population, the allele frequency for the normal (Hb^A) and sickle cell alleles (Hb^s) are 0.9 and 0.1 respectively. A group of 550 individuals migrate to a new population containing 10,000 individuals; in the recipient population, the allele frequencies are $Hb^A = 0.99$ & $Hb^S = 0.01$.

(a) Calculate the allele frequencies in the conglomerate population.

(b) Assuming that the donor and recipient populations are each in Hardy-Weinberg equilibrium, calculate the genotype frequencies in the conglomerate population prior to further mating between the donor and recipient population.

(c) What will be the genotype frequencies of the conglomerate population in the next generation assuming that it achieves Hardy-Weinberg equilibrium in one generation?

Solution:

Change of gene frequency in conglomerate population will occur after receiving the migrant population, i.e., ΔP_C with regard to sickle cell allele

(a) $\Delta P_C = m(P_D - P_R)$

[P_D = frequency of donor popular,
P_R = frequency of receiving popular]

$$\Delta P_C = \frac{550}{10550}(0.1 - 0.01)$$
$$= 0.052\,(0.09) = 0.00468 = 0.0047.$$

Frequency of conglomerate population is
$P_C = P_R + \Delta P_C = 0.01 + 0.0047 = 0.0147.$

(b) We have to calculate genotype frequency separately.

For 550 migrant individual $\frac{Hb^A}{Hb^A} = (0.9)^2 = 0.81$, i.e., 81%.

Among 550 individuals it will be (0.81 × 550) = 445.5 individuals have this genotype.

$\frac{Hb^A}{Hb^S}$ will be 2 × 0.9 × 0.1 = 0.18.

Among 550 individuals 0.18 × 550 = 99 individuals are heterozygous.

$\frac{Hb^S}{Hb^S}$ willl be $(0.1)^2 = 0.01$, in 550 it will be 0.01 × 550 = 5.5 have $\frac{Hb^S}{Hb^S}$ genotype.

Original recipient population is

$\frac{Hb^A}{Hb^A} = (0.99)^2 = 0.9801$, in 10,000 it will be 9801 individuals.

$\frac{Hb^A}{Hb^S} = 2 \times 0.99 \times 0.01 = 0.198$, in 10,000 it will be 198 individuals in this genotype.

$\frac{Hb^S}{Hb^S} = (0.01)^2 = 0.0001$, in 10,000 it will be individual.

In overall conglomerate population is

$$\frac{Hb^A}{Hb^A} = \frac{445.5 + 9801}{10,550} = \frac{10246.5}{10,550} = 0.971\left(\frac{Hb^A}{Hb^A}\right)$$
$$\frac{Hb^S}{Hb^S} = \frac{5.5 + 1}{10,550} = \frac{6.5}{10,550} = 0.000616 = 0.00062$$
$$\frac{Hb^A}{Hb^S} = \frac{99 + 198}{10,550} = \frac{297}{10,550} = 0.0281 = 0.028 \text{ heterozygous.}$$

After one round of mating, the allele frequency in the conglomerate population will be

Allele frequency $Hb^S = 0.0147$ and $Hb^A = 1 - 0.0147 = 0.985$.

$$\frac{Hb^A}{Hb^A} = (0.985)^2 = 0.97$$
$$\frac{Hb^A}{Hb^S} = 2 \times 0.985 \times 0.0147 = 0.0289 = 0.029$$
$$\frac{Hb^S}{Hb^S} = (0.0147)^2 = 0.0002.$$

(c) $(0.97 + 0.0002)^2$ = 94.09% + 0.0388% + 0.000004%

Problem 9

State the effect of gene flow on the amount of genetic variation in a population.

Solution:

(i) (a) In general it reduces genetic differences between population.

(b) It increases genetic variations within population.

(ii) If an allele is missing from a population, gene flow from a polymorphic population can introduce it to the population and therefore, increase genetic variation.

(iii) If the population is polymorphic, then gene flow from a monomorphic population can reduce the level of genetic variation.

Problem 10

How does genetic variation depends upon the level of gene flow and allele frequency in migrants?

Solution:

(i) A high level of gene flow can speed up these changes. It can rapidly change allelic frequencies.

(ii) Low level of gene flow may cause the changes to occur over many generation.

(iii) Gene flow is greater among more mobile organism.

Problem 11

Highlight gene flow and distribution pattern of population.

Solution:

(a) **Island model:**

(i) According to this model the population of a species is divided into a series of ideal subpopulation.

(ii) Each subpopulation contributes equal number of genes to a migrant pool.

(b) **Stepping stone model:**

(i) In natural population, migration is greater between the subpopulations that are closer to each other.

(ii) It rules out equal probabilities of gene exchange.

(c) **Continuously distribution model:**

(i) Same populations enjoy continuous distribution across large landscape, they are not subdivided.

(ii) These populations exhibit gene flow relatively at short distances.

Industrial melanism:

It is an classical example of directional selection from British peppered moth (*B. betularia* and *B. carbonaria*) in response to industrial revolution.

The effect of soot and pollution in industrial areas is increasing the frequency of darkly pigmented (*Biston carbonaria*) forms because of selection by predators against the nonpigmented or lightly pigmented forms.

A. In sooty environment

The melanic variety (*B. carbonaria*) could not survive. They were conspicuous on lichen covered trunk. As a result birds could detect them on lichens and were devoured by them. Therefore, they could not increase their population.

B. In sooty environment

With the elimination of lichen carbonaria varieties (melanic) because abundantly distributed. On the other hand, *B. bitularia* were easily detected on black environment by bird. Their populations were decreased.

Explanator:

(i) The difference between *betularia* and *carbonaria* is due to single dominant gene mutation from normal c to mutated C. Both homozygous (CC) and heterozygous (Cc) are black.

(ii) Selection coefficent (s): i.e., factor bird predator and industrialization (black envionment), so *bitularia* were lost and *carbonaria* survived.

(iii) Here directional selection leads to the change of the gene pool of peppered moth (*Biston*) tiggered by changes in the environment.

(iv) It is a kind of density dependent selection because camouflaged variety increases.

GENETIC DRIFT

Genetic drift:

Genetic drift may be defined as random variation of gene frequencies from generation to generation due to sampling error in population.

(i) It is one of the most nondirectional forces or factors arisen from variable sampling of the gene pool which influences the gene frequency.

(ii) Smaller population has more sampling errors, which may accumulate in successive generation. So that frequency of a variant may become higher or lower significantly.

(iii) A sample of population is always taken for an experiment. This sample of population although representative of the population, may not have exactly the same gene frequency as found in the population. Such deviation in gene frequency will be due to sampling errors.

(iv) On the other hand several populations are limited in size, genetic drift will cause changes in gene frequency because of sampling errors.

Steps:

(i) If a few parents are chosen to begin a population the gene frequency of this new population may deviate widely from the gene frequencies of the original population because in this case the sampling error will be high.

(ii) If the sample size is large, i.e., large number of parents are taken to begin a new population deviation in the gene frequencies may not be so large.

Calculation:

The extent of deviation for all sizes of population can be measured mathematically by the standard deviation.

$\sigma = \sqrt{\frac{pq}{2N}}$ where p is the frequency of one allele (A), q is the frequency of other allele (a) and N is the number of gene sampled.

For diploid organism (parents) each carrying two alleles

$\sigma = \sqrt{\frac{pq}{N}}$ where N is the number of actual parent

Illustration:

If there is a population in which A and a are alleles, assume that the frequency of A is equal to that of a. If p is the gene frequency of A and q is the gene frequency of a, then p = q = 0.5.

Now, take a large population of 5000 individuals in one hand and a small population of 2 individuals on the other.

(a) $\sigma = \sqrt{\frac{pq}{2N}} = \sqrt{\frac{0.5 \times 0.5}{2 \times 5000}} = \sqrt{\frac{0.25}{10,000}} = \sqrt{.000025}$

$= 0.005.$

The values of such population will, therefore, fluctuate mostly around 0.5 ± 0.005 or between 0.495 and 0.505.

(b) $\sigma = \sqrt{\frac{pq}{2N}} = \sqrt{\frac{0.5 \times 0.5}{2 \times 2}} = \sqrt{\frac{0.25}{4}} = \sqrt{.0625} = 0.25.$

The values of such population will, therefore, fluctuate mostly around 0.5 ± 0.25 or 0.25 to 0.75.

(c) Smaller population size can easily yield gene frequencies that depart considerably from the initial 0.5 value in a single generation. If the population remains small and the next generation begins with either of these extremes, i.e., gene frequencies of 0.25 or 0.75 for a particular allele, then

$\sigma = \sqrt{\frac{pq}{2N}} = \sqrt{\frac{0.25 \times 0.75}{2 \times 2}} = \sqrt{\frac{0.1875}{4}} = \sqrt{0.0468}$

$= 0.2163 = 0.22.$

The values of such population will, therefore, fluctuate mostly around 0.25 ± 0.22 (0.03 to 0.47) or 0.75 ± 0.22 (0.53 to 0.97).

Population bottleneck:

A population bottleneck (genetic bottleneck) is an evolutionary event in the form of genetic drift in which a significant percentage of a population is reduced in size. It represents a phenomenon of severe temporary drop of population. **Population**

crashes (species is either killed or prevented from reproduction) and later expands by steep increase, i.e., **population flushes** causing great fluctuation in allele frequencies. This change of population size resembles the neck of a bottle from wide to narrow, hence the name. The enlarge populations that may have gene frequencies are distinctly different from those before bottleneck.

Features:

(i) Population bottleneck leads to genetic bottleneck.

(ii) During genetic bottleneck drift in allele frequencies occurs. It reduces genetic variability in small population that persisted in the original population for generation.

(iii) The reduced variability in the gene pool of population makes its gene pool different from sister gene pool.

(iv) Fluctuation of population density may be annual long term or sporadic and provides a pivotal point for the change of gene frequencies (by genetic modification or gene elimination in the gene pool of the population).

(v) During **population flush** (too many individuals are present) the population spreads out far and wide and is exposed to different microenvironment. Therefore, it has more chances to accumulate mutation in its gene pool.

(vi) During **population crash** the number of indviduals may reduce to the extent that the small demes constituting the population become isolated and restricted in distribution. These are then exposed to random genetic drift resulting in the fixation of certain genes.

(vii) This period of crisis with low population density represents "**bottleneck**" and changes of gene frequencies during the bottleneck period constitute "**bottleneck effect**". It may result elimination of certain alleles.

Example: Cheetas population suffering from bottleneck, until 10,000 years ago they were widely spread out. But presently they are restricted to South and East Africa. Members of each populations are genetically so much alike. It is assumed that they have developed from a very few founder member.

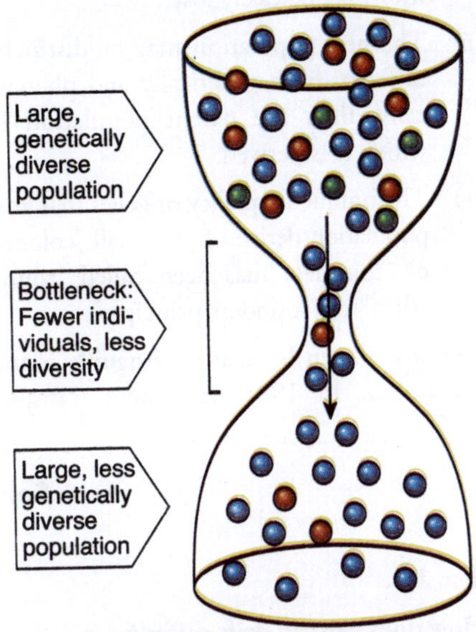

(A) Bottleneck effect

(B) An African cheetah

Fig. 18.3: Bottleneck effect

Founder effect:

A change in allele frequencies that occur when new population is establishd in a new area is called founder effect.

Features:

(i) The establishment of a new population by a few individuals or a few original founders that carry only a small fraction of the total genetic variation of the parental population or only a limited portion of parental gene pool.

(ii) In the small group, some genes may be absent or occur in such frequency that they may be easily lost.

(iii) The new population may be distinctively different both genetically and phenotypically, than the parent population from which it is derived.

(iv) The unique frequency of genes that arise in population derived from small "colonizers" or "founders" has been called "founders effect" or "founders principle".

Example: Green Iguanas on Anguilla Island in the Caribbean. In 1995 they were transported by hurricane from island of Guadeloupe. They multiplied and formed founder population on Anguila Island.

Significance of Founder effect:

(i) The isolated populations differ from source population in the frequencies of their alleles.

(ii) The genes or chromosomes arrangements, carried by the founders have chance of becoming established in new population.

(iii) Presence of high frequency of certain inherited disorders among isolated human population occurs one to the founder effect and inbreeding.

Solved Problems

Problem 1

How does genetic drift differ from natural selection?

Solution:

Genetic drift	Natural selection
(i) It operates in small population.	(i) It operates in large population.
(ii) It is nondirectional, random and unpredictable.	(ii) Its operation is directional and predictable.
(iii) Genetic drift does not support adaptive changes in the frequencies of alleles.	(iii) Natural selection favours in the frequency of those alleles that provide reproductive advantage and leads to adaptive evolution.
(iv) It is non-selective. Fluctuations in allele frequencies caused by genetic drift occur by chance.	(iv) It favours gene combinations that are most suited for the given environment. Gene frequency do not fluctuate, but change in a specific direction.
(v) Genetic drift leads to the loss of genetic variation and favours homozygosity by fixing or eliminating the other allele.	(v) Natural selection supports heterozygosity in the population because heterozygous superiority. It encourages and operates genetic variability.
(vi) It basically gives a head start to the founders of the isolated marginal population and accelerates speciations.	(vi) It favours good variant group and accelerates speciation though reproductive isolation.

Problem 2

A population of *Drosophila melanogaster* is polymorphic for two alleles A_1 and A_2. One thousand populations are derived and each is maintained by selecting random ten females and ten males in each generation as parents of the following generations. After many generations it is observed that 220 populations are fixed for allele A_1 and 780 for allele A_2. Estimate the allele frequencies in the original population, assuming that only genetic drift is involved.

Solution:

Here only the genetic drift is involved.

$$A_1 = p = \frac{220}{1000} = 0.220,\ A_2 = q = \frac{780}{1000} = 0.780.$$

Problem 3

A population of *D. melanogaster* is polymorphic for two alleles A_1 and A_2. One thousand populations are derived and each is maintained by selecting random ten females and ten males in each generations as parents of the following generations. After many generations it is observed that 220 populations are fixed for allele A_1 and 780 for allele A_2. What will be the range of allelic frequencies in the first generation among 1000 populations?

Solution:

$$A_1 = P_1 = \frac{220}{1000} = 0.220$$

$$A_2 = q = \frac{780}{1000} = 0.780;$$

Male	10
Female	10
Total (N)	20

$$\sigma = \sqrt{\frac{pq}{2N}} = \sqrt{\frac{0.220 \times 0.780}{2 \times 20}} = \sqrt{\frac{0.1716}{40}} = \sqrt{0.00429}$$

$$= 0.06549 = 0.0655.$$

Expected range: 0.220 ± 0.655

$0.780 \pm 0.655.$

Problem 4

Suppose that the alleles of the T genes are selectively neutral. In a population of 50 individuals, currently 34 are heterozygotes. Predict the frequency of heterozygotes in this population after ten generations in the future. Assume that the population size is constant and that mating is completely random (including the possibilities of self-fertilization).

Solution:

For a selectively neutral gene, evolution occurs by random genetic drift. The governing equation is

$$H_t = \left(1 - \frac{1}{12N}\right)^t H,$$

H_t = The frequency of heterozygote for t generations in future.

N = The population size

H = The frequency of population size now

$$N = 50, \quad H = \frac{34}{50} = 0.68, \quad t = 10$$

$$H_t = \left(1 - \frac{1}{2 \times 50}\right)^{10} \times 0.68 = \left(\frac{100-1}{100}\right)^{10} \times 0.68$$

$$= \left(\frac{99}{100}\right)^{10} \times 0.68 = (0.99)^{10} \times 0.68$$

$$= 0.9043 \times 0.68 = 0.6149 = 0.615.$$

Problem 5

Name four evolutionary processes that can change allele frequencies in natural populations? How do allele frequencies change in the absence of these processes?

Solution:

Mutation, migration, natural selection and random genetic drift. In the absence of these processes, the allele frequencies do not change.

Problem 6

What is random genetic drift and why does it occur? In the absence of any counteracting forces, what is the ultimate effect of random genetic drift on allele frequency? Are random changes in allele frequency from one generation to the next generation in small populations or in large ones and why?

Solution:

- Random genetic drift refers to changes in allele frequency that happen by chance in small populations.
- It occurs because the zygotes are actually formed but a small sample of a potentially infinite number and most samples do not exactly match the population from which they come to but differ to a greater or lesser extent because of chance.

  The ultimate fate of an allele when only random genetic drift is operating is either fixation or loss.

  Random changes in allele frequency are larger in small population because statistical fluctuations are greater in smaller samples.

Problem 7

If the genotype AA is an embryonic lethal and the genotype aa is fully viable and sterile, what genotype frequencies would be found in adults in an equilibrium population containing

the A and a alleles? Is it necessary to assume random mating?

Solution:

Adults $\frac{2}{3}$Aa and $\frac{1}{3}$aa (AA:Aa:aa, i.e., 1:2:1 but AA is lethal, therefore, $\frac{2}{3}$Aa and $\frac{1}{3}$aa).

It is not necessary to assume random mating, because the only fertile matings are Aa × Aa.

These produce $\frac{1}{4}$AA, $\frac{1}{2}$Aa and $\frac{1}{4}$aa zygotes, but the AA zygotes die leaving $\frac{2}{3}$Aa and $\frac{1}{3}$aa adults.

Problem 8

In a popultion of 100 individuals with q = 0.4, what are V and approximate value of Δq after one generation? How does the value of Δq compare to that for mutation?

Solution:

The expected variance (V) in allelic frequency over replicate population due to one generation of genetic drift is $V = \frac{pq}{2N}$.

Here q = 0.4, Therefore, p will be 0.6, N = 100

$$V = \frac{0.6 \times 0.4}{2 \times 100} = \frac{0.6 \times 0.2}{100} = \frac{0.12}{100} = 0.0012.$$

$$(\Delta q) = \sqrt{V} = \sqrt{.0012} = 0.0346 = 0.035.$$

Problem 9

Why genetic drift is more significant in small population? Why does it take longer for genetic drift to cause allele fixation in large population than in small one?

Solution:

(i) Genetic drift is due to sampling error and the degree of sampling error depends on the population size.

(ii) In small populations, the relative proportion of sampling error is much larger.

(iii) If genetic drift is moving an allele towards fixation, it will take longer in large population because the degree of sampling error is much smaller.

Problem 10

With regard to genetic drift, are the following statements true or false, if statement is false, explain why?

(a) Over the long run genetic drift will lead to allele fixation or loss. (b) When new mutation occurs within a population, genetic drift is more likely to cause the loss of less new allele rather than the fixation of the new allele. (c) Genetic drift is more significant in small population. (d) Genetic drift promotes genetic diversity in large population.

Solution:

(a) True, (b) True, (c) True, (d) False, It causes allele loss or fixation which results in less diversity.

Problem 11

Describe what happens to allele frequencies during the bottleneck effect. Discuss the relevance of the effect with regard to species that are approaching extinction.

Solution:

(i) During the bottleneck effect, allele frequencies are dramatically altered due to genetic drift.

(ii) In extreme cases, some alleles are lost, while others may become fixed at 100%.

(iii) The overall effect is to decrease genetic drift diversity within the population.

(iv) This may make it more difficult for the species to respopnd in a positive way to change in the environment.

(v) Species that are approaching extinction also face a bottleneck as their numbers decrease. The loss of genetic diversity may make it even more difficult for the species to rebound.

Problem 12

What is the effect of small population size of the amount of genetic variation in the population over a period of time? Discuss the effect on fixation of alleles and the level of homozygosity of the population.

Solution:

A small population size results in genetic drift, a factor that increases the fixation of alleles in a population and thereby reduces the heterozygosity.

EXERCISE

1. **Which of the following genotype frequencies of AA, Aa and aa respectively satisfy the Hardy-Weinberg principle?**

	AA	Aa	aa
(a)	0.29	0.42	0.29
(b)	0.64	0.27	0.09
(c)	0.49	0.42	0.09
(d)	0.36	0.55	0.09
(e)	0.25	0.50	0.25

2. **How many A and a alleles are present in a sample of organisms that consist of 10AA, 15Aa and 4aa individuals? What are the allele frequencies in this sample?**

3. **Hartnup disease is an autosomal recessive disorder of intenstinal and renal transport of amino acids. The frequency of affected newborn infants is about 1 in 14,000. Assuming random mating, what is the frequency of heterozygotes?**

4. **If in red-green colour blindness, a recessive sex-linked trait is present in 2 percent in males in a population with random mating, what is the frequency of carrier females?**

5. **In a population of *Drosophila*, an X-linked recessive allele causing yellow body colour is present in genotypes at frequencies typical of random mating, the frequency of recessive allele is 0.20. Among 1000 males and 1000 females, what are the expected numbers of the yellow and wild type phenotypes in each sex?**

6. **Among 35 individuals of the flowering plant *Phlox roemariana*, the following genotypes were observed for a gene determining electrophoretic forms of the enzyme phosphoglucose isomerase:**
2 AA, 13 AB, 20 BB.
   (a) What are the frequencies of the alleles A and B?
   (b) Assuming random mating, what are the expected number of the genotypes?

7. **A randomly mating popultion of dairy cattle contains an autosomal recessive allele causing dwarfism. If the frequency of dwarf calves is 10 percent, what is the frequency of heterozygous carriers of the allele in the entire herd? What is the frequency of heterozygotes among nondwarf cattle?**

8. **Allozyme phenotypes of alcohol dehydrogenase in the flowering plant *Phlox drummondii* are determined by codominant alleles of a single gene. In one sample of 35 plants, the following data were obtained:**

Genotype:	AA	AB	BB	BC	CC	AC
Number:	2	5	12	10	5	1

What are the frequencies of the alleles A, B and C in this sample?

9. **If the frequency of a homozygous dominant genotype in a randomly mating population is 0.09, what is the frequency of the dominant allele? What is the combined frequency of all other alleles of this gene?**

10. **In some regions of Ghatshila, the frequency of the Hb^s allele is 0.2. If this frequency is the result of a dynamic equilibrium due to the superior fitness of Hb^sHb^A heterozygotes, and if Hb^sHb^s homozygotes are essentially lethal what is the intensity of selection against the Hb^AHb^A homozygote.**

11. **Assume that a sex-linked recessive trait is observed in 25% of all the males. Apply Hardy-Weinberg calculations.**

12. **Assume that the gene frequency of the recessive i allele for the O blood type is 0.67 and that the frequency of the dominant Rh^+ allele is 0.60. What then would you expect for the frequency of persons having both blood type O and the Rh^+ blood factor?**

13. **Consider sex-influenced trait that is dominant in males (R) and recessive in females (r). In a population of 6000 males, 4850 were found to be r. How many females would be expected to be r in this population?**

14. Consider a hypothetical population in which you find 32.5% of the individuals to be homozygous ww. If the population is in Hardy-Weinberg equilibrium, what are the other genotypic frequencies?

15. About 90% of a population are tongue curlers, the allele for tongue curling is dominant. What are the gene and genotype frequencies of T and t?

16. The students of Baidyabati Charushila Bose Balika Vidyalaya are tested of phenylthiocarbamide (PTC). The ability to taste this chemical is dominant (T) over the inability to taste it (t). Among 1242 students, 67.8% were found to be tasters. What are the expected frequencies of the two alleles?

17. (a) If the frequency of individuals having a dominant trait in a population is $\frac{1}{100}$, what is the frequency of the dominant allele responsible for the trait?

(b) If the frequency for a certain autosomal recessive trait in a population is $\frac{1}{100}$, what is the frequency of the recessive allele involved in the occurrence of this trait?

18. From a sample of 361 American Indians, the following blood types were obtained. M 305, MN 52 and N 4, calculate gene frequency of M and N. Test data for H-W equilibrium.

19. The number of people in Arambagh town having the different MN allelic combinations for that blood systems were as follows:

MM 812, MN 1488, NN 664.

It was also determined that of the 1482 married couples, the following distribution existed.

(i) MM × MM	**=**	**116**
(ii) MM × MN	**=**	**404**
(iii) MN × MN	**=**	**380**
(iv) MM × NN	**=**	**176**
(v) MN × NN	**=**	**324**
(vi) NN × NN	**=**	**82**
	Total =	**1482**

(a) Show whether the population is in Hardy-Weinberg equilibrium with regard to the MN blood types.

(b) Show whether mating is occurring randomly among the married couples with regard to the MN blood types.

20. A particular abnormal trait in humans is suspected to be due to a recessive allele u, and the normal allele U is dominant. If the frequency of the u allele in a small country is 0.40,

(a) what is the probability of an individual having this recessive characteristic in this population?

(b) what is the probability of an individual who expresses dominant phenotype being heterozygous?

(c) what is the probability of both parents being heterozygous?

21. Observe the following hypothetical population:

Population	**DD**	**Dd**	**dd**
		Genotypes	
1.	**200**	**600**	**200**
2.	**0**	**1000**	**0**
3.	**498**	**2**	**498**
4.	**250**	**500**	**250**

(a) What are the gene frequencies of each of these population?

(b) Which of these populations, if any, is in Hardy-Weinberg equilibrium?

(c) Assume that no forces exist which would disturb these populations and that random mating now occurs, what will be genotypic frequencies among the progeny of the next generation?

22. The adult population of some villages of Orissa that have been heavily exposed to malaria have a composition in which 20% of the individuals are heterozygous for thalassaemia (Tt). The allele when homozygous, gives rise to Cooley's anaemia, this disease is lethal in the first years of life with practically 100% penetrance. Assuming random

mating, calculate the proportion of children in the next generation who will die from Cooley's anaemia.

23. Consider a recessive allele whose frequency in the general population is 1 in 10. You and your spouse do not have the trait, but you do not know whether either of you are the heterozygous for the allele.
    (a) What are the chances that you are a carrier for this allele?
    (b) What are the chances that both you and your spouse are carriers for this allele?
    (c) Since you do not know the genotypes of yourself and your spouse, what can you say about the chances that your child will actually have the trait?

24. Among 330 students of Serampore Girls College the following M, N blood types are obtained.

    MM 89, MN 162 and NN 79.

    Calculate gene frequency of M and N.

25. Phenylketonuria (PKU) is a biochemical anomaly in humans that can cause neurological problems. The genetic disorder is due to autosomal recessive inheritance and occurs in every 14,000 newborns approximately. A college professor asked his genetics class to calculate the number of heterozygous for this disorder among 62 students in his class.

26. The illustration at the top of the column at the right shows the gel patterns observed with a probe for a VNTR (variable number of tandem repeats) locus among four pairs of parents (1 – 8) and four children (A – D). One child comes from each pair of parents.
    (a) Why does each person has two bands?
    (b) How would it be possible for a person to have only one band?
    (c) What type of dominance is illustrated by the VNTR alleles?
    (d) Which pairs of people having the DNA in lanes 1 through 8 are the parents of each child?

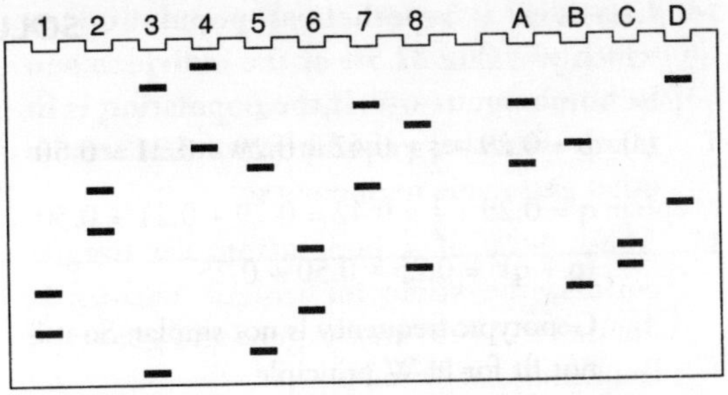

27. DNA from 100 unrelated people was digested with the restriction enzyme Hind III and the resulting fragments separated and probed with a sequence for a particular gene. Four fragment lengths that hybridized with the probe were observed—namely 5.7 kb, 6.0 kb, 6.2 kb, and 6.5 kb—each fragment defining a different restriction-fragment allele. The adjoining illustration shows the gel patterns observed, with the number of individuals with each combination shown across the top. Estimate the allele frequencies of the four restriction-fragment alleles.

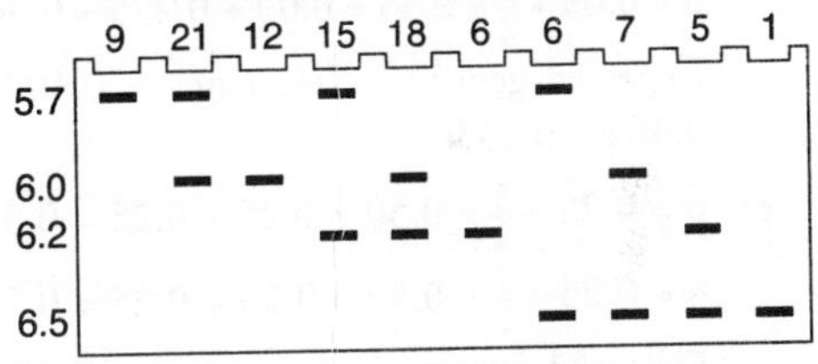

28. Xeroderma pigmentosum (XP) is often fatal skin cancer resulting from a recessive mutant allele that affects DNA repair. In north eastern region of India, the frequency of homozygous recessive affected people is approximately 1 in 250,000 (the mutant allele can be any one of about eight different genes).
    (a) What is the expected frequency of XP among the offsprings of first cousin matings.
    (b) What is the ratio of XP among the offsprings of first cousin matings to that among the offsprings of non-relatives?

SOLUTION

1. (a) $p = 0.29 + \frac{1}{2} \times 0.42 = 0.29 + 0.21 = 0.50$

$q = 0.29 + \frac{1}{2} \times 0.42 = 0.29 + 0.21 = 0.50$

$(p + q)^2 = 0.25 + 0.50 + 0.25$

Genotypic frequency is not similar. So it is not fit for H-W principle.

(b) $p = 0.64 + \frac{1}{2} \times 0.27 = 0.64 + 0.135 = 0.775$

$q = 0.09 + \frac{1}{2} \times 0.27 = 0.09 + 0.135 = 0.225$

Expected genotypic frequency = $(0.775 + 0.225)^2 = 0.601 + 0.349 + 0.051$.

It is fit for H-W principle.

(c) $p = 0.49 + \frac{1}{2} \times 0.42 = 0.49 + 0.21 = 0.7$

$q = 0.09 + \frac{1}{2} \times 0.42 = 0.09 + 0.21 = 0.3$

Expected genotypic frequency: $(0.7 + 0.3)^2 = 0.49 + 0.42 + 0.09$.

It is fit for H-W principle.

(d) $p = 0.36 + \frac{1}{2} \times 0.55 = 0.36 + 0.275 = 0.635$

$q = 0.09 + \frac{1}{2} \times 0.55 = 0.09 + 0.275 = 0.365$

Expected genotypic frequency = 0.403 + 0.464 + 0.133.

(e) $p = 0.25 + \frac{1}{2} \times 0.50 = 0.25 + 0.25 = 0.5$

$q = 0.25 + \frac{1}{2} \times 0.50 = 0.25 + 0.25 = 0.5$

Expected genotypic frequency = 0.25 + 0.50 + 0.25.

2.

	AA	Aa	aa	Total
	10	15	4	29
Genotypic frequency:	$\frac{10}{29}$ = 0.344	$\frac{15}{29}$ = 0.517	$\frac{4}{29}$ = 0.138	

$A = 0.344 + \frac{1}{2} \times 0.517 = 0.344 + 0.259 = 0.603 = 0.6$

$a = 0.138 + \frac{1}{2} \times 0.517 = 0.138 + 0.259 = 0.397 = 0.4$.

3. A genotypic frequency is $\frac{1}{14,000} = 0.000074$, i.e., $q^2 = 0.00074$

$\therefore$ $q = 0.00844 = 0.0085$;

$p = 1 - 0.0085 = 0.9915$.

The frequency of heterozygote is $2 \times 0.0085 \times 0.9915 = 0.01685 = 0.0168 = 0.017$ (about 1 in 60)

4. Let frequency of male (XY) = q and frequency of female (XY) = q^2.

$\therefore$ $q = \frac{2}{100} = 0.02$

The expected frequency of the trait in female is $(0.02)^2 = 0.0004$.

The expected frequency of heterozygotes females is $2pq = 2 \times 0.98 \times 0.02 = 0.0392$.

$(p = 1 - 0.02 = 0.98)$.

5. (i) The frequency of yellow allele y is 0.2.

In females, the expected frequencies of wild type $\left(\frac{+}{+}/\frac{y}{+}\right)$ and yellow $\left(\frac{y}{y}\right)$ are $(0.8)^2 + 2 \times 0.2 \times 0.8 = 0.64 + 0.32 = 0.96$ and $(0.2)^2 = 0.04$ respectively.

(ii)

	Wild type	Yellow body
Among 1000 females =	.96 × 1000 = 960	.04 × 1000 = 40

(iii) In males expected frequencies are very different. The expected frequencies of wild type (+) and yellow (y) are 0.8 and 0.2 respectively.

(iv)

	Wild type	Yellow body
Among 1000 males =	0.8 × 1000 = 800	0.2 × 1000 = 200

6. (a)

	AA	AB	BB	Total
	2	13	20	35
Genotypic frequency:	$\frac{2}{35}$ = 0.057	$\frac{13}{35}$ = 0.371	$\frac{20}{35}$ = 0.571	

Gene frequency of allele

$A = 0.057 + \frac{1}{2} \times 0.371 = 0.057 + 0.1855 = 0.057 + 0.185 = 0.242 = 0.24$

$a = 0.571 + \frac{1}{2} \times 0.371 = 0.571 + 0.185 = 0.756 = 0.76$.

(b) Expected number of genotypes

$(0.24 + 0.76)^2 = (0.24)^2 + 2 \times 0.24 \times 0.76 + 0.76^2$
$= 0.0576 + 0.3648 + 0.5776$
$= 0.058 + 0.365 + 0.8578.$

[AA = 0.058, AB = 0.365, BB = 0.578]

7. Frequency of homozygous recessive is 10%, i.e., 0.10 or, $q^2 = 0.10$.

   Therefore, the frequency of recessive allele is $q = \sqrt{0.10} = 0.316$; $p = 1 - 0.316 = 0.684$. So the frequency of heterozygotes is $2 \times 0.684 \times 0.316 = 0.432 = 0.43$. 90 percent of the individuals are nondwarf. Therefore, the frequency of heterozygous among nondwarf is $\frac{0.43}{0.90} = 0.4777 = 0.48$.

8. The number of alleles of each type are:

   $A = 2 + 2 + 5 + 1 = 10$

   $B = 5 + 12 + 12 + 10 = 39$

   $C = 10 + 5 + 5 + 1 = 21$

   Total $= 70$

Frequencies of alleles: A	B	C
$\frac{10}{70} = .142$	$\frac{39}{70} = .557$	$\frac{21}{70} = .3$
A = 0.14	B = 0.56	C = 0.3

9. Frequency of homozygous dominant is 0.9, i.e. $p^2 = 0.09$, $\therefore p = \sqrt{0.09} = 0.3$.

   All the other alleles have a combined frequency of $1 - 0.30 = 0.70$.

10. $p = 0.2$, at equilibrium $p = \frac{t}{1+t}$

    $\therefore p = \frac{t}{1+t}$, $\because s = 1$.

    $\therefore 0.2 + 0.2t = t$, or, $t - 0.2t = 0.2$, or, $0.8t = 0.2$

    $\therefore t = \frac{0.2}{0.8} = 0.25$.

11. $q = 0.25$ and $p = 1 - 0.25 = 0.75$.

    Therefore, these p and q frequencies represent both the gene frequencies for the population and the genotype frequencies for the males, with these values, the calculation for the female genotypes can be completed.

    $P^2 = (0.75)^2 = 0.56$

    $2pq = 2 \times 0.75 \times 0.25 = 0.375$

    $q^2 = (0.25)^2 = 0.0625 = 0.063$.

12. $q(i) = 0.67$ and $q^2(ii) = (0.67)^2 = 0.4489 = 0.449$

    For $Rh^+ = 0.60$

    $q(Rh^-) = 1 - 0.60 = 0.40$.

    Therefore, $q^2 = Rh– Rh– = (0.40)^2 = 0.16$.

    Now applying the product rule of probability,

    $0.449 \times 0.84 = 0.377$

    $0.377 \times 100 = 37.700 = 37.7\%$.

    37.7% of the individuals in this populations are expected to have O, Rh^+ blood types.

13. In the males, $q^2 = \frac{4850}{6000} = 0.808$.

    $\therefore q = \sqrt{0.808} = 0.899$.

    $p = 1 - 0.899 = 0.101$

    $2pq = 2 \times 0.899 \times 0.101 = 0.1815 = 0.182$

    In the females, q^2 and $2pq$ would have the r phenotype, because of the sex influenced mode of inheritance, hence $0.808 + 0.182 = 0.99$.

    Therefore, $0.99 \times 100 = 99\%$ of the females would be r.

14. $ww = q^2 = 32.5\% = 0.325$

    $\therefore q = \sqrt{0.325} = 0.570$.

    $\therefore p = 1 - 0.57 = 0.43$.

    $2pq = 2 \times 0.57 \times 0.43 = 0.490$.

    $p^2 = (0.43)^2 = 0.1849$.

    Therefore, $p^2 = 0.1849 = 0.1845$, $2pq = 0.490 = 0.490$.

15. The nontongue curlers (q^2) are equal to $1 - 0.90 = 0.10$.

    The gene frequency of non-curler (q^2) $= \sqrt{0.10} = 0.316$.

    $\therefore p = 1 - 0.316 = 0.684$, $2pq = 2 \times 0.684 \times 0.316 = 0.432$.

    $p^2 = (0.684)^2 = 0.4678 = 0.468$.

    The gene frequency of T(p) = 0.684 and t(q) = 0.316.

    The genotypic frequencies of

    $\left.\begin{array}{r} TT = 0.468 \\ Tt = 0.432 \end{array}\right\} 0.900 = 0.90 =$ Tongue curler

    $tt = 0.10 =$ Nontongue curler.

16. Taster (T) = 67.8%

    Non taster (t) = 100 – 67.8 = 32.2

    Non taster (t) $= q^2 = 0.322$, $q = \sqrt{0.322} = 0.5674 = 0.567$.

    $p = 1 - 0.57 = 0.43$

    The frequencies of the two alleles are $p = 0.433$ and $q = 0.567$.

17. The frequency $\frac{1}{100}$ that represents $p^2 + 2pq$. Therefore, $1 - 0.01 = 0.99$ represents the q^2 frequency.

    $q^2 = 0.99$, $\therefore q = \sqrt{0.99} = 0.9949 = 0.995$.

$\therefore$ p = 1 – 0.995 = 0.005 = 0.005 × 100 = 0.500 = 0.5%.

(b) $q^2 = \frac{1}{100} = 0.01$ or, $q = \sqrt{0.10} = 0.1$.

$\therefore$ 0.1 × 100 = 10.0 = 10%.

18.

Genotype/Phenotype	MM	MN	NN	Total
Number of individuals	305	52	4	361
Genotype frequency	$\frac{305}{100} = 0.84$	$\frac{52}{361} = 0.14$	$\frac{4}{361} = 0.01$	
Gene frequency of M = $0.845 + \frac{1}{2} \times 0.144 = 0.845 + 0.072 = 0.917$				
Gene frequency of N = $0.011 + \frac{1}{2} \times 0.144 = 0.011 + 0.072 = 0.083$				
$(0.917 + 0.083)^2$	0.8408	0.15222	0.00686	
Expected genotypic frequency	0.841	0.153	0.0069	
Observed genotypic frequency	0.845	0.144	0.011	

It is not exactly similar. So, apparently is not in equilibrium. But by chi-square test, it may be in equilibrium.

19.

Genotype/Phenotype	MM	MN	NN	Total
Number of individuals	812	1488	664	2964
Genotype frequency	$\frac{812}{2964} = 0.274$	$\frac{1488}{2964} = 0.0502$	$\frac{664}{2964} = 0.224$	
Gene frequency of M = $0.274 + \frac{1}{2} \times 0.502 = 0.274 + 0.251 = 0.525 = 0.52$				
Gene frequency of N = $0.224 + \frac{1}{2} \times 0.502 = 0.224 + 0.251 = 0.475 = 0.48$				
Expected results: $(0.525 + 0.475)^2$	0.2756	0.4987	0.2256	
Observed results:	0.2759	0.5020	0.2240	

The result appear to be very good fit.

	Mating	Calculation	Expected	Observed
i.	MM × MM =	0.274 × 0.274 =	0.075	$\frac{116}{1482} = 0.078$
ii.	MM × MN =	0.274 × 0.512 × 2 =	0.275	$\frac{404}{1482} = 0.272$
iii.	MN × MN =	0.502 × 0.502 =	0.252	$\frac{380}{1482} = 0.256$
iv.	MM × NN =	0.274 × 0.224 × 2 =	0.123	$\frac{176}{1482} = 0.119$
v.	MN × NN =	0.502 × 0.224 × 2 =	0.225	$\frac{324}{1482} = 0.218$
vi.	NN × NN =	0.224 × 0.224 =	0.050	$\frac{82}{1482} = 0.05$

The results appear to be good fit.

20. (a) Here q = u = 0.40 = $(0.40)^2$ = 0.16.

(b) p = 1 – 0.40 = 0.60

2pq = 2 × 0.60 × 0.40 = 0.48

$p^2 = (0.60)^2 = 0.36$.

$\therefore \frac{2pq}{p^2+2pq} = \frac{0.48}{0.36+0.48} = \frac{0.48}{0.48} = 0.571.$

(c) 0.48 × 0.48 = 0.230.

21. The gene frequencies of D = 0.5 and d = 0.5 in all cases.

(a) 1. $D = 0.2 + \frac{1}{2} \times 0.6 = 0.2 + 0.3 = 0.5$

$d = 0.2 + \frac{1}{2} \times 0.6 = 0.2 + 0.3 = 0.5.$

2. $D = 0 + \frac{1}{2} \times 1 = 0.5$

$d = 0 + \frac{1}{2} \times 1 = 0.5.$

3. $D = 0.499 + \frac{1}{2} \times 0.002 = 0.499 + 0.001 = 0.5$

$d = 0.499 + \frac{1}{2} \times 0.002 = 0.499 + 0.001 = 0.5.$

4. $D = 0.25 + \frac{1}{2} \times 0.5 = 0.25 + 0.25 = 0.5$

$d = 0.25 + \frac{1}{2} \times 0.5 = 0.25 + 0.25 = 0.5.$

(b) Only (4) is in equilibrium

$(0.5 + 0.5)^2 = 0.25 + 0.50 + 0.25.$

DD Dd dd

(c) $(0.5 + 0.5)^2 = 0.25 + 0.50 + 0.25$ in each case.

The progeny of each population would reach H-W equilibrium in one generation of random mating.

22. 20% are heterozygous for thalassaemia, i.e., 0.20.

Random mating results, $0.20 \times 0.20 = 0.04$.

And of the 0.04 frequency of progeny $\frac{1}{4}$ would be tt.

$\therefore 0.04 \times \frac{1}{4} = 0.01 = 1\%$.

Therefore, 1% of the progeny are expected to be affected with Colley's anaemia in the next generation.

23. (a) $q = \frac{1}{10} = 0.1$

$p = 1 - 0.1 = 0.9$

Carrier = $2pq = 2 \times 0.9 \times 0.1 = 0.18$

$0.18 \times 100 = 18\%$ population have the chance for carrier of these allele.

(b) Here product rule of probability is applied.

$0.18 \times 0.18 = 0.0324$

or, $0.0324 \times 100 = 3.2400 = 3.24\%$.

3.24% probability for both carriers of these disease on the basis of Mendelian relationship, i.e., heterozygous produces $\frac{1}{4}$ recessive trait.

Therefore, $0.0324 \times 0.25 = 0.0081 = 0.8100 = 81\%$.

0.81% chance to have the trait.

24.

Genotype/Phenotype	MM	MN	NN	Total
Number of individuals	89	162	79	330
Genotype frequency	$\frac{89}{330} = 0.2696$	$\frac{162}{330} = 0.4909$	$\frac{79}{330} = 0.2393$	

Gene frequency of M = $0.2696 + \frac{1}{2} \times 0.4909 = 0.2696 + 0.2454 = 0.515$

Gene frequency of N = $0.2393 + \frac{1}{2} \times 0.4909 = 0.2393 + 0.2454 = 0.4847 = 0.485$

25. $q^2 = \frac{1}{14000}$, $\therefore q = \sqrt{\frac{1}{14000}} = \sqrt{0.0000714}$.

$q = 0.008449 = 0.00845$

$p = 1 - 0.00845 = 0.9915 = 0.992$.

The frequency of heterozygote =

$2\,pq = 2 \times 0.792 \times 0.00845 = 0.0167 = 0.016$.

0.016 is approximately $= \frac{16}{1000} = \frac{1}{250} = \frac{1}{62.5}$.

Therefore, one student might be heterozygous for PKU.

26. (a) Two bands indicate heterozygosity for two different VNTR alleles.

(b) A person who is homozygous for VNTR allele will have a single VNTR band. Some VNTR alleles are present in small DNA fragments that migrate so fast that they are lost from the gel, a heterozygote for one of these alleles will have only one band – one corresponding to the smaller DNA fragment because the band corresponding to the smaller DNA fragment will not be detected.

(c) VNTR alleles are codominant because both are detected in heterozygotes.

(d) None of the bands in the four pairs of parents has the same electrophoretic mobility, the parents of each child can be identified as the persons who share a single VNTR band with the child.

Child A has parents 7 and 2, Child B has 4 and 1, Child C has 6 and 8 and Child D has 3 and 5.

27. Allele 5.7 frequency $= (2 \times 9 + 21 + 15 + 6)/200 = 0.3$

Allele 6.0 frequency $= (2 \times 12 + 21 + 18 + 7)/200 = 0.35$

Allele 6.2 frequency $= (2 \times 6 + 15 + 18 + 5)/200 = 0.25$

Allele 6.5 frequency $= (2 \times 1 + 6 + 7 + 5)/200 = 0.10$.

28. q is the frequency of homozygous recessive alleles.

$q^2 = \frac{1}{250{,}000}$

$q = \sqrt{\frac{1}{25 \times 10^4}}$, $q = \frac{1}{5 \times 10^2}$, $q = \frac{1}{500} = 0.002$.

(a) Among the offsprings of first cousin matings, the expected frequency of XP = $q^2(1-F) + qF$; where $F = \frac{1}{16}$ for offsprings of the first cousin. In this case, the formula yields $(0.002)^2 \times \frac{15}{16} + 0.002 \times \frac{1}{16}$ = 0.00012875 or about 1 in 7767.

(b) The ratio is given by the following formula, where q = 0.002 and $F = \frac{1}{16}$.

$$\text{Ratio} = \frac{q^2(1-F)+qF}{q^2} = \frac{\frac{1}{7767}}{\frac{1}{250{,}000}} = 32.2.$$

NOTES ON BEHAVIOURAL GENETICS

Behavioural genetics has emerged as important section/part within genetics because both genotype and phenotype have been found to have an immense impact in determining the animal's behaviour.

In studying genetic approach of behaviour, three approaches have been used.

- Use of closely related organisms from similar environments whose survival appears to be identical.
- Artificial selection of traits that can subsequently be transferred from one strain to another.
- Identification and isolation of single gene that have an impact in behaviour.

Example: 1. Geotaxis in *Drosophila*, 2. Alcohol preference and open field behaviour in Mice.

Extension and Application of Population Genetics 19

A. INBREEDING

INTRODUCTION

Mating between genetically related individuals such as selfing and sib mating is called inbreeding. As a result homozygosity increases among the offspring.

- **Features of Inbreeding:**
  - (i) The approach of inbreeding based on genotypic frequencies is useful for considering natural population.
  - (ii) Inbreeding is often practised in horticulture, animal husbandry. In human population inbreeding increases the frequency of recessive hereditary infirmities.
  - (iii) In human geneticist as well as in animal and plant breeding geneticist may wish to calculate the inbreeding coefficient of a particular individual whose pedigree is known inb order to determine the probability of each possible genotype for the individual.
  - (iv) During inbreeding, the gene pool is smaller because the parents are related genetically. The genetic consequence of inbreeding are measured by **coefficient of inbreeding** which is the probability of **autozygosity** or the probability that the two alleles in an individual at a given locus are **identical** by descent.

- **Primary key features of inbreeding**
  - (i) Increase in frequency of homozygotes.
  - (ii) Increase in variability between different inbred families.
  - (iii) Reduction in value of quantitative characters in the direction of recessive values.
  - (iv) The dependence of this reduction in value upon dominance.

- **Inbreeding coefficient(F):**

The probability that the two genes (alleles) at a locus in a diploid organism are identical to each other by descent from a common ancestor.

The coefficient of inbreeding is also called the **fixation coefficient (F)**. That is why the inbreeding coefficient is denoted by the letter F (for Fixation).

  - (i) If F = 1, all individuals in the population are homozygous and both alleles in every individuals are derived from the same ancestral copy.
  - (ii) If F = 0, no individual has two alleles derived from a common ancestral copy.

- **Estimation of F:**
  - (a) The method of estimating **F** for a population is based on the inverse relationship between inbreeding and the frequency of heterozygotes.
    - (i) As the level of inbreeding increases, the proportion of heterozygotes decreases.

$$F = \frac{H_e - H_0}{H_e}$$

[H_e = Expected heterozygosity based on H-W law
H_0 = Observed heterozygosity in a population]

    - (ii) In a completely random mating population, the expected and observed levels of heterozygosity will be equal and F = 0.
  - (b) Inbreeding coefficient (F) is defined as the proportionate reduction in heterozygosity (H_1) compared with the value of 2pq (expected heterozygosity) that would be expected with random mating.

$$F = \frac{2pq - H_1}{2pq}, \text{ or, } 2pq - H_1 = F \times 2pq$$

$$\text{or, } H_1 = 2pq - 2pqF = 2pq(1 - F).$$

The homozygous genotypes increase in frequency as the heterozygous decrease in frequency and the overall genotypic frequencies in the inbred population are

AA : $p^2(1 - F) + pF$

Aa : $2pq(1 - F)$

aa : $q^2(1 - F) + qF$.

- **Estimation of F through pedigree**

  (i) The inbreeding coefficient (F) can be computed by analyzing the degree of relatedness within a pedigree.

  (ii) The diagnostic criterion of inbreeding is a closed loop in the pedigree. Such closed loop is called path. An inbreeding path for an individual is the shortest path through the pedigree that includes both parents and common ancestors.

  (iii) In pedigree there is an inbreeding path for converting a pedigree path diagram by eliminating all extraneous individuals, those who cannot contribute to the inbreeding coefficient or in other words, a path diagram shows the direct line of descent from common ancestor.

  (iv) The common ancestor is any one who is an ancestor to both of an individual parents. The length of each inbreeding path is calculated by adding together all of the individuals in the path except the individual of interest.

$$F = \sum\left(\tfrac{1}{2}\right)^n(1+F_A)$$

F = Inbreeding coefficient of the individual of interest.

n = The number of individuals in the inbreeding path excluding the inbred offsprings.

F_A = The inbreeding coefficient of the common ancestor.

$\sum$ = Indicates that we add together $\left(\frac{1}{2}\right)^n(1+F_A)$ for each inbreeding path.

(i) The two dots in each individual represent the two copies of the particular gene and the lines indicate the mode of inheritance of genes from parent to offspring.

(ii) The two individuals in generation II, viz. A and B are half siblings. These individuals had a common father C but different mothers (D & E).

(iii) Mating between A and B produces an offspring F who is inbred. F inherits one gene copy from A and one gene copy from B. However, both of these copies might have originated from C, the common father of A and B. Thus, the two gene copies in F might be identical to each other by descent. The **identity of descent** is the important consequence of inbreeding.

(iv) Any individual whose gene copies are identical by descent must be homozygous for a particular allele of that gene. Thus, **consanguineous** matings are expected to produce relatively more homozygotes than matings between unrelated individuals.

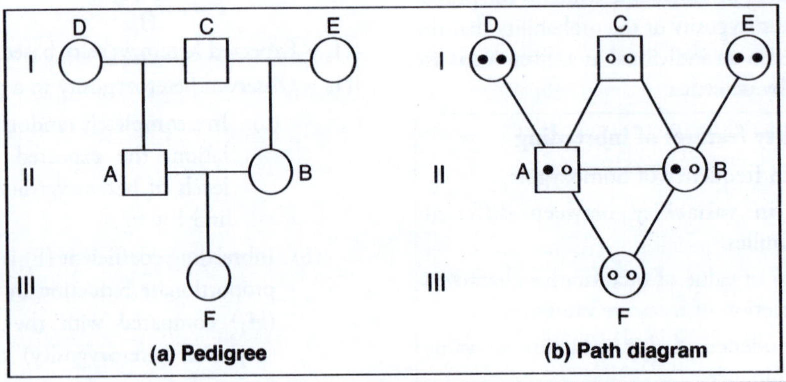

Fig. 19.1: Showing pedigree and Path diagram respectively

(v) In the above pedigree, C is referred to as the common ancestor of F because two paths of descent from C → A → F and C → B → F and together they form **inbreeding loop.**

Generation: **Case I** **Case 2**

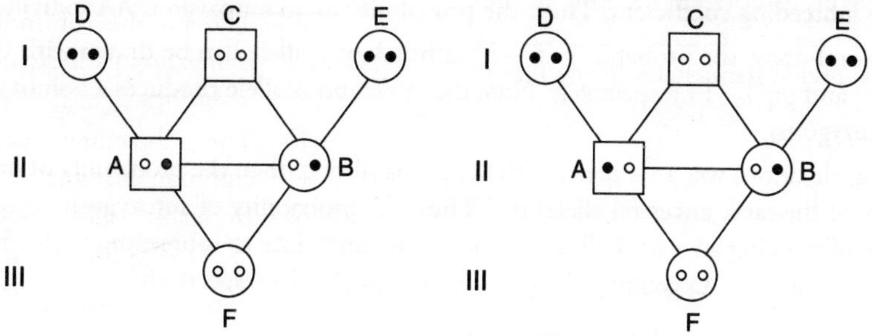

Probability: $\frac{1}{16}$ + $\frac{1}{16} = \frac{1}{8}$

Fig. 19.2: Showing inbreeding loop

In case I

(i) The chance of the gene copy (o) on the left in the common ancestor C is transmitted to offspring A is 1.

(ii) Once in A, that the chance that this gene copy is transmitted to F is 1. Thus, the probability that left gene (o) in C makes its way down to F through A is $\frac{1}{2} \times \frac{1}{2} = \frac{1}{4}$.

(iii) Similarly, the chance that left gene copy (o) makes its way down to I through B is $\frac{1}{2} \times \frac{1}{2} = \frac{1}{4}$.

(iv) The probability that the left gene copy (o) in C produces two identical gene copies in F, one transmitted through A and the other through B is $\frac{1}{4} \times \frac{1}{4} = \frac{1}{16}$.

In case II

(i) Similarly, the probability the right gene copy (•) in C produces two identical gene copies in F to be $\frac{1}{16}$. Thus, the probability that either left or right gene copies in C will produce two identical gene copies in F is $\frac{1}{16} + \frac{1}{16} = \frac{2}{16} = \frac{1}{8}$. It is as like as $\left(\frac{1}{2}\right)^3$ or, $\left(\frac{1}{2}\right)^n$.

Inbreeding coefficient

(i) In the case the common ancestors are not inbred, the inbreeding coefficient is as follows:

$$F = \sum\left(\frac{1}{2}\right)^{n_1+n_2+1}$$

n_1 = The number of generations from one parent back to the common ancestor.

n_2 = The number of generations from the other parent back to the same ancestor.

(ii) In the case, the common ancestors are inbred, the inbreeding coefficient is as follows:

$$F = \sum\left(\frac{1}{2}\right)^{n_1+n_1+1}(1+F \text{ ancestors})$$

(iii) The coefficient of inbreeding is also calculated by counting the number of arrows connecting the individual through one parent back to the common ancestor and back again to his other parent.

$$F = \sum\left(\frac{1}{2}\right)^{n}(1+F_A)$$

n = The number of arrows which connect the individual through one parent back to the common ancestor and back again to his other parent.

F_A = Inbreeding coefficient of the common ancestor.

- **Effects of Inbreeding on the Hardy-Weinberg equilibrium**
  (i) Let us assume that an allele of A type is drawn from the gene pool with the probability of p.
  (ii) On the second draw, the probability of autozygosity, i.e., the drawing a copy of the same type allele A, F is **the inbreeding coefficient**. Thus, the probability of an autozygous AA individual is pF.
  (iii) On the second draw, the probability (1 – F) either A or a allele can be drawn with the probabilities of $p^2(1 - F)$ and $pq(1 - F)$ respectively. Note that a second A allele produces a homozygote that is not inbred (**allozygous**).
  (iv) If the first allele drawn was an a allele, with the probability q, then the probability of drawing the same allele (copy of the same ancestral allele) is F. Thus, the probability of autozygosity is qF. However, the probability of drawing an a or A allele that does not contribute to inbreeding is (1 – F) and therefore, the probability of aa or Aa genotype is $q^2(1 - F)$ and $pq(1 - F)$ respectively.

Genotypic proportions in a population with inbreading:

Genotype	Due to Random mating (1 – F)		Due to Inbreeding (F)		Observed proportion
AA	$p^2(1 - F)$	+	pF	=	$p^2 + Fpq$
Aa	$2pq(1 - F)$				$2pq(1 - F)$
aa	$q^2(1 - F)$	+	qF	=	$q^2 + Fpq$
Total =	$(p^2 + 2pq + q^2)(1 - F)$	+	$(p + q)F$	=	$p^2 + 2Fpq + 2pq(1 - F) + q^2$
	$= 1 - F$	+	F	=	$p^2 + 2Fpq + 2pq - 2Fpq + q^2$
					$p^2 + 2pq + q^2 = 1$

- **Outcome:**
  (i) When the inbreeding coefficient (F) is zero (complete random mating), it reduces Hardy-Weinberg proportions.
  (ii) Inbreeding increases the proportion of homozygote in the population identity by descent implies homozygosity compared with Hardy-Weinberg proportion.
  (iii) With complete inbreeding (F = 1) only homozygotes will occur in the population.

- **Inbreeding affects allelic frequencies:**

  Lep p_{n+1} the allelic frequency of the allele A after one generation of inbreeding

  $$p_{n+1} = p^2(1 - F) + pF + \frac{1}{2}(2pq)(1 - F) = p^2 - p^2F + pF + pq - Fpq = p^2 + pq + pF - p^2F - Fpq$$
  $$= p(p + q) + pF(1 - p - q) = p\cdot 1 + pF(1 - p - q) = p + pF\cdot 0 = p.$$

  (i) Thus, inbreeding does not change allelic frequencies.
  (ii) Inbreeding affects genotypes (zygotic combinations). Although inbreeding may determine the genotypes of offspring, inbreeding does not change the number of each allele that an individual transmits into the next generation.

- **Relationship between Inbreeding and Genetic drift:**
  (i) Genetic drift causes changes in allelic frequencies, whereas inbreeding alters the proportions of genotypes, but does not alter allelic frequencies.
  (ii) Both inbreeding and genetic drift drive the population towards homozygosity.
  (iii) Inbreeding tends to increase the frequency of both homozygous classes at the expense of the heterozygotes. Genetic drift leads to fixation of one allele or other, thus producing a single homozygous class.
  (iv) Genetic drift is more common when the effective breeding size of the population is low. On the other hand, in the same condition inbreeding is much more.

- **Relationship between Inbreeding and Genetic drift through mathematical calculation**

Genetic drift causes changes in allelic frequencies, whereas inbreeding alters the proportions of genotypes, but does not alter allele frequencies.

Despite these differences inbreeding and genetic drift are closely related.

(i) Let a large random mating population becomes divided into **demes**. Within each deme, the probability that two randomly chosen gene copies are identical by descent increases with the passage of generations.

(ii) The probability of autozygosity (F) increases faster as the population becomes smaller. After 't' generations $F_t = 1 - \left(1 - \frac{1}{2N}\right)^t$.

(iii) Relationship between F (autozygosity) and H (heterozygosity). $H_t = \left(1 - \frac{1}{2N}\right)^t H_0$.

The quantity $\left(1 - \frac{1}{2N}\right)^t$ approaches zero as t becomes large, so ultimately F equals 1 and H equals 0. The population becomes completely inbred.

(iv) If individuals within each deme mate at random, mates are more closely related to each other than they would be if mating occured at random throughout the entire meta population. Thus, a group of isolated demes is conceptually similar to a population consisting of inbred lines propagated by mating between siblings. In both cases, as a whole exhibits an excess of all homozygous genotypes and deficiency of heterozygotes relative to Hardy-Weinberg proportion is expected.

Illustration:

(i) The frequencies p and q of alleles A_1 and A_2 are initially 0.5 in a large populations that is divided into demes.

(ii) If A_1 and A_2 each drift to fixation in half of the demes, the overall (mean) allele frequencies are still 0.5, but there is no heterozygous.

$$F = \frac{H_0 - H}{H_0} = \frac{0.5 - 0.0}{0.5} = 1.0.$$

(iii) Less extremely, suppose p has drifted to 0.75 in half of the demes and 0.25 in others. Then random mating within demes:

	A_1A_1	A_1A_2	A_2A_2
p = 0.75	0.5625	0.3750	0.0625
p = 0.25	0.0625	0.3750	0.5625
Overall (mean)	0.3125	0.3750	0.3125
H.W. expectation	0.25	0.50	0.25

$[0.3125 + \frac{1}{2} \times 0.3750 = 0.3125 + 0.1875 = 0.5$ $\left[\begin{matrix} p = 0.5 \\ q = 0.5 \end{matrix}\right]$

(iv) p in the population as a whole $(0.75 + \frac{0.25}{2} = 0.5)$ the frequency of heterozygote expected if all individuals mated at random (H_0) would be 2(0.5)(0.5) = 0.5, since observed frequency of heterozygotes is H = 0.375.

$$F = \frac{H_0 - H}{H_0} = \frac{(0.5 - 0.375)}{0.5} = \frac{0.125}{0.5} = 0.25.$$

(v) F represents the probability that two gene copies within a deme are the same relative to gene copies taken at random from entire population. The value of F which stems from divergence among demes by random genetic drift, i.e., Fst (fixation index).

If each deme has an effective size N, then after t generations $F_{ST} = 1 - \left(1 - \frac{1}{2N}\right)^t$.

F_{ST} often used as measure of the observed variation in allele frequencies among population

$$\left[F_{ST} = \frac{V_q}{\bar{q}(1 - \bar{q})}\right].$$

INBREEDING DEPRESSION:

A decrease or deterioration in population's viability, fitness or vigor or growth due to inbreeding (for several generation) of the normally outbreeding organism is termed as **inbreeding depression.**

(i) It refers to the reduction of fitness observed in population which are inbred.

(ii) It increases the frequencies of homozygous individual in a population and reduces genetic variability, which is necessary for genetic response to environmental change.

(iii) As deleterious genes become **homozygous**, more individuals are less fit in population.

- **Effect of inbreeding depression and selection**

(i) Let us examine the incidence of recessive genetic diseases among individuals from consanguineous matings.

(ii) Inbreeding increases the frequency of homozygous recessives which also increases the incidence of the recessive disease.

(iii) The genetic basis for this effect is that the proportion of recessive homozygotes in a noninbred population (θ, q^2) is compared to that for a given inbreeding coefficient ($\theta f = q^2 + 2fpq$),

$$\text{The ratio } \frac{\theta f}{\theta} = \frac{q^2+2fpq}{q^2} = \frac{q(q+2fp)}{q\times q} = \frac{q+2fp}{q}$$

(iv) Inbreeding can affect the mean fitness of the population.

(v) Assume that selection causes the relative fitness values of genotypes A_1A_1, A_1A_2 and A_2A_2 to be 1, 1 and 1 – s respectively. By using genotypic frequency for a given inbreeding population with mean fitness,

$$\begin{aligned}\text{i.e., } w &= 1(p^2+fpq)+1(2pq-2fpq)+(1-s)(q^2+fpq)\\ &= p^2+fpq+2pq-2fpq+(1-s)(q^2+fpq)\\ &= p^2+2pq-fpq+q^2+fpq-sq^2-sfpq\\ &= p^2+2pq+q^2-sq^2-sfpq\\ &= 1-sq^2-sfq\ [\because p^2+2pq+q^2=1].\end{aligned}$$

(The mean fitness for recessive gene when there is random mating is $1-sq^2$).

Remarks:

(i) The mean fitness is lowered from inbreeding by the amount sfpq.

(ii) The fitness is lowered (inbreeding depression increased) linearly as the inbreeding coefficient is increased.

- **Under what circumstances is inbreeding deleterious?**

(i) When it causes recessive deleterious alleles to become homozygous, this occurs in normally outbred populations of diploids that have built up these harmful alleles.

(ii) In species that normally inbred, these deleterious alleles are probably no longer present. They were either removed by selection long ago or cannot build up in the population because of the regular pattern of inbreeding.

- **Application of Inbreeding in measuring genetic relationship:**

(i) The inbreeding coefficient can also be used to measure the closeness of genetic relationship.

(ii) Full siblings are more closely related than half siblings.

(iii) Half siblings are more closely related than first cousins.

(iv) The coefficient of relationship is the fraction of genes that two individuals share by virtue of common ancestry.

- **Compare cause and effect of inbreeding depression and outbreeding depression.**

(i) Inbreeding depression reduces the level of heterozygosity.

(ii) Through the loss of heterozygosity, homozygosity increases and deleterious alleles are likely to become more of a load on population.

(iii) Out breeding depression occurs when there is a reduction in fitness of progeny from genetically diverse population.

It is usually attributed to the offspring being less well adapted to the local environmental conditions of the parents. Even though forced out breeding may be necessary to save a threatened species.

- **Coefficient of consanguinity:**

Consanguineous marriages are marriages between relatives.

The coefficient of consanguinity is the probability that two relatives will both have alleles which are identical by descent from a specific allele in a common ancestor.

- **Hybrid and Hybrid-vigour (heterosis)**

An offspring produced by mating two parents of different genotypes (differing in one or more genes) or unrelated strain is called hybrid.

The superiority of heterozygous genotypes (in terms of unusual growth survival, strength, health, fertility and phenotypic expression) over either two less vigorous homozygous genotypes of inbreed parents is called hybrid vigour.

Solved Problems

Problem 1

How does the frequency of heterozygotes in an inbred population compare with that in a randomly mating population with the same allele frequencies?

Solution:

The frequency of heterozygotes is smaller by the amount of 2pqF in which F is the inbreeding coefficient.

Problem 2

Galactosaemia is an autosomal-recessive condition associated with liver enlargement, cataracts and mental retardation. Among the offspring of unrelated individuals, the frequency of galactosaemia is 8.5×10^{-6}. What is the expected frequency among the offsprings of first cousins $\left(F=\frac{1}{16}\right)$ and second cousins $\left(F=\frac{1}{64}\right)$?

Solution:

The allele frequency is the genotypic frequency of affected offsprings among unrelated individuals, i.e., 8.5×10^{-6}.

Therefore, $q = \sqrt{8.5 \times 10^{-6}}$

$= (8.5 \times 10^{-6})^{\frac{1}{2}} = 2.9 \times 10^{-3}$

Among the offsprings first cousins.

(a) With the inbreeding $\left(F = \frac{1}{16}\right)$ the expected frequency of homozygous recessive is equal to $q^2(1-F) + qF$

$$= (2.9 \times 10^{-3})^2\left(1 - \frac{1}{16}\right) + 2.9 \times 10^{-3} \times \frac{1}{16}$$

$$= (2.9 \times 10^{-3})(2.9 \times 10^{-3}) \times \frac{15}{16} + 2.9 \times 10^{-3} \times \frac{1}{16}$$

$$= \frac{1}{16}(2.9 \times 10^{-3})\left\{(2.9 \times 10^{-3}) \times 15 + 1\right\}$$

$$= \frac{2.9 \times 10^{-3}}{16}(29 \times 10^{-4} \times 15 + 1)$$

$$= \frac{2.9 \times 10^{-3}}{16} \times \left(\frac{435}{10^4} + 1\right)$$

$$= \frac{2.9 \times 10^{-3}}{16} \times (0.0435 + 1)$$

$$= \frac{2.9 \times 10^{-3} \times 1.0435}{16}$$

$$= \frac{29 \times 1.0435 \times 10^{-3}}{16} = \frac{30.261}{16} \times 10^{-4}$$

$$= 1.891 \times 10^{-4} = 1.9 \times 10^{-4}.$$

(b) When $F = \frac{1}{64}$, the expected frequency among the offspring of second cousins is equal to $q^2(1 - F) + qF$

$$= (2.9 \times 10^{-3})^2\left(1 - \frac{1}{64}\right) + 2.9 \times 10^{-3} \times \frac{1}{64}$$

$$= (2.9 \times 10^{-3})(2.9 \times 10^{-3}) \times \frac{63}{64} + 2.9 \times 10^{-3} \times \frac{1}{64}$$

$$= \frac{1}{64}(2.9 \times 10^{-3})(2.9 \times 10^{-3} \times 63 + 1)$$

$$= \frac{1}{64}(2.9 \times 10^{-3})(29 \times 63 \times 10^{-4} + 1)$$

$$= \frac{1}{64}(2.9 \times 10^{-3})\left(\frac{1827}{10^4} + 1\right)$$

$$= \frac{1}{64}(2.9 \times 10^{-3})(0.1827 + 1)$$

$$= \frac{1}{64}(2.9 \times 10^{-3})(1.1827)$$

$$= \frac{2.9 \times 1.1827 \times 10^{-3}}{64} = \frac{3.42983}{64} \times 10^{-3}$$

$$= \frac{342.98 \times 10^{-5}}{64} = 5.359 \times 10^{-5} = 5.36 \times 10^{-5}.$$

Problem 3

Which of the following genotype frequencies of AA, Aa and aa respectively are suggestive of inbreeding?

	AA	Aa	aa
(a)	0.25	0.50	0.25
(b)	0.36	0.55	0.09
(c)	0.49	0.42	0.09
(d)	0.64	0.27	0.09
(e)	0.29	0.42	0.29

Solution:

Inbreeding is suggested by deficiency of heterozygotes relative to the 2pq expected with random mating.

	Expected frequency	Decision
(a) $p = 0.25 + \frac{1}{2} \times 0.50 = 0.25 + 0.25 = 0.5$	$(0.5 + 0.5)^2$	Random mating
$q = 25 + \frac{1}{2} \times 0.50 = 0.25 + 0.25 = 0.5$	$= 0.25 + 0.50 + 0.25$	
(b) $p = 0.36 + \frac{1}{2} \times 0.55 = 0.36 + 0.275 = 0.635$	$2pq = 2 \times 0.635 \times 0.365$	Inbreeding
$q = 0.09 + \frac{1}{2} \times 0.55 = 0.09 + 0.275 = 0.365$	$= 0.4635 = 0.47$	
(c) $p = 0.49 + \frac{1}{2} \times 0.42 = 0.49 + 0.21 = 0.70$	$0.49 + 0.42 + 0.09$	Random mating
$q = 0.09 + \frac{1}{2} \times 0.42 = 0.09 + 0.21 = 0.3$		
(d) $p = 0.64 + \frac{1}{2} \times 0.27 = 0.64 + 0.135 = 0.775$	$2pq = 2 \times 0.775 \times 0.225$	Inbreeding
$q = 0.09 + \frac{1}{2} \times 0.27 = 0.09 + 0.135 = 0.225$	$= 0.34875 = 0.349$	
(e) $p = 0.29 + \frac{1}{2} \times 0.42 = 0.29 + 0.21 = 0.50$	$2pq = 2 \times 0.5 \times 0.5 = 0.5$	Inbreeding
$q = 0.29 + \frac{1}{2} \times 0.42 = 0.29 + 0.21 = 0.5$		

Problem 4

Self-fertilization in the annual plant *Phlox cuspidata* results in an average inbreeding coefficient of F = 0.66.

(a) What frequencies of the genotypes for the enzyme phosphoglucose isomerase would be expected in a population with alleles A and B at frequencies 0.43 and 0.57 respectively?

(b) What frequencies of the genotypes would be expected with random mating?

Solution:

(a) F = 0.66

Frequencies of alleles A = 0.43 and B = 0.57 respectively.

The expected genotypic frequencies are

$p^2(1 - F) + pF$.

$AA = (0.43)^2(1 - 0.66) + 0.43 \times 0.66$

$= 0.43(0.43 \times 0.34) + 0.66$

$= 0.43(0.1462 + 0.66)$

$= 0.43 \times 0.8062 = 0.3466 = 0.347.$

$AB = 2pq(1 - F)$

$= 2 \times 0.43 \times 0.57(1 - 0.66)$

$= 2 \times 0.43 \times 0.57 \times 0.34$

$= 0.16666 = 0.167.$

$BB = q^2(1 - F) + qF$

$= (0.57)^2(1 - 0.66) + 0.57 \times 0.66$

$= (0.57)^2(0.34 + 0.57 \times 0.66)$

$= 0.57(0.57 \times 0.34) + 0.66$

$= 0.57(0.1938 + 0.66)$

$= 0.57 \times 8538$

$= 0.48666 = 0.487.$

(b) Random mating $p^2 + 2pq + q^2$

$= (0.43)+^2 + 2 \times 0.43 \times 0.57 + (0.57)^2.$

$AA = 0.1849 = 0.185$

$AB = 0.4902 = 0.490$

$BB = 0.3249 = 0.325.$

Problem 5

In a sample of 100 people of Serampore 14 MM, 32 MN and 44 NN are individuals. Calculate inbreeding coefficient.

Solution:

Genotype/Phenotype	MM	MN	NN	Total
Number of individuals	14	32	54	100
Genotype frequency	$\frac{14}{100} = 0.14$	$\frac{32}{100} = 0.32$	$\frac{54}{100} = 0.54$	

Gene frequency of M $= 0.14 + \frac{1}{2} \times 0.32 = 0.14 + 0.16 = 0.30$

Gene frequency of N $= 0.54 + \frac{1}{2} \times 0.32 = 0.54 + 0.16 = 0.70$

$$F = \frac{2pq - H_1}{2pq} = \frac{2 \times 0.3 \times 0.7 - 0.32}{0.42} = \frac{0.42 - 0.32}{0.42} = \frac{0.10}{0.42} = 0.2308$$

$[H_1 = 0.32;\ 2pq = 2 \times 0.3 \times 0.07 = 0.42]$

$\therefore F = 0.2308.$

Problem 6

If in a population with two alleles at an autosomal locus p = 0.8, q = 0.2 and the frequency of heterozygous is 0.20, what is the inbreeding coefficient?

Solution:

Here $H_1 = 0.20$

$$F = \frac{2pq - H_1}{2pq} = \frac{2 \times 0.8 \times 0.02 - 0.20}{2 \times 0.8 \times 0.2} = \frac{0.32 - 0.20}{0.32} = \frac{0.12}{0.32} = 0.375.$$

Problem 7

100 persons from a small town Uttarpara were tested for their MN blood types, viz. MM = 41, MN = 38, NN = 21. Calculate the inbreeding coefficient.

Solution:

Genotype/Phenotype	MM	MN	NN	Total
Number of individuals	41	38	21	100
Genotype frequency	$\frac{41}{100} = 0.41$	$\frac{38}{100} = 0.38$	$\frac{21}{100} = 0.21$	

Gene frequency of M $= 0.41 + \frac{1}{2} \times 0.38 = 0.41 + 0.19 = 0.60$

Gene frequency of N $= 0.21 + \frac{1}{2} \times 0.38 = 0.21 + 0.19 = 0.40$

Observed MN, i.e., $H_1 = 0.38$

$2pq = 2 \times 0.6 \times 0.4 = 0.48$

$$F = \frac{2pq - H_1}{2pq} = \frac{2 \times 0.6 \times 0.04 - 0.38}{2 \times 0.6 \times 0.4} = \frac{0.48 - 0.38}{0.48} = \frac{0.10}{0.48} = 0.208.$$

Problem 8

A sample of 140 plants are examined and the observed number of genotypes are 60 AA, 24 Aa and 56 aa. Calculate the inbreeding coefficient.

Solution:

	AA	Aa	aa
Genotypic frequency:	$\frac{60}{140} = 0.428$	$\frac{24}{140} = 0.171$	$\frac{56}{140} = 0.4$

Gene frequency of A $= 0.428 + \frac{1}{2} \times 0.171 = 0.428 + 0.085 = 0.513 = 0.51$

Gene frequency of a $= 0.4 + \frac{1}{2} \times 0.171 = 0.4 + 0.085 = 0.485 = 0.49$

$$F = \frac{2pq - H_1}{2pq} = \frac{0.499 - 0.171}{0.499} = \frac{0.328}{0.499} = 0.657 = 0.66 \qquad \left[\begin{matrix} 2pq = 2 \times 0.51 \times 0.49 = 0.499 \\ H_1 = 0.171 \end{matrix}\right]$$

Problem 9

Mention the basic difference between assortative mating and inbreeding.

Solution:

(i) Assortative mating is specific for a trait, where as inbreeding applies to the entire genome.

(ii) For example, individuals may mate assortatively with respect to height but at random with respect to blood group.

Problem 10

In a sample of 100 people of Lalgarh (W.B.) 14 MM, 32 MN and 54 NN blood types are found. Calculate the inbreeding coefficient.

Solution:

Genotypes/Phenotypes	MM	MN	NN	Total
Number of individuals	14	32	54	100
Genotypic frequencies	0.14	0.32	0.54	

Gene frequency of M $= 0.14 + \frac{1}{2} \times 0.32$
$= 0.14 + 0.16 = 0.3.$

Gene frequency of N $= 0.54 + \frac{1}{2} \times 0.32$
$= 0.54 + 0.16 = 0.7.$

Observed MN, i.e., $H_1 = 0.32$

$2pq = 2 \times 0.3 \times 0.7 = 0.42.$

$$F = \frac{2pq - H_1}{} = \frac{0.42 - 0.32}{} = \frac{0.10}{} = 0.238 = 0.24.$$

Problem 11

A population has an initial frequency of allele A at 0.7 and allele a at 0.3.

(a) What are the frequencies of genotypes AA, Aa and aa. When the inbreeding coefficient is 0.2?

(b) When the inbreeding coefficient is 0.5?

(c) When the inbreeding is complete (F = 1)?

Solution:

(a) $AA = p^2(1-F) + pF = (0.7)^2(1-0.2) + 0.7 \times 0.2$
$= 0.49 \times 0.8 + 0.14 = 0.392 + 0.14 = 0.532.$

$Aa = 2pq(1-F) = 2 \times 0.7 \times 0.3(1-0.2)$
$= 0.42 \times 0.8 = 0.336.$

$aa = q^2(1-F) + qF = (0.3)^2(1-0.2) + 0.3 \times 0.2$
$= 0.09 \times 0.8 + 0.06 = 0.072 + 0.06 = 0.132.$

(b) $AA = (0.7)^2(1-0.5) + 0.7 \times 0.5$
$= 0.49 \times 0.5 + 0.35 = 0.245 + 0.35 = 0.595.$

$Aa = 2 \times 0.7 \times 0.3(1-0.5) = 0.42 \times 0.5 = 0.21.$

$aa = (0.3)^2(1-0.5) = 0.09 \times 0.5 = 0.045.$

(c) $AA = (0.7)^2(1-1) + 0.7 \times 1 = 0.7 \times 0 + 0.7$
$= 0 + 0.7 = 0.7.$

$Aa = 2 \times 0.7 \times 0.3(1-1) = 0$

$aa = (0.3)^2(1-1) + 0.3 \times 1 = 0.3.$

2pq 0.42 0.42

Problem 12

A survey of three populations in North Bengal revealed that the observed frequencies of heterozygotes were 0.05, 0.07 and 0.03, while the expected frequencies of heterozygotes were 0.41, 0.44 and 0.32 respectively, (a) What are the estimated inbreeding coefficients in these three populations? (b) what do you think about this level of inbreeding in this species?

Solution:

(a) Observed heterozygotes (Ho): 0.05, 0.07 & 0.03

Expected heterozygotes (He): 0.41, 0.44 & 0.32

$$F = \frac{He - H_0}{He} = \frac{0.41 - 0.05}{0.41} = \frac{0.36}{0.41} = 0.878$$

$$= \frac{0.44 - 0.07}{0.44} = \frac{0.37}{0.44} = 0.8409$$

$$= \frac{0.32 - 0.03}{0.32} = \frac{0.29}{0.32} = 0.906.$$

(b) This species has a very high level of self-fertilization and low level of random mating, so that frequency of heterozygotes is very low.

Problem 13

An autosomal recessive polycystic kidney disease (AR-PKD) is a severe genetic disorder characterized by the growth of numerous cysts in the kidneys. Children with AR-PKD suffer from high blood pressure, urinary tract infections and frequent urination. They often develop kidney failure and require dialysis before adulthood. The frequency of affected children is about 1 per 10,000.

(a) Assuming Hardy-Weinberg proportions, what is the frequency of heterozygous carriers of AR-PKD.

(b) What is the expected frequency of AR-PKD among the children of first cousins?

(c) What is the ratio of AR-PKD among the offsprings of first cousins matings to that of among the offsprings of non-relatives?

Solution:

First we have to calculate the frequency of recessive allele q.

In the question, the frequency of homozygous recessive is 1 in 10,000, i.e., $q^2 = \frac{1}{10,000}$.

$$\therefore q = \sqrt{\frac{1}{10 \times 10^3}} = \sqrt{1 \times 10^{-4}} = 10^{-2} = \frac{1}{100} = 0.01.$$

(a) Assuming Hardy-Weinberg proportions the frequency of heterozygous carrier of AR-PKD will be

$2 \times 0.99 \times 0.01 = 0.0198$ [$P = 1 - 0.01 = 0.99$]

or, 0.02, i.e., 1 in 50 individuals.

(b) Among the offsprings of first cousins matings the expected frequency of AR-PKD equals to $q^2(1-F) + qF$.

F = inbreeding coefficient

$[F_1 = .5 \times .5 = .25; F_2 = .25 \times .25 - .0625 = \frac{1}{16}]$

Here, $\{(0.01)^2(1 - 0.0625) + 0.01 \times 0.0625\}$

$= 0.0001\{(0.9375 + 0.000625)\}$

$= 0.0001 \times 0.9375 + 0.000625$

$= 0.00009375 + 0.000628$

$= 0.00071875 = 0.0007186.$

(c) The ratio is given in the formula where F = 0.01 and F = 0625.

Ratio $= q^2(1-F) + qFq = q^2(1-F)q^2 + \frac{qF}{q^2}$

$= (1-F) + \frac{F}{q} = (1 - 0.0625) + \frac{0.0625}{0.01}$

$= 0.9375 + 6.25 = 7.1875.$

The offsprings of first cousins have more than 7-fold greater risk of being homozygous for a recessive nutant allele causing AR-PKD.

Problem 14

Galactosaemia is an autosomal recessive condition associated with liver enlargement, cataracts and mental retradation. Among the offsprings of unrelated individuals the frequency of galactosaemia is 8.5×10^{-6}. What is the expected freqency among offsprings of first cousins $\left(F = \frac{1}{16}\right)$ and among the offsprings of second cousins $\left(F = \frac{1}{64}\right)$?

Solution:

We have to calculate the frequency of glactosaemia (q), i.e., $q^2 = 8.5 \times 10^{-6}$.

$$\therefore q = \sqrt{8.5 \times 10^{-6}} = 2.9 \times 10^{-3}.$$

(a) Expected frequency among the offsprings of first cousins equals to $q^2(1-F) + qF$

i.e., $8.5 \times 10^{-6}\left(1 - \frac{1}{16}\right) + 2.9 \times 10^{-3}\left(\frac{1}{16}\right)$

$= 8.5 \times 10^{-6}\left(\frac{15}{16}\right) + 2.9 \times 10^{-3} \times 0.625$

$= 8.5 \times 0.9375 \times 10^{-6} + 2.9 \times 10^{-3} \times 0.625$

$= 7.969 \times 10^{-6} + 0.1812 \times 10^{-3}$

$= 10^{-6}(0.007969 + 0.182) + 3 \times 10^{-3} \times 0.19067$

$= 8.5 \times 10^{-6} \times \frac{63}{64} + 2.9 \times 10^{-3} \times \frac{1}{64}.$

$= 8.5 \times 0.984 \times 10^{-6} + 2.9 \times 0.0156 \times 10^{-3}$

$= 8.364 \times 10^{-6} + 0.0452 \times 10^{-3}.$

(b) Among the offsprings of second cousins

$= 8.5 \times 10^{-6}(1 - \frac{1}{64}) + 2.9 \times 10^{-3} \times \frac{1}{64}$

$= 8.5 \times 10^{-6} \times \frac{63}{64} + 2.9 \times 10^{-3} \times \frac{1}{64}$

$= 8.5 \times 0.984 \times 10^{-6} + 2.9 \times 0.156 \times 10^{-3}$

$= 8.364 \times 10^{-6} + 0.0452 \times 10^{-3}$

$= 10^{-3}(\frac{8.364}{10^3} + 0.0452)$

$= 10^{-3}(0.008364 + 0.0452)$

$= 10^{-6}(0.05356) = 0.0536 \times 10^{-3}$.

Problem 15

Suppose, the frequency of a recessive genetic disorder among the children of non-relatives in a population is I in 2.6 million.

(a) What is the estimated frequency of the harmful recessive allele?

(b) What is the expected frequency of homozygous recessive among the offsprings of first cousins?

(c) If first cousin matings make up 1 percent of all matings in the population and all other matings are random, what is the total frequency of homozygous recessive genotypes?

(d) What proportion of all homozygous recessive genotypes comes from the first cousins matings?

Solution:

Harmful recessive allele is 'q'. Therefore, from the question it is one in 2.6 million.

i.e., $\frac{1}{2.6 \times 1000000} = \frac{1}{2.6 \times 10^6} = \frac{0.03846}{106 \times 10^5}$

$= \frac{1}{2600000} = 0.000000384.$

$\therefore q = \sqrt{0.000000384} = 0.000620.$

(a) Expected frequency for homozygous recessive q = 0.000620.

(b) Expected frequency among the offspring of first cousins equals to $q^2F + qF$

$= 0.000000384 \times \frac{15}{16} + 0.000620 \times \frac{1}{16}$.

$= 1(5769 \times 10^{-9} + 6.20 \times 10^{-6})$

$= 1 \times 10^{-6}(\frac{5760}{10^3} + 620)$

$= \frac{625.76}{16} \times 10^{-6} = 39.11 \times 10^{-6}$

$= 3.912 \times 10^{-5}$ or, 1 in 25,561.

(c) If first cousins matings make up 1% of all matings is total frequency of homozygous recessive genotypes is

1% = 0.01 and others 1 – 0.01 = 0.99.

We know, $q^2 = 3.846 \times 10^{-7}$.

$\therefore q = 0.000620$

$= 0.99 \times 3{,}846 \times 10^{-7} + 0.01 \times 3.912 \times 10^{-5}$

($\because$ 1st cousin frequency 3.912×10^{-5})

$= 3.80754 \times 10^{-7} + 0.03912 + 10^{-5}$

$= 10^{-5}(0.0380754 + 0.03912)$

$= 0.0771954 \times 10^{-5} = 0.07720 \times 10^{-5}$

$= 7.720 \times 10^{-7}$.

(d) $\frac{0.01 \times 3.912 \times 10^{-5}}{7.720 \times 10^{-7}} = \frac{0.03912 \times 10^{-5}}{7.720 \times 10^{-7}}$

$= \frac{0.03912 \times 10^2}{7720} = \frac{3.912}{7720} = 0.5067 = 0.507.$

$\therefore 0.507 \times 100 = 50.7\%$.

Although only one percent of the matings are among first cousins, they account for more than half of the homozygous recessives for this gene.

Problem 16

(a) Inbreeding affects genetic variation in a different way than do the other factors. Describe the influence and tell how could you detect whether high breeding is present in a population.

(b) How might inbreeding be used in plant or animal breeding?

Solution:

(a) (i) Inbreeding increases the frequency of homozygotes and decreases the frequency of heterozygotes.

(ii) Calculating (examining) genotypic frequencies and compared it to Hardy-Weinberg expectation.

(b) Inbreeding in concert with selection may fix population for favourable homozygotes.

B. HERITABILITY

INTRODUCTION

It is a measure of resemblance between relatives and is estimated by evaluating the degree of resemblance, i.e., observed phenotypic differences (variance) of a trait is controlled by inheritance (genetic).

- **Broad sense Heritability:**

In quantitative genetics, the proportion of the total phenotypic variance or difference is due to genetic factors. This proportion is called broad sense heritability. It is symbolized H^2.

Phenotypic variance is due to the sum of three components, environmental variance (V_E), genetic variance (V_G) and variance resulting from the interaction of genetics and environments (V_{GE}).

Therefore, phenotypic variance (V_P) = $V_E + V_G + V_{GE}$ (V_{GE} is often negligible).

So, $V_P = V_E + V_G$

Broad sense heritability $(H^2) = \frac{V_G(\text{Genotypic variance})}{V_P(\text{Phenotypic variance})}$

If H^2 approaches 1.0, it indicates that environmental factors have had little impact on phenotyic variance. If H^2 value is close to 0, it indicates that variation in the natural population has almost solely responsible for phenotype variance.

- **Narrow sense variability:**

It is measured by the genetic variance due to the additive genes divided by the total phenotypic variance.

Genetic variance (V_G) is referred to additive variance (V_A). $V_G = V_A + V_D + V_I$

[V_D = Dominance variance, V_I = Interactive variance]

V_I is often negligible.

V_G is partioned into V_A and V_D.

In narrow sense (h^2) heritability

$$h^2 = \frac{V_A}{V_P} = \frac{V_A}{V_E+V_A+V_D}$$

we know, $V_P = V_E + V_G$ and $V_G = V_A + V_D$ (V_I negligible).

$$h^2 = \frac{V_A}{V_E+V_A+V_D}$$

- **Distinguish between broad sense and narrow sense heritability:**

(i) Broad sense heritability includes all the genetic variance as a fraction of the total phenotypic variance. $H^2 = \frac{V_G}{V_P}$.

(ii) The narrow sense heritability includes only the additive genetic variance as a fraction of the total phenotypic variance.

$$h^2 = \frac{V_A}{V_P} = \frac{V_A}{V_E+V_A+V_D}$$

- **Realized Heritability:**

Heritability measured by a response to selection

$$h^2N = \frac{R}{S} = \frac{Y_0-\bar{Y}}{Y_P-\bar{Y}} = \frac{\text{Gain}}{\text{Selection differential}}$$

[R = Response in the offspring to selection, S = selection differential in parent, h = heritability, Y_0 = offspring yield, $\bar{Y}$ means yield of the population].

$Y_0-\bar{Y}$ = Improvement over the population average

$Y_P-\bar{Y}$ = Amount of difference between the parents and the population average.

Calculation of Heritability:

(i) Several different methods are available for calculating heritability.

(ii) The environment and the gene are important in determining the phenotypic variance. Then closely related individuals should be more similar in phenotype because they have more genes in common. On the other hand, if environmental factors are responsible for determining differences in the trait, then related individuals should be no more similar in phenotype than unrelated individuals.

- **Heritability from Parent offspring regression:**

(i) If genetic component of variation is important in determining the differences among individuals, then we can expect that offsprings should resemble their parents.

(ii) To quantify the degree to which genes influence a trait, we can measure the phenotypes of parents and offsprings in a series of families.

(iii) To know the phenotyping relationship between parent and offspring, statistical analysis should be made through correlation and regression.

(iv) Here midparent value, i.e., the mean of the two parent's phenotypic value is calculated. If the variation between parents is due to completely additive genetic variation, then midparent value predicts the mean phenotype of the offspring.

(v) To know the relationship between offspring phenotype and parental phenotype, midparent values on X axis and mean of offspring phenotype on Y axis of a series of parent offsring are plotted. In the graph each point represents a family.

(vi) There are three of regression coefficients which are equal to h^2, i.e., heritability.

(vii) (a) When the slope of the parent offspring regression is 1, i.e., the regression coefficient the mean offspring phenotype is exactly intermediate to the phenotype of the two parents. Here the genes with additive effects determine all the phenotypic differences, where $h^2 = 1.0$.

(b) When the slope is less than 1 and greater than 0, the additive genes with dominance, or epistasis, and the environmental factors affects **Phenotypic variation.** Regression coefficient, i.e., $h^2 < 1.0$ (h^2 = between s and a).

(c) When slope is 0. The genetic differences do not contribute to the phenotypic differences among individuals.

Here the point will be scattered randomly, and gernerate G regressoin coefficient, i.e., $h^2 = 0$.

- **Heritability by elimination of variance components**

(i) We know, $V_P = V_G + V_E + V_{GE}$.

(ii) Elimination of environmental variance, i.e., $V_E = 0$, then V_{GE} is also zero, i.e., no genetic environment interaction can take place. Therefore, $V_P = V_G + 0 + 0$.

(iii) If $V_E = 0$, all individuals are developed in uniform environment. But it is not virtually possible.

(iv) In genetically identical trait the $V_G = 0$, then $V_P = V_E$.

- **Heritability and degrees of relatedness:**

(i) Heritability is calculated by comparing the phenotypes of individuals having different degrees of relatedness.

(ii) In monozygotic twins (MZ) 100% of this genes are common. On the other hand, in dizygotic twins (DZ) have only 50% common genes. $H^2 = 2(r_{M_z} - r_{D_z})$

[Correlation coefficient of monozygotic twin, i.e., r_{M_z} and dizygotic twins, i.e., r_{D_z}].

Solved Problems

Problem 1

The mean value of trait is 100 units, and the narrow sense heritability is 0.3. A male and female measuring 130 and 90 units respectively, mate and produce a large number of offsprings which are reared in randomized environments. What is the expected value of the trait among these offsprings?

Solution

(i) The midparent value (average of the two) is $\frac{130+90}{2} = 110$.

(ii) This value deviates from the population mean (100) by 10 units, i.e., (110 – 100).

(iii) If the narrow-sense heritability for the trait is 0.3, 30 percent of this deviation should be heritable.

(iv) Consequently, the predicted value of the trait for the offsprings of these two parent is 100 + (0.3 × 10) = 103.

Problem 2

A group of scientists of BHU genetics research laboratory studied variation in the number of abdominal bristles in female *Drosophila*. Two inbred strains that differed in bristle number were hybridzed to produce F_1 hybrids. The variance in bristle number among F_1 flies was 3.33. These F_1 flies were intercrossed and produced F_2 population in which the variance of bristle number is 5.44. Estimate the broad sense heritability for bristle number in F_2 population.

Solution

(i) By crossing two inbred strains the F_1 flies are produced. Therefore, they are genetically uniform.

(ii) The variance observed among those flies, estimates the environmental variance (V_E).

(iii) The variance observed among F_2 flies is V_T, which is the sum of the genetic variance (V_G) and environmental variance (V_E).

Here F_2 variance $V_T = 5.44$

From F_1 flies $V_E = 3.33$

$$V_T = V_G + V_E$$
$$V_G = V_T - V_E$$
$$= 5.44 - 3.33 = 2.11.$$

(iv) The broad sense heritability $H^2 = \frac{V_G}{V_P}$, or $\frac{V_G}{V_T}$.

$$H^2 = \frac{2.11}{5.44} = 0.38.$$

Problem 3

The narrow sense heritability for abdominal bristle number in a population of *Drosophila* is 0.3. The mean bristle number is 12. A male with 10 bristle number is mated to a female with 20 bristle number and a large number of progeny are scored for bristle number. What is the expected number of bristle among these progenies?

Solution

(i) The narrow sense variability is 0.3, i.e., 30% of this deviation should be heritable.

(ii) The midparent value (the average of the two parents) is $\frac{10+20}{2} = 15$. This value deviates from the population mean by 3, i.e., (15 – 12).

(iii) $H = \frac{Y_O - \bar{Y}}{Y_P - \bar{Y}}$ $[H = 0.3, \bar{Y} = 12, Y_P = 15, Y_O = ?]$

$$Y_O - \bar{Y} = Y_O - 12$$
$$Y_P - \bar{Y} = 15 - 12 = 3.$$
$$\therefore 0.3 = \frac{Y_O - 12}{15 - 12} = \frac{Y_O - 12}{3}, \text{ or, } Y_O = 12 + 0.9 = 12.9$$

The expected number of bristle among the progenies will be 12.9.

Problem 4

The narrow sense heritability for potato weight in the starting population of potato of Bidhan Chandra Agricultural University is 0.42 and the mean weight is 1.4 gram. An agriculture scientist crossed a two strains of potato with average potato weighing of 1.9 gram and 2.1 gram respectively. What is the expected average weight of potatoes in the offsprings?

Solution

The mean weight of the parental strains is 1.9 + 2.1 = $\frac{4.0}{2}$ = 2.0 gram.

$$H = \frac{Y_O - \bar{Y}}{Y_P - \bar{Y}} \quad [H = 0.42]$$

[H = heritability, Y_O = offspring yield, $\bar{Y}$ = mean yield of the population]

$$Y_O - \bar{Y} = Y_O - 1.4$$
$$Y_P - \bar{Y} = 2.0 - 1.4$$
$$\therefore H = \frac{Y_O - 1.4}{2.0 - 1.4} = \frac{Y_O - 1.4}{0.6}$$
$$0.42 \times 0.6 = Y_O - 1.4$$
$$Y_O - 1.4 = 0.252$$
$$Y_O = 0.252 + 1.4 = 1.65 \text{ gram.}$$

Problem 5

In the laboratory of Agriculture laboratory of Calcutta University, the mean and variance of corolla length in two highly inbred strains (short corolla and long corolla) of *Nicotiana* are observed.

Calculate the heritability (H^2) of corolla length of the plants.

Strain	Mean	Variance
P_1 (short)	40.47	3.12
P_2 (long)	93.75	3.87
F_1 ($P_1 \times P_2$)	63.90	4.74
F_2 ($F_2 \times F_1$)	68.72	47.70

Solution

$$\text{Heritability } (H^2) = \frac{V_G}{V_P} = \frac{\text{Genetic variability}}{\text{Phenotypic variability}}$$

Phenotypic variance (V_P): Combination of environmental (V_E) and genetic variance (V_G).

(i) Two parental strains are true in breeding. They are assumed to be homozygous and variances 3.12 and 3.87 are considered to be environmental influence. The average of the two values 3.12 + 3.87 = $\frac{6.99}{2}$ = 3.49 = 3.5.

(ii) F_1 is also genetically homogeneous and gives us an additional estimate of the impact of environmental factors.

(iii) By averaging this value along with that of the parent, $\frac{4.74 + 3.50}{2}$ = 4.12

(iv) The phenotypic variance in the F_2 is the sum of the genetic (V_G) and environmental (V_E) components, i.e., 4.12.

We know, $V_G = V_P - V_E$, i.e., 47.70 – 4.12 = 43.58.

$$\text{Heritability } (H^2) = \frac{V_G}{V_P} = \frac{43.58}{47.70} = 0.91 = 91\%.$$

So, 91% of the variation in corolla length is due to genetic influence.

Problem 6

A farmer of Haripal block wants to increase the average body weight in a herd of cattle. He begins with a herd having a mean weight of 595 kg and chooses individuals to breed that have a mean weight of 625 kg. 20 offsprings were obtained, having the following weights in kilograms 615, 587, 610, 611, 575, 612, 604, 589, 641, 598, 589, 601, 577, 620, 617, 609, 633, 599, 588, 611. Calculate the heritability in this herd with regard to body weight.

Solution:

The mean body weight of the starting herd is 625 kg.

$$\bar{X}_0 = \frac{\text{Sum of the offsprings' body weight}}{\text{Number of offsprings}}$$

$$\bar{X}_0 = \frac{615 + 587 + 610 + 611 + 575 + 612 + 604 + 589 + 641 + 598 + 589 + 601 + 577 + 620 + 617 + 609 + 633 + 599 + 588 + 611}{20}$$

$$\bar{X}_0 = \frac{12086}{20} = 604 \text{ kg}$$

$$\bar{X} = 595$$

$$\therefore H = \frac{\bar{X}_0 - \bar{X}}{\bar{X}_P - \bar{X}} = \frac{604 - 595}{625 - 595} = \frac{9}{30} = \frac{3}{10} = 0.3.$$

Problem 7

The mean and variance of plant height of two highly inbred strains (P_1 and P_2) and their progeny (F_1 & F_2) are shown here.

Strain	P_1	P_2	F_1	F_2
Mean (cm)	34.2	55.3	44.2	45.3
Variance	4.2	3.8	5.6	10.3

Calculate broad sense heritability.

Solution:

V_T or $V_{PH} = V_A + V_D + V_I + V_E$

$V_T = 4.0 + 1.8 + 0.5 + 2.5 = 8.8$

Heritability in the broad sense is the genetic variance divided by the total genetic variance.

$V_G = V_T - V_E = 8.8 - 2.5 = 6.3$

Broad sense heritability $(h^2) = \frac{V_G}{V_T} = \frac{6.3}{8.8}$

$= 0.7159 = 0.72.$

Narrow sense variability $(h^2) = \frac{V_A}{V_E + V_A + V_D}$

$= \frac{4}{2.5 + 4.0 + 1.8} = \frac{4}{8.3} = 2.075.$

Problem 8

A genetically heterogenous population of wheat has a variance in the number of days to maturation of 40, whereas two inbred populations derived from it have a variance in the number of days to maturation of 10.

(a) What is the genotypic variance $\sigma^2 g$, the environmental variance $\sigma^2 e$, and the broad-sense heritability H^2 of days to maturation in this population.

(b) If the inbred lines were crossed, what would be the predicted variance in days to maturation of the F generation?

Solution:

(a) (i) In the genetically heterogenous population, the phenotypic variance, i.e., $\sigma^2 p$ is the sum $\sigma^2 g + \sigma^2 e = 40$.

(ii) The variance in the inbred lines equals to the environmental variance (because $\sigma^2 g = 0$ in a genetically homogeneous population), hence $\sigma^2 e = 10$.

(iii) Therefore, $\sigma^2 g$ in the heterogenous population equals to $40 - 10 = 30$.

(iv) The broad-sense heritability (H^2) equals to $\frac{\sigma^2 g}{\sigma^2 g + \sigma^2 g} = \frac{30}{30 + 10} = \frac{30}{40} = \frac{3}{4} = 75.$

(b) (i) If the inbred lines were crossed, the F_1 generation would be genetically uniform and consequently the predicted variance would be $\sigma^2 e = 10$.

Problem 9

An animal breeder had a herd sheep with a mean weight of 254 pound at 3 years of age. He chooses animals with mean weight of 281 pound as parents for the next generation. When these offsprings reach 3 years of age, their mean weight is 269 pound. Calculate narrow-sense heritability for weight in this herd.

Solution:

Mean weight $(\bar{X})$ is 254 kg.

Mean weight of parents $\bar{X}_P$ is 281 kg.

The offspring mean weight $(\bar{X}_O)$ is 254 kg.

Narrow-sense heritability,

$h^2 = \frac{\bar{X}_O - \bar{X}}{\bar{X}_P - \bar{X}} = \frac{269 - 254}{281 - 254} = \frac{15}{27} = 0.555 = 0.56.$

Problem 10

Two inbred lines of *Drosophila* are crossed and the F_1 population has mean number of abdominal bristles of 20 and a standard deviation of 2. The F_2 generation has a mean of 20 and standard deviation of 3. What are the environmental variations, the genetic variance, the broadsense heritability of bristle number of this population.

Solution:

From the F data standard deviation $(\sigma^2 e) = 2$ variance $(\sigma^2 e) = (2)^2 = 4$.

From F_2 data standard deviation $(\sigma) = 3$, variance $(\sigma^2) = (3)^2 = 9$.

F_2 - phenotypic variance $(\sigma_P^2) = \sigma^2 g + \sigma^2 e$,
$9 = \sigma^2 g + 4$, or, $\sigma^2 g = 9 - 4 = 5$.

Broad-sense heritability $H^2 = \frac{\sigma^2 g}{\sigma_P} = \frac{5}{9} = 0.555 =$ 0.56.

Problem 11

In a population of 100 individuals where q = 0.4 and there are initial Hardy-Weinberg proportions, what are H_1 and H_2? If N = 10, what would H_1 and H_2 be?

Solution:

$H_1 = \left(1 - \frac{1}{2N}\right)H_0$

N = 100, q = 0.4, p = 0.6 (1 – 0.4 = 0.6)

2pq = 2 × 0.4 × 0.6, H_0 = 0.48

$$H_1 = \left(1 - \frac{1}{2\times 100}\right) \times 0.48 = \frac{199}{200} \times 0.48$$
$$= 0.995 \times 0.48 = 0.4776 = 0.478$$

$$H_2 = \left(1 - \frac{1}{2\times 100}\right) \times 0.478 = \frac{199}{200} \times 0.478 =$$
$$0.995 \times 0.478 = 0.475$$

If N = 10; H_0 = 0.48,

$$H_1 = \left(1 - \frac{1}{2\times 10}\right) \times 0.48 = \left(1 - \frac{1}{20}\right) \times 0.48 = \frac{19}{20} \times$$
$$0.48 = 0.95 \times 0.48 = 0.456.$$

$$H_2 = \left(1 - \frac{1}{20}\right) \times 0.456 = \frac{19}{20} \times 0.456 = 0.95 \times$$
$$0.456 = 0.433.$$

Problem 12

The amount of inbreeding depression is equal to sfpq. What is the expected breeding depression, when f = 0.25, s = 0.5 and q = 0.01. If q = 0.5? Why one of these values larger than the other?

Solution:

From the problems we know that inbreeding depression is equal to sfpq.

Here, s = 0.5, q = 0.01, f = 0.25

p will be 1 – 0.01 = 0.99.

So, sfpq = 0.5 × 0.25 × 0.99 × 0.01

$$= \frac{5\times 25\times 99\times 1}{10\times 100\times 100\times 100} = \frac{12375}{10^7} = 0.0012375$$
$$= 0.00124.$$

If q = 0.5 and p = 1 – 0.5 = 0.5.

So, sfpq = 0.5 × 0.25 × 0.5 × 0.5

$$= \frac{5\times 25\times 5\times 5}{10\times 100\times 10\times 10} = \frac{3125}{10^5} = 0.03125.$$

Problem 13

The genotype frequencies at one locus were found below:

f_{AA} = 0.58, f_{Aa} = 0.07, f_{aa} = 0.35.

Calculate inbreeding coefficient.

Solution:

$$\text{Inbreeding coefficient (F)} = \frac{2f_A f_a - f_{Aa}}{2f_A f_a}$$

Allele frequency $f_A = 0.58 + \frac{1}{2} \times 0.07$

$= 0.58 + 0.035 = 0.615$

$f_a = 0.35 + \frac{1}{2} \times 0.07$

$= 0.35 + 0.035 = 0.385$

$2f_{Aa} = 2 \times 0.615 \times 0.385$

$= 0.47355$

$$F = \frac{0.47355 - 0.07}{2 \times 0.615 \times 0.385} = \frac{0.40355}{0.47355} = 0.8525.$$

Problem 14

Whether natural population of animal (wolve) or domesticated population of animals (shephered) is more likely to have a higher heritability for the trait of size.

Solution:

1. A natural population of animals is more likely to have a higher genetic diversity compared to a domesticated population.

2. This is because domesticated populations have been subjected to many generations of selective breeding, which decreases the genetic diversity. Therefore, V_G is likely to be higher for the natural population.

3. On the other hand, it is difficult to say which group would have a more homogeneous environment.

4. The natural populations tend to have a more heterogeneous environment (not always). If the environment is more heterogeneous, this tends to cause more phenotypic variation, which makes V_E higher, i.e., $H^2 = V_G/V_T = V_G/(V_G + V_E)$, when V_G is high, heritability increases, when V_E is high, heritability decreases.

5. In the natural wolf population we can expect that V_G would be high, and we would guess that V_E might be high as well.

6. If the heritability of the wolf population might be similar to the domestic population due to high V_G in the wolf population is balanced by its high V_E.

7. On the other hand, if V_E is not that high in the wolf population, or if it is fairly high in the domestic population, then the wolf population would have a higher heritability for this trait.

C. CONSERVATION GENETICS

INTRODUCTION

Conservation genetics is concerned with the maintaning genetic variation in agricultural plants and animals as well as in rare and endangered species.

- **Features of Genetic diversity:**

(i) Genetic diversity includes interspecific differences and intraspecific differences within a population or between different populations. Genetic diversity can be measured by examining different phenotypes in the population or at the molecular structure, by using isozyme analysis or DNA profiling technique.

(ii) Loss of genetic diversity leads to risk of extinction of species population as a result of genetic drift, inbreeding or loss of gene flow, and the population suffer severe reduction in effective size. Reduced genetic diversity also results in greater levels of homozygosity in a population, often leading to an accumulation of deleterious alleles and inbreeding depression.

(iii) Conservation of genetic diversity depends on ex-situ methods, such as living collections, capture breeding programme and gene bank and in-situ approaches. Such as establishment of parks and preserves.

- **Genetic erosion:**

The loss of genetic diversity from a population or a species is known as genetic erosion.

- **Population augmentation:**

The process of boosting the numbers of declining population by transplanting and releasing individuals of the same species captured or collected from more numerous population elsewhere.

Effective population size:

(i) The effective population size (N_e) is almost smaller than the absolute population size (N).

(ii) In a sexually reproducing population that contains different numbers of males and females,

The effective population size $(N_e) = \frac{4(N_m N_f)}{N_m + N_f}$

N_m = Number of males

N_f = Number of females.

In-situ conservation:

(i) Attempts to preserve the population size and biological diversity of a species in its original habitat.

(ii) Identification of diversity hotspot which is an important tool for determining best places to establish parks and reserve forest, where plants and animals are protected from hunting or collecting.

Ex-situ conservation:

(i) It involves the removal of plants or animals from their original habitat to an artificially maintained location such as zoo or botanical garden.

(ii) Living collections have played a major role in the conservation and restoration of several species close to extinction.

(iii) Genetic management strategies, such as using DNA markers to identify the most genetically varied individuals and maintaianing careful pedigree records to avoid mating closely related animals have helped to maintain genetic diversity of the species.

(iv) In contrast to housing entire animals or plants establishment of gene bank, which provide long term storage of preservations for reproductive components (sperm, ova, frozen embryos, in case or animals seed pollen and culture tissues in case of plants). Many more individual genotypes can be preserved for a longer periods in a gene bank.

(v) Cryopreserved gametes or seeds can be used to reconstitute lost or endangered animals or plants after many years in storage.

(vi) A major problem with gene banks is that even large collections cannot contain all the genetic variation within a sperm.

(vii) Ex-situ conservation is that the artifitial condition under which a species is preserved in living condition and gene bank often creates their own selection preserve.

Solved Problems

Problem 1

Is a rare species found as several fragmented populations more vulnerable to extinction than an equally rare species found as one larger population? What factors should be considered, when managing fragmented populations of a rare species?

Solution:

(i) A rare species in which the remaining individuals are divided among smaller isolated subpopulations can appear to be less vulnerable.

(ii) If one subpopulation becomes extinct through local causes, such as disease or habitat loss, then the remaining subpopulations may still survive.

(iii) However, genetic drift, will cause smaller population to experience more rapidly increasing homozygosity over time compared with larger populations.

(iv) Even with random mating the change of heterozygosity from one generation to the next due to genetic drift can be calculated as

$H_1 = (1 - \frac{1}{2N})H_0.$

H_0 = the frequency of heterozygotes in the present generation,

H_1 = The frequency of heterozygotes in the next generation,

N = Number of individuals in the population.

(v) Let small populations of 50 individuals with an initial heterozygote frequency of 0.5, in just one generation, heterozygosity will decline to $H_1 = \left(1 - \frac{1}{2\times 50}\right) \times 0.5 = \left(1 - \frac{1}{100}\right) \times 0.5 =$ $0.99 \times 5 = 0.495$, a loss of 0.5%.

A larger population of 500 individuals with the same initial heterozygote frequency, i.e., 0.5, after one generation, heterozygosity will decline to $H_1 = \left(1 - \frac{1}{2\times 500}\right) \times 0.5 = \left(1 - \frac{1}{1000}\right) \times 0.5$ $= 0.999 \times 0.5 = 0.4995$, i.e., a loss of 0.05%.

(vi) This means smaller populations are likely to show the effects of homozygosity for deleterians alleles sooner than larger population, even with random mating.

(vii) If populations are fragmented, so movement of individuals or gametes among them is prevented. So management options could include transplanting individuals from one subpopulation to another to enable gene flow. Establishment of "wild life consider" of undisturbed habitat that will connect fragmented population could be considered.

(viii) In captive populations exchange of breeding adults or gametes or preserves sperms can be undertaken. Management for increased population to prevent to further gametic errosion.

Problem 2

Many populations of endangered species of Sunderban forests of West Bengal have been reduced to a small number. Assume one species with only ten individuals had a heterozygosity of 0.2 at a given gene, how much heterozygosity would be expected to remain after one generation and after twenty generations? Explain your answer.

Solution:

We know H_0 (initial heterozygosity) = 0.2.

N (number of individual) = 10

H_1 (Heterozygosity after the generation) = ?

$$H_1 = \left(1 - \frac{1}{2N}\right)H_0 = \left(1 - \frac{1}{2\times 10}\right) \times 0.2$$

$$= \left(1 - \frac{1}{20}\right) \times 0.2 = \frac{19}{20} \times \frac{1}{20} = 0.19.$$

So, after one generation the heterozygosity would be 0.19.

After 20 generation, i.e.,

$$H_{20} = \left(1 - \frac{1}{2N}\right)^{20} H_0 = \left(1 - \frac{1}{2\times 10}\right)^{20} \times 0.072.$$

(i) After one generation heterozygosity has hardly changed (0.2 → 0.19).

(ii) After 20 generations, it is only 36% of the initial value $\left(\frac{0.072\times 100}{0.2} = 36\%\right)$.

Problem 3

(a) A population of 100 males and 100 females. (b) A population of 180 males and 20 females. Calculate effective population size (N_e) from the followings:

	No. of Male	No. of Female
Pomplate (a)	100	100
Pomplate (b)	180	20

Solution:

(a) The effective population size $(N_e) = \frac{4(N_m N_f)}{N_m + N_f}$

$= \frac{4(100 \times 100)}{100+100} = \frac{4 \times 10^4}{200} = 2 \times 10^2 = 200.$

(b) The effective population size (N_e)

$= \frac{4(180 \times 20)}{180+20} = \frac{4 \times 36 \times 10^2}{200} = \frac{4 \times 36 \times 10^2}{2 \times 10^2} = 72.$

Problem 4

The effective population size N_e, can be estimated as $N_e = \frac{4(N_m N_f)}{N_m + N_f}$, where N_m and N_f are the numbers of males and females producing the following generation. If $N_m = N_f$, then $N_e = N_m + N_f$, otherwise N_e is smaller than sum of the two parents. Assume that in a cattle ranch 100 bulls, 400 cows are used to produce the following generation. What is the effective population size? A neighbouring rancher maintains 500 cows, all artificially inseminated with sperm of a single bull. What is the effective population size in this case?

Solution:

I. $N_e = \frac{4(N_m N_f)}{N_m + N_f}$

Here, $N_m = 100$ $N_f = 400$.

So, $N_e = \frac{4 \times (100 \times 400)}{100 + 400} = \frac{16 \times 10^4}{500} = \frac{16 \times 10^2}{5}$

$= 3.2 \times 10^2$.

II. $N_m = 1$, $N_f = 500$

$N_e = 4.1 \times \frac{500}{1} + 500 = \frac{2000}{50} = 4.$

Problem 5

(a) Calculate the effective population size (N_e) for a breeding population of cattle in the dairy farm (50 adult males and 50 adult females) at Mohanpur (Kalyani, W.B.) dairy of Veterinary College. (b) Calculate effective population size (N_e) for a breeding population of 60 adult males and 40 adult females. (c) Calculate effective population size (N_e) for a breeding population of 10 adult males and 90 adult females. (d) Calculate effective population size N_e for breeding population of 2 adult males and 98 adult females.

Solution:

(a) $N_m = 50$ $N_f = 50$

So, $N_e = \frac{4 \times 50 \times 50}{50 + 50} = \frac{4 \times 25 \times 100}{100} = 100.$

(b) $N_m = 60$ $N_f = 40$

So, $N_e = \frac{4 \times 60 \times 40}{60 + 40} = \frac{4 \times 24 \times 100}{100} = 96.$

(c) $N_m = 10$ $N_f = 90$

So, $N_e = \frac{4 \times 10 \times 90}{10 + 90} = \frac{4 \times 36 \times 100}{100} = 4 \times 36$

$= 144.$

(d) $N_m = 2$ $N_f = 98$

So, $N_e = \frac{4 \times 2 \times 98}{2 + 98} = \frac{8 \times 98}{2 + 98} = \frac{784}{100} = 78.4.$

Problem 6

Some animas, such as the black-footed ferret and North-American condor (vulture), now exist only in captivity. What kind of plan would you propose to conserve these species, considering such factors as inbreeding depression, loss of genetic variation, and reintroduction to the wild?

Solution:

(i) Avoid inbreeding.

(ii) Keep a large population.

(iii) Avoid unnatural selection.

- **Key Features of Conservation Genetics**
  - Genetic diversity includes both interspecific and intraspecific differences. It can be measured by examining different phenotypes in the population or at the molecular level, by using isozyme analysis or DNA profiling techniques.
  - Major declines in a species population numbers reduce genetic diversity and contribute to their risk of extinction as a result of genetic drift, inbreeding or loss of gene flow.
  - Population that suffer severe reduction in effective size and then recover are said to have passed through a population bottleneck and often show reduced genetic diversities.
  - Loss of genetic diversity reduces the capacity of a population to adapt to changing environmental condition. Reduced genetic diversities also results in greater level of homozygosity in a population. It often leads to an accumulation of deleterious alleles and inbreeding depression.
  - Conservation of genetic diversity depends on ex-situ methods, such as living collection, captive breeding programmes and gene banks in in-situ approaches as the establishment of parks and preserves.
  - Population augmentation, in which individuals are transplanted into a decline population for a more numerous population of the same species located elsewhere, can be used to increase numbers and genetic diversity.

- **Hybrid dysgenesis**

A syndrome of abnormal germline traits (including sterility mutation, chromosome breakage and male recombination) result from transposable elements activity in hybrid progeny of crosses between a certain laboratory and natural isolates of *Drosophila* is called hybrid dysgenesis.

- **Heterobeltiosis**

The superiority of a hybrid over its better parent is called heterobeltiosis.

$$\text{Heterobeltiosis} = \frac{F_1 - BP}{BP} \times 100.$$

BP = value of better parent.

- **Triple test cross**

Crossing of randomly selected F_2 plants with both the parents (P_1 and P_2) involved in the cross and with their F_1 hybrid is called triple test cross.

- Inbreeding: The mating of genetically related individuals.
- Outbreeding: The mating of genetically unrelated individuals.
- Autozygosity: Homozygosity in which the two alleles are identical by descent.
- Allozygosity: Homozygosity in which the two alleles are alike but unrelated.
- Assortative mating: Nonrandom selection of mates with respect to one or more traits.
- Positive assortative mating: Bias toward mating of like with like is called positive assortative mating.
- Negative assortative mating: Mating with unlike partners is called negative assortative mating.

Linkage Disequilibrium

20

INTRODUCTION

Linkage equilibrium is a state in which the alleles at the different loci are associated at **random,** i.e., any allelic combination in a gamete occurs as the product of the frequencies of each allele at its own locus.

- **Linkage disequilibrium:**

Linkage disequilibrium is a measure of the correlation in allelic frrequencies between two loci.

A state in which alleles at the different loci are associated at **non-random,** i.e., allelic combinations in a gamete do not follow the product rule of probability.

Example: Plant: European primrose = *Primula vulgaris*

Animal: land snail = *Cepaea nemoralis.*

When alleles at different loci are not associated at random (allele frequencies are non-randomly associated) and the gamete frequencies are not simply the product of the allele frequencies they are said to be linkage disequilibrium. On the other hand, it is a deviation from the random combinatorial proportion of haplotypes.

Here the alleles are not randomly associated, therefore, the product of the frequencies of the two coupling gametes and the product of the frequencies of the two repulsion gametes are different. This deviation or departure abbreviated by **D** or **d** is termed as **linkage disequilibrium.**

- **Explanation:**

Assume that the allele frequencies at loci are:

First locus: $A_1 = p$ $A_2 = q$

Second locus: $B_1 = r$ $B_2 = s$

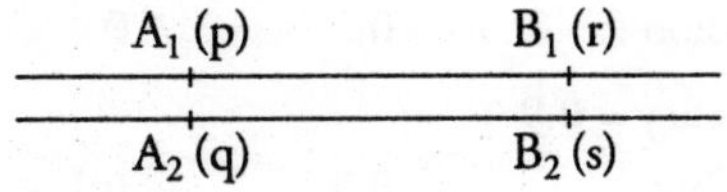

so that $p + q = 1$ and $r + s = 1$.

If the alleles of the two loci are associated at random, the four possible gametic classes (product of the frequencies of alleles) are expected:

$A_1B_1 = pr$; $A_2B_2 = qs$; $A_1B_2 = ps$; $A_2B_1 = qr$.

Because there are possible kinds of gametes, the sum of their frequencies must be 1.

$$\text{Indeed } pr + qs + ps + qr = pr + ps + qr + qs$$
$$= p(r + s) + q(r + s) = (p + q)(r + s)$$
$$= 1 \cdot 1 = 1.$$

If the alleles are associated at random, the product of the two coupling gametes ($pr \times qs = pqrs$) is the same as the product of the two repulsion gametes ($ps \times qr = pqrs$). However, if the alleles are not randomly associated, the two products will be different.

The extent of linkage disequilibrium (D) = Difference between the two products.

$$D = (\text{frequency of } A_1B_1)(\text{frequency of } A_2B_2) - (\text{frequency of } A_1B_2)(\text{frequency of } A_2B_1)$$

$$= pr \cdot qs - ps \cdot qr = pqrs - pqrs.$$

- Linkage equilibrium occurs when $D = 0$.
- Linkage disequilibrium is complete only when two gametic combinations exist—either only two coupling gametes (A_1B_1 and A_2B_2) or only the two repulsion gametes (A_1B_2 and A_2B_1).
- The actual numerical value of D depends on allelic frequencies at the two loci. The highest value of D is 0.25, if $p = q = r = s = 0.5$ and haplotype A_1B_2 and A_2B_1 are missing (complete linkage disequilibrium).

Frequency of gametic combination

Case:	A_1B_1	A_2B_2	A_1B_2	A_2B_1	D
1. Frequency = 0.5					
Coupling	0.5	0.5	0	0	$(0.5 \times 0.5) - (0 \times 0) = 0.25$
Repulsion	0	0	0.5	0.5	$(0 \times 0) - (0.5 \times 0.5) = -0.25$
2. Frequency = 0.6					
Coupling	0.6	0.4	0	0	$(0.6 \times 0.04) - (0.0) = 0.24$
Repulsion	0	0	0.6	0.4	$(0 \times 0) - (0.6 \times 0.4) = -0.24$
3. Frequency = 0.9					
Coupling	0.9	0.1	0	0	$(0.9 \times 0.1) - (0 \times 0) = 0.09$
Repulsion	0	0	0.9	0.1	$(0 \times 0) - (0.9 \times 0.1) = -0.09$

- The maximum absolute value of D = 0.25.
- The strength of linkage disequilibrium—the degree of association between alleles at different loci (*closely linked loci*).
- Maximum possible values of linkage disequilibrium D_1 for three different cases:
- **Coefficient of Linkage Disequilibrium (D_{AB})**

The coefficient of linkage disequilibrium (D) is a measure of how much the observed haplotype (gamete types) frequencies differ from their expected frequencies under random association.

The product of the coupling gamete frequencies minus the product of repulsion gamete frequencies is termed as **coefficient of linkage disequilibrium.**

Another two methods of LD are D′ and r^2 (squared correlation).

(a) (i) $D' = \dfrac{\text{Observed D value}}{D_{max}}$ (Maxmium D value i.e., D_{max})

(ii) It is always 1.

(b) (i) $r^2 = \dfrac{\text{Squared D value}}{\text{Product of all allelic frequencies}}$

$= \dfrac{D^2}{p_A\, p_B\, q_A\, q_B}$

(ii) r is always between 0 and 1.

Linkage disequilibrium will decay in each generation:

(i) Linkage disequilibrium "decays" at a rate proportion to the recombination rate (R) between the loci.

(ii) After 't' generations: $D^t_{AB} = D^0_{AB}(1-R)^t$, where D^0 is the original level. Eventually, D declines to zero unless other factors intervene.

- **Possible Reasons of Linkage Disequilibrium:**

1. Non-random mating can maintain linkage disequilibrium.
2. When new mutation arises, the single copy is necessarily associated with specific alleles at other loci on the chromosome, and therefore is in linkage disequilibrium with these alleles. Non-recombining region of the genome can form linkage disequilibrium (e.g., Human Y chromosome).
3. The population may have been formed recently by the union of two populations with different allele frequencies, and linkage disequilibrium has not yet decayed.
4. Recombination is very slow or nonexistent. **Recombination** breaks down non-random genic association. A number of generations are required for recombination to do its randomizing work.
5. Linkage disequilibrium is increased by reduction of the frequency of recombination between the loci involved which is accomplished through **translocation** and **inversion.**
6. Genetic drift in finite population may cause persistent not just transitory linkage disequilibrium.

7. **Gene flow** can produce significant levels of **disequilibrium** in population. This will occur when the frequencies of alleles at both loci differ between the population.
8. **Inbreeding** should slow the decay of linkage disequilibrium because it reduces the frequencies of double heterozygotes.
9. Natural selection may cause linkage disequilibrium by favouring individuals with particular combination. When selection occurs in one locus, it will influence gene frequencies at linked loci. The effect is called **hitchhiking.**
10. Evolution of **super gene—Natural selection** can create **linkage disequilibrium** between genes by favouring specific combinations of alleles at different loci. Recombination in the region of surrounding the super gene can also be reduced further to maintain the strong linkage disequilibrium between genes.

- **Difference between linkage equilibrium and linkage disequilibrium:**

Linkage equilibrium	Linkage disequilibrium
(i) Here alleles at different loci are associated at random (i.e. the proportion of their frequency).	(i) Here the alleles at different loci are not associated at random.
(ii) Linkage equilibrium occurs when D = 0 (D = measure of departure from linkage equilibrium).	(ii) Linkage disequilibrium occurs when $D \neq 0$. Maximum absolute value of D = 0.25.
(iii) It occurs when genotype present at one locus is independent of the genotype at second locus.	(iii) It occurs when the genotypes at the two loci are not independent of another.
(iv) Here alleles at two loci are combined independently. The two B alleles (B_1 and B_2) would be found with any one A alleles (A_1 and A_2) in the same frequency as they are found in the whole population.	(iv) It is complete when two gametic combinations exist either in the two coupling gametes (A_1B_1 and A_2B_2) or only two repulsion gametes (A_1B_2 and A_2B_1).
(v) Sexually reproducing populations (*Drosophila melanogaster*) pairs of polymorphic loci usually are found to be in linkage disequilibrium.	(v) It has been found in natural populations of European primrose (*Primula vulgaris*).

- **Linkage disequilibrium is misleading:**

(i) **Nonrandom association** of alleles at two loci can occur even if the two genes are unlinked.

(ii) Just because two loci are linked, this does not mean that they will be in linkage disequilibrium.

- **Linkage disequilibrium in different organism:**

(i) A high level of linkage disequilibrium occurs in *E. coli* bacteria, but low level in *Neisseria gonorrhoeae*.

(ii) In *Drosophila* low level of linkage disequilibrium is evident.

(iii) African swallow tail butterfly (*Papilio memmon*) exhibit high level of linkage disequilibrium.

- **Importance of linkage disequilibrium:**

1. To study genetic association in a isolated narrow population.
2. Genome-wide L. D. scans have been undertaken to locate simple Mendelian traits.
3. L. D. analysis is used to fine-map traits following initial localization to chromosomal regions by linkage analysis.
4. For studies complex disease, evidence for L. D. is sought in nuclear families.

Solved Problems

Problem 1

Two linked esterase loci were estimated in a population of *Drosophila montana*. At each locus there are two alleles, one exhibiting activity, the other being inactive (null). The observed numbers of each two-allele combination in a sample of 474 gametes are as follows:

Locus – 2	Locus – 1	
	Active	Null
Active	31	273
Null	97	73

Calculate linkage disequilibrium value, d.

Solution:

Locus – 2		Locus – 1	
		Active (A)	Null (a)
Active	(B)	31 (AB)	273 (aB) = 304
Null	(b)	97 (Ab)	73 (ab) = 170
		Total	474

$AB = \frac{31}{474} = 0.0654$, $aB = \frac{273}{474} = 0.5759$,

$Ab = \frac{97}{474} = 0.2046$, $ab = \frac{73}{474} = 0.1540$.

Linkage disequilibrium (d) = (AB × ab) – (Ab – aB)

= (0.0654 × 0.1540) – (0.2046 – 0.5759)

= 0.0100716 – 0.1178291

= –0.1077575 = –0.1078.

Problem 2

An experimental population of barley (*Hordeum vulgare*) was established by intercrossing 30 barley varieties from various parts of the world. The population was thereafter maintained by spontaneous self-fertilization. Two loci (A and B) coding for esterases were examined at various times, two alleles (1 and 2) were present at each locus. The genetic frequencies in three generations were as follows (several thousands gametes were examined in each generation):

Generation	A_1B_1	A_2B_2	A_1B_2	A_2B_1
4	0.453	0.019	0.076	0.452
14	0.407	0.004	0.098	0.491
26	0.354	0.003	0.256	0.387

Calculate the value of D for these three generations. What process(es) is(are) responsible for changing the linkage disequilibrium as the generations proceed?

Solution:

- Linkage disequilibrium for 4 generation, i.e., D_4
  = frequency of A_1B_1 × frequency of A_2B_2 – frequency of A_1B_2 × frequency of A_2B_1
  = 0.453 × 0.019 – 0.076 × 0.452
  = 0.008607 – 0.034352 = –0.025745.
- Linkage disequilibrium for 14 generation, i.e., D_{14}
  = 0.407 × 0.004 – 0.098 × 0.491
  = 0.001628 – 0.0048118
  = 0.04649 = 0.0465.
- Linkage disequilibrium for 26 generation i.e. D_{26}
  = 0.354 × 0.003 – 0.256 × 0.387
  = 0.001062 – 0.099072
  = –0.09801 = –0.0980.

Problem 3

Calculate coefficient of linkage disequilibrium (D) from the following:

$n_{AB} = 50$ $n_{Ab} = 20$ $n_{aB} = 10$ $n_{ab} = 20$ ($n = 100$)
$f_{AB} = 0.5$ $f_{Ab} = 0.2$ $f_{aB} = 0.1$ $f_{ab} = 0.2$

Solution:

Frequency of A, i.e., $f_A = f_{AB} + f_{Ab} = 0.5 + 0.2 = 0.7$

Frequency of B, i.e., $f_{AB} = f_B + f_{aB} = 0.5 + 0.1 = 0.6$

Frequency of a, i.e., $a = f_{aB} + f_{ab} = 0.1 + 0.2 = 0.3$

Frequency of b, i.e., $f_b = f_{Ab} + f_{ab} = 0.2 + 0.2 = 0.4$

Coefficient of linkage disequilibrium (D)

$D_{AB} = f_{AB} - f_A f_B = 0.5 - 0.7 \times 0.6 = 0.5 - 0.42$
$= +0.08.$

$D_{Ab} = f_{Ab} - f_A f_b = 0.2 - 0.7 \times 0.4 = 0.2 - 0.28$
$= -0.08.$

$D_{aB} = f_{aB} - f_a f_B = 0.1 - 0.3 \times 0.6 = 0.1 - 0.18$
$= -0.08.$

$D_{ab} = f_{ab} - f_a f_b = 0.2 - 0.3 \times 0.4 = 0.2 - 0.12$
$= +0.08.$

Problem 4

	n_{AB}	n_{Ab}	n_{aB}	n_{ab}	Total
Population A =	**70**	**0**	**10**	**20**	**100**
Population B =	**70**	**10**	**0**	**70**	**100**

(a) Determine haplotype frequency of each population.

(b) What are the coefficients of L.D in each population?

(c) If the two populations were mixed, what would be the additional L.D created by the two-locus Wahlun effect?

(d) Is the D′ larger or smaller in the max tune than in the two populations separated.

Solution:

(a)

	n_{AB}	n_{Ab}	n_{aB}	n_{ab}	Total
Population A:	70	0	10	20	100

Haplotype frequency of AB = f_{AB} = 0.7

Haplotype of A = f_A = 0.7 + 0 = 0.7

Haplotype of B = f_B = 0.7 + 0.1 = 0.8

	n_{AB}	n_{Ab}	n_{aB}	n_{ab}
Population B:	0.2	0.1	0	0.7

Haplotype of AB = f_{AB} = 0.2

Haplotype of A = f_A = 0.2 + 0.1 = 0.3

Haplotype of B = f_B = 0.2 + 0.2 = 0.2

Haplotype of ab = f_{ab} = 0.7.

(b) Coefficient of L.D of population A = $f_{AB} - f_A \times f_B$

$= 0.7 - 0.3 \times 0.2 = 0.7 - 0.7 \times 0.8 - 0.7 - 0.56 = 0.14$

Coefficient of L.D of population B = $f_{AB} - f_A \times f_B$

$= 0.2 - 0.3 \times 0.2 = 0.2 - 0.06 = 0.14.$

(c) When two populations were mixed:

	n_{AB}	n_A	n_B	n_{ab}
Population A:	70	0	10	20
Population B:	20	10	0	70
	$\frac{90}{200}$	$\frac{10}{200}$	$\frac{10}{200}$	$\frac{90}{200}$
	= 0.45	= 0.05	= 0.05	= 0.45

$f_A = 0.45 + 005 = 0.5$

$f_B = 0.45 + 0.05 = 0.5$

D = 0.45 – 0.05 × 0.05 = 0.45 – 0.25 = 0.20 = 0.2.

D in both the populations (a) and (b) is 0.14.

In the mixed population, D is 0.2. It is larger by 0.06 because of the population mixture.

Problem 5

Calculate D, D′ and r² from the following data.

AB = 22, Ab = 9, aB = 0, ab = 20

Solution:

Total = AB + Ab + aB + ab = 22 + 9 + 0 + 20 = 51.

$f_{AB} = \frac{90}{51} = 0.431$

$f_{Ab} = \frac{9}{51} = 0.176$

$f_{aB} = \frac{0}{51} = 0$

$f_{ab} = \frac{20}{51} = 0.392.$

Frequency of A = $f_A = f_{AB} + f_{Ab} = 0.431 + 0.176 = 0.607$

Frequency of B = $f_B = f_{AB} + f_{ab} = 0.431 + 0 = 0.431$

Frequency of a = $f_a = f_{aB} + f_{ab} = 0 + 0.392 = 0.392$

Frequency of b = $f_b = f_{Ab} + f_{ab} = 0.176 + 0.392 = 0.568.$

Coefficients of linkage disequilibrium:

D $= f_{AB} - f_A \times f_B = 0.431 - 0.607 \times 0.431$
$= 0.431 - 0.2616 = 0.169.$

D′ = 1, because there is no haplotype aB
(∵ aB = 0).

$$r^2 = \frac{D^2}{f_A \times f_B \times f_a \times f_b} = \frac{(0.169)^2}{0.607 \times 0.431 \times 0.392 \times 0.568}$$

$$= \frac{0.02856}{0.05825} = 0.490.$$

Problem 6

The frequencies of gametes AB, Ab, aB and ab are 0.1, 0.2, 0.3 and 0.4 respectively. What is D?

Solution:

D = (AB × ab) – (Ab × aB)
= (0.1 × 0.4) – (0.2 × 0.3)
= 0.04 – 0.06 = –0.02.

Problem 7

The disequilibrium value d or D is –0.02. What will be the value of D after four generations, if the recombination rate between the loci is 0.1?

Solution:

In the 1st generation disequilibrium value, i.e.,

$D_1 = (1 - r)D_0$. Recombination rate (r) = 0.1.

$D_1 = (1 - 0.1) \times -0.02 \quad D_0 = -0.02.$

For 4 generations, i.e., $D_4 = (1 - 0.1)^4 \times -0.02$

$D_4 = (0.9)^4 \times -0.02 = 0.6561 \times -0.02$

$= -0.013122 = -0.0131.$

Problem 8

The autosomal gametic disequilibrium in a population is expressed as d = 0.12. The two loci under consideration recombine with a frequency of 16%. Calculate the disequilibrium (d-value) which existed the gametic pool of (a) the previous generation and (b) the next generation.

Solution:

$d_0 = 0.12, \quad r = 16\% = 0.16$

(a) $dt = (1 - r) = 1 - 0.16 = 0.84$

In previous generation, dt = 0.84.

In present generation, $d_0 = 0.12$.

Disequilibrium value in gametic pool (previous generation) $\frac{d_0}{dt} = \frac{0.12}{0.84} = 0.143.$

(b) In the next generation, $dt = (1 - r)^1 d_0$

$= (1-0.16)^1 \times 0.12 = 0.84 \times 0.12 = 0.1008.$

Problem 9

If d = 0.25 initially and the two loci experience 20% recombination (i.e., the loci are 20 map units apart), calculate the disequilibrium (d value) after one generation.

Solution:

The disequilibrium value, i.e., d at any generation is expressed in the following equations:

$dt = (1-r)d_{t-1}$

$d = 0.25, \quad r = 20\% = 0.20$

$dt = (1-0.20)^1 d_0 = 0.8 \times 0.25 = 0.2.$

This represent $\frac{0.20}{0.25}$ or 80% of the maximum disequilibrium that could exist for a pair of alleles of linked loci.

Problem 10

A population of fish in lake is fixed at each of two unlinked loci for the dominant allele (AABB). A canal is built that connects the lake with a smaller one, where the same fish species is fixed for the recessive alleles (aabb). Assume that mating is random thereafter, the original population was ten times greater in the large lake than in small lake and that the two loci are not affected by natural selection. What is the linkage disequilibrium, d, immediately after the population of the two lakes become mixed and the five generations of random mating?

Solution:

AABB — aabb

The population of the large lake is ten times as large as the population of the small lake. Therefore, gametic frequencies immediately after mixing are

$AB = \frac{10}{11}$, $ab = \frac{1}{11}$, $Ab = 0$, $aB = 0$.

$\left(p + q = 1 = \frac{10}{11} + \frac{1}{11} = \frac{11}{11} = 1\right)$

The initial linkage disequilibrium is

d = AABB × aabb – AAbb × aaBB

$= \frac{10}{11} \times \frac{1}{11} - 0.0 = \frac{10}{121} = 0.0826.$

The value of d will decrease in each generation of random mating according to $d_1 = (1-r)d_0$, where r = 0.5, if the two loci assort independently. After five generations of random mating:

$d_5 = (1-r)^5 d_0 = (1-0.5)^5 \times 0.0826 = 0.03125 \times 0.0826 = 00025812 = 0.0026.$

Problem 11

If d = 0.0826 initially and the two loci are in the same chromosome with a recombination frequency 0.10, what will be the value of 'd' after five generations of random mating?

Solution:

$dt = (1-r)^t d_0, \quad r = 0.1, \quad d_0 = 0.0826.$

$d_5 = (1-r)^5 \times 0.0826 = (1-0.1)^5 \times 00826$
$= (0.9)^5 \times 0.0826 = 0.59049 \times 0.0826$
$= 0.4877447 = 0.488.$

Problem 12

Calculate 'd' and the equilibrium frequency of gametes for a population in which the frequencies of two unlinked gene pairs Aa and Bb are A = B = 0.6 and a = b = 0.4, and the initial genotypic frequencies are AABB = AAbb = aaBB = 0.30 and aabb = 0.10.

Solution:

Initial population	Type	Initial frequency	Equilibrium frequency	
30% AABB	AB	0.3	0.3 + d	coupling
30% AAbb	Ab	0.3	0.3 – d	repulsion
30% aaBB	aB	0.3	0.3 – d	repulsion
10% aabb	ab	0.1	0.1 + d	coupling

$D = (Ab \times aB) - (AB \times ab) = 0.3 \times 0.3 - 0.3 \times 0.1$
$= 0.09 - 0.03 = 0.06.$

Problem 13

In a haploid sexual population, size is affected by two loci, with two alleles at each locus. AB is larger than Ab or aB, which are both larger than ab. The four haplotypes are in linkage equilibrium. The fitness of AB, Ab, aB, are 1, 1 + S and 1 + t respectively. Will the linkage disequilibrium in the next generation be positive, zero or negative, if (a) the fitness of ab is (1 + S) (1 + t) (b) the fitness of ab is 1 + S + t?

Solution:

(a) Zero; (b) Negative

Let the initial frequencies of the haplotypes be p(AB), p(Ab), p(aB) and p(ab), where p(AB), p(ab) = p(Ab).p(aB).

In case (a), D after one generation is p(AB). p(aB) [(1 + s)(1 + t) – (1 + s) (1 + t)] = 0.

In case (b),

$D = p(AB).p(ab).[(1 + s + t) - (1 + s)(1 + t)]$
$= p(AB).p(ab).[1 + s + t - 1 - s - t - st]$
$= p(AB).p(ab).[-st]$
$= -st.p(AB).p(ab).$

(i) It is an important result due to Felsenstein, that if a population starts in linkage equilibrium and fitness are multiplicative (in case a), it remains in equilibrium.

(ii) But if population starts is exposed to directional selection and fitness increases less steeply than in the multiplicative case, then negative linkage disequilibrium is produced.

Problem 14

The observed number of haplotypes for SNPs at two polymorphic loci of *D. melanogaster* in the promote region.

locus 1 (A)	locus 2 (B)	Number	Frequency
C	A	40	$P_{11} = \frac{40}{172} = 0.2326$
C	G	42	$P_{12} = \frac{42}{172} = 0.2442$
G	A	14	$P_{21} = \frac{14}{172} = 0.0814$
G	G	76	$P_{22} = \frac{76}{172} = 0.4419$

Is this population in eqilibrium? If not, what is the coefficient of linkage equilibrium?

Solution:

We need to calculate the allele frequencies at each loci. From the given data we can calculate

$$p_A = \frac{40+42}{40+42+14+76} = \frac{82}{172} = 0.477$$

$$q_A = \frac{14+76}{40+42+14+76} = \frac{90}{172} = 0.523$$

$$p_B = \frac{40+14}{40+42+14+76} = \frac{54}{172} = 0.314$$

$$q_B = \frac{42+76}{40+42+14+76} = \frac{118}{172} = 0.686$$

It appears there are more CA and GG haplotypes and fewer CG and GA haplotypes.

The numerical value of $D = P_{11}P_{12} - P_{12}P_{21}$

$D = 0.2326 \times 0.449 - 0.2442 \times 0.0814$

$D = 0.10278 - 0.019877$

$D = 0.0829.$

Gene frequencies are $p_A = 0.477$, $q_A = 0.523$, $p_B = 0.314$, $q_B = 0.686$.

Expected frequency and Expected number		
$p_A p_B$	= 0.477 × 0.314 = 0.1478	0.1478 × 172 = 25.76
$p_A q_B$	= 0.477 × 0.686 = 0.3272	0.3272 × 172 = 56.28
$q_A p_B$	= 0.523 × 0.314 = 0.16422	0.16422 × 172 = 28.25
$q_A q_B$	= 0.523 × 0.686 = 0.35877	0.35877 × 172 = 61.71

Maximum expected frequency 0.1642

$\therefore D^1 = \frac{D}{D_{max}} = \frac{0.0829}{0.1642} = 0.5048 = 0.505$

D^1 is quite strong LD, 50% is theoretical maxima.

Alternatively $r^2 = \frac{D^2}{p_A p_B q_A q_B}$

$= \frac{0.0829 \times 0.0829}{0.477 \times 0.0523 \times 0.314 \times 0.686} = \frac{0.00687}{0.0054}$

$= 1.27266 = 0.128.$

Problem 15

From a population haplotypes are the following: AB = 30, Ab = 270, aB = 370, ab = 330. Calculate D.

Solution:

n_{AB}	n_{Ab}	n_{aB}	n_{ab}	Total
30	270	370	330	1000
0.03	0.27	0.37	0.33	

$t_{AB} = 0.03$

$f_A = 0.03 + 0.27 = 0.3$

$f_B = 0.03 + 0.37 = 0.4$

$f_{ab} = 0.33$

$D = f_{AB} = f_A \times f_B = 0.03 - 0.3 \times 0.4$

$= 0.03 - 0.12 = -0.09.$

- **Super gene:**

  Several closely related linked gene loci that affect a single trait or series of inherited trait and segregate as a single Mendelian locus are termed as super gene.

  In a super gene, the recombinants will often die and therefore, not be seen.

- **Hitchhiking:**

  When selection occurs in one locus—it will influence gene frequencies at linked loci. The effect is called hitchhiking.

- **Cytonuclear Disequilibrium:**

  A related form of disequilibrium which can arise between nuclear and mitochondrial gene is called cytonuclear disequilibrium.

Linkage disequilibrium of MHC gene in North-European population

(i) Two major histocompatibility (MHC) loci in human denoted as A and B.

(ii) The frequency of A1 allele at A locus is 17 percent, whereas the frequency of B8 allele is 11 percent. Therefore, the frequency of A1B8 chromosome (also called A1B8 haplotype) would be 0.17 × 0.11 = 0.0187 (1.8 for 2 percent).

(iii) Instead, the observed frequency of A_1B8 haplotype is 9 percent (0.09).

(iv) It results linkage disequilibrium (D) = observed frequency – expected frequency = 0.09 – 0.0187 = 0.0713 = 0.07.

- **Key features of Linkage disequilibrium:**

  A deviation from the random combinatorial proportions of haplotypes is called linkage disequilibrium.

  L.D can arise because of non-random mating, random sampling and natural selection.

  L.D is characteristics of very tightly linked loci.

Immuno Genetics

21

- **Immunogen, Antigen immunogenicity, Antigenicity:**

A substance capable of eliciting an immune response both humoral and cellular is called **immunogen.** All immunogens are antigen but some antigen, (e.g., haptens) are not immunogen.

Any substance (usually foreign) that binds specifically to an antibody or a T-cell receptor is called **antigen.**

The capacity of a substance to induce a detectable immune response under a given set of conditions is termed as **immunogenicity.**

The ability of a substance to combine specifically with antibody or T-cell receptor or MHC is known as antigenicity.

- **Antigenic determinant or Epitope:**

The restricted site on an antigen that is recognized and bound by a particular antibody, TCR-MHC component or TCR ligand-CDI complex is termed as epitope.

- **Superantigen:**
  (i) It is a class of antigen which binds to the V_β domain of the T-cell receptor and residues in the chain of class II MHC molecules.
  (ii) It produces activating signal that induces an activation of all T-cells that express T-cell receptors with a particular V_β domain. *Example*: Staphylococcal enterotoxins (food poisoning).

- **Affinity and Avidity:**

Affinity represents the strength of interaction between a single antigen binding site on an antibody and a single epitope of an antigen. It is represented quantitatively by the affinity constant κ_a, which is
  (i) The ratio of the forward rate constant (κ_1) to the reverse rate constant κ_{-1} in an antibody antigen reaction.
  (ii) Equivalent to the association constant in biochemical forms ($\kappa_a = \kappa_1/\kappa_{-1}$).

The strength of antigen-antibody binding when multiple epitopes of an antigen interact with multiple binding sites of an antibody is called avidity.

- **T-independent and T-dependent antigen:**

Immunogenic antigen which is able to elicit an antibody response without the help of T-cell is called T-independent antigen, and an antigen which requires helper T-cells (T_H) to elicit an antibody production is termed as T-dependent antigen.

Antigen (Immunogen)	Antibody (Immunoglobulin)
1. They are foreign materials such as protein or polysaccharide molecule.	1. These are protein molecule.
2. These tigger the formation of antibodies.	2. These are synthesized in the body to combat foreign materials.
3. They bind to macrophages C to reach helper T cells to initiate immune response.	3. They directly join to antigens to destroy them.
4. They produce diseases or allergic reaction.	4. These are protective and immobilize or lyse antigenic molecules.

Humoral Immunity	Cell-mediated Immunity
1. Mediated by B-lymphocyte in plasma.	1. Mediated by T-lymphocyte.
2. Operated through formation of antibodies.	2. Operated through T-cells.
3. It acts on pathogen that invades body fluids.	3. It operates against pathogen which invades body cells.
4. Do not react against transplants and cancerous cells.	4. React against transplants and cancerous cells.

IMMUNITY

A. **Innate immunity**: It is the non-specific natural and inborn immunity or resistance to infection, which is achieved by birth from natural side.

B. **Acquired immunity**: It is the specific immunity or resistance to infection mediated by B- or T-cells that an individual acquires during his/her life time following exposure to an antigen is called acquired immunity.

(a) ***Active immunity***:

(i) It is the resistance (immunity) of an individual by natural exposure to antigen (in the form of living or dead microbes or their protein) is called active immunity.

(ii) Here body cells manufacture antibody. It is long lasting.

- *Natural immunity*: It is obtained after exposure to antigen or infection to pathogen in natural environment.
- *Artificial immunity*: Here immunity developed after administration of antigen (by vaccination prepared by live attenuated or killed micro-organism).

(b) ***Passive immunity***: It is conferred by the transfer of specific immune product (readymade antibody or sensitized T-cells) from an immune individual to a non-immune one.

- *Natural immunity*: Conferred by transfer of maternal antibodies across placenta or breast milk.
- *Artificial immunity*: Conferred by administration of antibodies to combat infection. *Ex*: getting an injection of antisera.

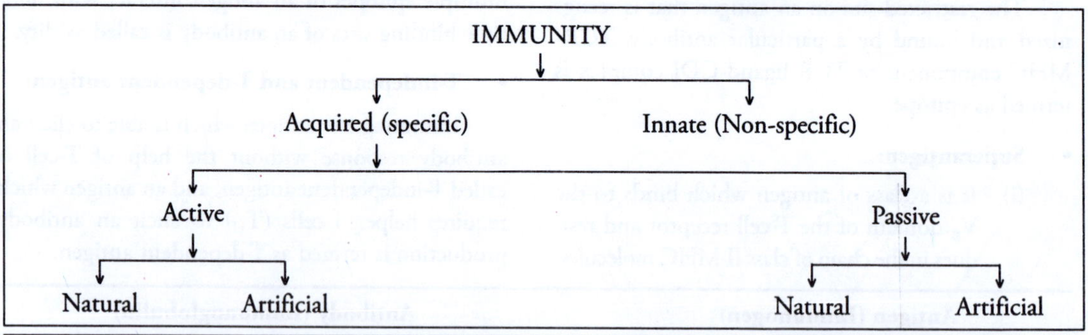

- **Severe combined immunodeficiency (SCID)**

(i) In this disorder there are very low number of circulating lymphocytes and both T- and B-cells are either absent or negligible at birth. Circulating new-born blood shows shortage of TRECs (T-cell receptor excision circles).

(ii) Since the body lack of commune system even a small infection can be fatal.

(iii) It is caused by a problem in a gene found in X chromosome and affect only patients who are carriers. The deficiency of enzyme **adenosine deaminase** (ADA) is found in the patients.

- **Hapten:**

A low molecular weight molecule which is not immunogenic, but can induce antibody production, when binds with carrier, provide epitopes recognized by the helper T-cells. It can react with an antibody of appropriate specificity.

STRUCTURE OF IMMUNOGLOBULIN

I. Each immunoglobulin molecule is made up of four polypeptide chains, two identical short light (L) chains (220 amino acid) and two identical heavy (H) chains. (440-450 amino acid) represented as H_2L_2.

II. Each of the four chains has a constant (c) region and a variable (v) region. Variable region is concerned with antigen binding and constant region concerned with complement fixation.

III. The variable region along with light chain is designated as V_L and along with heavy chain as V_H. The constant regions along with respective chains are designated as C_L and C_H.

IV. The variable regions (V) are specified by the amino acid sequences of the amino terminal portions of both heavy and light chains. The constant region (C) is specified by the carboxyl terminal portion of light and heavy chains.

V. The four polypeptide chains are held together by disulphide (-s-s-) bonds forming Y-shaped structure. Every chain (heavy and light) has an amino terminal variable region (Fab) within which amino acid sequence varies among the different kinds of antibodies.

VI. The light chain are two types, kappa (κ) and lambda (λ), distinguised by amino acid sequence. The top two tips of Y-shaped molecules are **antigen binding** sites. They recognise and bind to a specific antigen in a lock and key fashion, forming an **antigen-antibody complex**. Both the arms of Y have the same antigen binding sites and so recognise the same antigen.

VII. The light and heavy chains of an antibody are encoded by different loci in the genome. Each of these loci consists of a long array of gene segment. The genes in humans, coding for heavy chains are mapped on **chromosome 14**, and coding for light chain loci kappa (κ) on **chromosome 2**, and loci lambda (λ) on **chromosome 22.**

VIII. Three distinct gene segments code for light chains (C for constant, V for variable regions and J for joining the constant and variable regions).

IX. Four gene segments code for the heavy chains C, V and J for constant and variable and joining segment respectively, and D for the diversity region located between the joining and variable region.

X. An L_κ and V_κ gene segment encodes a leader peptide amino terminal 95 amino acid of the variable region of kappa light chain. The

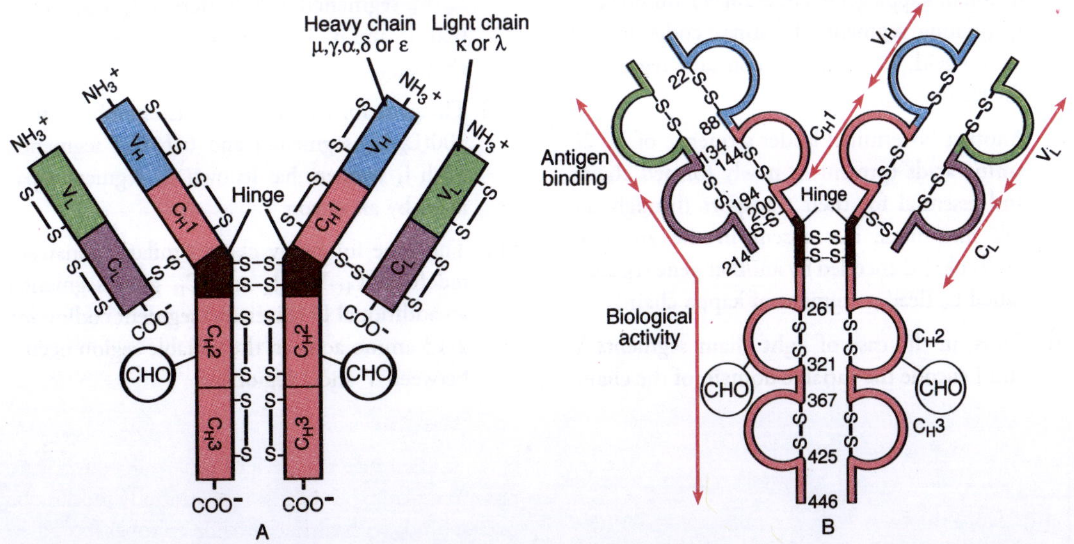

Fig. 21.1: Structure of Immunoglobulin

leader peptide is removed from kappa light chain by cleavage.

XI. The J_κ gene segment encodes the last 13 segments of the variable region of the kappa light chain. The symbol J_κ is used for the gene segment because the peptide it encodes joins the amino terminal peptide encoded by V_κ L_κ segment to carboxyl terminal peptide.

Immunoglobulin (antibody) genes:

I. Immunoglobulin (antibody) is a tetramere consisting of two identical light and two identical heavy chains.

II. The amino terminal end (100 amino acid) of both light and heavy chains constitutes their variable region designated as V_L and V_H respectively. The remaining region of both light and heavy chains are in each constant region (segmented as C_H and C_L).

III. The light chains have two families kappa and lambda. The type of light chain is determined by its constant region.

IV. The separate genes encoding kappa is located in human chromosome 2 and lambda in chromosome 22. The heavy chain is located in chromosome 14.

V. In the germline cell each gene consists of a large number of gene segments. V_κ (variable region of kappa gene) codes for 95 amino acid. J_κ (joining segment of kappa) codes for 13 amino acid. C_κ codes for constant regions of the chain.

VI. Another N-terminal leader sequence of 17-20 amino acids (present in newly formed chains and essential for their transport through the cell membrane but absent in the antibody molecules) is encoded by another gene segment called L_κ (leader segment of kappa chain).

VII. Thus, in the case of light chain segments V and J encode the variable domain of the chain. Therefore, together they are often termed as variable gene or V gene. Similarly the C segment encodes the constant region of chains, hence it is often termed as constant gene or C gene.

VIII. In case of heavy chain, V gene comprises V, D and J gene segments. Here the term gene is used to denote a DNA sequence that encodes a discrete domain of the polypeptide. Each chromosome 2 has about 76 different V_κ gene segments. Each V_κ gene segment has its own L_κ segments separated by an intron so that V_κ and L_κ segments alternate with each other.

IX. Five different J_κ gene segments (spread over 500-700 bp) are located next to V_κ region. V_κ and L_κ segments are separated by a large non-coding region.

X. In the end there is only C_κ gene segments located approximately 2-3 kb (noncoding intron) away from J_κ region.

XI. During the maturation of B-lymphocyte any one of the 76 L_κ-V_κ gene segments joins (by a process of somatic recombination) with any one of the 5 J_κ segments, the rest of the intervening DNA being deleted.

XII. The antibody specificity of the kappa chain produced depends on which of the L_κ-V_κ gene segments joins with which J_κ segments. Thus, there would be a total of about 76 (different L_κ-V_κ segments) × 5 (different J_κ segments), that is about 380 different classes of kappa chains.

XIII. The gene for lambda chains is assembled from 300(L_λ-V_λ segments) and 6(J_λ-C_λ) segments. Each J_λ segment has its own C_λ segment separated by an intron.

XIV. The gene for heavy chain similarly constructed from L_H-V_H, J_λ and C_H gene segments; an additional D (diversity) segment coding for 2-13 amino acids in the variable region occurs between V and J regions.

B and T Lymphocytes

I. Both B- and T-lymphocytes develop from the precursor cells, called **stem cells**, in the bone marrow. (These cells are pleuripotent in nature). The process of forming these B and T lymphocytes along with other blood cells in bone marrow is called **haematopoiesis.**

II. From the immature lymphocytes, i.e., stem cells destined to become thymocytes and migrate to thymus gland via blood. In thymus gland they become mature and are termed as **T-cells.**

III. The B-cells remain in bone marrow and mature in marrow itself.

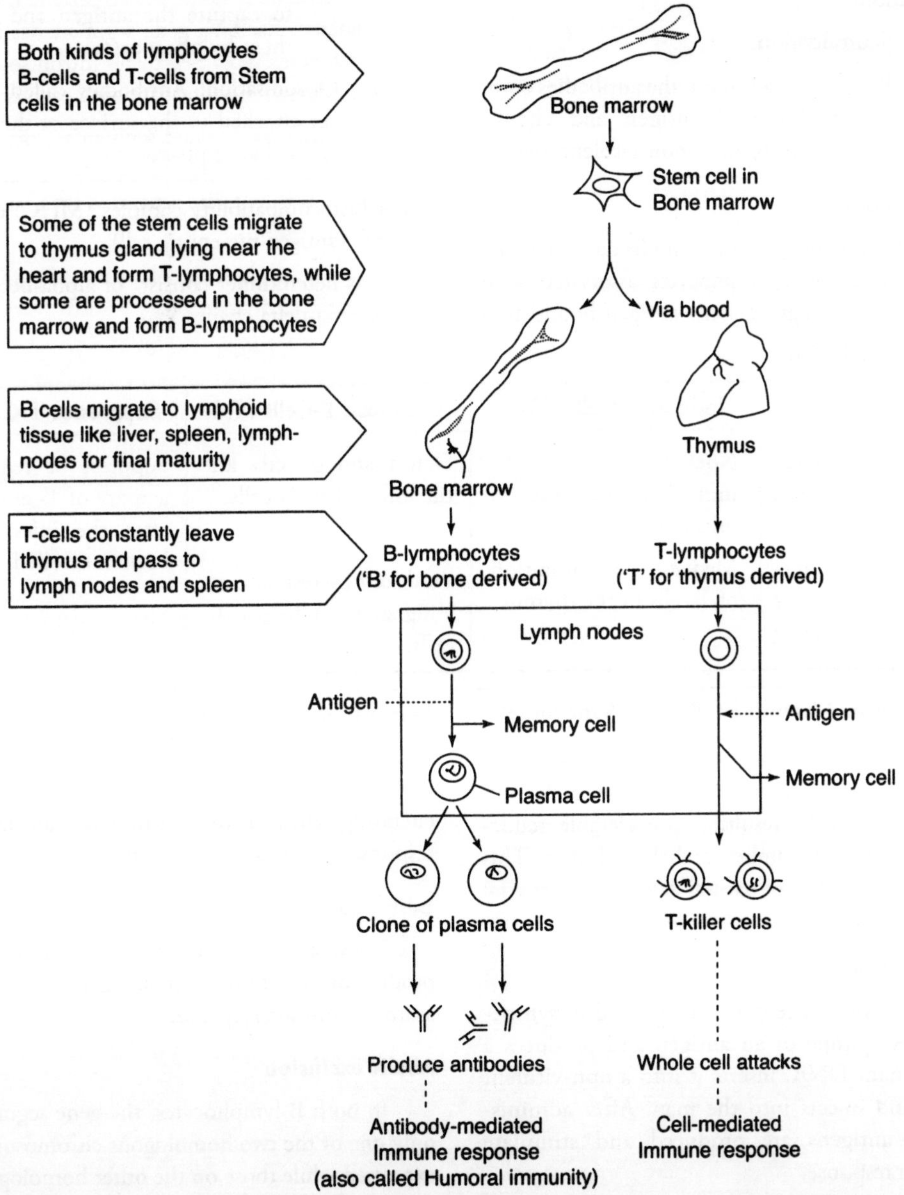

Fig. 21.2: Formation of B lymphocytes and T lymphocytes and their action in the immune response

- **Mechanism of Antigen and Antibody reaction**

I. Antibody binds more than one types of antigen on pathogen as on surface. It has more than one type of binding sites.

II. The free antibodies destroy or neutralise the leaving invading microbes by forming antigen-antibody complexes following three means: (1) Neutralisation, (2) Agglutination, (3) Opsonisation.

1. **Neutralisation:**

(i) In neutralization, the antibodies cover toxic site of antigen and change virulent form to non-virulent form.

(ii) Some antibodies neutralize toxins by acting as non-toxin.

2. **Agglutination:**

(i) Antibodies bind to antigen and cause clumping or agglutination.

(ii) Antibodies hold a clump of invading cells together.

(iii) Agglutination helps the macrophages to capture the antigen and destroys them by phagocytosis.

3. **Opsonisation:** Antibodies called opsonin get attached to the surface of the antigen and enhance phagocytosis.

T-Cell receptor

It is an antigen binding molecules expressed on the surface of T-lymphocytes associated with CD_3 and can recognize only antigen bound to a major histocompatibility complex (MHC) molecule on APC (antigen presenting cell).

It is heterodimer consist of alpha-beta (95%) and gamma-delta chain (5%).

- **Types of T-Cells**

Helper T-Cells (T_H)	Inducer T-Cells (T_I)	Cytotoxic T-Cells (T_C)	Suppressor T-Cells (T_S)
Bind with the antigen presenting cells and detect infection.	Not involved in the immediate response to infection.	Detect and directly kill the infected body cells.	Suppress or reduce the activity of T- and B-cells after the infection has been controlled.
Stimulate the production of both B-cells and cytotoxic T-cells.	Mediate the maturation of T-cells in the thymus.	Produced in response to a signal from helper T-cells.	

AIDS (Aquired Immunodeficiency Syndrome)

I. It is due to deficiency of immunosystem.

II. It is caused by retrovirus. Here HIV kills helper T-cells resulting considerable reduction in the number of helper T-cells. This results the loss of natural defence against viral infection.

DNA vaccine

It is a substance that is composed of synthetic DNA epitope of an antigen and produces a recombinant DNA, inserts it into a non-virulent vector and injects into the man. After administration, antigens are produced and stimulate immune response.

Recombinant DNA vaccine

It is a vaccine produced through recombinant DNA technology. This involves inserting the DNA encoding antigen that stimulates an immune response into bacterial or mammalian cells.

Edible vaccine

These are transgenic plant and animal based production or those contain agents that tiggers animals immune response.

Allelic exclusion

In both B lymphocytes, the gene segments of only one of the two homologous chromosomes are expressed while those on the other homologous are not expressed, this is known as allelic exclusion.

- **Location and function of different Immunoglobulins isotypes**

Types of Antibodies	Location	Function
IgA	Main form of antibodies in external secretion such as saliva and mothers milk. In secretory phase this Ig forms dimer.	Attack pathogen before they reach blood stream.
IgD	Present as receptors on the surface of B lymphocytes found in circulation in extremely low quantity.	Serve as receptors for antigen on the B lymphocytes (cells) – activation of B cells.
IgE	Present as membrane bound receptor on basophils in blood and mast cell in tissues.	Promote the release of histamines and other agents in autogenic reactions.
IgG	Most abandant antibody (about 75%) in blood-plasma in circulation – duly antibody that can cross placenta.	Stimulation of phagocytes. Attacks pathogens (viruses and bacteria and bacterial toxins).
IgM	Present as receptor on lymphocyte surface. In secretory form it form pentameric structure.	Provides passive immunity to foetus. (i) First type antibody to be serve during primary response. (ii) Causes activation of B cells and promotes clumping of cells. (iii) As a result antigen containing particles to stick together.

- **Antigen Presenting Cells (APCs)**

I. Those cells that process and present antigenic peptides in association with class II MHC molecules, and deliver a co-stimulatory signal required for TH Cell activation are called antigen presenting cells (APCs).

II. APCs are classified as (1) professional APCs and (2) nonprofessional APCs. Professional APCs are those cells that process and present antigens continually. They are three types: viz., dendritic cells, macrophages and B-cells.

III. The nonprofessional APCs can be induced to express MHC II or a co-stimulatory signal and many of them function as APCs only. All nucleated cells express MHC I molecules.

- **MHC protein**

It comprises a large cluster of closely linked polymorphic gene encoding cell surface glycoprotein(MHC antigen) associated with intracellular recognition and antigen presentation to T-lymphocyte. It is located on chromosomes 6 and 17 in (human and mice). On human, MHC is referred to as HLA complex and in mice as H-2 complex.

I. MHC loci are expressed co-dominantly and the genes are referred to as classes I, II and III.

II. The class I and II gene code impart for protein that presents antigen to T-cells. The class III gene code import for protein in the complex system.

- **TCR loci**

The TCR chains α, β, γ and δ are encoded by TCR loci.

I. TCRα locus is similar to the Ig light chain of κ locus (It has several V segment separated from a cluster of J segments).

II. They specify both alpha and delta chains. The loci encoding α and δ chains are in the same chromosome14 (human and mice). The entire α and δ gene segments span about 1100 kb.

III. TCRβ locus resembles the organization of heavy chain locus of immunoglobulin. It is on the chromosomes 6 in mice and 7 in human. It occupies 685 kb region.

IV. TCRγ locus has an organization similar to that of light chain of λ locus. It is on the chromosomes 13 in mice and 7 in human. The human gamma genes span about 150 kb.

Solved Problems

Problem 1

State briefly describe the three ways that antibody diversity is produced.

Solution:

I. Gene arrangement of V, D and J domains occurs within the light and heavy chains genes.

II. Within a given B-cell different combinations of light and heavy chains are possible.

III. Impressive fusion may occur between V, D and J domains.

Problem 2

Describe the functions of RAG-I and RAG-II proteins and NHEJ proteins.

Solution:

I. Site specific recombination (between V and J region of κ light chain of B-cells) event, is initiated by two proteins RAG I and RAG 2 (recombination activating gene). They are notable for having no introns and with opposite transcriptional orientation.

II. These proteins forms **recombinase** enzyme that are capable of recognizing recombination signals, sequences(12-23) and produces two double-strand breaks in DNA, at the junction of coding regions(V-J).

III. Recognition of the 12-23 signals is done with the help of HMG 1 and HMG 2 (high mobility group of protein) that bind DNA and bend it. It also enhances the activity of the RAG recombinase that brings together one of the variable regions (78) and one of the joining regions (2) and cut, link them in the process which generates a circle of DNA with intervening segment which is then lost.

IV. The connection phase of this process is catalysed by a group of proteins termed non homologous end joining (NHEJ) Protein.

V. The fusion process may not be entirely precise. So a few nucleotides can be added or lost at the junction between variable and joining domain.

Problem 3

What are the components of making an Ig light chain and heavy chain?

Solution:

I. The components of an immunoglobulin light chain are V, J, C regions.

II. The components of immunoglobulin heavy chains are V, D, J, C regions.

Problem 4

What are the nucleotide recognition signals in V-J joints?

Solution:

The V-J joints recognition signal is heptamer and nonamer separated by 23 and 12 bp.

Problem 5

Each Lk Vk gene segment in the kappa light chain locus on chromosome 2 consists of two coding exons, one for leader peptide and one for variable portion of the kappa light chain. Would you expect to find a stop codon at the end of the codon segment in the sequence of Vk exon?

Solution:

No, because the Vk coding sequence must be joined to the coding sequence of the constant region to encode a complete kappa light chain.

Problem 6

Why is an antibody molecule represented as H_2L_2?

Solution:

Each antibody molecule is formed of two heavy chains and two light chains. That is why antibody molecule is represented by H_2L_2.

Problem 7

What would happen to immune system if thymus gland is removed from the body of a person?

Solution:

I. In the thymus gland, immature lymphocytes differentiate into antigen sensitive lymphocytes.

II. If thymus gland is removed from the body, immune system of a person decreases. As a result person's body becomes prone to infectious diseases.

Problem 8

"Mother's milk is considered the most appropriate food for the newborn infant." Justify the statement.

Solution:

I. Mother's milk contains colostrums. It contains antibodies (IgA) and good bacteria.

II. It provides infant's immunity against infection and protects from illness.

Problem 9

"Frequent washing of eyes disadvantageous" Justify your statement.

Solution:

I. Lacrymal secretion that moistens the eyes, contains lysozyme.

II. It kills bacteria and prevent eye infection.

Problem 10

How can DNA be used to give an immunizing response?

Solution:

I. The plasmid DNA encoding antigenic proteins are injected directly into the muscle. In the recipient muscle cell, the DNA and encoded protein antigen are expressed.

II. It leads to both humoral antibody response and cell mediated response.

III. The DNA appears either to integrate into the chromosomal DNA or to be maintained for longer periods in an episomic form. The viral antigen is expressed not only by the muscle cell but also by dendritic cells in the area that take up the DNA and expresses. The muscle cell expresses low level of class I MHC molecule.

IV. DNA vaccines offer advantage over many of the existing vaccines. For example, the encoded protein is expressed in the host in its natural form, there is no denaturation or modification. So immune response is directed to antigen exactly as it is expressed by the pathogen.

V. An improved method for administering these vaccines entails coating microscopic gold beads with plasmid DNA and then delivering the coated particles through the skin into underlying muscle with an air gun (called gene gun). This will allow rapid delivery of a vaccine to large populations without requirement for huge supplies of needles and syringes.

Problem 11

Explain the mechanism of action of T-cell antigen.

Solution:

I. T-cells are antigen specific, i.e., T-cell recognises a specific antigen. Therefore, a body has separate T-cells for every antigen of the body.

II. When a T-cell comes in contact with an antigen the T-lymphocytes transform into lymphoblast which then divides to form a clone of T-cells.

III. A clone of T-cells has four types of cells: (a) Helper T-cells (b) Killer T-cells (c) Suppressor T-cells (d) Memory T-cells.

Steps:

1. Cytotoxic T-cell or killer cell (Tc) destroys specified cell having foreign antigen attached to their surface. They become active by receiving signal molecule from helper T-cell and antigen presenting cell or foreign antigen. It elaborates **perforins** or **granzymes**. Granzymes and perforins form pores and initiate apoptosis (cell death) of uninfected cell respectively.
2. Helper T-cell stimulate B-cell to produce antibodies and killer T-cell which destroy the cell.
3. Suppressor T-cell inhibits immune response by releasing cytokines and suppresses acti-

vity of other T-cell and B-cell. They provide immunological tolerance towards certain enzymes.

4. Memory T-cells are sensitized by antigen, retain sensitization for future and produce secondary response.

Problem 12

State briefly about Primary and Secondary immune response.

Solution:

Primary immune response: When our body encounters a pathogen for the first time, the response is called **Primary immune response** which is weak and of low intensity.

I. Here a pathogen invades the body, very few B- and T-cells have receptors which can recognize the pathogen.

II. During the first encounter, very few lymphocytes are exposed to an antigen.

III. After exposure, antigen triggers a reaction and two types of cells are formed: (a) Effector cells and (b) Memory cells.

(a) Effector cells produce antibodies that attack the infected cells to fight infection. They usually survive for a few days.

(b) Memory cells do not produce antibodies to kill infected cells. They remain in the lymph nodes and get activated by a secondary response to the same antigen. They can remain in the body for decades.

Secondary immune response: After primary immune response the subsequent exposure with the same pathogen (antigen) which elicits highly intensified and effective response is called **Secondary immune** or Anamnestic response. As a result of first infection, there is a large clone of memory cells (lymphocytes) in the body that can recognize that pathogen or antigen.

I. Second and subsequent exposure to the same antigen triggers a faster and a stronger response due to the presence of memory cells produced in the first encounter.

II. In the secondary response also, like the first response, there is a clone of effector cells and a clone of memory cells.

III. The effector cells produce antibodies or attack the foreign antigen for the secondary response.

IV. The memory cells remain in the body, capable of responding to the future exposures to the antigens

V. In certain cases, the memory cells give a life long immunity as in the case of childhood diseases like chicken pox and measles.

Problem 13

Define Vaccine/Vaccination and its types.

Solution:

A preparation of immunogenic material (attenuated or inactivated or dead pathogen) used to induce immunity against pathogen is termed as vaccine.

The process of introducing a vaccine in the body to induce a specific immune response that protect individual, later exposes to pathogen is called vaccination.

I. The vaccine can be introduced in the body mostly by ingestion and sometimes orally.

II. Immunization provides immunity without infection through the introduction of killed pathogen or its toxoid into a healthy person.

Types:

I. Killed pathogen

II. Attenuated living microorganism

III. Toxoid

IV. Cellular fraction

Problem 14

What do you mean by Autoimmune diseases?

Solution:

This is an abnormality which sometimes develops in the immune system of the body instead of destroying foreign molecules if starts attacking the bodies own cells leading to serious consequences. Such diseases are called autoimmune diseases.

Problem 15

Mention the two mechanisms of processing and presenting different types of antigens.

Solution:

I. Endogenous antigens are processed in the cytosolic pathway and presented on the membrane with class I MHC molecule.

II. Exogenous antigen are processed in the endocytic pathway and presented on membrane with class II MHC molecule.

Problem 16

How and where T-cells and B-cells are mature and differentiate?

Solution:

I. The progenitor T-cells are produced in the bone marrow. From there they migrate to the thymus.

II. During their maturation, positive and negative selections take place in the cirtical region of thymus and those cells sustain. The two selection, will be retained and other will be subjected to apoptosis.

III. The remaining T cells from thymus enter the blood circulation and from there they move to secondary lymphoid organs such as lymph node. There T cells come across antigens, when they come in contact with antigen they are activated and differentiated into plasma cells and memory cells.

IV. The progenitor B cells develop in the bone marrow. They mature in the bone marrow and undergo antigen induced activation and differentiation in the periphery. Activated B cells can give rise to antibody secreting plasma cells and memory cells.

Problem 17

What is stem cell? Mention its types:

Solution:

Stem cells are undifferentiated and specialized cells – that can multiply indefinitely by mitosis and also have potential to differentiate into diverse types of specialized cells with specialized function. They are two types, viz. embryonic stem cells and adult stem cell.

Problem 18

What are interferon?

Solution:

These are glycoproteins produced by different cells of immune system when exposed to foreign agents such as viruses, bacteria or parasites.

Problem 19

State briefly about the infection of HIV.

Solution:

HIV selectively infect and kills T helper cells resulting depletion of T helper cells and weakens the acquired immune response of the body. HIV utilises **reverse transcriptase** to produce cDNA (copy of RNA gene). cDNA integrates into the cells chromosomes before it reproduces and makes copies of itself.

I. HIV attaches itself and binds with plasma membrane of the T helper cells (macrophage).

II. HIV enters the cell and uses a special enzymes called reverse transcriptase, to make a DNA copy called cDNA of its genetic material.

III. The single strand replicates and produces the double-stranded cDNA.

IV. Now double-stranded viral DNA integrates into a host cell DNA, where it directs the production of viral RNA.

V. Each strand of viral RNA brings about synthesis of viral proteins.

VI. Assembly of RNA strand and viral proteins now gives rise to many viruses.

VII. These viruses are then released from the host cells before it dies, macrophages act like HIV factors.

Problem 20

What is molecular mimicry?

Solution:

This term is used to describe the sharing epitopes between an infectious agent or parasite and its host.

Problem 21

State briefly the mechanism of generating antibody diversity.

Solution:

1. **Multiple germline immunoglobulin gene:**

I. An individual has different (100 - 200) V segments, six different J segments and 30 different D segments located contiguously in his/her germline genes.

II. It allows for significant diversity.

2. **Somatic recombination:**

I. The process of joining a V gene to a C gene in lymphocyte to generate an immunoglobulin gene is known as somatic recombination.

II. During B lymphocyte maturation, a specific combination of single V and J segment is selected for light chain and another combination of V, J and D segments are selected for the heavy chain.

III. This is done by deleting the intervening DNA sequences between the V, J and D segments before they are transcribed into mRNA. This deletions are done by the enzyme **recombinases** encoded by RAG1 and RAG2 genes. This enzyme causes the double strand DNA break at the specific DNA sequences that flank the V and D gene segments.

IV. These non-deleted segments are joined by ligases. This cutting and pesting process is known as somatic recombination as against germ line recombination.

V. This results various types of rearranged DNA sequences that give rise many possible combination of V, J, D segments.

3. **Junctional Diversity:**

I. Variations occur in the V, D and J segments in the site where they are joined.

II. A number of nucleotides may be deleted or added at the junction joining the regions.

III. This adds to the capacity required to produce varied types of antibodies.

4. **Somatic Hypermutation:**

I. The induced increase (10^{-3} to 10^{-6} fold) of mutation in variable region of immunoglobulin is known as somatic hypermutation.

II. At the time of B-lymphocyte maturation, it undergoes somatic hypermutation.

III. This produces alteration in DNA sequence coding for immunoglobulins. Somatic hypermutation of the V, D and J genes results in significant antibody diversity.

5. **Random Multiple Combinations of Heavy and Light Chains:** A random combination of different heavy and light chains in immunoglobulin gives added variation to antibodies.

Problem 22

State briefly the mechanism of recombination of different segments of immunoglobulin genes.

Solution:

I. The mechanism joining of different gene segments during lymphocytes maturation is the same for lambda, kappa, and heavy chains.

II. It is based on the same consensus sequence found at the boundaries of the different gene segments in the germline. Consensus sequences consists of a heptamer (5′-CACAGTG-3′) and a nonamer (5′-CCAAAACC-3′) which are separated by 12 bp spacer to yield a signal sequence of 28 bp, while the other type of signal sequence (39 bp long) is separated by 23 bp spacer.

III. At the kappa locus, each V_κ segments is followed by a 28 bp signal sequence while each J_κ segment is preceded by a 39 bp signal sequence. But in case of lambda locus, each V_λ segment is followed by 39 bp signal sequence, while each J lambda is preceded by 28 bp consensus sequence. At heavy chain locus, each V_H segment is followed by a 39 bp signal sequence, each D segment is preceded and followed by 28 bp signal sequence while each J_H segment is preceded by 39 bp signal sequence. The orientation of the signal sequences differs with different segments and at different loci.

IV. According to the rule of joining reaction a 28 bp signal sequence can be joined only to a signal sequence of 39 bp and vice versa. The ends of the heptamer sequences of both the types of signalling sequences mark the ends of coding regions and they are recognized as the sites of double strand breaks. The two cleavages (one at the heptamer end of 28 bp sequence and other at the heptamer end of the 39 bp sequence) produce a central DNA fragment that is lost as circular molecule. The ends of Vk and Jk segments located at the cleavage site are joined together to give rise Vk-Jk combined sequence.

V. Proteins RAG1 and RAG2 are involved in rearangement. RAG1 recognizes the heptamer/nonamer consensus sequences of the signal sequences, and recruits RAG2 to the complex.

VI. The nonamer sequences serve for initial recognition, while the heptamer serve as the cleavage site. The RAG1 and RAG2 complexes then specifically nicks one strand at the junction of heptamer consensus sequence and the gene segment, e.g., V, J, etc. segments.

VII. The nick generates a free 3′ OH, which attacks the phosphate bond at the corresponding position in the other strand of DNA duplex; this tarnsesterification reaction generates hairpin at the ends of the V, J, etc segments. The hairpins at the ends of V, J, etc segments are removed and the two ends are joined together by the same mechanism that repairs double strands break in DNA.

VIII. Proteins Ku 70 and Ku 80 bind to the DNA ends and recruit a DNA dependent protein kinase(DNA-PKcs). Now DNA ligase IV, in the presence of protein XRCC4, joins the two ends. This end joining generates additional diversity in the variable regions of antibodies.

Problem 23

What components go into making an Ig light chain? A heavy chain?

Solution:

The components of an immunoglobulin light chain are V, J and C regions; the components of an immunoglobulin heavy chain are V, D, J and C regions.

Problem 24

What are NK cells?

Solution:

Natural Killer (NK) cells are a class of large, granular, cytotoxic lymphocytes that do not have T- or B-cell receptors. They are antibody-independent killers of tumour cells and also can participate in antibody-dependent cell-mediated cytotoxicity.

Problem 25

A disorder of the immune system is characterized by a complete lack of antibody production. Provide two possible molecular defects that would result in such a condition

Solution:

One explanation is a defect in the maturation process of B-cells. Another explanation is a defect in the process of V(D)J joining, which could involve five or more genes.

Problem 26

State briefly abougt Recombination Signal Sequence (RSS) of antibody.

Solution:

Highly conserved heptamer and nonamer nucleotide sequences that serve as signals for the gene rearrangement process and flank each germ line V, D and J segments is known as RSS.

I. One RSS is located 3′ to each V gene segment, 5′ to each J gene segment and on both sides of each D gene segment.

II. These sequences function as signal for recombination process:

1. Two-turn RSS:

```
CACAGTG–[23bp]–ACAAAAACC
GTGTCAC–[23bp]–TGTTTTTGG
 Heptamer        Nonamer
```

2. One-turn RSS:

```
GGTTTTTGT–[12bp]–CACTGTG
CCAAAAACA–[12bp]–GTGACAC
 Nonamer          Heptamer
```

Numerical Problems

Problem 1

Antibody molecules (Ig) composed of 4 polypeptide chains (two of one light chain type and two of one heavy chain type) held together by disulphide bonds.

(a) If for the light chain there were 300 different V_K segments and 4 J_K segments, how many different light chain combinations would be possible?

(b) If for the heavy chain there were 200 V_H segments, 12D segments and $4J_H$ segments, how many heavy chain combinations would be possible?

(c) Given the information in (a) and (b) what would be the number of possible types of IgG molecules (L + H chain combinations)?

Solution:

(a) Possible combinatorial association of light chain = 300 × 4= 1200.

(b) Possible combinatorial association of heavy chain = 200 × 4 × 12 = 9600.

(c) The number of possible type of Ig molecule (L+H) woud be 1200 × 96 = 15200.

Problem 2

Suppose that an antibody light chain gene is assembled from three different gene segments. How many different chains can be produced if the genome contains 5, 20, 200 copies of the three gene segments?

Solution:

If each gene is assembled using one copy of each gene segment, 5 × 20 × 200 = 20,000 = 2×10^4 different genes are possible.

Problem 3

Assume that an animal is capable of producing 100 million different antibodies and that each antibody contains a light chain 200 amino acid long and a heavy chain 500 amino acid long. How much genomic data would be needed to accommodate the coding sequences of three genes?

Solution:

I. If each antibody consists of one kind of light chain and one kind of heavy chains, it can combine freely and has the potential to produce 100 million different antibodies.

II. It produces 200 × 500 = 10000 light chain genes and 200 × 500 = 10^4 genes for heavy chain. Total = 10000 × 10000 = 100 million antibodies.

III. If each light chain is 200 amino acid long, then each light chain must comprises 3 × 200 = 600 nucleotides because each amino acid is specified by triplet of nucleotides. Similarly, each heavy chain must comprises 500 × 3 = 1500 nucleotides.

IV. Therefore, genome must contain 600 × 10^4 = 6 miillion nucleotides for light chain and 1500 × 10^4 = 15 million nucleotides for heavy chain. Altogether the genomic contains 6 + 15 = 21 million nucleotides dedicated to encode various antibody chains.

Problem 4

Considering only combinatorial joining of gene segments and association of light and heavy chains, how many different antibody molecules could be generated from germ line DNA containing 500 V_L and 4 J_L gene segments and $300V_H$, $15D_H$ and 4 J_H gene segments?

Solution:

Light chains = 500(V_L) × 4(J_L) = 2000 = 2×10^3.

Heavy chains = 300 (V_H) × 4(J_H) × 15 (D_H) = 18×10^3.

Antibody molecules = $2 \times 10^3 \times 18 \times 10^3 = 36 \times 10^6$

Problem 5

Considering only combinatorial joining of gene segments and association of light and heavy chains, how many different antibody molecules could be generated from a germ line DNA con-

taining 41 V_L snd 5 J_L gene segments and 48 V_H, 23 D_H and 6 J_H gene segments.

Solution:

Light chains = 41 × 5 = 205

Heavy chains = 48 × 23 × 6 = 6624

Antibody molecules = 205 × 6624 = 1357920.

Problem 6

Estimate the Combinatorial Diversity of human antibody from the following data:

Multiple germline segments	Heavy chain no. of segment	Light chain Kappa	Light chain Lambda
V	48	41	34
D	23	0	0
J	6	5	5

Solution:

Possible combinatorial association of light chains

(a) In kappa light chain = 41 × 5 = 205

(b) In lambda light chain = 34 × 5 = 170

Total = 205 + 170 = 375.

Combinatorial association of heavy chains = 48 × 23 × 6 = 6624.

The number of possible combinatorial association of light and heavy chains would be = 6624 × 375 = 2484000 = 2.48×10^6.

Problem 7

Estimate the Combinatorial diversity of mice antibody from the following data:

Multiple germline segments	Heavy chain no. of segment	Light chain Kappa	Light chain Lambda
V	101	85	2
D	13	0	0
J	4	4	3

Solution:

Possible combinatorial association of light chains.

(a) In kappa light chains = 85 × 4 = 340.

(b) In lambda light chains = 2 × 3 = 6.

Total = 340 + 6 = 346.

Combinatorial association of heavy chains = 101 × 13 × 4 = 5252.

The number of possible combinatorial association of light and heavy chains would be = 5252 × 346 = 1817192 = 1.817192×10^6 = 1.82×10^6.

Problem 8

If the genome of particular primate contains 300 V_κ, 5 J_κ, 1C_κ; 150 V_λ, 5 J_λ, C_λ; 300 V_H, 4 J_H, 50 D and 10 C_H gene segments and this can be joined in possible appropriate combination during the genome arrangement that occurs during the differentiation of B-lymphocytes, how many different antibodies could this mammal produce?

Solution:

Possible combinatorial association of light chains

(a) In kappa light chains = 300 × 5 × 1 = 1500

(b) In lambda light chains = 150 × 5 = 750

Total = 1500 +750 = 2250.

Combinatorial association of heavy chains = 300 × 4 × 50 × 10 = 6×10^5.

The number of combination of light chain and heavy chains would be = 2250 × 6×10^5 = 135×10^7.

Problem 9

If the genome of a particular mammal contains 200 V_κ, 4 J_κ, 1 C_κ; , 150 V_λ, 6 J_λ, 200 V_H, 4 J_H, 100 D, 8 C_H, gene segment and these can be joined in all possible appropriate combinations during the genome rearrangement that occur in the differentiation of B-lymphocytes, how many different antibody could this mammal produce?

Solution:

Combinatorial association of light chains:

(a) In kappa light chains 200 × 4 × 1=800

(b) In lambda light chains = 150 × 6

(c) Total = 800 + 900 = 1700.

Combinatorial association of heavy chains = 200 × 4 × 100 × 8 = 64×10^6.

The number of combination of light and heavy chains would be 1700 × 64 × 10000 = 1088×10^6.

Problem 10

In mouse the light chain has 350 L-V_κ gene segments, 4 functional J_κ segments, 1 C_κ gene segment. The heavy chain has 500 L-J_H segments 12 D segments and 4 J_H segments. Calculate combinatorial diversity of the antibody.

Solution:

I. The number of possible κ chain variable regions = 4 × 350 × 1 = 1400.

II. The number of possible heavy chain diversity = 500 × 12 × 4 = 24000.

The combinatorial diversity would be 24000 × 1400 = 33,600,000.

Problem 11

Estimate the combinatorial diversity from the following data:

Multiple germline segments	No. of segment	
	Heavy chain	Light chain
V	200	40
D	5	0
J	9	10

Solution:

The number of possible combinatorial association for light chains will be 40 × 10 = 400 and heavy chains will be 200 × 5 × 9 = 9000.

Final product = $9 \times 10^3 \times 4 \times 10^2 = 36 \times 10^5$ billion.

Problem 12

Estimate the combinatorial diversity of mice antibody from the following data:

Multiple germline segments	Heavy chain no. of segment	Light chain	
		κ	λ
V	300	300	2
D	13	0	0
J	4	4	3

Solution:

Possibe combinatorial association of light chains:

(a) In kappa light chain = 300 × 4 = 1200

(b) In lambda light chain = 2 × 3 = 6

Total = 1200 + 6 = 1206.

Combinatorial association of heavy chains = 300 × 13 × 4 = 15600.

The number of combination of light chain and heavy chain would be = 1206 × 15600 = 18813600 = $18.8136 \times 10^6 = 18.8 \times 10^6$.

Problem 13

Considering only combinatorial joining of gene segments and association of heavy chain and light chains, how many different antibody molecules could be generated from germ-line DNA containing 550 V_L and 4 J_L gene segments and 300 V_H, 5 D_H and 4 J_H gene sgments.

Solution:

Light chains = 550(V_L) × 4(J_L) = 2200 = 22×10^2

Heavy chains = 300 (V_H) × 4 J_H × 5 (D_H) = 6×10^3

Antibody molecules = $22 \times 10^2 \times 6 \times 10^3 = 132 \times 10^5$.

Bibliography

Name of the Authors	Name of the Books
1. Ayala F. J.	Population and Evolutionary Genetics: A Primer
2. Banerjee P. K.	Introduction to Bio-statistics A Text Book of Biometry (3rd edition)
3. Brown TA	Gene cloning and data analysis (5th edition, 2006)
4. Brooker Robert J.	Genetics, Analysis and Principles (3rd edition)
5. Chattopadhaya Sanjib	Evolution Adaptation Ethiology
6. Clark David P and Pazdernik Nanette	Biotechnology: Applying the Genetic Revolution
7. Dubey, R. C.; Maheshwari D. K.	A Text Book of Microbiology (1st edition)
8. Futuyama D.	Evolutionary Biology (3rd edition)
9. Fairbanks D and Anderson W. R.	Genetics
10. Gardner E. J.; Simmons M. J. and Snustad D. P.	Principles of Genetics (8th edition)
11. Goodenough Ursula	Genetics (3rd Edition)
12. Griffiths A. J. F., Gelbart W. M.; Miller J. H. and Lewontin R. C.	Modern Genetic Analysis
13. Hartwel, L. H.; Goldberg M. L.; Reynolds A. E.; Silver L. S. and Veres R. C.	Genetics: From Genes to Genomes
14. Hartl. Daniel L & Jones Elizabeth W	Genetics: Principles and Analysis (4th edition)
15. Kothari M. L.; Mehta L. A.; Roy Chowdhurys	Essentials of Human Genetics
16. Klug W. S.; Cummings, M. R. Spencer and Charlotte A	Concept of Genetics (8th edition)
17. Kowles Richord	Solving Problems in Genetics
18. Mandal S	Fundamentals of Human Genetics
19. Malacinski, George M	Freifelder's Essential of Molecular Biology (4th edition)
20. Murray R. K.; Granner D. K.; Mayes P. A. and Rodwell V. W.	Herper's Biochemistry (24th edition)
21. Nielsen Rasmus and Slatkin Montgomery	An Introduction to Population Genetics: Theory and Application
22. Pande, K.; Shukla J. P. and Trivedi S. P.	Fundamentals of Toxicology
23. Paul Ajoy	Textbook of Cell and Molecular Biology
24. Power C. B.	Genetics (Vol. II)

Name of the Authors	Name of the Books
25. Primrose S. B. and Twyman R. M.	Principles of Gene manipulation and Genomics
26. Ridley M.	Evolution (2nd edition)
27. Roy S. C. and De K. K.	Cell Biology (2nd edition)
28. Russell P. J.	Genetics (5th Edition)
29. Russell P. J.	Genetics: A molecular approach [International edition (2nd edition), 2006]
30. Sing B. D.	Fundamental of Genetics
31. Sastry K. S. and Padmanabhan G. Subramanyan	A Text Book of Molecular Biology (1st edition)
32. Stansfield W. D.	Schaum's Outline of Theory and Problems of Genetics (2nd edition)
33. Strachan Tom and Read Andrew	Human Molecular Genetics (4th edition)
34. Stern C.	Principles of Human Genetics (2nd edition)
35. Strickberger M. W.	Evolution (2nd edition)
36. Strickberger M. W.	Genetics (3rd edition)
37. Snustad D Peter and Simmons Michael S	Principles of Genetics (4th edition)
38. Tamarin R. H.	Principles of Genetics (7th edition)
39. Turnpenny, Peter D, Sian Ellard	Emery's Elements of Medical Genetics (13th edition)
40. Twyman R. M.	Advanced Molecular Biology
41. Watson J. D.; Baker T. A.; Bell S. P.; Gann A.; Levine M. and Losick R.	Molecular Biology of the Gene (5th edition)
42. Weaver Robert F and Hedrick Philip W.	Genetics (3rd edition)
43. Weaver F Robert	Molecular Biology (4th edition)
44. Winter P. C., Hickey G. J. and Fletcher H. L.	Genetics (2nd edition) (Instant notes)

Index

A

B

C

D

E

F

G

H

I

J

K

L

M

N

O

P

Q

R

S

T

U

V

W

Note

Note